Intermediate Algebra with Early Functions and Graphing

Seventh Edition

Margaret L. Lial
American River College

John Hornsby
University of New Orleans

Terry McGinnis

Boston San Francisco New York
London Toronto Sydney Tokyo Singapore Madrid
Mexico City Munich Paris Cape Town Hong Kong Montreal

Publisher	Greg Tobin
Editor-in-Chief	Maureen O'Connor
Editorial Project Management	Ruth Berry and Suzanne Alley
Editorial Assistants	Melissa Wright and Jolene Lehr
Managing Editor/Production Supervisor	Ron Hampton
Text and Cover Design	Dennis Schaefer
Supplements Production	Sheila C. Spinney
Production Services	Elm Street Publishing Services, Inc.
Media Producer	Lorie Reilly
Software Development	John O'Brien, InterAct Math; Marty Wright, TestGen-EQ
Marketing Manager	Dona Kenly
Marketing Coordinator	Heather Rosefsky
Prepress Services Buyer	Caroline Fell
Technical Art Supervisor	Joseph K. Vetere
Art Creation and Composition Services	Pre-Press Company, Inc.
First Print Buyer	Hugh Crawford
Cover Photo Credits	© Wolfgang Kaehler/Corbis; Walter Bibikow/Index Stock Imagery; Tina Buckman/Index Stock Imagery
Photo Credits	All photos from PhotoDisk except the following:

Bill Aron/PhotoEdit, p. 278; J. L. Atlan/Sygma, p. 87 left; Dick Blume/The Image Works, p. 196; Dick Blume/Syracuse Newspapers/The Image Works, p. 565; Cleve Bryant/PhotoEdit, p. 488 right; Nigel Cattlin, Holt Studios International/Photo Researchers, Inc., pp. 273, 285; Myrleen Ferguson Cate/PhotoEdit, pp. 1, 30; John Coletti/Stock Boston, p. 637; Bob Collins III/The Image Works, p. 542; Paul Conklin/PhotoEdit, pp. 89, 453 left; Gary A. Conner/PhotoEdit, pp. 7, 53, 88, 122, 463; Greg Crisp/SportsChrome USA, pp. 176, 183; Bob Daemmrich/Stock Boston, p. 734 top; Mary Kate Denny/PhotoEdit, pp. 29, 34; Michael Dwyer/Stock Boston, p. 440; Kathy Ferguson/PhotoEdit, p. 300; Eric Fowke/PhotoEdit, p. 189; Tony Freeman/PhotoEdit, pp. 128, 678, 683; Spencer Grant/PhotoEdit, pp. 24, 327, 696; Charles Gupton/Stock Boston, pp. 247, 286; Michael J. Howell/Stock Boston, p. 48; Iowa State University, p. 629; D. Jennings/The Image Works, p. 488 left; L. Kolvoord/The Image Works, p. 86; Jim Mahoney/The Image Works, p. 87 right; Joan Marcus & Marc Bryan-Brown/Photofest, p. 64; Doug Martin/Photo Researchers, Inc., pp. 587, 594; Mary Ellen Matthews/Photofest, p. 284; Tom McCarthy/PhotoEdit, p. 65; Courtesy, Terry McGinnis, p. 33; NASA, pp. 232, 272, 321 right, 705, 724, 734 bottom; Eric Neurath/Stock Boston, p. 621; Michael Newman/PhotoEdit, pp. 165, 543, 578, 593, 695, 695; Jonathan Nourok/PhotoEdit, p. 230; Richard Pasley/Stock Boston, pp. 113, 133; Charles Pefley/Stock Boston, p. 583; Photofest, pp. 157, 304, 635, 677; Phyllis Picardi/Stock Boston, p. 657; PIC Tommy Hindley/Professional Sport/Topham/The Image Works, p. 96; Powerball Lottery, pp. 307, 321 left; A. Ramey/PhotoEdit, pp. 333 right, 744 left; Randall/The Image Works, pp. 225, 233; Mark Richards/PhotoEdit, p. 282; Brian Spurlock/SportsChrome USA, pp. 73, 275; Jamie Squire/Allsport, p. 212; Rob Tringali, Jr., SportsChrome USA, pp. 49, 80, 127; Dana White/PhotoEdit, p. 159; David Young-Wolff/PhotoEdit, pp. 383, 388, 631, 743; Michael Zito/SportsChrome USA, pp. 138, 453 right; Elizabeth Zuckerman/PhotoEdit, p. 333 left.

Library of Congress Cataloging-in-Publication Data

Lial, Margaret L.

Intermediate algebra with early functions and graphing.—7 ed./Margaret L. Lial, John Hornsby, Terry McGinnis.

p. cm.

Includes index.

ISBN 0-321-06459-3 (Student Edition)

ISBN 0-321-08870-0 (Annotated Instructor's Edition)

1. Algebra. I. Title: Intermediate algebra. II. Hornsby, John. III. McGinnis, Terry. IV. Title.

QA152.2 .L528 2002

512.9—dc21 2001022977

2 3 4 5 6 7 8 9 10 WC 04 03 02

Contents

List of Applications

Miscellaneous

Physics

Sports/Entertainment

Statistics/Demographics

Technology

Transportation

List of Focus on Real-Data Applications

Preface

The seventh edition of *Intermediate Algebra with Early Functions and Graphing* continues our ongoing commitment to provide the best possible text and supplements package to help instructors teach and students succeed. To that end, we have tried to address the diverse needs of today's students through an attractive design, updated figures and graphs, helpful features, careful explanations of topics, and an expanded package of supplements and study aids. We have also taken special care to respond to the suggestions of users and reviewers and have added many new examples and exercises based on their feedback. Students who have completed a course in introductory algebra—as well as those who require further review of basic algebraic concepts before taking additional courses in mathematics, business, science, nursing, or other fields—will benefit from the text's student-oriented approach.

This text is part of a series that also includes the following books:

- *Essential Mathematics*, by Lial and Salzman
- *Basic College Mathematics*, Sixth Edition, by Lial, Salzman, and Hestwood
- *Prealgebra*, Second Edition, by Lial and Hestwood
- *Introductory Algebra*, Seventh Edition, by Lial, Hornsby, and McGinnis
- *Introductory and Intermediate Algebra*, Second Edition, by Lial, Hornsby, and McGinnis.

WHAT'S NEW IN THIS EDITION?

We believe students and instructors will welcome the following new features.

New, Real-Life Applications We are always on the lookout for interesting data to use in real-life applications. As a result, we have included many new or updated examples and exercises throughout the text that focus on real-life applications of mathematics. These applied problems provide a modern flavor that will appeal to and motivate students. (See pp. 7, 196, and 261.) A comprehensive List of Applications appears at the beginning of the text.

New Figures and Photos Today's students are more visually oriented than ever. Thus, we have made a concerted effort to add mathematical figures, diagrams, tables, and graphs whenever possible. (See pp. 34, 127, and 183.) Many of the graphs use a style similar to that seen by students in today's print and electronic media. Photos have also been incorporated to enhance applications in examples and exercises.

Increased Emphasis on Problem Solving Introduced in Chapter 2, our six-step problem-solving method has been refined and integrated throughout the text. The six steps, *Read*, *Assign a Variable*, *Write an Equation*, *Solve*, *State the Answer*, and *Check*, are emphasized in boldface type and repeated in examples and exercises to reinforce the problem-solving process for students. (See pp. 79, 91, and 275.)

Study Skills Component A desk-light icon at key points in the text directs students to a separate *Study Skills Workbook* containing activities correlated directly to the text. (See pp. 43, 61, and 357.) This unique workbook explains *how* the brain actually learns, so students understand *why* the study tips presented will help them succeed in the course. Students are introduced to the workbook in an updated To the Student section at the beginning of the text.

Focus on Real-Data Applications These one-page activities present a relevant and in-depth look at how mathematics is used in the real world. Designed to help instructors answer the often-asked question, "When will I ever use this stuff?," these activities ask students to read and interpret data from newspaper articles, the Internet, and other familiar, real sources. (See pp. 84 and 156.) The activities are well-suited to collaborative work and can also be completed by individuals or used for open-ended class discussions. Instructor teaching notes and extensions for the activities are provided in the *Printed Test Bank and Instructor's Resource Guide*.

Diagnostic Pretest A diagnostic pretest is now included on p. xxix and covers all the material in the book, much like a sample final exam. This pretest can be used to facilitate student placement in the correct chapter according to skill level.

Chapter Openers New chapter openers feature real-world applications of mathematics that are relevant to students and tied to specific material within the chapters. Examples of topics include higher education costs, personal computers, and political affiliation. (See pp. 113, 165, and 247—Chapters 3, 4, and 5.)

Calculator Tips These optional tips, marked with calculator icons, offer basic information and instruction for students using calculators in the course. (See pp. 249, 290, and 316.) In addition, a new Introduction to Calculators has been included at the beginning of the text.

Test Your Word Power To help students understand and master mathematical vocabulary, this new feature has been incorporated in each chapter summary. Key terms from the chapter are presented along with four possible definitions in a multiple-choice format. Answers and examples illustrating each term are provided. (See pp. 236, 295, and 390.)

What Familiar Features Have Been Retained?

We have retained the popular features of previous editions of the text, some of which follow.

Learning Objectives Each section begins with clearly stated, numbered objectives, and the included material is directly keyed to these objectives so that students know exactly what is covered in each section. (See pp. 177, 248, and 323.)

Cautions and Notes One of the most popular features of previous editions, Caution and Note boxes warn students about common errors and emphasize important ideas throughout the exposition. (See pp. 142, 170, and 217.) There are more of these in the seventh edition than in the sixth, and the new text design makes them easier to spot; Cautions are highlighted in bright yellow and Notes are highlighted in green.

Margin Problems Margin problems, with answers immediately available at the bottom of the page, are found in every section of the text. (See pp. 29, 117, and 252.) This key feature allows students to immediately practice the material covered in the examples in preparation for the exercise sets. Based on reviewer feedback, we have added more margin exercises to the seventh edition.

Ample and Varied Exercise Sets The text contains a wealth of exercises to provide students with opportunities to practice, apply, connect, and extend the algebraic skills they are learning. Numerous illustrations, tables, graphs, and photos have been added to the exercise sets to help students visualize the problems they are solving. Problem types include writing, estimation, and calculator exercises as well as applications and multiple-choice, matching, true/false, and fill-in-the-blank problems. In the *Annotated Instructor's Edition* of the text, writing exercises are marked with icons so that instructors may assign these problems at their discretion. Exercises suitable for calculator work are marked in both the student and instructor editions with calculator icons. (See pp. 85–90, 135–138, and 257–262.)

Relating Concepts Exercises Formerly titled Mathematical Connections, these sets of exercises help students tie together topics and develop problem-solving skills as they compare and contrast ideas, identify and describe patterns, and extend concepts to new situations. (See pp. 71, 176, and 234.) These exercises make great collaborative activities for pairs or small groups of students.

Summary Exercises Four sets of in-chapter summary exercises on problem solving, linear and absolute value equations and inequalities, factoring, and rational expressions provide students with the all-important *mixed* practice problems they need to master these typically difficult topics. (See pp. 99, 151, 377, and 441.)

Ample Opportunity for Review Each chapter concludes with a Chapter Summary that features Key Terms with definitions and helpful graphics, New Symbols, Test Your Word Power, and a Quick Review of each section's content with additional examples. A comprehensive set of Chapter Review Exercises, keyed to individual sections, is included, as are Mixed Review Exercises and a Chapter Test. Beginning with Chapter 2, each chapter concludes with a set of Cumulative Review Exercises that cover material going back to Chapter 1. (See pp. 235, 295, and 389.)

WHAT CONTENT CHANGES HAVE BEEN MADE?

We have worked hard to fine-tune and polish presentations of topics throughout the text based on user and reviewer feedback. Some of the content changes include the following:

- Review material on inequalities, formerly in Section 1.2, has been redistributed between Sections 1.1 and 3.1.
- A new set of summary exercises on solving applied problems is included in Chapter 2.
- Topics on linear inequalities and absolute value are covered in a separate Chapter 3.
- Solving linear systems of equations by matrix methods is included in a new Section 5.4.
- The material on integer exponents is consolidated in Section 6.1.
- Functions are introduced early, as in the previous edition, but are more fully integrated throughout the text. We now introduce graphs of simple polynomial functions in Chapter 6, rational functions in Chapter 7, and radical functions in Chapter 8.
- Quadratic functions and graphs are presented earlier in the text when quadratic equations are solved (Chapter 9).
- The presentation on solving quadratic and rational inequalities in Section 9.7 has been rewritten to build on the graphing concepts presented in the preceding sections.
- The material on exponential and logarithmic functions (Chapter 10) is now placed before conic sections (Chapter 11) instead of after it.
- Two new appendices have been included. Appendix A provides a review of fractions, and Appendix B covers determinants and Cramer's Rule, formerly included in the chapter on systems of linear equations.

WHAT SUPPLEMENTS ARE AVAILABLE?

Our extensive supplements package includes an *Annotated Instructor's Edition*, testing materials, solutions manuals, tutorial software, videotapes, and a state-of-the-art Web site. For more information about any of the following supplements, please contact your Addison-Wesley sales consultant.

FOR THE STUDENT

***Student's Solutions Manual* (ISBN 0-321-09204-X)** The *Student's Solutions Manual* provides detailed solutions to the odd-numbered section exercises and to all margin, Relating Concepts, Summary, Chapter Review, Chapter Test, and Cumulative Review exercises.

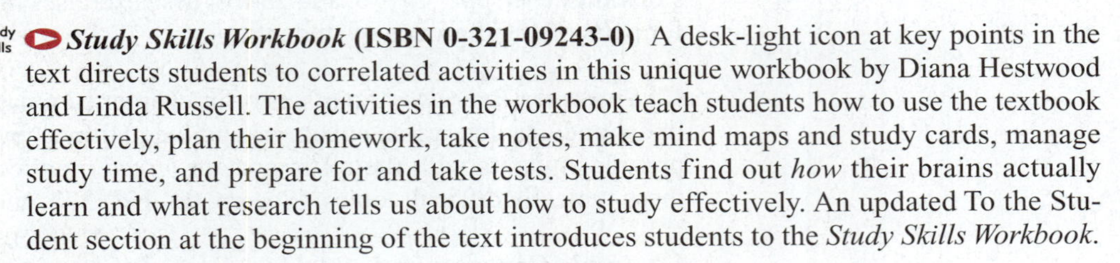

***Study Skills Workbook* (ISBN 0-321-09243-0)** A desk-light icon at key points in the text directs students to correlated activities in this unique workbook by Diana Hestwood and Linda Russell. The activities in the workbook teach students how to use the textbook effectively, plan their homework, take notes, make mind maps and study cards, manage study time, and prepare for and take tests. Students find out *how* their brains actually learn and what research tells us about how to study effectively. An updated To the Student section at the beginning of the text introduces students to the *Study Skills Workbook*.

Addison-Wesley Math Tutor Center The Addison-Wesley Math Tutor Center is staffed by qualified college mathematics instructors who tutor students on examples and exercises from the textbook. Tutoring is provided via toll-free telephone, toll-free fax, e-mail, and the Internet. White Board technology allows tutors and students to actually see problems being worked while they "talk" in real time over the Internet during tutoring sessions. The Math Tutor Center is accessed through a registration number that can be bundled free with a new textbook or purchased separately.

***InterAct Math® Tutorial Software* (ISBN 0-321-09210-4)** This interactive CD-ROM tutorial software provides algorithmically generated practice exercises that are correlated at the objective level to the content of the text. Every exercise in the program is accompanied by an example and a guided solution designed to involve students in the solution process. Selected problems also include a video clip to help students visualize concepts. The software tracks student activity and scores and can generate printed summaries of students' progress. Instructors can use the InterAct Math® Plus course-management software to create, administer, and track tests and monitor student performance during practice sessions. (See For the Instructor.)

***InterAct MathXL:* www.mathxl.com** InterAct MathXL is a Web-based tutorial system that enables students to take practice tests and receive personalized study plans based on their results. Practice tests are correlated directly to the section objectives in the text, and once a student has taken a practice test, the software scores the test and generates a study plan that identifies strengths, pinpoints topics where more review is needed, and links directly to InterAct Math® tutorial software for additional practice and review. A course-management feature allows instructors to create and administer tests and view students' test results, study plans, and practice work. Students gain access to the InterAct MathXL Web site through a password-protected subscription, which can either be bundled free with a new copy of the text or purchased separately.

***Real-to-Reel Videotape Series* (ISBN 0-321-09212-0)** This series of videotapes, created specifically for *Intermediate Algebra with Early Functions and Graphing*, Seventh Edition, features an engaging team of lecturers who provide comprehensive lessons on every objective in the text. The videos include a stop-the-tape feature that encourages students to pause the video, work through the example presented on their own, and then resume play to watch the video instructor go over the solution.

***Digital Video Tutor* (ISBN 0-321-09213-9)** This supplement provides the entire set of Real-to-Reel videotapes for the text in digital format on CD-ROM, making it easy and convenient for students to watch video segments from a computer, either at home or on campus. Available for purchase with the text at minimal cost, the Digital Video Tutor is ideal for distance learning and supplemental instruction.

MyMathLab.com

***Web Site:* www.MyMathLab.com** Ideal for lecture-based, lab-based, and on-line courses, MyMathLab.com provides students with a centralized point of access to the wide variety of on-line resources available with this text. The pages of the actual book are loaded into MyMathLab.com, and as students work through a section of the on-line text, they can link directly from the pages to supplementary resources (such as tutorial software, interactive animations, and audio and video clips) that provide instruction, exploration, and practice beyond what is offered in the printed book. MyMathLab.com generates personalized study plans for students and allows instructors to track all student work on tutorials, quizzes, and tests.

For the Instructor

***Annotated Instructor's Edition* (ISBN 0-321-08870-0)** The *Annotated Instructor's Edition* provides answers to all text exercises in color next to the corresponding problems. To assist instructors in assigning homework problems, icons identify writing and calculator exercises.

***Instructor's Solutions Manual* (ISBN 0-321-09205-8)** The *Instructor's Solutions Manual* provides complete solutions to all even-numbered section exercises.

***Answer Book* (ISBN 0-321-09207-4)** The *Answer Book* provides answers to all the exercises in the text.

***Printed Test Bank and Instructor's Resource Guide* (ISBN 0-321-09206-6)** The *Printed Test Bank* portion of this manual contains two diagnostic pretests, six free-response and two multiple-choice test forms per chapter, and two final exams. The *Instructor's Resource Guide* portion of the manual contains teaching suggestions for each chapter, additional practice exercises for every objective of every section, a correlation guide from the sixth to the seventh edition, phonetic spellings for all key terms in the text, and teaching notes and extensions for the Focus on Real-Data Applications in the text.

***TestGen-EQ with QuizMaster-EQ* (ISBN 0-321-09208-2)** This fully networkable software enables instructors to create, edit, and administer tests using a computerized test bank of questions organized according to the chapter content of the text. Six question formats are available, and a built-in question editor allows the user to create graphs, import graphics, and insert mathematical symbols and templates, variables, or text. An "Export to HTML" feature allows practice tests to be posted to the Internet, and instructors can use QuizMaster-EQ to post quizzes to a local computer network so that students can take them on-line and have their results tracked automatically.

***Web Site:* www.MyMathLab.com** In addition to providing a wealth of resources for lecture-based courses, MyMathLab.com gives instructors a quick and easy way to create a complete on-line course based on *Intermediate Algebra with Early Functions and Graphing*, Seventh Edition. MyMathLab.com is hosted nationally at no cost to instructors, students, or schools, and it provides access to an interactive learning environment where all content is keyed directly to the text. Using a customized version of Blackboard™ as the course-management platform, MyMathLab.com lets instructors administer preexisting tests and quizzes or create their own. It provides detailed tracking of all student work as well as a wide array of communication tools for course participants. Within MyMathLab.com, students link directly from on-line pages of their text to supplementary resources such as tutorial software, interactive animations, and audio and video clips.

Acknowledgments

The comments, criticisms, and suggestions of users, nonusers, instructors, and students have positively shaped this textbook over the years, and we are most grateful for the many responses we have received. The feedback gathered for this revision of the text was particularly helpful, and we especially wish to thank the following individuals who provided invaluable suggestions:

Mary Kay Abbey, *Montgomery College*
Randall Allbritton, *Daytona Beach Community College*
Sonya Armstrong, *West Virginia State College*
Linda Beller, *Brevard Community College*
Dawn Cox, *Cochise College*
Julie Dewan, *Mohawk Valley Community College*
Lucy Edwards, *Las Positas College*
Rob Farinelli, *Community College of Allegheny—Boyce Campus*
Anthony Hearn, *Community College of Philadelphia*
Jeffrey Kroll, *Brazosport College*
Barbara Krueger, *Cochise College*
Sandy Lofstock, *California Lutheran University*
Janice Rech, *University of Nebraska at Omaha*
Dwight Smith, *Prestonburg Community College*
Theresa Stalder, *University of Illinois—Chicago*
Mark Tom, *College of the Sequoias*

Our sincere thanks go to these dedicated individuals at Addison-Wesley who worked long and hard to make this revision a success: Maureen O'Connor, Ruth Berry, Ron Hampton, Dennis Schaefer, Dona Kenly, Suzanne Alley, and Jolene Lehr.

While Kitty Pellissier did her usual outstanding job checking the answers to all the exercises, she also reviewed the entire manuscript and provided invaluable content suggestions during both the writing and production processes. Steven Pusztai of Elm Street Publishing Services provided his customary excellent production work. We are most grateful to Peg Crider for researching and writing the Focus on Real-Data Applications feature; Paul Van Erden for his accurate and useful index; Becky Troutman for preparing the comprehensive List of Applications; Abby Tanenbaum for writing the new Diagnostic Pretest; and Valerie Maley and Sharon Testone for accuracy checking the manuscript.

As an author team, we are committed to the goal stated earlier in this Preface—to provide the best possible text and supplements package to help instructors teach and students succeed. We are most grateful to all those over the years who have aspired to this goal with us. As we continue to work toward it, we would welcome any comments or suggestions you might have. Please feel free to send your comments via e-mail to math@awl.com.

Margaret L. Lial
John Hornsby
Terry McGinnis

Feature Walk-Through

New! Chapter Openers New chapter openers feature real-world applications of mathematics that are relevant to students and tied to specific material within the chapters. (page 53)

Linear Equations and Applications

2

In 1998, 995,000 college-bound students took the ACT exam. Of these students, 57% were female and 43% were male. Average composite scores on the ACT exam rose from 20.6 in 1990 to 21.0 in 1998. (*Source:* The ACT, Inc., Iowa City, IA, *High School Profile Report.*) In Section 2.3 we discuss the use of percent—one of the most common everyday applications of mathematics—and find the percent increases in numbers and scores of college-bound students taking the ACT exam.

53

Section 2.3 87

Use the six-step problem-solving method to solve each problem. See Examples 3 and 4.

29. The John Hancock Center in Chicago has a rectangular base. The length of the base measures 65 ft less than twice the width. The perimeter of this base is 860 ft. What are the dimensions of the base?

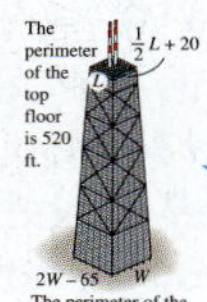

30. The John Hancock Center (Exercise 29) tapers as it rises. The top floor is rectangular and has perimeter 520 ft. The width of the top floor measures 20 ft more than one-half its length. What are the dimensions of the top floor?

31. The Bermuda Triangle supposedly causes trouble for aircraft pilots. It has a perimeter of 3075 mi. The shortest side measures 75 mi less than the middle side, and the longest side measures 375 mi more than the middle side. Find the lengths of the three sides.

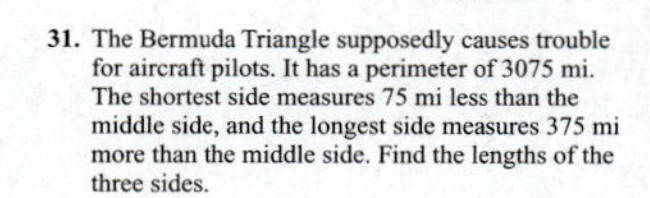

32. The Vietnam Veterans Memorial in Washington, D.C., is in the shape of two sides of an isosceles triangle. If the two walls of equal length were joined by a straight line of 438 ft, the perimeter of the resulting triangle would be 931.5 ft. Find the lengths of the two walls. (*Source:* Pamphlet obtained at Vietnam Veterans Memorial.)

33. In a recent year, the two U.S. industrial corporations with the highest profits were General Motors and General Electric. Their profits together totaled \$13.5 billion. General Electric profits were \$.3 billion less than General Motors profits. What were the profits for each corporation? (*Source: The Universal Almanac,* 1997.)

34. In a recent year, video rental revenue was \$.27 billion more than twice video sales revenue. Together, these revenues amounted to \$9.81 billion. What was the revenue from each of these sources? (*Source:* Paul Kagan Associates, Inc.)

35. In the 1996 presidential election, Bill Clinton and Bob Dole together received 538 electoral votes. Clinton received 220 more votes than Dole. How many votes did each candidate receive? (*Source:* Congressional Quarterly, Inc.)

36. Ted Williams and Rogers Hornsby were two great hitters. Together they got 5584 hits in their careers. Hornsby got 276 more hits than Williams. How many base hits did each get? (*Source:* Neft, D. S. and Cohen, R. M., *The Sports Encyclopedia: Baseball,* St. Martins Griffin; New York, 1997.)

Figures and Photos Today's students are more visually oriented than ever. Thus, a concerted effort has been made to add mathematical figures, diagrams, tables, and graphs whenever possible. Many of the graphs use a style similar to that seen by students in today's print and electronic media. Photos have been incorporated to enhance applications in examples and exercises. (page 87)

Relating Concepts Formerly titled *Mathematical Connections,* these sets of exercises help students tie together topics and develop problem-solving skills as they compare and contrast ideas, identify and describe patterns, and extend concepts to new situations. These exercises make great collaborative activities for pairs or small groups of students. (page 262)

262 Chapter 5 Systems of Linear Equations

Use the graph given at the beginning of this section (repeated here) to work Exercises 59–62.

59. For which years was Hewlett-Packard's share less than Packard Bell, NEC's share?

60. Estimate the year in which market share for Hewlett-Packard and Packard Bell, NEC was the same. About what was this share?

61. If $x = 0$ represents 1995 and $x = 3$ represents 1998, the market shares y (in percent) of these companies are closely modeled by the linear equations in the following system.

$$y = -.5x + 7 \quad \text{Packard Bell, NEC}$$
$$y = x + 3.5 \quad \text{Hewlett-Packard}$$

Solve this system. Round values to the nearest tenth as necessary.

62. Using your solution from Exercise 61, in what month and year did these companies have the same market share? What was that share?

MARKET SHARE

Packard Bell, NEC

Hewlett-Packard

Market Share (in percent): 0, 3.5, 4.0, 4.5, 5.0, 5.5, 6.0, 6.5, 7.0

Year: 1995, 1996, 1997, 1998

Source: Intelliquest; IDC.

RELATING CONCEPTS (Exercises 63–66) FOR INDIVIDUAL OR GROUP WORK

Work Exercises 63–66 in order *to see the connections between systems of linear equations and the graphs of linear functions.*

63. Use elimination or substitution to solve the system.

$$3x + y = 6 \quad (1)$$
$$-2x + 3y = 7 \quad (2)$$

64. For equation (1) in the system of Exercise 63, solve for y and rename it $f(x)$. What special kind of function is f?

65. For equation (2) in the system of Exercise 63, solve for y and rename it $g(x)$. What special kind of function is g?

66. Use the result of Exercise 63 to fill in the blanks with the appropriate responses:

Because the graphs of f and g are straight lines that are neither parallel nor coincide, they intersect in exactly ——— point. The coordinates of the point are (———, ———). Using function notation, this is given by f(———) = ——— and g(———) = ———.

Focus on

Real-Data Applications

Comparing Long-Distance Costs

Cellular phones are becoming popular tools for both local and long-distance phone calls. Frequently, rate plans include long-distance telephoning as an option if the calls are made from within the home area. The plans vary among different companies and often offer a limited number of "anytime" minutes. Information about the rate plans offered by different cellular phone companies is readily available on the Internet.

Using the Internet, you have found the following pricing schemes for both regular and the cellular phones.

- The long-distance plan for the *in-home* phone costs $6.95 per month plus $.05 per min for long-distance calls both within your state or between states, with no limit to the number of minutes of call time.

- One option for the *cellular* phone is a flat monthly fee of $59.99 that includes 450 min of "anytime" local or long-distance calls.

Note: Basic phone rates are *not* included in the in-home plan, but since you intend to have an in-home phone anyway, you can disregard those costs. Also, calls in excess of the limits for the cellular plan are expensive: $.35 per minute over the maximum. You do *not* expect to exceed the number of minutes included in the basic cellular rate plan, so do not worry about those extra charges.

For Group Discussion

The question is "Which plan is more economical?" Of course, economy is only one of the criteria that you will use when deciding whether to use your in-home phone or a cellular phone for long-distance calls, but it is one of the most important issues.

Let x represent the number of minutes of long-distance calls in a month.

1. Write an expression that represents the monthly costs for the in-home rate plan.
2. Write the expression that represents the monthly cost for the cellular rate plan.
3. The question is: "How many minutes of long-distance calls would you have to make in one month with the in-home phone to exceed the cost of the cellular phone plan?" Write a linear inequality that states that the in-home rate plan costs more than the cellular rate plan.
4. Solve the linear inequality and answer the question posed in Problem 3. What does your answer mean in terms of comparing phone costs?
5. Suppose you use the cellular phone plan for 450 min (the maximum number of minutes without incurring excess charges). How much more money would you pay compared to the in-home plan?

134

New! Focus on Real-Data Applications These one-page activities found throughout the text present even more relevant and in-depth looks at how mathematics is used in the real world. Designed to help instructors answer the often-asked question, "When will I ever use this stuff?," these activities ask students to read and interpret data from newspaper articles, the Internet, and other familiar, real sources. The activities are well suited to collaborative work and can also be completed by individuals or used for open-ended class discussions. (page 134)

Calculator Tip Figure 9 shows how a graphing calculator displays the preceding two matrices. Work with matrices is made much easier by using technology when available. Consult your owner's manual for details.

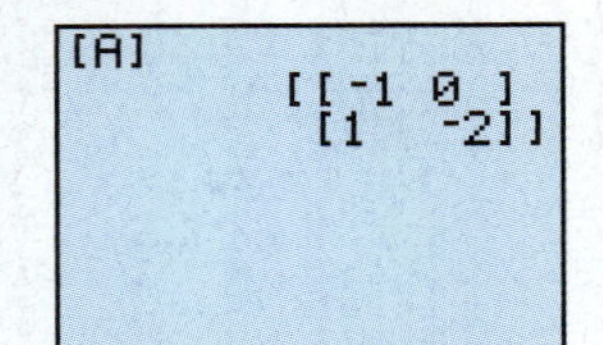

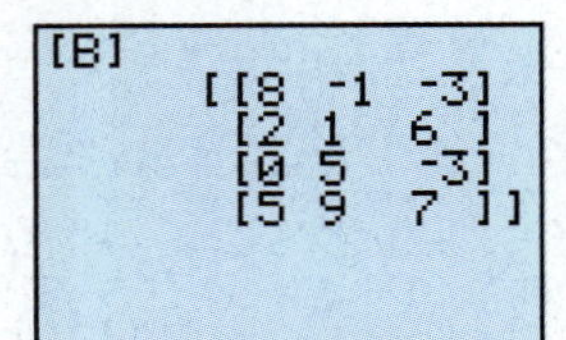

Figure 9

New! Calculator Tips These optional tips, marked with calculator icons, offer basic information and instruction for students using calculators in the course. (page 287)

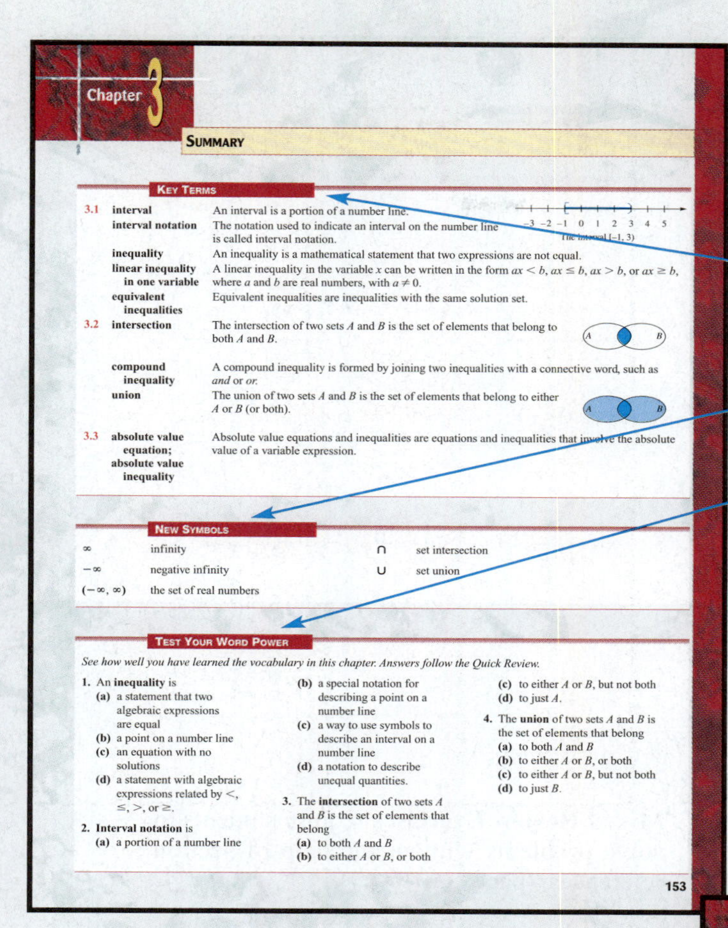

Chapter 3

SUMMARY

KEY TERMS

3.1	interval	An interval is a portion of a number line.
	interval notation	The notation used to indicate an interval on the number line is called interval notation.
	inequality	An inequality is a mathematical statement that two expressions are not equal.
	linear inequality in one variable	A linear inequality in the variable x can be written in the form $ax < b$, $ax \leq b$, $ax > b$, or $ax \geq b$, where a and b are real numbers, with $a \neq 0$.
	equivalent inequalities	Equivalent inequalities are inequalities with the same solution set.
3.2	intersection	The intersection of two sets A and B is the set of elements that belong to both A and B.
	compound inequality	A compound inequality is formed by joining two inequalities with a connective word, such as *and* or *or*.
	union	The union of two sets A and B is the set of elements that belong to either A or B (or both).
3.3	absolute value equation; absolute value inequality	Absolute value equations and inequalities are equations and inequalities that involve the absolute value of a variable expression.

NEW SYMBOLS

∞	infinity	$\cap$	set intersection
$-\infty$	negative infinity	$\cup$	set union
$(-\infty, \infty)$	the set of real numbers		

TEST YOUR WORD POWER

See how well you have learned the vocabulary in this chapter. Answers follow the Quick Review.

1. An **inequality** is
   (a) a statement that two algebraic expressions are equal
   (b) a point on a number line
   (c) an equation with no solutions
   (d) a statement with algebraic expressions related by $<$, $\leq$, $>$, or $\geq$.
2. **Interval notation** is
   (a) a portion of a number line
   (b) a special notation for describing a point on a number line
   (c) a way to use symbols to describe an interval on a number line
   (d) a notation to describe unequal quantities.
3. The **intersection** of two sets A and B is the set of elements that belong
   (a) to both A and B
   (b) to either A or B, or both
   (c) to either A or B, but not both
   (d) to just A.
4. The **union** of two sets A and B is the set of elements that belong
   (a) to both A and B
   (b) to either A or B, or both
   (c) to either A or B, but not both
   (d) to just B.

153

End-of-Chapter Material One of the most admired features of the Lial textbooks is the extensive and well-thought-out end-of-chapter material. At the end of each chapter, students will find:

Key Terms are listed, defined, and referenced back to the appropriate section number. (page 153)

New Symbols are listed for easy reference and study.

New! Test Your Word Power To help students understand and master mathematical vocabulary, Test Your Word Power has been incorporated in each Chapter Summary. Students are quizzed on Key Terms from the chapter in a multiple-choice format. Answers and examples illustrating each term are provided.

A Chapter Test helps students practice for the real thing. (page 161)

New! Study Skills Component A desk-light icon at key points in the text directs students to a separate *Study Skills Workbook* containing activities correlated directly to the text. This unique workbook explains how the brain actually learns, so students understand *why* the study tips presented will help them succeed in the course.

161

Chapter 3 TEST

Study Skills Workbook Activity 9

1. What is the special rule that must be remembered when multiplying or dividing each side of an inequality by a negative number? 1. ______

Solve each inequality. Give the solution set in both interval and graph forms.

2. $4 - 6(x + 3) \leq -2 - 3(x + 6) + 3x$ 2. ______
3. $-\frac{4}{7}x > -16$ 3. ______
4. $-6 \leq \frac{4}{3}x - 2 \leq 2$ 4. ______
5. Which one of the following inequalities is equivalent to $x < -3$? 5. ______
   A. $-3x < 9$ B. $-3x > -9$ C. $-3x > 9$ D. $-3x < -9$
6. The graph shows the number (in millions) of U.S. citizen departures to Europe. During which years were departures to Europe 6. (a) ______ (b) ______ (c) ______
   (a) at least 8 million, (b) less than 7 million,
   (c) between 7 million and 9 million?

154 Chapter 3 Linear Inequalities and Absolute Value

QUICK REVIEW

Concepts	Examples
3.1 Linear Inequalities in One Variable **Solving Linear Inequalities in One Variable**	Solve $3(x + 2) - 5x \leq 12$.
Step 1 Simplify each side of the inequality by clearing parentheses and combining like terms.	$3x + 6 - 5x \leq 12$; $-2x + 6 \leq 12$
Step 2 Use the addition property of inequality to get all terms with variables on one side and all terms without variables on the other side.	$-2x \leq 6$
Step 3 Use the multiplication property of inequality to write the inequality in the form $x < k$ or $x > k$.	$\frac{-2x}{-2} \geq \frac{6}{-2}$; $x \geq -3$
If an inequality is multiplied or divided by a *negative* number, the inequality symbol *must be reversed.*	The solution set $[-3, \infty)$ is graphed below. −3 −2 −1 0 1 2 3
3.2 Set Operations and Compound Inequalities **Solving a Compound Inequality**	Solve $x + 1 > 2$ and $2x < 6$.
Step 1 Solve each inequality in the compound inequality individually.	$x + 1 > 2$ and $2x < 6$; $x > 1$ and $x < 3$
Step 2 If the inequalities are joined with *and*, the solution set is the intersection of the two individual solution sets.	The solution set is $(1, 3)$. −1 0 1 2 3

Quick Review sections give students not only the main concepts from the chapter (referenced back to the appropriate section), but also an adjacent example of each concept. (page 154)

Review Exercises are keyed to the appropriate sections so that students can refer to examples of that type of problem if they need help. (page 157)

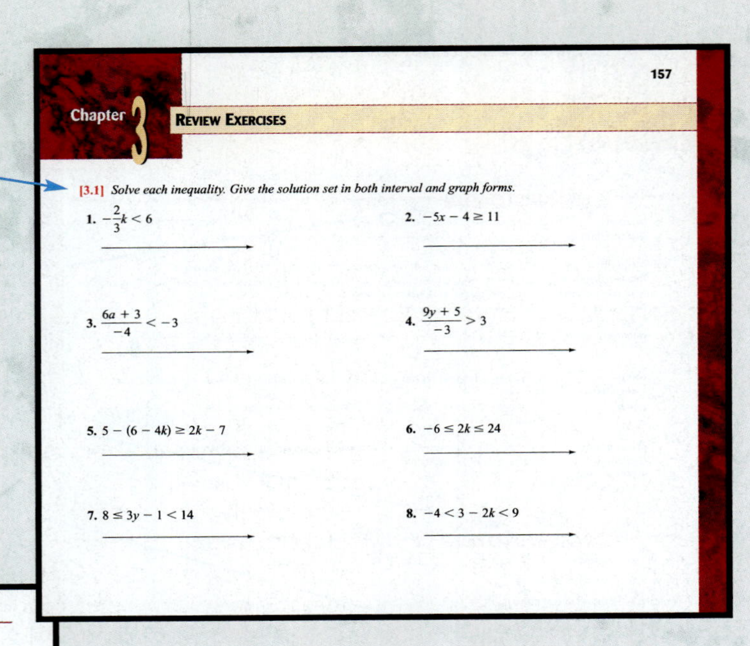

157

Chapter 3 REVIEW EXERCISES

[3.1] *Solve each inequality. Give the solution set in both interval and graph forms.*

1. $-\frac{2}{3}k < 6$
2. $-5x - 4 \ge 11$
3. $\frac{6a + 3}{-4} < -3$
4. $\frac{9y + 5}{-3} > 3$
5. $5 - (6 - 4k) \ge 2k - 7$
6. $-6 \le 2k \le 24$
7. $8 \le 3y - 1 < 14$
8. $-4 < 3 - 2k < 9$

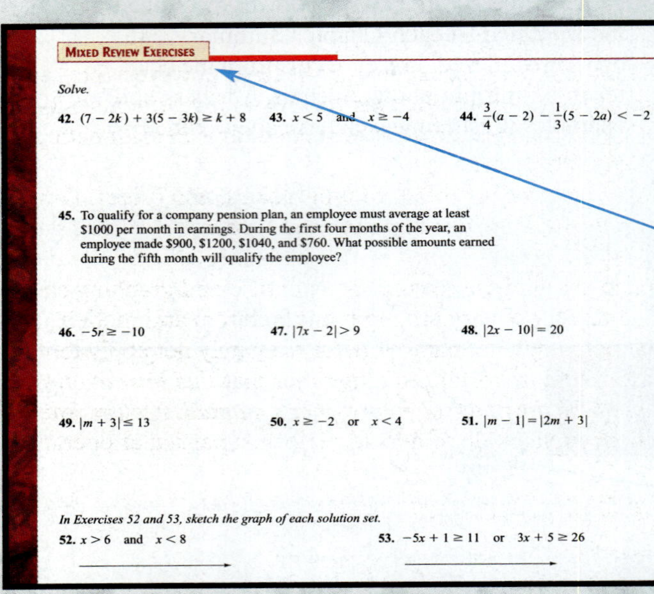

MIXED REVIEW EXERCISES

Solve.

42. $(7 - 2k) + 3(5 - 3k) \ge k + 8$
43. $x < 5$ and $x \ge -4$
44. $\frac{3}{4}(a - 2) - \frac{1}{3}(5 - 2a) < -2$
45. To qualify for a company pension plan, an employee must average at least \$1000 per month in earnings. During the first four months of the year, an employee made \$900, \$1200, \$1040, and \$760. What possible amounts earned during the fifth month will qualify the employee?
46. $-5r \ge -10$
47. $|7x - 2| > 9$
48. $|2x - 10| = 20$
49. $|m + 3| \le 13$
50. $x \ge -2$ or $x < 4$
51. $|m - 1| = |2m + 3|$

In Exercises 52 and 53, sketch the graph of each solution set.

52. $x > 6$ and $x < 8$
53. $-5x + 1 \ge 11$ or $3x + 5 \ge 26$

Mixed Review Exercises require students to solve problems without the help of section references. (page 160)

Cumulative Review Exercises gather various types of exercises from preceding chapters to help students remember and retain what they are learning throughout the course. (page 163)

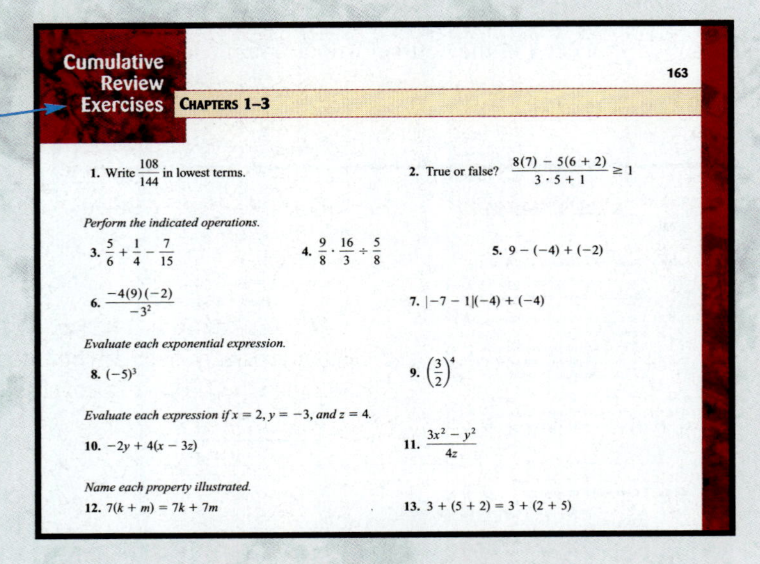

163

Cumulative Review Exercises CHAPTERS 1–3

1. Write $\frac{108}{144}$ in lowest terms.
2. True or false? $\frac{8(7) - 5(6 + 2)}{3 \cdot 5 + 1} \ge 1$

Perform the indicated operations.

3. $\frac{5}{6} + \frac{1}{4} - \frac{7}{15}$
4. $\frac{9}{8} \cdot \frac{16}{3} \div \frac{5}{8}$
5. $9 - (-4) + (-2)$
6. $\frac{-4(9)(-2)}{-3^2}$
7. $|-7 - 1|(-4) + (-4)$

Evaluate each exponential expression.

8. $(-5)^3$
9. $\left(\frac{3}{2}\right)^4$

Evaluate each expression if $x = 2$, $y = -3$, and $z = 4$.

10. $-2y + 4(x - 3z)$
11. $\frac{3x^2 - y^2}{4z}$

Name each property illustrated.

12. $7(k + m) = 7k + 7m$
13. $3 + (5 + 2) = 3 + (2 + 5)$

An Introduction to Calculators

There is little doubt that the appearance of handheld calculators three decades ago and the later development of scientific and graphing calculators have changed the methods of learning and studying mathematics forever. For example, computations with tables of logarithms and slide rules made up an important part of mathematics courses prior to 1970. Today, with the widespread availability of calculators, these topics are studied only for their historical significance.

Most consumer models of calculators are inexpensive. At first, however, they were costly. One of the first consumer models available was the Texas Instruments SR-10, which sold for about $150 in 1973. It could perform the four operations of arithmetic and take square roots, but could do very little more.

Today, calculators come in a large array of different types, sizes, and prices. *For the course for which this textbook is intended, the most appropriate type is the scientific calculator*, which costs $10–$20.

In this introduction, we explain some of the features of scientific and graphing calculators. However, remember that calculators vary among manufacturers and models, and that while the methods explained here apply to many of them, they may not apply to your specific calculator. For this reason, it is important to remember that *this introduction is only a guide and is not intended to take the place of your owner's manual.* Always refer to the manual in the event you need an explanation of how to perform a particular operation.

Scientific Calculators

Scientific calculators are capable of much more than the typical four-function calculator that you might use for balancing your checkbook. Most scientific calculators use *algebraic logic*. (Models sold by Texas Instruments, Sharp, Casio, and Radio Shack, for example, use algebraic logic.) A notable exception is Hewlett-Packard, a company whose calculators use *Reverse Polish Notation* (RPN). In this introduction, we explain the use of calculators with algebraic logic.

Arithmetic Operations To perform an operation of arithmetic, simply enter the first number, press the operation key (+, −, ×, or ÷), enter the second number, and then press the = key. For example, to add 4 and 3, use the following keystrokes.

Change Sign Key The key marked +/− allows you to change the sign of a display. This is particularly useful when you wish to enter a negative number. For example, to enter −3, use the following keystrokes.

Memory Key Scientific calculators can hold a number in memory for later use. The label of the memory key varies among models; two of these are [M] and [STO]. The [M+] and [M−] keys allow you to add to or subtract from the value currently in memory. The memory recall key, labeled [MR], [RM], or [RCL], allows you to retrieve the value stored in memory.

Suppose that you wish to store the number 5 in memory. Enter 5, then press the key for memory. You can then perform other calculations. When you need to retrieve the 5, press the key for memory recall.

If a calculator has a constant memory feature, the value in memory will be retained even after the power is turned off. Some advanced calculators have more than one memory. It is best to read the owner's manual for your model to see exactly how memory is activated.

Clearing/Clear Entry Keys These keys allow you to clear the display or clear the last entry entered into the display. They are usually marked [C] and [CE]. In some models, pressing the [C] key once will clear the last entry, while pressing it twice will clear the entire operation in progress.

Second Function Key This key is used in conjunction with another key to activate a function that is printed *above* an operation key (and not on the key itself). It is usually marked [2nd]. For example, suppose you wish to find the square of a number, and the squaring function (explained in more detail later) is printed above another key. You would need to press [2nd] before the desired squaring function can be activated.

Square Root Key Pressing the square root key, [$\sqrt{x}$], will give the square root (or an approximation of the square root) of the number in the display. For example, to find the square root of 36, use the following keystrokes.

The square root of 2 is an example of an irrational number (Chapter 8). The calculator will give an approximation of its value, since the decimal for $\sqrt{2}$ never terminates and never repeats. The number of digits shown will vary among models. To find an approximation of $\sqrt{2}$, use the following keystrokes.

Squaring Key The [x^2] key allows you to square the entry in the display. For example, to square 35.7, use the following keystrokes.

The squaring key and the square root key are often found on the same key, with one of them being a second function (that is, activated by the second function key previously described).

Reciprocal Key The key marked [1/x] is the reciprocal key. (When two numbers have a product of 1, they are called *reciprocals*. See Chapter 1.) Suppose that you wish to find the reciprocal of 5. Use the following keystrokes.

Inverse Key Some calculators have an inverse key, marked [INV]. Inverse operations are operations that "undo" each other. For example, the operations of squaring and taking the square root are inverse operations. The use of the [INV] key varies among different models of calculators, so read your owner's manual carefully.

Exponential Key The key marked x^y or y^x allows you to raise a number to a power. For example, if you wish to raise 4 to the fifth power (that is, find 4^5, as explained in Chapter 1), use the following keystrokes.

Root Key Some calculators have this key specifically marked $\sqrt[y]{x}$ or $\sqrt[x]{y}$; with others, the operation of taking roots is accomplished by using the inverse key in conjunction with the exponential key. Suppose, for example, your calculator is of the latter type and you wish to find the fifth root of 1024. Use the following keystrokes.

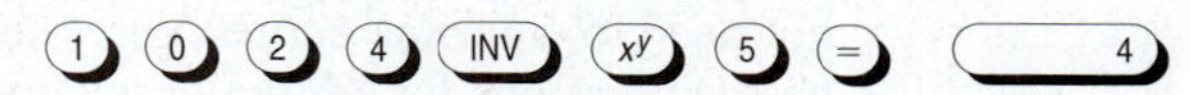

Notice how this "undoes" the operation explained in the exponential key discussion.

Pi Key The number π is an important number in mathematics. It occurs, for example, in the area and circumference formulas for a circle. By pressing the π key, you can display the first few digits of π. (Because π is irrational, the display shows only an approximation.) One popular model gives the following display when the π key is pressed.

3.1415927 An approximation for π

Methods of Display When decimal approximations are shown on scientific calculators, they are either *truncated* or *rounded*. To see how a particular model is programmed, evaluate 1/18 as an example. If the display shows .0555555 (last digit 5), it truncates the display. If it shows .0555556 (last digit 6), it rounds the display.

When very large or very small numbers are obtained as answers, scientific calculators often express these numbers in scientific notation (Chapter 6). For example, if you multiply 6,265,804 by 8,980,591, the display might look like this:

The 13 at the far right means that the number on the left is multiplied by 10^{13}. This means that the decimal point must be moved 13 places to the right if the answer is to be expressed in its usual form. Even then, the value obtained will only be an approximation: 56,270,623,000,000.

Graphing Calculators

Graphing calculators are becoming increasingly popular in mathematics classrooms. While you are not expected to have a graphing calculator to study from this book, we include the following as background information and reference should your course or future courses require the use of graphing calculators.

Basic Features

Graphing calculators provide many features beyond those found on scientific calculators. In addition to the typical keys found on scientific calculators, they have keys that can be used to create graphs, make tables, analyze data, and change settings. One of the major differences between graphing and scientific calculators is that a graphing calculator has a larger viewing screen with graphing capabilities. The screens below illustrate the graphs of $y = x$ and $y = x^2$.

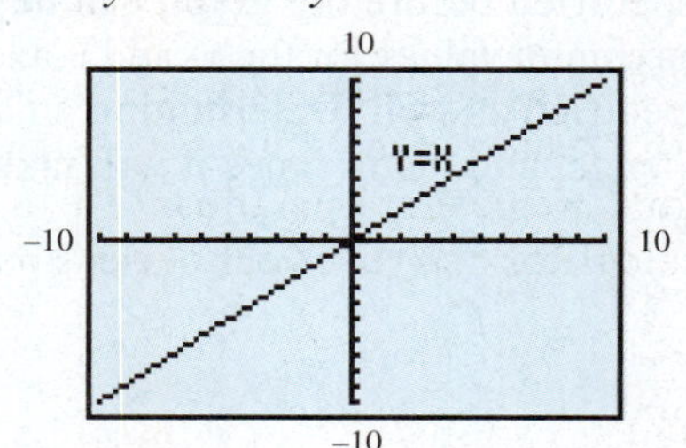

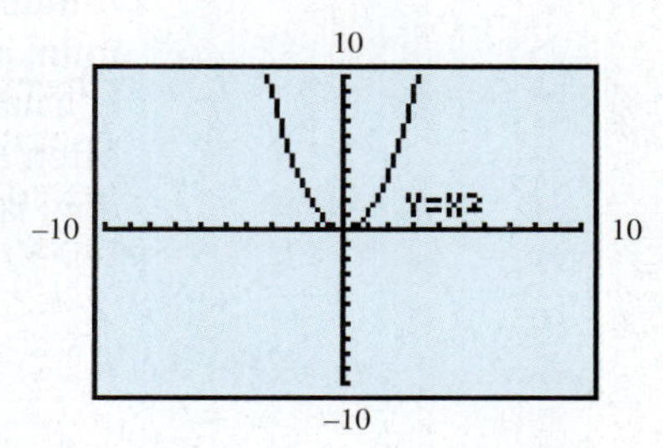

If you look closely at the screens, you will see that the graphs appear to be jagged rather than smooth, as they should be. The reason for this is that graphing calculators have much lower resolution than computer screens. Because of this, graphs generated by graphing calculators must be interpreted carefully.

Editing Input

The screen of a graphing calculator can display several lines of text at a time. This feature allows you to view both previous and current expressions. If an incorrect expression is entered, an error message is displayed. The erroneous expression can be viewed and corrected by using various editing keys, much like a word-processing program. You do not need to enter the entire expression again. Many graphing calculators can also recall past expressions for editing or updating. The screen on the left below shows how two expressions are evaluated. The final line is entered incorrectly, and the resulting error message is shown in the screen on the right.

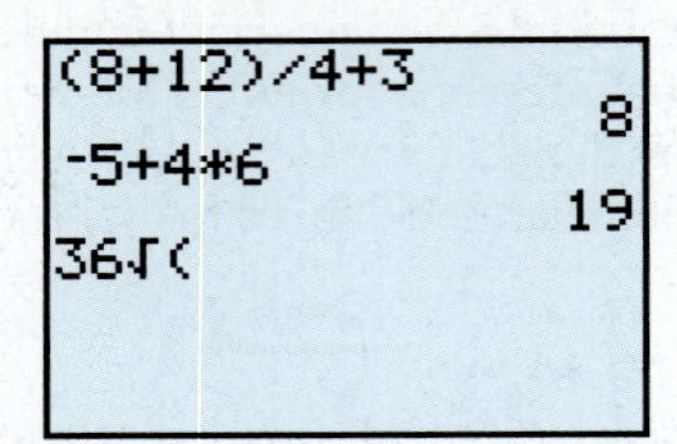

Order of Operations

Arithmetic operations on graphing calculators are usually entered as they are written in mathematical expressions. For example, to evaluate $\sqrt{36}$ on a typical scientific calculator, you would first enter 36 and then press the square root key. As seen above, this is not the correct syntax for a graphing calculator. To find this root, you would first press the square root key, and then enter 36. See the screen on the left below. The order of operations on a graphing calculator is also important, and current models assist the user by inserting parentheses when typical errors might occur. The open parenthesis that follows the square root symbol is automatically entered by the calculator so that an expression such as $\sqrt{2 \times 8}$ will not be calculated incorrectly as $\sqrt{2} \times 8$. Compare the two entries and their results in the screen on the right.

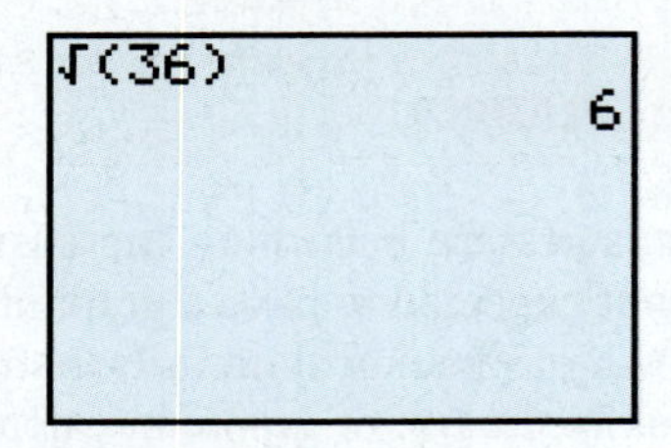

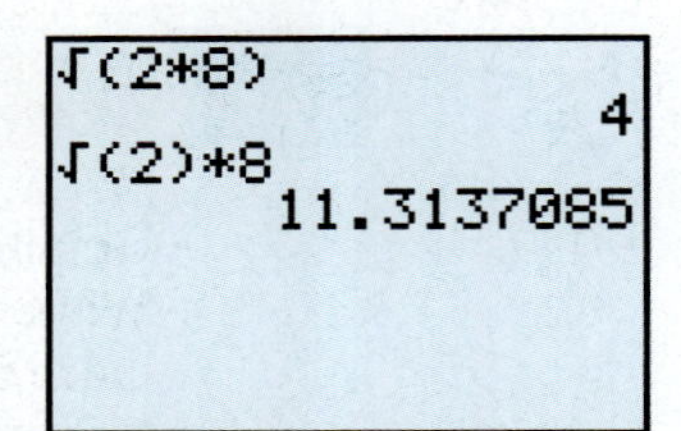

Viewing Windows

The viewing window for a graphing calculator is similar to the viewfinder in a camera. A camera usually cannot take a photograph of an entire view of a scene. The camera must be centered on some object and can capture only a portion of the available scenery. A camera with a zoom lens can photograph different views of the same scene by zooming in and out. Graphing calculators have similar capabilities. The *xy*-coordinate plane is infinite. The calculator screen can only show a finite, rectangular region in the plane, and it must be specified before the graph can be drawn. This is done by setting both minimum and maximum values for the *x*- and *y*-axes. The scale (distance between tick marks) is usually specified as well. Determining an appropriate viewing window for a graph is often a challenge, and many times it will take a few attempts before a satisfactory window is found.

The screen on the left shows a standard viewing window, and the graph of $y = 2x + 1$ is shown on the right. Using a different window would give a different view of the line.

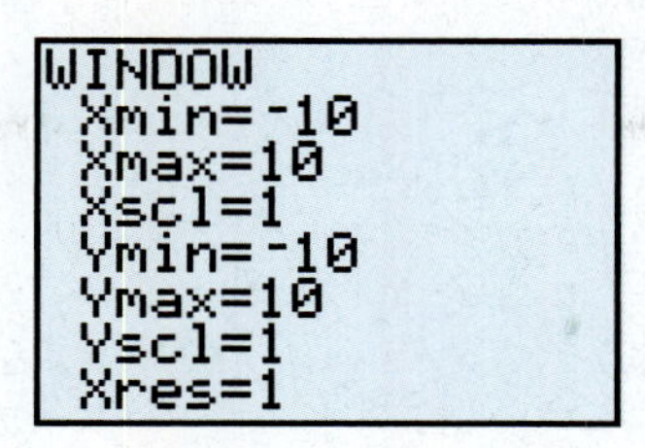

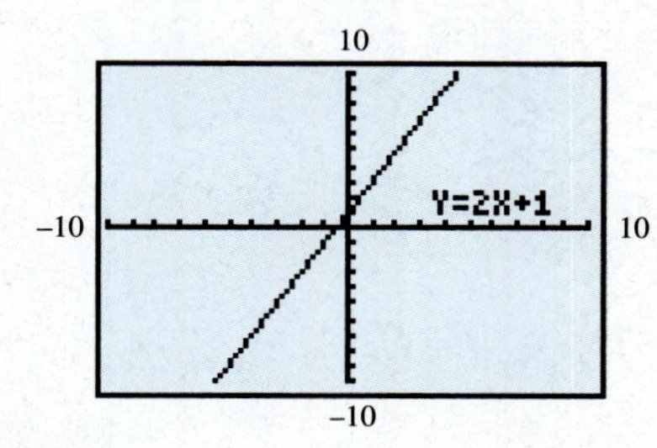

Locating Points on a Graph: Tracing and Tables

Graphing calculators allow you to trace along the graph of an equation and display the coordinates of points on the graph. See the screen on the left below, which indicates that the point (2, 5) lies on the graph of $y = 2x + 1$. Tables for equations can also be displayed. The screen on the right shows a partial table for this same equation. Note the middle of the screen, which indicates that when $x = 2, y = 5$.

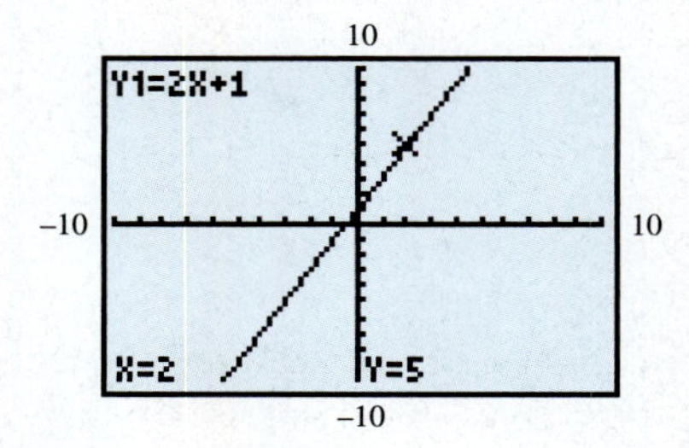

X	Y1
-1	-1
0	1
1	3
2	5
3	7
4	9
5	11

Y1=2X+1

Additional Features

There are many features of graphing calculators that go far beyond the scope of this book. These calculators can be programmed, much like computers. Many of them can solve equations at the stroke of a key, analyze statistical data, and perform symbolic algebraic manipulations. Mathematicians from the past would have been amazed by today's calculators. Many important equations in mathematics cannot be solved by hand. However, their solutions can often be approximated using a calculator. Calculators also provide the opportunity to ask "What if . . . ?" more easily. Values in algebraic expressions can be altered and conjectures tested quickly.

Final Comments

Despite the power of today's calculators, they cannot replace human thought. ***In the entire problem-solving process, your brain is the most important component.*** Calculators are only tools and, like any tool, they must be used appropriately in order to enhance our ability to understand mathematics. Mathematical insight may often be the quickest and easiest way to solve a problem; a calculator may neither be needed nor appropriate. By applying mathematical concepts, you can make the decision whether or not to use a calculator.

To the Student: Success in Algebra

There are two main reasons students have difficulty with mathematics:

- Students start in a course for which they do not have the necessary background knowledge.
- Students don't know how to study mathematics effectively.

Your instructor can help you decide whether this is the right course for you. We can give you some study tips.

Studying mathematics *is* different from studying subjects like English and history. The key to success is regular practice. This should not be surprising. After all, can you learn to play the piano or ski well without a lot of regular practice? The same is true for learning mathematics. Working problems nearly every day is the key to becoming successful. Here is a list of things that will help you succeed in studying algebra.

1. *Attend class regularly.* Pay attention to what your instructor says and does in class, and take careful notes. In particular, note the problems the instructor works on the board and copy the complete solutions. Keep these notes separate from your homework to avoid confusion when you review them later.
2. Don't hesitate to *ask questions in class.* It is not a sign of weakness but of strength. There are always other students with the same question who are too shy to ask.
3. *Read your text carefully.* Many students read only enough to get by, usually only the examples. Reading the complete section will help you solve the homework problems. Most exercises are keyed to specific examples or objectives that will explain the procedures for working them.
4. Before you start on your homework assignment, *rework the problems the teacher worked in class.* This will reinforce what you have learned. Many students say, "I understand it perfectly when you do it, but I get stuck when I try to work the problem myself."
5. Do your homework assignment only *after reading the text* and reviewing your notes from class. Check your work against the answers in the back of the book. If you get a problem wrong and are unable to understand why, mark that problem and ask your instructor about it. Then practice working additional problems of the same type to reinforce what you have learned.
6. *Work as neatly as you can.* Write your symbols clearly, and make sure the problems are clearly separated from each other. Working neatly will help you to think clearly and also make it easier to review the homework before a test.
7. After you complete a homework assignment, *look over the text again.* Try to identify the main ideas that are in the lesson. Often they are clearly highlighted or boxed in the text.
8. *Use the chapter test at the end of each chapter as a practice test.* Work through the problems under test conditions, without referring to the text or the answers until you are finished. You may want to time yourself to see how long it takes you. When you finish, check your answers against those in the back of the book, and study the problems you missed.

9. *Keep all quizzes and tests that are returned to you,* and use them when you study for future tests and the final exam. These quizzes and tests indicate what concepts your instructor considers to be most important. Be sure to correct any problems on these tests that you missed, so you will have the corrected work to study.

10. *Don't worry if you do not understand a new topic right away.* As you read more about it and work through the problems, you will gain understanding. Each time you review a topic you will understand it a little better. Few people understand each topic completely right from the start.

Reading a list of study tips is a good start, but you may need some help actually *applying* the tips to your work in this math course.

Study Skills Watch for this icon as you work in this textbook, particularly in the first few chapters. It will direct you to one of 12 activities in the *Study Skills Workbook* that comes with this text. Each activity helps you to actually *use* a study skills technique. These techniques will greatly improve your chances for success in this course.

- Find out *how your brain learns new material*. Then use that information to set up effective ways to learn math.
- Find out *why short-term memory is so short* and what you can do to help your brain remember new material weeks and months later.
- Find out *what happens when you "blank out" on a test* and simple ways to prevent it from happening.

All the activities in the *Study Skills Workbook* are practical ways to enjoy and succeed at math. Whether you need help with note taking, managing homework, taking tests, or preparing for a final exam, you'll find specific, clearly explained ideas that really work because they're based on research about how the brain learns and remembers.

Diagnostic Pretest

Study Skills Workbook
Activity 1

[Chapter 1]

1. Perform the indicated operations.

$$\frac{5 \cdot (-4)^2 - \sqrt{121}}{-4[-8 - (-3 - 9)] + (-7)}$$

1. ________

2. The table below shows the heights of some selected mountains and the depths of some selected trenches.

2. ________

Mountain	Height (in feet)	Trench	Depth (in feet)
Kennedy	16,239	Tonga	−35,433
Hood	11,239	Palau	−26,424
Washington	6,288	Verna	−21,004

Source: World Almanac and Book of Facts, 2000.

What is the difference between the height of Mt. Hood and the depth of the Verna Trench?

3. Evaluate $\frac{6m + 3n^3}{p + 4}$ if $m = 5, n = -1$, and $p = -7$.

3. ________

[Chapter 2]

4. Solve $.09x + .11(x + 8) = -.12$.

4. ________

5. Two trains leave from the same point at the same time, traveling in opposite directions. One travels 13 mph faster than the other. After 4 hr, they are 484 mi apart. Find the rate of each train.

5. ________

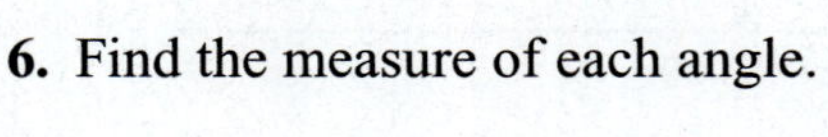

6. Find the measure of each angle.

6. ________

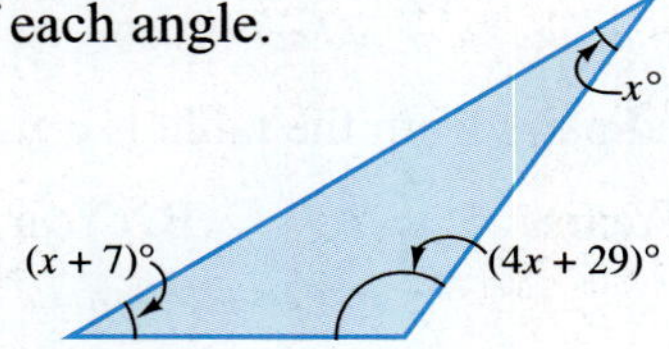

[Chapter 3]

Solve each inequality. Give the solution set in both interval and graph forms.

7. $3x - 2(x - 7) \geq 4(3x + 2) + x$

7.

8. $-8t < 32$ and $3t - 5 \leq 10$

8.

9. Solve $|4 - 5y| = |3y + 9|$.

9. ________

[Chapter 4]

10. ______________________

10. Find the slope of the line through $(5, -12)$ and $(-9, 10)$.

11. x-intercept: ______________

y-intercept: ______________

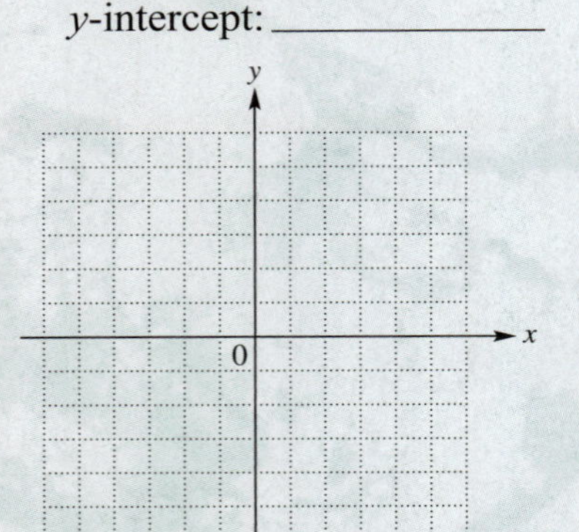

11. Find the x- and y-intercepts and graph the equation $3x - 5y = -15$.

12.

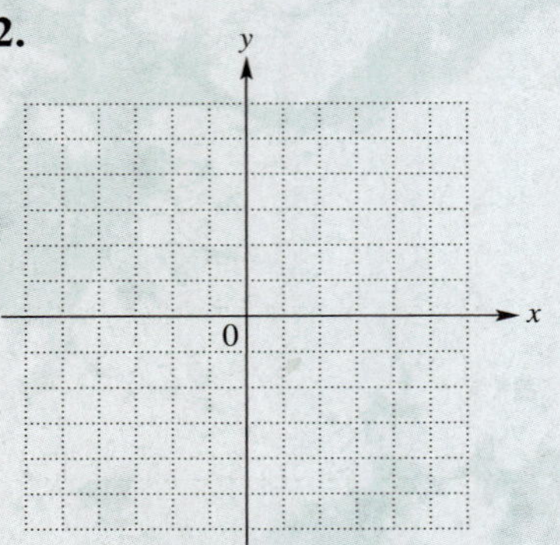

12. Graph $4x + 3y < -12$.

13. ______________________

13. The median ages for first marriage for women in the United States from 1994–1998 are shown in the table.

Marriage Age	24.5	24.5	24.8	25.0	25.0
Year	1994	1995	1996	1997	1998

Source: World Almanac and Book of Facts, 2000.

Which set of ordered pairs from the table is a function? Explain.

A. (Marriage Age, Year) **B.** (Year, Marriage Age)

[Chapter 5]

Solve each system.

14. ______________________

14. $-2x + 5y = 11$
$y = x + 7$

15. ______________________

15. $2x + 3y = 1$
$3x - 4y = 27$

16. ______________________

16. $3x + y - z = -2$
$x - 2y + z = 8$
$-2x + 3y + 3z = 1$

17. A party mix is made by combining nuts that sell for \$3.50 per lb with raisins that sell for \$1.50 per lb. How much of each should be used to get 32 lb of a mix that will sell for \$2.75 per lb?

17. ____________

[Chapter 6]

18. Simplify $(7s^0t^{-5})^2(t^{-3}s^2)^{-4}$ and write your answer with only positive exponents.

18. ____________

19. Divide $(2x^3 + 5x^2 - 13x + 7) \div (2x - 1)$.

19. ____________

20. Factor $y^3 + 3y^2 - 4y - 12$.

20. ____________

21. Solve $4x^2 + 17x - 15 = 0$.

21. ____________

[Chapter 7]

22. Multiply $\dfrac{z^2 - 16}{z^2 - 4z - 5} \cdot \dfrac{z^2 - 10z + 25}{z^2 - 9z + 20}$.

22. ____________

23. Subtract $\dfrac{5x}{x - 3} - \dfrac{4}{x + 3}$.

23. ____________

24. Simplify the complex fraction.

$$\frac{16 - \dfrac{1}{x^2}}{\dfrac{4}{x} - \dfrac{1}{x^2}}$$

24. ____________

[Chapter 8]

Simplify each expression. Assume that all variables represent positive real numbers.

25. $(-64m^{15}n^{-9})^{2/3}$

25. ____________

26. $\sqrt[3]{250y^7z^{11}}$

26. ____________

27. Solve $\sqrt{2x - 5} + 4 = x$.

27. ____________

28. Multiply $(8 - 5i)(8 + 5i)$.

28. ____________

[Chapter 9]

Solve each equation.

29. $(3x + 8)^2 = 49$

29. ____________

30. $3y^2 + 5y - 1 = 0$

30. ____________

31. Two cars left the same intersection at the same time, one heading due east and the other heading due south. Some time later, they were exactly 75 mi apart. The car headed east had gone 15 mi farther than the car headed south. How far had each car traveled?

31. ____________

32. vertex: ____________
domain: ____________
range: ____________

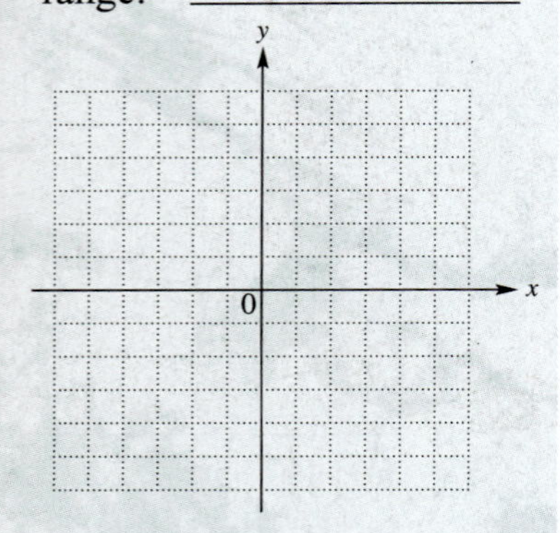

32. Graph $f(x) = -x^2 + 6x - 4$. Give the vertex, domain, and range.

[Chapter 10]

33. ____________

33. Find $f^{-1}(x)$ for the one-to-one function with $f(x) = x^3 - 8$.

Solve.

34. ____________

34. $5^{3x+2} = 25^{2x}$

35. ____________

35. $\log_9 x = \dfrac{3}{2}$

36. ____________

36. $\log_2 x + \log_2(x - 6) = 4$

[Chapter 11]

37. (a) ____________
(b) ____________
(c) ____________
(d) ____________

37. Let $f(x) = x^2 + 3$ and $g(x) = 2x - 1$. Find each of the following.

(a) $(f \circ g)(2)$ **(b)** $(g \circ f)(2)$
(c) $(f \circ g)(x)$ **(d)** $(g \circ f)(x)$

Graph.

38.

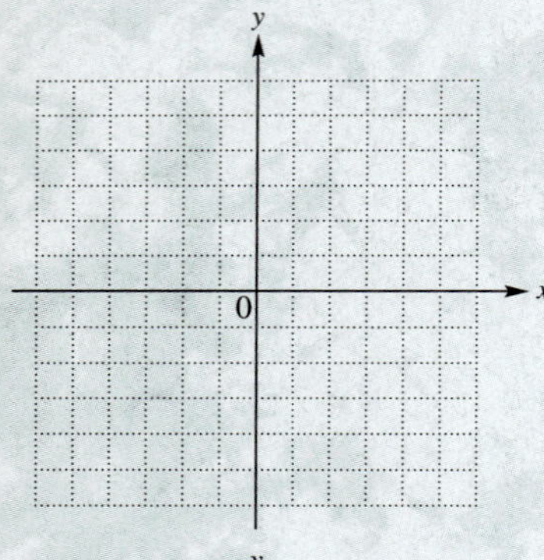

38. $25x^2 + 4y^2 = 100$

39.

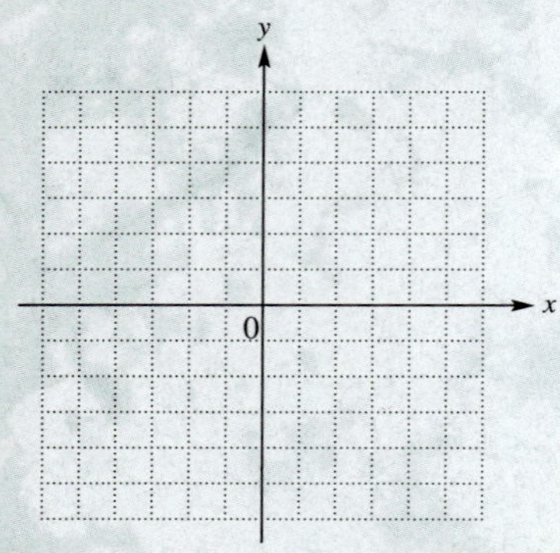

39. $4x^2 - 9y^2 = 36$

40. ____________

40. Solve the system.

$$\begin{aligned} y &= x^2 - 4 \\ 2x - y &= 1 \end{aligned}$$

Review of the Real Number System

Social Security is the largest source of income for elderly Americans. It is projected that in about 30 years, however, there will be twice as many older Americans as there are today, and the excess revenues now accumulating in Social Security's trust funds will be exhausted. (*Source:* Social Security Administration.) To supplement their retirement incomes, more and more Americans have begun investing in mutual funds, pension plans, and other means of savings. In Section 1.3, we relate the concepts of this chapter to the percent of U.S. households investing in mutual funds.

1.1 Basic Concepts

Objectives

1. Write sets using set notation.
2. Use number lines.
3. Know the common sets of numbers.
4. Find additive inverses.
5. Use absolute value.
6. Use inequality symbols.

Study Skills Workbook **Activity 2**

1 Consider the set

$$\left\{0, 10, \frac{3}{10}, 52, 98.6\right\}.$$

(a) Which elements of the set are natural numbers?

(b) Which elements of the set are whole numbers?

2 List the elements in each set.

(a) $\{x \mid x$ is a whole number less than $5\}$

(b) $\{y \mid y$ is a whole number greater than $12\}$

Answers
1. **(a)** 10 and 52 **(b)** 0, 10, and 52
2. **(a)** $\{0, 1, 2, 3, 4\}$ **(b)** $\{13, 14, 15, \ldots\}$

In this chapter we review some of the basic symbols and rules of algebra.

1 Write sets using set notation. A **set** is a collection of objects called the **elements** or **members** of the set. In algebra, the elements of a set are usually numbers. Set braces, $\{\ \}$, are used to enclose the elements. For example, 2 is an element of the set $\{1, 2, 3\}$. Since we can count the number of elements in the set $\{1, 2, 3\}$, it is a *finite set.*

In our study of algebra, we refer to certain sets of numbers by name. The set

$$N = \{1, 2, 3, 4, 5, 6, \ldots\}$$

is called the **natural numbers** or the **counting numbers.** The three dots show that the list continues in the same pattern indefinitely. We cannot list all of the elements of the set of natural numbers, so it is an *infinite set.*

When 0 is included with the set of natural numbers, we have the set of **whole numbers,** written

$$W = \{0, 1, 2, 3, 4, 5, 6, \ldots\}.$$

A set containing no elements, such as the set of whole numbers less than 0, is called the **empty set,** or **null set,** usually written $\emptyset$.

CAUTION

Do not write $\{\emptyset\}$ for the empty set; $\{\emptyset\}$ is a set with one element, $\emptyset$. Use only the notation $\emptyset$ for the empty set.

Work Problem 1 at the Side.

In algebra, letters called **variables** are often used to represent numbers or to define sets of numbers. For example,

$$\{x \mid x \text{ is a natural number between 3 and 15}\}$$

(read "the set of all elements x such that x is a natural number between 3 and 15") defines the set

$$\{4, 5, 6, 7, \ldots, 14\}.$$

The notation $\{x \mid x$ is a natural number between 3 and 15$\}$ is an example of **set-builder notation.**

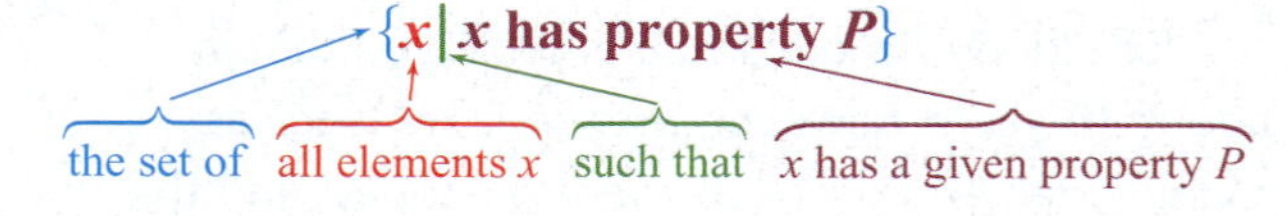

Example 1 Listing the Elements in Sets

List the elements in each set.

(a) $\{x \mid x$ is a natural number less than $4\}$
The natural numbers less than 4 are 1, 2, and 3. This set is $\{1, 2, 3\}$.

(b) $\{y \mid y$ is one of the first five even natural numbers$\} = \{2, 4, 6, 8, 10\}$

(c) $\{z \mid z$ is a natural number greater than or equal to $7\}$
The set of natural numbers greater than or equal to 7 is an infinite set, written with three dots as $\{7, 8, 9, 10, \ldots\}$.

Work Problem 2 at the Side.

Example 2 Using Set-Builder Notation to Describe Sets

Use set-builder notation to describe each set.

(a) $\{1, 3, 5, 7, 9\}$

There are often several ways to describe a set with set-builder notation. One way to describe this set is

$$\{y \mid y \text{ is one of the first five odd natural numbers}\}.$$

(b) $\{5, 10, 15, \ldots\}$

This set can be described as $\{d \mid d \text{ is a multiple of 5 greater than 0}\}$.

Work Problem 3 at the Side.

2 **Use number lines.** A good way to get a picture of a set of numbers is by using a **number line.** To construct a number line, choose any point on a horizontal line and label it 0. Next, choose a point to the right of 0 and label it 1. The distance from 0 to 1 establishes a scale that can be used to locate more points, with positive numbers to the right of 0 and negative numbers to the left of 0. The number 0 is neither positive nor negative. A number line is shown in Figure 1.

−5 −4 −3 −2 −1 0 1 2 3 4 5

Figure 1

The set of numbers identified on the number line in Figure 1, including positive and negative numbers and 0, is part of the set of **integers,** written

$$I = \{\ldots, -3, -2, -1, 0, 1, 2, 3, \ldots\}.$$

Each number on a number line is called the **coordinate** of the point that it labels, while the point is the **graph** of the number. Figure 2 shows a number line with several selected points graphed on it.

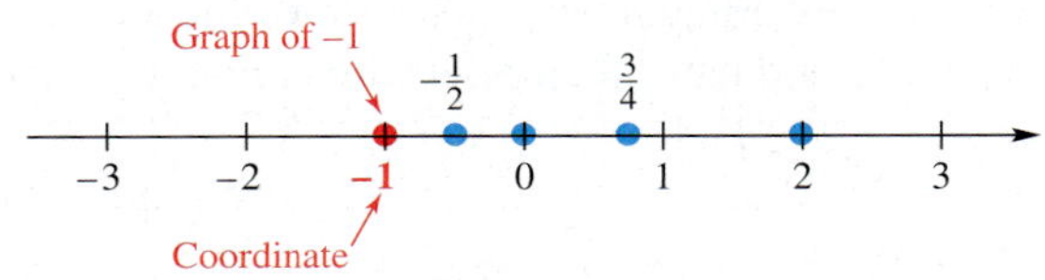

Figure 2

Work Problem 4 at the Side.

The fractions $-\frac{1}{2}$ and $\frac{3}{4}$, graphed on the number line in Figure 2, are examples of **rational numbers.** Rational numbers can be written in decimal form, either as terminating decimals such as $\frac{3}{5} = .6$, $\frac{1}{8} = .125$, or $\frac{11}{4} = 2.75$, or as repeating decimals such as $\frac{1}{3} = .33333\ldots$ or $\frac{3}{11} = .272727\ldots$. A repeating decimal is often written with a bar over the repeating digit(s). Using this notation, $.2727\ldots$ is written $.\overline{27}$.

Decimal numbers that neither terminate nor repeat are *not* rational, and thus are called **irrational numbers.** Many square roots are irrational numbers; for example, $\sqrt{2} = 1.4142136\ldots$ and $-\sqrt{7} = -2.6457513\ldots$ repeat indefinitely without pattern. (Some square roots *are* rational: $\sqrt{16} = 4$, $\sqrt{100} = 10$, and so on.) Another irrational number is π, the ratio of the circumference of a circle to its diameter.

Some of the rational and irrational numbers discussed above are graphed on the number line in Figure 3 on the next page. The rational numbers together with the irrational numbers make up the set of **real numbers.** Every point on a number line corresponds to a real number, and every real number corresponds to a point on the number line.

3 Use set-builder notation to describe each set.

(a) $\{0, 1, 2, 3, 4, 5\}$

(b) $\{7, 14, 21, 28, \ldots\}$

4 Graph the elements of each set.

(a) $\{-4, -2, 0, 2, 4, 6\}$

(b) $\left\{-1, 0, \frac{2}{3}, \frac{5}{2}\right\}$

(c) $\left\{5, \frac{16}{3}, 6, \frac{13}{2}, 7, \frac{29}{4}\right\}$

ANSWERS

3. (a) One answer is $\{x \mid x \text{ is a whole number less than } 6\}$. **(b)** One answer is $\{x \mid x \text{ is a multiple of 7 greater than } 0\}$.

4. (a) −4 −2 0 2 4 6

(b) −2 −1 0 1 2 3

(c) 4 5 6 7 8

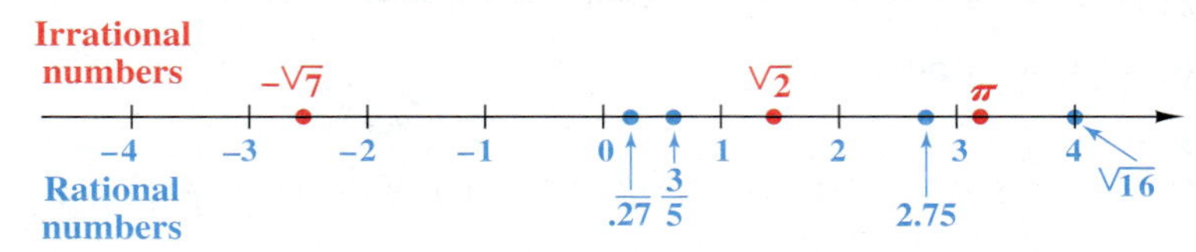

Figure 3

3 Know the common sets of numbers. The sets of numbers listed below will be used throughout the rest of this text.

Sets of Numbers

Natural numbers or counting numbers	$\{1, 2, 3, 4, 5, 6, \ldots\}$
Whole numbers	$\{0, 1, 2, 3, 4, 5, 6, \ldots\}$
Integers	$\{\ldots, -3, -2, -1, 0, 1, 2, 3, \ldots\}$
Rational numbers	$\left\{\frac{p}{q} \middle\vert p \text{ and } q \text{ are integers, } q \neq 0\right\}$ *Examples:* $\frac{4}{1}$, 1.3, $-\frac{9}{2}$, $\frac{16}{8}$ or 2, $\sqrt{9}$ or 3, $.\overline{6}$
Irrational numbers	$\{x \mid x \text{ is a real number that is not rational}\}$ *Examples:* $\sqrt{3}, -\sqrt{2}, \pi$
Real numbers	$\{x \mid x$ is represented by a point on a number line$\}$*

The relationships among these various sets of numbers are shown in Figure 4; in particular, the figure shows that the set of real numbers includes both the rational and irrational numbers. Every real number is either rational or irrational. Also, notice that the integers are elements of the set of rational numbers and that whole numbers and natural numbers are elements of the set of integers.

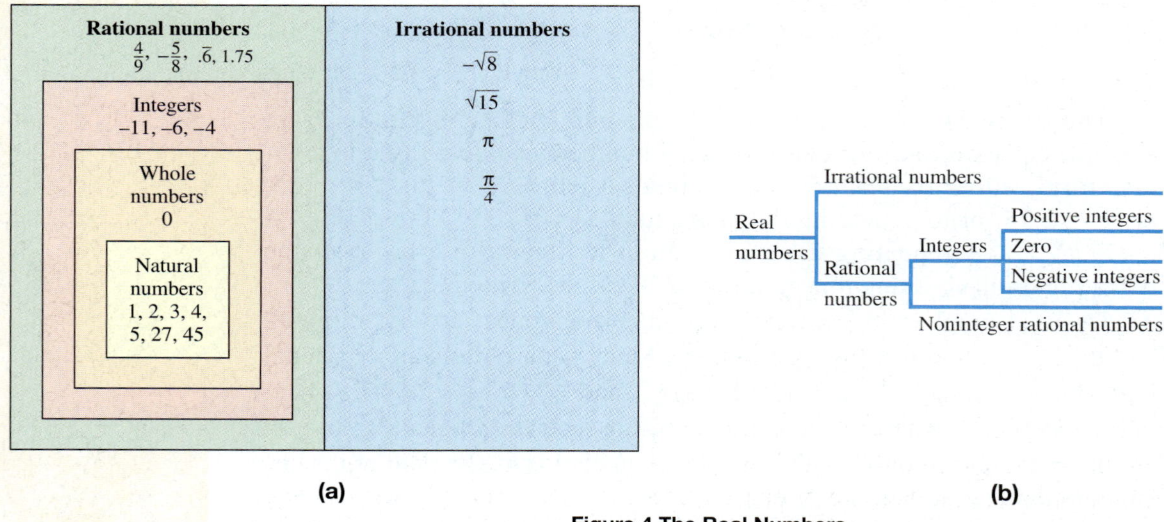

Figure 4 The Real Numbers

* An example of a number that is not a coordinate of a point on a number line is $\sqrt{-1}$. This number, called an *imaginary number,* is discussed in Chapter 8.

Example 3 **Identifying Examples of Number Sets**

Which numbers in

$$\left\{-8, -\sqrt{2}, -\frac{9}{64}, 0, .5, \frac{2}{3}, 1.\overline{12}, \sqrt{3}, 2\right\}$$

are elements of each set?

(a) Integers
-8, 0, and 2 are integers.

(b) Rational numbers
$-8, -\frac{9}{64}, 0, .5, \frac{2}{3}, 1.\overline{12}$, and 2 are rational numbers.

(c) Irrational numbers
$-\sqrt{2}$ and $\sqrt{3}$ are irrational numbers.

(d) Real numbers
All the numbers in the given set are real numbers.

Work Problem 5 at the Side.

Example 4 **Determining Relationships between Sets of Numbers**

Decide whether each statement is *true* or *false.*

(a) All irrational numbers are real numbers.
This is true. As shown in Figure 4, the set of real numbers includes all irrational numbers.

(b) Every rational number is an integer.
This statement is false. Although some rational numbers are integers, other rational numbers, such as $\frac{2}{3}$ and $-\frac{1}{4}$, are not.

Work Problem 6 at the Side.

4 **Find additive inverses.** Look again at the number line in Figure 1. For each positive number, there is a negative number on the opposite side of 0 that lies the same distance from 0. These pairs of numbers are called **additive inverses, negatives,** or **opposites** of each other. For example, 5 is the additive inverse of -5, and -5 is the additive inverse of 5.

Additive Inverse

For any real number a, the number $-a$ is the additive inverse of a.

Change the sign of a number to get its additive inverse. The sum of a number and its additive inverse is always 0.

The symbol "$-$" can be used to indicate any of the following:

1. a negative number, such as -9 or -15;
2. the additive inverse of a number, as in "-4 is the additive inverse of 4";
3. subtraction, as in $12 - 3$.

In the expression $-(-5)$, the symbol "$-$" is being used in two ways: the first $-$ indicates the additive inverse of -5, and the second indicates a negative number, -5. Since the additive inverse of -5 is 5, then $-(-5) = 5$. This example suggests the following property.

For any real number a, $\quad -(-a) = a.$

5 Select all the words from the following list that apply to each number.
Whole number
Rational number
Irrational number
Real number

(a) -6

(b) 12

(c) $.\overline{3}$

(d) $-\sqrt{15}$

(e) π

(f) $\frac{22}{7}$

(g) 3.14

6 Decide whether the statement is *true* or *false*. If *false*, tell why.

(a) All whole numbers are integers.

(b) Some integers are whole numbers.

(c) Every real number is irrational.

ANSWERS

5. **(a)** rational, real
(b) whole, rational, real
(c) rational, real
(d) irrational, real
(e) irrational, real
(f) rational, real
(g) rational, real

6. **(a)** true **(b)** true
(c) False; some real numbers are irrational, but others are rational numbers.

7 Give the additive inverse of each number.

(a) 9

(b) -12

(c) $-\frac{6}{5}$

(d) 0

(e) 1.5

Numbers written with positive or negative signs, such as $+4$, $+8$, -9, and -5, are called **signed numbers.** A positive number can be called a signed number even though the positive sign is usually left off. The following table shows the additive inverses of several signed numbers.

Number	Additive Inverse
6	-6
-4	4
$\frac{2}{3}$	$-\frac{2}{3}$
-8.7	8.7

Note that 0 is its own additive inverse.

Work Problem 7 at the Side.

5 Use absolute value. Geometrically, the **absolute value** of a number a, written $|a|$, is the distance on the number line from 0 to a. For example, the absolute value of 5 is the same as the absolute value of -5 because each number lies five units from 0. See Figure 5. That is,

$$|5| = 5 \quad \text{and} \quad |-5| = 5.$$

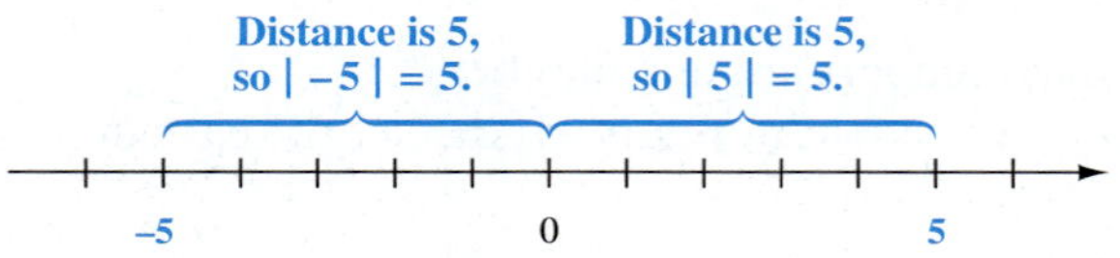

Distance is 5, so $|-5| = 5$.
Distance is 5, so $|5| = 5$.

-5 0 5

Figure 5

CAUTION

Because absolute value represents distance, and distance is always positive (or 0), ***the absolute value of a number is always positive (or 0).***

The formal definition of absolute value follows.

Absolute Value

$$|a| = \begin{cases} a & \text{if } a \text{ is positive or } 0 \\ -a & \text{if } a \text{ is negative} \end{cases}$$

The second part of this definition, $|a| = -a$ if a is negative, requires careful thought. If a is a *negative* number, then $-a$, the additive inverse or opposite of a, is a positive number, so $|a|$ is positive. For example, if $a = -3$, then

$$|a| = |-3| = -(-3) = 3. \qquad |a| = -a \text{ if } a \text{ is negative.}$$

Example 5 **Evaluating Absolute Value Expressions**

Find the value of each expression.

(a) $|13| = 13$

(b) $|-2| = -(-2) = 2$

(c) $|0| = 0$

Continued on Next Page

ANSWERS

7. (a) -9 **(b)** 12 **(c)** $\frac{6}{5}$ **(d)** 0 **(e)** -1.5

(d) $-|8|$
Evaluate the absolute value first. Then find the additive inverse.

$$-|8| = -(8) = -8$$

(e) $-|-8|$
Work as in part (d): $|-8| = 8$, so

$$-|-8| = -(8) = -8.$$

(f) $|-2| + |5|$
Evaluate each absolute value first, then add.

$$|-2| + |5| = 2 + 5 = 7$$

(g) $-|5 - 2| = -|3| = -3$

Work Problem 8 at the Side.

Absolute value is useful in applications comparing size without regard to sign.

Example 6 Comparing Rates of Change in Industries

The projected annual rates of employment change (in percent) in some of the fastest growing and most rapidly declining industries from 1994 through 2005 are shown in the table.

Industry (1994–2005)	Percent Rate of Change
Health services	5.7
Computer and data processing services	4.9
Child day care services	4.3
Footware, except rubber and plastic	−6.7
Household audio and video equipment	−4.2
Luggage, handbags, and leather products	−3.3

Source: U.S. Bureau of Labor Statistics.

What industry in the list is expected to see the greatest change? the least change?

We want the greatest *change*, without regard to whether the change is an increase or a decrease. Look for the number in the list with the largest absolute value. That number is found in footware, since $|-6.7| = 6.7$. Similarly, the least change is in the luggage, handbags, and leather products industry: $|-3.3| = 3.3$.

Work Problem 9 at the Side.

6 Use inequality symbols. The statement $4 + 2 = 6$ is an **equation;** it states that two quantities are equal. The statement $4 \neq 6$ (read "4 is not equal to 6") is an **inequality,** a statement that two quantities are *not* equal. When two numbers are not equal, one must be less than the other. The symbol $<$ means "is less than." For example,

$$8 < 9, \quad -6 < 15, \quad -6 < -1, \quad \text{and} \quad 0 < \frac{4}{3}.$$

The symbol $>$ means "is greater than." For example,

$$12 > 5, \quad 9 > -2, \quad -4 > -6, \quad \text{and} \quad \frac{6}{5} > 0.$$

Notice that in each case, the symbol "points" toward the smaller number.

8 Find the value of each expression.

(a) $|6|$

(b) $|-3|$

(c) $-|5|$

(d) $-|-2|$

(e) $-|-7|$

(f) $|-6| + |-3|$

(g) $|-9| - |-4|$

(h) $-|9 - 4|$

9 Refer to the table in Example 6. Of the household audio/video equipment industry and computer/data processing services, which will show the greater change (without regard to sign)?

ANSWERS

8. **(a)** 6 **(b)** 3 **(c)** −5 **(d)** −2 **(e)** −7 **(f)** 9 **(g)** 5 **(h)** −5

9. Computer/data processing services

10 Insert $<$ or $>$ in each blank to make a true statement.

(a) 3 ______ 7

(b) 9 ______ 2

(c) -4 ______ -8

(d) -2 ______ -1

(e) 0 ______ -3

ANSWERS
10. **(a)** $<$ **(b)** $>$ **(c)** $>$ **(d)** $<$ **(e)** $>$

The number line in Figure 6 shows the numbers 4 and 9, and we know that $4 < 9$. On the graph, 4 is to the left of 9. The smaller of two numbers is always to the left of the other on a number line.

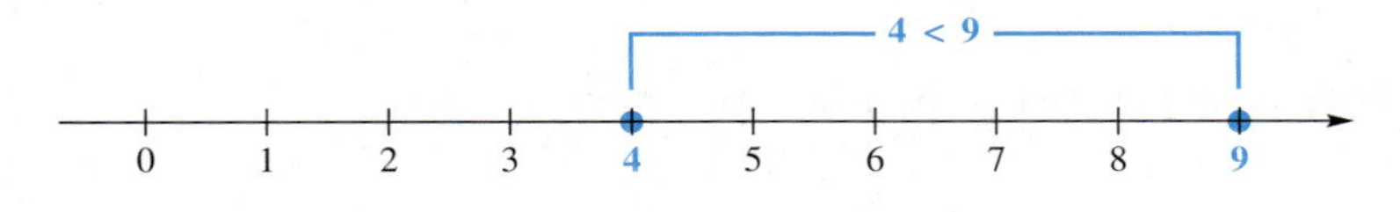

Figure 6

Inequalities on a Number Line

On a number line,

$a < b$ if a is to the left of b; $a > b$ if a is to the right of b.

We can use a number line to determine order. As shown on the number line in Figure 7, -6 is located to the left of 1. For this reason, $-6 < 1$. Also, $1 > -6$. From the same number line, $-5 < -2$, or $-2 > -5$.

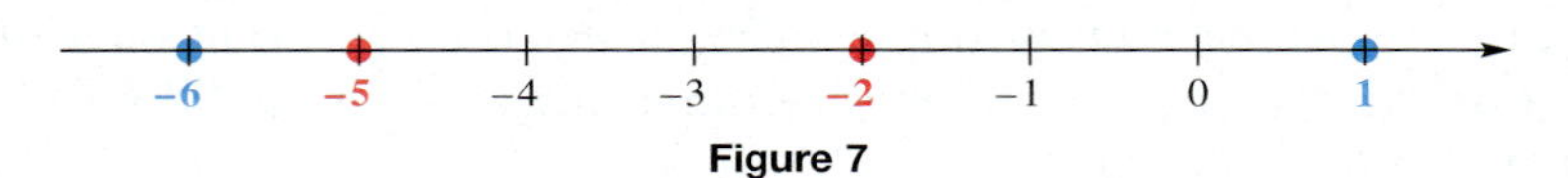

Figure 7

CAUTION

Be careful when ordering negative numbers. Since -5 is to the left of -2 on the number line in Figure 7, $-5 < -2$, or $-2 > -5$. In each case, the symbol points to -5, the smaller number.

Work Problem 10 at the Side.

The following table summarizes results about positive and negative numbers in both words and symbols.

Words	Symbols
Every negative number is less than 0.	If a is negative, then $a < 0$.
Every positive number is greater than 0.	If a is positive, then $a > 0$.
0 is neither positive nor negative.	

In addition to the symbols $\neq$, $<$, and $>$, the symbols $\leq$ and $\geq$ are often used.

INEQUALITY SYMBOLS

Symbol	Meaning	Example
$\neq$	is not equal to	$3 \neq 7$
$<$	is less than	$-4 < -1$
$>$	is greater than	$3 > -2$
$\leq$	is less than or equal to	$6 \leq 6$
$\geq$	is greater than or equal to	$-8 \geq -10$

The following table shows several inequalities and why each is true.

Inequality	Why It Is True
$6 \leq 8$	$6 < 8$
$-2 \leq -2$	$-2 = -2$
$-9 \geq -12$	$-9 > -12$
$-3 \geq -3$	$-3 = -3$
$6 \cdot 4 \leq 5(5)$	$24 < 25$

Notice the reason why $-2 \leq -2$ is true. With the symbol $\leq$, if *either* the $<$ part or the $=$ part is true, then the inequality is true. This is also the case with the $\geq$ symbol.

In the last line, recall that the dot in $6 \cdot 4$ indicates the product 6×4, or 24, and $5(5)$ means 5×5, or 25. Thus, the inequality $6 \cdot 4 \leq 5(5)$ becomes $24 \leq 25$, which is true.

Work Problem 11 at the Side.

11 Answer *true* or *false*.

(a) $-2 \leq -3$

(b) $8 \leq 8$

(c) $-9 \geq -1$

(d) $5 \cdot 8 \leq 7 \cdot 7$

(e) $3(4) > 2(6)$

ANSWERS

11. **(a)** false **(b)** true **(c)** false **(d)** true **(e)** false

Focus on **Real-Data Applications**

Stock Splits

Buying stock in a corporation makes you part owner, and you can earn **dividends** from the profits. You can track the status of your holdings by reading stock exchange reports in daily papers such as the *Wall Street Journal.* Stock prices are also reported on Internet sites such as *Yahoo! Finance* or *Morningside.com*. In fact, many reputable Internet sites have on-line "education centers" to help you learn about investing in stocks, bonds, and mutual funds.

- Your **equity,** or the value of your stock holdings, is the product of the price per share times the number of shares. For example, if you owned 25.400 shares of Washington Mutual stock, worth \$36.59 per share on May 23, 2001, then your equity to the nearest cent would be

$$\$36.59 \times 25.400 = \$929.39.$$

- The **number of shares** owned is typically given to three-decimal-place accuracy (nearest thousandth).
- The **price per share** is rounded to the nearest cent.

Occasionally, corporations announce a **stock split,** resulting in an increase in the number of shares, a decrease in the price per share, and a constant equity. For example, in a **2 for 1 split** the number of shares that you own doubles, the price per share is halved, and the equity remains the same. Proposed stock splits are published in advance.

Suppose that your 25.400 shares of Washington Mutual, priced at \$36.59, split 3 for 2 on May 23, 2001.

1. The increased number of shares becomes $\frac{3}{2} \times 25.400 = 38.100$.
2. The decreased price per share becomes $\frac{2}{3} \times \$36.59 = \$24.39\overline{3} = \$24.39$ (nearest cent).

Note: Because the price per share was rounded down, the number of shares would be adjusted slightly to maintain constant equity.

For Group Discussion

1. For the purpose of discussion only, assume that the stock prices in the table are those quoted on May 23, 2001. For each of the announced stock splits, calculate the *new* number of shares, price per share, and value of equity. To check your work, show that the equity is the same before and after the stock split.

	Before Split			After Split		
Stock	Number of Shares	Price per Share	Split Ratio	Number of Shares	Price per share	Equity
(a) Chico's FAS, Inc.	25	\$34.26	3–2	________	________	________
(b) Ivax Corp	100	\$29.50	5–4	________	________	________
(c) State Street	52.406	\$114.68	2–1	________	________	________

2. What are some reasons why a corporation institutes a stock split?

1.1 EXERCISES

FOR EXTRA HELP

Student's Solutions Manual

MyMathLab.com

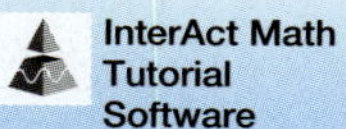
InterAct Math Tutorial Software

AW Math Tutor Center

www.mathxl.com

Digital Video Tutor CD 1 Videotape 1

Study Skills *Study Skills Workbook* **Activity 3**

Write each set by listing its elements. See Example 1.

1. $\{x \mid x$ is a natural number less than $6\}$

2. $\{m \mid m$ is a natural number less than $9\}$

3. $\{z \mid z$ is an integer greater than $4\}$

4. $\{y \mid y$ is an integer greater than $8\}$

5. $\{a \mid a$ is an even integer greater than $8\}$

6. $\{k \mid k$ is an odd integer less than $1\}$

7. $\{x \mid x$ is an irrational number that is also rational$\}$

8. $\{r \mid r$ is a number that is both positive and negative$\}$

9. $\{p \mid p$ is a number whose absolute value is $4\}$

10. $\{w \mid w$ is a number whose absolute value is $7\}$

Write each set using set-builder notation. See Example 2. (More than one description is possible.)

11. $\{2, 4, 6, 8\}$

12. $\{11, 12, 13, 14\}$

13. $\{4, 8, 12, 16, \ldots\}$

14. $\{\ldots, -6, -3, 0, 3, 6, \ldots\}$

Graph the elements of each set on a number line.

15. $\{-3, -1, 0, 4, 6\}$

16. $\{-4, -2, 0, 3, 5\}$

17. $\left\{-\frac{2}{3}, 0, \frac{4}{5}, \frac{12}{5}, \frac{9}{2}, 4.8\right\}$

18. $\left\{-\frac{6}{5}, -\frac{1}{4}, 0, \frac{5}{6}, \frac{13}{4}, 5.2, \frac{11}{2}\right\}$

*Which elements of each set are **(a)** natural numbers, **(b)** whole numbers, **(c)** integers, **(d)** rational numbers, **(e)** irrational numbers, **(f)** real numbers? See Example 3.*

19. $\left\{-8, -\sqrt{5}, -.6, 0, \frac{3}{4}, \sqrt{3}, \pi, 5, \frac{13}{2}, 17, \frac{40}{2}\right\}$

20. $\left\{-9, -\sqrt{6}, -.7, 0, \frac{6}{7}, \sqrt{7}, 4.\overline{6}, 8, \frac{21}{2}, 13, \frac{75}{5}\right\}$

Decide whether each statement is true *or* false. *If* false, *tell why. See Example 4.*

21. Every rational number is an integer.

22. Every natural number is an integer.

23. Every irrational number is an integer.

24. Every integer is a rational number.

25. Every natural number is a whole number.

26. Some rational numbers are irrational.

27. Some rational numbers are whole numbers.

28. Some real numbers are integers.

29. The absolute value of any number is the same as the absolute value of its additive inverse.

30. The absolute value of any nonzero number is positive.

Give **(a)** *the additive inverse and* **(b)** *the absolute value of each number. See the discussion of additive inverses and Example 5.*

31. 6

32. 8

33. -12

34. -15

35. $\frac{6}{5}$

36. $.13$

Find the value of each expression. See Example 5.

37. $|-8|$

38. $|-11|$

39. $\left|\frac{3}{2}\right|$

40. $\left|\frac{7}{4}\right|$

41. $-|5|$

42. $-|17|$

43. $-|-2|$

44. $-|-8|$

45. $-|4.5|$

46. $-|12.6|$

47. $|-2| + |3|$

48. $|-16| + |12|$

49. $|-9| - |-3|$

50. $|-10| - |-5|$

51. $|-1| + |-2| - |-3|$

52. $|-6| + |-4| - |-10|$

Solve each problem. See Example 6.

53. The table shows the percent change in population from 1990 through 1999 for some of the largest cities in the United States.

City	Percent Change
New York	1.4
Los Angeles	4.2
Chicago	.6
Philadelphia	−10.6
Houston	8.7
Detroit	−6.1

Source: U.S. Bureau of the Census.

(a) Which city had the greatest change in population? What was this change? Was it an increase or a decline?

(b) Which city had the smallest change in population? What was this change? Was it an increase or a decline?

54. The table gives the net trade balance, in millions of dollars, for selected U.S. trade partners for August 2000.

Country	Trade Balance (in millions of dollars)
Germany	−2721
China	−8600
Netherlands	1081
Venezuela	−1016
Panama	96
Australia	504

Source: U.S. Bureau of the Census.

A negative balance means that imports exceeded exports, while a positive balance means that exports exceeded imports.

(a) Which country had the greatest discrepancy between exports and imports? Explain.

(b) Which country had the smallest discrepancy between exports and imports? Explain.

Sea level refers to the surface of the ocean. The depth of a body of water such as an ocean or sea can be expressed as a negative number, representing average depth in feet below sea level. On the other hand, the altitude of a mountain can be expressed as a positive number, indicating its height in feet above sea level. The table gives selected depths and heights.

Body of Water	Average Depth in Feet (as a negative number)	Mountain	Altitude in Feet (as a positive number)
Pacific Ocean	−12,925	McKinley	20,320
South China Sea	−4,802	Point Success	14,158
Gulf of California	−2,375	Matlalcueyetl	14,636
Caribbean Sea	−8,448	Rainier	14,410
Indian Ocean	−12,598	Steele	16,644

Source: World Almanac and Book of Facts, 2000.

55. List the bodies of water in order, starting with the deepest and ending with the shallowest.

56. List the mountains in order, starting with the shortest and ending with the tallest.

57. *True* or *false*: The absolute value of the depth of the Pacific Ocean is greater than the absolute value of the depth of the Indian Ocean.

58. *True* or *false*: The absolute value of the depth of the Gulf of California is greater than the absolute value of the depth of the Caribbean Sea.

Use the number line to answer true *or* false *to each statement.*

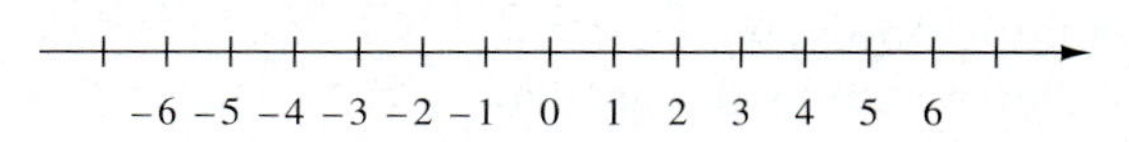

59. $-6 < -2$

60. $-4 < -3$

61. $-4 > -3$

62. $-2 > -1$

63. $3 > -2$

64. $5 > -3$

65. $-3 \geq -3$

66. $-4 \leq -4$

Use an inequality symbol to write each statement.

67. 7 is greater than y.

68. -4 is less than 12.

69. 5 is greater than or equal to 5.

70. -3 is less than or equal to -3.

71. $3t - 4$ is less than or equal to 10.

72. $5x + 4$ is greater than or equal to 19.

73. $5x + 3$ is not equal to 0.

74. $6x + 7$ is not equal to -3.

First simplify each side of the inequality. Then tell whether the resulting statement is true *or* false.

75. $-6 < 7 + 3$

76. $-7 < 4 + 2$

77. $2 \cdot 5 \geq 4 + 6$

78. $8 + 7 \leq 3 \cdot 5$

79. $-|-3| \geq -3$

80. $-|-5| \leq -5$

81. $-8 > -|-6|$

82. $-9 > -|-4|$

The graph shows egg production in millions of eggs in selected states for 1998 and 1999. Use this graph to work Exercises 83–87.

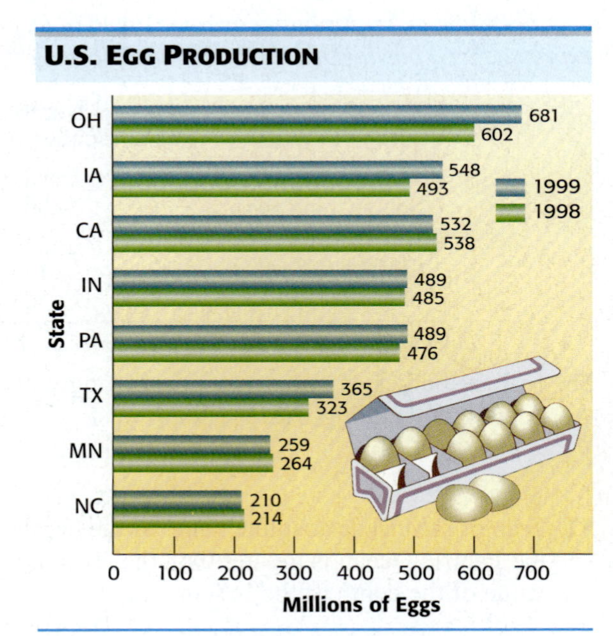

Source: Iowa Agricultural Statistics.

83. In 1998, was egg production in Iowa (IA) less than or greater than egg production in California (CA)?

84. In 1999, which states had production greater than 500 million eggs?

85. In which states was 1999 egg production less than 1998 egg production?

86. If x represents 1999 egg production for Texas (TX) and y represents 1999 egg production for Ohio (OH), which is true: $x < y$ or $x > y$?

87. If x represents 1999 egg production for Indiana (IN) and y represents 1999 egg production for Pennsylvania (PA), write an equation or inequality that compares the production in these two states.

1.2 Operations on Real Numbers

In this section we review the rules for adding, subtracting, multiplying, and dividing real numbers.

Objectives

1. Add real numbers.
2. Subtract real numbers.
3. Multiply real numbers.
4. Find the reciprocal of a number.
5. Divide real numbers.

1 Add real numbers. Number lines can be used to illustrate addition and subtraction of real numbers. To add two real numbers on a number line, start at 0. Move right (the *positive* direction) to add a positive number or left (the *negative* direction) to add a negative number. See Figure 8.

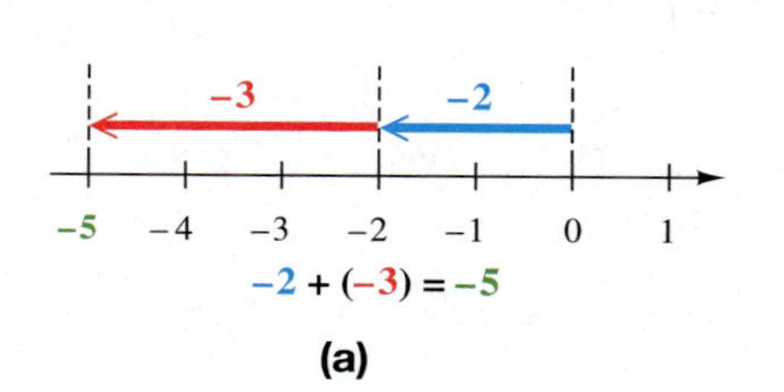

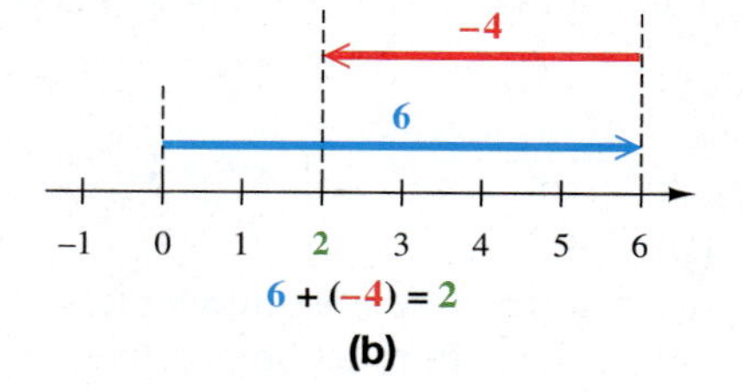

Figure 8

This procedure for adding real numbers can be generalized in the following rules.

Adding Real Numbers

Like signs To add two numbers with the *same* sign, add their absolute values. The sign of the answer (either + or −) is the same as the sign of the two numbers.

Unlike signs To add two numbers with *different* signs, subtract the smaller absolute value from the larger. The sign of the answer is the same as the sign of the number with the larger absolute value.

Recall that the answer to an addition problem is called the **sum.**

Example 1 Adding Two Negative Numbers

Find each sum.

(a) $-12 + (-8)$
First find the absolute values.

$$|-12| = 12 \qquad \text{and} \qquad |-8| = 8$$

Because −12 and −8 have the *same* sign, add their absolute values. Both numbers are negative, so the answer is negative.

$$-12 + (-8) = -(12 + 8) = -(20) = -20$$

(b) $-6 + (-3) = -(|-6| + |-3|) = -(6 + 3) = -9$

(c) $-1.2 + (-.4) = -(1.2 + .4) = -1.6$

(d) $-\frac{5}{6} + \left(-\frac{1}{3}\right) = -\left(\frac{5}{6} + \frac{1}{3}\right) = -\left(\frac{5}{6} + \frac{2}{6}\right) = -\frac{7}{6}$

Work Problem 1 at the Side.

1 Find each sum.

(a) $-2 + (-7)$

(b) $-15 + (-6)$

(c) $-1.1 + (-1.2)$

(d) $-\frac{3}{4} + \left(-\frac{1}{2}\right)$

Answers

1. (a) −9 (b) −21 (c) −2.3 (d) $-\frac{5}{4}$

2 Find each sum.

(a) $12 + (-1)$

(b) $3 + (-7)$

(c) $-17 + 5$

(d) $-\frac{3}{4} + \frac{1}{2}$

(e) $-1.5 + 3.2$

ANSWERS

2. (a) 11 **(b)** −4 **(c)** −12 **(d)** $-\frac{1}{4}$
(e) 1.7

Example 2 Adding Numbers with Different Signs

Find each sum.

(a) $-17 + 11$

First find the absolute values.

$$|-17| = 17 \quad \text{and} \quad |11| = 11$$

Because -17 and 11 have *different* signs, subtract their absolute values.

$$17 - 11 = 6$$

The number -17 has a larger absolute value than 11, so the answer is negative.

$$-17 + 11 = -6$$

Negative because $|-17| > |11|$

(b) $4 + (-1)$

Subtract the absolute values, 4 and 1. Because 4 has the larger absolute value, the sum must be positive.

$$4 + (-1) = 4 - 1 = 3$$

Positive because $|4| > |-1|$

(c) $-9 + 17 = 17 - 9 = 8$

(d) $-16 + 12$

The absolute values are 16 and 12. Subtract the absolute values. The negative number has the larger absolute value, so the answer is negative.

$$-16 + 12 = -(16 - 12) = -4$$

(e) $-\frac{4}{5} + \frac{2}{3}$

Write each number with a common denominator.

$$\frac{4}{5} = \frac{4 \cdot 3}{5 \cdot 3} = \frac{12}{15} \quad \text{and} \quad \frac{2}{3} = \frac{2 \cdot 5}{3 \cdot 5} = \frac{10}{15}$$

$$-\frac{4}{5} + \frac{2}{3} = -\frac{12}{15} + \frac{10}{15}$$

$$= -\left(\frac{12}{15} - \frac{10}{15}\right) \quad -\tfrac{12}{15} \text{ has the larger absolute value.}$$

$$= -\frac{2}{15} \quad \text{Subtract.}$$

(f) $-2.3 + 5.6 = 3.3$

Work Problem 2 at the Side.

2 **Subtract real numbers.** Recall that the answer to a subtraction problem is called the **difference.** Thus, the difference between 6 and 4 is 2. To see how subtraction should be defined, compare the following two statements.

$$6 - 4 = 2$$
$$6 + (-4) = 2$$

The second statement is pictured on the number line in Figure 8(b) at the beginning of this section. Similarly, $9 - 3 = 6$ and $9 + (-3) = 6$ so that $9 - 3 = 9 + (-3)$. These examples suggest the following rule for subtraction.

Subtraction

For all real numbers a and b,

$$a - b = a + (-b).$$

In words, change the sign of the second number and add.

Example 3 **Subtracting Real Numbers**

Find each difference.

Change to addition.
Change sign of second number.

(a) $6 - 8 = 6 + (-8) = -2$

Changed
Sign changed

(b) $-12 - 4 = -12 + (-4) = -16$

(c) $-10 - (-7) = -10 + [-(-7)]$ This step is often omitted.

$= -10 + 7$

$= -3$

(d) $-2.4 - (-8.1) = -2.4 + 8.1 = 5.7$

(e) $\frac{8}{3} - \left(-\frac{5}{3}\right) = \frac{8}{3} + \frac{5}{3} = \frac{13}{3}$

Work Problem 3 at the Side.

When working a problem that involves both addition and subtraction, add and subtract in order from left to right. Work inside brackets or parentheses first.

Example 4 **Adding and Subtracting Real Numbers**

Perform the indicated operations.

(a) $-8 + 5 - 6 = (-8 + 5) - 6$ Work from left to right.

$= -3 - 6$

$= -3 + (-6)$

$= -9$

(b) $15 - (-3) - 5 - 12 = (15 + 3) - 5 - 12$

$= 18 - 5 - 12$

$= 13 - 12$

$= 1$

(c) $-4 - (-6) + 7 - 1 = (-4 + 6) + 7 - 1$

$= 2 + 7 - 1$

$= 9 - 1$

$= 8$

Continued on Next Page

3 Find each difference.

(a) $9 - 12$

(b) $-7 - 2$

(c) $-8 - (-2)$

(d) $-6.3 - (-11.5)$

(e) $12 - (-5)$

ANSWERS

3. **(a)** -3 **(b)** -9 **(c)** -6 **(d)** 5.2 **(e)** 17

(d) $-9 - [-8 - (-4)] + 6 = -9 - [-8 + 4] + 6$
$= -9 - [-4] + 6$
$= -9 + 4 + 6$
$= -5 + 6$
$= 1$

Work Problem 4 at the Side.

3 **Multiply real numbers.** The answer to a multiplication problem is called the **product.** For example, 24 is the product of 8 and 3. The rules for finding signs of products of real numbers are given below.

Multiplying Real Numbers

Like signs The product of two numbers with the *same* sign is positive.

Unlike signs The product of two numbers with *different* signs is negative.

Example 5 Multiplying Real Numbers

Find each product.

(a) $-3(-9) = 27$ Same sign; product is positive.

(b) $-.5(-.4) = .2$

(c) $-\frac{3}{4}\left(-\frac{5}{3}\right) = \frac{5}{4}$

(d) $6(-9) = -54$ Different signs; product is negative.

(e) $-.05(.3) = -.015$

(f) $\frac{2}{3}(-3) = -2$

(g) $-\frac{5}{8}\left(\frac{12}{13}\right) = -\frac{15}{26}$

Work Problem 5 at the Side.

4 **Find the reciprocal of a number.** Earlier, subtraction was defined in terms of addition. Now, division is defined in terms of multiplication. The definition of division depends on the idea of a **multiplicative inverse** or *reciprocal*; two numbers are *reciprocals* if they have a product of 1.

Reciprocal

The **reciprocal** of a nonzero number a is $\frac{1}{a}$.

Calculator Tip Reciprocals (in decimal form) can be found with a calculator that has a key labeled $1/x$ or x^{-1}. For example, a calculator shows that the reciprocal of 25 is .04.

4 Perform the indicated operations.

(a) $-6 + 9 - 2$

(b) $12 - (-4) + 8$

(c) $-6 - (-2) - 8 - 1$

(d) $-3 - [(-7) + 15] + 6$

5 Find each product.

(a) $-7(-5)$

(b) $-.9(-15)$

(c) $-\frac{4}{7}\left(-\frac{14}{3}\right)$

(d) $7(-2)$

(e) $-.8(.006)$

(f) $\frac{5}{8}(-16)$

(g) $-\frac{2}{3}(12)$

ANSWERS

4. **(a)** 1 **(b)** 24 **(c)** -13 **(d)** -5

5. **(a)** 35 **(b)** 13.5 **(c)** $\frac{8}{3}$ **(d)** -14 **(e)** $-.0048$ **(f)** -10 **(g)** -8

The table gives several numbers and their reciprocals.

Number	Reciprocal	
$-\frac{2}{5}$	$-\frac{5}{2}$	$-\frac{2}{5}\left(-\frac{5}{2}\right) = 1$
-6	$-\frac{1}{6}$	$-6\left(-\frac{1}{6}\right) = 1$
$\frac{7}{11}$	$\frac{11}{7}$	$\frac{7}{11}\left(\frac{11}{7}\right) = 1$
.05	20	$.05(20) = 1$
0	None	

There is no reciprocal for 0 because there is no number that can be multiplied by 0 to give a product of 1.

CAUTION

A number and its additive inverse have *opposite* signs; however, a number and its reciprocal always have the *same* sign.

Work Problem 6 at the Side.

5 Divide real numbers. The result of dividing one number by another is called the **quotient.** For example, when 45 is divided by 3, the quotient is 15. To define division of real numbers, we first write the quotient of 45 and 3 as $\frac{45}{3}$, which equals 15. The same answer will be obtained if 45 and $\frac{1}{3}$ are multiplied, as follows.

$$45 \div 3 = \frac{45}{3} = 45 \cdot \frac{1}{3} = 15$$

This suggests the following definition of division of real numbers.

Division

For all real numbers a and b (where $b \neq 0$),

$$a \div b = \frac{a}{b} = a \cdot \frac{1}{b}.$$

In words, multiply the first number by the reciprocal of the second number.

There is no reciprocal for the number 0, so ***division by 0 is undefined.*** For example,

$$\frac{15}{0} \text{ is undefined} \quad \text{and} \quad -\frac{1}{0} \text{ is undefined.}$$

CAUTION

Division by 0 is undefined. However, dividing 0 by a nonzero number gives the quotient 0. For example,

$$\frac{6}{0} \text{ is undefined,} \quad \text{but} \quad \frac{0}{6} = 0 \quad (\text{since } 6 \cdot 0 = 0).$$

Be careful when 0 is involved in a division problem.

Work Problem 7 at the Side.

Since division is defined as multiplication by the reciprocal, the rules for signs of quotients are the same as those for signs of products.

6 Give the reciprocal of each number.

(a) 15

(b) -7

(c) $\frac{8}{9}$

(d) $-\frac{1}{3}$

(e) .125

7 Divide where possible.

(a) $\frac{9}{0}$

(b) $\frac{0}{9}$

(c) $\frac{-9}{0}$

(d) $\frac{0}{-9}$

Answers

6. (a) $\frac{1}{15}$ **(b)** $-\frac{1}{7}$ **(c)** $\frac{9}{8}$ **(d)** -3 **(e)** 8

7. (a) undefined **(b)** 0 **(c)** undefined **(d)** 0

8 Find each quotient.

(a) $\dfrac{-16}{4}$

(b) $\dfrac{8}{-2}$

(c) $\dfrac{-15}{-3}$

(d) $\dfrac{\frac{3}{8}}{\frac{11}{16}}$

9 Which of the following fractions are equal to $\frac{-3}{5}$?

A. $\dfrac{3}{5}$ **B.** $\dfrac{3}{-5}$

C. $-\dfrac{3}{5}$ **D.** $\dfrac{-3}{-5}$

ANSWERS

8. (a) -4 **(b)** -4 **(c)** 5 **(d)** $\frac{6}{11}$

9. B, C

Dividing Real Numbers

Like signs The quotient of two nonzero real numbers with the *same* sign is positive.

Unlike signs The quotient of two nonzero real numbers with *different* signs is negative.

Example 6 Dividing Real Numbers

Find each quotient.

(a) $\dfrac{-12}{4} = -12 \cdot \dfrac{1}{4} = -3$ $\quad \frac{a}{b} = a \cdot \frac{1}{b}$

(b) $\dfrac{6}{-3} = 6\left(-\dfrac{1}{3}\right) = -2$ $\quad$ The reciprocal of -3 is $-\frac{1}{3}$.

(c) $\dfrac{-30}{-2} = -30\left(-\dfrac{1}{2}\right) = 15$

(d) $\dfrac{\frac{2}{3}}{\frac{5}{9}}$

This is a *complex fraction,* a fraction that has a fraction in the numerator, the denominator, or both. Find the quotient as in parts (a)–(c)—multiply the first number (the numerator) by the reciprocal of the second number (the denominator).

$$\frac{\frac{2}{3}}{\frac{5}{9}} = \frac{2}{3} \cdot \frac{9}{5} = \frac{6}{5} \qquad \text{The reciprocal of } \tfrac{5}{9} \text{ is } \tfrac{9}{5}.$$

Work Problem 8 at the Side.

The rules for multiplication and division suggest the following results.

Equivalent Forms of a Fraction

The fractions $\dfrac{-x}{y}$, $-\dfrac{x}{y}$, and $\dfrac{x}{-y}$ are equal. (Assume $y \neq 0$.)

Example: $\dfrac{-4}{7}$, $-\dfrac{4}{7}$, and $\dfrac{4}{-7}$ are equal.

The fractions $\dfrac{x}{y}$ and $\dfrac{-x}{-y}$ are equal.

Example: $\dfrac{4}{7}$ and $\dfrac{-4}{-7}$ are equal.

The forms $\frac{x}{-y}$ and $\frac{-x}{-y}$ are not used very often.

Every fraction has three signs: the sign of the numerator, the sign of the denominator, and the sign of the fraction itself. Changing any two of these three signs does not change the value of the fraction. Changing only one sign, or changing all three, *does* change the value.

Work Problem 9 at the Side.

1.2 Exercises

FOR EXTRA HELP

Student's Solutions Manual

MyMathLab.com

InterAct Math Tutorial Software

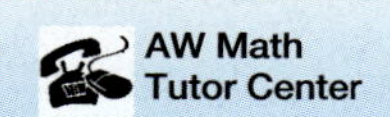
AW Math Tutor Center

www.mathxl.com

Digital Video Tutor CD 1 Videotape 1

Complete each statement and give an example.

1. The sum of a positive number and a negative number is 0 if ___________.

2. The sum of two positive numbers is a ___________ number.

3. The sum of two negative numbers is a ___________ number.

4. The sum of a positive number and a negative number is negative if ___________.

5. The sum of a positive number and a negative number is positive if ___________.

6. The difference between two positive numbers is negative if ___________.

7. The difference between two negative numbers is negative if ___________.

8. The product of two numbers with like signs is ___________.

9. The product of two numbers with unlike signs is ___________.

10. The quotient formed by any nonzero number divided by 0 is ___________, and the quotient formed by 0 divided by any nonzero number is ___________.

Add or subtract as indicated. See Examples 1–3.

11. $13 + (-4)$

12. $19 + (-13)$

13. $-6 + (-13)$

14. $-8 + (-15)$

15. $-\frac{7}{3} + \frac{3}{4}$

16. $-\frac{5}{6} + \frac{3}{8}$

17. $-2.3 + .45$

18. $-.238 + 4.55$

19. $-6 - 5$ **20.** $-8 - 13$ **21.** $8 - (-13)$ **22.** $13 - (-22)$

23. $-16 - (-3)$ **24.** $-21 - (-8)$ **25.** $-12.31 - (-2.13)$ **26.** $-15.88 - (-9.22)$

27. $\frac{9}{10} - \left(-\frac{4}{3}\right)$ **28.** $\frac{3}{14} - \left(-\frac{1}{4}\right)$ **29.** $-2 - |-4|$ **30.** $9 - |-13|$

Multiply. See Example 5.

31. $5(-7)$ **32.** $6(-6)$ **33.** $-8(-5)$ **34.** $-10(-4)$

35. $-10\left(-\frac{1}{5}\right)$ **36.** $-\frac{1}{2}(-12)$ **37.** $\frac{3}{4}(-16)$ **38.** $\frac{4}{5}(-35)$

39. $-\frac{5}{2}\left(-\frac{12}{25}\right)$ **40.** $-\frac{9}{7}\left(-\frac{35}{36}\right)$ **41.** $-\frac{3}{8}\left(-\frac{24}{9}\right)$ **42.** $-\frac{2}{11}\left(-\frac{99}{4}\right)$

43. $-2.4(-2.45)$ **44.** $-3.45(-2.14)$ **45.** $3.4(-3.14)$ **46.** $5.66(-2.1)$

Give the reciprocal of each number.

47. 6 **48.** 8 **49.** -7 **50.** -11

51. $-\frac{2}{3}$ **52.** $-\frac{7}{8}$ **53.** $\frac{1}{5}$ **54.** $\frac{1}{4}$

55. $.02$ **56.** $.45$ **57.** $-.001$ **58.** $-.0003$

Divide where possible. See Example 6.

59. $\dfrac{-14}{2}$ **60.** $\dfrac{-26}{13}$ **61.** $\dfrac{-24}{-4}$ **62.** $\dfrac{-36}{-9}$ **63.** $\dfrac{100}{-25}$

64. $\dfrac{300}{-60}$ **65.** $\dfrac{0}{-8}$ **66.** $\dfrac{0}{-10}$ **67.** $\dfrac{5}{0}$ **68.** $\dfrac{12}{0}$

69. $-\dfrac{10}{17} \div \left(-\dfrac{12}{5}\right)$ **70.** $-\dfrac{22}{23} \div \left(-\dfrac{33}{4}\right)$ **71.** $\dfrac{\frac{12}{13}}{-\frac{4}{3}}$ **72.** $\dfrac{\frac{5}{6}}{-\frac{1}{30}}$

73. $-\dfrac{27.72}{13.2}$ **74.** $\dfrac{-126.7}{36.2}$ **75.** $\dfrac{-100}{-.01}$ **76.** $\dfrac{-50}{-.05}$

Perform the indicated operations. Work inside parentheses or brackets first. Remember to add and subtract in order, working from left to right. See Example 4.

77. $-7 + 5 - 9$ **78.** $-12 + 13 - 19$ **79.** $6 - (-2) + 8$

80. $7 - (-3) + 12$ **81.** $-9 - 4 - (-3) + 6$ **82.** $-10 - 5 - (-12) + 8$

83. $-4 - [(-4 - 6) + 12] - 13$ **84.** $-10 - [(-2 + 3) - 4] - 17$

Solve each problem.

85. The highest temperature ever recorded in Juneau, Alaska, was 90°F. The lowest temperature ever recorded there was −22°F. What is the difference between these two temperatures? (*Source: World Almanac and Book of Facts*, 2000.)

86. On August 10, 1936, a temperature of 120°F was recorded in Arkansas. On February 13, 1905, Arkansas recorded a temperature of −29°F. What is the difference between these two temperatures? (*Source: World Almanac and Book of Facts*, 2000.)

87. When George W. Bush took office in January 2001, the U.S. federal budget was at a record surplus of \$236 billion. It was at a record deficit of −\$255 billion when his father, George H. W. Bush, left office in January 1993. Find the difference between these two amounts. (*Source: Economic Report of the President,* 2001.)

88. The Standard and Poor's 500, an index measuring the performance of 500 leading stocks, had an annual return of 37.58% in 1995. For 2000, its annual return was −9.10%. Find the difference between these two percents. (*Source:* Legg Mason Wood Walker, Inc.)

The table shows Social Security finances (in billions of dollars). Use this table to work Exercises 89 and 90.

Year	Tax Revenue	Cost of Benefits
2000	538	409
2010*	916	710
2020*	1479	1405
2030*	2041	2542

*Projected
Source: Social Security Board of Trustees.

89. Find the difference between Social Security tax revenue and cost of benefits for each year shown in the table.

90. Interpret your answer for 2030.

Use the graph of California exports to work Exercises 91–94.

91. What is the difference between the January and February changes?

92. What is the difference between the changes in April and May?

93. Which of the following is the best estimate of the difference between the October and November changes?

A. −6000 **B.** −6500 **C.** −7000 **D.** −7500

94. Which of the following is the best estimate of the difference between the November and December changes?

A. 17,000 **B.** 18,000 **C.** 19,000 **D.** 20,000

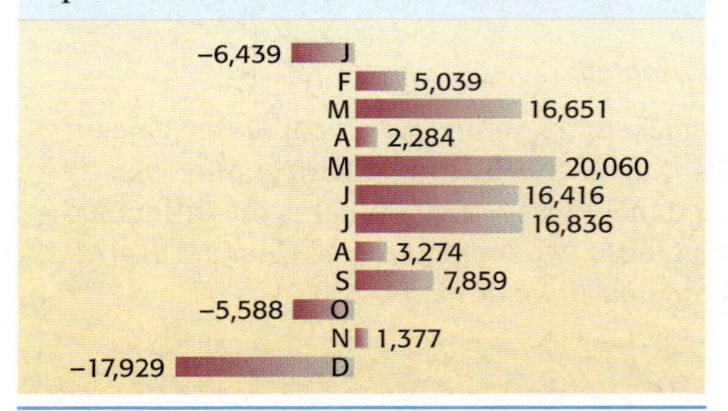

Source: USA Today, June 23, 1998.

1.3 EXPONENTS, ROOTS, AND ORDER OF OPERATIONS

OBJECTIVES

1. Use exponents.
2. Identify exponents and bases.
3. Find square roots.
4. Use the order of operations.
5. Evaluate expressions for given values of variables.

Two or more numbers whose product is a third number are **factors** of that third number. For example, 2 and 6 are factors of 12 since $2 \cdot 6 = 12$. Other factors of 12 are 1, 3, 4, 12, −1, −2, −3, −4, −6, and −12.

1 Use exponents. In algebra, we use *exponents* as a way of writing products of repeated factors. For example, the product $2 \cdot 2 \cdot 2 \cdot 2 \cdot 2$ is written

$$\underbrace{2 \cdot 2 \cdot 2 \cdot 2 \cdot 2}_{\text{5 factors of 2}} = 2^5.$$

The number 5 shows that 2 is used as a factor 5 times. The number 5 is the **exponent,** and 2 is the **base.**

$$2^5 \longleftarrow \text{Exponent}$$
Base (2)

Read 2^5 as "2 to the fifth power" or simply "2 to the fifth." Multiplying out the five 2s gives

$$2^5 = 2 \cdot 2 \cdot 2 \cdot 2 \cdot 2 = 32.$$

Exponential Expression

If a is a real number and n is a natural number,

$$a^n = \underbrace{a \cdot a \cdot a \cdots a}_{n \text{ factors of } a},$$

where n is the **exponent,** a is the **base,** and a^n is an **exponential expression.** Exponents are also called **powers.**

Example 1 Using Exponential Notation

Write each expression using exponents.

(a) $4 \cdot 4 \cdot 4$
Here, 4 is used as a factor 3 times, so

$$\underbrace{4 \cdot 4 \cdot 4}_{\text{3 factors of 4}} = 4^3.$$

Read 4^3 as "4 cubed."

(b) $\frac{3}{5} \cdot \frac{3}{5} = \left(\frac{3}{5}\right)^2$ 2 factors of $\frac{3}{5}$

Read $\left(\frac{3}{5}\right)^2$ as "$\frac{3}{5}$ squared."

(c) $(-6)(-6)(-6)(-6) = (-6)^4$

(d) $x \cdot x \cdot x \cdot x \cdot x \cdot x = x^6$

Work Problem 1 at the Side.

1 Write each expression using exponents.

(a) $3 \cdot 3 \cdot 3 \cdot 3 \cdot 3$

(b) $\frac{2}{7} \cdot \frac{2}{7} \cdot \frac{2}{7} \cdot \frac{2}{7}$

(c) $(-10)(-10)(-10)$

(d) $(.5)(.5)$

(e) $y \cdot y \cdot y \cdot y \cdot y \cdot y \cdot y \cdot y$

ANSWERS

1. (a) 3^5 (b) $\left(\frac{2}{7}\right)^4$ (c) $(-10)^3$ (d) $(.5)^2$ (e) y^8

2 Write each expression without exponents.

(a) 5^3

(b) 3^4

(c) $(-4)^5$

(d) $(-3)^4$

(e) $(.75)^3$

(f) $\left(\frac{2}{5}\right)^4$

ANSWERS

2. **(a)** 125 **(b)** 81 **(c)** -1024 **(d)** 81 **(e)** .421875 **(f)** $\frac{16}{625}$

Example 2 Evaluating Exponential Expressions

Write each expression without exponents.

(a) $5^2 = 5 \cdot 5 = 25$ 5 is used as a factor 2 times.

(b) $\left(\frac{2}{3}\right)^3 = \frac{2}{3} \cdot \frac{2}{3} \cdot \frac{2}{3} = \frac{8}{27}$ $\frac{2}{3}$ is used as a factor 3 times.

(c) $2^6 = 2 \cdot 2 \cdot 2 \cdot 2 \cdot 2 \cdot 2 = 64$

(d) $(-2)^4 = (-2)(-2)(-2)(-2) = 16$

(e) $(-3)^5 = (-3)(-3)(-3)(-3)(-3) = -243$

Parts (d) and (e) of Example 2 suggest the following generalization.

The product of an *even* number of negative factors is positive.
The product of an *odd* number of negative factors is negative.

Calculator Tip Most calculators have a key labeled x^y or y^x that can be used to raise a number to a power. See "An Introduction to Calculators" at the beginning of this book for more information.

Work Problem 2 at the Side.

In parts (a) and (b) of Example 1, we used the terms *squared* and *cubed* to refer to powers of 2 and 3, respectively. The term *squared* comes from the figure of a square, which has the same measure for both length and width, as shown in Figure 9(a). Similarly, the term *cubed* comes from the figure of a cube. As shown in Figure 9(b), the length, width, and height of a cube have the same measure.

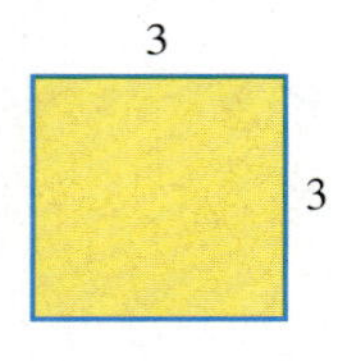

(a) $3 \cdot 3 = 3$ squared, or 3^2

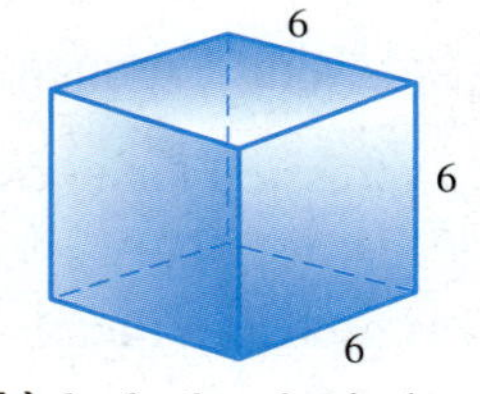

(b) $6 \cdot 6 \cdot 6 = 6$ cubed, or 6^3

Figure 9

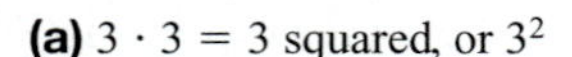

2 Identify exponents and bases.

Example 3 Identifying Exponents and Bases

Identify the exponent and the base. Then evaluate each expression.

(a) 3^6 The exponent is 6, and the base is 3.

$$3^6 = 3 \cdot 3 \cdot 3 \cdot 3 \cdot 3 \cdot 3 = 729$$

(b) 5^4 The exponent is 4, and the base is 5.

$$5^4 = 5 \cdot 5 \cdot 5 \cdot 5 = 625$$

Continued on Next Page

(c) $(-2)^6$

The exponent 6 applies to the number -2, so the base is -2.

$$(-2)^6 = (-2)(-2)(-2)(-2)(-2)(-2) = 64 \quad \text{The base is } -2.$$

(d) -2^6

Since there are no parentheses, the exponent 6 applies *only* to the number 2, not to -2; the base is 2.

$$-2^6 = -(2 \cdot 2 \cdot 2 \cdot 2 \cdot 2 \cdot 2) = -64 \quad \text{The base is } 2.$$

CAUTION

As shown in Examples 3(c) and (d), it is important to distinguish between $-a^n$ and $(-a)^n$.

$$-a^n = -1\underbrace{(a \cdot a \cdot a \cdots a)}_{n \text{ factors of } a} \quad \text{The base is } a.$$

$$(-a)^n = \underbrace{(-a)(-a) \cdots (-a)}_{n \text{ factors of } -a} \quad \text{The base is } -a.$$

Work Problem 3 at the Side.

3 Find square roots. As we saw in Example 2(a), $5^2 = 5 \cdot 5 = 25$, so 5 squared is 25. The opposite of squaring a number is called taking its **square root.** For example, a square root of 25 is 5. Another square root of 25 is -5 since $(-5)^2 = 25$; thus, 25 has two square roots, 5 and -5.

We write the positive or *principal* square root of a number with the symbol $\sqrt{}$, called a **radical sign.** For example, the positive or principal square root of 25 is written $\sqrt{25} = 5$. The negative square root of 25 is written $-\sqrt{25} = -5$. Since the square of any nonzero real number is positive, *the square root of a negative number, such as* $\sqrt{-25}$, *is not a real number.*

Example 4 Finding Square Roots

Find each square root that is a real number.

(a) $\sqrt{36} = 6$ since 6 is positive and $6^2 = 36$.

(b) $\sqrt{0} = 0$ since $0^2 = 0$.

(c) $\sqrt{\frac{9}{16}} = \frac{3}{4}$ since $\left(\frac{3}{4}\right)^2 = \frac{9}{16}$.

(d) $\sqrt{.16} = .4$ since $(.4)^2 = .16$.

(e) $\sqrt{100} = 10$ since $10^2 = 100$.

(f) $-\sqrt{100} = -10$ since the negative sign is outside the radical sign.

(g) $\sqrt{-100}$ is not a real number because the negative sign is inside the radical sign. No *real number* squared equals -100.

Notice the difference among the expressions in parts (e), (f), and (g). Part (e) is the positive or principal square root of 100, part (f) is the negative square root of 100, and part (g) is the square root of -100, which is not a real number.

3 Identify the exponent and the base. Then evaluate each expression.

(a) 7^3

(b) $(-5)^4$

(c) -5^4

(d) $-(.9)^5$

ANSWERS

3. (a) 3; 7; 343 **(b)** 4; -5; 625 **(c)** 4; 5; -625 **(d)** 5; .9; $-.59049$

4 Find each square root that is a real number.

(a) $\sqrt{9}$

(b) $\sqrt{49}$

(c) $-\sqrt{81}$

(d) $\sqrt{\frac{121}{81}}$

(e) $\sqrt{.25}$

(f) $\sqrt{-9}$

(g) $-\sqrt{-169}$

5 Simplify.

(a) $5 \cdot 9 + 2 \cdot 4$

(b) $4 - 12 \div 4 \cdot 2$

ANSWERS

4. (a) 3 (b) 7 (c) -9 (d) $\frac{11}{9}$ (e) .5 (f) not a real number (g) not a real number

5. (a) 53 (b) -2

CAUTION

The symbol $\sqrt{}$ is used only for the *positive* square root, except that $\sqrt{0} = 0$. The symbol $-\sqrt{}$ is used for the negative square root.

Work Problem 4 at the Side.

Calculator Tip Most calculators have a square root key, usually labeled $\sqrt{x}$, that allows us to find the square root of a number. On some models, the square root key must be used in conjunction with the key marked INV or 2nd.

4 Use the order of operations. To simplify an expression such as $5 + 2 \cdot 3$, what should we do first—add 5 and 2, or multiply 2 and 3? When an expression involves more than one operation symbol, we use the following **order of operations.**

Order of Operations

1. Work separately above and below any fraction bar.
2. If **parentheses** or **square brackets** are present, start with the innermost set and work outward.
3. Evaluate all **powers, roots,** and **absolute values.**
4. Do any **multiplications** or **divisions** in order, working from left to right.
5. Do any **additions** or **subtractions** in order, working from left to right.

Example 5 Using the Order of Operations

Simplify.

(a) $5 + 2 \cdot 3$

First multiply and then add.

$$5 + 2 \cdot 3 = 5 + 6 \quad \text{Multiply.}$$
$$= 11 \quad \text{Add.}$$

(b) $24 \div 3 \cdot 2 + 6$

Multiplications and divisions are done *in the order in which they appear from left to right,* so divide first.

$$24 \div 3 \cdot 2 + 6 = 8 \cdot 2 + 6 \quad \text{Divide.}$$
$$= 16 + 6 \quad \text{Multiply.}$$
$$= 22 \quad \text{Add.}$$

Work Problem 5 at the Side.

Example 6 Using the Order of Operations

Simplify.

(a) $4 \cdot 3^2 + 7 - (2 + 8)$

Work inside the parentheses first.

Continued on Next Page

$$4 \cdot 3^2 + 7 - (2 + 8) = 4 \cdot 3^2 + 7 - 10 \quad \text{Add inside parentheses.}$$
$$= 4 \cdot 9 + 7 - 10 \quad \text{Evaluate powers.}$$
$$= 36 + 7 - 10 \quad \text{Multiply.}$$
$$= 43 - 10 \quad \text{Add.}$$
$$= 33 \quad \text{Subtract.}$$

(b) $\frac{1}{2} \cdot 4 + (6 \div 3 - 7)$

Work inside the parentheses, dividing before subtracting.

$$\frac{1}{2} \cdot 4 + (6 \div 3 - 7) = \frac{1}{2} \cdot 4 + (2 - 7) \quad \text{Divide inside parentheses.}$$
$$= \frac{1}{2} \cdot 4 + (-5) \quad \text{Subtract inside parentheses.}$$
$$= 2 + (-5) \quad \text{Multiply.}$$
$$= -3 \quad \text{Add.}$$

Work Problem **6** at the Side.

6 Simplify.

(a) $(4 + 2) - 3^2 - (8 - 3)$

(b) $6 + \frac{2}{3}(-9) - \frac{5}{8} \cdot 16$

Example 7 Using the Order of Operations

Simplify $\dfrac{5 + 2^4}{6\sqrt{9} - 9 \cdot 2}$.

$$\frac{5 + 2^4}{6\sqrt{9} - 9 \cdot 2} = \frac{5 + 16}{6 \cdot 3 - 9 \cdot 2} \quad \text{Evaluate powers and roots.}$$
$$= \frac{5 + 16}{18 - 18} \quad \text{Multiply.}$$
$$= \frac{21}{0} \quad \text{Add and subtract.}$$

Because division by 0 is undefined, the given expression is undefined.

Work Problem **7** at the Side.

7 Simplify $\dfrac{10 - 6 + 2\sqrt{9}}{11 \cdot 2 - 3(2)^2}$.

Calculator Tip Most calculators follow the order of operations given in this section. You may want to try some of the examples to see whether your calculator gives the same answers. Be sure to use the parentheses keys to insert parentheses where they are needed. To work Example 7 with a calculator, you must put parentheses around the numerator and the denominator.

5 Evaluate expressions for given values of variables.

Any collection of numbers, variables, operation symbols, and grouping symbols, such as

$$6ab, \quad 5m - 9n, \quad \text{and} \quad -2(x^2 + 4y),$$

is called an **algebraic expression.** Algebraic expressions have different numerical values for different values of the variables. We can evaluate such expressions by *substituting* given values for the variables.

Algebraic expressions are used in problem solving. For example, if movie tickets cost \$7 each, the amount in dollars you pay for x tickets can be represented by the algebraic expression $7x$. We can substitute different numbers of tickets to get the costs to purchase those tickets.

ANSWERS
6. **(a)** -8 **(b)** -10
7. 1

8 Evaluate each expression if $w = 4, x = -12, y = 64$, and $z = -3$.

(a) $5x - 2w$

(b) $-6(x - \sqrt{y})$

(c) $\dfrac{5x - 3 \cdot \sqrt{y}}{x - 1}$

(d) $w^2 + 2z^3$

9 Use the expression in Example 9 to approximate the percent of U.S. households investing in mutual funds in 1990 and 1999. Round answers to the nearest tenth.

ANSWERS

8. **(a)** -68 **(b)** 120 **(c)** $\dfrac{84}{13}$ **(d)** -38

9. 1990: 25.8%; 1999: 45.0%

Example 8 Evaluating Expressions

Evaluate each expression if $m = -4$, $n = 5$, $p = -6$, and $q = 25$.

(a) $5m - 9n$

Replace m with -4 and n with 5.

$$5m - 9n = 5(-4) - 9(5) = -20 - 45 = -65$$

(b) $\dfrac{m + 2n}{4p} = \dfrac{-4 + 2(5)}{4(-6)} = \dfrac{-4 + 10}{-24} = \dfrac{6}{-24} = -\dfrac{1}{4}$

(c)

$$\begin{aligned} -3m^3 - n^2(\sqrt{q}) &= -3(-4)^3 - (5)^2(\sqrt{25}) && \text{Substitute; } m = -4, n = 5, \text{ and } q = 25. \\ &= -3(-64) - 25(5) && \text{Evaluate powers and roots.} \\ &= 192 - 125 && \text{Multiply.} \\ &= 67 && \text{Subtract.} \end{aligned}$$

CAUTION

To avoid errors when evaluating expressions, it is a good idea to use parentheses around any negative numbers that are substituted for variables.

Work Problem **8** at the Side.

Example 9 Evaluating an Expression to Approximate Mutual Fund Investors

An approximation of the percent of U.S. households investing in mutual funds during the years 1980–1999 can be obtained by substituting a given year for x in the expression

$$2.1331x - 4219.1$$

and then evaluating. (*Source:* Investment Company Institute.)

(a) Approximate the percent of U.S. households investing in mutual funds in 1980.

Substitute 1980 for x in the given expression.

$$\begin{aligned} 2.1331x - 4219.1 &= 2.1331(1980) - 4219.1 && \text{Let } x = 1980. \\ &\approx 4.4 && \text{Use a calculator.} \end{aligned}$$

Recall that the symbol $\approx$ means "is approximately equal to." In 1980, about 4.4% of U.S. households invested in mutual funds.

Work Problem **9** at the Side.

(b) Give the results found above and in Problem 9 at the side in a table. How has the percent of households investing in mutual funds changed during these years?

The table follows. The percent of U.S. households investing in mutual funds increased dramatically during these years.

Year	Percent of U.S. Households Investing in Mutual Funds
1980	4.4
1990	25.8
1999	45.0

Percent in 1999 is more than ten times percent in 1980.

1.3 EXERCISES

FOR EXTRA HELP

Student's Solutions Manual

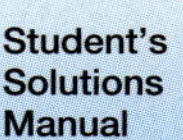
MyMathLab.com

InterAct Math Tutorial Software

AW Math Tutor Center

www.mathxl.com

Digital Video Tutor CD 1 Videotape 1

Decide whether each statement is true *or* false. *If* false, *correct the statement so it is* true.

1. $-4^6 = (-4)^6$

2. $-4^7 = (-4)^7$

3. $\sqrt{16}$ is a positive number.

4. $3 + 5 \cdot 6 = 3 + (5 \cdot 6)$

5. $(-2)^7$ is a negative number.

6. $(-2)^8$ is a positive number.

7. The product of 8 positive factors and 8 negative factors is positive.

8. The product of 3 positive factors and 3 negative factors is positive.

9. In the exponential expression -3^5, -3 is the base.

10. $\sqrt{a}$ is positive for all positive numbers a.

11. Evaluate each exponential expression.
(a) 8^2 **(b)** -8^2 **(c)** $(-8)^2$ **(d)** $-(-8)^2$

12. Evaluate each exponential expression.
(a) 4^3 **(b)** -4^3 **(c)** $(-4)^3$ **(d)** $-(-4)^3$

Write each expression using exponents. See Example 1.

13. $8 \cdot 8 \cdot 8$

14. $10 \cdot 10 \cdot 10 \cdot 10$

15. $\frac{1}{2} \cdot \frac{1}{2}$

16. $\frac{3}{4} \cdot \frac{3}{4} \cdot \frac{3}{4} \cdot \frac{3}{4} \cdot \frac{3}{4}$

17. $(-4)(-4)(-4)(-4)$

18. $(-9)(-9)(-9)$

19. $z \cdot z \cdot z \cdot z \cdot z \cdot z \cdot z$

20. $a \cdot a \cdot a \cdot a \cdot a \cdot a$

Evaluate each expression. See Examples 2 and 3.

21. 4^2

22. 2^4

23. $.28^3$

24. $.91^3$

25. $\left(\frac{1}{5}\right)^3$

26. $\left(\frac{1}{6}\right)^4$

27. $\left(\frac{7}{10}\right)^3$

28. $\left(\frac{4}{5}\right)^4$

29. $(-5)^3$ **30.** $(-3)^5$ **31.** $(-2)^8$ **32.** $(-3)^6$

33. -3^6 **34.** -4^6 **35.** -8^4 **36.** -10^3

Identify the exponent and the base in each expression. Do not evaluate. See Example 3.

37. $(-4.1)^7$ **38.** $(-3.4)^9$ **39.** -4.1^7 **40.** -3.4^9

Find each square root. If it is not a real number, say so. See Example 4.

41. $\sqrt{81}$ **42.** $\sqrt{64}$ **43.** $\sqrt{169}$ **44.** $\sqrt{225}$

45. $-\sqrt{400}$ **46.** $-\sqrt{900}$ **47.** $\sqrt{\frac{100}{121}}$ **48.** $\sqrt{\frac{225}{169}}$

49. $-\sqrt{.49}$ **50.** $-\sqrt{.64}$ **51.** $\sqrt{-36}$ **52.** $\sqrt{-121}$

53. Match each square root with the appropriate value or description.

(a) $\sqrt{144}$ **A.** -12

(b) $\sqrt{-144}$ **B.** 12

(c) $-\sqrt{144}$ **C.** Not a real number

54. Explain why $\sqrt{-900}$ is not a real number.

55. If a is a positive number, is $-\sqrt{-a}$ positive, negative, or not a real number?

56. If a is a positive number, is $-\sqrt{a}$ positive, negative, or not a real number?

Simplify each expression. Use the order of operations. See Examples 5–7.

57. $12 + 3 \cdot 4$ **58.** $15 + 5 \cdot 2$ **59.** $2[-5 - (-7)]$ **60.** $3[-8 - (-2)]$

61. $-12\left(-\frac{3}{4}\right) - (-5)$ **62.** $-7\left(-\frac{2}{14}\right) - (-8)$ **63.** $6 \cdot 3 - 12 \div 4$ **64.** $9 \cdot 4 - 8 \div 2$

65. $10 + 30 \div 2 \cdot 3$ **66.** $12 + 24 \div 3 \cdot 2$ **67.** $-3(5)^2 - (-2)(-8)$ **68.** $-9(2)^2 - (-3)(-2)$

69. $5 - 7 \cdot 3 - (-2)^3$ **70.** $-4 - 3 \cdot 5 + 6^2$ **71.** $-7(\sqrt{36}) - (-2)(-3)$ **72.** $-8(\sqrt{64}) - (-3)(-7)$

73. $\dfrac{-8 + (-16)}{-3}$ **74.** $\dfrac{-9 + (-11)}{-2}$ **75.** $\dfrac{(-5 + \sqrt{4})(-2^2)}{-5 - 1}$

76. $\dfrac{(-9 + \sqrt{16})(-3^2)}{-4 - 1}$ **77.** $\dfrac{2(-5) + (-3)(-2)}{-8 + 3^2 - 1}$ **78.** $\dfrac{3(-4) + (-5)(-8)}{2^3 - 2 - 6}$

Evaluate each expression if $a = -3$, $b = 64$, *and* $c = 6$. *See Example 8.*

79. $3a + \sqrt{b}$ **80.** $-2a - \sqrt{b}$ **81.** $\sqrt{b} + c - a$ **82.** $\sqrt{b} - c + a$

83. $4a^3 + 2c$ **84.** $-3a^4 - 3c$ **85.** $\dfrac{2c + a^3}{4b + 6a}$ **86.** $\dfrac{3c + a^2}{2b - 6c}$

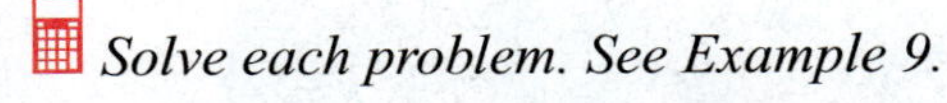

Solve each problem. See Example 9.

Residents of Linn County, Iowa in the Cedar Rapids Community School District can use the expression

$$(v \times .5485 - 4850) \div 1000 \times 31.44$$

to determine their property taxes, where v is home value. (*Source: The Gazette,* August 19, 2000.) Use the expression to calculate the amount of property taxes to the nearest dollar that the owner of a home with each of the following values would pay. Follow the order of operations.

87. \$100,000 **88.** \$150,000 **89.** \$200,000

The Blood Alcohol Concentration (BAC) of a person who has been drinking is given by the expression

$$\text{number of oz} \times \%\ \text{alcohol} \times .075 \div \text{body weight in lb} - \text{hr of drinking} \times .015.$$

(*Source:* Lawlor, J., *Auto Math Handbook: Mathematical Calculations, Theory, and Formulas for Automotive Enthusiasts,* HP Books, 1991.)

90. Suppose a policeman stops a 190-lb man who, in 2 hr, has ingested four 12-oz beers (48 oz), each having a 3.2% alcohol content.

(a) Substitute the values in the formula, and write the expression for the man's BAC.

(b) Calculate the man's BAC to the nearest thousandth. Follow the order of operations.

91. Find the BAC to the nearest thousandth for a 135-lb woman who, in 3 hr, has drunk three 12-oz beers (36 oz), each having a 4.0% alcohol content.

92. Calculate the BACs in Exercises 90 and 91 if each person weighs 25 lb more and the rest of the variables stay the same. How does increased weight affect a person's BAC?

93. Predict how decreased weight would affect the BAC of each person in Exercises 90 and 91. Calculate the BACs if each person weighs 25 lb less and the rest of the variables stay the same.

94. An approximation of federal spending on education in billions of dollars from 1996 through 2001 can be obtained using the expression

$$3.31714x - 6597.86,$$

where x represents the year. (*Source:* U.S. Department of Education.)

(a) Use this expression to complete the following table. Round answers to the nearest tenth.

Year	Education Spending (in billions of dollars)
1997	26.5
1998	29.8
1999	______
2000	______
2001	______

(b) Describe the trend in the amount of federal spending on education during these years.

1.4 Properties of Real Numbers

Objectives

1. Use the distributive property.
2. Use the inverse properties.
3. Use the identity properties.
4. Use the commutative and associative properties.
5. Use the multiplication property of 0.

The study of any object is simplified when we know the properties of the object. For example, a property of water is that it freezes when cooled to 0°C. Knowing this helps us to predict the behavior of water.

The study of numbers is no different. The basic properties of addition and multiplication of real numbers studied in this section will be used in later work in algebra. These properties reflect results that occur consistently in work with numbers, so they have been generalized to apply to expressions with variables as well.

1 Use the distributive property. Notice that

$$2(3 + 5) = 2 \cdot 8 = 16$$

and

$$2 \cdot 3 + 2 \cdot 5 = 6 + 10 = 16,$$

so

$$2(3 + 5) = 2 \cdot 3 + 2 \cdot 5.$$

This idea is illustrated by the divided rectangle in Figure 10.

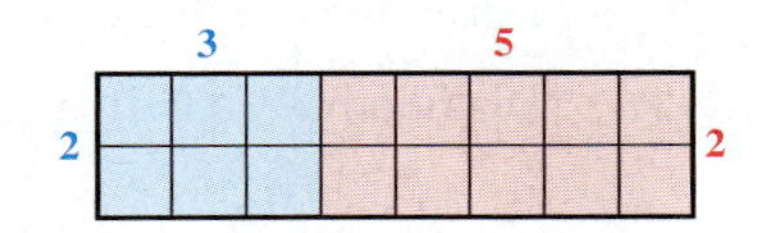

Area of left part is $2 \cdot 3 = 6$.
Area of right part is $2 \cdot 5 = 10$.
Area of total rectangle is $2(3 + 5) = 16$.

Figure 10

Similarly,

$$-4[5 + (-3)] = -4(2) = -8$$

and

$$-4(5) + (-4)(-3) = -20 + 12 = -8,$$

so

$$-4[5 + (-3)] = -4(5) + (-4)(-3).$$

These arithmetic examples are generalized to *all* real numbers as the **distributive property of multiplication with respect to addition,** or simply the **distributive property.**

Distributive Property

For any real numbers a, b, and c,

$$a(b + c) = ab + ac \quad \text{and} \quad (b + c)a = ba + ca.$$

The distributive property can also be written

$$ab + ac = a(b + c) \quad \text{and} \quad ba + ca = (b + c)a.$$

It can be extended to more than two numbers as well.

$$a(b + c + d) = ab + ac + ad$$

This property is important because it provides a way to rewrite a *product* $a(b + c)$ as a sum $ab + ac$, or a *sum* as a product.

1 Use the distributive property to rewrite each expression.

(a) $8(m + n)$

(b) $-4(p - 5)$

(c) $3k + 6k$

(d) $-6m + 2m$

(e) $2r + 3s$

(f) $5(4p - 2q + r)$

2 Use the distributive property to calculate each expression.

(a) $14 \cdot 5 + 14 \cdot 85$

(b) $78 \cdot 33 + 22 \cdot 33$

ANSWERS

1. **(a)** $8m + 8n$ **(b)** $-4p + 20$ **(c)** $9k$ **(d)** $-4m$ **(e)** cannot be rewritten **(f)** $20p - 10q + 5r$

2. **(a)** 1260 **(b)** 3300

NOTE

When we rewrite $a(b + c)$ as $ab + ac$, we sometimes refer to the process as "removing parentheses."

Example 1 Using the Distributive Property

Use the distributive property to rewrite each expression.

(a) $3(x + y)$

Use the first form of the property.

$$3(x + y) = 3x + 3y$$

(b)
$$-2(5 + k) = -2(5) + (-2)(k)$$
$$= -10 - 2k$$

(c) $4x + 8x$

Use the second form of the property.

$$4x + 8x = (4 + 8)x = 12x$$

(d)
$$3r - 7r = 3r + (-7r) \quad \text{Definition of subtraction}$$
$$= [3 + (-7)]r \quad \text{Distributive property}$$
$$= -4r$$

(e) $5p + 7q$

Because there is no common number or variable here, we cannot use the distributive property to rewrite the expression.

(f)
$$6(x + 2y - 3z) = 6x + 6(2y) + 6(-3z)$$
$$= 6x + 12y - 18z$$

As illustrated in Example 1(d), the distributive property can also be used for subtraction, so

$$a(b - c) = ab - ac.$$

Work Problem 1 at the Side.

The distributive property can be used to mentally perform calculations.

Example 2 Using the Distributive Property for Calculation

Calculate $38 \cdot 17 + 38 \cdot 3$.

$$38 \cdot 17 + 38 \cdot 3 = 38(17 + 3) \quad \text{Distributive property}$$
$$= 38(20)$$
$$= 760$$

Work Problem 2 at the Side.

2 **Use the inverse properties.** In Section 1.1 we saw that the additive inverse of a number a is $-a$ and that the sum of a number and its additive inverse is 0. For example, 3 and -3 are additive inverses, as are -8 and 8. The number 0 is its own additive inverse. In Section 1.2, we saw that two numbers with a product of 1 are reciprocals. As mentioned there, another name for reciprocal is *multiplicative inverse.* This is similar to the idea of an additive inverse. Thus, 4 and $\frac{1}{4}$ are multiplicative inverses, as are $-\frac{2}{3}$ and $-\frac{3}{2}$. (Recall that reciprocals have the same sign.) We can extend these properties of arithmetic, the **inverse properties** of addition and multiplication, to the real numbers of algebra.

Inverse Properties

For any real number a, there is a single real number $-a$, such that

$$a + (-a) = 0 \quad \text{and} \quad -a + a = 0.$$

The inverse "undoes" addition with the result 0.

For any *nonzero* real number a, there is a single real number $\frac{1}{a}$ such that

$$a \cdot \frac{1}{a} = 1 \quad \text{and} \quad \frac{1}{a} \cdot a = 1.$$

The inverse "undoes" multiplication with the result 1.

Work Problem 3 at the Side.

3 Use the identity properties. The numbers 0 and 1 each have a special property. Zero is the only number that can be added to any number to get that number. That is, adding 0 leaves the identity of a number unchanged. For this reason, 0 is called the **identity element for addition** or the **additive identity.** In a similar way, multiplying by 1 leaves the identity of any number unchanged, so 1 is the **identity element for multiplication** or the **multiplicative identity.** The following **identity properties** summarize this discussion and extend these properties from arithmetic to algebra.

Identity Properties

For any real number a,

$$a + 0 = 0 + a = a.$$

Start with a number a; add 0. The answer is "identical" to a.

Also, $$a \cdot 1 = 1 \cdot a = a.$$

Start with a number a; multiply by 1. The answer is "identical" to a.

Example 3 Using the Identity Property $1 \cdot a = a$

Simplify each expression.

(a) $12m + m = 12m + 1m$ Identity property

$= (12 + 1)m$ Distributive property

$= 13m$ Add inside parentheses.

(b) $y + y = 1y + 1y$ Identity property

$= (1 + 1)y$ Distributive property

$= 2y$ Add inside parentheses.

(c) $-(m - 5n) = -1(m - 5n)$ Identity property

$= -1(m) + (-1)(-5n)$ Distributive property

$= -m + 5n$ Multiply.

Work Problem 4 at the Side.

3 Complete each statement.

(a) $4 +$ ______ $= 0$

(b) $-7.1 +$ ______ $= 0$

(c) $-9 + 9 =$ ______

(d) $5 \cdot$ ______ $= 1$

(e) $-\frac{3}{4} \cdot$ ______ $= 1$

(f) $7 \cdot \frac{1}{7} =$ ______

4 Simplify each expression.

(a) $p - 3p$

(b) $r + r + r$

(c) $-(3 + 4p)$

(d) $-(k - 2)$

Answers

3. **(a)** -4 **(b)** 7.1 **(c)** 0 **(d)** $\frac{1}{5}$ **(e)** $-\frac{4}{3}$ **(f)** 1

4. **(a)** $-2p$ **(b)** $3r$ **(c)** $-3 - 4p$ **(d)** $-k + 2$

Expressions such as $12m$ and $5n$ from Example 3 are examples of *terms.* A **term** is a number or the product of a number and one or more variables. Terms with exactly the same variables raised to exactly the same powers are called **like terms.** Some examples of like terms are

$$5p \text{ and } -21p \qquad -6x^2 \text{ and } 9x^2. \qquad \text{Like terms}$$

Some examples of unlike terms are

$$3m \text{ and } 16x \qquad 7y^3 \text{ and } -3y^2. \qquad \text{Unlike terms}$$

The numerical factor in a term is called the **numerical coefficient,** or just the **coefficient.** For example, in the term $9x^2$, the coefficient is 9.

4 Use the commutative and associative properties. Simplifying expressions as in parts (a) and (b) of Example 3 is called **combining like terms.** Only like terms may be combined. To combine like terms in an expression such as

$$-2m + 5m + 3 - 6m + 8,$$

we need two more properties. From arithmetic, we know that

$$3 + 9 = 12 \quad \text{and} \quad 9 + 3 = 12.$$

Also,

$$3 \cdot 9 = 27 \quad \text{and} \quad 9 \cdot 3 = 27.$$

Furthermore, notice that

$$(5 + 7) + (-2) = 12 + (-2) = 10$$

and

$$5 + [7 + (-2)] = 5 + 5 = 10.$$

Also,

$$(5 \cdot 7)(-2) = 35(-2) = -70$$

and

$$(5)[7 \cdot (-2)] = 5(-14) = -70.$$

These arithmetic examples can now be extended to algebra.

Commutative and Associative Properties

For any real numbers a, b, and c,

$$a + b = b + a$$

and

$$ab = ba. \qquad \text{Commutative properties}$$

There are two terms or factors; reverse the order.

Also, $$a + (b + c) = (a + b) + c$$

and

$$a(bc) = (ab)c. \qquad \text{Associative properties}$$

There are three terms or factors; same order, parentheses shifted.

The commutative properties are used to change the *order* of the terms or factors in an expression. Think of commuting from home to work and then from work to home. The associative properties are used to *regroup* the terms or factors of an expression. Remember, to *associate* is to be part of a group.

Example 4 Using the Commutative and Associative Properties

Simplify $-2m + 5m + 3 - 6m + 8$.

$$-2m + 5m + 3 - 6m + 8$$
$= (-2m + 5m) + 3 - 6m + 8$ Order of operations
$= (-2 + 5)m + 3 - 6m + 8$ Distributive property
$= 3m + 3 - 6m + 8$

By the order of operations, the next step would be to add $3m$ and 3, but they are unlike terms. To get $3m$ and $-6m$ together, use the associative and commutative properties. Begin by inserting parentheses and brackets according to the order of operations.

$$[(3m + 3) - 6m] + 8$$
$= [3m + (3 - 6m)] + 8$ Associative property
$= [3m + (-6m + 3)] + 8$ Commutative property
$= [(3m + [-6m]) + 3] + 8$ Associative property
$= (-3m + 3) + 8$ Combine like terms.
$= -3m + (3 + 8)$ Associative property
$= -3m + 11$ Add.

In practice, many of the steps are not written down, but you should realize that the commutative and associative properties are used whenever the terms in an expression are rearranged to combine like terms.

Example 5 Using the Properties of Real Numbers

Simplify each expression.

(a) $5y - 8y - 6y + 11y$
$= (5 - 8 - 6 + 11)y$ Distributive property
$= 2y$ Combine like terms.

(b) $3x + 4 - 5(x + 1) - 8$
$= 3x + 4 - 5x - 5 - 8$ Distributive property
$= 3x - 5x + 4 - 5 - 8$ Commutative property
$= -2x - 9$ Combine like terms.

(c) $8 - (3m + 2) = 8 - 1(3m + 2)$ Identity property
$= 8 - 3m - 2$ Distributive property
$= 6 - 3m$ Combine like terms.

(d) $(3x)(5)(y) = [(3x)(5)]y$ Order of operations
$= [3(x \cdot 5)]y$ Associative property
$= [3(5x)]y$ Commutative property
$= [(3 \cdot 5)x]y$ Associative property
$= (15x)y$ Multiply.
$= 15(xy)$ Associative property
$= 15xy$

As previously mentioned, many of these steps are not usually written out.

Work Problem 5 at the Side.

5 Simplify each expression.

(a) $12b - 9b + 4b - 7b + b$

(b) $-3w + 7 - 8w - 2$

(c) $-3(6 + 2t)$

(d) $9 - 2(a - 3) + 4 - a$

(e) $(4m)(2n)$

ANSWERS

5. **(a)** b **(b)** $-11w + 5$ **(c)** $-18 - 6t$ **(d)** $19 - 3a$ **(e)** $8mn$

CAUTION

Be careful. Notice that the distributive property does not apply in Example 5(d), because there is no addition involved.

$$(3x)(5)(y) \neq (3x)(5) \cdot (3x)(y)$$

5 **Use the multiplication property of 0.** The additive identity property gives a special property of 0, namely that $a + 0 = a$ for any real number a. The **multiplication property of 0** gives a special property of 0 that involves multiplication: The product of any real number and 0 is 0.

Multiplication Property of 0

For any real number a,

$$a \cdot 0 = 0 \quad \text{and} \quad 0 \cdot a = 0.$$

Work Problem 6 at the Side.

6 Complete each statement.

(a) $197 \cdot 0 =$ ______

(b) $0\left(-\frac{8}{9}\right) =$ ______

(c) $0 \cdot$ ______ $= 0$

ANSWERS

6. (a) 0 **(b)** 0 **(c)** any real number

1.4 Exercises

FOR EXTRA HELP

 Student's Solutions Manual

 MyMathLab.com

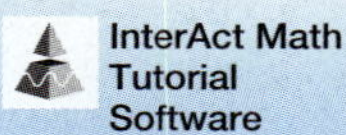 InterAct Math Tutorial Software

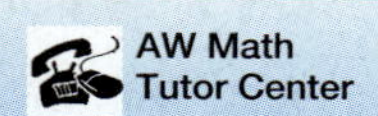 AW Math Tutor Center

 www.mathxl.com

Digital Video Tutor CD 1 Videotape 1

Choose the correct response in Exercises 1–4.

1. The identity element for addition is

A. $-a$ **B.** 0 **C.** 1 **D.** $\frac{1}{a}$.

2. The identity element for multiplication is

A. $-a$ **B.** 0 **C.** 1 **D.** $\frac{1}{a}$.

3. The additive inverse of a is

A. $-a$ **B.** 0 **C.** 1 **D.** $\frac{1}{a}$.

4. The multiplicative inverse of a, where $a \neq 0$, is

A. $-a$ **B.** 0 **C.** 1 **D.** $\frac{1}{a}$.

Complete each statement.

5. The multiplication property of 0 says that the ________ of 0 and any real number is ________.

6. The commutative property is used to change the ________ of two terms or factors.

7. The associative property is used to change the ________ of three terms or factors.

8. Like terms are terms with the ________ variables raised to the ________ powers.

9. When simplifying an expression, only ________ terms can be combined.

10. The coefficient in the term $-8yz^2$ is ________.

Use the properties of real numbers to simplify each expression. See Examples 1 and 3.

11. $5k + 3k$

12. $6a + 5a$

13. $-9r + 7r$

14. $-4n + 6n$

15. $-8z + 4w$

16. $-12k + 3r$

17. $-a + 7a$

18. $-s + 9s$

19. $2(m + p)$

20. $3(a + b)$

21. $-5(2d - f)$

22. $-2(3m - n)$

Use the distributive property to calculate each value mentally. See Example 2.

23. $96 \cdot 19 + 4 \cdot 19$

24. $27 \cdot 60 + 27 \cdot 40$

25. $58 \cdot \frac{3}{2} - 8 \cdot \frac{3}{2}$

26. $\frac{8}{5} \cdot 17 + \frac{8}{5} \cdot 13$

27. $4.31(69) + 4.31(31)$

28. $\frac{4}{5}(17) + \frac{4}{5}(23)$

Simplify each expression by removing parentheses and combining like terms. See Examples 1 and 3–5.

29. $-12y + 4y + 3 + 2y$

30. $-5r - 9r + 8r - 5$

31. $-6p + 11p - 4p + 6 + 5$

32. $-8x - 5x + 3x - 12 + 9$

33. $3(k + 2) - 5k + 6 + 3$

34. $5(r - 3) + 6r - 2r + 4$

35. $-2(m + 1) + 3(m - 4)$

36. $6(a - 5) - 4(a + 6)$

37. $.25(8 + 4p) - .5(6 + 2p)$

38. $.4(10 - 5x) - .8(5 + 10x)$

39. $-(2p + 5) + 3(2p + 4) - 2p$

40. $-(7m - 12) - 2(4m + 7) - 8m$

41. $2 + 3(2z - 5) - 3(4z + 6) - 8$

42. $-4 + 4(4k - 3) - 6(2k + 8) + 7$

Complete each statement so that the indicated property is illustrated. Simplify each answer, if possible.

43. $5x + 8x =$ ____________________ (distributive property)

44. $9y - 6y =$ ____________________ (distributive property)

45. $5(9r) =$ ____________________ (associative property)

46. $-4 + (12 + 8) =$ ____________________ (associative property)

47. $5x + 9y =$ ____________________ (commutative property)

48. $-5(7) =$ ____________________ (commutative property)

49. $1 \cdot 7 =$ ____________________ (identity property)

50. $-12x + 0 =$ ____________________ (identity property)

51. $8(-4 + x) =$ ____________________ (distributive property)

52. $3(x - y + z) =$ ____________________ (distributive property)

53. Give an "everyday" example of a commutative operation.

54. Give an "everyday" example of inverse operations.

RELATING CONCEPTS (Exercises 55–60) FOR INDIVIDUAL OR GROUP WORK

While it may seem that simplifying the expression $3x + 4 + 2x + 7$ *to* $5x + 11$ *is fairly easy, there are several important steps that require mathematical justification. These steps are usually done mentally.* ***Work Exercises 55–60 in order,*** *providing the property that justifies each statement in the given simplification. (These steps could be done in other orders.)*

55. $3x + 4 + 2x + 7 = (3x + 4) + (2x + 7)$ ____________________

56. $= 3x + (4 + 2x) + 7$ ____________________

57. $= 3x + (2x + 4) + 7$ ____________________

58. $= (3x + 2x) + (4 + 7)$ ____________________

59. $= (3 + 2)x + (4 + 7)$ ____________________

60. $= 5x + 11$ ____________________

Chapter 1

SUMMARY

Study Skills Workbook
Activity 5

KEY TERMS

1.1	**set**	A set is a collection of objects.
	elements	The elements (members) of a set are the numbers or objects that make up the set.
	empty set	The set with no elements is called the empty (null) set.
	variable	A variable is a letter used to represent a number or a set of numbers.
	set-builder notation	Set-builder notation is used to describe a set of numbers without listing them.
	number line	A number line is a line with a scale to indicate the set of real numbers.
	coordinate	The number that corresponds to a point on the number line is its coordinate.
	graph	The point on the number line that corresponds to a number is its graph.
	additive inverse	The additive inverse (**negative, opposite**) of a number a is $-a$.
	signed numbers	Positive and negative numbers are signed numbers.
	absolute value	The absolute value of a number is its distance from 0 on a number line.
	equation	An equation is a mathematical statement that two quantities are equal.
	inequality	An inequality is a mathematical statement that two quantities are not equal.
1.2	**sum**	The answer to an addition problem is called the sum.
	difference	The answer to a subtraction problem is called the difference.
	product	The answer to a multiplication problem is called the product.
	reciprocals	Two numbers whose product is 1 are reciprocals (**multiplicative inverses**).
	quotient	The answer to a division problem is called the quotient.
1.3	**factors**	Two (or more) numbers whose product is a third number are factors of that third number.
	exponent	An exponent (**power**) is a number that shows how many times a factor is repeated in a product.
	base	The base is the number that is a repeated factor in a product.
	exponential expression	A base with an exponent is called an exponential expression.
	square root	A square root of a number r is a number that can be squared to get r.
	algebraic expression	Any collection of numbers, variables, operation symbols, and grouping symbols is an algebraic expression.
1.4	**term**	A term is a number or the product of a number and one or more variables.
	like terms	Like terms are terms with the same variables raised to the same powers.
	coefficient	A coefficient (**numerical coefficient**) is the numerical factor of a term.
	combining like terms	Combining like terms is a method of adding or subtracting like terms by using the properties of real numbers.

Graph of -1

$-3 \quad -2 \quad -1 \quad 0 \quad 1 \quad 2 \quad 3$

Coordinate

2^5 ← Exponent

Base

New Symbols

$\{a, b\}$	set containing the elements a and b
$\emptyset$	empty (null) set
$\{x \mid x$ **has property** $P\}$	set-builder notation
$\lvert x \rvert$	absolute value of x
$\neq$	is not equal to
$<$	is less than
$\leq$	is less than or equal to
$>$	is greater than
$\geq$	is greater than or equal to
a^m	m factors of a
$\sqrt{}$	radical sign
$\sqrt{a}$	the positive (or principal) square root of a
$\approx$	is approximately equal to

Test Your Word Power

See how well you have learned the vocabulary in this chapter. Answers follow the Quick Review.

1. The **empty set** is a set
- **(a)** with 0 as its only element
- **(b)** with an infinite number of elements
- **(c)** with no elements
- **(d)** of ideas.

2. A **variable** is
- **(a)** a symbol used to represent an unknown number
- **(b)** a value that makes an equation true
- **(c)** a solution of an equation
- **(d)** the answer in a division problem.

3. The **absolute value** of a number is
- **(a)** the graph of the number
- **(b)** the reciprocal of the number
- **(c)** the opposite of the number
- **(d)** the distance between 0 and the number on a number line.

4. The **reciprocal** of a nonzero number a is
- **(a)** a
- **(b)** $\frac{1}{a}$
- **(c)** $-a$
- **(d)** 1.

5. A **factor** is
- **(a)** the answer in an addition problem
- **(b)** the answer in a multiplication problem
- **(c)** one of two or more numbers that are added to get another number
- **(d)** any number that divides evenly into a given number.

6. An **exponential expression** is
- **(a)** a number that is a repeated factor in a product
- **(b)** a number or a variable written with an exponent
- **(c)** a number that shows how many times a factor is repeated in a product
- **(d)** an expression that involves addition.

7. A **term** is
- **(a)** a numerical factor
- **(b)** a number or a product of numbers and variables raised to powers
- **(c)** one of several variables with the same exponents
- **(d)** a sum of numbers and variables raised to powers.

8. A **numerical coefficient** is
- **(a)** the numerical factor in a term
- **(b)** the number of terms in an expression
- **(c)** a variable raised to a power
- **(d)** the variable factor in a term.

Quick Review

Concepts	*Examples*
1.1 Basic Concepts	
Sets of Numbers	
Natural Numbers $\{1, 2, 3, 4, \ldots\}$	$10, 25, 143$
Whole Numbers $\{0, 1, 2, 3, 4, \ldots\}$	$0, 8, 47$
Integers $\{\ldots, -2, -1, 0, 1, 2, \ldots\}$	$-22, -7, 0, 4, 9$
Rational Numbers $\left\{\frac{p}{q} \middle\| p \text{ and } q \text{ are integers}, q \neq 0\right\}$ (all terminating or repeating decimals)	$-\frac{2}{3}, -.14, 0, 6, \frac{5}{8}, .33333\ldots$
Irrational Numbers $\{x \| x \text{ is a real number that is not rational}\}$ (all nonterminating, nonrepeating decimals)	$\pi, .125469\ldots, \sqrt{3}, -\sqrt{22}$
Real Numbers $\{x \| x \text{ is represented by a point on a number line}\}$ (all rational and irrational numbers)	$-3, .7, \pi, -\frac{2}{3}$
Absolute Value $\lvert a \rvert = \begin{cases} a & \text{if } a \text{ is positive or } 0 \\ -a & \text{if } a \text{ is negative} \end{cases}$	$\lvert 12 \rvert = 12$ $\lvert -12 \rvert = 12$
1.2 Operations on Real Numbers	
Addition	
Like Signs: To add two numbers with the same sign, add the absolute values. The answer has the same sign as the two numbers.	$-2 + (-7) = -(2 + 7) = -9$
Unlike Signs: To add two numbers with different signs, subtract the smaller absolute value from the larger. The answer has the sign of the number with the larger absolute value.	$-5 + 8 = 8 - 5 = 3$ $-12 + 4 = -(12 - 4) = -8$
Subtraction Change the sign of the second number and add.	$-5 - (-3) = -5 + 3 = -2$
Multiplication and Division	
Like Signs: The answer is positive when multiplying or dividing two numbers with the same sign.	$-3(-8) = 24$ $\quad \frac{-15}{-5} = 3$
Unlike Signs: The answer is negative when multiplying or dividing two numbers with different signs.	$-7(5) = -35$ $\quad \frac{-24}{12} = -2$
1.3 Exponents, Roots, and Order of Operations	
The product of an even number of negative factors is positive. The product of an odd number of negative factors is negative.	$(-5)^2$ is positive: $(-5)^2 = (-5)(-5) = 25$ $(-5)^3$ is negative: $(-5)^3 = (-5)(-5)(-5) = -125$
Order of Operations	
1. Work separately above and below any fraction bar.	$\frac{12 + 3}{5 \cdot 2} = \frac{15}{10} = \frac{3}{2}$
2. If parentheses or brackets are present, start with the innermost set and work outward.	$(-6)[2^2 - (3 + 4)] + 3 = (-6)[2^2 - 7] + 3$
3. Evaluate all exponents, roots, and absolute values.	$= (-6)[4 - 7] + 3$ $= (-6)[-3] + 3$
4. Multiply or divide in order from left to right.	$= 18 + 3$
5. Add or subtract in order from left to right.	$= 21$

Concepts	*Examples*
1.4 Properties of Real Numbers	
Distributive Property	
$a(b + c) = ab + ac$ (Remove parentheses.)	$12(4 + 2) = 12 \cdot 4 + 12 \cdot 2$
Inverse Properties	
$a + (-a) = 0$ and $-a + a = 0$ (The additive inverse "undoes" addition to give 0.)	$5 + (-5) = 0$
$a \cdot \frac{1}{a} = 1$ and $\frac{1}{a} \cdot a = 1$ (The multiplicative inverse "undoes" multiplication to give 1.)	$-\frac{1}{3}(-3) = 1$
Identity Properties	
$a + 0 = 0 + a = a$ (Start with a number a, add 0; the answer is identical to a.)	$-32 + 0 = -32$
$a \cdot 1 = 1 \cdot a = a$ (Start with a number a, multiply by 1; the answer is identical to a.)	$17.5 \cdot 1 = 17.5$
Commutative Properties	
$a + b = b + a$ and $ab = ba$ (Two terms or factors; reverse the order.)	$9 + (-3) = -3 + 9$ $6(-4) = (-4)6$
Associative Properties	
$a + (b + c) = (a + b) + c$ and $a(bc) = (ab)c$ (Three terms or factors; same order, parentheses shifted.)	$7 + (5 + 3) = (7 + 5) + 3$ $-4(6 \cdot 3) = (-4 \cdot 6)3$
Multiplication Property of 0	
$a \cdot 0 = 0$ and $0 \cdot a = 0$ (Multiplying any number by 0 gives 0.)	$4 \cdot 0 = 0$ $\quad 0(-3) = 0$

ANSWERS TO TEST YOUR WORD POWER

1. (c) *Example:* The set of whole numbers less than 0 is the empty set, written $\emptyset$. **2. (a)** *Examples:* a, b, c **3. (d)** *Examples:* $|2| = 2$ and $|-2| = 2$ **4. (b)** *Examples*: 3 is the reciprocal of $\frac{1}{3}$; $-\frac{5}{2}$ is the reciprocal of $-\frac{2}{5}$. **5. (d)** *Examples:* 2 and 5 are factors of 10 since both divide evenly (without remainder) into 10; other factors of 10 are $-10, -5, -2, -1, 1$, and 10. **6. (b)** *Examples:* 3^4 and x^{10} **7. (b)** *Examples:* $6, \frac{x}{2}, -4ab^2$ **8. (a)** *Examples:* The term $8z$ has numerical coefficient 8, and $-10x^3y$ has numerical coefficient -10.

Chapter 1 Review Exercises

If you need help with any of these Review Exercises, look in the section indicated in brackets.

[1.1] *Graph the elements of each set on a number line.*

1. $\left\{-4, -1, 2, \frac{9}{4}, 4\right\}$

2. $\left\{-5, -\frac{11}{4}, -.5, 0, 3, \frac{13}{3}\right\}$

Find the value of each expression.

3. $|-16|$

4. $|23|$

5. $-|-4|$

6. $|-8| - |-3|$

Let set $S = \{-9, -\frac{4}{3}, -\sqrt{4}, -.25, 0, .\overline{35}, \frac{5}{3}, \sqrt{7}, \sqrt{-9}, \frac{12}{3}\}$. *Simplify the elements of S as necessary, and then list the elements that belong to the specified set.*

7. Whole numbers

8. Integers

9. Rational numbers

10. Real numbers

Write each set by listing its elements.

11. $\{x \mid x$ is a natural number between 3 and 9$\}$

12. $\{y \mid y$ is a whole number less than 4$\}$

Write true *or* false *for each inequality.*

13. $4 \cdot 2 \leq |12 - 4|$

14. $2 + |-2| > 4$

15. $4(3 + 7) > -|40|$

The graph shows the percent change in car sales from January 2000 to January 2001 for various automakers. Use this graph to work Exercises 16–19.

16. Which automaker had the greatest change in sales? What was that change?

17. Which automaker had the smallest change in sales? What was that change?

18. *True* or *false:* The absolute value of the percent change for Honda was greater than the absolute value of the percent change for Toyota.

19. *True* or *false:* The percent change for Hyundai was more than four times greater than the percent change for Honda.

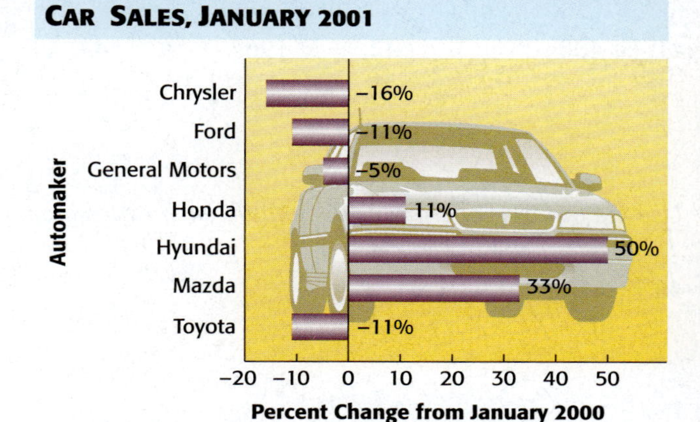

Source: Automakers.

[1.2] *Add or subtract as indicated.*

20. $-\frac{5}{8} - \left(-\frac{7}{3}\right)$

21. $-\frac{4}{5} - \left(-\frac{3}{10}\right)$

22. $-5 + (-11) + 20 - 7$

23. $-9.42 + 1.83 - 7.6 - 1.9$

24. $-15 + (-13) + (-11)$

25. $-1 - 3 - (-10) + (-7)$

26. $\frac{3}{4} - \left(\frac{1}{2} - \frac{9}{10}\right)$

27. $-\frac{2}{3} - \left(\frac{1}{6} - \frac{5}{9}\right)$

28. $-|-12| - |-9| + (-4) - |10|$

29. Telescope Peak, altitude 11,049 ft, is next to Death Valley, 282 ft below sea level. Find the difference between these altitudes. (*Source: World Almanac and Book of Facts*, 2000.)

Find each product or quotient.

30. $2(-5)(-3)(-3)$

31. $-\frac{3}{7}\left(-\frac{14}{9}\right)$

32. $-4.6(2.48)$

33. $\frac{75}{-5}$

34. $\frac{-2.3754}{-.74}$

35. Which one of the following is undefined: $\frac{5}{7-7}$ or $\frac{7-7}{5}$?

[1.3] *Evaluate each expression.*

36. 10^4

37. $\left(\frac{3}{7}\right)^3$

38. $(-5)^3$

39. -5^3

40. $(1.7)^2$

Find each square root. If it is not a real number, say so.

41. $\sqrt{400}$

42. $-\sqrt{196}$

43. $\sqrt{\frac{64}{121}}$

44. $-\sqrt{.81}$

45. $\sqrt{-64}$

Use the order of operations to simplify each expression.

46. $-14\left(\frac{3}{7}\right) + 6 \div 3$

47. $-\frac{2}{3}[5(-2) + 8 - 4^3]$

48. $\frac{-5(3^2) + 9(\sqrt{4}) - 5}{6 - 5(-2)}$

Evaluate each expression if $k = -4$, $m = 2$, and $n = 16$.

49. $4k - 7m$

50. $-3\sqrt{n} + m + 5k$

51. $\dfrac{4m^3 - 3n}{7k^2 - 10}$

52. The following expression for *body mass index* (BMI) can help determine ideal body weight.

$$704 \times (\text{weight in pounds}) \div (\text{height in inches})^2$$

A BMI of 19 to 25 corresponds to a healthy weight. (*Source: Washington Post.*)

(a) Derek Jeter is 6 ft 3 in. tall and weighs 195 lb. (*Source:* www.mlb.com) Find his BMI (to the nearest whole number).

(b) Calculate your BMI.

[1.4] *Use the properties of real numbers to simplify each expression.*

53. $2q + 19q$

54. $13z - 17z$

55. $-m + 6m$

56. $5p - p$

57. $-2(k + 3)$

58. $6(r + 3)$

59. $9(2m + 3n)$

60. $-(-p + 6q) - (2p - 3q)$

61. $-3y + 6 - 5 + 4y$

62. $2a + 3 - a - 1 - a - 2$

63. $-3(4m - 2) + 2(3m - 1) - 4(3m + 1)$

Complete each statement so that the indicated property is illustrated. Simplify each answer, if possible.

64. $2x + 3x =$ ________________ (distributive property)

65. $-4 \cdot 1 =$ ________________ (identity property)

66. $2(4x) =$ ________________ (associative property)

67. $-3 + 13 =$ ________________ (commutative property)

68. $-3 + 3 =$ ________________ (inverse property)

69. $5(x + z) =$ ________________ (distributive property)

70. $0 + 7 =$ ________________ (identity property)

71. $8 \cdot \dfrac{1}{8} =$ ________________ (inverse property)

72. $3a + 5a + 6a =$ ________________ (distributive property)

73. $\dfrac{9}{28} \cdot 0 =$ ________________ (multiplication property of 0)

MIXED REVIEW EXERCISES*

The table gives revenue and expenditures (both in millions of dollars) for Yahoo, Inc., for three different years.

Year	Revenue	Expenditures
1996	19.07	21.39
1997	67.41	90.29
1998	203.20	177.61

Source: www.quote.com

Determine the absolute value of the difference between revenue and expenditures for each year, and tell whether the company made a profit (i.e., was "in the black") or experienced a loss (i.e., was "in the red"). (These descriptions go back to the days when bookkeepers used black ink to represent gains and red ink to represent losses. This convention is still used. For example, Yahoo and on-line brokers display stock gains in black and losses in red.)

74. 1996

75. 1997

76. 1998

Perform the indicated operations.

77. $\left(-\frac{4}{5}\right)^4$

78. $-\frac{5}{8}(-40)$

79. $-25\left(-\frac{4}{5}\right) + 3^3 - 32 \div \sqrt{4}$

80. $-8 + |-14| + |-3|$

81. $\dfrac{6 \cdot \sqrt{4} - 3 \cdot \sqrt{16}}{-2 \cdot 5 + 7(-3) - 10}$

82. $-\sqrt{25}$

83. $-\frac{10}{21} \div \left(-\frac{5}{14}\right)$

84. $.8 - 4.9 - 3.2 + 1.14$

85. -3^2

86. $\dfrac{-38}{-19}$

87. $-2(k - 1) + 3k - k$

88. $-\sqrt{-100}$

89. Evaluate $-m(3k^2 + 5m)$ if $k = -4$ and $m = 2$.

90. To evaluate $(3 + 2)^2$, should you work within the parentheses first, or should you square 3 and square 2 and then add?

* The order of exercises in this final group does not correspond to the order in which topics occur in the chapter. This random ordering should help you prepare for the chapter test in yet another way.

Chapter 1 TEST

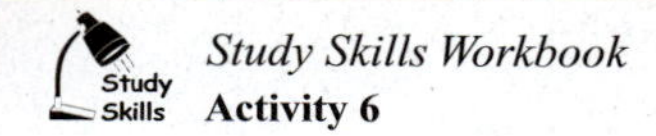

Study Skills Workbook
Activity 6

1. Graph $\{-3, .75, \frac{5}{3}, 5, 6.3\}$ on the number line.

Let $A = \{-\sqrt{6}, -1, -.5, 0, 3, \sqrt{25}, 7.5, \frac{24}{2}, \sqrt{-4}\}$. *First simplify each element as needed, and then list the elements from A that belong to each set.*

2. Whole numbers
3. Integers
4. Rational numbers
5. Real numbers

Perform the indicated operations.

6. $-6 + 14 + (-11) - (-3)$
7. $10 - 4 \cdot 3 + 6(-4)$
8. $7 - 4^2 + 2(6) + (-4)^2$
9. $\dfrac{10 - 24 + (-6)}{\sqrt{16}(-5)}$
10. $\dfrac{-2[3 - (-1 - 2) + 2]}{\sqrt{9}(-3) - (-2)}$
11. $\dfrac{8 \cdot 4 - 3^2 \cdot 5 - 2(-1)}{-3 \cdot 2^3 + 1}$

The table shows the heights in feet of some selected mountains and the depths in feet (as negative numbers) of some selected ocean trenches.

Mountain	Height	Trench	Depth
Foraker	17,400	Philippine	−32,995
Wilson	14,246	Cayman	−24,721
Pikes Peak	14,110	Java	−23,376

Source: World Almanac and Book of Facts, 2000.

12. What is the difference between the height of Mt. Foraker and the depth of the Philippine Trench?
13. What is the difference between the height of Pikes Peak and the depth of the Java Trench?
14. How much deeper is the Cayman Trench than the Java Trench?

1. ___
2. ___
3. ___
4. ___
5. ___
6. ___
7. ___
8. ___
9. ___
10. ___
11. ___
12. ___
13. ___
14. ___

15. ____________

16. ____________

17. ____________

18. (a) ____________

(b) ____________

(c) ____________

19. ____________

20. ____________

21. ____________

22. ____________

23. ____________

24. ____________

25. ____________

26. ____________

27. ____________

28. ____________

29. ____________

30. ____________

Find each square root. If it is not a real number, say so.

15. $\sqrt{196}$

16. $-\sqrt{225}$

17. $\sqrt{-16}$

18. For the expression $\sqrt{a}$, under what conditions will its value be **(a)** positive, **(b)** not real, **(c)** 0?

Evaluate each expression if $k = -3$, $m = -3$, *and* $r = 25$.

19. $\sqrt{r} + 2k - m$

20. $\dfrac{8k + 2m^2}{r - 2}$

21. Use the properties of real numbers to simplify

$$-3(2k - 4) + 4(3k - 5) - 2 + 4k.$$

22. How does the subtraction sign affect the terms $-4r$ and 6 when simplifying $(3r + 8) - (-4r + 6)$? What is the simplified form?

Match each statement in Column I with the appropriate property in Column II. Answers may be used more than once.

I	II
23. $6 + (-6) = 0$	**A.** Distributive property
24. $4 + 5 = 5 + 4$	**B.** Inverse property
25. $-2 + (3 + 6) = (-2 + 3) + 6$	**C.** Identity property
26. $5x + 15x = (5 + 15)x$	**D.** Associative property
27. $13 \cdot 0 = 0$	**E.** Commutative property
28. $-9 + 0 = -9$	**F.** Multiplication property of 0
29. $4 \cdot 1 = 4$	
30. $(a + b) + c = (b + a) + c$	

Linear Equations and Applications

In 1998, 995,000 college-bound students took the ACT exam. Of these students, 57% were female and 43% were male. Average composite scores on the ACT exam rose from 20.6 in 1990 to 21.0 in 1998. (*Source:* The ACT, Inc., Iowa City, IA, *High School Profile Report.*) In Section 2.3 we discuss the use of percent—one of the most common everyday applications of mathematics—and find the percent increases in numbers and scores of college-bound students taking the ACT exam.

2.1 Linear Equations in One Variable

Objectives

1. Decide whether a number is a solution of a linear equation.
2. Solve linear equations using the addition and multiplication properties of equality.
3. Solve linear equations using the distributive property.
4. Solve linear equations with fractions or decimals.
5. Identify conditional equations, contradictions, and identities.

Study Skills Workbook
Activity 2

1 Are the given numbers solutions of the given equations?

(a) $3k = 15$; 5

(b) $r + 5 = 4$; 1

(c) $-8m = 12$; $\frac{3}{2}$

In the previous chapter we began to use *algebraic expressions.* Some examples of algebraic expressions are

$$8x + 9, \quad y - 4, \quad \text{and} \quad \frac{x^3y^8}{z}. \qquad \text{Algebraic expressions}$$

Equations and inequalities compare algebraic expressions, just as a balance scale compares the weights of two quantities. Many applications of mathematics lead to *equations*, statements that two algebraic expressions are equal. A *linear equation in one variable* involves only real numbers and one variable raised to the first power. Examples are

$$x + 1 = -2, \quad x - 3 = 5, \quad \text{and} \quad 2k + 5 = 10. \qquad \text{Linear equations}$$

It is important to be able to distinguish between algebraic expressions and equations. *An equation always contains an equals sign, while an expression does not.*

Linear Equation in One Variable

A **linear equation in one variable** can be written in the form

$$ax = b,$$

where a and b are real numbers, with $a \neq 0$.

A linear equation is also called a **first-degree equation** since the highest power on the variable is one. Some examples of equations that are not linear (that is, *nonlinear*) are

$$x^2 + 3y = 5, \quad \frac{8}{x} = -22, \quad \text{and} \quad \sqrt{x} = 6. \qquad \text{Nonlinear equations}$$

1 Decide whether a number is a solution of a linear equation. If the variable in an equation can be replaced by a real number that makes the statement true, then that number is a **solution** of the equation. For example, 8 is a solution of the equation $x - 3 = 5$, since replacing x with 8 gives a true statement. An equation is *solved* by finding its **solution set,** the set of all solutions. The solution set of the equation $x - 3 = 5$ is $\{8\}$.

Work Problem 1 at the Side.

Equivalent equations are equations that have the same solution set. To solve an equation, we usually start with the given equation and replace it with a series of simpler equivalent equations. For example,

$$5x + 2 = 17, \quad 5x = 15, \quad \text{and} \quad x = 3$$

are all equivalent since each has the solution set $\{3\}$.

2 Solve linear equations using the addition and multiplication properties of equality. Two important properties that are used in producing equivalent equations are the **addition** and **multiplication properties of equality.**

Answers
1. (a) yes (b) no (c) no

Addition and Multiplication Properties of Equality

Addition Property of Equality
For all real numbers a, b, and c, the equations

$$a = b \quad \text{and} \quad a + c = b + c$$

are equivalent.

In words, the same number may be added to each side of an equation without changing the solution set.

Multiplication Property of Equality
For all real numbers a and b, and for $c \neq 0$, the equations

$$a = b \quad \text{and} \quad ac = bc$$

are equivalent.

In words, each side of an equation may be multiplied by the same nonzero number without changing the solution set.

Because subtraction and division are defined in terms of addition and multiplication, respectively, these properties can be extended: The same number may be subtracted from each side of an equation, and each side of an equation may be divided by the same nonzero number, without changing the solution set.

Example 1 Using the Addition and Multiplication Properties to Solve a Linear Equation

Solve $4x - 2x - 5 = 4 + 6x + 3$.

The goal is to use the addition and multiplication properties to get x alone on one side of the equation. First, combine like terms on each side of the equation to get

$$2x - 5 = 7 + 6x.$$

Next, use the addition property to get the terms with x on the same side of the equation and the remaining terms (the numbers) on the other side. One way to do this is to first add 5 to each side.

$$2x - 5 + 5 = 7 + 6x + 5 \qquad \text{Add 5.}$$
$$2x = 12 + 6x$$
$$2x - 6x = 12 + 6x - 6x \qquad \text{Subtract } 6x.$$
$$-4x = 12 \qquad \text{Combine like terms.}$$
$$\frac{-4x}{-4} = \frac{12}{-4} \qquad \text{Divide by } -4.$$
$$x = -3$$

To be sure that -3 is the solution, check by substituting for x in the *original* equation.

Check:

$$4x - 2x - 5 = 4 + 6x + 3 \qquad \text{Original equation}$$
$$4(-3) - 2(-3) - 5 = 4 + 6(-3) + 3 \quad ? \qquad \text{Let } x = -3.$$
$$-12 + 6 - 5 = 4 - 18 + 3 \quad ? \qquad \text{Multiply.}$$
$$-11 = -11 \qquad \text{True}$$

The true statement indicates that $\{-3\}$ is the solution set.

2 Solve and check.

(a) $3p + 2p + 1 = -24$

(b) $3p = 2p + 4p + 5$

(c) $4x + 8x = 17x - 9 - 1$

(d) $-7 + 3t - 9t = 12t - 5$

NOTE

Notice that in Example 1 the equality symbols are aligned in a column. Do not use more than one equality symbol in a horizontal line of work when solving an equation.

Work Problem 2 at the Side.

The steps needed to solve a linear equation in one variable are as follows. (Some equations may not require all of these steps.)

Solving a Linear Equation in One Variable

Step 1 **Clear fractions.** Eliminate any fractions by multiplying each side by the least common denominator.

Step 2 **Simplify each side separately.** Simplify each side of the equation as much as possible by using the distributive property to clear parentheses and by combining like terms as needed.

Step 3 **Isolate the variable terms on one side.** Use the addition property to get all terms with variables on one side of the equation and all numbers on the other.

Step 4 **Isolate the variable.** Use the multiplication property to get an equation with just the variable (with coefficient 1) on one side.

Step 5 **Check.** Check by substituting back into the original equation.

3 Solve linear equations using the distributive property. In Example 1 we did not use Step 1 or the distributive property in Step 2 as given in the box. Many equations, however, will require one or both of these steps, as shown in the next examples.

Example 2 **Using the Distributive Property to Solve a Linear Equation**

Solve $2(k - 5) + 3k = k + 6$.

Step 1 Since there are no fractions in this equation, Step 1 does not apply.

Step 2 Use the distributive property to simplify and combine terms on the left side of the equation.

$$2(k - 5) + 3k = k + 6$$
$$2k - 10 + 3k = k + 6 \quad \text{Distributive property}$$
$$5k - 10 = k + 6 \quad \text{Combine like terms.}$$

Step 3 Next, use the addition property of equality.

$$5k - 10 + 10 = k + 6 + 10 \quad \text{Add 10.}$$
$$5k = k + 16 \quad \text{Combine like terms.}$$
$$5k - k = k + 16 - k \quad \text{Subtract } k.$$
$$4k = 16 \quad \text{Combine like terms.}$$

Step 4 Use the multiplication property of equality to get just k on the left.

$$\frac{4k}{4} = \frac{16}{4} \quad \text{Divide by 4.}$$
$$k = 4$$

Step 5 Check that the solution set is $\{4\}$ by substituting 4 for k in the original equation.

ANSWERS

2. (a) $\{-5\}$ (b) $\left\{-\frac{5}{3}\right\}$ (c) $\{2\}$ (d) $\left\{-\frac{1}{9}\right\}$

NOTE

Because of space limitations, we will not always show the check when solving an equation. To be sure that your solution is correct, you should *always* check your work.

Work Problem 3 at the Side.

4 Solve linear equations with fractions or decimals. When fractions or decimals appear as coefficients in equations, our work can be made easier if we multiply each side of the equation by the least common denominator (LCD) of all the fractions. This is an application of the multiplication property of equality, and it produces an equivalent equation with integer coefficients.

Example 3 Solving a Linear Equation with Fractions

Solve $\frac{x+7}{6} + \frac{2x-8}{2} = -4$.

Step 1 Start by eliminating the fractions. Multiply each side by the LCD, 6.

$$6\left(\frac{x+7}{6} + \frac{2x-8}{2}\right) = 6(-4)$$

Step 2

$$6\left(\frac{x+7}{6}\right) + 6\left(\frac{2x-8}{2}\right) = 6(-4) \quad \text{Distributive property}$$

$$(x+7) + 3(2x-8) = -24 \quad \text{Multiply.}$$

$$x + 7 + 6x - 24 = -24 \quad \text{Distributive property}$$

$$7x - 17 = -24 \quad \text{Combine like terms.}$$

Step 3

$$7x - 17 + 17 = -24 + 17 \quad \text{Add 17.}$$

$$7x = -7$$

Step 4

$$\frac{7x}{7} = \frac{-7}{7} \quad \text{Divide by 7.}$$

$$x = -1$$

Step 5 Check by substituting -1 for x in the original equation.

$$\frac{x+7}{6} + \frac{2x-8}{2} = -4$$

$$\frac{-1+7}{6} + \frac{2(-1)-8}{2} = -4 \quad ? \quad \text{Let } x = -1.$$

$$\frac{6}{6} + \frac{-10}{2} = -4 \quad ?$$

$$1 - 5 = -4 \quad ?$$

$$-4 = -4 \quad \text{True}$$

The solution checks, so the solution set is $\{-1\}$.

Work Problem 4 at the Side.

3 Solve and check.

(a) $5p + 4(3 - 2p) = 2 + p - 10$

(b) $3(z - 2) + 5z = 2$

(c) $-2 + 3(x + 4) = 8x$

(d) $6 - (4 + m) = 8m - 2(3m + 5)$

4 Solve and check.

(a) $\frac{2p}{7} - \frac{p}{2} = -3$

(b) $\frac{k+1}{2} + \frac{k+3}{4} = \frac{1}{2}$

ANSWERS

3. (a) $\{5\}$ **(b)** $\{1\}$ **(c)** $\{2\}$ **(d)** $\{4\}$

4. (a) $\{14\}$ **(b)** $\{-1\}$

5 Solve and check.

(a) $.04x + .06(20 - x) = .05(50)$

(b) $.10(x - 6) + .05x = .06(50)$

In later sections we solve problems involving interest rates and concentrations of solutions. These problems involve percents that are converted to decimals. The equations that are used to solve such problems involve decimal coefficients. We can clear these decimals by multiplying by a power of ten that will allow us to obtain integer coefficients.

Example 4 Solving a Linear Equation with Decimals

Solve $.06x + .09(15 - x) = .07(15)$.

Because each decimal number is given in hundredths, multiply each side of the equation by 100. (Move the decimal points two places to the right.)

$$.06x + .09(15 - x) = .07(15)$$

$$.06x + .09(15 - x) = .07(15) \quad \text{Multiply by 100.}$$

$$6x + 9(15 - x) = 7(15)$$

$$6x + 135 - 9x = 105 \quad \text{Distributive property}$$

$$-3x + 135 = 105 \quad \text{Combine like terms.}$$

$$-3x + 135 - 135 = 105 - 135 \quad \text{Subtract 135.}$$

$$-3x = -30$$

$$\frac{-3x}{-3} = \frac{-30}{-3} \quad \text{Divide by } -3.$$

$$x = 10$$

Check to verify that the solution set is $\{10\}$.

Work Problem 5 at the Side.

5 Identify conditional equations, contradictions, and identities. All of the preceding equations had solution sets containing one element; for example, $2(k - 5) + 3k = k + 6$ has solution set $\{4\}$. Some linear equations, however, have no solutions, while others have an infinite number of solutions. The table below gives the names of these types of equations.

Type of Linear Equation	Number of Solutions	Indication When Solving
Conditional	One	Final line is $x =$ a number. (See Example 5(a).)
Contradiction	None; solution set $\emptyset$	Final line is false, such as $0 = 1$. (See Example 5(c).)
Identity	Infinite; solution set {all real numbers}	Final line is true, such as $0 = 0$. (See Example 5(b).)

Example 5 Recognizing Conditional Equations, Identities, and Contradictions

Solve each equation. Decide whether it is a *conditional equation*, an *identity*, or a *contradiction*.

(a) $5x - 9 = 4(x - 3)$

$$5x - 9 = 4x - 12 \quad \text{Distributive property}$$

$$5x - 9 - 4x = 4x - 12 - 4x \quad \text{Subtract } 4x.$$

$$x - 9 = -12 \quad \text{Combine like terms.}$$

$$x - 9 + 9 = -12 + 9 \quad \text{Add 9.}$$

$$x = -3$$

Continued on Next Page

ANSWERS

5. (a) $\{-65\}$ (b) $\{24\}$

The solution set, $\{-3\}$, has only one element, so $5x - 9 = 4(x - 3)$ is a conditional equation.

(b) $5x - 15 = 5(x - 3)$

Use the distributive property to clear parentheses on the right side.

$$5x - 15 = 5x - 15$$

$$0 = 0 \qquad \text{Subtract } 5x \text{ and add 15.}$$

The final line, $0 = 0$, indicates that the solution set is {all real numbers}, and the equation $5x - 15 = 5(x - 3)$ is an identity. (*Note:* The first step yielded $5x - 15 = 5x - 15$, which is true for all values of x. We could have identified the equation as an identity at that point.)

(c) $5x - 15 = 5(x - 4)$

$$5x - 15 = 5x - 20 \qquad \text{Distributive property}$$

$$5x - 15 - 5x = 5x - 20 - 5x \qquad \text{Subtract } 5x.$$

$$-15 = -20 \qquad \text{False}$$

Since the result, $-15 = -20$, is *false*, the equation has no solution. The solution set is $\emptyset$, so the equation $5x - 15 = 5(x - 4)$ is a contradiction.

Work Problem 6 at the Side.

6 Solve each equation. Decide whether it is a *conditional equation*, an *identity*, or a *contradiction*. Give the solution set.

(a) $5(x + 2) - 2(x + 1) = 3x + 1$

(b) $\dfrac{x + 1}{3} + \dfrac{2x}{3} = x + \dfrac{1}{3}$

(c) $5(3x + 1) = x + 5$

ANSWERS

6. (a) contradiction; $\emptyset$

(b) identity; {all real numbers}

(c) conditional; $\{0\}$

Focus on Real-Data Applications

International Time Zones

Companies that operate globally, such as Coca-Cola and BP Amoco, must contend with time-zone differences between offices located in different countries. To ensure that everyone experiences daylight during morning and afternoon hours, the world is divided into 24 time zones. (Why 24?) Longitudinal lines are drawn between the North Pole and South Pole and are measured in degrees of longitude between 0° and 360°. The 0° longitudinal meridian passes through Greenwich, England (a suburb of London) and is called the **prime meridian.** The time at the prime meridian is called **Greenwich Mean Time (GMT).** The longitudinal meridians increase going west, and the 180° longitudinal meridian is called the **International Date Line.** When the International Date Line is crossed going west, the date is advanced one day; when it is crossed going east, the date becomes one day earlier. A map of the International Time Zones is shown here.

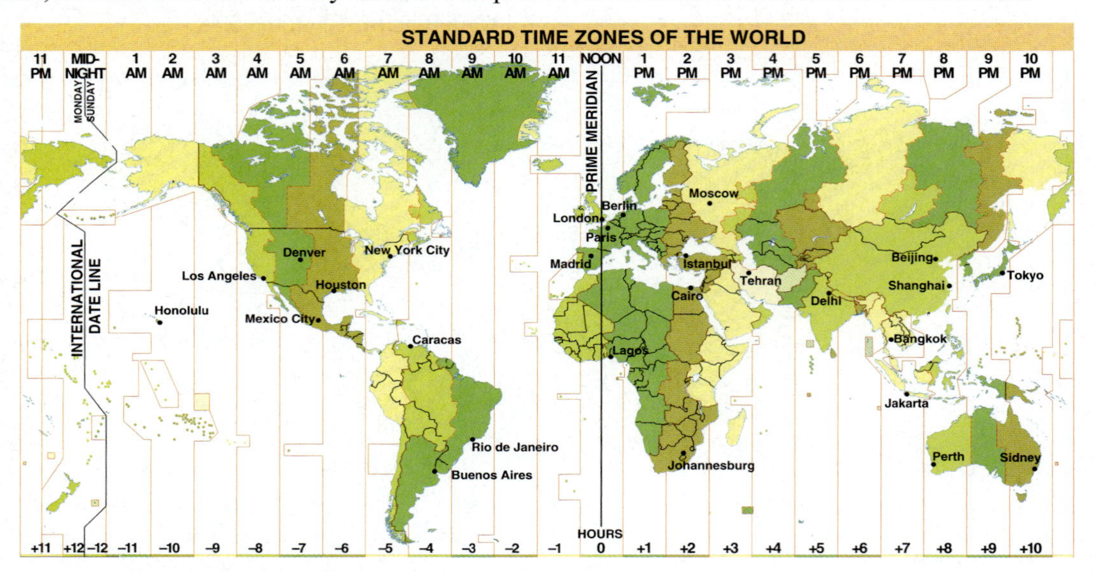

For Group Discussion

A London-based company employs a courier to deliver important documents between international offices. The courier adjusts her watch to local times as she travels between cities. For example, traveling from London to New York and then to Paris, she would have to adjust her watch $-5 + 6$ hr. In Paris, the time would be one hour ahead of GMT $(+1)$. Write a similar integer expression to describe the changes in clock settings at each local time zone for the courier's trip. By how many hours does the time at the final location differ from GMT time?

1. London to Tokyo to Cairo to Houston

2. London to Los Angeles to Caracas to Johannesburg to Bangkok to Paris

3. London to Mexico City to Honolulu to Tokyo (Remember the International Date Line.)

4. Determine one possible route, starting in London, that matches the time changes given by the integer expression $-4 - 4 + 5 + 5 + 6$.

2.1 Exercises

FOR EXTRA HELP

Student's Solutions Manual

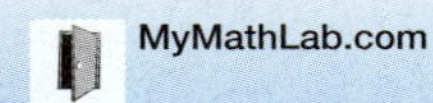
MyMathLab.com

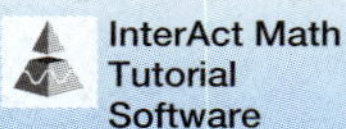
InterAct Math Tutorial Software

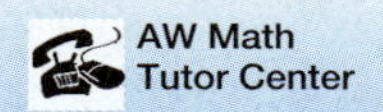
AW Math Tutor Center

www.mathxl.com

Digital Video Tutor CD 1 Videotape 2

Study Skills Workbook **Activity 3**

1. Which equations are linear equations in x?

A. $3x + x - 1 = 0$ **B.** $8 = x^2$

C. $6x + 2 = 9$ **D.** $\frac{1}{2}x - \frac{1}{x} = 0$

2. Which of the equations in Exercise 1 are nonlinear equations in x? Explain why.

3. Decide whether 6 is a solution of $3(x + 4) = 5x$ by substituting 6 for x. If it is not a solution, explain why.

4. Use substitution to decide whether -2 is a solution of $5(x + 4) - 3(x + 6) = 9(x + 1)$. If it is not a solution, explain why.

5. The equation $4[x + (2 - 3x)] = 2(4 - 4x)$ is an identity. Let x represent the number of letters in your last name. Is this number a solution of this equation? Check your answer.

6. The expression $.06(10 - x)(100)$ is equivalent to which of the following?

A. $.06 - .06x$ **B.** $60 - 6x$

C. $6 - 6x$ **D.** $6 - .06x$

7. Identify each as an *expression* or an *equation.*

(a) $3x = 6$

(b) $3x + 6$

(c) $5x + 6(x - 3) = 12x + 6$

(d) $5x + 6(x - 3) - (12x + 6)$

8. Explain why $6x + 9 = 6x + 8$ cannot have a solution. (No work is necessary.)

Solve and check each equation. See Examples 1 and 2.

9. $7x + 8 = 1$

10. $5x - 4 = 21$

11. $7x - 5x + 15 = x + 8$

12. $2x + 4 - x = 4x - 5$

13. $12w + 15w - 9 + 5 = -3w + 5 - 9$

14. $-4t + 5t - 8 + 4 = 6t - 4$

15. $2(x + 3) = -4(x + 1)$

16. $4(t - 9) = 8(t + 3)$

17. $3(2w + 1) - 2(w - 2) = 5$

18. $4(x - 2) + 2(x + 3) = 6$

19. $2x + 3(x - 4) = 2(x - 3)$

20. $6x - 3(5x + 2) = 4(1 - x)$

21. $6p - 4(3 - 2p) = 5(p - 4) - 10$

22. $-2k - 3(4 - 2k) = 2(k - 3) + 2$

23. $-[2z - (5z + 2)] = 2 + (2z + 7)$

24. $-[6x - (4x + 8)] = 9 + (6x + 3)$

25. $-(9 - 3x) - (4 + 2x) = -(2 - 5x) + (-x) + 4$

26. $-(-2 + 4x) - (3 - 4x) + 4 = -(-3 + 6x) + x$

27. $-3(x + 2) + 4(3x - 8) = 2(4x + 7) + 2(3x - 6)$

28. $-7(2x + 1) + 5(3x + 2) = 6(2x - 4) - (12x + 3)$

29. In order to solve the linear equation

$$\frac{8x}{3} - \frac{2x}{4} = -13,$$

we are allowed to multiply each side by the least common denominator of all the fractions in the equation. What is this least common denominator?

30. In order to solve the linear equation

$$.05x + .12(x + 5000) = 940,$$

we are allowed to multiply each side by a power of ten so that all coefficients are integers. What is the smallest power of ten that will accomplish this goal?

31. Suppose that in solving the equation

$$\frac{1}{3}x + \frac{1}{2}x = \frac{1}{6}x,$$

you begin by multiplying each side by 12, rather than the *least* common denominator, 6. Would you get the correct solution anyway? Explain.

32. What is the final line in the check for the solution of the equation in Example 4?

Solve and check each equation. See Examples 3 and 4.

33. $\frac{3x}{4} + \frac{5x}{2} = 13$

34. $\frac{8x}{3} - \frac{2x}{4} = -13$

35. $\frac{x - 8}{5} + \frac{8}{5} = -\frac{x}{3}$

36. $\frac{2r - 3}{7} + \frac{3}{7} = -\frac{r}{3}$

37. $\frac{4t + 1}{3} = \frac{t + 5}{6} + \frac{t - 3}{6}$

38. $\frac{2x + 5}{5} = \frac{3x + 1}{2} + \frac{-x + 7}{2}$

39. $.05x + .12(x + 5000) = 940$

40. $.09k + .13(k + 300) = 61$

41. $.02(50) + .08r = .04(50 + r)$

42. $.20(14{,}000) + .14t = .18(14{,}000 + t)$

43. $.05x + .10(200 - x) = .45x$

44. $.08x + .12(260 - x) = .48x$

45. Explain the distinction between a conditional equation, an identity, and a contradiction.

46. A student tried to solve the equation $8x = 7x$ by dividing each side by x, obtaining $8 = 7$. He gave the solution set as $\emptyset$. Why is this incorrect?

Decide whether each equation is a conditional equation, *an* identity, *or a* contradiction. *Give the solution set. See Example 5.*

47. $-2x + 5x - 9 = 3(x - 4) - 5$

48. $-6x + 2x - 11 = -2(2x - 3) + 4$

49. $-11x + 4(x - 3) + 6x = 4x - 12$

50. $3x - 5(x + 4) + 9 = -11 + 15x$

51. $7[2 - (3 + 4x)] - 2x = -9 + 2(1 - 15x)$

52. $4[6 - (1 + 2x)] + 10x = 2(10 - 3x) + 8x$

53. If two equations are equivalent, they have the same ________ ________.

54. Which equation is equivalent to $3x = 9$?

A. $x = -3$ **B.** $x^2 = 9$

C. $4x = -1$ **D.** $5x = 15$

Two equations with the same solution set are equivalent. In Exercises 55–58, the given pair of equations are not *equivalent. Explain why this is so.*

55. $3x + 2 = 2x + 2$
$3x + 6 = 3x + 7$

56. $2x - 3 = -1$
$2x - 1 = -3$

57. $k = 4$
$k^2 = 16$

58. $x = -6$
$x^2 = 36$

Solve each problem.

59. This graph indicates the number of tickets sold for Broadway shows in the 1990s.

Source: League of American Theaters and Producers.

For the period shown in the graph, between which two seasons did ticket sales increase the most?

60. The linear equation

$$y = .55x - 42.5$$

models the number of Broadway tickets sold from 1990 through 1997. In the equation, $x = 90$ corresponds to the 1990–1991 season, $x = 91$ corresponds to the 1991–1992 season, and so on, and y is the number of tickets sold in millions.

(a) Based on this model, how many tickets were sold in 1995–1996? (*Hint:* 1995–1996 corresponds to $x = 95$.)

(b) Based on the model, in what season did the number of tickets sold reach or exceed 7.9 million? Round up to the nearest year. (*Hint:* The number of tickets sold in millions corresponds to y.)

61. Use the equation in Exercise 60 to find the number of tickets sold in 1994–1995. Is your answer a good approximation of the number shown in the graph in Exercise 59?

62. The table shows the number of tickets (in millions) sold to theatergoers under age 18.

Season	Millions of Tickets
1990–1991	.5
1996–1997	1.1

Source: League of American Theaters and Producers.

If $x = 90$ corresponds to the 1990–1991 season, and so on, the linear equation

$$y = .1x - 8.5$$

approximates the number under age 18 attending the theater.

(a) Use the equation to estimate the number of these youths attending in 1993–1994.

(b) In what season did ticket sales reach or exceed .75 million? Round up to the nearest year.

2.2 FORMULAS

Models for many applied problems already exist; they are called *formulas.* A **formula** is a mathematical equation in which variables are used to describe a relationship. Some formulas that we will be using are

$$d = rt, \quad I = prt, \quad \text{and} \quad P = 2L + 2W.$$

A list of some common formulas used in algebra is given inside the covers of the book.

OBJECTIVES

1. Solve a formula for a specified variable.
2. Solve applied problems using formulas.
3. Solve percent problems.

1 Solve a formula for a specified variable. In some applications, the appropriate formula may be solved for a different variable than the one to be found. For example, the formula $I = prt$ says that interest on a loan or investment equals principal (amount borrowed or invested) times rate (percent) times time at interest (in years). To determine how long it will take for an investment at a stated interest rate to earn a predetermined amount of interest, it would help to first solve the formula for t. This process is called **solving for a specified variable.**

The steps used in the following examples are very similar to those used in solving linear equations. When you are solving for a specified variable, the key is to treat that variable as if it were the only one; treat all other variables like numbers (constants).

Example 1 Solving for a Specified Variable

Solve the formula $I = prt$ for t.

We solve this formula for t by assuming that I, p, and r are constants (having fixed values) and that t is the variable. We first write the formula so that the variable for which we are solving, t, is on the left side. Then we use the properties of the previous section as follows.

$$prt = I$$

$$(pr)t = I \qquad \text{Associative property}$$

$$\frac{(pr)t}{pr} = \frac{I}{pr} \qquad \text{Divide by } pr.$$

$$t = \frac{I}{pr}$$

The result is a formula for t, time in years.

Work Problem 1 at the Side.

1 Solve $I = prt$ for each given variable.

(a) p

(b) r

ANSWERS

1. (a) $p = \frac{I}{rt}$ (b) $r = \frac{I}{pt}$

While the process of solving for a specified variable uses the same steps as solving a linear equation from Section 2.1, the following additional suggestions may be helpful.

Solving for a Specified Variable

Step 1 Get all terms containing the specified variable on one side of the equation and all terms without that variable on the other side.

Step 2 If necessary, use the distributive property to combine the terms with the specified variable.* The result should be the product of a sum or difference and the variable.

Step 3 Divide each side by the factor that is the coefficient of the specified variable.

Example 2 **Solving for a Specified Variable**

Solve the formula $P = 2L + 2W$ for W.

This formula gives the relationship between the perimeter of (distance around) a rectangle, P, the length of the rectangle, L, and the width of the rectangle, W. See Figure 1.

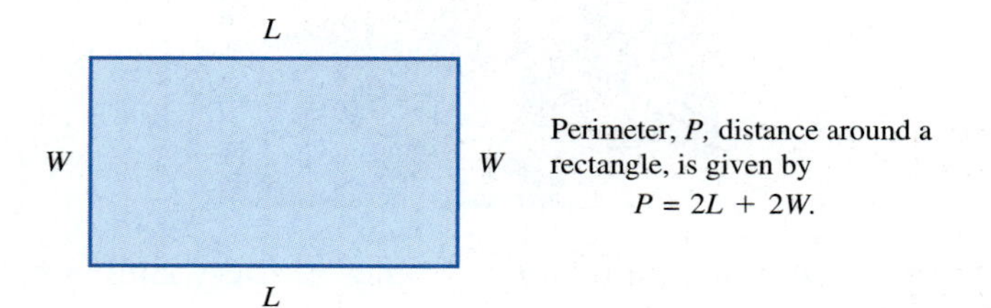

Figure 1

Solve the formula for W by getting W alone on one side of the equals sign. To begin, subtract $2L$ from each side.

$$P = 2L + 2W$$

Step 1 $$P - 2L = 2L + 2W - 2L \quad \text{Subtract } 2L.$$

$$P - 2L = 2W$$

Step 2 is not needed here.

Step 3 $$\frac{P - 2L}{2} = \frac{2W}{2} \quad \text{Divide by 2.}$$

$$\frac{P - 2L}{2} = W$$

$$W = \frac{P - 2L}{2}$$

*Using the distributive property to write $ab + ac$ as $a(b + c)$ is called *factoring*. See Chapter 6.

CAUTION

In Step 3 of Example 2, you cannot simplify the fraction by dividing 2 into the term $2L$. The subtraction in the numerator must be done before the division.

$$\frac{P - 2L}{2} \neq P - L$$

Work Problem 2 at the Side.

A rectangular solid has the shape of a box, but is solid. See Figure 2. The labels H, W, and L represent the height, width, and length of the figure, respectively.

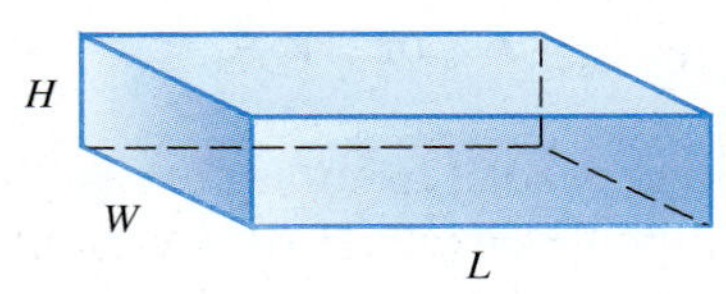

Figure 2

The surface area of any solid three-dimensional figure is the total area of its surface. For a rectangular solid, the surface area A is

$$A = 2HW + 2LW + 2LH.$$

Example 3 **Using the Distributive Property to Solve for a Specified Variable**

Given the surface area, height, and width of a rectangular solid, write a formula for the length.

To solve for the length L, treat L as the only variable and treat all other variables as constants.

$$A = 2HW + 2LW + 2LH$$

$$A - 2HW = 2LW + 2LH \quad \text{Subtract } 2HW.$$

$$A - 2HW = L(2W + 2H) \quad \text{Use the distributive property on the right side.}$$

$$\frac{A - 2HW}{2W + 2H} = L \quad \text{Divide by } 2W + 2H.$$

$$L = \frac{A - 2HW}{2W + 2H}$$

CAUTION

The most common error in working a problem like Example 3 is not using the distributive property correctly. We must write the expression so that the specified variable is a *factor*; then we can divide by its coefficient in the final step.

Work Problem 3 at the Side.

2 **(a)** Solve the formula

$$P = a + b + c$$

for a.

(b) Solve the formula

$$m = 2k + 3b$$

for k.

3 Solve the formula

$$A = 2HW + 2LW + 2LH$$

for W.

ANSWERS

2. **(a)** $a = P - b - c$ **(b)** $k = \frac{m - 3b}{2}$

3. $W = \frac{A - 2LH}{2H + 2L}$

4 (a) A triangle has an area of 36 in.^{2} (square inches) and a base of 12 in. Find its height.

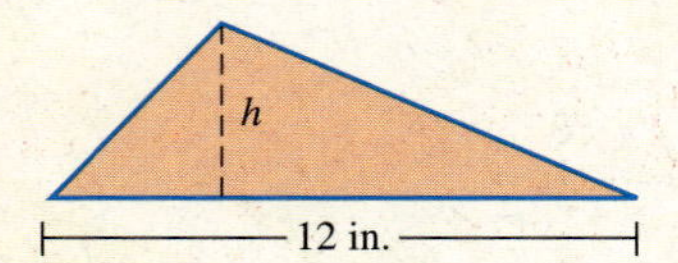

(b) In 1997 Mark Martin won the Talledega 500 (mile) race with a speed of 188.345 mph. (*Source: Sports Illustrated 2000 Sports Almanac.*) Find his time to the nearest thousandth.

ANSWERS
4. (a) 6 in. **(b)** 2.655 hr

2 Solve applied problems using formulas. The next example uses the distance formula, $d = rt$, which relates d, the distance traveled, r, the rate or speed, and t, the travel time.

Example 4 Finding Average Speed

Janet Branson found that on average it took her $\frac{3}{4}$ hr each day to drive a distance of 15 mi to work. What was her average speed?

Find the speed r by solving $d = rt$ for r.

$$d = rt$$

$$\frac{d}{t} = \frac{rt}{t} \quad \text{Divide by } t.$$

$$\frac{d}{t} = r \quad \text{or} \quad r = \frac{d}{t}$$

Notice that only Step 3 was needed to solve for r in this example. Now find the speed by substituting the given values of d and t into this formula.

$$r = \frac{15}{\frac{3}{4}} \quad \text{Let } d = 15, t = \tfrac{3}{4}.$$

$$r = 15 \cdot \frac{4}{3} \quad \text{Multiply by the reciprocal of } \tfrac{3}{4}.$$

$$r = 20$$

Her average speed was 20 mph.

Work Problem 4 at the Side.

3 Solve percent problems. An important everyday use of mathematics involves the concept of percent. Percent is written with the symbol %. The word **percent** means "per one hundred." One percent means "one per one hundred" or "one one-hundredth."

$$1\% = .01 \quad \text{or} \quad 1\% = \frac{1}{100}$$

Solving a Percent Problem

Let a represent a partial amount of b, the base, or whole amount. Then the following formula can be used in solving a percent problem.

$$\frac{\text{amount}}{\text{base}} = \frac{a}{b} = \text{percent (represented as a decimal)}$$

For example, if a class consists of 50 students and 32 are males, then the percent of males in the class is

$$\frac{\text{amount}}{\text{base}} = \frac{a}{b}$$

$$= \frac{32}{50} \quad \text{Let } a = 32, b = 50.$$

$$= .64 = 64\%.$$

Example 5 **Solving Percent Problems**

(a) A 50-L mixture of acid and water contains 10 L of acid. What is the percent of acid in the mixture?

The given amount of the mixture is 50 L, and the part that is acid (percentage) is 10 L. Let x represent the percent of acid. Then, the percent of acid in the mixture is

$$x = \frac{10}{50}$$

$$x = .20 = 20\%.$$

(b) If a savings account balance of $3550 earns 8% interest in one year, how much interest is earned?

Let x represent the amount of interest earned (that is, the part of the whole amount invested). Since 8% = .08, the equation is

$$\frac{x}{3550} = .08 \qquad \frac{a}{b} = \text{percent}$$

$$x = .08(3550) \qquad \text{Multiply by 3550.}$$

$$x = 284.$$

The interest earned is $284.

Work Problem 5 at the Side.

Graphs sometimes represent the percents of a whole amount that satisfy certain conditions, as shown in the next example.

Example 6 **Interpreting Percents from a Graph**

In 1998, the United States consumed 94.2 quadrillion Btu of energy. Use the graph in Figure 3 to determine how many Btu were produced using nuclear electric power.

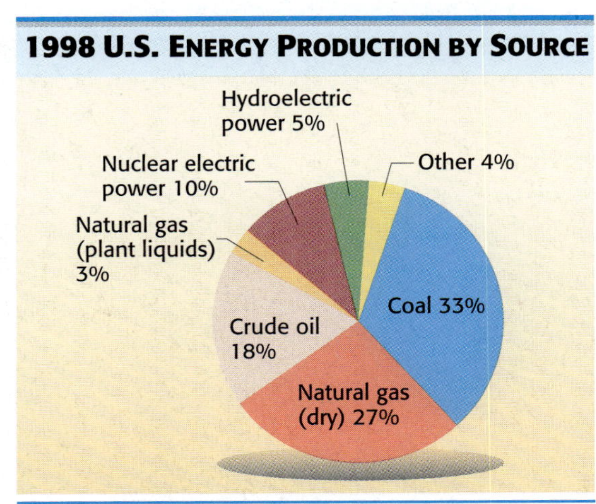

Source: U.S. Energy Department.

Figure 3

According to the graph, 10% of the energy was produced using nuclear electric power. Let x represent the required number of Btu.

Continued on Next Page

5 **(a)** A mixture of gasoline and oil contains 20 oz, 1 oz of which is oil. What percent of the mixture is oil?

(b) An automobile salesman earns an 8% commission on every car he sells. How much does he earn on a car that sells for $12,000?

ANSWERS

5. **(a)** 5% **(b)** $960

6 Refer to Figure 3. How much energy was produced by natural gas (plant liquids)?

$$\frac{x}{94.2} = .10$$

$$x = .10(94.2) \quad \text{Multiply by 94.2.}$$

$$x = 9.42$$

Therefore, 9.42 quadrillion Btu were produced using nuclear methods.

Work Problem 6 at the Side.

ANSWERS

6. 2.826 quadrillion Btu

2.2 EXERCISES

FOR EXTRA HELP

Student's Solutions Manual

MyMathLab.com

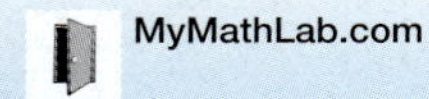
InterAct Math Tutorial Software

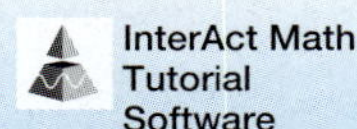
AW Math Tutor Center

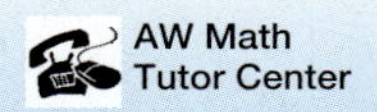
www.mathxl.com

Digital Video Tutor CD 1 Videotape 2

RELATING CONCEPTS (Exercises 1–6) FOR INDIVIDUAL OR GROUP WORK

Consider the following equations:

First Equation	*Second Equation*
$x = \frac{5x + 8}{3}$	$t = \frac{bt + k}{c} \quad (c \neq 0).$

Solving the second equation for t requires the same logic as solving the first equation for x. When solving for t, we treat all other variables as though they were constants. ***Work Exercises 1–6 in order,*** *to see the "parallel logic" of solving for x and solving for t.*

1. (a) Clear the first equation of fractions by multiplying each side by 3.

(b) Clear the second equation of fractions by multiplying each side by c.

2. (a) Get the terms involving x on the left side of the first equation by subtracting $5x$ from each side.

(b) Get the terms involving t on the left side of the second equation by subtracting bt from each side.

3. (a) Combine like terms on the left side of the first equation. What property allows us to write $3x - 5x$ as $(3 - 5)x = -2x$?

(b) Write the expression on the left side of the second equation so that t is a factor. What property allows us to do this?

4. (a) Divide each side of the first equation by the coefficient of x.

(b) Divide each side of the second equation by the coefficient of t.

5. Look at your answer for the second equation. What restriction must be placed on the variables? Why is this necessary?

6. Write a short paragraph summarizing what you have learned in this group of exercises.

Solve each formula for the specified variable. See Examples 1 and 2.

7. $I = prt$ for r (simple interest)

8. $d = rt$ for t (distance)

9. $P = 2L + 2W$ for L
(perimeter of a rectangle)

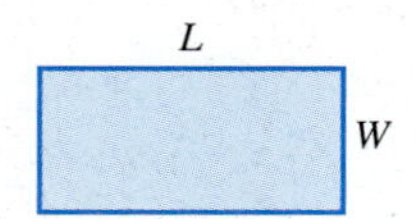

10. $A = bh$ for b
(area of a parallelogram)

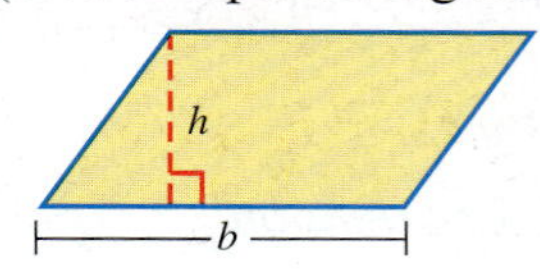

11. $V = LWH$ for W
(volume of a rectangular solid)

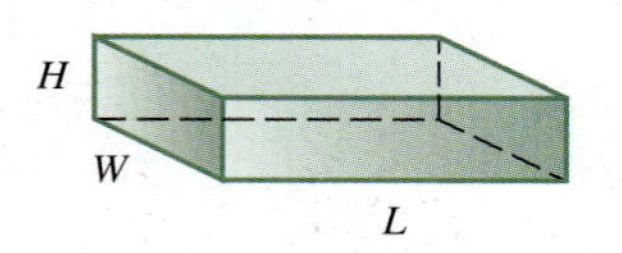

12. $P = a + b + c$ for b
(perimeter of a triangle)

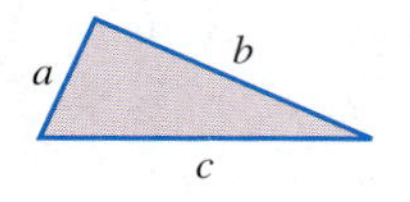

13. $C = 2\pi r$ for r (circumference of a circle)

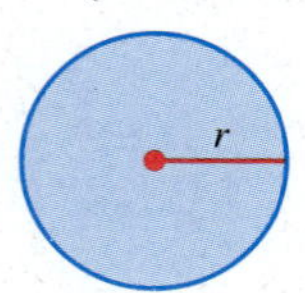

14. $A = \frac{1}{2}bh$ for h (area of a triangle)

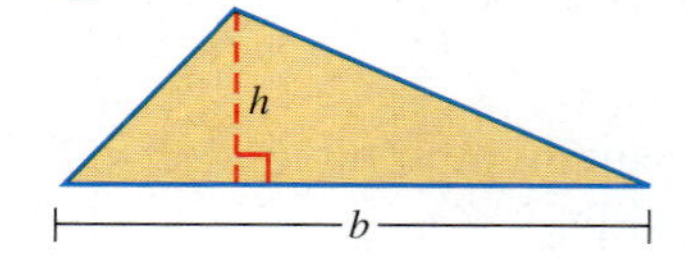

15. $A = \frac{1}{2}h(B + b)$ for B (area of a trapezoid)

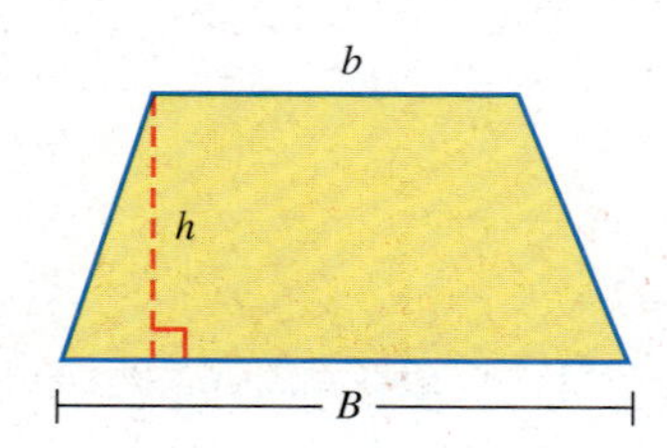

16. $S = 2\pi rh + 2\pi r^2$ for h
(surface area of a right circular cylinder)

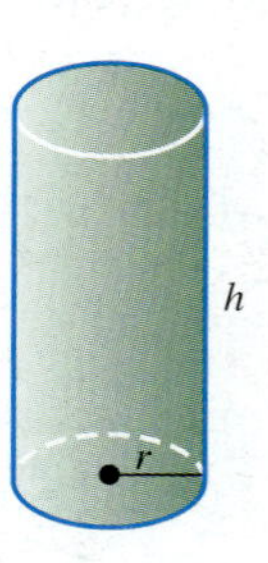

17. $F = \frac{9}{5}C + 32$ for C (Celsius to Fahrenheit)

18. $C = \frac{5}{9}(F - 32)$ for F (Fahrenheit to Celsius)

19. When a formula is solved for a particular variable, several different equivalent forms may be possible. If we solve $A = \frac{1}{2}bh$ for h, one possible correct answer is

$$h = \frac{2A}{b}.$$

Which one of the following is *not* equivalent to this?

A. $h = 2\left(\frac{A}{b}\right)$ **B.** $h = 2A\left(\frac{1}{b}\right)$

C. $h = \frac{A}{\frac{1}{2}b}$ **D.** $h = \frac{\frac{1}{2}A}{b}$

20. Suppose the formula

$$A = 2HW + 2LW + 2LH$$

is solved for L as follows.

$$A = 2HW + 2LW + 2LH$$
$$A - 2LW - 2HW = 2LH$$
$$\frac{A - 2LW - 2HW}{2H} = L$$

While there are no algebraic errors here, what is wrong with the final equation, if we are interested in solving for L?

Solve each equation for the specified variable. Use the distributive property to factor as necessary. See Example 3.

21. $2k + ar = r - 3y$ for r

22. $4s + 7p = tp - 7$ for p

23. $w = \frac{3y - x}{y}$ for y

24. $c = \frac{-2t + 4}{t}$ for t

Solve each problem. See Example 4.

25. In 1997 Jeff Gordon won the World 600 (mile) race with a speed of 136.745 mph. Find his time to the nearest thousandth. (*Source: Sports Illustrated 1998 Sports Almanac.*)

26. In 1975, rain shortened the Indianapolis 500 race to 435 mi. It was won by Bobby Unser, who averaged 149.213 mph. What was his time to the nearest thousandth? (*Source: Sports Illustrated 1998 Sports Almanac.*)

27. The highest temperature ever recorded in Chicago was 40°C. Find the corresponding Fahrenheit temperature. (*Source: World Almanac and Book of Facts*, 2000.)

28. The lowest temperature recorded in Salt Lake City in 1997 was 8°F. Find the corresponding Celsius temperature. (*Source: World Almanac and Book of Facts*, 1999.)

29. The base of the Great Pyramid of Cheops is a square whose perimeter is 920 m. What is the length of each side of this square? (*Source: Atlas of Ancient Archaeology*, 1994.)

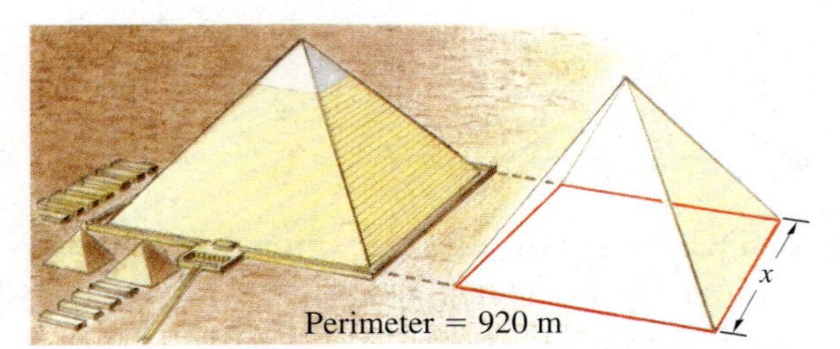

30. The Peachtree Plaza Hotel in Atlanta is in the shape of a cylinder with radius 46 m and height 220 m. Find its volume to the nearest tenth. (*Hint:* Use the π key on your calculator.)

31. The circumference of a circle is 480π in. What is its radius? What is its diameter?

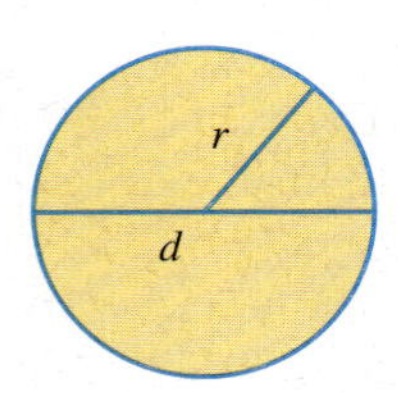

32. The radius of a circle is 2.5 in. What is its diameter? What is its circumference?

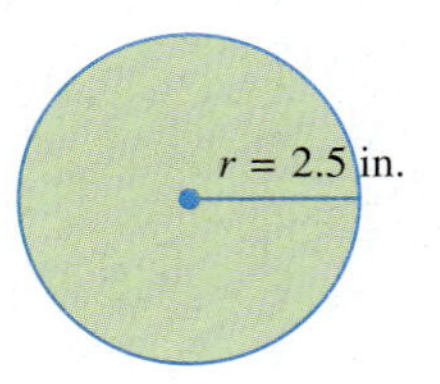

33. A cord of wood contains 128 ft^3 (cubic feet) of wood. If a stack of wood is 4 ft wide and 4 ft high, how long must it be if it contains exactly 1 cord?

34. Give one set of possible dimensions for a stack of wood that contains 1.5 cords. (See Exercise 33.)

Solve each problem. See Example 5.

35. A mixture of alcohol and water contains a total of 36 oz of liquid. There are 9 oz of pure alcohol in the mixture. What percent of the mixture is water? What percent is alcohol?

36. A mixture of acid and water is 35% acid. If the mixture contains a total of 40 L, how many liters of pure acid are in the mixture? How many liters of pure water are in the mixture?

37. A real estate agent earned $6300 commission on a property sale of $210,000. What is her rate of commission?

38. A certificate of deposit for one year pays $221 simple interest on a principal of $3400. What is the interest rate being paid on this deposit?

When a consumer loan is paid off ahead of schedule, the finance charge is smaller than if the loan were paid off over its scheduled life. By one method, called the rule of 78, *the amount of* unearned interest *(finance charge that need not be paid) is given by*

$$u = f \cdot \frac{k(k + 1)}{n(n + 1)},$$

where u is the amount of unearned interest (money saved) when a loan scheduled to run n payments is paid off k payments ahead of schedule. The total scheduled finance charge is f. Use this formula to solve Exercises 39–42.

39. Rhonda Alessi bought a new Ford and agreed to pay it off in 36 monthly payments. The total finance charge is $700. Find the unearned interest if she pays the loan off 4 payments ahead of schedule.

40. Charles Vosburg bought a car and agreed to pay it off in 36 monthly payments. The total finance charge on the loan was $600. With 12 payments remaining, Charles decided to pay the loan in full. Find the amount of unearned interest.

41. The finance charge on a loan taken out by Vic Denicola is $380.50. If there were 24 equal monthly installments needed to repay the loan, and the loan is paid in full with 8 months remaining, find the amount of unearned interest.

42. Adrian Ortega is scheduled to repay a loan in 24 equal monthly installments. The total finance charge on the loan is $450. With 9 payments remaining, he decides to repay the loan in full. Find the amount of unearned interest.

Exercises 43 and 44 deal with winning percentage in the standings of sports teams.

43. At the start of play on September 15, 2000, the standings of the Central Division of the American League were as shown. Winning percentage (Pct.) is commonly expressed as a decimal rounded to the nearest thousandth. To find the winning percentage of a team, divide the number of wins (W) by the total number of games played (W + L). Find the winning percentage of each team.

(a) Chicago **(b)** Cleveland **(c)** Detroit

	W	L	Pct.
Chicago	87	58	______
Cleveland	77	65	______
Detroit	71	74	______
Kansas City	68	78	.466
Minnesota	63	82	.434

44. Repeat Exercise 43 for the following standings for the Eastern Division of the National League.

(a) Atlanta **(b)** New York **(c)** Florida

	W	L	Pct.
Atlanta	86	60	______
New York	84	62	______
Florida	69	76	______
Montreal	61	84	.421
Philadelphia	60	85	.414

Television networks have been losing viewers to cable programming since 1982, as the two graphs show. Use these graphs to answer Exercises 45–48. See Example 6.

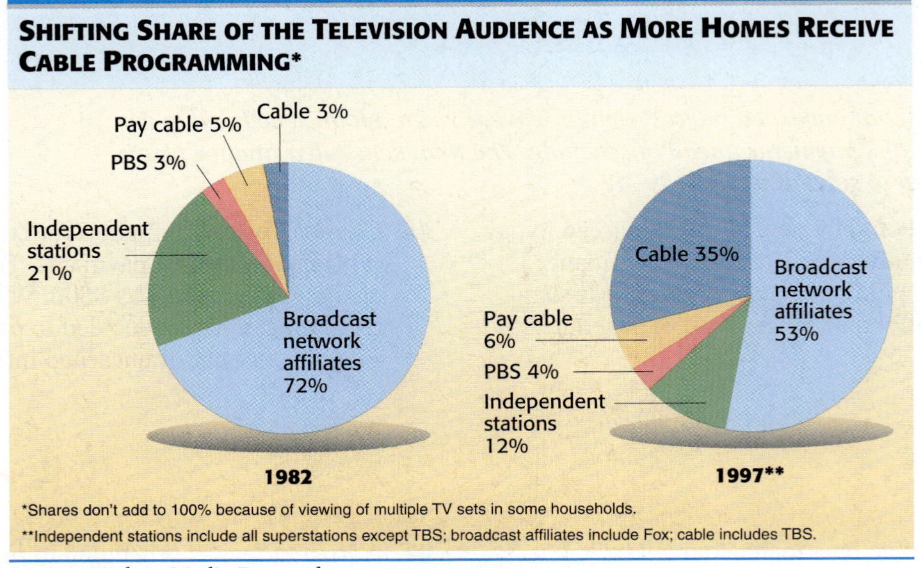

Source: Nielsen Media Research.

45. In a typical group of 50,000 television viewers, how many would have watched cable in 1982?

46. In 1982, how many of a typical group of 110,000 viewers watched independent stations?

47. How many of a typical group of 35,000 viewers watched cable in 1997?

48. In a typical group of 65,000 viewers, how many watched independent stations in 1997?

An average middle-income family will spend \$160,140 *to raise a child born in 1999 from birth to age 17. The graph shows the percentage spent for various categories. Use the graph to answer Exercises 49 and 50. See Example 6.*

49. To the nearest dollar, how much will be spent to provide housing for the child?

50. To the nearest dollar, how much will be spent for healthcare?

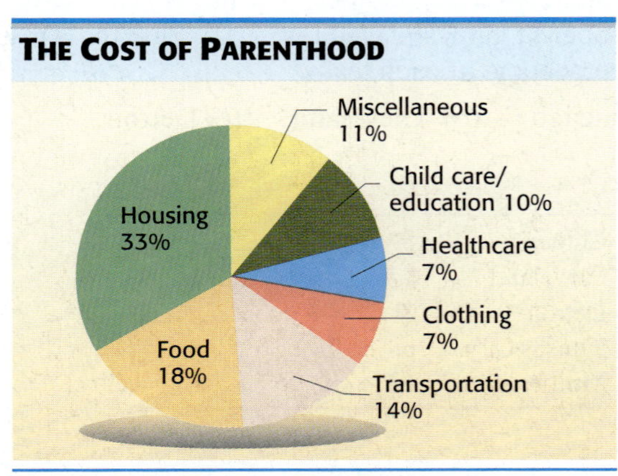

Source: U.S. Department of Agriculture.

2.3 Applications of Linear Equations

Objectives

1. Translate from words to mathematical expressions.
2. Write equations from given information.
3. Distinguish between expressions and equations.
4. Use the six steps in solving an applied problem.
5. Solve percent problems.
6. Solve investment problems.
7. Solve mixture problems.

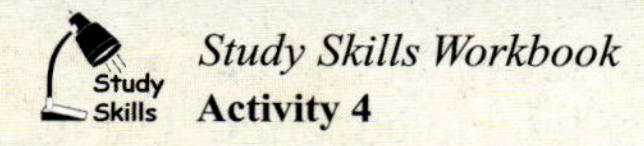

Study Skills Workbook
Activity 4

1 Translate from words to mathematical expressions. Producing a mathematical model of a real situation often involves translating verbal statements into mathematical statements. Although the problems we will be working with are simple ones, the methods we use will also apply to more difficult problems later.

Problem Solving

Usually there are key words and phrases in a verbal problem that translate into mathematical expressions involving addition, subtraction, multiplication, and division. Translations of some commonly used expressions follow.

TRANSLATING FROM WORDS TO MATHEMATICAL EXPRESSIONS

Verbal Expression	Mathematical Expression (where x and y are numbers)
Addition	
The **sum** of a number and 7	$x + 7$
6 **more than** a number	$x + 6$
3 **plus** a number	$3 + x$
24 **added to** a number	$x + 24$
A number **increased by** 5	$x + 5$
The **sum** of two numbers	$x + y$
Subtraction	
2 **less than** a number	$x - 2$
12 **minus** a number	$12 - x$
A number **decreased by** 12	$x - 12$
A number **subtracted from** 10	$10 - x$
The **difference between** two numbers	$x - y$
Multiplication	
16 **times** a number	$16x$
A number **multiplied by** 6	$6x$
$\frac{2}{3}$ **of** a number (used with fractions and percent)	$\frac{2}{3}x$
Twice (2 times) a number	$2x$
The **product** of two numbers	xy
Division	
The **quotient** of 8 and a number	$\frac{8}{x}$ $(x \neq 0)$
A number **divided by** 13	$\frac{x}{13}$
The **ratio** of two numbers or the **quotient** of two numbers	$\frac{x}{y}$ $(y \neq 0)$

1 Translate each verbal expression as a mathematical expression. Use x as the variable.

(a) 9 added to a number

(b) The difference between 7 and a number

(c) Four times a number

(d) The quotient of 7 and a nonzero number

2 Translate each verbal sentence into an equation. Use x as the variable.

(a) The sum of a number and 6 is 28.

(b) If twice a number is decreased by 3, the result is 17.

(c) The product of a number and 7 is twice the number plus 12.

(d) The quotient of a number and 6, added to twice the number, is 7.

3 Decide whether each is an expression or an equation.

(a) $5x - 3(x + 2) = 7$

(b) $5x - 3(x + 2)$

CAUTION

Because subtraction and division are not commutative operations, it is important to correctly translate expressions involving them. For example, "2 less than a number" is translated as $x - 2$, *not* $2 - x$. "A number subtracted from 10" is expressed as $10 - x$, *not* $x - 10$.

For division, it is understood that the number by which we are dividing is the denominator, and the number that is divided is the numerator. For example, "a number divided by 13" and "13 divided into x" both translate as $\frac{x}{13}$. Similarly, "the quotient of x and y" is translated as $\frac{x}{y}$.

Work Problem 1 at the Side.

2 Write equations from given information. The symbol for equality, =, is often indicated by the word *is*. In fact, because equal mathematical expressions represent names for the same number, any words that indicate the idea of "sameness" translate to =.

Example 1 Translating Words into Equations

Translate each verbal sentence into an equation.

Verbal Sentence	Equation
Twice a number, **decreased by** 3, **is** 42.	$2x - 3 = 42$
If the **product of a number and 12** is decreased by 7, the result **is** 105.	$12x - 7 = 105$
The **quotient of a number and the number plus 4 is** 28.	$\frac{x}{x + 4} = 28$
The **quotient of a number and 4,** plus the number, **is** 10.	$\frac{x}{4} + x = 10$

Work Problem 2 at the Side.

3 Distinguish between expressions and equations. It is important to be able to distinguish between algebraic expressions and equations. An expression translates as a phrase; an equation includes the = symbol and translates as a sentence.

Example 2 Distinguishing between Expressions and Equations

Decide whether each is an expression or an equation.

(a) $2(3 + x) - 4x + 7$

There is no equals sign, so this is an expression.

(b) $2(3 + x) - 4x + 7 = -1$

Because of the equals sign, this is an equation. Note that the expression in part (a) simplifies to the expression $-2x + 13$, and the equation in part (b) has solution 7.

Work Problem 3 at the Side.

4 Use the six steps in solving an applied problem. Throughout this book we will be solving different types of applications. While there is no one method that will allow us to solve all types of applied problems, the following six steps are helpful.

ANSWERS

1. **(a)** $9 + x$ or $x + 9$ **(b)** $7 - x$ **(c)** $4x$ **(d)** $\frac{7}{x}$

2. **(a)** $x + 6 = 28$ **(b)** $2x - 3 = 17$ **(c)** $7x = 2x + 12$ **(d)** $\frac{x}{6} + 2x = 7$

3. **(a)** equation **(b)** expression

Solving an Applied Problem

Step 1 **Read** the problem carefully until you understand what is given and what is to be found.

Step 2 **Assign a variable** to represent the unknown value, using diagrams or tables as needed. Write down what the variable represents. If necessary, express any other unknown values in terms of the variable.

Step 3 **Write an equation** using the variable expression(s).

Step 4 **Solve** the equation.

Step 5 **State the answer** to the problem. Does it seem reasonable?

Step 6 **Check** the answer in the words of the original problem.

We now see how these steps are applied.

Example 3 Solving a Geometry Problem

The length of a rectangle is 1 cm more than twice the width. The perimeter of the rectangle is 110 cm. Find the length and the width of the rectangle.

Step 1 **Read** the problem. What must be found? The length and width of the rectangle. What is given? The length is 1 cm more than twice the width; the perimeter is 110 cm.

Step 2 **Assign a variable.** Make a sketch, as in Figure 4.

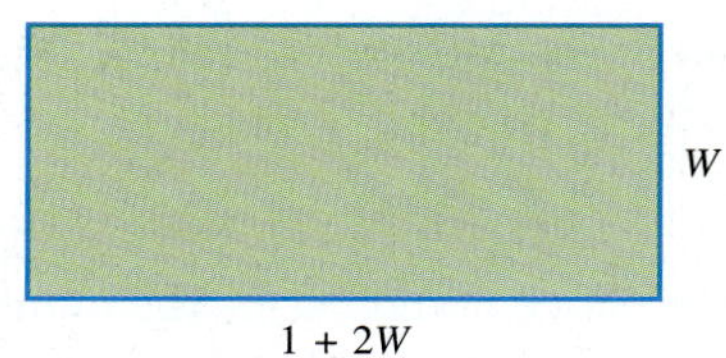

Figure 4

Choose a variable: let $W =$ the width; then $1 + 2W =$ the length.

Step 3 **Write an equation.** The perimeter of a rectangle is given by the formula $P = 2L + 2W$.

$$P = 2L + 2W$$

$$110 = 2(1 + 2W) + 2W \quad \text{Let } L = 1 + 2W \text{ and } P = 110.$$

Step 4 **Solve** the equation obtained in Step 3.

$$110 = 2(1 + 2W) + 2W$$

$$110 = 2 + 4W + 2W \quad \text{Distributive property}$$

$$110 = 2 + 6W \quad \text{Combine like terms.}$$

$$110 - 2 = 2 + 6W - 2 \quad \text{Subtract 2.}$$

$$108 = 6W$$

$$\frac{108}{6} = \frac{6W}{6} \quad \text{Divide by 6.}$$

$$18 = W$$

Step 5 **State the answer.** The width of the rectangle is 18 cm and the length is $1 + 2(18) = 37$ cm.

Step 6 **Check** the answer by substituting these dimensions into the words of the original problem.

Work Problem 4 at the Side.

4 Solve the problem.

The length of a rectangle is 5 cm more than its width. The perimeter is five times the width. What are the dimensions of the rectangle?

ANSWERS

4. width: 10 cm; length: 15 cm

5 Solve the problem.

At the end of the 1999 baseball season, Sammy Sosa and Mark McGwire had a lifetime total of 858 home runs. McGwire had 186 more than Sosa. How many home runs did each player have? (*Source: Who's Who in Baseball,* 2000.)

Answers

5. Sosa: 336; McGwire: 522

Example 4 Finding Unknown Numerical Quantities

Two outstanding major league pitchers in recent years are Roger Clemens and Greg Maddux. Between 1984 and 1999, they pitched a total of 916 games. Clemens pitched 44 more games than Maddux. How many games did each player pitch? (*Source: Who's Who in Baseball,* 2000.)

Step 1 **Read** the problem. We are asked to find the number of games each player pitched.

Step 2 **Assign a variable** to represent the number of games of one of the men.

Let m = the number of games for Maddux.

We must also find the number of games for Clemens. Since he pitched 44 more games than Maddux,

$m + 44$ = Clemens' number of games.

Step 3 **Write an equation.** The sum of the numbers of games is 916, so

$$\underbrace{m}_{\text{Maddux's games}} + \underbrace{(m + 44)}_{\text{Clemens' games}} = \underbrace{916}_{\text{Total}}.$$

Step 4 **Solve** the equation.

$$\begin{aligned} m + (m + 44) &= 916 \\ 2m + 44 &= 916 && \text{Combine like terms.} \\ 2m &= 872 && \text{Subtract 44.} \\ m &= 436 && \text{Divide by 2.} \end{aligned}$$

Step 5 **State the answer.** Since m represents the number of Maddux's games, Maddux pitched 436 games. Also, $m + 44 = 436 + 44 = 480$ is the number of games pitched by Clemens.

Step 6 **Check.** 480 is 44 more than 436, and the sum of 436 and 480 is 916.

The conditions of the problem are satisfied, and our solution checks.

CAUTION

A common error in solving applied problems is forgetting to answer all the questions asked in the problem. In Example 4, we were asked for the number of games for *each* player, so there was an extra step at the end in order to find Clemens' number.

Work Problem 5 at the Side.

5 **Solve percent problems.** Recall from the previous section that percent means "per one hundred," so 5% means .05, 14% means .14, and so on.

Example 5 Solving a Percent Problem

In 1998 there were 212 long distance area codes in the United States. This was an increase of 147% over the number when the area code plan originated in 1947. How many area codes were there in 1947? (*Source:* Pacific Bell Telephone Directory.)

Continued on Next Page

Step 1 **Read** the problem. We are given that the number of area codes increased by 147% from 1947 to 1998, and there were 212 area codes in 1998. We must find the original number of area codes.

Step 2 **Assign a variable.** Let x represent the number of area codes in 1947. Since $147\% = 1.47$, $1.47x$ represents the number of codes added since then.

Step 3 **Write an equation** from the given information.

$$\begin{array}{ccccc} \text{the number in 1947} & + & \text{the increase} & = & 212 \\ \downarrow & & \downarrow & & \downarrow \\ x & + & 1.47x & = & 212 \end{array}$$

Step 4 **Solve** the equation.

$$\begin{aligned} 1x + 1.47x &= 212 && \text{Identity property} \\ 2.47x &= 212 && \text{Combine like terms.} \\ x &\approx 86 && \text{Divide by 2.47. (Use a calculator if you wish.) Round to the nearest whole number.} \end{aligned}$$

Step 5 **State the answer.** There were 86 area codes in 1947.

Step 6 **Check** that the increase, $212 - 86 = 126$, is 147% of 86.

6 Solve each problem.

(a) A number increased by 15% is 287.5. Find the number.

(b) Michelle Raymond was paid $162 for a week's work at her part-time job after 10% deductions for taxes. How much did she make before the deductions were made?

CAUTION

Watch for two common errors that occur in solving problems like the one in Example 5.

1. Do not try to find 147% of 212 and subtract that amount from 212. The 147% should be applied to *the amount in 1947, not the amount in 1998.*
2. Do not write the equation as

$$x + 1.47 = 212.$$

The percent must be multiplied by some amount; in this case, the amount is the number of area codes in 1947, giving $1.47x$.

Work Problem 6 at the Side.

6 Solve investment problems. We can use linear equations to solve certain types of investment problems. The investment problems in this chapter deal with *simple interest.* In most real-world applications, *compound interest* is used. However, more advanced methods (covered in a later chapter) are needed for compound interest problems, so we will deal only with simple interest until then.

Example 6 Solving an Investment Problem

After winning the state lottery, Mark LeBeau has $40,000 to invest. He will put part of the money in an account paying 4% interest and the remainder into stocks paying 6% interest. His accountant tells him that the total annual income from these investments should be $2040. How much should he invest at each rate?

Step 1 **Read** the problem again.

Continued on Next Page

ANSWERS
6. (a) 250 (b) $180

7 Solve each problem.

(a) A woman invests $72,000 in two ways—some at 5% and some at 3%. Her total annual interest income is $3160. Find the amount she invests at each rate.

(b) A man has $34,000 to invest. He invests some at 5% and the balance at 4%. His total annual interest income is $1545. Find the amount he invests at each rate.

Step 2 **Assign a variable.**

Let $x =$ the amount to invest at 4%;

$40{,}000 - x =$ the amount to invest at 6%.

The formula for interest is $I = prt$. Here the time, t, is 1 year. Make a table to organize the given information.

Rate (as a Decimal)	Principal	Interest
.04	x	$.04x$
.06	$40{,}000 - x$	$.06(40{,}000 - x)$
		2040

Step 3 **Write an equation.** The last column of the table gives the equation.

$$\underset{\text{interest at 4\%}}{.04x} + \underset{\text{interest at 6\%}}{.06(40{,}000 - x)} = \underset{\text{total interest}}{2040}$$

Step 4 **Solve** the equation. We do so without clearing decimals.

$$\begin{aligned} .04x + .06(40{,}000) - .06x &= 2040 && \text{Distributive property} \\ -.02x + 2400 &= 2040 && \text{Combine like terms.} \\ -.02x &= -360 && \text{Subtract 2400.} \\ x &= 18{,}000 && \text{Divide by } -.02. \end{aligned}$$

Step 5 **State the answer.** Mark should invest $18,000 at 4% and $40,000 − $18,000 = $22,000 at 6%.

Step 6 **Check** by finding the annual interest at each rate; they should total $2040.

Work Problem 7 at the Side.

NOTE

In Example 6, we chose to let the variable represent the amount invested at 4%. Students often ask, "Can I let the variable represent the other unknown?" The answer is yes. The equation will be different, but in the end the two answers will be the same.

7 **Solve mixture problems.** Mixture problems involving rates of concentration can be solved with linear equations.

Example 7 Solving a Mixture Problem

A chemist must mix 8 L of a 40% acid solution with some 70% solution to get a 50% solution. How much of the 70% solution should be used?

Step 1 **Read** the problem. The problem asks for the amount of 70% solution to be used.

Step 2 **Assign a variable.** The information in the problem is illustrated in Figure 5.

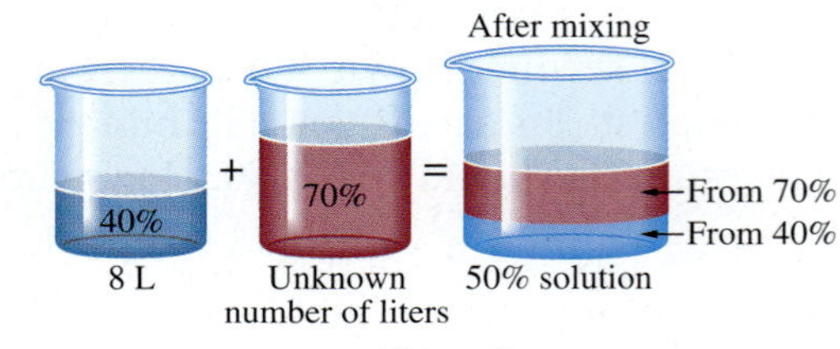

Figure 5

Continued on Next Page

Answers

7. **(a)** $50,000 at 5%; $22,000 at 3%
(b) $18,500 at 5%; $15,500 at 4%

Let x = the number of liters of 70% solution to be used. Use the given information to complete the following table.

Percent (as a Decimal)	Number of Liters	Liters of Pure Acid
.40	8	$.40(8) = 3.2$
.70	x	$.70x$
.50	$8 + x$	$.50(8 + x)$

Sum must equal

The numbers in the right column were found by multiplying the strengths and the numbers of liters. The number of liters of pure acid in the 40% solution plus the number of liters in the 70% solution must equal the number of liters in the 50% solution.

Step 3 **Write an equation.**

$$3.2 + .70x = .50(8 + x)$$

Step 4 **Solve.**

$$3.2 + .70x = 4 + .50x \quad \text{Distributive property}$$

$$.20x = .8 \quad \text{Subtract 3.2 and } .50x.$$

$$x = 4 \quad \text{Divide by .20.}$$

Step 5 **State the answer.** The chemist should use 4 L of the 70% solution.

Step 6 **Check.** (See margin problem 8(c).)

Work Problem 8 at the Side.

NOTE

When pure water is added to a solution, remember that water is 0% of the chemical (acid, alcohol, etc.).

8 Solve each problem.

(a) How many liters of a 10% solution should be mixed with 60 L of a 25% solution to get a 15% solution?

(b) How many pounds of candy worth \$8 per lb should be mixed with 100 lb of candy worth \$4 per lb to get a mixture that can be sold for \$7 per lb?

(c) Verify that the answer in Example 7 is correct.

ANSWERS

8. (a) 120 L **(b)** 300 lb **(c)** The total amount of pure acid is 6 L in both cases.

Focus on Real-Data Applications

Same (International) Time, Second Place

Employees of companies that operate globally must be able to converse by telephone with their colleagues around the world. It is sometimes challenging to find the common working hours at different global locations. Refer to the map of the International Time Zones. The time zone map is explored in the Focus-on application *International Time Zones* on p. 60.

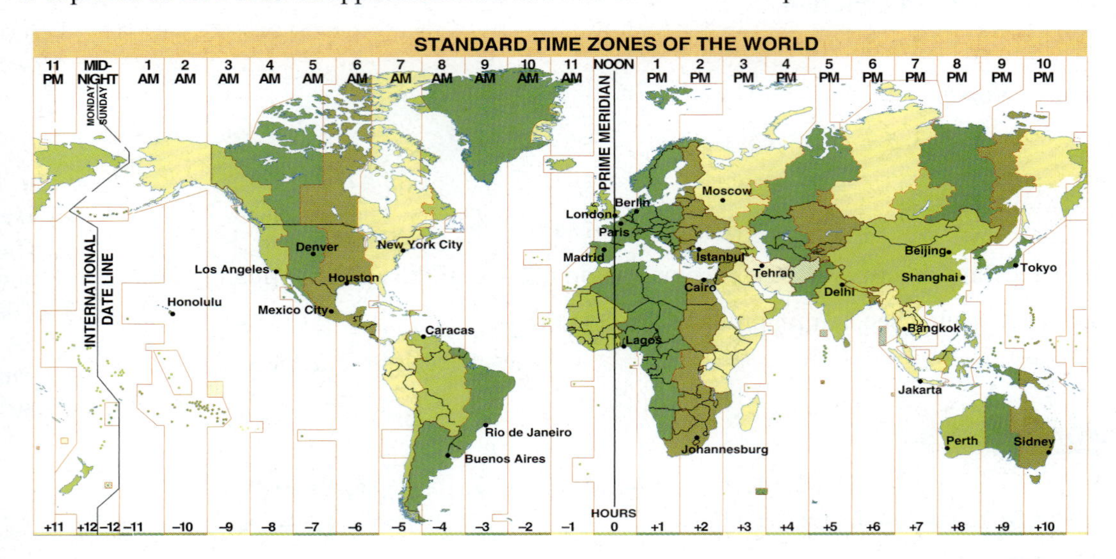

For Group Discussion

Europeans use a 24-hr clock, or military time. For example, 03:45 is 3:45 A.M. and 15:45 is 3:45 P.M. At 15:45 in London, the time would be 8 hr earlier in Los Angeles or 7:45 A.M., since 15:45 − 8:00 is 07:45. Assume that it is common for workdays to last from 8:00 A.M. to 6:00 P.M. worldwide (08:00–18:00). For each pair of cities, find the time interval during the workday and the workdays during the week that a person in the first city can talk by telephone with a colleague in the second city.

1. Moscow and Houston

   Moscow is ___________ hr ahead of Houston.

   08:00–18:00 Moscow time corresponds to ___________ Houston time.

   The overlap in time is ___________ hr between ___________ Moscow time and ___________ Houston time.

   Monday–Friday in Moscow overlaps _____________ in Houston.

2. Los Angeles and Tokyo

3. London and Los Angeles and London and Tokyo

2.3 EXERCISES

FOR EXTRA HELP

Student's Solutions Manual 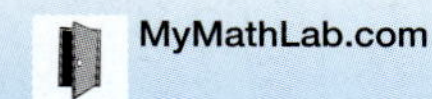

MyMathLab.com 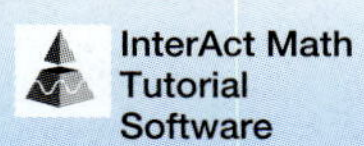

InterAct Math Tutorial Software 

AW Math Tutor Center 

www.mathxl.com

Digital Video Tutor CD 1 Videotape 2 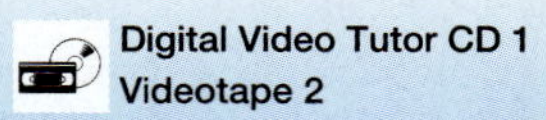

In each of the following, ***(a)*** *translate as an expression and* ***(b)*** *translate as an equation or inequality. Use x to represent the number.*

1. (a) 12 more than a number ________

(b) 12 is more than a number. ________

2. (a) 3 less than a number ________

(b) 3 is less than a number. ________

3. (a) 4 smaller than a number ________

(b) 4 is smaller than a number. ________

4. (a) 6 greater than a number ________

(b) 6 is greater than a number. ________

5. Which one of the following is *not* a valid translation of "20% of a number"?

A. $.20x$ **B.** $.2x$ **C.** $\frac{x}{5}$ **D.** $20x$

6. Explain why $13 - x$ is *not* a correct translation of "13 less than a number."

Translate each verbal phrase into a mathematical expression. Use x to represent the unknown number. See Example 1.

7. Twice a number, decreased by 13

8. The product of 6 and a number, decreased by 12

9. 12 increased by three times a number

10. 12 more than one-half of a number

11. The product of 8 and 12 less than a number

12. The product of 9 more than a number and 6 less than the number

13. The quotient of three times a number and 7

14. The quotient of 6 and five times a nonzero number

Use the variable x for the unknown, and write an equation representing the verbal sentence. Then solve the problem. See Example 1.

15. The sum of a number and 6 is -31. Find the number.

16. The sum of a number and -4 is 12. Find the number.

17. If the product of a number and -4 is subtracted from the number, the result is 9 more than the number. Find the number.

18. If the quotient of a number and 6 is added to twice the number, the result is 8 less than the number. Find the number.

19. When $\frac{2}{3}$ of a number is subtracted from 12, the result is 10. Find the number.

20. When 75% of a number is added to 6, the result is 3 more than the number. Find the number.

Decide whether each is an expression *or an* equation. *See Example 2.*

21. $5(x + 3) - 8(2x - 6)$

22. $-7(y + 4) + 13(y - 6)$

23. $5(x + 3) - 8(2x - 6) = 12$

24. $-7(y + 4) + 13(y - 6) = 18$

25. $\frac{r}{2} - \frac{r + 9}{6} - 8$

26. $\frac{r}{2} - \frac{r + 9}{6} = 8$

27. In your own words, list the six steps suggested for solving an applied problem.

Step 1 ________________________

Step 2 ________________________

Step 3 ________________________

Step 4 ________________________

Step 5 ________________________

Step 6 ________________________

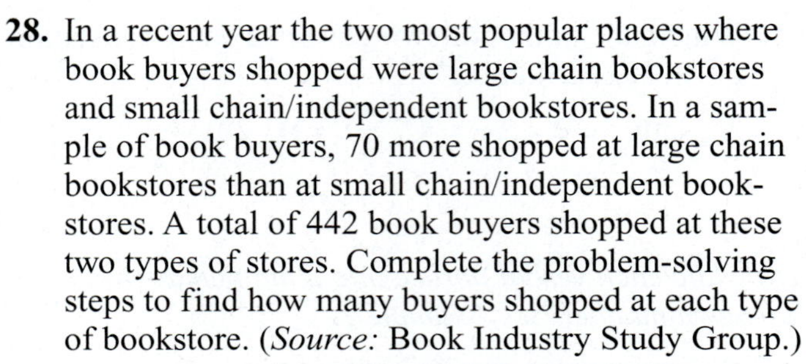

28. In a recent year the two most popular places where book buyers shopped were large chain bookstores and small chain/independent bookstores. In a sample of book buyers, 70 more shopped at large chain bookstores than at small chain/independent bookstores. A total of 442 book buyers shopped at these two types of stores. Complete the problem-solving steps to find how many buyers shopped at each type of bookstore. (*Source:* Book Industry Study Group.)

Step 1 Read the problem carefully. What are you asked to find?

____

____

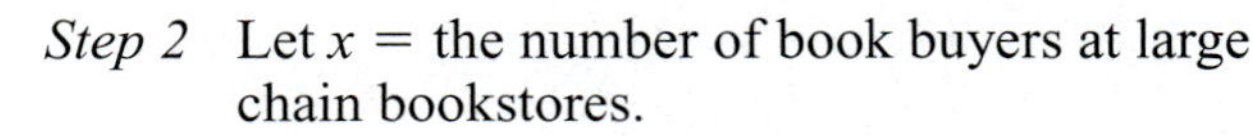

Step 2 Let $x =$ the number of book buyers at large chain bookstores.

Then $x - 70 =$ ________________________

____

________________________.

Step 3 __________ + __________ $= 442$

Step 4 $x =$ __________

Step 5 There were __________ large chain bookstore shoppers and __________ small chain/independent shoppers.

Step 6 The number of ________________________ was ________ more than the number of ________________________, and the total number of these shoppers was ________.

Use the six-step problem-solving method to solve each problem. See Examples 3 and 4.

29. The John Hancock Center in Chicago has a rectangular base. The length of the base measures 65 ft less than twice the width. The perimeter of this base is 860 ft. What are the dimensions of the base?

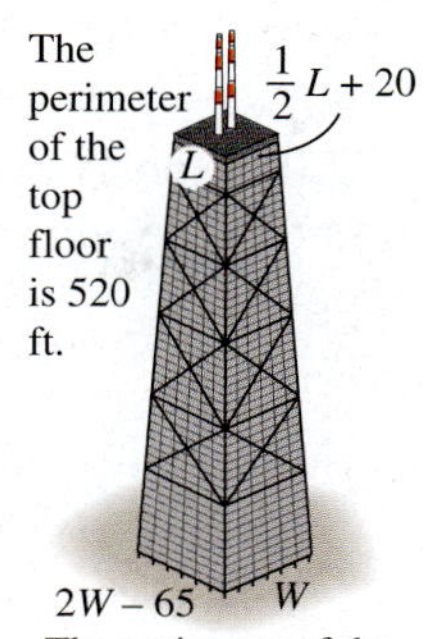

The perimeter of the base is 860 ft.

30. The John Hancock Center (Exercise 29) tapers as it rises. The top floor is rectangular and has perimeter 520 ft. The width of the top floor measures 20 ft more than one-half its length. What are the dimensions of the top floor?

31. The Bermuda Triangle supposedly causes trouble for aircraft pilots. It has a perimeter of 3075 mi. The shortest side measures 75 mi less than the middle side, and the longest side measures 375 mi more than the middle side. Find the lengths of the three sides.

32. The Vietnam Veterans Memorial in Washington, D.C., is in the shape of two sides of an isosceles triangle. If the two walls of equal length were joined by a straight line of 438 ft, the perimeter of the resulting triangle would be 931.5 ft. Find the lengths of the two walls. (*Source:* Pamphlet obtained at Vietnam Veterans Memorial.)

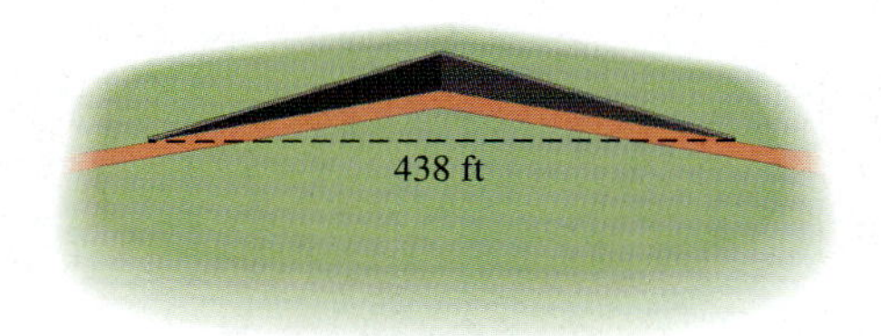

33. In a recent year, the two U.S. industrial corporations with the highest profits were General Motors and General Electric. Their profits together totaled $13.5 billion. General Electric profits were $.3 billion less than General Motors profits. What were the profits for each corporation? (*Source: The Universal Almanac,* 1997.)

34. In a recent year, video rental revenue was $.27 billion more than twice video sales revenue. Together, these revenues amounted to $9.81 billion. What was the revenue from each of these sources? (*Source:* Paul Kagan Associates, Inc.)

35. In the 1996 presidential election, Bill Clinton and Bob Dole together received 538 electoral votes. Clinton received 220 more votes than Dole. How many votes did each candidate receive? (*Source:* Congressional Quarterly, Inc.)

36. Ted Williams and Rogers Hornsby were two great hitters. Together they got 5584 hits in their careers. Hornsby got 276 more hits than Williams. How many base hits did each get? (*Source:* Neft, D. S. and Cohen, R. M., *The Sports Encyclopedia: Baseball*, St. Martins Griffin; New York, 1997.)

Solve each percent problem. See Example 5.

37. Composite scores on the ACT exam rose from 20.6 in 1990 to 21.0 in 1998. What percent increase was this? (*Source:* The American College Testing Program.)

38. In 1998, the number of participants in the ACT exam was 995,000. Earlier, in 1990, a total of 817,000 took the exam. What percent increase was this? (*Source:* The American College Testing Program.)

39. In 1990, the population of Palm Beach county in Florida was 863,500. The 1997 population was 117.95% of the 1990 population. What was the approximate 1997 population? (*Source: The World Book Almanac,* 1999.)

40. The consumer price index (CPI) in 1998 was 163.0. This represented a 7.03% increase from 1995. What was the CPI in 1995? (*Source:* U.S. Bureau of Labor Statistics.)

41. At the end of a day, Jeff Hornsby found that the total cash register receipts at the motel where he works amounted to $2725. This included the 9% sales tax charged. Find the amount of the tax.

42. Fino Roverato sold his house for $159,000. He got this amount knowing that he would have to pay a 6% commission to his agent. What amount did he have after the agent was paid?

Solve each investment problem. See Example 6.

43. Carter Fenton earned $12,000 last year by giving tennis lessons. He invested part at 3% simple interest and the rest at 4%. He earned a total of $440 in interest. How much did he invest at each rate?

Rate (as a Decimal)	Principal	Interest
.03	x	
.04		

44. Melissa Wright won $60,000 on a slot machine in Las Vegas. She invested part at 2% simple interest and the rest at 3%. She earned a total of $1600 in interest. How much was invested at each rate?

Rate (as a Decimal)	Principal	Interest
.02	x	

45. Michael Pellissier invested some money at 4.5% simple interest and $1000 less than twice this amount at 3%. His total annual income from the interest was $1020. How much was invested at each rate?

46. Holly Rioux invested some money at 3.5% simple interest, and $5000 more than 3 times this amount at 4%. She earned $1440 in interest. How much did she invest at each rate?

47. Jerry and Lucy Keefe have \$29,000 invested in stocks paying 5%. How much additional money should they invest in certificates of deposit paying 2% so that the average return on the two investments is 3%?

48. Ron Hampton placed \$15,000 in an account paying 6%. How much additional money should he deposit at 4% so that the average return on the two investments is 5.5%?

Solve each problem involving rates of concentration and mixtures. See Example 7.

49. Ten liters of a 4% acid solution must be mixed with a 10% solution to get a 6% solution. How many liters of the 10% solution are needed?

Percent (as a Decimal)	Liters of Solution	Liters of Pure Acid
.04	10	
.10	x	
.06		

50. How many liters of a 14% alcohol solution must be mixed with 20 L of a 50% solution to get a 30% solution?

Percent (as a Decimal)	Liters of Solution	Liters of Pure Alcohol
.14	x	
.50		

51. In a chemistry class, 12 L of a 12% alcohol solution must be mixed with a 20% solution to get a 14% solution. How many liters of the 20% solution are needed?

52. How many liters of a 10% alcohol solution must be mixed with 40 L of a 50% solution to get a 40% solution?

53. How much pure dye must be added to 4 gal of a 25% dye solution to increase the solution to 40%? (*Hint:* Pure dye is 100% dye.)

54. How much water must be added to 6 gal of a 4% insecticide solution to reduce the concentration to 3%? (*Hint:* Water is 0% insecticide.)

55. Randall Albritton wants to mix 50 lb of nuts worth \$2 per lb with some nuts worth \$6 per lb to make a mixture worth \$5 per lb. How many pounds of \$6 nuts must he use?

56. Lee Ann Spahr wants to mix tea worth 2¢ per oz with 100 oz of tea worth 5¢ per oz to make a mixture worth 3¢ per oz. How much 2¢ tea should be used?

57. Why is it impossible to add two mixtures of candy worth \$4 per lb and \$5 per lb to obtain a final mixture worth \$6 per lb?

58. Write an equation based on the following problem, solve the equation, and explain why the problem has no solution.

How much 30% acid should be mixed with 15 L of 50% acid to obtain a mixture that is 60% acid?

RELATING CONCEPTS (Exercises 59–63) FOR INDIVIDUAL OR GROUP WORK

Consider each problem.

Problem A
Jack has \$800 invested in two accounts. One pays 5% interest per year and the other pays 10% interest per year. The amount of yearly interest is the same as he would get if the entire \$800 was invested at 8.75%. How much does he have invested at each rate?

Problem B
Jill has 800 L of acid solution. She obtained it by mixing some 5% acid with some 10% acid. Her final mixture of 800 L is 8.75% acid. How much of each of the 5% and 10% solutions did she use to get her final mixture?

In Problem A, let x represent the amount invested at 5% interest, and in Problem B, let y represent the amount of 5% acid used. ***Work Exercises 59–63 in order.***

59. (a) Write an expression in x that represents the amount of money Jack invested at 10% in Problem A.

(b) Write an expression in y that represents the amount of 10% acid solution Jill used in Problem B.

60. (a) Write expressions that represent the amount of interest Jack earns per year at 5% and at 10%.

(b) Write expressions that represent the amount of pure acid in Jill's 5% and 10% acid solutions.

61. (a) The sum of the two expressions in part (a) of Exercise 60 must equal the total amount of interest earned in one year. Write an equation representing this fact.

(b) The sum of the two expressions in part (b) of Exercise 60 must equal the amount of pure acid in the final mixture. Write an equation representing this fact.

62. (a) Solve Problem A.

(b) Solve Problem B.

63. Explain the similarities between the processes used in solving Problems A and B.

2.4 Further Applications of Linear Equations

There are three common applications of linear equations that we did not discuss in Section 2.3: money problems, uniform motion problems, and problems involving the angles of a triangle.

Objectives

1. Solve problems about different denominations of money.
2. Solve problems about uniform motion.
3. Solve problems involving the angles of a triangle.

1 Solve problems about different denominations of money. These problems are very similar to the simple interest problems in Section 2.3.

Problem Solving

In problems involving money, use the fact that

$$\text{denomination} \times \begin{array}{c}\text{number of monetary}\\\text{units of the same kind}\end{array} = \begin{array}{c}\text{total monetary}\\\text{value}\end{array}.$$

For example, 30 dimes have a monetary value of $\$.10(30) = \$3$. Fifteen five-dollar bills have a value of $\$5(15) = \$75$.

1 Solve the problem.

At the end of a day, a cashier had 26 dimes and half-dollars. The total value of these coins was \$8.60. How many of each type did he have?

Example 1 Solving a Money Denomination Problem

For a bill totaling \$5.65, a cashier received 25 coins consisting of nickels and quarters. How many of each type of coin did the cashier receive?

Step 1 **Read** the problem. The problem asks that we find the number of nickels and the number of quarters the cashier received.

Step 2 **Assign a variable.**

Let x represent the number of nickels;
then $25 - x$ represents the number of quarters.

We can organize the information in a table as we did with investment problems.

Denomination	Number of Coins	Total Value
\$.05	x	$.05x$
\$.25	$25 - x$	$.25(25 - x)$
		5.65

Step 3 **Write an equation.** From the last column of the table,

$$.05x + .25(25 - x) = 5.65.$$

Step 4 **Solve.**

$$\begin{aligned} 5x + 25(25 - x) &= 565 && \text{Multiply by 100.} \\ 5x + 625 - 25x &= 565 && \text{Distributive property} \\ -20x &= -60 && \text{Subtract 625; combine terms.} \\ x &= 3 && \text{Divide by } -20. \end{aligned}$$

Step 5 **State the answer.** The cashier has 3 nickels and $25 - 3 = 22$ quarters.

Step 6 **Check.** The cashier has $3 + 22 = 25$ coins, and the value of the coins is $\$.05(3) + \$.25(22) = \$5.65$, as required.

Work Problem 1 at the Side.

Answers

1. 11 dimes, 15 half-dollars

CAUTION

Be sure that your answer is reasonable when working problems like Example 1. Because you are dealing with a number of coins, the correct answer can neither be negative nor a fraction.

2 Solve problems about uniform motion.

Problem Solving

Uniform motion problems use the distance formula, $d = rt$. In this formula, when rate (or speed) is given in miles per hour, time must be given in hours. To solve such problems, draw a sketch to illustrate what is happening in the problem, and make a table to summarize the given information.

Example 2 Solving a Motion Problem (Motion in Opposite Directions)

Two cars leave the same place at the same time, one going east and the other west. The eastbound car averages 40 mph, while the westbound car averages 50 mph. In how many hours will they be 300 mi apart?

Step 1 **Read** the problem. We are looking for the time it takes for the two cars to be 300 mi apart.

Step 2 **Assign a variable.** A sketch shows what is happening in the problem: The cars are going in *opposite* directions. See Figure 6.

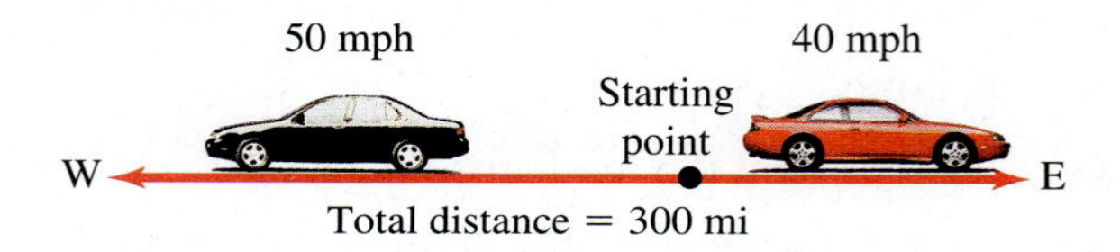

Figure 6

Let x represent the time traveled by each car. Summarize the information of the problem in a table. When the expressions for rate and time are entered, *fill in each distance by multiplying rate by time* using the formula $d = rt$.

	Rate	Time	Distance
Eastbound Car	40	x	$40x$
Westbound Car	50	x	$50x$

From the sketch in Figure 6, the sum of the two distances is 300.

Step 3 **Write an equation.** $40x + 50x = 300$

Step 4 **Solve.**

$$90x = 300 \quad \text{Combine like terms.}$$

$$x = \frac{300}{90} \quad \text{Divide by 90.}$$

$$x = \frac{10}{3} \quad \text{Lowest terms}$$

Step 5 **State the answer.** The cars travel $\frac{10}{3} = 3\frac{1}{3}$ hr, or 3 hr and 20 min.

Step 6 **Check.** The eastbound car traveled $40\left(\frac{10}{3}\right) = \frac{400}{3}$ mi, and the westbound car traveled $50\left(\frac{10}{3}\right) = \frac{500}{3}$ mi, for a total of $\frac{400}{3} + \frac{500}{3} = \frac{900}{3} = 300$ mi, as required.

CAUTION

It is a common error to write 300 as the distance for each car in Example 2. Three hundred miles is the *total* distance traveled.

As in Example 2, in general, the equation for a problem involving motion in opposite directions is of the form

partial distance + partial distance = total distance.

Work Problem 2 at the Side.

Example 3 **Solving a Motion Problem (Motion in the Same Direction)**

Jeff Bezzone can bike to work in $\frac{3}{4}$ hr. When he takes the bus, the trip takes $\frac{1}{4}$ hr. If the bus travels 20 mph faster than Jeff rides his bike, how far is it to his workplace?

Step 1 **Read** the problem. We must find the distance between Jeff's home and his workplace.

Step 2 **Assign a variable.** Although the problem asks for a distance, it is easier here to let x be Jeff's speed when he rides his bike to work. Then the speed of the bus is $x + 20$. For the trip by bike,

$$d = rt = x \cdot \frac{3}{4} = \frac{3}{4}x,$$

and by bus,

$$d = rt = (x + 20) \cdot \frac{1}{4} = \frac{1}{4}(x + 20).$$

Summarize this information in a table.

	Rate	Time	Distance
Bike	x	$\frac{3}{4}$	$\frac{3}{4}x$
Bus	$x + 20$	$\frac{1}{4}$	$\frac{1}{4}(x + 20)$

(Distances: Same)

Step 3 **Write an equation.** The key to setting up the correct equation is to understand that the distance in each case is the same. See Figure 7.

Figure 7

Since the distance is the same in each case,

$$\frac{3}{4}x = \frac{1}{4}(x + 20).$$

Step 4 **Solve** the equation. First multiply each side by 4.

$$4\left(\frac{3}{4}x\right) = 4\left(\frac{1}{4}\right)(x + 20) \quad \text{Multiply by 4.}$$

$$3x = x + 20 \quad \text{Multiply; distributive property}$$

$$2x = 20 \quad \text{Subtract } x.$$

$$x = 10 \quad \text{Divide by 2.}$$

Continued on Next Page

2 Solve the problem.

Two cars leave the same location at the same time. One travels north at 60 mph and the other south at 45 mph. In how many hours will they be 420 mi apart?

Answers

2. 4 hr

Step 5 **State the answer.** The required distance is given by

$$d = \frac{3}{4}x = \frac{3}{4}(10) = \frac{30}{4} = 7.5.$$

Step 6 **Check.** Check by finding the distance using

$$d = \frac{1}{4}(x + 20) = \frac{1}{4}(10 + 20) = \frac{30}{4} = 7.5,$$

the same result.

As in Example 3, the equation for a problem involving motion in the same direction is often of the form

one distance = other distance.

NOTE

In Example 3 it was easier to let the variable represent a quantity other than the one that we were asked to find. This is the case in some problems. It takes practice to learn when this approach is the best, and practice means working lots of problems!

Work Problem 3 at the Side.

3 **Solve problems involving the angles of a triangle.** An important result of Euclidean geometry (the geometry of the Greek mathematician Euclid) is that the sum of the angle measures of any triangle is 180°. This property is used in the next example.

Example 4 Finding Angle Measures

Find the value of x, and determine the measure of each angle in Figure 8.

Step 1 **Read** the problem. We are asked to find the measure of each angle.

Step 2 **Assign a variable.** Let x represent the measure of one angle.

Step 3 **Write an equation.** The sum of the three measures shown in the figure must be 180°.

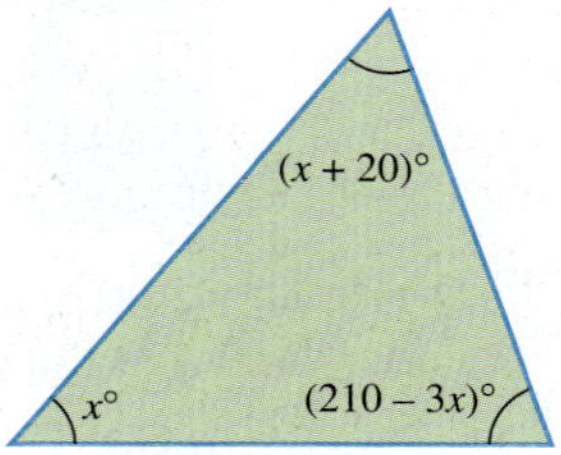

$(x + 20)°$

$x°$ $(210 - 3x)°$

Figure 8

$$x + (x + 20) + (210 - 3x) = 180$$

Step 4 **Solve.**

$$-x + 230 = 180 \quad \text{Combine like terms.}$$
$$-x = -50 \quad \text{Subtract 230.}$$
$$x = 50 \quad \text{Divide by } -1.$$

Step 5 **State the answer.** One angle measures 50°, another measures $x + 20 = 50 + 20 = 70°$, and the third measures $210 - 3x = 210 - 3(50) = 60°$.

Step 6 **Check.** Since $50° + 70° + 60° = 180°$, the answers are correct.

Work Problem 4 at the Side.

3 Solve the problem.

Elayn begins jogging at 5:00 A.M., averaging 3 mph. Clay leaves at 5:30 A.M., following her, averaging 5 mph. How long will it take him to catch up to her? (*Hint:* 30 min = $\frac{1}{2}$ hr.)

4 Solve the problem.

One angle in a triangle is 15° larger than a second angle. The third angle is 25° larger than twice the second angle. Find the measure of each angle.

Answers

3. $\frac{3}{4}$ hr or 45 min

4. 35°, 50°, and 95°

2.4 EXERCISES

FOR EXTRA HELP — Student's Solutions Manual — MyMathLab.com — InterAct Math Tutorial Software — AW Math Tutor Center — www.mathxl.com — Digital Video Tutor CD 2 Videotape 2

Solve each problem.

1. What amount of money is found in a coin hoard containing 38 nickels and 26 dimes?

2. The distance between Cape Town, South Africa, and Miami is 7700 mi. If a jet averages 480 mph between the two cities, what is its travel time in hours?

3. Tri Phong traveled from Louisville to Kansas City, a distance of 520 mi, in 10 hr. What was his rate in miles per hour?

4. A square has perimeter 40 in. What would be the perimeter of an equilateral triangle whose sides each measure the same length as the side of the square?

Write a short explanation in Exercises 5 and 6.

5. Read over Example 3 in this section. The solution of the equation is 10. Why is *10 mph* not the answer to the problem?

6. Suppose that you know that two angles of a triangle have equal measures, and the third angle measures 36°. Explain in a few words the strategy you would use to find the measures of the equal angles without actually writing an equation.

Solve each problem. See Example 1.

7. Otis Taylor has a box of coins that he uses when playing poker with his friends. The box currently contains 44 coins, consisting of pennies, dimes, and quarters. The number of pennies is equal to the number of dimes, and the total value is $4.37. How many of each denomination of coin does he have in the box?

Denomination	Number of Coins	Total Value
.01	x	$.01x$
	x	
.25		

8. Nana Nantambu found some coins while looking under her sofa pillows. There were equal numbers of nickels and quarters, and twice as many half-dollars as quarters. If she found $2.60 in all, how many of each denomination of coin did she find?

Denomination	Number of Coins	Total Value
.05	x	$.05x$
	x	
.50	$2x$	

9. Kim Falgout's daughter, Madeline, has a piggy bank with 47 coins. Some are quarters, and the rest are half-dollars. If the total value of the coins is $17.00, how many of each denomination does she have?

10. John Joslyn has a jar in his office that contains 39 coins. Some are pennies, and the rest are dimes. If the total value of the coins is $2.64, how many of each denomination does he have?

11. Dave Bowers collects U.S. gold coins. He has a collection of 41 coins. Some are \$10 coins, and the rest are \$20 coins. If the face value of the coins is \$540, how many of each denomination does he have?

12. In the nineteenth century, the United States minted two-cent and three-cent pieces. Frances Steib has three times as many three-cent pieces as two-cent pieces, and the face value of these coins is \$2.42. How many of each denomination does she have?

13. A total of 550 people attended a Kenny Loggins concert. Floor tickets cost \$40 each, while balcony tickets cost \$28 each. If a total of \$20,800 was collected, how many of each type of ticket were sold?

14. The U.N.O. production of *The Music Man* was a big success. For opening night, 410 tickets were sold. Students paid \$3 each, while nonstudents paid \$7 each. If a total of \$1650 was collected, how many students and how many nonstudents attended?

In Exercises 15–18, find the rate based on the information provided. Use a calculator and round your answers to the nearest hundredth. All events were at the Sydney Olympics. (*Source:* http://espn.go.com/oly/summer00)

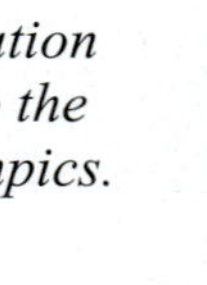

	Event and Year	Participant	Distance	Time
15.	100-m hurdles, Women	Olga Shishigina, Kazakhstan	100 m	12.65 sec
16.	400-m hurdles, Women	Irina Privalova, Russia	400 m	53.02 sec
17.	400-m hurdles, Men	Angelo Taylor, USA	400 m	47.50 sec
18.	400-m dash, Men	Michael Johnson, USA	400 m	43.84 sec

Solve each problem. See Examples 2 and 3.

19. Two steamers leave a port on a river at the same time, traveling in opposite directions. Each is traveling 22 mph. How long will it take for them to be 110 mi apart?

	Rate	Time	Distance
First Steamer		t	
Second Steamer	22		

20. A train leaves Kansas City, Kansas, and travels north at 85 km per hr. Another train leaves at the same time and travels south at 95 km per hr. How long will it take before they are 315 km apart?

	Rate	Time	Distance
First Train	85	t	
Second Train			

21. Agents Mulder and Scully are driving to Georgia to investigate "Big Blue," a giant aquatic reptile reported to inhabit one of the local lakes. Mulder leaves Washington at 8:30 A.M. and averages 65 mph. His partner, Scully, leaves at 9:00 A.M., following the same path and averaging 68 mph. At what time will Scully catch up with Mulder?

	Rate	Time	Distance
Mulder			
Scully			

22. Lois and Clark are covering separate stories and have to travel in opposite directions. Lois leaves the *Daily Planet* at 8:00 A.M. and travels at 35 mph. Clark leaves at 8:15 A.M. and travels at 40 mph. At what time will they be 140 mi apart?

	Rate	Time	Distance
Lois			
Clark			

23. Latrella can get to school in 15 min if she rides her bike. It takes her 45 min if she walks. Her speed when walking is 10 mph slower than her speed when riding. What is her speed when she rides?

	Rate	Time	Distance
Riding			
Walking			

24. When Dewayne drives his car to work, the trip takes 30 min. When he rides the bus, it takes 45 min. The average speed of the bus is 12 mph less than his speed when driving. Find the distance he travels to work.

	Rate	Time	Distance
Car			
Bus			

25. Johnny leaves Memphis to visit his cousin, Anne Hoffman, in the town of Hornsby, TN, 80 mi away. He travels at an average speed of 50 mph. One-half hour later, Anne leaves to visit Johnny, traveling at an average speed of 60 mph. How long after Anne leaves will it be before they meet?

26. On an automobile trip, Amy Cardella maintained a steady speed for the first two hours. Rush-hour traffic slowed her speed by 25 mph for the last part of the trip. The entire trip, a distance of 125 mi, took $2\frac{1}{2}$ hr. What was her speed during the first part of the trip?

Find the measure of each angle in the triangles shown. (Be sure to substitute your value of x into each angle expression.) See Example 4.

27.

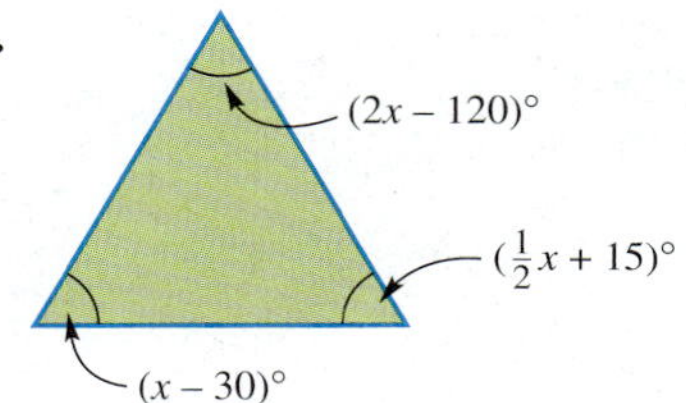

28.

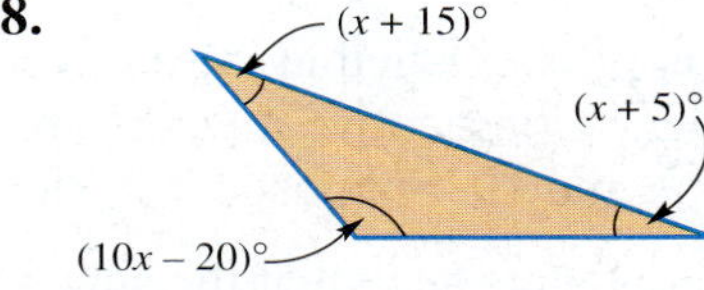

29.

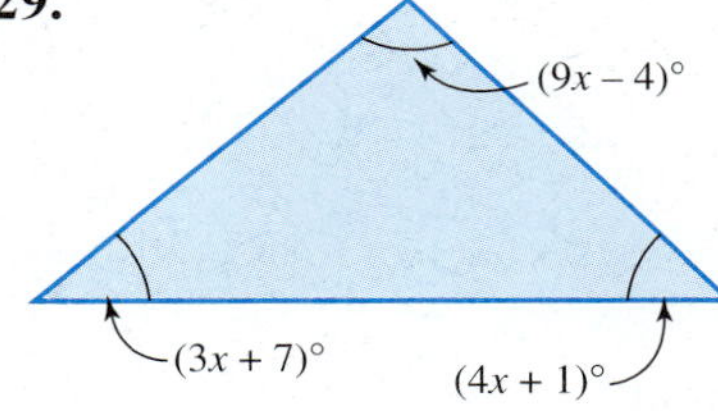

30.

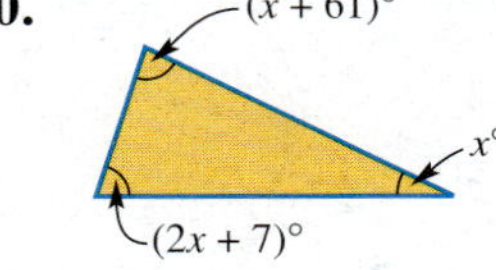

Relating Concepts (Exercises 31–34) For Individual or Group Work

Consider the following two figures. ***Work Exercises 31–34 in order.***

31. Solve for the measures of the unknown angles in Figure A.

32. Solve for the measure of the unknown angle marked $y°$ in Figure B.

33. Add the measures of the two angles you found in Exercise 31. How does the sum compare to the measure of the angle you found in Exercise 32?

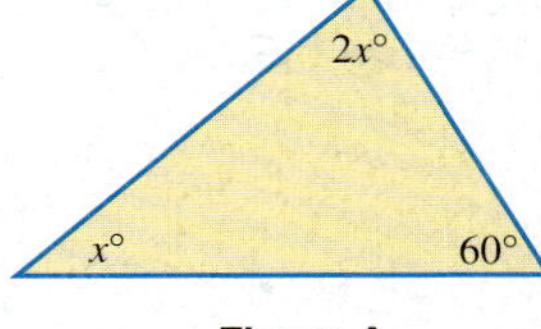

Figure A

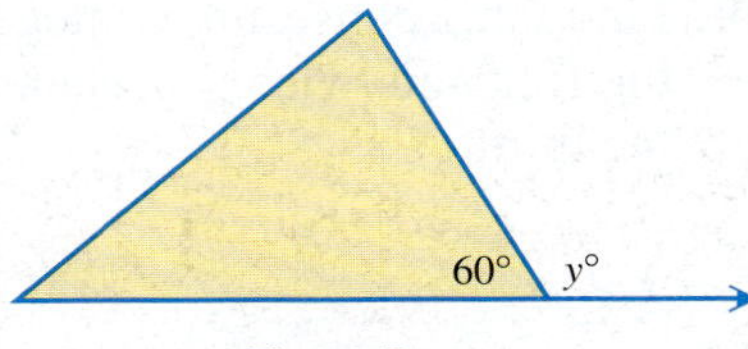

Figure B

34. From Exercises 31–33, make a conjecture (an educated guess) about the relationship among the angles marked ①, ②, and ③ in the figure shown here.

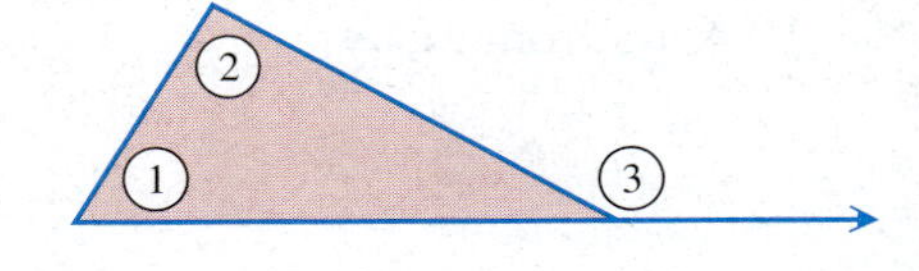

In Exercises 35 and 36, the angles marked with variable expressions are called ***vertical angles.*** *It is shown in geometry that vertical angles have equal measures. Find the measure of each angle.*

35.

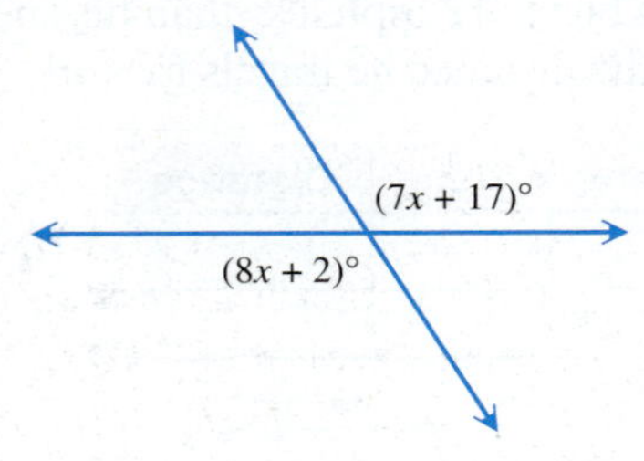

36.

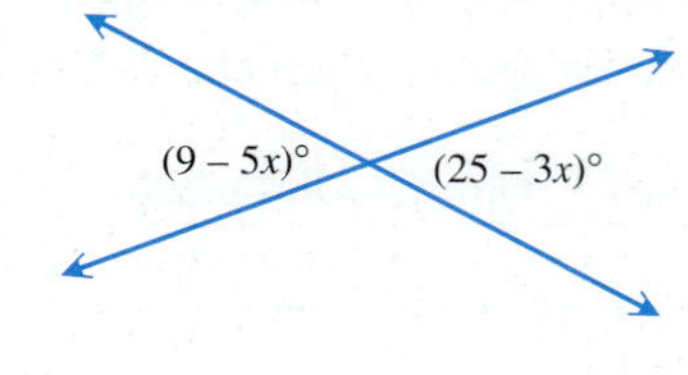

37. Two angles whose sum is 90° are called **complementary angles.** Find the measures of the complementary angles shown in the figure.

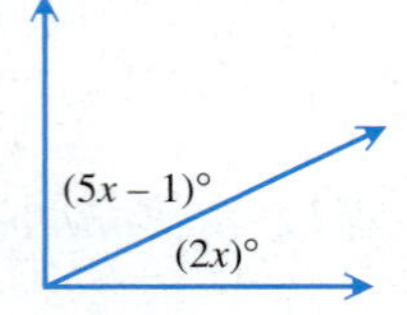

38. Two angles whose sum is 180° are called **supplementary angles.** Find the measures of the supplementary angles shown in the figure.

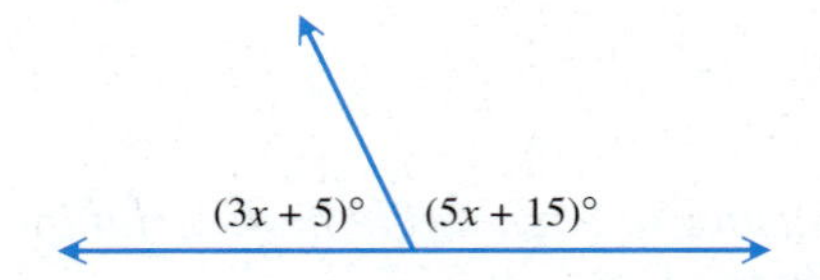

Another type of application often studied in introductory and intermediate algebra courses involves consecutive integers. Consecutive integers are integers that follow each other in counting order, such as 8, 9, and 10. Suppose we wish to solve the following problem:

Find three consecutive integers such that the sum of the first and third, increased by 3, is 50 more than the second.

Let x represent the first of the unknown integers. Then $x + 1$ will be the second, and $x + 2$ will be the third. The equation we need can be found by going back to the words of the original problem.

Sum of the first and third	increased by 3	is	50 more than the second.
↓	↓	↓	↓
$x + (x + 2)$	$+ 3$	$=$	$(x + 1) + 50$

The solution of this equation is 46, meaning that the first integer is $x = 46$, the second is $x + 1 = 47$, and the third is $x + 2 = 48$. The three integers are 46, 47, and 48. Check by substituting these numbers back into the words of the original problem.

Solve each problem involving consecutive integers.

39. Find three consecutive integers such that the sum of the first and twice the second is 17 more than twice the third.

40. Find four consecutive integers such that the sum of the first three is 54 more than the fourth.

41. If I add my current age to the age I will be next year on this date, the sum is 103 years. How old will I be 10 years from today?

42. Two pages facing each other in this book have 189 as the sum of their page numbers. What are the two page numbers?

Summary Exercises on SOLVING APPLIED PROBLEMS

The applications that follow are of the various types introduced in this chapter. Use the strategies you have developed to solve each problem.

1. The length of a rectangle is 3 in. more than its width. If the length were decreased by 2 in. and the width were increased by 1 in., the perimeter of the resulting rectangle would be 24 in. Find the dimensions of the original rectangle.

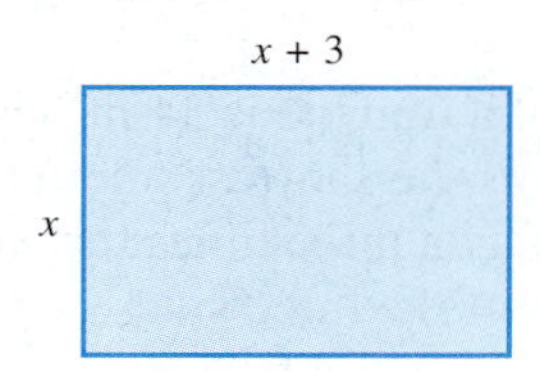

2. A farmer wishes to enclose a rectangular region with 210 m of fencing in such a way that the length is twice the width and the region is divided into two equal parts, as shown in the figure. What length and width should be used?

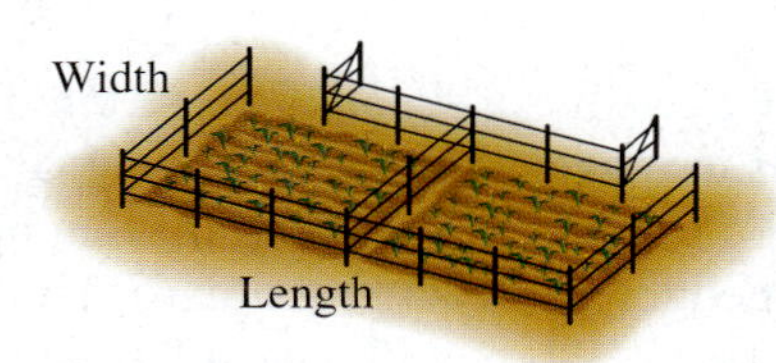

3. An electronics store offered a videodisc player for $255. This was the sale price, after the regular price had been discounted 40%. What was the regular price?

4. After a discount of 30%, the sale price of *The Parents' Guide to Kids' Sports* was $6.27. What was the regular price of the book? (Give your answer to the nearest 5¢.)

5. An amount of money is invested at 4% annual simple interest, and twice that amount is invested at 5%. The total annual interest is $112. How much is invested at each rate?

6. An amount of money is invested at 3% annual simple interest, and $2000 more than that amount is invested at 4%. The total annual interest is $920. How much is invested at each rate?

7. In the 1940 presidential election, Franklin Roosevelt and Wendell Willkie together received 531 electoral votes. Roosevelt received 367 more votes than Willkie in the landslide. How many votes did each man receive? (*Source:* Congressional Quarterly, Inc.)

8. Two of the highest paid business executives in a recent year were Mike Eisner, chairman of Disney, and Ed Horrigan, vice chairman of RJR Nabisco. Together their salaries totaled $61.8 million. Eisner earned $18.4 million more than Horrigan. What was the salary for each executive?

9. Moses Tanui from Kenya won the 1996 men's Boston marathon with a rate of 12.19 mph. The women's race was won by Uta Pippig from Germany, who ran at 10.69 mph. Pippig's time was .3 hr longer than Tanui's. Find their winning times. (*Source: The Universal Almanac*, 1997.)

10. A newspaper recycling collection bin is in the shape of a box, 1.5 ft wide and 5 ft long. If the volume of the bin is 75 ft^3, find the height.

11. Joshua Rogers has a sheet of tin 12 cm by 16 cm. He plans to make a box by cutting equal squares out of each of the four corners and folding up the remaining edges. How large a square should he cut so that the finished box will have a length that is 5 cm less than twice the width?

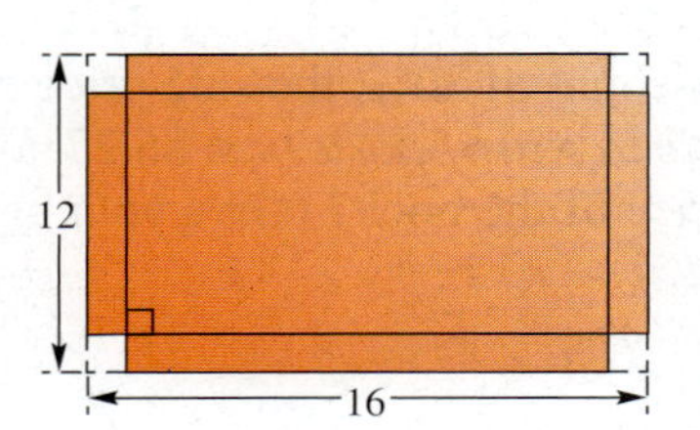

12. The perimeter of a triangle is 34 in. The middle side is twice as long as the shortest side. The longest side is 2 in. less than three times the shortest side. Find the lengths of the three sides.

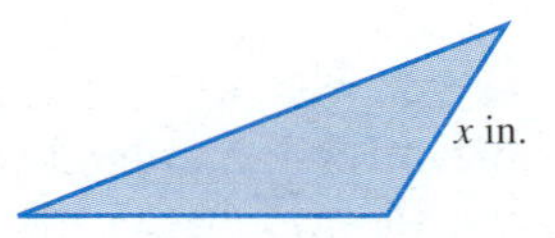

13. According to the Wilderness Society, in the early 1990s there were 1631 breeding pairs of northern spotted owls in California and Washington. California had 289 more pairs than Washington. How many pairs were there in each state?

14. The hit movie *Titanic* earned more in Europe than in the United States. As of September 1998, the movie had earned about $686 million in Europe. At that time, the average cost for a movie ticket in the United States was $4.59. In London, it cost $10.59 and in Frankfurt, $8.42. If 74,000,000 tickets were sold in London and Frankfurt, how many would have to be sold in each country to earn $686,000,000? Round answers to the nearest million. (*Source: Parade* magazine, September 13, 1998.)

15. The sum of the smallest and largest of three consecutive integers is 32 more than the middle integer. What are the three integers?

16. Find the measure of each angle.

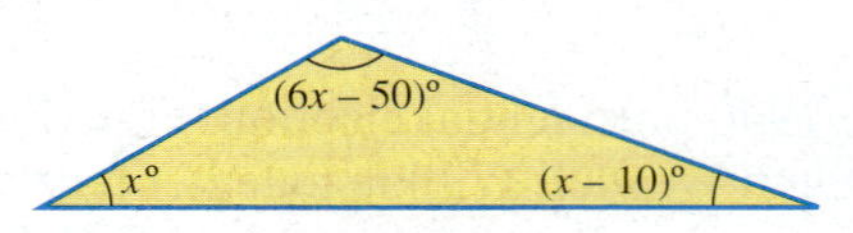

Chapter 2

Summary

Study Skills Workbook
Activity 5

Key Terms

2.1

linear (first-degree) equation in one variable A linear equation in one variable can be written in the form $ax = b$, where a and b are real numbers, with $a \neq 0$.

solution A solution of an equation is a number that makes the equation true when substituted for the variable.

solution set The solution set of an equation is the set of all its solutions.

equivalent equations Equivalent equations are equations that have the same solution set.

conditional equation An equation that has a finite (but nonzero) number of elements in its solution set is called a conditional equation.

contradiction An equation that has no solution (that is, its solution set is $\emptyset$) is called a contradiction.

identity An equation that is satisfied by every number is called an identity.

2.2

formula A formula is a mathematical equation in which variables are used to describe a relationship.

percent One percent means "one per hundred."

2.4

vertical angles Angles (1) and (2) shown in the figure are called vertical angles. They have equal measures.

complementary angles Two angles whose sum is 90° are called complementary angles.

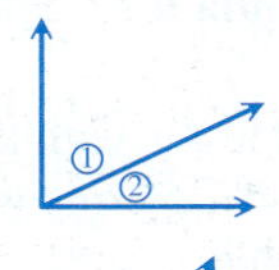

supplementary angles Two angles whose sum is 180° are called supplementary angles.

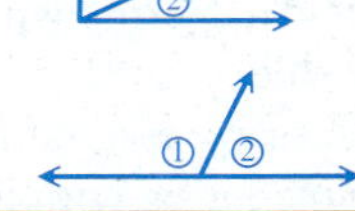

Test Your Word Power

See how well you have learned the vocabulary in this chapter. Answers follow the Quick Review.

1. An **algebraic expression** is
   - **(a)** an expression that uses any of the four basic operations or the operation of taking roots on any collection of variables and numbers
   - **(b)** an expression that contains fractions
   - **(c)** an equation that uses any of the four basic operations or the operation of taking roots on any collection of variables and numbers
   - **(d)** an equation in algebra.

2. An **equation** is
   - **(a)** an algebraic expression
   - **(b)** an expression that contains fractions
   - **(c)** an expression that uses any of the four basic operations or the operation of taking roots on any collection of variables and numbers
   - **(d)** a statement that two algebraic expressions are equal.

3. A **solution set** is the set of numbers that
   - **(a)** make an expression undefined
   - **(b)** make an equation false
   - **(c)** make an equation true
   - **(d)** make an expression equal to 0.

QUICK REVIEW

Concepts	Examples
2.1 Linear Equations in One Variable **Addition and Multiplication Properties of Equality** The same number may be added to (or subtracted from) each side of an equation to obtain an equivalent equation. Similarly, the same nonzero number may be multiplied by or divided into each side of an equation to obtain an equivalent equation.	
Solving a Linear Equation in One Variable *Step 1* Clear fractions. *Step 2* Simplify each side separately. *Step 3* Isolate the variable terms on one side. *Step 4* Isolate the variable. *Step 5* Check.	Solve the equation. $4(8 - 3t) = 32 - 8(t + 2)$ $32 - 12t = 32 - 8t - 16$ $32 - 12t = 16 - 8t$ $32 - 12t + 12t = 16 - 8t + 12t$ $32 = 16 + 4t$ $32 - 16 = 16 + 4t - 16$ $16 = 4t$ $\frac{16}{4} = \frac{4t}{4}$ $4 = t$ The solution set is $\{4\}$. This can be checked by substituting 4 for t in the original equation.
2.2 Formulas **Solving a Formula for a Specified Variable** *Step 1* Get all terms with the specified variable on one side and all terms without that variable on the other side. *Step 2* If necessary, use the distributive property to combine terms with the specified variable. *Step 3* Divide each side by the factor that is the coefficient of the specified variable.	Solve for h: $A = \frac{1}{2}bh$. $A = \frac{1}{2}bh$ $2A = 2\left(\frac{1}{2}bh\right)$ $2A = bh$ $\frac{2A}{b} = h$
2.3 Applications of Linear Equations **Solving an Applied Problem** *Step 1* Read the problem. *Step 2* Assign a variable.	How many liters of 30% alcohol solution and 80% alcohol solution must be mixed to obtain 100 L of 50% alcohol solution? Let x = number of liters of 30% solution needed; $100 - x$ = number of liters of 80% solution needed. Summarize the information of the problem in a table.

Percent (as a Decimal)	Liters of Solution	Liters of Pure Alcohol
.30	x	$.30x$
.80	$100 - x$	$.80(100 - x)$
.50	100	$.50(100)$

Concepts	*Examples*
2.3 Applications of Linear Equations (continued)	
Step 3 Write an equation.	The equation is $$.30x + .80(100 - x) = .50(100).$$
Step 4 Solve the equation.	The solution of the equation is 60.
Step 5 State the answer.	60 L of 30% solution and $100 - 60 = 40$ L of 80% solution are needed.
Step 6 Check.	$.30(60) + .80(100 - 60) = 50$ is true.

2.4 Further Applications of Linear Equations

To solve a uniform motion problem, draw a sketch and make a table. Use the formula $d = rt$.

Two cars start from towns 400 mi apart and travel toward each other. They meet after 4 hr. Find the speed of each car if one travels 20 mph faster than the other.

Let x = speed of the slower car in miles per hour;
$x + 20$ = speed of the faster car.

Use the information in the problem and $d = rt$ to complete the table.

	Rate	Time	Distance
Slower Car	x	4	$4x$
Faster Car	$x + 20$	4	$4(x + 20)$

A sketch shows that the sum of the distances, $4x$ and $4(x + 20)$, must be 400.

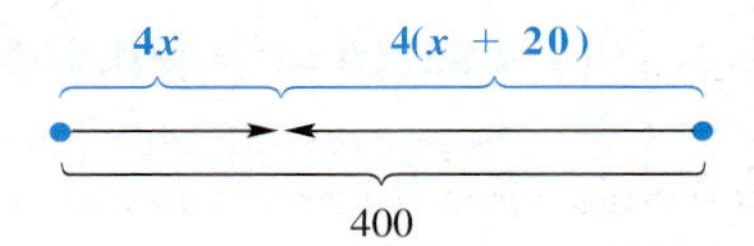

The equation is

$$4x + 4(x + 20) = 400.$$

Solving this equation gives $x = 40$. The slower car travels 40 mph, and the faster car travels $40 + 20 = 60$ mph.

Problems involving denominations of money and mixture problems are solved using methods similar to the one used for the mixture problem shown in the example for Section 2.3.

ANSWERS TO TEST YOUR WORD POWER

1. (a) *Examples:* $\frac{3y - 1}{2}, 6 + \sqrt{2x}, 4a^3b - c$ **2. (d)** *Examples:* $2a + 3 = 7; 3y = -8, x^2 = 4$

3. (c) *Example:* $\{8\}$ is the solution set of $2x + 5 = 21$.

Focus on Real-Data Applications

Currency Exchange and Best Buys

When you travel to a foreign country, you can exchange dollars (in cash) for the local currency (in cash), or you can obtain cash using an ATM card, or you can charge purchases using a credit card. There are advantages and disadvantages to each method of currency exchange.

- ***Currency Exchange Agency.*** Travel agencies, banks, airport and rail stations, and some stores will house a currency exchange agency. You give cash in dollars in exchange for the local currency, but the agency charges different exchange rates for buying and selling dollars and will often impose commission fees.
- ***ATM Cards.*** You may need cash for routine purchases. You can use your ATM card to get local currency in cash from almost any bank. Typically, both your bank and the local bank will charge a fee for using the ATM. The exchange rate for that day is applied.
- ***Credit Cards.*** Credit cards are simple to use, accepted by many businesses, and readily provide records of your purchases. Typically, if you purchase goods in another currency, such as British pounds, then the exchange rate for the day that the transaction is recorded is applied. The credit card company will charge a fee for the currency exchange. It can be as little as 1.5%, but the company can charge wire fees that are very expensive. You must read the fine print of the card agreement.

For Group Discussion

Suppose you had planned a two-week trip to England in summer 2000 to see Wimbledon and to tour London. You budgeted $1000 for souvenirs, touring, and lunches for the entire two weeks.

The official exchange rate on June 30, 2000, was $1 = £.6614.

You discovered in London that the currency exchange agency advertised the following rates for June 30, 2000.

"[The agency will] Buy pounds at 1.4509. Sell pounds at 1.6141. Fee: 2% to buy, 3% to sell, £2 minimum fee per transaction."

1. If you decided to exchange $1000 (U.S.) into British pounds at the *exchange agency*, how many British pounds would you receive (after the transaction fee)? How many British pounds did you anticipate that $1000 would have purchased on June 30, 2000, based on the exchange rate of $1 = £.6614?
2. If you decided to use your *ATM card* and withdrew £600, how much money was deducted from your U.S. account in dollars? Assume that the local bank charged a £1.50 fee and your bank charged a $1.25 fee.
3. You decide to buy an antique map for £600. You are trying to decide whether to use your *credit card* or get cash using your *ATM card*. The credit card charges a fee of 1.5% for currency exchange. Which is more economical?
4. The currency exchange agency makes a profit when changing money. Calculate the total amount (in dollars) that the agency would receive on the transaction to change $100 (U.S.) into British pounds and then to change the proceeds (after the transaction fee) back into dollars. What percent of the original $100 does the agency receive?

Chapter 2 Review Exercises

[2.1] *Solve each equation.*

1. $-(8 + 3x) + 5 = 2x + 6$

2. $-(r + 5) - (2 + 7r) + 8r = 3r - 8$

3. $\dfrac{m - 2}{4} + \dfrac{m + 2}{2} = 8$

4. $\dfrac{2q + 1}{3} - \dfrac{q - 1}{4} = 0$

5. $5(2x - 3) = 6(x - 1) + 4x$

6. $-3x + 2(4x + 5) = 10$

7. $-\dfrac{3}{4}x = -12$

8. $.05x + .03(1200 - x) = 42$

9. Which equation has $\{0\}$ as its solution set?

   A. $x - 5 = 5$ **B.** $4x = 5x$

   C. $x + 3 = -3$ **D.** $6x - 6 = 6$

10. Give the steps you would use to solve the equation $-2x + 5 = 7$.

Decide whether each equation is a conditional equation, *an* identity, *or a* contradiction. *Give the solution set.*

11. $7r - 3(2r - 5) + 5 + 3r = 4r + 20$

12. $8p - 4p - (p - 7) + 9p + 13 = 12p$

13. $-2r + 6(r - 1) + 3r - (4 - r) = -(r + 5) - 5$

[2.2] *Solve each formula for the indicated variable.*

14. $V = LWH$ for H

15. $A = \dfrac{1}{2}h(B + b)$ for h

16. $C = \pi d$ for d

Solve each problem.

17. A rectangular solid has a volume of 180 ft^3. Its length is 6 ft and its width is 5 ft. Find its height.

18. The total number of AIDS cases reported in 1997 was 58,443. In 1998, this figure had decreased to 48,269. What approximate percent decrease did this represent? (*Source:* U.S. Centers for Disease Control.)

19. Find the simple interest rate that Francesco Castellucio is earning, if a principal of \$30,000 earns \$7800 interest in 4 years.

20. If the Fahrenheit temperature is 77°, what is the corresponding Celsius temperature?

21. The circle graph shows the projected racial composition of the U.S. workforce in the year 2006. The projected total number of people in the workforce for that year is 148,847,000. How many of these will be in the Hispanic category? (*Source:* U.S. Bureau of Labor Statistics.)

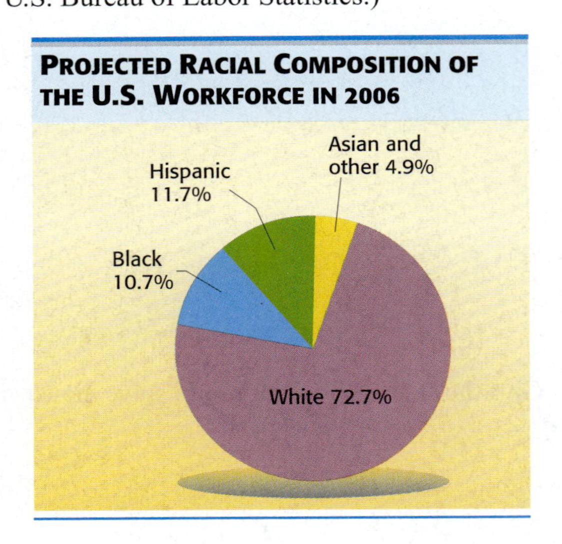

22. The Sioux drum that Wade purchased has a circumference of 200π mm. Find the measure of its radius.

[2.3] *Write each phrase as a mathematical expression, using x as the variable.*

23. One-third of a number, subtracted from 9

24. The product of 4 and a number, divided by 9 more than the number

Solve each problem.

25. The length of a rectangle is 3 m less than twice the width. The perimeter of the rectangle is 42 m. Find the length and width of the rectangle.

26. In a triangle with two sides of equal length, the third side measures 15 in. less than the sum of the two equal sides. The perimeter of the triangle is 53 in. Find the lengths of the three sides.

27. A candy clerk has three times as many kilograms of chocolate creams as peanut clusters. The clerk has 48 kg of the two candies altogether. How many kilograms of peanut clusters does the clerk have?

28. How many liters of a 20% solution of a chemical should be mixed with 15 L of a 50% solution to get a 30% mixture?

29. How much water should be added to 30 L of a 40% acid solution to reduce it to a 30% solution?

Percent (as a Decimal)	Liters of Solution	Liters of Pure Acid
.40		
.00	x	
.30		

30. Anna Mae Wood invested some money at 6% and $4000 less than this amount at 4%. Find the amount invested at each rate if her total annual interest income is $840.

Rate (as a Decimal)	Principal	Interest
.06	x	
.04		

[2.4]

31. Which choice is the best *estimate* for the average speed of a trip of 405 mi that lasted 8.2 hr?

A. 50 mph

B. 30 mph

C. 60 mph

D. 40 mph

32. (a) A driver averaged 53 mph and took 10 hr to travel from Memphis to Chicago. What is the distance between Memphis and Chicago?

(b) A small plane traveled from Warsaw to Rome, averaging 164 mph. The trip took 2 hr. What is the distance from Warsaw to Rome?

33. A passenger train and a freight train leave a town at the same time and go in opposite directions. They travel at 60 mph and 75 mph, respectively. How long will it take for them to be 297 mi apart?

	Rate	Time	Distance
Passenger Train	60	x	
Freight Train	75	x	

34. Two cars leave towns 230 km apart at the same time, traveling directly toward one another. One car travels 15 km per hr slower than the other. They pass one another 2 hr later. What are their speeds?

	Rate	Time	Distance
Faster Car	x	2	
Slower Car	$x - 15$	2	

35. An automobile averaged 45 mph for the first part of a trip and 50 mph for the second part. If the entire trip took 4 hr and covered 195 mi, for how long was the rate 45 mph?

36. An 85-mi trip to the beach took the Valenzuela family 2 hr. During the second hour, a rainstorm caused them to average 7 mph less than they traveled during the first hour. Find their average rate for the first hour.

Mixed Review Exercises

Solve.

37. $(7 - 2k) + 3(5 - 3k) = k + 8$

38. $\frac{4x + 2}{4} + \frac{3x - 1}{8} = \frac{x + 6}{16}$

39. $-5(6p + 4) - 2p = -32p + 14$

40. The perimeter of a triangle is 68 in. The middle side is twice as long as the shortest side. The longest side is 4 in. less than three times the shortest side. Find the lengths of the three sides.

41. $ak + bt = 6t - sk$ for k

42. $5(2r - 3) + 7(2 - r) = 3(r + 2) - 7$

43. A square is such that if each side were increased by 4 in., the perimeter would be 8 in. less than twice the perimeter of the original square. Find the length of a side of the original square.

44. In the 2000 presidential election, Al Gore received five fewer electoral votes than George W. Bush. A total of 537 electoral votes were cast. How many electoral votes did each candidate receive? (*Source:* www.cnn.com)

45. Two cars start from the same point and travel in opposite directions. The car traveling west leaves 1 hr later than the car traveling east. The eastbound car travels 40 mph, and the westbound car travels 60 mph. When they are 240 mi apart, how long had each car traveled?

46. Some money is invested at 4% simple annual interest and $500 more than that amount is invested at 5%. After 1 year, a total of $142 interest was earned. How much was invested at each rate?

47. $.08x + .04(x + 200) = 188$

48. $P = 2L + 2W$ for W

Chapter 2 TEST

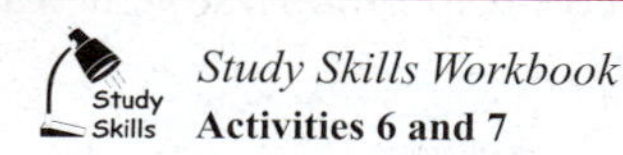

Study Skills Workbook
Activities 6 and 7

Solve each equation.

1. $3(2x - 2) - 4(x + 6) = 4x + 8$

1. ______________________

2. $.08x + .06(x + 9) = 1.24$

2. ______________________

3. $\frac{x + 6}{10} + \frac{x - 4}{15} = 1$

3. ______________________

4. Decide whether the equation

$$3x - (2 - x) + 4x + 2 = 8x + 3$$

is a *conditional equation*, an *identity*, or a *contradiction*. Give its solution set.

4. ______________________

5. Solve for v: $-16t^2 + vt - S = 0$.

5. ______________________

6. Solve for r: $ar + 2 = 3r - 6t$.

6. ______________________

Solve each problem.

7. The 1997 Daytona 500 (mile) race was won by Jeff Gordon, who averaged 148.295 mph. What was Gordon's time?

7. ______________________

8. A certificate of deposit pays \$2281.25 in simple interest for 1 year on a principal of \$36,500. What is the rate of interest?

8. ______________________

9. Of the 38,159 offices, stations, and branches of the U.S. Postal Service in 1998, 27,952 were actually classified as post offices. What percent to the nearest tenth were classified as post offices? (*Source:* U.S. Postal Service.)

9. ______________________

10. ______________________

11. ______________________

12. ______________________

13. ______________________

14. ______________________

15. ______________________

10. Tyler McGinnis invested some money at 3% simple interest and some at 5% simple interest. The total amount of his investments was $28,000, and the interest he earned during the first year was $1240. How much did he invest at each rate?

11. Two cars leave from the same point at the same time, traveling in opposite directions. One travels 15 mph slower than the other. After 6 hr, they are 630 mi apart. Find the rate of each car.

12. Find the measure of each angle.

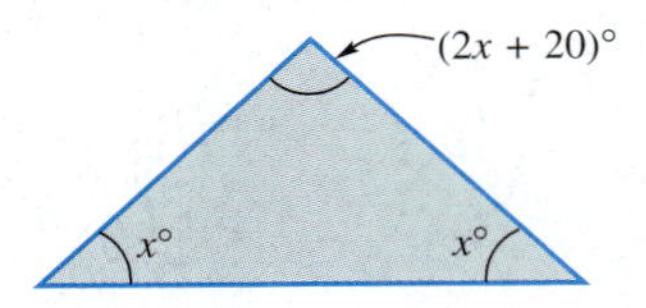

The formula

$$A = \frac{24f}{b(p + 1)}$$

gives the approximate annual interest rate for a consumer loan paid off with monthly payments. Here f is the finance charge on the loan, p is the number of payments, and b is the original amount of the loan. Use this formula to solve Problems 13 and 14.

13. Find the approximate annual interest rate for an installment loan to be repaid in 24 monthly installments. The finance charge on the loan is $200, and the original loan balance is $1920.

14. Find the approximate annual interest rate for an automobile loan to be repaid in 36 monthly installments. The finance charge on the loan is $740, and the amount financed is $3600. (Round to the nearest hundredth of a percent.)

15. The circle graph shows the percents of various occupations in a representative sample of stockholders. Based on the figure, in a group of 5000 stockholders, how many would you expect to be white-collar workers? (*Source: Study by Peter D. Hart Research Associates for the Nasdaq Stock Market.*)

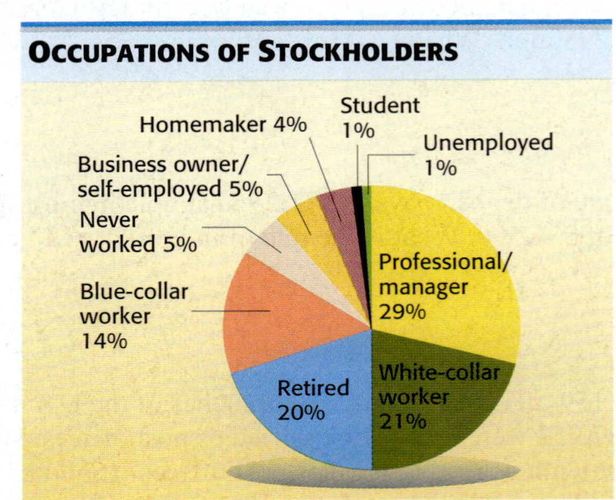

Cumulative Review Exercises

Chapters 1–2

From now on, each chapter will conclude with a set of cumulative review exercises designed to cover the major topics from the beginning of the course. This allows you to constantly review topics that have been introduced up to that point.

Let $A = \{-8, -\frac{2}{3}, -\sqrt{6}, 0, \frac{4}{5}, 9, \sqrt{36}\}$. *Simplify the elements of A as necessary, and then list the elements that belong to each set listed in Exercises 1–6.*

1. Natural numbers

2. Whole numbers

3. Integers

4. Rational numbers

5. Irrational numbers

6. Real numbers

Add or subtract, as indicated.

7. $-\frac{4}{3} - \left(-\frac{2}{7}\right)$

8. $|-4.2| + |5.6| - |-1.9|$

9. $(-2)^4 + (-2)^3$

10. $\sqrt{25} - \frac{\sqrt{100}}{2}$

Evaluate each expression.

11. $(-3)^5$

12. $\left(\frac{6}{7}\right)^3$

13. 4^6

14. -4^6

15. Which one of the following is not a real number: $-\sqrt{36}$ or $\sqrt{-36}$?

16. Which one of the following is undefined: $\frac{4-4}{4+4}$ or $\frac{4+4}{4-4}$?

Evaluate each expression if $a = 2$, $b = -3$, *and* $c = 4$.

17. $-3a + 2b - c$

18. $-2b^2 - c^2$

19. $-8(a^2 + b^3)$

20. $\frac{3a^3 - b}{4 + 3c}$

Use the properties of real numbers to simplify each expression.

21. $-7r + 5 - 13r + 12$

22. $-(3k + 8) - 2(4k - 7) + 3(8k + 12)$

Identify the property of real numbers illustrated in each equation.

23. $(a + b) + 4 = 4 + (a + b)$

24. $4x + 12x = (4 + 12)x$

25. $-9 + 9 = 0$

Solve each equation.

26. $-4x + 7(2x + 3) = 7x + 36$

27. $-\frac{3}{5}x + \frac{2}{3}x = 2$

28. $.06x + .03(100 + x) = 4.35$

29. $P = a + b + c$ for c

30. $4(2x - 6) + 3(x - 2) = 11x + 1$

31. $\frac{2}{3}x + \frac{5}{8}x = \frac{31}{24}x$

Solve each problem.

32. How much pure alcohol should be added to 7 L of 10% alcohol to increase the concentration to 30% alcohol?

33. A coin collection contains 29 coins. It consists of pennies, nickels, and quarters. The number of quarters is 4 less than the number of nickels, and the face value of the collection is \$2.69. How many of each denomination are there in the collection?

34. Linda Casse invested some money at 5% simple annual interest and \$2000 more than that amount at 6%. Her annual interest from the two investments totaled \$670. How much did she invest at each rate?

35. Eric and Trish are running in the Fresh Water Fun Run. Eric runs at 7 mph and Trish runs at 5 mph. If they start at the same time, how long will it be before Eric is $\frac{1}{4}$ mi ahead of Trish?

36. Refer to the circle graph in Problem 15 of the Chapter 2 Test. In a group of 3000 stockholders, how many would you expect to be blue-collar workers?

37. The body mass index, or BMI, of a person is given by the formula

$$\text{BMI} = \frac{704 \times (\text{weight in pounds})}{(\text{height in inches})^2}.$$

Ken Griffey, Jr., is listed as being 6 ft, 3 in. tall and weighing 205 lb. What is his BMI (to the nearest tenth)? (*Source: Reader's Digest*, October 1993.)

38. Clark's rule, a formula used in reducing drug dosage according to weight from the recommended adult dosage to a child dosage, is

$$\frac{\text{weight of child in pounds}}{150} \times \text{adult dose} = \text{child's dose}.$$

Find a child's dosage if the child weighs 55 lb and the recommended adult dosage is 120 mg.

Linear Inequalities and Absolute Value

The cost of a college education has risen rapidly in the last decade. Average higher education tuition and fees increased by 50.2% from the 1990–1991 school year to the 1998–1999 school year. (*Source: World Almanac and Book of Facts,* 2000.) In Example 8 of Section 3.2, we apply the concepts of this chapter to college student expenses.

3.1 Linear Inequalities in One Variable

Objectives

1. Graph intervals on a number line.
2. Solve linear inequalities using the addition property.
3. Solve linear inequalities using the multiplication property.
4. Solve linear inequalities with three parts.
5. Solve applied problems using linear inequalities.

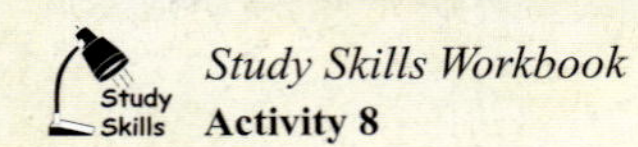
Study Skills Workbook
Activity 8

Solving inequalities is closely related to solving equations. In this section we introduce properties for solving inequalities.

Inequalities are algebraic expressions related by

$<$ "is less than,"
$\leq$ "is less than or equal to,"
$>$ "is greater than,"
$\geq$ "is greater than or equal to."

We solve an inequality by finding all real number solutions for it. For example, the solution set of $x \leq 2$ includes *all* real numbers that are less than or equal to 2, not just the integers less than or equal to 2. For example, $-2.5, -1.7, -1, \frac{1}{2}, \sqrt{2}, \frac{7}{4}$, and 2, are real numbers less than or equal to 2 and are therefore solutions of $x \leq 2$.

1 Graph intervals on a number line. A good way to show the solution set of an inequality is by graphing. We graph all the real numbers satisfying $x \leq 2$ by placing a square bracket at 2 on a number line and drawing an arrow extending from the bracket to the left (to represent the fact that all numbers less than 2 are also part of the graph). The graph is shown in Figure 1.

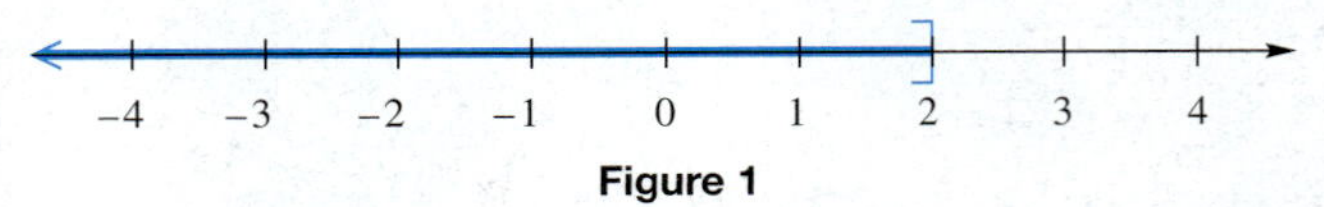

Figure 1

The set of numbers less than or equal to 2 is an example of an **interval** on the number line. To write intervals, we use **interval notation.** For example, using this notation, the interval of all numbers less than or equal to 2 is written $(-\infty, 2]$. The negative infinity symbol $-\infty$ does not indicate a number. It is used to show that the interval includes all real numbers less than 2. As on the number line, the square bracket indicates that 2 is included in the solution set. A parenthesis is always used next to the infinity symbol. The set of real numbers is written in interval notation as $(-\infty, \infty)$.

Example 1 **Graphing Intervals Written in Interval Notation on Number Lines**

Write each inequality in interval notation and graph it.

(a) $x > -5$

The statement $x > -5$ says that x can represent any number greater than -5, but x cannot equal -5. This interval is written $(-5, \infty)$. We show this solution set on a number line by placing a parenthesis at -5 and drawing an arrow to the right, as in Figure 2. The parenthesis at -5 shows that -5 is *not* part of the graph.

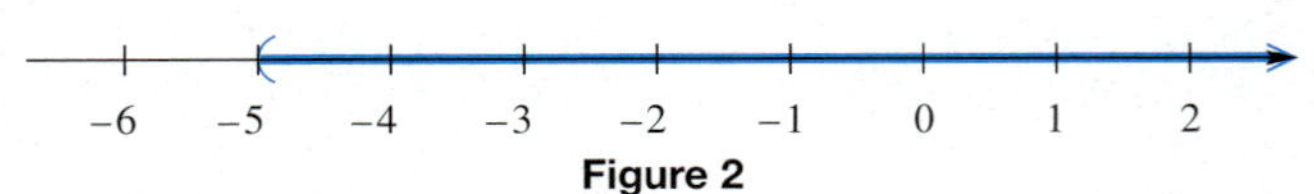

Figure 2

Continued on Next Page

(b) $-1 \leq x < 3$

This statement is read "-1 is less than or equal to x *and* x is less than 3." Thus, we want the set of numbers that are *between* -1 and 3, with -1 included and 3 excluded. In interval notation, we write the solution set as $[-1, 3)$, using a square bracket at -1 because it is part of the graph and a parenthesis at 3 because it is not part of the graph. The graph is shown in Figure 3.

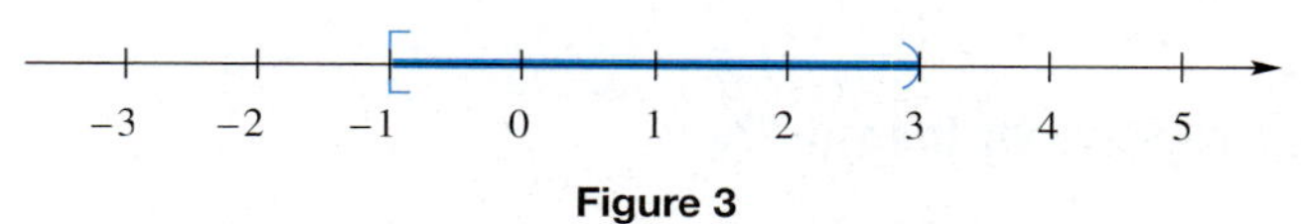

Figure 3

Work Problem 1 at the Side.

We now summarize the various types of intervals.

Type of Interval	Set	Interval Notation	Graph
Open interval	$\{x \mid a < x\}$	(a, ∞)	a
	$\{x \mid a < x < b\}$	(a, b)	a b
	$\{x \mid x < b\}$	$(-\infty, b)$	b
	$\{x \mid x$ is a real number$\}$	$(-\infty, \infty)$	
Half-open interval	$\{x \mid a \leq x\}$	$[a, \infty)$	a
	$\{x \mid a < x \leq b\}$	$(a, b]$	a b
	$\{x \mid a \leq x < b\}$	$[a, b)$	a b
	$\{x \mid x \leq b\}$	$(-\infty, b]$	b
Closed interval	$\{x \mid a \leq x \leq b\}$	$[a, b]$	a b

An **inequality** says that two expressions are *not* equal. Solving inequalities is similar to solving equations.

Linear Inequality

A **linear inequality in one variable** can be written in the form

$$ax < b,$$

where a and b are real numbers, with $a \neq 0$.

Examples of linear inequalities include

$$x + 5 < 2, \quad y - 3 \geq 5, \quad \text{or} \quad 2k + 5 \leq 10.$$

(Throughout this section we give definitions and rules only for $<$, but they are also valid for $>$, $\leq$, and $\geq$.)

1 Write each inequality in interval notation and graph it.

(a) $x < -1$

(b) $x \geq -3$

(c) $-4 \leq x < 2$

ANSWERS

1. **(a)** $(-\infty, -1)$

−4 −3 −2 −1 0 1 2

(b) $[-3, \infty)$

−4 −3 −2 −1 0

(c) $[-4, 2)$

−6 −4 −2 0 2 4

2 Solve each inequality, check your solutions, and graph the solution set.

(a) $p + 6 < 8$

(b) $8y < 7y - 6$

2 Solve linear inequalities using the addition property.

We solve an inequality by finding all numbers that make the inequality true. Usually, an inequality has an infinite number of solutions. These solutions, like solutions of equations, are found by producing a series of simpler equivalent inequalities. **Equivalent inequalities** are inequalities with the same solution set. We use the addition and multiplication properties of inequality to produce equivalent inequalities.

Addition Property of Inequality

For all real numbers a, b, and c, the inequalities

$$a < b \quad \text{and} \quad a + c < b + c$$

are equivalent.

In words, adding the same number to each side of an inequality does not change the solution set.

As with equations, the addition property can be used to *subtract* the same number from each side of an inequality.

Example 2 Using the Addition Property of Inequality

Solve $x - 7 < -12$.

Add 7 to each side.

$$x - 7 + 7 < -12 + 7 \qquad \text{Add 7.}$$

$$x < -5$$

Check: Substitute -5 for x in the equation $x - 7 = -12$. The result should be a true statement.

$$x - 7 = -12$$

$$-5 - 7 = -12 \quad ? \qquad \text{Let } x = -5.$$

$$-12 = -12 \qquad \text{True}$$

This shows that -5 is the boundary point. Now we test a number on each side of -5 to verify that numbers *less than* -5 make the inequality true. We choose -4 and -6.

$$x - 7 < -12$$

$-4 - 7 < -12$? Let $x = -4$.	$-6 - 7 < -12$? Let $x = -6$.
$-11 < -12$ False	$-13 < -12$ True
-4 is not in the solution set.	-6 is in the solution set.

The check confirms that $(-\infty, -5)$, graphed in Figure 4, is the correct solution set.

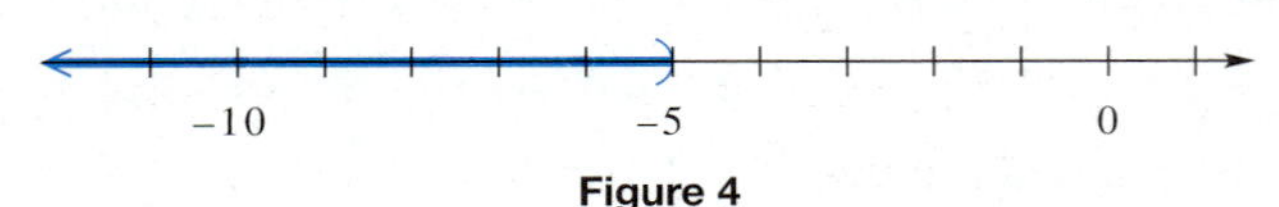

Figure 4

Work Problem **2** at the Side.

Example 3 Using the Addition Property of Inequality

Solve the inequality $14 + 2m \leq 3m$, and graph the solution set.

Continued on Next Page

ANSWERS

2. (a) $(-\infty, 2)$

−3 −2 −1 0 1 2 3

(b) $(-\infty, -6)$

−10 −9 −8 −7 −6 −5 −4

First, subtract $2m$ from each side.

$$14 + 2m \leq 3m$$

$$14 + 2m - \mathbf{2m} \leq 3m - \mathbf{2m} \quad \text{Subtract } 2m.$$

$$14 \leq m \quad \text{Combine like terms.}$$

The inequality $14 \leq m$ (14 is less than or equal to m) can also be written $m \geq 14$ (m is greater than or equal to 14). Notice that in each case, the inequality symbol points to the smaller number, 14.

Check:

$$14 + 2\mathbf{m} = 3\mathbf{m}$$

$$14 + 2(\mathbf{14}) = 3(\mathbf{14}) \quad ? \quad \text{Let } m = 14.$$

$$42 = 42 \quad \text{True}$$

So 14 satisfies the equality part of $\leq$. Choose 10 and 15 as test points.

$$14 + 2\mathbf{m} < 3\mathbf{m}$$

$14 + 2(\mathbf{10}) < 3(\mathbf{10})$? Let $m = 10$.	$14 + 2(\mathbf{15}) < 3(\mathbf{15})$? Let $m = 15$.
$34 < 30$ False	$44 < 45$ True
10 is not in the solution set.	15 is in the solution set.

The check confirms that $[14, \infty)$ is the correct solution set. See Figure 5.

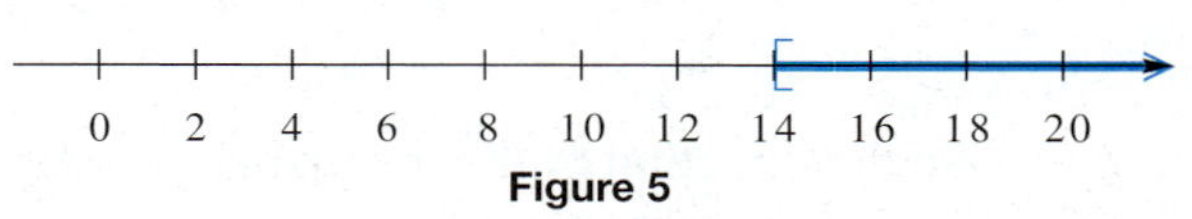

Figure 5

CAUTION

Errors often occur in graphing inequalities when the variable term is on the right side. (This is probably due to the fact that we read from left to right.) To guard against such errors, it is a good idea to rewrite these inequalities so that the variable is on the left, as discussed in Example 3.

Work Problem 3 at the Side.

3 Solve linear inequalities using the multiplication property. Solving an inequality such as $3x \leq 15$ requires dividing each side by 3, using the *multiplication property of inequality*, which is a little more involved than the multiplication property of *equality*. To see how the multiplication property of inequality works, start with the true statement

$$-2 < 5.$$

Multiply each side by, say, 8.

$$-2(\mathbf{8}) < 5(\mathbf{8}) \quad \text{Multiply by 8.}$$

$$-16 < 40 \quad \text{True}$$

This gives a true statement. Start again with $-2 < 5$, and multiply each side by -8.

$$-2(\mathbf{-8}) < 5(\mathbf{-8}) \quad \text{Multiply by } -8.$$

$$16 < -40 \quad \text{False}$$

The result, $16 < -40$, is false. To make it true, we must change the direction of the inequality symbol to get

$$16 \mathbf{>} -40. \quad \text{True}$$

Work Problem 4 at the Side.

3 Solve $2k - 5 \geq 1 + k$, check, and graph the solution set.

4 Multiply both sides of each inequality by -5. Then insert the correct symbol, either $<$ or $>$, in the first blank, and fill in the other blank in part (b).

(a) $7 < 8$

-35 ________ -40

(b) $-1 > -4$

5 ________ ________

ANSWERS

3. $[6, \infty)$

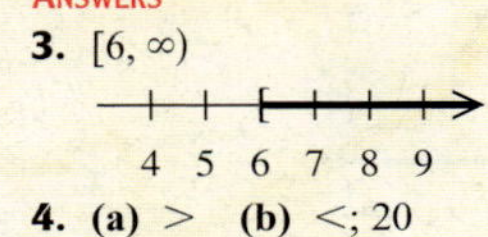

4. (a) $>$ **(b)** $<$; 20

5 Solve each inequality, check, and graph the solution set.

(a) $2y < -10$

(b) $-7k \geq 8$

(c) $-9m < -81$

As these examples suggest, multiplying each side of an inequality by a *negative* number reverses the direction of the inequality symbol. The same is true for dividing by a negative number since division is defined in terms of multiplication.

Multiplication Property of Inequality

For all real numbers a, b, and c, with $c \neq 0$,
(a) the inequalities

$$a < b \quad \text{and} \quad ac < bc$$

are equivalent **if $c > 0$**;
(b) the inequalities

$$a < b \quad \text{and} \quad ac > bc$$

are equivalent **if $c < 0$**.

In words, each side of an inequality may be multiplied by a *positive* number without changing the direction of the inequality symbol. ***Multiplying or dividing by a* negative *number requires that we reverse the inequality symbol.***

CAUTION

Remember to reverse the direction of the inequality symbol when multiplying or dividing by a *negative* number.

Example 4 Using the Multiplication Property of Inequality

Solve each inequality, and graph the solution set.

(a) $5m \leq -30$

Use the multiplication property to divide each side by 5. Since $5 > 0$, do *not* reverse the inequality symbol.

$$5m \leq -30$$

$$\frac{5m}{5} \leq \frac{-30}{5} \qquad \text{Divide by 5.}$$

$$m \leq -6$$

Check that the solution set is the interval $(-\infty, -6]$, graphed in Figure 6.

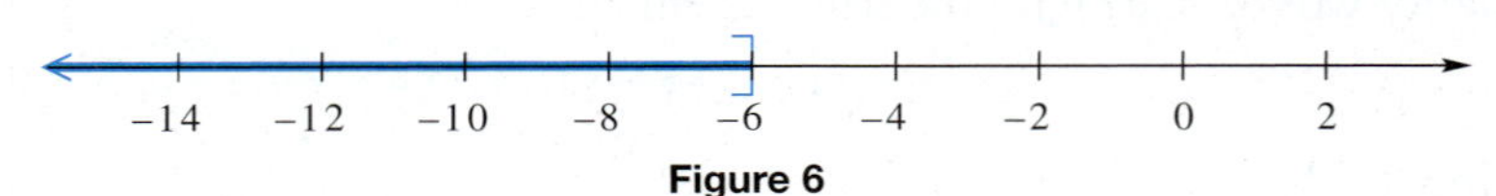

Figure 6

(b) $-4k \leq 32$

Divide each side by -4. Since $-4 < 0$, reverse the inequality symbol.

$$-4k \leq 32$$

$$\frac{-4k}{-4} \geq \frac{32}{-4} \qquad \text{Divide by } -4 \text{ and reverse the symbol.}$$

$$k \geq -8$$

Check the solution set. Figure 7 shows the graph of the solution set, $[-8, \infty)$.

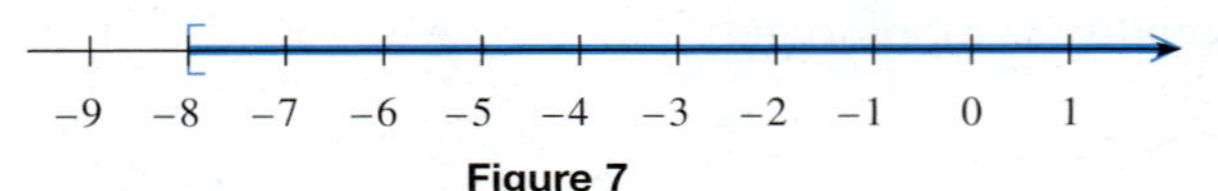

Figure 7

Work Problem 5 at the Side.

ANSWERS

5. (a) $(-\infty, -5)$

−8 −7 −6 −5 −4

(b) $\left(-\infty, -\frac{8}{7}\right]$

$-\frac{8}{7}$

−3 −2 −1 0 1 2

(c) $(9, \infty)$

7 8 9 10 11 12

The steps used in solving a linear inequality are given below.

Solving a Linear Inequality

Step 1 **Simplify each side separately.** Simplify each side of the inequality as much as possible by using the distributive property to clear parentheses and by combining like terms as needed.

Step 2 **Isolate the variable terms on one side.** Use the addition property of inequality to get all terms with variables on one side of the inequality and all numbers on the other side.

Step 3 **Isolate the variable.** Use the multiplication property of inequality to change the inequality to the form $x < k$ or $x > k$.

Remember: Reverse the direction of the inequality symbol ***only*** when ***multiplying or dividing each side of an inequality by a* negative *number.***

Example 5 **Solving a Linear Inequality Using the Distributive Property**

Solve $-3(x + 4) + 2 \geq 7 - x$, and graph the solution set.

$$-3x - 12 + 2 \geq 7 - x \quad \text{Distributive property}$$

$$-3x - 10 \geq 7 - x$$

$$-3x - 10 + x \geq 7 - x + x \quad \text{Add } x.$$

$$-2x - 10 \geq 7$$

$$-2x - 10 + 10 \geq 7 + 10 \quad \text{Add 10.}$$

$$-2x \geq 17$$

$$\frac{-2x}{-2} \leq \frac{17}{-2} \quad \text{Divide by } -2\text{; change } \geq \text{ to } \leq.$$

$$x \leq -\frac{17}{2}$$

Figure 8 shows the graph of the solution set, $(-\infty, -\frac{17}{2}]$.

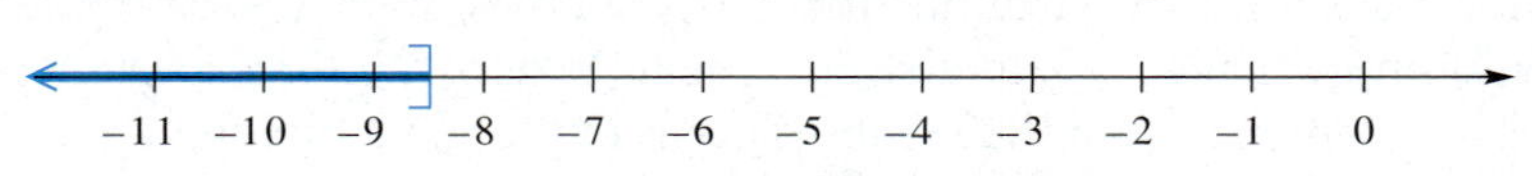

Figure 8

Example 6 **Solving a Linear Inequality with Fractions**

Solve $-\frac{2}{3}(r - 3) - \frac{1}{2} < \frac{1}{2}(5 - r)$, and graph the solution set.

To clear fractions, multiply each side by the least common denominator, 6.

$$-\frac{2}{3}(r - 3) - \frac{1}{2} < \frac{1}{2}(5 - r)$$

$$-4(r - 3) - 3 < 3(5 - r) \quad \text{Multiply by 6.}$$

Step 1

$$-4r + 12 - 3 < 15 - 3r \quad \text{Distributive property}$$

$$-4r + 9 < 15 - 3r$$

Continued on Next Page

Step 2

$$3r - 4r + 9 < 3r + 15 - 3r \quad \text{Add } 3r.$$
$$-r + 9 < 15$$
$$-r + 9 - 9 < 15 - 9 \quad \text{Subtract 9.}$$
$$-r < 6$$

Step 3 To solve for r, multiply each side of the inequality by -1. Since -1 is negative, change the direction of the inequality symbol.

$$-1(-r) > -1(6) \quad \text{Multiply by } -1, \text{ change } < \text{ to } >.$$
$$r > -6$$

Check that the solution set is $(-6, \infty)$. See Figure 9.

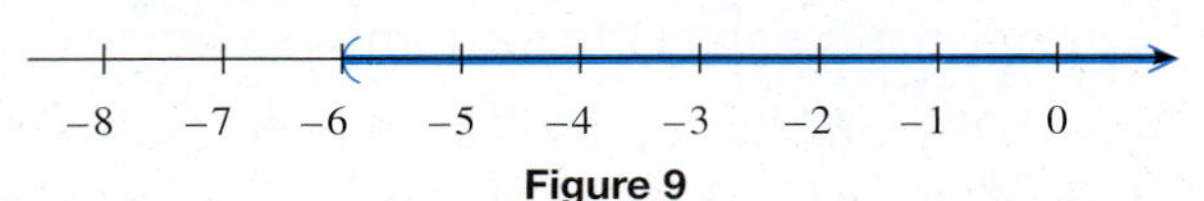

Figure 9

Work Problem 6 at the Side.

6 Solve, check, and graph the solution set of each inequality.

(a) $5 - 3(m - 1) \leq 2(m + 3) + 1$

(b) $\frac{1}{4}(m + 3) + 2 \leq \frac{3}{4}(m + 8)$

4 Solve linear inequalities with three parts.

For some applications, it is necessary to work with an inequality such as

$$3 < x + 2 < 8,$$

where $x + 2$ is *between* 3 and 8. To solve this inequality, we subtract 2 from each of the three parts of the inequality, giving

$$3 - 2 < x + 2 - 2 < 8 - 2$$
$$1 < x < 6.$$

Thus, x must be between 1 and 6, so $x + 2$ will be between 3 and 8. The solution set, $(1, 6)$, is graphed in Figure 10.

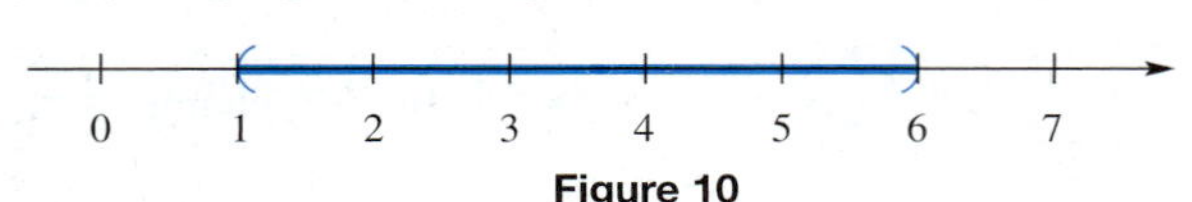

Figure 10

CAUTION

When inequalities have three parts, the order of the parts is important. It would be *wrong* to write an inequality as $8 < x + 2 < 3$, since this would imply that $8 < 3$, a false statement. In general, three-part inequalities are written so that the symbols point in the same direction, and both point toward the smaller number.

Example 7 Solving a Three-Part Inequality

Solve the inequality $-2 \leq -3k - 1 \leq 5$, and graph the solution set.

Begin by adding 1 to each of the three parts to isolate the variable term in the middle.

$$-2 + 1 \leq -3k - 1 + 1 \leq 5 + 1 \quad \text{Add 1 to each part.}$$
$$-1 \leq -3k \leq 6$$
$$\frac{-1}{-3} \geq \frac{-3k}{-3} \geq \frac{6}{-3} \quad \text{Divide each part by } -3; \text{ reverse the inequality symbols.}$$
$$\frac{1}{3} \geq k \geq -2$$
$$-2 \leq k \leq \frac{1}{3} \quad \text{Rewrite in the order on the number line.}$$

Continued on Next Page

ANSWERS

6. (a) $\left[\frac{1}{5}, \infty\right)$

$\frac{1}{5}$

0 1 2 3 4 5

(b) $\left[-\frac{13}{2}, \infty\right)$

$-\frac{13}{2}$

−9 −8 −7 −6 −5 −4

Check that the solution set is $\left[-2, \frac{1}{3}\right]$, as shown in Figure 11.

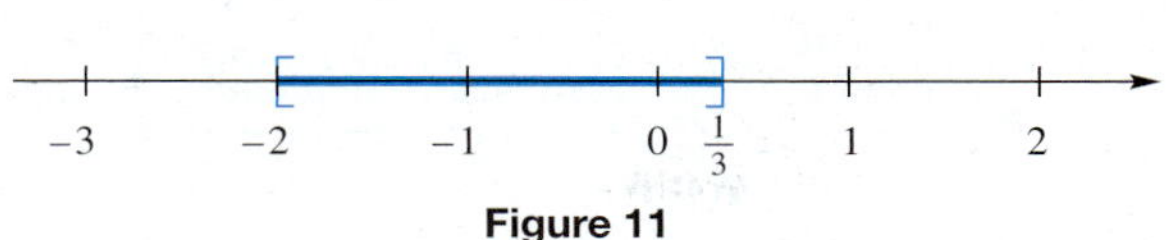

Figure 11

Work Problem 7 at the Side.

Examples of the types of solution sets to be expected from solving linear equations or linear inequalities are shown below.

SOLUTIONS OF LINEAR EQUATIONS AND INEQUALITIES

Equation or Inequality	Typical Solution Set	Graph of Solution Set
Linear equation $5x + 4 = 14$	$\{2\}$	2
Linear inequality $5x + 4 < 14$	$(-\infty, 2)$	2
$5x + 4 > 14$	$(2, \infty)$	2
Three-part inequality $-1 \leq 5x + 4 \leq 14$	$[-1, 2]$	−1 2

5 **Solve applied problems using linear inequalities.** In addition to the familiar "is less than" and "is greater than," the expressions "is no more than" and "is at least" also indicate inequalities. Expressions for inequalities sometimes appear in applied problems. The table below shows how to interpret these expressions.

Word Expression	Interpretation
a **is at least** b	$a \geq b$
a **is no less than** b	$a \geq b$
a **is at most** b	$a \leq b$
a **is no more than** b	$a \leq b$

In Examples 8 and 9, we show how to solve applied problems with inequalities. We use the six problem-solving steps from Chapter 2, changing Step 3 to "Write an inequality" instead of "Write an equation."

Example 8 Using a Linear Inequality to Solve a Rental Problem

A rental company charges $15 to rent a chain saw, plus $2 per hr. Al Ghandi can spend no more than $35 to clear some logs from his yard. What is the *maximum* amount of time he can use the rented saw?

Step 1 **Read** the problem again.

Step 2 **Assign a variable.** Let h = the number of hours he can rent the saw.

Continued on Next Page

7 Solve, check, and graph the solution set of each inequality.

(a) $-3 \leq x - 1 \leq 7$

(b) $5 < 3x - 4 < 9$

ANSWERS

7. **(a)** $[-2, 8]$

−4 −2 0 2 4 6 8 10

(b) $\left(3, \frac{13}{3}\right)$

13/3

2 3 4 5

Step 3 **Write an inequality.** He must pay \$15, plus \$2h, to rent the saw for h hours, and this amount must be *no more than* \$35.

$$\underbrace{15 + 2h}_{\text{Cost of renting}} \quad \underbrace{\leq}_{\text{is no more than}} \quad \underbrace{35}_{\text{35 dollars.}}$$

Step 4 **Solve.**

$$2h \leq 20 \quad \text{Subtract 15.}$$

$$h \leq 10 \quad \text{Divide by 2.}$$

Step 5 **State the answer.** He can use the saw for a maximum of 10 hr. (Of course, he may use it for less time, as indicated by the inequality $h \leq 10$.)

Step 6 **Check.** If Al uses the saw for **10** hr, he will spend $15 + 2(\mathbf{10}) = 35$ dollars, the maximum amount.

Work Problem 8 at the Side.

8 Solve the problem.

Maureen O'Connor can rent a car from Ames for \$48 per day plus 10¢ per mile, or from Hughes at \$40 per day plus 15¢ per mile. She plans to use the car for 3 days. What number of miles would make Hughes cost at most as much as Ames?

Example 9 Finding an Average Test Score

Martha has scores of 88, 86, and 90 on her first three algebra tests. An average score of at least 90 will earn an A in the class. What possible scores on her fourth test will earn her an A average?

Let x represent the score on the fourth test. Her average score must be at least 90. To find the average of four numbers, add them and then divide by 4.

$$\underbrace{\frac{88 + 86 + 90 + x}{4}}_{\text{Average}} \quad \underbrace{\geq}_{\text{is at least}} \quad \underbrace{90}_{\text{90.}}$$

$$\frac{264 + x}{4} \geq 90 \quad \text{Add the scores.}$$

$$264 + x \geq 360 \quad \text{Multiply by 4.}$$

$$x \geq 96 \quad \text{Subtract 264.}$$

She must score **96** or more on her fourth test.

Check: $$\frac{88 + 86 + 90 + \mathbf{96}}{4} = \frac{360}{4} = 90$$

A score of 96 or more will give an average of at least 90, as required.

Work Problem 9 at the Side.

9 Solve the problem.

Michael has grades of 92, 90, and 84 on his first three tests. What grade must he make on his fourth test in order to keep an average of at least 90?

Answers

8. 480 mi or less

9. at least 94

3.1 EXERCISES

FOR EXTRA HELP

 Student's Solutions Manual

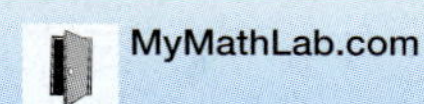 MyMathLab.com

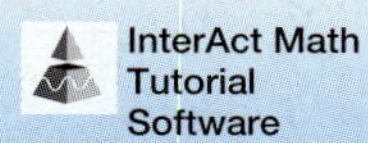 InterAct Math Tutorial Software

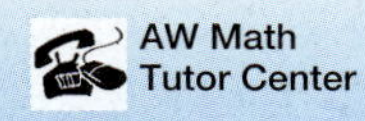 AW Math Tutor Center

 www.mathxl.com

Digital Video Tutor CD 2 Videotape 3

Match each inequality with the correct graph or interval notation.

1. $x \leq 3$	**A.** (number line: bracket at 3, shaded to the right; labels 0, 3)
2. $x > 3$	**B.** (number line: parenthesis at 3, shaded to the left; labels 0, 3)
3. $x < 3$	**C.** $(3, \infty)$
4. $x \geq 3$	**D.** $(-\infty, 3]$
5. $-3 \leq x \leq 3$	**E.** $(-3, 3)$
6. $-3 < x < 3$	**F.** $[-3, 3]$

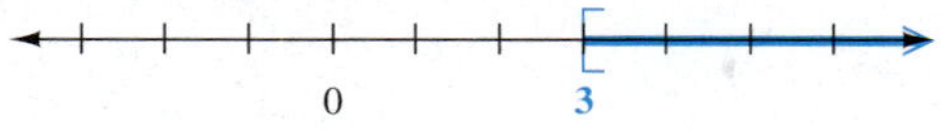

7. Explain how you will determine whether to use parentheses or brackets when graphing the solution set of an inequality.

8. Describe the steps used to solve a linear inequality. Explain when it is necessary to reverse the inequality symbol.

Solve each inequality, giving its solution set in both interval and graph forms. Check your answers. See Examples 1–6.

9. $4x + 1 \geq 21$

10. $5t + 2 \geq 52$

11. $\dfrac{3k - 1}{4} > 5$

12. $\dfrac{5z - 6}{8} < 8$

13. $-4x < 16$

14. $-2m > 10$

15. $-\frac{3}{4}r \geq 30$

16. $-\frac{2}{3}y \leq 12$

17. $-1.3m \geq -5.2$

18. $-2.5y \leq -1.25$

19. $\frac{2k - 5}{-4} > 5$

20. $\frac{3z - 2}{-5} < 6$

21. $y + 4(2y - 1) \geq y$

22. $m - 2(m - 4) \leq 3m$

23. $-(4 + r) + 2 - 3r < -14$

24. $-(9 + k) - 5 + 4k \geq 4$

25. $-3(z - 6) > 2z - 2$

26. $-2(y + 4) \leq 6y + 16$

27. $\frac{2}{3}(3k - 1) \geq \frac{3}{2}(2k - 3)$

28. $\frac{7}{5}(10m - 1) < \frac{2}{3}(6m + 5)$

29. $-\frac{1}{4}(p + 6) + \frac{3}{2}(2p - 5) < 10$

30. $\frac{3}{5}(k - 2) - \frac{1}{4}(2k - 7) \leq 3$

RELATING CONCEPTS (Exercises 31–35) FOR INDIVIDUAL OR GROUP WORK

Work Exercises 31–35 in order.

31. Solve the linear equation

$$5(x + 3) - 2(x - 4) = 2(x + 7),$$

and graph the solution set on a number line.

32. Solve the linear inequality

$$5(x + 3) - 2(x - 4) > 2(x + 7),$$

and graph the solution set on a number line.

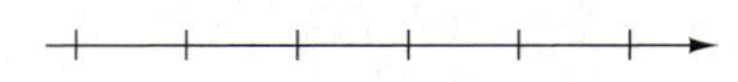

33. Solve the linear inequality

$$5(x + 3) - 2(x - 4) < 2(x + 7),$$

and graph the solution set on a number line.

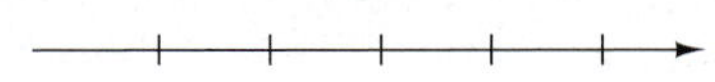

34. Graph all the solution sets of the equation and inequalities in Exercises 31–33 on the same number line. What set do you obtain?

35. Based on the results of Exercises 31–33, complete the following using a conjecture (educated guess): The solution set of

$$-3(x + 2) = 3x + 12$$

is $\{-3\}$, and the solution set of

$$-3(x + 2) < 3x + 12$$

is $(-3, \infty)$. Therefore the solution set of

$$-3(x + 2) > 3x + 12$$

is __________.

36. Which is the graph of $-2 < x$?

A. 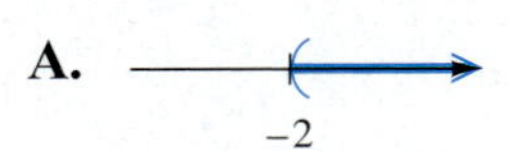 **B.** **C.** **D.**

Solve each inequality, giving its solution set in both interval and graph forms. Check your answers. See Example 7.

37. $-4 < x - 5 < 6$

38. $-1 < x + 1 < 8$

39. $-9 \le k + 5 \le 15$

40. $-4 \le m + 3 \le 10$

41. $-6 \le 2(z + 2) \le 16$

42. $-15 < 3(p + 2) < -12$

43. $-16 < 3t + 2 < -10$

44. $-1 \le \frac{2x - 5}{6} \le 5$

45. $-3 \le \frac{3m + 1}{4} \le 3$

The July 14th weather forecast by time of day for the 2000 U.S. Olympic Track and Field Trials, held July 14–23, 2000, in Sacramento, California, is shown in the figure. Use this graph to work Exercises 46–49.

TRACKING THE HEAT

The forecast for the U.S. Olympic Track and Field Trials July 14–23, by time of day. (Average temperature this time of year is a high of 93.5, low of 60.5.)

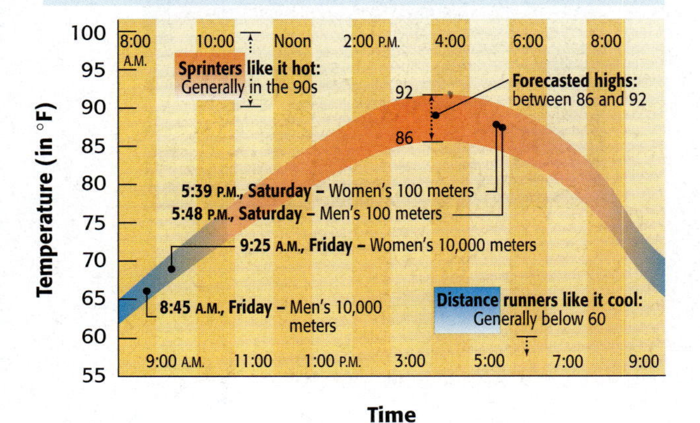

Source: Accuweather, Bee research.

46. Sprinters prefer Fahrenheit temperatures in the 90s. Using the upper boundary of the forecast, in what time period is the temperature expected to be at least 90°F?

47. Distance runners prefer cool temperatures. During what time period are temperatures predicted to be no more than 70°F? Use the lower forecast boundary.

48. What range of temperatures is predicted for the Women's 100-m event?

49. What range of temperatures is forecast for the Men's 10,000-m event?

Solve each problem. See Examples 8 and 9.

50. Margaret Westmoreland earned scores of 90 and 82 on her first two tests in English Literature. What score must she make on her third test to keep an average of 84 or greater?

51. Jacques d'Hemecourt scored 92 and 96 on his first two tests in Methods in Teaching Mathematics. What score must he make on his third test to keep an average of 90 or greater?

52. A couple wishes to rent a car for one day while on vacation. Ford Automobile Rental wants \$15.00 per day and 14¢ per mi, while Chevrolet-For-A-Day wants \$14.00 per day and 16¢ per mi. After how many miles would the price to rent the Chevrolet exceed the price to rent a Ford?

53. Jane and Terry Brandsma went to Mobile, Alabama, for a week. They needed to rent a car, so they checked out two rental firms. Avis wanted \$28 per day, with no mileage fee. Downtown Toyota wanted \$108 per week and 14¢ per mi. How many miles would they have to drive before the Avis price is less than the Toyota price?

A product will produce a profit only when the revenue (R) from selling the product exceeds the cost (C) of producing it. Find the smallest whole number of units x that must be sold for each business to show a profit for the item described.

54. Peripheral Visions, Inc. finds that the cost to produce x studio-quality videotapes is

$$C = 20x + 100,$$

while the revenue produced from them is $R = 24x$ (C and R in dollars).

55. Speedy Delivery finds that the cost to make x deliveries is

$$C = 3x + 2300,$$

while the revenue produced from them is $R = 5.50x$ (C and R in dollars).

56. A BMI (body mass index) between 19 and 25 is considered healthy. Use the formula

$$\text{BMI} = \frac{704 \times (\text{weight in pounds})}{(\text{height in inches})^2}$$

to find the weight range w, to the nearest pound, that gives a healthy BMI for each height. (*Source: Washington Post.*)

(a) 72 in. **(b)** Your height in inches

57. To achieve the maximum benefit from exercising, the heart rate in beats per minute should be in the target heart rate zone (THR). For a person aged A, the formula is

$$.7(220 - A) \leq THR \leq .85(220 - A).$$

Find the THR to the nearest whole number for each age. (*Source:* Hockey, Robert V., *Physical Fitness: The Pathway to Healthful Living,* Times Mirror/Mosby College Publishing, 1989.)

(a) 35 **(b)** Your age

3.2 Set Operations and Compound Inequalities

The table shows symptoms of an overactive thyroid and an underactive thyroid.

Underactive Thyroid	Overactive Thyroid
Sleepiness, s	Insomnia, i
Dry hands, d	Moist hands, m
Intolerance of cold, c	Intolerance of heat, h
Goiter, g	Goiter, g

Source: The Merck Manual of Diagnosis and Therapy, 16th Edition, Merck Research Laboratories, 1992.

Let N be the set of symptoms for an underactive thyroid, and let O be the set of symptoms for an overactive thyroid. Suppose we are interested in the set of symptoms that are found in *both* sets N *and* O. In this section we discuss the use of the words *and* and *or* as they relate to sets and inequalities.

Objectives

1. Find the intersection of two sets.
2. Solve compound inequalities with the word *and*.
3. Find the union of two sets.
4. Solve compound inequalities with the word *or*.

1 Find the intersection of two sets. The intersection of two sets is defined using the word *and*.

Intersection of Sets

For any two sets A and B, the **intersection** of A and B, symbolized $A \cap B$, is defined as follows:

$$A \cap B = \{x \mid x \text{ is an element of } A \textbf{ and } x \text{ is an element of } B\}.$$

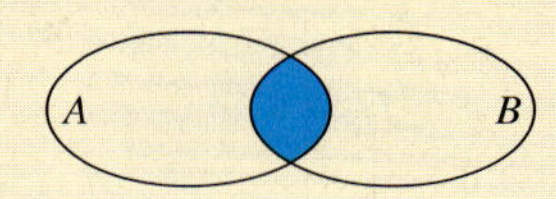

Example 1 Finding the Intersection of Two Sets

Let $A = \{1, 2, 3, 4\}$ and $B = \{2, 4, 6\}$. Find $A \cap B$.

The set $A \cap B$ contains those elements that belong to both A *and* B: the numbers 2 and 4. Therefore,

$$A \cap B = \{1, 2, 3, 4\} \cap \{2, 4, 6\}$$
$$= \{2, 4\}.$$

Work Problem 1 at the Side.

A **compound inequality** consists of two inequalities linked by a connective word such as *and* or *or*. Examples of compound inequalities are

$$x + 1 \leq 9 \quad \textbf{and} \quad x - 2 \geq 3$$

and

$$2x > 4 \quad \textbf{or} \quad 3x - 6 < 5.$$

2 Solve compound inequalities with the word *and*. Use the following steps.

Solving a Compound Inequality with *and*

Step 1 Solve each inequality in the compound inequality individually.

Step 2 Since the inequalities are joined with *and*, the solution set of the compound inequality will include all numbers that satisfy both inequalities in Step 1 (the intersection of the solution sets).

---

1 List the elements in each set.

(a) $A \cap B$, if $A = \{3, 4, 5, 6\}$ and $B = \{5, 6, 7\}$

(b) $N \cap O$ (Refer to the thyroid table.)

Answers

1. (a) $\{5, 6\}$ **(b)** $\{g\}$

2 Solve each compound inequality, and graph the solution set.

(a) $x < 10$ and $x > 2$

(b) $x + 3 \leq 1$ and $x - 4 \geq -12$

3 Solve

$2x \geq x - 1$ and $3x \geq 3 + 2x$,

and graph the solution set.

Answers

2. (a) (2, 10)

(b) [−8, −2]

3. $[3, \infty)$

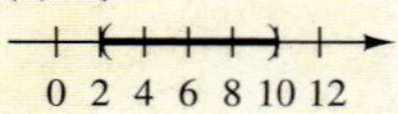

Example 2 Solving a Compound Inequality with *and*

Solve the compound inequality

$$x + 1 \leq 9 \quad \text{and} \quad x - 2 \geq 3.$$

Step 1 Solve each inequality in the compound inequality individually.

$$\begin{aligned} x + 1 &\leq 9 &\quad \text{and} \quad x - 2 &\geq 3 \\ x + 1 - 1 &\leq 9 - 1 &\quad \text{and} \quad x - 2 + 2 &\geq 3 + 2 \\ x &\leq 8 &\quad \text{and} \quad x &\geq 5 \end{aligned}$$

Step 2 Because the inequalities are joined with the word *and*, the solution set will include all numbers that satisfy both inequalities in Step 1 at the same time. Thus, the compound inequality is true whenever $x \leq 8$ and $x \geq 5$ are both true. The top graph in Figure 12 shows $x \leq 8$, and the bottom graph shows $x \geq 5$.

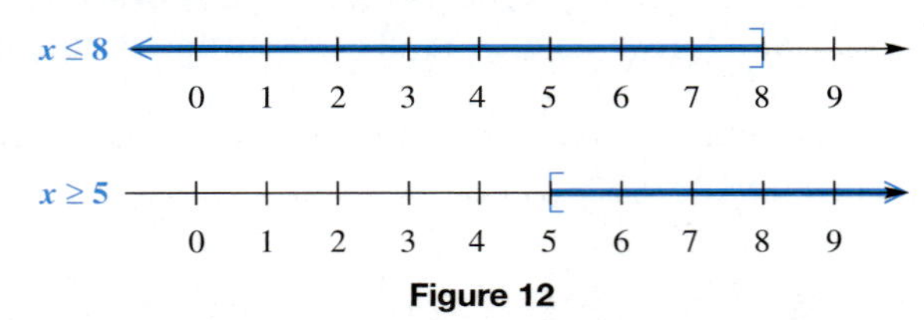

Figure 12

Find the intersection of the two graphs in Figure 12 to get the solution set of the compound inequality. The intersection of the two graphs in Figure 13 shows that the solution set in interval notation is [5, 8].

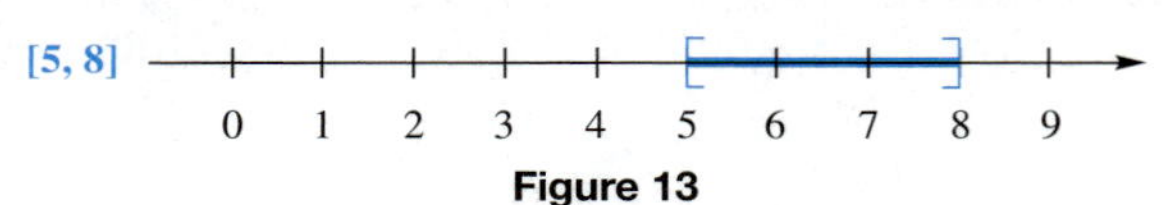

Figure 13

Work Problem 2 at the Side.

Example 3 Solving a Compound Inequality with *and*

Solve the compound inequality

$$-3x - 2 > 5 \quad \text{and} \quad 5x - 1 \leq -21.$$

Step 1 Solve each inequality separately.

$$\begin{aligned} -3x - 2 &> 5 &\quad \text{and} \quad 5x - 1 &\leq -21 \\ -3x &> 7 &\quad \text{and} \quad 5x &\leq -20 \\ x &< -\frac{7}{3} &\quad \text{and} \quad x &\leq -4 \end{aligned}$$

The graphs of $x < -\frac{7}{3}$ and $x \leq -4$ are shown in Figure 14.

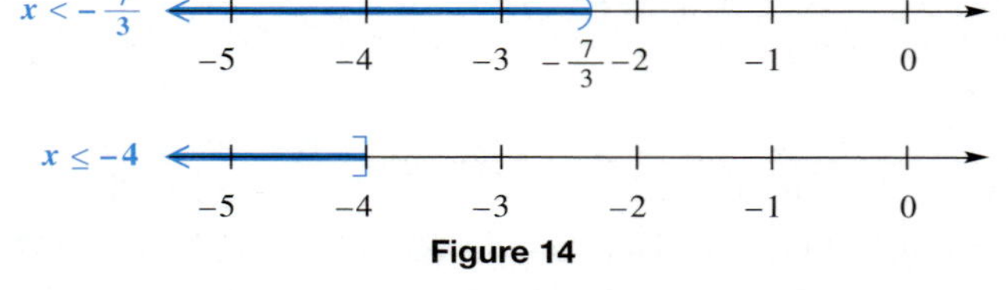

Figure 14

Step 2 Now find all values of x that satisfy both conditions; that is, the real numbers that are less than $-\frac{7}{3}$ and also less than or equal to -4. As shown by the graph in Figure 15, the solution set is $(-\infty, -4]$.

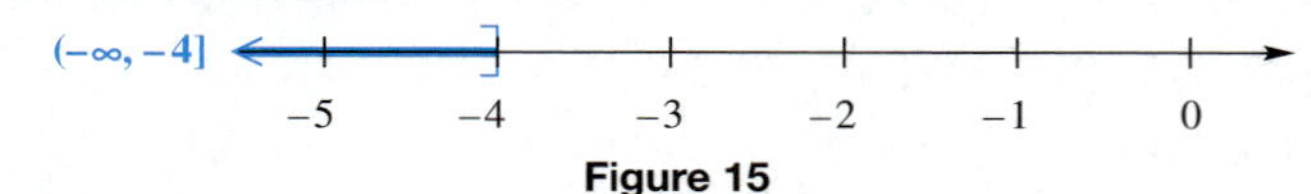

Figure 15

Work Problem 3 at the Side.

Example 4 Solving a Compound Inequality with *and*

Solve $x + 2 < 5$ and $x - 10 > 2$.

First solve each inequality separately.

$$x + 2 < 5 \quad \text{and} \quad x - 10 > 2$$
$$x < 3 \quad \text{and} \quad x > 12$$

The graphs of $x < 3$ and $x > 12$ are shown in Figure 16.

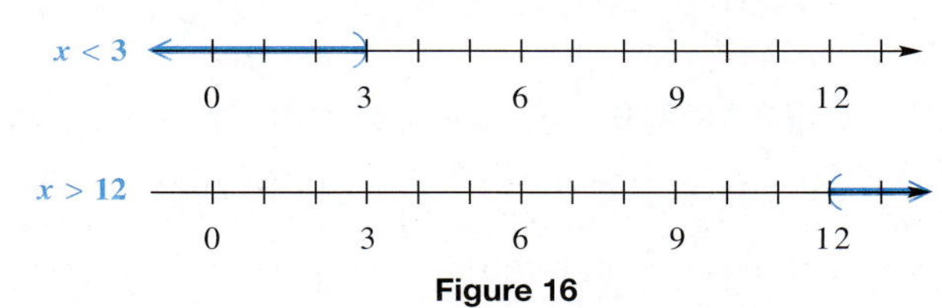

Figure 16

There is no number that is both less than 3 *and* greater than 12, so the given compound inequality has no solution. The solution set is $\emptyset$. See Figure 17.

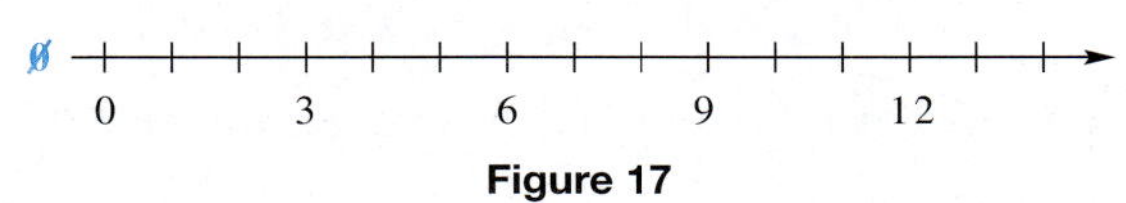

Figure 17

Work Problem 4 at the Side.

4 Solve.

(a) $x < 5$ and $x > 5$

(b) $x + 2 > 3$ and $2x + 1 < -3$

3 **Find the union of two sets.** The union of two sets is defined using the word *or*.

Union of Sets

For any two sets A and B, the **union** of A and B, symbolized $A \cup B$, is defined as follows:

$$A \cup B = \{x \mid x \text{ is an element of } A \textbf{ or } x \text{ is an element of } B\}.$$

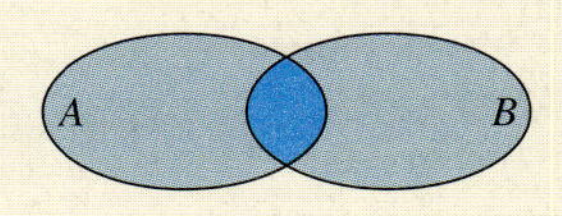

5 List the elements in each set.

(a) $A \cup B$, if $A = \{3, 4, 5, 6\}$ and $B = \{5, 6, 7\}$

(b) $N \cup O$ from the thyroid table at the beginning of this section

Example 5 Finding the Union of Two Sets

Let $A = \{1, 2, 3, 4\}$ and $B = \{2, 4, 6\}$. Find $A \cup B$.

Begin by listing all the elements of set A: 1, 2, 3, 4. Then list any additional elements from set B. In this case the elements 2 and 4 are already listed, so the only additional element is 6. Therefore,

$$A \cup B = \{1, 2, 3, 4\} \cup \{2, 4, 6\}$$
$$= \{1, 2, 3, 4, 6\}.$$

The union consists of all elements in either A *or* B (or both).

In Example 5, notice that although the elements 2 and 4 appeared in both sets A and B, they are written only once in $A \cup B$.

Work Problem 5 at the Side.

Answers

4. **(a)** $\emptyset$ **(b)** $\emptyset$

5. **(a)** $\{3, 4, 5, 6, 7\}$ **(b)** $\{s, d, c, g, i, m, h\}$

6 Give each solution set in both interval and graph forms.

(a) $x + 2 > 3$ or $2x + 1 < -3$

(b) $y - 1 > 2$ or $3y + 5 < 2y + 6$

4 **Solve compound inequalities with the word *or*.** Use the following steps.

Solving a Compound Inequality with *or*

Step 1 Solve each inequality in the compound inequality individually.

Step 2 Since the inequalities are joined with *or,* the solution set includes all numbers that satisfy either one of the two inequalities in Step 1 (the union of the solution sets).

Example 6 Solving a Compound Inequality with *or*

Solve $6x - 4 < 2x$ or $-3x \leq -9$.

Step 1 Solve each inequality separately.

$$\begin{aligned} 6x - 4 &< 2x \quad \text{or} \quad -3x \leq -9 \\ 4x &< 4 \\ x &< 1 \quad \text{or} \quad x \geq 3 \end{aligned}$$

The graphs of these two inequalities are shown in Figure 18.

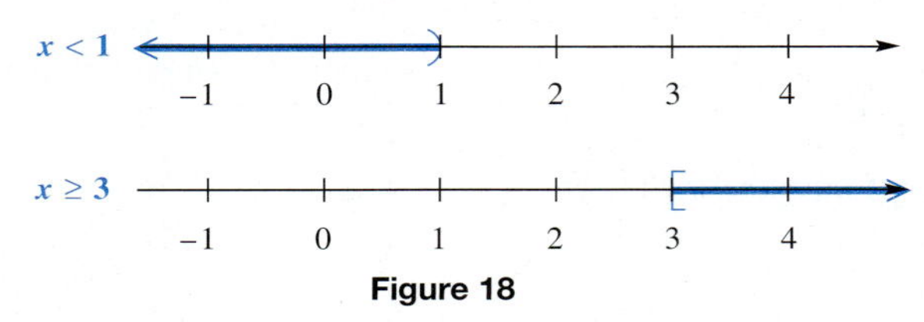

Figure 18

Step 2 Since the inequalities are joined with *or*, find the union of the two solution sets. The union is shown in Figure 19 and is written $(-\infty, 1) \cup [3, \infty)$.

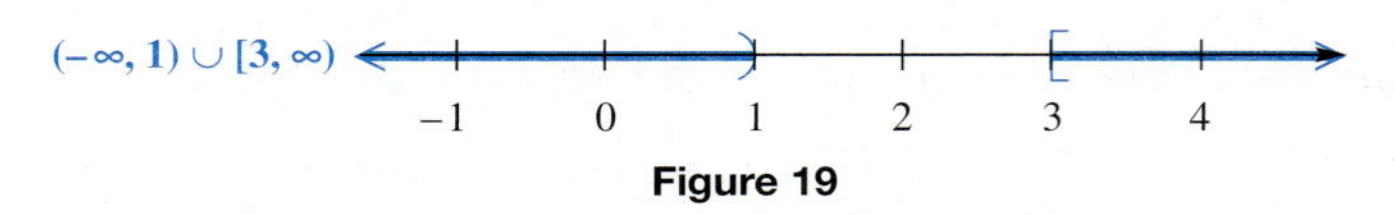

Figure 19

CAUTION

When inequalities are used to write the solution set in Example 6, it *should* be written as

$$x < 1 \quad \text{or} \quad x \geq 3,$$

which keeps the numbers 1 and 3 in their order on the number line. Writing $3 \leq x < 1$ would imply that $3 \leq 1$, which is ***FALSE.*** There is no other way to write the solution set of such a union.

Work Problem 6 at the Side.

Example 7 Solving a Compound Inequality with *or*

Solve $-4x + 1 \geq 9$ or $5x + 3 \geq -12$.

First, solve each inequality separately.

$$\begin{aligned} -4x + 1 &\geq 9 \quad \text{or} \quad 5x + 3 \geq -12 \\ -4x &\geq 8 \quad \text{or} \quad 5x \geq -15 \\ x &\leq -2 \quad \text{or} \quad x \geq -3 \end{aligned}$$

Continued on Next Page

ANSWERS

6. (a) $(-\infty, -2) \cup (1, \infty)$

–4 –3 –2 –1 0 1 2 3

(b) $(-\infty, 1) \cup (3, \infty)$

–1 0 1 2 3 4 5

The graphs of these two inequalities are shown in Figure 20.

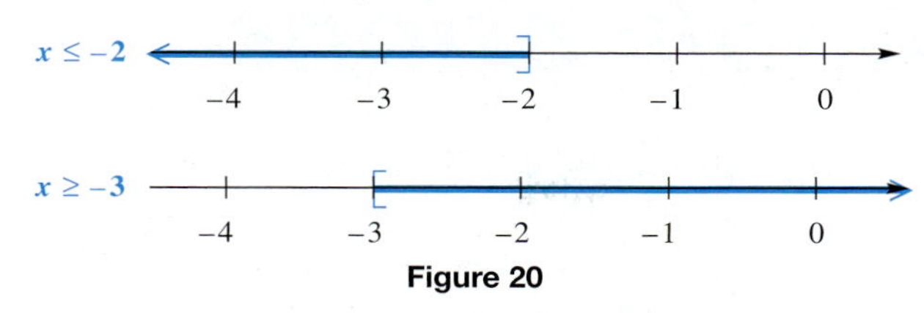

Figure 20

By taking the union, we obtain every real number as a solution, since every real number satisfies at least one of the two inequalities. The set of all real numbers is written in interval notation as $(-\infty, \infty)$ and graphed as in Figure 21.

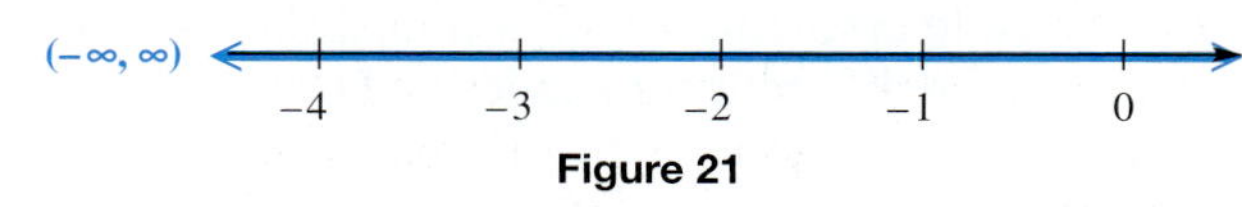

Figure 21

Work Problem 7 at the Side.

Example 8 Applying Intersection and Union

Average expenses for full-time college students during the 1997–1998 academic year are shown in the table.

COLLEGE EXPENSES IN 1997–1998 (IN DOLLARS)

Type of Expense	Public Schools	Private Schools
Tuition and fees	2365	13,013
Board rates	2180	2742
Dormitory charges	2243	2990

Source: U.S. National Center for Education Statistics, *Digest of Education Statistics,* annual.

Use the table to list the elements of each set.

(a) The set of expenses that are less than \$2500 for public schools *and* greater than \$3000 for private schools

The only expense that satisfies both conditions is for tuition and fees, so the set is

$$\{\text{Tuition and fees}\}.$$

(b) The set of expenses that are less than \$2200 for public schools *or* are greater than \$3000 for private schools

Here, an expense that satisfies at least one of the conditions is in the set. Only the public school expense for board rates is less than \$2200, and only the private school expense for tuition and fees is greater than \$3000, so the set is

$$\{\text{Tuition and fees, Board rates}\}.$$

Work Problem 8 at the Side.

7 Solve.

(a) $2x + 1 \leq 9$ or $2x + 3 \leq 5$

(b) $3x - 2 \leq 13$ or $x + 5 \geq 7$

8 Refer to the table in Example 8. List the elements in each set.

(a) The set of expenses less than \$2500 for public schools and less than \$3000 for private schools

(b) The set of expenses greater than \$10,000 or less than \$2000

Answers

7. **(a)** $(-\infty, 4]$ **(b)** $(-\infty, \infty)$

8. **(a)** {Board rates, Dormitory charges}
(b) {Tuition and fees}

Focus on

Real-Data Applications

Comparing Long-Distance Costs

Cellular phones are becoming popular tools for both local and long-distance phone calls. Frequently, rate plans include long-distance telephoning as an option if the calls are made from within the home area. The plans vary among different companies and often offer a limited number of "anytime" minutes. Information about the rate plans offered by different cellular phone companies is readily available on the Internet.

Using the Internet, you have found the following pricing schemes for both regular and the cellular phones.

- The long-distance plan for the *in-home* phone costs \$6.95 per month plus \$.05 per min for long-distance calls both within your state or between states, with no limit to the number of minutes of call time.

- One option for the *cellular* phone is a flat monthly fee of \$59.99 that includes 450 min of "anytime" local or long-distance calls.

Note: Basic phone rates are *not* included in the in-home plan, but since you intend to have an in-home phone anyway, you can disregard those costs. Also, calls in excess of the limits for the cellular plan are expensive: \$.35 per minute over the maximum. You do *not* expect to exceed the number of minutes included in the basic cellular rate plan, so do not worry about those extra charges.

For Group Discussion

The question is "Which plan is more economical?" Of course, economy is only one of the criteria that you will use when deciding whether to use your in-home phone or a cellular phone for long-distance calls, but it is one of the most important issues.

Let x represent the number of minutes of long-distance calls in a month.

1. Write an expression that represents the monthly costs for the in-home rate plan.
2. Write the expression that represents the monthly cost for the cellular rate plan.
3. The question is: "How many minutes of long-distance calls would you have to make in one month with the in-home phone to exceed the cost of the cellular phone plan?" Write a linear inequality that states that the in-home rate plan costs more than the cellular rate plan.
4. Solve the linear inequality and answer the question posed in Problem 3. What does your answer mean in terms of comparing phone costs?
5. Suppose you use the cellular phone plan for 450 min (the maximum number of minutes without incurring excess charges). How much more money would you pay compared to the in-home plan?

3.2 EXERCISES

Student's Solutions Manual | MyMathLab.com | InterAct Math Tutorial Software |

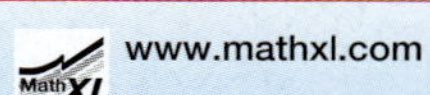

Digital Video Tutor CD 2 Videotape 3

Decide whether each statement is true *or* false. *If it is* false, *explain why.*

1. The union of the solution sets of $2x + 1 = 3$, $2x + 1 > 3$, and $2x + 1 < 3$ is $(-\infty, \infty)$.

2. The intersection of the sets $\{x \mid x \geq 5\}$ and $\{x \mid x \leq 5\}$ is $\emptyset$.

3. The union of the sets $(-\infty, 6)$ and $(6, \infty)$ is $\{6\}$.

4. The intersection of the sets $[6, \infty)$ and $(-\infty, 6]$ is $\{6\}$.

Let $A = \{1, 2, 3, 4, 5, 6\}$, $B = \{1, 3, 5\}$, $C = \{1, 6\}$, *and* $D = \{4\}$. *Specify each set. See Examples 1 and 5.*

5. $A \cap D$

6. $B \cap C$

7. $B \cap \emptyset$

8. $A \cap \emptyset$

9. $A \cup B$

10. $B \cup D$

11. $B \cup C$

12. $C \cup B$

Two sets are specified by graphs. Graph the intersection of the two sets.

13.

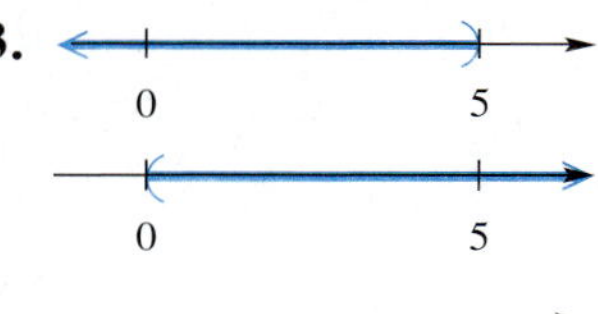

14.

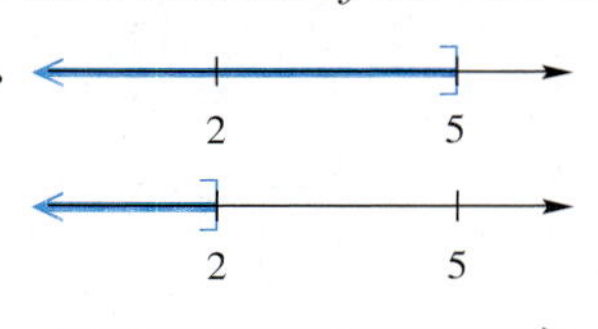

15.

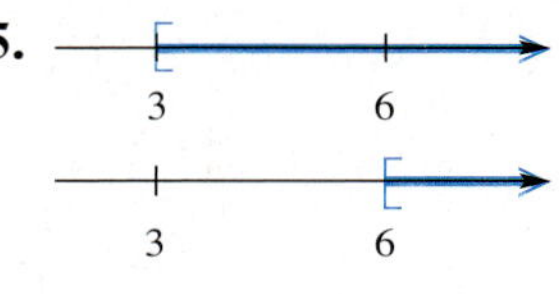

Two sets are specified by graphs. Graph the union of the two sets.

16.

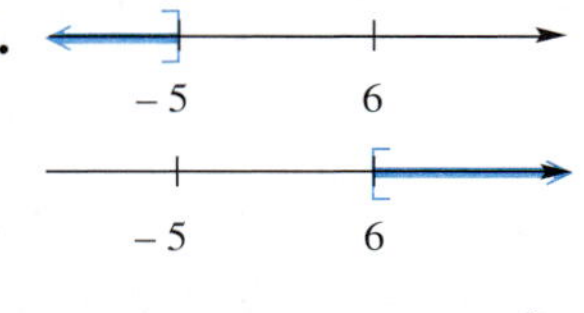

17.

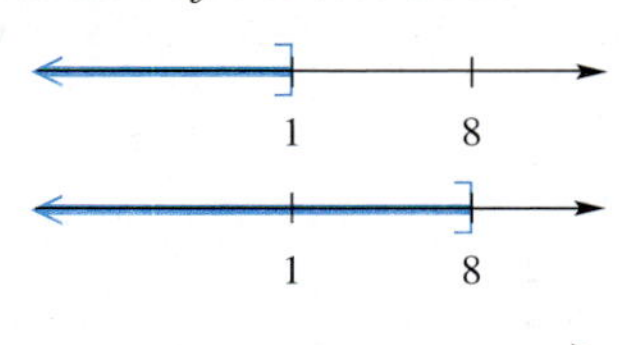

18.

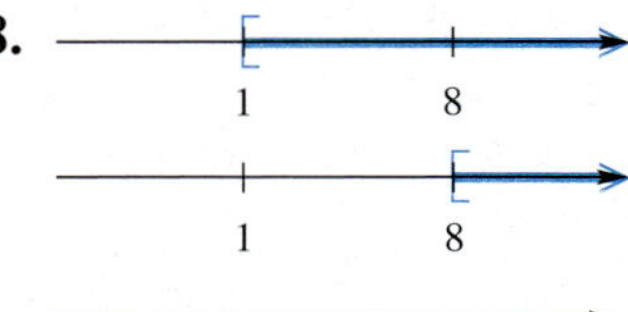

19. Give an example of intersection applied to a real-life situation.

20. A compound inequality uses one of the words *and* or *or*. Explain how you will determine whether to use *intersection* or *union* when graphing the solution set.

For each compound inequality, give the solution set in both interval and graph forms. See Examples 2–4.

21. $x < 2$ and $x > -3$

22. $x < 5$ and $x > 0$

23. $x \leq 2$ and $x \leq 5$

24. $x \geq 3$ and $x \geq 6$

25. $x \leq 3$ and $x \geq 6$

26. $x \leq -1$ and $x \geq 3$

27. $x - 3 \leq 6$ and $x + 2 \geq 7$

28. $x + 5 \leq 11$ and $x - 3 \geq -1$

29. $3x - 4 \leq 8$ and $4x - 1 \leq 15$

30. $7x + 6 \leq 48$ and $-4x \geq -24$

For each compound inequality, give the solution set in both interval and graph forms. See Examples 6 and 7.

31. $x \leq 1$ or $x \leq 8$

32. $x \geq 1$ or $x \geq 8$

33. $x \geq -2$ or $x \geq 5$

34. $x \leq -2$ or $x \leq 6$

35. $x \geq -2$ or $x \leq 4$

36. $x \geq 5$ or $x \leq 7$

37. $x + 2 > 7$ or $x - 1 < -6$

38. $x + 1 > 3$ or $x + 4 < 2$

39. $4x - 8 > 0$ or $4x - 1 < 7$

40. $3x < x + 12$ or $3x - 8 > 10$

Express each set in the simplest interval form.

41. $(-\infty, -1] \cap [-4, \infty)$

42. $[-1, \infty) \cap (-\infty, 9]$

43. $(-\infty, -6] \cap [-9, \infty)$

44. $(5, 11] \cap [6, \infty)$

45. $(-\infty, 3) \cup (-\infty, -2)$

46. $[-9, 1] \cup (-\infty, -3)$

47. $[3, 6] \cup (4, 9)$

48. $[-1, 2] \cup (0, 5)$

For each compound inequality, state whether intersection or union should be used. Then give the solution set in both interval and graph forms. See Examples 2, 3, 4, 6, and 7.

49. $x < -1$ and $x > -5$

50. $x > -1$ and $x < 7$

51. $x < 4$ or $x < -2$

52. $x < 5$ or $x < -3$

53. $x + 1 \geq 5$ and $x - 2 \leq 10$

54. $2x - 6 \leq -18$ and $2x \geq -18$

55. $-3x \leq -6$ or $-3x \geq 0$

56. $-8x \leq -24$ or $-5x \geq 15$

RELATING CONCEPTS (Exercises 57–60) FOR INDIVIDUAL OR GROUP WORK

The figures represent the backyards of neighbors Luigi, Mario, Than, and Joe. Find the area and the perimeter of each yard. Suppose that each resident has 150 *ft of fencing and enough sod to cover* 1400 ft^2 *of lawn.*

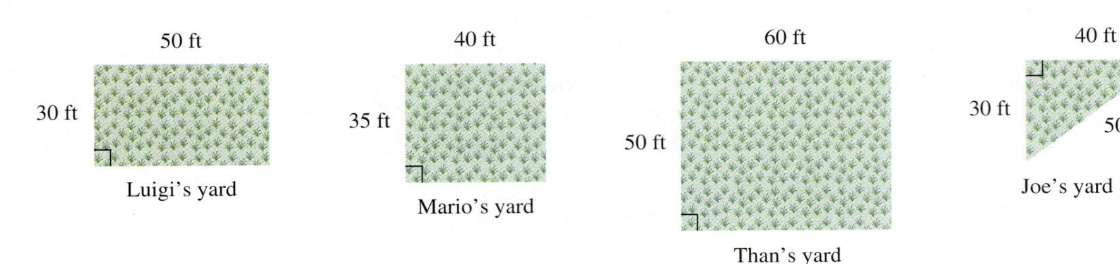

Give the name or names of the residents whose yards satisfy each description.

57. The yard can be fenced *and* the yard can be sodded.

58. The yard can be fenced *and* the yard cannot be sodded.

59. The yard cannot be fenced *and* the yard can be sodded.

60. The yard cannot be fenced *and* the yard cannot be sodded.

Use the graphs to answer Exercises 61 and 62. See Example 8.

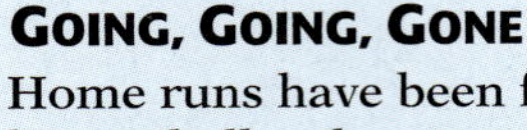

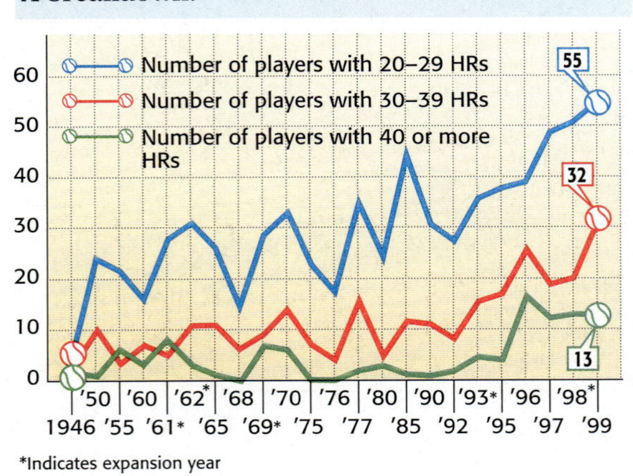

Source: Bee research.

61. In which years did the number of players with 30–39 home runs exceed 20 *and* the number of players with 40 or more home runs exceed 45?

62. In which years was the number of players with 20–29 home runs less than 20 *or* the number of players with 30–39 home runs at least 20?

3.3 Absolute Value Equations and Inequalities

In a production line, quality is controlled by randomly choosing items from the line and checking to see how selected measurements vary from the optimum measure. These differences are sometimes positive and sometimes negative, so they are expressed with absolute value. For example, a machine that fills quart milk cartons might be set to release 1 qt plus or minus 2 oz per carton. Then the number of ounces in each carton should satisfy the *absolute value inequality* $|x - 32| \leq 2$, where x is the number of ounces.

Objectives

1. Use the distance definition of absolute value.
2. Solve equations of the form $|ax + b| = k$, for $k > 0$.
3. Solve inequalities of the form $|ax + b| < k$ and of the form $|ax + b| > k$, for $k > 0$.
4. Solve absolute value equations that involve rewriting.
5. Solve equations of the form $|ax + b| = |cx + d|$.
6. Solve special cases of absolute value equations and inequalities.

1 Use the distance definition of absolute value. In Chapter 1 we saw that the absolute value of a number x, written $|x|$, represents the distance from x to 0 on the number line. For example, the solutions of $|x| = 4$ are 4 and -4, as shown in Figure 22.

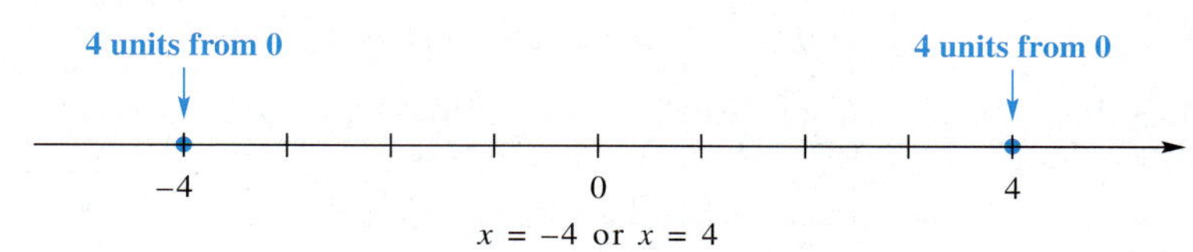

Figure 22

Because absolute value represents distance from 0, it is reasonable to interpret the solutions of $|x| > 4$ to be all numbers that are *more* than 4 units from 0. The set $(-\infty, -4) \cup (4, \infty)$ fits this description. Figure 23 shows the graph of the solution set of $|x| > 4$. Because the graph consists of two separate intervals, the solution set is described using *or* as

$$x < -4 \quad \text{or} \quad x > 4.$$

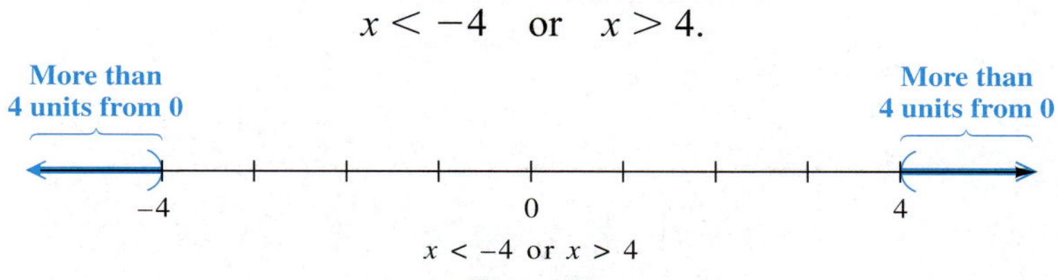

Figure 23

The solution set of $|x| < 4$ consists of all numbers that are *less* than 4 units from 0 on the number line. Another way of thinking of this is to think of all numbers *between* -4 and 4. This set of numbers is given by $(-4, 4)$, as shown in Figure 24. Here, the graph shows that $-4 < x < 4$, which means $x > -4$ *and* $x < 4$.

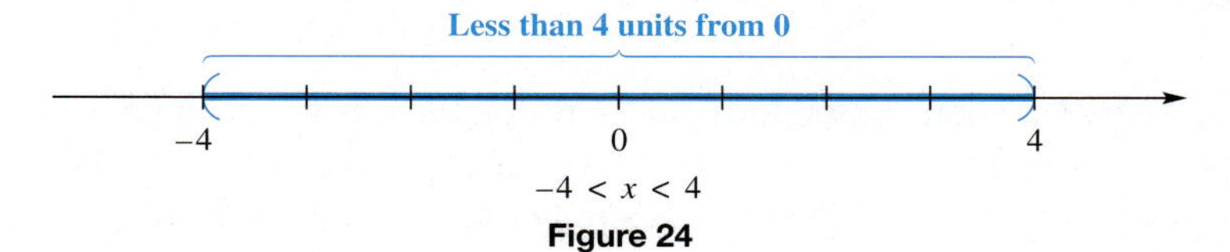

Figure 24

Work Problem 1 at the Side.

1 Graph the solution set of each equation or inequality.

(a) $|x| = 3$

(b) $|x| > 3$

(c) $|x| < 3$

The equation and inequalities just described are examples of **absolute value equations and inequalities.** They involve the absolute value of a variable expression and generally take the form

$$|ax + b| = k, \qquad |ax + b| > k, \qquad \text{or} \qquad |ax + b| < k,$$

where k is a positive number. From Figures 22–24, we see that

$|x| = 4$ has the same solution set as $x = -4$ **or** $x = 4$,

$|x| > 4$ has the same solution set as $x < -4$ **or** $x > 4$,

$|x| < 4$ has the same solution set as $x > -4$ **and** $x < 4$.

ANSWERS

1. (a), (b)

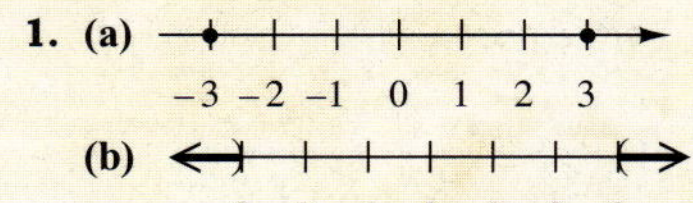

(c)

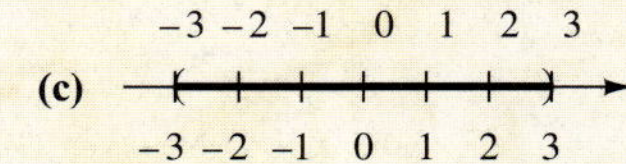

Thus, we can solve an absolute value equation or inequality by solving the appropriate compound equation or inequality.

Solving Absolute Value Equations and Inequalities

Let k be a positive real number, and p and q be real numbers.

1. To solve $|ax + b| = k$, solve the compound equation

$$ax + b = k \quad \text{or} \quad ax + b = -k.$$

The solution set is usually of the form $\{p, q\}$, which includes two numbers.

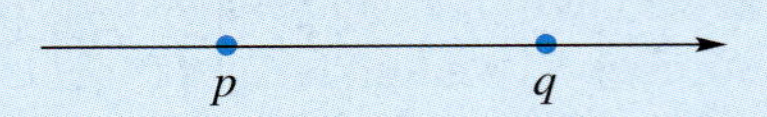

2. To solve $|ax + b| > k$, solve the compound inequality

$$ax + b > k \quad \text{or} \quad ax + b < -k.$$

The solution set is of the form $(-\infty, p) \cup (q, \infty)$, which consists of two separate intervals.

3. To solve $|ax + b| < k$, solve the compound inequality

$$-k < ax + b < k.$$

The solution set is of the form (p, q), a single interval.

NOTE

Some people prefer to write the compound statements in parts 1 and 2 of the summary as

$$ax + b = k \quad \text{or} \quad -(ax + b) = k$$

and

$$ax + b > k \quad \text{or} \quad -(ax + b) > k.$$

These forms are equivalent to those we give in the summary and produce the same results.

2 **Solve equations of the form $|ax + b| = k$, for $k > 0$.** The next example shows how we use a compound equation to solve a typical absolute value equation. Remember that because absolute value refers to distance from the origin, each absolute value equation will have two parts.

Example 1 Solving an Absolute Value Equation

Solve $|2x + 1| = 7$.

For $|2x + 1|$ to equal 7, $2x + 1$ must be 7 units from 0 on the number line. This can happen only when $2x + 1 = 7$ or $2x + 1 = -7$. This is the first case in the preceding summary. Solve this compound equation as follows.

$$\begin{aligned} 2x + 1 &= 7 \quad \text{or} \quad & 2x + 1 &= -7 \\ 2x &= 6 \quad \text{or} \quad & 2x &= -8 \\ x &= 3 \quad \text{or} \quad & x &= -4 \end{aligned}$$

Continued on Next Page

Check by substitution in the original absolute value equation to verify that the solution set is $\{-4, 3\}$. The graph is shown in Figure 25.

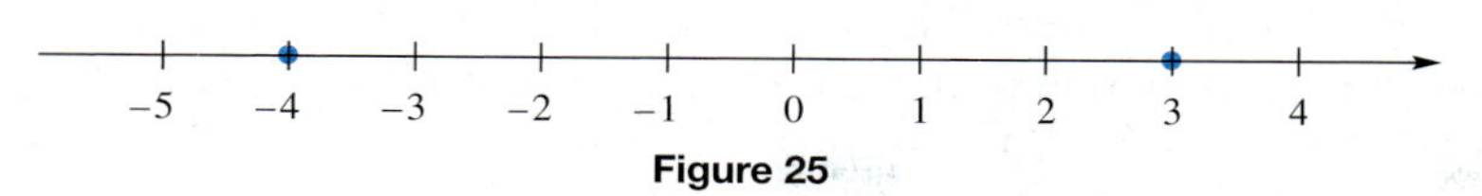

Figure 25

Work Problem 2 at the Side.

3 Solve inequalities of the form $|ax + b| < k$ and of the form $|ax + b| > k$, for $k > 0$.

Example 2 Solving an Absolute Value Inequality with >

Solve $|2x + 1| > 7$.

By part 2 of the summary, this absolute value inequality is rewritten as

$$2x + 1 > 7 \quad \text{or} \quad 2x + 1 < -7,$$

because $2x + 1$ must represent a number that is *more* than 7 units from 0 on either side of the number line. Now, solve the compound inequality.

$$\begin{aligned} 2x + 1 > 7 \quad &\text{or} \quad 2x + 1 < -7 \\ 2x > 6 \quad &\text{or} \quad 2x < -8 \\ x > 3 \quad &\text{or} \quad x < -4 \end{aligned}$$

Check these solutions. The solution set is $(-\infty, -4) \cup (3, \infty)$. See Figure 26. Notice that the graph consists of two intervals.

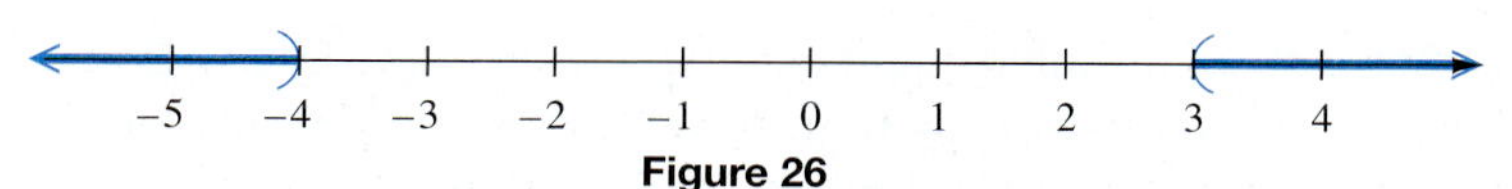

Figure 26

Work Problem 3 at the Side.

Example 3 Solving an Absolute Value Inequality with <

Solve $|2x + 1| < 7$.

The expression $2x + 1$ must represent a number that is less than 7 units from 0 on either side of the number line. Another way of thinking of this is to realize that $2x + 1$ must be between -7 and 7. As part 3 of the summary shows, this is written as the three-part inequality

$$-7 < 2x + 1 < 7.$$

We solved such inequalities in Section 3.1 by working with all three parts at the same time.

$$\begin{aligned} -7 < 2x + 1 < 7& \\ -8 < 2x < 6& \qquad \text{Subtract 1 from each part.} \\ -4 < x < 3& \qquad \text{Divide each part by 2.} \end{aligned}$$

Check that the solution set is $(-4, 3)$, so the graph consists of the single interval shown in Figure 27.

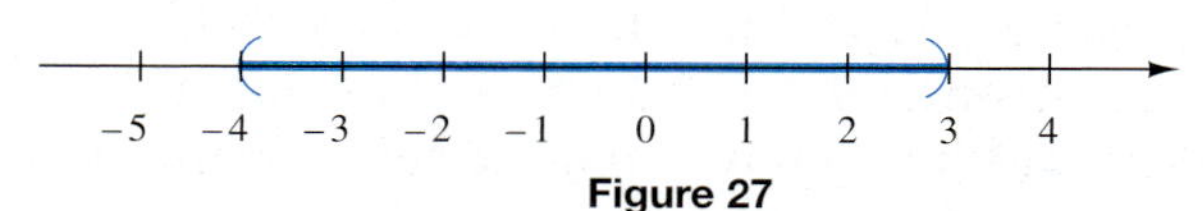

Figure 27

Work Problem 4 at the Side.

2 Solve each equation, check, and graph the solution set.

(a) $|x + 2| = 3$

(b) $|3x - 4| = 11$

3 Solve each inequality, check, and graph the solution set.

(a) $|x + 2| > 3$

(b) $|3x - 4| \geq 11$

4 Solve each inequality, check, and graph the solution set.

(a) $|x + 2| < 3$

(b) $|3x - 4| \leq 11$

Answers

2. (a) $\{-5, 1\}$

−5 −4 −3 −2 −1 0 1

(b) $\left\{-\frac{7}{3}, 5\right\}$

$-\frac{7}{3}$ 0 2 4 5

3. (a) $(-\infty, -5) \cup (1, \infty)$

−5 −4 −3 −2 −1 0 1

(b) $\left(-\infty, -\frac{7}{3}\right] \cup [5, \infty)$

$-\frac{7}{3}$

−2 0 2 4 5

4. (a) $(-5, 1)$

−5 −4 −3 −2 −1 0 1

(b) $\left[-\frac{7}{3}, 5\right]$

$-\frac{7}{3}$ 0 2 4 5

Look back at Figures 25, 26, and 27, with the graphs of $|2x + 1| = 7$, $|2x + 1| > 7$, and $|2x + 1| < 7$. If we find the union of the three sets, we get the set of all real numbers. This is because for any value of x, $|2x + 1|$ will satisfy one and only one of the following: it is equal to 7, greater than 7, or less than 7.

CAUTION

When solving absolute value equations and inequalities of the types in Examples 1, 2, and 3, remember the following.

1. The methods described apply when the constant is alone on one side of the equation or inequality and is *positive.*
2. Absolute value equations and absolute value inequalities in the form $|ax + b| > k$ translate into "or" compound statements.
3. Absolute value inequalities in the form $|ax + b| < k$ translate into "and" compound statements, which may be written as three-part inequalities.
4. An "or" statement *cannot* be written in three parts. It would be incorrect to use

$$-7 > 2x + 1 > 7$$

in Example 2, because this would imply that $-7 > 7$, which is *false.*

4 Solve absolute value equations that involve rewriting. Sometimes an absolute value equation or inequality requires some rewriting before it can be set up as a compound statement, as shown in the next example.

Example 4 Solving an Absolute Value Equation That Requires Rewriting

Solve the equation $|x + 3| + 5 = 12$.

First get the absolute value alone on one side of the equals sign by subtracting 5 from each side.

$$|x + 3| + 5 - 5 = 12 - 5 \quad \text{Subtract 5.}$$
$$|x + 3| = 7$$

Now use the method shown in Example 1.

$$x + 3 = 7 \quad \text{or} \quad x + 3 = -7$$
$$x = 4 \quad \text{or} \quad x = -10$$

Check that the solution set is $\{4, -10\}$ by substituting into the original equation.

We use a similar method to solve an absolute value *inequality* that requires rewriting.

Work Problem 5 at the Side.

5 Solve equations of the form $|ax + b| = |cx + d|$. By definition, for two expressions to have the same absolute value, they must either be equal or be negatives of each other.

5 **(a)** Solve $|5a + 2| - 9 = -7$.

(b) Solve $|m + 2| - 3 > 2$, and graph the solution set.

(c) Solve, and graph the solution set.

$|3a + 2| + 4 \leq 15$

Answers

5. (a) $\left\{-\frac{4}{5}, 0\right\}$

(b) $(-\infty, -7) \cup (3, \infty)$

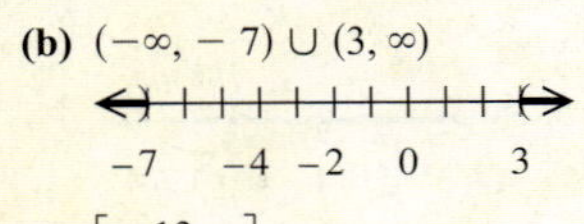

(c) $\left[-\frac{13}{3}, 3\right]$

$-\frac{13}{3}$ -2 0 2 3

Solving $|ax + b| = |cx + d|$

To solve an absolute value equation of the form

$$|ax + b| = |cx + d|,$$

solve the compound equation

$$ax + b = cx + d \quad \textbf{or} \quad ax + b = -(cx + d).$$

Example 5 Solving an Equation with Two Absolute Values

Solve the equation $|z + 6| = |2z - 3|$.

This equation is satisfied either if $z + 6$ and $2z - 3$ are equal to each other, or if $z + 6$ and $2z - 3$ are negatives of each other. Thus,

$$z + 6 = 2z - 3 \quad \text{or} \quad z + 6 = -(2z - 3).$$

Solve each equation.

$$\begin{aligned} z + 6 &= 2z - 3 \quad &\text{or} \quad z + 6 &= -2z + 3 \\ 9 &= z & 3z &= -3 \\ & & z &= -1 \end{aligned}$$

Check that the solution set is $\{9, -1\}$.

Work Problem 6 at the Side.

6 Solve special cases of absolute value equations and inequalities. When a typical absolute value equation or inequality involves a *negative* constant or *0* alone on one side, use the properties of absolute value to solve. Keep the following in mind.

1. The absolute value of an expression can never be negative: $|a| \geq 0$ for all real numbers a.
2. The absolute value of an expression equals 0 only when the expression is equal to 0.

The next two examples illustrate these special cases.

Example 6 Solving Special Cases of Absolute Value Equations

Solve each equation.

(a) $|5r - 3| = -4$

Since the absolute value of an expression can never be negative, there are no solutions for this equation. The solution set is $\emptyset$.

(b) $|7x - 3| = 0$

The expression $7x - 3$ will equal 0 *only* if

$$7x - 3 = 0$$

$$x = \frac{3}{7}.$$

Checking shows that the solution set is $\{\frac{3}{7}\}$, with just one element.

Work Problem 7 at the Side.

6 Solve each equation.

(a) $|k - 1| = |5k + 7|$

(b) $|4r - 1| = |3r + 5|$

7 Solve each equation.

(a) $|6x + 7| = -5$

(b) $\left|\frac{1}{4}x - 3\right| = 0$

ANSWERS

6. (a) $\{-1, -2\}$ **(b)** $\left\{-\frac{4}{7}, 6\right\}$

7. (a) $\emptyset$ **(b)** $\{12\}$

8 Solve.

(a) $|x| > -1$

(b) $|y| < -5$

(c) $|k + 2| \leq 0$

Answers

8. (a) $(-\infty, \infty)$ **(b)** $\emptyset$ **(c)** $\{-2\}$

Example 7 Solving Special Cases of Absolute Value Inequalities

Solve each inequality.

(a) $|x| \geq -4$

The absolute value of a number is always greater than or equal to 0. For this reason, $|x| \geq -4$ is true for *all* real numbers. The solution set is $(-\infty, \infty)$.

(b) $|k + 6| - 3 < -5$

Add 3 to each side to get the absolute value expression alone on one side.

$$|k + 6| < -2$$

There is no number whose absolute value is less than -2, so this inequality has no solution. The solution set is $\emptyset$.

(c) $|m - 7| + 4 \leq 4$

Subtracting 4 from each side gives

$$|m - 7| \leq 0.$$

The value of $|m - 7|$ will never be less than 0. However, $|m - 7|$ will equal 0 when $m = 7$. Therefore, the solution set is $\{7\}$.

Work Problem 8 at the Side.

3.3 EXERCISES

FOR EXTRA HELP: Student's Solutions Manual · MyMathLab.com · InterAct Math Tutorial Software · AW Math Tutor Center · www.mathxl.com · Digital Video Tutor CD 2 Videotape 3

Match each absolute value equation or inequality in Column I with the graph of its solution set in Column II.

1.

I

$|x| = 5$

$|x| < 5$

$|x| > 5$

$|x| \leq 5$

$|x| \geq 5$

II

A. 

B. 

C. 

D. 

E. 

2.

I

$|x| = 9$

$|x| > 9$

$|x| \geq 9$

$|x| < 9$

$|x| \leq 9$

II

A. 

B. 

C. 

D. 

E. 

3. Explain when to use *and* and when to use *or* if you are solving an absolute value equation or inequality of the form $|ax + b| = k$, $|ax + b| < k$, or $|ax + b| > k$, where k is a positive number.

4. How many solutions will $|ax + b| = k$ have if **(a)** $k = 0$; **(b)** $k > 0$; **(c)** $k < 0$?

Solve each equation. See Example 1.

5. $|x| = 12$

6. $|k| = 14$

7. $|4x| = 20$

8. $|5x| = 30$

9. $|y - 3| = 9$

10. $|p - 5| = 13$

11. $|2x + 1| = 7$

12. $|2y + 3| = 19$

13. $|4r - 5| = 17$

14. $|5t - 1| = 21$

15. $|2y + 5| = 14$

16. $|2x - 9| = 18$

17. $\left|\frac{1}{2}x + 3\right| = 2$

18. $\left|\frac{2}{3}q - 1\right| = 5$

19. $\left|1 - \frac{3}{4}k\right| = 7$

20. $\left|2 - \frac{5}{2}m\right| = 14$

Solve each inequality, and graph the solution set. See Example 2.

21. $|x| > 3$

22. $|y| > 5$

23. $|k| \geq 4$

24. $|r| \geq 6$

25. $|t + 2| > 10$

26. $|r + 5| > 20$

27. $|3x - 1| \geq 8$

28. $|4x + 1| \geq 21$

29. $|3 - x| > 5$

30. $|5 - x| > 3$

31. The graph of the solution set of $|2x + 1| = 9$ is given here.

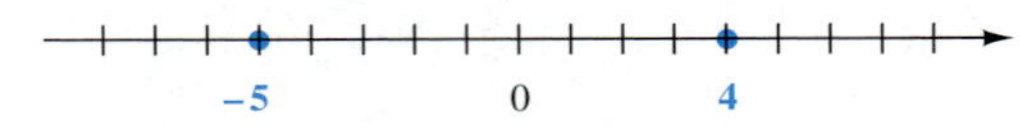

Without actually doing the algebraic work, graph the solution set of each inequality, referring to the graph above.

(a) $|2x + 1| < 9$

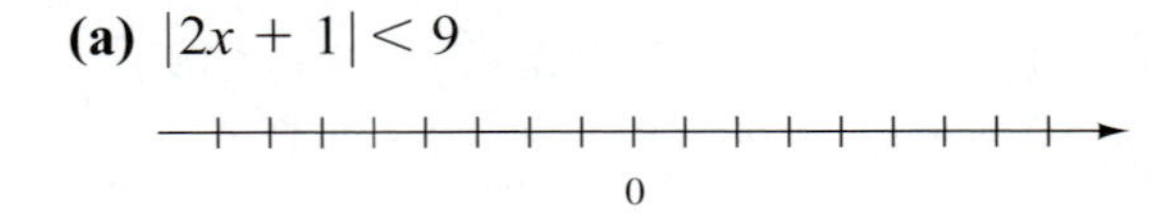

(b) $|2x + 1| > 9$

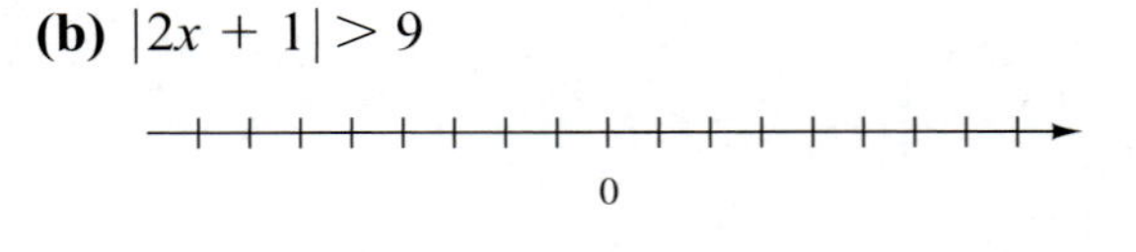

32. The graph of the solution set of $|3y - 4| < 5$ is given here.

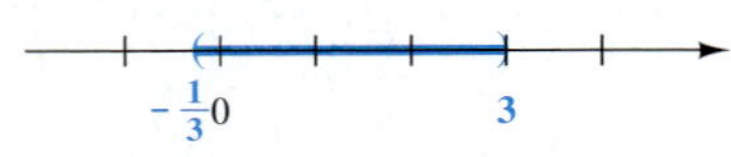

Without actually doing the algebraic work, graph the solution set of the equation and the inequality, referring to the graph above.

(a) $|3y - 4| = 5$

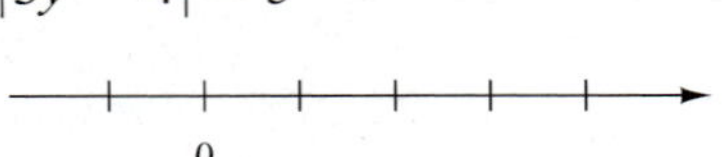

(b) $|3y - 4| > 5$

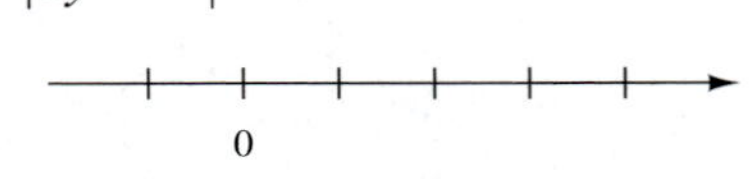

Solve each inequality, and graph the solution set. See Example 3. (Hint: Compare your answers to those in Exercises 21–30.)

33. $|x| \leq 3$

34. $|y| \leq 5$

35. $|k| < 4$

36. $|r| < 6$

37. $|t + 2| \leq 10$

38. $|r + 5| \leq 20$

39. $|3x - 1| < 8$

40. $|4x + 1| < 21$

41. $|3 - x| \leq 5$

42. $|5 - x| \leq 3$

Exercises 43–50 represent a sampling of the various types of absolute value equations and inequalities covered in Exercises 1–42. Decide which method of solution applies, find the solution set, and graph. See Examples 1–3.

43. $|-4 + k| > 9$

44. $|-3 + t| > 8$

45. $|7 + 2z| = 5$

46. $|9 - 3p| = 3$

47. $|3r - 1| \leq 11$

48. $|2s - 6| \leq 6$

49. $|-6x - 6| \leq 1$

50. $|-2x - 6| \leq 5$

Solve each equation or inequality. Give the solution set in set notation for equations and in interval notation for inequalities. See Example 4.

51. $|x| - 1 = 4$

52. $|y| + 3 = 10$

53. $|x + 4| + 1 = 2$

54. $|y + 5| - 2 = 12$

55. $|2x + 1| + 3 > 8$

56. $|6x - 1| - 2 > 6$

57. $|x + 5| - 6 \leq -1$

58. $|r - 2| - 3 \leq 4$

Solve each equation. See Example 5.

59. $|3x + 1| = |2x + 4|$

60. $|7x + 12| = |x - 8|$

61. $\left|m - \frac{1}{2}\right| = \left|\frac{1}{2}m - 2\right|$

62. $\left|\frac{2}{3}r - 2\right| = \left|\frac{1}{3}r + 3\right|$

63. $|6x| = |9x + 1|$

64. $|13y| = |2y + 1|$

65. $|2p - 6| = |2p + 11|$

66. $|3x - 1| = |3x + 9|$

Solve each equation or inequality. See Examples 6 and 7.

67. $|12t - 3| = -8$

68. $|13w + 1| = -3$

69. $|4x + 1| = 0$

70. $|6r - 2| = 0$

71. $|2q - 1| < -6$

72. $|8n + 4| < -4$

73. $|x + 5| > -9$

74. $|x + 9| > -3$

75. $|7x + 3| \leq 0$

76. $|4x - 1| \leq 0$

77. $|5x - 2| \geq 0$

78. $|4 + 7x| \geq 0$

79. $|10z + 7| > 0$

80. $|4x + 1| > 0$

81. The 1998 recommended daily intake (RDI) of calcium for females aged 19–50 is 1000 mg/day. (*Source: World Almanac and Book of Facts,* 2000.) Actual vitamin needs vary from person to person. Write an absolute value inequality to express the RDI plus or minus 100 mg and solve it.

82. The average clotting time of blood is 7.45 sec with a variation of plus or minus 3.6 sec. Write this statement as an absolute value inequality and solve it.

RELATING CONCEPTS (Exercises 83–86) FOR INDIVIDUAL OR GROUP WORK

The ten tallest buildings in Kansas City, Missouri, are listed along with their heights.

Building	Height (in feet)
One Kansas City Place	632
AT&T Town Pavilion	590
Hyatt Regency	504
Kansas City Power and Light	476
City Hall	443
Fidelity Bank and Trust Building	433
1201 Walnut	427
Federal Office Building	413
Commerce Tower	407
City Center Square	404

Source: World Almanac and Book of Facts, 2001.

Use this information to ***work Exercises 83–86 in order.***

83. To find the average of a group of numbers, we add the numbers and then divide by the number of items added. Use a calculator to find the average of the heights.

84. Let k represent the average height of these buildings. If a height x satisfies the inequality

$$|x - k| < t,$$

then the height is said to be within t ft of the average. Using your result from Exercise 83, list the buildings that are within 50 ft of the average.

85. Repeat Exercise 84, but find the buildings that are within 75 ft of the average.

86. (a) Write an absolute value inequality that describes the height of a building that is *not* within 75 ft of the average.

(b) Solve the inequality you wrote in part (a).

(c) Use the result of part (b) to find the buildings that are not within 75 ft of the average.

(d) Confirm that your answer to part (c) makes sense by comparing it with your answer to Exercise 85.

Summary Exercises on Solving Linear and Absolute Value Equations and Inequalities

Students often have difficulty distinguishing between the various types of equations and inequalities introduced in Chapters 2 and 3. This section of miscellaneous equations and inequalities provides practice in solving all such types. You might wish to refer to the boxes in these chapters that summarize the various methods of solution.

Solve each equation or inequality.

1. $4z + 1 = 49$

2. $|m - 1| = 6$

3. $6q - 9 = 12 + 3q$

4. $3p + 7 = 9 + 8p$

5. $|a + 3| = -4$

6. $2m + 1 \leq m$

7. $8r + 2 \geq 5r$

8. $4(a - 11) + 3a = 20a - 31$

9. $2q - 1 = -7$

10. $|3q - 7| - 4 = 0$

11. $6z - 5 \leq 3z + 10$

12. $|5z - 8| + 9 \geq 7$

13. $9y - 3(y + 1) = 8y - 7$

14. $|y| \geq 8$

15. $9y - 5 \geq 9y + 3$

16. $13p - 5 > 13p - 8$

17. $|q| < 5.5$

18. $4z - 1 = 12 + z$

19. $\frac{2}{3}y + 8 = \frac{1}{4}y$

20. $-\frac{5}{8}y \geq -20$

21. $\frac{1}{4}p < -6$

22. $7z - 3 + 2z = 9z - 8z$

23. $\frac{3}{5}q - \frac{1}{10} = 2$

24. $|r - 1| < 7$

25. $r + 9 + 7r = 4(3 + 2r) - 3$

26. $6 - 3(2 - p) < 2(1 + p) + 3$

27. $|2p - 3| > 11$

28. $\frac{x}{4} - \frac{2x}{3} = -10$

29. $|5a + 1| \leq 0$

30. $5z - (3 + z) \geq 2(3z + 1)$

31. $-2 \leq 3x - 1 \leq 8$

32. $-1 \leq 6 - x \leq 5$

33. $|7z - 1| = |5z + 3|$

34. $|p + 2| = |p + 4|$

35. $|1 - 3x| \geq 4$

36. $\frac{1}{2} \leq \frac{2}{3}r \leq \frac{5}{4}$

37. $-(m + 4) + 2 = 3m + 8$

38. $\frac{p}{6} - \frac{3p}{5} = p - 86$

39. $-6 \leq \frac{3}{2} - x \leq 6$

40. $|5 - y| < 4$

41. $|y - 1| \geq -6$

42. $|2r - 5| = |r + 4|$

43. $8q - (1 - q) = 3(1 + 3q) - 4$

44. $8y - (y + 3) = -(2y + 1) - 12$

45. $|r - 5| = |r + 9|$

46. $|r + 2| < -3$

47. $2x + 1 > 5$ or $3x + 4 < 1$

48. $1 - 2x \geq 5$ and $7 + 3x \geq -2$

Chapter 3

Summary

Key Terms

3.1

interval An interval is a portion of a number line.

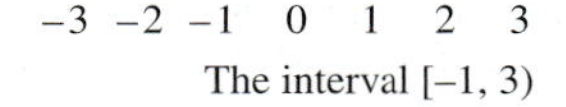

The interval [−1, 3)

interval notation The notation used to indicate an interval on the number line is called interval notation.

inequality An inequality is a mathematical statement that two expressions are not equal.

linear inequality in one variable A linear inequality in the variable x can be written in the form $ax < b$, $ax \leq b$, $ax > b$, or $ax \geq b$, where a and b are real numbers, with $a \neq 0$.

equivalent inequalities Equivalent inequalities are inequalities with the same solution set.

3.2

intersection The intersection of two sets A and B is the set of elements that belong to both A and B.

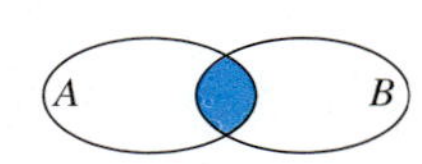

compound inequality A compound inequality is formed by joining two inequalities with a connective word, such as *and* or *or*.

union The union of two sets A and B is the set of elements that belong to either A or B (or both).

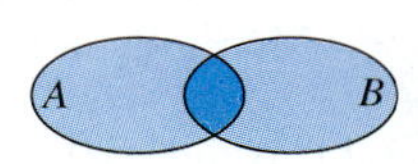

3.3

absolute value equation; absolute value inequality Absolute value equations and inequalities are equations and inequalities that involve the absolute value of a variable expression.

New Symbols

∞	infinity	$\cap$	set intersection
$-\infty$	negative infinity	$\cup$	set union
$(-\infty, \infty)$	the set of real numbers		

Test Your Word Power

See how well you have learned the vocabulary in this chapter. Answers follow the Quick Review.

1. An **inequality** is
   - **(a)** a statement that two algebraic expressions are equal
   - **(b)** a point on a number line
   - **(c)** an equation with no solutions
   - **(d)** a statement with algebraic expressions related by $<$, $\leq$, $>$, or $\geq$.

2. **Interval notation** is
   - **(a)** a portion of a number line
   - **(b)** a special notation for describing a point on a number line
   - **(c)** a way to use symbols to describe an interval on a number line
   - **(d)** a notation to describe unequal quantities.

3. The **intersection** of two sets A and B is the set of elements that belong
   - **(a)** to both A and B
   - **(b)** to either A or B, or both
   - **(c)** to either A or B, but not both
   - **(d)** to just A.

4. The **union** of two sets A and B is the set of elements that belong
   - **(a)** to both A and B
   - **(b)** to either A or B, or both
   - **(c)** to either A or B, but not both
   - **(d)** to just B.

Quick Review

Concepts

Examples

3.1 Linear Inequalities in One Variable

Solving Linear Inequalities in One Variable

Step 1 Simplify each side of the inequality by clearing parentheses and combining like terms.

Step 2 Use the addition property of inequality to get all terms with variables on one side and all terms without variables on the other side.

Step 3 Use the multiplication property of inequality to write the inequality in the form $x < k$ or $x > k$.

If an inequality is multiplied or divided by a *negative* number, the inequality symbol *must be reversed.*

Solve $3(x + 2) - 5x \leq 12$.

$$3x + 6 - 5x \leq 12$$
$$-2x + 6 \leq 12$$
$$-2x \leq 6$$
$$\frac{-2x}{-2} \geq \frac{6}{-2}$$
$$x \geq -3$$

The solution set $[-3, \infty)$ is graphed below.

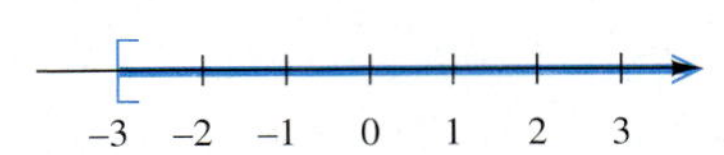

3.2 Set Operations and Compound Inequalities

Solving a Compound Inequality

Step 1 Solve each inequality in the compound inequality individually.

Step 2 If the inequalities are joined with *and*, the solution set is the intersection of the two individual solution sets.

If the inequalities are joined with *or*, the solution set is the union of the two individual solution sets.

Solve $x + 1 > 2$ and $2x < 6$.

$$x + 1 > 2 \quad \text{and} \quad 2x < 6$$
$$x > 1 \quad \text{and} \quad x < 3$$

The solution set is $(1, 3)$.

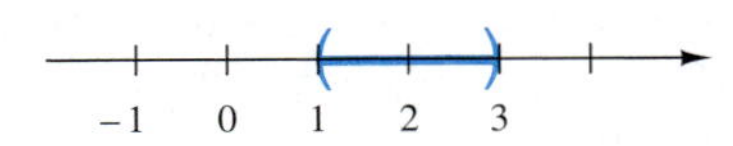

Solve $x \geq 4$ or $x \leq 0$.

The solution set is $(-\infty, 0] \cup [4, \infty)$.

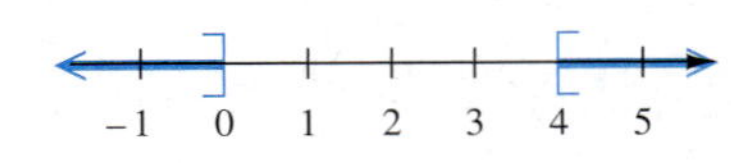

Concepts	*Examples*

3.3 Absolute Value Equations and Inequalities

Let k be a positive number.

To solve $|ax + b| = k$, solve the compound equation

$$ax + b = k \quad \text{or} \quad ax + b = -k.$$

Solve $|x - 7| = 3$.

$$x - 7 = 3 \quad \text{or} \quad x - 7 = -3$$
$$x = 10 \quad \text{or} \quad x = 4$$

The solution set is $\{4, 10\}$.

To solve $|ax + b| > k$, solve the compound inequality

$$ax + b > k \quad \text{or} \quad ax + b < -k.$$

Solve $|x - 7| > 3$.

$$x - 7 > 3 \quad \text{or} \quad x - 7 < -3$$
$$x > 10 \quad \text{or} \quad x < 4$$

The solution set is $(-\infty, 4) \cup (10, \infty)$.

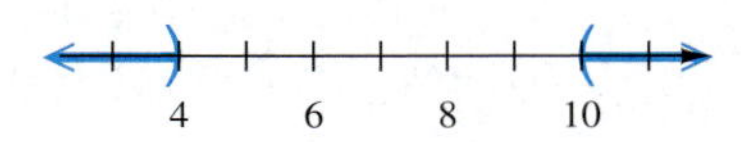

To solve $|ax + b| < k$, solve the compound inequality

$$-k < ax + b < k.$$

Solve $|x - 7| < 3$.

$$-3 < x - 7 < 3$$
$$4 < x < 10$$

The solution set is $(4, 10)$.

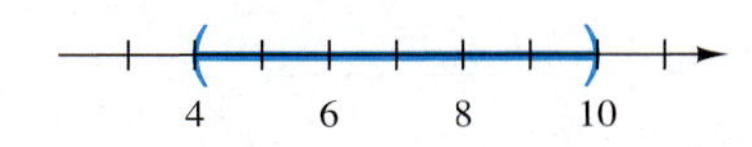

To solve an absolute value equation of the form

$$|ax + b| = |cx + d|,$$

solve the compound equation

$$ax + b = cx + d \quad \text{or} \quad ax + b = -(cx + d).$$

Solve $|x + 2| = |2x - 6|$.

$$x + 2 = 2x - 6 \quad \text{or} \quad x + 2 = -(2x - 6)$$
$$x = 8 \qquad\qquad x + 2 = -2x + 6$$
$$3x = 4$$
$$x = \frac{4}{3}$$

The solution set is $\{\frac{4}{3}, 8\}$.

ANSWERS TO TEST YOUR WORD POWER

1. (d) *Examples:* $x < 5$, $7 + 2y \geq 11$, $-5 < 2z - 1 \leq 3$ **2. (c)** *Examples:* $(-\infty, 5]$, $(1, \infty)$, $[-3, 3)$ **3. (a)** *Example:* If $A = \{2, 4, 6, 8\}$ and $B = \{1, 2, 3\}$, $A \cap B = \{2\}$. **4. (b)** *Example:* Using the preceding sets A and B, $A \cup B = \{1, 2, 3, 4, 6, 8\}$.

Focus on **Real-Data Applications**

What Do I Have to Average on My Tests to Get the Grade I Want?

On the first day of class, you are typically given a syllabus that describes the course requirements. If the syllabus includes a grading scale for homework, tests, projects, and final exam, then you should be able to predict the points you need on the final exam to earn a specific grade.

One intermediate algebra teacher bases final grades on points earned for three major exams, a comprehensive final exam, a daily activities grade (scaled), and lab participation and completion. The number of points available for each activity is given in the Graded Classwork table on the left. The teacher strictly adheres to the point ranges given in the Grade Distribution table on the right. A grade of IP (In Progress) is given to a student who participates fully but fails to achieve the course objectives.

GRADED CLASSWORK

Activity	Points Available
Homework and vocabulary	45
Daily activities (scaled)	55
Lab participation and completion	100
Major exams (3 at 100 pt)	300
Final exam	150
Total points	650

GRADE DISTRIBUTION

Grade	Points Required
A	585–650
B	520–584
C	455–519
IP	< 455 and active
F	< 455 and inactive

Notice that exams account for 450 of the possible 650 points. The remaining 200 points should be fairly easy to earn by keeping up with the day-to-day course requirements.

Assumption: You earn a "baseline" number of points based on the following criteria.

1. You earn *all* of the homework and vocabulary points.
2. You earn a minimum of 50 points based on daily activities.
3. You earn a minimum of 90 lab participation and completion points.

For Group Discussion

1. Assume that you earn the baseline number of points. Let x represent the test points to be earned. Write and solve linear inequalities to find the minimum number of points that you need in test scores to earn grades no lower than A, B, and C. What "test average" is each minimum score? Round *up* to the nearest whole percent.

2. To keep your scholarship, you must earn a B in the course. Write a compound inequality to find the range of points that you need in test scores to earn a B average. Solve the inequality. What range of "test averages" are those minimum scores? Round *up* to the nearest whole percent.

3. Mark does not like to do the homework or participate in labs. Assume that Mark earns only 15 points in homework and vocabulary, 40 points in daily activities, and 50 points in lab participation. Write and solve linear inequalities to find the minimum number of points that Mark needs in test scores to earn grades no lower than A, B, and C. What "test average" is each minimum score? Round *up* to the nearest whole percent.

Chapter 3 Review Exercises

[3.1] *Solve each inequality. Give the solution set in both interval and graph forms.*

1. $-\frac{2}{3}k < 6$

2. $-5x - 4 \geq 11$

3. $\frac{6a + 3}{-4} < -3$

4. $\frac{9y + 5}{-3} > 3$

5. $5 - (6 - 4k) \geq 2k - 7$

6. $-6 \leq 2k \leq 24$

7. $8 \leq 3y - 1 < 14$

8. $-4 < 3 - 2k < 9$

9. The perimeter of a rectangular playground must be no greater than 120 m. The width of the playground must be 22 m. Find the possible lengths of the playground.

10. The hit movie *Titanic* earned more in Europe than in the United States. The average movie ticket in London, for example, costs the equivalent of \$10.59. (*Source: Parade* magazine, September 13, 1998.) A student group from the United States is touring London and wishes to see the movie there. If \$1000 is available to purchase tickets and the group receives a \$50 discount from the tour company, how many tickets can be purchased?

11. To pass algebra, a student must have an average of at least 70% on five tests. On the first four tests, a student has grades of 75%, 79%, 64%, and 71%. What possible grades on the fifth test would guarantee a passing grade in the class?

12. While solving the inequality

$$10x + 2(x - 4) < 12x - 13,$$

a student did all the work correctly and obtained the statement $-8 < -13$. The student did not know what to do at this point, because the variable "disappeared." How would you explain to the student the interpretation of this result?

[3.2] *Let* $A = \{a, b, c, d\}$, $B = \{a, c, e, f\}$, *and* $C = \{a, e, f, g\}$. *Find each set.*

13. $A \cap B$

14. $A \cap C$

15. $B \cup C$

16. $A \cup C$

Solve each compound inequality. Give the solution set in both interval and graph forms.

17. $x > 6$ and $x < 9$

18. $x + 4 > 12$ and $x - 2 < 12$

19. $x > 5$ or $x \leq -3$

20. $x \geq -2$ or $x < 2$

21. $x - 4 > 6$ and $x + 3 \leq 10$

22. $-5x + 1 \geq 11$ or $3x + 5 \geq 26$

Express each union or intersection in simplest interval form.

23. $(-3, \infty) \cap (-\infty, 4)$

24. $(-\infty, 6) \cap (-\infty, 2)$

25. $(4, \infty) \cup (9, \infty)$

26. $(1, 2) \cup (1, \infty)$

27. The table shows the median weekly earnings of full-time workers by occupation for men and women.

Occupation	Men	Women
Managerial and professional specialty	\$ 852	\$616
Mathematical and computer scientists	\$1005	\$754
Waiters and waitresses	\$ 300	\$264
Bus drivers	\$ 482	\$354

Source: U.S. Bureau of Labor Statistics.

Give the occupation that satisfies each description.

(a) The median earnings for men are less than \$900 *and* for women are greater than \$500.

(b) The median earnings for men are greater than \$900 *or* for women are greater than \$600.

[3.3] *Solve each absolute value equation.*

28. $|x| = 7$

29. $|y + 2| = 9$

30. $|3k - 7| = 8$

31. $|z - 4| = -12$

32. $|2k - 7| + 4 = 11$

33. $|4a + 2| - 7 = -3$

34. $|3p + 1| = |p + 2|$

35. $|2m - 1| = |2m + 3|$

Solve each absolute value inequality. Give the solution set in both interval and graph forms.

36. $|p| < 14$

37. $|-y + 6| \leq 7$

38. $|2p + 5| \leq 1$

39. $|x + 1| \geq -3$

40. $|5r - 1| > 9$

41. $|3k + 6| \geq 0$

Mixed Review Exercises

Solve.

42. $(7 - 2k) + 3(5 - 3k) \geq k + 8$

43. $x < 5$ and $x \geq -4$

44. $\frac{3}{4}(a - 2) - \frac{1}{3}(5 - 2a) < -2$

45. To qualify for a company pension plan, an employee must average at least \$1000 per month in earnings. During the first four months of the year, an employee made \$900, \$1200, \$1040, and \$760. What possible amounts earned during the fifth month will qualify the employee?

46. $-5r \geq -10$

47. $|7x - 2| > 9$

48. $|2x - 10| = 20$

49. $|m + 3| \leq 13$

50. $x \geq -2$ or $x < 4$

51. $|m - 1| = |2m + 3|$

In Exercises 52 and 53, sketch the graph of each solution set.

52. $x > 6$ and $x < 8$

53. $-5x + 1 \geq 11$ or $3x + 5 \geq 26$

54. If $k < 0$, what is the solution set of

(a) $|5x + 3| < k$, **(b)** $|5x + 3| > k$, **(c)** $|5x + 3| = k$?

Chapter 3 Test

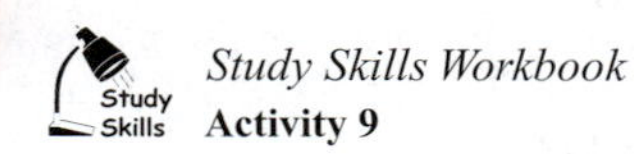

1. What is the special rule that must be remembered when multiplying or dividing each side of an inequality by a negative number?

1. ______

Solve each inequality. Give the solution set in both interval and graph forms.

2. $4 - 6(x + 3) \leq -2 - 3(x + 6) + 3x$

2. ______

3. $-\frac{4}{7}x > -16$

3. ______

4. $-6 \leq \frac{4}{3}x - 2 \leq 2$

4. ______

5. Which one of the following inequalities is equivalent to $x < -3$?

A. $-3x < 9$ **B.** $-3x > -9$ **C.** $-3x > 9$ **D.** $-3x < -9$

5. ______

6. The graph shows the number (in millions) of U.S. citizen departures to Europe. During which years were departures to Europe

(a) at least 8 million, **(b)** less than 7 million,

(c) between 7 million and 9 million?

6. (a) ______

(b) ______

(c) ______

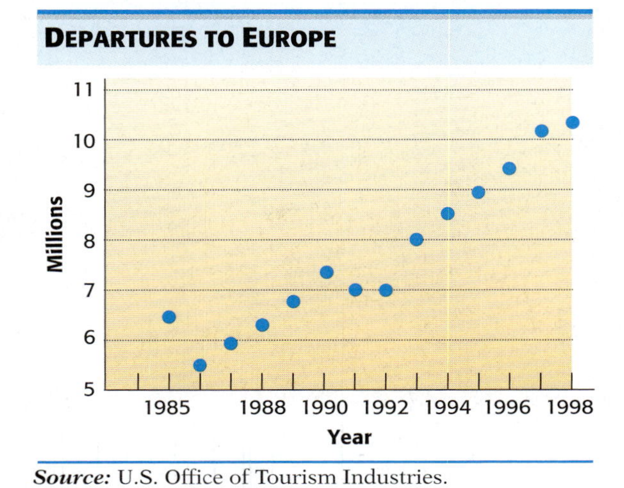

Source: U.S. Office of Tourism Industries.

Solve each problem.

7. A student must have an average of at least 80% on the four tests in a course to get a B. The student had 83%, 76%, and 79% on the first three tests. What minimum percent on the fourth test would guarantee a B in the course?

7. ______

8. ______________

8. A product will break even or produce a profit only if the revenue R (in dollars) from selling the product is at least the cost C (in dollars) of producing it. Suppose that the cost to produce x units of carpet is $C = 50x + 5000$, while the revenue is $R = 60x$. For what values of x is R at least equal to C?

9. (a) ______________

(b) ______________

9. Let $A = \{1, 2, 5, 7\}$ and $B = \{1, 5, 9, 12\}$. Find

(a) $A \cap B$, **(b)** $A \cup B$.

10. ______________

10. Solve $x \leq 2$ and $x \geq 2$.

Solve each compound or absolute value inequality. For Exercises 11–14, give the solution set in both interval and graph forms.

11. ______________

11. $3k \geq 6$ and $k - 4 < 5$

12. ______________

12. $-4x \leq -24$ or $4x - 2 < 10$

13. ______________

13. $|4x + 3| \leq 7$

14. ______________

14. $|5 - 6x| > 12$

15. ______________

15. $|7 - x| \leq -1$

Solve each absolute value equation.

16. ______________

16. $|3k - 2| + 1 = 8$

17. ______________

17. $|3 - 5x| = |2x + 8|$

Cumulative Review Exercises CHAPTERS 1–3

1. Write $\frac{108}{144}$ in lowest terms.

2. True or false? $\frac{8(7) - 5(6 + 2)}{3 \cdot 5 + 1} \geq 1$

Perform the indicated operations.

3. $\frac{5}{6} + \frac{1}{4} - \frac{7}{15}$

4. $\frac{9}{8} \cdot \frac{16}{3} \div \frac{5}{8}$

5. $9 - (-4) + (-2)$

6. $\frac{-4(9)(-2)}{-3^2}$

7. $|-7 - 1|(-4) + (-4)$

Evaluate each exponential expression.

8. $(-5)^3$

9. $\left(\frac{3}{2}\right)^4$

Evaluate each expression if $x = 2$, $y = -3$, and $z = 4$.

10. $-2y + 4(x - 3z)$

11. $\frac{3x^2 - y^2}{4z}$

Name each property illustrated.

12. $7(k + m) = 7k + 7m$

13. $3 + (5 + 2) = 3 + (2 + 5)$

14. Simplify $-4(k + 2) + 3(2k - 1)$ by combining terms.

Solve each equation, and check the solution.

15. $4 - 5(a + 2) = 3(a + 1) - 1$

16. $\frac{2}{3}y + \frac{3}{4}y = -17$

17. $\frac{2x + 3}{5} = \frac{x - 4}{2}$

18. $|3m - 5| = |m + 2|$

19. $3x + 4y = 24$ for y

20. $A = P(1 + ni)$ for n

Solve each inequality. Give the solution set in both interval and graph forms.

21. $3 - 2(x + 7) \leq -x + 3$

22. $-4 < 5 - 3x \leq 0$

23. $2x + 1 > 5$ or $2 - x > 2$

24. $|-7k + 3| \geq 4$

Solve each problem.

25. Kathy Manley invested some money at 7% interest and the same amount at 10%. Her total interest for the year was $150 less than one-tenth of the total amount she invested. How much did she invest at each rate?

26. A dietician must use three foods, A, B, and C, in a diet. He must include twice as many grams of food A as food C, and 5 g of food B. The three foods must total at most 24 g. What is the largest amount of food C that the dietician can use?

27. Lorie Reilly got scores of 88 and 78 on her first two tests. What score must she make on her third test to keep an average of 80 or greater?

28. Two cars are 400 mi apart. Both start at the same time and travel toward one another. They meet 4 hr later. If the speed of one car is 20 mph faster than the other, what is the speed of each car?

29. Since 1975, the number of daily newspapers has steadily declined.

Year	Number of Daily Newspapers
1975	1756
1980	1745
1985	1676
1990	1611
1995	1533
1996	1520
1997	1509
1998	1489

Source: Statistical Abstract of the United States, 1999.

According to the table,

(a) by how much did the number of daily newspapers decrease between 1990 and 1998?

(b) by what *percent* did the number of daily newspapers decrease from 1990 to 1998?

30. For a woven hanging, Miguel Hidalgo needs three pieces of yarn, which he will cut from a 40 cm piece. The longest piece is to be 3 times as long as the middle-sized piece, and the shortest piece is to be 5 cm shorter than the middle-sized piece. What lengths should he cut?

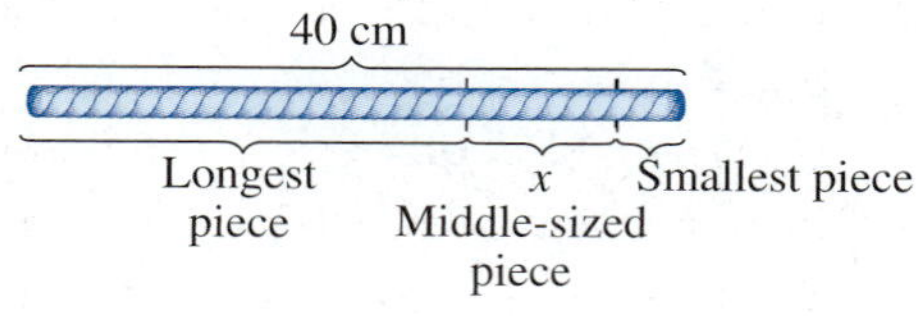

Graphs, Linear Equations, and Functions

Graphs are widely used in the media because they present a lot of information in an easy-to-understand form. As the saying goes, "A picture is worth a thousand words." It is important to be able to read graphs correctly and understand how to use the data they provide. In Section 4.2, Example 8, we use a graph to find the average rate of change each year from 1997 to 2001 in the number of U.S. households owning more than one personal computer.

4.1 THE RECTANGULAR COORDINATE SYSTEM

OBJECTIVES

1. Plot ordered pairs.
2. Find ordered pairs that satisfy a given equation.
3. Graph lines.
4. Find x- and y-intercepts.
5. Recognize equations of vertical and horizontal lines.

There are many ways to present information graphically. The circle graph (or pie chart) in Figure 1 shows the cost breakdown (in cents) for a gallon of regular unleaded gasoline (in California), selling for about \$1.54, based on a crude oil cost of \$17.16 per barrel. What contributes most to the cost?

Figure 1 also shows a bar graph in which the heights of the bars represent the Btu (British thermal units) required to cool different sized rooms. How many Btu are needed to cool a 1400 ft^2 room?

The line graph in Figure 1 shows CNBC profits from 1994 through 1999. CNBC, a fast-growing television channel, dominates the financial-news category on cable and is preparing to enter the Internet news market. By how many million dollars did profits increase from 1998 to 1999?

WHAT GOES INTO THE PRICE OF A GALLON OF GAS?

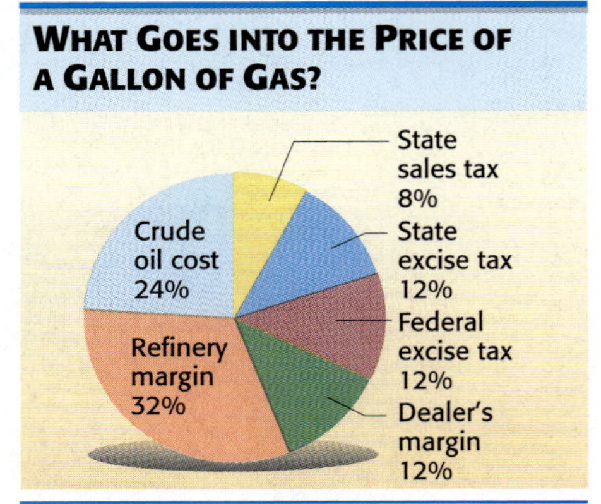

Source: California Energy Commission, 1999.

CHOOSE THE RIGHT AIR CONDITIONER

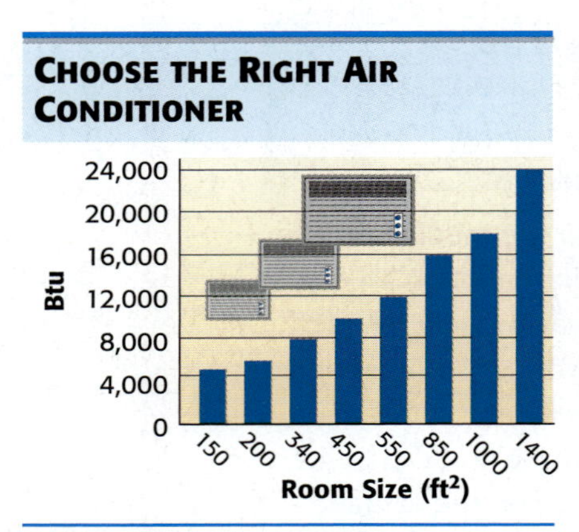

Source: Carey, Morris and James, *Home Improvement for Dummies*, IDG Books.

CNBC PROFITS

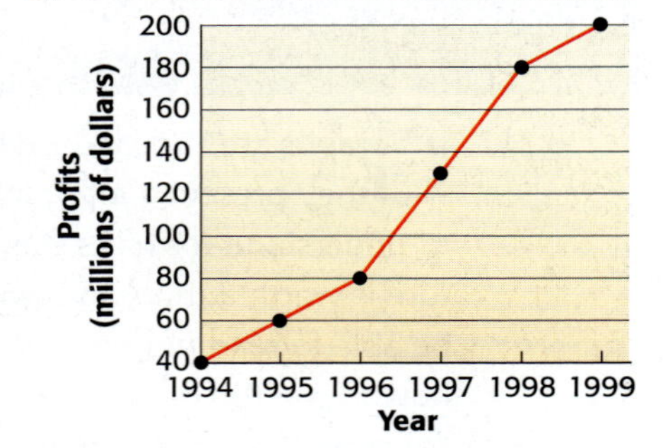

Source: *Fortune*, May 24, 1999, p.142.

Figure 1

René Descartes, a 17th-century French mathematician, is credited with giving us an indispensable method of locating a point on a plane. It seems Descartes, who was lying in bed ill, was watching a fly crawl about on the ceiling near a corner of the room. It occurred to him that the location of the fly on the ceiling could be described by determining its distances from the two adjacent walls. In this chapter we use this insight to plot points and graph linear equations in two variables whose graphs are straight lines.

1 **Plot ordered pairs.** Each of the pairs of numbers (3, 1), (−5, 6), and (4, −1) is an example of an **ordered pair;** that is, a pair of numbers written within parentheses in which the order of the numbers is important. We graph an ordered pair using two perpendicular number lines that intersect at their 0 points, as shown in Figure 2. The common 0 point is called the **origin.** The position of any point in this plane is determined by referring to the horizontal number line, the ***x*-axis,** and the vertical number line, the ***y*-axis.** The first number in the ordered pair indicates the position relative to the x-axis, and the second number indicates the position relative to the y-axis. The x-axis and the y-axis make up a **rectangular** (or **Cartesian,** for Descartes) **coordinate system.**

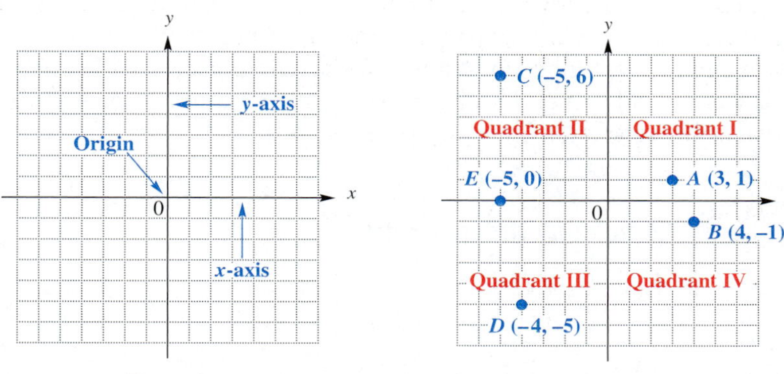

Figure 2 Figure 3

To locate, or **plot,** the point on the graph that corresponds to the ordered pair (3, 1), we move three units from 0 to the right along the x-axis, and then one unit up parallel to the y-axis. The point corresponding to the ordered pair (3, 1) is labeled A in Figure 3. The point (4, −1) is labeled B, (−5, 6) is labeled C, and (−4, −5) is labeled D. Point E corresponds to (−5, 0). The phrase "the point corresponding to the ordered pair (3, 1)" is often abbreviated as "the point (3, 1)." The numbers in the ordered pairs are called the **coordinates** of the corresponding point.

CAUTION

The parentheses used to represent an ordered pair are also used to represent an open interval (introduced in Chapter 3). The context of the discussion tells whether ordered pairs or open intervals are being represented.

The four regions of the graph, shown in Figure 3, are called **quadrants I, II, III,** and **IV,** reading counterclockwise from the upper right quadrant. The points on the x-axis and y-axis do not belong to any quadrant. For example, point E in Figure 3 belongs to no quadrant.

Work Problem 1 at the Side.

2 **Find ordered pairs that satisfy a given equation.** Each solution to an equation with two variables will include two numbers, one for each variable. To keep track of which number goes with which variable, we write the solutions as ordered pairs. (If x and y are used as the variables, the x-value is given first.) For example, we can show that (6, −2) is a solution of $2x + 3y = 6$ by substitution.

1 Plot each point. Name the quadrant (if any) in which each point is located.

(a) (−4, 2)

(b) (3, −2)

(c) (−5, −6)

(d) (4, 6)

(e) (−3, 0)

(f) (0, −5)

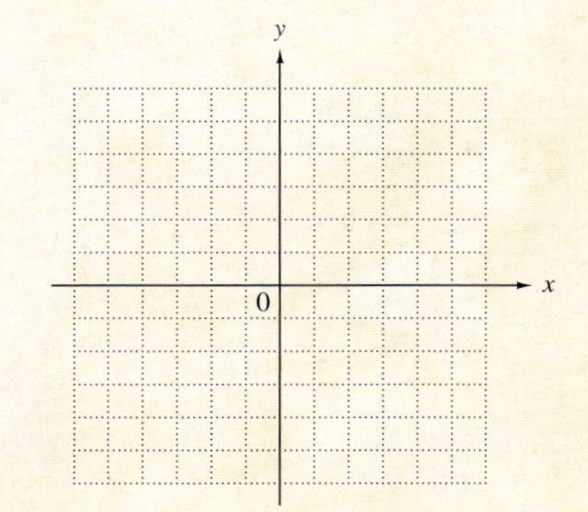

ANSWERS

1.

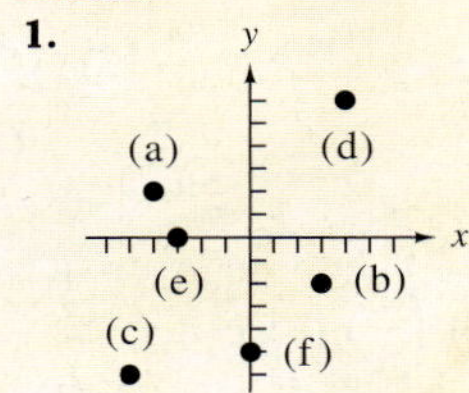

(a) II **(b)** IV **(c)** III **(d)** I
(e) no quadrant **(f)** no quadrant

2 (a) Complete each ordered pair for $3x - 4y = 12$: $(0,\ \ )$, $(\ \ , 0)$, $(\ \ , -2)$, $(-4,\ \ )$.

(b) Find one other ordered pair that satisfies the equation.

$$\begin{aligned} 2x + 3y &= 6 \\ 2(6) + 3(-2) &= 6 \quad ? \quad \text{Let } x = 6, y = -2. \\ 12 - 6 &= 6 \quad ? \\ 6 &= 6 \quad \text{True} \end{aligned}$$

Because the pair of numbers $(6, -2)$ makes the equation true, it is a solution. On the other hand, because

$$2(5) + 3(1) = 10 + 3 = 13 \neq 6,$$

$(5, 1)$ is not a solution of the equation $2x + 3y = 6$.

To find ordered pairs that satisfy an equation, select any number for one of the variables, substitute it into the equation for that variable, and then solve for the other variable. Two other ordered pairs satisfying $2x + 3y = 6$ are $(0, 2)$ and $(3, 0)$. Since any real number could be selected for one variable and would lead to a real number for the other variable, linear equations in two variables have an infinite number of solutions.

Example 1 Completing Ordered Pairs

Complete each ordered pair for $2x + 3y = 6$.

(a) $(-3,\ \ )$

We are given $x = -3$. We substitute into the equation to find y.

$$\begin{aligned} 2x + 3y &= 6 \\ 2(-3) + 3y &= 6 \quad \text{Let } x = -3. \\ -6 + 3y &= 6 \\ 3y &= 12 \\ y &= 4 \end{aligned}$$

The ordered pair is $(-3, 4)$.

(b) $(\ \ , -4)$

Replace y with -4 in the equation to find x.

$$\begin{aligned} 2x + 3y &= 6 \\ 2x + 3(-4) &= 6 \quad \text{Let } y = -4. \\ 2x - 12 &= 6 \\ 2x &= 18 \\ x &= 9 \end{aligned}$$

The ordered pair is $(9, -4)$.

Work Problem 2 at the Side.

3 Graph lines. The **graph of an equation** is the set of points corresponding to all ordered pairs that satisfy the equation. It gives a "picture" of the equation. Most equations in two variables are satisfied by an infinite number of ordered pairs, so their graphs include an infinite number of points. To graph an equation, we plot a number of ordered pairs that satisfy the equation until we have enough points to suggest the shape of the graph. For example, to graph $2x + 3y = 6$, plot all the ordered pairs found in Objective 2 and Example 1. These points, shown in Figure 4(a), appear to lie on

ANSWERS

2. (a) $(0, -3)$, $(4, 0)$, $\left(\frac{4}{3}, -2\right)$, $(-4, -6)$

(b) Many answers are possible; for example, $\left(-6, -\frac{15}{2}\right)$.

a straight line. If all the ordered pairs that satisfy the equation $2x + 3y = 6$ were graphed, they would form a straight line. The graph of $2x + 3y = 6$ is the line shown in Figure 4(b).

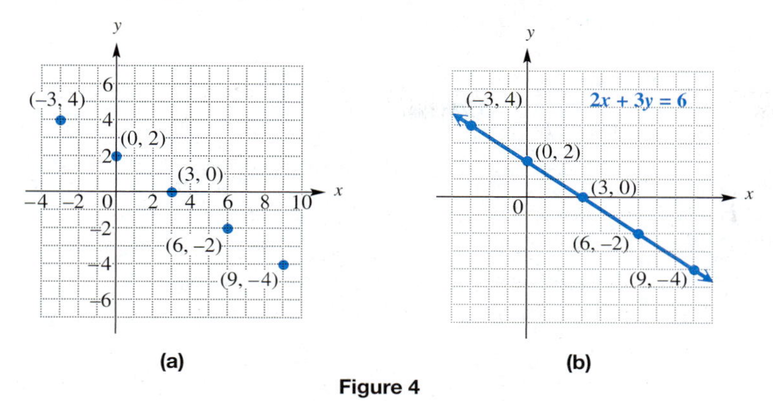

Figure 4

❸ Graph $3x - 4y = 12$. Use the points from Problem 2.

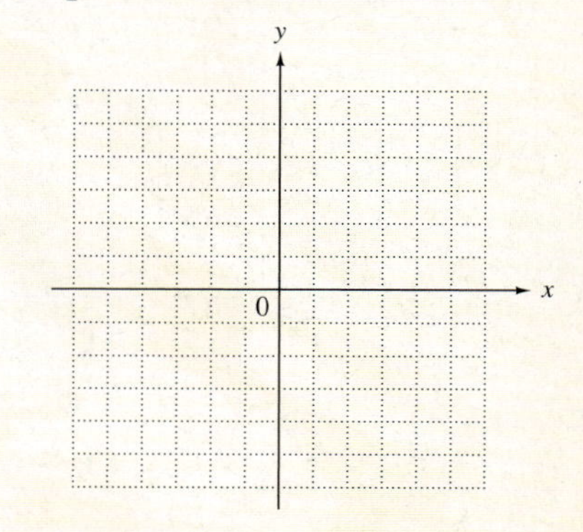

Work Problem ❸ at the Side.

The equation $2x + 3y = 6$ is called a **first-degree equation** because it has no term with a variable to a power greater than one.

The graph of any first-degree equation in two variables is a straight line.

Since first-degree equations with two variables have straight-line graphs, they are called *linear equations in two variables.* (We discussed linear equations in one variable in Chapter 2.)

Linear Equation in Two Variables

A **linear equation in two variables** can be written in the form

$$Ax + By = C \quad (A \text{ and } B \text{ not both } 0).$$

This form is called **standard form.**

4 **Find *x*- and *y*-intercepts.** A straight line is determined if any two different points on the line are known, so finding two different points is enough to graph the line. Two useful points for graphing are the x- and y-intercepts. The ***x*-intercept** is the point (if any) where the line intersects the x-axis; likewise, the ***y*-intercept** is the point (if any) where the line intersects the y-axis.* In Figure 4(b), the y-value of the point where the line intersects the x-axis is 0. Similarly, the x-value of the point where the line intersects the y-axis is 0. This suggests a method for finding the x- and y-intercepts.

Intercepts

In the equation of a line, let $y = 0$ to find the x-intercept; let $x = 0$ to find the y-intercept.

* Some texts define an intercept as a number, not a point.

ANSWERS

3.

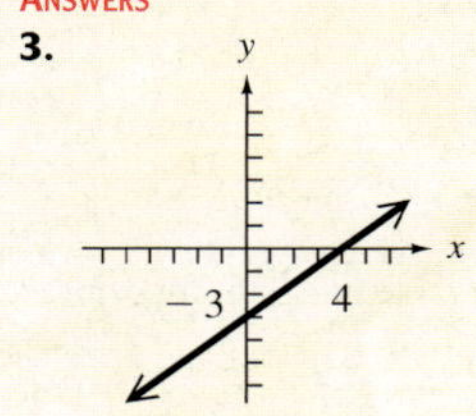

4 Find the intercepts, and graph $2x - y = 4$.

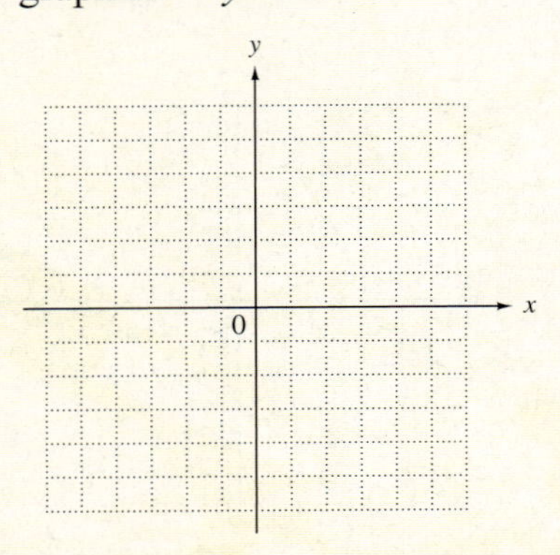

Example 2 Finding Intercepts

Find the x- and y-intercepts of $4x - y = -3$, and graph the equation.

We find the x-intercept by letting $y = 0$.

$$4x - \mathbf{0} = -3 \quad \text{Let } y = 0.$$
$$4x = -3$$
$$x = -\frac{3}{4} \quad x\text{-intercept is } \left(-\tfrac{3}{4}, 0\right).$$

For the y-intercept, let $x = 0$.

$$4(\mathbf{0}) - y = -3 \quad \text{Let } x = 0.$$
$$-y = -3$$
$$y = 3 \quad y\text{-intercept is } (0, 3).$$

The intercepts are the two points $\left(-\frac{3}{4}, 0\right)$ and $(0, 3)$. We show these ordered pairs in the table next to Figure 5 and use these points to draw the graph.

x	y
$-\frac{3}{4}$	0
0	3

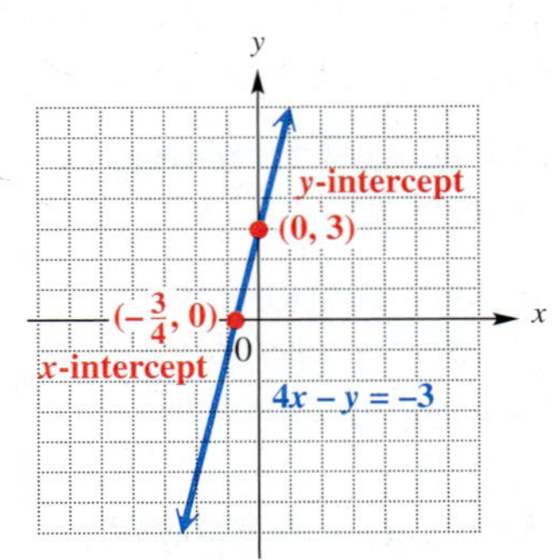

Figure 5

NOTE

While two points, such as the two intercepts in Figure 5, are sufficient to graph a straight line, it is a good idea to use a third point to guard against errors. Verify by substitution that $(-1, -1)$ also lies on the graph of $4x - y = -3$.

Work Problem 4 at the Side.

5 Recognize equations of vertical and horizontal lines. A graph can fail to have an x-intercept or a y-intercept, which is why the phrase "if any" was added when discussing intercepts.

Example 3 Graphing a Horizontal Line

Graph $y = 2$.

Since y is always 2, there is no value of x corresponding to $y = 0$, so the graph has no x-intercept. The y-intercept is $(0, 2)$. The graph in Figure 6, shown with a table of ordered pairs, is a horizontal line.

Continued on Next Page

Answers

4. x-intercept is $(2, 0)$; y-intercept is $(0, -4)$.

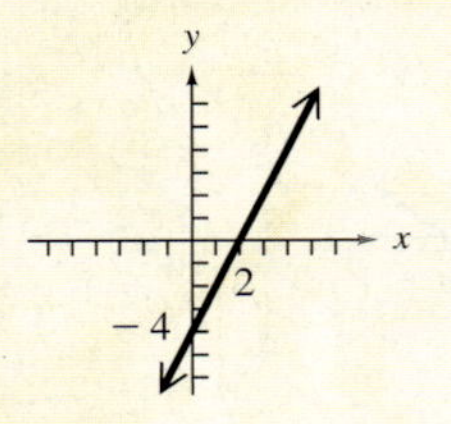

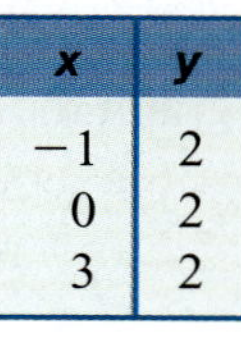

x	y
-1	2
0	2
3	2

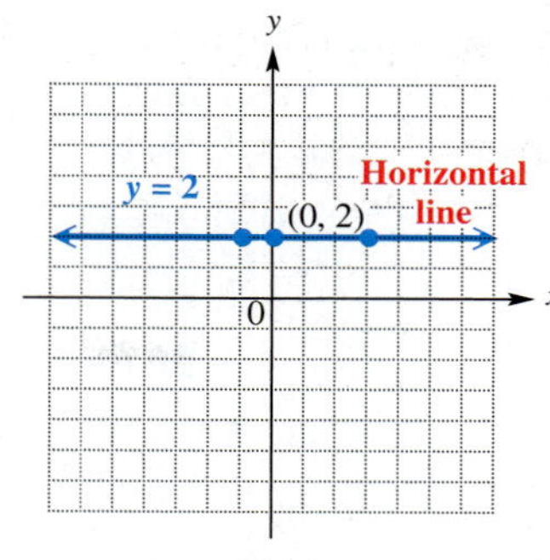

Figure 6

Work Problem 5 at the Side.

Example 4 Graphing a Vertical Line

Graph $x + 1 = 0$.

The x-intercept is $(-1, 0)$. The standard form $1x + 0y = -1$ shows that every value of y leads to $x = -1$, so no value of y makes $x = 0$. The only way a straight line can have no y-intercept is if it is vertical, as in Figure 7.

x	y
-1	-4
-1	0
-1	5

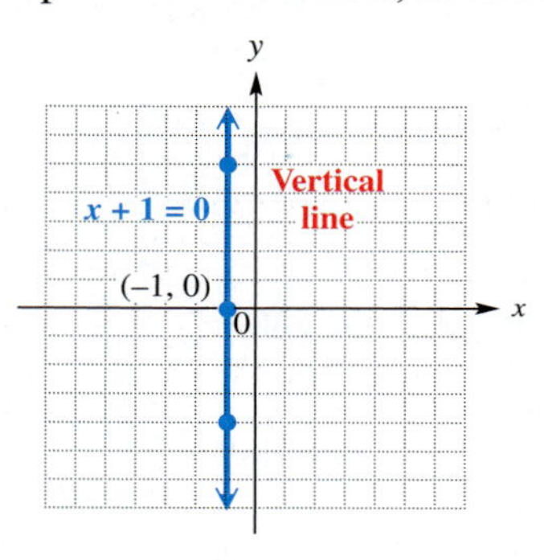

Figure 7

Work Problem 6 at the Side.

Some lines have both the x- and y-intercepts at the origin.

Example 5 Graphing a Line That Passes through the Origin

Graph $x + 2y = 0$.

Find the x-intercept by letting $y = 0$.

$$x + 2y = 0$$
$$x + 2(0) = 0 \quad \text{Let } y = 0.$$
$$x + 0 = 0$$
$$x = 0 \quad x\text{-intercept is } (0, 0).$$

To find the y-intercept, let $x = 0$.

$$x + 2y = 0$$
$$0 + 2y = 0 \quad \text{Let } x = 0.$$
$$y = 0 \quad y\text{-intercept is } (0, 0).$$

Continued on Next Page

5 Find the intercepts, and graph $y + 4 = 0$.

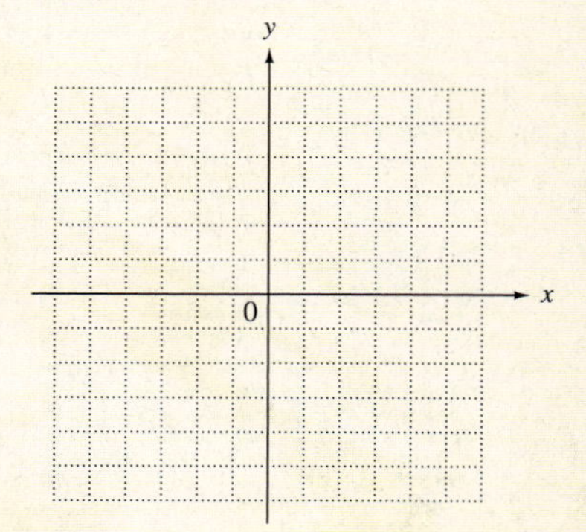

6 Find the intercepts, and graph the line $x = 2$.

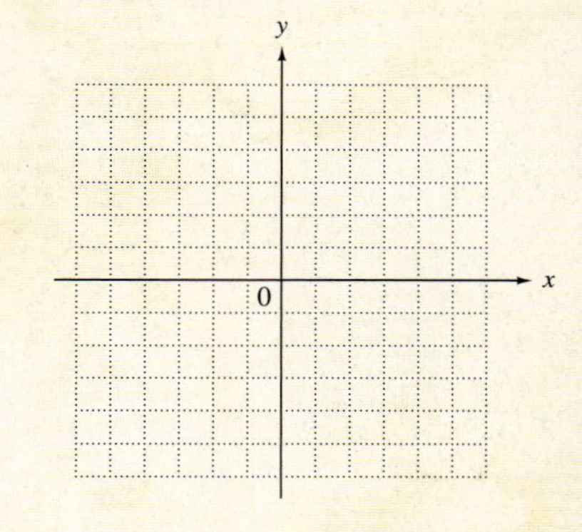

Answers

5. no x-intercept; y-intercept is $(0, -4)$.

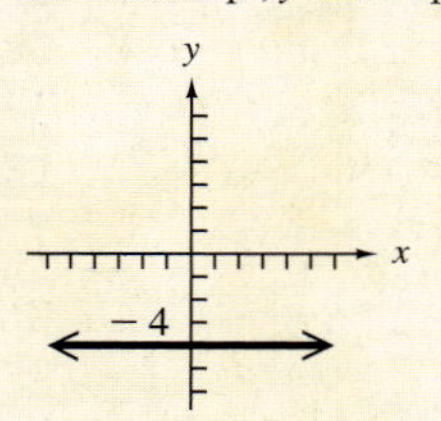

6. no y-intercept; x-intercept is $(2, 0)$.

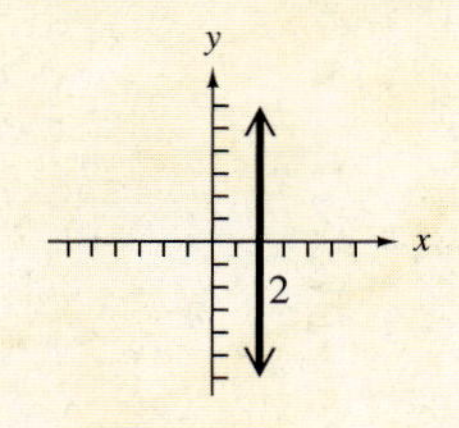

7 Find the intercepts, and graph the line $3x - y = 0$.

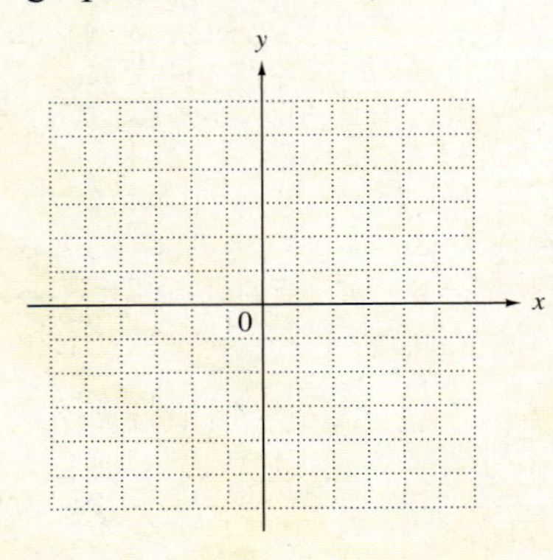

Both intercepts are the same ordered pair, (0, 0). (This means that the graph goes through the origin.) To find another point to graph the line, choose any nonzero number for x, say $x = 4$, and solve for y.

$$x + 2y = 0$$
$$4 + 2y = 0 \quad \text{Let } x = 4.$$
$$2y = -4$$
$$y = -2$$

This gives the ordered pair $(4, -2)$. These two points lead to the graph shown in Figure 8. As a check, verify that $(-2, 1)$ also lies on the line.

x	y
-2	1
0	0
4	-2

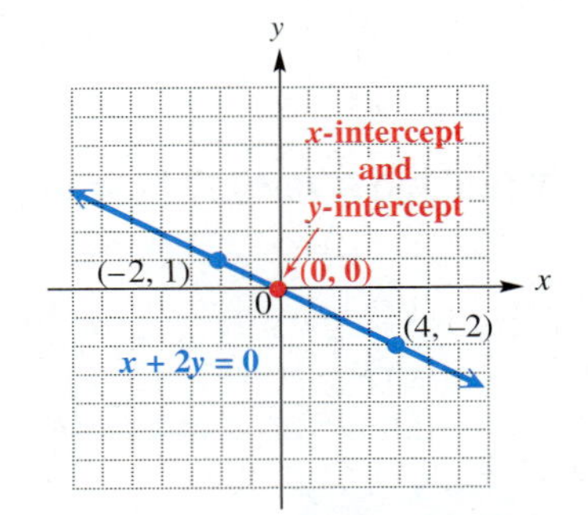

Figure 8

To find the additional point, we could have chosen any number (except 0) for y instead of x.

Work Problem 7 at the Side.

ANSWERS

7. Both intercepts are (0, 0).

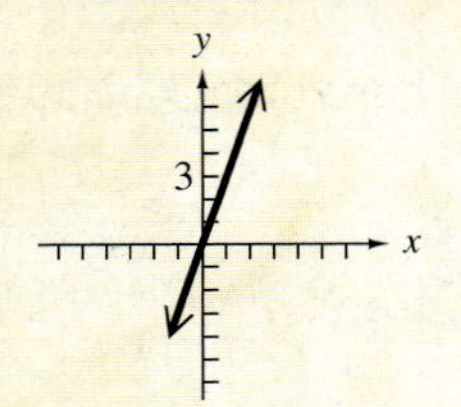

4.1 EXERCISES

FOR EXTRA HELP: Student's Solutions Manual; MyMathLab.com; InterAct Math Tutorial Software; AW Math Tutor Center; www.mathxl.com; Digital Video Tutor CD 2 Videotape 4

In Exercises 1 and 2, answer each question by locating ordered pairs on the graphs.

1. The graph shows the percent of women in math or computer science professions since 1970.

(a) If (x, y) represents a point on the graph, what does x represent? What does y represent?

(b) In what decade (10-year period) did the percent of women in math or computer science professions decrease?

(c) When did the percent of women in math or computer science professions reach a maximum?

(d) In what year was the percent of women in math or computer science professions about 27%?

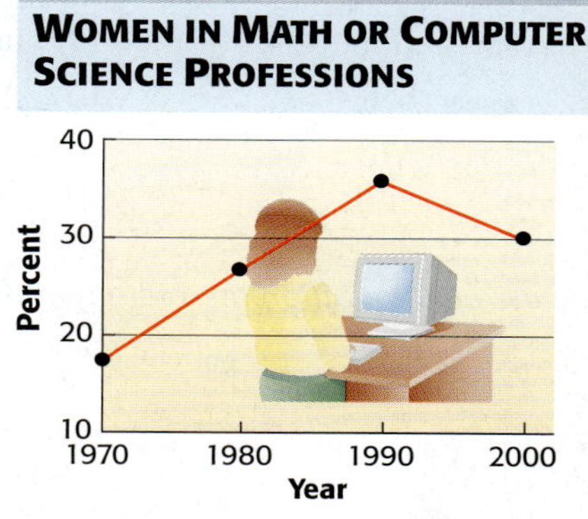

Source: U.S. Bureau of the Census and Bureau of Labor Statistics.

2. The graph indicates federal government tax revenues in billions of dollars.

(a) If (x, y) represents a point on the graph, what does x represent? What does y represent?

(b) What was the revenue in 1996?

(c) In what year was revenue about $1700 billion?

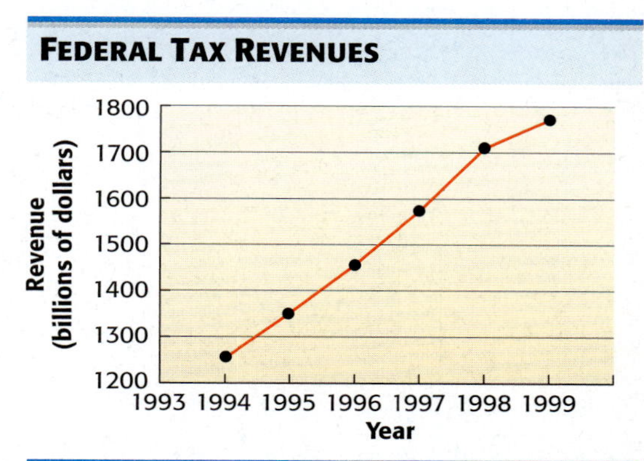

Source: U.S. Office of Management and Budget.

3. Observe the graphs in Exercises 1 and 2. If you were to use one of the other types of graphs mentioned in the opening paragraphs of this section to depict the information given there, which one would you choose?

4. What is another name for the rectangular coordinate system? After whom is it named?

Fill in each blank with the correct response.

5. The point with coordinates (0, 0) is called the _______ of a rectangular coordinate system.

6. For any value of x, the point $(x, 0)$ lies on the _______-axis.

7. To find the x-intercept of a line, we let _______ equal 0 and solve for _______.

8. The equation _______ $(x$ or $y)$ $= 4$ has a horizontal line as its graph.

9. To graph a straight line, we must find a minimum of _______ points.

10. The point (_______, 4) is on the graph of $2x - 3y = 0$.

Name the quadrant, if any, in which each point is located.

11. **(a)** $(1, 6)$ **(b)** $(-4, -2)$
(c) $(-3, 6)$ **(d)** $(7, -5)$
(e) $(-3, 0)$

12. **(a)** $(-2, -10)$ **(b)** $(4, 8)$
(c) $(-9, 12)$ **(d)** $(3, -9)$
(e) $(0, -8)$

13. Use the given information to determine the possible quadrants in which the point (x, y) must lie. (*Hint:* Consider the signs of the coordinates in each quadrant, and the signs of their product and quotient.)

(a) $xy > 0$ **(b)** $xy < 0$

(c) $\frac{x}{y} < 0$ **(d)** $\frac{x}{y} > 0$

14. What must be true about the coordinates of any point that lies on an axis?

Locate each point on the rectangular coordinate system.

15. $(2, 3)$ **16.** $(-1, 2)$ **17.** $(-3, -2)$ **18.** $(1, -4)$

19. $(0, 5)$ **20.** $(-2, -4)$ **21.** $(-2, 4)$ **22.** $(3, 0)$

23. $(-2, 0)$ **24.** $(3, -3)$

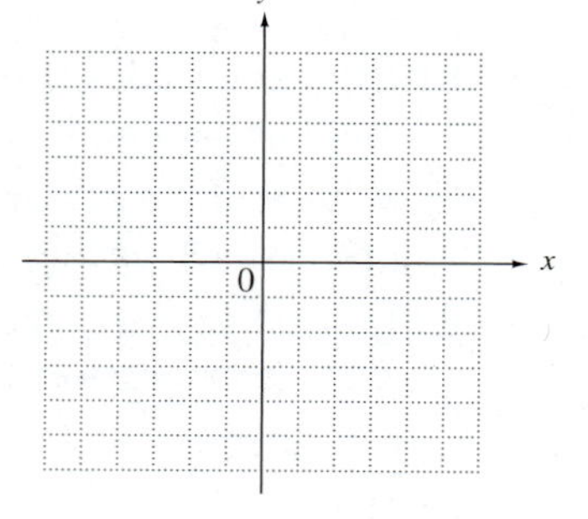

In each exercise, complete the given ordered pairs for the equation, and then graph the equation. See Example 1.

25. $x - y = 3$

(0,), (, 0)
(5,), (2,)

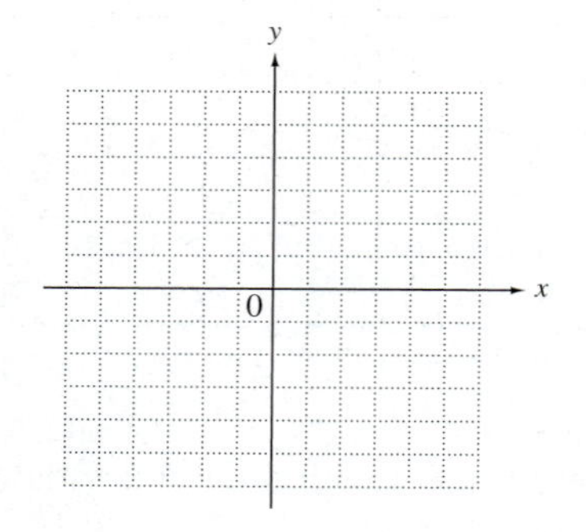

26. $x - y = 5$

(0,), (, 0)
(1,), (3,)

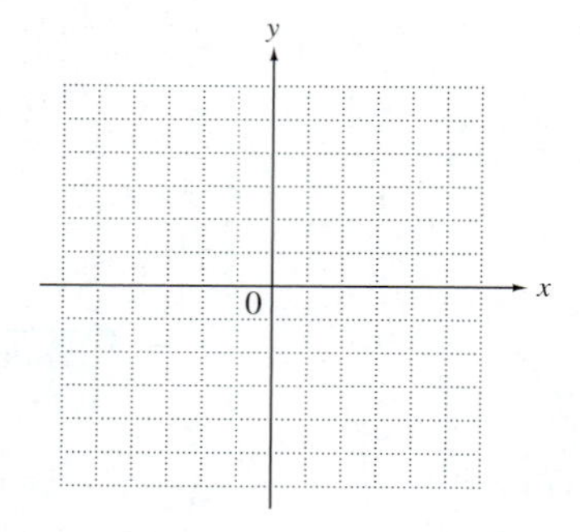

27. $x + 2y = 5$

(0,), (, 0)
(2,), (, 2)

28. $x + 3y = -5$

(0,), (, 0)
(1,), (, −1)

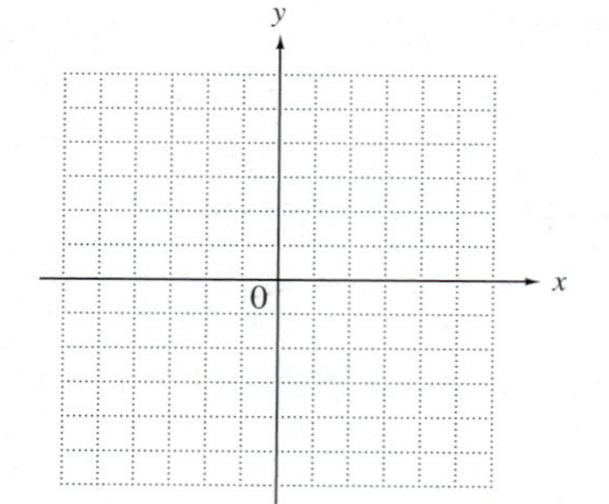

29. $4x - 5y = 20$

(0,), (, 0)
(2,), (, −3)

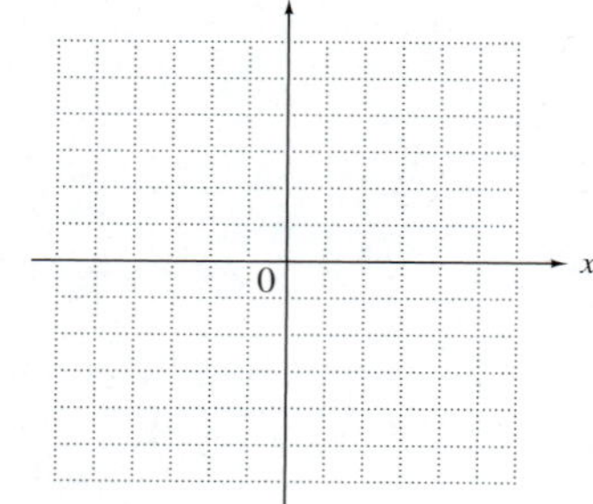
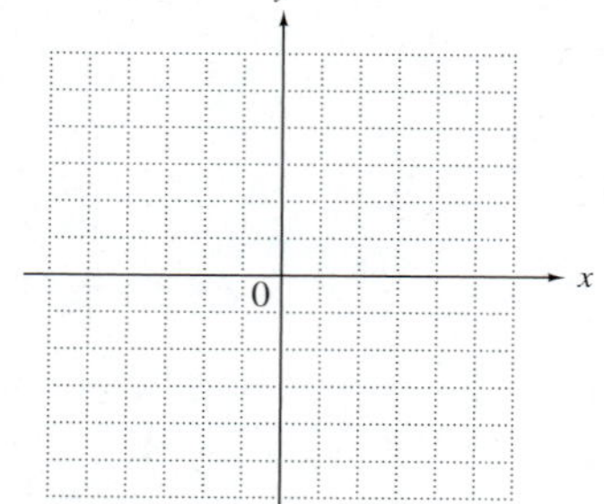

30. $6x - 5y = 30$

(0,), (, 0)
(3,), (, −2)

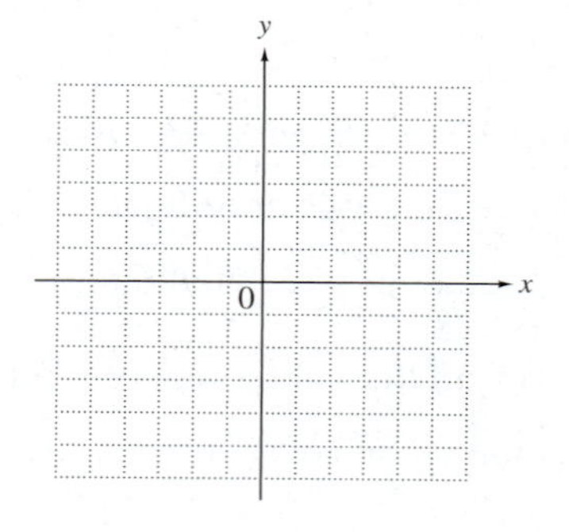

31. Explain why the graph of $x + y = k$ cannot pass through quadrant III if $k > 0$.

32. What is the equation of a line that coincides with the x-axis? the y-axis?

Find the x- and y-intercepts. Then graph each equation. See Examples 2–5.

33. $2x + 3y = 12$

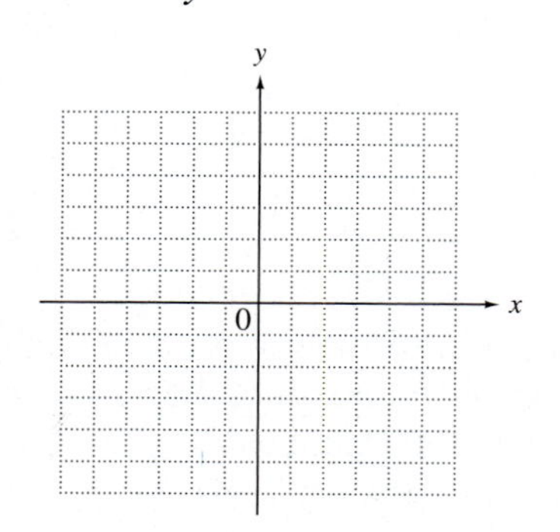

34. $5x + 2y = 10$

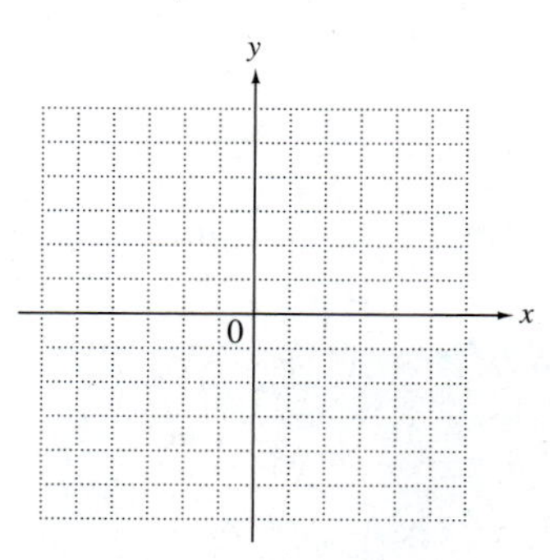

35. $x - 3y = 6$

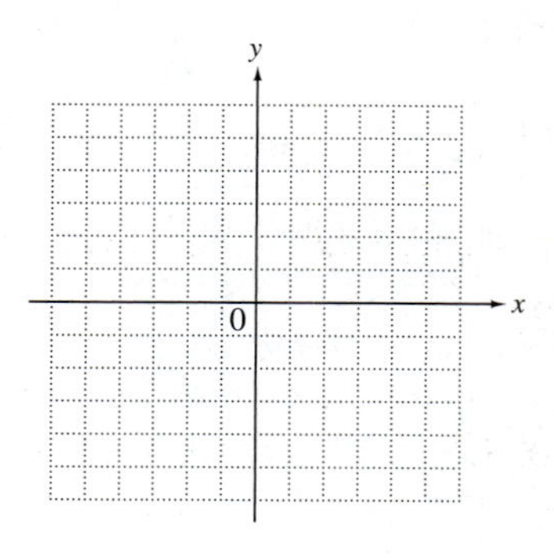

36. $x - 2y = -4$

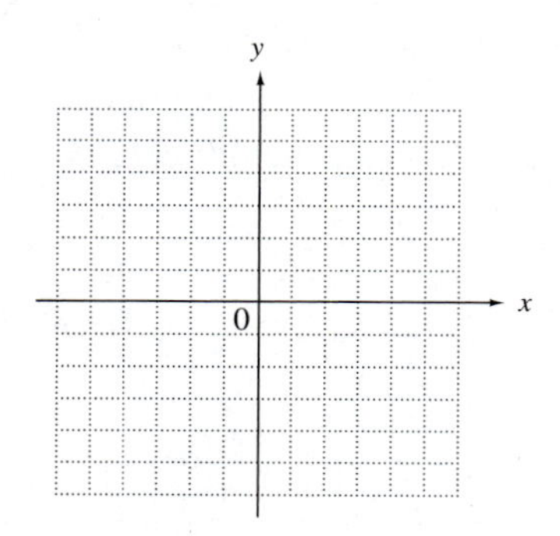

37. $3x - 7y = 9$

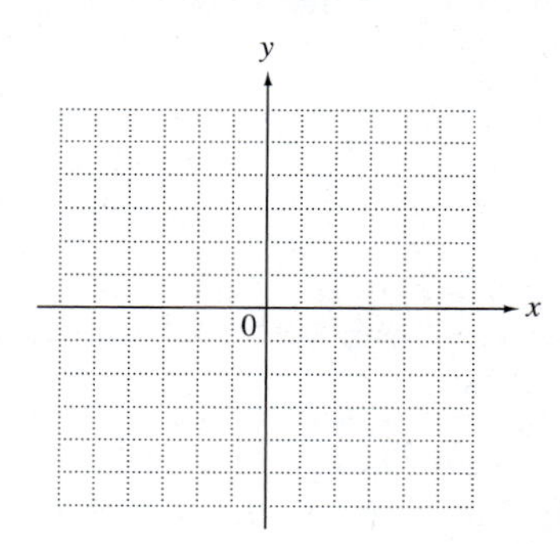

38. $5x + 6y = -10$

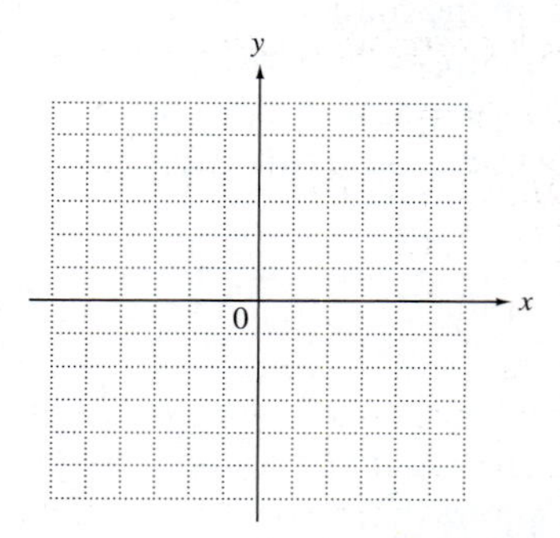

39. $y = 5$

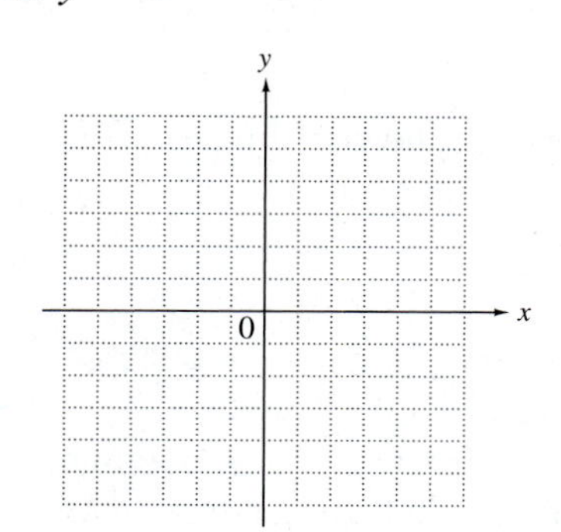

40. $y = -3$

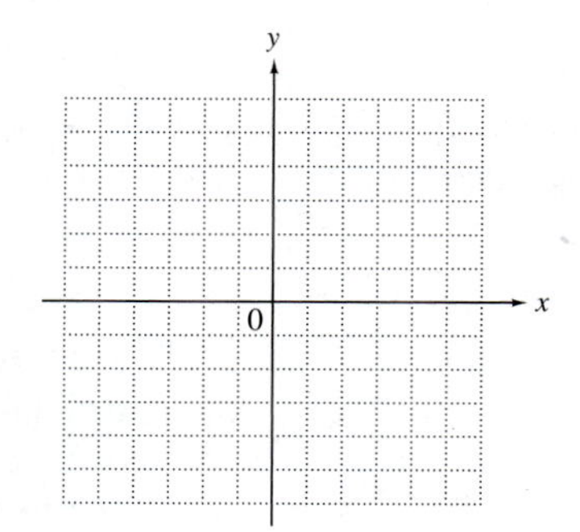

41. $x = 2$

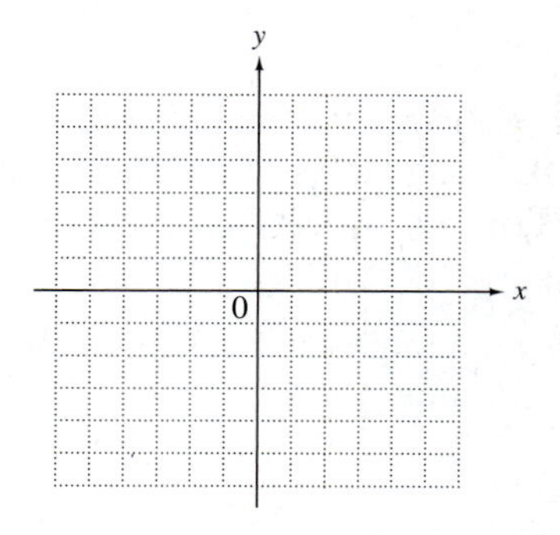

42. $x = -3$

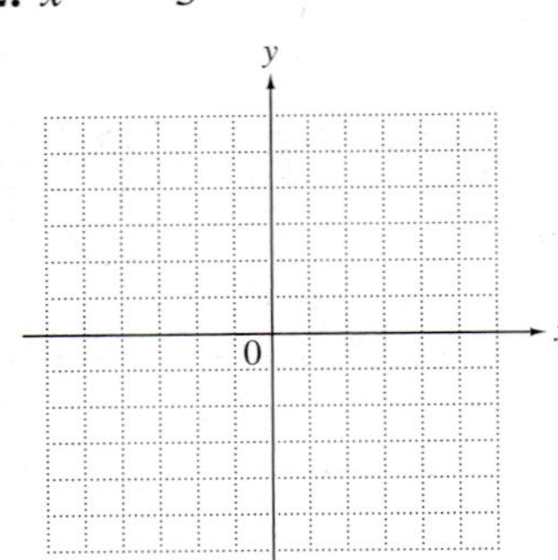

43. $x + 5y = 0$

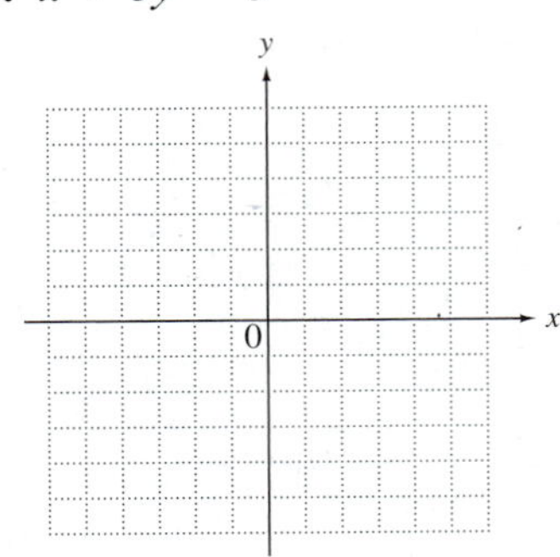

44. $x - 3y = 0$

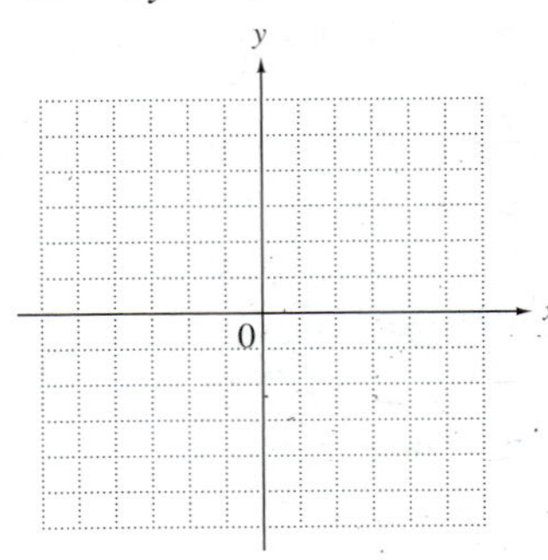

A linear equation can be used as a model to describe data in some cases. Exercises 45 and 46 are based on this idea.

45. Track qualifying records at North Carolina Motor Speedway from 1965 through 1998 are approximated by the linear equation

$$y = 1.22x + 118,$$

where y is the speed (in mph) in year x. In the model, $x = 0$ corresponds to 1965, $x = 10$ corresponds to 1975, and so on. Use the model, to approximate the speed of the 1995 winner, Hut Stricklin. (*Source:* NASCAR.)

46. According to information provided by Families USA Foundation, the national average annual family health care cost in dollars between 1980 and 2000 can be approximated by the linear model

$$y = 382.75x + 1742,$$

where $x = 0$ corresponds to 1980 and $x = 20$ corresponds to 2000. Based on this model, what was the expected national average health care cost in 2000?

RELATING CONCEPTS (Exercises 47–52) FOR INDIVIDUAL OR GROUP WORK

If the endpoints of a line segment are known, the coordinates of the midpoint of the segment can be found. The figure shows the coordinates of the points P and Q. Let PQ represent the line segment with endpoints at P and Q. To derive a formula for the midpoint of PQ, **work Exercises 47–52 in order.**

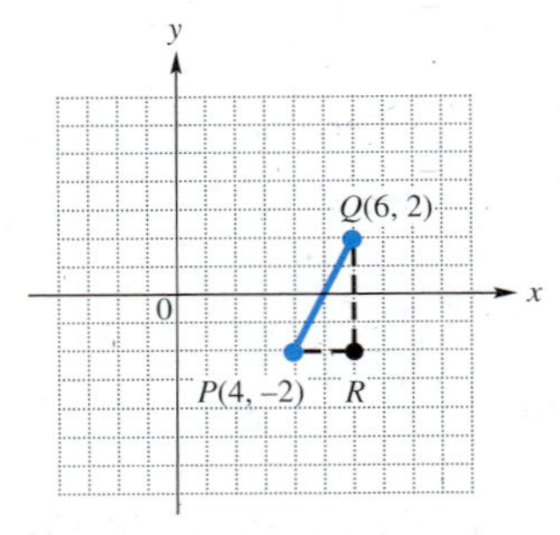

47. In the figure, R is the point with the same x-coordinate as Q and the same y-coordinate as P. Write the ordered pair that corresponds to R.

48. From the graph, determine the coordinates of the midpoint of PR.

49. From the graph, determine the coordinates of the midpoint of QR.

50. The x-coordinate of the midpoint M of PQ is the x-coordinate of the midpoint of PR and the y-coordinate is the y-coordinate of the midpoint of QR. Write the ordered pair that corresponds to M.

51. The average of two numbers is found by dividing their sum by 2. Find the average of the x-coordinates of points P and Q. Find the average of the y-coordinates of points P and Q.

52. Compare your answers to Exercises 50 and 51. What connection is there between the coordinates of P and Q and the coordinates of M?

4.2 Slope

Slope (steepness) is used in many practical ways. The slope of a highway (sometimes called the *grade*) is often given as a percent. For example, a 10% (or $\frac{10}{100} = \frac{1}{10}$) slope means the highway rises 1 unit for every 10 horizontal units. Stairs and roofs have slopes too, as shown in Figure 9.

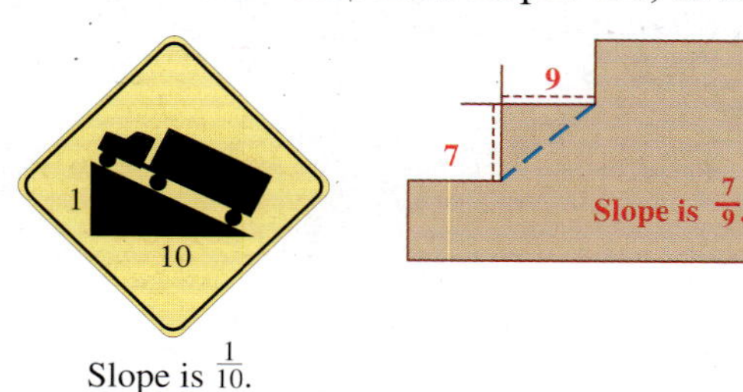

Slope is $\frac{1}{10}$.

Figure 9

In each example mentioned, slope is the ratio of vertical change, or **rise**, to horizontal change, or **run**. A simple way to remember this is to think "slope is rise over run."

OBJECTIVES

1. Find the slope of a line given two points on the line.
2. Find the slope of a line given an equation of the line.
3. Graph a line given its slope and a point on the line.
4. Use slopes to determine whether two lines are parallel, perpendicular, or neither.
5. Solve problems involving average rate of change.

1 Find the slope of a line given two points on the line. To get a formal definition of the slope of a line, we designate two different points on the line. To differentiate between the points, we write them as (x_1, y_1) and (x_2, y_2). See Figure 10. (The small numbers 1 and 2 in these ordered pairs are called *subscripts*. Read (x_1, y_1) as "x-sub-one, y-sub-one.")

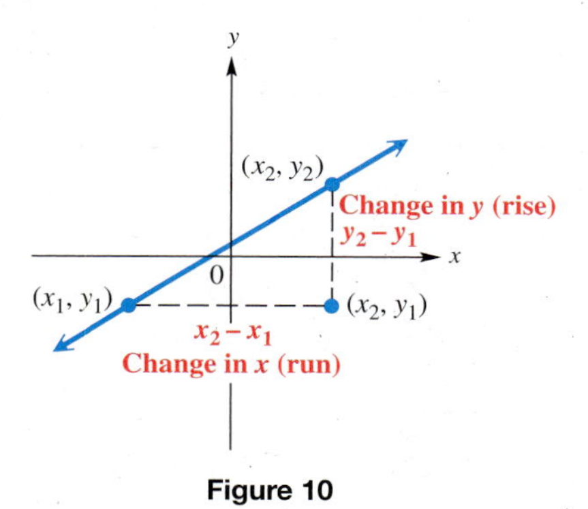

Figure 10

As we move along the line in Figure 10 from (x_1, y_1) to (x_2, y_2), the y-value changes (vertically) from y_1 to y_2, an amount equal to $y_2 - y_1$. As y changes from y_1 to y_2, the value of x changes (horizontally) from x_1 to x_2 by the amount $x_2 - x_1$. The ratio of the change in y to the change in x (the rise over the run) is called the **slope** of the line, with the letter m traditionally used for slope.

Slope Formula

The slope of the line through the distinct points (x_1, y_1) and (x_2, y_2) is

$$m = \frac{\text{rise}}{\text{run}} = \frac{\text{change in } y}{\text{change in } x} = \frac{y_2 - y_1}{x_2 - x_1} \quad (x_1 \neq x_2).$$

Work Problem **1** at the Side.

1 Use the information given for the walkway in the figure to find the following.

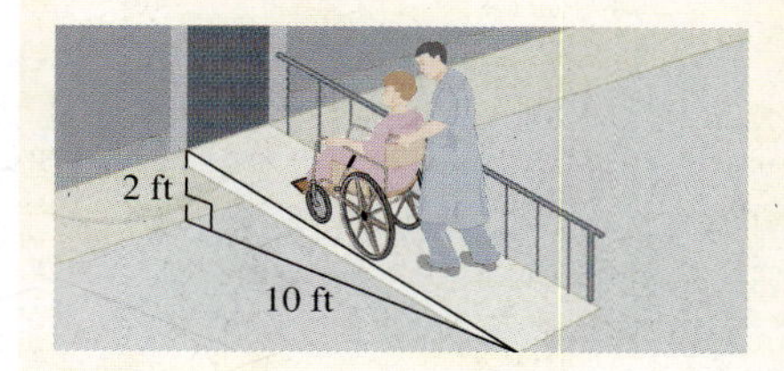

(a) The rise

(b) The run

(c) The slope

ANSWERS

1. (a) 2 ft **(b)** 10 ft **(c)** $\frac{2}{10}$ or $\frac{1}{5}$

2 Find the slope of the line through each pair of points.

(a) $(-2, 7), (4, -3)$

(b) $(1, 2), (8, 5)$

(c) $(8, -4), (3, -2)$

Example 1 Finding the Slope of a Line

Find the slope of the line through the points $(2, -1)$ and $(-5, 3)$.

If $(2, -1) = (x_1, y_1)$ and $(-5, 3) = (x_2, y_2)$, then

$$m = \frac{y_2 - y_1}{x_2 - x_1} = \frac{3 - (-1)}{-5 - 2} = \frac{4}{-7} = -\frac{4}{7}.$$

See Figure 11. On the other hand, if the pairs are reversed so that $(2, -1) = (x_2, y_2)$ and $(-5, 3) = (x_1, y_1)$, the slope is

$$m = \frac{-1 - 3}{2 - (-5)} = \frac{-4}{7} = -\frac{4}{7},$$

the same answer.

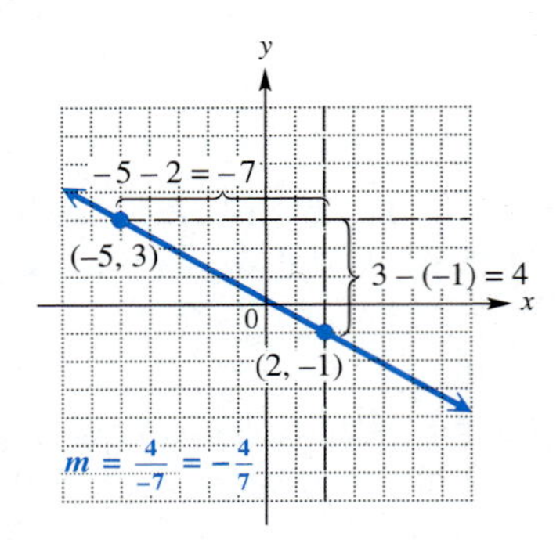

Figure 11

Example 1 suggests that the slope is the same no matter which point we consider first. Also, using similar triangles from geometry, we can show that the slope is the same no matter which two different points on the line we choose.

CAUTION

In calculating the slope, be careful to subtract the y-values and the x-values in the *same* order.

Correct		Incorrect	
$\frac{y_2 - y_1}{x_2 - x_1}$ or	$\frac{y_1 - y_2}{x_1 - x_2}$	$\frac{y_2 - y_1}{x_1 - x_2}$ or	$\frac{y_1 - y_2}{x_2 - x_1}$

Also, remember that the change in y is the *numerator* and the change in x is the *denominator*.

Work Problem 2 at the Side.

2 Find the slope of a line given an equation of the line. When an equation of a line is given, one way to find the slope is to use the definition of slope by first finding two different points on the line.

ANSWERS

2. (a) $-\frac{5}{3}$ (b) $\frac{3}{7}$ (c) $-\frac{2}{5}$

Example 2 **Finding the Slope of a Line**

Find the slope of the line $4x - y = 8$.

The intercepts can be used as the two different points needed to find the slope. Let $y = 0$ to find that the x-intercept is $(2, 0)$. Then let $x = 0$ to find that the y-intercept is $(0, -8)$. Use these two points in the slope formula. The slope is

$$m = \frac{\text{rise}}{\text{run}} = \frac{-8 - 0}{0 - 2} = \frac{-8}{-2} = 4.$$

Example 3 **Finding the Slopes of Vertical and Horizontal Lines**

Find the slope of each line.

(a) $x = -1$

As shown in Figure 7 (Section 4.1), the graph of $x = -1$ or $x + 1 = 0$ is a vertical line. Two points that satisfy the equation $x = -1$ are $(-1, 5)$ and $(-1, -4)$. Use these two points to find the slope.

$$m = \frac{\text{rise}}{\text{run}} = \frac{-4 - 5}{-1 - (-1)} = \frac{-9}{0}$$

Since division by 0 is undefined, the slope is undefined. This is why the definition of slope includes the restriction $x_1 \neq x_2$.

(b) $y = 2$

Figure 6 in Section 4.1 shows that the graph of $y = 2$ is a horizontal line. To find the slope, select two different points on the line, such as $(3, 2)$ and $(-1, 2)$, and use the slope formula.

$$m = \frac{\text{rise}}{\text{run}} = \frac{2 - 2}{3 - (-1)} = \frac{0}{4} = 0$$

In this case, the *rise* is 0, so the slope is 0.

Generalizing from Example 3, we can make the following statements about vertical and horizontal lines.

Slopes of Vertical and Horizontal Lines

The slope of a vertical line is undefined; the slope of a horizontal line is 0.

Work Problem 3 at the Side.

The slope of a line can also be found directly from its equation. Look again at the equation $4x - y = 8$ from Example 2. Solve this equation for y.

$4x - y = 8$	Equation from Example 2
$-y = -4x + 8$	Subtract $4x$.
$y = 4x - 8$	Multiply by -1.

Notice that the slope, 4, found using the slope formula in Example 2 is the same number as the coefficient of x in the equation $y = 4x - 8$. We will

3 Find the slope of each line.

(a) $2x + y = 6$

(b) $3x - 4y = 12$

(c) $x = -6$

(d) $y + 5 = 0$

ANSWERS

3. (a) -2 **(b)** $\frac{3}{4}$ **(c)** undefined **(d)** 0

see in the next section that this always happens, *as long as the equation is solved for y.*

4 Find the slope of the graph of $2x - 5y = 8$.

Example 4 **Finding the Slope from an Equation**

Find the slope of the graph of $3x - 5y = 8$.

Solve the equation for y.

$$3x - 5y = 8$$

$$-5y = -3x + 8 \quad \text{Subtract } 3x.$$

$$y = \frac{3}{5}x - \frac{8}{5} \quad \text{Divide by } -5.$$

The slope is given by the coefficient of x, so the slope is $\frac{3}{5}$.

Work Problem 4 at the Side.

3 **Graph a line given its slope and a point on the line.** Example 5 shows how to graph a straight line by using the slope and one point on the line.

Example 5 **Using the Slope and a Point to Graph Lines**

Graph each line.

(a) With slope $\frac{2}{3}$ through the point $(-1, 4)$

First locate the point $P(-1, 4)$ on a graph as shown in Figure 12. Then use the slope to find a second point. From the slope formula,

$$m = \frac{\text{change in } y}{\text{change in } x} = \frac{2}{3},$$

so move *up* 2 units and then 3 units to the *right* to locate another point on the graph (labeled R). The line through $(-1, 4)$ and R is the required graph.

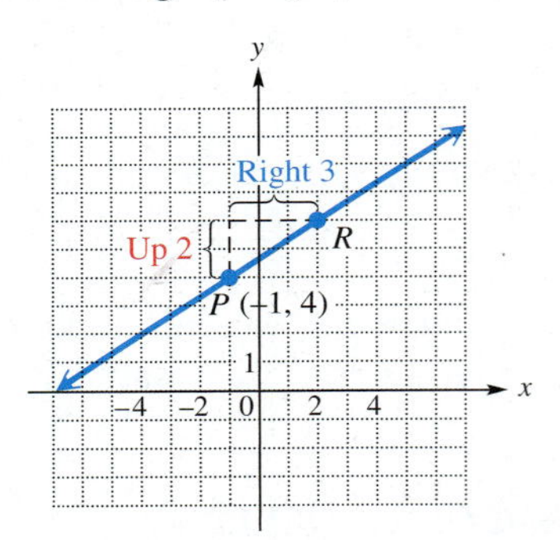

Figure 12

(b) Through $(3, 1)$ with slope -4

Start by locating the point $P(3, 1)$ on a graph. Find a second point R on the line by writing -4 as $\frac{-4}{1}$ and using the slope formula.

$$m = \frac{\text{change in } y}{\text{change in } x} = \frac{-4}{1}$$

Move *down* 4 units from $(3, 1)$, and then move 1 unit to the *right*. Draw a line through this second point R and $(3, 1)$, as shown in Figure 13.

Continued on Next Page

ANSWERS

4. $\frac{2}{5}$

The slope also could be written as

$$m = \frac{\text{change in } y}{\text{change in } x} = \frac{4}{-1}.$$

In this case the second point R is located *up* 4 units and 1 unit to the *left*. Verify that this approach produces the same line.

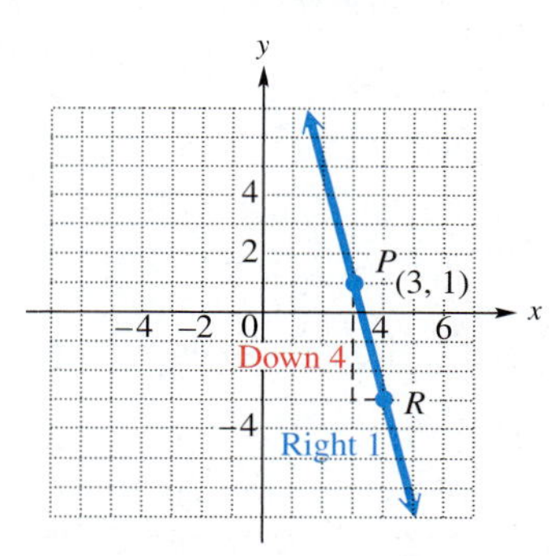

Figure 13

Work Problem 5 at the Side.

In Example 5(a), the slope of the line is the *positive* number $\frac{2}{3}$. The graph of the line in Figure 12 goes up (rises) from left to right. The line in Example 5(b) has a *negative* slope, −4. As Figure 13 shows, its graph goes down (falls) from left to right. These facts suggest the following generalization.

A positive slope indicates that the line goes *up* from left to right; a negative slope indicates that the line goes *down* from left to right.

Figure 14 shows lines of positive, 0, negative, and undefined slopes.

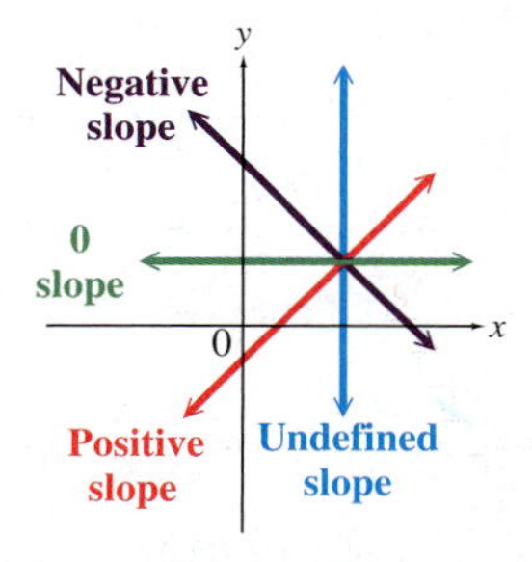

Figure 14

4 **Use slopes to determine whether two lines are parallel, perpendicular, or neither.** The slopes of a pair of parallel or perpendicular lines are related in a special way. The slope of a line measures the steepness of the line. Since parallel lines have equal steepness, their slopes must be equal; also, lines with the same slope are parallel.

Slopes of Parallel Lines

Two nonvertical lines with the same slope are parallel; two nonvertical parallel lines have the same slope.

5 Graph each line.

(a) Through (1, −3); $m = -\frac{3}{4}$

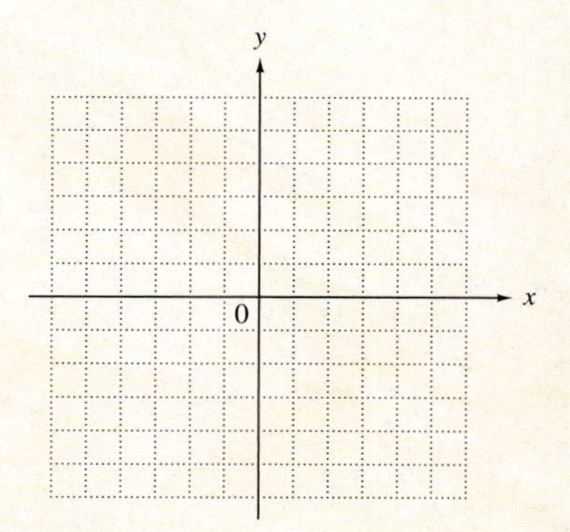

(b) Through (−1, −4); $m = 2$

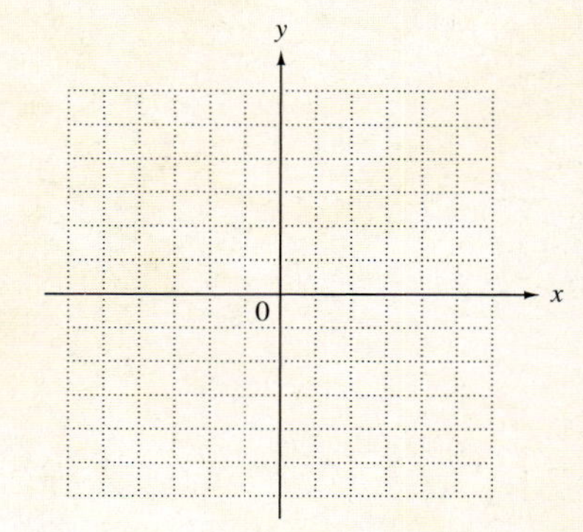

ANSWERS

5. (a)

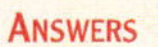

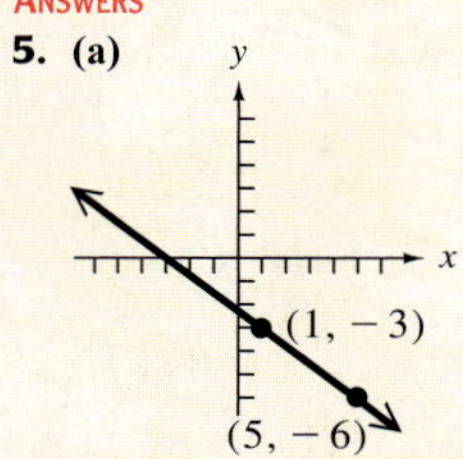

(b)

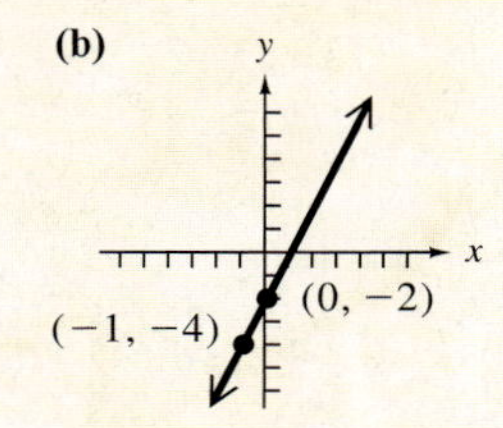

Example 6 **Determining whether Two Lines Are Parallel**

Are the lines L_1, through $(-2, 1)$ and $(4, 5)$, and L_2, through $(3, 0)$ and $(0, -2)$, parallel?

The slope of L_1 is

$$m_1 = \frac{5 - 1}{4 - (-2)} = \frac{4}{6} = \frac{2}{3}.$$

The slope of L_2 is

$$m_2 = \frac{-2 - 0}{0 - 3} = \frac{-2}{-3} = \frac{2}{3}.$$

Because the slopes are equal, the two lines are parallel.

To see how the slopes of perpendicular lines are related, consider a nonvertical line with slope $\frac{a}{b}$. If this line is rotated 90°, the vertical change and the horizontal change are reversed and the slope is $-\frac{b}{a}$, since the horizontal change is now negative. See Figure 15. Thus, the slopes of perpendicular lines have product -1 and are negative reciprocals of each other. For example, if the slopes of two lines are $\frac{3}{4}$ and $-\frac{4}{3}$, then the lines are perpendicular because $\frac{3}{4}(-\frac{4}{3}) = -1$.

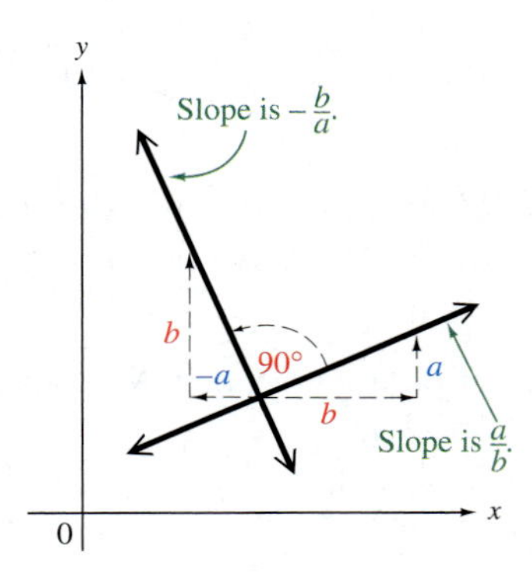

Figure 15

Slopes of Perpendicular Lines

If neither is vertical, perpendicular lines have slopes that are negative reciprocals; that is, their product is -1. Also, lines with slopes that are negative reciprocals are perpendicular.

Example 7 **Determining whether Two Lines Are Perpendicular**

Are the lines with equations $2y = 3x - 6$ and $2x + 3y = -6$ perpendicular?

Find the slope of each line by first solving each equation for y.

$$2y = 3x - 6$$
$$y = \frac{3}{2}x - 3$$

↑ Slope

$$2x + 3y = -6$$
$$3y = -2x - 6$$
$$y = -\frac{2}{3}x - 2$$

↑ Slope

Continued on Next Page

Since the product of the slopes of the two lines is $\frac{3}{2}(-\frac{2}{3}) = -1$, the lines are perpendicular.

NOTE

In Example 7, alternatively, we could have found the slope of each line by using intercepts and the slope formula.

Work Problem 6 at the Side.

5 Solve problems involving average rate of change. We know that the slope of a line is the ratio of the change in y (vertical) to the change in x (horizontal). This idea can be applied to real-life situations. The slope gives the average rate of change in y per unit of change in x, where the value of y depends on the value of x. The next example further illustrates this idea of average rate of change. We assume a linear relationship between x and y.

Example 8 Interpreting Slope as Average Rate of Change

The graph in Figure 16 approximates the percent of U.S. households owning multiple personal computers in the years 1997–2001. Find the average rate of change in percent per year.

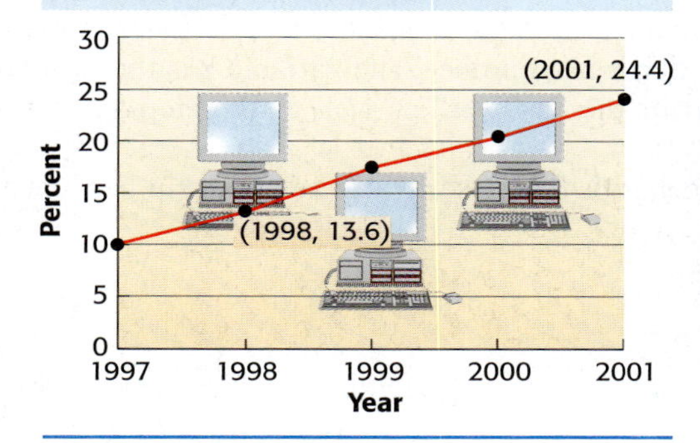

Source: The Yankee Group.

Figure 16

To use the slope formula, we need two pairs of data. From the graph, if $x = 1998$, then $y = 13.6$ and if $x = 2001$, then $y = 24.4$, so we have the ordered pairs (1998, 13.6) and (2001, 24.4). By the slope formula,

$$\text{average rate of change} = \frac{y_2 - y_1}{x_2 - x_1} = \frac{24.4 - 13.6}{2001 - 1998} = \frac{10.8}{3} = 3.6.$$

This means that the number of U.S. households owning multiple computers *increased* by 3.6% each year in the period from 1997 to 2001.

Work Problem 7 at the Side.

6 Write *parallel, perpendicular,* or *neither* for each pair of two distinct lines.

(a) The line through $(-1, 2)$ and $(3, 5)$ and the line through $(4, 7)$ and $(8, 10)$

(b) The line through $(5, -9)$ and $(3, 7)$ and the line through $(0, 2)$ and $(8, 3)$

(c) $2x - y = 4$ and $2x + y = 6$

(d) $3x + 5y = 6$ and $5x - 3y = 2$

7 Use the ordered pairs (1997, 10) and (2000, 20.8), which are plotted in Figure 16, to find the average rate of change. How does it compare to the average rate of change found in Example 8?

ANSWERS

6. **(a)** parallel **(b)** perpendicular **(c)** neither **(d)** perpendicular

7. 3.6; It is the same.

Focus on Real-Data Applications

"Ins and Outs" of Algebraic Expressions

An algebraic expression such as $3x + 2$ represents both a *process* (what to do) and a *concept* (the result). To illustrate this idea, choose a number. Multiply by 3, then add 2. What is your answer? Now choose a different number and repeat this procedure. Did you get the same answer?

The examples in the table imitate the activity that you just finished. Compare the examples to the generalization.

	Example 1	Example 2	Example 3	Generalization
Choose a number.	5	-3.1	$\frac{2}{5}$	x
Multiply by 3.	$3 \times 5 = 15$	$-3.1 \times 3 = -9.3$	$\frac{2}{5} \times 3 = \frac{6}{5}$	$3x$
Add 2.	$3 \times 5 + 2 = 15 + 2$	$-3.1 \times 3 + 2 = -9.3 + 2$	$\frac{2}{5} \times 3 + 2 = \frac{6}{5} + 2$	$3x + 2$
What is the answer?	17	-7.3	$\frac{16}{5}$	$3x + 2$

As these examples illustrate, $3x + 2$ represents both the arithmetic procedure and the answer. Remember that $3x + 2$ is a number in the same way that x is a number. The variable x is the **input** and the expression $3x + 2$ represents the **output** or the "answer" as well as the step-by-step process for calculating the output.

Algebraic expressions are useful in real-life scenarios that describe the relationship between two variables, the input and the output.

For Group Discussion

For each scenario, define the input and the output, and write an expression that represents the relationship between those two variables.

1. R. Conniff claims that a hummingbird's heart beats about 1200 times per minute in flight. (*Source: Smithsonian.*)

What is the input?

What is the output?

Write an expression for the number of times that a hummingbird's heart beats during x minutes of flight.

2. D. Webster claims that malaria kills one child every 30 seconds. (*Source: Smithsonian.*)

What is the input?

What is the output?

Write an expression for the number of children who die of malaria after x seconds.

4.2 EXERCISES

FOR EXTRA HELP

Student's Solutions Manual | MyMathLab.com | InterAct Math Tutorial Software | AW Math Tutor Center | 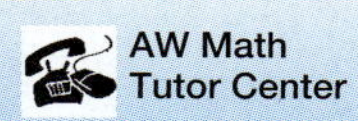 www.mathxl.com | Digital Video Tutor CD 2 Videotape 4

1. A ski slope drops 30 ft for every horizontal 100 ft.

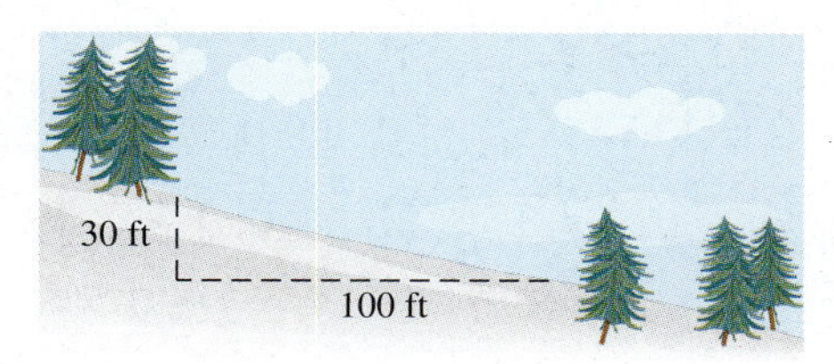

Which of the following express its slope? (There are several correct choices.)

A. $-.3$ **B.** $-\frac{3}{10}$ **C.** $-3\frac{1}{3}$ **D.** $-\frac{30}{100}$ **E.** $-\frac{10}{3}$

2. A hill has a slope of $-.05$. How many feet in the vertical direction correspond to a run of 50 ft?

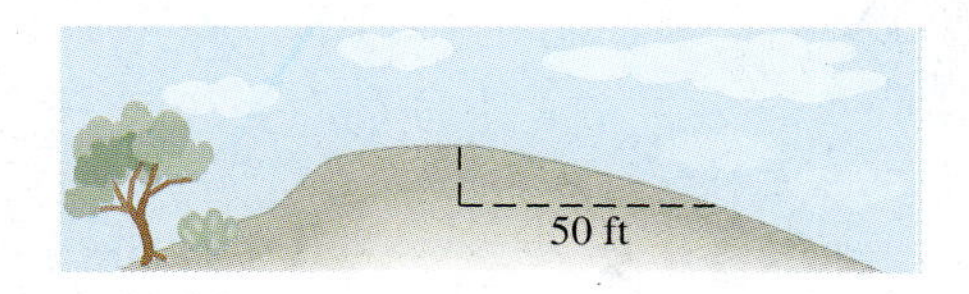

Use the given figure to determine the slope of the line segment described, by counting the number of units of "rise," the number of units of "run," and then finding the quotient.

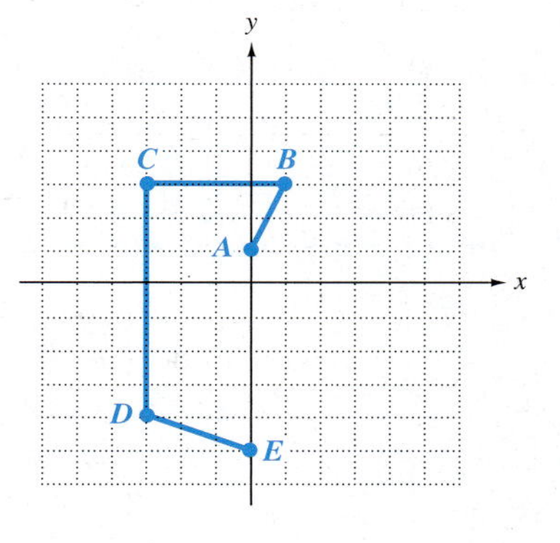

3. AB **4.** BC **5.** CD **6.** DE

Find each slope using the slope formula.

7. $m = \frac{6 - 2}{5 - 3}$ **8.** $m = \frac{5 - 7}{-4 - 2}$ **9.** $m = \frac{4 - (-1)}{-3 - (-5)}$

10. $m = \frac{-6 - 0}{0 - (-3)}$ **11.** $m = \frac{-5 - (-5)}{3 - 2}$ **12.** $m = \frac{7 - (-2)}{-3 - (-3)}$

Find the slope of the line through each pair of points using the slope formula. See Example 1.

13. $(-2, -3)$ and $(-1, 5)$ **14.** $(-4, 3)$ and $(-3, -4)$ **15.** $(-4, 1)$ and $(2, 6)$

16. $(-3, -3)$ and $(5, 6)$ **17.** $(2, 4)$ and $(-4, 4)$ **18.** $(-6, 3)$ and $(2, 3)$

Find the slope of each line.

19.

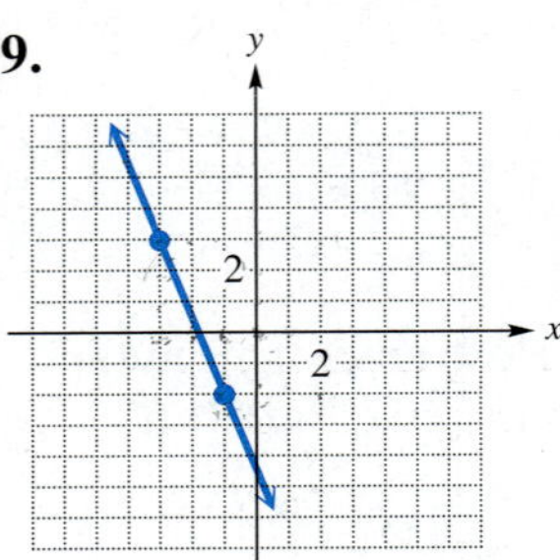

20.

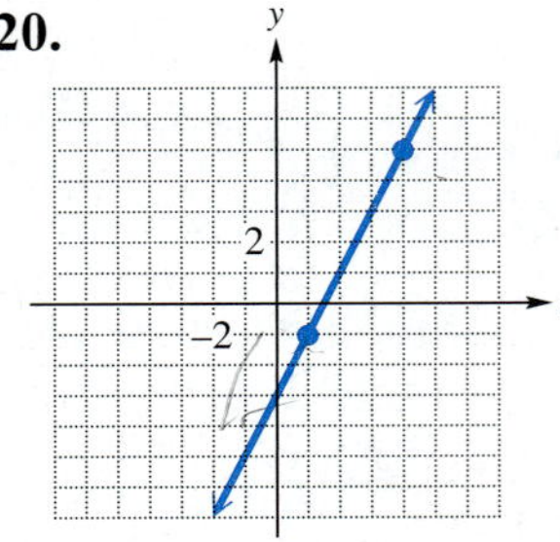

21.

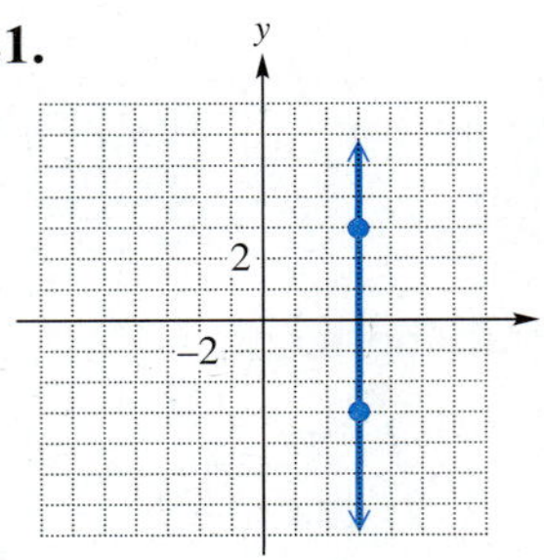

22.

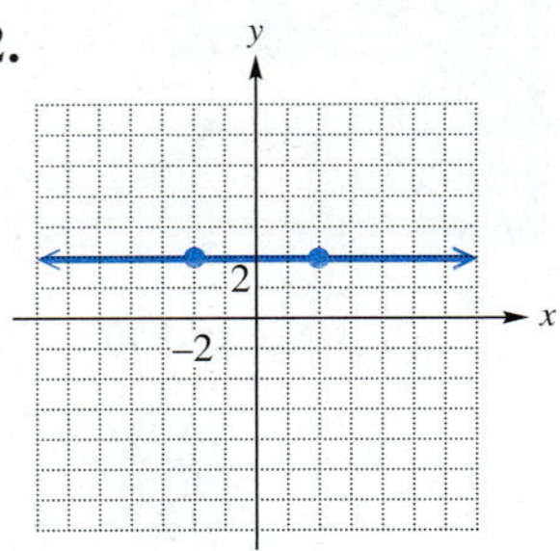

Based on the figure shown here, determine which line satisfies the given description.

23. The line has positive slope.

24. The line has negative slope.

25. The line has slope 0.

26. The line has undefined slope.

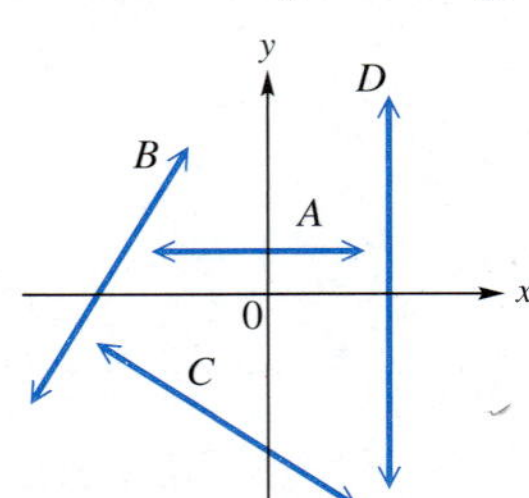

Find the slope of each line, and sketch the graph. See Examples 2–4.

27. $x + 2y = 4$

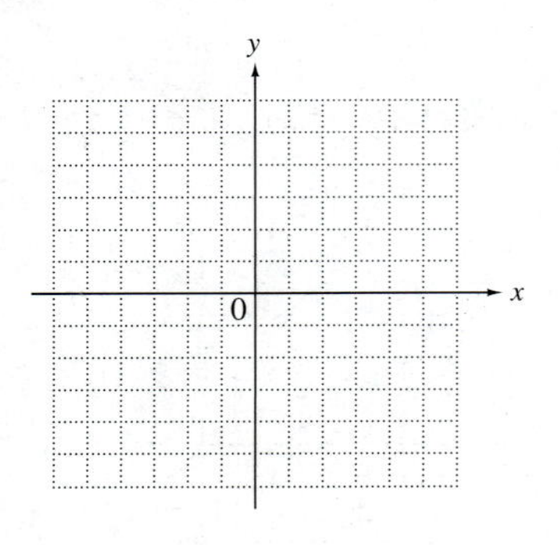

28. $x + 3y = -6$

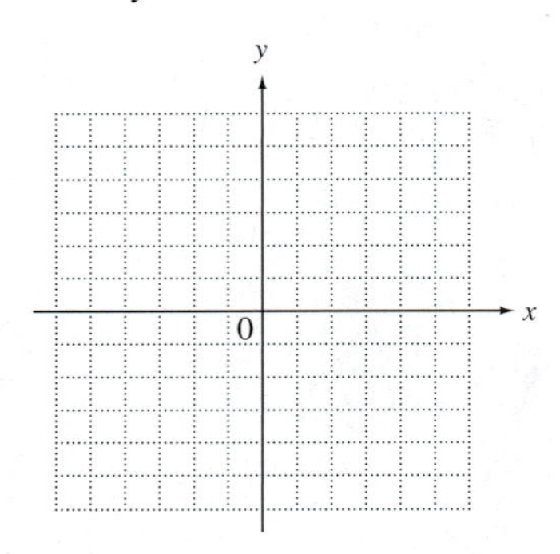

29. $-x + y = 4$

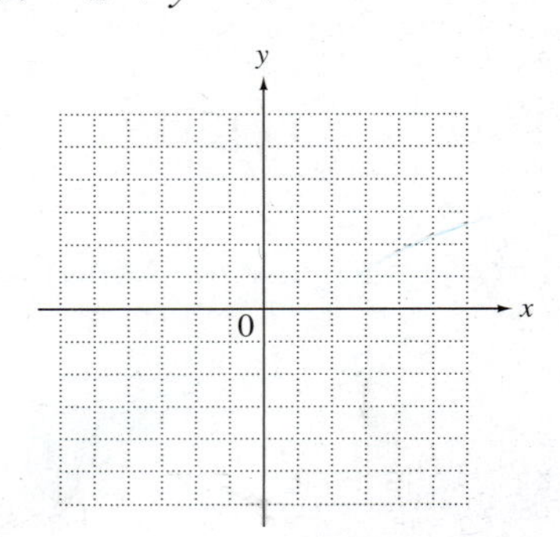

30. $-x + y = 6$

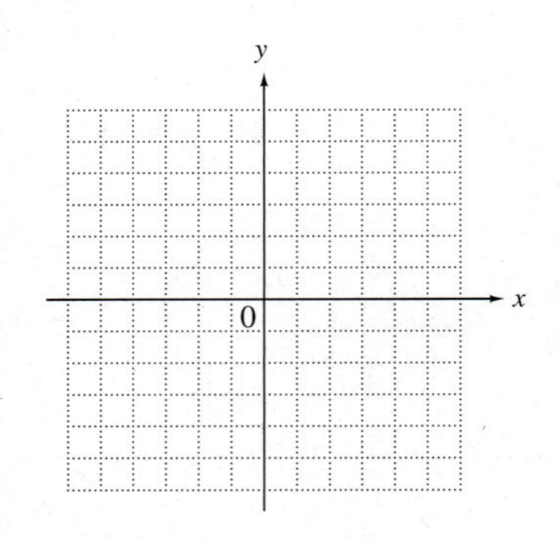

31. $6x + 5y = 30$

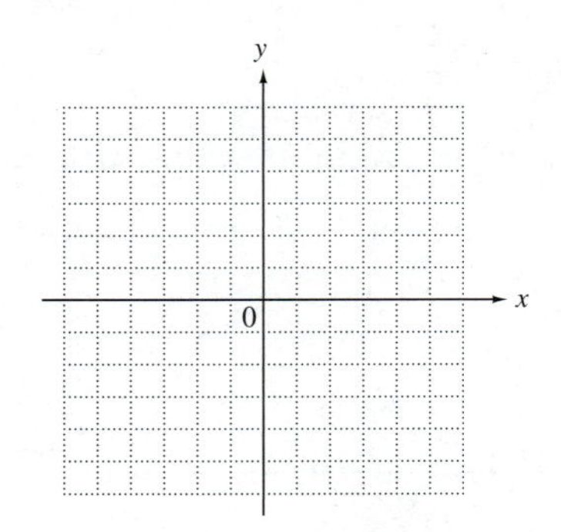

32. $3x + 4y = 12$

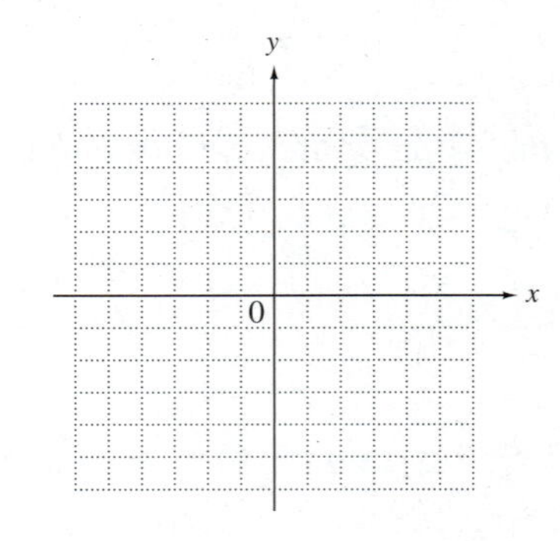

33. $x + 2 = 0$

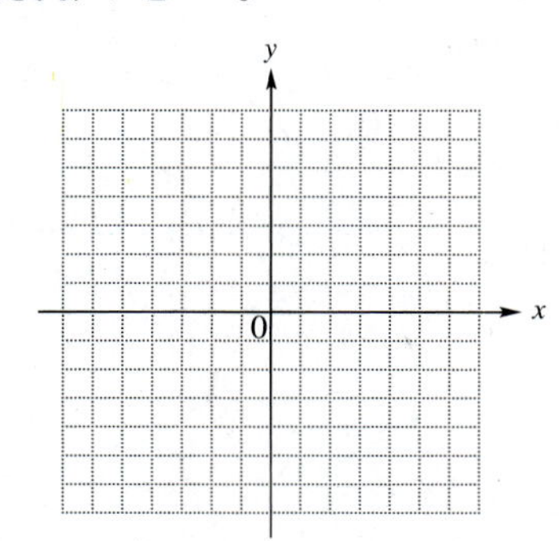

34. $x - 4 = 0$

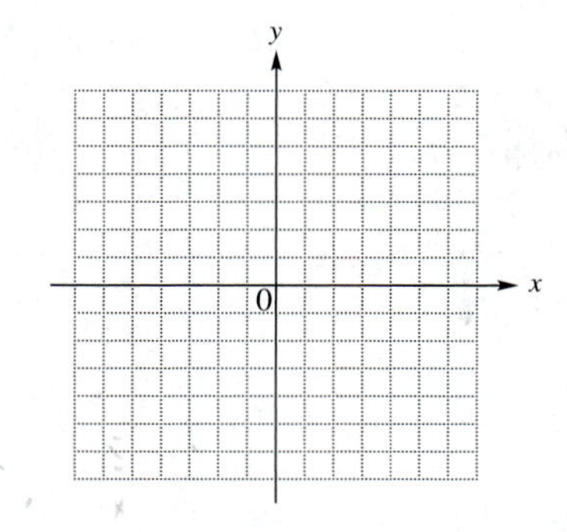

35. $y = 4x$

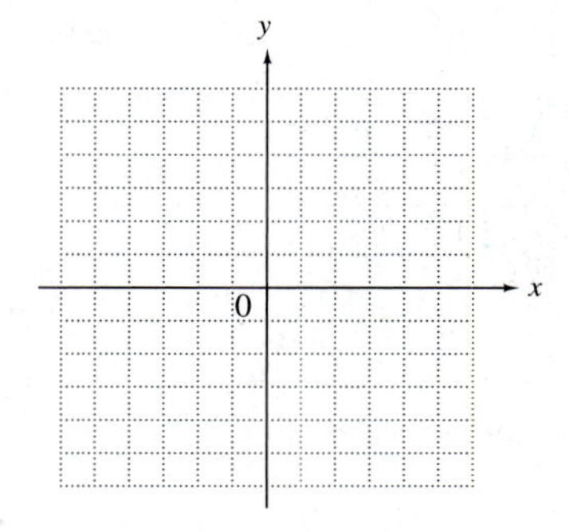

36. $y = -3x$

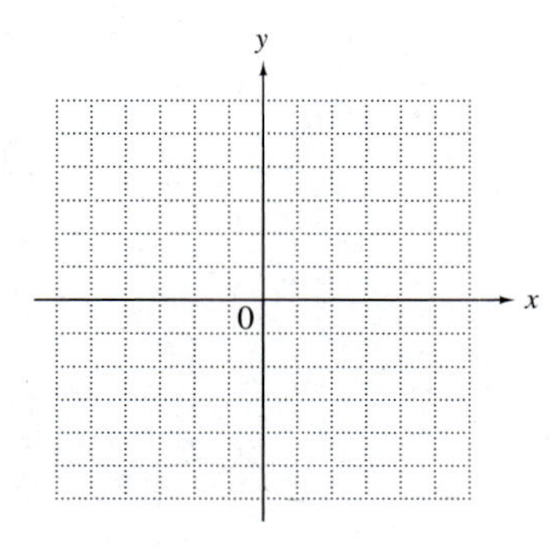

37. $y - 3 = 0$

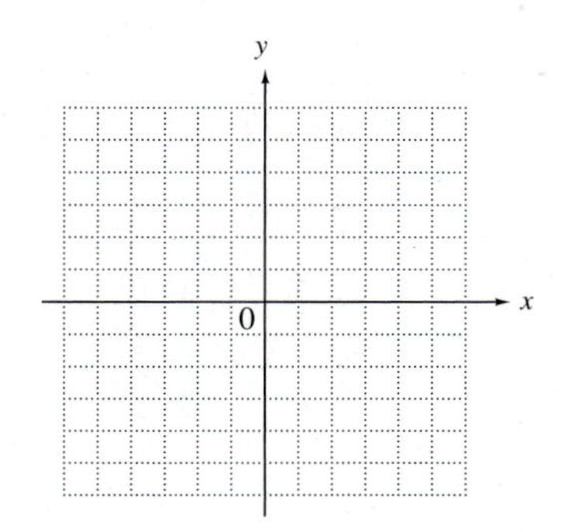

38. $y + 5 = 0$

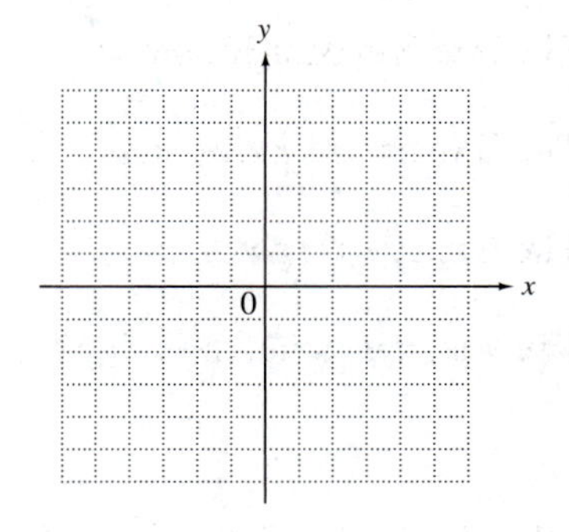

Use the method shown in Example 5 to graph each line.

39. Through $(-4, 2)$; $m = \frac{1}{2}$

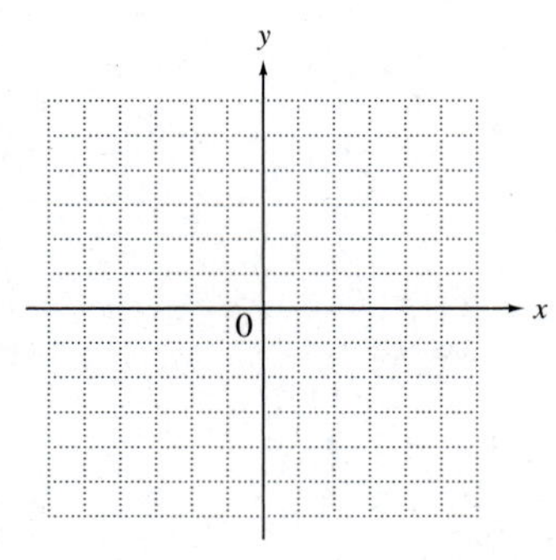

40. Through $(-2, -3)$; $m = \frac{5}{4}$

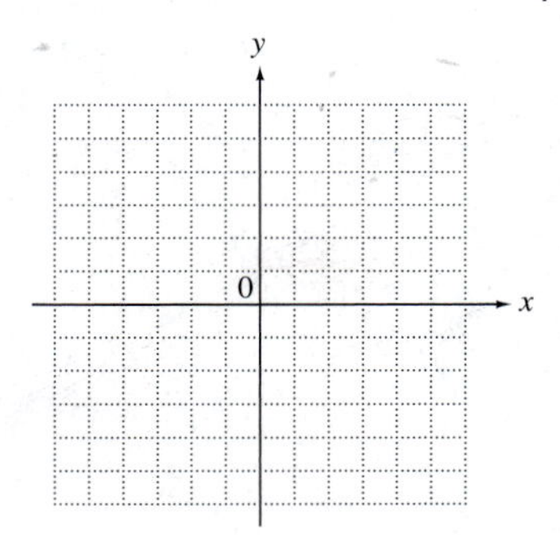

41. Through $(0, -2)$; $m = -\frac{2}{3}$

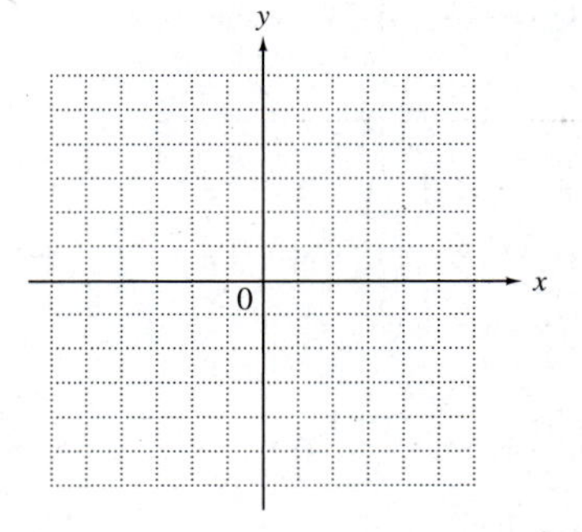

42. Through $(0, -4)$; $m = -\frac{3}{2}$

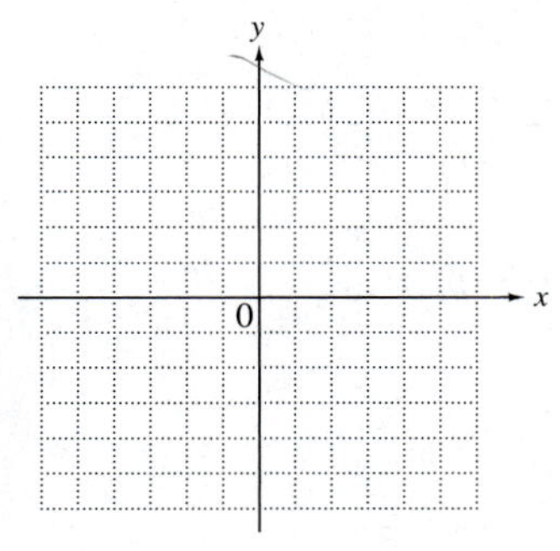

43. Through $(-1, -2)$; $m = 3$

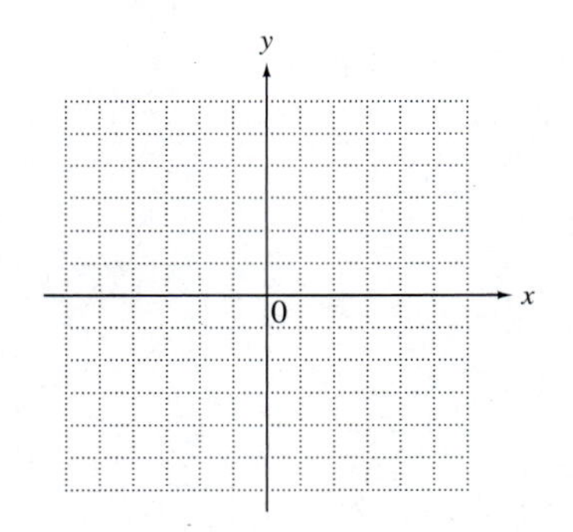

44. Through $(-2, -4)$; $m = 4$

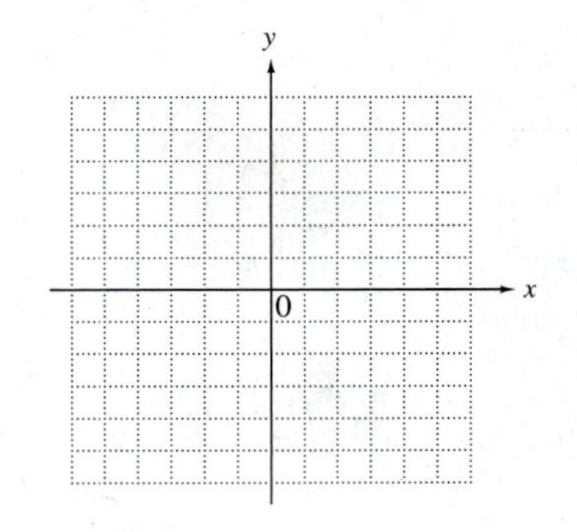

Decide whether each pair of lines is parallel, perpendicular, *or* neither. *See Examples 6 and 7.*

45. $2x + 5y = -7$ and $5x - 2y = 1$

46. $x + 4y = 7$ and $4x - y = 3$

47. The line through $(4, 6)$ and $(-8, 7)$ and the line through $(-5, 5)$ and $(7, 4)$

48. The line through $(15, 9)$ and $(12, -7)$ and the line through $(8, -4)$ and $(5, -20)$

49. $2x + y = 6$ and $x - y = 4$

50. $4x - 3y = 6$ and $3x - 4y = 2$

51. $3x = y$ and $2y - 6x = 5$

52. $x = 6$ and $6 - x = 8$

53. $2x + 5y = -8$ and $6 + 2x = 5y$

54. $4x + y = 0$ and $5x - 8 = 2y$

55. $4x - 3y = 8$ and $4y + 3x = 12$

56. $2x = y + 3$ and $2y + x = 3$

Use the concept of slope to solve each problem.

57. The upper deck at Comiskey Park in Chicago has produced, among other complaints, displeasure with its steepness. It's been compared to a ski jump. It is 160 ft from home plate to the front of the upper deck and 250 ft from home plate to the back. The top of the upper deck is 63 ft above the bottom. What is its slope?

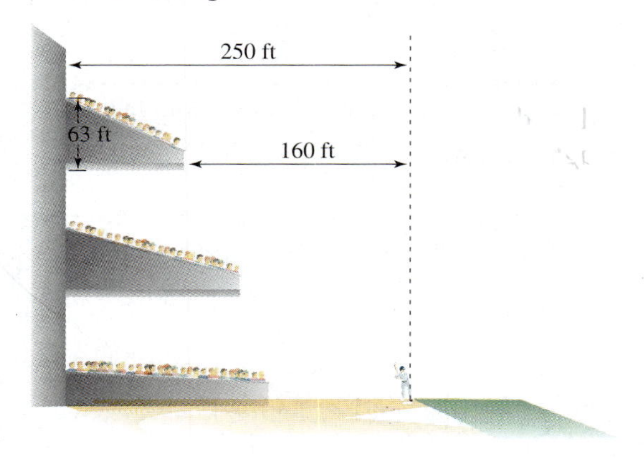

58. When designing the new FleetCenter arena in Boston to replace the old Boston Garden, architects were careful to design the ramps leading up to the entrances so that circus elephants would be able to march up the ramps. The maximum grade (or slope) that an elephant will walk on is 13%. Suppose that such a ramp was constructed with a horizontal run of 150 ft. What would be the maximum vertical rise the architects could use?

Use the idea of average rate of change to solve each problem. See Example 8.

59. Merck pharmaceutical company research and development expenditures (in millions of dollars) in recent years are closely approximated by the graph.

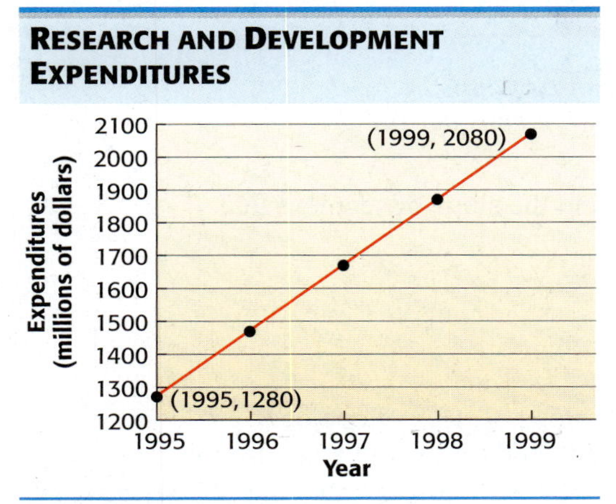

Source: Merck & Co., Inc. 1999 Annual Report.

(a) Use the given ordered pairs to determine the average rate of change in these expenditures per year.

(b) Explain how a positive rate of change is interpreted in this situation.

60. The graph provides a good approximation of the number of food stamp recipients (in millions) during 1994–1998.

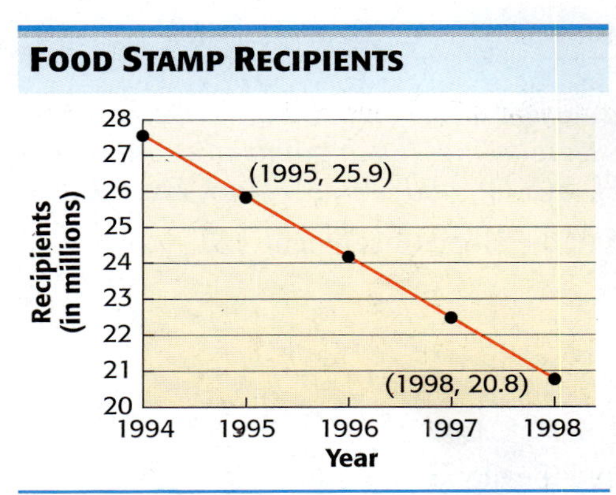

Source: U.S. Bureau of the Census.

(a) Use the given ordered pairs to find the average rate of change in food stamp recipients per year during this period.

(b) Interpret what a negative slope means in this situation.

61. The table gives book publishers' approximate net dollar sales (in millions) from 1995 through 2000.

BOOK PUBLISHERS' SALES

Year	Sales (in millions)
1995	19,000
1996	20,000
1997	21,000
1998	22,000
1999	23,000
2000	24,000

Source: Book Industry Study Group.

(a) Find the average rate of change for 1995–1996, 1995–1999, and 1998–2000. What do you notice about your answers? What does this tell you?

(b) Calculate the rates of change in part (a) as percents. What do you notice?

62. The table gives the number of cellular telephone subscribers (in thousands) from 1994 through 1999.

CELLULAR TELEPHONE SUBSCRIBERS

Year	Subscribers (in thousands)
1994	24,134
1995	33,786
1996	44,043
1997	55,312
1998	69,209
1999	86,047

Source: Cellular Telecommunications Industry Association, Washington, D.C. *State of the Cellular Industry* (Annual).

(a) Find the average rate of change in subscribers for 1994–1995, 1995–1996, and so on.

(b) Is the average rate of change in successive years approximately the same? If the ordered pairs in the table were plotted, could an approximately straight line be drawn through them?

RELATING CONCEPTS (Exercises 63–68) FOR INDIVIDUAL OR GROUP WORK

In these exercises we investigate a method of determining whether three points lie on the same straight line. (Such points are said to be **collinear.***) The points we consider are* $A(3, 1)$, $B(6, 2)$, *and* $C(9, 3)$. ***Work Exercises 63–68 in order.***

63. Find the slope of segment AB.

64. Find the slope of segment BC.

65. Find the slope of segment AC.

66. If slope of AB = slope of BC = slope of AC, then A, B, and C are collinear. Use the results of Exercises 63–65 to show that this statement is satisfied.

67. Use the slope formula to determine whether the points $(1, -2)$, $(3, -1)$, and $(5, 0)$ are collinear.

68. Repeat Exercise 67 for the points $(0, 6)$, $(4, -5)$, and $(-2, 12)$.

4.3 Linear Equations in Two Variables

Objectives

1. Write the equation of a line given its slope and a point on the line.
2. Write the equation of a line given two points on the line.
3. Write the equation of a line given its slope and y-intercept.
4. Find the slope and y-intercept of a line given its equation.
5. Write the equation of a line parallel or perpendicular to a given line.
6. Apply concepts of linear equations to real data.

Many real-world situations can be described by straight-line graphs. This section shows how to write a linear equation that satisfies given conditions.

1 Write the equation of a line given its slope and a point on the line. A straight line is a set of points in the plane such that the slope of the line between any two of the points is the same. In Figure 17, point P is on the line through P_1 and P_2 if the slope of the line through points P_1 and P equals the slope of the line through points P and P_2.

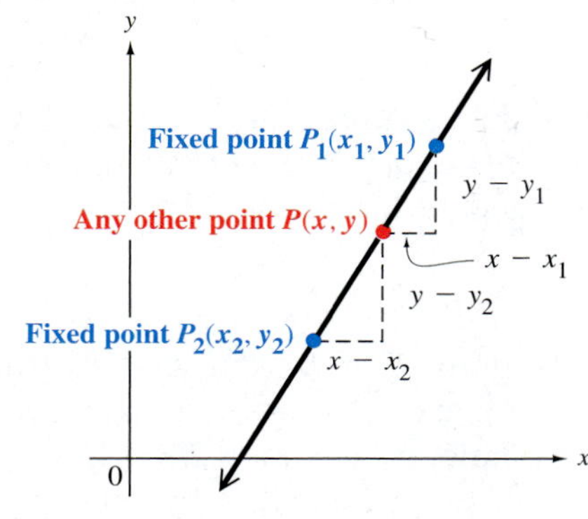

Figure 17

To write an equation for the line through the point (x, y), given the slope m and one other point on the line, say (x_1, y_1), we use the slope formula as follows.

$$\frac{y - y_1}{x - x_1} = m$$

$$y - y_1 = m(x - x_1) \quad \text{Multiply each side by } x - x_1.$$

This last equation is the *point-slope form* of the equation of the line. To use this form to write the equation of a line, we need to know the coordinates of a point (x_1, y_1) and the slope of the line, m.

Point-Slope Form

The **point-slope form** of the equation of a line is

Slope ↓

$$y - y_1 = m(x - x_1).$$

↑ Given point ↑

Example 1 Using the Point-Slope Form

Find an equation of the line with slope $\frac{1}{3}$ going through the point $(-2, 5)$.

Use the point-slope form of the equation of a line, with $(x_1, y_1) = (-2, 5)$ and $m = \frac{1}{3}$.

Continued on Next Page

$$y - y_1 = m(x - x_1) \quad \text{Point-slope form}$$

$$y - 5 = \frac{1}{3}[x - (-2)] \quad y_1 = 5, m = \tfrac{1}{3}, x_1 = -2$$

$$y - 5 = \frac{1}{3}(x + 2)$$

$$3y - 15 = x + 2 \quad \text{Multiply by 3.}$$

$$-x + 3y = 17 \quad \text{Subtract } x\text{; add 15.}$$

In Section 4.1, we defined *standard form* for a linear equation as

$$Ax + By = C.$$

In addition, from now on, let us agree that A, B, and C will be integers with no common factor (except 1) and $A \geq 0$. For example, the final equation in Example 1 is written in standard form as $x - 3y = -17$.

NOTE

The definition of "standard form" is not standard from one text to another. Any linear equation can be written in many different (all equally correct) forms. For example, the equation $2x + 3y = 8$ can be written as $2x = 8 - 3y$, $3y = 8 - 2x$, $x + \frac{3}{2}y = 4$, $4x + 6y = 16$, and so on. In addition to writing it in the form $Ax + By = C$ with $A \geq 0$, let us agree that the form $2x + 3y = 8$ is preferred over any multiples of each side, such as $4x + 6y = 16$. (To write $4x + 6y = 16$ in standard form, divide each side by 2.)

Work Problem 1 at the Side.

2 Write the equation of a line given two points on the line. To find an equation of a line when two points on the line are known, first use the slope formula to find the slope of the line. Then use the slope with either of the given points and the point-slope form of the equation of a line.

Example 2 Finding an Equation of a Line Given Two Points

Find an equation of the line through the points $(-4, 3)$ and $(5, -7)$.

First find the slope by using the slope formula.

$$m = \frac{-7 - 3}{5 - (-4)} = -\frac{10}{9}$$

Use either $(-4, 3)$ or $(5, -7)$ as (x_1, y_1) in the point-slope form of the equation of a line. If you choose $(-4, 3)$, then $-4 = x_1$ and $3 = y_1$.

$$y - y_1 = m(x - x_1) \quad \text{Point-slope form}$$

$$y - 3 = -\frac{10}{9}[x - (-4)] \quad y_1 = 3, m = -\tfrac{10}{9}, x_1 = -4$$

$$y - 3 = -\frac{10}{9}(x + 4)$$

$$9y - 27 = -10x - 40 \quad \text{Multiply by 9.}$$

$$10x + 9y = -13 \quad \text{Standard form}$$

Verify that if $(5, -7)$ were used, the same equation would result.

Work Problem 2 at the Side.

---

1 Write an equation of each line in standard form.

(a) Through $(-2, 7)$; $m = 3$

(b) Through $(1, 3)$; $m = -\frac{5}{4}$

2 Write an equation in standard form for each line.

(a) Through $(-1, 2)$ and $(5, 7)$

(b) Through $(-2, 6)$ and $(1, 4)$

ANSWERS

1. **(a)** $3x - y = -13$ **(b)** $5x + 4y = 17$

2. **(a)** $5x - 6y = -17$ **(b)** $2x + 3y = 14$

Notice that the point-slope form does not apply to a vertical line, since the slope of a vertical line is undefined. A vertical line through the point (c, d) has equation $x = c$.

A horizontal line has slope 0. From the point-slope form, the equation of a horizontal line through the point (c, d) is

$$y - y_1 = m(x - x_1)$$
$$y - d = 0(x - c) \qquad y_1 = d,\ m = 0,\ x_1 = c$$
$$y - d = 0$$
$$y = d.$$

In summary, horizontal and vertical lines have the following special equations.

Equations of Vertical and Horizontal Lines

The vertical line through (c, d) has equation $x = c$.
The horizontal line through (c, d) has equation $y = d$.

Work Problem 3 at the Side.

3 Write the equation of a line given its slope and *y*-intercept. Suppose a line has slope m and y-intercept $(0, b)$. Using the point-slope form, the equation of the line is

$$y - y_1 = m(x - x_1)$$
$$y - b = m(x - 0) \qquad x_1 = 0,\ y_1 = b$$
$$y = mx + b. \qquad \text{Add } b.$$

When the equation is solved for y, the coefficient of x is the slope, m, and the constant b is the y-value of the y-intercept. Because this form of the equation shows the slope and the y-intercept, it is called the *slope-intercept form.*

Slope-Intercept Form

The equation of a line with slope m and y-intercept $(0, b)$ is written in **slope-intercept form** as

$$y = mx + b.$$

↑ Slope ↑ y-intercept is $(0, b)$.

Example 3 Using the Slope-Intercept Form to Find the Equation of a Line

Find an equation of the line with slope $-\frac{4}{5}$ and y-intercept $(0, -2)$.

Here $m = -\frac{4}{5}$ and $b = -2$. Substitute these values into the slope-intercept form.

$$y = mx + b \qquad \text{Slope-intercept form}$$
$$y = -\frac{4}{5}x - 2 \qquad m = -\frac{4}{5};\ b = -2$$

Work Problem 4 at the Side.

3 Write an equation for each line.

(a) Through $(8, -2)$; $m = 0$

(b) The vertical line through $(3, 5)$

4 Write an equation in slope-intercept form for each line with the given slope and y-intercept.

(a) Slope 2; y-intercept $(0, -3)$

(b) Slope $-\frac{2}{3}$; y-intercept $(0, 0)$

(c) Slope 0; y-intercept $(0, 3)$

ANSWERS

3. **(a)** $y = -2$ **(b)** $x = 3$

4. **(a)** $y = 2x - 3$ **(b)** $y = -\frac{2}{3}x$ **(c)** $y = 3$

5 Find the slope and the y-intercept of each line.

(a) $x + y = 2$

(b) $2x - 5y = 1$

4 **Find the slope and y-intercept of a line given its equation.** If the equation of a line is written in slope-intercept form, the coefficient of x is the slope and the constant leads to the y-intercept.

Example 4 Find the Slope and y-Intercept from the Equation

Find the slope and y-intercept of the graph of $3y + 2x = 9$.

Write the equation in slope-intercept form by solving for y.

$$3y + 2x = 9$$
$$3y = -2x + 9$$
$$y = -\frac{2}{3}x + 3 \quad \text{Slope-intercept form}$$

Slope ↑ (the coefficient $-\frac{2}{3}$); ↑ y-intercept is $(0, 3)$.

From the slope-intercept form, the slope is $-\frac{2}{3}$ and the y-intercept is $(0, 3)$.

NOTE

The slope-intercept form of a linear equation is the most useful for several reasons. Every linear equation (of a nonvertical line) has a *unique* (one and only one) slope-intercept form. In Section 4.5 we will study *linear functions,* which are defined by the slope-intercept form. Also, this is the form we must use when graphing a line with a graphing calculator.

Work Problem 5 at the Side.

5 **Write the equation of a line parallel or perpendicular to a given line.** As mentioned in the previous section, parallel lines have the same slope and perpendicular lines have slopes with product -1.

Example 5 Finding Equations of Parallel or Perpendicular Lines

Find the equation in slope-intercept form of the line passing through the point $(-4, 5)$ and **(a)** parallel to the line $2x + 3y = 6$; **(b)** perpendicular to the line $2x + 3y = 6$.

(a) The slope of the line $2x + 3y = 6$ can be found by solving for y.

$$2x + 3y = 6$$
$$3y = -2x + 6 \quad \text{Subtract } 2x.$$
$$y = -\frac{2}{3}x + 2 \quad \text{Divide by 3.}$$

└ Slope

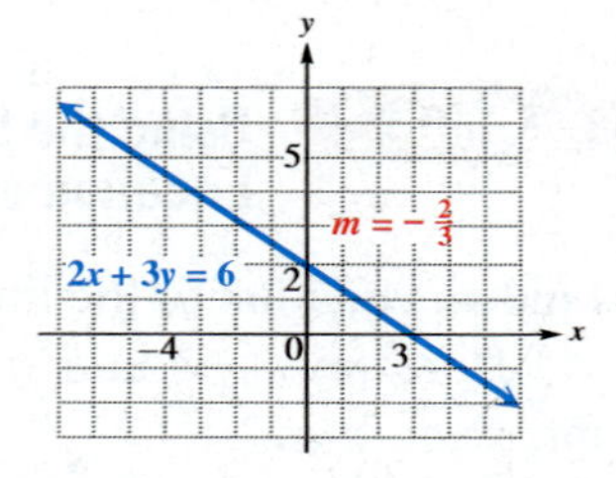

The slope is given by the coefficient of x, so $m = -\frac{2}{3}$. See the figure. The required equation of the line through $(-4, 5)$ and parallel to $2x + 3y = 6$ must also have slope $-\frac{2}{3}$. To find this equation, use the point-slope form, with $(x_1, y_1) = (-4, 5)$ and $m = -\frac{2}{3}$.

Continued on Next Page

ANSWERS

5. (a) -1; $(0, 2)$ (b) $\frac{2}{5}$; $\left(0, -\frac{1}{5}\right)$

$$y - 5 = -\frac{2}{3}[x - (-4)] \qquad y_1 = 5, m = -\tfrac{2}{3}, x_1 = -4$$

$$y - 5 = -\frac{2}{3}(x + 4)$$

$$y - 5 = -\frac{2}{3}x - \frac{8}{3} \qquad \text{Distributive property}$$

$$y = -\frac{2}{3}x - \frac{8}{3} + \frac{15}{3} \qquad \text{Add } 5 = \tfrac{15}{3}.$$

$$y = -\frac{2}{3}x + \frac{7}{3} \qquad \text{Combine like terms.}$$

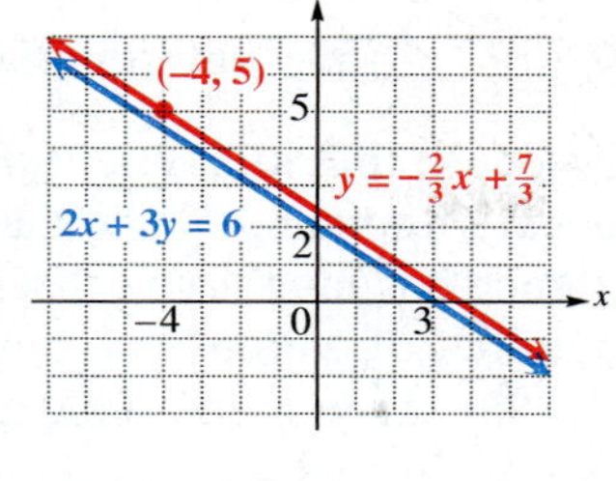

We did not clear fractions after the substitution step here because we want the equation in slope-intercept form—that is, solved for y. Both lines are shown in the figure.

(b) To be perpendicular to the line $2x + 3y = 6$, a line must have a slope that is the negative reciprocal of $-\frac{2}{3}$, which is $\frac{3}{2}$. Use the point $(-4, 5)$ and slope $\frac{3}{2}$ in the point-slope form to get the equation of the perpendicular line shown in the figure.

$$y - 5 = \frac{3}{2}[x - (-4)] \qquad y_1 = 5, m = \tfrac{3}{2}, x_1 = -4$$

$$y - 5 = \frac{3}{2}(x + 4)$$

$$y - 5 = \frac{3}{2}x + 6 \qquad \text{Distributive property}$$

$$y = \frac{3}{2}x + 11 \qquad \text{Add 5.}$$

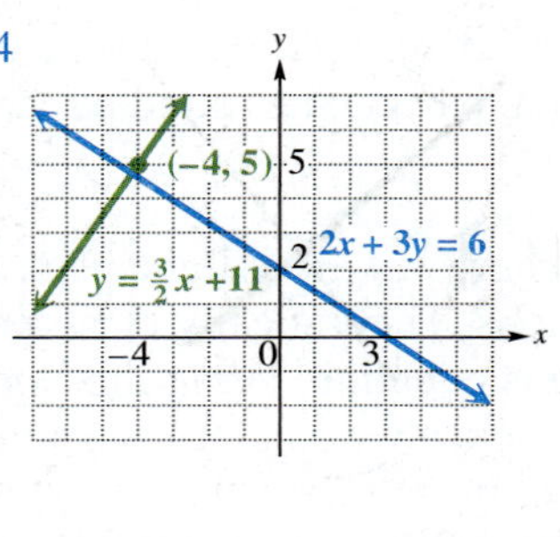

Work Problem 6 at the Side.

A summary of the various forms of linear equations follows.

Summary of Forms of Linear Equations

$y - y_1 = m(x - x_1)$	**Point-Slope Form** Slope is m. Line passes through (x_1, y_1).
$y = mx + b$	**Slope-Intercept Form** Slope is m. y-intercept is $(0, b)$.
$Ax + By = C$	**Standard Form** $(A \geq 0)$ Slope is $-\frac{A}{B}$ $(B \neq 0)$. x-intercept is $(\frac{C}{A}, 0)$ $(A \neq 0)$. y-intercept is $(0, \frac{C}{B})$ $(B \neq 0)$.
$x = c$	**Vertical Line** Slope is undefined. x-intercept is $(c, 0)$.
$y = d$	**Horizontal Line** Slope is 0. y-intercept is $(0, d)$.

6 Write an equation in slope-intercept form of the line passing through the point $(-8, 3)$ and

(a) parallel to the line $2x - 3y = 10$.

(b) perpendicular to the line $2x - 3y = 10$.

ANSWERS

6. **(a)** $y = \frac{2}{3}x + \frac{25}{3}$ **(b)** $y = -\frac{3}{2}x - 9$

7 Suppose it costs \$.10 per minute to make a long-distance call. Write an equation to describe the cost y to make an x-minute call.

6 Apply concepts of linear equations to real data.

Example 6 Determining a Linear Equation to Describe Real Data

Suppose that it is time to fill up your car with gasoline. You drive into your local station and notice that 89-octane gas is selling for \$1.60 per gal. Experience has taught you that the final price you pay can be determined by the number of gallons you buy multiplied by the price per gallon (in this case, \$1.60). As you pump the gas you observe two sets of numbers spinning by: one is the number of gallons you have pumped, and the other is the price you pay for that number of gallons.

The table uses ordered pairs to illustrate this situation.

Number of Gallons Pumped	Price for This Number of Gallons
0	0(\$1.60) = \$0.00
1	1(\$1.60) = \$1.60
2	2(\$1.60) = \$3.20
3	3(\$1.60) = \$4.80
4	4(\$1.60) = \$6.40

8 Suppose there is a flat rate of \$.20 plus a charge of \$.10 per minute to make a call.

(a) Write an equation that gives the cost y for a call of x minutes.

If we let x denote the number of gallons pumped, then the price y that you pay can be found by the linear equation

$$y = 1.60x,$$

where y is in dollars. This is a simple, realistic application of linear equations. Theoretically, there are infinitely many ordered pairs (x, y) that satisfy this equation, but in this application we are limited to nonnegative values for x, since we cannot have a negative number of gallons. There is also a practical maximum value for x in this situation, which varies from one car to another. What do you think determines this maximum value?

Work Problem 7 at the Side.

(b) Interpret the ordered pair (15, 1.7) in relation to the equation from part (a).

In Example 6, the ordered pair (0, 0) satisfied the equation, so the equation has the form $y = mx$, where $b = 0$. If a realistic situation involves an initial charge plus a charge per unit, the equation will have the form $y = mx + b$, where $b \neq 0$.

Example 7 Determining an Equation and Interpreting Ordered Pairs That Satisfy It

Suppose that you can get a car wash at the gas station in Example 6 if you pay an additional \$3.00.

(a) Write an equation that defines the price you will pay.

Since an additional \$3.00 will be charged, you will pay $1.60x + 3.00$ dollars for x gallons of gas and a car wash. Thus, if y represents the price, the equation is $y = 1.6x + 3$. (We delete the unnecessary 0s.)

(b) Interpret the ordered pairs (5, 11) and (10, 19) in relation to the equation from part (a).

The ordered pair (5, 11) indicates that the price of 5 gal of gas and a car wash is \$11.00. Similarly, (10, 19) indicates that the price of 10 gal of gas and a car wash is \$19.00.

Work Problem 8 at the Side.

ANSWERS

7. $y = .1x$

8. **(a)** $y = .1x + .2$ **(b)** The ordered pair (15, 1.7) indicates that the price of a 15-min call is \$1.70.

4.3 EXERCISES

FOR EXTRA HELP

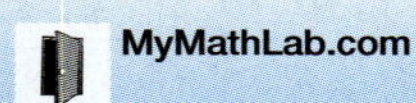
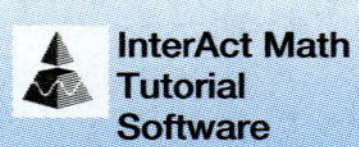

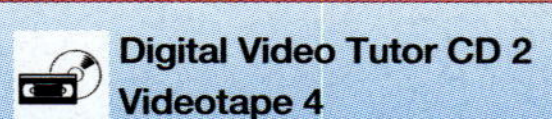

Student's Solutions Manual | MyMathLab.com | InterAct Math Tutorial Software | AW Math Tutor Center | www.mathxl.com | Digital Video Tutor CD 2 Videotape 4

1. The following equations all represent the same line. Which one is in standard form as defined in the text?

A. $3x - 2y = 5$ **B.** $2y = 3x - 5$

C. $\frac{3}{5}x - \frac{2}{5}y = 1$ **D.** $3x = 2y + 5$

2. Which equation is in point-slope form?

A. $y = 6x + 2$ **B.** $4x + y = 9$

C. $y - 3 = 2(x - 1)$ **D.** $2y = 3x - 7$

3. Which equation in Exercise 2 is in slope-intercept form?

4. Write the equation $y + 2 = -3(x - 4)$ in slope-intercept form.

5. Write the equation from Exercise 4 in standard form.

6. Write the equation $10x - 7y = 70$ in slope-intercept form.

Match each equation with the graph that it most closely resembles. (Hint: Determining the signs of m and b will help you make your decision.)

A.
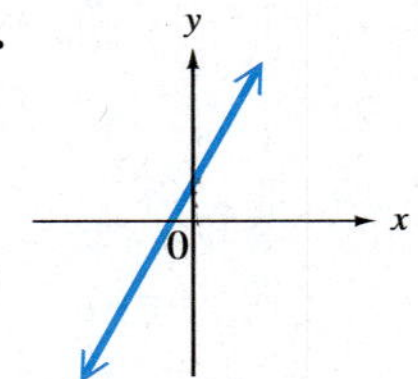

B.
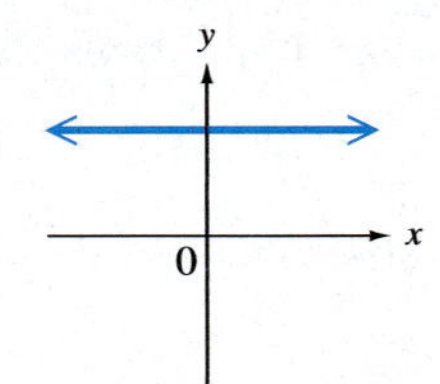

C.
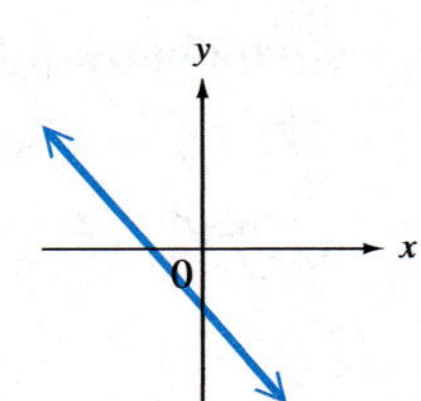

D.
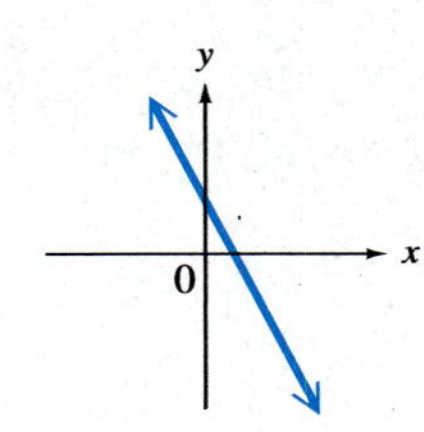

E.
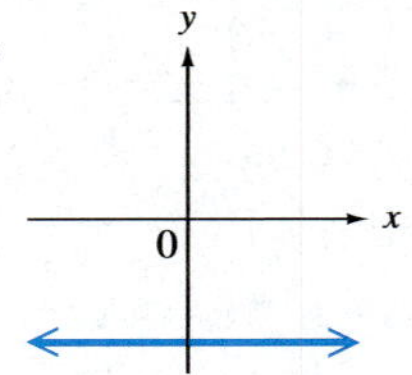

F.
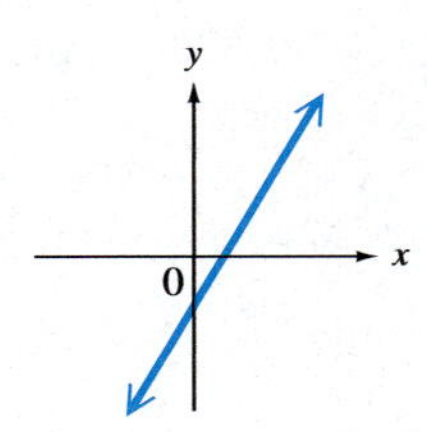

G.
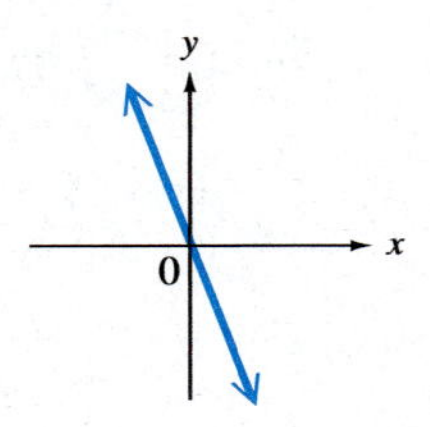

H.
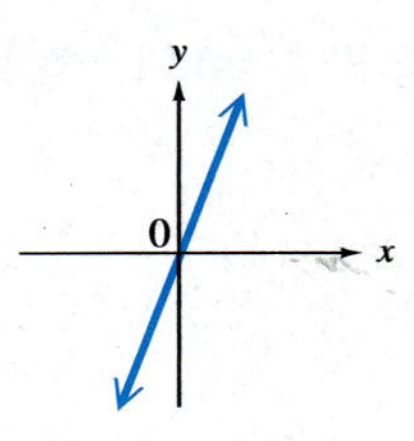

7. $y = 2x + 3$ **8.** $y = -2x + 3$ **9.** $y = -2x - 3$ **10.** $y = 2x - 3$

11. $y = 2x$ **12.** $y = -2x$ **13.** $y = 3$ **14.** $y = -3$

Write the equation in standard form of the line satisfying the given conditions. See Example 1.

15. Through $(-2, 4)$; $m = -\frac{3}{4}$

16. Through $(-1, 6)$; $m = -\frac{5}{6}$

17. Through $(5, 8)$; $m = -2$

18. Through $(12, 10)$; $m = 1$

19. Through $(-5, 4)$; $m = \frac{1}{2}$

20. Through $(7, -2)$; $m = \frac{1}{4}$

21. Through $(-4, 12)$; horizontal

22. Through $(1, 5)$; horizontal

Write an equation that satisfies the given conditions.

23. Through $(9, 10)$; undefined slope

24. Through $(-2, 8)$; 0 slope

25. Through $(.5, .2)$; horizontal

26. Through $\left(\frac{5}{8}, \frac{2}{9}\right)$; vertical

Write the equation in standard form of the line through the given points. See Example 2.

27. $(3, 4)$ and $(5, 8)$

28. $(5, -2)$ and $(-3, 14)$

29. $(6, 1)$ and $(-2, 5)$

30. $(-2, 5)$ and $(-8, 1)$

31. $\left(-\frac{2}{5}, \frac{2}{5}\right)$ and $\left(\frac{4}{3}, \frac{2}{3}\right)$

32. $\left(\frac{3}{4}, \frac{8}{3}\right)$ and $\left(\frac{2}{5}, \frac{2}{3}\right)$

33. $(2, 5)$ and $(1, 5)$

34. $(-2, 2)$ and $(4, 2)$

35. $(7, 6)$ and $(7, -8)$

36. $(13, 5)$ and $(13, -1)$

Write the equation in slope-intercept form of the line satisfying the given conditions. See Example 3.

37. $m = 5; b = 15$

38. $m = -2; b = 12$

39. $m = -\frac{2}{3}$; through $\left(0, \frac{4}{5}\right)$

40. $m = -\frac{5}{8}$; through $\left(0, -\frac{1}{3}\right)$

41. Slope $\frac{2}{5}$; y-intercept $(0, 5)$

42. Slope $-\frac{3}{4}$; y-intercept $(0, 7)$

Write each equation in slope-intercept form; then give the slope of the line and the y-intercept. See Example 4.

	Equation	Slope	y-Intercept
43. $5x + 2y = 20$	__________	__________	__________
44. $6x + 5y = 40$	__________	__________	__________
45. $2x - 3y = 10$	__________	__________	__________
46. $4x - 3y = 7$	__________	__________	__________

Write the equation in slope-intercept form of the line satisfying the given conditions. See Example 5.

47. Through $(7, 2)$; parallel to $3x - y = 8$

48. Through $(4, 1)$; parallel to $2x + 5y = 10$

49. Through $(-2, -2)$; parallel to $-x + 2y = 10$

50. Through $(-1, 3)$; parallel to $-x + 3y = 12$

51. Through $(8, 5)$; perpendicular to $2x - y = 7$

52. Through $(2, -7)$; perpendicular to $5x + 2y = 18$

53. Through $(-2, 7)$; perpendicular to $x = 9$

54. Through $(8, 4)$; perpendicular to $x = -3$

Write an equation in the form $y = mx$ for each situation. Then give the three ordered pairs associated with the equation for x-values of 0, 5, *and* 10. *See Example 6.*

55. x represents the number of hours traveling at 45 mph, and y represents the distance traveled (in miles).

56. x represents the number of compact discs sold at $16 each, and y represents the total cost of the discs (in dollars).

57. x represents the number of gallons of gas sold at $1.50 per gal, and y represents the total cost of the gasoline (in dollars).

58. x represents the number of days a videocassette is rented at $2.50 per day, and y represents the total charge for the rental (in dollars).

For each situation, **(a)** *write an equation in the form $y = mx + b$;* **(b)** *find and interpret the ordered pair associated with the equation for $x = 5$; and* **(c)** *answer the question. See Example 7.*

59. A membership to the Midwest Athletic Club costs $99 plus $39 per month. (*Source:* Midwest Athletic Club.) Let x represent the number of months selected. How much does the first year's membership cost?

60. For a family membership, the athletic club in Exercise 59 charges a membership fee of $159 plus $60 for each additional family member after the first. Let x represent the number of additional family members. What is the membership fee for a four-person family?

61. A rental car costs $50 plus $.20 per mile. Let x represent the number of miles driven, so y represents the total charge to the renter. How many miles was the car driven if the renter paid $84.60?

62. There is a $30 fee to rent a chain saw, plus $6 per day. Let x represent the number of days the saw is rented and y represent the charge to the user in dollars. If the total charge is $138, for how many days is the saw rented?

In Exercises 63 and 64, assume that the situation described in the figure can be modeled by a straight-line graph, and use the information to find the $y = mx + b$ form of the equation of the line.

63. The number of post offices in the United States has been declining in recent years. Use the information given on the bar graph for the years 1994 and 1998, letting $x = 0$ represent the year 1990 and y represent the number of post offices.

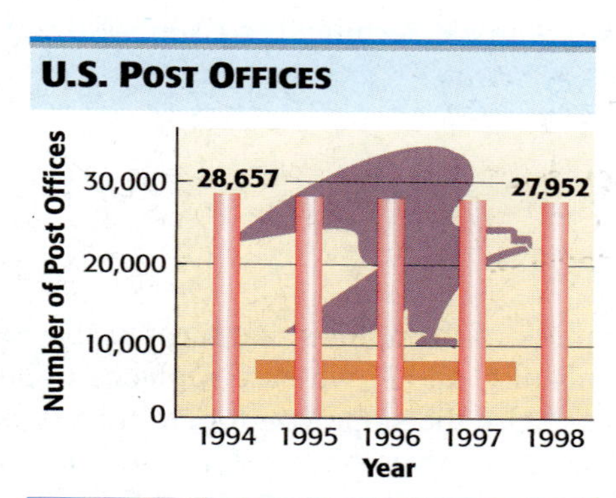

Source: U.S. Postal Service, *Annual Report of the Postmaster General.*

64. When selecting a room air conditioner, it is important to match its cooling capacity (in Btu*) to the size of the room to be cooled. The graph shows the recommended Btu size for selected room sizes in square feet. Use the information given for rooms of 150 ft^2 and 1400 ft^2. Let x represent the number of square feet and y represent the corresponding Btu size air conditioner.

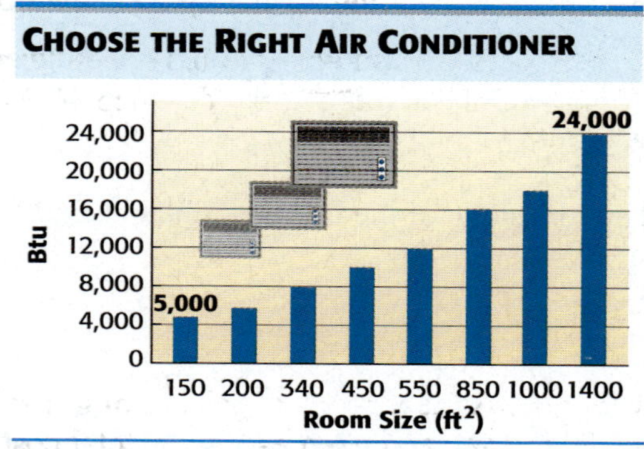

Source: Carey, Morris and James, *Home Improvement for Dummies,* IDG Books.

Solve each problem. In part (a), give equations in slope-intercept form.

65. Median household income of African Americans increased in recent years, as shown in the bar graph.

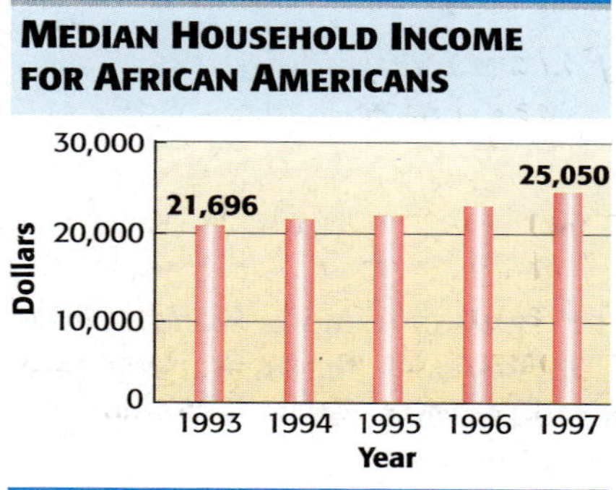

Source: U.S. Bureau of the Census.

(a) Use the information given for the years 1993 and 1997, letting $x = 3$ represent 1993, $x = 7$ represent 1997, and y represent the median income, to write an equation that models median household income.

(b) Use the equation to approximate the median income for 1995. How does your result compare to the actual value, \$23,583?

66. The bar graph shows median household income for Hispanics.

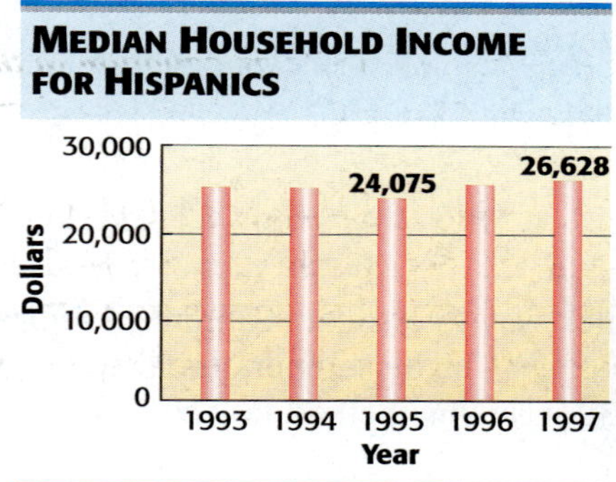

Source: U.S. Bureau of the Census.

(a) Use the information for the years 1995 and 1997 to write an equation. Let $x = 5$ represent 1995, $x = 7$ represent 1997, and y represent the median income.

(b) Looking at the graph, would you expect the equation from part (a) to give good approximations for 1993 and 1994 incomes? Would the equation give a reasonable approximation for 1998? Explain.

---

* British thermal units

RELATING CONCEPTS (Exercises 67–76) FOR INDIVIDUAL OR GROUP WORK

In Section 2.2 we learned how formulas can be applied to problem solving. In Exercises 67–76, we will see how the formula that relates Celsius and Fahrenheit temperatures is derived. ***Work Exercises 67–76 in order.***

67. There is a linear relationship between Celsius and Fahrenheit temperatures. When $C = 0°$, $F =$ ________°, and when $C = 100°$, $F =$ ________°.

68. Think of ordered pairs of temperatures (C, F), where C and F represent corresponding Celsius and Fahrenheit temperatures. The equation that relates the two scales has a straight-line graph that contains the two points determined in Exercise 67. What are these two points?

69. Find the slope of the line described in Exercise 68.

70. Now think of the point-slope form of the equation in terms of C and F, where C replaces x and F replaces y. Use the slope you found in Exercise 69 and one of the two points determined earlier, and find the equation that gives F in terms of C.

71. To obtain another form of the formula, use the equation you found in Exercise 70 and solve for C in terms of F.

72. For what temperature does $F = C$?

73. A quick way to estimate Fahrenheit temperature for a given Celsius temperature is to double C and add 30. Use this method to find F if $C = 15$.

74. Use the equation found in Exercise 70 to find F if $C = 15$. How does that answer compare with your answer to Exercise 73?

75. Use the method given in Exercise 73 to estimate the Fahrenheit temperature given $C = 30$. Then use the equation from Exercise 70 to find F when $C = 30$. How do the temperatures compare?

76. Explain why the method given in Exercise 73 to estimate Fahrenheit temperature gives a good approximation of $F = \frac{9}{5}C + 32$.

4.4 Linear Inequalities in Two Variables

Objectives

1. Graph linear inequalities in two variables.
2. Graph the intersection of two linear inequalities.
3. Graph the union of two linear inequalities.

1 Graph linear inequalities in two variables. In Chapter 3 we graphed linear inequalities in one variable on the number line. In this section we will graph linear inequalities in two variables on a rectangular coordinate system.

Linear Inequality in Two Variables

An inequality that can be written as

$$Ax + By < C \quad \text{or} \quad Ax + By > C,$$

where A, B, and C are real numbers and A and B are not both 0, is a **linear inequality in two variables.**

Also, $\leq$ and $\geq$ may replace $<$ and $>$ in the definition.

A line divides the plane into three regions: the line itself and the two half-planes on either side of the line. Recall that graphs of linear inequalities in one variable are intervals on the number line that sometimes include endpoints. The graphs of linear inequalities in two variables are *regions* in the real number plane and may include *boundary lines*. The **boundary line** for the inequality $Ax + By < C$ or $Ax + By > C$ is the graph of the *equation* $Ax + By = C$. To graph a linear inequality in two variables, follow these steps.

Graphing a Linear Inequality

Step 1 **Draw the boundary.** Draw the graph of the straight line that is the boundary. Make the line solid if the inequality involves $\leq$ or $\geq$; make the line dashed if the inequality involves $<$ or $>$.

Step 2 **Choose a test point.** Choose any point not on the line as a test point.

Step 3 **Shade the appropriate region.** Shade the region that includes the test point if it satisfies the original inequality; otherwise, shade the region on the other side of the boundary line.

Example 1 Graphing a Linear Inequality

Graph $3x + 2y \geq 6$.

Step 1 First graph the line $3x + 2y = 6$. The graph of this line, the boundary of the graph of the inequality, is shown in Figure 18.

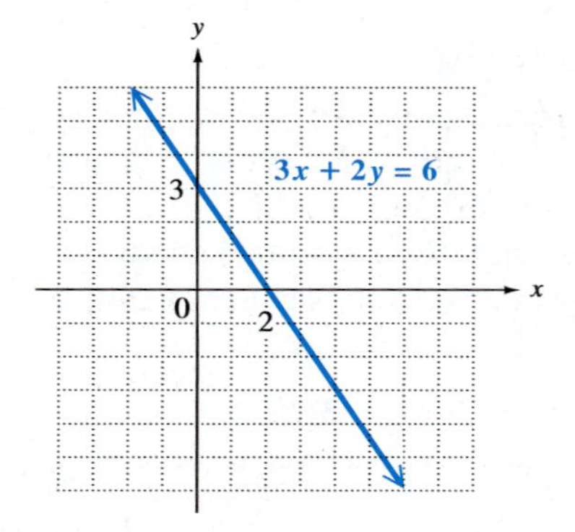

Figure 18

Continued on Next Page

1 Graph each inequality.

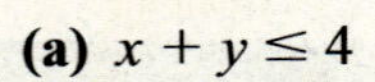

(a) $x + y \leq 4$

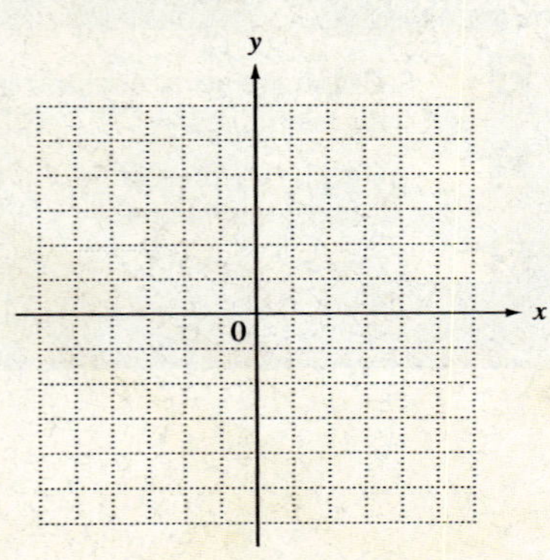

(b) $3x + y \geq 6$

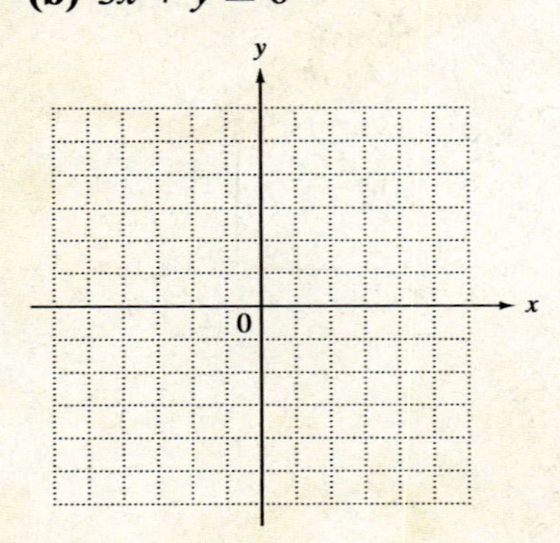

Step 2 The graph of the inequality $3x + 2y \geq 6$ includes the points of the line $3x + 2y = 6$ and either the points *above* the line $3x + 2y = 6$ or the points *below* that line. To decide which, select any point not on the line $3x + 2y = 6$ as a test point. The origin, (0, 0), is often a good choice. Substitute the values from the test point (0, 0) for x and y in the inequality $3x + 2y > 6$.

$$3(0) + 2(0) > 6 \quad ?$$
$$0 > 6 \quad \text{False}$$

Step 3 Because the result is false, (0, 0) does *not* satisfy the inequality, and so the solution set includes all points on the other side of the line. This region is shaded in Figure 19.

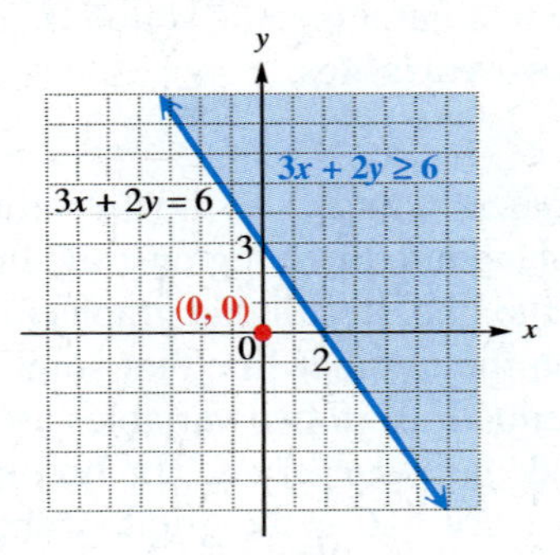

Figure 19

Work Problem 1 at the Side.

If the inequality is written in the form $y > mx + b$ or $y < mx + b$, the inequality symbol indicates which half-plane to shade.

If $y > mx + b$, shade **above** the boundary line;
if $y < mx + b$, shade **below** the boundary line.

This method works *only* if the inequality is solved for y.

Example 2 Graphing a Linear Inequality

Graph $x - 3y < 4$.

First graph the boundary line, shown in Figure 20. The points of the boundary line do not belong to the inequality $x - 3y < 4$ (because the inequality symbol is $<$, not $\leq$). For this reason, the line is dashed. Now solve the inequality for y.

$$x - 3y < 4$$
$$-3y < -x + 4$$
$$y > \frac{x}{3} - \frac{4}{3} \quad \text{Multiply by } -\tfrac{1}{3}\text{; change } < \text{ to } >.$$

Because of the *is greater than* symbol, shade *above* the line. As a check, choose a test point not on the line, say (1, 2), and substitute for x and y in the original inequality.

$$1 - 3(2) < 4 \quad ?$$
$$-5 < 4 \quad \text{True}$$

Continued on Next Page

ANSWERS

1. (a) **(b)**

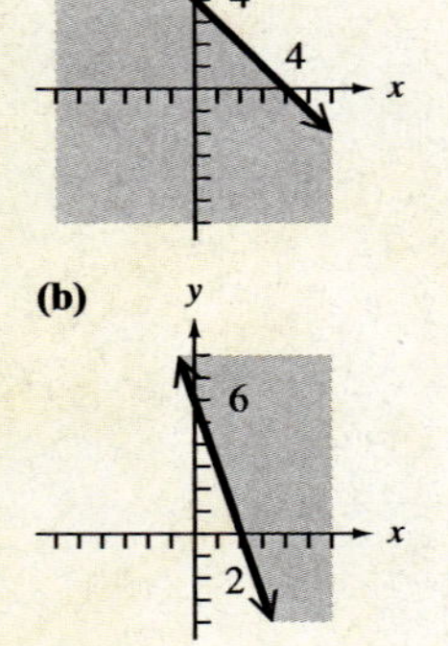

This result agrees with the decision to shade above the line. The solution set, graphed in Figure 20, includes only those points in the shaded half-plane (not those on the line).

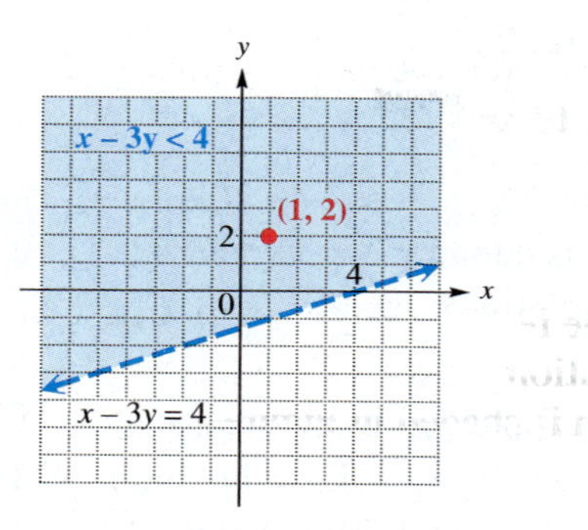

Figure 20

Work Problem **2** at the Side.

2 Graph each inequality.

(a) $x - y > 2$

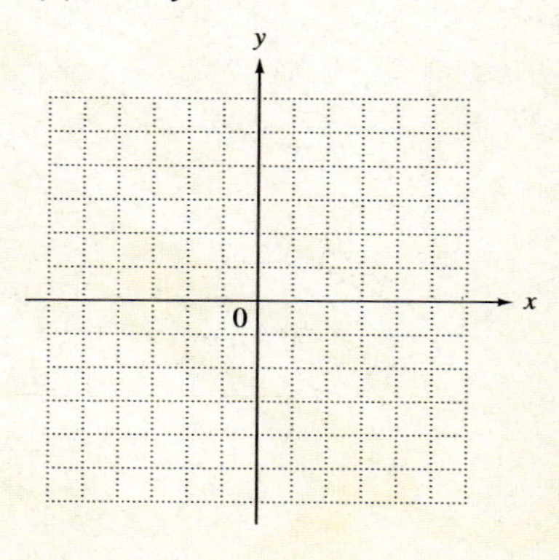

(b) $3x + 4y < 12$

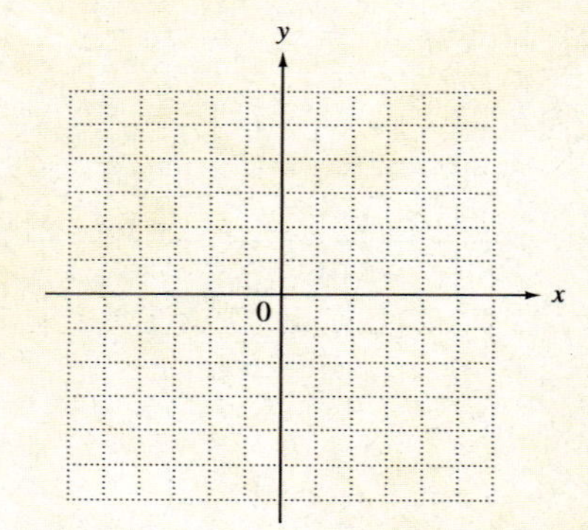

2 Graph the intersection of two linear inequalities. In Section 3.2 we discussed how the words *and* and *or* are used with compound inequalities. In that section, the inequalities had one variable. Those ideas can be extended to include inequalities in two variables. A pair of inequalities joined with the word *and* is interpreted as the intersection of the solution sets of the inequalities. The graph of the intersection of two or more inequalities is the region of the plane where all points satisfy all of the inequalities at the same time.

Example 3 Graphing the Intersection of Two Inequalities

Graph $2x + 4y \geq 5$ and $x \geq 1$.

To begin, we graph each of the two inequalities $2x + 4y \geq 5$ and $x \geq 1$ separately. The graph of $2x + 4y \geq 5$ is shown in Figure 21(a), and the graph of $x \geq 1$ is shown in Figure 21(b).

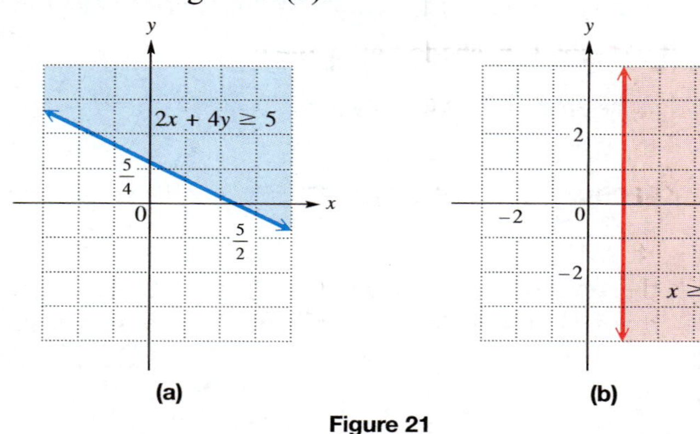

Figure 21

In practice, the two graphs in Figure 21 are graphed on the same axes. Then we use heavy shading to identify the intersection of the graphs, as shown in Figure 22.

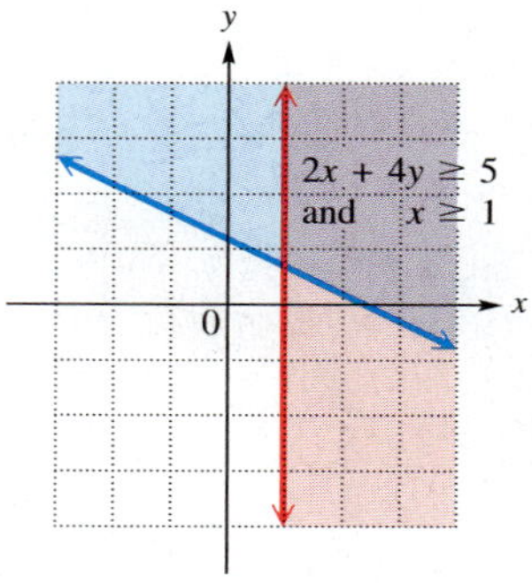

Figure 22

Continued on Next Page

ANSWERS

2. (a) **(b)**

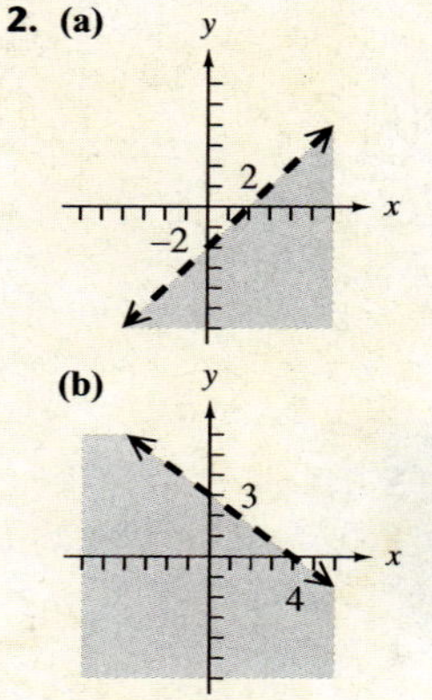

3 Graph $x - y \leq 4$ and $x \geq -2$.

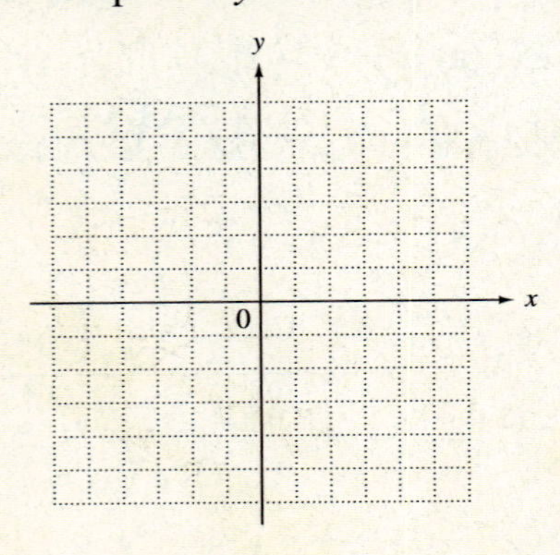

4 Graph $7x - 3y < 21$ or $x > 2$.

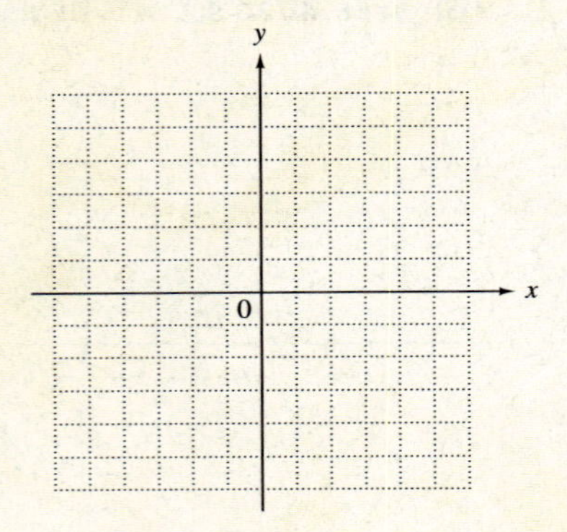

To check, we can use a test point from each of the four regions formed by the intersection of the boundary lines. Verify that only ordered pairs in the heavily shaded region satisfy both inequalities.

Work Problem 3 at the Side.

Calculator Tip Graphing calculators can graph inequalities in two variables, as well as equations in two variables. Refer to your owner's manual for specific directions.

3 Graph the union of two linear inequalities. When two inequalities are joined by the word *or*, we must find the union of the graphs of the inequalities. The graph of the union of two inequalities includes all of the points that satisfy either inequality.

Example 4 Graphing the Union of Two Inequalities

Graph $2x + 4y \geq 5$ or $x \geq 1$.

The graphs of the two inequalities are shown in Figure 21 on the previous page. The graph of the union is shown in Figure 23.

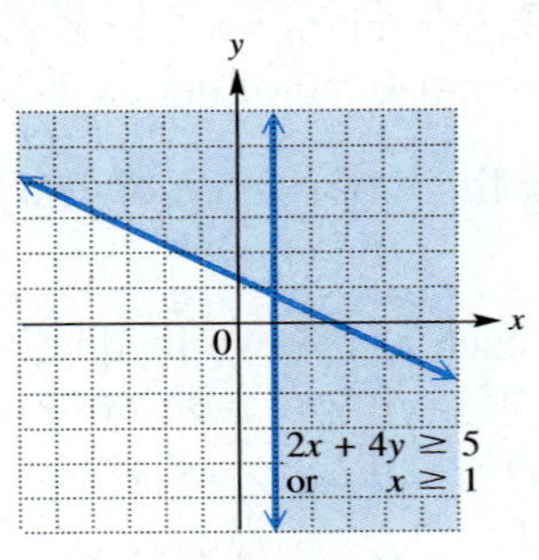

Figure 23

Work Problem 4 at the Side.

ANSWERS

3.

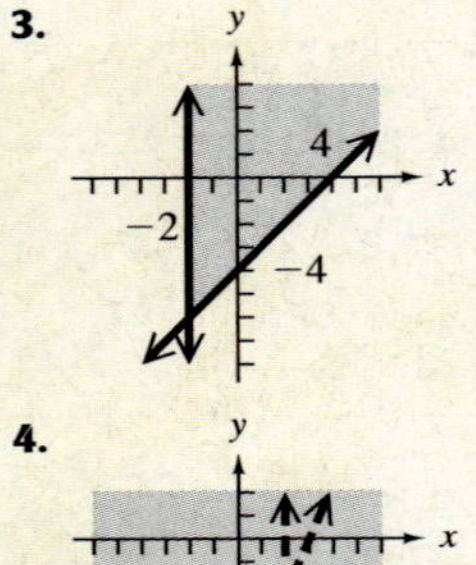

4.

4.4 EXERCISES

FOR EXTRA HELP

Student's Solutions Manual

MyMathLab.com

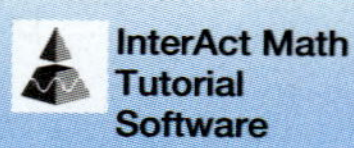
InterAct Math Tutorial Software

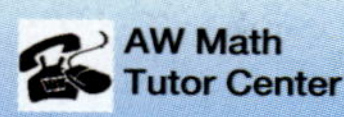
AW Math Tutor Center

www.mathxl.com

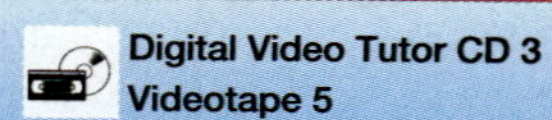
Digital Video Tutor CD 3 Videotape 5

In each statement, fill in the first blank with either solid *or* dashed. *Fill in the second blank with* above *or* below.

1. The boundary of the graph of $y \leq -x + 2$ will be a ________ line, and the shading will be ________ the line.

2. The boundary of the graph of $y < -x + 2$ will be a ________ line, and the shading will be ________ the line.

3. The boundary of the graph of $y > -x + 2$ will be a ________ line, and the shading will be ________ the line.

4. The boundary of the graph of $y \geq -x + 2$ will be a ________ line, and the shading will be ________ the line.

5. How is the boundary line $Ax + By = C$ used in graphing either $Ax + By < C$ or $Ax + By > C$?

6. Describe the two methods discussed in the text for deciding which region is the solution set of a linear inequality in two variables.

Graph each linear inequality. See Examples 1 and 2.

7. $x + y \leq 2$

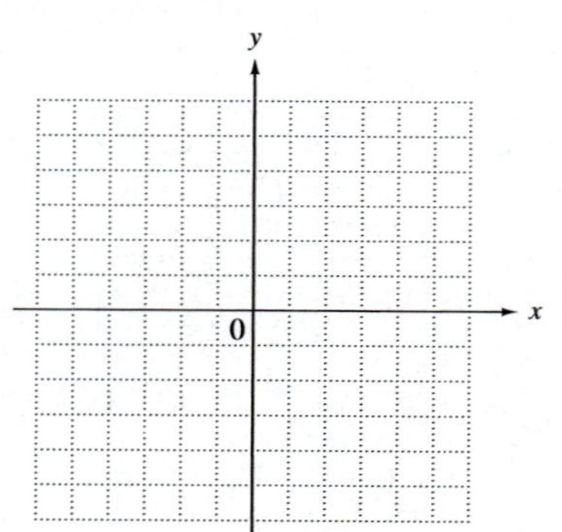

8. $x + y \leq -3$

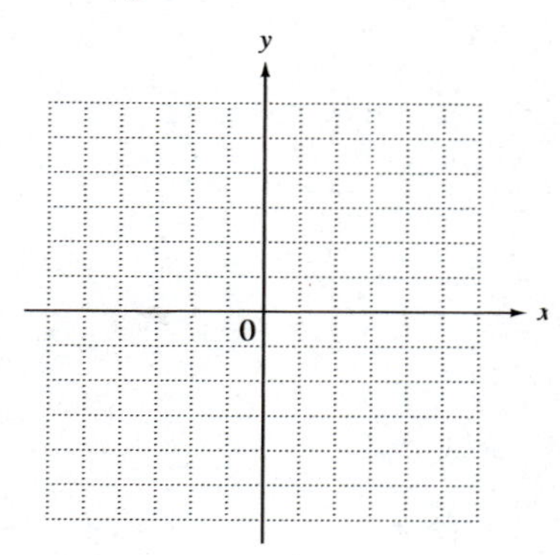

9. $4x - y < 4$

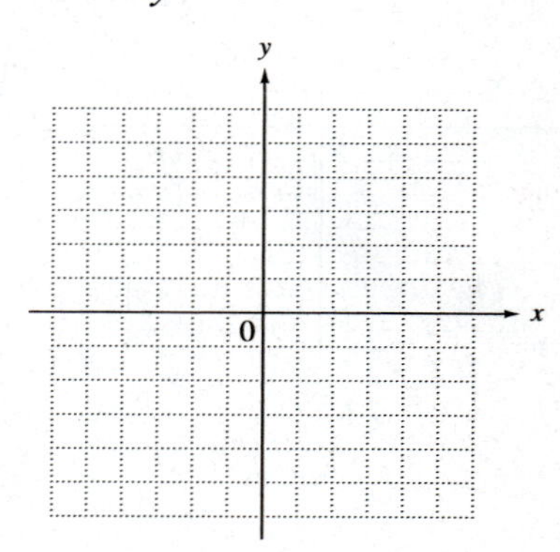

10. $3x - y < 3$

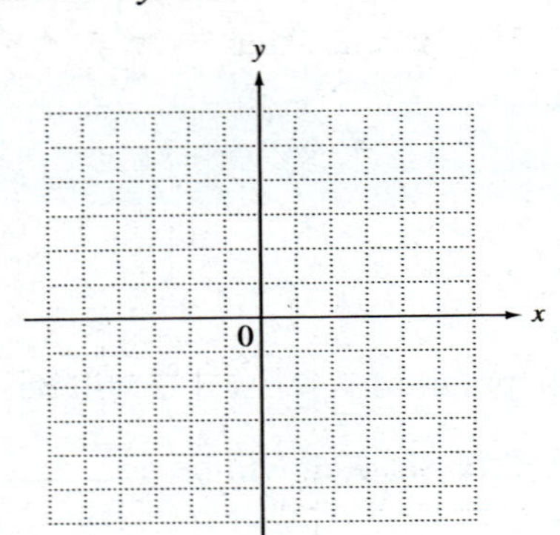

11. $x + 3y \geq -2$

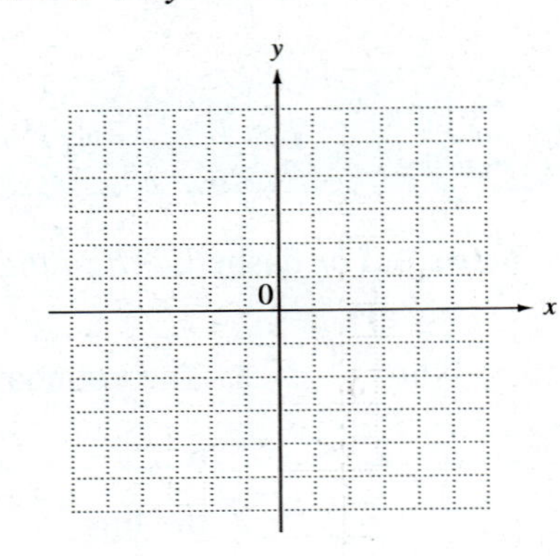

12. $x + 4y \geq -3$

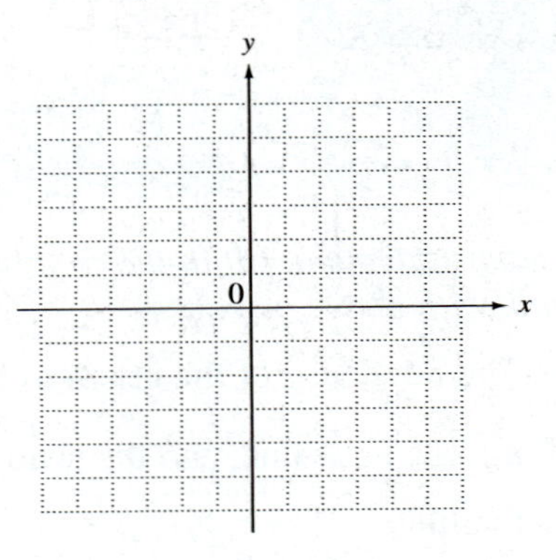

13. $x + y > 0$

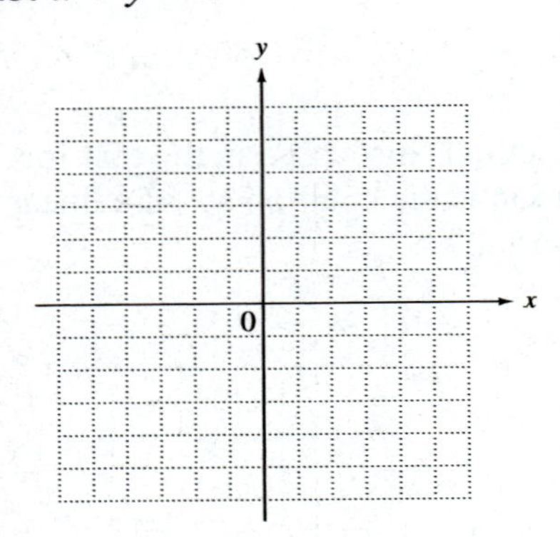

14. $x + 2y > 0$

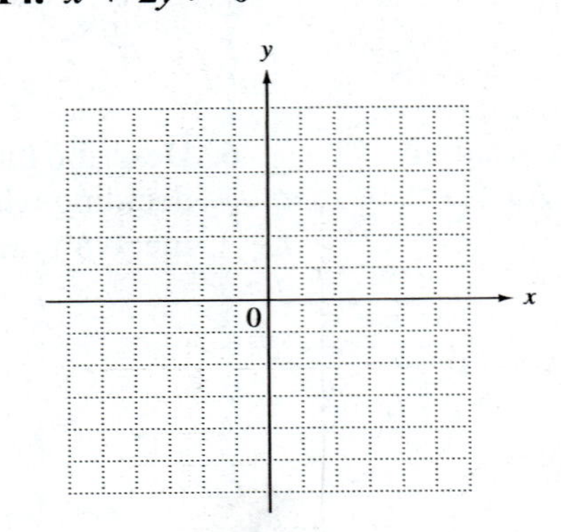

15. $x - 3y \leq 0$

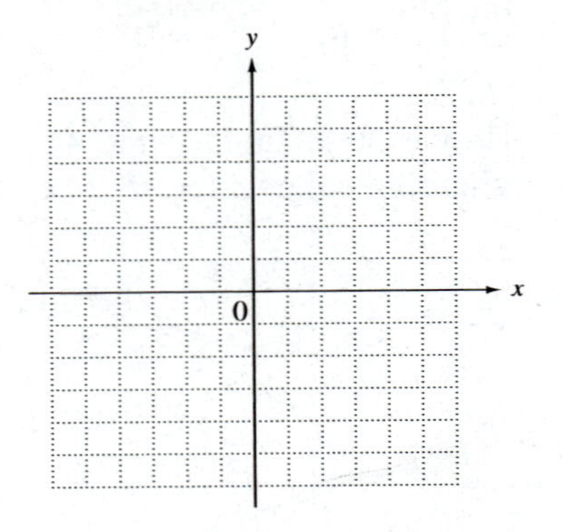

16. $x - 5y \leq 0$

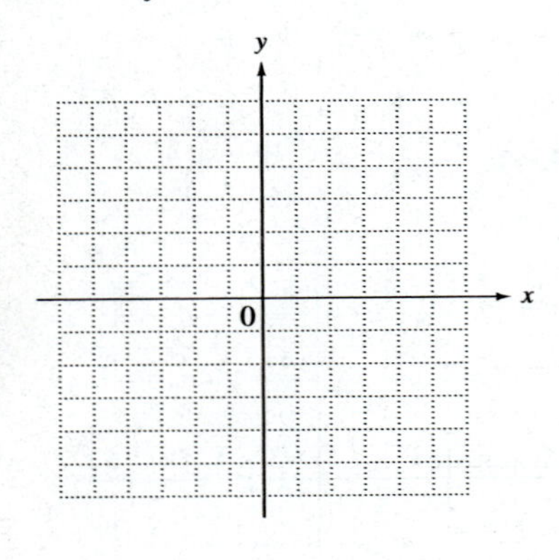

17. $y < x$

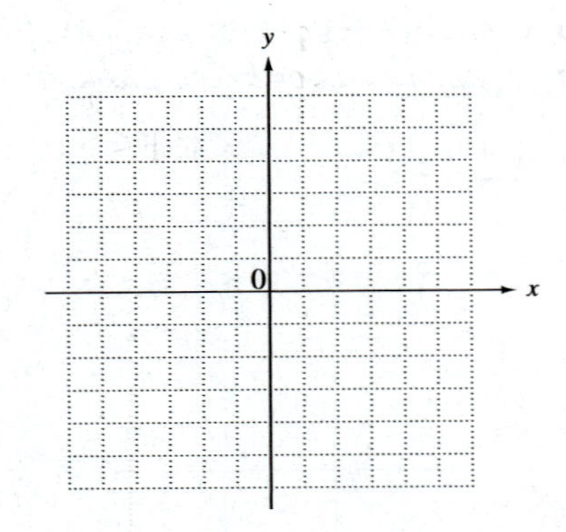

18. $y \leq 4x$

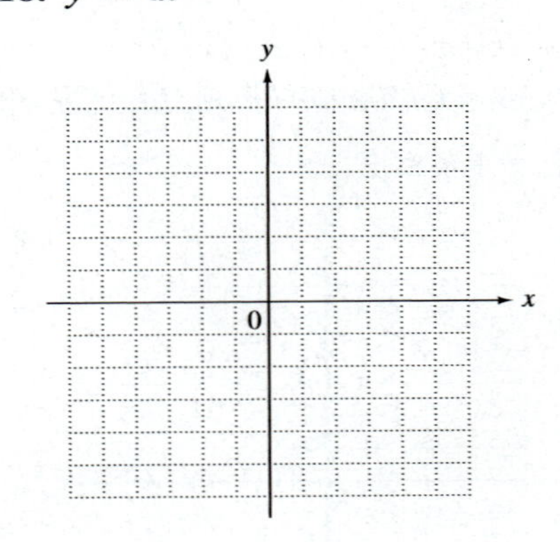

Graph the intersection of each pair of inequalities. See Example 3.

19. $x + y \leq 1$ and $x \geq 1$

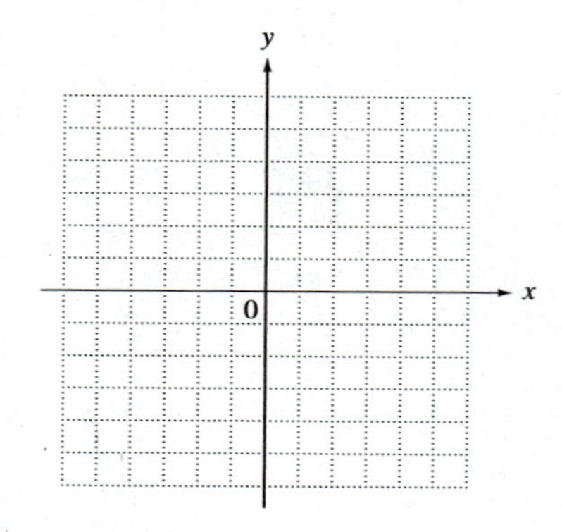

20. $x - y \geq 2$ and $x \geq 3$

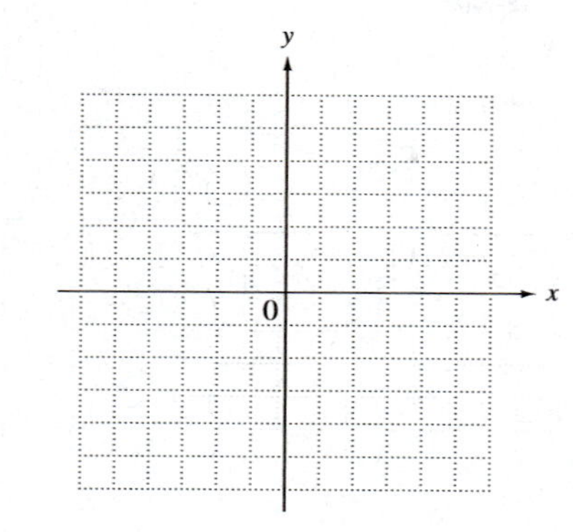

21. $2x - y \geq 2$ and $y < 4$

22. $3x - y \geq 3$ and $y < 3$

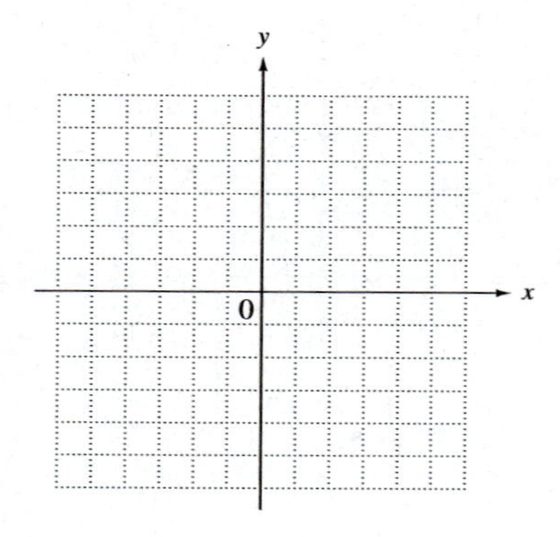

23. $x + y > -5$ and $y < -2$

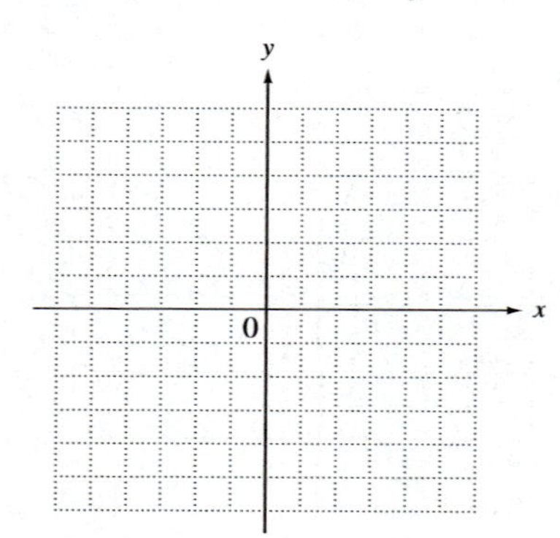

24. $6x - 4y < 10$ and $y > 2$

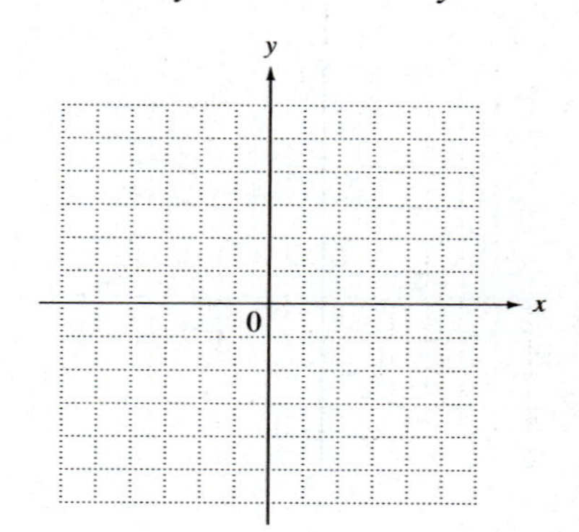

Use the method described in Section 3.3 to write each inequality as a compound inequality, and graph its solution set in the rectangular coordinate plane.

25. $|x| \geq 3$

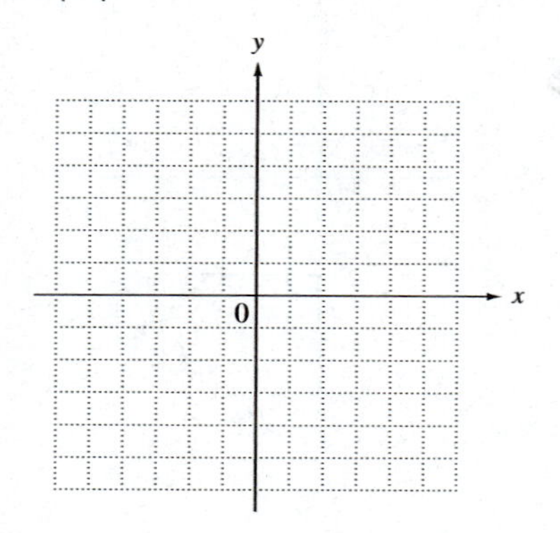

26. $|y| < 5$

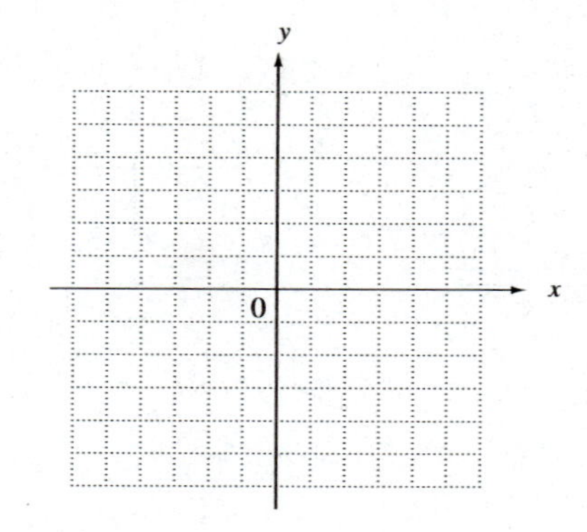

27. $|y + 1| < 2$

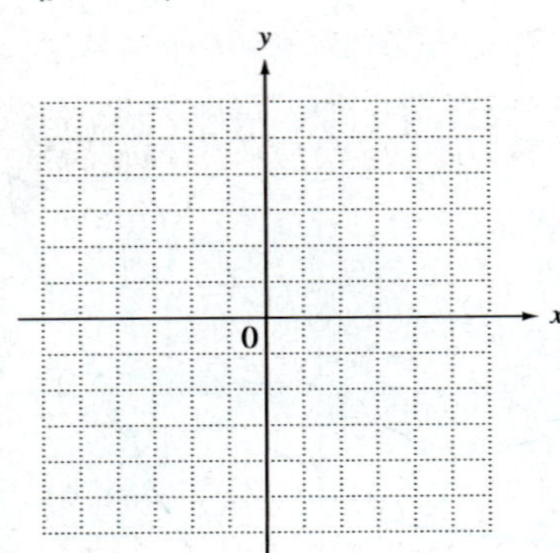

28. $|x - 2| \geq 1$

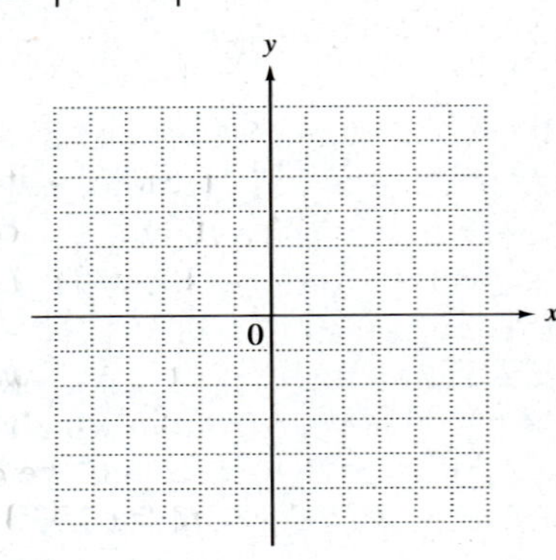

Graph the union of each pair of inequalities. See Example 4.

29. $x - y \geq 1$ or $y \geq 2$

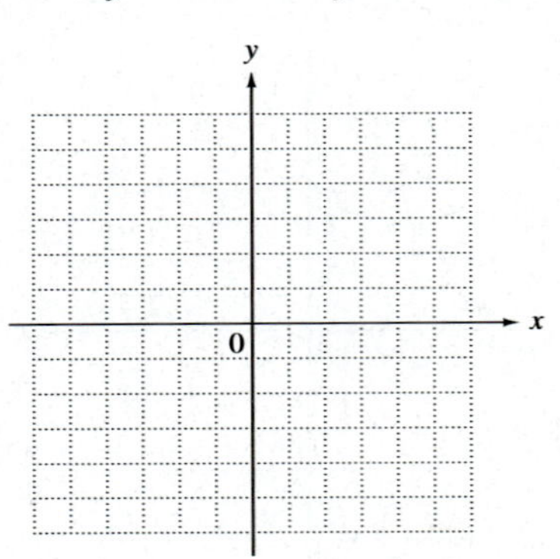

30. $x + y \leq 2$ or $y \geq 3$

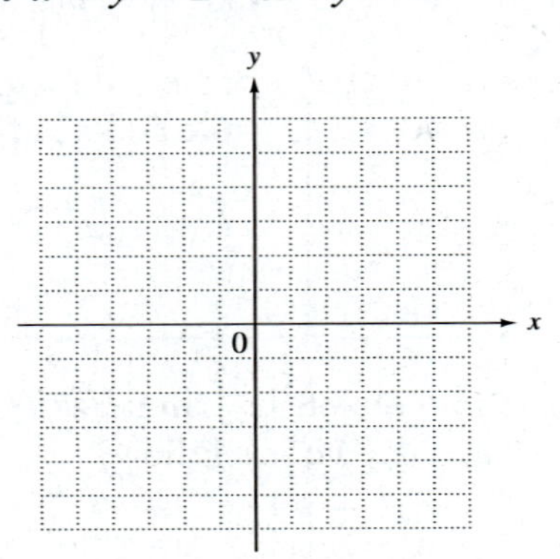

31. $x - 2 > y$ or $x < 1$

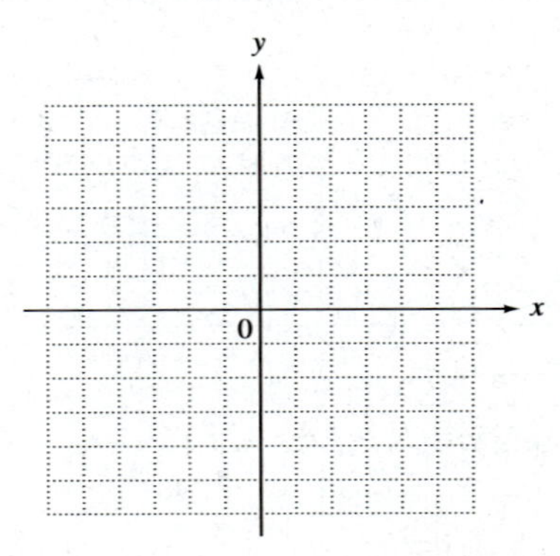

32. $x + 3 < y$ or $x > 3$

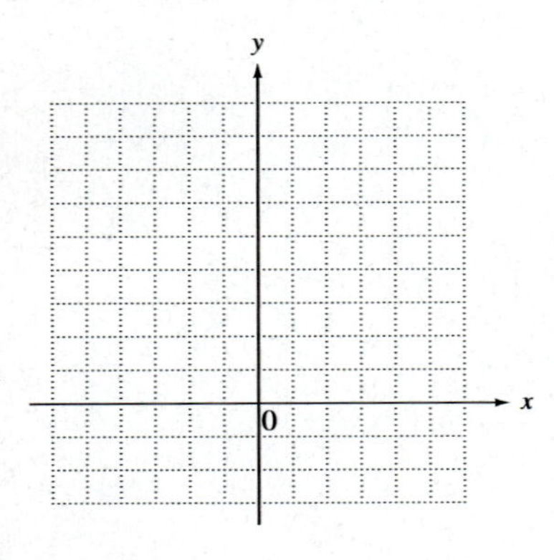

33. $3x + 2y < 6$ or $x - 2y > 2$

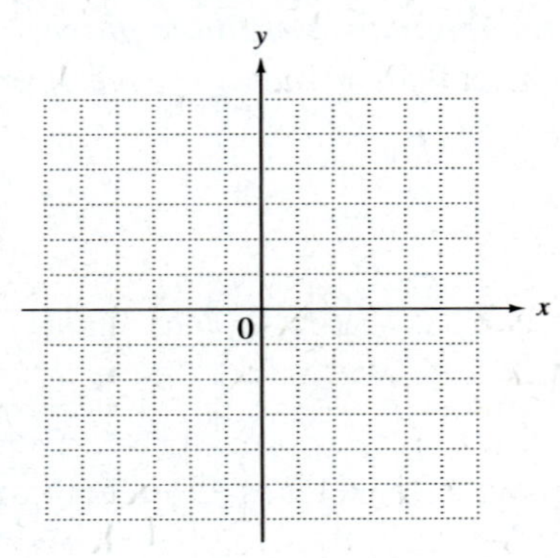

34. $x - y \geq 1$ or $x + y \leq 4$

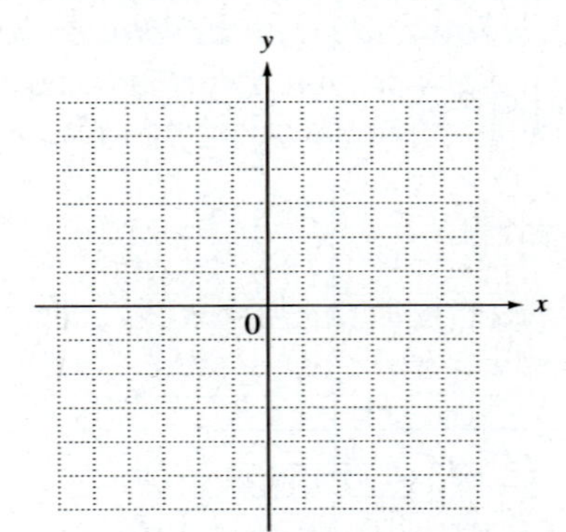

4.5 Introduction to Functions

Objectives

1. Define relation and function.
2. Find domain and range.
3. Identify functions.
4. Use function notation.
5. Identify linear functions.

It is often useful to describe one quantity in terms of another; for example, the growth of a plant is related to the amount of light it receives, the demand for a product is related to the price of the product, the cost of a trip is related to the distance traveled, and so on. To represent these corresponding quantities, it is helpful to use ordered pairs.

For example, we can indicate the relationship between the demand for a product and its price by writing ordered pairs in which the first number represents the price and the second number represents the demand. Then the ordered pair (5, 1000) would indicate a demand for 1000 items when the price of the item is \$5. Since the demand depends on the price charged, we place the price first and the demand second. The ordered pair is an abbreviation for the sentence "If the price is 5 (dollars), then the demand is for 1000 (items)." Similarly, the ordered pairs (3, 5000) and (10, 250) show that a price of \$3 produces a demand for 5000 items, and a price of \$10 produces a demand for 250 items.

In this example, the demand depends on the price of the item. For this reason, demand is called the *dependent variable,* and price is called the *independent variable.* Generalizing, if the value of the variable y depends on the value of the variable x, then y is the **dependent variable** and x is the **independent variable.**

Independent variable → x, Dependent variable → y

$$(x, y)$$

1 Define relation and function. Since related quantities can be written using ordered pairs, the concept of *relation* can be defined as follows.

Relation

A **relation** is a set of ordered pairs.

For example, the sets

$$F = \{(1, 2), (-2, 5), (3, -1)\} \quad \text{and} \quad G = \{(-4, 1), (-2, 1), (-2, 0)\}$$

are both relations. A special kind of relation, called a *function,* is very important in mathematics and its applications.

Function

A **function** is a relation in which, for each value of the first component of the ordered pairs, there is *exactly one value* of the second component.

Of the two examples of a relation just given, only set F is a function, because for each x-value, there is exactly one y-value. In set G, the last two ordered pairs have the same x-value paired with two different y-values, so G is a relation, but not a function.

$$F = \{(1, 2), (-2, 5), (3, -1)\} \qquad \text{Function}$$

Different x-values (1, −2, 3)

$$G = \{(-4, 1), (-2, 1), (-2, 0)\} \qquad \text{Not a function}$$

Same x-value (−2)

In a function, there is *exactly one* value of the dependent variable, the second component, for each value of the independent variable, the first component. This is what makes functions so important in applications. It would not be as useful, for example, to know a price/demand relationship that gave more than one demand for a given price.

Another way to think of a functional relationship is to think of the independent variable as an input and the dependent variable as an output. A calculator is an input-output machine, for example. To find 8^2, we input 8, press the squaring key, and see that the output is 64. Inputs and outputs can also be determined from a graph or a table.

A third way to describe a function is to give a rule that tells how to determine the dependent variable for a specific value of the independent variable. The rule may be given in words: the dependent variable is twice the independent variable. Usually the rule is an equation:

$$y = 2x.$$

Dependent variable (y) Independent variable (x)

This is the most efficient way to define a function.

Example 1 **Determining Independent and Dependent Variables of Functions**

Determine the independent and dependent variables for each function. Give an example of an ordered pair belonging to the function.

(a) The 2000 Summer Olympics medal winners in men's basketball were {(gold, United States), (silver, France), (bronze, Lithuania)}. (*Source:* espn.go.com/oly/summer00/)

The independent variable (the first component in each ordered pair) is the type of medal; the dependent variable (the second component) is the recipient. Any of the three ordered pairs could be given as an example.

(b) A calculator that finds square roots

The independent variable (the input) is a nonnegative real number, since the square root of a negative number is not a real number. The dependent variable is the nonnegative square root. For example, (81, 9) belongs to this function.

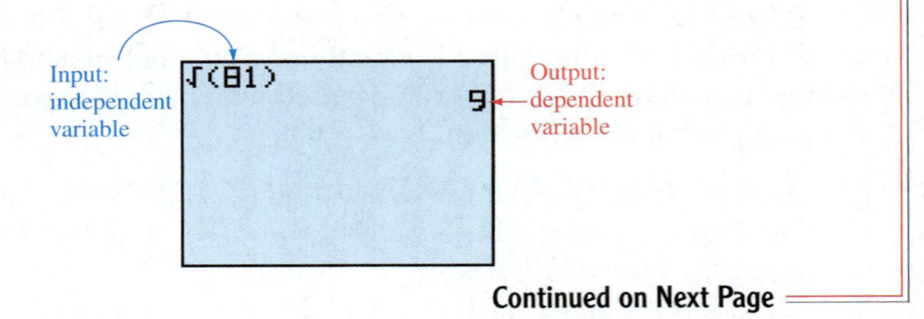

Continued on Next Page

(c) The graph of the relationship between the number of gallons of water in a small swimming pool and time in hours

The independent variable is the number of hours, and the dependent variable is the number of gallons of water in the pool. One ordered pair is (25, 3000).

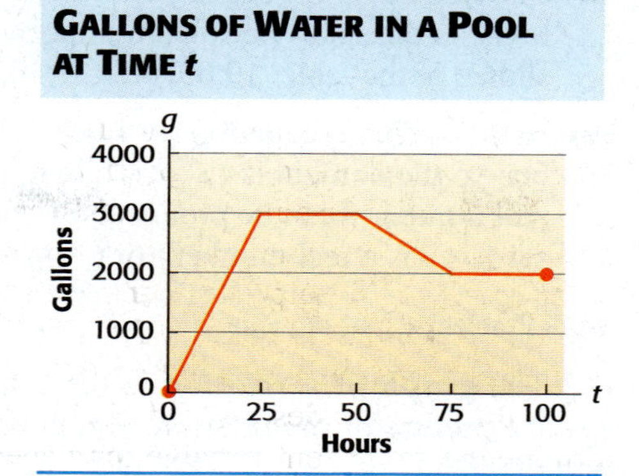

(d) Petroleum imports in millions of barrels per day for selected years given in the table

The independent variable is the year; the dependent variable is the number of barrels of petroleum. An example of an ordered pair is (1994, 9.00).

U.S. PETROLEUM IMPORTS

Year	Barrels Per Day (in millions)
1994	9.00
1995	8.84
1996	9.40
1997	10.16
1998	10.71

Source: World Almanac and Book of Facts, 2000.

(e) $y = 3x + 4$

The independent variable is x, and the dependent variable is y. One ordered pair is $(\frac{1}{3}, 5)$.

Work Problem 1 at the Side.

2 Find domain and range.

Domain and Range

In a relation, the set of all values of the independent variable (x) is the **domain;** the set of all values of the dependent variable (y) is the **range.**

Example 2 Determining Domains and Ranges of Relations

Give the domain and range of each function in Example 1.

(a) The domain is the type of medal, {gold, silver, bronze}, and the range is the set of winning countries, {United States, France, Lithuania}.

(b) Here, the domain is restricted to nonnegative numbers: $[0, \infty)$. The range is also $[0, \infty)$.

(c) The domain includes all possible values of t, the time in hours, which is the interval $[0, 100]$. The range is the set of the number of gallons at time t, the interval $[0, 3000]$.

Continued on Next Page

1 Determine the independent variable, the dependent variable, and an ordered pair for each function.

(a) The reciprocal key on a calculator

(b) The graph given in Section 4.1, Exercise 2 (See page 173.)

(c) The table given in Section 4.2, Exercise 62 and repeated here

Year	Subscribers (in thousands)
1994	24,134
1995	33,786
1996	44,043
1997	55,312
1998	69,209
1999	86,047

Source: Cellular Telecommunications Industry Association, Washington, D.C., *State of the Cellular Industry* (Annual).

(d) $y = \frac{1}{2}x$

ANSWERS

1. (a) independent variable: any nonzero real number; dependent variable: any nonzero real number; $\left(-\frac{2}{3}, -\frac{3}{2}\right)$ **(b)** independent variable: the year; dependent variable: the tax revenue; any ordered pair corresponding to a point on the graph **(c)** independent variable: the year; dependent variable: the number of subscribers; any ordered pair from the table **(d)** independent variable: x; dependent variable: y; $(-10, -5)$ is one example.

(d) The domain is the set of years, {1994, 1995, 1996, 1997, 1998}; the range is the set of petroleum imports (in millions of barrels per day) shown in the table, {9.00, 8.84, 9.40, 10.16, 10.71}.

(e) In the defining equation (or rule), $y = 3x + 4$, x can be any real number, so the domain is $\{x \mid x \text{ is a real number}\}$ or $(-\infty, \infty)$. Since every real number y can be produced by some value of x, the range is also the set $\{y \mid y \text{ is a real number}\}$ or $(-\infty, \infty)$.

Work Problem 2 at the Side.

The **graph of a relation** is the graph of its ordered pairs. The graph gives a picture of the relation, which can be used to determine its domain and range.

Example 3 Finding Domains and Ranges from Graphs

Give the domain and range of each relation.

(a)

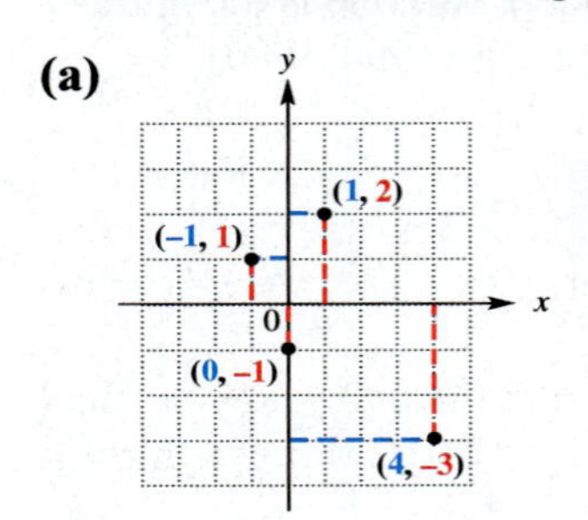

The domain is the set of x-values,

$$\{-1, 0, 1, 4\}.$$

The range is the set of y-values,

$$\{-3, -1, 1, 2\}.$$

(b)

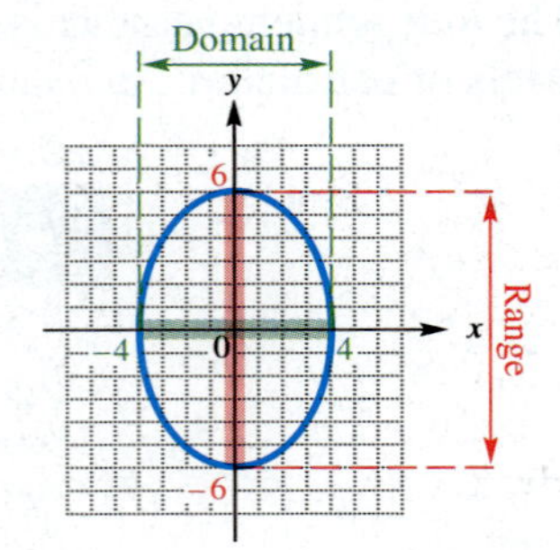

The x-values of the points on the graph include all numbers between -4 and 4, inclusive. The y-values include all numbers between -6 and 6, inclusive. Using interval notation,

the domain is $[-4, 4]$;

the range is $[-6, 6]$.

(c)

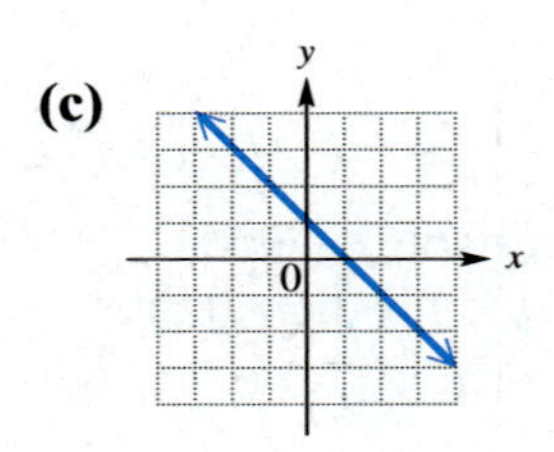

The arrowheads indicate that the line extends indefinitely left and right, as well as up and down. Therefore, both the domain and the range include all real numbers, written $(-\infty, \infty)$.

(d)

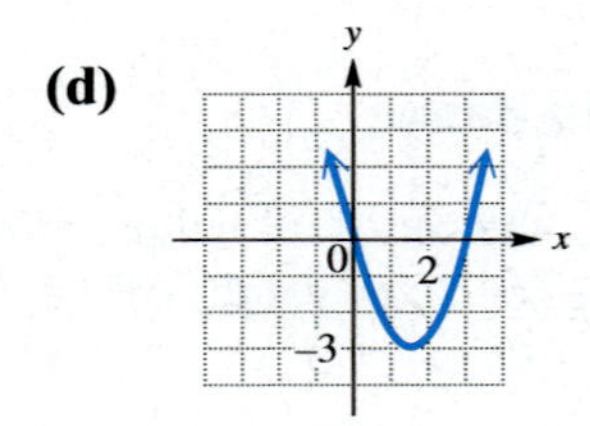

The arrowheads indicate that the graph extends indefinitely left and right, as well as upward. The domain is $(-\infty, \infty)$. Because there is a least y-value, -3, the range includes all numbers greater than or equal to -3, written $[-3, \infty)$.

Work Problem 3 at the Side.

2 Give the domain and range for each function.

(a) The reciprocal key on a calculator

(b) The graph given in Section 4.1, Exercise 1 (See page 173.)

(c) The table given in Section 4.2, Exercise 62 and repeated in margin Problem 1(c)

(d) $y = \frac{1}{2}x$

3 Give the domain and range of each relation.

(a)

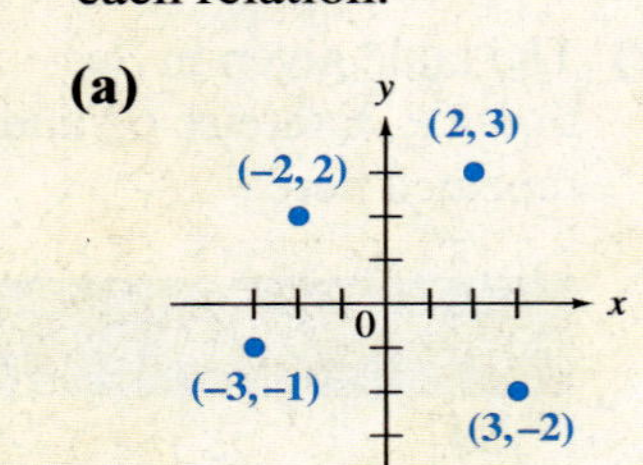

(b)

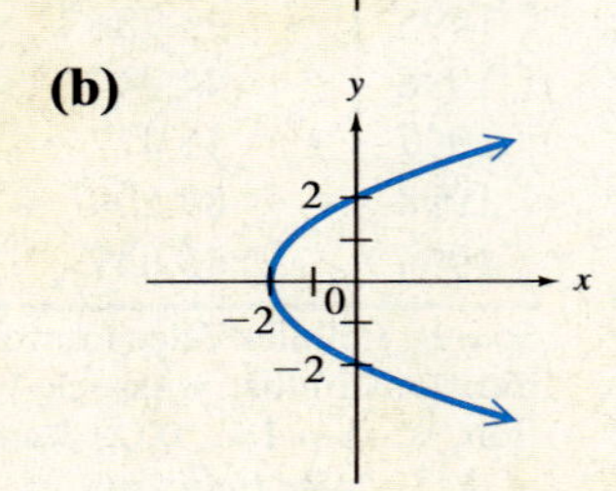

(c)

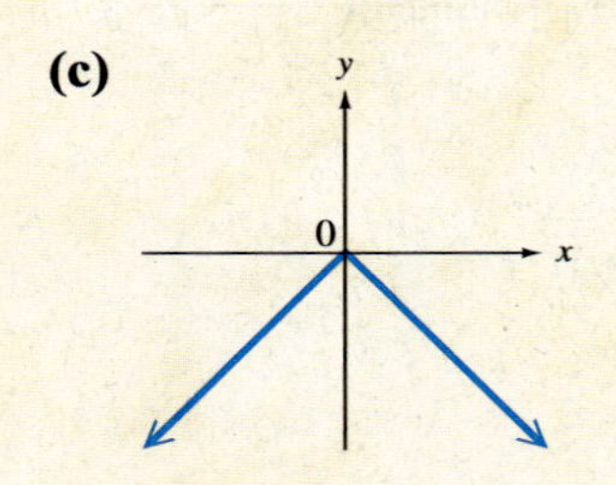

ANSWERS

2. **(a)** domain: $(-\infty, 0) \cup (0, \infty)$; range: $(-\infty, 0) \cup (0, \infty)$
(b) domain: $[1970, 2000]$; range: approximately $[17, 36]$
(c) domain: {1994, 1995, 1996, 1997, 1998, 1999}; range: {24,134, 33,786, 44,043, 55,312, 69,209, 86,047}
(d) domain: $(-\infty, \infty)$; range: $(-\infty, \infty)$

3. **(a)** domain: $\{-3, -2, 2, 3\}$; range: $\{-2, -1, 2, 3\}$
(b) domain: $[-2, \infty)$; range: $(-\infty, \infty)$
(c) domain: $(-\infty, \infty)$; range: $(-\infty, 0]$

Relations are often defined by equations, such as $y = 2x + 3$ and $y^2 = x$. It is sometimes necessary to determine the domain of a relation from its equation. In this book, the following agreement on the domain of a relation is assumed.

Agreement on Domain

The domain of a relation is assumed to be all real numbers that produce real numbers when substituted for the independent variable.

To illustrate this agreement, since any real number can be used as a replacement for x in $y = 2x + 3$, the domain of this function is the set of real numbers. As another example, the function defined by $y = \frac{1}{x}$ has all real numbers except 0 as domain, since y is undefined if $x = 0$. In general, the domain of a function defined by an algebraic expression is all real numbers, except those numbers that lead to division by 0 or an even root of a negative number.

3 **Identify functions.** Most of the relations we have seen in the examples are functions—that is, each x-value corresponds to exactly one y-value. Now we look at ways to determine whether a given relation, defined graphically, is a function.

In a function each value of x leads to only one value of y, so any vertical line drawn through the graph of a function must intersect the graph in at most one point. This is the *vertical line test for a function.*

Vertical Line Test

If every vertical line intersects the graph of a relation in no more than one point, then the relation represents a function.

For example, the graph shown in Figure 24(a) is not the graph of a function since a vertical line intersects the graph in more than one point. The graph in Figure 24(b) does represent a function.

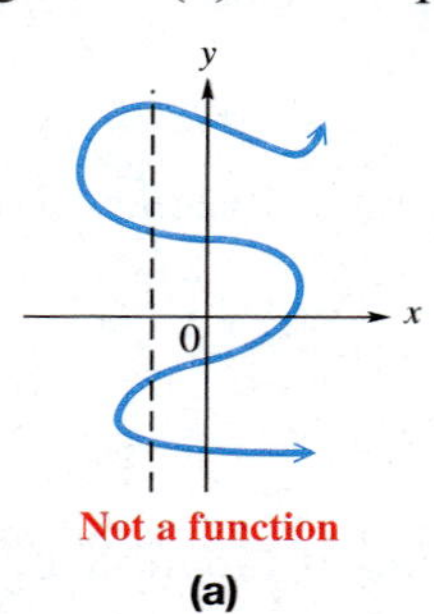

Not a function

(a)

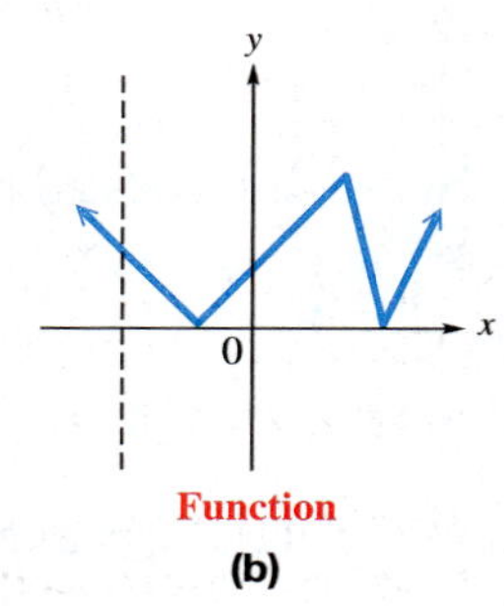

Function

(b)

Figure 24

Work Problem 4 at the Side.

The vertical line test is a simple method for identifying a function defined by a graph. It is more difficult to decide whether a relation defined by an equation is a function. The next example gives some hints that may help.

4 Use the vertical line test to decide which graphs represent functions.

A.

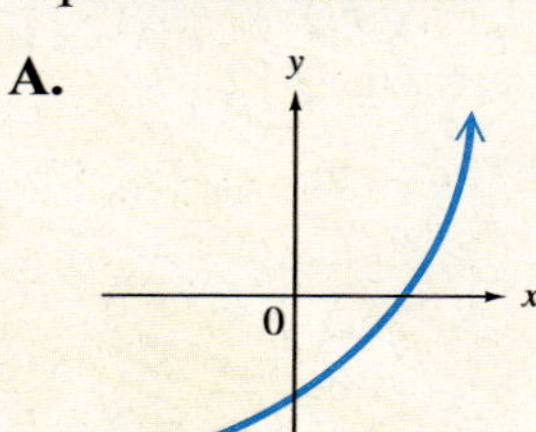

B.

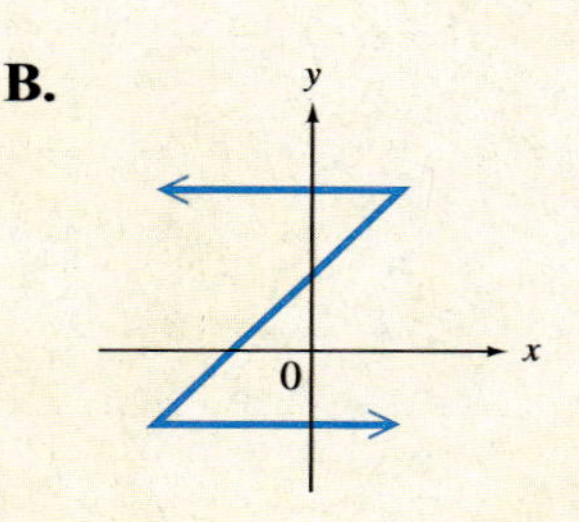

C.

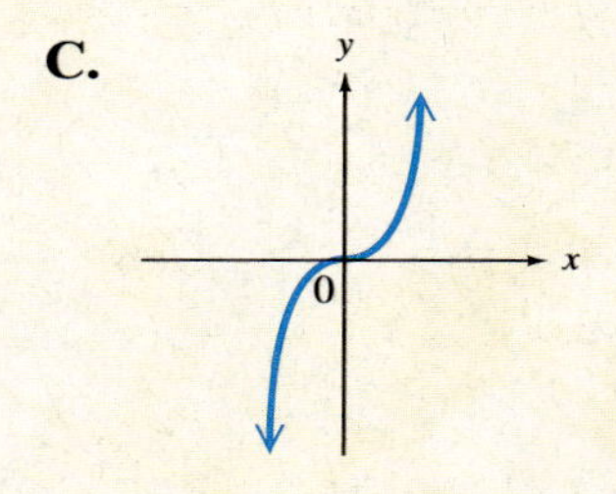

ANSWERS

4. A and C are graphs of functions.

5 Decide whether each relation defines a function, and give the domain.

(a) $y = 6x + 12$

(b) $y \leq 4x$

(c) $y = -\sqrt{3x - 2}$

(d) $y^2 = 25x$

ANSWERS

5. **(a)** yes; $(-\infty, \infty)$ **(b)** no; $(-\infty, \infty)$ **(c)** yes; $\left[\frac{2}{3}, \infty\right)$ **(d)** no; $[0, \infty)$

Example 4 Identifying Functions

Decide whether each relation defines a function, and give the domain.

(a) $y = \sqrt{2x - 1}$

For any choice of x in the domain, there is exactly one corresponding value for y (the radical is a nonnegative number), so this equation defines a function. Since the quantity under the radical sign cannot be negative,

$$2x - 1 \geq 0$$
$$2x \geq 1$$
$$x \geq \frac{1}{2}.$$

The domain is $[\frac{1}{2}, \infty)$.

(b) $y^2 = x$

The ordered pairs $(16, 4)$ and $(16, -4)$ both satisfy this equation. Since one value of x, 16, corresponds to two values of y, 4 and -4, this equation does not define a function. Because x is equal to the square of y, the values of x must always be nonnegative. The domain of the relation is $[0, \infty)$.

(c) $y \leq x - 1$

By definition, y is a function of x if every value of x leads to exactly one value of y. In this example, a particular value of x, say 1, corresponds to many values of y. The ordered pairs $(1, 0)$, $(1, -1)$, $(1, -2)$, $(1, -3)$, and so on, all satisfy the inequality. For this reason, an inequality does not define a function. Any number can be used for x, so the domain is the set of real numbers, $(-\infty, \infty)$.

(d) $y = \dfrac{5}{x - 1}$

Given any value of x in the domain, we find y by subtracting 1, then dividing the result into 5. This process produces exactly one value of y for each value in the domain, so this equation defines a function. The domain includes all real numbers except those that make the denominator 0. We find these numbers by setting the denominator equal to 0 and solving for x.

$$x - 1 = 0$$
$$x = 1$$

Thus, the domain includes all real numbers except 1. In interval notation this is written as

$$(-\infty, 1) \cup (1, \infty).$$

Work Problem 5 at the Side.

In summary, three variations of the definition of function are given here.

Variations of the Definition of Function

1. A **function** is a relation in which, for each value of the first component of the ordered pairs, there is exactly one value of the second component.
2. A **function** is a set of ordered pairs in which no first component is repeated.
3. A **function** is a rule or correspondence that assigns exactly one range value to each domain value.

4 Use function notation. When a function f is defined with a rule or an equation using x and y for the independent and dependent variables, we say "y is a function of x" to emphasize that y *depends on* x. We use the notation

$$y = f(x)$$

to express this. (In this special notation the parentheses do not indicate multiplication.) The letter f stands for *function*. For example, if $y = 2x - 7$, we write

$$f(x) = 2x - 7.$$

When you see the notation $f(x)$, remember that it is just another name for the dependent variable y. This **function notation** is useful for simplifying certain statements. For example, if $y = f(x) = 9x - 5$, then replacing x with 2 gives

$$\begin{aligned} y &= f(2) \\ &= 9 \cdot 2 - 5 \\ &= 18 - 5 \\ &= 13. \end{aligned}$$

The statement "if $x = 2$, then $y = 13$" is abbreviated with function notation as

$$f(2) = 13.$$

Read $f(2)$ as "f of 2" or "f at 2." Also,

$$f(0) = 9 \cdot 0 - 5 = -5, \quad \text{and} \quad f(-3) = 9(-3) - 5 = -32.$$

These ideas and the symbols used to represent them can be illustrated as follows.

Name of the function

Defining expression

$$y = f(x) = 9x - 5$$

Value of the function

Name of the independent variable

CAUTION

The symbol $f(x)$ *does not* indicate "f times x," but represents the y-value for the indicated x-value. As just shown, $f(2)$ is the y-value that corresponds to the x-value 2.

Example 5 Using Function Notation

Let $f(x) = -x^2 + 5x - 3$. Find the following.

(a) $f(2)$

Replace x with 2.

$$\begin{aligned} f(2) &= -2^2 + 5 \cdot 2 - 3 \\ &= -4 + 10 - 3 \\ &= 3 \end{aligned}$$

Continued on Next Page

6 Find $f(-3)$, $f(p)$, and $f(m + 1)$.

(a) $f(x) = 6x - 2$

(b) $f(x) = \dfrac{-3x + 5}{2}$

(c) $f(x) = \dfrac{1}{6}x - 1$

(b) $$f(-1) = -(-1)^2 + 5(-1) - 3$$
$$= -1 - 5 - 3$$
$$= -9$$

(c) $f(q)$
Replace x with q.

$$f(q) = -q^2 + 5q - 3$$

The replacement of one variable with another is important in later courses.

Sometimes letters other than f, such as g, h, or capital letters F, G, and H are used to name functions.

Example 6 Using Function Notation

Let $g(x) = 2x + 3$. Find and simplify the following.

(a) $g(a + 1)$
Replace x with $a + 1$.

$$g(a + 1) = 2(a + 1) + 3$$
$$= 2a + 2 + 3$$
$$= 2a + 5$$

(b) $$g\left(\frac{1}{b + 4}\right) = 2\left(\frac{1}{b + 4}\right) + 3$$
$$= \frac{2}{b + 4} + 3$$

Work Problem 6 at the Side.

If a function is defined by an equation with x and y, not with function notation, use the following steps to find $f(x)$.

Finding an Expression for $f(x)$

Step 1 Solve the equation for y.

Step 2 Replace y with $f(x)$.

Example 7 Writing Equations Using Function Notation

Rewrite each equation using function notation. Then find $f(-2)$ and $f(a)$.

(a) $y = x^2 + 1$
This equation is already solved for y. Since $y = f(x)$,

$$f(x) = x^2 + 1.$$

To find $f(-2)$, let $x = -2$.

$$f(-2) = (-2)^2 + 1$$
$$= 4 + 1$$
$$= 5$$

Find $f(a)$ by letting $x = a$: $f(a) = a^2 + 1$.

Continued on Next Page

ANSWERS

6. (a) -20; $6p - 2$; $6m + 4$

(b) 7; $\dfrac{-3p + 5}{2}$; $\dfrac{-3m + 2}{2}$

(c) $-\dfrac{3}{2}$; $\dfrac{1}{6}p - 1$; $\dfrac{1}{6}(m + 1) - 1$ or $\dfrac{1}{6}m - \dfrac{5}{6}$

(b) $x - 4y = 5$

First solve $x - 4y = 5$ for y. Then replace y with $f(x)$.

$$x - 4y = 5$$
$$x - 5 = 4y$$
$$y = \frac{x-5}{4} \quad \text{so} \quad f(x) = \frac{1}{4}x - \frac{5}{4}$$

Now find $f(-2)$ and $f(a)$.

$$f(-2) = \frac{1}{4}(-2) - \frac{5}{4} = -\frac{7}{4}$$
$$f(a) = \frac{1}{4}a - \frac{5}{4}$$

Work Problem 7 at the Side.

7 Rewrite each equation using function notation. Then find $f(-1)$.

(a) $y = \sqrt{x + 2}$

(b) $x^2 - 4y = 3$

5 Identify linear functions. Our first two-dimensional graphing was of straight lines. Linear equations (except for $x = c$) define *linear functions.*

Linear Function

A function that can be defined by

$$f(x) = mx + b$$

for real numbers m and b is a **linear function.**

Recall from Section 4.3 that m is the slope of the line and $(0, b)$ is the y-intercept. A linear function defined by $f(x) = d$ (whose graph is a horizontal line) is sometimes called a **constant function.** The domain of any linear function is $(-\infty, \infty)$. The range of a nonconstant linear function is $(-\infty, \infty)$, while the range of the constant function with $f(x) = d$ is $\{d\}$.

In later chapters of this book, we will learn about several other types of functions.

ANSWERS

7. (a) $f(x) = \sqrt{x + 2}$; 1

(b) $f(x) = \frac{x^2 - 3}{4}$ or $f(x) = \frac{1}{4}x^2 - \frac{3}{4}$; $-\frac{1}{2}$

Focus on Real-Data Applications

Linear or Nonlinear? That Is the Question about College Tuition

The data used in this table was taken from the fee schedule for North Harris Montgomery Community College for fall semester, 2000.

Credit Hours, *x*	Change in *x*	Resident Tuition Costs, *y*	Change in *y*
0		12	
3	$3 - 0 = 3$	102	$102 - 12 = 90$
6	$6 - 3 = 3$	192	$192 - 102 = 90$
9	$9 - 6 = 3$	282	$282 - 192 = 90$
12		372	
15		462	

Source: Credit schedule for North Harris Montgomery Community College District.

The data represents the relationship between the number of credit hours enrolled (*input*, or *domain*) and the tuition costs (*output*, or *range*). Is this relationship linear?

For Group Discussion

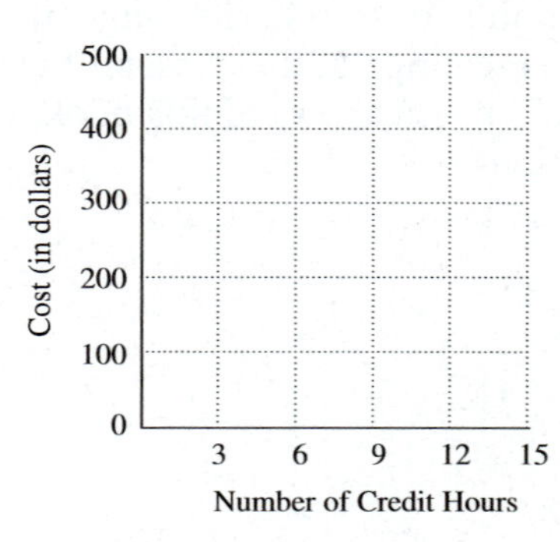

1. Complete the change in x and change in y columns of the table.

2. Graph the ordered-pair data (x, y) and observe the shape of the graph. If the graph approximates a line, then the data represents a linear relationship. Is the relationship linear?

3. **(a)** Linear relationships change at a constant rate. For example,

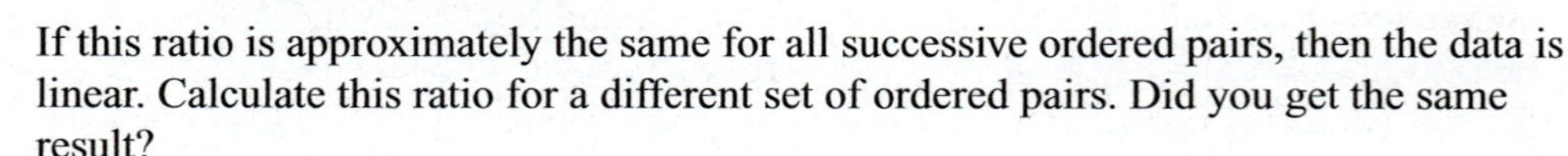

$$\frac{\text{change in fees, } y}{\text{change in credit hours, } x} = \frac{192 - 102}{6 - 3} = \frac{90}{3} = \underline{\qquad\qquad}.$$

If this ratio is approximately the same for all successive ordered pairs, then the data is linear. Calculate this ratio for a different set of ordered pairs. Did you get the same result?

(b) What is the rate of change, or slope, of the line?

4. The y-intercept is the range value (*output*) associated with a domain value (*input*) of 0. From the table, predict the cost to enroll in 0 credit hours. What might the y-intercept represent in this situation?

5. Write an equation for a linear function that represents the relationship between number of credit hours and tuition costs. Use function notation.

4.5 EXERCISES

FOR EXTRA HELP

Student's Solutions Manual

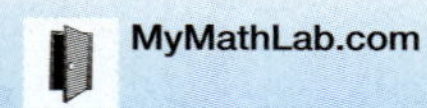
MyMathLab.com

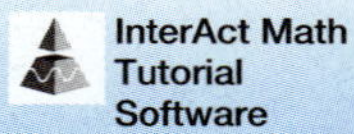
InterAct Math Tutorial Software

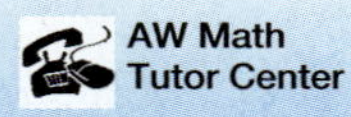
AW Math Tutor Center

www.mathxl.com

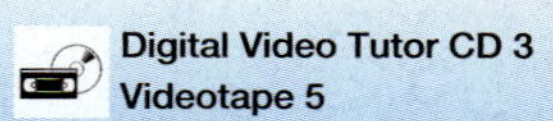
Digital Video Tutor CD 3 Videotape 5

1. In an ordered pair of a relation, is the first element the independent or the dependent variable?

2. Give an example of a relation that is not a function, having domain $\{-3, 2, 6\}$ and range $\{4, 6\}$. (There are many possible correct answers.)

3. Explain what is meant by each term.

(a) Relation **(b)** Domain of a relation

(c) Range of a relation **(d)** Function

Decide whether each relation is a function, and give the domain and the range. Use the vertical line test in Exercises 11–13. See Examples 1–3.

4. $\{(5, 1), (3, 2), (4, 9), (7, 3)\}$

5. $\{(8, 0), (5, 4), (9, 3), (3, 9)\}$

6. $\{(2, 4), (0, 2), (2, 6)\}$

7. $\{(9, -2), (-3, 5), (9, 1)\}$

8. The set containing certain countries and their predicted life expectancy estimates for persons born in 2050 is {(U.S., 83.9), (Japan, 90.91), (Canada, 85.26), (Britain, 83.79), (France, 87.01), (Germany, 83.12), (Italy, 82.26)}. (*Source:* Shripad Tuljapurkar, Mountain View Research, Los Altos, California.)

9. An input-output machine accepts positive real numbers as input, and outputs both their positive and negative square roots.

10.

U.S. Voting-Age Population in 2000 (in millions)	
Hispanic	21.3
Native American	1.6
Asian American	8.2
African American	24.6
White	152.0

Source: U.S. Bureau of the Census.

11.

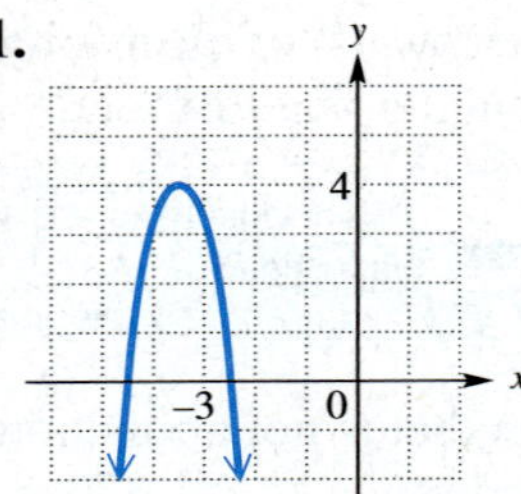

12.

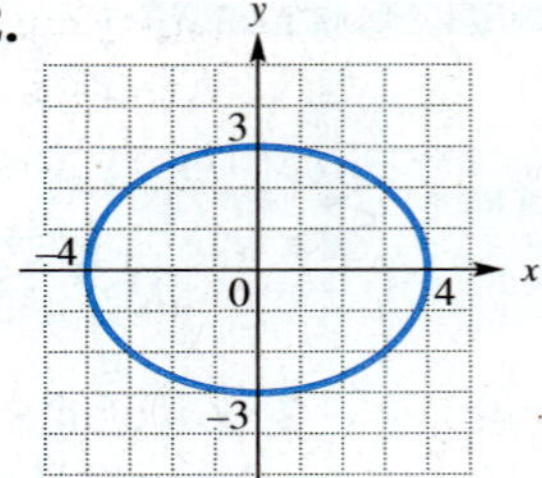

13.

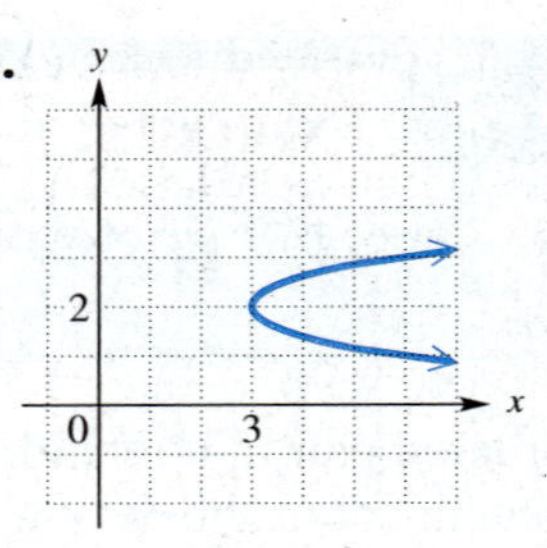

14. Describe the use of the vertical line test.

Decide whether each relation defines y as a function of x. Give the domain. See Example 4.

15. $y = x^2$

16. $y = x^3$

17. $x = y^6$

18. $x = y^4$

19. $x + y < 4$

20. $x - y < 3$

21. $y = \sqrt{x}$

22. $y = -\sqrt{x}$

23. $xy = 1$

24. $xy = -3$

25. $y = 2x - 6$

26. $y = -6x + 8$

27. $y = \sqrt{4x + 2}$

28. $y = \sqrt{9 - 2x}$

29. $y = \dfrac{2}{x - 9}$

30. $y = \dfrac{-7}{x - 16}$

31. Refer to the graph to answer the questions.

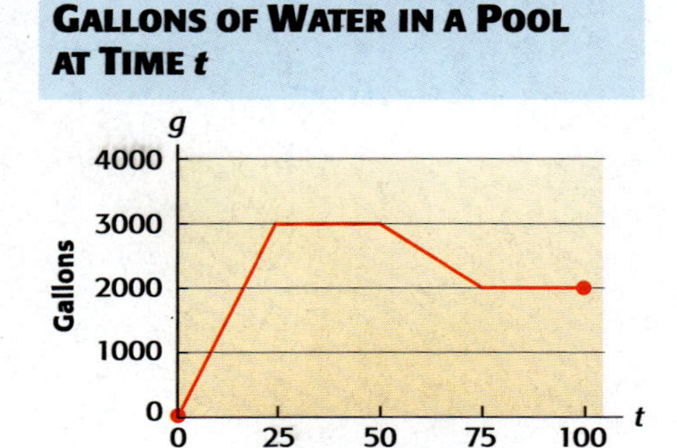

(a) What numbers are possible values of the dependent variable?

(b) For how long is the water level increasing? decreasing?

(c) How many gallons are in the pool after 90 hr?

(d) Call this function f. What is $f(0)$? What does it mean in this example?

32. The graph shows the daily megawatts of electricity used on a record-breaking summer day in Sacramento, California.

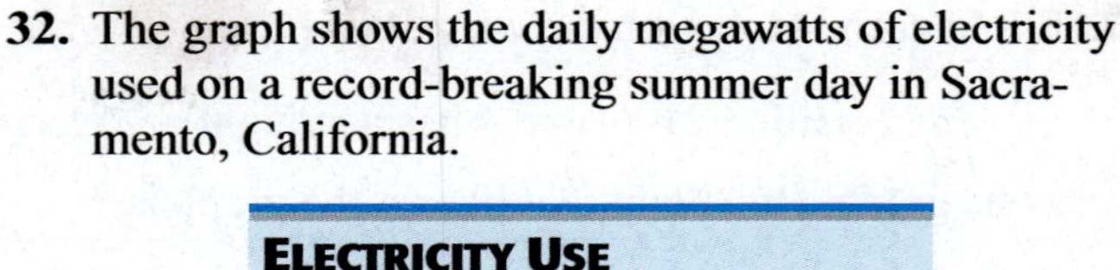

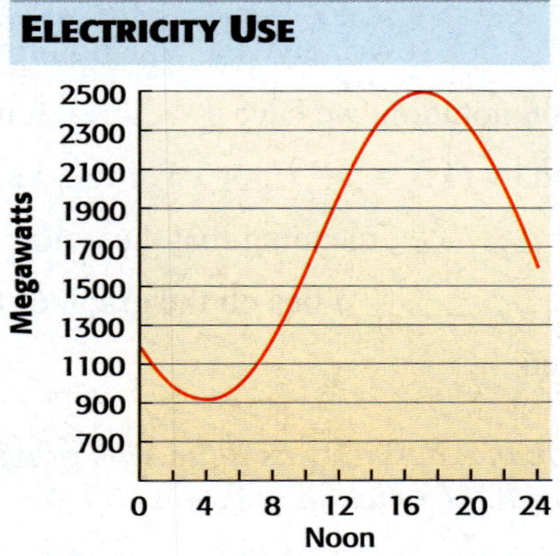

Source: Sacramento Municipal Utility District.

(a) Is this the graph of a function?

(b) What is the domain?

(c) Estimate the number of megawatts used at 8 A.M.

(d) At what time was the most electricity used? the least electricity?

33. Give an example of a function from everyday life. (*Hint:* Fill in the blanks: _______ depends on _______, so _______ is a function of _______.)

34. Choose the correct response: The notation $f(3)$ means

A. the variable f times 3 or $3f$

B. the value of the dependent variable when the independent variable is 3

C. the value of the independent variable when the dependent variable is 3

D. f equals 3.

Let $f(x) = -3x + 4$ and $g(x) = -x^2 + 4x + 1$. Find the following. See Examples 5 and 6.

35. $f(0)$

36. $f(-3)$

37. $g(-2)$

38. $g(10)$

39. $f(p)$

40. $g(k)$

41. $f(-x)$

42. $g(-x)$

43. $f(x + 2)$

44. $g\left(-\frac{1}{x}\right)$

45. $g\left(\frac{p}{3}\right)$

46. $f(3t - 2)$

47. Fill in each blank with the correct response.

The equation $2x + y = 4$ has a straight ______ as its graph. One point that lies on the graph is (3, ______). If we solve the equation for y and use function notation, we have a ______ function defined by $f(x) =$ ______. For this function, $f(3) =$ ______, meaning that the point (______, ______) lies on the graph of the function.

48. Which of the following defines a linear function?

A. $y = \dfrac{x - 5}{4}$ **B.** $y = \dfrac{1}{x}$

C. $y = x^2$ **D.** $y = \sqrt{x}$

An equation that defines y as a function of x is given. ***(a)*** *Solve for y in terms of x, and replace y with the function notation f(x).* ***(b)*** *Find f(3). See Example 7.*

49. $x + 3y = 12$

50. $x - 4y = 8$

51. $y + 2x^2 = 3$

52. $y - 3x^2 = 2$

53. $4x - 3y = 8$

54. $-2x + 5y = 9$

Solve each problem.

55. Suppose that a taxicab driver charges \$1.50 per mi.

(a) Fill in the table with the correct response for the price $f(x)$ she charges for a trip of x miles.

x	f(x)
0	
1	
2	
3	

(b) The linear function that gives a rule for the amount charged is $f(x) =$ ______.

(c) Graph this function for the domain {0, 1, 2, 3}.

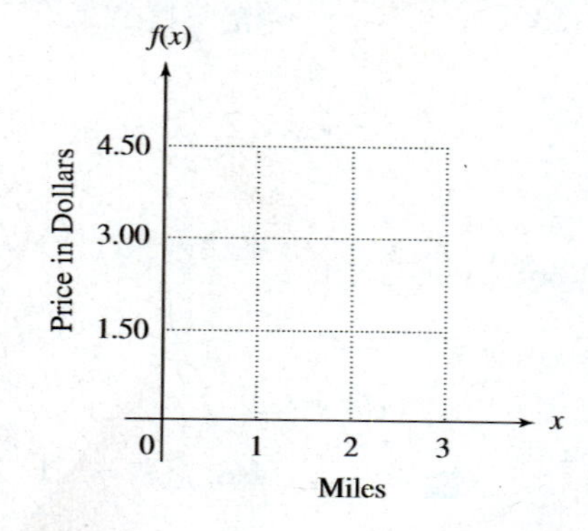

56. Suppose that a package weighing x pounds costs $f(x)$ dollars to mail to a given location, where

$$f(x) = 2.75x.$$

(a) What is the value of $f(3)$?

(b) In your own words, describe what 3 and the value $f(3)$ mean in part (a), using the terms *independent variable* and *dependent variable*.

(c) How much would it cost to mail a 5-lb package? Write the answer using function notation.

4.6 Variation

Objectives

1. Write an equation expressing direct variation.
2. Find the constant of variation, and solve direct variation problems.
3. Solve inverse variation problems.
4. Solve joint variation problems.
5. Solve combined variation problems.

Certain types of functions are very common, especially in business and the physical sciences. These are functions where y depends on a multiple of x, or y depends on a number divided by x. In such situations, y is said to *vary directly as x* (in the first case) or *vary inversely as x* (in the second case). For example, by the distance formula, the distance traveled varies directly as the rate (or speed) and the time. The simple interest formula and the formulas for area and volume are other familiar examples of *direct variation.*

On the other hand, the force required to keep a car from skidding on a curve varies inversely as the radius of the curve. Other examples of *inverse variation* are how travel time is inversely proportional to rate or speed and how, for a predetermined gain, principal is inversely proportional to the length of time the amount is invested.

1 Write an equation expressing direct variation. The circumference of a circle is given by the formula $C = 2\pi r$, where r is the radius of the circle. As the formula shows, the circumference is always a constant multiple of the radius. (C is always found by multiplying r by the constant 2π.) Because of this, the circumference is said to *vary directly* as the radius.

Direct Variation

y varies directly as x if there exists some constant k such that

$$y = kx.$$

Also, y is said to be **proportional to** x. The number k is called the **constant of variation.** In direct variation, for $k > 0$, as the value of x increases, the value of y also increases. Similarly, as x decreases, y decreases.

2 Find the constant of variation, and solve direct variation problems. The direct variation equation defines a linear function. In applications, functions are often defined by variation equations. For example, if Tom earns $8 per hour, his wages vary directly as, or are proportional to, the number of hours he works. If y represents his total wages and x the number of hours he has worked, then

$$y = 8x.$$

Here k, the constant of variation, is 8.

Example 1 Finding the Constant of Variation and the Variation Equation

Steven Hidalgo is paid an hourly wage. One week he worked 43 hr and was paid $795.50. How much does he earn per hour?

Let h represent the number of hours he works and P represent his corresponding pay. Then, P **varies directly as** h, so

$$P = kh.$$

Continued on Next Page

Here k represents Steven's hourly wage. Since $P = 795.50$ when $h = 43$,

$$795.50 = 43k$$

$$k = 18.50. \quad \text{Use a calculator.}$$

His hourly wage is \$18.50, and P and h are related by

$$P = 18.50h.$$

Work Problem 1 at the Side.

1 Find the constant of variation, and write a direct variation equation.

(a) Suzanne Alley is paid a daily wage. One month she worked 17 days and earned \$1334.50.

(b) Distance varies directly as time (at a constant speed). A car travels 100 mi at a constant speed in 2 hr.

Example 2 Solving a Direct Variation Problem

Hooke's law for an elastic spring states that the distance a spring stretches is proportional to the force applied. If a force of 150 newtons* stretches a certain spring 8 cm, how much will a force of 400 newtons stretch the spring? See Figure 25.

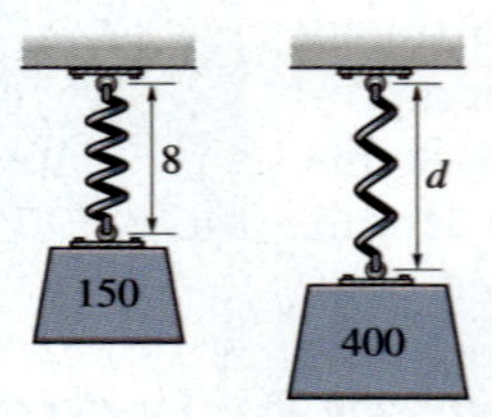

Figure 25

If d is the distance the spring stretches and f is the force applied, then $d = kf$ for some constant k. Since a force of 150 newtons stretches the spring 8 cm, we can use these values to find k.

$$d = kf \quad \text{Variation equation}$$

$$8 = k \cdot 150 \quad \text{Let } d = 8 \text{ and } f = 150.$$

$$k = \frac{8}{150} \quad \text{Find } k.$$

$$k = \frac{4}{75}$$

Substitute $\frac{4}{75}$ for k in the variation equation $d = kf$ to get

$$d = \frac{4}{75}f.$$

For a force of 400 newtons,

$$d = \frac{4}{75}(400) \quad \text{Let } f = 400.$$

$$= \frac{64}{3}.$$

The spring will stretch $\frac{64}{3}$ cm if a force of 400 newtons is applied.

Work Problem 2 at the Side.

2 The charge (in dollars) to customers for electricity (in kilowatt-hours) varies directly as the number of kilowatt-hours used. It costs \$52 to use 800 kilowatt-hours. Find the cost to use 1000 kilowatt-hours.

In summary, use the following steps to solve a variation problem.

* A newton is a unit of measure of force used in physics.

ANSWERS

1. (a) $k = 78.50$; Let E represent her earnings for d days. Then $E = 78.50d$. (b) $k = 50$; Let d represent the distance traveled in h hours. Then $d = 50h$.
2. \$65

Solving a Variation Problem

Step 1 Write the variation equation.

Step 2 Substitute the initial values and solve for k.

Step 3 Rewrite the variation equation with the value of k from Step 2.

Step 4 Substitute the remaining values, solve for the unknown, and find the required answer.

The direct variation equation $y = kx$ is a linear equation. However, other kinds of variation involve other types of equations. For example, one variable can be proportional to a power of another variable.

Direct Variation as a Power

y varies directly as the nth power of x if there exists a real number k such that

$$y = kx^n.$$

An example of direct variation as a power is the formula for the area of a circle, $A = \pi r^2$. Here, π is the constant of variation, and the area varies directly as the square of the radius.

Example 3 Solving a Direct Variation Problem

The distance a body falls from rest varies directly as the square of the time it falls (disregarding air resistance). If a skydiver falls 64 ft in 2 sec, how far will she fall in 8 sec?

Step 1 If d represents the distance the skydiver falls and t the time it takes to fall, then d is a function of t, and

$$d = kt^2$$

for some constant k.

Step 2 To find the value of k, use the fact that the skydiver falls 64 ft in 2 sec.

$$d = kt^2 \quad \text{Variation equation}$$
$$64 = k(2)^2 \quad \text{Let } d = 64 \text{ and } t = 2.$$
$$k = 16 \quad \text{Find } k.$$

Step 3 Using 16 for k, the variation equation becomes

$$d = 16t^2.$$

Step 4 Now let $t = 8$ to find the number of feet the skydiver will fall in 8 sec.

$$d = 16(8)^2 \quad \text{Let } t = 8.$$
$$= 1024$$

The skydiver will fall 1024 ft in 8 sec.

Work Problem 3 at the Side.

3 Solve inverse variation problems. In direct variation, where $k > 0$, as x increases, y increases. Similarly, as x decreases, y decreases. Another type of variation is *inverse variation.* With inverse variation, where $k > 0$, as one variable increases, the other variable decreases. For example, in a closed

3 The area of a circle varies directly as the square of its radius. A circle with radius 3 in. has area 28.278 in.^{2}.

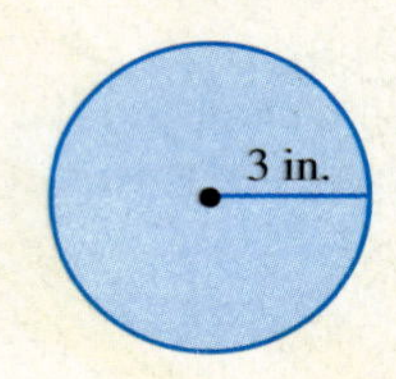

(a) Write a variation equation and give the value of k.

(b) What is the area of a circle with radius 4.1 in.?

ANSWERS

3. (a) $A = kr^2$; 3.142
(b) 52.817 in.^{2} (to the nearest thousandth)

space, volume decreases as pressure increases, as illustrated by a trash compactor. See Figure 26. As the compactor presses down, the pressure on the trash increases; in turn, the trash occupies a smaller space.

Figure 26

Inverse Variation

y varies inversely as x if there exists a real number k such that

$$y = \frac{k}{x}.$$

Also, **y varies inversely as the nth power of x** if there exists a real number k such that

$$y = \frac{k}{x^n}.$$

Notice that the inverse variation equation also defines a function. Since x is in the denominator, these functions are called *rational functions.* (Rational functions are discussed in Chapter 7.) Another example of inverse variation can be found by looking at the formula for the area of a parallelogram. In its usual form, the formula is

$$A = bh.$$

Dividing each side by b gives

$$h = \frac{A}{b}.$$

Here, h (height) varies inversely as b (base), with A (the area) serving as the constant of variation. For example, if a parallelogram has an area of 72 in.^{2}, the values of b and h might be any of the following.

$$\left.\begin{aligned} b &= 2, h = 36 \\ b &= 3, h = 24 \\ b &= 4, h = 18 \end{aligned}\right\} \text{As } b \text{ increases, } h \text{ decreases.} \qquad \left.\begin{aligned} b &= 12, h = 6 \\ b &= 9, \; h = 8 \\ b &= 8, \; h = 9 \end{aligned}\right\} \text{As } b \text{ decreases, } h \text{ increases.}$$

Example 4 **Solving an Inverse Variation Problem**

The weight of an object above Earth varies inversely as the square of its distance from the center of Earth. A space shuttle in an elliptical orbit has a maximum distance from the center of Earth (apogee) of 6700 mi. Its minimum distance from the center of Earth (perigee) is 4090 mi. See Figure 27. If an astronaut in the shuttle weighs 57 lb at its apogee, what does the astronaut weigh at its perigee?

Continued on Next Page

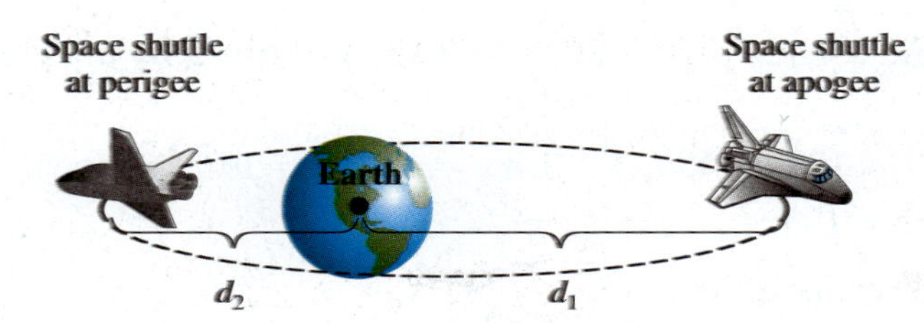

Figure 27

If w is the weight and d is the distance from the center of Earth, then

$$w = \frac{k}{d^2}$$

for some constant k. At the apogee the astronaut weighs 57 lb, and the distance from the center of Earth is 6700 mi. Use these values to find k.

$$57 = \frac{k}{(6700)^2} \quad \text{Let } w = 57 \text{ and } d = 6700.$$

$$k = 57(6700)^2$$

Then the weight at the perigee with $d = 4090$ mi is

$$w = \frac{57(6700)^2}{(4090)^2} \approx 153 \text{ lb.} \quad \text{Use a calculator.}$$

Work Problem 4 at the Side.

4 **Solve joint variation problems.** It is common for one variable to depend on several others. If one variable varies directly as the *product* of several other variables (perhaps raised to powers), the first variable is said to **vary jointly** as the others.

CAUTION

Note that *and* in the expression "y varies directly as m *and* n" translates as the product

$$y = kmn.$$

The word *and* does not indicate addition here.

Example 5 Solving a Joint Variation Problem

The interest on a loan or an investment is given by the formula $I = prt$. Here, for a given principal p, the interest earned I varies jointly as the interest rate r and the time t the principal is left at interest. If an investment earns \$100 interest at 5% for 2 yr, how much interest will the same principal earn at 4.5% for 3 yr?

We use the formula $I = prt$, where p is the constant of variation because it is the same for both investments. For the first investment, we have $I = 100$, $r = .05$, and $t = 2$, so

$$I = prt$$

$$100 = p(.05)(2) \quad \text{Let } I = 100, r = .05, \text{ and } t = 2.$$

$$100 = .1p$$

$$\frac{100}{.1} = p$$

$$p = 1000.$$

Continued on Next Page

4 If the temperature is constant, the volume of a gas varies inversely as the pressure. For a certain gas, the volume is 10 cm^3 when the pressure is 6 kg per cm^2.

(a) Find the variation equation.

(b) Find the volume when the pressure is 12 kg per cm^2.

ANSWERS

4. (a) $V = \frac{60}{P}$ (b) 5 cm^3

5 The volume of a rectangular box of a given height is proportional to its width and length. A box with width 2 ft and length 4 ft has volume 12 ft^3. Find the volume of a box with the same height that is 3 ft wide and 5 ft long.

6 The maximum load that a cylindrical column with a circular cross section can hold varies directly as the fourth power of the diameter of the cross section and inversely as the square of the height. A 9-m column 1 m in diameter will support 8 metric tons. How many metric tons can be supported by a column 12 m high and $\frac{2}{3}$ m in diameter?

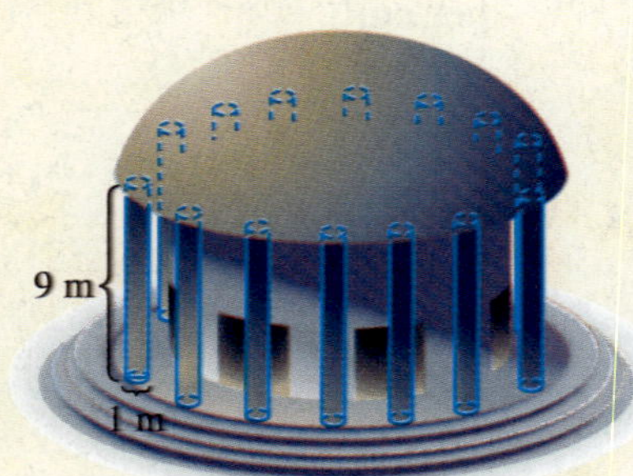

Load = 8 metric tons

Answers

5. 22.5 ft^3

6. $\frac{8}{9}$ metric ton

Now we find I when $p = 1000$, $r = .045$, and $t = 3$ by substituting into $I = prt$.

$$I = 1000(.045)(3) \qquad \text{Let } p = 1000,\ r = .045, \text{ and } t = 3.$$
$$I = 135$$

The interest will be \$135.

Work Problem 5 at the Side.

5 Solve combined variation problems. There are many combinations of direct and inverse variation. Example 6 shows a typical **combined variation** problem.

Example 6 Solving a Combined Variation Problem

Body mass index, or BMI, is used by physicians to assess a person's level of fatness. A BMI from 19 through 25 is considered desirable. BMI varies directly as an individual's weight in pounds and inversely as the square of the individual's height in inches. A person who weighs 118 lb and is 64 in. tall has a BMI of 20. (The BMI is rounded to the nearest whole number.) Find the BMI of a person who weighs 165 lb with a height of 70 in.

Let B represent the BMI, w the weight, and h the height. Then

$$B = \frac{kw}{h^2}. \qquad \begin{array}{l}\leftarrow \text{BMI varies directly as the weight.} \\ \leftarrow \text{BMI varies inversely as the square of the height.}\end{array}$$

To find k, let $B = 20$, $w = 118$, and $h = 64$.

$$20 = \frac{k(118)}{64^2}$$

$$k = \frac{20(64^2)}{118} \qquad \text{Multiply by } 64^2\text{; divide by } 118.$$

$$k \approx 694 \qquad \text{Use a calculator.}$$

Now find B when $k = 694$, $w = 165$, and $h = 70$.

$$B = \frac{694(165)}{70^2} \approx 23 \qquad \text{Nearest whole number}$$

The person's BMI is 23.

Work Problem 6 at the Side.

4.6 EXERCISES

FOR EXTRA HELP

Student's Solutions Manual

MyMathLab.com

InterAct Math Tutorial Software

AW Math Tutor Center

www.mathxl.com

Digital Video Tutor CD 3 Videotape 5

Determine whether each equation represents direct, inverse, joint, *or* combined *variation.*

1. $y = \frac{3}{x}$

2. $y = \frac{8}{x}$

3. $y = 10x^2$

4. $y = 2x^3$

5. $y = 3xz^4$

6. $y = 6x^3z^2$

7. $y = \frac{4x}{wz}$

8. $y = \frac{6x}{st}$

Solve each problem. See Examples 2–5.

9. If x varies directly as y, and $x = 9$ when $y = 3$, find x when $y = 12$.

10. If x varies directly as y, and $x = 10$ when $y = 7$, find y when $x = 50$.

11. If z varies inversely as w, and $z = 10$ when $w = .5$, find z when $w = 8$.

12. If t varies inversely as s, and $t = 3$ when $s = 5$, find s when $t = 5$.

13. p varies jointly as q and r^2, and $p = 200$ when $q = 2$ and $r = 3$. Find p when $q = 5$ and $r = 2$.

14. f varies jointly as g^2 and h, and $f = 50$ when $g = 4$ and $h = 2$. Find f when $g = 3$ and $h = 6$.

15. For $k > 0$, if y varies directly as x, when x increases, y ________, and when x decreases, y ________.

16. For $k > 0$, if y varies inversely as x, when x increases, y ________, and when x decreases, y ________.

17. Explain the difference between inverse variation and direct variation.

18. What is meant by the constant of variation in a direct variation problem? If you were to graph the linear equation $y = kx$ for some constant k, what role would the value of k play in the graph?

Solve each problem involving variation. See Examples 1–6.

19. Todd bought 8 gal of gasoline and paid \$13.59. To the nearest tenth of a cent, what is the price of gasoline per gallon?

20. Melissa gives horseback rides at Shadow Mountain Ranch. A 2.5-hr ride costs \$50.00. What is the price per hour?

21. The volume of a can of tomatoes is proportional to the height of the can. If the volume of the can is 300 cm^3 when its height is 10.62 cm, find the volume of a can with height 15.92 cm.

22. The weight of an object on Earth is directly proportional to the weight of that same object on the moon. A 200-lb astronaut would weigh 32 lb on the moon. How much would a 50-lb dog weigh on the moon?

23. A large federally funded research project at the University of Michigan's Institute for Social Research indicated that the higher a woman's BMI (see Example 6) in late middle age, the lower her net worth. That is, her BMI varies inversely as her net worth. This was not true for men. (*Source: Sacramento Bee,* December 13, 2000.) Suppose such a woman with a BMI of 35 has a net worth of \$10,000. According to the study results, what would be the net worth of a woman with a BMI of 38?

24. The frequency (number of vibrations per second) of a vibrating guitar string varies inversely as its length. That is, a longer string vibrates fewer times in a second than a shorter string. Suppose a guitar string .65 m long vibrates 4.3 times per sec. What frequency would a string .5 m long have?

25. The amount of light (measured in foot-candles) produced by a light source varies inversely as the square of the distance from the source. If the illumination produced 1 m from a light source is 768 foot-candles, find the illumination produced 6 m from the same source.

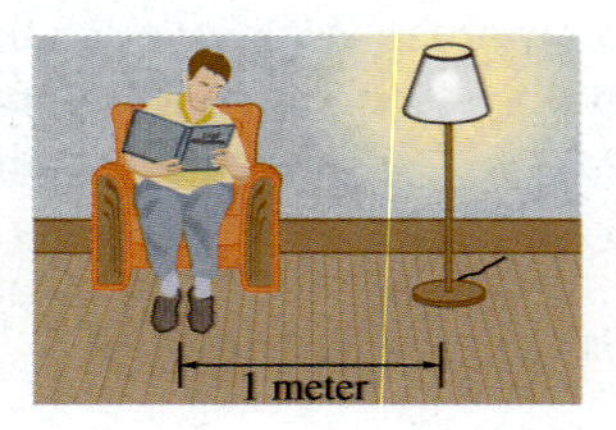

26. The current in a simple electrical circuit is inversely proportional to the resistance. If the current is 20 amperes (an *ampere* is a unit for measuring current) when the resistance is 5 ohms, find the current when the resistance is 7.5 ohms.

27. For a given interest rate, simple interest varies jointly as principal and time. If $2000 left in an account for 4 yr earned interest of $280, how much interest would be earned in 6 yr?

28. The collision impact of an automobile varies jointly as its mass and the square of its speed. Suppose a 2000-lb car traveling at 55 mph has a collision impact of 6.1. What is the collision impact of the same car at 65 mph?

29. The force needed to keep a car from skidding on a curve varies inversely as the radius of the curve and jointly as the weight of the car and the square of the speed. If 242 lb of force keep a 2000-lb car from skidding on a curve of radius 500 ft at 30 mph, what force would keep the same car from skidding on a curve of radius 750 ft at 50 mph?

30. Natural gas provides 35.8% of U.S. energy. (*Source:* U.S. Energy Department.) The volume of gas varies inversely as the pressure and directly as the temperature. (Temperature must be measured in *Kelvin* (K), a unit of measurement used in physics.) If a certain gas occupies a volume of 1.3 L at 300 K and a pressure of 18 newtons per cm^2, find the volume at 340 K and a pressure of 24 newtons per cm^2.

31. The number of long-distance phone calls between two cities in a certain time period varies jointly as the populations of the cities, p_1 and p_2, and inversely as the distance between them. If 80,000 calls are made between two cities 400 mi apart, with populations of 70,000 and 100,000, how many calls are made between cities with populations of 50,000 and 75,000 that are 250 mi apart?

32. A body mass index from 27 through 29 carries a slight risk of weight-related health problems, while one of 30 or more indicates a great increase in risk. Use your own height and weight and the information in Example 6 to determine whether you are at risk.

Exercises 33 and 34 describe weight-estimation formulas that fishermen have used over the years. Girth *is the distance around the body of the fish.* (*Source: Sacramento Bee,* November 9, 2000.)

33. The weight of a bass varies jointly as its girth and the square of its length. A prize-winning bass weighed in at 22.7 lb and measured 36 in. long with 21 in. girth. How much would a bass 28 in. long with 18 in. girth weigh?

34. The weight of a trout varies jointly as its length and the square of its girth. One angler caught a trout that weighed 10.5 lb and measured 26 in. long with 18 in. girth. Find the weight of a trout that is 22 in. long with 15 in. girth.

RELATING CONCEPTS (Exercises 35–40) FOR INDIVIDUAL OR GROUP WORK

A routine activity such as pumping gasoline can be related to many of the concepts studied in this chapter. Suppose that premium unleaded costs $1.75 *per gal.* ***Work Exercises 35–40 in order.***

35. 0 gal of gasoline cost $0.00, while 1 gal costs $1.75. Represent these two pieces of information as ordered pairs of the form (gallons, price).

36. Use the information from Exercise 35 to find the slope of the line on which the two points lie.

37. Write the slope-intercept form of the equation of the line on which the two points lie.

38. Using function notation, if $f(x) = ax + b$ represents the line from Exercise 37, what are the values of a and b?

39. How does the value of a from Exercise 38 relate to gasoline in this situation? With relationship to the line, what do we call this number?

40. Why does the equation from Exercise 38 satisfy the conditions for direct variation? In the context of variation, what do we call the value of a?

Chapter 4

Summary

Key Terms

	Term	Definition
4.1	**ordered pair**	An ordered pair is a pair of numbers written in parentheses in which the order of the numbers matters.
	origin	When two number lines intersect at a right angle, the origin is the common 0 point.
	***x*-axis**	The horizontal number line in a rectangular coordinate system is called the x-axis.
	***y*-axis**	The vertical number line in a rectangular coordinate system is called the y-axis.
	rectangular (Cartesian) coordinate system	Two number lines that intersect at a right angle at their 0 points form a rectangular coordinate system, also called the Cartesian coordinate system.
	plot	To plot an ordered pair is to locate it on a rectangular coordinate system.
	coordinate	Each number in an ordered pair is a coordinate of the corresponding point.
	quadrant	A quadrant is one of the four regions in the plane determined by a rectangular coordinate system.
	graph of an equation	The graph of an equation is the set of points corresponding to all ordered pairs that satisfy the equation.
	first-degree equation	A first-degree equation has no term with a variable to a power greater than one.
	linear equation in two variables	A first-degree equation with two variables is a linear equation in two variables.
	***x*-intercept**	The point where a line intersects the x-axis is the x-intercept.
	***y*-intercept**	The point where a line intersects the y-axis is the y-intercept.
4.2	**rise**	The rise of a line is the vertical change between two points on the line.
	run	The run of a line is the horizontal change between two points on the line.
	slope	The ratio of the change in y compared to the change in x (rise/run) along a line is the slope of the line.
4.4	**linear inequality in two variables**	A linear inequality in two variables is a first-degree inequality with two variables.
	boundary line	In the graph of a linear inequality, the boundary line separates the region that satisfies the inequality from the region that does not satisfy the inequality.
4.5	**dependent variable**	If the quantity y depends on x, then y is called the dependent variable in a relation between x and y.
	independent variable	If y depends on x, then x is the independent variable in a relation between x and y.
	relation	A relation is a set of ordered pairs of real numbers.
	function	A function is a set of ordered pairs in which each value of the first component, x, corresponds to exactly one value of the second component, y.
	domain	The domain of a relation is the set of first components (x-values) of the ordered pairs of the relation.
	range	The range of a relation is the set of second components (y-values) of the ordered pairs of the relation.
	graph of a relation	The graph of a relation is the graph of the ordered pairs of the relation.
	function notation	The function notation $f(x)$ is another way to represent the dependent variable y for the function f.
	linear function	A function that is defined by $f(x) = mx + b$ is a linear function.

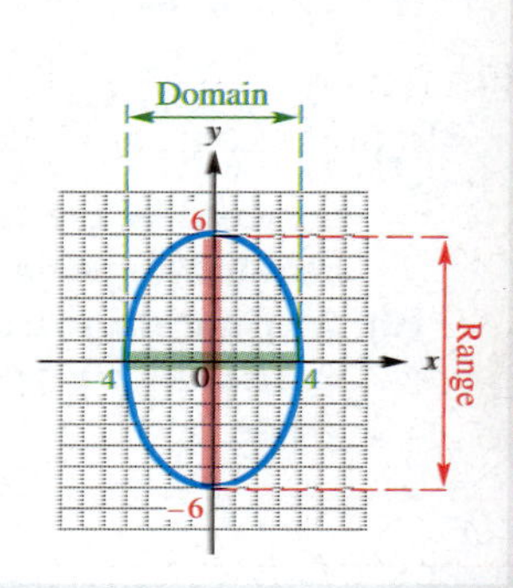

	constant function	A constant function is a linear function of the form $f(x) = d$, for a real number d.
4.6	**variation equation**	A variation equation describes how a dependent variable varies with respect to the corresponding independent variable.

New Symbols

(a, b) ordered pair

x_1 a specific value of the variable x (read "x sub one")

m slope

$f(x)$ function of x (read "f of x")

Test Your Word Power

See how well you have learned the vocabulary in this chapter. Answers follow the Quick Review.

1. An **ordered pair** is a pair of numbers written
   - **(a)** in numerical order between brackets
   - **(b)** between parentheses or brackets
   - **(c)** between parentheses in which order is important
   - **(d)** between parentheses in which order does not matter.

2. The **coordinates** of a point are
   - **(a)** the numbers in the corresponding ordered pair
   - **(b)** the solution of an equation
   - **(c)** the values of the x- and y-intercepts
   - **(d)** the graph of the point.

3. A **linear equation in two variables** is an equation that can be written in the form
   - **(a)** $Ax + By < C$
   - **(b)** $ax = b$
   - **(c)** $y = x^2$
   - **(d)** $Ax + By = C$.

4. An **intercept** is
   - **(a)** the point where the x-axis and y-axis intersect
   - **(b)** a pair of numbers written between parentheses in which order matters
   - **(c)** one of the four regions determined by a rectangular coordinate system
   - **(d)** the point where a graph intersects the x-axis or the y-axis.

5. The **slope** of a line is
   - **(a)** the measure of the run over the rise of the line
   - **(b)** the distance between two points on the line
   - **(c)** the ratio of the change in y to the change in x along the line
   - **(d)** the horizontal change compared to the vertical change of two points on the line.

6. In a relationship between two variables x and y, the **independent variable** is
   - **(a)** x, if x depends on y
   - **(b)** x, if y depends on x
   - **(c)** either x or y
   - **(d)** the larger of x and y.

7. In a relationship between two variables x and y, the **dependent variable** is
   - **(a)** y, if y depends on x
   - **(b)** y, if x depends on y
   - **(c)** either x or y
   - **(d)** the smaller of x and y.

8. A **relation** is
   - **(a)** a set of ordered pairs
   - **(b)** the ratio of the change in y to the change in x along a line
   - **(c)** the set of all possible values of the independent variable
   - **(d)** all the second elements of a set of ordered pairs.

9. A **function** is
   - **(a)** the numbers in an ordered pair
   - **(b)** a set of ordered pairs in which each x-value corresponds to exactly one y-value
   - **(c)** a pair of numbers written between parentheses in which order matters
   - **(d)** the set of all ordered pairs that satisfy an equation.

10. The **domain** of a function is
    - **(a)** the set of all possible values of the dependent variable y
    - **(b)** a set of ordered pairs
    - **(c)** the difference between the x-values
    - **(d)** the set of all possible values of the independent variable x.

11. The **range** of a function is
    - **(a)** the set of all possible values of the dependent variable y
    - **(b)** a set of ordered pairs
    - **(c)** the difference between the y-values
    - **(d)** the set of all possible values of the independent variable x.

Quick Review

Concepts	*Examples*
4.1 The Rectangular Coordinate System	
Finding Intercepts To find the x-intercept, let $y = 0$. To find the y-intercept, let $x = 0$.	The graph of $2x + 3y = 12$ has x-intercept $(6, 0)$ and y-intercept $(0, 4)$.
4.2 Slope	
If $x_1 \neq x_2$, then $m = \dfrac{\text{rise}}{\text{run}} = \dfrac{y_2 - y_1}{x_2 - x_1}$.	For $2x + 3y = 12$, $m = \dfrac{4 - 0}{0 - 6} = -\dfrac{2}{3}$.
A vertical line has undefined slope.	$x = 3$ has undefined slope.
A horizontal line has 0 slope.	$y = -5$ has $m = 0$.
Parallel lines have equal slopes.	$y = 2x + 5$, $m = 2$ $4x - 2y = 6$ $-2y = -4x + 6$ $y = 2x - 3$ $m = 2$ These lines are **parallel**.
The slopes of perpendicular lines are negative reciprocals (with a product of -1).	$y = 3x - 1$, $m = 3$ $x + 3y = 4$ $3y = -x + 4$ $y = -\dfrac{1}{3}x + \dfrac{4}{3}$ $m = -\dfrac{1}{3}$ These lines are **perpendicular**.
4.3 Linear Equations in Two Variables	
Point-Slope Form $y - y_1 = m(x - x_1)$	$y - 3 = 4(x - 5)$ $(5, 3)$ is on the line, $m = 4$.
Slope-Intercept Form $y = mx + b$	$y = 2x + 3$ $m = 2$, y-intercept is $(0, 3)$.
Standard Form $Ax + By = C$	$2x - 5y = 8$
Vertical Line $x = c$	$x = -1$
Horizontal Line $y = d$	$y = 4$

Concepts	*Examples*
4.4 Linear Inequalities in Two Variables	
Graphing a Linear Inequality	
Step 1 Draw the graph of the line that is the boundary. Make the line solid if the inequality involves $\leq$ or $\geq$; make the line dashed if the inequality involves $<$ or $>$.	Graph $2x - 3y \leq 6$. Draw the graph of $2x - 3y = 6$. Use a solid line because the symbol $\leq$ is used.
Step 2 Choose any point not on the line as a test point.	Choose (1, 2). $2(1) - 3(2) = 2 - 6 \leq 6$ True
Step 3 Shade the region that includes the test point if the test point satisfies the original inequality; otherwise, shade the region on the other side of the boundary line.	Shade the side of the line that includes (1, 2). 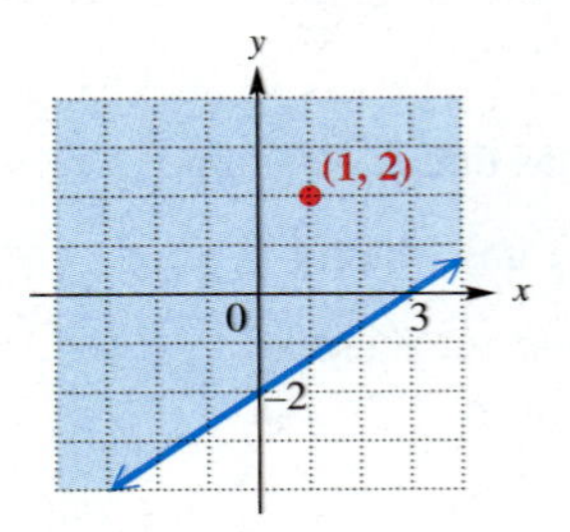
4.5 Introduction to Functions	
To evaluate a function using function notation (that is, $f(x)$ notation) for a given value of x, substitute the value wherever x appears.	If $f(x) = x^2 - 7x + 12$, then $f(1) = 1^2 - 7(1) + 12 = 6.$
To write the equation that defines a function in function notation, solve the equation for y.	Write $2x + 3y = 12$ in function notation. $3y = -2x + 12$; $y = -\frac{2}{3}x + 4$
Then replace y with $f(x)$.	$f(x) = -\frac{2}{3}x + 4$
4.6 Variation	
If there is some constant k such that: $y = kx^n$, then y varies directly as, or is proportional to, x^n; $y = \frac{k}{x^n}$, then y varies inversely as x^n.	The area of a circle **varies directly as** the square of the radius. $A = kr^2$ Pressure **varies inversely as** volume. $P = \frac{k}{V}$

Answers to Test Your Word Power

1. (c) *Examples:* (0, 3), (3, 8), (4, 0) **2. (a)** *Example:* The point associated with the ordered pair (1, 2) has x-coordinate 1 and y-coordinate 2. **3. (d)** *Examples:* $3x + 2y = 6$, $x = y - 7$, $4x = y$ **4. (d)** *Example:* In Figure 4(b) of Section 4.1, the x-intercept is (3, 0) and the y-intercept is (0, 2). **5. (c)** *Example:* The line through (3, 6) and (5, 4) has slope $\frac{4-6}{5-3} = \frac{-2}{2} = -1$. **6. (b)** *Example:* See Answer 7, which follows. **7. (a)** *Example:* When borrowing money, the amount you borrow (independent variable) determines the size of your payments (dependent variable). **8. (a)** *Example:* The set $\{(2, 0), (4, 3), (6, 6), (8, 9)\}$ defines a relation. **9. (b)** *Example:* The relation given in Answer 8 is a function since each x-value corresponds to exactly one y-value. **10. (d)** *Example:* In the function in Answer 8, the domain is the set of x-values, $\{2, 4, 6, 8\}$. **11. (a)** *Example:* In the function in Answer 8, the range is the set of y-values, $\{0, 3, 6, 9\}$.

Chapter 4 Review Exercises

[4.1] *Complete the given ordered pairs for each equation, and then graph the equation.*

1. $3x + 2y = 6$
   (0,), (, 0), (, −2)

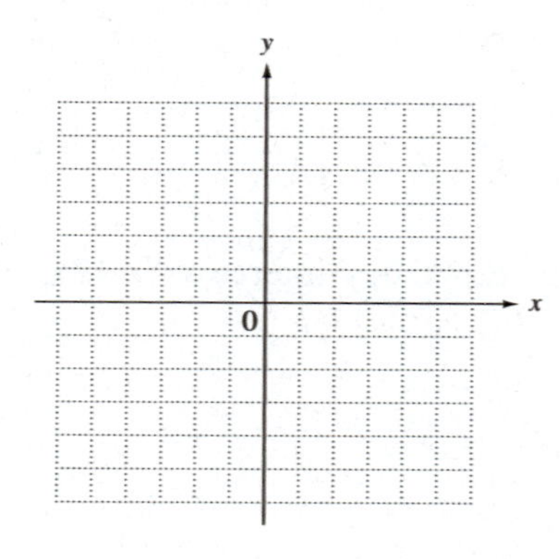

2. $x - y = 6$
   (2,), (, −3), (1,), (, −2)

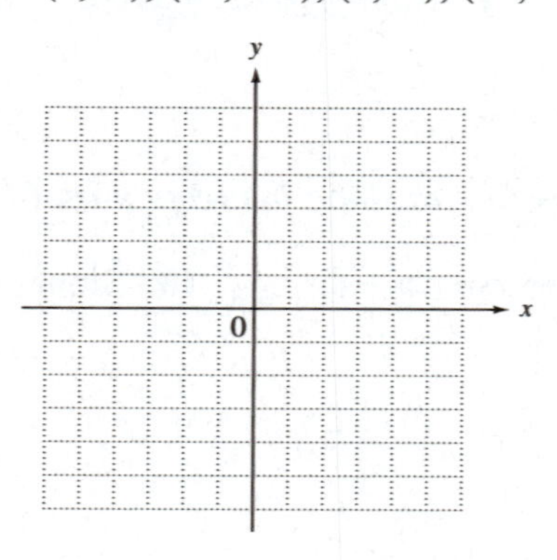

Find the x- and y-intercepts, and then graph each equation.

3. $4x + 3y = 12$

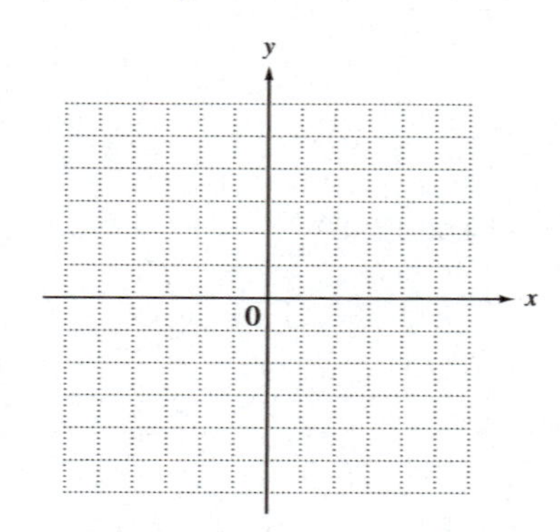

4. $5x + 7y = 15$

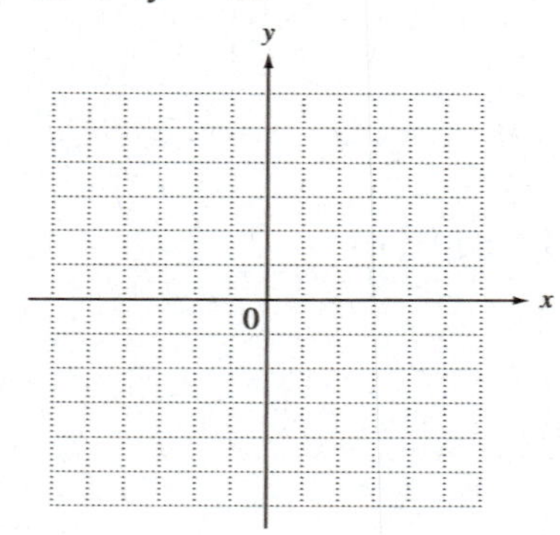

[4.2] *Find the slope of each line.*

5. Through (−1, 2) and (4, −6)

6. $y = 2x + 3$

7. $-3x + 4y = 5$

8. $y = 4$

9. A line parallel to $3y = -2x + 5$

10. A line perpendicular to $3x - y = 6$

Tell whether the line has positive, negative, 0, *or* undefined *slope.*

11.

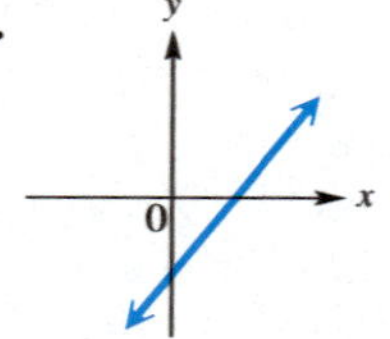

12.

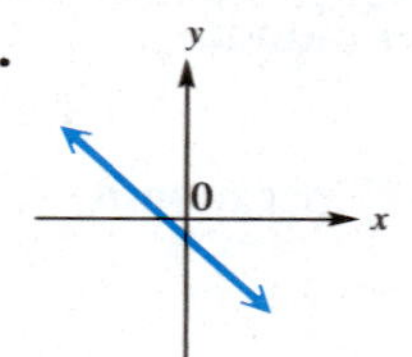

13.

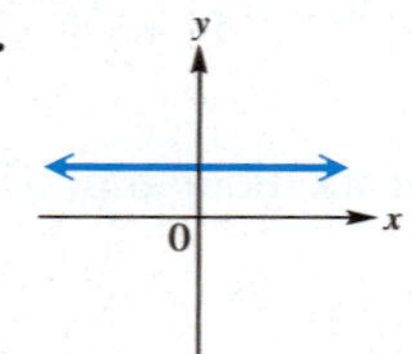

14.

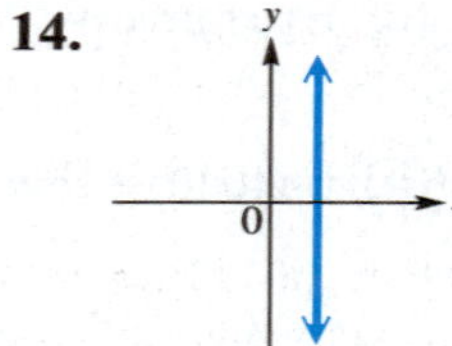

15. If the pitch of a roof is $\frac{1}{4}$, how many feet in the horizontal direction correspond to a rise of 3 ft?

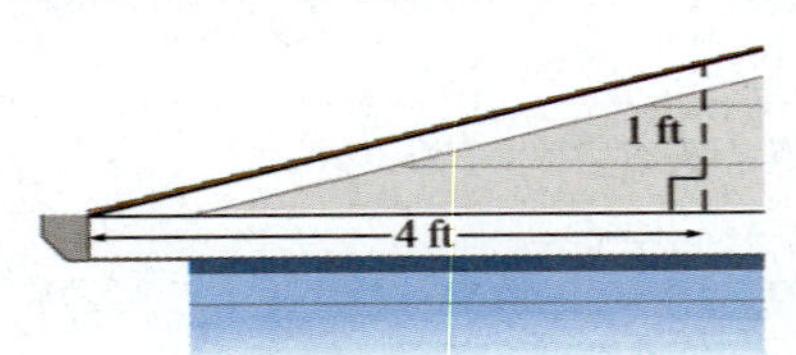

16. Family income in the United States has steadily increased for many years (primarily due to inflation). In 1970 the median family income was about $10,000 per yr. In 1998 it was about $47,000 per yr. Find the average rate of change of median family income over that period. (*Source:* U.S. Bureau of the Census.)

[4.3] *Write an equation in slope-intercept form (if possible) for each line.*

17. Slope $\frac{3}{5}$; y-intercept $(0, -8)$

18. Slope $-\frac{1}{3}$; y-intercept $(0, 5)$

19. Slope 0; y-intercept $(0, 12)$

20. Undefined slope; through $(2, 7)$

21. Horizontal; through $(-1, 4)$

22. Vertical; through $(.3, .6)$

23. Through $(2, -5)$ and $(1, 4)$

24. Through $(-3, -1)$ and $(2, 6)$

25. Parallel to $4x - y = 3$ and through $(6, -2)$

26. Perpendicular to $2x - 5y = 7$ and through $(0, 1)$

27. The Midwest Athletic Club (Section 4.3, Exercises 59 and 60) offers two special membership plans. (*Source:* Midwest Athletic Club.) For each plan, write a linear equation and give the cost y of a year's membership. Let x represent the number of months.

(a) Executive VIP/Gold membership: $159 fee plus $57 per month

(b) Executive Regular/Silver membership: $159 fee plus $47 per month

[4.4] *Graph each inequality.*

28. $3x - 2y \leq 12$

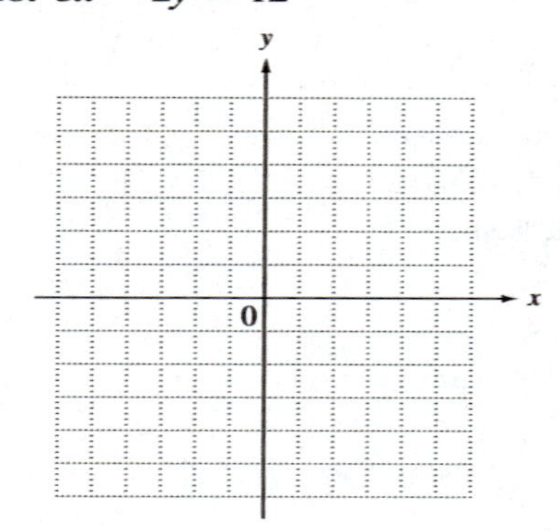

29. $5x - y > 6$

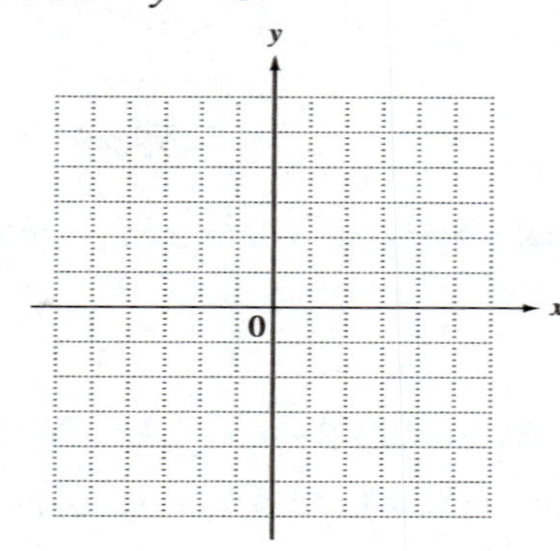

30. $x \geq 2$ or $y \geq 2$

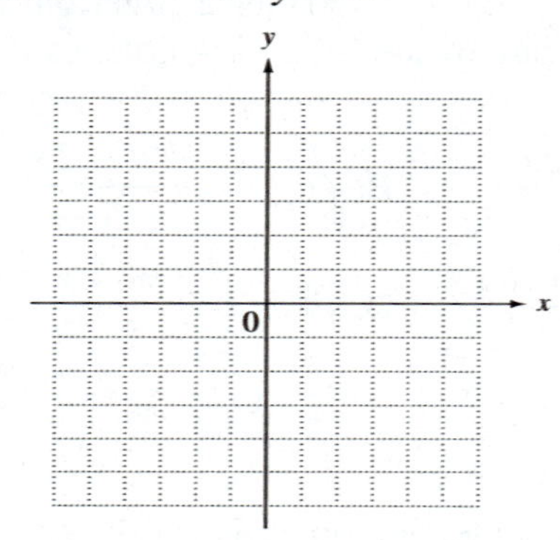

31. $2x + y \leq 1$ and $x \geq 2y$

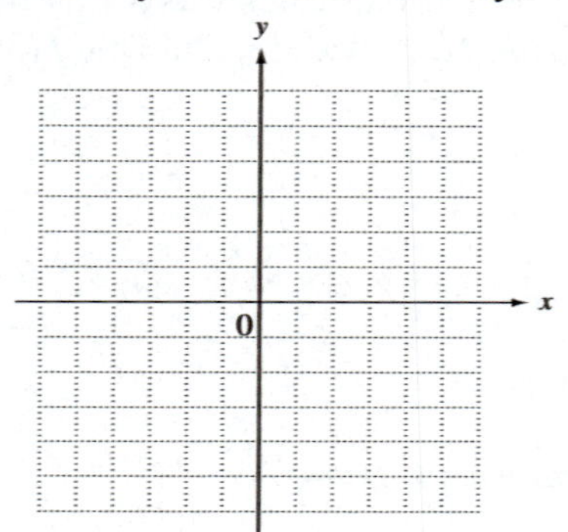

[4.5] *Give the domain and range of each relation. Identify any functions.*

32. $\{(-4, 2), (-4, -2), (1, 5), (1, -5)\}$

33. The number of small offices/home offices in 1996 for the top five states were {(California, 71,266), (New York, 50,101), (Texas, 48,010), (Pennsylvania, 42,142), (Washington, 38,240)}. (*Source:* Dunn & Bradstreet's Cottage Industry File.)

34.

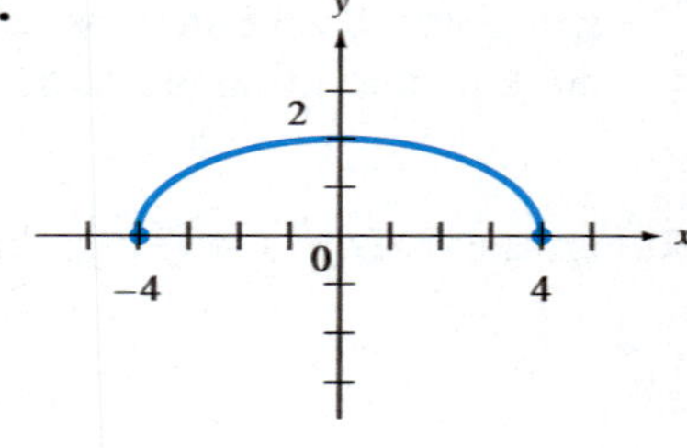

Determine whether each relation defines y as a function of x. Identify any linear functions. Give the domain in each case.

35. $y = 3x - 3$

36. $y < x + 2$

37. $y = |x - 4|$

38. $y = \sqrt{4x + 7}$

39. $x = y^2$

40. $y = \dfrac{7}{x - 36}$

41. Explain the test that allows us to determine whether a graph is that of a function.

Given $f(x) = -2x^2 + 3x - 6$, *find each of the following.*

42. $f(0)$

43. $f(3)$

44. $f(p)$

45. $f(-k)$

46. The equation $2x^2 - y = 0$ defines y as a function of x. Rewrite it using $f(x)$ notation, and find $f(3)$.

47. Suppose that $2x - 5y = 7$ defines a function. If $y = f(x)$, which one of the following defines the same function?

A. $f(x) = \dfrac{7 - 2x}{5}$

B. $f(x) = \dfrac{-7 - 2x}{5}$

C. $f(x) = \dfrac{-7 + 2x}{5}$

D. $f(x) = \dfrac{7 + 2x}{5}$

[4.6] *Solve each variation problem.*

48. The amount of water emptied by a pipe varies directly as the square of the diameter of the pipe. For a certain constant water flow, a pipe emptying into a canal will allow 200 gal of water to escape in an hour. The diameter of the pipe is 6 in. How much water would a 12-in. pipe empty into the canal in an hour, assuming the same water flow?

49. For the subject in a photograph to appear in the same perspective in the photograph as in real life, the viewing distance must be properly related to the amount of enlargement. For a particular camera, the viewing distance varies directly as the amount of enlargement. A picture taken with this camera that is enlarged 5 times should be viewed from a distance of 250 mm. Suppose a print 8.6 times the size of the negative is made. From what distance should it be viewed?

50. The force with which Earth attracts an object above Earth's surface varies inversely with the square of the object's distance from the center of Earth. If an object 4000 mi from the center of Earth is attracted with a force of 160 lb, find the force of attraction on an object 6000 mi from the center of Earth.

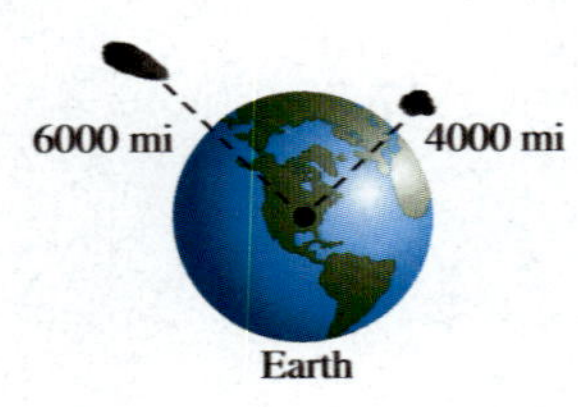

51. The period of a pendulum varies directly as the square root of the length of the pendulum and inversely as the square root of the acceleration due to gravity. Find the period when the length is 4 ft and the acceleration due to gravity is 32 ft per sec^2, if the period is 1.06π sec when the length is 9 ft and the acceleration due to gravity is 32 ft per sec^2.

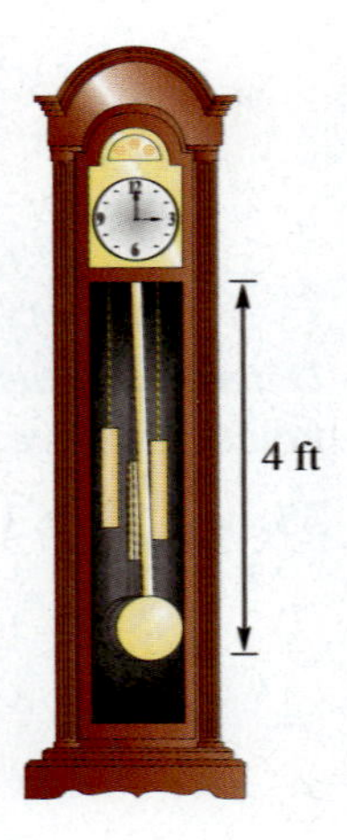

Chapter 4 Test

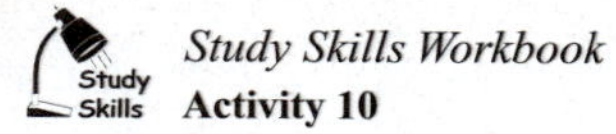

Study Skills Workbook
Activity 10

1. Find the slope of the line through (6, 4) and $(-4, -1)$.

1. ________________

For each line, find the slope and the x- and y-intercepts.

2. $3x - 2y = 13$

3. $y = 5$

4. Describe the graph of a line with undefined slope in a rectangular coordinate system.

2. ________________

3. ________________

4. ________________

5. ________________

Find the x- and y-intercepts, and graph each equation.

5. $4x - 3y = -12$

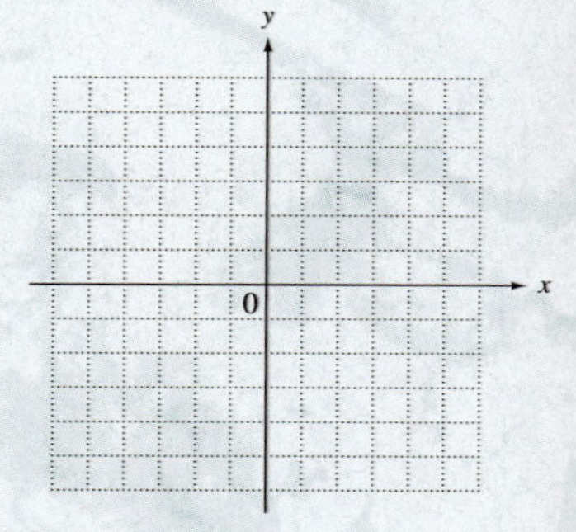

6. $y - 2 = 0$

6. ________________

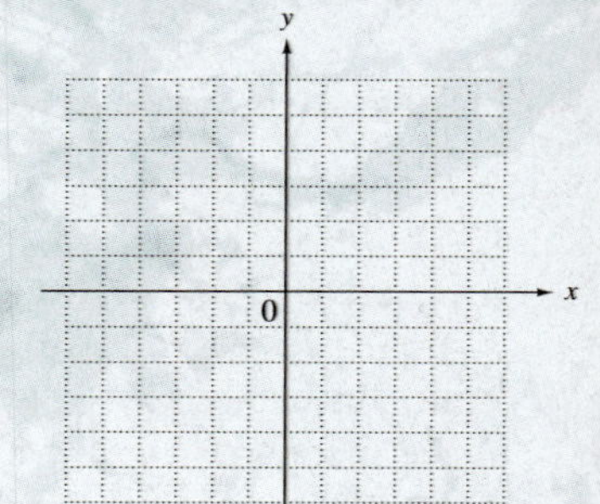

7. $y = -2x$

7. ________________

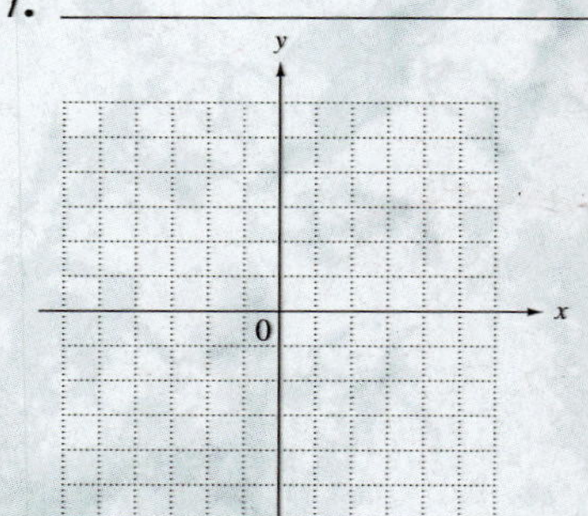

8. Graph $3x - 2y > 6$.

8.

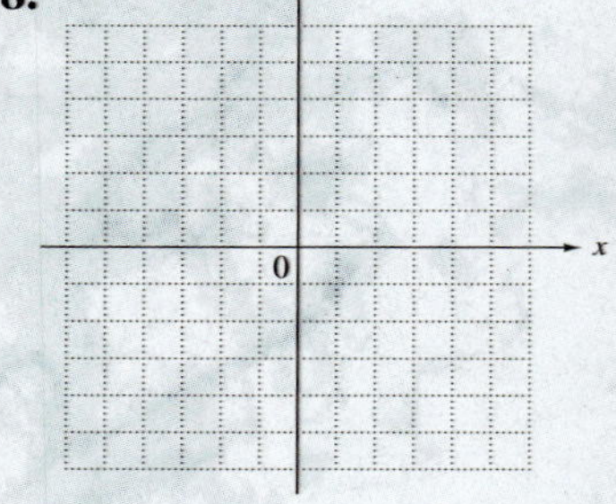

Write the equation of each line in slope-intercept form.

9. ____________

9. Through $(-3, 14)$; horizontal

10. ____________

10. Through $(4, -1)$; $m = -5$

11. (a) ____________

(b) ____________

11. Through $(-7, 2)$;

(a) parallel to $3x + 5y = 6$

(b) perpendicular to $y = 2x$

12. ____________

12. Which of the following is the graph of a function?

A.

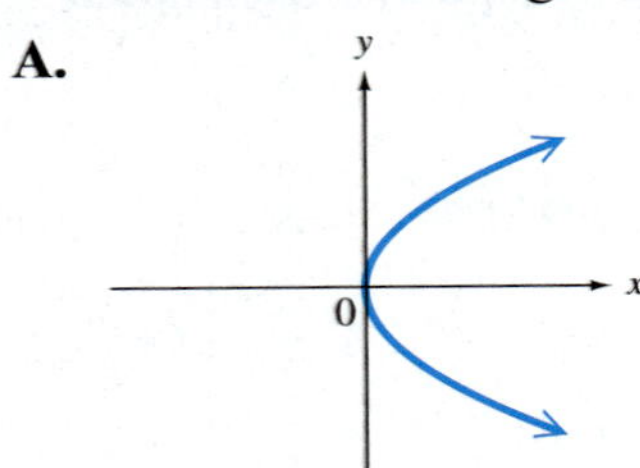

B.

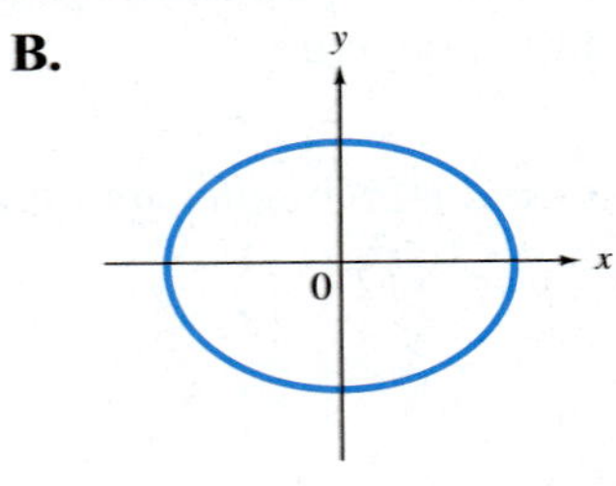

C.

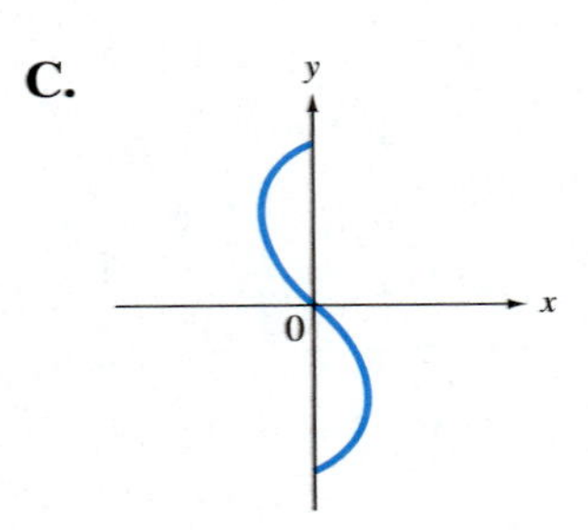

D.

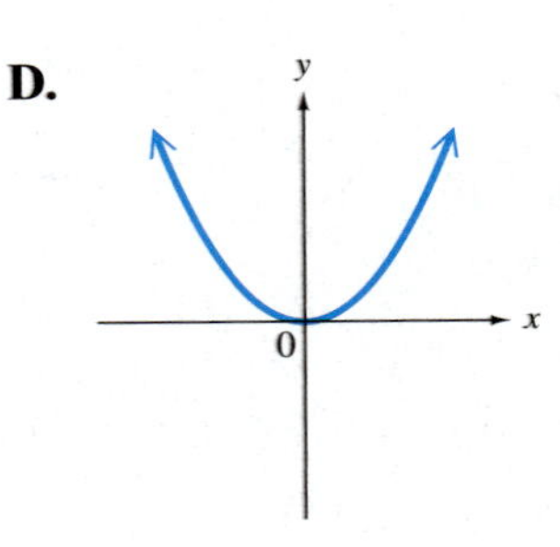

13. (a) ____________

(b) ____________

13. For the function defined by $f(x) = \sqrt{x - 3}$,

(a) give the domain, **(b)** find $f(7)$.

14. Deaths per 1000 population from 1993 through 1997 are shown in the table.

Death Rate	8.8	8.8	8.8	8.7	8.6
Year	1993	1994	1995	1996	1997

Source: U.S. National Center for Health Statistics.

Which set of ordered pairs from the table is a function? Explain.

14. ____________

A. {Death Rate, Year} **B.** {Year, Death Rate}

15. ____________

15. For a body falling freely from rest (disregarding air resistance), the distance the body falls varies directly as the square of the time. If an object is dropped from the top of a tower 576 ft high and hits the ground in 6 sec, how far did it fall in the first 4 sec?

16. ____________

16. The force of the wind blowing on a vertical surface varies jointly as the area of the surface and the square of the velocity. If a wind blowing at 40 mph exerts a force of 50 lb on a surface of 500 ft^2, how much force will a wind of 80 mph place on a surface of 2 ft^2?

Cumulative Review Exercises CHAPTERS 1–4

Decide whether each statement is always true, sometimes true, *or* never true. *If the statement is* sometimes true, *give examples where it is true and where it is false.*

1. The absolute value of a negative number equals the additive inverse of the number.

2. The quotient of two integers with nonzero denominator is a rational number.

3. The sum of two negative numbers is positive.

4. The sum of a positive number and a negative number is 0.

Perform each operation.

5. $-|-2| - 4 + |-3| + 7$

6. $(-.8)^2$

7. $\sqrt{-64}$

8. $-\frac{2}{3}\left(-\frac{12}{5}\right)$

Simplify.

9. $-(-4m + 3)$

10. $3x^2 - 4x + 4 + 9x - x^2$

11. $\dfrac{3\sqrt{16} - (-1)7}{4 + (-6)}$

12. Write $-3 < x \leq 5$ in interval notation.

13. Is $\sqrt{\dfrac{-2 + 4}{-5}}$ a real number?

Evaluate if $p = -4$, $q = -2$, *and* $r = 5$.

14. $-3(2q - 3p)$

15. $|p|^3 - |q^3|$

16. $\dfrac{\sqrt{r}}{-p + 2q}$

Solve.

17. $2z - 5 + 3z = 4 - (z + 2)$

18. $\dfrac{3a - 1}{5} + \dfrac{a + 2}{2} = -\dfrac{3}{10}$

19. $V = \frac{1}{3}\pi r^2 h$ for h

20. Two planes leave the Dallas–Fort Worth airport at the same time. One travels east at 550 mph, and the other travels west at 500 mph. Assuming no wind, how long will it take for the planes to be 2100 mi apart?

21. Ms. Bell must take at least 30 units of a certain medication each day. She can get the medication from white pills or yellow pills, each of which contains 3 units of the drug. To provide other benefits, she needs to take twice as many of the yellow pills as white pills. Find the smallest number of white pills that will satisfy these requirements.

22. If each side of a square were increased by 4 in., the perimeter would be 8 in. less than twice the perimeter of the original square. Find the length of a side of the original square.

23. How are the solution sets of a linear equation and the two associated inequalities related?

Solve.

24. $3 - 2(m + 3) < 4m$

25. $2k + 4 < 10$ and $3k - 1 > 5$

26. $2k + 4 > 10$ or $3k - 1 < 5$

27. $|5x + 3| = 13$

28. $|x + 2| < 9$

29. $|2y - 5| \geq 9$

30. Complete the ordered pairs (0,), (, 0), and (2,) for the equation $3x - 4y = 12$.

31. Graph $-4x + 2y = 8$, and give the intercepts.

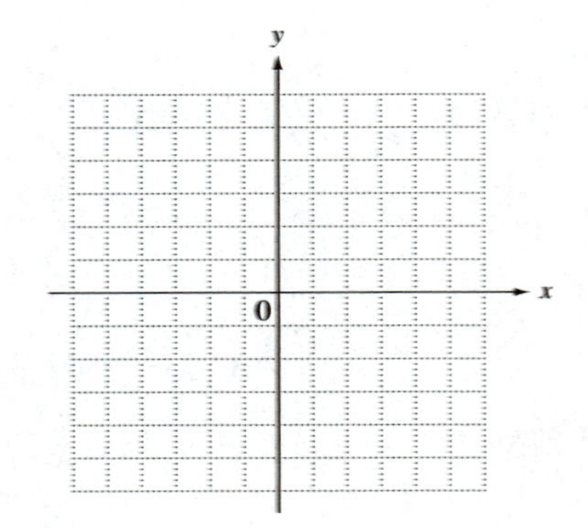

Find the slope of each line.

32. Through $(-5, 8)$ and $(-1, 2)$

33. Parallel to $y = -\frac{1}{2}x + 5$

34. Perpendicular to $4x - 3y = 12$

Write an equation in slope-intercept form for each line.

35. Slope $-\frac{3}{4}$; y-intercept $(0, -1)$

36. Horizontal; through $(2, -2)$

37. Through $(4, -3)$ and $(1, 1)$

38. For the function defined by $f(x) = -4x + 10$,

(a) what is the domain?

(b) what is $f(-3)$?

Use the graph to answer Exercises 39 and 40.

39. What is the slope of the line segment joining the points for 1992 and 2000?

40. Which one of the two line segments shown has a greater slope?

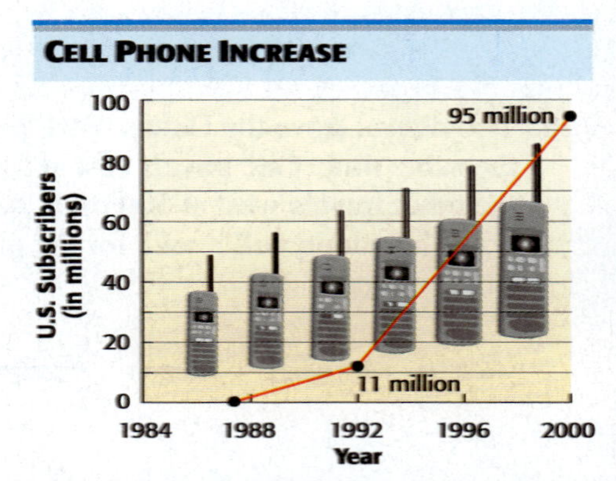

Source: Cellular Telecommunications Industry Association, Intel Corp.

Systems of Linear Equations

On November 7, 2000, in what was to become the most hotly contested presidential election in U.S. history, over 100,000,000 Americans went to the polls to vote. Although Al Gore won the popular vote by .5%, George W. Bush carried the Electoral College by 271 to 267 and became the 43rd president. (*Source: The Gazette,* January 18, 2001.) In Exercise 43 of Section 5.3, we determine the political affiliations of Americans using the concepts of this chapter.

5.1 Linear Systems of Equations in Two Variables

Objectives

1. Solve linear systems by graphing.
2. Decide whether an ordered pair is a solution of a linear system.
3. Solve linear systems by substitution.
4. Solve linear systems by elimination.
5. Solve special systems.

The worldwide personal computer market share for different manufacturers has varied, with first one, then another obtaining a larger share. As shown in Figure 1, Hewlett-Packard's share rose from 1995–1998, while Packard Bell, NEC saw its share decline. The graphs intersect at the point when the two companies had the same market share.

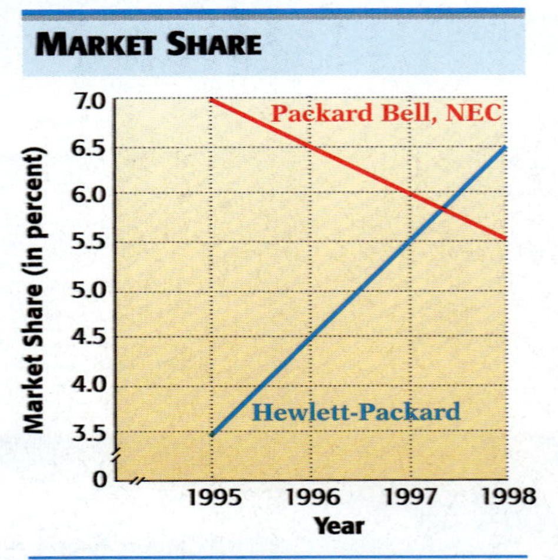

Source: Intelliquest; IDC.

Figure 1

We could use a linear equation to model the graph of Hewlett-Packard's market share and another linear equation to model the graph of Packard Bell, NEC's market share. Such a set of equations is called a **system of equations,** in this case a **linear system of equations.** The point where the graphs in Figure 1 intersect is a solution of each of the individual equations. It is also the solution of the linear system of equations.

1 Solve linear systems by graphing. The **solution set of a system of equations** contains all ordered pairs that satisfy all the equations of the system *at the same time.* An example of a linear system is

$$x + y = 5$$
$$2x - y = 4.$$

One way to find the solution set of a linear system of equations is to graph each equation and find the point where the graphs intersect.

Example 1 **Solving a System by Graphing**

Solve the system of equations by graphing.

$$x + y = 5 \quad (1)$$
$$2x - y = 4 \quad (2)$$

When we graph these linear equations as shown in Figure 2, the graph suggests that the point of intersection is the ordered pair (3, 2).

Continued on Next Page

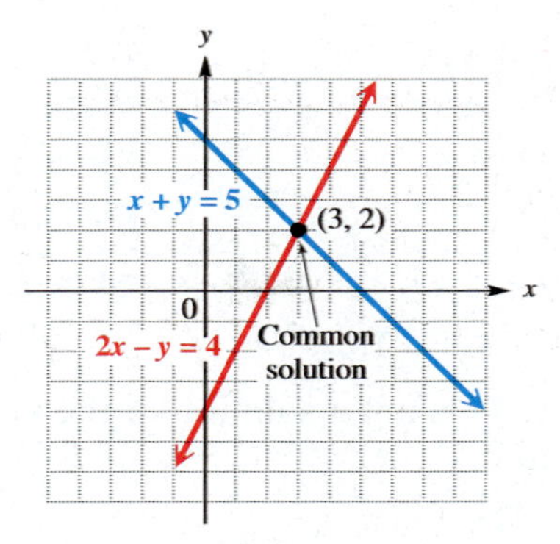

Figure 2

To be sure that (3, 2) is a solution of *both* equations, we check by substituting 3 for x and 2 for y in each equation.

$x + y = 5$	(1)	$2x - y = 4$	(2)
$3 + 2 = 5$?		$2(3) - 2 = 4$?	
$5 = 5$	True	$6 - 2 = 4$?	
		$4 = 4$	True

Since (3, 2) makes both equations true, $\{(3, 2)\}$ is the solution set of the system.

Work Problem 1 at the Side.

Calculator Tip A graphing calculator can be used to solve a system of equations. Each equation must be solved for y before being entered in the calculator. The coordinates of the point of intersection of the graphs, which is the solution of the system, can then be displayed. Consult your owner's manual for details.

2 Decide whether an ordered pair is a solution of a linear system. To do this, we substitute the ordered pair in both equations of the system, just as we did when we checked the solution in Example 1.

Example 2 Deciding whether an Ordered Pair Is a Solution

Decide whether the given ordered pair is a solution of the given system.

(a) $\begin{aligned} x + y &= 6 \\ 4x - y &= 14 \end{aligned}$; (4, 2)

Replace x with 4 and y with 2 in each equation of the system.

$x + y = 6$		$4x - y = 14$	
$4 + 2 = 6$?		$4(4) - 2 = 14$?	
$6 = 6$	True	$14 = 14$	True

Since (4, 2) makes both equations true, (4, 2) is a solution of the system.

Continued on Next Page

1 Solve each system by graphing.

(a) $x - y = 3$ (1)
$2x - y = 4$ (2)

(b) $2x + y = -5$ (1)
$-x + 3y = 6$ (2)

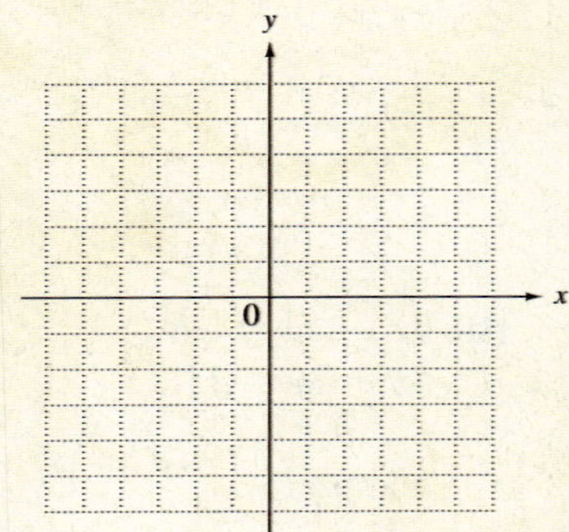

ANSWERS

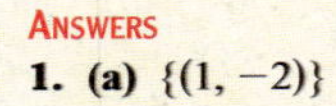

1. (a) $\{(1, -2)\}$

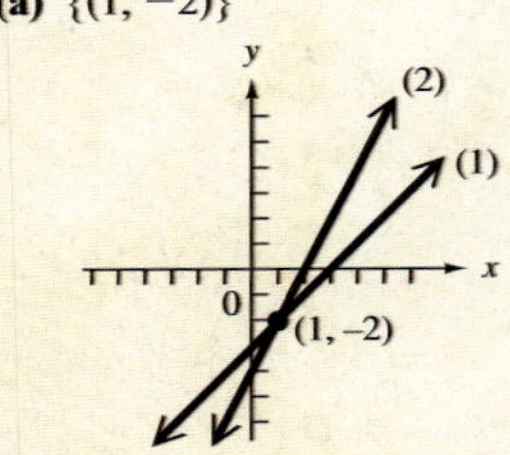

(b) $\{(-3, 1)\}$

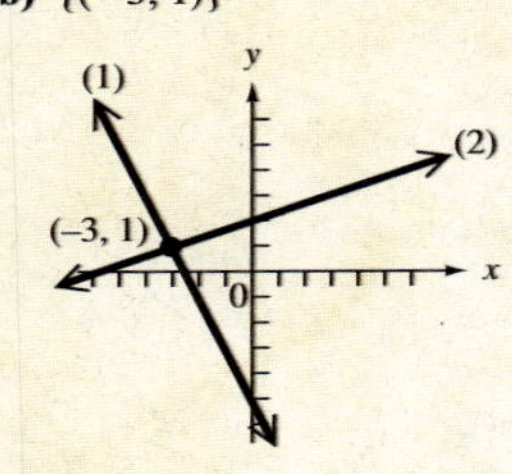

2 Are the given ordered pairs solutions of the given systems?

(a) $\begin{aligned} 2x + y &= -6 \\ x + 3y &= 2 \end{aligned}$; $(-4, 2)$

(b) $\begin{aligned} 9x - y &= -4 \\ 4x + 3y &= 11 \end{aligned}$; $(-1, 5)$

(b) $\begin{aligned} 3x + 2y &= 11 \\ x + 5y &= 36 \end{aligned}$; $(-1, 7)$

$3x + 2y = 11$		$x + 5y = 36$	
$3(-1) + 2(7) = 11$	?	$-1 + 5(7) = 36$	?
$-3 + 14 = 11$	?	$-1 + 35 = 36$	?
$11 = 11$	True	$34 = 36$	**False**

The ordered pair $(-1, 7)$ is not a solution of the system, since it does not make *both* equations true.

Work Problem 2 at the Side.

Since the graph of a linear equation is a straight line, there are three possibilities for the solution set of a linear system in two variables.

Graphs of Linear Systems in Two Variables

1. The two graphs intersect in a single point. The coordinates of this point give the only solution of the system. In this case the system is **consistent,** and the equations are **independent.** This is the most common case. See Figure 3(a).
2. The graphs are parallel lines. In this case the system is **inconsistent;** that is, there is no solution common to both equations of the system, and the solution set is $\emptyset$. See Figure 3(b).
3. The graphs are the same line. In this case the equations are **dependent,** since any solution of one equation of the system is also a solution of the other. The solution set is an infinite set of ordered pairs representing the points on the line. See Figure 3(c).

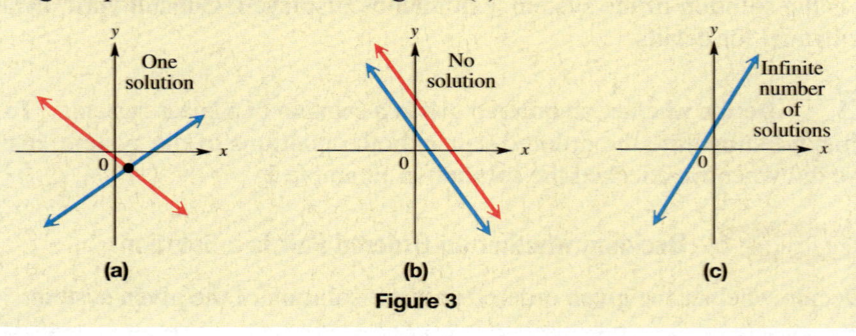

Figure 3

3 Solve linear systems by substitution. While it is possible to find the solution of a system of equations by graphing, it can be difficult to read exact coordinates, especially if they are not integers, from a graph. Because of this, we usually use algebraic methods to solve systems. One such method is called the **substitution method.** This method is most useful for solving linear systems in which one equation is solved or can be easily solved for one variable in terms of the other.

Example 3 Solving a System by Substitution

Solve the system.

$$2x - y = 6 \quad (1)$$

$$x = y + 2 \quad (2)$$

Since equation (2) is solved for x, substitute $y + 2$ for x in equation (1) to find y.

Continued on Next Page

Answers

2. (a) yes (b) no

$$\begin{aligned} 2x - y &= 6 && (1) \\ 2(y+2) - y &= 6 && \text{Let } x = y + 2. \\ 2y + 4 - y &= 6 && \text{Distributive property} \\ y + 4 &= 6 && \text{Combine terms.} \\ y &= 2 && \text{Subtract 4.} \end{aligned}$$

Now find x by substituting 2 for y in equation (2).

$$x = y + 2 = 2 + 2 = 4$$

Thus, $x = 4$ and $y = 2$, giving the ordered pair (4, 2). Check this solution in both equations of the original system. The solution set is $\{(4, 2)\}$.

CAUTION

Be careful when you write the ordered pair solution of a system. Even though we found y first in Example 3, the x-coordinate is *always* written first in the ordered pair.

Work Problem 3 at the Side.

The substitution method is summarized as follows.

Solving a Linear System by Substitution

Step 1 **Solve for one variable in terms of the other.** Solve one of the equations for either variable. (If one of the variable terms has coefficient 1 or −1, choose it, since the substitution method is usually easier this way.)

Step 2 **Substitute.** Substitute for that variable in the other equation. The result should be an equation with just one variable.

Step 3 **Solve.** Solve the equation from Step 2.

Step 4 **Find the other value.** Substitute the result from Step 3 into the equation from Step 1 to find the value of the other variable.

Step 5 **Check.** Check the solution in both of the original equations. Then write the solution set.

Example 4 Solving a System by Substitution

Solve the system.

$$\begin{aligned} 3x + 2y &= 13 && (1) \\ 4x - y &= -1 && (2) \end{aligned}$$

Step 1 To use the substitution method, first solve one of the equations for either x or y. Since the coefficient of y in equation (2) is −1, it is easiest to solve for y in equation (2).

$$\begin{aligned} 4x - y &= -1 && (2) \\ -y &= -1 - 4x && \text{Subtract } 4x. \\ y &= 1 + 4x && \text{Multiply by } -1. \end{aligned}$$

Step 2 Substitute $1 + 4x$ for y in equation (1).

$$\begin{aligned} 3x + 2y &= 13 && (1) \\ 3x + 2(1 + 4x) &= 13 && \text{Let } y = 1 + 4x. \end{aligned}$$

Continued on Next Page

3 Solve by substitution.

(a) $7x - 2y = -2$
$y = 3x$

(b) $5x - 3y = -6$
$x = 2 - y$

ANSWERS

3. (a) $\{(-2, -6)\}$ **(b)** $\{(0, 2)\}$

4 Solve by substitution.

(a) $3x - y = 10$
$2x + 5y = 1$

(b) $4x - 5y = -11$
$x + 2y = 7$

Step 3 Solve for x.

$$3x + 2 + 8x = 13 \quad \text{Distributive property}$$
$$11x = 11 \quad \text{Combine terms; subtract 2.}$$
$$x = 1 \quad \text{Divide by 11.}$$

Step 4 Now solve for y. Since $y = 1 + 4x$,

$$y = 1 + 4(1) = 5. \quad \text{Let } x = 1.$$

Step 5 Check the solution (1, 5) in both equations (1) and (2).

$3x + 2y = 13$	(1)	$4x - y = -1$	(2)
$3(1) + 2(5) = 13$	?	$4(1) - 5 = -1$	?
$3 + 10 = 13$	?	$4 - 5 = -1$	?
$13 = 13$	True	$-1 = -1$	True

The solution set is $\{(1, 5)\}$.

Work Problem 4 at the Side.

4 **Solve linear systems by elimination.** Another algebraic method for solving linear systems of equations is the **elimination method.** This method involves combining the two equations in the system so that one variable is eliminated. This is done using the following logic:

$$\text{If } a = b \text{ and } c = d, \text{ then } a + c = b + d.$$

Example 5 Solving a System by Elimination

Solve the system.

$$2x + 3y = -6 \quad (1)$$
$$4x - 3y = 6 \quad (2)$$

Notice that adding the equations together will eliminate the variable y.

$$\begin{aligned} 2x + 3y &= -6 \quad (1)\\ 4x - 3y &= 6 \quad (2)\\ \hline 6x &= 0 \quad \text{Add.}\\ x &= 0 \quad \text{Solve for } x. \end{aligned}$$

To find y, substitute 0 for x in either equation (1) or equation (2). Choosing equation (1) gives

$$2x + 3y = -6 \quad (1)$$
$$2(0) + 3y = -6 \quad \text{Let } x = 0.$$
$$0 + 3y = -6$$
$$3y = -6$$
$$y = -2.$$

The solution of the system is $(0, -2)$. Check by substituting 0 for x and -2 for y in both equations of the original system. The solution set is $\{(0, -2)\}$.

Work Problem 5 at the Side.

5 Solve by elimination.

(a) $3x - y = -7$
$2x + y = -3$

(b) $-2x + 3y = -10$
$2x + 2y = 5$

By adding the equations in Example 5, we eliminated the variable y because the coefficients of the y-terms were opposites. In many cases the coefficients will *not* be opposites. In these cases it is necessary to transform one or both equations so that the coefficients of one pair of variable terms are opposites. The general procedure for solving a system by the elimination method is summarized as follows.

ANSWERS
4. (a) $\{(3, -1)\}$ (b) $\{(1, 3)\}$
5. (a) $\{(-2, 1)\}$ (b) $\left\{\left(\frac{7}{2}, -1\right)\right\}$

Solving a Linear System by Elimination

Step 1 **Write in standard form.** Write both equations in the form

$$Ax + By = C.$$

Step 2 **Make the coefficients of one pair of variable terms opposites.** Multiply one or both equations by appropriate numbers so that the sum of the coefficients of either the x- or y-terms is 0.

Step 3 **Add.** Add the new equations to eliminate a variable. The sum should be an equation with just one variable.

Step 4 **Solve.** Solve the equation from Step 3 for the remaining variable.

Step 5 **Find the other value.** Substitute the result of Step 4 into either of the original equations and solve for the other variable.

Step 6 **Check.** Check the solution in both of the original equations. Then write the solution set.

6 Solve by elimination.

(a) $x + 3y = 8$
$2x - 5y = -17$

(b) $6x - 2y = -21$
$-3x + 4y = 36$

(c) $2x + 3y = 19$
$3x - 7y = -6$

Example 6 Solving a System by Elimination

Solve the system.

$$5x - 2y = 4 \quad (1)$$
$$2x + 3y = 13 \quad (2)$$

Step 1 Both equations are in standard form.

Step 2 Suppose that you wish to eliminate the variable x. One way to do this is to multiply equation (1) by 2 and equation (2) by -5.

$$10x - 4y = 8 \quad \text{2 times each side of equation (1)}$$
$$-10x - 15y = -65 \quad \text{−5 times each side of equation (2)}$$

Step 3 Now add.

$$\begin{aligned} 10x - 4y &= 8 \\ -10x - 15y &= -65 \\ \hline -19y &= -57 \end{aligned}$$

Step 4 Solve for y. $y = 3$ Divide by -19.

Step 5 To find x, substitute 3 for y in either equation (1) or (2). Substituting in equation (2) gives

$$\begin{aligned} 2x + 3y &= 13 && (2) \\ 2x + 3(3) &= 13 && \text{Let } y = 3. \\ 2x + 9 &= 13 \\ 2x &= 4 && \text{Subtract 9.} \\ x &= 2. && \text{Divide by 2.} \end{aligned}$$

Step 6 The solution is (2, 3). To check, substitute 2 for x and 3 for y in both equations (1) and (2).

$$\begin{aligned} 5x - 2y &= 4 && (1) \\ 5(2) - 2(3) &= 4 && ? \\ 10 - 6 &= 4 && ? \\ 4 &= 4 && \text{True} \end{aligned}$$

$$\begin{aligned} 2x + 3y &= 13 && (2) \\ 2(2) + 3(3) &= 13 && ? \\ 4 + 9 &= 13 && ? \\ 13 &= 13 && \text{True} \end{aligned}$$

The solution set is $\{(2, 3)\}$.

Work Problem **6** at the Side.

ANSWERS

6. (a) $\{(-1, 3)\}$ (b) $\left\{\left(-\frac{2}{3}, \frac{17}{2}\right)\right\}$ (c) $\{(5, 3)\}$

7 Solve each system.

(a) $\frac{1}{3}x - \frac{1}{2}y = \frac{1}{6}$
$3x - 2y = 9$

(b) $\frac{x}{5} + \frac{2y}{3} = -\frac{8}{5}$
$3x - y = 9$

Example 7 **Solving a System with Fractional Coefficients**

Solve the system.

$$5x - 2y = 4 \quad (1)$$
$$\frac{1}{2}x + \frac{3}{4}y = \frac{13}{4} \quad (2)$$

If an equation in a system has fractional coefficients, as in equation (2), first multiply by the least common denominator to clear the fractions.

$$4\left(\frac{1}{2}x + \frac{3}{4}y\right) = 4 \cdot \frac{13}{4} \quad \text{Multiply equation (2) by the LCD, 4.}$$
$$4 \cdot \frac{1}{2}x + 4 \cdot \frac{3}{4}y = 4 \cdot \frac{13}{4} \quad \text{Distributive property}$$
$$2x + 3y = 13 \quad \text{Equivalent to equation (2)}$$

The system of equations becomes

$$5x - 2y = 4 \quad (1)$$
$$2x + 3y = 13, \quad \text{Equation (2) with fractions cleared}$$

which is identical to the system we solved in Example 6. The solution set is $\{(2, 3)\}$. To confirm this, check the solution in both equations (1) and (2).

Work Problem 7 at the Side.

NOTE

If an equation in a system contains decimal coefficients, it is best to first clear the decimals by multiplying by 10, 100, or 1000, depending on the number of decimal places. Then solve the system. For example, we multiply *each side* of the equation

$$.5x + .75y = 3.25$$

by 100 to get the equivalent equation

$$50x + 75y = 325.$$

5 **Solve special systems.** As we saw in Figures 3(b) and (c), some systems of linear equations have no solution or an infinite number of solutions. Examples 8 and 9 show how to recognize these systems when solving algebraically.

Example 8 **Solving a System of Dependent Equations**

Solve the system.

$$2x - y = 3 \quad (1)$$
$$6x - 3y = 9 \quad (2)$$

We multiply equation (1) by -3, and then add the result to equation (2).

$$-6x + 3y = -9 \quad -3 \text{ times each side of equation (1)}$$
$$6x - 3y = 9 \quad (2)$$
$$0 = 0 \quad \text{True}$$

Continued on Next Page

ANSWERS
7. (a) $\{(5, 3)\}$ (b) $\{(2, -3)\}$

Adding these equations gives the true statement $0 = 0$. In the original system, we could get equation (2) from equation (1) by multiplying equation (1) by 3. Because of this, equations (1) and (2) are equivalent and have the same graph, as shown in Figure 4. The equations are dependent. The solution set is the set of all points on the line with equation $2x - y = 3$, written

$$\{(x, y) \mid 2x - y = 3\}$$

and read "the set of all ordered pairs (x, y), such that $2x - y = 3$."

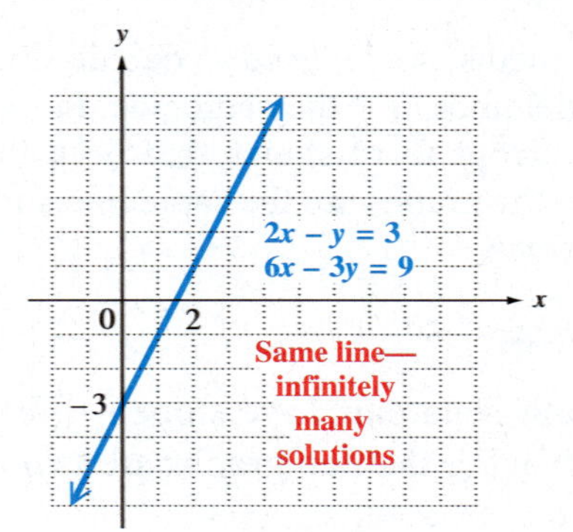

Figure 4

NOTE

When a system has dependent equations and an infinite number of solutions, as in Example 8, either equation of the system could be used to write the solution set. We prefer to use the equation (in standard form) with coefficients that are integers having no common factor (except 1). Other texts may express such solutions differently.

Work Problem 8 at the Side.

Example 9 Solving an Inconsistent System

Solve the system.

$$x + 3y = 4 \quad (1)$$
$$-2x - 6y = 3 \quad (2)$$

Multiply equation (1) by 2, and then add the result to equation (2).

$$2x + 6y = 8 \quad \text{Equation (1) multiplied by 2}$$
$$-2x - 6y = 3 \quad (2)$$
$$0 = 11 \quad \text{False}$$

The result of the addition step is a false statement, which indicates that the system is inconsistent. As shown in Figure 5, the graphs of the equations of the system are parallel lines. There are no ordered pairs that satisfy both equations, so there is no solution for the system; the solution set is $\emptyset$.

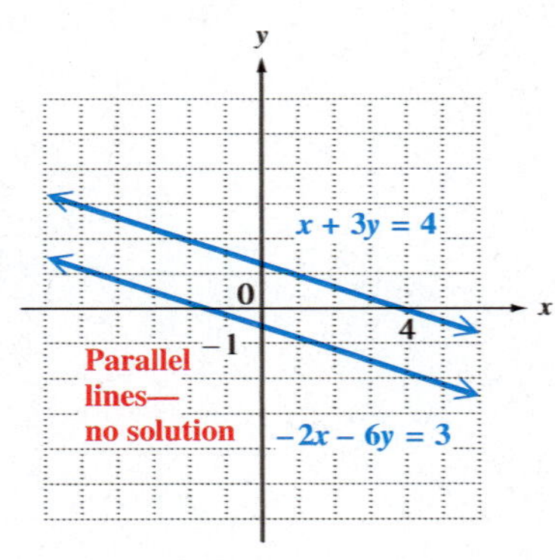

Figure 5

Work Problem 9 at the Side.

8 Solve the system. Then graph both equations.

$$2x + y = 6 \quad (1)$$
$$-8x - 4y = -24 \quad (2)$$

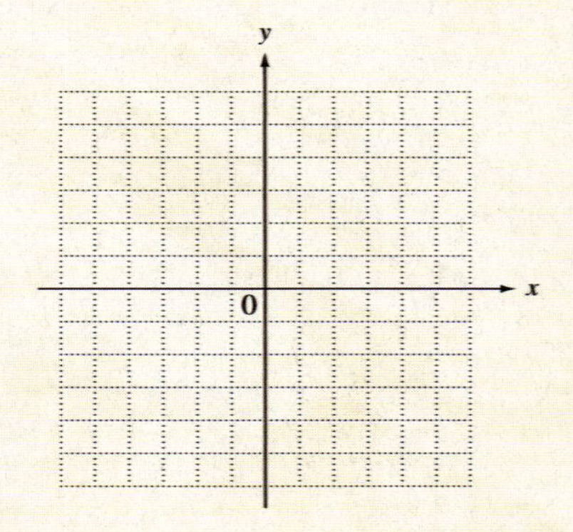

9 Solve the system. Then graph both equations.

$$4x - 3y = 8 \quad (1)$$
$$8x - 6y = 14 \quad (2)$$

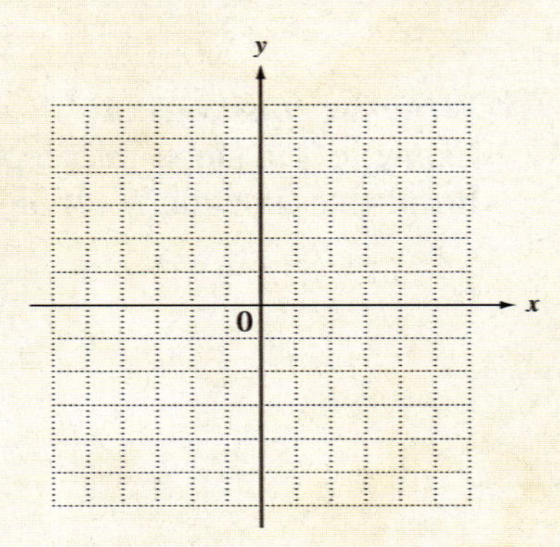

ANSWERS

8. $\{(x, y) \mid 2x + y = 6\}$

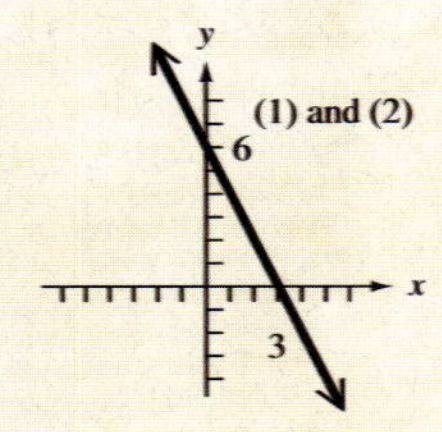

9. $\emptyset$

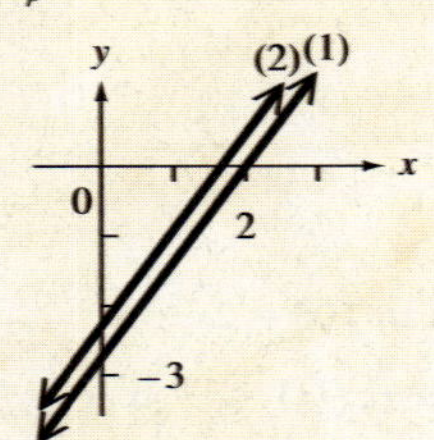

The results of Examples 8 and 9 are generalized as follows.

Special Cases of Linear Systems

If both variables are eliminated when a system of linear equations is solved,

1. there is no solution if the resulting statement is *false;*
2. there are infinitely many solutions if the resulting statement is *true.*

Slopes and y-intercepts can be used to decide if the graphs of a system of equations are parallel lines or if they coincide. In Example 8, writing each equation in slope-intercept form shows that both lines have slope 2 and y-intercept $(0, -3)$, so the graphs are the same line and the system has an infinite number of solutions.

Work Problem 10 at the Side.

In Example 9, both equations have slope $-\frac{1}{3}$, but the y-intercepts are $(0, \frac{4}{3})$ and $(0, -\frac{1}{2})$, showing that the graphs are two distinct parallel lines. Thus, the system has $\emptyset$ as its solution set.

Work Problem 11 at the Side.

10 Write the equations of Example 8 in slope-intercept form. Use function notation.

11 Write the equations of Example 9 in slope-intercept form. Use function notation.

ANSWERS

10. Both equations are $f(x) = 2x - 3$.

11. $f(x) = -\frac{1}{3}x + \frac{4}{3}$; $f(x) = -\frac{1}{3}x - \frac{1}{2}$

5.1 EXERCISES

FOR EXTRA HELP

 Student's Solutions Manual  MyMathLab.com 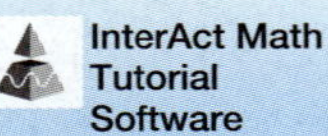 InterAct Math Tutorial Software  AW Math Tutor Center  www.mathxl.com  Digital Video Tutor CD 3 Videotape 6

Fill in the blanks with the correct responses.

1. If $(3, -6)$ is a solution of a linear system in two variables, then substituting ________ for x and ________ for y leads to true statements in *both* equations.

2. A solution of a system of independent linear equations in two variables is a(n) ____________.

3. If the solution process leads to a false statement such as $0 = 5$ when solving a system, the solution set is ________.

4. If the solution process leads to a true statement such as $0 = 0$ when solving a system, the system has __________ equations.

5. If the two lines forming a system have the same slope and different y-intercepts, the system has __________ solution(s).
(how many?)

6. If the two lines forming a system have different slopes, the system has __________ solution(s).
(how many?)

7. Which ordered pair could possibly be a solution of the graphed system of equations? Why?

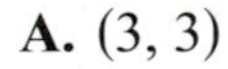
A. $(3, 3)$
B. $(-3, 3)$
C. $(-3, -3)$
D. $(3, -3)$

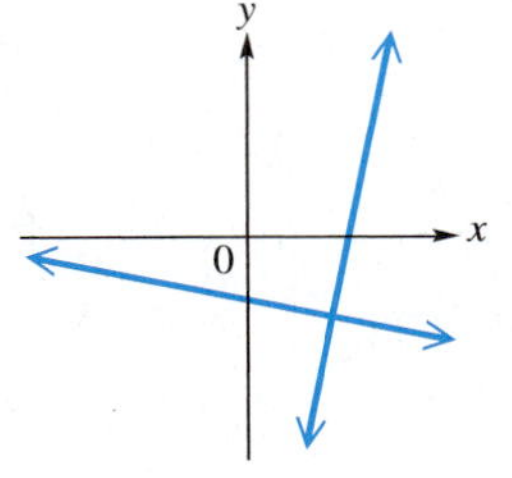

8. Which ordered pair could possibly be a solution of the graphed system of equations? Why?

A. $(3, 0)$
B. $(-3, 0)$
C. $(0, 3)$
D. $(0, -3)$

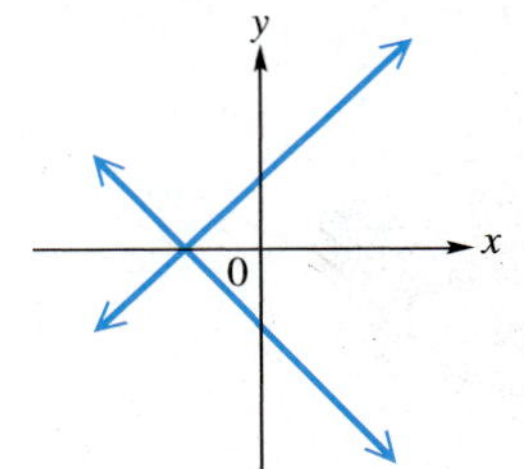

9. Match each system with the correct graph.

(a) $x + y = 6$
$x - y = 0$

(b) $x + y = -6$
$x - y = 0$

(c) $x + y = 0$
$x - y = -6$

(d) $x + y = 0$
$x - y = 6$

A.
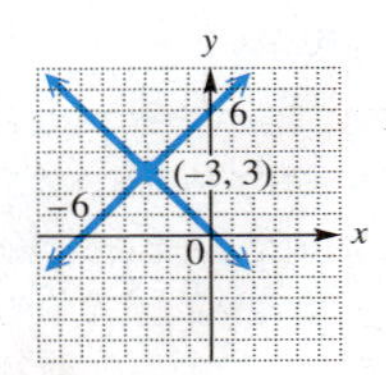

B.
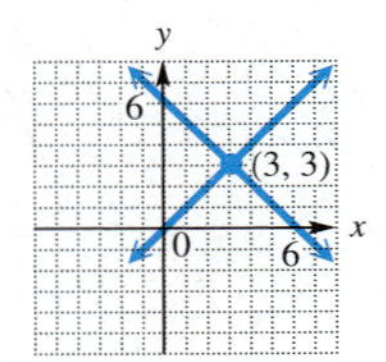

C.
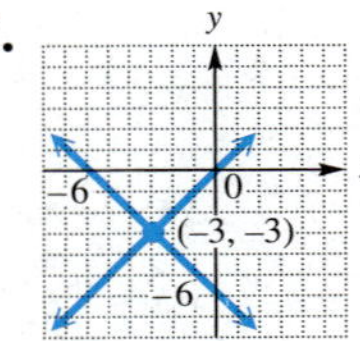

D.
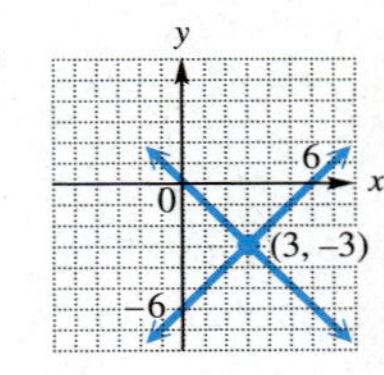

Solve each system by graphing. See Example 1.

10. $x + y = 4$
$2x - y = 2$

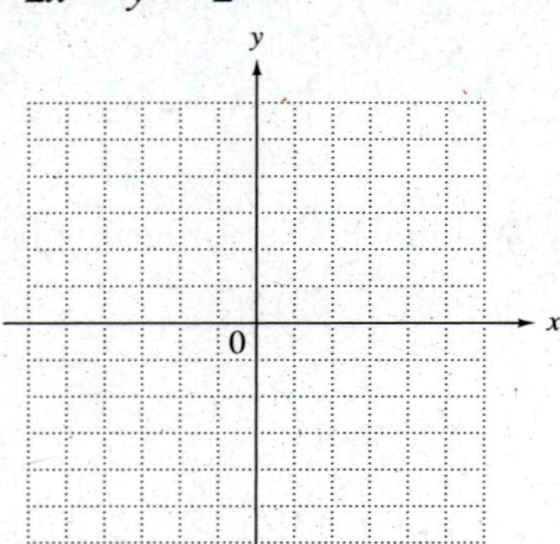

11. $x + y = -5$
$-2x + y = 1$

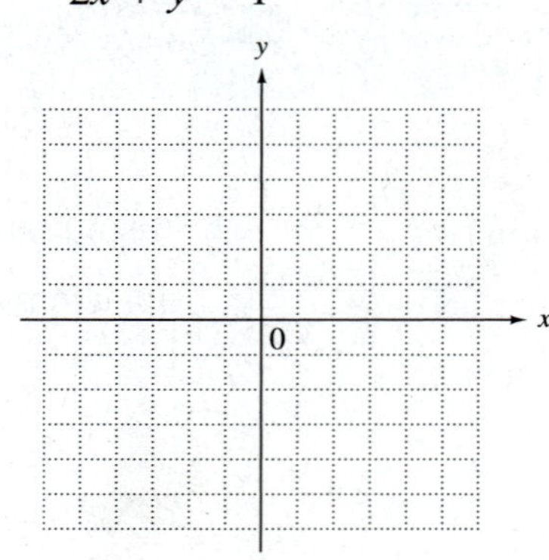

12. $x - 4y = -4$
$3x + y = 1$

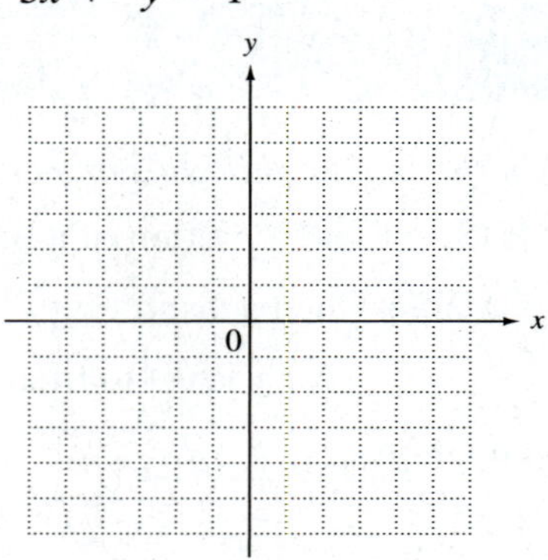

Decide whether the given ordered pair is a solution of the given system. See Example 2.

13. $x + y = 6$
$x - y = 4$; $(5, 1)$

14. $x - y = 17$
$x + y = -1$; $(8, -9)$

15. $2x - y = 8$
$3x + 2y = 20$; $(5, 2)$

16. $3x - 5y = -12$
$x - y = 1$; $(-1, 2)$

Solve each system by substitution. See Examples 3, 4, and 7.

17. $4x + y = 6$
$y = 2x$

18. $2x - y = 6$
$y = 5x$

19. $3x - 4y = -22$
$-3x + y = 0$

20. $-3x + y = -5$
$x + 2y = 0$

21. $-x - 4y = -14$
$2x = y + 1$

22. $-3x - 5y = -17$
$4x = y - 8$

23. $5x - 4y = 9$
$3 - 2y = -x$

24. $6x - y = -9$
$4 + 7x = -y$

25. $x = 3y + 5$
$x = \frac{3}{2}y$

26. $x = 6y - 2$
$x = \frac{3}{4}y$

27. $\frac{1}{2}x + \frac{1}{3}y = 3$
$y = 3x$

28. $\frac{1}{4}x - \frac{1}{5}y = 9$
$y = 5x$

Solve each system by elimination. If the system is inconsistent or has dependent equations, say so. See Examples 5–9.

29. $2x - 5y = 11$
$3x + y = 8$

30. $-2x + 3y = 1$
$-4x + y = -3$

31. $3x + 4y = -6$
$5x + 3y = 1$

32. $4x + 3y = 1$
$3x + 2y = 2$

33. $3x + 3y = 0$
$4x + 2y = 3$

34. $8x + 4y = 0$
$4x - 2y = 2$

35. $7x + 2y = 6$
$-14x - 4y = -12$

36. $x - 4y = 2$
$4x - 16y = 8$

37. $\frac{x}{2} + \frac{y}{3} = -\frac{1}{3}$
$\frac{x}{2} + 2y = -7$

38. $\frac{x}{5} + y = \frac{6}{5}$
$\frac{x}{10} + \frac{y}{3} = \frac{5}{6}$

39. $5x - 5y = 3$
$x - y = 12$

40. $2x - 3y = 7$
$-4x + 6y = 14$

Write each equation in slope-intercept form, and then tell how many solutions the system has. Do not actually solve.

41. $3x + 7y = 4$
$6x + 14y = 3$

42. $-x + 2y = 8$
$4x - 8y = 1$

43. $2x = -3y + 1$
$6x = -9y + 3$

44. $5x = -2y + 1$
$10x = -4y + 2$

45. Assuming you want to minimize the amount of work required, tell whether you would use the substitution or elimination method to solve each system. Explain your answers. *Do not actually solve.*

(a) $6x - y = 5$
$y = 11x$

(b) $3x + y = -7$
$x - y = -5$

(c) $3x - 2y = 0$
$9x + 8y = 7$

Solve each system by the method of your choice. (For Exercises 46–48, see your answers for Exercise 45.)

46. $6x - y = 5$
$y = 11x$

47. $3x + y = -7$
$x - y = -5$

48. $3x - 2y = 0$
$9x + 8y = 7$

49. $2x + 3y = 10$
$-3x + y = 18$

50. $3x - 5y = 7$
$2x + 3y = 30$

51. $\frac{1}{2}x - \frac{1}{8}y = -\frac{1}{4}$
$-4x + y = 2$

52. $\frac{1}{6}x + \frac{1}{3}y = 8$
$\frac{1}{4}x + \frac{1}{2}y = 12$

53. $.3x + .2y = .4$
$.5x + .4y = .7$

54. $.2x + .5y = 6$
$.4x + y = 9$

Answer the questions in Exercises 55–58 by observing the graphs provided.

55. Eboni Perkins compared the monthly payments she would incur for two types of mortgages: fixed-rate and variable-rate. Her observations led to the following graphs.

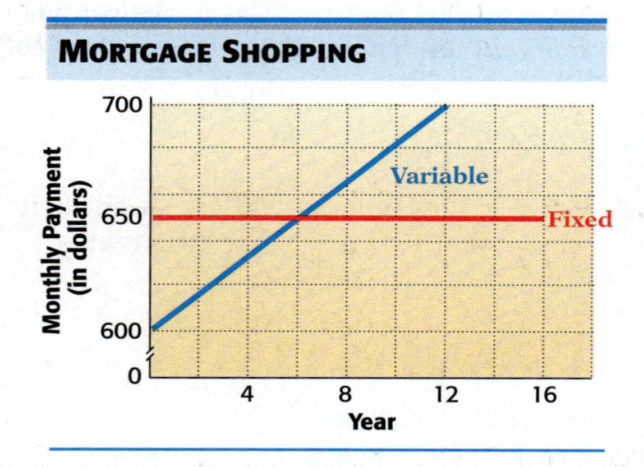

(a) For which years would the monthly payment be more for the fixed-rate mortgage than for the variable-rate mortgage?

(b) In what year would the payments be the same, and what would those payments be?

56. The figure shows graphs that represent supply and demand for a certain brand of low-fat frozen yogurt at various prices per half-gallon (in dollars).

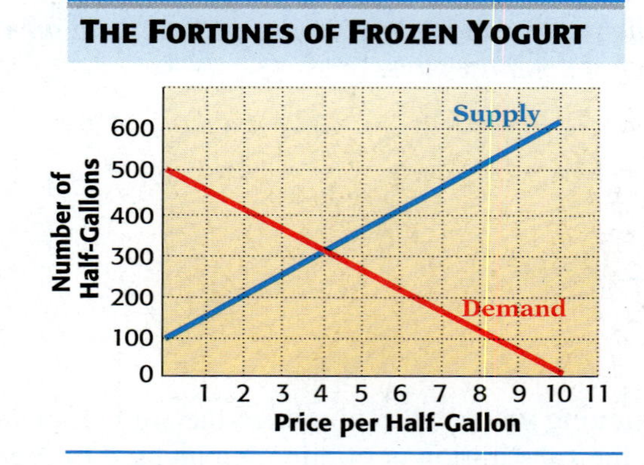

(a) At what price does supply equal demand?

(b) For how many half-gallons does supply equal demand?

(c) What are the supply and demand at a price of $2 per half-gallon?

57. The graph shows network share (the percentage of TV sets in use) for the early evening news programs for the three major broadcast networks from 1986–2000.

Source: Nielson Media Research.

(a) Between what years did the ABC early evening news dominate?

(b) During what year did ABC's dominance end? Which network equaled ABC's share that year? What was that share?

(c) During what years did ABC and CBS have equal network share? What was the share for each of these years?

(d) Which networks most recently had equal share? Write their share as an ordered pair of the form (year, share).

(e) Describe the general trend in viewership for the three major networks during these years.

58. The graph shows how the production of vinyl LPs, audiocassettes, and compact discs (CDs) changed over the years from 1986 through 1998.

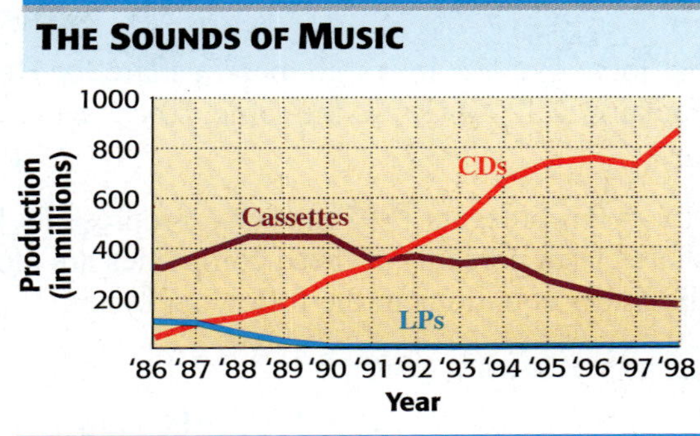

Source: Recording Industry Association of America.

(a) In what year did cassette production and CD production reach equal levels? What was that level?

(b) Express the point of intersection of the graphs of LP production and CD production as an ordered pair of the form (year, production level).

(c) Between what years did cassette production first stabilize and remain fairly constant?

(d) Describe the trend in CD production from 1986 through 1998. If a straight line were used to approximate its graph, would the line have positive, negative, or 0 slope?

(e) If a straight line were used to approximate the graph of cassette production from 1990 through 1998, would the line have positive, negative, or 0 slope? Explain.

Use the graph given at the beginning of this section (repeated here) to work Exercises 59–62.

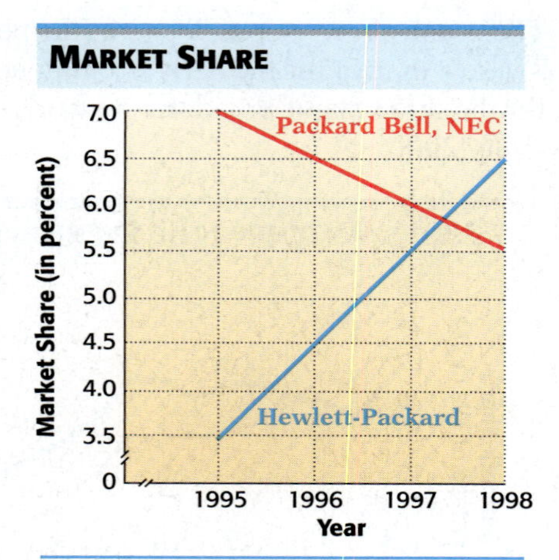

Source: Intelliquest; IDC.

59. For which years was Hewlett-Packard's share less than Packard Bell, NEC's share?

60. Estimate the year in which market share for Hewlett-Packard and Packard Bell, NEC was the same. About what was this share?

61. If $x = 0$ represents 1995 and $x = 3$ represents 1998, the market shares y (in percent) of these companies are closely modeled by the linear equations in the following system.

$$y = -.5x + 7 \quad \text{Packard Bell, NEC}$$

$$y = x + 3.5 \quad \text{Hewlett-Packard}$$

Solve this system. Round values to the nearest tenth as necessary.

62. Using your solution from Exercise 61, in what month and year did these companies have the same market share? What was that share?

RELATING CONCEPTS (Exercises 63–66) FOR INDIVIDUAL OR GROUP WORK

Work Exercises 63–66 in order *to see the connections between systems of linear equations and the graphs of linear functions.*

63. Use elimination or substitution to solve the system.

$$3x + y = 6 \quad (1)$$

$$-2x + 3y = 7 \quad (2)$$

64. For equation (1) in the system of Exercise 63, solve for y and rename it $f(x)$. What special kind of function is f?

65. For equation (2) in the system of Exercise 63, solve for y and rename it $g(x)$. What special kind of function is g?

66. Use the result of Exercise 63 to fill in the blanks with the appropriate responses:

Because the graphs of f and g are straight lines that are neither parallel nor coincide, they intersect in exactly ______ point. The coordinates of the point are (______, ______). Using function notation, this is given by $f($______$) =$ ______ and $g($______$) =$ ______.

5.2 Linear Systems of Equations in Three Variables

Objectives

1. Understand the geometry of systems of three equations in three variables.
2. Solve linear systems by elimination.
3. Solve linear systems where some of the equations have missing terms.
4. Solve special systems.

A solution of an equation in three variables, such as

$$2x + 3y - z = 4,$$

is called an **ordered triple** and is written (x, y, z). For example, the ordered triple $(0, 1, -1)$ is a solution of the equation, because

$$2(0) + 3(1) - (-1) = 0 + 3 + 1 = 4.$$

Verify that another solution of this equation is $(10, -3, 7)$.

In the rest of this chapter, the term *linear equation* is extended to equations of the form

$$Ax + By + Cz + \ldots + Dw = K,$$

where not all the coefficients $A, B, C, \ldots, D$ equal 0. For example,

$$2x + 3y - 5z = 7 \quad \text{and} \quad x - 2y - z + 3u - 2w = 8$$

are linear equations, the first with three variables and the second with five variables.

1 Understand the geometry of systems of three equations in three variables. In this section, we discuss the solution of a system of linear equations in three variables, such as

$$\begin{aligned} 4x + 8y + z &= 2 \\ x + 7y - 3z &= -14 \\ 2x - 3y + 2z &= 3. \end{aligned}$$

Theoretically, a system of this type can be solved by graphing. However, the graph of a linear equation with three variables is a *plane,* not a line. Since the graph of each equation of the system is a plane, which requires three-dimensional graphing, this method is not practical. However, it does illustrate the number of solutions possible for such systems, as shown in Figure 6.

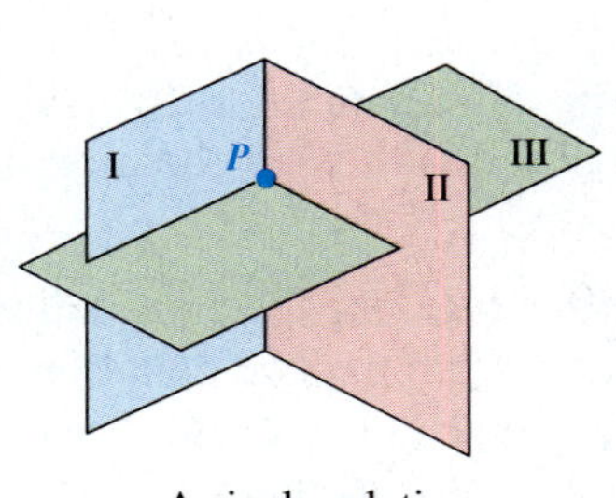

A single solution
(a)

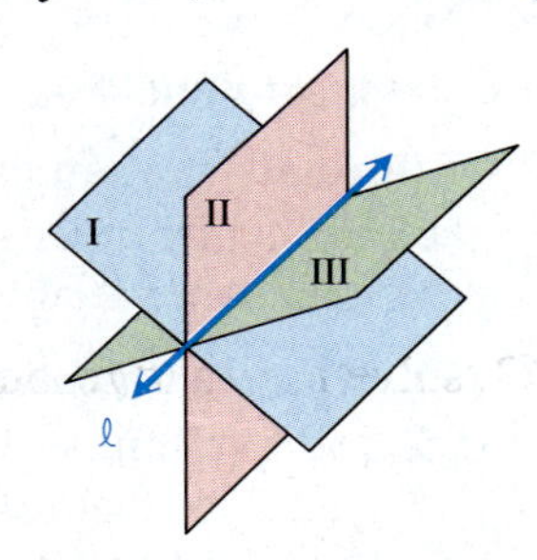

Points of a line in common
(b)

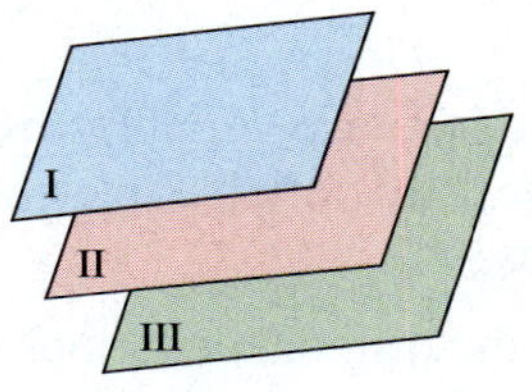

No points in common
(c)

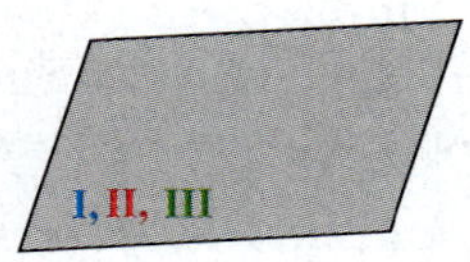

All points in common
(d)

Figure 6

Figure 6 illustrates the following cases.

Graphs of Linear Systems in Three Variables

1. The three planes may meet at a single, common point that is the solution of the system. See Figure 6(a).
2. The three planes may have the points of a line in common so that the infinite set of points that satisfy the equation of the line is the solution of the system. See Figure 6(b).
3. The planes may have no points common to all three so that there is no solution of the system. See Figure 6(c).
4. The three planes may coincide so that the solution of the system is the set of all points on a plane. See Figure 6(d).

NOTE

There are other illustrations of these cases. For example, two of the planes might intersect in a line, while the third plane is parallel to one of these planes, again resulting in no points common to all three planes. We give only one example of each case in Figure 6.

2 **Solve linear systems by elimination.** Since graphing to find the solution set of a system of three equations in three variables is impractical, these systems are solved with an extension of the elimination method, summarized as follows.

Solving a Linear System in Three Variables

Step 1 **Eliminate a variable.** Use the elimination method to eliminate any variable from any two of the original equations. The result is an equation in two variables.

Step 2 **Eliminate the same variable again.** Eliminate the *same* variable from any *other* two equations. The result is an equation in the same two variables as in Step 1.

Step 3 **Eliminate a different variable and solve.** Use the elimination method to eliminate a second variable from the two equations in two variables that result from Steps 1 and 2. The result is an equation in one variable that gives the value of that variable.

Step 4 **Find a second value.** Substitute the value of the variable found in Step 3 into either of the equations in two variables to find the value of the second variable.

Step 5 **Find a third value.** Use the values of the two variables from Steps 3 and 4 to find the value of the third variable by substituting into any of the original equations.

Step 6 **Check.** Check the solution in all of the original equations. Then write the solution set.

Example 1 Solving a System in Three Variables

Solve the system.

$$
\begin{aligned}
4x + 8y + z &= 2 \qquad (1)\\
x + 7y - 3z &= -14 \qquad (2)\\
2x - 3y + 2z &= 3 \qquad (3)
\end{aligned}
$$

Step 1 As before, the elimination method involves eliminating a variable from the sum of two equations. The choice of which variable to eliminate is arbitrary. Suppose we decide to begin by eliminating z. To do this, we multiply equation (1) by 3 and then add the result to equation (2).

$$
\begin{aligned}
12x + 24y + 3z &= 6 \qquad \text{Multiply each side of (1) by 3.}\\
x + 7y - 3z &= -14 \qquad (2)\\
\hline
13x + 31y &= -8 \qquad \text{Add.} \quad (4)
\end{aligned}
$$

Step 2 Equation (4) has only two variables. To get another equation without z, we multiply equation (1) by -2 and add the result to equation (3). It is essential at this point to *eliminate the same variable, z*.

$$
\begin{aligned}
-8x - 16y - 2z &= -4 \qquad \text{Multiply each side of (1) by } -2.\\
2x - 3y + 2z &= 3 \qquad (3)\\
\hline
-6x - 19y &= -1 \qquad \text{Add.} \quad (5)
\end{aligned}
$$

Step 3 Now we solve the system of equations (4) and (5) for x and y. This step is possible only if the *same* variable is eliminated in Steps 1 and 2.

$$
\begin{aligned}
78x + 186y &= -48 \qquad \text{Multiply each side of (4) by 6.}\\
-78x - 247y &= -13 \qquad \text{Multiply each side of (5) by 13.}\\
\hline
-61y &= -61 \qquad \text{Add.}\\
y &= 1
\end{aligned}
$$

Step 4 Now we substitute 1 for y in either equation (4) or (5). Choosing (5) gives

$$
\begin{aligned}
-6x - 19y &= -1 \qquad (5)\\
-6x - 19(1) &= -1 \qquad \text{Let } y = 1.\\
-6x - 19 &= -1\\
-6x &= 18\\
x &= -3.
\end{aligned}
$$

Step 5 We substitute -3 for x and 1 for y in any one of the three original equations to find z. Choosing (1) gives

$$
\begin{aligned}
4x + 8y + z &= 2 \qquad (1)\\
4(-3) + 8(1) + z &= 2 \qquad \text{Let } x = -3 \text{ and } y = 1.\\
-4 + z &= 2\\
z &= 6.
\end{aligned}
$$

Continued on Next Page

1 Check that the solution $(-3, 1, 6)$ satisfies equations (2) and (3) of Example 1.

(a) $x + 7y - 3z = -14$ (2)

Does the solution satisfy equation (2)?

(b) $2x - 3y + 2z = 3$ (3)

Does the solution satisfy equation (3)?

2 Solve each system.

(a)
$$\begin{aligned} x + y + z &= 2 \\ x - y + 2z &= 2 \\ -x + 2y - z &= 1 \end{aligned}$$

(b)
$$\begin{aligned} 2x + y + z &= 9 \\ -x - y + z &= 1 \\ 3x - y + z &= 9 \end{aligned}$$

Step 6 It appears that the ordered triple $(-3, 1, 6)$ is the only solution of the system. We must check that the solution satisfies all three equations of the system. For equation (1),

$$\begin{aligned} 4x + 8y + z &= 2 && (1) \\ 4(-3) + 8(1) + 6 &= 2 && ? \\ -12 + 8 + 6 &= 2 && ? \\ 2 &= 2. && \text{True} \end{aligned}$$

Work Problem 1 at the Side.

Because $(-3, 1, 6)$ also satisfies equations (2) and (3), the solution set is $\{(-3, 1, 6)\}$.

Work Problem 2 at the Side.

3 **Solve linear systems where some of the equations have missing terms.** When this happens, one elimination step can be omitted.

Example 2 Solving a System of Equations with Missing Terms

Solve the system.

$$\begin{aligned} 6x - 12y &= -5 && (1) \\ 8y + z &= 0 && (2) \\ 9x - z &= 12 && (3) \end{aligned}$$

Since equation (3) is missing the variable y, a good way to begin the solution is to eliminate y again using equations (1) and (2).

$$\begin{aligned} 12x - 24y \quad &= -10 && \text{Multiply each side of (1) by 2.} \\ 24y + 3z &= 0 && \text{Multiply each side of (2) by 3.} \\ \hline 12x \quad + 3z &= -10 && \text{Add.} \quad (4) \end{aligned}$$

Use this result, together with equation (3), to eliminate z. Multiply equation (3) by 3. This gives

$$\begin{aligned} 27x - 3z &= 36 && \text{Multiply each side of (3) by 3.} \\ 12x + 3z &= -10 && (4) \\ \hline 39x \quad &= 26 && \text{Add.} \end{aligned}$$

$$x = \frac{26}{39} = \frac{2}{3}.$$

Substituting into equation (3) gives

$$\begin{aligned} 9x - z &= 12 && (3) \\ 9\left(\frac{2}{3}\right) - z &= 12 && \text{Let } x = \tfrac{2}{3}. \\ 6 - z &= 12 \\ z &= -6. \end{aligned}$$

Continued on Next Page

Answers
1. (a) yes (b) yes
2. (a) $\{(-1, 1, 2)\}$ (b) $\{(2, 1, 4)\}$

Substituting -6 for z in equation (2) gives

$$\begin{aligned} 8y + z &= 0 && (2) \\ 8y - 6 &= 0 && \text{Let } z = -6. \\ 8y &= 6 \\ y &= \frac{3}{4}. \end{aligned}$$

Check in each of the original equations of the system to verify that the solution set of the system is $\{(\frac{2}{3}, \frac{3}{4}, -6)\}$.

Work Problem 3 at the Side.

3 Solve each system.

(a)
$$\begin{aligned} x - y &= 6 \\ 2y + 5z &= 1 \\ 3x - 4z &= 8 \end{aligned}$$

(b)
$$\begin{aligned} 5x - y &= 26 \\ 4y + 3z &= -4 \\ x + z &= 5 \end{aligned}$$

4 Solve special systems. Linear systems with three variables may be inconsistent or may include dependent equations. The next examples illustrate these cases.

Example 3 Solving an Inconsistent System with Three Variables

Solve the system.

$$\begin{aligned} 2x - 4y + 6z &= 5 && (1) \\ -x + 3y - 2z &= -1 && (2) \\ x - 2y + 3z &= 1 && (3) \end{aligned}$$

Eliminate x by adding equations (2) and (3) to get the equation

$$y + z = 0.$$

Now, *eliminate x again,* using equations (1) and (3).

$$\begin{aligned} -2x + 4y - 6z &= -2 && \text{Multiply each side of (3) by } -2. \\ 2x - 4y + 6z &= 5 && (1) \\ \hline 0 &= 3 && \text{False} \end{aligned}$$

The resulting false statement indicates that equations (1) and (3) have no common solution. Thus, the system is inconsistent and the solution set is $\emptyset$. The graph of this system would show these two planes parallel to one another.

4 Solve each system.

(a)
$$\begin{aligned} 3x - 5y + 2z &= 1 \\ 5x + 8y - z &= 4 \\ -6x + 10y - 4z &= 5 \end{aligned}$$

(b)
$$\begin{aligned} 7x - 9y + 2z &= 0 \\ y + z &= 0 \\ 8x - z &= 0 \end{aligned}$$

NOTE

If you get a false statement when adding as in Example 3, you do not need to go any further with the solution. Since two of the three planes are parallel, it is not possible for the three planes to have any common points.

Work Problem 4 at the Side.

ANSWERS

3. **(a)** $\{(4, -2, 1)\}$ **(b)** $\{(5, -1, 0)\}$

4. **(a)** $\emptyset$ **(b)** $\{(0, 0, 0)\}$

5 Solve the system.

$$\begin{aligned} x - y + z &= 4 \\ -3x + 3y - 3z &= -12 \\ 2x - 2y + 2z &= 8 \end{aligned}$$

Example 4 Solving a System of Dependent Equations with Three Variables

Solve the system.

$$2x - 3y + 4z = 8 \quad (1)$$

$$-x + \frac{3}{2}y - 2z = -4 \quad (2)$$

$$6x - 9y + 12z = 24 \quad (3)$$

Multiplying each side of equation (1) by 3 gives equation (3). Multiplying each side of equation (2) by -6 also gives equation (3). Because of this, the equations are dependent. All three equations have the same graph, as illustrated in Figure 6(d). The solution set is written

$$\{(x, y, z) \mid 2x - 3y + 4z = 8\}.$$

Although any one of the three equations could be used to write the solution set, we use the equation with coefficients that are integers with no common factor (except 1), as we did in Section 5.1.

Work Problem 5 at the Side.

ANSWERS

5. $\{(x, y, z) \mid x - y + z = 4\}$

5.2 EXERCISES

FOR EXTRA HELP

Student's Solutions Manual

MyMathLab.com

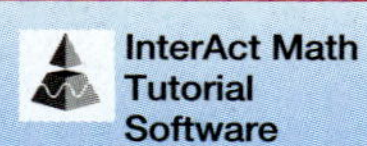
InterAct Math Tutorial Software

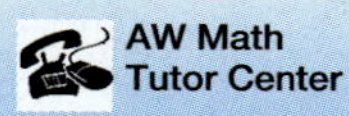
AW Math Tutor Center

www.mathxl.com

Digital Video Tutor CD 3 Videotape 6

1. Explain what the following statement means: The solution set of the system

$$\begin{aligned} 2x + y + z &= 3 \\ 3x - y + z &= -2 \\ 4x - y + 2z &= 0 \end{aligned}$$

is $\{(-1, 2, 3)\}$.

2. The two equations

$$\begin{aligned} x + y + z &= 6 \\ 2x - y + z &= 3 \end{aligned}$$

have a common solution of $(1, 2, 3)$. Which equation would complete a system of three linear equations in three variables having solution set $\{(1, 2, 3)\}$?

A. $3x + 2y - z = 1$ **B.** $3x + 2y - z = 4$

C. $3x + 2y - z = 5$ **D.** $3x + 2y - z = 6$

Solve each system of equations. See Example 1.

3. $$\begin{aligned} 2x - 5y + 3z &= -1 \\ x + 4y - 2z &= 9 \\ x - 2y - 4z &= -5 \end{aligned}$$

4. $$\begin{aligned} x + 3y - 6z &= 7 \\ 2x - y + z &= 1 \\ x + 2y + 2z &= -1 \end{aligned}$$

5. $$\begin{aligned} 3x + 2y + z &= 8 \\ 2x - 3y + 2z &= -16 \\ x + 4y - z &= 20 \end{aligned}$$

6. $$\begin{aligned} -3x + y - z &= -10 \\ -4x + 2y + 3z &= -1 \\ 2x + 3y - 2z &= -5 \end{aligned}$$

7. $$\begin{aligned} -x + 2y + 6z &= 2 \\ 3x + 2y + 6z &= 6 \\ x + 4y - 3z &= 1 \end{aligned}$$

8. $$\begin{aligned} 2x + y + 2z &= 1 \\ x + 2y + z &= 2 \\ x - y - z &= 0 \end{aligned}$$

9. $$\begin{aligned} 2x + 5y + 2z &= 0 \\ 4x - 7y - 3z &= 1 \\ 3x - 8y - 2z &= -6 \end{aligned}$$

10. $$\begin{aligned} 5x - 2y + 3z &= -9 \\ 4x + 3y + 5z &= 4 \\ 2x + 4y - 2z &= 14 \end{aligned}$$

11. $$\begin{aligned} x + y - z &= -2 \\ 2x - y + z &= -5 \\ -x + 2y - 3z &= -4 \end{aligned}$$

12. $$\begin{aligned} x + 2y + 3z &= 1 \\ -x - y + 3z &= 2 \\ -6x + y + z &= -2 \end{aligned}$$

Solve each system of equations. See Example 2.

13. $$\begin{aligned} 2x - 3y + 2z &= -1 \\ x + 2y + z &= 17 \\ 2y - z &= 7 \end{aligned}$$

14. $$\begin{aligned} 2x - y + 3z &= 6 \\ x + 2y - z &= 8 \\ 2y + z &= 1 \end{aligned}$$

15. $$\begin{aligned} 4x + 2y - 3z &= 6 \\ x - 4y + z &= -4 \\ -x + 2z &= 2 \end{aligned}$$

16. $$\begin{aligned} 2x + 3y - 4z &= 4 \\ x - 6y + z &= -16 \\ -x + 3z &= 8 \end{aligned}$$

17. $$\begin{aligned} 2x + y &= 6 \\ 3y - 2z &= -4 \\ 3x - 5z &= -7 \end{aligned}$$

18. $$\begin{aligned} 4x - 8y &= -7 \\ 4y + z &= 7 \\ -8x + z &= -4 \end{aligned}$$

19. Using your immediate surroundings, give an example of three planes that

(a) intersect in a single point;

(b) do not intersect;

(c) intersect in infinitely many points.

20. Suppose that a system has infinitely many ordered triple solutions of the form (x, y, z) such that

$$x + y + 2z = 1.$$

Give three specific ordered triples that are solutions of the system.

Solve each system of equations. See Examples 1, 3, and 4.

21. $$\begin{aligned} 2x + 2y - 6z &= 5 \\ -3x + y - z &= -2 \\ -x - y + 3z &= 4 \end{aligned}$$

22. $$\begin{aligned} -2x + 5y + z &= -3 \\ 5x + 14y - z &= -11 \\ 7x + 9y - 2z &= -5 \end{aligned}$$

23. $$\begin{aligned} -5x + 5y - 20z &= -40 \\ x - y + 4z &= 8 \\ 3x - 3y + 12z &= 24 \end{aligned}$$

24. $$\begin{aligned} x + 4y - z &= 3 \\ -2x - 8y + 2z &= -6 \\ 3x + 12y - 3z &= 9 \end{aligned}$$

25. $$\begin{aligned} 2x + y - z &= 6 \\ 4x + 2y - 2z &= 12 \\ -x - \frac{1}{2}y + \frac{1}{2}z &= -3 \end{aligned}$$

26. $$\begin{aligned} 2x - 8y + 2z &= -10 \\ -x + 4y - z &= 5 \\ \frac{1}{8}x - \frac{1}{2}y + \frac{1}{8}z &= -\frac{5}{8} \end{aligned}$$

27. $$\begin{aligned} x + y - 2z &= 0 \\ 3x - y + z &= 0 \\ 4x + 2y - z &= 0 \end{aligned}$$

28. $$\begin{aligned} 2x + 3y - z &= 0 \\ x - 4y + 2z &= 0 \\ 3x - 5y - z &= 0 \end{aligned}$$

RELATING CONCEPTS (Exercises 29–36) FOR INDIVIDUAL OR GROUP WORK

Suppose that on a distant planet a function of the form

$$f(x) = ax^2 + bx + c \quad (a \neq 0)$$

describes the height in feet of a projectile x seconds after it has been projected upward. ***Work Exercises 29–36 in order*** *to see how this can be related to a system of three equations in three variables a, b, and c.*

29. After 1 sec, the height of a certain projectile is 128 ft. Thus, $f(1) = 128$. Use this information to find one equation in the variables a, b, and c. (*Hint:* Substitute 1 for x and 128 for $f(x)$.)

30. After 1.5 sec, the height is 140 ft. Find a second equation in a, b, and c.

31. After 3 sec, the height is 80 ft. Find a third equation in a, b, and c.

32. Write a system of three equations in a, b, and c, based on your answers in Exercises 29–31. Solve the system.

33. What is the function f for this particular projectile?

34. In the function f written in Exercise 33, the ______ of the projectile is a function of the ______ elapsed since it was projected.

35. What was the initial height of the projectile? (*Hint:* Find $f(0)$.)

36. The projectile reaches its maximum height in 1.625 sec. Find its maximum height.

5.3 Applications of Linear Systems of Equations

Objectives

1. Solve problems using two variables.
2. Solve money problems using two variables.
3. Solve mixture problems using two variables.
4. Solve distance-rate-time problems using two variables.
5. Solve problems with three variables using a system of three equations.

Many applied problems involve more than one unknown quantity. Although some problems with two unknowns can be solved using just one variable, it is often easier to use two variables. To solve a problem with two unknowns, we must write two equations that relate the unknown quantities. The system formed by the pair of equations can then be solved using the methods of this chapter.

Problems that can be solved by writing a system of equations have been of interest historically. The following problem, which is given in the exercises for this section, first appeared in a Hindu work that dates back to about A.D. 850.

> The mixed price of 9 citrons [a lemonlike fruit shown in the photo] and 7 fragrant wood apples is 107; again, the mixed price of 7 citrons and 9 fragrant wood apples is 101. O you arithmetician, tell me quickly the price of a citron and the price of a wood apple here, having distinctly separated those prices well.

The following steps, based on the six-step problem-solving method first introduced in Chapter 2, give a strategy for solving applied problems using more than one variable.

Solving an Applied Problem by Writing a System of Equations

Step 1 **Read** the problem carefully until you understand what is given and what is to be found.

Step 2 **Assign variables** to represent the unknown values, using diagrams or tables as needed. *Write down* what each variable represents.

Step 3 **Write a system of equations** that relates the unknowns.

Step 4 **Solve** the system of equations.

Step 5 **State the answer** to the problem. Does it seem reasonable?

Step 6 **Check** the answer in the words of the original problem.

1 Solve problems using two variables. Problems about the perimeter of a geometric figure often involve two unknowns and can be solved using systems of equations.

Example 1 Finding the Dimensions of a Soccer Field

Unlike football, where the dimensions of a playing field cannot vary, a rectangular soccer field may have a width between 50 and 100 yd and a length between 50 and 100 yd. Suppose that one particular field has a perimeter of 320 yd. Its length measures 40 yd more than its width. What are the dimensions of this field? (*Source: Microsoft Encarta Encyclopedia 2000.*)

Step 1 **Read** the problem again. We are asked to find the dimensions of the field.

Continued on Next Page

1 Solve the problem.

The length of the foundation of a rectangular house is to be 6 m more than its width. Find the length and width of the house if the perimeter must be 48 m.

Step 2 **Assign variables.** Let $L =$ the length and $W =$ the width. Figure 7 shows a soccer field with the length labeled L and the width labeled W.

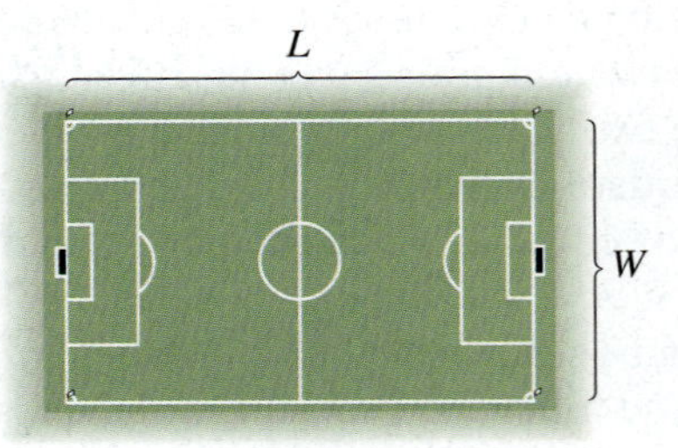

Figure 7

Step 3 **Write a system of equations.** Because the perimeter is 320 yd, we find one equation by using the perimeter formula:

$$2L + 2W = 320.$$

Because the length is 40 yd more than the width, we have

$$L = W + 40.$$

The system is, therefore,

$$2L + 2W = 320 \quad (1)$$
$$L = W + 40. \quad (2)$$

Step 4 **Solve** the system of equations. Since equation (2) is solved for L, we can use the substitution method. We substitute $W + 40$ for L in equation (1), and solve for W.

$$\begin{aligned} 2L + 2W &= 320 && (1) \\ 2(W + 40) + 2W &= 320 && \text{Let } L = W + 40. \\ 2W + 80 + 2W &= 320 && \text{Distributive property} \\ 4W + 80 &= 320 && \text{Combine terms.} \\ 4W &= 240 && \text{Subtract 80.} \\ W &= 60 && \text{Divide by 4.} \end{aligned}$$

Let $W = 60$ in the equation $L = W + 40$ to find L.

$$L = 60 + 40 = 100$$

Step 5 **State the answer.** The length is 100 yd, and the width is 60 yd. Both dimensions are within the ranges given in the problem.

Step 6 **Check.** The perimeter of this soccer field is

$$2(100) + 2(60) = 320 \text{ yd},$$

and the length, 100 yd, is indeed 40 yd more than the width, since

$$100 - 40 = 60.$$

The answer is correct.

Work Problem 1 at the Side.

2 Solve money problems using two variables. Professional sport ticket prices increase annually. Average per-ticket prices in three of the four major sports (football, basketball, and hockey) now exceed $30.00.

ANSWERS

1. length: 15 m; width: 9 m

Example 2 Solving a Problem about Ticket Prices

During recent National Hockey League and National Basketball Association seasons, two hockey tickets and one basketball ticket purchased at their average prices would have cost $110.40. One hockey ticket and two basketball tickets would have cost $106.32. What were the average ticket prices for the two sports? (*Source:* Team Marketing Report, Chicago.)

Step 1 **Read** the problem again. There are two unknowns.

Step 2 **Assign variables.** Let h represent the average price for a hockey ticket and b represent the average price for a basketball ticket.

Step 3 **Write a system of equations.** Because two hockey tickets and one basketball ticket cost a total of $110.40, one equation for the system is

$$2h + b = 110.40.$$

By similar reasoning, the second equation is

$$h + 2b = 106.32.$$

Therefore, the system is

$$2h + b = 110.40 \quad (1)$$
$$h + 2b = 106.32. \quad (2)$$

Step 4 **Solve** the system of equations. To eliminate h, multiply equation (2) by -2 and add.

$$2h + b = 110.40 \quad (1)$$
$$-2h - 4b = -212.64 \quad \text{Multiply each side of (2) by } -2.$$
$$-3b = -102.24 \quad \text{Add.}$$
$$b = 34.08 \quad \text{Divide by } -3.$$

To find the value of h, let $b = 34.08$ in equation (2).

$$h + 2b = 106.32 \quad (2)$$
$$h + 2(34.08) = 106.32 \quad \text{Let } b = 34.08.$$
$$h + 68.16 = 106.32 \quad \text{Multiply.}$$
$$h = 38.16 \quad \text{Subtract 68.16.}$$

Step 5 **State the answer.** The average price for one basketball ticket was $34.08. For one hockey ticket, the average price was $38.16.

Step 6 **Check** that these values satisfy the conditions stated in the problem.

Work Problem 2 at the Side.

3 Solve mixture problems using two variables. We solved mixture problems earlier using one variable. For many mixture problems it seems more natural to use more than one variable and a system of equations.

Example 3 Solving a Mixture Problem

How many ounces each of 5% hydrochloric acid and 20% hydrochloric acid must be combined to get 10 oz of solution that is 12.5% hydrochloric acid?

Continued on Next Page

2 Solve the problem.

For recent Major League Baseball and National Football League seasons, based on average ticket prices, three baseball tickets and two football tickets would have cost $105.05, while two baseball tickets and one football ticket would have cost $58.12. What were the average ticket prices for the two sports? (*Source:* Team Marketing Report, Chicago.)

ANSWERS

2. baseball: $11.19; football: $35.74

3 Solve each problem.

(a) A grocer has some \$4 per lb coffee and some \$8 per lb coffee, which he will mix to make 50 lb of \$5.60 per lb coffee. How many pounds of each should be used?

(b) Some 40% ethyl alcohol solution is to be mixed with some 80% solution to get 200 L of a 50% solution. How many liters of each should be used?

Step 1 **Read** the problem. Two solutions of different strengths are being mixed together to get a specific amount of a solution with an "in-between" strength.

Step 2 **Assign variables.** Let x represent the number of ounces of 5% solution and y represent the number of ounces of 20% solution. Use a table to summarize the information from the problem.

Percent (as a Decimal)	Ounces of Solution	Ounces of Pure Acid
$5\% = .05$	x	$.05x$
$20\% = .20$	y	$.20y$
$12.5\% = .125$	10	$(.125)10$

Figure 8 also illustrates what is happening in the problem.

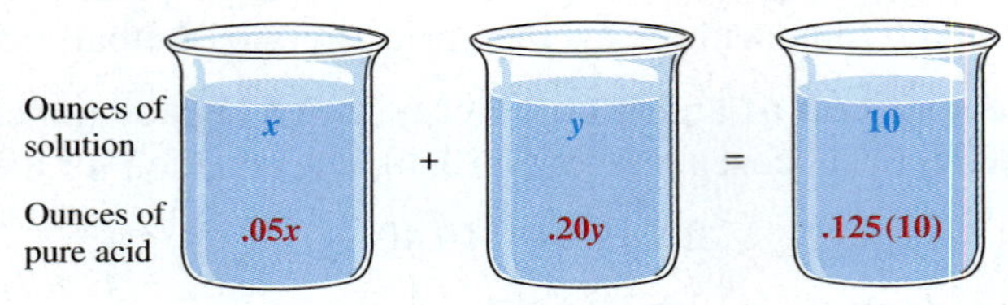

Figure 8

Step 3 **Write a system of equations.** When the x ounces of 5% solution and the y ounces of 20% solution are combined, the total number of ounces is 10, so

$$x + y = 10. \qquad (1)$$

The ounces of acid in the 5% solution ($.05x$) plus the ounces of acid in the 20% solution ($.20y$) should equal the total ounces of acid in the mixture, which is $(.125)10$, or 1.25. That is,

$$.05x + .20y = 1.25. \qquad (2)$$

Notice that these equations can be quickly determined by reading down in the table or using the labels in Figure 8.

Step 4 **Solve** the system of equations (1) and (2). Eliminate x by first multiplying equation (2) by 100 to clear it of decimals and then multiplying equation (1) by -5.

$$\begin{aligned} 5x + 20y &= 125 && \text{Multiply each side of (2) by 100.} \\ -5x - 5y &= -50 && \text{Multiply each side of (1) by } -5. \\ \hline 15y &= 75 && \text{Add.} \\ y &= 5 \end{aligned}$$

Because $y = 5$ and $x + y = 10$, x is also 5.

Step 5 **State the answer.** The desired mixture will require 5 oz of the 5% solution and 5 oz of the 20% solution.

Step 6 **Check** that these values satisfy both equations of the system.

Work Problem 3 at the Side.

4 Solve distance-rate-time problems using two variables. Motion problems require the distance formula, $d = rt$, where d is distance, r is rate (or speed), and t is time. These applications often lead to systems of equations, as in the next example.

ANSWERS

3. (a) 30 lb of \$4; 20 lb of \$8
(b) 150 L of 40%; 50 L of 80%

Example 4 Solving a Motion Problem

A car travels 250 km in the same time that a truck travels 225 km. If the speed of the car is 8 km per hr faster than the speed of the truck, find both speeds.

Step 1 **Read** the problem again. Given the distances traveled, you need to find the speed of each vehicle.

Step 2 **Assign variables.**

Let $x =$ the speed of the car

and $y =$ the speed of the truck.

As in Example 3, a table helps organize the information. Fill in the given information for each vehicle (in this case, distance) and use the assigned variables for the unknown speeds (rates).

	d	*r*	*t*
Car	250	x	
Truck	225	y	

The table shows nothing about time. To get an expression for time, solve the distance formula, $d = rt$, for t.

$$\frac{d}{r} = t$$

The two times can be written as $\frac{250}{x}$ and $\frac{225}{y}$.

Step 3 **Write a system of equations.** The problem states that the car travels 8 km per hr faster than the truck. Since the two speeds are x and y,

$$x = y + 8.$$

Both vehicles travel for the same time, so from the table

$$\frac{250}{x} = \frac{225}{y}.$$

This is not a linear equation. However, multiplying each side by xy gives

$$250y = 225x,$$

which is linear. The system is

$$x = y + 8 \qquad (1)$$
$$250y = 225x. \qquad (2)$$

Step 4 **Solve** the system of equations by substitution. Replace x with $y + 8$ in equation (2).

$$250y = 225\mathbf{x} \qquad (2)$$
$$250y = 225(\mathbf{y + 8}) \qquad \text{Let } x = y + 8.$$
$$250y = 225y + 1800 \qquad \text{Distributive property}$$
$$25y = 1800 \qquad \text{Subtract } 225y.$$
$$y = 72 \qquad \text{Divide by 25.}$$

Because $x = y + 8$, the value of x is $72 + 8 = 80$.

Continued on Next Page

Step 5 **State the answer.** The car's speed is 80 km per hr, and the truck's speed is 72 km per hr.

Step 6 **Check.** This is especially important since one of the equations had variable denominators.

$$\text{Car: } t = \frac{d}{r} = \frac{250}{80} = 3.125$$

$$\text{Truck: } t = \frac{d}{r} = \frac{225}{72} = 3.125$$

Times are equal.

Since $80 - 72 = 8$, the conditions of the problem are satisfied.

Work Problem 4 at the Side.

4 Solve the problem.

A train travels 600 mi in the same time that a truck travels 520 mi. Find the speed of each vehicle if the train's average speed is 8 mph faster than the truck's.

5 Solve problems with three variables using a system of three equations. To solve such problems, we extend the method used for two unknowns. Since three variables are used, three equations are necessary to find a solution.

Example 5 Solving a Problem Involving Prices

At Panera Bread, a loaf of honey wheat bread costs \$2.40, a loaf of pumpernickel bread costs \$3.35, and a loaf of French bread costs \$2.10. On a recent day, three times as many loaves of honey wheat were sold as pumpernickel. The number of loaves of French bread sold was 5 less than the number of loaves of honey wheat sold. Total receipts for these breads were \$56.90. How many loaves of each type of bread were sold? (*Source:* Panera Bread menu.)

Step 1 **Read** the problem again. There are three unknowns in this problem.

Step 2 **Assign variables** to represent the three unknowns.

Let x = the number of loaves of honey wheat,

y = the number of loaves of pumpernickel,

and z = the number of loaves of French bread.

Step 3 **Write a system of three equations** using the information in the problem. Since three times as many loaves of honey wheat were sold as pumpernickel,

$$x = 3y, \quad \text{or} \quad x - 3y = 0. \quad (1)$$

Also,

Number of loaves of French bread	equals	5 less than the number of loaves of honey wheat.
↓	↓	↓
z	$=$	$x - 5$,

so $\quad x - z = 5. \quad (2)$

Multiplying the cost of a loaf of each kind of bread by the number of loaves of that kind sold and adding gives the total receipts.

$$2.40x + 3.35y + 2.10z = 56.90$$

Multiply each side of this equation by 100 to clear it of decimals.

$$240x + 335y + 210z = 5690 \quad (3)$$

Step 4 **Solve** the system of three equations using the method shown in Section 5.2.

Work Problem 5 at the Side.

5 Solve the system of equations from Example 5.

$$\begin{aligned} x - 3y &= 0 \quad (1) \\ x - z &= 5 \quad (2) \\ 240x + 335y + 210z &= 5690 \quad (3) \end{aligned}$$

Answers

4. train: 60 mph; truck: 52 mph

5. (12, 4, 7)

Continued on Next Page

Step 5 **State the answer.** The solution is (12, 4, 7), so 12 loaves of honey wheat, 4 loaves of pumpernickel, and 7 loaves of French bread were sold.

Step 6 **Check.** Since $12 = 3 \cdot 4$, the number of loaves of honey wheat is three times the number of loaves of pumpernickel. Also, $12 - 7 = 5$, so the number of loaves of French bread is 5 less than the number of loaves of honey wheat. Multiply the appropriate cost per loaf by the number of loaves sold and add the results to check that total receipts were $56.90.

Work Problem 6 at the Side.

6 Solve the problem.

A department store has three kinds of perfume: cheap, better, and best. It has 10 more bottles of cheap than better, and 3 fewer bottles of best than better. Each bottle of cheap costs $8, better costs $15, and best costs $32. The total value of all the perfume is $589. How many bottles of each are there?

Example 6 Solving a Business Production Problem

A company produces three color television sets, models X, Y, and Z. Each model X set requires 2 hr of electronics work, 2 hr of assembly time, and 1 hr of finishing time. Each model Y requires 1, 3, and 1 hr of electronics, assembly, and finishing time, respectively. Each model Z requires 3, 2, and 2 hr of the same work, respectively. There are 100 hr available for electronics, 100 hr available for assembly, and 65 hr available for finishing per week. How many of each model should be produced each week if all available time must be used?

Step 1 **Read** the problem again. There are three unknowns.

Step 2 **Assign variables.**

Let x = the number of model X produced per week,

y = the number of model Y produced per week,

and z = the number of model Z produced per week.

Organize the information in a table.

	Each Model X	Each Model Y	Each Model Z	Totals
Hours of Electronics Work	2	1	3	100
Hours of Assembly Time	2	3	2	100
Hours of Finishing Time	1	1	2	65

Step 3 **Write a system of three equations.** The x model X sets require $2x$ hr of electronics, the y model Y sets require $1y$ (or y) hr of electronics, and the z model Z sets require $3z$ hr of electronics. Since 100 hr are available for electronics,

$$2x + y + 3z = 100. \quad (1)$$

Similarly, from the fact that 100 hr are available for assembly,

$$2x + 3y + 2z = 100, \quad (2)$$

and the fact that 65 hr are available for finishing leads to the equation

$$x + y + 2z = 65. \quad (3)$$

Again, notice the advantage of setting up a table. By reading across, we can easily determine the coefficients and constants in the equations of the system.

Continued on Next Page

ANSWERS

6. 21 bottles of cheap; 11 of better; 8 of best

Step 4 **Solve** the system

$$
\begin{aligned}
2x + y + 3z &= 100 \\
2x + 3y + 2z &= 100 \\
x + y + 2z &= 65
\end{aligned}
$$

to find $x = 15$, $y = 10$, and $z = 20$.

Step 5 **State the answer.** The company should produce 15 model X, 10 model Y, and 20 model Z sets per week.

Step 6 **Check** that these values satisfy the conditions of the problem.

Work Problem 7 at the Side.

7 Solve the problem.

During the 1999–2000 National Hockey League regular season, the eventual Stanley Cup winners, the New Jersey Devils, played 82 games. Twice the number of ties added to the number of losses was equal to the number of wins. The difference between the numbers of wins and ties was 8 more than the number of losses. How many wins, losses, and ties did the New Jersey Devils have? (*Source: The World Almanac and Book of Facts* 2001.)

ANSWERS

7. 45 wins, 29 losses, 8 ties

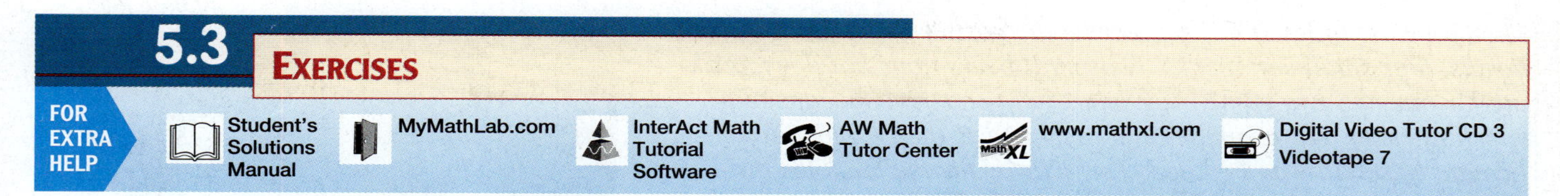

Solve each problem. See Example 1.

1. During the 1999–2000 Major League Baseball regular season, the St. Louis Cardinals played 162 games. They won 28 more games than they lost. What was their win–loss record that year?

2. Refer to Exercise 1. During the same 162-game season, the Chicago Cubs lost 32 more games than they won. What was the team's win–loss record?

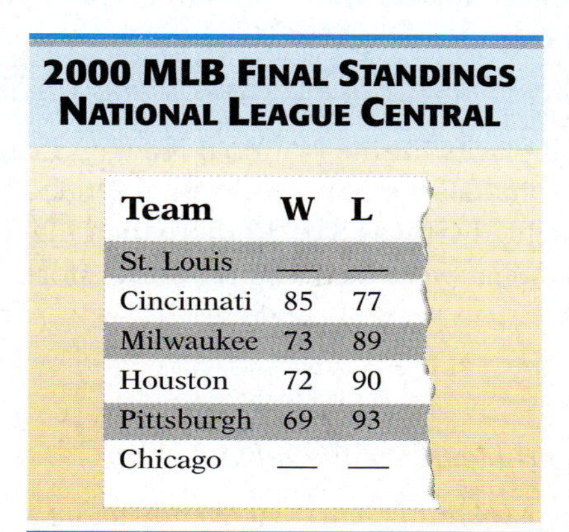

2000 MLB Final Standings
National League Central

Team	W	L
St. Louis	___	___
Cincinnati	85	77
Milwaukee	73	89
Houston	72	90
Pittsburgh	69	93
Chicago	___	___

Source: www.mlb.com

3. Venus and Serena measured a tennis court and found that it was 42 ft longer than it was wide and had a perimeter of 228 ft. What were the length and the width of the tennis court?

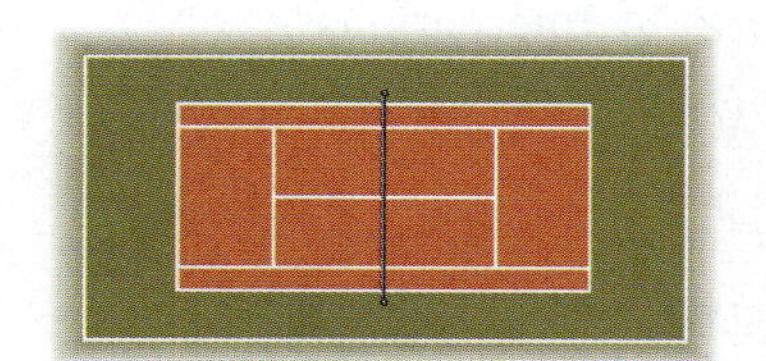

4. Shaq and Kobe found that the width of their basketball court was 44 ft less than the length. If the perimeter was 288 ft, what were the length and the width of their court?

5. The two biggest U.S. companies in terms of revenue in 2000 were ExxonMobil and General Motors. ExxonMobil's revenue was \$29 billion more than that of General Motors. Total revenue for the two companies was \$399 billion. What was the revenue for each company? (*Source:* Bridge News, MarketGuide.com)

6. The top two U.S. trading partners during the first four months of 2000 were Canada and Mexico. Exports and imports with Mexico were \$57 billion less than those with Canada. Total exports and imports involving these two countries were \$211 billion. How much were U.S. exports and imports with each country? (*Source:* U.S. Bureau of the Census.)

In Exercises 7 and 8, find the measures of the angles marked x and y. Remember that (1) *the sum of the measures of the angles of a triangle is* 180°, (2) *supplementary angles have a sum of* 180°, *and* (3) *vertical angles have equal measures.*

7.

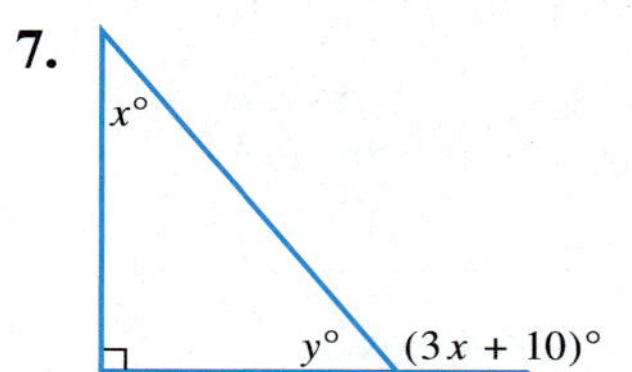

8.

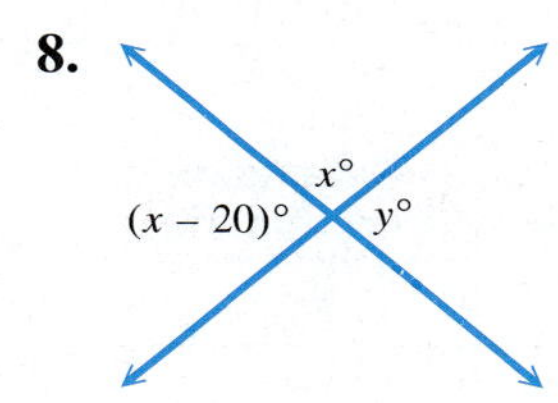

The Fan Cost Index (FCI) represents the cost of four average-price tickets, four small soft drinks, two small beers, four hot dogs, parking for one car, two game programs, and two souvenir caps to a sporting event. For example, in a recent year, the FCI for Major League Baseball was \$105.63. This was by far the least for the four major professional sports. (*Source:* Team Marketing Report, Chicago.)

Use the concept of FCI in Exercises 9 and 10. See Example 2.

9. The FCI prices for the National Hockey League and the National Basketball Association totaled \$423.12. The hockey FCI was \$16.36 more than that of basketball. What were the FCIs for these sports?

10. The FCI prices for Major League Baseball and the National Football League totaled \$311.03. The football FCI was \$105.87 more than that of baseball. What were the FCIs for these sports?

Solve each problem. See Example 2.

11. Andrew McGinnis works at Wendy's Old Fashioned Hamburgers. During one particular lunch hour, he sold 15 single hamburgers and 10 double hamburgers, totaling \$63.25. Another lunch hour, he sold 30 singles and 5 doubles, totaling \$78.65. How much did each type of burger cost? (*Source:* Wendy's Old Fashioned Hamburgers menu.)

12. Tokyo and New York are among the most expensive cities worldwide for business travelers. Using average costs per day for each city (which includes room, meals, laundry, and two taxi fares), 2 days in Tokyo and 3 days in New York cost \$2015. Four days in Tokyo and 2 days in New York cost \$2490. What is the average cost per day for each city? (*Source:* ECA International.)

The formulas $p = br$ (percentage = base × rate) and $I = prt$ (simple interest = principal × rate × time) are used in the applications in Exercises 17–24. To prepare to use these formulas, answer the questions in Exercises 13 and 14.

13. If a container of liquid contains 60 oz of solution, what is the number of ounces of pure acid if the given solution contains the following acid concentrations?

(a) 10% **(b)** 25% **(c)** 40% **(d)** 50%

14. If \$5000 is invested in an account paying simple annual interest, how much interest will be earned during the first year at the following rates?

(a) 2% **(b)** 3% **(c)** 4% **(d)** 3.5%

15. If a pound of turkey costs \$.99, how much will x pounds cost?

16. If a ticket to the movie *Pearl Harbor* costs \$8 and y tickets are sold, how much is collected from the sale?

Solve each problem. See Example 3.

17. How many gallons each of 25% alcohol and 35% alcohol should be mixed to get 20 gal of 32% alcohol?

Percent (as a Decimal)	Gallons of Solution	Gallons of Pure Alcohol
25% = .25	x	
35% = .35	y	
32% =	20	

18. How many liters each of 15% acid and 33% acid should be mixed to get 120 L of 21% acid?

Percent (as a Decimal)	Liters of Solution	Liters of Pure Acid
15% = .15	x	
33% =	y	
21% =	120	

19. Pure acid is to be added to a 10% acid solution to obtain 54 L of a 20% acid solution. What amounts of each should be used?

20. A truck radiator holds 36 L of fluid. How much pure antifreeze must be added to a mixture that is 4% antifreeze to fill the radiator with a mixture that is 20% antifreeze?

21. A party mix is made by adding nuts that sell for \$2.50 per kg to a cereal mixture that sells for \$1 per kg. How much of each should be added to get 30 kg of a mix that will sell for \$1.70 per kg?

	Price per Kilogram	Number of Kilograms	Value
Nuts	2.50	x	
Cereal	1.00	y	
Mixture	1.70		

22. A popular fruit drink is made by mixing fruit juices. Such a drink with 50% juice is to be mixed with another drink that is 30% juice to get 200 L of a drink that is 45% juice. How much of each should be used?

	Percent (as a Decimal)	Liters of Drink	Liters of Pure Juice
50% Juice	.50	x	
30% Juice	.30	y	
Mixture	.45		

23. A total of \$3000 is invested, part at 2% simple interest and part at 4%. If the total annual return from the two investments is \$100, how much is invested at each rate?

Rate (as a Decimal)	Principal	Interest
.02	x	$.02x$
.04	y	$.04y$
	3000	100

24. An investor must invest a total of \$15,000 in two accounts, one paying 4% annual simple interest, and the other 3%. If he wants to earn \$550 annual interest, how much should he invest at each rate?

Rate (as a Decimal)	Principal	Interest
.04	x	
.03	y	
	15,000	

The formula $d = rt$ (distance = rate × time) is used in the applications in Exercises 27–30. To prepare to use this formula, answer the questions in Exercises 25 and 26.

25. If the speed of a killer whale is 25 mph and the whale swims for y hr, how many miles does the whale travel?

26. If the speed of a boat in still water is 10 mph, and the speed of the current of a river is x mph, what is the speed of the boat

(a) going upstream (that is, against the current);

(b) going downstream (that is, with the current)?

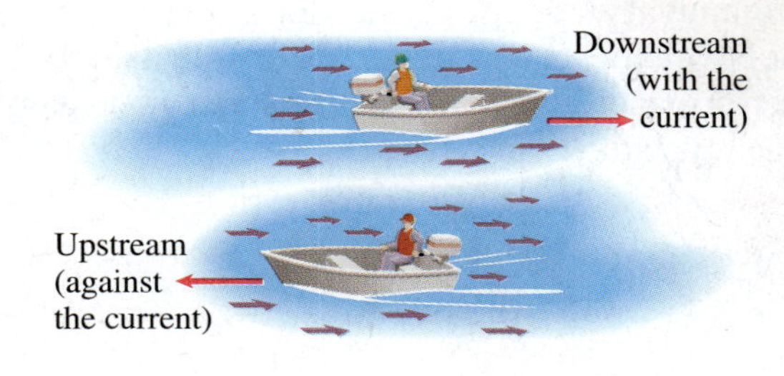

Solve each problem. See Example 4.

27. A freight train and an express train leave towns 390 km apart, traveling toward one another. The freight train travels 30 km per hr slower than the express train. They pass one another 3 hr later. What are their speeds?

	r	t	d
Freight Train	x	3	
Express Train	y	3	

28. A train travels 150 km in the same time that a plane covers 400 km. If the speed of the plane is 20 km per hr less than 3 times the speed of the train, find both speeds.

29. In his motorboat, Bill Ruhberg travels upstream at top speed to his favorite fishing spot, a distance of 36 mi, in 2 hr. Returning, he finds that the trip downstream, still at top speed, takes only 1.5 hr. Find the speed of Bill's boat and the speed of the current.

	r	t	d
Upstream	$x - y$	2	
Downstream	$x + y$		

30. Traveling for 3 hr into a steady headwind, a plane flies 1650 mi. The pilot determines that flying *with* the same wind for 2 hr, he could make a trip of 1300 mi. Find the speed of the plane and the speed of the wind.

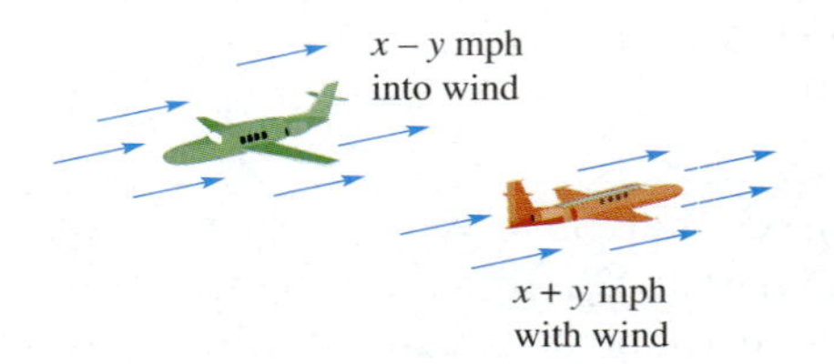

Use the problem-solving techniques of this section to solve each problem with two unknowns. See Examples 1–6.

31. At age 61, rock icon Tina Turner generated the most revenue on the concert circuit in 2000. Turner and second-place 'N Sync together took in \$157 million from ticket sales. If 'N Sync took in \$3.8 million less than Turner, how much did each generate? (*Source:* Pollstar.)

32. Carol Britz plans to mix pecan clusters that sell for \$3.60 per lb with chocolate truffles that sell for \$7.20 per lb to get a mixture that she can sell in Valentine boxes for \$4.95 per lb. How much of the \$3.60 clusters and the \$7.20 truffles should she use to create 80 lb of the mix?

	Price per Pound	Number of Pounds	Value
Pecan Clusters		x	
Chocolate Truffles		y	
Valentine Mixture		80	

33. Tickets to a production of *King Lear* at Cape Fear Community College cost \$5 for general admission or \$4 with a student ID. If 184 people paid to see a performance and \$812 was collected, how many of each type of ticket were sold?

34. At a business meeting at Panera Bread, the bill for two cappuccinos and three house lattes was \$10.95. At another table, the bill for one cappuccino and two house lattes was \$6.65. How much did each type of beverage cost? (*Source:* Panera Bread menu.)

35. The mixed price of 9 citrons and 7 fragrant wood apples is 107; again, the mixed price of 7 citrons and 9 fragrant wood apples is 101. O you arithmetician, tell me quickly the price of a citron and the price of a wood apple here, having distinctly separated those prices well. (*Source:* Hindu work, A.D. 850.) (*Hint:* "Mixed price" refers to the price of a mixture of the two fruits.)

36. Braving blizzard conditions on the planet Hoth, Luke Skywalker sets out at top speed in his snow speeder for a rebel base 4800 mi away. He travels into a steady headwind and makes the trip in 3 hr. Returning, he finds that the trip back, still at top speed but now with a tailwind, takes only 2 hr. Find the top speed of Luke's snow speeder and the speed of the wind.

	r	t	d
Into Headwind			
With Tailwind			

Solve each problem involving three unknowns. See Examples 5 and 6. (In Exercises 37–40, remember that the sum of the measures of the angles of a triangle is 180°.)

37. In the figure, $z = x + 10$ and $x + y = 100$. Determine a third equation involving x, y, and z, and then find the measures of the three angles.

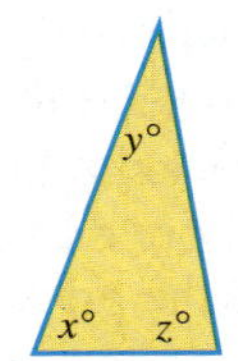

38. In the figure, x is 10 less than y and 20 less than z. Write a system of equations and find the measures of the three angles.

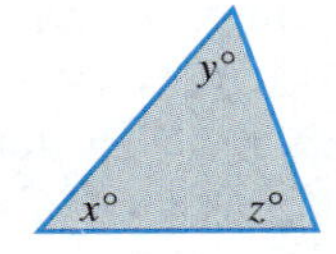

39. In a certain triangle, the measure of the second angle is 10° more than three times the first. The third angle measure is equal to the sum of the measures of the other two. Find the measures of the three angles.

40. The measure of the largest angle of a triangle is 12° less than the sum of the measures of the other two. The smallest angle measures 58° less than the largest. Find the measures of the angles.

41. The perimeter of a triangle is 70 cm. The longest side is 4 cm less than the sum of the other two sides. Twice the shortest side is 9 cm less than the longest side. Find the length of each side of the triangle.

42. The perimeter of a triangle is 56 in. The longest side measures 4 in. less than the sum of the other two sides. Three times the shortest side is 4 in. more than the longest side. Find the lengths of the three sides.

43. In a random sample of 100 Americans of voting age, 10 more Americans identify themselves as Independents than Republicans. Six fewer Americans identify themselves as Republicans than Democrats. Assuming that all of those sampled are Republican, Democrat, or Independent, how many of those in the sample identify themselves with each political affiliation? (*Source:* The Gallop Organization.)

44. In the 2000 Summer Olympics in Sydney, Australia, the United States earned 14 more gold medals than silver. The number of bronze medals earned was 17 less than twice the number of silver medals. The United States earned a total of 97 medals. How many of each kind of medal did the United States earn? (*Source: The Gazette,* October 2, 2000.)

45. Tickets for one show on the Harlem Globetrotters' 75th Anniversary Tour cost \$10, \$18, or, for VIP seats, \$30. So far, five times as many \$18 tickets have been sold as VIP tickets. The number of \$10 tickets equals the number of \$18 tickets plus twice the number of VIP tickets. Sales of these tickets total \$9500. How many of each kind of ticket have been sold? (*Source:* www.ticketmaster.com)

46. Three kinds of tickets are available for a Green Day concert: "up close," "in the middle," and "far out." "Up close" tickets cost \$10 more than "in the middle" tickets, while "in the middle" tickets cost \$10 more than "far out" tickets. Twice the cost of an "up close" ticket is \$20 more than 3 times the cost of a "far out" ticket. Find the price of each kind of ticket.

47. A hardware supplier manufactures three kinds of clamps, types A, B, and C. Production restrictions require it to make 10 units more type C clamps than the total of the other types and twice as many type B clamps as type A. The shop must produce a total of 490 units of clamps per day. How many units of each type can be made per day?

48. A Mardi Gras trinket manufacturer supplies three wholesalers, A, B, and C. The output from a day's production is 320 cases of trinkets. She must send wholesaler A three times as many cases as she sends B, and she must send wholesaler C 160 cases less than she provides A and B together. How many cases should she send to each wholesaler to distribute the entire day's production to them?

5.4 Solving Linear Systems of Equations by Matrix Methods

Objectives

1. Define a matrix.
2. Write the augmented matrix for a system.
3. Use row operations to solve a system with two equations.
4. Use row operations to solve a system with three equations.
5. Use row operations to solve special systems.

1 Define a matrix. An ordered array of numbers such as

$$\text{Rows}\rightarrow\overset{\text{Columns}}{\begin{bmatrix} 2 & 3 & 5 \\ 7 & 1 & 2 \end{bmatrix}}$$

is called a **matrix.** The numbers are called **elements** of the matrix. Matrices (the plural of *matrix*) are named according to the number of **rows** and **columns** they contain. The rows are read horizontally, and the columns are read vertically. For example, the first row in the preceding matrix is 2 3 5 and the first column is $\begin{matrix} 2 \\ 7 \end{matrix}$. This matrix is a 2×3 (read "two by three") matrix because it has 2 rows and 3 columns. The number of rows is given first, and then the number of columns. Two other examples follow.

$$\begin{bmatrix} -1 & 0 \\ 1 & -2 \end{bmatrix} \quad 2 \times 2 \text{ matrix} \qquad \begin{bmatrix} 8 & -1 & -3 \\ 2 & 1 & 6 \\ 0 & 5 & -3 \\ 5 & 9 & 7 \end{bmatrix} \quad 4 \times 3 \text{ matrix}$$

A **square matrix** is one that has the same number of rows as columns. The 2×2 matrix is a square matrix.

Calculator Tip Figure 9 shows how a graphing calculator displays the preceding two matrices. Work with matrices is made much easier by using technology when available. Consult your owner's manual for details.

```
[A]
      [[-1 0 ]
       [1  -2]]
```

```
[B]
      [[8 -1 -3]
       [2 1  6 ]
       [0 5  -3]
       [5 9  7 ]]
```

Figure 9

In this section, we discuss a method of solving linear systems that uses matrices. This method is really just a very structured way of using the elimination method to solve a linear system. The advantage of this new method is that it can be done by a graphing calculator or a computer, allowing large systems of equations to be solved easily.

2 Write the augmented matrix for a system. To begin, we write an *augmented matrix* for the system. An **augmented matrix** has a vertical bar that separates the columns of the matrix into two groups. For example, to solve the system

$$\begin{aligned} x - 3y &= 1 \\ 2x + y &= -5, \end{aligned}$$

start with the augmented matrix

$$\left[\begin{array}{cc|c} 1 & -3 & 1 \\ 2 & 1 & -5 \end{array}\right].$$

Place the coefficients of the variables to the left of the bar, and the constants to the right. The bar separates the coefficients from the constants. The matrix is just a shorthand way of writing the system of equations, so the rows of the augmented matrix can be treated the same as the equations of a system of equations.

We know that exchanging the position of two equations in a system does not change the system. Also, multiplying any equation in a system by a nonzero number does not change the system. Comparable changes to the augmented matrix of a system of equations produce new matrices that correspond to systems with the same solutions as the original system.

The following **row operations** produce new matrices that lead to systems having the same solutions as the original system.

Matrix Row Operations

1. Any two rows of the matrix may be interchanged.
2. The numbers in any row may be multiplied by any nonzero real number.
3. Any row may be changed by adding to the numbers of the row the product of a real number and the corresponding numbers of another row.

Examples of these row operations follow.

Row operation 1:

$$\begin{bmatrix} 2 & 3 & 9 \\ 4 & 8 & -3 \\ 1 & 0 & 7 \end{bmatrix} \text{ becomes } \begin{bmatrix} 1 & 0 & 7 \\ 4 & 8 & -3 \\ 2 & 3 & 9 \end{bmatrix}.$$

Interchange row 1 and row 3.

Row operation 2:

$$\begin{bmatrix} 2 & 3 & 9 \\ 4 & 8 & -3 \\ 1 & 0 & 7 \end{bmatrix} \text{ becomes } \begin{bmatrix} 6 & 9 & 27 \\ 4 & 8 & -3 \\ 1 & 0 & 7 \end{bmatrix}.$$

Multiply the numbers in row 1 by 3.

Row operation 3:

$$\begin{bmatrix} 2 & 3 & 9 \\ 4 & 8 & -3 \\ 1 & 0 & 7 \end{bmatrix} \text{ becomes } \begin{bmatrix} 0 & 3 & -5 \\ 4 & 8 & -3 \\ 1 & 0 & 7 \end{bmatrix}.$$

Multiply the numbers in row 3 by -2; add them to the corresponding numbers in row 1.

The third row operation corresponds to the way we eliminated a variable from a pair of equations in the previous sections.

3 Use row operations to solve a system with two equations. Row operations can be used to rewrite a matrix until it is the matrix of a system where the solution is easy to find. The goal is a matrix in the form

$$\left[\begin{array}{cc|c} 1 & a & b \\ 0 & 1 & c \end{array}\right] \quad \text{or} \quad \left[\begin{array}{ccc|c} 1 & a & b & c \\ 0 & 1 & d & e \\ 0 & 0 & 1 & f \end{array}\right]$$

for systems with two or three equations, respectively. Notice that there are 1s down the diagonal from upper left to lower right and 0s below the 1s. A matrix written this way is said to be in **row echelon form.** When these matrices are rewritten as systems of equations, the value of one variable is known, and the rest can be found by substitution. The following examples illustrate this method.

Example 1 Using Row Operations to Solve a System with Two Variables

Use row operations to solve the system.

$$\begin{aligned} x - 3y &= 1 \\ 2x + y &= -5 \end{aligned}$$

We start with the augmented matrix of the system.

$$\left[\begin{array}{rr|r} 1 & -3 & 1 \\ 2 & 1 & -5 \end{array}\right]$$

Now we use the various row operations to change this matrix into one that leads to a system that is easier to solve.

It is best to work by columns. We start with the first column and make sure that there is a 1 in the first row, first column position. There is already a 1 in this position. Next, we get 0 in every position below the first. To get a 0 in row two, column one, we use the third row operation and add to the numbers in row two the result of multiplying each number in row one by -2. (We abbreviate this as $-2R_1 + R_2$.) Row one remains unchanged.

$$\left[\begin{array}{cc|c} 1 & -3 & 1 \\ 2 + \mathbf{1(-2)} & 1 + \mathbf{-3(-2)} & -5 + \mathbf{1(-2)} \end{array}\right]$$

Original number from row two ↑   ↑ -2 times number from row one

$$\left[\begin{array}{rr|r} 1 & -3 & 1 \\ 0 & 7 & -7 \end{array}\right] \quad -2R_1 + R_2$$

The matrix now has a 1 in the first position of column one, with 0 in every position below the first.

Now we go to column two. A 1 is needed in row two, column two. We get this 1 by using the second row operation, multiplying each number of row two by $\frac{1}{7}$.

$$\left[\begin{array}{rr|r} 1 & -3 & 1 \\ 0 & 1 & -1 \end{array}\right] \quad \tfrac{1}{7}R_2$$

This augmented matrix leads to the system of equations

$$\begin{aligned} 1x - 3y &= 1 \\ 0x + 1y &= -1 \end{aligned} \quad \text{or} \quad \begin{aligned} x - 3y &= 1 \\ y &= -1. \end{aligned}$$

From the second equation, $y = -1$. We substitute -1 for y in the first equation to get

$$\begin{aligned} x - 3y &= 1 \\ x - 3(-1) &= 1 \\ x + 3 &= 1 \\ x &= -2. \end{aligned}$$

The solution set of the system is $\{(-2, -1)\}$. Check this solution by substitution in both equations of the system.

Work Problem 1 at the Side.

1 Use row operations to solve the system.

$$\begin{aligned} x - 2y &= 9 \\ 3x + y &= 13 \end{aligned}$$

ANSWERS

1. $\{(5, -2)\}$

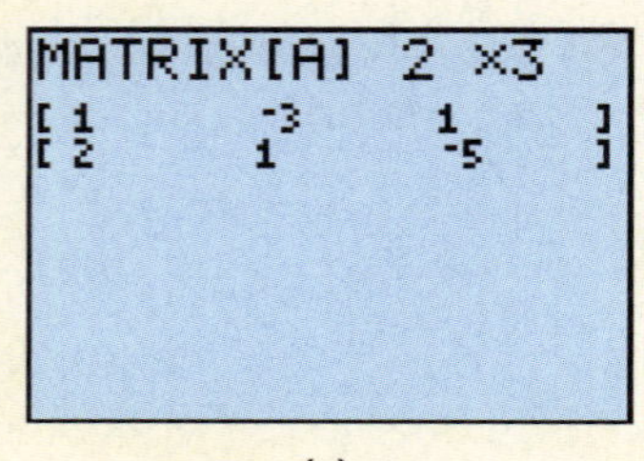

(a)

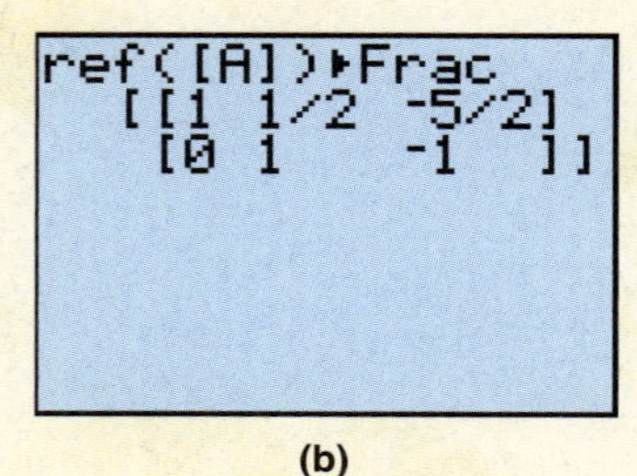

(b)

Figure 10

Calculator Tip If the augmented matrix of the system in Example 1 is entered as matrix A in a graphing calculator (Figure 10(a)) and the row echelon form of the matrix is found (Figure 10(b)), the system becomes

$$x + \frac{1}{2}y = -\frac{5}{2}$$
$$y = -1.$$

While this system looks different from the one we obtained in Example 1, it is equivalent, since its solution set is also $\{(-2, -1)\}$.

4 Use row operations to solve a system with three equations. A linear system with three equations is solved in a similar way. We use row operations to get 1s down the diagonal from left to right and all 0s below each 1.

Example 2 Using Row Operations to Solve a System with Three Variables

Use row operations to solve the system.

$$\begin{aligned} x - y + 5z &= -6 \\ 3x + 3y - z &= 10 \\ x + 3y + 2z &= 5 \end{aligned}$$

Start by writing the augmented matrix of the system.

$$\left[\begin{array}{rrr|r} 1 & -1 & 5 & -6 \\ 3 & 3 & -1 & 10 \\ 1 & 3 & 2 & 5 \end{array}\right]$$

This matrix already has 1 in row one, column one. Next get 0s in the rest of column one. First, add to row two the results of multiplying each number of row one by -3. This gives the matrix

$$\left[\begin{array}{rrr|r} 1 & -1 & 5 & -6 \\ 0 & 6 & -16 & 28 \\ 1 & 3 & 2 & 5 \end{array}\right]. \qquad -3R_1 + R_2$$

Now add to the numbers in row three the results of multiplying each number of row one by -1.

$$\left[\begin{array}{rrr|r} 1 & -1 & 5 & -6 \\ 0 & 6 & -16 & 28 \\ 0 & 4 & -3 & 11 \end{array}\right] \qquad -1R_1 + R_3$$

Get 1 in row two, column two by multiplying each number in row two by $\frac{1}{6}$.

$$\left[\begin{array}{rrr|r} 1 & -1 & 5 & -6 \\ 0 & 1 & -\frac{8}{3} & \frac{14}{3} \\ 0 & 4 & -3 & 11 \end{array}\right] \qquad \tfrac{1}{6}R_2$$

Get 0 in row three, column two by adding to row three the results of multiplying each number in row two by -4.

$$\left[\begin{array}{rrr|r} 1 & -1 & 5 & -6 \\ 0 & 1 & -\frac{8}{3} & \frac{14}{3} \\ 0 & 0 & \frac{23}{3} & -\frac{23}{3} \end{array}\right] \qquad -4R_2 + R_3$$

Continued on Next Page

Finally, get 1 in row three, column three by multiplying each number in row three by $\frac{3}{23}$.

$$\left[\begin{array}{ccc|c} 1 & -1 & 5 & -6 \\ 0 & 1 & -\frac{8}{3} & \frac{14}{3} \\ 0 & 0 & 1 & -1 \end{array}\right] \quad \frac{3}{23}R_3$$

This final matrix gives the system of equations

$$\begin{aligned} x - y + 5z &= -6 \\ y - \frac{8}{3}z &= \frac{14}{3} \\ z &= -1. \end{aligned}$$

Substitute -1 for z in the second equation, $y - \frac{8}{3}z = \frac{14}{3}$, to get $y = 2$. Finally, substitute 2 for y and -1 for z in the first equation, $x - y + 5z = -6$, to get $x = 1$. The solution set of the original system is $\{(1, 2, -1)\}$. Check by substitution in the original system.

Work Problem 2 at the Side.

2 Use row operations to solve the system.

$$\begin{aligned} 2x - y + z &= 7 \\ x - 3y - z &= 7 \\ -x + y - 5z &= -9 \end{aligned}$$

5 Use row operations to solve special systems. In the final example we show how to recognize inconsistent systems or systems with dependent equations when solving these systems with row operations.

Example 3 Recognizing Inconsistent Systems or Dependent Equations

Use row operations to solve each system.

(a) $2x - 3y = 8$
$-6x + 9y = 4$

$$\left[\begin{array}{cc|c} 2 & -3 & 8 \\ -6 & 9 & 4 \end{array}\right] \quad \text{Write the augmented matrix.}$$

$$\left[\begin{array}{cc|c} 1 & -\frac{3}{2} & 4 \\ -6 & 9 & 4 \end{array}\right] \quad \frac{1}{2}R_1$$

$$\left[\begin{array}{cc|c} 1 & -\frac{3}{2} & 4 \\ \mathbf{0} & \mathbf{0} & \mathbf{28} \end{array}\right] \quad 6R_1 + R_2$$

The corresponding system of equations is

$$\begin{aligned} x - \frac{3}{2}y &= 4 \\ 0 &= 28, \quad \textbf{False} \end{aligned}$$

which has no solution and is inconsistent. The solution set is $\emptyset$.

(b) $-10x + 12y = 30$
$5x - 6y = -15$

$$\left[\begin{array}{cc|c} -10 & 12 & 30 \\ 5 & -6 & -15 \end{array}\right] \quad \text{Write the augmented matrix.}$$

$$\left[\begin{array}{cc|c} 1 & -\frac{6}{5} & -3 \\ 5 & -6 & -15 \end{array}\right] \quad -\frac{1}{10}R_1$$

$$\left[\begin{array}{cc|c} 1 & -\frac{6}{5} & -3 \\ \mathbf{0} & \mathbf{0} & \mathbf{0} \end{array}\right] \quad -5R_1 + R_2$$

Continued on Next Page

ANSWERS
2. $\{(2, -2, 1)\}$

The corresponding system is

$$x - \frac{6}{5}y = -3$$
$$0 = 0, \qquad \text{True}$$

which has dependent equations. Using the second equation of the original system, we write the solution set as

$$\{(x, y) \mid 5x - 6y = -15\}.$$

Work Problem 3 at the Side.

3 Use row operations to solve each system.

(a)
$$\begin{aligned} x - y &= 2 \\ -2x + 2y &= 2 \end{aligned}$$

(b)
$$\begin{aligned} x - y &= 2 \\ -2x + 2y &= -4 \end{aligned}$$

ANSWERS

3. **(a)** $\emptyset$ **(b)** $\{(x, y) \mid x - y = 2\}$

5.4 EXERCISES

FOR EXTRA HELP

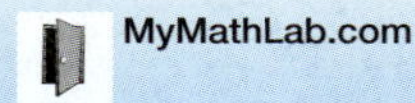
Student's Solutions Manual

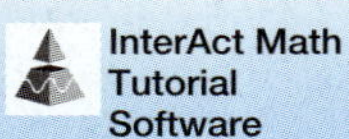
MyMathLab.com

InterAct Math Tutorial Software

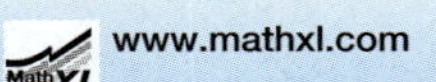
AW Math Tutor Center

www.mathxl.com

Digital Video Tutor CD 3 Videotape 7

1. Consider the matrix $\begin{bmatrix} -2 & 3 & 1 \\ 0 & 5 & -3 \\ 1 & 4 & 8 \end{bmatrix}$, and answer the following.

(a) What are the elements of the second row?

(b) What are the elements of the third column?

(c) Is this a square matrix? Explain.

(d) Give the matrix obtained by interchanging the first and third rows.

(e) Give the matrix obtained by multiplying the first row by $-\frac{1}{2}$.

(f) Give the matrix obtained by multiplying the third row by 3 and adding to the first row.

2. Give the dimensions of each matrix.

(a) $\begin{bmatrix} 3 & -7 \\ 4 & 5 \\ -1 & 0 \end{bmatrix}$ **(b)** $\begin{bmatrix} 4 & 9 & 0 \\ -1 & 2 & -4 \end{bmatrix}$ **(c)** $\begin{bmatrix} 6 & 3 \\ -2 & 5 \\ 4 & 10 \\ 1 & -11 \end{bmatrix}$

Complete the steps in the matrix solution of each system by filling in the blanks. Give the final system and the solution set. See Example 1.

3. $4x + 8y = 44$
$2x - y = -3$

$\left[\begin{array}{cc|c} 4 & 8 & 44 \\ 2 & -1 & -3 \end{array}\right]$

$\left[\begin{array}{cc|c} 1 & \underline{\quad} & \underline{\quad} \\ 2 & -1 & -3 \end{array}\right]$ $\quad \frac{1}{4}R_1$

$\left[\begin{array}{cc|c} 1 & 2 & 11 \\ 0 & \underline{\quad} & \underline{\quad} \end{array}\right]$ $\quad -2R_1 + R_2$

$\left[\begin{array}{cc|c} 1 & 2 & 11 \\ 0 & 1 & \underline{\quad} \end{array}\right]$ $\quad -\frac{1}{5}R_2$

4. $2x - 5y = -1$
$3x + y = 7$

$\left[\begin{array}{cc|c} 2 & -5 & -1 \\ 3 & 1 & 7 \end{array}\right]$

$\left[\begin{array}{cc|c} 1 & -\frac{5}{2} & \underline{\quad} \\ 3 & 1 & 7 \end{array}\right]$ $\quad \frac{1}{2}R_1$

$\left[\begin{array}{cc|c} 1 & -\frac{5}{2} & -\frac{1}{2} \\ 0 & \underline{\quad} & \underline{\quad} \end{array}\right]$ $\quad -3R_1 + R_2$

$\left[\begin{array}{cc|c} 1 & -\frac{5}{2} & -\frac{1}{2} \\ 0 & 1 & \underline{\quad} \end{array}\right]$ $\quad \frac{2}{17}R_2$

Use row operations to solve each system. See Examples 1 and 3.

5. $x + y = 5$
$x - y = 3$

6. $x + 2y = 7$
$x - y = -2$

7. $2x + 4y = 6$
$3x - y = 2$

8. $4x + 5y = -7$
$x - y = 5$

9. $3x + 4y = 13$
$2x - 3y = -14$

10. $5x + 2y = 8$
$3x - y = 7$

11. $-4x + 12y = 36$
$x - 3y = 9$

12. $2x - 4y = 8$
$-3x + 6y = 5$

Complete the steps in the matrix solution of each system by filling in the blanks. Give the final system and the solution set. See Example 2.

13.
$$\begin{aligned} x + y - z &= -3 \\ 2x + y + z &= 4 \\ 5x - y + 2z &= 23 \end{aligned}$$

$$\left[\begin{array}{ccc|c} 1 & 1 & -1 & -3 \\ 2 & 1 & 1 & 4 \\ 5 & -1 & 2 & 23 \end{array}\right]$$

$$\left[\begin{array}{ccc|c} 1 & 1 & -1 & -3 \\ 0 & \underline{\quad} & \underline{\quad} & \underline{\quad} \\ 0 & \underline{\quad} & \underline{\quad} & \underline{\quad} \end{array}\right] \begin{array}{l} \\ -2R_1 + R_2 \\ -5R_1 + R_3 \end{array}$$

$$\left[\begin{array}{ccc|c} 1 & 1 & -1 & -3 \\ 0 & 1 & \underline{\quad} & \underline{\quad} \\ 0 & -6 & 7 & 38 \end{array}\right] \quad -1R_2$$

$$\left[\begin{array}{ccc|c} 1 & 1 & -1 & -3 \\ 0 & 1 & -3 & -10 \\ 0 & 0 & \underline{\quad} & \underline{\quad} \end{array}\right] \quad 6R_2 + R_3$$

$$\left[\begin{array}{ccc|c} 1 & 1 & -1 & -3 \\ 0 & 1 & -3 & -10 \\ 0 & 0 & 1 & \underline{\quad} \end{array}\right] \quad -\tfrac{1}{11}R_3$$

14.
$$\begin{aligned} 2x + y + 2z &= 11 \\ 2x - y - z &= -3 \\ 3x + 2y + z &= 9 \end{aligned}$$

$$\left[\begin{array}{ccc|c} 2 & 1 & 2 & 11 \\ 2 & -1 & -1 & -3 \\ 3 & 2 & 1 & 9 \end{array}\right]$$

$$\left[\begin{array}{ccc|c} 1 & \underline{\quad} & \underline{\quad} & \underline{\quad} \\ 2 & -1 & -1 & -3 \\ 3 & 2 & 1 & 9 \end{array}\right] \quad \tfrac{1}{2}R_1$$

$$\left[\begin{array}{ccc|c} 1 & \frac{1}{2} & 1 & \frac{11}{2} \\ 0 & \underline{\quad} & \underline{\quad} & \underline{\quad} \\ 0 & \underline{\quad} & \underline{\quad} & \underline{\quad} \end{array}\right] \begin{array}{l} \\ -2R_1 + R_2 \\ -3R_1 + R_3 \end{array}$$

$$\left[\begin{array}{ccc|c} 1 & \frac{1}{2} & 1 & \frac{11}{2} \\ 0 & 1 & \underline{\quad} & \underline{\quad} \\ 0 & \frac{1}{2} & -2 & -\frac{15}{2} \end{array}\right] \quad -\tfrac{1}{2}R_2$$

$$\left[\begin{array}{ccc|c} 1 & \frac{1}{2} & 1 & \frac{11}{2} \\ 0 & 1 & \frac{3}{2} & 7 \\ 0 & 0 & \underline{\quad} & \underline{\quad} \end{array}\right] \quad -\tfrac{1}{2}R_2 + R_3$$

$$\left[\begin{array}{ccc|c} 1 & \frac{1}{2} & 1 & \frac{11}{2} \\ 0 & 1 & \frac{3}{2} & 7 \\ 0 & 0 & 1 & \underline{\quad} \end{array}\right] \quad -\tfrac{4}{11}R_3$$

Use row operations to solve each system. See Examples 2 and 3.

15. $\begin{aligned} x + y - 3z &= 1 \\ 2x - y + z &= 9 \\ 3x + y - 4z &= 8 \end{aligned}$

16. $\begin{aligned} 2x + 4y - 3z &= -18 \\ 3x + y - z &= -5 \\ x - 2y + 4z &= 14 \end{aligned}$

17. $\begin{aligned} x + y - z &= 6 \\ 2x - y + z &= -9 \\ x - 2y + 3z &= 1 \end{aligned}$

18. $\begin{aligned} x + 3y - 6z &= 7 \\ 2x - y + 2z &= 0 \\ x + y + 2z &= -1 \end{aligned}$

19. $\begin{aligned} x - y &= 1 \\ y - z &= 6 \\ x + z &= -1 \end{aligned}$

20. $\begin{aligned} x + y &= 1 \\ 2x - z &= 0 \\ y + 2z &= -2 \end{aligned}$

21. $\begin{aligned} x - 2y + z &= 4 \\ 3x - 6y + 3z &= 12 \\ -2x + 4y - 2z &= -8 \end{aligned}$

22. $\begin{aligned} 4x + 8y + 4z &= 9 \\ x + 3y + 4z &= 10 \\ 5x + 10y + 5z &= 12 \end{aligned}$

Solve each problem by first setting up a system of equations. Use row operations.

23. The manager of a small company deposits some money in a bank account paying 5% per yr. He uses additional money, amounting to $\frac{1}{3}$ the amount placed in the bank, to buy bonds paying 6% per yr. With the balance of funds he buys an 8% certificate of deposit. The first year his investments earn $690. If the total investment is $10,000, how much is invested at each rate?

Rate (as a Decimal)	Amount Invested	Annual Interest
.05	x	$.05x$
.06		
.08		
		690

24. A small company took out three loans totaling $25,000. The company was able to borrow some of the money at 8%. It borrowed $2000 more than $\frac{1}{2}$ the amount of the 8% loan at 10%, and the rest at 9%. The total annual interest was $2220. How much did the company borrow at each rate?

Rate (as a Decimal)	Amount Borrowed	Annual Interest
	x	
.10		
.09		
		2220

Chapter 5

Summary

Key Terms

5.1	**system of equations**	Two or more equations that are to be solved at the same time form a system of equations.
	linear system	A linear system is a system of equations that contains only linear equations.
	solution set of a system	All ordered pairs that satisfy all the equations of a system at the same time make up the solution set of the system.
	consistent system	A system is consistent if it has a solution.
	independent equations	Independent equations are equations whose graphs are different lines.
	inconsistent system	A system is inconsistent if it has no solution.
	dependent equations	Dependent equations are equations whose graphs are the same line.
5.4	**matrix**	A matrix is a rectangular array of numbers, consisting of horizontal **rows** and vertical **columns.**
	elements of a matrix	The numbers in a matrix are its elements.
	square matrix	A square matrix is a matrix that has the same number of rows as columns.
	augmented matrix	An augmented matrix is a matrix that has a vertical bar that separates the columns of the matrix into two groups.
	row echelon form	If a matrix is written with 1s down the diagonal from upper left to lower right and 0s below the 1s, it is said to be in row echelon form.

New Symbols

(x, y, z) ordered triple

$\begin{bmatrix} a & b \\ c & d \end{bmatrix}$ 2×2 matrix

Test Your Word Power

See how well you have learned the vocabulary in this chapter. Answers follow the Quick Review.

1. A **system of equations** consists of
   - **(a)** at least two equations with different variables
   - **(b)** two or more equations that have an infinite number of solutions
   - **(c)** two or more equations that are to be solved at the same time
   - **(d)** two or more inequalities that are to be solved.

2. The **solution set of a system of equations** is
   - **(a)** all ordered pairs that satisfy one equation of the system
   - **(b)** all ordered pairs that satisfy all the equations of the system at the same time
   - **(c)** any ordered pair that satisfies one or more equations of the system
   - **(d)** the set of values that make all the equations of the system false.

3. An **inconsistent system** is a system of equations
   - **(a)** with one solution
   - **(b)** with no solution
   - **(c)** with an infinite number of solutions
   - **(d)** that have the same graph.

4. **Dependent equations**
   - **(a)** have different graphs
   - **(b)** have no solution
   - **(c)** have one solution
   - **(d)** are different forms of the same equation.

5. A **matrix** is
   - **(a)** an ordered pair of numbers
   - **(b)** an array of numbers with the same number of rows and columns
   - **(c)** a pair of numbers written between brackets
   - **(d)** a rectangular array of numbers.

QUICK REVIEW

Concepts	*Examples*
5.1 Linear Systems of Equations in Two Variables **Solving a Linear System by Substitution**	Solve by substitution. $4x - y = 7$ (1) $3x + 2y = 30$ (2)
Step 1 Solve one of the equations for either variable.	Solve for y in equation (1). $y = 4x - 7$
Step 2 Substitute for that variable in the other equation. The result should be an equation with just one variable. *Step 3* Solve the equation from Step 2.	Substitute $4x - 7$ for y in equation (2), and solve for x. $3x + 2(4x - 7) = 30$; $3x + 8x - 14 = 30$; $11x - 14 = 30$; $11x = 44$; $x = 4$
Step 4 Find the value of the other variable by substituting the result from Step 3 into the equation from Step 1.	Substitute 4 for x in the equation $y = 4x - 7$ to find that $y = 9$.
Step 5 Check the solution in both of the original equations. Then write the solution set.	Check to see that $\{(4, 9)\}$ is the solution set.
Solving a Linear System by Elimination	Solve by elimination.
Step 1 Write both equations in standard form.	$5x + y = 2$ (1) $2x - 3y = 11$ (2)
Step 2 Make the coefficients of one pair of variable terms opposites.	To eliminate y, multiply equation (1) by 3, and add the result to equation (2).
Step 3 Add the new equations. The sum should be an equation with just one variable.	$15x + 3y = 6$; $2x - 3y = 11$ (2); $17x = 17$
Step 4 Solve the equation from Step 3.	$x = 1$
Step 5 Find the value of the other variable by substituting the result of Step 4 into either of the original equations.	Let $x = 1$ in equation (1), and solve for y. $5(1) + y = 2$; $y = -3$
Step 6 Check the solution in both of the original equations. Then write the solution set.	Check to verify that $\{(1, -3)\}$ is the solution set.

Concepts	*Examples*
5.2 *Linear Systems of Equations in Three Variables* **Solving a Linear System in Three Variables** *Step 1* Use the elimination method to eliminate any variable from any two of the original equations.	Solve the system. $x + 2y - z = 6$ (1) $x + y + z = 6$ (2) $2x + y - z = 7$ (3) Add equations (1) and (2); z is eliminated and the result is $2x + 3y = 12$.
Step 2 Eliminate the *same* variable from any *other* two equations.	Eliminate z again by adding equations (2) and (3) to get $3x + 2y = 13$. Now solve the system $2x + 3y = 12$ (4) $3x + 2y = 13.$ (5)
Step 3 Eliminate a second variable from the two equations in two variables that result from Steps 1 and 2. The result is an equation in one variable that gives the value of that variable.	To eliminate x, multiply equation (4) by -3 and equation (5) by 2. $-6x - 9y = -36$ $6x + 4y = 26$ $-5y = -10$ $y = 2$
Step 4 Substitute the value of the variable found in Step 3 into either of the equations in two variables to find the value of the second variable.	Let $y = 2$ in equation (4). $2x + 3(2) = 12$ $2x + 6 = 12$ $2x = 6$ $x = 3$
Step 5 Use the values of the two variables from Steps 3 and 4 to find the value of the third variable by substituting into any of the original equations.	Let $y = 2$ and $x = 3$ in any of the original equations to find $z = 1$.
Step 6 Check the solution in all of the original equations. Then write the solution set.	Check. The solution set is $\{(3, 2, 1)\}$.
5.3 *Applications of Linear Systems of Equations* Use the six-step problem-solving method. *Step 1* Read the problem carefully. *Step 2* Assign variables. *Step 3* Write a system of equations that relates the unknowns. *Step 4* Solve the system. *Step 5* State the answer. *Step 6* Check.	The perimeter of a rectangle is 18 ft. The length is 3 ft more than twice the width. What are the dimensions of the rectangle? Let x represent the length and y represent the width. From the perimeter formula, one equation is $2x + 2y = 18$. From the problem, another equation is $x = 3 + 2y$. Solve the system $2x + 2y = 18$ $x = 3 + 2y$ to get $x = 7$ and $y = 2$. The length is 7 ft, and the width is 2 ft. Since the perimeter is $2(7) + 2(2) = 18,$ and $3 + 2(2) = 7,$ the solution checks.

Concepts	Examples
5.4 Solving Linear Systems of Equations by Matrix Methods	
Matrix Row Operations	
1. Any two rows of the matrix may be interchanged.	$\begin{bmatrix} 1 & 5 & 7 \\ 3 & 9 & -2 \\ 0 & 6 & 4 \end{bmatrix}$ becomes $\begin{bmatrix} 3 & 9 & -2 \\ 1 & 5 & 7 \\ 0 & 6 & 4 \end{bmatrix}$ Interchange R_1 and R_2.
2. The numbers in any row may be multiplied by any nonzero real number.	$\begin{bmatrix} 1 & 5 & 7 \\ 3 & 9 & -2 \\ 0 & 6 & 4 \end{bmatrix}$ becomes $\begin{bmatrix} 1 & 5 & 7 \\ 1 & 3 & -\frac{2}{3} \\ 0 & 6 & 4 \end{bmatrix}$ $\frac{1}{3}R_2$
3. Any row may be changed by adding to the numbers of the row the product of a real number and the numbers of another row.	$\begin{bmatrix} 1 & 5 & 7 \\ 3 & 9 & -2 \\ 0 & 6 & 4 \end{bmatrix}$ becomes $\begin{bmatrix} 1 & 5 & 7 \\ 0 & -6 & -23 \\ 0 & 6 & 4 \end{bmatrix}$ $-3R_1 + R_2$
A system can be solved by matrix methods. Write the augmented matrix, and use row operations to obtain a matrix in row echelon form.	Solve using row operations. $\begin{aligned} x + 3y &= 7 \\ 2x + y &= 4 \end{aligned}$ $\left[\begin{array}{cc\|c} 1 & 3 & 7 \\ 2 & 1 & 4 \end{array}\right]$ Augmented matrix $\left[\begin{array}{cc\|c} 1 & 3 & 7 \\ 0 & -5 & -10 \end{array}\right]$ $-2R_1 + R_2$ $\left[\begin{array}{cc\|c} 1 & 3 & 7 \\ 0 & 1 & 2 \end{array}\right]$ $-\frac{1}{5}R_2$ $x + 3y = 7$ $y = 2$ When $y = 2$, $x + 3(2) = 7$, so $x = 1$. The solution set is $\{(1, 2)\}$.

Answers to Test Your Word Power

1. (c) *Example:* $\begin{aligned} 3x - y &= 3 \\ 2x + y &= 7 \end{aligned}$ **2. (b)** *Example:* The ordered pair (2, 3) satisfies both equations of the system in Answer 1, so $\{(2, 3)\}$ is the solution set of the system. **3. (b)** *Example:* The equations of two parallel lines form an inconsistent system; their graphs never intersect, so the system has no solution. **4. (d)** *Example:* The equations $4x - y = 8$ and $8x - 2y = 16$ are dependent because their graphs are the same line. **5. (d)** *Examples:* $\begin{bmatrix} 3 & -1 & 0 \\ 4 & 2 & 1 \end{bmatrix}$, $\begin{bmatrix} 1 & 2 \\ 4 & 3 \end{bmatrix}$

Chapter 5 Review Exercises

[5.1] **1.** Solve the system by graphing.

$$x + 3y = 8$$
$$2x - y = 2$$

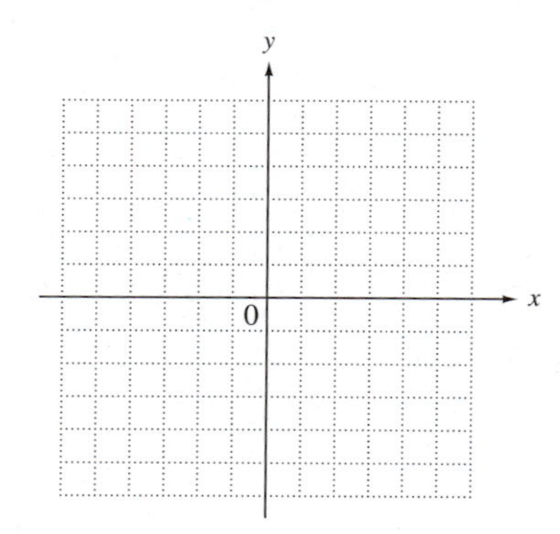

2. The graph shows the trends during the years 1974–1996 relating to bachelor's degrees awarded in the United States.

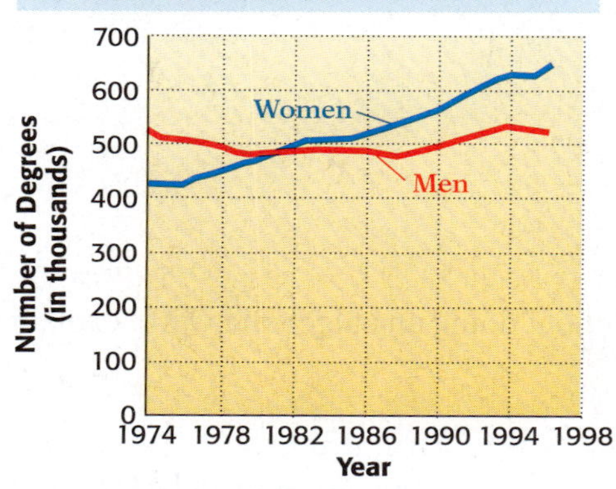

Source: U.S. National Center for Education Statistics, *Digest of Education Statistics*, annual.

(a) Between what years shown on the horizontal axis did the number of degrees for men and women reach equal numbers?

(b) When the number of degrees for men and women reached equal numbers, what was that number (approximately)?

Solve each system using the substitution method.

3. $3x + y = -4$
$x = \frac{2}{3}y$

4. $9x - y = -4$
$y = x + 4$

5. $-5x + 2y = -2$
$x + 6y = 26$

Solve each system using the elimination method.

6. $6x + 5y = 4$
$-4x + 2y = 8$

7. $\frac{x}{6} + \frac{y}{6} = -\frac{1}{2}$
$x - y = -9$

8. $4x + 5y = 9$
$3x + 7y = -1$

9. $-3x + y = 6$
$2y = 12 + 6x$

10. $5x - 4y = 2$
$-10x + 8y = 7$

Suppose that two linear equations are graphed on the same set of coordinate axes. Sketch what the graph might look like if the system has the given description.

11. The system has a single solution.

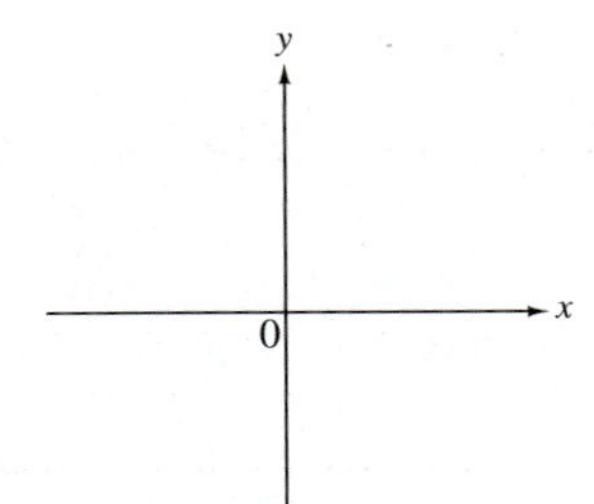

12. The system has no solution.

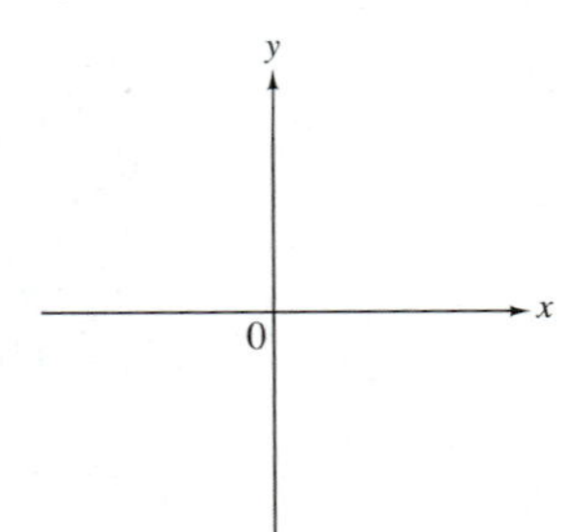

13. The system has infinitely many solutions.

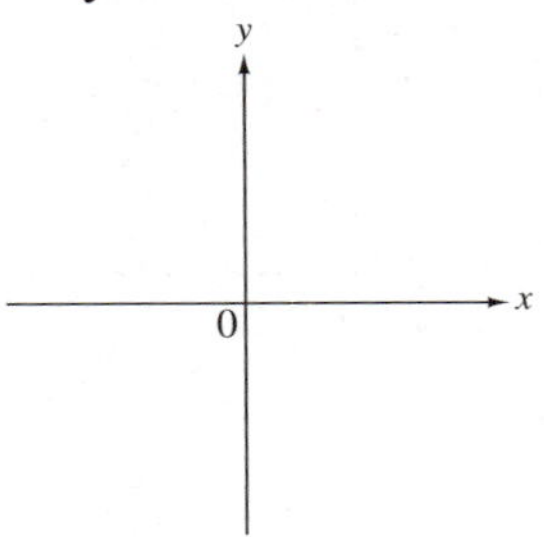

14. Without doing any algebraic work, explain why the system

$$y = 3x + 2$$
$$y = 3x - 4$$

has $\emptyset$ as its solution set. Answer based only on your knowledge of the graphs of the two lines.

[5.2] *Solve each system of equations.*

15. $\begin{aligned} 2x + 3y - z &= -16 \\ x + 2y + 2z &= -3 \\ -3x + y + z &= -5 \end{aligned}$

16. $\begin{aligned} 3x - y - z &= -8 \\ 4x + 2y + 3z &= 15 \\ -6x + 2y + 2z &= 10 \end{aligned}$

17. $\begin{aligned} 4x - y &= 2 \\ 3y + z &= 9 \\ x + 2z &= 7 \end{aligned}$

[5.3] *Solve each problem using a system of equations.*

18. A regulation National Hockey League ice rink has perimeter 570 ft. The length is 30 ft longer than twice the width. What are the dimensions of an NHL ice rink? (*Source: Microsoft Encarta Encyclopedia 2000.*)

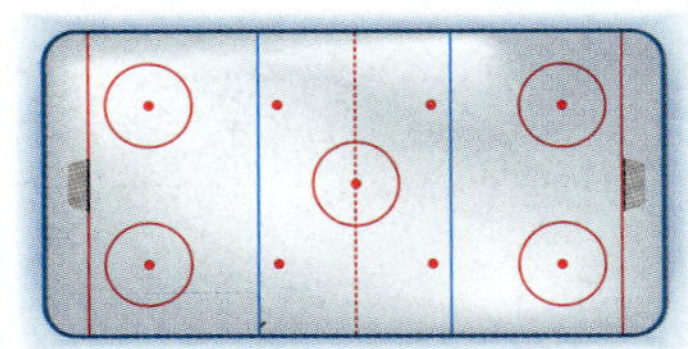

19. On a 6-day business trip, Todd Hall rented a car for \$53 per day at weekday rates and \$35 per day at weekend rates. If his total rental bill was \$264, how many days did he rent at each rate? (*Source:* Enterprise.)

20. A plane flies 560 mi in 1.75 hr traveling with the wind. The return trip later against the same wind takes the plane 2 hr. Find the speed of the plane and the speed of the wind.

	r	t	d
With Wind	$x + y$	1.75	
Against Wind		2	

21. Sweet's Candy Store is offering a special mix for Valentine's Day. Ms. Sweet will mix some \$2 per lb nuts with some \$1 per lb chocolate candy to get 100 lb of mix, which she will sell at \$1.30 per lb. How many pounds of each should she use?

	Price per Pound	Number of Pounds	Value
Nuts		x	
Chocolate		y	
Mixture		100	

22. A biologist wants to grow two types of algae, green and brown. She has 15 kg of nutrient X and 26 kg of nutrient Y. A vat of green algae needs 2 kg of nutrient X and 3 kg of nutrient Y, while a vat of brown algae needs 1 kg of nutrient X and 2 kg of nutrient Y. How many vats of each type of algae should she grow in order to use all the nutrients?

23. The sum of the measures of the angles of a triangle is 180°. The largest angle measures 10° less than the sum of the other two. The measure of the middle-sized angle is the average of the other two. Find the measures of the three angles.

24. How many liters each of 8%, 10%, and 20% hydrogen peroxide should be mixed together to get 8 L of 12.5% solution, if the amount of 8% solution used must be 2 L more than the amount of 20% solution used?

25. In the great baseball year of 1961, Yankee teammates Mickey Mantle, Roger Maris, and John Blanchard combined for 136 home runs. Mantle hit 7 fewer than Maris. Maris hit 40 more than Blanchard. What were the home run totals for each player? (*Source:* Neft, David S. and Richard M. Cohen, *The Sports Encyclopedia: Baseball 1997.*)

[5.4] *Solve each system using row operations.*

26. $2x + 5y = -4$
$4x - y = 14$

27. $6x + 3y = 9$
$-7x + 2y = 17$

28. $x + 2y - z = 1$
$3x + 4y + 2z = -2$
$-2x - y + z = -1$

Mixed Review Exercises

Solve by any method.

29. $\frac{2}{3}x + \frac{1}{6}y = \frac{19}{2}$
$\frac{1}{3}x - \frac{2}{9}y = 2$

30. $2x - 5y = 8$
$3x + 4y = 10$

31. $x = 7y + 10$
$2x + 3y = 3$

32. $x + 4y = 17$
$-3x + 2y = -9$

33. $-7x + 3y = 12$
$5x + 2y = 8$

34. $2x + 5y - z = 12$
$-x + y - 4z = -10$
$-8x - 20y + 4z = 31$

35. To make a 10% acid solution for chemistry class, Xavier wants to mix some 5% solution with 10 L of 20% solution. How many liters of 5% solution should he use?

Percent (as a Decimal)	Liters of Solution	Liters of Pure Acid

36. In the 2000 Summer Olympics in Sydney, Australia, the top three medal-winning countries were the United States, Russia, and China, with a combined total of 244 medals. The United States won 9 more medals than Russia, while China won 29 fewer medals than Russia. How many medals did each country win? (*Source: The Gazette,* October 2, 2000.)

Relating Concepts (Exercises 37–41) For Individual or Group Work

Thus far in this text we have studied only linear *equations. In later chapters we will study the graphs of other kinds of equations. One such graph is a* circle, *which has an equation of the form*

$$x^2 + y^2 + ax + by + c = 0.$$

It is a fact from geometry that given three noncollinear *points (that is, points that do not all lie on the same straight line), there will be a circle that contains them. For example, the points* (4, 2), (−5, −2), *and* (0, 3) *lie on the circle whose equation is shown in the figure.* ***Work Exercises 37–41 in order*** *to find an equation of the circle passing through the points* (2, 1), (−1, 0), *and* (3, 3).

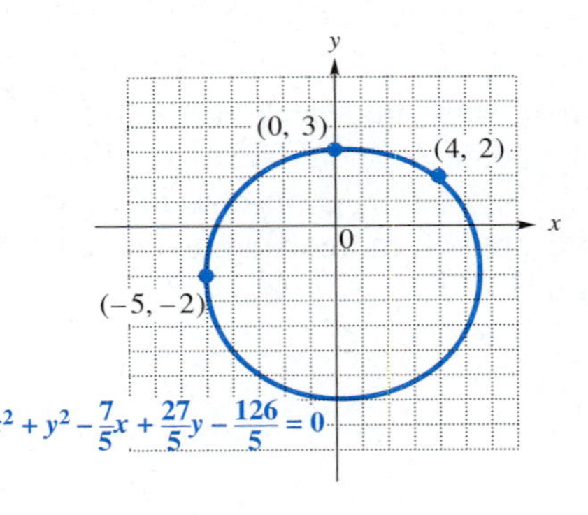

37. Let $x = 2$ and $y = 1$ in the equation $x^2 + y^2 + ax + by + c = 0$ to find an equation in a, b, and c.

38. Let $x = -1$ and $y = 0$ to find a second equation in a, b, and c.

39. Let $x = 3$ and $y = 3$ to find a third equation in a, b, and c.

40. Solve the system of equations formed by your answers in Exercises 37–39 to find the values of a, b, and c. What is the equation of the circle?

41. Explain why the relation whose graph is a circle is not a function.

Chapter 5 Test

Hank Aaron and Babe Ruth are the all-time home run leaders in the major leagues. Use the graphs to answer Problems 1 and 2.

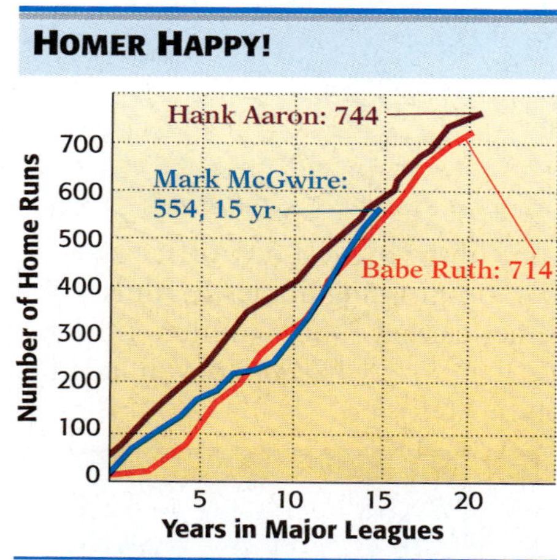

Source: *ESPN 1998 Sports Almanac*; CNN/ *Sports Illustrated* and Major League Baseball web sites.

1. Was there any year in Babe Ruth's career that he had more home runs than Hank Aaron in the same year of Aaron's career? Explain.

2. After 15 yr, which player had the most home runs? Who had the fewest?

3. Use a graph to solve the system.

$$x + y = 7$$
$$x - y = 5$$

1. ____________

2. ____________

3. ____________

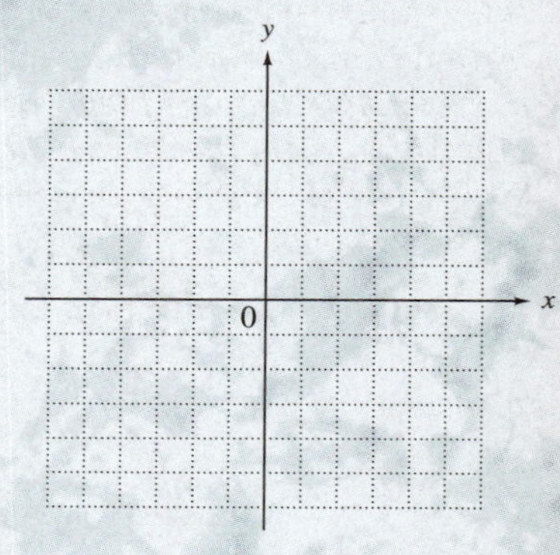

Solve each system using substitution.

4. $2x - 3y = 24$

$$y = -\frac{2}{3}x$$

5. $12x - 5y = 8$

$$3x = \frac{5}{4}y + 2$$

6. $3x - y = -8$
$2x + 6y = 3$

4. ____________

5. ____________

6. ____________

7. ____________________

8. ____________________

9. ____________________

10. ____________________

11. ____________________

Solve each system using elimination.

7. $3x + y = 12$
$2x - y = 3$

8. $-5x + 2y = -4$
$6x + 3y = -6$

9. $3x + 4y = 8$
$8y = 7 - 6x$

10. $3x + 5y + 3z = 2$
$6x + 5y + z = 0$
$3x + 10y - 2z = 6$

11. $4x + y + z = 11$
$x - y - z = 4$
$y + 2z = 0$

Solve each problem using a system of equations.

12. ____________________

12. Julia Roberts is one of the biggest box-office stars in Hollywood. As of July 2001, her two top-grossing domestic films, *Pretty Woman* and *Runaway Bride,* together earned $330.7 million. If *Runaway Bride* grossed $26.1 million less than *Pretty Woman,* how much did each film gross? (*Source:* ACNielsen EDI.)

13. ____________________

13. Two cars start from points 420 mi apart and travel toward each other. They meet after 3.5 hr. Find the average speed of each car if one travels 30 mph slower than the other.

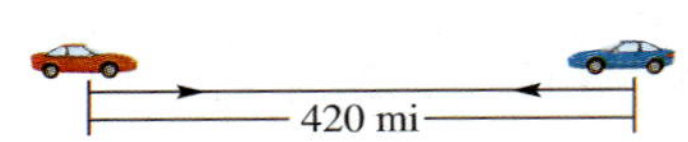

14. ____________________

14. A chemist needs 12 L of a 40% alcohol solution. She must mix a 20% solution and a 50% solution. How many liters of each will be required to obtain what she needs?

15. ____________________

15. A local electronics store will sell 7 AC adaptors and 2 rechargeable flashlights for $86, or 3 AC adaptors and 4 rechargeable flashlights for $84. What is the price of a single AC adaptor and a single rechargeable flashlight?

16. ____________________

16. The owner of a tea shop wants to mix three kinds of tea to make 100 oz of a mixture that will sell for $.83 per oz. He uses Orange Pekoe, which sells for $.80 per oz, Irish Breakfast, for $.85 per oz, and Earl Grey, for $.95 per oz. If he wants to use twice as much Orange Pekoe as Irish Breakfast, how much of each kind of tea should he use?

Solve each system using row operations.

17. ____________________

18. ____________________

17. $3x + 2y = 4$
$5x + 5y = 9$

18. $x + 3y + 2z = 11$
$3x + 7y + 4z = 23$
$5x + 3y - 5z = -14$

Cumulative Review Exercises CHAPTERS 1–5

Evaluate.

1. $(-3)^4$
2. -3^4
3. $-(-3)^4$
4. $\sqrt{.49}$
5. $-\sqrt{.49}$
6. $\sqrt{-.49}$

Evaluate if $x = -4$, $y = 3$, and $z = 6$.

7. $|2x| + y^2 - z^3$
8. $-5(x^3 - y^3)$
9. $\dfrac{2x^2 - x + z}{y^2 - z}$

Solve each equation.

10. $7(2x + 3) - 4(2x + 1) = 2(x + 1)$
11. $.04x + .06(x - 1) = 1.04$
12. $ax + by = cx + d$ for x
13. $|6x - 8| = 4$

Solve each inequality.

14. $\dfrac{2}{3}y + \dfrac{5}{12}y \leq 20$
15. $|3x + 2| \leq 4$
16. $|12t + 7| \geq 0$

17. A recent survey measured public recognition of the most popular contemporary advertising slogans. Complete the results shown in the table if 2500 people were surveyed.

Slogan (product or company)	Percent Recognition (nearest tenth of a percent)	Actual Number Who Recognized Slogan (nearest whole number)
Please Don't Squeeze the . . . (Charmin)	80.4%	______
The Breakfast of Champions (Wheaties)	72.5%	______
The King of Beers (Budweiser)	______	1570
Like a Good Neighbor (State Farm)	______	1430

(Other slogans included "You're in Good Hands" (Allstate), "Snap, Crackle, Pop" (Rice Krispies), and "The Un-Cola" (7-Up).)
Source: Department of Integrated Marketing Communications, Northwestern University.

Solve each problem.

18. On February 12, 1999, the U.S. Senate voted to acquit William Jefferson Clinton on both counts of impeachment (perjury and obstruction of justice). Of the 200 votes cast that day, there were 10 more "not guilty" votes than "guilty" votes. How many of each vote were there? (*Source:* MSNBC Web site, February 13, 1999.)

19. Two angles of a triangle have the same measure. The measure of the third angle is 4° less than twice the measure of each of the equal angles. Find the measures of the three angles.

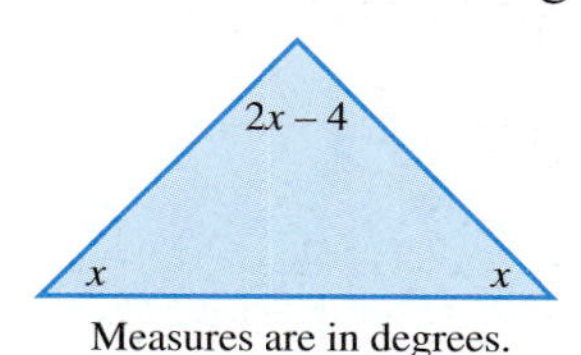

Measures are in degrees.

In Exercises 20–25, point A has coordinates $(-2, 6)$ *and point B has coordinates* $(4, -2)$.

20. What is the equation of the horizontal line through A?

21. What is the equation of the vertical line through B?

22. What is the slope of AB?

23. What is the slope of a line perpendicular to line AB?

24. What is the standard form of the equation of line AB?

25. Write the equation of the line in the form of a linear function.

26. Graph the linear function whose graph has slope $\frac{2}{3}$ and passes through the point $(-1, -3)$.

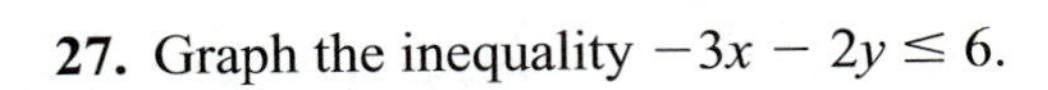

27. Graph the inequality $-3x - 2y \le 6$.

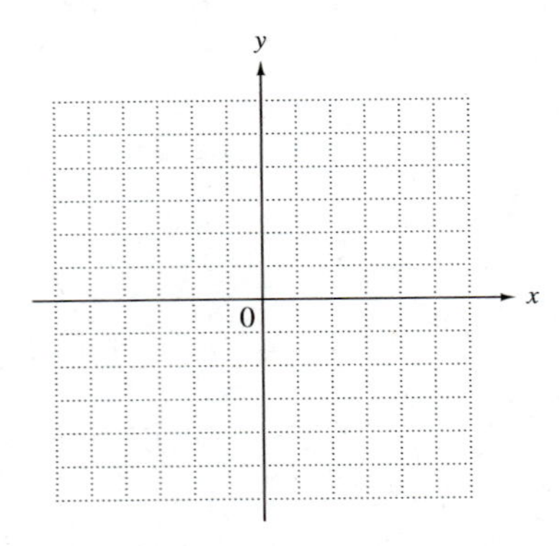

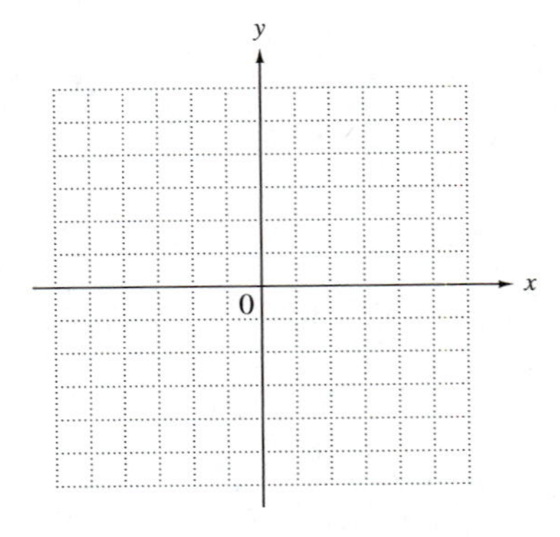

Solve by any method.

28. $$\begin{aligned} -2x + 3y &= -15 \\ 4x - y &= 15 \end{aligned}$$

29. $$\begin{aligned} x + y + z &= 10 \\ x - y - z &= 0 \\ -x + y - z &= -4 \end{aligned}$$

Solve each problem using a system of equations.

30. Two of the best-selling toys of 1996 were Tickle Me Elmo and Snacktime Kid. Based on their average retail prices, Elmo cost \$8.63 less than Kid, and together they cost \$63.89. What was the average retail price for each toy? (*Source:* NPD Group, Inc.)

31. A grocer plans to mix candy that sells for \$1.20 per lb with candy that sells for \$2.40 per lb to get a mixture that he plans to sell for \$1.65 per lb. How much of the \$1.20 and \$2.40 candy should he use if he wants 80 lb of the mix?

The graph shows a company's costs to produce computer parts and the revenue from the sale of those parts.

32. At what production level does the cost equal the revenue? What is the revenue at that point?

33. Profit is revenue less cost. Estimate the profit on the sale of 1100 parts.

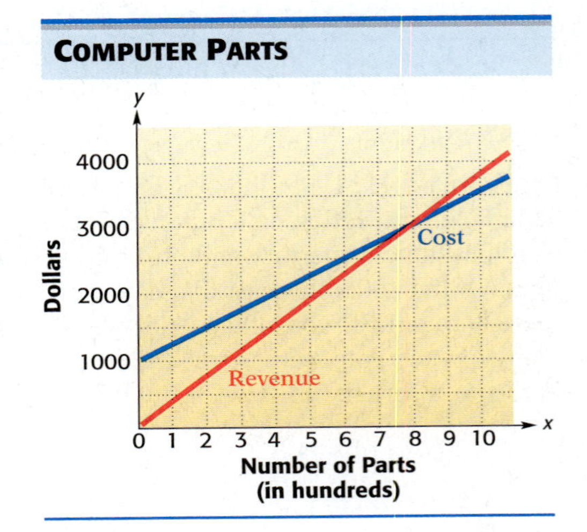

Exponents and Polynomials

In the Powerball lottery, a player must choose five numbers from 1 through 49 and one number from 1 through 42. It can be shown that there are about 8.009×10^7 different ways to do this. Suppose that a group of 2000 people decide to purchase tickets for all these numbers and each ticket costs $1.00. (*Source:* www.powerball.com) Do you think it would be worth your while to participate in such an activity if the Powerball lottery jackpot were $50 million? $100 million? See Exercise 117 in Section 6.1 to help you decide.

6.1 Integer Exponents and Scientific Notation

Objectives

1. Use the product rule for exponents.
2. Define 0 and negative exponents.
3. Use the quotient rule for exponents.
4. Use the power rules for exponents.
5. Simplify exponential expressions.
6. Use the rules for exponents with scientific notation.

Recall that we use exponents to write products of repeated factors. For example,

$$2^5 \text{ is defined as } 2 \cdot 2 \cdot 2 \cdot 2 \cdot 2 = 32.$$

The number 5, the *exponent*, shows that the *base* 2 appears as a factor 5 times. The quantity 2^5 is called an *exponential* or a *power*. We read 2^5 as "2 to the fifth power" or "2 to the fifth."

1 Use the product rule for exponents. There are several useful rules that simplify work with exponents. For example, the product $2^5 \cdot 2^3$ can be simplified as follows.

$$5 + 3 = 8$$

$$2^5 \cdot 2^3 = (2 \cdot 2 \cdot 2 \cdot 2 \cdot 2)(2 \cdot 2 \cdot 2) = 2^8$$

This result, that products of exponential expressions with the *same base* are found by adding exponents, is generalized as the **product rule for exponents.**

Product Rule for Exponents

If m and n are natural numbers and a is any real number, then

$$a^m \cdot a^n = a^{m+n}.$$

To see that the product rule is true, use the definition of an exponent as follows.

$$a^m = \underbrace{a \cdot a \cdot a \cdots a}_{a \text{ appears as a factor } m \text{ times.}} \qquad a^n = \underbrace{a \cdot a \cdot a \cdots a}_{a \text{ appears as a factor } n \text{ times.}}$$

From this,

$$a^m \cdot a^n = \underbrace{a \cdot a \cdot a \cdots a}_{m \text{ factors}} \cdot \underbrace{a \cdot a \cdot a \cdots a}_{n \text{ factors}}$$

$$= \underbrace{a \cdot a \cdot a \cdots a}_{(m+n) \text{ factors}}$$

$$a^m \cdot a^n = a^{m+n}.$$

Example 1 Using the Product Rule for Exponents

Apply the product rule for exponents, if possible, in each case.

(a) $3^4 \cdot 3^7 = 3^{4+7} = 3^{11}$

(b) $5^3 \cdot 5 = 5^3 \cdot 5^1 = 5^{3+1} = 5^4$

(c) $y^3 \cdot y^8 \cdot y^2 = y^{3+8+2} = y^{13}$

(d) $(5y^2)(-3y^4)$ Use the associative and commutative properties as necessary to multiply the numbers and multiply the variables.

$$\begin{aligned}(5y^2)(-3y^4) &= 5(-3)y^2y^4 \\ &= -15y^{2+4} \\ &= -15y^6\end{aligned}$$

(e) $(7p^3q)(2p^5q^2) = 7(2)p^3p^5qq^2 = 14p^8q^3$

(f) $x^2 \cdot y^4$

Because the bases are not the same, the product rule does not apply.

Work Problem 1 at the Side.

---

1 Apply the product rule for exponents, if possible, in each case.

(a) $m^8 \cdot m^6$

(b) $r^7 \cdot r$

(c) $k^4k^3k^6$

(d) $m^5 \cdot p^4$

(e) $(-4a^3)(6a^2)$

(f) $(-5p^4)(-9p^5)$

Answers

1. (a) m^{14} (b) r^8 (c) k^{13} (d) The product rule does not apply. (e) $-24a^5$ (f) $45p^9$

CAUTION

Be careful in problems like Example 1(a) not to multiply the bases. Notice that $3^4 \cdot 3^7 \neq 9^{11}$. Remember to keep the *same* base and add the exponents.

2 Define 0 and negative exponents. So far we have discussed only positive exponents. Let us consider how we might define a 0 exponent. Suppose we multiply 4^2 by 4^0. By the product rule,

$$4^2 \cdot 4^0 = 4^{2+0} = 4^2.$$

For the product rule to hold true, 4^0 must equal 1, and so we define a^0 this way for any nonzero real number a.

Zero Exponent

If a is any nonzero real number, then

$$a^0 = 1.$$

The expression 0^0 is undefined.*

Example 2 Using 0 as an Exponent

Evaluate each expression.

(a) $12^0 = 1$

(b) $(-6)^0 = 1$ Base is -6.

(c) $-6^0 = -(6^0) = -1$ Base is 6.

(d) $5^0 + 12^0 = 1 + 1 = 2$

(e) $(8k)^0 = 1, \quad k \neq 0$

Work Problem 2 at the Side.

How should we define a negative exponent? Using the product rule again,

$$8^2 \cdot 8^{-2} = 8^{2+(-2)} = 8^0 = 1.$$

This indicates that 8^{-2} is the reciprocal of 8^2. But $\frac{1}{8^2}$ is the reciprocal of 8^2, and a number can have only one reciprocal. Therefore, it is reasonable to conclude that $8^{-2} = \frac{1}{8^2}$. We can generalize and make the following definition.

Negative Exponent

For any natural number n and any nonzero real number a,

$$a^{-n} = \frac{1}{a^n}.$$

With this definition, the expression a^n is meaningful for any integer exponent n and any nonzero real number a.

* In advanced treatments, 0^0 is called an *indeterminant form*.

2 Evaluate each expression.

(a) 46^0

(b) $(-29)^0$

(c) -29^0

(d) $8^0 - 15^0$

(e) $(-15p^5)^0, \quad p \neq 0$

ANSWERS

2. **(a)** 1 **(b)** 1 **(c)** -1 **(d)** 0 **(e)** 1

3 In parts (a)–(f), write the expressions with only positive exponents. In parts (g) and (h), simplify each expression.

(a) 6^{-3}

(b) 8^{-1}

(c) $(2x)^{-4}, \quad x \neq 0$

(d) $7r^{-6}, \quad r \neq 0$

(e) $-q^{-4}, \quad q \neq 0$

(f) $(-q)^{-4}, \quad q \neq 0$

(g) $3^{-1} + 5^{-1}$

(h) $4^{-1} - 2^{-1}$

CAUTION

A negative exponent does not indicate a negative number; negative exponents lead to reciprocals.

Expression	Example	
a^{-m}	$3^{-2} = \frac{1}{3^2} = \frac{1}{9}$	**Not negative**
$-a^{-m}$	$-3^{-2} = -\frac{1}{3^2} = -\frac{1}{9}$	**Negative**

Example 3 Using Negative Exponents

In parts (a)–(f), write the expressions with only positive exponents. In parts (g) and (h), simplify each expression.

(a) $2^{-3} = \frac{1}{2^3}$

(b) $6^{-1} = \frac{1}{6^1} = \frac{1}{6}$

(c) $(5z)^{-3} = \frac{1}{(5z)^3}, \quad z \neq 0$

(d) $5z^{-3} = 5\left(\frac{1}{z^3}\right) = \frac{5}{z^3}, \quad z \neq 0$

(e) $-m^{-2} = -\frac{1}{m^2}, \quad m \neq 0$

(f) $(-m)^{-2} = \frac{1}{(-m)^2}, \quad m \neq 0$

(g) $3^{-1} + 4^{-1} = \frac{1}{3} + \frac{1}{4} = \frac{4}{12} + \frac{3}{12} = \frac{7}{12}$

(h) $5^{-1} - 2^{-1} = \frac{1}{5} - \frac{1}{2} = \frac{2}{10} - \frac{5}{10} = -\frac{3}{10}$

CAUTION

In Example 3(g), note that $3^{-1} + 4^{-1} \neq (3 + 4)^{-1}$. The expression on the left is equal to $\frac{7}{12}$, as shown in the solution, while the expression on the right is $7^{-1} = \frac{1}{7}$. Similar reasoning can be applied to part (h).

Work Problem 3 at the Side.

Example 4 Using Negative Exponents

Evaluate each expression.

(a) $\frac{1}{2^{-3}} = \frac{1}{\frac{1}{2^3}} = 1 \div \frac{1}{2^3} = 1 \cdot \frac{2^3}{1} = 2^3 = 8$

(b) $\frac{2^{-3}}{3^{-2}} = \frac{\frac{1}{2^3}}{\frac{1}{3^2}} = \frac{1}{2^3} \cdot \frac{3^2}{1} = \frac{3^2}{2^3} = \frac{9}{8}$

Example 4 suggests the following generalizations.

Answers

3. (a) $\frac{1}{6^3}$ **(b)** $\frac{1}{8}$ **(c)** $\frac{1}{(2x)^4}$ **(d)** $\frac{7}{r^6}$
(e) $-\frac{1}{q^4}$ **(f)** $\frac{1}{(-q)^4}$ **(g)** $\frac{8}{15}$ **(h)** $-\frac{1}{4}$

Special Rules for Negative Exponents

If $a \neq 0$ and $b \neq 0$, $\dfrac{1}{a^{-n}} = a^n$ and $\dfrac{a^{-n}}{b^{-m}} = \dfrac{b^m}{a^n}$.

Work Problem 4 at the Side.

3 Use the quotient rule for exponents. A quotient, such as a^8/a^3, can be simplified in much the same way as a product. (In all quotients of this type, assume that the denominator is not 0.) Using the definition of an exponent,

$$\frac{a^8}{a^3} = \frac{a \cdot a \cdot a \cdot a \cdot a \cdot a \cdot a \cdot a}{a \cdot a \cdot a} = a \cdot a \cdot a \cdot a \cdot a = a^5.$$

Notice that $8 - 3 = 5$. In the same way,

$$\frac{a^3}{a^8} = \frac{a \cdot a \cdot a}{a \cdot a \cdot a \cdot a \cdot a \cdot a \cdot a \cdot a} = \frac{1}{a^5}.$$

Here again, $8 - 3 = 5$, but this time the exponent 5 is in the denominator. These examples suggest the following **quotient rule for exponents.**

Quotient Rule for Exponents

If a is any nonzero real number and m and n are integers, then

$$\frac{a^m}{a^n} = a^{m-n}.$$

Example 5 Using the Quotient Rule for Exponents

Apply the quotient rule for exponents, if possible, and write each result using only positive exponents.

Numerator exponent
Denominator exponent

(a) $\dfrac{3^7}{3^2} = 3^{7-2} = 3^5$ (Minus sign)

(b) $\dfrac{p^6}{p^2} = p^{6-2} = p^4, \quad p \neq 0$

(c) $\dfrac{k^7}{k^{12}} = k^{7-12} = k^{-5} = \dfrac{1}{k^5}, \quad k \neq 0$

(d) $\dfrac{2^7}{2^{-3}} = 2^{7-(-3)} = 2^{7+3} = 2^{10}$

(e) $\dfrac{8^{-2}}{8^5} = 8^{-2-5} = 8^{-7} = \dfrac{1}{8^7}$

(f) $\dfrac{6}{6^{-1}} = \dfrac{6^1}{6^{-1}} = 6^{1-(-1)} = 6^2$

(g) $\dfrac{z^{-5}}{z^{-8}} = z^{-5-(-8)} = z^3, \quad z \neq 0$

(h) $\dfrac{a^3}{b^4}, \quad b \neq 0$

The quotient rule does not apply because the bases are different.

CAUTION

As seen in Example 5, be very careful when working with quotients that involve negative exponents in the denominator. Always be sure to write the numerator exponent, then a minus sign, and then the denominator exponent.

Work Problem 5 at the Side.

4 Evaluate each expression.

(a) $\dfrac{1}{4^{-3}}$

(b) $\dfrac{3^{-3}}{9^{-1}}$

5 Apply the quotient rule for exponents, if possible, and write each result using only positive exponents.

(a) $\dfrac{4^8}{4^6}$

(b) $\dfrac{x^{12}}{x^3}, \quad x \neq 0$

(c) $\dfrac{r^5}{r^8}, \quad r \neq 0$

(d) $\dfrac{2^8}{2^{-4}}$

(e) $\dfrac{6^{-3}}{6^4}$

(f) $\dfrac{8}{8^{-1}}$

(g) $\dfrac{t^{-4}}{t^{-6}}, \quad t \neq 0$

(h) $\dfrac{x^3}{y^5}, \quad y \neq 0$

Answers

4. **(a)** 64 **(b)** $\dfrac{1}{3}$

5. **(a)** 4^2 **(b)** x^9 **(c)** $\dfrac{1}{r^3}$ **(d)** 2^{12} **(e)** $\dfrac{1}{6^7}$ **(f)** 8^2 **(g)** t^2
(h) The quotient rule does not apply.

6 Use one or more power rules to simplify each expression.

(a) $(r^5)^4$

(b) $\left(\frac{3}{4}\right)^3$

(c) $(9x)^3$

(d) $(5r^6)^3$

(e) $\left(\frac{-3n^4}{m}\right)^3, \quad m \neq 0$

ANSWERS

6. (a) r^{20} **(b)** $\frac{27}{64}$ **(c)** $729x^3$ **(d)** $125r^{18}$ **(e)** $\frac{-27n^{12}}{m^3}$

4 Use the power rules for exponents. The expression $(3^4)^2$ can be simplified as

$$(3^4)^2 = 3^4 \cdot 3^4 = 3^{4+4} = 3^8,$$

where $4 \cdot 2 = 8$. This example suggests the first **power rule for exponents.** The other two power rules can be demonstrated with similar examples.

Power Rules for Exponents

If a and b are real numbers and m and n are integers, then

$$(a^m)^n = a^{mn}, \quad (ab)^m = a^m b^m, \quad \text{and} \quad \left(\frac{a}{b}\right)^m = \frac{a^m}{b^m} \quad (b \neq 0).$$

Example 6 Using the Power Rules for Exponents

Use one or more power rules to simplify each expression.

(a) $(p^8)^3 = p^{8 \cdot 3} = p^{24}$

(b) $\left(\frac{2}{3}\right)^4 = \frac{2^4}{3^4} = \frac{16}{81}$

(c) $(3y)^4 = 3^4y^4 = 81y^4$

(d) $(6p^7)^2 = 6^2p^{7 \cdot 2} = 6^2p^{14} = 36p^{14}$

(e) $\left(\frac{-2m^5}{z}\right)^3 = \frac{(-2)^3m^{5 \cdot 3}}{z^3} = \frac{(-2)^3m^{15}}{z^3} = \frac{-8m^{15}}{z^3}, \quad z \neq 0$

Work Problem 6 at the Side.

The reciprocal of a^n is $\frac{1}{a^n} = \left(\frac{1}{a}\right)^n$. Also, by definition, a^n and a^{-n} are reciprocals since

$$a^n \cdot a^{-n} = a^n \cdot \frac{1}{a^n} = 1.$$

Thus, since both are reciprocals of a^n,

$$a^{-n} = \left(\frac{1}{a}\right)^n.$$

Some examples of this result are

$$6^{-3} = \left(\frac{1}{6}\right)^3 \quad \text{and} \quad \left(\frac{1}{3}\right)^{-2} = 3^2.$$

This discussion can be generalized as follows.

More Special Rules for Negative Exponents

Any nonzero number raised to the negative nth power is equal to the reciprocal of that number raised to the nth power. That is, if $a \neq 0$ and $b \neq 0$ and n is an integer,

$$a^{-n} = \left(\frac{1}{a}\right)^n \quad \text{and} \quad \left(\frac{a}{b}\right)^{-n} = \left(\frac{b}{a}\right)^n.$$

Example 7 Using Negative Exponents with Fractions

Write each expression with only positive exponents and then evaluate.

(a) $\left(\frac{3}{7}\right)^{-2} = \left(\frac{7}{3}\right)^{2} = \frac{49}{9}$ **(b)** $\left(\frac{4}{5}\right)^{-3} = \left(\frac{5}{4}\right)^{3} = \frac{125}{64}$

Work Problem 7 at the Side.

The definitions and rules of this section are summarized here.

Definitions and Rules for Exponents

For all integers m and n and all real numbers a and b, the following rules apply.

Product Rule $a^m \cdot a^n = a^{m+n}$

Quotient Rule $\frac{a^m}{a^n} = a^{m-n} \quad (a \neq 0)$

Zero Exponent $a^0 = 1 \quad (a \neq 0)$

Negative Exponent $a^{-n} = \frac{1}{a^n} \quad (a \neq 0)$

Power Rules $(a^m)^n = a^{mn}$ $\quad (ab)^m = a^m b^m$

$\left(\frac{a}{b}\right)^m = \frac{a^m}{b^m} \quad (b \neq 0)$

Special Rules $\frac{1}{a^{-n}} = a^n \quad (a \neq 0)$ $\quad \frac{a^{-n}}{b^{-m}} = \frac{b^m}{a^n} \quad (a, b \neq 0)$

$a^{-n} = \left(\frac{1}{a}\right)^n \quad (a \neq 0)$ $\quad \left(\frac{a}{b}\right)^{-n} = \left(\frac{b}{a}\right)^n \quad (a, b \neq 0)$

5 **Simplify exponential expressions.** With the rules of exponents developed so far in this section, we can simplify expressions that involve one or more rules.

Example 8 Using the Definitions and Rules for Exponents

Simplify each expression so that no negative exponents appear in the final result. Assume all variables represent nonzero real numbers.

(a) $3^2 \cdot 3^{-5} = 3^{2+(-5)} = 3^{-3} = \frac{1}{3^3}$ or $\frac{1}{27}$

(b) $x^{-3} \cdot x^{-4} \cdot x^2 = x^{-3+(-4)+2} = x^{-5} = \frac{1}{x^5}$

(c) $(4^{-2})^{-5} = 4^{(-2)(-5)} = 4^{10}$

(d) $(x^{-4})^6 = x^{(-4)6} = x^{-24} = \frac{1}{x^{24}}$

(e)
$$\frac{x^{-4}y^2}{x^2y^{-5}} = \frac{x^{-4}}{x^2} \cdot \frac{y^2}{y^{-5}}$$
$$= x^{-4-2} \cdot y^{2-(-5)}$$
$$= x^{-6}y^7$$
$$= \frac{y^7}{x^6}$$

(f)
$$(2^3x^{-2})^{-2} = (2^3)^{-2} \cdot (x^{-2})^{-2}$$
$$= 2^{-6}x^4$$
$$= \frac{x^4}{2^6} \text{ or } \frac{x^4}{64}$$

---

7 Write each expression with only positive exponents and then evaluate.

(a) $\left(\frac{3}{4}\right)^{-3}$

(b) $\left(\frac{5}{6}\right)^{-2}$

Answers

7. **(a)** $\left(\frac{4}{3}\right)^3; \frac{64}{27}$ **(b)** $\left(\frac{6}{5}\right)^2; \frac{36}{25}$

8 Simplify each expression so that no negative exponents appear in the final result. Assume all variables represent nonzero real numbers.

(a) $5^4 \cdot 5^{-6}$

(b) $x^{-4} \cdot x^{-6} \cdot x^8$

(c) $(5^{-3})^{-2}$

(d) $(y^{-2})^7$

(e) $\dfrac{a^{-3}b^5}{a^4b^{-2}}$

(f) $(3^2k^{-4})^{-1}$

NOTE

There is often more than one way to simplify expressions like those in Example 8. For instance, we could simplify Example 8(e) as follows.

$$\frac{x^{-4}y^2}{x^2y^{-5}} = \frac{y^5y^2}{x^4x^2} \qquad \text{Use } \frac{a^{-n}}{b^{-m}} = \frac{b^m}{a^n}.$$

$$= \frac{y^7}{x^6} \qquad \text{Product rule}$$

Work Problem 8 at the Side.

6 **Use the rules for exponents with scientific notation.** Scientists often need to use numbers that are very large or very small. For example, the number of one-celled organisms that will sustain a whale for a few hours is 400,000,000,000,000, and the shortest wavelength of visible light is approximately .0000004 m. It is simpler to write these numbers using *scientific notation.*

In scientific notation, a number is written with the decimal point after the first nonzero digit and multiplied by a power of 10, as indicated in the following definition.

Scientific Notation

A number is written in **scientific notation** when it is expressed in the form

$$a \times 10^n$$

where $1 \leq |a| < 10$, and n is an integer.

For example, in scientific notation,

$$8000 = 8 \times 1000 = 8 \times 10^3.$$

The following numbers are not in scientific notation.

$.230 \times 10^4$ — .230 is less than 1.

46.5×10^{-3} — 46.5 is greater than 10.

To write a number in scientific notation, use the following steps. (If the number is negative, ignore the negative sign, go through these steps, and then attach a negative sign to the result.)

Converting to Scientific Notation

Step 1 **Position the decimal point.** Place a caret, ^, to the right of the first nonzero digit, where the decimal point will be placed.

Step 2 **Determine the numeral for the exponent.** Count the number of digits from the decimal point to the caret. This number gives the absolute value of the exponent on 10.

Step 3 **Determine the sign for the exponent.** Decide whether multiplying by 10^n should make the result of Step 1 larger or smaller. The exponent should be positive to make the result larger; it should be negative to make the result smaller.

It is helpful to remember that for $n \geq 1$, $10^{-n} < 1$ and $10^n \geq 10$.

ANSWERS

8. **(a)** $\frac{1}{5^2}$ or $\frac{1}{25}$ **(b)** $\frac{1}{x^2}$ **(c)** 5^6 **(d)** $\frac{1}{y^{14}}$ **(e)** $\frac{b^7}{a^7}$ **(f)** $\frac{k^4}{3^2}$ or $\frac{k^4}{9}$

Example 9 **Writing Numbers in Scientific Notation**

Write each number in scientific notation.

(a) 820,000

Place a caret to the right of the 8 (the first nonzero digit) to mark the new location of the decimal point.

$$8_{\wedge}20{,}000$$

Count from the decimal point, which is understood to be after the last 0, to the caret.

$$8.\underbrace{20{,}000}_{\text{Count 5 places.}}. \leftarrow \text{Decimal point}$$

Since the number 8.2 is to be made larger, the exponent on 10 is positive.

$$820{,}000 = 8.2 \times 10^5$$

(b) .0000072

Count from left to right.

$$.\underbrace{000007}_{\text{6 places}}.2$$

Since the number 7.2 is to be made smaller, the exponent on 10 is negative.

$$.0000072 = 7.2 \times 10^{-6}$$

Work Problem 9 at the Side.

To convert a number written in scientific notation to standard notation, just work in reverse.

Converting from Scientific Notation

Multiplying a number by a positive power of 10 makes the number larger, so move the decimal point to the right if n is positive in 10^n.

Multiplying by a negative power of 10 makes a number smaller, so move the decimal point to the left if n is negative.

If n is 0, leave the decimal point where it is.

Example 10 **Converting from Scientific Notation to Standard Notation**

Write each number in standard notation.

(a) 6.93×10^7

$$6.\underbrace{9300000}_{\text{7 places}} \quad \text{Attach 0s as necessary.}$$

We moved the decimal point 7 places to the right. (It was necessary to attach five 0s.)

$$6.93 \times 10^7 = 69{,}300{,}000$$

(b) 4.7×10^{-6}

$$\underbrace{000004}_{\text{6 places}}.7$$

We moved the decimal point 6 places to the left.

$$4.7 \times 10^{-6} = .0000047$$

(c) $1.083 \times 10^0 = 1.083$

Work Problem 10 at the Side.

9 Write each number in scientific notation.

(a) 400,000

(b) 29,800,000

(c) −6083

(d) .00172

(e) .0000000503

10 Write each number in standard notation.

(a) 4.98×10^5

(b) 6.8×10^{-7}

(c) -5.372×10^0

ANSWERS

9. **(a)** 4×10^5 **(b)** 2.98×10^7 **(c)** -6.083×10^3 **(d)** 1.72×10^{-3} **(e)** 5.03×10^{-8}

10. **(a)** 498,000 **(b)** .00000068 **(c)** −5.372

11 Evaluate

$$\frac{200{,}000 \times .0003}{.06 \times 4{,}000{,}000}.$$

12 The distance to the sun is 9.3×10^7 mi. How long would it take a rocket, traveling at 3.2×10^3 mph, to reach the sun?

When problems require operations with numbers that are very large and/or very small, it is often advantageous to write the numbers in scientific notation first, and then perform the calculations using the rules for exponents.

Example 11 Using Scientific Notation in Computation

Evaluate $\dfrac{1{,}920{,}000 \times .0015}{.000032 \times 45{,}000}$.

First, express all numbers in scientific notation.

$$\frac{1{,}920{,}000 \times .0015}{.000032 \times 45{,}000} = \frac{1.92 \times 10^6 \times 1.5 \times 10^{-3}}{3.2 \times 10^{-5} \times 4.5 \times 10^4}$$

Next, use the commutative and associative properties and the rules for exponents to simplify the expression.

$$\begin{aligned}\frac{1{,}920{,}000 \times .0015}{.000032 \times 45{,}000} &= \frac{1.92 \times 1.5 \times 10^6 \times 10^{-3}}{3.2 \times 4.5 \times 10^{-5} \times 10^4}\\ &= \frac{1.92 \times 1.5 \times 10^3}{3.2 \times 4.5 \times 10^{-1}}\\ &= \frac{1.92 \times 1.5}{3.2 \times 4.5} \times 10^4\\ &= .2 \times 10^4\\ &= (2 \times 10^{-1}) \times 10^4\\ &= 2 \times 10^3\\ &= 2000\end{aligned}$$

Work Problem 11 at the Side.

Calculator Tip To enter numbers in scientific notation, you can use the EE or EXP key on a scientific calculator. For instance, to work Example 11 using a calculator with an EE key, enter the following symbols.

1.92 EE 6 × 1.5 EE 3 +/− ÷ (3.2 EE 5 +/− × 4.5 EE 4) =

The EXP key is used in exactly the same way. Notice that the negative exponent −3 is entered by pressing 3, then +/−. (Keystrokes vary among different models of calculators, so you should refer to your owner's manual if this sequence does not apply to your particular model.)

Example 12 Using Scientific Notation to Solve Problems

In 1985, the national health care expenditure was 428.2 billion dollars. By 1995, this figure had risen by a factor of 2.3; that is, it more than doubled in only 10 yr. (*Source:* U.S. Health Care Financing Administration.)

(a) Write the 1985 health care expenditure using scientific notation.

$$\begin{aligned}428.2 \text{ billion} = 428.2 \times 10^9 &= (4.282 \times 10^2) \times 10^9\\ &= 4.282 \times 10^{11} \quad \text{Product rule}\end{aligned}$$

In 1985, the expenditure was $\$4.282 \times 10^{11}$.

(b) What was the expenditure in 1995?
Multiply the result in part (a) by 2.3.

$$\begin{aligned}(4.282 \times 10^{11}) \times 2.3 &= (2.3 \times 4.282) \times 10^{11} \quad \text{Commutative and associative properties}\\ &= 9.849 \times 10^{11} \quad \text{Round to three decimal places.}\end{aligned}$$

The 1995 expenditure was \$984,900,000,000.

Work Problem 12 at the Side.

ANSWERS
11. 2.5×10^{-4} or .00025
12. approximately 2.9×10^4 hr

6.1 EXERCISES

Student's Solutions Manual

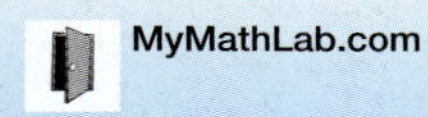
MyMathLab.com

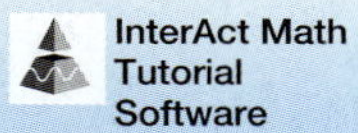
InterAct Math Tutorial Software

AW Math Tutor Center

www.mathxl.com

Digital Video Tutor CD 4 Videotape 8

Decide whether each expression has been simplified correctly. If not, correct it.

1. $(ab)^2 = ab^2$

2. $(5x)^3 = 5^3x^3$

3. $\left(\frac{4}{a}\right)^3 = \frac{4^3}{a} \quad (a \neq 0)$

4. $y^2 \cdot y^6 = y^{12}$

5. $x^3 \cdot x^4 = x^7$

6. $xy^0 = 0 \quad (y \neq 0)$

Apply the product rule for exponents, if possible, in each case. See Example 1.

7. $13^4 \cdot 13^8$

8. $9^6 \cdot 9^4$

9. $x^3 \cdot x^5 \cdot x^9$

10. $y^4 \cdot y^5 \cdot y^6$

11. $(-3w^5)(9w^3)$

12. $(-5x^2)(3x^4)$

13. $(2x^2y^5)(9xy^3)$

14. $(8s^4t)(3s^3t^5)$

15. $r^2 \cdot s^4$

16. $p^3 \cdot q^2$

Evaluate. Assume all variables represent nonzero numbers. See Example 2.

17. 25^0

18. 14^0

19. -7^0

20. -10^0

21. $(-15)^0$

22. $(-20)^0$

23. $-4^0 - m^0$

24. $-8^0 - k^0$

Write each expression with only positive exponents. Assume all variables represent nonzero numbers. In Exercises 37–40, simplify each expression. See Example 3.

25. 5^{-4}

26. 7^{-2}

27. 8^{-1}

28. 12^{-1}

29. $(4x)^{-2}$ **30.** $(5t)^{-3}$ **31.** $4x^{-2}$ **32.** $5t^{-3}$

33. $-a^{-3}$ **34.** $-b^{-4}$ **35.** $(-a)^{-4}$ **36.** $(-b)^{-6}$

37. $5^{-1} + 6^{-1}$ **38.** $2^{-1} + 8^{-1}$ **39.** $8^{-1} - 3^{-1}$ **40.** $6^{-1} - 4^{-1}$

Evaluate each expression. See Examples 4 and 7.

41. $\frac{1}{4^{-2}}$ **42.** $\frac{1}{3^{-3}}$ **43.** $\frac{2^{-2}}{3^{-3}}$ **44.** $\frac{3^{-3}}{2^{-2}}$

45. $\left(\frac{2}{3}\right)^{-3}$ **46.** $\left(\frac{3}{2}\right)^{-3}$ **47.** $\left(\frac{4}{5}\right)^{-2}$ **48.** $\left(\frac{5}{4}\right)^{-2}$

Apply the quotient rule for exponents, if applicable, and write each result using only positive exponents. Assume all variables represent nonzero numbers. See Example 5.

49. $\frac{4^8}{4^6}$ **50.** $\frac{5^9}{5^7}$ **51.** $\frac{x^{12}}{x^8}$ **52.** $\frac{y^{14}}{y^{10}}$

53. $\frac{r^7}{r^{10}}$ **54.** $\frac{y^8}{y^{12}}$ **55.** $\frac{6^4}{6^{-2}}$ **56.** $\frac{7^5}{7^{-3}}$

57. $\frac{6^{-3}}{6^7}$ **58.** $\frac{5^{-4}}{5^2}$ **59.** $\frac{7}{7^{-1}}$ **60.** $\frac{8}{8^{-1}}$

61. $\dfrac{r^{-3}}{r^{-6}}$ **62.** $\dfrac{s^{-4}}{s^{-8}}$ **63.** $\dfrac{x^3}{y^2}$ **64.** $\dfrac{y^5}{t^3}$

Use one or more power rules to simplify each expression. Assume all variables represent nonzero numbers. See Example 6.

65. $(x^3)^6$ **66.** $(y^5)^4$ **67.** $\left(\dfrac{3}{5}\right)^3$ **68.** $\left(\dfrac{4}{3}\right)^2$

69. $(4t)^3$ **70.** $(5t)^4$ **71.** $(-6x^2)^3$ **72.** $(-2x^5)^5$

73. $\left(\dfrac{-4m^2}{t}\right)^3$ **74.** $\left(\dfrac{-5n^4}{r^2}\right)^3$

Simplify each expression so that no negative exponents appear in the final result. Assume all variables represent nonzero numbers. See Example 8.

75. $3^5 \cdot 3^{-6}$ **76.** $4^4 \cdot 4^{-6}$ **77.** $a^{-3}a^2a^{-4}$ **78.** $k^{-5}k^{-3}k^4$

79. $(k^2)^{-3}k^4$ **80.** $(x^3)^{-4}x^5$ **81.** $-4r^{-2}(r^4)^2$ **82.** $-2m^{-1}(m^3)^2$

83. $(5a^{-1})^4(a^2)^{-3}$ **84.** $(3p^{-4})^2(p^3)^{-1}$ **85.** $(z^{-4}x^3)^{-1}$ **86.** $(y^{-2}z^4)^{-3}$

87. $\dfrac{(p^{-2})^3}{5p^4}$ **88.** $\dfrac{(m^4)^{-1}}{9m^3}$ **89.** $\dfrac{4a^5(a^{-1})^3}{(a^{-2})^{-2}}$ **90.** $\dfrac{12k^{-2}(k^{-3})^{-4}}{6k^5}$

91. $\dfrac{(-y^{-4})^2}{6(y^{-5})^{-1}}$

92. $\dfrac{2(-m^{-1})^{-4}}{9(m^{-3})^2}$

93. $\dfrac{(2k)^2m^{-5}}{(km)^{-3}}$

94. $\dfrac{(3rs)^{-2}}{3^2r^2s^{-4}}$

Write each number in scientific notation. See Example 9.

95. 530

96. 1600

97. .830

98. .0072

99. .00000692

100. .875

101. $-38{,}500$

102. $-976{,}000{,}000$

Write each number in standard notation. See Example 10.

103. 7.2×10^4

104. 8.91×10^2

105. 2.54×10^{-3}

106. 5.42×10^{-4}

107. -6×10^4

108. -9×10^3

109. 1.2×10^{-5}

110. 2.7×10^{-6}

Use the rules for exponents to find each value. See Example 11.

111. $\dfrac{3 \times 10^{-2}}{12 \times 10^3}$

112. $\dfrac{5 \times 10^{-3}}{25 \times 10^2}$

113. $\dfrac{.05 \times 1600}{.0004}$

114. $\dfrac{.003 \times 40{,}000}{.00012}$

Solve each problem. See Example 12.

115. The projected budget of NASA (National Aeronautics and Space Administration) for 2003 is \$13,757,400,000. Write this number in scientific notation. (*Source:* U.S. National Aeronautics and Space Administration.)

116. In 1997, there were 207,754,000 motor vehicle registrations in the United States. Write this number in scientific notation. (*Source:* U.S. Federal Highway Administration.)

117. In the Powerball lottery, a player must choose five numbers from 1 through 49 and one number from 1 through 42. It can be shown that there are about 8.009×10^7 different ways to do this. Suppose that a group of 2000 people decide to purchase tickets for all these numbers and each ticket costs \$1.00. How much should each person expect to pay? (*Source:* www.powerball.com)

118. The average distance from Earth to the sun is 9.3×10^7 mi. How long would it take a rocket, traveling at 2.9×10^3 mph, to reach the sun?

119. A *light-year* is the distance that light travels in one year. Find the number of miles in a light-year if light travels 1.86×10^5 mi per sec.

120. Use the information given in the previous two exercises to find the number of minutes necessary for light from the sun to reach Earth.

121. (a) The planet Mercury has an average distance from the sun of 3.6×10^7 mi, while the average distance of Venus to the sun is 6.7×10^7 mi. How long would it take a spacecraft traveling at 1.55×10^3 mph to travel from Venus to Mercury? (Give your answer in hours, in standard notation.)

(b) Use the information from part (a) to find the number of days it would take the spacecraft to travel from Venus to Mercury. Round your answer to the nearest whole number of days.

122. When the distance between the centers of the moon and Earth is 4.60×10^8 m, an object on the line joining the centers of the moon and Earth exerts the same gravitational force on each when it is 4.14×10^8 m from the center of Earth. How far is the object from the center of the moon at that point?

The graph shows the estimated annual number of Americans, by age and sex, experiencing heart attacks. Use scientific notation to represent the number of people for the following categories.

123. Males between 29 and 44

124. Females between 29 and 44

125. Males between 45 and 64

126. Females between 45 and 64

127. Males 65 or older

128. Females 65 or older

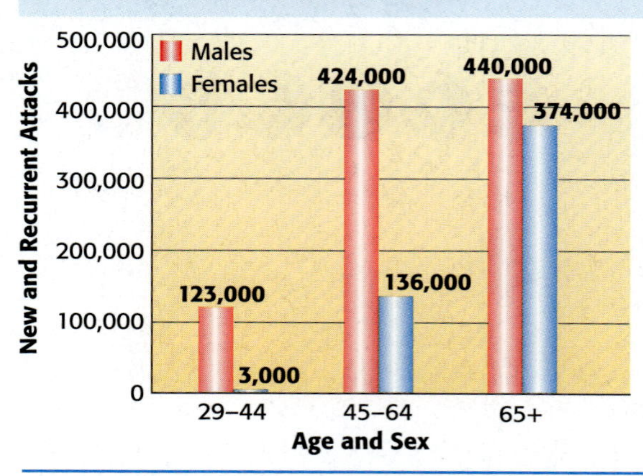

Source: American Heart Association and Framingham Heart Study, 24-year follow-up.

129. In some cases, $-a^n$ and $(-a)^n$ do give the same result. Using $a = 2$ and $n = 2, 3, 4,$ and 5, draw a conclusion as to when they are equal and when they are opposites.

130. Your friend evaluated $4^5 \cdot 4^2$ as 16^7. Explain to him why his answer is incorrect.

131. In your own words, describe how to rewrite a fraction raised to a negative power as a fraction raised to a positive power.

132. Explain in your own words how to raise a power to a power.

6.2 ADDING AND SUBTRACTING POLYNOMIALS; POLYNOMIAL FUNCTIONS

OBJECTIVES

1. Know the basic definitions for polynomials.
2. Find the degree of a polynomial.
3. Add and subtract polynomials.
4. Recognize and evaluate polynomial functions.
5. Use a polynomial function to model data.
6. Add and subtract polynomial functions.
7. Graph basic polynomial functions.

1 Know the basic definitions for polynomials. Just as whole numbers are the basis of arithmetic, *polynomials* are fundamental in algebra. To understand polynomials, we must review several words from Chapter 1. A *term* is a number, a variable, or the product or quotient of a number and one or more variables raised to powers. Examples of terms include

$$4x, \quad \frac{1}{2}m^5\left(\text{or } \frac{m^5}{2}\right), \quad -7z^9, \quad 6x^2z, \quad \frac{5}{3x^2}, \quad \text{and} \quad 9. \qquad \text{Terms}$$

The number in the product is called the *numerical coefficient,* or just the *coefficient.** In the term $8x^3$, the coefficient is 8. In the term $-4p^5$, it is -4. The coefficient of the term k is understood to be 1. The coefficient of $-r$ is -1.

Work Problem 1 at the Side.

Recall that any combination of variables or constants (numerical values) joined by the basic operations of addition, subtraction, multiplication, and division (except by 0), or taking roots is called an *algebraic expression*. The simplest kind of algebraic expression is a *polynomial.*

Polynomial

A **polynomial** is a term or a finite sum of terms in which all variables have whole number exponents and no variables appear in denominators.

Examples of polynomials include

$$3x - 5, \quad 4m^3 - 5m^2p + 8, \quad \text{and} \quad -5t^2s^3. \qquad \text{Polynomials}$$

Even though the expression $3x - 5$ involves subtraction, it is a sum of terms since it could be written as $3x + (-5)$.

Some examples of expressions that are not polynomials are

$$x^{-1} + 3x^{-2}, \quad \sqrt{9 - x}, \quad \text{and} \quad \frac{1}{x}. \qquad \text{Not polynomials}$$

The first of these is not a polynomial because it has negative integer exponents, the second because it involves a variable under a radical, and the third because it contains a variable in the denominator.

Most of the polynomials used in this book contain only one variable. A polynomial containing only the variable x is called a **polynomial in x.** A polynomial in one variable is written in **descending powers** of the variable if the exponents on the variable decrease from left to right. For example,

$$x^5 - 6x^2 + 12x - 5$$

is a polynomial in descending powers of x. The term -5 in this polynomial can be thought of as $-5x^0$, since $-5x^0 = -5(1) = -5$.

Work Problem 2 at the Side.

* More generally, any factor in a term is the coefficient of the product of the remaining factors. For example, $3x^2$ is the coefficient of y in the term $3x^2y$, and $3y$ is the coefficient of x^2 in $3x^2y$.

1 Identify each coefficient.

(a) $-9m^5$

(b) $12y^2x$

(c) x

(d) $-y$

2 Write each polynomial in descending powers.

(a) $-4 + 9y + y^3$

(b) $-3z^4 + 2z^3 + z^5 - 6z$

(c) $-12m^{10} + 8m^9 + 10m^{12}$

ANSWERS

1. (a) -9 **(b)** 12 **(c)** 1 **(d)** -1

2. (a) $y^3 + 9y - 4$ **(b)** $z^5 - 3z^4 + 2z^3 - 6z$ **(c)** $10m^{12} - 12m^{10} + 8m^9$

3 Identify each polynomial as a *trinomial, binomial, monomial,* or *none of these.*

(a) $12m^4 - 6m^2$

(b) $-6y^3 + 2y^2 - 8y$

(c) $3a^5$

(d) $-2k^{10} + 2k^9 - 8k^5 + 2k$

4 Give the degree of each polynomial.

(a) $9y^4 + 8y^3 - 6$

(b) $-12m^7 + 11m^3 + m^9$

(c) $-2k$

(d) 10

(e) $3mn^2 + 2m^3n$

ANSWERS

3. **(a)** binomial **(b)** trinomial **(c)** monomial **(d)** none of these

4. **(a)** 4 **(b)** 9 **(c)** 1 **(d)** 0 **(e)** 4

Some polynomials with a specific number of terms are so common that they are given special names. A polynomial with exactly three terms is a **trinomial,** and a polynomial with exactly two terms is a **binomial.** A single-term polynomial is a **monomial.** The table that follows gives examples.

Type of Polynomial	Examples
Monomial	$5x$, $7m^9$, -8, x^2y^2
Binomial	$3x^2 - 6$, $11y + 8$, $5a^2b + 3a$
Trinomial	$y^2 + 11y + 6$, $8p^3 - 7p + 2m$, $-3 + 2k^5 + 9z^4$
None of these	$p^3 - 5p^2 + 2p - 5$, $-9z^3 + 5c^3 + 2m^5 + 11r^2 - 7r$

Work Problem **3** at the Side.

2 Find the degree of a polynomial. The **degree of a term** with one variable is the exponent on the variable. For example, the degree of $2x^3$ is 3, the degree of $-x^4$ is 4, and the degree of $17x$ is 1. The degree of a term in more than one variable is defined to be the sum of the exponents on the variables. For example, the degree of $5x^3y^7$ is 10, because $3 + 7 = 10$.

The greatest degree of any term in a polynomial is called the **degree of the polynomial.** In most cases, we will be interested in finding the degree of a polynomial in one variable. For example, $4x^3 - 2x^2 - 3x + 7$ has degree 3, because the greatest degree of any term is 3 (the degree of $4x^3$).

The table shows several polynomials and their degrees.

Polynomial	Degree
$9x^2 - 5x + 8$	2
$17m^9 + 18m^{14} - 9m^3$	14
$5x$	1, because $5x = 5x^1$
-2	0, because $-2 = -2x^0$ (Any nonzero constant has degree 0.)
$5a^2b^5$	7, because $2 + 5 = 7$
$x^3y^9 + 12xy^4 + 7xy$	12, because $3 + 9 = 12$, and $12 > 5$ and $12 > 2$

NOTE

The number 0 has no degree, since 0 times a variable to any power is 0.

Work Problem **4** at the Side.

3 Add and subtract polynomials. We use the distributive property to simplify polynomials by combining terms. For example,

$$x^3 + 4x^2 + 5x^2 - 1 = x^3 + (4 + 5)x^2 - 1 \quad \text{Distributive property}$$
$$= x^3 + 9x^2 - 1.$$

On the other hand, the terms in the polynomial $4x + 5x^2$ cannot be combined. As these examples suggest, only terms containing exactly the same variables to the same powers may be combined. As mentioned in Chapter 1, such terms are called *like terms.*

CAUTION

Remember that only *like terms* can be combined.

Example 1 **Combining Like Terms**

Combine terms.

(a) $-5y^3 + 8y^3 - y^3 = (-5 + 8 - 1)y^3 = 2y^3$

(b) $6x + 5y - 9x + 2y = 6x - 9x + 5y + 2y$ Associative and commutative properties

$= -3x + 7y$ Combine like terms.

Since $-3x$ and $7y$ are unlike terms, no further simplification is possible.

(c) $5x^2y - 6xy^2 + 9x^2y + 13xy^2 = 5x^2y + 9x^2y - 6xy^2 + 13xy^2$

$= 14x^2y + 7xy^2$

Work Problem 5 at the Side.

We use the following rule to add two polynomials.

Adding Polynomials

To add two polynomials, combine like terms.

Polynomials can be added horizontally or vertically, as seen in the next example.

Example 2 **Adding Polynomials**

Add.

$(3a^5 - 9a^3 + 4a^2) + (-8a^5 + 8a^3 + 2)$

$= 3a^5 - 8a^5 - 9a^3 + 8a^3 + 4a^2 + 2$

$= -5a^5 - a^3 + 4a^2 + 2$ Combine like terms.

Add these same two polynomials vertically by placing like terms in columns.

$$\begin{array}{r} 3a^5 - 9a^3 + 4a^2 \\ -8a^5 + 8a^3 + 2 \\ \hline -5a^5 - a^3 + 4a^2 + 2 \end{array}$$

Work Problem 6 at the Side.

In Chapter 1, we defined subtraction of real numbers as

$$a - b = a + (-b).$$

That is, we add the first number and the negative (or opposite) of the second. We can give a similar definition for subtraction of polynomials by defining the **negative of a polynomial** as that polynomial with the sign of every coefficient changed.

Subtracting Polynomials

To subtract two polynomials, add the first polynomial and the negative of the *second* polynomial.

5 Combine terms.

(a) $11x + 12x - 7x - 3x$

(b) $11p^5 + 4p^5 - 6p^3 + 8p^3$

(c) $2y^2z^4 + 3y^4 + 5y^4 - 9y^4z^2$

6 Add, using both the horizontal and vertical methods.

(a) $(12y^2 - 7y + 9) + (-4y^2 - 11y + 5)$

(b)
$$\begin{array}{r} -6r^5 + 2r^3 - r^2 \\ 8r^5 - 2r^3 + 5r^2 \\ \hline \end{array}$$

Answers

5. **(a)** $13x$ **(b)** $15p^5 + 2p^3$ **(c)** $2y^2z^4 + 8y^4 - 9y^4z^2$

6. **(a)** $8y^2 - 18y + 14$ **(b)** $2r^5 + 4r^2$

7 Subtract, using both the horizontal and vertical methods.

(a) $(6y^3 - 9y^2 + 8) - (2y^3 + y^2 + 5)$

(b)
$$\begin{array}{r} 6y^3 - 2y^2 + 5y \\ \underline{-2y^3 + 8y^2 - 11y} \end{array}$$

Example 3 Subtracting Polynomials

Subtract.

$$(-6m^2 - 8m + 5) - (-5m^2 + 7m - 8)$$

Change every sign in the second polynomial and add.

$$\begin{aligned} &(-6m^2 - 8m + 5) - (-5m^2 + 7m - 8) \\ &\qquad = -6m^2 - 8m + 5 + 5m^2 - 7m + 8 \\ &\qquad = -6m^2 + 5m^2 - 8m - 7m + 5 + 8 && \text{Rearrange terms.} \\ &\qquad = -m^2 - 15m + 13 && \text{Combine like terms.} \end{aligned}$$

Check by adding the sum, $-m^2 - 15m + 13$, to the second polynomial. The result should be the first polynomial.

To subtract these two polynomials vertically, write the first polynomial above the second, lining up like terms in columns.

$$\begin{array}{r} -6m^2 - 8m + 5 \\ \underline{-5m^2 + 7m - 8} \end{array}$$

Change all the signs in the second polynomial, and add.

$$\begin{array}{rl} -6m^2 - 8m + 5 & \\ \underline{+\,5m^2 - 7m + 8} & \text{Change all signs.} \\ -m^2 - 15m + 13 & \text{Add in columns.} \end{array}$$

Work Problem 7 at the Side.

4 Recognize and evaluate polynomial functions. In Chapter 4 we studied linear (first-degree polynomial) functions, defined as $f(x) = mx + b$. Now we consider more general polynomial functions.

Polynomial Function

A **polynomial function of degree *n*** is defined by

$$f(x) = a_n x^n + a_{n-1} x^{n-1} + \cdots + a_1 x + a_0,$$

for real numbers $a_n, a_{n-1}, \ldots, a_1$, and a_0, where $a_n \neq 0$ and n is a whole number.

Another way of describing a polynomial function is to say that it is a function defined by a polynomial in one variable, consisting of one or more terms. It is usually written in descending powers of the variable, and its degree is the degree of the polynomial that defines it.

Suppose that we consider the polynomial $3x^2 - 5x + 7$, so

$$f(x) = 3x^2 - 5x + 7.$$

If $x = -2$, then $f(x) = 3x^2 - 5x + 7$ takes on the value

$$\begin{aligned} f(-2) &= 3(-2)^2 - 5(-2) + 7 && \text{Let } x = -2. \\ &= 3 \cdot 4 + 10 + 7 \\ &= 29. \end{aligned}$$

Example 4 shows additional illustrations of this idea.

ANSWERS

7. **(a)** $4y^3 - 10y^2 + 3$ **(b)** $8y^3 - 10y^2 + 16y$

Example 4 **Evaluating Polynomial Functions**

Let $f(x) = 4x^3 - x^2 + 5$. Find each value.

(a) $f(3)$

$$\begin{aligned} f(x) &= 4x^3 - x^2 + 5 \\ f(3) &= 4 \cdot 3^3 - 3^2 + 5 && \text{Substitute 3 for } x. \\ &= 4 \cdot 27 - 9 + 5 && \text{Order of operations} \\ &= 108 - 9 + 5 \\ &= 104 \end{aligned}$$

(b) $f(-4) = 4 \cdot (-4)^3 - (-4)^2 + 5$ Let $x = -4$.

$$\begin{aligned} &= 4 \cdot (-64) - 16 + 5 \\ &= -267 \end{aligned}$$

While f is the most common letter used to represent functions, recall that other letters such as g and h are also used. The capital letter P is often used for polynomial functions. Note that the function defined as $P(x) = 4x^3 - x^2 + 5$ yields the same ordered pairs as the function f in Example 4.

Work Problem 8 at the Side.

5 **Use a polynomial function to model data.** Polynomial functions can be used to approximate data. They are usually valid for small intervals, and they allow us to predict (with caution) what might happen for values just outside the intervals. These intervals are often periods of years, as shown in Example 5.

Example 5 **Using a Polynomial Model to Approximate Data**

Average annual pay for registered nurses during the years 1990–1996 can be modeled by the polynomial function defined by

$$P(x) = -128.57x^2 + 2000x + 30{,}100,$$

where $x = 0$ corresponds to the year 1990, $x = 1$ corresponds to 1991, and so on, and $P(x)$ is in dollars. Use this function to approximate the pay for the year 1991. (*Source:* Watson Wyandotte Worldwide.)

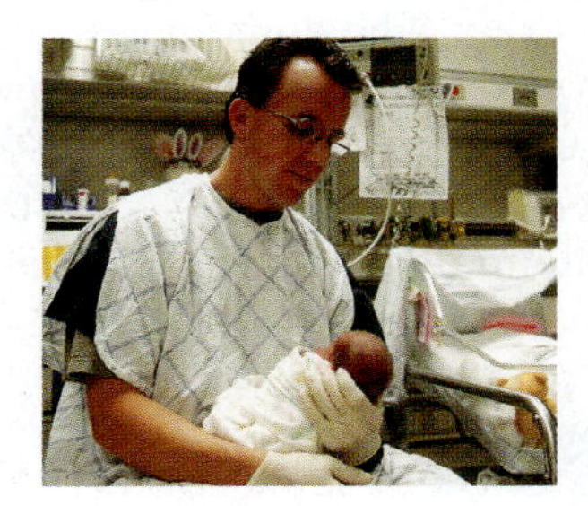

Since $x = 1$ corresponds to 1991, we must find $P(1)$:

$$\begin{aligned} P(1) &= -128.57(1)^2 + 2000(1) + 30{,}100 && \text{Let } x = 1. \\ &= 31{,}971.43. && \text{Evaluate.} \end{aligned}$$

Therefore, in 1991 a registered nurse earned on the average approximately $32,000.

Work Problem 9 at the Side.

8 Let $f(x) = -x^2 + 5x - 11$. Find each value.

(a) $f(1)$

(b) $f(-4)$

(c) $f(0)$

9 Use the function in Example 5 to approximate pay in 1998.

ANSWERS

8. **(a)** -7 **(b)** -47 **(c)** -11

9. $37,872.52, or approximately $38,000

10 Let
$f(x) = 3x^2 + 8x - 6$ and
$g(x) = -4x^2 + 4x - 8$.
Find each function.

(a) $(f + g)(x)$

(b) $(f - g)(x)$

6 **Add and subtract polynomial functions.** The operations of addition, subtraction, multiplication, and division are also defined for functions. We now consider addition and subtraction of functions, using polynomial functions as examples.

Adding and Subtracting Functions

If $f(x)$ and $g(x)$ define functions, their sum and difference are defined as follows.

$$(f + g)(x) = f(x) + g(x)$$
$$(f - g)(x) = f(x) - g(x)$$

In both cases, the domain of the new function is the intersection of the domains of $f(x)$ and $g(x)$.

Example 6 Adding and Subtracting Polynomial Functions

For the functions defined by

$$f(x) = x^2 - 3x + 7 \quad \text{and} \quad g(x) = -3x^2 - 7x + 7,$$

find **(a)** the sum and **(b)** the difference.

(a) $(f + g)(x) = f(x) + g(x)$

$= (x^2 - 3x + 7) + (-3x^2 - 7x + 7)$ Substitution

$= -2x^2 - 10x + 14$ Add the polynomials.

(b) $(f - g)(x) = f(x) - g(x)$

$= (x^2 - 3x + 7) - (-3x^2 - 7x + 7)$ Substitution

$= (x^2 - 3x + 7) + (3x^2 + 7x - 7)$ Change subtraction to addition.

$= 4x^2 + 4x$ Add.

Work Problem 10 at the Side.

7 **Graph basic polynomial functions.** Functions were introduced in Section 4.5. Recall that each input (or x-value) of a function results in one output (or y-value). The simplest polynomial function is the **identity function,** defined by $f(x) = x$. The domain (set of x-values) of this function is all real numbers, $(-\infty, \infty)$, and it pairs each real number with itself. Therefore, the range (set of y-values) is also $(-\infty, \infty)$. Its graph is a straight line, as first seen in Chapter 4. (Notice that a *linear function* is a specific kind of polynomial function.) Figure 1 shows its graph and a table of selected ordered pairs.

x	f(x) = x
−2	−2
−1	−1
0	0
1	1
2	2

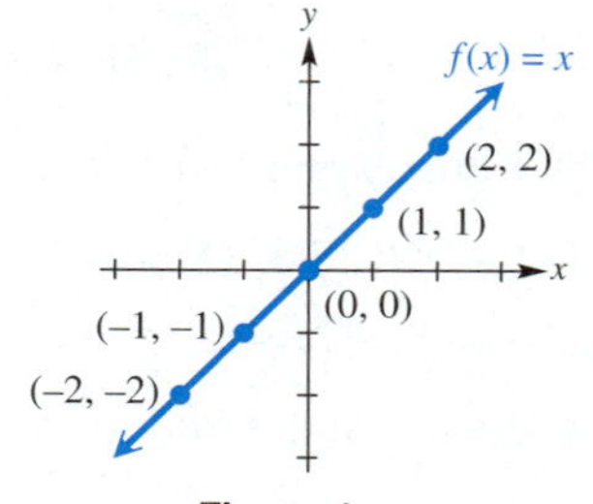

Figure 1

ANSWERS
10. (a) $-x^2 + 12x - 14$ **(b)** $7x^2 + 4x + 2$

Another polynomial function, defined by $f(x) = x^2$, is the **squaring function.** For this function, every real number is paired with its square. The input can be any real number, so the domain is $(-\infty, \infty)$. Since the square of any real number is nonnegative, the range is $[0, \infty)$. Its graph is a *parabola.* Figure 2 shows the graph and a table of selected ordered pairs.

x	$f(x) = x^2$
-2	4
-1	1
0	0
1	1
2	4

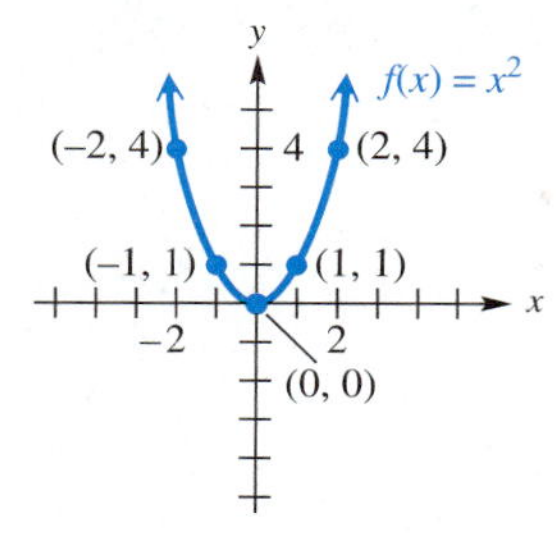

Figure 2

The **cubing function** is defined by $f(x) = x^3$. Every real number is paired with its cube. The domain and the range are both $(-\infty, \infty)$. Its graph is neither a line nor a parabola. See Figure 3 and the table of ordered pairs. (Polynomial functions of degree 3 and greater are studied in detail in more advanced courses.)

x	$f(x) = x^3$
-2	-8
-1	-1
0	0
1	1
2	8

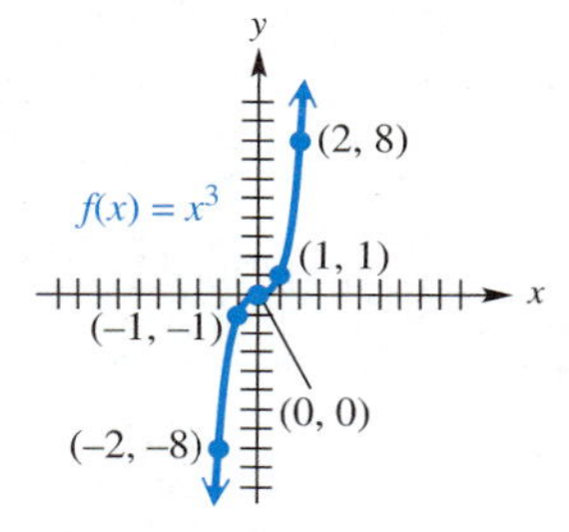

Figure 3

Example 7 **Graphing Variations of the Identity, Squaring, and Cubing Functions**

Graph each function by creating a table of ordered pairs. Give the domain and the range of each function by observing the graphs.

(a) $f(x) = 2x$

To find each range value, multiply the domain value by 2. Plot the points and join them with a straight line. See Figure 4. Both the domain and the range are $(-\infty, \infty)$.

x	$f(x) = 2x$
-2	-4
-1	-2
0	0
1	2
2	4

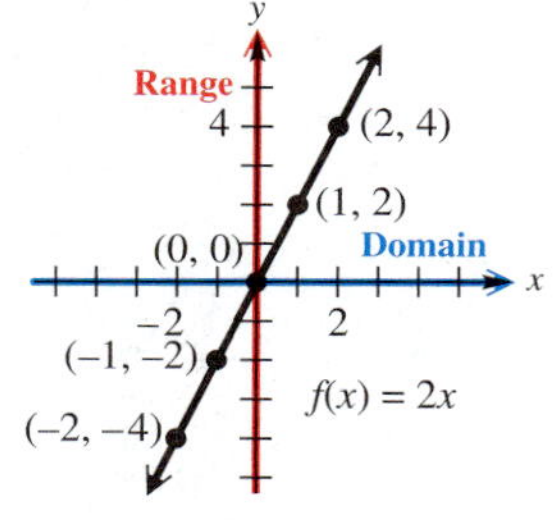

Figure 4

Continued on Next Page

11 Graph $f(x) = -2x^2$. Give the domain and the range.

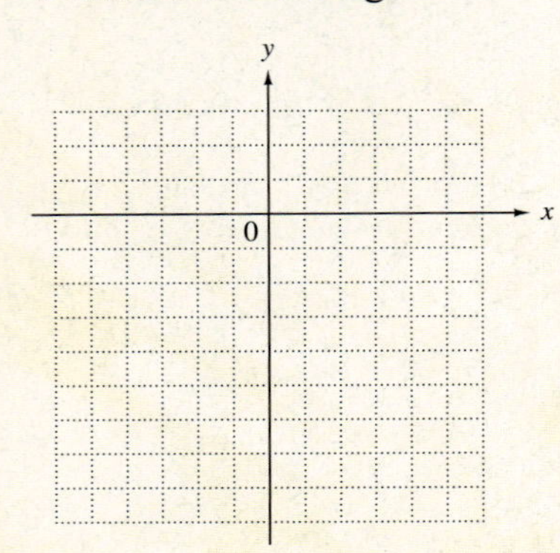

(b) $f(x) = -x^2$

For each input x, square it and then take its opposite. Plotting and joining the points gives a parabola that opens down. See the table and Figure 5. The domain is $(-\infty, \infty)$, and the range is $(-\infty, 0]$.

x	$f(x) = -x^2$
-2	-4
-1	-1
0	0
1	-1
2	-4

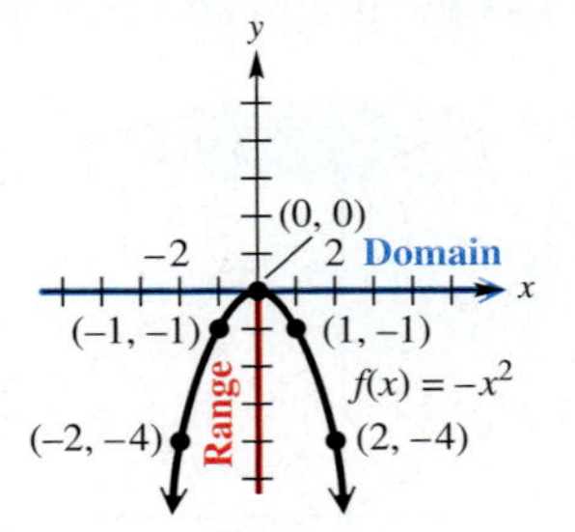

Figure 5

(c) $f(x) = x^3 - 2$

For this function, cube the input and then subtract 2 from the result. The graph is that of the cubing function *shifted* 2 units down. See the table and Figure 6. The domain and the range are both $(-\infty, \infty)$.

x	$f(x) = x^3 - 2$
-2	-10
-1	-3
0	-2
1	-1
2	6

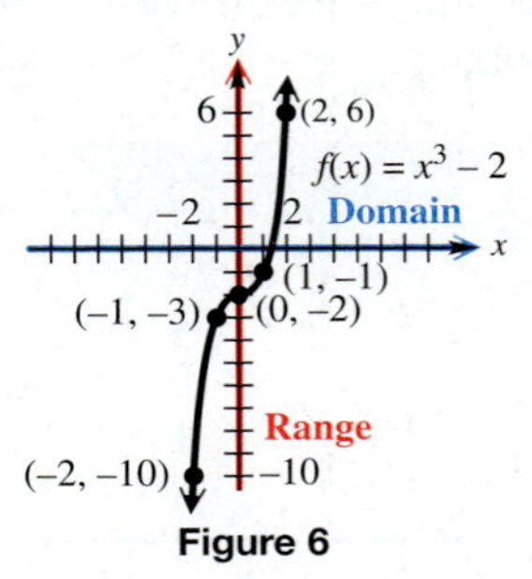

Figure 6

Work Problem 11 at the Side.

ANSWERS

11.

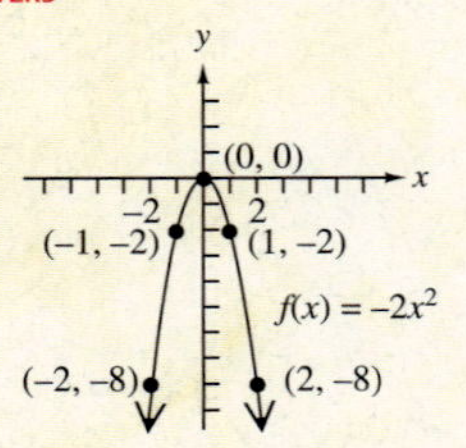

domain: $(-\infty, \infty)$; range: $(-\infty, 0]$

6.2 EXERCISES

FOR EXTRA HELP

Student's Solutions Manual

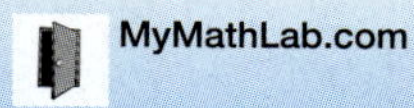
MyMathLab.com

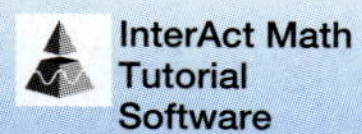
InterAct Math Tutorial Software

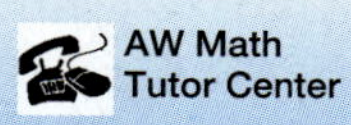
AW Math Tutor Center

www.mathxl.com

Digital Video Tutor CD 4 Videotape 8

We defined a polynomial written in descending powers in the text. Sometimes we write a polynomial in ascending powers, *with the degree of the terms increasing from left to right. Decide whether each polynomial is written in* descending *powers,* ascending *powers, or* neither.

1. $2x^3 + x - 3x^2$

2. $3x^5 + x^4 - 2x^3 + x$

3. $4p^3 - 8p^5 + p^7$

4. $q^2 + 3q^4 - 2q + 1$

5. $-m^3 + 5m^2 + 3m + 10$

6. $4 - x + 3x^2$

Give the coefficient and the degree of each term.

7. $7z$

8. $3r$

9. $-15p^2$

10. $-27k^3$

11. x^4

12. y^6

13. $-mn^5$

14. $-a^5b$

Identify each polynomial as a monomial, binomial, trinomial, *or* none of these. *Give the degree of each.*

Polynomial	Type	Degree
15. 24		
17. $7m - 21$		
19. $2r^3 + 3r^2 + 5r$		
21. $-6p^4q - 3p^3q^2 + 2pq^3 - q^4$		

Polynomial	Type	Degree
16. 5		
18. $-x^2 + 3x^5$		
20. $5z^2 - 5z + 7$		
22. $8s^3t - 3s^2t^2 + 2st^3 + 9$		

Combine terms. See Example 1.

23. $5z^4 + 3z^4$

24. $8r^5 - 2r^5$

25. $-m^3 + 2m^3 + 6m^3$

26. $3p^4 + 5p^4 - 2p^4$

27. $x + x + x + x + x$

28. $z - z - z + z$

29. $m^4 - 3m^2 + m$

30. $5a^5 + 2a^4 - 9a^3$

31. $y^2 + 7y - 4y^2$

32. $2c^2 - 4 + 8 - c^2$

33. $2k + 3k^2 + 5k^2 - 7$

34. $4x^2 + 2x - 6x^2 - 6$

35. $n^4 - 2n^3 + n^2 - 3n^4 + n^3$

36. $2q^3 + 3q^2 - 4q - q^3 + 5q^2$

Add or subtract as indicated. See Examples 2 and 3.

37. Add.
$$\begin{array}{r} -12p^2 + 4p - 1 \\ 3p^2 + 7p - 8 \\ \hline \end{array}$$

38. Add.
$$\begin{array}{r} -6y^3 + 8y + 5 \\ 9y^3 + 4y - 6 \\ \hline \end{array}$$

39. Subtract.
$$\begin{array}{r} 12a + 15 \\ 7a - 3 \\ \hline \end{array}$$

40. Subtract.
$$\begin{array}{r} -3b + 6 \\ 2b - 8 \\ \hline \end{array}$$

41. Subtract.
$$\begin{array}{r} 6m^2 - 11m + 5 \\ -8m^2 + 2m - 1 \\ \hline \end{array}$$

42. Subtract.
$$\begin{array}{r} -4z^2 + 2z - 1 \\ 3z^2 - 5z + 2 \\ \hline \end{array}$$

43. Add.
$$\begin{array}{r} 12z^2 - 11z + 8 \\ 5z^2 + 16z - 2 \\ -4z^2 + 5z - 9 \\ \hline \end{array}$$

44. Add.
$$\begin{array}{r} -6m^3 + 2m^2 + 5m \\ 8m^3 + 4m^2 - 6m \\ -3m^3 + 2m^2 - 7m \\ \hline \end{array}$$

45. Add.
$$\begin{array}{r} 6y^3 - 9y^2 \qquad + 8 \\ 4y^3 + 2y^2 + 5y \qquad \\ \hline \end{array}$$

46. Add.
$$\begin{array}{r} -7r^8 + 2r^6 - r^5 \qquad \\ 3r^6 \qquad\qquad + 5 \\ \hline \end{array}$$

47. Subtract.
$$\begin{array}{r} -5a^4 \qquad + 8a^2 - 9 \\ 6a^3 - a^2 + 2 \\ \hline \end{array}$$

48. Subtract.
$$\begin{array}{r} -2m^3 + 8m^2 \qquad \\ m^4 - m^3 \qquad\qquad + 2m \\ \hline \end{array}$$

49. $(3r + 8) - (2r - 5)$

50. $(2d + 7) - (3d - 1)$

51. $(5x^2 + 7x - 4) + (3x^2 - 6x + 2)$

52. $(4k^3 + k^2 + k) + (2k^3 - 4k^2 - 3k)$

53. $(2a^2 + 3a - 1) - (4a^2 + 5a + 6)$

54. $(q^4 - 2q^2 + 10) - (3q^4 + 5q^2 - 5)$

55. $(z^5 + 3z^2 + 2z) - (4z^5 + 2z^2 - 5z)$

56. $(5t^3 - 3t^2 + 2t) - (4t^3 + 2t^2 + 3t)$

For each polynomial function, find ***(a)*** *$f(-1)$ and* ***(b)*** *$f(2)$. See Example 4.*

57. $f(x) = 6x - 4$

58. $f(x) = -2x + 5$

59. $f(x) = x^2 - 3x + 4$

60. $f(x) = 3x^2 + x - 5$

61. $f(x) = 5x^4 - 3x^2 + 6$

62. $f(x) = -4x^4 + 2x^2 - 1$

63. $f(x) = -x^2 + 2x^3 - 8$

64. $f(x) = -x^2 - x^3 + 11x$

Solve each problem. See Example 5.

65. The number of airports in the United States during the period from 1970 through 1997 can be approximated by the polynomial function defined by

$$f(x) = -6.77x^2 + 445.34x + 11{,}279.82,$$

where $x = 0$ represents 1970, $x = 1$ represents 1971, and so on. Use this function to approximate the number of airports in each given year. (*Source:* U.S. Federal Aviation Administration.)

(a) 1970 **(b)** 1985 **(c)** 1997

66. The number of cases commenced by U.S. Courts of Appeals during the period from 1990 through 1998 can be approximated by the polynomial function defined by

$$f(x) = -145.32x^2 + 2610.84x + 41{,}341.13,$$

where $x = 0$ represents 1990, $x = 1$ represents 1991, and so on. Use this function to approximate the number of cases commenced in each given year. (*Source:* Administrative Office of the U.S. Courts, *Statistical Tables for the Federal Judiciary*, annual.)

(a) 1993 **(b)** 1995 **(c)** 1997

67. The number of people, in millions, enrolled in Health Maintenance Organizations (HMOs) during the period 1990–1995 can be modeled by the polynomial function defined by

$$f(x) = .39x^2 + .71x + 33.0,$$

where $x = 0$ corresponds to 1990, $x = 1$ corresponds to 1991, and so on. Use this model to approximate the number of people enrolled in each given year. (*Source:* Interstudy; U.S. National Center for Health Statistics.)

(a) 1990 **(b)** 1991 **(c)** 1992

68. The number of medical doctors, in thousands, in the United States during the period 1990–1995 can be modeled by the polynomial function defined by

$$f(x) = 1.23x^2 + 13.9x + 616.7,$$

where $x = 0$ corresponds to 1990, $x = 1$ corresponds to 1991, and so on. Use this model to approximate the number of doctors in each given year. (*Source:* American Medical Association.)

(a) 1993 **(b)** 1994 **(c)** 1995

For each pair of functions, find **(a)** $(f + g)(x)$ *and* **(b)** $(f - g)(x)$. *See Example 6.*

69. $f(x) = 5x - 10, g(x) = 3x + 7$

70. $f(x) = -4x + 1, g(x) = 6x + 2$

71. $f(x) = 4x^2 + 8x - 3, g(x) = -5x^2 + 4x - 9$

72. $f(x) = 3x^2 - 9x + 10, g(x) = -4x^2 + 2x + 12$

73. Construct two polynomial functions defined by $f(x)$, a polynomial of degree 3, and $g(x)$, a polynomial of degree 4. Find $(f - g)(x)$ and $(g - f)(x)$. Use your answers to decide whether subtraction of polynomial functions is a commutative operation.

74. Construct two polynomial functions defined by $f(x)$ and $g(x)$ such that $(f + g)(x) = 3x^3 - x + 3$.

Graph each function by creating a table of ordered pairs. Give the domain and the range. See Example 7.

75. $f(x) = -2x + 1$

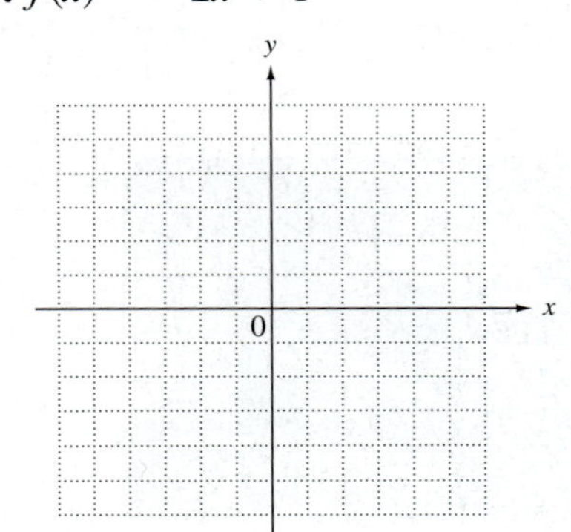

76. $f(x) = 3x + 2$

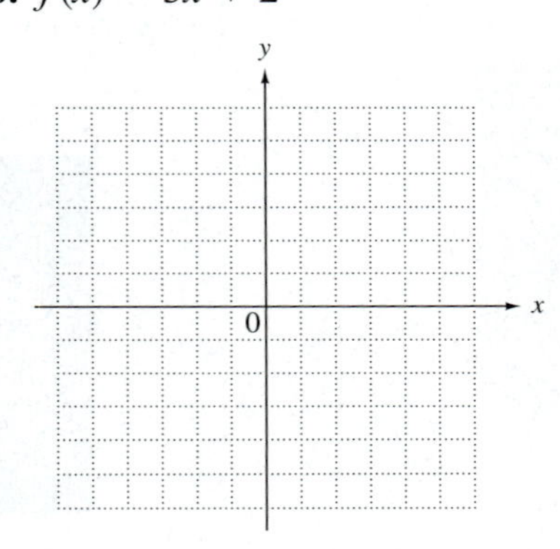

77. $f(x) = -3x^2$

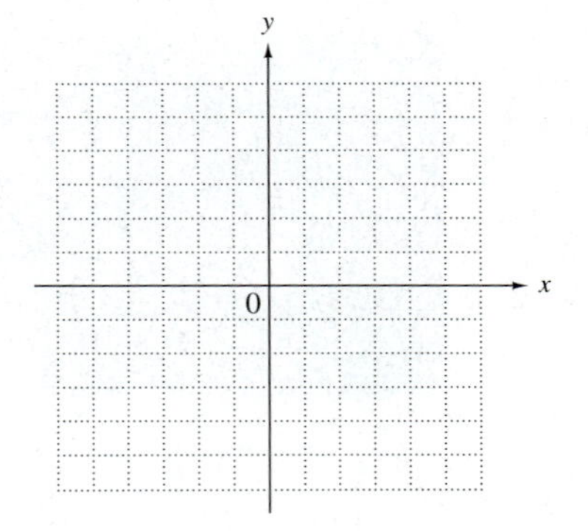

78. $f(x) = \frac{1}{2}x^2$

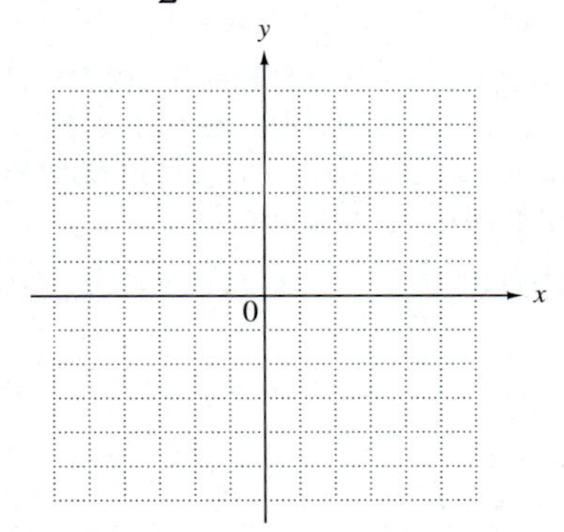

79. $f(x) = x^3 + 1$

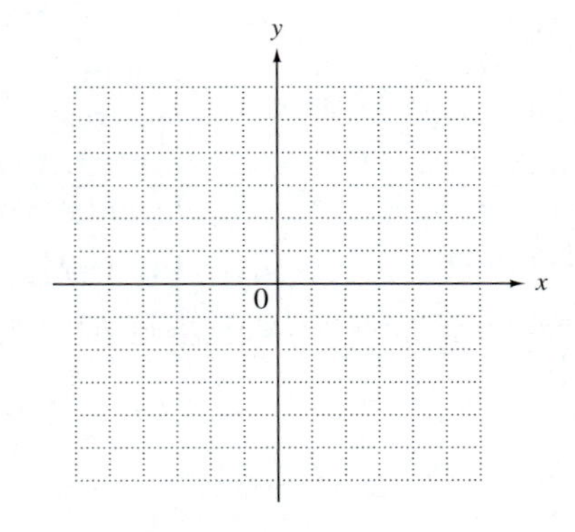

80. $f(x) = -x^3 + 2$

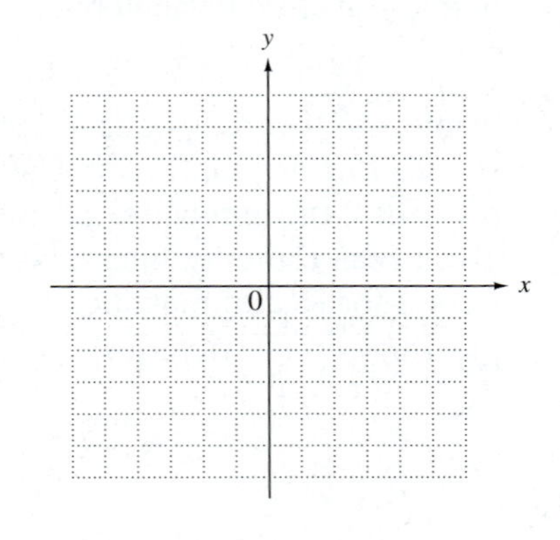

6.3 MULTIPLYING POLYNOMIALS

OBJECTIVES

1. Multiply terms.
2. Multiply any two polynomials.
3. Multiply binomials.
4. Find the product of the sum and difference of two terms.
5. Find the square of a binomial.
6. Multiply polynomial functions.

1 Multiply terms. Recall that the product of the two terms $3x^4$ and $5x^3$ is found by using the commutative and associative properties, along with the rules for exponents.

$$\begin{aligned}(3x^4)(5x^3) &= 3 \cdot 5 \cdot x^4 \cdot x^3\\ &= 15x^{4+3}\\ &= 15x^7\end{aligned}$$

Example 1 Multiplying Monomials

Find each product.

(a) $-4a^3(3a^5) = -4(3)a^3 \cdot a^5 = -12a^8$

(b) $2m^2z^4(8m^3z^2) = 2(8)m^2 \cdot m^3 \cdot z^4 \cdot z^2 = 16m^5z^6$

Work Problem 1 at the Side.

1 Find each product.

(a) $-6m^5(2m^4)$

(b) $8k^3y(9ky^3)$

2 Multiply any two polynomials. We use the distributive property to extend this process to find the product of any two polynomials.

Example 2 Multiplying Polynomials

Find each product.

(a) $-2(8x^3 - 9x^2)$

$$\begin{aligned}-2(8x^3 - 9x^2) &= -2(8x^3) - 2(-9x^2) \quad \text{Distributive property}\\ &= -16x^3 + 18x^2\end{aligned}$$

(b) $$\begin{aligned}5x^2(-4x^2 + 3x - 2) &= 5x^2(-4x^2) + 5x^2(3x) + 5x^2(-2)\\ &= -20x^4 + 15x^3 - 10x^2\end{aligned}$$

(c) $(3x - 4)(2x^2 + x)$

Use the distributive property to multiply each term of $2x^2 + x$ by $3x - 4$.

$$(3x - 4)(2x^2 + x) = (3x - 4)(2x^2) + (3x - 4)(x)$$

Here $3x - 4$ has been treated as a single expression so that the distributive property could be used. Now use the distributive property two more times.

$$\begin{aligned}&= 3x(2x^2) + (-4)(2x^2) + (3x)(x) + (-4)(x)\\ &= 6x^3 - 8x^2 + 3x^2 - 4x\\ &= 6x^3 - 5x^2 - 4x\end{aligned}$$

(d) $$\begin{aligned}2x^2(x + 1)(x - 3) &= 2x^2[(x + 1)(x) + (x + 1)(-3)]\\ &= 2x^2[x^2 + x - 3x - 3]\\ &= 2x^2(x^2 - 2x - 3)\\ &= 2x^4 - 4x^3 - 6x^2\end{aligned}$$

Work Problem 2 at the Side.

2 Find each product.

(a) $-2r(9r - 5)$

(b) $3p^2(5p^3 + 2p^2 - 7)$

(c) $(4a - 5)(3a + 6)$

(d) $3x^3(x + 4)(x - 6)$

ANSWERS

1. (a) $-12m^9$ (b) $72k^4y^4$
2. (a) $-18r^2 + 10r$ (b) $15p^5 + 6p^4 - 21p^2$ (c) $12a^2 + 9a - 30$ (d) $3x^5 - 6x^4 - 72x^3$

3 Find each product.

(a)
$$\begin{array}{r} 2m - 5 \\ \underline{3m + 4} \end{array}$$

(b)
$$\begin{array}{r} 5a^3 - 6a^2 + 2a - 3 \\ \underline{2a - 5} \end{array}$$

It is often easier to multiply polynomials by writing them vertically.

Example 3 Multiplying Polynomials Vertically

Find each product.

(a) $(5a - 2b)(3a + b)$

$$\begin{array}{rrrl} & 5a & -2b & \\ & 3a & +\ b & \\ \hline & 5ab & -2b^2 & \leftarrow b(5a - 2b) \\ 15a^2 & -6ab & & \leftarrow 3a(5a - 2b) \\ \hline 15a^2 & -ab & -2b^2 & \text{Combine like terms.} \end{array}$$

(b) $(3m^3 - 2m^2 + 4)(3m - 5)$

$$\begin{array}{rrrrrl} & 3m^3 & -2m^2 & +\ 4 & & \\ & & & 3m & -5 & \\ \hline & -15m^3 & +10m^2 & & -20 & -5(3m^3 - 2m^2 + 4) \\ 9m^4 & -6m^3 & & +12m & & 3m(3m^3 - 2m^2 + 4) \\ \hline 9m^4 & -21m^3 & +10m^2 & +12m & -20 & \text{Combine like terms.} \end{array}$$

Work Problem 3 at the Side.

3 Multiply binomials. When working with polynomials, the product of two binomials occurs repeatedly. There is a shortcut method for finding these products. Recall that a binomial has just two terms, such as $3x - 4$ or $2x + 3$. We can find the product of these binomials using the distributive property as follows.

$$\begin{aligned} (3x - 4)(2x + 3) &= 3x(2x + 3) - 4(2x + 3) \\ &= 3x(2x) + 3x(3) - 4(2x) - 4(3) \\ &= 6x^2 + 9x - 8x - 12 \end{aligned}$$

Before combining like terms to find the simplest form of the answer, let us check the origin of each of the four terms in the sum. First, $6x^2$ is the product of the two *first* terms.

$$(3x - 4)(2x + 3) \qquad 3x(2x) = 6x^2 \qquad \text{First terms}$$

To get $9x$, the *outer* terms are multiplied.

$$(3x - 4)(2x + 3) \qquad 3x(3) = 9x \qquad \text{Outer terms}$$

The term $-8x$ comes from the *inner* terms.

$$(3x - 4)(2x + 3) \qquad -4(2x) = -8x \qquad \text{Inner terms}$$

Finally, -12 comes from the *last* terms.

$$(3x - 4)(2x + 3) \qquad -4(3) = -12 \qquad \text{Last terms}$$

The product is found by combining these four results.

$$\begin{aligned} (3x - 4)(2x + 3) &= 6x^2 + 9x - 8x - 12 \\ &= 6x^2 + x - 12 \end{aligned}$$

To keep track of the order of multiplying these terms, we use the initials FOIL (**F**irst, **O**uter, **I**nner, **L**ast). All the steps of the FOIL method can be done as follows. Try to do as many of these steps as possible mentally.

ANSWERS
3. (a) $6m^2 - 7m - 20$
(b) $10a^4 - 37a^3 + 34a^2 - 16a + 15$

Example 4 **Using the FOIL Method**

Use the FOIL method to find each product.

(a) $(4m - 5)(3m + 1)$

First terms $(4m - 5)(3m + 1)$ $4m(3m) = 12m^2$

Outer terms $(4m - 5)(3m + 1)$ $4m(1) = 4m$

Inner terms $(4m - 5)(3m + 1)$ $-5(3m) = -15m$

Last terms $(4m - 5)(3m + 1)$ $-5(1) = -5$

Simplify by combining the four terms.

$$\begin{aligned}(4m - 5)(3m + 1) &= \overset{F}{12m^2} + \overset{O}{4m} - \overset{I}{15m} - \overset{L}{5} \\ &= 12m^2 - 11m - 5\end{aligned}$$

The procedure can be written in compact form as follows.

Combine these four results to get $12m^2 - 11m - 5$.

(b) $$\begin{aligned}(6a - 5b)(3a + 4b) &= \overset{\text{First}}{18a^2} + \overset{\text{Outer}}{24ab} - \overset{\text{Inner}}{15ab} - \overset{\text{Last}}{20b^2} \\ &= 18a^2 + 9ab - 20b^2\end{aligned}$$

(c) $(2k + 3z)(5k - 3z) = 10k^2 + 9kz - 9z^2$ FOIL

Work Problem 4 at the Side.

4 **Find the product of the sum and difference of two terms.** Some types of binomial products occur frequently. The product of the sum and difference of the same two terms, x and y, is

$$\begin{aligned}(x + y)(x - y) &= x^2 - xy + xy - y^2 \quad \text{FOIL} \\ &= x^2 - y^2.\end{aligned}$$

Product of the Sum and Difference of Two Terms

The **product of the sum and difference of the two terms x and y** is the difference of the squares of the terms, or

$$(x + y)(x - y) = x^2 - y^2.$$

4 Use the FOIL method to find each product.

(a) $(3z + 2)(z + 1)$

(b) $(5r - 3)(2r - 5)$

(c) $(4p + 5q)(3p - 2q)$

(d) $(4y - z)(2y + 3z)$

(e) $(8r + 1)(8r - 1)$

Answers

4. **(a)** $3z^2 + 5z + 2$ **(b)** $10r^2 - 31r + 15$ **(c)** $12p^2 + 7pq - 10q^2$ **(d)** $8y^2 + 10yz - 3z^2$ **(e)** $64r^2 - 1$

5 Find each product.

(a) $(m + 5)(m - 5)$

(b) $(x - 4y)(x + 4y)$

(c) $(7m - 2n)(7m + 2n)$

(d) $4y^2(y + 7)(y - 7)$

Example 5 Multiplying the Sum and Difference of Two Terms

Find each product.

(a) $(p + 7)(p - 7) = p^2 - 7^2$
$= p^2 - 49$

(b) $(2r + 5)(2r - 5) = (2r)^2 - 5^2$
$= 2^2r^2 - 25$
$= 4r^2 - 25$

(c) $(6m + 5n)(6m - 5n) = (6m)^2 - (5n)^2$
$= 36m^2 - 25n^2$

(d) $2x^3(x + 3)(x - 3) = 2x^3(x^2 - 9)$
$= 2x^5 - 18x^3$

Work Problem **5** at the Side.

5 Find the square of a binomial. Another special binomial product is the *square of a binomial.* To find the square of $x + y$, or $(x + y)^2$, multiply $x + y$ by itself.

$$(x + y)(x + y) = x^2 + xy + xy + y^2$$
$$= x^2 + 2xy + y^2$$

A similar result is true for the square of a difference.

Square of a Binomial

The **square of a binomial** is the sum of the square of the first term, twice the product of the two terms, and the square of the last term.

$$(x + y)^2 = x^2 + 2xy + y^2$$
$$(x - y)^2 = x^2 - 2xy + y^2$$

Example 6 Squaring Binomials

Find each product.

(a) $(m + 7)^2 = m^2 + 2 \cdot m \cdot 7 + 7^2$
$= m^2 + 14m + 49$

(b) $(p - 5)^2 = p^2 - 2 \cdot p \cdot 5 + 5^2$
$= p^2 - 10p + 25$

(c) $(2p + 3v)^2 = (2p)^2 + 2(2p)(3v) + (3v)^2$
$= 4p^2 + 12pv + 9v^2$

(d) $(3r - 5s)^2 = (3r)^2 - 2(3r)(5s) + (5s)^2$
$= 9r^2 - 30rs + 25s^2$

ANSWERS
5. (a) $m^2 - 25$ **(b)** $x^2 - 16y^2$
(c) $49m^2 - 4n^2$ **(d)** $4y^4 - 196y^2$

CAUTION

As the products in the formula for the square of a binomial show,

$$(x + y)^2 \neq x^2 + y^2.$$

More generally,

$$(x + y)^n \neq x^n + y^n.$$

Work Problem 6 at the Side.

We can use the patterns for the special products with more complicated products, as the following example shows.

Example 7 Multiplying More Complicated Binomials

Use special products to find each product.

(a) $[(3p - 2) + 5q][(3p - 2) - 5q]$

$= (3p - 2)^2 - (5q)^2$ Product of sum and difference of terms

$= 9p^2 - 12p + 4 - 25q^2$ Square both quantities.

(b) $[(2z + r) + 1]^2 = (2z + r)^2 + 2(2z + r)(1) + 1^2$ Square of a binomial

$= 4z^2 + 4zr + r^2 + 4z + 2r + 1$ Square again; use the distributive property.

(c) $(x + y)^3 = (x + y)^2(x + y)$

$= (x^2 + 2xy + y^2)(x + y)$ Square $x + y$.

$= x^3 + 2x^2y + xy^2 + x^2y + 2xy^2 + y^3$

$= x^3 + 3x^2y + 3xy^2 + y^3$

(d) $(2a + b)^4 = (2a + b)^2(2a + b)^2$

$= (4a^2 + 4ab + b^2)(4a^2 + 4ab + b^2)$ Square $2a + b$.

$= 16a^4 + 16a^3b + 4a^2b^2 + 16a^3b + 16a^2b^2 + 4ab^3 + 4a^2b^2 + 4ab^3 + b^4$

$= 16a^4 + 32a^3b + 24a^2b^2 + 8ab^3 + b^4$

Work Problem 7 at the Side.

6 Multiply polynomial functions. In the previous section we saw how functions can be added and subtracted. Functions can also be multiplied.

Multiplying Functions

If $f(x)$ and $g(x)$ define functions, their product is defined as follows.

$$(fg)(x) = f(x) \cdot g(x)$$

The domain of the product function is the intersection of the domains of $f(x)$ and $g(x)$.

6 Find each product.

(a) $(a + 2)^2$

(b) $(2m - 5)^2$

(c) $(y + 6z)^2$

(d) $(3k - 2n)^2$

7 Find each product.

(a) $[(m - 2n) - 3] \cdot [(m - 2n) + 3]$

(b) $[(k - 5h) + 2]^2$

(c) $(p + 2q)^3$

(d) $(x + 2)^4$

Answers

6. **(a)** $a^2 + 4a + 4$
(b) $4m^2 - 20m + 25$
(c) $y^2 + 12yz + 36z^2$
(d) $9k^2 - 12kn + 4n^2$

7. **(a)** $m^2 - 4mn + 4n^2 - 9$
(b) $k^2 - 10kh + 25h^2 + 4k - 20h + 4$
(c) $p^3 + 6p^2q + 12pq^2 + 8q^3$
(d) $x^4 + 8x^3 + 24x^2 + 32x + 16$

8 For

$$f(x) = 2x + 7 \quad \text{and}$$
$$g(x) = x^2 - 4,$$

find $(fg)(x)$.

Example 8 Multiplying Polynomial Functions

For $f(x) = 3x + 4$ and $g(x) = 2x^2 + x$, find $(fg)(x)$.

$$
\begin{aligned}
(fg)(x) &= f(x) \cdot g(x) && \text{Use the definition.} \\
&= (3x + 4)(2x^2 + x) \\
&= 6x^3 + 3x^2 + 8x^2 + 4x && \text{FOIL} \\
&= 6x^3 + 11x^2 + 4x && \text{Combine terms.}
\end{aligned}
$$

Work Problem 8 at the Side.

ANSWERS

8. $2x^3 + 7x^2 - 8x - 28$

6.3 EXERCISES

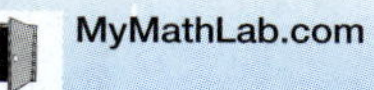

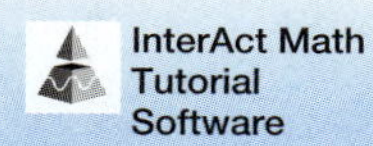

AW Math Tutor Center | www.mathxl.com | Digital Video Tutor CD 4 Videotape 8

Find each product. See Examples 1–3.

1. $-8m^3(3m^2)$

2. $4p^2(-5p^4)$

3. $3x(-2x + 5)$

4. $5y(-6y - 1)$

5. $-q^3(2 + 3q)$

6. $-3a^4(4 - a)$

7. $6k^2(3k^2 + 2k + 1)$

8. $5r^3(2r^2 - 3r - 4)$

9. $(2m + 3)(3m^2 - 4m - 1)$

10. $(4z - 2)(z^2 + 3z + 5)$

11. $4x^3(x - 3)(x + 2)$

12. $2y^5(y - 8)(y + 2)$

13. $(2y + 3)(3y - 4)$

14. $(5m - 3)(2m + 6)$

15. $$\begin{array}{r} -b^2 + 3b + 3 \\ \underline{2b + 4} \end{array}$$

16. $$\begin{array}{r} -r^2 - 4r + 8 \\ \underline{3r - 2} \end{array}$$

17. $$\begin{array}{r} 5m - 3n \\ \underline{5m + 3n} \end{array}$$

18. $$\begin{array}{r} 2k + 6q \\ \underline{2k - 6q} \end{array}$$

19. $$\begin{array}{r} 2z^3 - 5z^2 + 8z - 1 \\ \underline{4z + 3} \end{array}$$

20. $$\begin{array}{r} 3z^4 - 2z^3 + z - 5 \\ \underline{2z - 5} \end{array}$$

21. $$\begin{array}{r} 2p^2 + 3p + 6 \\ \underline{3p^2 - 4p - 1} \end{array}$$

22. $$\begin{array}{r} 5y^2 - 2y + 4 \\ \underline{2y^2 + y + 3} \end{array}$$

Use the FOIL method to find each product. See Example 4.

23. $(m + 5)(m - 8)$

24. $(p - 6)(p + 4)$

25. $(4k + 3)(3k - 2)$

26. $(5w + 2)(2w + 5)$

27. $(z - w)(3z + 4w)$

28. $(s + t)(2s - 5t)$

29. $(6c - d)(2c + 3d)$

30. $(2m - n)(3m + 5n)$

31. $(.2x + 1.3)(.5x - .1)$

32. $(.5y - .4)(.1y + 2.1)$

33. $\left(3r + \frac{1}{4}y\right)(r - 2y)$

34. $\left(5w - \frac{2}{3}z\right)(w + 5z)$

35. Describe the FOIL method in your own words.

36. Explain why the product of the sum and difference of two terms is not a trinomial.

Find each product. See Example 5.

37. $(2p - 3)(2p + 3)$

38. $(3x - 8)(3x + 8)$

39. $(5m - 1)(5m + 1)$

40. $(6y + 3)(6y - 3)$

41. $(3a + 2c)(3a - 2c)$

42. $(5r - 4s)(5r + 4s)$

43. $\left(4x - \frac{2}{3}\right)\left(4x + \frac{2}{3}\right)$

44. $\left(3t + \frac{5}{4}\right)\left(3t - \frac{5}{4}\right)$

45. $(4m + 7n^2)(4m - 7n^2)$

46. $(2k^2 + 6h)(2k^2 - 6h)$

47. $(5y^3 + 2)(5y^3 - 2)$

48. $(3x^3 + 4)(3x^3 - 4)$

Find each square. See Example 6.

49. $(y - 5)^2$

50. $(a - 3)^2$

51. $(2p + 7)^2$

52. $(3z + 8)^2$

53. $(4n - 3m)^2$

54. $(5r - 7s)^2$

55. $\left(k - \frac{5}{7}p\right)^2$

56. $\left(q - \frac{3}{4}r\right)^2$

57. Explain how the expressions $(x + y)^2$ and $x^2 + y^2$ differ.

58. Explain how you can find the product $101 \cdot 99$ using the special product

$$(a + b)(a - b) = a^2 - b^2.$$

Find each product. See Example 7.

59. $[(5x + 1) + 6y]^2$

60. $[(3m - 2) + p]^2$

61. $[(2a + b) - 3][(2a + b) + 3]$

62. $[(m + p) + 5][(m + p) - 5]$

63. $[(2h - k) + j][(2h - k) - j]$

64. $[(3m - y) + z][(3m - y) - z]$

65. $(5r - s)^3$

66. $(x + 3y)^3$

67. $(m - p)^4$

68. Expand the following: $(x + y)^2$, $(x + y)^3$, and $(x + y)^4$. Now, evaluate 11^2, 11^3, and 11^4. Compare the coefficients of x and y in your expansions to these powers of 11. What do you notice?

RELATING CONCEPTS (Exercises 69–76) FOR INDIVIDUAL OR GROUP WORK

Consider the figure. ***Work Exercises 69–76 in order.***

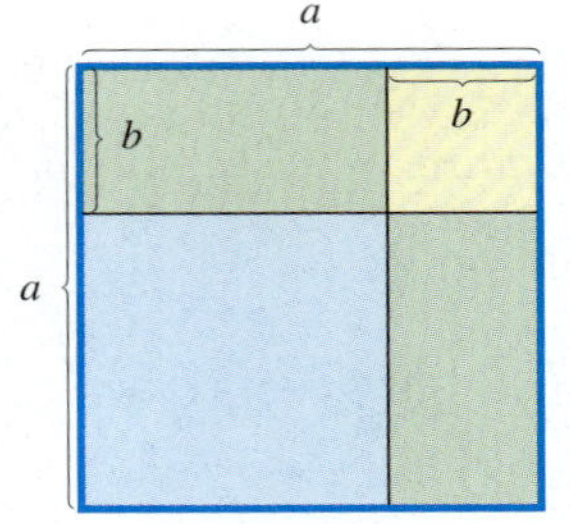

69. What is the length of each side of the blue square in terms of a and b?

70. What is the formula for the area of a square? Use the formula to write an expression, in the form of a product, for the area of the blue square.

71. Each green rectangle has an area of ________________. Therefore, the total area in green is represented by the polynomial ________.

72. The yellow square has an area of ________.

73. The area of the entire colored region is represented by ________, because each side of the entire colored region has length ________.

74. The area of the blue square is equal to the area of the entire colored region minus the total area of the green squares minus the area of the yellow square. Write this as a simplified polynomial in a and b.

75. What must be true about the expressions for the area of the blue square you found in Exercises 70 and 74?

76. Write a statement of equality based on your answer in Exercise 75. How does this reinforce one of the main ideas of this section?

For each pair of functions, find the product $(fg)(x)$. See Example 8.

77. $f(x) = 2x,\ g(x) = 5x - 1$

78. $f(x) = 3x,\ g(x) = 6x - 8$

79. $f(x) = x + 1,\ g(x) = 2x - 3$

80. $f(x) = x - 7,\ g(x) = 4x + 5$

81. $f(x) = 2x - 3,\ g(x) = 4x^2 + 6x + 9$

82. $f(x) = 3x + 4,\ g(x) = 9x^2 - 12x + 16$

Show that each statement is false by replacing x with 2 and y with 3. Then, rewrite each statement with the correct product.

83. $(x + y)^2 = x^2 + y^2$

84. $(x + y)^3 = x^3 + y^3$

85. $(x + y)^4 = x^4 + y^4$

6.4 Dividing Polynomials

OBJECTIVES

1. Divide a polynomial by a monomial.
2. Divide a polynomial by a polynomial of two or more terms.
3. Divide polynomial functions.

1 Divide a polynomial by a monomial. In the previous two sections, we added, subtracted, and multiplied polynomials. We now discuss polynomial division, beginning with division by a monomial. (Recall that a monomial is a single term, such as $8x$, $-9m^4$, or $11y^2$.)

Dividing by a Monomial

To divide a polynomial by a monomial, divide each term in the polynomial by the monomial, and then write each quotient in lowest terms.

Example 1 Dividing a Polynomial by a Monomial

Divide.

(a)
$$\frac{15x^2 - 12x + 6}{3} = \frac{15x^2}{3} - \frac{12x}{3} + \frac{6}{3} \quad \text{Divide each term by 3.}$$
$$= 5x^2 - 4x + 2 \quad \text{Write in lowest terms.}$$

Check this answer by multiplying it by the divisor, 3. You should get $15x^2 - 12x + 6$ as the result.

$$\underset{\text{Divisor}}{3}\underbrace{(5x^2 - 4x + 2)}_{\text{Quotient}} = \underbrace{15x^2 - 12x + 6}_{\text{Original polynomial}}$$

(b)
$$\frac{5m^3 - 9m^2 + 10m}{5m^2} = \frac{5m^3}{5m^2} - \frac{9m^2}{5m^2} + \frac{10m}{5m^2} \quad \text{Divide each term by } 5m^2.$$
$$= m - \frac{9}{5} + \frac{2}{m} \quad \text{Write in lowest terms.}$$

This result is not a polynomial. (Why?) The quotient of two polynomials need not be a polynomial.

(c)
$$\frac{8xy^2 - 9x^2y + 6x^2y^2}{x^2y^2} = \frac{8xy^2}{x^2y^2} - \frac{9x^2y}{x^2y^2} + \frac{6x^2y^2}{x^2y^2}$$
$$= \frac{8}{x} - \frac{9}{y} + 6$$

Work Problem 1 at the Side.

1 Divide.

(a) $\dfrac{12p + 30}{6}$

(b) $\dfrac{9y^3 - 4y^2 + 8y}{2y^2}$

(c) $\dfrac{8a^2b^2 - 20ab^3}{4a^3b}$

2 Divide a polynomial by a polynomial of two or more terms. The process for dividing one polynomial by another polynomial that is not a monomial is similar to that for dividing whole numbers.

Example 2 Dividing a Polynomial by a Polynomial

Divide $2m^2 + m - 10$ by $m - 2$.

Write the problem, making sure that both polynomials are written in descending powers of the variables.

$$m - 2\overline{)2m^2 + m - 10}$$

Continued on Next Page

ANSWERS

1. (a) $2p + 5$ (b) $\dfrac{9y}{2} - 2 + \dfrac{4}{y}$ (c) $\dfrac{2b}{a} - \dfrac{5b^2}{a^2}$

2 Divide.

(a) $\dfrac{2r^2 + r - 21}{r - 3}$

(b) $\dfrac{2k^2 + 17k + 30}{2k + 5}$

Divide the first term of $2m^2 + m - 10$ by the first term of $m - 2$. Since $\frac{2m^2}{m} = 2m$, place this result above the division line.

$$m - 2\overline{)2m^2 + m - 10} \quad \text{quotient: } 2m \leftarrow \text{Result of } \tfrac{2m^2}{m}$$

Multiply $m - 2$ and $2m$, and write the result below $2m^2 + m - 10$.

$$\begin{array}{r} 2m \\ m - 2\overline{)2m^2 + m - 10} \\ \underline{2m^2 - 4m} \end{array} \quad \leftarrow 2m(m - 2) = 2m^2 - 4m$$

Now subtract $2m^2 - 4m$ from $2m^2 + m$. Do this by mentally changing the signs on $2m^2 - 4m$ and *adding*.

$$\begin{array}{r} 2m \\ m - 2\overline{)2m^2 + m - 10} \\ \underline{2m^2 - 4m} \\ 5m \end{array} \quad \leftarrow \text{Subtract. The difference is } 5m.$$

Bring down -10 and continue by dividing $5m$ by m.

$$\begin{array}{rl} 2m + 5 & \leftarrow \tfrac{5m}{m} = 5 \\ m - 2\overline{)2m^2 + m - 10} & \\ \underline{2m^2 - 4m} & \\ 5m - 10 & \leftarrow \text{Bring down } -10. \\ \underline{5m - 10} & \leftarrow 5(m - 2) = 5m - 10 \\ 0 & \leftarrow \text{Subtract. The difference is 0.} \end{array}$$

Finally, $(2m^2 + m - 10) \div (m - 2) = 2m + 5$. Check by multiplying $m - 2$ and $2m + 5$. The result should be $2m^2 + m - 10$.

Work Problem 2 at the Side.

Example 3 Dividing a Polynomial with a Missing Term

Divide $3x^3 - 2x + 5$ by $x - 3$.

Make sure that $3x^3 - 2x + 5$ is in descending powers of the variable. Add a term with 0 coefficient as a placeholder for the missing x^2-term.

$$x - 3\overline{)3x^3 + \mathbf{0x^2} - 2x + 5} \quad \leftarrow \text{Missing term } (0x^2)$$

Start with $\frac{3x^3}{x} = 3x^2$.

$$\begin{array}{rl} 3x^2 & \leftarrow \tfrac{3x^3}{x} = 3x^2 \\ x - 3\overline{)3x^3 + 0x^2 - 2x + 5} & \\ \underline{3x^3 - 9x^2} & \leftarrow 3x^2(x - 3) \end{array}$$

Subtract by mentally changing the signs on $3x^3 - 9x^2$ and adding.

$$\begin{array}{rl} 3x^2 & \\ x - 3\overline{)3x^3 + 0x^2 - 2x + 5} & \\ \underline{3x^3 - 9x^2} & \\ 9x^2 & \leftarrow \text{Subtract.} \end{array}$$

Bring down the next term.

$$\begin{array}{rl} 3x^2 & \\ x - 3\overline{)3x^3 + 0x^2 - 2x + 5} & \\ \underline{3x^3 - 9x^2} & \\ 9x^2 - 2x & \leftarrow \text{Bring down } -2x. \end{array}$$

Continued on Next Page

ANSWERS

2. (a) $2r + 7$ (b) $k + 6$

In the next step, $\frac{9x^2}{x} = 9x$.

$$\begin{array}{r} 3x^2 + 9x \phantom{{}+5} \\ x - 3\overline{)3x^3 + 0x^2 - 2x + 5} \\ \underline{3x^3 - 9x^2} \phantom{{}- 2x + 5} \\ 9x^2 - 2x \phantom{{}+5} \\ \underline{9x^2 - 27x} \phantom{{}+5} \\ 25x + 5 \end{array} \quad \begin{array}{l} \leftarrow \frac{9x^2}{x} = 9x \\ \\ \\ \\ \leftarrow 9x(x-3) \\ \leftarrow \text{Subtract; bring down 5.} \end{array}$$

Finally, $\frac{25x}{x} = 25$.

$$\begin{array}{r} 3x^2 + 9x + 25 \\ x - 3\overline{)3x^3 + 0x^2 - 2x + 5} \\ \underline{3x^3 - 9x^2} \phantom{{}- 2x + 5} \\ 9x^2 - 2x \phantom{{}+5} \\ \underline{9x^2 - 27x} \phantom{{}+5} \\ 25x + 5 \\ \underline{25x - 75} \\ 80 \end{array} \quad \begin{array}{l} \leftarrow \frac{25x}{x} = 25 \\ \\ \\ \\ \\ \\ \leftarrow 25(x-3) \\ \leftarrow \text{Remainder} \end{array}$$

Write the remainder, 80, as the numerator of the fraction $\frac{80}{x-3}$. In summary,

$$\frac{3x^3 - 2x + 5}{x - 3} = 3x^2 + 9x + 25 + \frac{80}{x - 3}.$$

Check by multiplying $x - 3$ and $3x^2 + 9x + 25$ and adding 80 to the result. You should get $3x^3 - 2x + 5$.

CAUTION

Remember to write $\frac{\text{remainder}}{\text{divisor}}$ as part of the quotient.

Work Problem 3 at the Side.

Example 4 **Performing a Division with a Fractional Coefficient in the Quotient**

Divide $2p^3 + 5p^2 + p - 2$ by $2p + 2$.

$$\begin{array}{r} p^2 + \frac{3}{2}p - 1 \\ 2p + 2\overline{)2p^3 + 5p^2 + p - 2} \\ \underline{2p^3 + 2p^2} \phantom{{}+p-2} \\ 3p^2 + p \phantom{{}-2} \\ \underline{3p^2 + 3p} \phantom{{}-2} \\ -2p - 2 \\ \underline{-2p - 2} \\ 0 \end{array} \quad \frac{3p^2}{2p} = \frac{3}{2}p$$

Since the remainder is 0, the quotient is $p^2 + \frac{3}{2}p - 1$.

Work Problem 4 at the Side.

3 Divide.

$$\frac{3k^3 + 9k - 14}{k - 2}$$

4 Divide $2p^3 + 7p^2 + 9p + 2$ by $2p + 2$.

Answers

3. $3k^2 + 6k + 21 + \frac{28}{k - 2}$

4. $p^2 + \frac{5}{2}p + 2 + \frac{-2}{2p + 2}$

5 Divide.

(a) $\dfrac{3r^5-15r^4-2r^3+19r^2-7}{3r^2-2}$

(b) $\dfrac{4x^4-7x^2+x+5}{2x^2-x}$

6 For

$$f(x)=2x^2+17x+30$$
$$\text{and } g(x)=2x+5,$$

find $\left(\frac{f}{g}\right)(x)$.

Answers

5. (a) $r^3-5r^2+3+\dfrac{-1}{3r^2-2}$

(b) $2x^2+x-3+\dfrac{-2x+5}{2x^2-x}$

6. $x+6, \quad x\neq-\dfrac{5}{2}$

Example 5 **Dividing by a Polynomial with a Missing Term**

Divide $6r^4+9r^3+2r^2-8r+7$ by $3r^2-2$.

The polynomial $3r^2-2$ has a missing term. Write it as $3r^2+0r-2$ and divide as usual.

$$\begin{array}{r} 2r^2+3r+2 \\ 3r^2+0r-2\overline{)6r^4+9r^3+2r^2-8r+7} \\ \underline{6r^4+0r^3-4r^2} \qquad\qquad \\ 9r^3+6r^2-8r \qquad \\ \underline{9r^3+0r^2-6r} \qquad \\ 6r^2-2r+7 \\ \underline{6r^2+0r-4} \\ -2r+11 \end{array}$$

Since the degree of the remainder, $-2r+11$, is less than the degree of $3r^2-2$, the division process is now finished. The result is written

$$2r^2+3r+2+\frac{-2r+11}{3r^2-2}.$$

Work Problem **5** at the Side.

CAUTION

Remember the following steps when dividing a polynomial by a polynomial of two or more terms.

1. Be sure the terms in both polynomials are in descending powers.
2. Write any missing terms with 0 placeholders.

3 **Divide polynomial functions.** Earlier in this chapter, we saw how polynomial functions are added, subtracted, and multiplied. We now give the definition for the quotient of two functions.

Dividing Functions

If $f(x)$ and $g(x)$ define functions, their quotient is defined as follows.

$$\left(\frac{f}{g}\right)(x)=\frac{f(x)}{g(x)}$$

The domain of the quotient function is the intersection of the domains of $f(x)$ and $g(x)$, excluding any values of x where $g(x)=0$.

Example 6 **Dividing Polynomial Functions**

For $f(x)=2x^2+x-10$ and $g(x)=x-2$, find $\left(\frac{f}{g}\right)(x)$.

$$\left(\frac{f}{g}\right)(x)=\frac{f(x)}{g(x)}=\frac{2x^2+x-10}{x-2}$$

This quotient was found in Example 2, with m replacing x. The result here is $2x+5$, so

$$\left(\frac{f}{g}\right)(x)=2x+5, \quad x\neq 2.$$

Work Problem **6** at the Side.

6.4 EXERCISES

FOR EXTRA HELP

Student's Solutions Manual

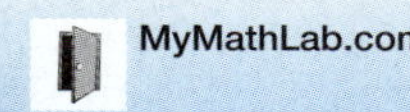
MyMathLab.com

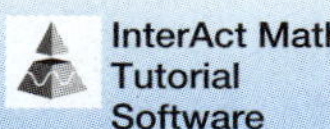
InterAct Math Tutorial Software

AW Math Tutor Center

www.mathxl.com

Digital Video Tutor CD 4 Videotape 9

Divide. See Example 1.

1. $\dfrac{15x^3 - 10x^2 + 5}{5}$

2. $\dfrac{27m^4 - 18m^3 + 9m}{9}$

3. $\dfrac{9y^2 + 12y - 15}{3y}$

4. $\dfrac{80r^2 - 40r + 10}{10r}$

5. $\dfrac{15m^3 + 25m^2 + 30m}{5m^2}$

6. $\dfrac{64x^3 - 72x^2 + 12x}{8x^3}$

7. $\dfrac{14m^2n^2 - 21mn^3 + 28m^2n}{14m^2n}$

8. $\dfrac{24h^2k + 56hk^2 - 28hk}{16h^2k^2}$

Divide. See Examples 2–5.

9. $\dfrac{y^2 + 3y - 18}{y + 6}$

10. $\dfrac{q^2 + 4q - 32}{q - 4}$

11. $\dfrac{3t^2 + 17t + 10}{3t + 2}$

12. $\dfrac{2k^2 - 3k - 20}{2k + 5}$

13. $(2z^3 - 5z^2 + 6z - 15) \div (2z - 5)$

14. $(3p^3 + p^2 + 18p + 6) \div (3p + 1)$

15. $(4x^3 + 9x^2 - 10x + 3) \div (4x + 1)$

16. $(10z^3 - 26z^2 + 17z - 13) \div (5z - 3)$

17. $\dfrac{14x + 6x^3 - 15 - 19x^2}{3x^2 - 2x + 4}$

18. $\dfrac{37m - 18m^2 - 13 + 8m^3}{2m^2 - 3m + 6}$

19. $\dfrac{4k^4 + 6k^3 + 3k - 1}{2k^2 + 1}$

20. $\dfrac{6y^4 + 9y^3 + 10y^2 + 6y + 4}{3y^2 + 2}$

21. $(9z^4 - 13z^3 + 23z^2 - 10z + 8) \div (z^2 - z + 2)$

22. $(2q^4 + 5q^3 - 11q^2 + 11q - 20) \div (2q^2 - q + 2)$

23. $\left(2x^2 - \frac{7}{3}x - 1\right) \div (3x + 1)$

24. $\left(m^2 + \frac{7}{2}m + 3\right) \div (2m + 3)$

25. $\left(3a^2 - \frac{23}{4}a - 5\right) \div (4a + 3)$

26. $\left(3q^2 + \frac{19}{5}q - 3\right) \div (5q - 2)$

For each pair of functions, find the quotient $\left(\frac{f}{g}\right)(x)$ *and give any x-values that are not in the domain of the quotient function. See Example 6.*

27. $f(x) = 10x^2 - 2x,\ g(x) = 2x$

28. $f(x) = 18x^2 - 24x,\ g(x) = 3x$

29. $f(x) = 2x^2 - x - 3,\ g(x) = x + 1$

30. $f(x) = 4x^2 - 23x - 35,\ g(x) = x - 7$

31. $f(x) = 8x^3 - 27,\ g(x) = 2x - 3$

32. $f(x) = 27x^3 + 64,\ g(x) = 3x + 4$

Solve each problem.

33. The volume of a box is $2p^3 + 15p^2 + 28p$ ft. The height is p ft and the length is $p + 4$ ft. Find an expression for the width.

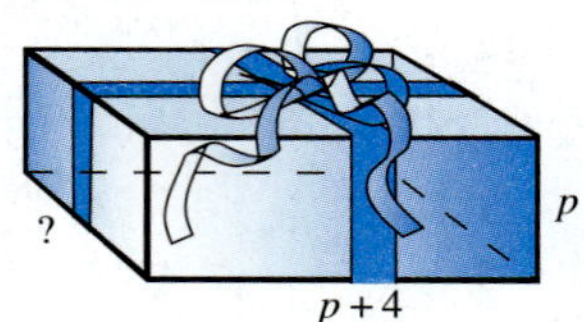

34. Suppose a car goes $2m^3 + 15m^2 + 13m - 63$ km in $2m + 9$ hr. Find an expression for its rate.

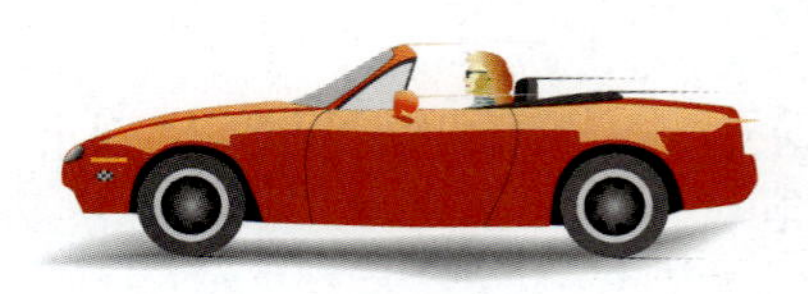

RELATING CONCEPTS (Exercises 35–40) FOR INDIVIDUAL OR GROUP WORK

Let $10 = t$. *Then* $100 = t^2$, $1000 = t^3$, *and so on. We can write integers using t as follows:*
$23 = 20 + 3 = 2t + 3$, $547 = 500 + 40 + 7 = 5t^2 + 4t + 7$, $1080 = 1000 + 80 = t^3 + 8t$.
To see the close connection between division of polynomials and division of integers, ***work Exercises 35–40 in order.***

35. Write 1654 using t.

36. Write 14 using t.

37. Divide your answer to Exercise 35 by your answer to Exercise 36. What is the quotient?

38. Divide 1654 by 14. (Use a calculator if you like, but write any decimal as a common fraction.)

39. Write your answer to Exercise 37 without t, substituting $t = 10$, $t^2 = 100$, and so on.

40. What do you observe about the quotients in Exercises 38 and 39?

6.5 Synthetic Division

Objectives

1. Use synthetic division to divide by a polynomial of the form $x - k$.
2. Use the remainder theorem to evaluate a polynomial.
3. Decide whether a given number is a solution of an equation.

1 Use synthetic division to divide by a polynomial of the form $x - k$. Often when one polynomial is divided by a second, the second polynomial is of the form $x - k$, where the coefficient of the x-term is 1. There is a shortcut method for doing these divisions. To see how it works, look at the left below, where the division of $3x^3 - 2x + 5$ by $x - 3$ is shown. Notice that 0 was inserted for the missing x^2-term.

$$\begin{array}{r|l}
 & 3x^2 + 9x + 25 \\ \hline
x - 3 & 3x^3 + 0x^2 - 2x + 5 \\
 & 3x^3 - 9x^2 \\
 & \quad 9x^2 - 2x \\
 & \quad 9x^2 - 27x \\
 & \qquad 25x + 5 \\
 & \qquad 25x - 75 \\
 & \qquad\qquad 80
\end{array}
\qquad
\begin{array}{r|rrrr}
 & & 3 & 9 & 25 \\ \hline
1 - 3 & 3 & 0 & -2 & 5 \\
 & 3 & -9 & & \\
 & & 9 & -2 & \\
 & & 9 & -27 & \\
 & & & 25 & 5 \\
 & & & 25 & -75 \\
 & & & & 80
\end{array}$$

On the right, the same division is shown written without the variables. This is why it is *essential* to use 0 as a placeholder in synthetic division. All the numbers in color on the right are repetitions of the numbers directly above them, so they may be omitted, as shown on the left below.

$$\begin{array}{r|rrrr}
 & & 3 & 9 & 25 \\ \hline
1 - 3 & 3 & 0 & -2 & 5 \\
 & & -9 & & \\
 & & 9 & -2 & \\
 & & & -27 & \\
 & & & 25 & 5 \\
 & & & & -75 \\
 & & & & 80
\end{array}
\qquad
\begin{array}{r|rrrr}
 & & 3 & 9 & 25 \\ \hline
1 - 3 & 3 & 0 & -2 & 5 \\
 & & -9 & & \\
 & & 9 & & \\
 & & & -27 & \\
 & & & 25 & \\
 & & & & -75 \\
 & & & & 80
\end{array}$$

The numbers in color on the left are again repetitions of the numbers directly above them; they too may be omitted, as shown on the right above.

Now the problem can be condensed. If the 3 in the dividend is brought down to the beginning of the bottom row, the top row can be omitted since it duplicates the bottom row.

$$\begin{array}{r|rrrr}
1 - 3 & 3 & 0 & -2 & 5 \\
 & & -9 & -27 & -75 \\ \hline
 & 3 & 9 & 25 & 80
\end{array}$$

The 1 at the upper left can be omitted, since it represents $1x$, which will *always* be the first term in the divisor. Also, to simplify the arithmetic, subtraction in the second row is replaced by addition. We compensate for this by changing the -3 at the upper left to its additive inverse, 3. The result of doing all this is shown below.

$$\begin{array}{r|rrrrl}
\text{Additive inverse} \longrightarrow 3 & 3 & 0 & -2 & 5 & \\
 & & 9 & 27 & 75 & \longleftarrow \text{Change signs.} \\ \hline
 & 3 & 9 & 25 & 80 & \longleftarrow \text{Remainder} \\
 & \downarrow & \downarrow & \downarrow & \downarrow &
\end{array}$$

The quotient is read from the bottom row.

$$3x^2 + 9x + 25 + \frac{80}{x - 3}$$

The first three numbers in the bottom row are the coefficients of the quotient polynomial with degree 1 less than the degree of the dividend. The last number gives the remainder.

1 Divide, using synthetic division.

(a) $\dfrac{3z^2 + 10z - 8}{z + 4}$

(b) $(2x^2 + 3x - 5) \div (x + 1)$

2 Divide, using synthetic division.

(a) $\dfrac{3a^3 - 2a + 21}{a + 2}$

(b) $(-4x^4 + 3x^3 + 18x + 2) \div (x - 2)$

ANSWERS

1. (a) $3z - 2$ (b) $2x + 1 + \dfrac{-6}{x + 1}$

2. (a) $3a^2 - 6a + 10 + \dfrac{1}{a + 2}$

(b) $-4x^3 - 5x^2 - 10x - 2 + \dfrac{-2}{x - 2}$

Synthetic Division

This shortcut method is called **synthetic division.** It is used *only* when dividing a polynomial by a binomial of the form $x - k$.

Example 1 Using Synthetic Division

Use synthetic division to divide $5x^2 + 16x + 15$ by $x + 2$.

As mentioned above, use synthetic division only when dividing by a polynomial of the form $x - k$. Get $x + 2$ in this form by writing it as

$$x + 2 = x - (-2),$$

where $k = -2$. Now write the coefficients of $5x^2 + 16x + 15$, placing -2 to the left.

$x + 2$ leads to -2. ⟶ $-2\overline{)5 \quad 16 \quad 15}$ ⟵ Coefficients

Bring down the 5, and multiply: $-2 \cdot 5 = -10$.

$$\begin{array}{r|rrr} -2 & 5 & 16 & 15 \\ & & -10 & \\ \hline & 5 & & \end{array}$$

Add 16 and -10, getting 6. Multiply 6 and -2 to get -12.

$$\begin{array}{r|rrr} -2 & 5 & 16 & 15 \\ & & -10 & -12 \\ \hline & 5 & 6 & \end{array}$$

Add 15 and -12, getting 3.

$$\begin{array}{r|rrr} -2 & 5 & 16 & 15 \\ & & -10 & -12 \\ \hline & 5 & 6 & 3 \end{array}$$

⟵ Remainder

Read the result from the bottom row.

$$\frac{5x^2 + 16x + 15}{x + 2} = 5x + 6 + \frac{3}{x + 2}$$

Work Problem 1 at the Side.

Example 2 Using Synthetic Division with a Missing Term

Use synthetic division to find $(-4x^5 + x^4 + 6x^3 + 2x^2 + 50) \div (x - 2)$.

Use the steps given above, inserting a 0 for the missing x-term.

$$\begin{array}{r|rrrrrr} 2 & -4 & 1 & 6 & 2 & 0 & 50 \\ & & -8 & -14 & -16 & -28 & -56 \\ \hline & -4 & -7 & -8 & -14 & -28 & -6 \end{array}$$

Read the result from the bottom row.

$$\frac{-4x^5 + x^4 + 6x^3 + 2x^2 + 50}{x - 2} = -4x^4 - 7x^3 - 8x^2 - 14x - 28 + \frac{-6}{x - 2}$$

Work Problem 2 at the Side.

2 Use the remainder theorem to evaluate a polynomial. We can use synthetic division to evaluate polynomials. For example, in the synthetic division of Example 2, where the polynomial was divided by $x - 2$, the remainder was -6.

Replacing x in the polynomial with 2 gives

$$\begin{aligned} -4x^5 + x^4 + 6x^3 + 2x^2 + 50 &= -4 \cdot 2^5 + 2^4 + 6 \cdot 2^3 + 2 \cdot 2^2 + 50 \\ &= -4 \cdot 32 + 16 + 6 \cdot 8 + 2 \cdot 4 + 50 \\ &= -128 + 16 + 48 + 8 + 50 \\ &= -6, \end{aligned}$$

the same number as the remainder; that is, dividing by $x - 2$ produced a remainder equal to the result when x is replaced with 2. This always happens, as the following remainder theorem states.

Remainder Theorem

If the polynomial $P(x)$ is divided by $x - k$, then the remainder is equal to $P(k)$.

This result is proved in more advanced courses.

Example 3 Using the Remainder Theorem

Let $P(x) = 2x^3 - 5x^2 - 3x + 11$. Find $P(-2)$.

Use the remainder theorem; divide $P(x)$ by $x - (-2)$.

$$\begin{array}{r|rrrr} \text{Value of } x \rightarrow -2 & 2 & -5 & -3 & 11 \\ & & -4 & 18 & -30 \\ \hline & 2 & -9 & 15 & -19 \end{array} \leftarrow \text{Remainder}$$

By this result, $P(-2) = -19$.

Work Problem 3 at the Side.

3 Decide whether a given number is a solution of an equation. The remainder theorem can also be used to show that a given number is a solution of an equation.

Example 4 Using the Remainder Theorem

Show that -5 is a solution of the equation

$$2x^4 + 12x^3 + 6x^2 - 5x + 75 = 0.$$

One way to show that -5 is a solution is to substitute -5 for x in the equation. However, an easier way is to use synthetic division and the remainder theorem.

$$\begin{array}{r|rrrrr} \text{Proposed solution} \rightarrow -5 & 2 & 12 & 6 & -5 & 75 \\ & & -10 & -10 & 20 & -75 \\ \hline & 2 & 2 & -4 & 15 & 0 \end{array} \leftarrow \text{Remainder}$$

Since the remainder is 0, the polynomial has a value of 0 when $x = -5$, so -5 is a solution of the given equation.

3 Let $P(x) = x^3 - 5x^2 + 7x - 3$. Use synthetic division to find each value.

(a) $P(1)$ (Divide by $x - 1$.)

(b) $P(-2)$

ANSWERS

3. **(a)** 0 **(b)** -45

4 Use synthetic division to decide whether 2 is a solution of each equation.

(a) $3x^3 - 11x^2 + 17x - 14 = 0$

(b) $4x^5 - 7x^4 - 11x^2 + 2x + 6 = 0$

ANSWERS

4. **(a)** yes **(b)** no

Work Problem 4 at the Side.

The synthetic division in Example 4 also shows that $x - (-5)$ divides the polynomial with 0 remainder. Thus $x - (-5) = x + 5$ is a *factor* of the polynomial and

$$2x^4 + 12x^3 + 6x^2 - 5x + 75 = (x + 5)(2x^3 + 2x^2 - 4x + 15).$$

The second factor is the quotient polynomial found in the last row of the synthetic division.

6.5 EXERCISES

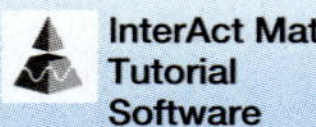

www.mathxl.com

Digital Video Tutor CD 4
Videotape 9

Choose the letter of the correct setup to perform synthetic division on the indicated quotient.

1. $\dfrac{x^2 + 3x - 6}{x - 2}$

A. $-2\overline{)1 \quad 3 \quad -6}$ **B.** $-2\overline{)-1 \quad -3 \quad 6}$

C. $2\overline{)1 \quad 3 \quad -6}$ **D.** $2\overline{)-1 \quad -3 \quad 6}$

2. $\dfrac{x^3 - 3x^2 + 2}{x - 1}$

A. $1\overline{)1 \quad -3 \quad 2}$ **B.** $-1\overline{)1 \quad -3 \quad 2}$

C. $1\overline{)1 \quad -3 \quad 0 \quad 2}$ **D.** $1\overline{)-1 \quad 3 \quad 0 \quad -2}$

Use synthetic division to find each quotient. See Examples 1 and 2.

3. $\dfrac{x^2 - 6x + 5}{x - 1}$

4. $\dfrac{x^2 - 4x - 21}{x + 3}$

5. $\dfrac{4m^2 + 19m - 5}{m + 5}$

6. $\dfrac{3k^2 - 5k - 12}{k - 3}$

7. $\dfrac{2a^2 + 8a + 13}{a + 2}$

8. $\dfrac{4y^2 - 5y - 20}{y - 4}$

9. $(p^2 - 3p + 5) \div (p + 1)$

10. $(z^2 + 4z - 6) \div (z - 5)$

11. $\dfrac{4a^3 - 3a^2 + 2a - 3}{a - 1}$

12. $\dfrac{5p^3 - 6p^2 + 3p + 14}{p + 1}$

13. $(x^5 - 2x^3 + 3x^2 - 4x - 2) \div (x - 2)$

14. $(2y^5 - 5y^4 - 3y^2 - 6y - 23) \div (y - 3)$

15. $(-4r^6 - 3r^5 - 3r^4 + 5r^3 - 6r^2 + 3r) \div (r - 1)$

16. $(-3t^5 + 2t^4 - 5t^3 + 6t^2 - 3t - 2) \div (t - 2)$

17. $(-3y^5 + 2y^4 - 5y^3 - 6y^2 - 1) \div (y + 2)$

18. $(m^6 + 2m^4 - 5m + 11) \div (m - 2)$

19. $\dfrac{y^3 + 1}{y - 1}$

20. $\dfrac{z^4 + 81}{z - 3}$

Use the remainder theorem to find P(k). See Example 3.

21. $P(x) = 2x^3 - 4x^2 + 5x - 3;\ k = 2$

22. $P(x) = x^3 + 3x^2 - x + 5;\ k = -1$

23. $P(r) = -r^3 - 5r^2 - 4r - 2;\ k = -4$

24. $P(z) = -z^3 + 5z^2 - 3z + 4;\ k = 3$

25. $P(x) = 2x^3 - 4x^2 + 5x - 33;\ k = 3$

26. $P(x) = x^3 - 3x^2 + 4x - 4;\ k = 2$

Use synthetic division to decide whether the given number is a solution of each equation. See Example 4.

27. $x^3 - 2x^2 - 3x + 10 = 0;\ x = -2$

28. $x^3 - 3x^2 - x + 10 = 0;\ x = -2$

29. $m^4 + 2m^3 - 3m^2 + 8m - 8 = 0;\ m = -2$

30. $r^4 - r^3 - 6r^2 + 5r + 10 = 0;\ r = -2$

31. $3x^3 + 2x^2 - 2x + 11 = 0;\ x = -2$

32. $3z^3 + 10z^2 + 3z - 9 = 0;\ z = -2$

33. Explain why it is important to insert 0s as placeholders for missing terms before performing synthetic division.

34. Explain why a 0 remainder in synthetic division of $P(x)$ by k indicates that k is a solution of the equation $P(x) = 0$.

6.6 GREATEST COMMON FACTORS; FACTORING BY GROUPING

OBJECTIVES

1. Factor out the greatest common factor.
2. Factor by grouping.

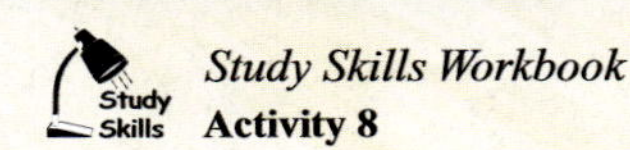

Study Skills Workbook
Activity 8

Writing a polynomial as the product of two or more simpler polynomials is called **factoring** the polynomial. For example, the product of $3x$ and $5x - 2$ is $15x^2 - 6x$, and $15x^2 - 6x$ can be factored as the product $3x(5x - 2)$.

$$3x(5x - 2) = 15x^2 - 6x \quad \text{Multiplying}$$
$$15x^2 - 6x = 3x(5x - 2) \quad \text{Factoring}$$

Notice that both multiplying and factoring use the distributive property, but in opposite directions. Factoring "undoes" or reverses multiplying.

1 Factor out the greatest common factor. The first step in factoring a polynomial is to find the *greatest common factor* for the terms of the polynomial. The **greatest common factor (GCF)** is the largest term that is a factor of all terms in the polynomial. For example, the greatest common factor for $8x + 12$ is 4, since 4 is the largest term that is a factor of both $8x$ and 12. Using the distributive property,

$$8x + 12 = 4(2x) + 4(3) = 4(2x + 3).$$

As a check, multiply 4 and $2x + 3$. The result should be $8x + 12$. Using the distributive property this way is called *factoring out the greatest common factor.*

Example 1 Factoring Out the Greatest Common Factor

Factor out the greatest common factor.

(a) $9z - 18$
Since 9 is the GCF, factor 9 from each term.

$$9z - 18 = 9 \cdot z - 9 \cdot 2 = 9(z - 2)$$

(b) $56m + 35p = 7(8m + 5p)$

(c) $2y + 5$ There is no common factor other than 1.

(d) $12 + 24z = 12 \cdot 1 + 12 \cdot 2z$
$= 12(1 + 2z)$ 12 is the GCF.

CAUTION

In Example 1(d), remember to write the factor 1. Always check answers by multiplying.

Work Problem 1 at the Side.

1 Factor out the greatest common factor.

(a) $7k + 28$

(b) $32m + 24$

(c) $8a - 9$

(d) $5z + 5$

Example 2 Factoring Out the Greatest Common Factor

Factor out the greatest common factor.

(a) $9x^2 + 12x^3$
The numerical part of the GCF is 3, the largest number that divides into both 9 and 12. For the variable parts, x^2 and x^3, use the least exponent that appears on x; here the least exponent is 2. The GCF is $3x^2$.

$$9x^2 + 12x^3 = 3x^2(3) + 3x^2(4x)$$
$$= 3x^2(3 + 4x)$$

Continued on Next Page

ANSWERS
1. (a) $7(k + 4)$ (b) $8(4m + 3)$
(c) cannot be factored (d) $5(z + 1)$

2 Factor out the greatest common factor.

(a) $16y^4 + 8y^3$

(b) $14p^2 - 9p^3 + 6p^4$

(c) $15z^2 + 45z^5 - 60z^6$

(d) $4x^2z - 2xz + 8z^2$

(e) $12y^5x^2 + 8y^3x^3$

(f) $5m^4x^3 + 15m^5x^6 - 20m^4x^6$

3 Factor out the greatest common factor.

(a) $(a + 2)(a - 3) + (a + 2)(a + 6)$

(b) $(y - 1)(y + 3) - (y - 1)(y + 4)$

(c) $k^2(a + 5b) + m^2(a + 5b)$

(d) $r^2(y + 6) + r^2(y + 3)$

ANSWERS

2. **(a)** $8y^3(2y + 1)$ **(b)** $p^2(14 - 9p + 6p^2)$
(c) $15z^2(1 + 3z^3 - 4z^4)$
(d) $2z(2x^2 - x + 4z)$ **(e)** $4y^3x^2(3y^2 + 2x)$
(f) $5m^4x^3(1 + 3mx^3 - 4x^3)$

3. **(a)** $(a + 2)(2a + 3)$
(b) $(y - 1)(-1)$, or $-y + 1$
(c) $(a + 5b)(k^2 + m^2)$ **(d)** $r^2(2y + 9)$

(b) $32p^4 - 24p^3 + 40p^5 = 8p^3(4p) + 8p^3(-3) + 8p^3(5p^2)$ GCF $= 8p^3$

$$= 8p^3(4p - 3 + 5p^2)$$

(c) $3k^4 - 15k^7 + 24k^9 = 3k^4(1 - 5k^3 + 8k^5)$

(d) $24m^3n^2 - 18m^2n + 6m^4n^3$

The numerical part of the GCF is 6. Here 2 is the least exponent that appears on m, while 1 is the least exponent on n. The GCF is $6m^2n$.

$$24m^3n^2 - 18m^2n + 6m^4n^3 = 6m^2n(4mn) + 6m^2n(-3) + 6m^2n(m^2n^2)$$
$$= 6m^2n(4mn - 3 + m^2n^2)$$

(e) $25x^2y^3 + 30y^5 - 15x^4y^7 = 5y^3(5x^2 + 6y^2 - 3x^4y^4)$

Work Problem 2 at the Side.

A greatest common factor need not be a monomial. The next example shows a binomial greatest common factor.

Example 3 Factoring Out a Binomial Factor

Factor out the greatest common factor.

(a) $(x - 5)(x + 6) + (x - 5)(2x + 5)$

The greatest common factor here is $x - 5$.

$$(x - 5)(x + 6) + (x - 5)(2x + 5) = (x - 5)[(x + 6) + (2x + 5)]$$
$$= (x - 5)(x + 6 + 2x + 5)$$
$$= (x - 5)(3x + 11)$$

(b) $z^2(m + n) + x^2(m + n) = (m + n)(z^2 + x^2)$

(c) $p(r + 2s) - q^2(r + 2s) = (r + 2s)(p - q^2)$

(d) $(p - 5)(p + 2) - (p - 5)(3p + 4)$

$$= (p - 5)[(p + 2) - (3p + 4)]$$
$$= (p - 5)[p + 2 - 3p - 4] \quad \text{Be careful with signs.}$$
$$= (p - 5)[-2p - 2]$$
$$= (p - 5)[-2(p + 1)] \text{ or } -2(p - 5)(p + 1) \quad \text{Look for a common factor.}$$

Work Problem 3 at the Side.

When the coefficient of the term of greatest degree is negative, it is sometimes preferable to factor out the -1 that is understood along with the GCF.

Example 4 Factoring Out a Negative Common Factor

Factor $-a^3 + 3a^2 - 5a$ in two ways.

First, a could be used as the common factor, giving

$$-a^3 + 3a^2 - 5a = a(-a^2) + a(3a) + a(-5)$$
$$= a(-a^2 + 3a - 5).$$

Because of the leading negative sign, $-a$ could be used as the common factor.

$$-a^3 + 3a^2 - 5a = -a(a^2) + (-a)(-3a) + (-a)(5)$$
$$= -a(a^2 - 3a + 5)$$

Either answer is correct.

NOTE

Example 4 showed two ways of factoring a polynomial. Sometimes there may be a reason to prefer one of these forms over the other, but both are correct. The answer section in this book will *usually* give the form where the common factor has a positive coefficient.

Work Problem 4 at the Side.

2 Factor by grouping. Sometimes a polynomial has a greatest common factor of 1, but it still may be possible to factor the polynomial by using a process called *factoring by grouping.* We usually factor by grouping when a polynomial has more than three terms. For example, to factor the polynomial

$$ax - ay + bx - by,$$

group the terms as follows.

Terms with common factors

$$(ax - ay) + (bx - by)$$

Then factor $ax - ay$ as $a(x - y)$ and factor $bx - by$ as $b(x - y)$ to get

$$(ax - ay) + (bx - by) = a(x - y) + b(x - y).$$

On the right, the common factor is $x - y$. The final factored form is

$$ax - ay + bx - by = (x - y)(a + b).$$

Example 5 Factoring by Grouping

Factor $3x - 3y - ax + ay$.

Grouping terms gives

$$(3x - 3y) + (-ax + ay) = 3(x - y) + a(-x + y).$$

There is no simple common factor here. However, if we factor out $-a$ instead of a in the second group of terms, we get

$$3(x - y) - a(x - y),$$

which equals

$$(x - y)(3 - a).$$

Check by multiplying.

NOTE

In Example 5, different grouping would lead to the product

$$(a - 3)(y - x).$$

Verify by multiplying that this is also correct.

4 Factor each polynomial in two ways.

(a) $-k^2 + 3k$

(b) $-6r^2 + 5r$

ANSWERS

4. **(a)** $k(-k + 3)$ or $-k(k - 3)$
(b) $r(-6r + 5)$ or $-r(6r - 5)$

5 Factor $xy + 2y - 4x - 8$.

6 Factor.

(a) $mn + 6 + 2n + 3m$

(b) $4y - zx + yx - 4z$

ANSWERS
5. $(x + 2)(y - 4)$
6. (a) $(m + 2)(n + 3)$ (b) $(y - z)(x + 4)$

The steps used in factoring by grouping are listed here.

Factoring by Grouping

Step 1 **Group terms.** Collect the terms into groups so that each group has a common factor.

Step 2 **Factor within the groups.** Factor out the common factor in each group.

Step 3 **Factor the entire polynomial.** If each group now has a common factor, factor it out. If not, try a different grouping.

Example 6 Factoring by Grouping

Factor $6ax + 12bx + a + 2b$ by grouping.

$$6ax + 12bx + a + 2b = (6ax + 12bx) + (a + 2b) \quad \text{Group terms.}$$

Now factor $6x$ from the first group, and use the identity property of multiplication to introduce the factor 1 in the second group.

$$(6ax + 12bx) + (a + 2b) = 6x(a + 2b) + 1(a + 2b)$$
$$= (a + 2b)(6x + 1) \quad \text{Factor out } a + 2b.$$

Again, as in Example 1(d), remember to write the 1.

Work Problem 5 at the Side.

Example 7 Rearranging Terms before Factoring by Grouping

Factor $p^2q^2 - 10 - 2q^2 + 5p^2$.

Neither the first two terms nor the last two terms have a common factor except 1. Rearrange and group the terms as follows.

$$(p^2q^2 - 2q^2) + (5p^2 - 10) \quad \text{Rearrange and group the terms.}$$
$$= q^2(p^2 - 2) + 5(p^2 - 2) \quad \text{Factor out the common factors.}$$
$$= (p^2 - 2)(q^2 + 5) \quad \text{Factor out } p^2 - 2.$$

CAUTION

In Example 7, do not stop at the step

$$q^2(p^2 - 2) + 5(p^2 - 2).$$

This expression is *not in factored form* because it is a *sum* of two terms, $q^2(p^2 - 2)$ and $5(p^2 - 2)$, not a product.

Work Problem 6 at the Side.

6.6 EXERCISES

MyMathLab.com

1. Explain in your own words what it means to factor a polynomial.

2. What is the first step in attempting to factor a polynomial?

3. What is the GCF of the following terms?
$7z^2(m + n)^4$, $9z^3(m + n)^5$

4. Which one of the following is an example of a polynomial in factored form?

A. $3x^2y^3 + 6x^2(2x + y)$

B. $5(x + y)^2 - 10(x + y)^3$

C. $(-2 + 3x)(5y^2 + 4y + 3)$

D. $(3x + 4)(5x - y) - (3x + 4)(2x - 1)$

Find the greatest common factor for each list of terms.

5. $9m^3$, $3m^2$, $15m$

6. $4a^2$, $6ab$, $2a^3$

7. $6m(r + t)^2$, $3p(r + t)^4$

8. $7z^2(m + n)^4$, $9z^3(m + n)^5$

9. Which one of the following has the greatest common factor of $6x^3y^4 - 12x^5y^2 + 24x^4y^8$ as one of the factors?

A. $6x^3y^2(y^2 - 2x^2 + 4xy^6)$

B. $6xy(x^2y^3 - 2x^4y + 4x^3y^7)$

C. $2x^3y^2(3y^2 - 6x^2 + 12xy^6)$

D. $6x^2y^2(xy^2 - 2x^3 + 4x^2y^6)$

10. When directed to factor the polynomial $4x^2y^5 - 8xy^3$ completely, a student responded with $2xy^3(2xy^2 - 4)$. When the teacher did not give him full credit, he complained because when his factors are multiplied, the product is the original polynomial. Was the teacher justified in her grading? Why or why not?

Factor out the greatest common factor. See Examples 1–4.

11. $8k^3 + 24k$

12. $9z^4 + 27z$

13. $3xy - 5xy^2$

14. $5h^2j + 7hj$

15. $-4p^3q^4 - 2p^2q^5$

16. $-3z^5w^2 - 18z^3w^4$

17. $21x^5 + 35x^4 - 14x^3$

18. $18k^3 - 36k^4 + 48k^5$

19. $15a^2c^3 - 25ac^2 + 5a^2c$

20. $15y^3z^3 + 27y^2z^4 - 36yz^5$

21. $-27m^3p^5 + 5r^4s^3 - 8x^5z^4$

22. $-50r^4t^2 + 81x^3y^3 - 49p^2q^4$

23. $(m-4)(m+2)+(m-4)(m+3)$

24. $(z-5)(z+7)+(z-5)(z+9)$

25. $(2z-1)(z+6)-(2z-1)(z-5)$

26. $(3x+2)(x-4)-(3x+2)(x+8)$

27. $5(2-x)^2-(2-x)^3+4(2-x)$

28. $3(5-x)^4+2(5-x)^3-(5-x)^2$

Factor each polynomial twice. First use a common factor with a positive coefficient, and then use a common factor with a negative coefficient. See Example 4.

29. $-r^3+3r^2+5r$

30. $-t^4+8t^3-12t$

31. $-12s^5+48s^4$

32. $-16y^4+64y^3$

33. $-2x^5+6x^3+4x^2$

34. $-5a^3+10a^4-15a^5$

Factor by grouping. See Examples 5–7.

35. $mx+3qx+my+3qy$

36. $2k+2h+jk+jh$

37. $10m+2n+5mk+nk$

38. $3ma+3mb+2ab+2b^2$

39. $m^2-3m-15+5m$

40. $z^2-6z-54+9z$

41. $p^2-4zq+pq-4pz$

42. $r^2-9tw+3rw-3rt$

43. $3a^2+15a-10-2a$

44. $7k+2k^2-6k-21$

45. $-15p^2+5pq-6pq+2q^2$

46. $-6r^2+9rs+8rs-12s^2$

47. $-3a^3-3ab^2+2a^2b+2b^3$

48. $-16m^3+4m^2p^2-4mp+p^3$

49. $4+xy-2y-2x$

50. $2ab^2-4-8b^2+a$

51. $8+9y^4-6y^3-12y$

52. $x^3y^2-3-3y^2+x^3$

Factor out the variable that is raised to the smaller exponent. (For example, in Exercise 53, factor out m^{-5}.)

53. $3m^{-5}+m^{-3}$

54. $k^{-2}+2k^{-4}$

55. $3p^{-3}+2p^{-2}$

6.7 Factoring Trinomials

Objectives

1. Factor trinomials when the coefficient of the squared term is 1.
2. Factor trinomials when the coefficient of the squared term is not 1.
3. Use an alternative method for factoring trinomials.
4. Factor by substitution.

1 Factor trinomials when the coefficient of the squared term is 1. We begin by finding the product of $x + 3$ and $x - 5$.

$$\begin{aligned}(x + 3)(x - 5) &= x^2 - 5x + 3x - 15 \\ &= x^2 - 2x - 15\end{aligned}$$

By this result, the factored form of $x^2 - 2x - 15$ is $(x + 3)(x - 5)$.

Multiplying →

Factored form → $(x + 3)(x - 5) = x^2 - 2x - 15$ ← Product

← Factoring

Since multiplying and factoring are operations that "undo" each other, factoring trinomials involves using FOIL backwards. As shown here, the x^2-term comes from multiplying x and x, and -15 comes from multiplying 3 and -5.

Product of x and x is x^2.

$$(x + 3)(x - 5) = x^2 - 2x - 15$$

Product of 3 and -5 is -15.

We find the $-2x$ in $x^2 - 2x - 15$ by multiplying the outer terms, and then the inner terms, and adding.

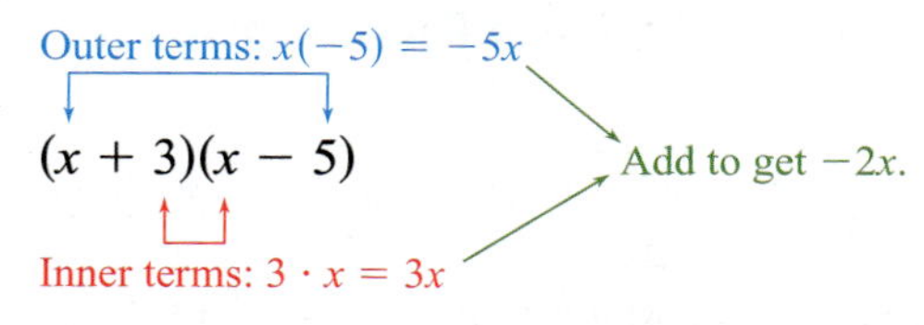

Based on this example, follow these steps to factor a trinomial $x^2 + bx + c$, with 1 as the coefficient of the squared term.

Factoring $x^2 + bx + c$

Step 1 **Find pairs whose product is *c*.** Find all pairs of integers whose product is the third term of the trinomial, c.

Step 2 **Find pairs whose sum is *b*.** Choose the pair whose sum is the coefficient of the middle term, b.

If there are no such integers, the polynomial cannot be factored. A polynomial that cannot be factored with integer coefficients is **prime**.

1 Factor each polynomial.

(a) $p^2 + 6p + 5$

(b) $a^2 + 9a + 20$

(c) $k^2 - k - 6$

(d) $b^2 - 7b + 10$

(e) $y^2 - 8y + 6$

2 Factor each polynomial.

(a) $m^2 + 2mn - 8n^2$

(b) $z^2 - 7zx + 9x^2$

ANSWERS

1. **(a)** $(p + 1)(p + 5)$ **(b)** $(a + 5)(a + 4)$ **(c)** $(k - 3)(k + 2)$ **(d)** $(b - 5)(b - 2)$ **(e)** prime
2. **(a)** $(m - 2n)(m + 4n)$ **(b)** prime

Example 1 Factoring Trinomials in $x^2 + bx + c$ Form

Factor each polynomial.

(a) $y^2 + 2y - 35$

Step 1 Find pairs of numbers whose product is -35.

Step 2 Write sums of those numbers.

Step 1	Step 2
$-35(1)$	$-35 + 1 = -34$
$35(-1)$	$35 + (-1) = 34$
$7(-5)$	$7 + (-5) = 2$ ← Coefficient of the middle term
$5(-7)$	$5 + (-7) = -2$

The required numbers are 7 and -5, so

$$y^2 + 2y - 35 = (y + 7)(y - 5).$$

Check by finding the product of $y + 7$ and $y - 5$.

(b) $r^2 + 8r + 12$

Look for two numbers with a product of 12 and a sum of 8. Of all pairs of numbers having a product of 12, only the pair 6 and 2 has a sum of 8. Therefore,

$$r^2 + 8r + 12 = (r + 6)(r + 2).$$

Because of the commutative property, it would be equally correct to write $(r + 2)(r + 6)$. Check by multiplying.

Example 2 Recognizing a Prime Polynomial

Factor $m^2 + 6m + 7$.

Look for two numbers whose product is 7 and whose sum is 6. Only two pairs of integers, 7 and 1 and -7 and -1, give a product of 7. Neither of these pairs has a sum of 6, so $m^2 + 6m + 7$ cannot be factored with integer coefficients and is prime.

Work Problem 1 at the Side.

Factoring a trinomial that has more than one variable uses a similar process.

Example 3 Factoring a Trinomial in Two Variables

Factor $p^2 + 6ap - 16a^2$.

Look for two expressions whose product is $-16a^2$ and whose sum is $6a$. The quantities $8a$ and $-2a$ have the necessary product and sum, so

$$p^2 + 6ap - 16a^2 = (p + 8a)(p - 2a).$$

Work Problem 2 at the Side.

Sometimes a trinomial will have a common factor that should be factored out first.

Example 4 Factoring a Trinomial with a Common Factor

Factor $16y^3 - 32y^2 - 48y$.

Start by factoring out the greatest common factor, $16y$.

$$16y^3 - 32y^2 - 48y = 16y(y^2 - 2y - 3)$$

Continued on Next Page

To factor $y^2 - 2y - 3$, look for two integers whose product is -3 and whose sum is -2. The necessary integers are -3 and 1, so

$$16y^3 - 32y^2 - 48y = 16y(y - 3)(y + 1).$$

CAUTION

When factoring, always look for a common factor first. Remember to write the common factor as part of the answer.

Work Problem 3 at the Side.

3 Factor $5m^4 - 5m^3 - 100m^2$.

2 **Factor trinomials when the coefficient of the squared term is not 1.** We can use a generalization of the method shown in Objective 1 to factor a trinomial of the form $ax^2 + bx + c$, where $a \neq 1$. To factor $3x^2 + 7x + 2$, for example, we first identify the values of a, b, and c.

$$\begin{array}{ccc} ax^2 & +\, bx & +\, c \\ \downarrow & \downarrow & \downarrow \\ 3x^2 & +\, 7x & +\, 2 \end{array}$$

$$a = 3, \quad b = 7, \quad c = 2$$

The product ac is $3 \cdot 2 = 6$, so we must find integers having a product of 6 and a sum of 7 (since the middle term has coefficient 7). The necessary integers are 1 and 6, so we write $7x$ as $1x + 6x$, or $x + 6x$, giving

$$3x^2 + 7x + 2 = 3x^2 + \underbrace{x + 6x}_{x + 6x = 7x} + 2.$$

Now we factor by grouping.

$$\begin{aligned} (3x^2 + x) + (6x + 2) &= x(3x + 1) + 2(3x + 1) \\ &= (3x + 1)(x + 2) \end{aligned}$$

4 Factor each trinomial.

(a) $3y^2 - 11y - 4$

(b) $6k^2 - 19k + 10$

Example 5 Factoring a Trinomial in $ax^2 + bx + c$ Form

Factor $12r^2 - 5r - 2$.

Since $a = 12$, $b = -5$, and $c = -2$, the product ac is -24. The two integers whose product is -24 and whose sum is -5 are 3 and -8.

$12r^2 - 5r - 2 = 12r^2 + 3r - 8r - 2$	Write $-5r$ as $3r - 8r$.
$= 3r(4r + 1) - 2(4r + 1)$	Factor by grouping.
$= (4r + 1)(3r - 2)$	Factor out the common factor.

Work Problem 4 at the Side.

3 **Use an alternative method for factoring trinomials.** An alternative approach, the method of trying repeated combinations and using FOIL, is especially helpful when the product ac is large. This method is shown using the two trinomials we just factored.

Example 6 Factoring Trinomials in $ax^2 + bx + c$ Form

Factor each polynomial.

(a) $3x^2 + 7x + 2$

To factor this trinomial, find the correct numbers to put in the blanks.

$$3x^2 + 7x + 2 = (\_\_\_x + \_\_\_)(\_\_\_x + \_\_\_)$$

Continued on Next Page

ANSWERS

3. $5m^2(m - 5)(m + 4)$

4. (a) $(y - 4)(3y + 1)$ **(b)** $(2k - 5)(3k - 2)$

5 Use the method of Example 6 to factor each trinomial.

(a) $10x^2 + 17x + 3$

(b) $16y^2 - 34y - 15$

(c) $8t^2 - 13t + 5$

Addition signs are used since all the signs in the trinomial indicate addition. The first two expressions have a product of $3x^2$, so they must be $3x$ and x.

$$3x^2 + 7x + 2 = (3x + \_\_\_\_)(x + \_\_\_\_)$$

The product of the two last terms must be 2, so the numbers must be 2 and 1. There is a choice. The 2 could be used with the $3x$ or with the x. Only one of these choices can give the correct middle term, $7x$. Use the FOIL method to try each one.

$(3x + 2)(x + 1)$: outer $3x$, inner $2x$

$3x + 2x = 5x$
Wrong middle term

$(3x + 1)(x + 2)$: outer $6x$, inner x

$6x + x = 7x$
Correct middle term

Therefore, $3x^2 + 7x + 2 = (3x + 1)(x + 2)$.

(b) $12r^2 - 5r - 2$

To reduce the number of trials, we note that the trinomial has no common factor (except 1). This means that neither of its factors can have a common factor. We should keep this in mind as we choose factors. We try 4 and 3 for the two first terms.

$$12r^2 - 5r - 2 = (4r\_\_\_\_)(3r\_\_\_\_)$$

We do not know what signs to use yet. The factors of -2 are -2 and 1 or 2 and -1. Try both possibilities.

$(4r - 2)(3r + 1)$
Wrong: $4r - 2$ has a common factor of 2.

$(4r - 1)(3r + 2)$: outer $8r$, inner $-3r$

$8r - 3r = 5r$
Wrong middle term

The middle term on the right is $5r$, instead of the $-5r$ that is needed. We get $-5r$ by interchanging the signs in the factors.

$(4r + 1)(3r - 2)$: outer $-8r$, inner $3r$

$-8r + 3r = -5r$
Correct middle term

Thus, $12r^2 - 5r - 2 = (4r + 1)(3r - 2)$.

NOTE

As shown in Example 6(b), if the terms of a polynomial have no common factor (except 1), then none of the terms of its factors can have a common factor. Remembering this will eliminate some potential factors.

Work Problem 5 at the Side.

ANSWERS

5. (a) $(5x + 1)(2x + 3)$ (b) $(8y + 3)(2y - 5)$ (c) $(8t - 5)(t - 1)$

This alternative method of factoring a trinomial $ax^2 + bx + c$, $a \neq 1$, is summarized here.

Factoring $ax^2 + bx + c$

Step 1 **Find pairs whose product is *a*.** Write all pairs of integer factors of the coefficient of the squared term, a.

Step 2 **Find pairs whose product is *c*.** Write all pairs of integer factors of the last term, c.

Step 3 **Choose inner and outer terms.** Use FOIL and various combinations of the factors from Steps 1 and 2 until the necessary middle term is found.

If no such combinations exist, the trinomial is prime.

Example 7 Factoring a Trinomial in Two Variables

Factor $18m^2 - 19mx - 12x^2$.

There is no common factor (except 1). Follow the steps to factor the trinomial. There are many possible factors of both 18 and -12. Try 6 and 3 for 18 and -3 and 4 for -12.

$(6m - 3x)(3m + 4x)$ Wrong: common factor

$(6m + 4x)(3m - 3x)$ Wrong: common factors

Since 6 and 3 do not work in this situation, try 9 and 2 instead, with -4 and 3 as factors of -12.

$(9m + 3x)(2m - 4x)$ Wrong: common factors

$(9m - 4x)(2m + 3x)$ (outer: $27mx$; inner: $-8mx$)

$27mx + (-8mx) = 19mx$

The result on the right differs from the correct middle term only in sign, so interchange the signs in the factors. Check by multiplying.

$$18m^2 - 19mx - 12x^2 = (9m + 4x)(2m - 3x)$$

Work Problem 6 at the Side.

Example 8 Factoring $ax^2 + bx + c$, $a < 0$

Factor $-3x^2 + 16x + 12$.

While it is possible to factor this trinomial directly, it is helpful to first factor out -1. Then proceed as in the earlier examples.

$$\begin{aligned} -3x^2 + 16x + 12 &= -1(3x^2 - 16x - 12) \\ &= -1(3x + 2)(x - 6) \\ &= -(3x + 2)(x - 6) \end{aligned}$$

This factored form can be written in other ways. Two of them are

$$(-3x - 2)(x - 6) \quad \text{and} \quad (3x + 2)(-x + 6).$$

Verify that these both give the original trinomial when multiplied.

Work Problem 7 at the Side.

6 Factor each trinomial.

(a) $7p^2 + 15pq + 2q^2$

(b) $6m^2 + 7mn - 5n^2$

(c) $12z^2 - 5zy - 2y^2$

(d) $8m^2 + 18mx - 5x^2$

7 Factor each trinomial.

(a) $-6r^2 + 13r + 5$

(b) $-8x^2 + 10x - 3$

ANSWERS

6. (a) $(7p + q)(p + 2q)$
(b) $(3m + 5n)(2m - n)$
(c) $(3z - 2y)(4z + y)$
(d) $(4m - x)(2m + 5x)$

7. (a) $-(2r - 5)(3r + 1)$
(b) $-(4x - 3)(2x - 1)$

8 Factor each trinomial.

(a) $2m^3 - 4m^2 - 6m$

(b) $12r^4 + 6r^3 - 90r^2$

(c) $30y^5 - 55y^4 - 50y^3$

9 Factor each polynomial.

(a) $6(a - 1)^2 + (a - 1) - 2$

(b) $8(z + 5)^2 - 2(z + 5) - 3$

(c) $15(m - 4)^2 - 11(m - 4) + 2$

10 Factor each trinomial.

(a) $y^4 + y^2 - 6$

(b) $2p^4 + 7p^2 - 15$

(c) $6r^4 - 13r^2 + 5$

ANSWERS

8. **(a)** $2m(m + 1)(m - 3)$
(b) $6r^2(r + 3)(2r - 5)$
(c) $5y^3(2y - 5)(3y + 2)$

9. **(a)** $(2a - 3)(3a - 1)$
(b) $(4z + 17)(2z + 11)$
(c) $(3m - 13)(5m - 22)$

10. **(a)** $(y^2 - 2)(y^2 + 3)$
(b) $(2p^2 - 3)(p^2 + 5)$
(c) $(3r^2 - 5)(2r^2 - 1)$

Example 9 **Factoring a Trinomial with a Common Factor**

Factor $16y^3 + 24y^2 - 16y$.

$$16y^3 + 24y^2 - 16y = 8y(2y^2 + 3y - 2) \quad \text{GCF} = 8y$$

$$= 8y(2y - 1)(y + 2) \quad \text{Remember the common factor.}$$

Work Problem 8 at the Side.

4 **Factor by substitution.** Sometimes we can factor a more complicated polynomial by making a substitution of one variable for an expression.

Example 10 **Factoring a Polynomial Using Substitution**

Factor $2(x + 3)^2 + 5(x + 3) - 12$.

Since the binomial $x + 3$ appears to powers 2 and 1, we let the substitution variable represent $x + 3$. We may choose any letter we wish except x. We choose y to equal $x + 3$.

$$2(x + 3)^2 + 5(x + 3) - 12 = 2y^2 + 5y - 12 \quad \text{Let } y = x + 3.$$

$$= (2y - 3)(y + 4) \quad \text{Factor.}$$

Now we replace y with $x + 3$ to get

$$2(x + 3)^2 + 5(x + 3) - 12 = [2(x + 3) - 3][(x + 3) + 4]$$

$$= (2x + 6 - 3)(x + 7)$$

$$= (2x + 3)(x + 7).$$

CAUTION

Remember to make the final substitution of $x + 3$ for y in Example 10.

Work Problem 9 at the Side.

Example 11 **Factoring a Trinomial in $ax^4 + bx^2 + c$ Form**

Factor $6y^4 + 7y^2 - 20$.

The variable y appears to powers in which the larger exponent is twice the smaller exponent. In a case such as this, let a substitution variable equal the smaller power. Here, let $m = y^2$. Since $y^4 = (y^2)^2 = m^2$, the given trinomial becomes

$$6m^2 + 7m - 20,$$

which is factored as

$$6m^2 + 7m - 20 = (3m - 4)(2m + 5).$$

Since $m = y^2$,

$$6y^4 + 7y^2 - 20 = (3y^2 - 4)(2y^2 + 5).$$

NOTE

Some students feel comfortable enough about factoring to factor polynomials like the one in Example 11 directly, without using the substitution method.

Work Problem 10 at the Side.

6.7 EXERCISES

FOR EXTRA HELP

 Student's Solutions Manual

 MyMathLab.com

 InterAct Math Tutorial Software

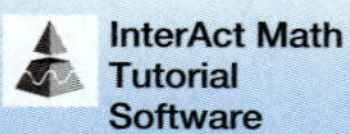 AW Math Tutor Center

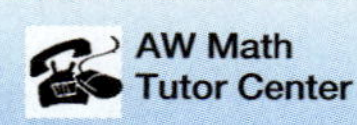 www.mathxl.com

 Digital Video Tutor CD 5 Videotape 10

1. Which one of the following is *not* a valid way of starting the process of factoring $12x^2 + 29x + 10$?

A. $(12x \quad)(x \quad)$ **B.** $(4x \quad)(3x \quad)$

C. $(6x \quad)(2x \quad)$ **D.** $(8x \quad)(4x \quad)$

2. Which one of the following is the completely factored form of $2x^6 - 5x^5 - 3x^4$?

A. $x^4(2x + 1)(x - 3)$ **B.** $x^4(2x - 1)(x + 3)$

C. $(2x^5 + x^4)(x - 3)$ **D.** $x^3(2x^2 + x)(x - 3)$

3. Which one of the following is *not* a factored form of $-x^2 + 16x - 60$?

A. $(x - 10)(-x + 6)$ **B.** $(-x - 10)(x + 6)$

C. $(-x + 10)(x - 6)$ **D.** $-1(x - 10)(x - 6)$

4. Which one of the following is the completely factored form of $4x^2 - 4x - 24$?

A. $4(x - 2)(x + 3)$ **B.** $4(x + 2)(x + 3)$

C. $4(x + 2)(x - 3)$ **D.** $4(x - 2)(x - 3)$

Factor each trinomial. See Examples 1–9.

5. $y^2 + 7y - 30$

6. $z^2 + 2z - 24$

7. $p^2 - p - 56$

8. $k^2 - 11k + 30$

9. $-m^2 + 16m - 60$

10. $-p^2 + 6p + 27$

11. $a^2 - 2ab - 35b^2$

12. $z^2 + 8zw + 15w^2$

13. $y^2 - 3yq - 15q^2$

14. $k^2 - 11hk + 28h^2$

15. $x^2y^2 + 11xy + 18$

16. $p^2q^2 - 5pq - 18$

17. $-6m^2 - 13m + 15$

18. $-15y^2 + 17y + 18$

19. $10x^2 + 3x - 18$

20. $8k^2 + 34k + 35$

21. $20k^2 + 47k + 24$

22. $27z^2 + 42z - 5$

23. $15a^2 - 22ab + 8b^2$

24. $15p^2 + 24pq + 8q^2$

25. $36m^2 - 60m + 25$

26. $25r^2 - 90r + 81$

27. $40x^2 + xy + 6y^2$

28. $14c^2 - 17cd - 6d^2$

29. $6x^2z^2 + 5xz - 4$

30. $8m^2n^2 - 10mn + 3$

31. $24x^2 + 42x + 15$

32. $36x^2 + 18x - 4$

33. $-15a^2 - 70a + 120$

34. $-12a^2 - 10a + 42$

35. $-11x^3 + 110x^2 - 264x$

36. $-9k^3 - 36k^2 + 189k$

37. $2x^3y^3 - 48x^2y^4 + 288xy^5$

38. $6m^3n^2 - 24m^2n^3 - 30mn^4$

Factor each trinomial. See Example 10.

39. $10(k + 1)^2 - 7(k + 1) + 1$

40. $4(m - 5)^2 - 4(m - 5) - 15$

41. $3(m + p)^2 - 7(m + p) - 20$

42. $4(x - y)^2 - 23(x - y) - 6$

43. $a^2(a + b)^2 - ab(a + b)^2 - 6b^2(a + b)^2$

44. $m^2(m - p) + mp(m - p) - 2p^2(m - p)$

Factor each trinomial. See Example 11.

45. $2x^4 - 9x^2 - 18$

46. $6z^4 + z^2 - 1$

47. $16x^4 + 16x^2 + 3$

48. $9r^4 + 9r^2 + 2$

49. $12p^6 - 32p^3r + 5r^2$

50. $2y^6 + 7xy^3 + 6x^2$

RELATING CONCEPTS (Exercises 51–56) FOR INDIVIDUAL OR GROUP WORK

Refer to the note following Example 6 in this section. Then ***work Exercises 51–56 in order.***

51. Is 2 a factor of the composite number 45?

52. List all positive integer factors of 45. Is 2 a factor of any of these factors?

53. Is 5 a factor of $10x^2 + 29x + 10$?

54. Factor $10x^2 + 29x + 10$. Is 5 a factor of either of its factors?

55. Suppose that k is an odd integer and you are asked to factor $2x^2 + kx + 8$. Why is $2x + 4$ not a possible choice in factoring this polynomial?

56. The polynomial $12y^2 - 11y - 15$ can be factored using the methods of this section. Explain why $3y + 15$ cannot be one of its factors.

6.8 SPECIAL FACTORING

1 Factor a difference of squares. The special products introduced in Section 6.3 are used in reverse when factoring. Recall that the product of the sum and difference of two terms leads to a **difference of squares,** a pattern that occurs often when factoring.

Difference of Squares

$$x^2 - y^2 = (x + y)(x - y)$$

Example 1 Factoring Differences of Squares

Factor each polynomial.

(a) $4a^2 - 64$

There is a common factor of 4.

$$4a^2 - 64 = 4(a^2 - 16) \quad \text{Factor out the common factor.}$$
$$= 4(a + 4)(a - 4) \quad \text{Factor the difference of squares.}$$

(b)
$$\begin{array}{ccccccc} A^2 & - & B^2 & = & (A + B) & (A - B) \\ \downarrow & & \downarrow & & \downarrow \ \downarrow & \downarrow \ \downarrow \end{array}$$
$$16m^2 - 49p^2 = (4m)^2 - (7p)^2 = (4m + 7p)(4m - 7p)$$

(c)
$$\begin{array}{ccccccc} A^2 & - & B^2 & = & (A + B) & (A - B) \\ \downarrow & & \downarrow & & \downarrow \ \downarrow & \downarrow \ \downarrow \end{array}$$
$$81k^2 - (a + 2)^2 = (9k)^2 - (a + 2)^2 = (9k + a + 2)(9k - [a + 2])$$
$$= (9k + a + 2)(9k - a - 2)$$

We could have used the method of substitution here.

(d) $x^4 - 81 = (x^2 + 9)(x^2 - 9)$ Factor the difference of squares.

$= (x^2 + 9)(x + 3)(x - 3)$ Factor $x^2 - 9$.

Work Problem 1 at the Side.

CAUTION

Assuming no greatest common factor except 1, it is not possible to factor (with real numbers) a *sum* of squares such as $x^2 + 25$. In particular, $x^2 + y^2 \neq (x + y)^2$, as shown next.

2 Factor a perfect square trinomial. Two other special products from Section 6.3 lead to the following rules for factoring.

Perfect Square Trinomial

$$x^2 + 2xy + y^2 = (x + y)^2$$
$$x^2 - 2xy + y^2 = (x - y)^2$$

Because the trinomial $x^2 + 2xy + y^2$ is the square of $x + y$, it is called a **perfect square trinomial.** In this pattern, both the first and the last terms of the trinomial must be perfect squares. In the factored form, twice the

OBJECTIVES

1. Factor a difference of squares.
2. Factor a perfect square trinomial.
3. Factor a difference of cubes.
4. Factor a sum of cubes.

1 Factor each polynomial.

(a) $2x^2 - 18$

(b) $9a^2 - 16b^2$

(c) $(m + 3)^2 - 49z^2$

(d) $y^4 - 16$

ANSWERS

1. (a) $2(x + 3)(x - 3)$
(b) $(3a - 4b)(3a + 4b)$
(c) $(m + 3 + 7z)(m + 3 - 7z)$
(d) $(y^2 + 4)(y + 2)(y - 2)$

product of the first and the last terms must give the middle term of the trinomial. It is important to understand these patterns in terms of words, since they occur with many different symbols (other than x and y).

$4m^2 + 20m + 25$ — Perfect square trinomial

$p^2 - 8p + 64$ — Not a perfect square trinomial; middle term should be $\pm 16p$.

Work Problem 2 at the Side.

2 Identify any perfect square trinomials.

(a) $z^2 + 12z + 36$

(b) $2x^2 - 4x + 4$

(c) $9a^2 + 12ab + 16b^2$

Example 2 Factoring Perfect Square Trinomials

Factor each polynomial.

(a) $144p^2 - 120p + 25$

Here $144p^2 = (12p)^2$ and $25 = 5^2$. The sign on the middle term is $-$, so if $144p^2 - 120p + 25$ is a perfect square trinomial, the factored form will have to be

$$(12p - 5)^2.$$

Take twice the product of the two terms to see if this is correct.

$$2(12p)(-5) = -120p$$

This is the middle term of the given trinomial, so

$$144p^2 - 120p + 25 = (12p - 5)^2.$$

(b) $4m^2 + 20mn + 49n^2$

If this is a perfect square trinomial, it will equal $(2m + 7n)^2$. By the pattern in the box, if multiplied out, this squared binomial has a middle term of $2(2m)(7n) = 28mn$, which *does not equal* $20mn$. Verify that this trinomial cannot be factored by the methods of the previous section either. It is prime.

(c) $(r + 5)^2 + 6(r + 5) + 9 = [(r + 5) + 3]^2$
$= (r + 8)^2,$

since $2(r + 5)(3) = 6(r + 5)$, the middle term.

(d) $m^2 - 8m + 16 - p^2$

Since there are four terms, we will use factoring by grouping. The first three terms here are a perfect square trinomial. Group them together, and factor as follows.

$$(m^2 - 8m + 16) - p^2 = (m - 4)^2 - p^2$$

The result is the difference of squares. Factor again to get

$$(m - 4)^2 - p^2 = (m - 4 + p)(m - 4 - p).$$

Work Problem 3 at the Side.

3 Factor each polynomial.

(a) $49z^2 - 14zk + k^2$

(b) $9a^2 + 48ab + 64b^2$

(c) $(k + m)^2 - 12(k + m) + 36$

(d) $x^2 - 2x + 1 - y^2$

Perfect square trinomials, of course, can be factored using the general methods shown earlier for other trinomials. The patterns given here provide "shortcuts."

3 **Factor a difference of cubes.** A **difference of cubes,** $x^3 - y^3$, can be factored as follows.

Difference of Cubes

$$x^3 - y^3 = (x - y)(x^2 + xy + y^2)$$

We could check this pattern by finding the product of $x - y$ and $x^2 + xy + y^2$.

ANSWERS

2. (a) square (b) not square (c) not square
3. (a) $(7z - k)^2$ (b) $(3a + 8b)^2$
(c) $[(k + m) - 6]^2$ or $(k + m - 6)^2$
(d) $(x - 1 + y)(x - 1 - y)$

Example 3 Factoring Differences of Cubes

Factor each polynomial.

(a) $m^3 - 8 = m^3 - 2^3 = (m - 2)(m^2 + 2m + 2^2) = (m - 2)(m^2 + 2m + 4)$

Check: $(m - 2)(m^2 + 2m + 4)$

m^3, -8, $-2m$

Opposite of the product of the cube roots gives the middle term.

(b) $27x^3 - 8y^3 = (3x)^3 - (2y)^3$

$= (3x - 2y)[(3x)^2 + (3x)(2y) + (2y)^2]$

$= (3x - 2y)(9x^2 + 6xy + 4y^2)$

(c) $1000k^3 - 27n^3 = (10k)^3 - (3n)^3$

$= (10k - 3n)[(10k)^2 + (10k)(3n) + (3n)^2]$

$= (10k - 3n)(100k^2 + 30kn + 9n^2)$

Work Problem 4 at the Side.

4 **Factor a sum of cubes.** While an expression of the form $x^2 + y^2$ (a sum of squares) cannot be factored with real numbers, a **sum of cubes** is factored as follows.

Sum of Cubes

$$x^3 + y^3 = (x + y)(x^2 - xy + y^2)$$

To verify this result, find the product of $x + y$ and $x^2 - xy + y^2$. Compare this pattern with the pattern for a difference of cubes.

NOTE

The sign of the second term in the binomial factor of a sum or difference of cubes is *always the same* as the sign in the original polynomial. In the trinomial factor, the first and last terms are *always positive;* the sign of the middle term is *the opposite of* the sign of the second term in the binomial factor.

Example 4 Factoring Sums of Cubes

Factor each polynomial.

(a) $r^3 + 27 = r^3 + 3^3 = (r + 3)(r^2 - 3r + 3^2)$

$= (r + 3)(r^2 - 3r + 9)$

(b) $27z^3 + 125 = (3z)^3 + 5^3 = (3z + 5)[(3z)^2 - (3z)(5) + 5^2]$

$= (3z + 5)(9z^2 - 15z + 25)$

(c) $125t^3 + 216s^6 = (5t)^3 + (6s^2)^3$

$= (5t + 6s^2)[(5t)^2 - (5t)(6s^2) + (6s^2)^2]$

$= (5t + 6s^2)(25t^2 - 30ts^2 + 36s^4)$

4 Factor each polynomial.

(a) $x^3 - 1000$

(b) $8k^3 - y^3$

(c) $27m^3 - 64$

ANSWERS

4. (a) $(x - 10)(x^2 + 10x + 100)$
(b) $(2k - y)(4k^2 + 2ky + y^2)$
(c) $(3m - 4)(9m^2 + 12m + 16)$

5 Factor each polynomial.

(a) $2x^3 + 2000$

(b) $8p^3 + 125$

(c) $27m^3 + 125n^3$

ANSWERS

5. (a) $2(x + 10)(x^2 - 10x + 100)$
(b) $(2p + 5)(4p^2 - 10p + 25)$
(c) $(3m + 5n)(9m^2 - 15mn + 25n^2)$

CAUTION

A common error is to think that the xy-term has a coefficient of 2 when factoring the sum or difference of cubes. Since there is no coefficient of 2, expressions of the form $x^2 + xy + y^2$ and $x^2 - xy + y^2$ cannot be factored further.

Work Problem 5 at the Side.

The special types of factoring in this section are summarized here. *These should be memorized.*

Special Types of Factoring

Difference of Squares	$x^2 - y^2 = (x + y)(x - y)$
Perfect Square Trinomial	$x^2 + 2xy + y^2 = (x + y)^2$ $x^2 - 2xy + y^2 = (x - y)^2$
Difference of Cubes	$x^3 - y^3 = (x - y)(x^2 + xy + y^2)$
Sum of Cubes	$x^3 + y^3 = (x + y)(x^2 - xy + y^2)$

6.8 EXERCISES

FOR EXTRA HELP

Student's Solutions Manual

MyMathLab.com

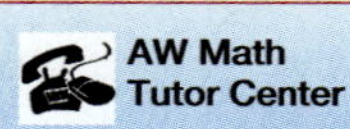
InterAct Math Tutorial Software

AW Math Tutor Center

www.mathxl.com

Digital Video Tutor CD 5 Videotape 10

1. Which of the following binomials are differences of squares?

   A. $64 - m^2$ B. $2x^2 - 25$

   C. $k^2 + 9$ D. $4z^4 - 49$

2. Which of the following binomials are sums or differences of cubes?

   A. $64 + y^3$ B. $125 - p^6$

   C. $9x^3 + 125$ D. $(x + y)^3 - 1$

3. Which of the following trinomials are perfect squares?

   A. $x^2 - 8x - 16$ B. $4m^2 + 20m + 25$

   C. $9z^4 + 30z^2 + 25$ D. $25a^2 - 45a + 81$

4. Of the twelve polynomials listed in Exercises 1–3, which ones can be factored using the methods of this section?

5. The binomial $9x^2 + 81$ is an example of the sum of two squares that can be factored. Under what conditions can the sum of two squares be factored?

6. Insert the correct signs in the blanks.

   (a) $8 + t^3 = (2 \_\_ t)(4 \_\_ 2t \_\_ t^2)$

   (b) $z^3 - 1 = (z \_\_ 1)(z^2 \_\_ z \_\_ 1)$

Factor each polynomial. See Examples 1–4.

7. $p^2 - 16$
8. $k^2 - 9$
9. $25x^2 - 4$
10. $36m^2 - 25$
11. $18a^2 - 98b^2$
12. $32c^2 - 98d^2$
13. $64m^4 - 4y^4$
14. $243x^4 - 3t^4$
15. $(y + z)^2 - 81$
16. $(h + k)^2 - 9$
17. $16 - (x + 3y)^2$
18. $64 - (r + 2t)^2$
19. $(p + q)^2 - (p - q)^2$
20. $(a + b)^2 - (a - b)^2$
21. $k^2 - 6k + 9$
22. $x^2 + 10x + 25$
23. $4z^2 + 4zw + w^2$
24. $9y^2 + 6yz + z^2$
25. $16m^2 - 8m + 1 - n^2$
26. $25c^2 - 20c + 4 - d^2$
27. $4r^2 - 12r + 9 - s^2$
28. $9a^2 - 24a + 16 - b^2$
29. $x^2 - y^2 + 2y - 1$
30. $-k^2 - h^2 + 2kh + 4$
31. $98m^2 + 84mn + 18n^2$
32. $80z^2 - 40zw + 5w^2$
33. $(p + q)^2 + 2(p + q) + 1$

34. $(x + y)^2 + 6(x + y) + 9$

35. $(a - b)^2 + 8(a - b) + 16$

36. $(m - n)^2 + 4(m - n) + 4$

37. $8x^3 - y^3$

38. $z^3 + 125p^3$

39. $64g^3 + 27h^3$

40. $27a^3 - 8b^3$

41. $24n^3 + 81p^3$

42. $250x^3 - 16y^3$

43. $(y + z)^3 - 64$

44. $(p - q)^3 + 125$

45. $m^6 - 125$

46. $k^6 + (k + 3)^3$

47. $(a + b)^3 - (a - b)^3$

RELATING CONCEPTS (Exercises 48–53) FOR INDIVIDUAL OR GROUP WORK

The binomial $x^6 - y^6$ may be considered either as a difference of squares or a difference of cubes. ***Work Exercises 48–53 in order.***

48. Factor $x^6 - y^6$ by first factoring as a difference of squares. Then factor further by considering one of the factors as a sum of cubes and the other factor as a difference of cubes.

49. Based on your answer in Exercise 48, fill in the blank with the correct factors so that $x^6 - y^6$ is factored completely:

$$x^6 - y^6 = (x - y)(x + y)\ \underline{\hspace{3cm}}.$$

50. Factor $x^6 - y^6$ by first factoring as a difference of cubes. Then factor further by considering one of the factors as a difference of squares.

51. Based on your answer in Exercise 50, fill in the blank with the correct factor so that $x^6 - y^6$ is factored:

$$x^6 - y^6 = (x - y)(x + y)\ \underline{\hspace{3cm}}.$$

52. Notice that the factor you wrote in the blank in Exercise 51 is a fourth-degree polynomial, while the two factors you wrote in the blank in Exercise 49 are both second-degree polynomials. What must be true about the product of the two factors you wrote in the blank in Exercise 49? Verify this.

53. If you have a choice of factoring as a difference of squares or a difference of cubes, how should you start to more easily obtain the factored form of the polynomial? Base the answer on your results in Exercises 48–52 and the methods of factoring explained in this section.

Summary Exercises on FACTORING

A polynomial is completely factored when the polynomial is in the form described below.

1. The polynomial is written as a product of prime polynomials with integer coefficients.
2. None of the polynomial factors can be factored further, except that a monomial factor need not be factored completely.

Factoring a Polynomial

Step 1 **Factor out any common factor.**

Step 2 **If the polynomial is a binomial,** check to see if it is the difference of squares, the difference of cubes, or the sum of cubes.

If the polynomial is a trinomial, check to see if it is a perfect square trinomial. If it is not, factor as in Section 6.7.

If the polynomial has more than three terms, try to factor by grouping.

Factor each polynomial.

1. $100a^2 - 9b^2$

2. $10r^2 + 13r - 3$

3. $18p^5 - 24p^3 + 12p^6$

4. $15x^2 - 20x$

5. $x^2 + 2x - 35$

6. $9 - a^2 + 2ab - b^2$

7. $225p^2 + 256$

8. $x^3 - 100$

9. $6b^2 - 17b - 3$

10. $k^2 - 6k + 16$

11. $18m^3n + 3m^2n^2 - 6mn^3$

12. $6t^2 + 19tu - 77u^2$

13. $2p^2 + 11pq + 15q^2$

14. $9m^2 - 45m + 18m^3$

15. $4k^2 + 28kr + 49r^2$

16. $54m^3 - 2000$

17. $mn - 2n + 5m - 10$

18. $9m^2 - 30mn + 25n^2 - p^2$

19. $x^3 + 3x^2 - 9x - 27$

20. $56k^3 - 875$

21. $9r^2 + 100$

22. $8p^3 - 125$

23. $6k^2 - k - 1$

24. $27m^2 + 144mn + 192n^2$

25. $x^4 - 625$

26. $125m^6 + 216$

27. $ab + 6b + ac + 6c$

28. $p^3 + 64$

29. $4y^2 - 8y$

30. $6a^4 - 11a^2 - 10$

31. $14z^2 - 3zk - 2k^2$

32. $12z^3 - 6z^2 + 18z$

33. $256b^2 - 400c^2$

34. $z^2 - zp + 20p^2$

35. $1000z^3 + 512$

36. $64m^2 - 25n^2$

37. $10r^2 + 23rs - 5s^2$

38. $12k^2 - 17kq - 5q^2$

39. $32x^2 + 16x^3 - 24x^5$

40. $48k^4 - 243$

41. $14x^2 - 25xq - 25q^2$

42. $5p^2 - 10p$

43. $y^2 + 3y - 10$

44. $b^2 - 7ba - 18a^2$

45. $2a^3 + 6a^2 - 4a$

46. $12m^2rx + 4mnrx + 40n^2rx$

47. $18p^2 + 53pr - 35r^2$

48. $21a^2 - 5ab - 4b^2$

49. $(x - 2y)^2 - 4$

50. $(3m - n)^2 - 25$

51. $(5r + 2s)^2 - 6(5r + 2s) + 9$

52. $(p + 8q)^2 - 10(p + 8q) + 25$

53. $z^4 - 9z^2 + 20$

54. $21m^4 - 32m^2 - 5$

6.9 SOLVING EQUATIONS BY FACTORING

OBJECTIVES

1. Learn and use the zero-factor property.
2. Solve applied problems that require the zero-factor property.

The equations that we have solved so far in this book have been linear equations. Recall that in a linear equation, the greatest power of the variable is 1. To solve equations of degree greater than 1, other methods must be developed. One of these methods involves factoring.

1 Learn and use the zero-factor property. Some equations can be solved by factoring. Solving equations by factoring depends on a special property of the number 0, called the **zero-factor property.**

Zero-Factor Property

If two numbers have a product of 0, then at least one of the numbers must be 0. That is, if $ab = 0$, then either $a = 0$ or $b = 0$.

To prove the zero-factor property, we first assume $a \neq 0$. (If a does equal 0, then the property is proved already.) If $a \neq 0$, then $\frac{1}{a}$ exists, and each side of $ab = 0$ can be multiplied by $\frac{1}{a}$ to get

$$\frac{1}{a} \cdot ab = \frac{1}{a} \cdot 0$$

$$b = 0.$$

Thus, if $a \neq 0$, then $b = 0$, and the property is proved.

CAUTION

If $ab = 0$, then $a = 0$ or $b = 0$. However, if $ab = 6$, for example, it is not necessarily true that $a = 6$ or $b = 6$; in fact, it is very likely that *neither* $a = 6$ *nor* $b = 6$. *The zero-factor property works only for a product equal to 0.*

Example 1 Using the Zero-Factor Property to Solve an Equation

Solve the equation $(x + 6)(2x - 3) = 0$.

Here the product of $x + 6$ and $2x - 3$ is 0. By the zero-factor property, this can be true only if

$$x + 6 = 0 \quad \text{or} \quad 2x - 3 = 0.$$

Solve these two equations.

$$x + 6 = 0 \quad \text{or} \quad 2x - 3 = 0$$

$$x = -6 \qquad\qquad 2x = 3$$

$$x = \frac{3}{2}$$

Continued on Next Page

1 Solve each equation.

(a) $(3x + 5)(x + 1) = 0$

(b) $(3x + 11)(5x - 2) = 0$

Check these two solutions by substitution in the original equation.

If $x = -6$, then

$$(x + 6)(2x - 3) = 0$$
$$(-6 + 6)[2(-6) - 3] = 0 \quad ?$$
$$0(-15) = 0. \quad \text{True}$$

If $x = \frac{3}{2}$, then

$$(x + 6)(2x - 3) = 0$$
$$\left(\frac{3}{2} + 6\right)\left(2 \cdot \frac{3}{2} - 3\right) = 0 \quad ?$$
$$\frac{15}{2}(0) = 0. \quad \text{True}$$

Both solutions check; the solution set is $\{-6, \frac{3}{2}\}$.

Work Problem 1 at the Side.

Since the product $(x + 6)(2x - 3)$ equals $2x^2 + 9x - 18$, the equation of Example 1 has a squared term and is an example of a *quadratic equation*. A quadratic equation has degree 2.

Quadratic Equation

An equation that can be written in the form

$$ax^2 + bx + c = 0,$$

where $a \neq 0$, is a **quadratic equation**. This form is called **standard form.**

Quadratic equations are discussed in more detail in Chapter 9.

The steps involved in solving a quadratic equation by factoring are summarized below.

Solving a Quadratic Equation by Factoring

Step 1 **Write in standard form.** Rewrite the equation if necessary so that one side is 0.

Step 2 **Factor.** Factor the polynomial.

Step 3 **Use the zero-factor property.** Set each variable factor equal to 0.

Step 4 **Find the solution(s).** Solve each equation formed in Step 3.

Step 5 **Check.** Check each solution in the *original* equation.

Example 2 Solving a Quadratic Equation by Factoring

Solve the equation $2x^2 + 3x = 2$.

Step 1 $\quad 2x^2 + 3x = 2$

$2x^2 + 3x - 2 = 0 \quad$ Standard form

Step 2 $\quad (2x - 1)(x + 2) = 0 \quad$ Factor.

Step 3 $\quad 2x - 1 = 0 \quad$ or $\quad x + 2 = 0 \quad$ Zero-factor property

Step 4 $\quad 2x = 1 \qquad x = -2 \quad$ Solve each equation.

$x = \frac{1}{2}$

Continued on Next Page

ANSWERS

1. (a) $\left\{-\frac{5}{3}, -1\right\}$ (b) $\left\{-\frac{11}{3}, \frac{2}{5}\right\}$

Step 5 Check each solution in the original equation.

If $x =$, then

$$2\left(\frac{1}{2}\right)^2 + 3\left(\frac{1}{2}\right) = 2 \quad ?$$

$$2\left(\frac{1}{4}\right) + \frac{3}{2} = 2 \quad ?$$

$$\frac{1}{2} + \frac{3}{2} = 2 \quad ?$$

$$2 = 2. \quad \text{True}$$

If $x = -2$, then

$$2(-2)^2 + 3(-2) = 2 \quad ?$$

$$2(4) - 6 = 2 \quad ?$$

$$8 - 6 = 2 \quad ?$$

$$2 = 2. \quad \text{True}$$

Because both solutions check, the solution set is $\{\frac{1}{2}, -2\}$.

Work Problem 2 at the Side.

2 Solve each equation.

(a) $3x^2 - x = 4$

(b) $15x^2 + 7x = 2$

Example 3 Solving a Quadratic Equation with a Missing Term

Solve $5z^2 - 25z = 0$.

This quadratic equation has a missing term. Comparing it with the standard form $ax^2 + bx + c = 0$ shows that $c = 0$. The zero-factor property can still be used.

$$5z^2 - 25z = 0$$

$$5z(z - 5) = 0 \quad \text{Factor.}$$

$$5z = 0 \quad \text{or} \quad z - 5 = 0 \quad \text{Zero-factor property}$$

$$z = 0 \quad \text{or} \quad z = 5$$

The solutions are 0 and 5, as can be verified by substituting in the original equation. The solution set is $\{0, 5\}$.

CAUTION

Remember to include 0 as a solution of the equation in Example 3.

Work Problem 3 at the Side.

3 Solve each equation.

(a) $x^2 = -12x$

(b) $t^2 - 16 = 0$

Example 4 Solving an Equation That Requires Rewriting

Solve $(2q + 1)(q + 1) = 2(1 - q) + 6$.

Write the equation in standard form $ax^2 + bx + c = 0$ by first multiplying on each side.

$$(2q + 1)(q + 1) = 2(1 - q) + 6$$

$$2q^2 + 3q + 1 = 2 - 2q + 6$$

$$2q^2 + 5q - 7 = 0 \quad \text{Standard form}$$

$$(2q + 7)(q - 1) = 0 \quad \text{Factor.}$$

$$2q + 7 = 0 \quad \text{or} \quad q - 1 = 0 \quad \text{Zero-factor property}$$

$$2q = -7 \qquad q = 1$$

$$q = -\frac{7}{2}$$

Check that the solution set is $\{-\frac{7}{2}, 1\}$.

Work Problem 4 at the Side.

4 Solve.

$(x + 6)(x - 2) = -8 + x$

ANSWERS

2. (a) $\left\{\frac{4}{3}, -1\right\}$ **(b)** $\left\{\frac{1}{5}, -\frac{2}{3}\right\}$

3. (a) $\{0, -12\}$ **(b)** $\{4, -4\}$

4. $\{-4, 1\}$

5 Solve.

$3x^3 + x^2 = 4x$

The zero-factor property can be extended to solve certain polynomial equations of degree 3 or higher, as shown in the next example.

Example 5 **Solving an Equation of Degree 3**

Solve $-x^3 + x^2 = -6x$.

Start by adding $6x$ to each side to get 0 on the right side.

$$-x^3 + x^2 + 6x = 0$$

To make the factoring step easier, multiply each side by -1.

$$x^3 - x^2 - 6x = 0$$

$$x(x^2 - x - 6) = 0 \quad \text{Factor out } x.$$

$$x(x - 3)(x + 2) = 0 \quad \text{Factor the trinomial.}$$

Use the zero-factor property, extended to include the three variable factors.

$$x = 0 \quad \text{or} \quad x - 3 = 0 \quad \text{or} \quad x + 2 = 0$$

$$x = 3 \qquad\qquad x = -2$$

Check that the solution set is $\{0, 3, -2\}$.

Work Problem 5 at the Side.

2 **Solve applied problems that require the zero-factor property.** The next example shows an application that leads to a quadratic equation. We continue to use the six-step problem-solving method introduced in Chapter 2.

Example 6 **Using a Quadratic Equation in an Application**

Some surveyors are surveying a lot that is in the shape of a parallelogram. They find that the longer sides of the parallelogram are each 8 m longer than the distance between them. The area of the lot is 48 m^2. Find the length of the longer sides and the distance between them.

Step 1 **Read** the problem again. There will be two answers.

Step 2 **Assign a variable.** Let x represent the distance between the longer sides. Then $x + 8$ is the length of each longer side. See Figure 7.

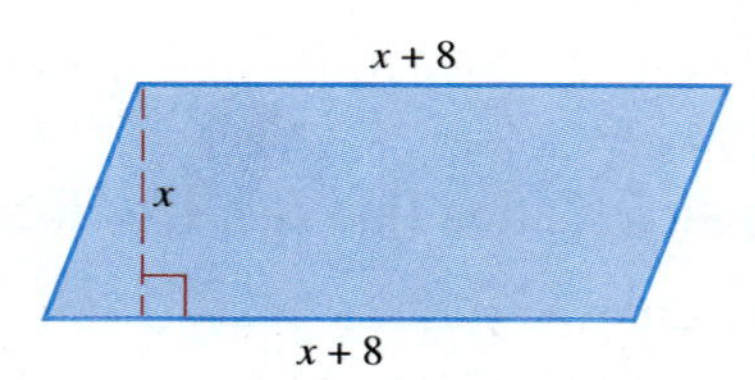

Figure 7

Step 3 **Write an equation.** The area of a parallelogram is given by $A = bh$, where b is the length of the longer side and h is the distance between the longer sides. Here $b = x + 8$ and $h = x$.

$$A = bh$$

$$48 = (x + 8)x \quad \text{Let } A = 48,\ b = x + 8,\ h = x.$$

Step 4 **Solve.**

$$48 = x^2 + 8x \quad \text{Distributive property}$$

$$0 = x^2 + 8x - 48 \quad \text{Standard form}$$

$$0 = (x + 12)(x - 4) \quad \text{Factor.}$$

$$x + 12 = 0 \quad \text{or} \quad x - 4 = 0 \quad \text{Zero-factor property}$$

$$x = -12 \quad \text{or} \quad x = 4$$

Continued on Next Page

Answers

5. $\left\{-\frac{4}{3}, 0, 1\right\}$

Step 5 **State the answer.** A distance cannot be negative, so reject -12 as a solution. The only possible solution is 4, so the distance between the longer sides is 4 m. The length of the longer sides is $4 + 8 = 12$ m.

Step 6 **Check.** The length of the longer sides is 8 m more than the distance between them, and the area is $4 \cdot 12 = 48$ m^2 as required, so the answers check.

CAUTION

When applications lead to quadratic equations, a solution of the equation may not satisfy the physical requirements of the problem, as in Example 6. Reject such solutions.

Work Problem 6 at the Side.

A function defined by a quadratic polynomial is called a *quadratic function*. In Chapter 9 we will investigate quadratic functions in detail. The next example uses such a function.

Example 7 Using a Quadratic Function in an Application

Quadratic functions are used to describe the height a falling object or a propelled object reaches in a specific time. For example, if a toy rocket is launched vertically upward from ground level with an initial velocity of 128 ft per sec, then its height in feet after t seconds is a function defined by

$$h(t) = -16t^2 + 128t,$$

if air resistance is neglected. After how many seconds will the rocket be 220 ft above the ground?

We must let $h(t) = 220$ and solve for t.

$220 = -16t^2 + 128t$	Let $h(t) = 220$.
$16t^2 - 128t + 220 = 0$	Standard form
$4t^2 - 32t + 55 = 0$	Divide by 4.
$(2t - 11)(2t - 5) = 0$	Factor.
$2t - 11 = 0$ or $2t - 5 = 0$	Zero-factor property
$t = 5.5$ or $t = 2.5$	

The rocket will reach a height of 220 ft twice: on its way up at 2.5 sec and again on its way down at 5.5 sec.

Work Problem 7 at the Side.

6 Solve the problem.

Carl is planning to build a rectangular deck along the back of his house. He wants the area of the deck to be 60 m^2, and the width to be 1 m less than half the length. What length and width should he use?

7 Solve the problem.

How long will it take the rocket in Example 7 to reach a height of 256 ft?

ANSWERS
6. length: 12 m; width: 5 m
7. 4 sec

Focus on

Real-Data Applications

Long Division: Arithmetic versus Algebra

In arithmetic, you learn to add, subtract, multiply, and divide numbers. In algebra, you learn to do the same with polynomials. This activity shows that the *process* used to divide polynomials mimics that used in long division of numbers. The step-by-step operations are identical.

Arithmetic Problem	Expanded Arithmetic Problem	Polynomial Division
21 $23\overline{)485}$ $\underline{46}$ 25 $\underline{23}$ 2	Long division in the expanded arithmetic problem is shown. $2 \cdot 10 + 1$ $2 \cdot 10 + 3\overline{)4 \cdot 10^2 + 8 \cdot 10 + 5}$ $\underline{4 \cdot 10^2 + 6 \cdot 10}$ $2 \cdot 10 + 5$ $\underline{2 \cdot 10 + 3}$ 2	The algebraic problem mimics the arithmetic problem written in expanded form. For $x = 10$, the problems are identical. $2x + 1$ $2x + 3\overline{)4x^2 + 8x + 5}$ $\underline{4x^2 + 6x}$ $2x + 5$ $\underline{2x + 3}$ 2

One difference between polynomial and numerical long division is that the polynomial coefficients may not be integers.

For Group Discussion

(a) Compute the long division of each polynomial given.

(b) Compute the problem in expanded form for $x = 10$.

(c) Determine the corresponding arithmetic problem, and verify that the computations are equivalent. (*Hint:* In Problem 1, when $x = 10$, the expression $3x - 1$ has value 29.)

	Polynomial Division	Expanded Arithmetic Problem	Arithmetic Problem
1.	**(a)** $3x - 1\overline{)6x^2 - 5x + 1}$	**(b)** $3 \cdot 10 - 1\overline{)6 \cdot 10^2 - 5 \cdot 10 + 1}$	**(c)** $29\overline{)}$
2.	**(a)** $x + 3\overline{)2x^2 \qquad - 8}$	**(b)** $10 + 3\overline{)2 \cdot 10^2 \qquad - 8}$	**(c)**

6.9 EXERCISES

FOR EXTRA HELP

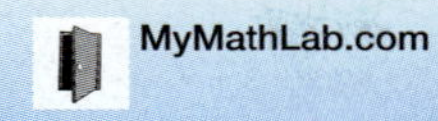

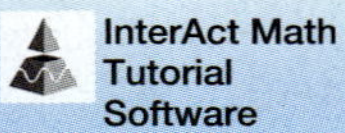

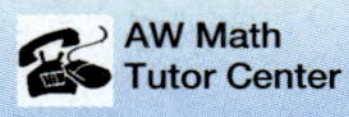

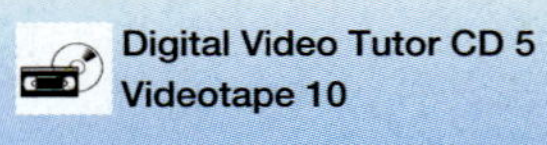

Student's Solutions Manual | MyMathLab.com | InterAct Math Tutorial Software | AW Math Tutor Center | www.mathxl.com | Digital Video Tutor CD 5 Videotape 10

1. Explain in your own words how the zero-factor property is used in solving a quadratic equation.

2. One of the following equations is *not* in proper form for using the zero-factor property. Which one is it? Explain why it is not in proper form.

A. $(x + 2)(x - 6) = 0$

B. $x(3x - 7) = 0$

C. $3t(t + 8)(t - 9) = 0$

D. $y(y - 3) + 6(y - 3) = 0$

Solve each equation using the zero-factor property. See Example 1.

3. $(x - 5)(x + 10) = 0$

4. $(y + 3)(y + 7) = 0$

5. $(2k - 5)(3k + 8) = 0$

6. $(3q - 4)(2q + 5) = 0$

Find all solutions by factoring. See Examples 2–4.

7. $m^2 - 3m - 10 = 0$

8. $x^2 + x - 12 = 0$

9. $z^2 + 9z + 18 = 0$

10. $x^2 - 18x + 80 = 0$

11. $2x^2 = 7x + 4$

12. $2x^2 = 3 - x$

13. $15k^2 - 7k = 4$

14. $3c^2 + 3 = -10c$

15. $2y^2 - 12 - 4y = y^2 - 3y$

16. $3p^2 + 9p + 30 = 2p^2 - 2p$

17. $(5y + 1)(y + 3) = -2(5y + 1)$

18. $(3x + 1)(x - 3) = 2 + 3(x + 5)$

19. $6m^2 - 36m = 0$

20. $-3m^2 + 27m = 0$

21. $-3m^2 + 27 = 0$

22. $-2a^2 + 8 = 0$

23. $4p^2 - 16 = 0$

24. $9x^2 - 81 = 0$

25. $(x - 3)(x + 5) = -7$

26. $(x + 8)(x - 2) = -21$

27. $(2x + 1)(x - 3) = 6x + 3$

28. $(3x + 2)(x - 3) = 7x - 1$

29. $(x + 3)(x - 6) = (2x + 2)(x - 6)$

30. $(2x + 1)(x + 5) = (x + 11)(x + 3)$

Solve each equation using the zero-factor property. See Example 5.

31. $2x^3 - 9x^2 - 5x = 0$

32. $6x^3 - 13x^2 - 5x = 0$

33. $9t^3 = 16t$

34. $25y^3 = 64y$

35. $2r^3 + 5r^2 - 2r - 5 = 0$

36. $2p^3 + p^2 - 98p - 49 = 0$

37. A student tried to solve the equation in Exercise 33 by first dividing each side by t, obtaining $9t^2 = 16$. She then solved the resulting equation by the zero-factor property to get the solution set $\left\{-\frac{4}{3}, \frac{4}{3}\right\}$. What was incorrect about her procedure?

38. Without actually solving each equation, determine which one of the following has 0 in its solution set.

A. $4x^2 - 25 = 0$

B. $x^2 + 2x - 3 = 0$

C. $6x^2 + 9x + 1 = 0$

D. $x^3 + 4x^2 = 3x$

Solve each problem by writing a quadratic equation and then solving it using the zero-factor property. See Examples 6 and 7.

39. A garden has an area of 320 ft^2. Its length is 4 ft more than its width. What are the dimensions of the garden?

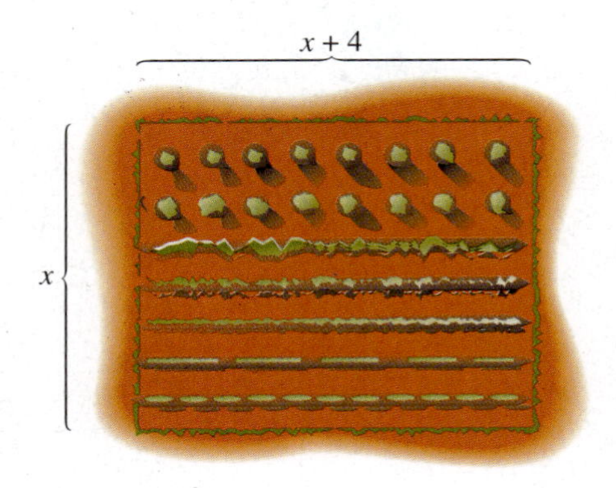

40. A square mirror has sides measuring 2 ft less than the sides of a square painting. If the difference between their areas is 32 ft^2, find the lengths of the sides of the mirror and the painting.

41. A sign has the shape of a triangle. The length of the base is 3 m less than the height. What are the measures of the base and the height, if the area is 44 m^2?

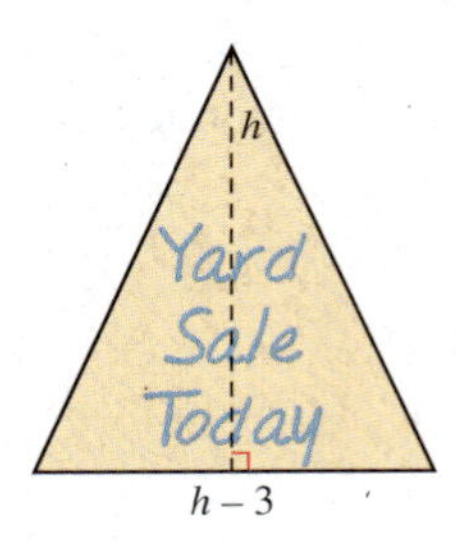

42. The base of a parallelogram is 7 ft more than the height. If the area of the parallelogram is 60 ft^2, what are the measures of the base and the height?

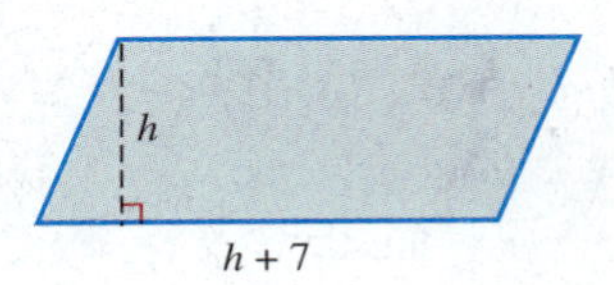

43. A farmer has 300 ft of fencing and wants to enclose a rectangular area of 5000 ft^2. What dimensions should she use?

44. A rectangular landfill has an area of 30,000 ft^2. Its length is 200 ft more than its width. What are the dimensions of the landfill?

45. A box with no top is to be constructed from a piece of cardboard whose length measures 6 in. more than its width. The box is to be formed by cutting squares that measure 2 in. on each side from the four corners and then folding up the sides. If the volume of the box will be 110 in.^{3}, what are the dimensions of the piece of cardboard?

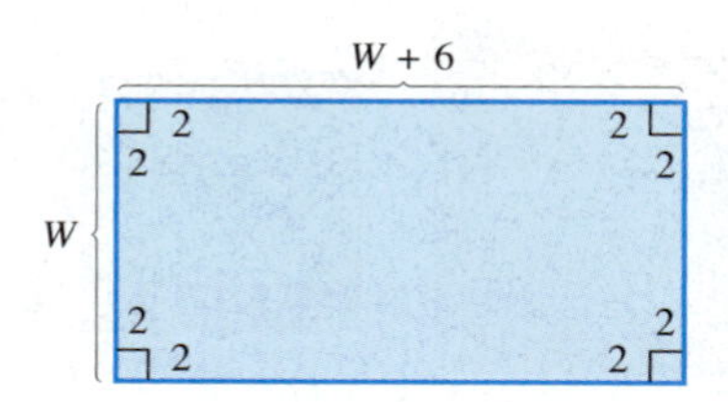

46. The surface area of the box with open top shown in the figure is 161 in.^{2}. Find the dimensions of the base. (*Hint:* The surface area is a function defined by $S(x) = x^2 + 16x$.)

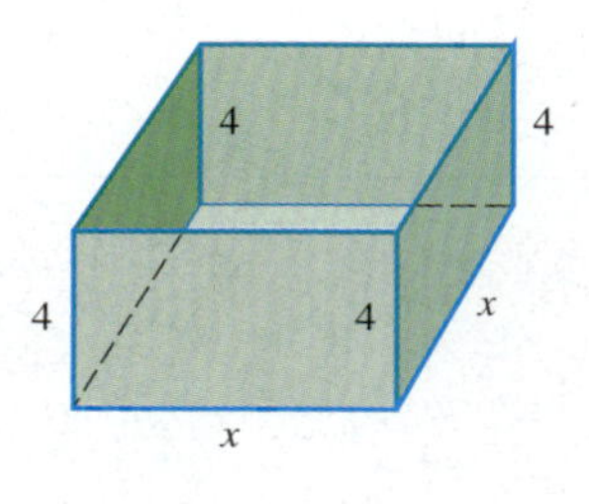

47. Refer to Example 7. After how many seconds will the rocket be 240 ft above the ground? 112 ft above the ground?

48. If an object is propelled upward with an initial velocity of 64 ft per sec from a height of 80 ft, then its height in feet t seconds after it is propelled is a function defined by

$$f(t) = -16t^2 + 64t + 80.$$

How long after it is propelled will it hit the ground? (*Hint:* When it hits the ground, its height is 0 ft.)

49. If a baseball is dropped from a helicopter 625 ft above the ground, then its distance in feet from the ground t seconds later is a function defined by

$$f(t) = -16t^2 + 625.$$

How long after it is dropped will it hit the ground?

50. If a rock is dropped from a building 576 ft high, then its distance in feet from the ground t seconds later is a function defined by

$$f(t) = -16t^2 + 576.$$

How long after it is dropped will it hit the ground?

Summary

Key Terms

6.2	**term**	A term is a number, a variable, or the product or quotient of a number and one or more variables raised to powers.
	coefficient (numerical coefficient)	A coefficient is a factor in a term (usually used for the numerical factor).
	algebraic expression	An algebraic expression is any combination of variables or constants (numerical values) joined by the basic operations of addition, subtraction, multiplication, and division (except by 0), or taking roots.
	polynomial	A polynomial is a term or a finite sum of terms in which all variables have whole number exponents and no variables appear in denominators.
	polynomial in *x*	A polynomial in x is a polynomial containing only the variable x.
	descending powers	A polynomial in one variable is written in descending powers if the exponents on the variable in the terms decrease from left to right.
	trinomial	A trinomial is a polynomial with exactly three terms.
	binomial	A binomial is a polynomial with exactly two terms.
	monomial	A monomial is a polynomial with exactly one term.
	degree of a term	The degree of a term with one variable is the exponent on that variable.
	degree of a polynomial	The degree of a polynomial is the greatest degree of any of the terms in the polynomial.
	negative of a polynomial	The negative of a polynomial is obtained by changing the sign of every coefficient in the polynomial.
	polynomial function of degree *n*	A function defined by $f(x) = a_nx^n + a_{n-1}x^{n-1} + \cdots + a_1x + a_0$ is a polynomial function of degree n.
6.5	**synthetic division**	Synthetic division is a shortcut procedure that can be used when dividing a polynomial by a binomial of the form $x - k$.
6.6	**greatest common factor**	The product of the largest common numerical factor and each variable factor of least degree common to every term in a polynomial is the greatest common factor of the terms of the polynomial.
6.7	**prime polynomial**	A polynomial that cannot be factored with integer coefficients is a prime polynomial.
6.9	**standard form of a quadratic equation**	An equation that can be written in the form $ax^2 + bx + c = 0$, where $a \neq 0$, is a quadratic equation. This form is called standard form.

Test Your Word Power

See how well you have learned the vocabulary in this chapter. Answers follow the Quick Review.

1. A **polynomial** is an algebraic expression made up of
(a) a term or a finite product of terms with positive coefficients and exponents
(b) the sum of two or more terms with whole number coefficients and exponents
(c) the product of two or more terms with positive exponents
(d) a term or a finite sum of terms with real coefficients and whole number exponents.

2. A **binomial** is a polynomial with
(a) only one term
(b) exactly two terms
(c) exactly three terms
(d) more than three terms.

3. A **trinomial** is a polynomial with
(a) only one term
(b) exactly two terms
(c) exactly three terms
(d) more than three terms.

4. FOIL is a method for
(a) adding two binomials
(b) adding two trinomials
(c) multiplying two binomials
(d) multiplying two trinomials.

5. Factoring is
(a) a method of multiplying polynomials
(b) the process of writing a polynomial as a product
(c) the answer in a multiplication problem
(d) a way to add the terms of a polynomial.

6. A **difference of squares** is a binomial
(a) that can be factored as the difference of two cubes
(b) that cannot be factored
(c) that is squared
(d) that can be factored as the product of the sum and difference of two terms.

7. A **perfect square trinomial** is a trinomial
(a) that can be factored as the square of a binomial
(b) that cannot be factored
(c) that is multiplied by a binomial
(d) where all terms are perfect squares.

8. A **quadratic equation** is a polynomial equation of
(a) degree one
(b) degree two
(c) degree three
(d) degree four.

Quick Review

Concepts	*Examples*
6.1 Integer Exponents and Scientific Notation **Definitions and Rules for Exponents**	Apply the rules of exponents.
Product Rule: $a^m \cdot a^n = a^{m+n}$	$3^4 \cdot 3^2 = 3^6$
Quotient Rule: $\frac{a^m}{a^n} = a^{m-n}$	$\frac{2^5}{2^3} = 2^2$
Negative Exponent: $a^{-n} = \frac{1}{a^n}$	$5^{-2} = \frac{1}{5^2}$
Zero Exponent: $a^0 = 1$	$27^0 = 1, \quad (-5)^0 = 1$

Concepts	*Examples*
6.1 Integer Exponents and Scientific Notation (continued)	
Power Rules: $(a^m)^n = a^{mn}$	$(6^3)^4 = 6^{12}$
$(ab)^m = a^m b^m$	$(5p)^4 = 5^4 p^4$
$\left(\frac{a}{b}\right)^n = \frac{a^n}{b^n}$	$\left(\frac{2}{3}\right)^5 = \frac{2^5}{3^5}$
$\frac{1}{a^{-n}} = a^n$	$\frac{1}{x^{-3}} = x^3$
$\frac{a^{-n}}{b^{-m}} = \frac{b^m}{a^n}$	$\frac{r^{-3}}{t^{-4}} = \frac{t^4}{r^3}$
$a^{-n} = \left(\frac{1}{a}\right)^n$	$4^{-3} = \left(\frac{1}{4}\right)^3$
$\left(\frac{a}{b}\right)^{-n} = \left(\frac{b}{a}\right)^n$	$\left(\frac{4}{7}\right)^{-2} = \left(\frac{7}{4}\right)^2$
Scientific Notation A number is in scientific notation when it is written as a product of a number between 1 and 10 (inclusive of 1) and an integer power of 10.	Write 23,500,000,000 in scientific notation. $23{,}500{,}000{,}000 = 2.35 \times 10^{10}$ Write 4.3×10^{-6} in standard notation. $4.3 \times 10^{-6} = .0000043$
6.2 Adding and Subtracting Polynomials; Polynomial Functions Add or subtract polynomials by combining like terms.	$(x^2 - 2x + 3) + (2x^2 - 8) = 3x^2 - 2x - 5$ $(5x^4 + 3x^2) - (7x^4 + x^2 - x) = -2x^4 + 2x^2 + x$
The graph of $f(x) = x$ is a line, and the graph of $f(x) = x^2$ is a parabola. The graph of $f(x) = x^3$ is neither of these. They define the identity, squaring, and cubing functions, respectively.	Graph the identity, squaring, and cubing functions. 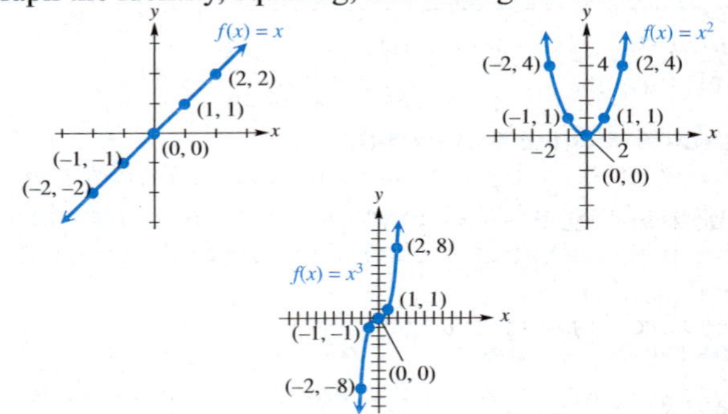
6.3 Multiplying Polynomials To multiply two polynomials, multiply each term of one by each term of the other.	$(x^3 + 3x)(4x^2 - 5x + 2)$ $= 4x^5 + 12x^3 - 5x^4 - 15x^2 + 2x^3 + 6x$ $= 4x^5 - 5x^4 + 14x^3 - 15x^2 + 6x$
To multiply two binomials, use the FOIL method. Multiply the First terms, the Outer terms, the Inner terms, and the Last terms. Then add these products.	$(2x + 3)(x - 7) = 2x(x) + 2x(-7) + 3x + 3(-7)$ $= 2x^2 - 14x + 3x - 21$ $= 2x^2 - 11x - 21$
Special Products $(x + y)(x - y) = x^2 - y^2$ $(x + y)^2 = x^2 + 2xy + y^2$ $(x - y)^2 = x^2 - 2xy + y^2$	$(3m + 8)(3m - 8) = 9m^2 - 64$ $(5a + 3b)^2 = 25a^2 + 30ab + 9b^2$ $(2k - 1)^2 = 4k^2 - 4k + 1$

Concepts	*Examples*
6.4 Dividing Polynomials **Dividing by a Monomial** To divide a polynomial by a monomial, divide each term in the polynomial by the monomial, and then write each fraction in lowest terms.	$\frac{2x^3 - 4x^2 + 6x - 8}{2x} = \frac{2x^3}{2x} - \frac{4x^2}{2x} + \frac{6x}{2x} - \frac{8}{2x}$ $= x^2 - 2x + 3 - \frac{4}{x}$
Dividing by a Polynomial Use the "long division" process.	Divide $\frac{m^3 - m^2 + 2m + 5}{m + 1}$. $\begin{array}{r} m^2 - 2m + 4 \\ m + 1\overline{)m^3 - m^2 + 2m + 5} \\ \underline{m^3 + m^2} \qquad\qquad \\ -2m^2 + 2m \qquad \\ \underline{-2m^2 - 2m} \qquad \\ 4m + 5 \\ \underline{4m + 4} \\ 1 \end{array}$ ← Remainder The quotient is $m^2 - 2m + 4 + \frac{1}{m + 1}$.
6.5 Synthetic Division To use synthetic division, the divisor must be of the form $x - k$.	Divide $\frac{x^3 - x^2 + 2x + 5}{x + 1}$ by synthetic division. $\begin{array}{r\|rrrr} -1 & 1 & -1 & 2 & 5 \\ & & -1 & 2 & -4 \\ \hline & 1 & -2 & 4 & 1 \end{array}$ $x^2 - 2x + 4$ Remainder The quotient is $x^2 - 2x + 4 + \frac{1}{x + 1}$.
6.6 Greatest Common Factors; Factoring by Grouping **The Greatest Common Factor** The product of the largest common numerical factor and each variable of lowest degree common to every term in a polynomial is the greatest common factor of the terms of the polynomial.	Factor $4x^2y - 50xy^2 = 2^2x^2y - 2 \cdot 5^2xy^2$. The greatest common factor is $2xy$. $4x^2y - 50xy^2 = 2xy(2x - 25y)$
Factoring by Grouping Group the terms so that each group has a common factor. Factor out the common factor in each group. If the groups now have a common factor, factor it out. If not, try a different grouping.	Factor by grouping. $5a - 5b - ax + bx = (5a - 5b) + (-ax + bx)$ $= 5(a - b) - x(a - b)$ $= (a - b)(5 - x)$

Concepts	Examples
6.7 Factoring Trinomials To factor a trinomial, choose factors of the first term and factors of the last term. Then, place them in a pair of parentheses of this form: ( )( ). Try various combinations of the factors until the correct middle term of the trinomial is found.	Factor $15x^2 + 14x - 8$. The factors of 15 are 5 and 3, and 15 and 1. The factors of -8 are -4 and 2, 4 and -2, -1 and 8, and 1 and -8. Various combinations of these factors lead to the correct factorization. $15x^2 + 14x - 8 = (5x - 2)(3x + 4)$. Check by multiplying, using the FOIL method.
6.8 Special Factoring **Difference of Squares** $x^2 - y^2 = (x + y)(x - y)$	$4m^2 - 25n^2 = (2m)^2 - (5n)^2$ $= (2m + 5n)(2m - 5n)$
Perfect Square Trinomials $x^2 + 2xy + y^2 = (x + y)^2$ $x^2 - 2xy + y^2 = (x - y)^2$	$9y^2 + 6y + 1 = (3y + 1)^2$ $16p^2 - 56p + 49 = (4p - 7)^2$
Difference of Cubes $x^3 - y^3 = (x - y)(x^2 + xy + y^2)$	$8 - 27a^3 = (2 - 3a)(4 + 6a + 9a^2)$
Sum of Cubes $x^3 + y^3 = (x + y)(x^2 - xy + y^2)$	$64z^3 + 1 = (4z + 1)(16z^2 - 4z + 1)$
6.9 Solving Equations by Factoring *Step 1* Rewrite the equation if necessary so that one side is 0.	Solve. $2x^2 + 5x = 3$ $2x^2 + 5x - 3 = 0$ Standard form
Step 2 Factor the polynomial.	$(2x - 1)(x + 3) = 0$
Step 3 Set each factor equal to 0.	$2x - 1 = 0$ or $x + 3 = 0$
Step 4 Solve each equation from Step 3.	$2x = 1$ $x = -3$ $x = \frac{1}{2}$
Step 5 Check each solution.	A check verifies that the solution set is $\{-3, \frac{1}{2}\}$.

ANSWERS TO TEST YOUR WORD POWER

1. (d) *Example:* $5x^3 + 2x^2 - 7$ **2. (b)** *Example:* $3t^3 + 5t$ **3. (c)** *Example:* $2a^2 - 3ab + b^2$ **4. (c)** *Example:* $(m + 4)(m - 3) = \overset{F}{m(m)} \overset{O}{- 3m} \overset{I}{+ 4m} + \overset{L}{4(-3)} = m^2 + m - 12$ **5. (b)** *Example:* $x^2 - 5x - 14 = (x - 7)(x + 2)$ **6. (d)** *Example:* $b^2 - 49$ is the difference of the squares b^2 and 7^2. It can be factored as $(b + 7)(b - 7)$. **7. (a)** *Example:* $a^2 + 2a + 1$ is a perfect square trinomial; its factored form is $(a + 1)^2$. **8. (b)** *Examples:* $y^2 - 3y + 2 = 0$, $x^2 - 9 = 0$, $2m^2 = 6m + 8$

Focus on **Real-Data Applications**

Reply All: The Computer Network Manager's Nightmare

Most businesses, including colleges, rely on e-mail messaging for internal communications. The computer network manager is responsible for ensuring that e-mail is a reliable and efficient system for maintaining that communication. One of the biggest problems involves the Reply All option in most e-mail software. If a message was sent to 10 addresses, the Reply All option sends a copy of the answer to each address in the list. In comparison, the Reply option sends an answer to only one address. A computer virus creates an even worse problem.

For Group Discussion

Assume that the discussion about e-mail messages relates only to those described in the following problem.

1. A committee of 10 faculty members is formed to recommend faculty salary increases for the next year. One member is selected as the committee chair. The chair requests that e-mail discussions be sent to all 10 committee members (using the Reply All option).

(a) The committee chair sends an e-mail message to every committee member, including himself, to which all other committee members reply. How many relevant e-mail messages are in each person's mailbox? How many e-mail messages has the computer network processed? (*Hint:* Think about the number of e-mails that would be sent if the committee had only 2 members, 3 members, 4 members, etc.)

(b) The committee chair forwards an e-mail message from the chancellor to the committee members suggesting that summer salaries must be reduced for the faculty to receive a pay raise for the next year. A heated debate ensues. Counting the original e-mail message about the chancellor's comment, the chairman counts eight relevant e-mail messages in his mailbox. How many e-mail messages has the computer network processed?

(c) Suppose the committee chair inadvertently sent the original message to the "All College Faculty" group, a list of 150 names set up by the computer manager primarily for administrators to notify the faculty about policy. The chairman counted eight relevant e-mail messages in his mailbox, each sent using Reply All. How many e-mail messages would the computer network have processed?

2. The "Love Bug" computer virus spreads by sending copies of itself to each e-mail address in the recipient's address book. Suppose that each person has 50 e-mail addresses in his or her address book and that it takes a virus 10 sec to process. Also assume that each recipient's computer automatically receives and processes the computer virus. (It is more reasonable to assume that a virus is processed only after the recipient "opens" the message.) How many messages has the computer network processed after 1 hr?

3. It is illegal to send a chain letter using the U.S. postal service. A chain letter requests that you send copies of it to, say, 10 people. Discuss why you think chain letters are outlawed.

Chapter 6 Review Exercises

[6.1] *Simplify. Write answers with only positive exponents. Assume that all variables represent positive real numbers.*

1. 4^3

2. $\left(\frac{1}{3}\right)^4$

3. $(-5)^3$

4. $\frac{2}{(-3)^{-2}}$

5. $\left(\frac{2}{3}\right)^{-4}$

6. $\left(\frac{5}{4}\right)^{-2}$

7. $5^{-1} + 6^{-1}$

8. $-3^0 + 3^0$

9. $(-3x^4y^3)(4x^{-2}y^5)$

10. $\frac{6m^{-4}n^3}{-3mn^2}$

11. $\frac{(5p^{-2}q)(4p^5q^{-3})}{2p^{-5}q^5}$

12. $\frac{x^{-2}y^{-4}}{x^{-4}y^{-2}}$

13. $(3^{-4})^2$

14. $(x^{-4})^{-2}$

15. $(xy^{-3})^{-2}$

16. $(z^{-3})^3z^{-6}$

17. $(5m^{-3})^2(m^4)^{-3}$

18. $\frac{(3r)^2r^4}{r^{-2}r^{-3}}(9r^{-3})^{-2}$

19. $\left(\frac{5z^{-3}}{z^{-1}}\right)\left(\frac{5}{z^2}\right)$

20. $\left(\frac{6m^{-4}}{m^{-9}}\right)^{-1}\left(\frac{m^{-2}}{16}\right)$

21. $\left(\frac{3r^5}{5r^{-3}}\right)^{-2}\left(\frac{9r^{-1}}{2r^{-5}}\right)^3$

Write in scientific notation.

22. 13,450

23. .0000000765

24. .138

Write without scientific notation.

25. 1.21×10^6

26. 5.8×10^{-3}

27. In 1999, maquiladoras (low-wage export factories) in Mexico exported \$63,750,000,000 worth of goods. Write this number in scientific notation. (*Source:* John Christman, CIEMEX-WEFA.)

Use scientific notation to compute. Give answers in both scientific notation and standard form.

28. $\dfrac{16 \times 10^4}{8 \times 10^8}$

29. $\dfrac{6 \times 10^{-2}}{4 \times 10^{-5}}$

30. $\dfrac{.0000000164}{.0004}$

31. $\dfrac{.0009 \times 12{,}000{,}000}{400{,}000}$

[6.2] *Give the numerical coefficient for each term.*

32. $14p^5$

33. $-z$

For each polynomial, ***(a)*** *write in descending powers,* ***(b)*** *identify as* monomial, binomial, trinomial, *or* none of these, *and* ***(c)*** *give the degree.*

34. $9k + 11k^3 - 3k^2$

35. $14m^6 + 9m^7$

36. $-7q^5r^3$

37. Give an example of a polynomial in the variable x such that it has degree 5, is lacking a third-degree term, and is in descending powers of the variable.

Add or subtract as indicated.

38. Add.

$$\begin{array}{r} 3x^2 - 5x + 6 \\ \underline{-4x^2 + 2x - 5} \end{array}$$

39. Subtract.

$$\begin{array}{r} -5y^3 \qquad + 8y - 3 \\ \underline{4y^2 + 2y + 9} \end{array}$$

40. $(4a^3 - 9a + 15) - (-2a^3 + 4a^2 + 7a)$

41. $(3y^2 + 2y - 1) + (5y^2 - 11y + 6)$

42. Find the perimeter of the triangle.

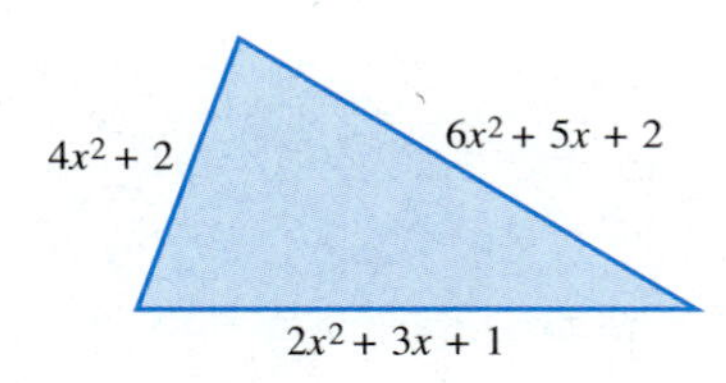

43. For the polynomial function defined by $f(x) = -2x^2 + 5x + 7$, find each value.

(a) $f(-2)$ **(b)** $f(3)$

44. For $f(x) = 2x + 3$ and $g(x) = 5x^2 - 3x + 2$, find each of the following.

(a) $(f + g)(x)$ **(b)** $(f - g)(x)$

45. The number of people employed in health service industries, in thousands, is approximated by the polynomial function defined by

$$f(x) = 22x^2 + 243x + 8992,$$

where $x = 0$ represents 1994, $x = 1$ represents 1995, and $x = 2$ represents 1996. Find the number of employees for each of these years. (*Source:* U.S. Bureau of Labor Statistics.)

Graph each polynomial function defined as follows.

46. $f(x) = -2x + 5$

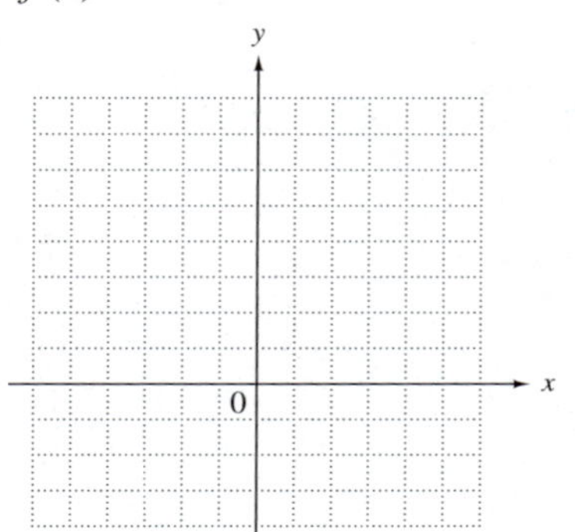

47. $f(x) = x^2 - 6$

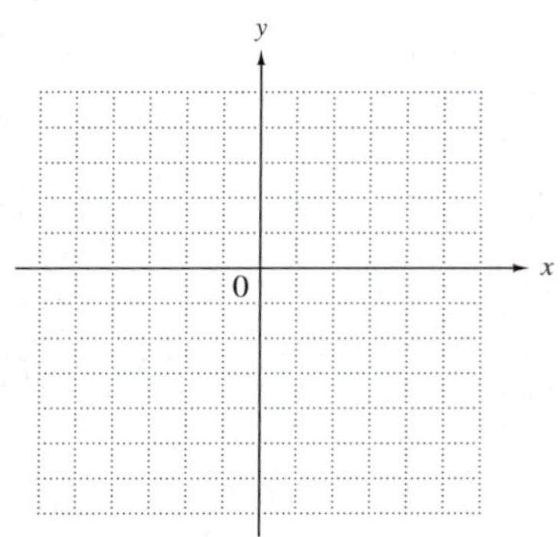

48. $f(x) = -x^3 + 1$

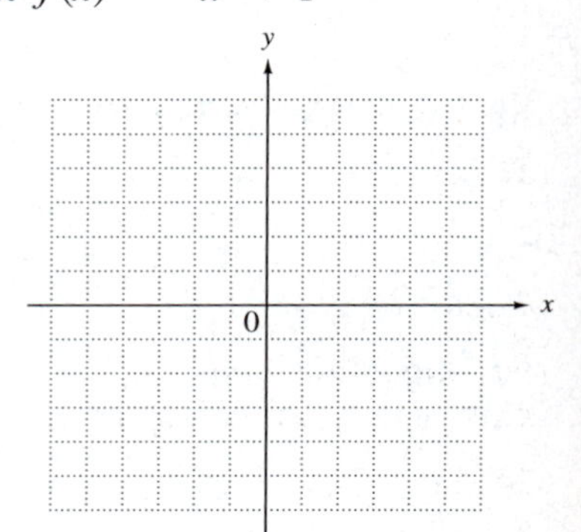

[6.3] *Find each product.*

49. $-6k(2k^2 + 7)$

50. $(7y - 8)(2y + 3)$

51. $(3w - 2t)(2w - 3t)$

52. $(2p^2 + 6p)(5p^2 - 4)$

53. $(3z^3 - 2z^2 + 4z - 1)(3z - 2)$

54. $(6r^2 - 1)(6r^2 + 1)$

55. $\left(z + \frac{3}{5}\right)\left(z - \frac{3}{5}\right)$

56. $(4m + 3)^2$

57. $(2x + 5)^3$

[6.4] *Divide.*

58. $\dfrac{4y^3 - 12y^2 + 5y}{4y}$

59. $\dfrac{2p^3 + 9p^2 + 27}{2p - 3}$

60. $\dfrac{5p^4 + 15p^3 - 33p^2 - 9p + 18}{5p^2 - 3}$

[6.5] *Use synthetic division to perform each division.*

61. $\dfrac{3p^2 - p - 2}{p - 1}$

62. $(2k^3 - 5k^2 + 12) \div (k - 3)$

Use synthetic division to decide whether -5 *is a solution of each equation.*

63. $2w^3 + 8w^2 - 14w - 20 = 0$

64. $-3q^4 + 2q^3 + 5q^2 - 9q + 1 = 0$

Use synthetic division to evaluate $P(k)$ *for the given value of* k.

65. $P(x) = 3x^3 - 5x^2 + 4x - 1;\ k = -1$

66. $P(z) = z^4 - 2z^3 - 9z - 5;\ k = 3$

[6.6] *Factor out the greatest common factor.*

67. $21y^2 + 35y$

68. $12q^2b + 8qb^2 - 20q^3b^2$

69. $(x + 3)(4x - 1) - (x + 3)(3x + 2)$

70. $(z + 1)(z - 4) + (z + 1)(2z + 3)$

Factor by grouping.

71. $4m + nq + mn + 4q$

72. $x^2 + 5y + 5x + xy$

73. $2m + 6 - am - 3a$

74. $2am - 2bm - ap + bp$

[6.7] *Factor completely.*

75. $3p^2 - p - 4$

76. $12r^2 - 5r - 3$

77. $10m^2 + 37m + 30$

78. $10k^2 - 11kh + 3h^2$

79. $9x^2 + 4xy - 2y^2$

80. $24x - 2x^2 - 2x^3$

81. $2k^4 - 5k^2 - 3$

82. $p^2(p + 2)^2 + p(p + 2)^2 - 6(p + 2)^2$

83. When asked to factor $x^2y^2 - 6x^2 + 5y^2 - 30$, a student gave the following incorrect answer: $x^2(y^2 - 6) + 5(y^2 - 6)$. Why is this answer incorrect? What is the correct answer?

84. If the area of this rectangle is represented by $4p^2 + 3p - 1$, what is the width in terms of p?

$4p - 1$

[6.8] *Factor completely.*

85. $16x^2 - 25$

86. $9t^2 - 49$

87. $x^2 + 14x + 49$

88. $9k^2 - 12k + 4$

89. $r^3 + 27$

90. $125x^3 - 1$

91. $m^6 - 1$

92. $x^8 - 1$

93. $x^2 + 6x + 9 - 25y^2$

[6.9] *Use the zero-factor property to solve each equation.*

94. $(5x + 2)(x + 1) = 0$

95. $p^2 - 5p + 6 = 0$

96. $6z^2 = 5z + 50$

97. $6r^2 + 7r = 3$

98. $-4m^2 + 36 = 0$

99. $6y^2 + 9y = 0$

100. $(2x + 1)(x - 2) = -3$

101. $x^2 - 8x + 16 = 0$

102. $2x^3 - x^2 - 28x = 0$

Solve each problem.

103. A triangular wall brace creates the shape of a right triangle. One of the perpendicular sides is 1 ft longer than twice the other. The area enclosed by the triangle is 10.5 ft^2. Find the shorter of the perpendicular sides.

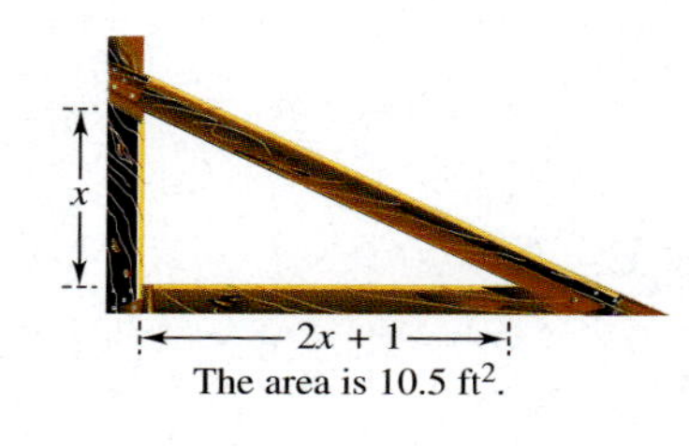

The area is 10.5 ft^2.

104. A rectangular parking lot has a length 20 ft more than its width. Its area is 2400 ft^2. What are the dimensions of the lot?

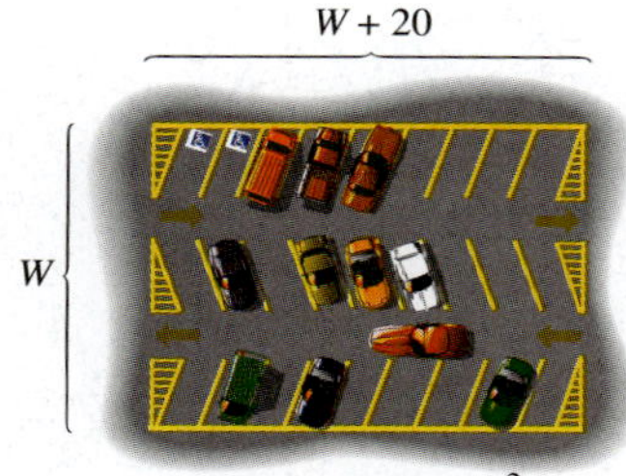

The area is 2400 ft^2.

A rock is propelled directly upward from ground level. After t seconds, its height in feet is given by $f(t) = -16t^2 + 256t$ (if air resistance is neglected).

105. When will the rock return to the ground?

106. After how many seconds will it be 240 ft above the ground?

107. Why does the question in Exercise 106 have two answers?

Mixed Review Exercises

Perform the indicated operations, then simplify. Write answers with only positive exponents. Assume all variables represent nonzero real numbers.

108. $(4x + 1)(2x - 3)$

109. $\dfrac{6^{-1}y^3(y^2)^{-2}}{6y^{-4}(y^{-1})}$

110. $(y^6)^{-5}(2y^{-3})^{-4}$

111. $(-5 + 11w) + (6 + 5w) + (-15 - 8w^2)$

112. $7p^5(3p^4 + p^3 + 2p^2)$

113. $\dfrac{(-z^{-2})^3}{5(z^{-3})^{-1}}$

114. $\dfrac{x^3 + 7x^2 + 7x - 15}{x + 5}$

115. $(2k - 1) - (3k^2 - 2k + 6)$

Factor completely.

116. $30a + am - am^2$

117. $8 - a^3$

118. $9x^2 + 13xy - 3y^2$

119. $15y^3 + 20y^2$

Solve.

120. $5x^2 - 17x - 12 = 0$

121. $x^3 - x = 0$

122. When Europeans arrived in America, many native Americans of the Northeast lived in *longhouses* that sheltered several related families. The rectangular floor area of a typical Huron longhouse was about 2750 ft^2. The length was 85 ft greater than the width. What were the dimensions of the floor?

Chapter 6 TEST

Simplify. Write answers with only positive exponents. Assume that all variables represent nonzero real numbers.

1. $(3x^{-2}y^3)^{-2}(4x^3y^{-4})$

2. $\dfrac{36r^{-4}(r^2)^{-3}}{6r^4}$

3. $\left(\dfrac{4p^2}{q^4}\right)^3\left(\dfrac{6p^8}{q^{-8}}\right)^{-2}$

4. $(-2x^4y^{-3})^0(-4x^{-3}y^{-8})^2$

5. (a) Write 9.1×10^{-7} without using scientific notation.

(b) Use scientific notation to simplify $\dfrac{2{,}500{,}000 \times .00003}{.05 \times 5{,}000{,}000}$.

Write the answer in both scientific notation and standard form.

6. If $f(x) = -2x^2 + 5x - 6$ and $g(x) = 7x - 3$, find each of the following.

(a) $f(4)$ **(b)** $(f + g)(x)$ **(c)** $(f - g)(x)$

7. Graph the function defined by $f(x) = -2x^2 + 3$.

Perform the indicated operations.

8. $(4x^3 - 3x^2 + 2x - 5) - (3x^3 + 11x + 8) + (x^2 - x)$

9. $(5x - 3)(2x + 1)$

10. $(2m - 5)(3m^2 + 4m - 5)$

11. $(6x + y)(6x - y)$

12. $(3k + q)^2$

13. $[2y + (3z - x)][2y - (3z - x)]$

14. $(x^3 + 3x^2 - 4) \div (x - 1)$

15. (a) Use synthetic division to divide $\dfrac{2x^2 + 3x - 6}{x - 2}$.

(b) If $f(x) = 2x^2 + 3x - 6$, what is $f(2)$ based on your answer in part (a)?

1. ______
2. ______
3. ______
4. ______
5. (a) ______
   (b) ______
6. (a) ______
   (b) ______
   (c) ______
7.

y

0 x

8. ______
9. ______
10. ______
11. ______
12. ______
13. ______
14. ______
15. (a) ______
    (b) ______

16. ______________________

17. ______________________

18. ______________________

19. ______________________

20. ______________________

21. ______________________

22. ______________________

23. ______________________

24. ______________________

25. ______________________

26. ______________________

27. ______________________

28. ______________________

29. ______________________

30. ______________________

Factor.

16. $11z^2 - 44z$

17. $3x + by + bx + 3y$

18. $4p^2 + 3pq - q^2$

19. $16a^2 + 40ab + 25b^2$

20. $y^3 - 216$

21. $9k^2 - 121j^2$

22. $6k^4 - k^2 - 35$

23. $27x^6 + 1$

24. $-x^2 - x + 12$

25. $(t^2 + 3)^2 + 4(t^2 + 3) - 5$

26. Explain why $(x^2 + 2y)p + 3(x^2 + 2y)$ is not in factored form. Then factor the polynomial.

Solve each equation using the zero-factor property.

27. $3x^2 + 8x + 4 = 0$

28. $10x^2 = 17x - 3$

Solve each problem.

29. The area of the rectangle shown is 40 in.^{2}. Find the length and the width of the rectangle.

$x + 7$

$2x + 3$

The area is 40 in.^{2}.

30. A ball is propelled upward from ground level. After t seconds, its height in feet is described by a function defined by $f(t) = -16t^2 + 96t$. After how many seconds will it reach a height of 128 ft?

Cumulative Review Exercises CHAPTERS 1–6

Use the properties of real numbers to simplify each expression.

1. $-2(m - 3)$
2. $-(-4m + 3)$
3. $3x^2 - 4x + 4 + 9x - x^2$

Evaluate if $p = -4$, $q = -2$, and $r = 5$.

4. $-3(2q - 3p)$
5. $8r^2 + q^2$
6. $\dfrac{\sqrt{r}}{-p + 2q}$
7. $\dfrac{rp + 6r^2}{p^2 + q - 1}$

Solve.

8. $2z - 5 + 3z = 4 - (z + 2)$
9. $\dfrac{3a - 1}{5} + \dfrac{a + 2}{2} = -\dfrac{3}{10}$
10. $-\dfrac{4}{3}d \geq -5$
11. $3 - 2(m + 3) < 4m$
12. $2k + 4 < 10$ and $3k - 1 > 5$
13. $2k + 4 > 10$ or $3k - 1 < 5$
14. $|5x + 3| - 10 = 3$
15. $|x + 2| < 9$
16. $|2y - 5| \geq 9$
17. $V = lwh$ for h

18. Two planes leave the Dallas-Fort Worth airport at the same time. One travels east at 550 mph, and the other travels west at 500 mph. Assuming no wind, how long will it take for the planes to be 2100 mi apart?

	r	t	d
Eastbound plane	550	x	
Westbound plane	500	x	

19. Graph $4x + 2y = -8$.

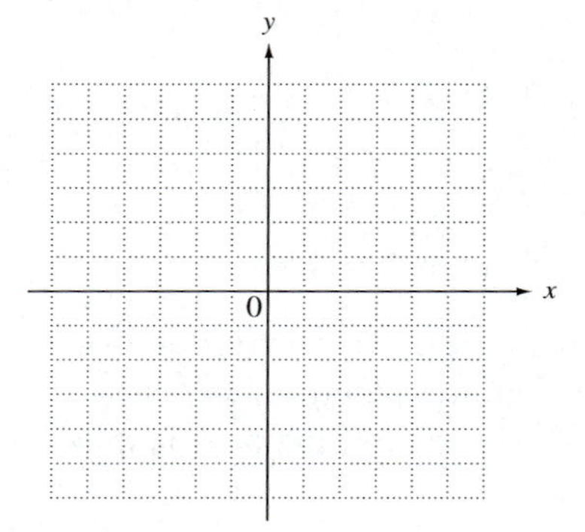

20. Find the slope of the line through the points $(-4, 8)$ and $(-2, 6)$.

21. What is the slope of the line shown here?

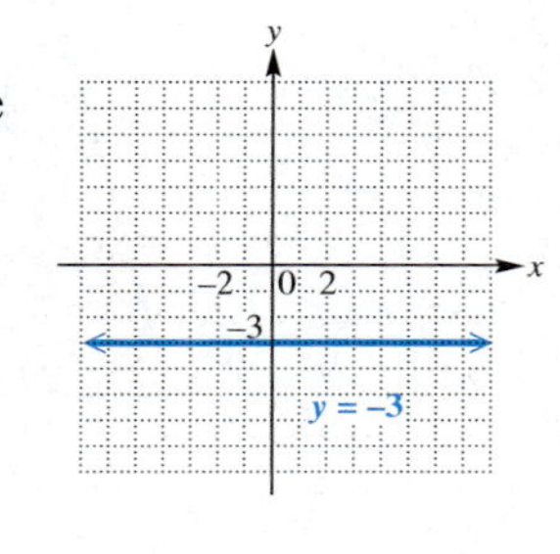

Use the function defined by $f(x) = 2x + 7$ to find the following.

22. $f(-4)$

23. The x-intercept of its graph

24. The y-intercept of its graph

Solve each system.

25. $3x - 2y = -7$
$2x + 3y = 17$

26. $2x + 3y - 6z = 5$
$8x - y + 3z = 7$
$3x + 4y - 3z = 7$

Perform the indicated operations. Assume variables represent nonzero real numbers.

27. $(3x^2y^{-1})^{-2}(2x^{-3}y)^{-1}$

28. $\dfrac{5m^{-2}y^3}{3m^{-3}y^{-1}}$

Perform the indicated operations.

29. $(3x^3 + 4x^2 - 7) - (2x^3 - 8x^2 + 3x)$

30. $(7x + 3y)^2$

31. $(2p + 3)(5p^2 - 4p - 8)$

Factor.

32. $16w^2 + 50wz - 21z^2$

33. $4x^2 - 4x + 1 - y^2$

34. $4y^2 - 36y + 81$

35. $100x^4 - 81$

36. $8p^3 + 27$

Solve.

37. $(p - 1)(2p + 3)(p + 4) = 0$

38. $9q^2 = 6q - 1$

39. A sign is to have the shape of a triangle with a height 3 ft greater than the length of the base. How long should the base be if the area is to be 14 ft^2?

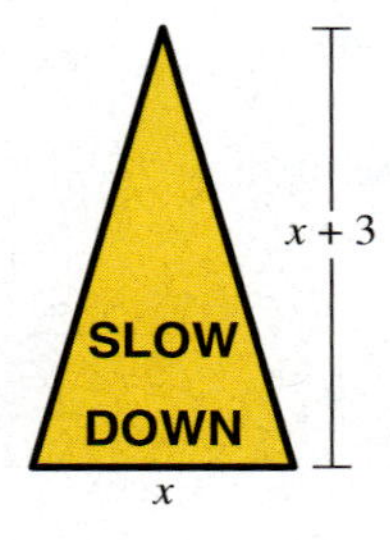

40. A game board has the shape of a rectangle. The longer sides are each 2 in. longer than the distance between them. The area of the board is 288 in.^{2}. Find the length of the longer sides and the distance between them.

Rational Expressions

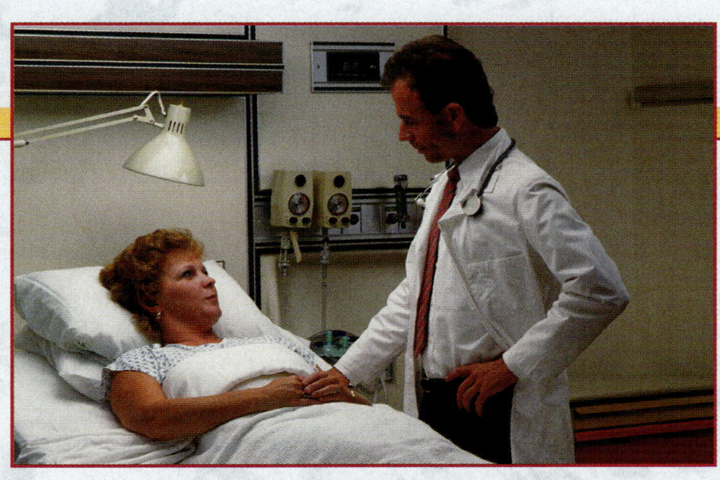

In 1998, 16 of every 100 Americans had no health insurance coverage. (If you happen to be one such person, you should be very concerned, considering the exorbitant cost of health care today.) The population at that time was about 227,000,000 Americans. (*Source*: U.S. Bureau of the Census.) In Example 4 of Section 7.5, we use proportions to determine the number of Americans without health insurance.

7.1 Rational Expressions and Functions; Multiplying and Dividing

Objectives

1. Define rational expressions.
2. Define rational functions and describe their domains.
3. Write rational expressions in lowest terms.
4. Multiply rational expressions.
5. Find reciprocals for rational expressions.
6. Divide rational expressions.

1 Define rational expressions. In arithmetic, a rational number is the quotient of two integers, with the denominator not 0. In algebra, a **rational expression** or *algebraic fraction* is the quotient of two polynomials, again with the denominator not 0. For example,

$$\frac{x}{y}, \quad \frac{-a}{4}, \quad \frac{m+4}{m-2}, \quad \frac{8x^2-2x+5}{4x^2+5x}, \quad \text{and} \quad x^5\left(\text{or } \frac{x^5}{1}\right)$$

are all rational expressions. In other words, rational expressions are the elements of the set

$$\left\{\frac{P}{Q} \middle| P, Q \text{ polynomials, with } Q \neq 0\right\}.$$

2 Define rational functions and describe their domains. A function that is defined by a rational expression is called a **rational function** and has the form

$$f(x) = \frac{P(x)}{Q(x)},$$

where $Q(x) \neq 0$.

The domain of a rational function includes all real numbers except those that make $Q(x)$, that is, the denominator, equal to 0. For example, the domain of

$$f(x) = \frac{2}{x-5}$$

includes all real numbers except 5, because 5 would make the denominator equal to 0.

Figure 1 shows a graph of the function defined by $f(x) = \frac{2}{x-5}$. Notice that the graph does not exist when $x = 5$. It does not intersect the dashed vertical line whose equation is $x = 5$. This line is an *asymptote*. We will discuss graphs of rational functions in more detail in Section 7.4.

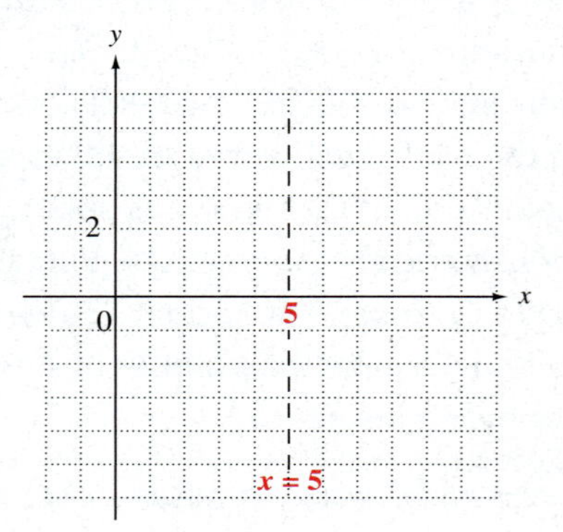

Figure 1

Example 1 **Finding Numbers That Are Not in the Domains of Rational Functions**

Find all numbers that are not in the domain of each rational function.

(a) $f(x) = \frac{3}{7x-14}$

Continued on Next Page

The only values that cannot be used are those that make the denominator 0. To find these values, set the denominator equal to 0 and solve the resulting equation.

$$
\begin{aligned}
7x - 14 &= 0 \\
7x &= 14 \qquad \text{Add 14.} \\
x &= 2 \qquad \text{Divide by 7.}
\end{aligned}
$$

The number 2 cannot be used as a replacement for x; the domain of f includes all real numbers except 2.

(b) $g(x) = \dfrac{3 + x}{x^2 - 4x + 3}$

Set the denominator equal to 0, and solve the equation.

$$
\begin{aligned}
x^2 - 4x + 3 &= 0 \\
(x - 3)(x - 1) &= 0 \qquad \text{Factor.} \\
x - 3 = 0 \quad &\text{or} \quad x - 1 = 0 \qquad \text{Zero-factor property} \\
x = 3 \quad &\text{or} \quad x = 1
\end{aligned}
$$

The domain of g includes all real numbers except 3 and 1.

(c) $h(x) = \dfrac{8x + 2}{3}$

The denominator, 3, can never be 0, so the domain includes all real numbers.

(d) $f(x) = \dfrac{2}{x^2 + 4}$

Setting $x^2 + 4$ equal to 0 leads to $x^2 = -4$. There is no real number whose square is -4. Therefore, any real number can be used, and as in part (c), the domain includes all real numbers.

Work Problem 1 at the Side.

1 Find all numbers that are not in the domain of each rational function.

(a) $f(x) = \dfrac{x + 4}{x - 6}$

(b) $f(x) = \dfrac{x + 6}{x^2 - x - 6}$

(c) $f(x) = \dfrac{3 + 2x}{5}$

(d) $f(x) = \dfrac{2}{x^2 + 1}$

3 Write rational expressions in lowest terms. In arithmetic, we write the fraction $\frac{15}{20}$ in lowest terms by dividing the numerator and denominator by 5 to get $\frac{3}{4}$. We write rational expressions in lowest terms in a similar way, using the **fundamental property of rational numbers.**

Fundamental Property of Rational Numbers

If $\frac{a}{b}$ is a rational number and if c is any nonzero real number, then

$$\frac{a}{b} = \frac{ac}{bc}.$$

In words, the numerator and denominator of a rational number may either be multiplied or divided by the same nonzero number without changing the value of the rational number.

Since $\frac{c}{c}$ is equivalent to 1, the fundamental property is based on the identity property of multiplication.

A rational expression is a quotient of two polynomials. Since the value of a polynomial is a real number for every value of the variable for which it is defined, any statement that applies to rational numbers will also apply to rational expressions. We use the following steps to write rational expressions in lowest terms.

ANSWERS

1. **(a)** 6 **(b)** $-2, 3$
(c) none (The domain includes all real numbers.)
(d) none (The domain includes all real numbers.)

Writing a Rational Expression in Lowest Terms

Step 1 **Factor.** Factor both numerator and denominator to find their greatest common factor (GCF).

Step 2 **Apply the fundamental property.**

Example 2 Writing Rational Expressions in Lowest Terms

Write each rational expression in lowest terms.

(a) $\frac{8k}{16} = \frac{k \cdot 8}{2 \cdot 8} = \frac{k}{2} \cdot 1 = \frac{k}{2}$

Here, the GCF of the numerator and denominator is 8. We then applied the fundamental property.

(b) $\frac{8 + k}{16}$

The numerator cannot be factored, so this expression cannot be simplified further and is in lowest terms.

(c) $\frac{a^2 - a - 6}{a^2 + 5a + 6} = \frac{(a - 3)(a + 2)}{(a + 3)(a + 2)}$ Factor the numerator and the denominator.

$= \frac{a - 3}{a + 3} \cdot 1$ $\quad \frac{a + 2}{a + 2} = 1$

$= \frac{a - 3}{a + 3}$ Lowest terms

(d) $\frac{y^2 - 4}{2y + 4} = \frac{(y + 2)(y - 2)}{2(y + 2)} = \frac{y - 2}{2}$

(e) $\frac{x^3 - 27}{x - 3} = \frac{(x - 3)(x^2 + 3x + 9)}{x - 3}$ Factor the difference of cubes.

$= x^2 + 3x + 9$ Lowest terms

(f) $\frac{pr + qr + ps + qs}{pr + qr - ps - qs} = \frac{(pr + qr) + (ps + qs)}{(pr + qr) - (ps + qs)}$ Group terms.

$= \frac{r(p + q) + s(p + q)}{r(p + q) - s(p + q)}$ Factor within groups.

$= \frac{(p + q)(r + s)}{(p + q)(r - s)}$ Factor by grouping.

$= \frac{r + s}{r - s}$ Lowest terms

CAUTION

Be careful! When using the fundamental property of rational numbers, only common *factors* may be divided. For example,

$$\frac{y - 2}{2} \neq y \quad \text{and} \quad \frac{y - 2}{2} \neq y - 1$$

because the 2 in $y - 2$ is not a *factor* of the numerator. Remember to *factor* before writing a fraction in lowest terms.

Work Problem 2 at the Side.

In the rational expression from Example 2(c),

$$\frac{a^2 - a - 6}{a^2 + 5a + 6}, \quad \text{or} \quad \frac{(a-3)(a+2)}{(a+3)(a+2)},$$

a can take any value except -3 or -2 since these values make the denominator 0. In the simplified rational expression

$$\frac{a-3}{a+3},$$

a cannot equal -3. Because of this,

$$\frac{a^2 - a - 6}{a^2 + 5a + 6} = \frac{a-3}{a+3}$$

for all values of a except -3 or -2. From now on such statements of equality will be made with the understanding that they apply only for those real numbers that make neither denominator equal 0. We will no longer state such restrictions.

Example 3 Writing Rational Expressions in Lowest Terms

Write each rational expression in lowest terms.

(a) $\dfrac{m-3}{3-m}$

In this rational expression, the numerator and denominator are opposites. The given expression can be written in lowest terms by writing the denominator as $-1(m-3)$, giving

$$\frac{m-3}{3-m} = \frac{m-3}{-1(m-3)} = \frac{1}{-1} = -1.$$

The numerator could have been rewritten instead to get the same result.

(b) $\dfrac{r^2-16}{4-r} = \dfrac{(r+4)(r-4)}{4-r}$

$= \dfrac{(r+4)(r-4)}{-1(r-4)}$ Write $4 - r$ as $-1(r-4)$.

$= \dfrac{r+4}{-1}$ Fundamental property

$= -(r+4)$ or $-r-4$ Lowest terms

As shown in Examples 3(a) and (b), the quotient

$$\frac{a}{-a} \quad (a \neq 0)$$

can be simplified as

$$\frac{a}{-a} = \frac{a}{-1(a)} = \frac{1}{-1} = -1.$$

The following statement summarizes this result.

In general, if the numerator and the denominator of a rational expression are opposites, the expression equals -1.

2 Write each rational expression in lowest terms.

(a) $\dfrac{y^2+2y-3}{y^2-3y+2}$

(b) $\dfrac{3y+9}{y^2-9}$

(c) $\dfrac{y+2}{y^2+4}$

(d) $\dfrac{1+p^3}{1+p}$

(e) $\dfrac{3x+3y+rx+ry}{5x+5y-rx-ry}$

ANSWERS

2. (a) $\dfrac{y+3}{y-2}$ **(b)** $\dfrac{3}{y-3}$

(c) already in lowest terms

(d) $1-p+p^2$ **(e)** $\dfrac{3+r}{5-r}$

3 Write each rational expression in lowest terms.

(a) $\dfrac{y-2}{2-y}$

(b) $\dfrac{8-b}{8+b}$

(c) $\dfrac{p-2}{4-p^2}$

Based on this result,

$$\frac{q-7}{7-q}=-1 \quad \text{and} \quad \frac{-5a+2b}{5a-2b}=-1.$$

However,

$$\frac{r-2}{r+2}$$

cannot be simplified further since the numerator and the denominator are *not* opposites.

Work Problem 3 at the Side.

4 Multiply rational expressions. To multiply rational expressions, follow these steps. (In practice, we usually simplify before multiplying.)

Multiplying Rational Expressions

Step 1 **Factor.** Factor all numerators and denominators as completely as possible.

Step 2 **Apply the fundamental property.**

Step 3 **Multiply.** Multiply remaining factors in the numerator and remaining factors in the denominator. Leave the denominator in factored form.

Step 4 **Check.** Check to be sure the product is in lowest terms.

Example 4 Multiplying Rational Expressions

Multiply.

(a)
$$\frac{5p-5}{p}\cdot\frac{3p^2}{10p-10}=\frac{5(p-1)}{p}\cdot\frac{3p\cdot p}{2\cdot 5(p-1)} \quad \text{Factor.}$$
$$=\frac{1}{1}\cdot\frac{3p}{2} \quad \text{Lowest terms}$$
$$=\frac{3p}{2} \quad \text{Multiply.}$$

(b)
$$\frac{k^2+2k-15}{k^2-4k+3}\cdot\frac{k^2-k}{k^2+k-20}=\frac{(k+5)(k-3)}{(k-3)(k-1)}\cdot\frac{k(k-1)}{(k+5)(k-4)}$$
$$=\frac{k}{k-4}$$

(c)
$$(p-4)\cdot\frac{3}{5p-20}=\frac{p-4}{1}\cdot\frac{3}{5p-20} \quad \text{Write } p-4 \text{ as } \frac{p-4}{1}.$$
$$=\frac{p-4}{1}\cdot\frac{3}{5(p-4)} \quad \text{Factor.}$$
$$=\frac{3}{5}$$

Continued on Next Page

ANSWERS

3. (a) -1 (b) already in lowest terms (c) $\dfrac{-1}{2+p}$

(d) $$\frac{x^2+2x}{x+1}\cdot\frac{x^2-1}{x^3+x^2}=\frac{x(x+2)}{x+1}\cdot\frac{(x+1)(x-1)}{x^2(x+1)}$$ Factor.

$$=\frac{(x+2)(x-1)}{x(x+1)}$$ Multiply; lowest terms.

(e) $$\frac{x-6}{x^2-12x+36}\cdot\frac{x^2-3x-18}{x^2+7x+12}=\frac{x-6}{(x-6)^2}\cdot\frac{(x+3)(x-6)}{(x+3)(x+4)}$$ Factor.

$$=\frac{1}{x+4}$$ Lowest terms

Remember to include 1 in the numerator when all other factors are eliminated using the fundamental property.

Work Problem 4 at the Side.

5 Find reciprocals for rational expressions. The rational numbers $\frac{a}{b}$ and $\frac{c}{d}$ are reciprocals of each other if they have a product of 1. The **reciprocal** of a rational expression is defined in the same way: Two rational expressions are reciprocals of each other if they have a product of 1. Recall that 0 has no reciprocal. The table shows several rational expressions and their reciprocals. In the first two cases, check that the product of the rational expression and its reciprocal is 1.

Rational Expression	Reciprocal
$\frac{5}{k}$	$\frac{k}{5}$
$\frac{m^2-9m}{2}$	$\frac{2}{m^2-9m}$
$\frac{0}{4}$	undefined

The examples in the table suggest the following procedure.

Finding the Reciprocal

To find the reciprocal of a nonzero rational expression, invert the rational expression.

Work Problem 5 at the Side.

6 Divide rational expressions. Dividing rational expressions is like dividing rational numbers.

Dividing Rational Expressions

To divide two rational expressions, *multiply* the first by the reciprocal of the second.

4 Multiply.

(a) $\frac{2r+4}{5r}\cdot\frac{3r}{5r+10}$

(b) $\frac{c^2+2c}{c^2-4}\cdot\frac{c^2-4c+4}{c^2-c}$

(c) $\frac{m^2-16}{m+2}\cdot\frac{1}{m+4}$

(d) $\frac{x-3}{x^2+2x-15}\cdot\frac{x^2-25}{x^2+3x-40}$

5 Find each reciprocal.

(a) $\frac{-3}{r}$

(b) $\frac{7}{y+8}$

(c) $\frac{a^2+7a}{2a-1}$

(d) $\frac{0}{-5}$

ANSWERS

4. (a) $\frac{6}{25}$ **(b)** $\frac{c-2}{c-1}$ **(c)** $\frac{m-4}{m+2}$ **(d)** $\frac{1}{x+8}$

5. (a) $\frac{r}{-3}$ **(b)** $\frac{y+8}{7}$ **(c)** $\frac{2a-1}{a^2+7a}$

(d) There is no reciprocal.

6 Divide.

(a) $\dfrac{16k^2}{5} \div \dfrac{3k}{10}$

(b) $\dfrac{5p+2}{6} \div \dfrac{15p+6}{5}$

(c)

$\dfrac{y^2-2y-3}{y^2+4y+4} \div \dfrac{y^2-1}{y^2+y-2}$

Example 5 Dividing Rational Expressions

Divide.

(a) $\dfrac{2z}{9} \div \dfrac{5z^2}{18} = \dfrac{2z}{9} \cdot \dfrac{18}{5z^2}$ Multiply by the reciprocal of the divisor.

$= \dfrac{2z}{9} \cdot \dfrac{2 \cdot 9}{5z^2}$ Factor.

$= \dfrac{4}{5z}$ Multiply; lowest terms

(b) $\dfrac{8k-16}{3k} \div \dfrac{3k-6}{4k^2} = \dfrac{8k-16}{3k} \cdot \dfrac{4k^2}{3k-6}$ Multiply by the reciprocal.

$= \dfrac{8(k-2)}{3k} \cdot \dfrac{4k^2}{3(k-2)}$ Factor.

$= \dfrac{32k}{9}$ Multiply; lowest terms

(c) $\dfrac{5m^2+17m-12}{3m^2+7m-20} \div \dfrac{5m^2+2m-3}{15m^2-34m+15}$

$= \dfrac{5m^2+17m-12}{3m^2+7m-20} \cdot \dfrac{15m^2-34m+15}{5m^2+2m-3}$ Definition of division

$= \dfrac{(5m-3)(m+4)}{(m+4)(3m-5)} \cdot \dfrac{(3m-5)(5m-3)}{(5m-3)(m+1)}$ Factor.

$= \dfrac{5m-3}{m+1}$ Lowest terms

Work Problem **6** at the Side.

ANSWERS

6. (a) $\dfrac{32k}{3}$ (b) $\dfrac{5}{18}$ (c) $\dfrac{y-3}{y+2}$

7.1 EXERCISES

FOR EXTRA HELP

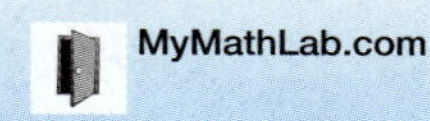
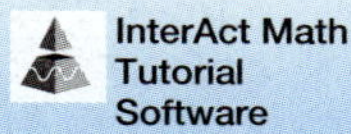
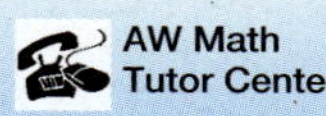

Student's Solutions Manual | MyMathLab.com | InterAct Math Tutorial Software | AW Math Tutor Center | www.mathxl.com | Digital Video Tutor CD 5 Videotape 11

Study Skills Workbook **Activity 11**

Rational expressions can often be written in lowest terms in seemingly different ways. For example,

$$\frac{y-3}{-5} \quad and \quad \frac{-y+3}{5}$$

look different, but we get the second expression by multiplying the first by -1 in both the numerator and denominator. To practice recognizing equivalent rational expressions, match the expressions in Exercises 1–6 with their equivalents in Choices A–F.

1. $\frac{x-3}{x+4}$ **2.** $\frac{x+3}{x-4}$ **3.** $\frac{x-3}{x-4}$ **4.** $\frac{x+3}{x+4}$ **5.** $\frac{3-x}{x+4}$ **6.** $\frac{x+3}{4-x}$

A. $\frac{-x-3}{4-x}$ **B.** $\frac{-x-3}{-x-4}$ **C.** $\frac{3-x}{-x-4}$ **D.** $\frac{-x+3}{-x+4}$ **E.** $\frac{x-3}{-x-4}$ **F.** $\frac{-x-3}{x-4}$

7. In Example 1(a), we showed that the domain of the rational function defined by $f(x) = \frac{3}{7x-14}$ does not include 2. Explain in your own words why this is so. In general, how do we find the value or values excluded from the domain of a rational function?

8. The domain of the rational function defined by $g(x) = \frac{x+1}{x^2+3}$ includes all real numbers. Explain.

Find all numbers that are not in the domain of each function. See Example 1.

9. $f(x) = \frac{x}{x-7}$

10. $f(x) = \frac{x}{x+3}$

11. $f(x) = \frac{6x-5}{7x+1}$

12. $f(x) = \frac{8x-3}{2x+7}$

13. $f(x) = \frac{12x+3}{x}$

14. $f(x) = \frac{9x+8}{x}$

15. $f(x) = \frac{3x+1}{2x^2+x-6}$

16. $f(x) = \frac{2x+4}{3x^2+11x-42}$

17. $f(x) = \frac{x+2}{14}$

18. $f(x) = \frac{x-9}{26}$

19. $f(x) = \frac{2x^2-3x+4}{3x^2+8}$

20. $f(x) = \frac{9x^2-8x+3}{4x^2+1}$

21. (a) Identify the two *terms* in the numerator and the two *terms* in the denominator of the rational expression $\frac{x^2 + 4x}{x + 4}$.

(b) Describe the steps you would use to write this rational expression in lowest terms. (*Hint:* It simplifies to x.)

22. Only one of the following rational expressions can be simplified. Which one is it?

A. $\frac{x^2 + 2}{x^2}$ **B.** $\frac{x^2 + 2}{2}$

C. $\frac{x^2 + y^2}{y^2}$ **D.** $\frac{x^2 - 5x}{x}$

23. Only one of the following rational expressions is *not* equivalent to $\frac{x - 3}{4 - x}$. Which one is it?

A. $\frac{3 - x}{x - 4}$ **B.** $\frac{x + 3}{4 + x}$

C. $-\frac{3 - x}{4 - x}$ **D.** $-\frac{x - 3}{x - 4}$

24. Which two of the following rational expressions equal -1?

A. $\frac{2x + 3}{2x - 3}$ **B.** $\frac{2x - 3}{3 - 2x}$

C. $\frac{2x + 3}{3 + 2x}$ **D.** $\frac{2x + 3}{-2x - 3}$

Write each rational expression in lowest terms. See Example 2.

25. $\frac{x^2(x + 1)}{x(x + 1)}$ **26.** $\frac{y^3(y - 4)}{y^2(y - 4)}$ **27.** $\frac{(x + 4)(x - 3)}{(x + 5)(x + 4)}$

28. $\frac{(2x + 7)(x - 1)}{(2x + 3)(2x + 7)}$ **29.** $\frac{4x(x + 3)}{8x^2(x - 3)}$ **30.** $\frac{5y^2(y + 8)}{15y(y - 8)}$

31. $\frac{3x + 7}{3}$ **32.** $\frac{4x - 9}{4}$ **33.** $\frac{6m + 18}{7m + 21}$

34. $\frac{5r - 20}{3r - 12}$ **35.** $\frac{3z^2 + z}{18z + 6}$ **36.** $\frac{2x^2 - 5x}{16x - 40}$

37. $\frac{2t + 6}{t^2 - 9}$ **38.** $\frac{5s - 25}{s^2 - 25}$ **39.** $\frac{x^2 + 2x - 15}{x^2 + 6x + 5}$

40. $\frac{y^2 - 5y - 14}{y^2 + y - 2}$

41. $\frac{8x^2 - 10x - 3}{8x^2 - 6x - 9}$

42. $\frac{12x^2 - 4x - 5}{8x^2 - 6x - 5}$

43. $\frac{a^3 + b^3}{a + b}$

44. $\frac{r^3 - s^3}{r - s}$

45. $\frac{2c^2 + 2cd - 60d^2}{2c^2 - 12cd + 10d^2}$

46. $\frac{3s^2 - 9st - 54t^2}{3s^2 - 6st - 72t^2}$

47. $\frac{ac - ad + bc - bd}{ac - ad - bc + bd}$

48. $\frac{2xy + 2xw + y + w}{2xy + y - 2xw - w}$

Write each rational expression in lowest terms. See Example 3.

49. $\frac{7 - b}{b - 7}$

50. $\frac{r - 13}{13 - r}$

51. $\frac{x^2 - y^2}{y - x}$

52. $\frac{m^2 - n^2}{n - m}$

53. $\frac{(a - 3)(x + y)}{(3 - a)(x - y)}$

54. $\frac{(8 - p)(x + 2)}{(p - 8)(x - 2)}$

55. $\frac{5k - 10}{20 - 10k}$

56. $\frac{7x - 21}{63 - 21x}$

57. $\frac{a^2 - b^2}{a^2 + b^2}$

58. $\frac{p^2 + q^2}{p^2 - q^2}$

Multiply or divide as indicated. See Examples 4 and 5.

59. $\frac{(x + 2)(x + 1)}{(x + 3)(x - 2)} \cdot \frac{(x + 3)(x + 4)}{(x + 2)(x + 1)}$

60. $\frac{(x + 3)(x - 4)}{(x - 4)(x + 2)} \cdot \frac{(x + 5)(x - 6)}{(x + 3)(x - 6)}$

61. $\frac{(2x + 3)(x - 4)}{(x + 8)(x - 4)} \div \frac{(x - 4)(x + 2)}{(x - 4)(x + 8)}$

62. $\frac{(6x + 5)(x - 3)}{(x + 9)(x - 1)} \div \frac{(x - 3)(2x + 7)}{(x - 1)(x + 9)}$

63. $\dfrac{7t+7}{-6} \div \dfrac{4t+4}{15}$

64. $\dfrac{8z-16}{-20} \div \dfrac{3z-6}{40}$

65. $\dfrac{4x}{8x+4} \cdot \dfrac{14x+7}{6}$

66. $\dfrac{12x-20}{5x} \cdot \dfrac{6}{9x-15}$

67. $\dfrac{p^2-25}{4p} \cdot \dfrac{2}{5-p}$

68. $\dfrac{a^2-1}{4a} \cdot \dfrac{2}{1-a}$

69. $\dfrac{m^2-49}{m+1} \div \dfrac{7-m}{m}$

70. $\dfrac{k^2-4}{3k^2} \div \dfrac{2-k}{11k}$

71. $\dfrac{12x-10y}{3x+2y} \cdot \dfrac{6x+4y}{10y-12x}$

72. $\dfrac{9s-12t}{2s+2t} \cdot \dfrac{3s+3t}{4t-3s}$

73. $\dfrac{x^2-25}{x^2+x-20} \cdot \dfrac{x^2+7x+12}{x^2-2x-15}$

74. $\dfrac{t^2-49}{t^2+4t-21} \cdot \dfrac{t^2+8t+15}{t^2-2t-35}$

75. $\dfrac{6x^2+5xy-6y^2}{12x^2-11xy+2y^2} \div \dfrac{4x^2-12xy+9y^2}{8x^2-14xy+3y^2}$

76. $\dfrac{8a^2-6ab-9b^2}{6a^2-5ab-6b^2} \div \dfrac{4a^2+11ab+6b^2}{9a^2+12ab+4b^2}$

77. $\dfrac{3k^2+17kp+10p^2}{6k^2+13kp-5p^2} \div \dfrac{6k^2+kp-2p^2}{6k^2-5kp+p^2}$

78. $\dfrac{16c^2+24cd+9d^2}{16c^2-16cd+3d^2} \div \dfrac{16c^2-9d^2}{16c^2-24cd+9d^2}$

79. $\left(\dfrac{6k^2-13k-5}{k^2+7k} \div \dfrac{2k-5}{k^3+6k^2-7k}\right) \cdot \dfrac{k^2-5k+6}{3k^2-8k-3}$

80. $\left(\dfrac{2x^3+3x^2-2x}{3x-15} \div \dfrac{2x^3-x^2}{x^2-3x-10}\right) \cdot \dfrac{5x^2-10x}{3x^2+12x+12}$

7.2 Adding and Subtracting Rational Expressions

Objectives

1. Add and subtract rational expressions with the same denominator.
2. Find a least common denominator.
3. Add and subtract rational expressions with different denominators.

1 Add and subtract rational expressions with the same denominator. The following steps, used to add or subtract rational numbers, are also used to add or subtract rational expressions.

Adding or Subtracting Rational Expressions

Step 1 **If the denominators are the same,** add or subtract the numerators. Place the result over the common denominator.

If the denominators are different, first find the least common denominator. Write all rational expressions with this LCD, and then add or subtract the numerators. Place the result over the common denominator.

Step 2 **Simplify.** Write all answers in lowest terms.

Example 1 **Adding and Subtracting Rational Expressions with the Same Denominator**

Add or subtract as indicated.

(a) $\frac{3y}{5} + \frac{x}{5} = \frac{3y + x}{5}$

The denominators of these rational expressions are the same, so just add the numerators, and place the sum over the common denominator.

(b) $\frac{7}{2r^2} - \frac{11}{2r^2} = \frac{7 - 11}{2r^2} = \frac{-4}{2r^2} = -\frac{2}{r^2}$ Lowest terms

Subtract the numerators since the denominators are the same, and keep the common denominator.

(c) $\frac{m}{m^2 - p^2} + \frac{p}{m^2 - p^2} = \frac{m + p}{m^2 - p^2}$ Add the numerators; keep the common denominator.

$= \frac{m + p}{(m + p)(m - p)}$ Factor.

$= \frac{1}{m - p}$ Lowest terms

(d) $\frac{4}{x^2 + 2x - 8} + \frac{x}{x^2 + 2x - 8} = \frac{4 + x}{x^2 + 2x - 8}$

$= \frac{4 + x}{(x - 2)(x + 4)}$

$= \frac{1}{x - 2}$

Work Problem 1 at the Side.

1 Add or subtract.

(a) $\frac{3m}{8} + \frac{5n}{8}$

(b) $\frac{7}{3a} + \frac{10}{3a}$

(c) $\frac{2}{y^2} - \frac{5}{y^2}$

(d) $\frac{a}{a + b} + \frac{b}{a + b}$

(e) $\frac{2y - 1}{y^2 + y - 2} - \frac{y}{y^2 + y - 2}$

Answers

1. (a) $\frac{3m + 5n}{8}$ **(b)** $\frac{17}{3a}$ **(c)** $-\frac{3}{y^2}$ **(d)** 1 **(e)** $\frac{1}{y + 2}$

2 Find the LCD for each pair of denominators.

(a) $5k^3s, \quad 10ks^4$

(b) $3 - x, \quad 9 - x^2$

(c) $z, \quad z + 6$

(d) $2y^2 - 3y - 2,$
$2y^2 + 3y + 1$

Answers

2. (a) $10k^3s^4$ **(b)** $(3 + x)(3 - x)$
(c) $z(z + 6)$ **(d)** $(y - 2)(2y + 1)(y + 1)$

2 **Find a least common denominator.** We add or subtract rational expressions with different denominators by first writing them with a common denominator, usually the **least common denominator (LCD).**

Finding the Least Common Denominator

Step 1 **Factor.** Factor each denominator.

Step 2 **Find the least common denominator.** The LCD is the product of all different factors from each denominator, with each factor raised to the *greatest* power that occurs in any denominator.

Example 2 Finding Least Common Denominators

Assume that the given expressions are denominators of two fractions. Find the LCD for each pair.

(a) $5xy^2, \quad 2x^3y$

Each denominator is already factored.

$$5xy^2 = 5 \cdot x \cdot y^2$$
$$2x^3y = 2 \cdot x^3 \cdot y$$

Greatest exponent on x is 3.

$$\text{LCD} = 5 \cdot 2 \cdot x^3 \cdot y^2 \quad \leftarrow \text{Greatest exponent on } y \text{ is 2.}$$
$$= 10x^3y^2$$

(b) $k - 3, \quad k$

Each denominator is already factored. The LCD, an expression divisible by *both* $k - 3$ and k, is

$$k(k - 3).$$

It is usually best to leave a least common denominator in factored form.

(c) $y^2 - 2y - 8, \quad y^2 + 3y + 2$

Factor the denominators.

$$\left.\begin{aligned} y^2 - 2y - 8 &= (y - 4)(y + 2) \\ y^2 + 3y + 2 &= (y + 2)(y + 1) \end{aligned}\right\} \text{Factor.}$$

The LCD, divisible by both polynomials, is

$$(y - 4)(y + 2)(y + 1).$$

(d) $8z - 24, \quad 5z^2 - 15z$

$$\left.\begin{aligned} 8z - 24 &= 8(z - 3) \\ 5z^2 - 15z &= 5z(z - 3) \end{aligned}\right\} \text{Factor.}$$

The LCD is $8 \cdot 5z \cdot (z - 3) = 40z(z - 3)$.

Work Problem 2 at the Side.

3 Add and subtract rational expressions with different denominators. Before adding or subtracting two rational expressions, we write each expression with the least common denominator by multiplying its numerator and denominator by the factors needed to get the LCD. This procedure is valid because we are multiplying each rational expression by a form of 1, the identity element for multiplication.

Adding or subtracting rational expressions follows the same procedure as that used for rational numbers. Consider the sum $\frac{7}{15} + \frac{5}{12}$. The LCD for 15 and 12 is 60. Multiply $\frac{7}{15}$ by $\frac{4}{4}$ (a form of 1) and multiply $\frac{5}{12}$ by $\frac{5}{5}$ so that each fraction has denominator 60, and then add the numerators.

$$\frac{7}{15} + \frac{5}{12} = \frac{7 \cdot 4}{15 \cdot 4} + \frac{5 \cdot 5}{12 \cdot 5} \quad \text{Fundamental property}$$

$$= \frac{28}{60} + \frac{25}{60}$$

$$= \frac{28 + 25}{60} \quad \text{Add the numerators.}$$

$$= \frac{53}{60}$$

Example 3 Adding and Subtracting Rational Expressions with Different Denominators

Add or subtract as indicated.

(a) $\frac{5}{2p} + \frac{3}{8p}$

The LCD for $2p$ and $8p$ is $8p$. To write the first rational expression with a denominator of $8p$, multiply by $\frac{4}{4}$.

$$\frac{5}{2p} + \frac{3}{8p} = \frac{5 \cdot 4}{2p \cdot 4} + \frac{3}{8p} \quad \text{Fundamental property}$$

$$= \frac{20}{8p} + \frac{3}{8p}$$

$$= \frac{20 + 3}{8p} \quad \text{Add the numerators.}$$

$$= \frac{23}{8p}$$

(b) $\frac{6}{r} - \frac{5}{r - 3}$

The LCD is $r(r - 3)$. Rewrite each rational expression with this denominator.

$$\frac{6}{r} - \frac{5}{r - 3} = \frac{6(r - 3)}{r(r - 3)} - \frac{r \cdot 5}{r(r - 3)} \quad \text{Fundamental property}$$

$$= \frac{6r - 18}{r(r - 3)} - \frac{5r}{r(r - 3)} \quad \text{Distributive and commutative properties}$$

$$= \frac{6r - 18 - 5r}{r(r - 3)} \quad \text{Subtract the numerators.}$$

$$= \frac{r - 18}{r(r - 3)} \quad \text{Combine terms in the numerator.}$$

Work Problem 3 at the Side.

3 Add or subtract.

(a) $\frac{6}{7} + \frac{1}{5}$

(b) $\frac{8}{3k} - \frac{2}{9k}$

(c) $\frac{2}{y} - \frac{1}{y + 4}$

Answers

3. (a) $\frac{37}{35}$ **(b)** $\frac{22}{9k}$ **(c)** $\frac{y + 8}{y(y + 4)}$

4 Subtract.

(a) $\dfrac{5x + 7}{2x + 7} - \dfrac{-x - 14}{2x + 7}$

(b) $\dfrac{2}{r - 2} - \dfrac{r}{r - 1}$

CAUTION

One of the most common sign errors in algebra occurs when a rational expression with two or more terms in the numerator is being subtracted. Remember that in this situation, the subtraction sign must be distributed to *every* term in the numerator of the fraction that follows it. Study Example 4 carefully to see how this is done.

Example 4 **Using the Distributive Property When Subtracting Rational Expressions**

Subtract.

(a) $\dfrac{7x}{3x + 1} - \dfrac{x - 2}{3x + 1}$

The denominators are the same for both rational expressions. The subtraction sign must be applied to *both* terms in the numerator of the second rational expression. Notice the careful use of the distributive property here.

$$\frac{7x}{3x+1} - \frac{x-2}{3x+1} = \frac{7x - (x-2)}{3x+1} \qquad \text{Write as a single rational expression.}$$

$$= \frac{7x - x + 2}{3x+1} \qquad \text{Distributive property; be careful with signs.}$$

$$= \frac{6x + 2}{3x+1} \qquad \text{Combine terms in the numerator.}$$

$$= \frac{2(3x+1)}{3x+1} \qquad \text{Factor the numerator.}$$

$$= 2 \qquad \text{Lowest terms}$$

(b) $\dfrac{1}{q - 1} - \dfrac{1}{q + 1}$

$$= \frac{1(q+1)}{(q-1)(q+1)} - \frac{1(q-1)}{(q+1)(q-1)} \qquad \text{Fundamental property}$$

$$= \frac{(q+1) - (q-1)}{(q-1)(q+1)} \qquad \text{Subtract.}$$

$$= \frac{q + 1 - q + 1}{(q-1)(q+1)} \qquad \text{Distributive property}$$

$$= \frac{2}{(q-1)(q+1)} \qquad \text{Combine terms in the numerator.}$$

Work Problem 4 at the Side.

In some problems, rational expressions to be added or subtracted have denominators that are opposites of each other. The next example illustrates how to proceed in such a problem.

ANSWERS

4. **(a)** 3 **(b)** $\dfrac{-r^2 + 4r - 2}{(r - 2)(r - 1)}$

Example 5 Adding Rational Expressions with Denominators That Are Opposites

Add.

$$\frac{y}{y-2}+\frac{8}{2-y}$$

To get a common denominator of $y-2$, multiply the second expression by -1 in both the numerator and the denominator.

$$\frac{y}{y-2}+\frac{8}{2-y}=\frac{y}{y-2}+\frac{8(-1)}{(2-y)(-1)}$$

$$=\frac{y}{y-2}+\frac{-8}{y-2}$$

$$=\frac{y-8}{y-2} \qquad \text{Add the numerators.}$$

Work Problem 5 at the Side.

The next example illustrates addition and subtraction involving more than two rational expressions.

Example 6 Adding and Subtracting Three Rational Expressions

Add and subtract as indicated.

$$\frac{3}{x-2}+\frac{5}{x}-\frac{6}{x^2-2x}$$

The denominator of the third rational expression factors as $x(x-2)$, which is the LCD for the three rational expressions.

$$\frac{3}{x-2}+\frac{5}{x}-\frac{6}{x^2-2x}$$

$$=\frac{3x}{x(x-2)}+\frac{5(x-2)}{x(x-2)}-\frac{6}{x(x-2)} \qquad \text{Fundamental property}$$

$$=\frac{3x+5(x-2)-6}{x(x-2)} \qquad \text{Add and subtract the numerators.}$$

$$=\frac{3x+5x-10-6}{x(x-2)} \qquad \text{Distributive property}$$

$$=\frac{8x-16}{x(x-2)} \qquad \text{Combine terms in the numerator.}$$

$$=\frac{8(x-2)}{x(x-2)} \qquad \text{Factor the numerator.}$$

$$=\frac{8}{x} \qquad \text{Lowest terms}$$

Work Problem 6 at the Side.

5 Add or subtract as indicated.

(a) $\frac{8}{x-4}+\frac{2}{4-x}$

(b) $\frac{9}{2x-9}-\frac{4}{9-2x}$

6 Add and subtract as indicated.

$$\frac{4}{x-5}+\frac{-2}{x}-\frac{10}{x^2-5x}$$

ANSWERS

5. **(a)** $\frac{6}{x-4}$ or $\frac{-6}{4-x}$ **(b)** $\frac{13}{2x-9}$ or $\frac{-13}{9-2x}$

6. $\frac{2}{x-5}$

7 Subtract.

$$\frac{-a}{a^2+3a-4}-\frac{4a}{a^2+7a+12}$$

Example 7 **Subtracting Rational Expressions**

Subtract.

$$\frac{m+4}{m^2-2m-3}-\frac{2m-3}{m^2-5m+6}$$

$$=\frac{m+4}{(m-3)(m+1)}-\frac{2m-3}{(m-3)(m-2)}$$ Factor each denominator.

The LCD is $(m-3)(m+1)(m-2)$.

$$=\frac{(m+4)(m-2)}{(m-3)(m+1)(m-2)}-\frac{(2m-3)(m+1)}{(m-3)(m-2)(m+1)}$$ Fundamental property

$$=\frac{(m+4)(m-2)-(2m-3)(m+1)}{(m-3)(m+1)(m-2)}$$ Subtract.

$$=\frac{m^2+2m-8-(2m^2-m-3)}{(m-3)(m+1)(m-2)}$$ Multiply in the numerator.

$$=\frac{m^2+2m-8-2m^2+m+3}{(m-3)(m+1)(m-2)}$$ Distributive property; be careful with signs.

$$=\frac{-m^2+3m-5}{(m-3)(m+1)(m-2)}$$ Combine terms in the numerator.

If we try to factor the numerator, we find that this rational expression is in lowest terms.

Work Problem 7 at the Side.

ANSWERS

7. $\frac{-5a^2+a}{(a+4)(a-1)(a+3)}$

7.2 EXERCISES

FOR EXTRA HELP

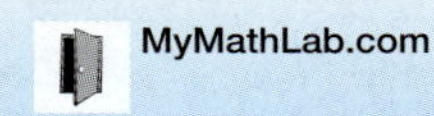
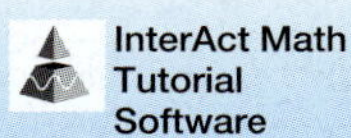

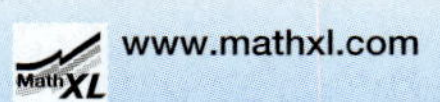
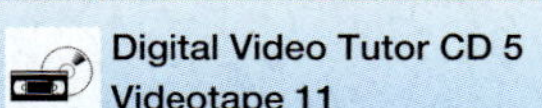

1. Write an explanation for adding or subtracting rational expressions that have a common denominator.

2. Write an explanation for adding or subtracting rational expressions that have different denominators.

Add or subtract as indicated. Write all answers in lowest terms. See Example 1.

3. $\frac{7}{t} + \frac{2}{t}$

4. $\frac{5}{r} + \frac{9}{r}$

5. $\frac{11}{5x} - \frac{1}{5x}$

6. $\frac{7}{4y} - \frac{3}{4y}$

7. $\frac{5x + 4}{6x + 5} + \frac{x + 1}{6x + 5}$

8. $\frac{6y + 12}{4y + 3} + \frac{2y - 6}{4y + 3}$

9. $\frac{x^2}{x + 5} - \frac{25}{x + 5}$

10. $\frac{y^2}{y + 6} - \frac{36}{y + 6}$

11. $\frac{4}{p^2 + 7p + 12} + \frac{p}{p^2 + 7p + 12}$

12. $\frac{5}{x^2 + x - 20} + \frac{x}{x^2 + x - 20}$

13. $\frac{a^3}{a^2 + ab + b^2} - \frac{b^3}{a^2 + ab + b^2}$

14. $\frac{p^3}{p^2 - pq + q^2} + \frac{q^3}{p^2 - pq + q^2}$

Assume that the expressions given are denominators of fractions. Find the least common denominator (LCD) for each group. See Example 2.

15. $18x^2y^3, \quad 24x^4y^5$

16. $24a^3b^4, \quad 18a^5b^2$

17. $z - 2, \quad z$

18. $k + 3, \quad k$

19. $2y + 8, \quad y + 4$

20. $3r - 21, \quad r - 7$

21. $x^2 - 81, \quad x^2 + 18x + 81$

22. $y^2 - 16, \quad y^2 - 8y + 16$

23. $m + n, \quad m - n, \quad m^2 - n^2$

24. $r + s, \quad r - s, \quad r^2 - s^2$

25. $x^2 - 3x - 4, \quad x + x^2$

26. $y^2 - 8y + 12, \quad y^2 - 6y$

27. $2t^2 + 7t - 15, \quad t^2 + 3t - 10$

28. $s^2 - 3s - 4, \quad 3s^2 + s - 2$

29. $2y + 6, \quad y^2 - 9, \quad y$

30. $9x + 18, \quad x^2 - 4, \quad x$

31. One student added two rational expressions and obtained the answer $\dfrac{3}{5-y}$. Another student obtained the answer $\dfrac{-3}{y-5}$ for the same problem. Is it possible that both answers are correct? Explain.

32. What is *wrong* with the following work?

$$\frac{x}{x+2} - \frac{4x-1}{x+2} = \frac{x-4x-1}{x+2} = \frac{-3x-1}{x+2}$$

Add or subtract as indicated. Write all answers in lowest terms. See Examples 3–7.

33. $\dfrac{8}{t} + \dfrac{7}{3t}$

34. $\dfrac{5}{x} + \dfrac{9}{4x}$

35. $\dfrac{5}{12x^2y} - \dfrac{11}{6xy}$

36. $\dfrac{7}{18a^3b^2} - \dfrac{2}{9ab}$

37. $\dfrac{1}{x-1} - \dfrac{1}{x}$

38. $\dfrac{3}{x-3} - \dfrac{1}{x}$

39. $\dfrac{3a}{a+1} + \dfrac{2a}{a-3}$

40. $\dfrac{2x}{x+4} + \dfrac{3x}{x-7}$

41. $\dfrac{17y+3}{9y+7} - \dfrac{-10y-18}{9y+7}$

42. $\dfrac{7x+8}{3x+2} - \dfrac{x+4}{3x+2}$

43. $\dfrac{2}{4-x} + \dfrac{5}{x-4}$

44. $\dfrac{3}{2-t} + \dfrac{1}{t-2}$

45. $\dfrac{w}{w-z} - \dfrac{z}{z-w}$

46. $\dfrac{a}{a-b} - \dfrac{b}{b-a}$

47. $\dfrac{5}{12+4x} - \dfrac{7}{9+3x}$

48. $\dfrac{3}{10x+15} - \dfrac{8}{12x+18}$

49. $\dfrac{4x}{x-1} - \dfrac{2}{x+1} - \dfrac{4}{x^2-1}$

50. $\dfrac{4}{x+3} - \dfrac{x}{x-3} - \dfrac{18}{x^2-9}$

51. $\dfrac{15}{y^2 + 3y} + \dfrac{2}{y} + \dfrac{5}{y + 3}$

52. $\dfrac{7}{t - 2} - \dfrac{6}{t^2 - 2t} - \dfrac{3}{t}$

53. $\dfrac{5}{x - 2} + \dfrac{1}{x} + \dfrac{2}{x^2 - 2x}$

54. $\dfrac{5x}{x - 3} + \dfrac{2}{x} + \dfrac{6}{x^2 - 3x}$

55. $\dfrac{3x}{x + 1} + \dfrac{4}{x - 1} - \dfrac{6}{x^2 - 1}$

56. $\dfrac{5x}{x + 3} + \dfrac{x + 2}{x} - \dfrac{6}{x^2 + 3x}$

57. $\dfrac{4}{x + 1} + \dfrac{1}{x^2 - x + 1} - \dfrac{12}{x^3 + 1}$

58. $\dfrac{5}{x + 2} + \dfrac{2}{x^2 - 2x + 4} - \dfrac{60}{x^3 + 8}$

59. $\dfrac{2x + 4}{x + 3} + \dfrac{3}{x} - \dfrac{6}{x^2 + 3x}$

60. $\dfrac{4x + 1}{x + 5} - \dfrac{2}{x} + \dfrac{10}{x^2 + 5x}$

61. $\dfrac{5x}{x^2 + xy - 2y^2} - \dfrac{3x}{x^2 + 5xy - 6y^2}$

62. $\dfrac{6x}{6x^2 + 5xy - 4y^2} - \dfrac{2y}{9x^2 - 16y^2}$

A concours d'elegance *is a competition in which a maximum of 100 points is awarded to a car based on its general attractiveness. The function defined by the rational expression*

$$c(x) = \frac{1010}{49(101 - x)} - \frac{10}{49}$$

approximates the cost, in thousands of dollars, of restoring a car so that it will win x points.

Use this information to work Exercises 63 and 64.

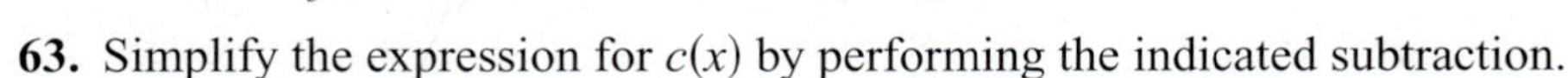

63. Simplify the expression for $c(x)$ by performing the indicated subtraction.

64. Use the simplified expression to determine how much it would cost to win 95 points.

RELATING CONCEPTS (Exercises 65–70) FOR INDIVIDUAL OR GROUP WORK

In Example 6 we showed that

$$\frac{3}{x-2}+\frac{5}{x}-\frac{6}{x^2-2x}$$

is equal to $\frac{8}{x}$. *Algebra is, in a sense, a generalized form of arithmetic.* ***Work Exercises 65–70 in order,*** *to see how the algebra in this example is related to the arithmetic of common fractions.*

65. Perform the following operations, and express your answer in lowest terms.

$$\frac{3}{7}+\frac{5}{9}-\frac{6}{63}$$

66. Substitute 9 for x in the given problem from Example 6. Compare this problem to the one given in Exercise 65. What do you notice?

67. Now substitute 9 for x in the answer given in Example 6. Do your results agree with the result you obtained in Exercise 65?

68. Replace x in the problem from Example 6 with the number of letters in your last name, assuming that this number is not 2. If your last name has two letters, let $x = 3$. Now predict the answer to your problem. Verify that your prediction is correct.

69. Why will $x = 2$ not work for the problem from Example 6?

70. What other value of x is not allowed in the problem given from Example 6?

7.3 Complex Fractions

A **complex fraction** is an expression having a fraction in the numerator, denominator, or both. Examples of complex fractions include

$$\frac{1+\frac{1}{x}}{2}, \quad \frac{\frac{4}{y}}{6-\frac{3}{y}}, \quad \text{and} \quad \frac{\frac{m^2-9}{m+1}}{\frac{m+3}{m^2-1}}.$$

Objectives

1. Simplify complex fractions by simplifying the numerator and denominator. (Method 1)
2. Simplify complex fractions by multiplying by a common denominator. (Method 2)
3. Compare the two methods of simplifying complex fractions.
4. Simplify rational expressions with negative exponents.

1 Simplify complex fractions by simplifying the numerator and denominator. (Method 1) There are two different methods for simplifying complex fractions.

Simplifying a Complex Fraction: Method 1

Step 1 Simplify the numerator and denominator separately.

Step 2 Divide by multiplying the numerator by the reciprocal of the denominator.

Step 3 Simplify the resulting fraction, if possible.

In Step 2, we are treating the complex fraction as a quotient of two rational expressions and dividing. Before performing this step, be sure that both the numerator and denominator are single fractions.

Example 1 Simplifying Complex Fractions by Method 1

Use Method 1 to simplify each complex fraction.

(a) $\dfrac{\frac{x+1}{x}}{\frac{x-1}{2x}}$

Both the numerator and the denominator are already simplified, so divide by multiplying the numerator by the reciprocal of the denominator.

$$\frac{\frac{x+1}{x}}{\frac{x-1}{2x}} = \frac{x+1}{x} \div \frac{x-1}{2x} \qquad \text{Write as a division problem.}$$

$$= \frac{x+1}{x} \cdot \frac{2x}{x-1} \qquad \text{Reciprocal of } \tfrac{x-1}{2x}$$

$$= \frac{2(x+1)}{x-1} \qquad \text{Multiply and simplify.}$$

Continued on Next Page

1 Use Method 1 to simplify each complex fraction.

(a) $\dfrac{\dfrac{a+2}{5a}}{\dfrac{a-3}{7a}}$

(b) $\dfrac{2+\dfrac{1}{k}}{2-\dfrac{1}{k}}$

(c) $\dfrac{\dfrac{r^2-4}{4}}{1+\dfrac{2}{r}}$

(b) $\dfrac{2+\dfrac{1}{y}}{3-\dfrac{2}{y}} = \dfrac{\dfrac{2y}{y}+\dfrac{1}{y}}{\dfrac{3y}{y}-\dfrac{2}{y}}$

$= \dfrac{\dfrac{2y+1}{y}}{\dfrac{3y-2}{y}}$ Simplify the numerator and denominator.

$= \dfrac{2y+1}{y} \cdot \dfrac{y}{3y-2}$ Reciprocal of $\frac{3y-2}{y}$

$= \dfrac{2y+1}{3y-2}$

Work Problem 1 at the Side.

2 Simplify complex fractions by multiplying by a common denominator. (Method 2) The second method for simplifying complex fractions uses the identity property of multiplication.

Simplifying a Complex Fraction: Method 2

Step 1 Multiply the numerator and denominator of the complex fraction by the least common denominator of the fractions in the numerator and the fractions in the denominator of the complex fraction.

Step 2 Simplify the resulting fraction, if possible.

Example 2 **Simplifying Complex Fractions by Method 2**

Use Method 2 to simplify each complex fraction.

(a) $\dfrac{2+\dfrac{1}{y}}{3-\dfrac{2}{y}}$

Multiply the numerator and denominator by the LCD of all the fractions in the numerator and denominator of the complex fraction. (This is the same as multiplying by 1.) Here the LCD is y.

$\dfrac{2+\dfrac{1}{y}}{3-\dfrac{2}{y}} = \dfrac{2+\dfrac{1}{y}}{3-\dfrac{2}{y}} \cdot 1 = \dfrac{\left(2+\dfrac{1}{y}\right)\cdot y}{\left(3-\dfrac{2}{y}\right)\cdot y}$ Multiply the numerator and denominator by y, since $\frac{y}{y}=1$.

$= \dfrac{2\cdot y+\dfrac{1}{y}\cdot y}{3\cdot y-\dfrac{2}{y}\cdot y}$ Distributive property

$= \dfrac{2y+1}{3y-2}$

Compare this method with that used in Example 1(b).

Continued on Next Page

ANSWERS

1. (a) $\dfrac{7(a+2)}{5(a-3)}$ (b) $\dfrac{2k+1}{2k-1}$ (c) $\dfrac{r(r-2)}{4}$

(b) $\dfrac{2p + \dfrac{5}{p-1}}{3p - \dfrac{2}{p}}$

The LCD is $p(p-1)$.

$$\frac{2p + \dfrac{5}{p-1}}{3p - \dfrac{2}{p}} = \frac{\left(2p + \dfrac{5}{p-1}\right) \cdot p(p-1)}{\left(3p - \dfrac{2}{p}\right) \cdot p(p-1)}$$

Multiply the numerator and denominator by the LCD.

$$= \frac{2p[p(p-1)] + \dfrac{5}{p-1} \cdot p(p-1)}{3p[p(p-1)] - \dfrac{2}{p} \cdot p(p-1)}$$

Distributive property

$$= \frac{2p[p(p-1)] + 5p}{3p[p(p-1)] - 2(p-1)}$$

$$= \frac{2p^3 - 2p^2 + 5p}{3p^3 - 3p^2 - 2p + 2}$$

This rational expression is in lowest terms.

Work Problem 2 at the Side.

3 Compare the two methods of simplifying complex fractions. Choosing whether to use Method 1 or Method 2 to simplify a complex fraction is usually a matter of preference. Some students prefer one method over the other, while other students feel comfortable with both methods and rely on practice with many examples to determine which method they will use on a particular problem. In the next example, we illustrate how to simplify a complex fraction using both methods so that you can observe the processes and decide for yourself the pros and cons of each method.

Example 3 Simplifying Complex Fractions Using Both Methods

Use both Method 1 and Method 2 to simplify each complex fraction.

Method 1

(a) $\dfrac{\dfrac{2}{x-3}}{\dfrac{5}{x^2-9}}$

$$= \frac{\dfrac{2}{x-3}}{\dfrac{5}{(x-3)(x+3)}}$$

$$= \frac{2}{x-3} \div \frac{5}{(x-3)(x+3)}$$

$$= \frac{2}{x-3} \cdot \frac{(x-3)(x+3)}{5}$$

$$= \frac{2(x+3)}{5}$$

Method 2

(a) $\dfrac{\dfrac{2}{x-3}}{\dfrac{5}{x^2-9}}$

$$= \frac{\dfrac{2}{x-3} \cdot (x-3)(x+3)}{\dfrac{5}{(x-3)(x+3)} \cdot (x-3)(x+3)}$$

$$= \frac{2(x+3)}{5}$$

Continued on Next Page

2 Use Method 2 to simplify each complex fraction.

(a) $\dfrac{\dfrac{5}{y} + 6}{\dfrac{8}{3y} - 1}$

(b) $\dfrac{\dfrac{1}{y} + \dfrac{1}{y-1}}{\dfrac{1}{y} - \dfrac{2}{y-1}}$

Answers

2. (a) $\dfrac{15 + 18y}{8 - 3y}$ **(b)** $\dfrac{2y-1}{-y-1}$ or $\dfrac{1-2y}{y+1}$

3 Use both methods to simplify each complex fraction.

(a) $\dfrac{\dfrac{5}{y+2}}{\dfrac{-3}{y^2-4}}$

(b) $\dfrac{\dfrac{1}{a}-\dfrac{1}{b}}{\dfrac{1}{a^2}-\dfrac{1}{b^2}}$

4 Simplify each expression, using only positive exponents in the answer.

(a) $\dfrac{r^{-2}-s^{-1}}{4r^{-1}+s^{-2}}$

(b) $\dfrac{b^{-4}}{b^{-5}+2}$

ANSWERS

3. (Both methods give the same answers.)

(a) $\dfrac{5(y-2)}{-3}$ (b) $\dfrac{ab}{b+a}$

4. (a) $\dfrac{s^2-r^2s}{4rs^2+r^2}$ (b) $\dfrac{b}{1+2b^5}$

Method 1

(b) $\dfrac{\dfrac{1}{x}+\dfrac{1}{y}}{\dfrac{1}{x^2}-\dfrac{1}{y^2}}$

$$=\frac{\dfrac{y}{xy}+\dfrac{x}{xy}}{\dfrac{y^2}{x^2y^2}-\dfrac{x^2}{x^2y^2}}$$

$$=\frac{\dfrac{y+x}{xy}}{\dfrac{y^2-x^2}{x^2y^2}}$$

$$=\frac{y+x}{xy}\div\frac{y^2-x^2}{x^2y^2}$$

$$=\frac{y+x}{xy}\cdot\frac{x^2y^2}{(y-x)(y+x)}$$

$$=\frac{xy}{y-x}$$

Method 2

(b) $\dfrac{\dfrac{1}{x}+\dfrac{1}{y}}{\dfrac{1}{x^2}-\dfrac{1}{y^2}}$

$$=\frac{\left(\dfrac{1}{x}+\dfrac{1}{y}\right)\cdot x^2y^2}{\left(\dfrac{1}{x^2}-\dfrac{1}{y^2}\right)\cdot x^2y^2}$$

$$=\frac{xy^2+x^2y}{y^2-x^2}$$

$$=\frac{xy(y+x)}{(y+x)(y-x)}$$

$$=\frac{xy}{y-x}$$

Work Problem 3 at the Side.

4 Simplify rational expressions with negative exponents. Rational expressions and complex fractions sometimes involve negative exponents. To simplify such expressions, we begin by rewriting the expressions with only positive exponents.

Example 4 Simplifying a Rational Expression with Negative Exponents

Simplify $\dfrac{m^{-1}+p^{-2}}{2m^{-2}-p^{-1}}$, using only positive exponents in the answer.

First write the expression with only positive exponents using the definition of a negative exponent.

$$\frac{m^{-1}+p^{-2}}{2m^{-2}-p^{-1}}=\frac{\dfrac{1}{m}+\dfrac{1}{p^2}}{\dfrac{2}{m^2}-\dfrac{1}{p}}$$

Note that the 2 in $2m^{-2}$ is not raised to the -2 power, so $2m^{-2}=\dfrac{2}{m^2}$. Simplify the complex fraction using Method 2, multiplying numerator and denominator by the LCD, m^2p^2.

$$\frac{\dfrac{1}{m}+\dfrac{1}{p^2}}{\dfrac{2}{m^2}-\dfrac{1}{p}}=\frac{m^2p^2\cdot\dfrac{1}{m}+m^2p^2\cdot\dfrac{1}{p^2}}{m^2p^2\cdot\dfrac{2}{m^2}-m^2p^2\cdot\dfrac{1}{p}}$$

$$=\frac{mp^2+m^2}{2p^2-m^2p}\qquad\text{Lowest terms}$$

Work Problem 4 at the Side.

7.3 EXERCISES

FOR EXTRA HELP

Student's Solutions Manual

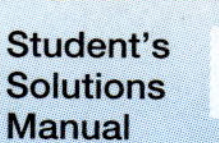
MyMathLab.com

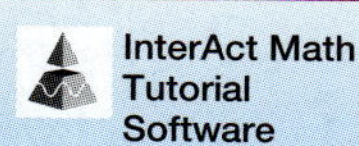
InterAct Math Tutorial Software

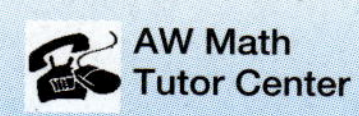
AW Math Tutor Center

www.mathxl.com

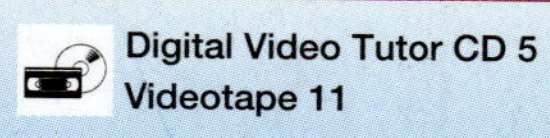
Digital Video Tutor CD 5 Videotape 11

1. Explain in your own words Method 1 for simplifying complex fractions.

2. Method 2 for simplifying complex fractions says that we can multiply both the numerator and the denominator of the complex fraction by the same nonzero expression. What property of real numbers from Section 1.4 justifies this method?

Use either method to simplify each complex fraction. See Examples 1–3.

3. $\dfrac{\dfrac{12}{x-1}}{\dfrac{6}{x}}$

4. $\dfrac{\dfrac{24}{t+4}}{\dfrac{6}{t}}$

5. $\dfrac{\dfrac{k+1}{2k}}{\dfrac{3k-1}{4k}}$

6. $\dfrac{\dfrac{1-r}{4r}}{\dfrac{-1-r}{8r}}$

7. $\dfrac{\dfrac{4z^2x^4}{9}}{\dfrac{12x^2z^5}{15}}$

8. $\dfrac{\dfrac{3y^2x^3}{8}}{\dfrac{9y^3x^4}{16}}$

9. $\dfrac{\dfrac{1}{x}+1}{-\dfrac{1}{x}+1}$

10. $\dfrac{\dfrac{2}{k}-1}{\dfrac{2}{k}+1}$

11. $\dfrac{\dfrac{3}{x}+\dfrac{3}{y}}{\dfrac{3}{x}-\dfrac{3}{y}}$

12. $\dfrac{\dfrac{4}{t}-\dfrac{4}{s}}{\dfrac{4}{t}+\dfrac{4}{s}}$

13. $\dfrac{\dfrac{8x-24y}{10}}{\dfrac{x-3y}{5x}}$

14. $\dfrac{\dfrac{10x-5y}{12}}{\dfrac{2x-y}{6y}}$

15. $\dfrac{\dfrac{x^2-16y^2}{xy}}{\dfrac{1}{y}-\dfrac{4}{x}}$

16. $\dfrac{\dfrac{2}{s}-\dfrac{3}{t}}{\dfrac{4t^2-9s^2}{st}}$

17. $\dfrac{y-\dfrac{y-3}{3}}{\dfrac{4}{9}+\dfrac{2}{3y}}$

18. $\dfrac{p - \dfrac{p+2}{4}}{\dfrac{3}{4} - \dfrac{5}{2p}}$

19. $\dfrac{\dfrac{x+2}{x} + \dfrac{1}{x+2}}{\dfrac{5}{x} + \dfrac{x}{x+2}}$

20. $\dfrac{\dfrac{y+3}{y} - \dfrac{4}{y-1}}{\dfrac{y}{y-1} + \dfrac{1}{y}}$

RELATING CONCEPTS (Exercises 21–26) FOR INDIVIDUAL OR GROUP WORK

Simplifying a complex fraction by Method 1 is a good way to review the methods of adding, subtracting, multiplying, and dividing rational expressions. Method 2 gives a good review of the fundamental property of rational expressions. Refer to the following complex fraction, and ***work Exercises 21–26 in order.***

$$\frac{\dfrac{4}{m} + \dfrac{m+2}{m-1}}{\dfrac{m+2}{m} - \dfrac{2}{m-1}}$$

21. Add the fractions in the numerator.

22. Subtract as indicated in the denominator.

23. Divide your answer from Exercise 21 by your answer from Exercise 22.

24. Go back to the original complex fraction and find the least common denominator of all denominators.

25. Multiply the numerator and denominator of the complex fraction by your answer from Exercise 24.

26. Your answers for Exercises 23 and 25 should be the same. Write an explanation comparing the two methods. Which method do you prefer? Explain why.

Simplify each expression, using only positive exponents in the answer. See Example 4.

27. $\dfrac{1}{x^{-2} + y^{-2}}$

28. $\dfrac{1}{p^{-2} - q^{-2}}$

29. $\dfrac{x^{-2} + y^{-2}}{x^{-1} + y^{-1}}$

30. $\dfrac{x^{-1} - y^{-1}}{x^{-2} - y^{-2}}$

31. $\dfrac{x^{-1} + 2y^{-1}}{2y + 4x}$

32. $\dfrac{a^{-2} - 4b^{-2}}{3b - 6a}$

7.4 Graphs and Equations with Rational Expressions

Objectives

1. Recognize the graph of a rational function.
2. Determine the domain of a rational equation.
3. Solve rational equations.

1 Recognize the graph of a rational function. As we saw in Section 7.1, one or more values of x may be excluded from the domain of some rational functions. As a result, the graph of a rational function is often *discontinuous*. That is, there will be one or more breaks in the graph. For example, we use point plotting and observing the domain to graph the simple rational function defined by

$$f(x) = \frac{1}{x}.$$

The domain of this function includes all real numbers except 0. Thus, there will be no point on the graph with $x = 0$. The vertical line with equation $x = 0$ is called a **vertical asymptote** of the graph. We show some typical ordered pairs in the table for both negative and positive x-values.

x	-3	-2	-1	$-.5$	$-.25$	$-.1$	$.1$	$.25$	$.5$	1	2	3
y	$-\frac{1}{3}$	$-\frac{1}{2}$	-1	-2	-4	-10	10	4	2	1	$\frac{1}{2}$	$\frac{1}{3}$

Notice that the closer positive values of x are to 0, the larger y is. Similarly, the closer negative values of x are to 0, the smaller (more negative) y is. Using this observation, excluding 0 from the domain, and plotting the points in the table, we obtain the graph in Figure 2.

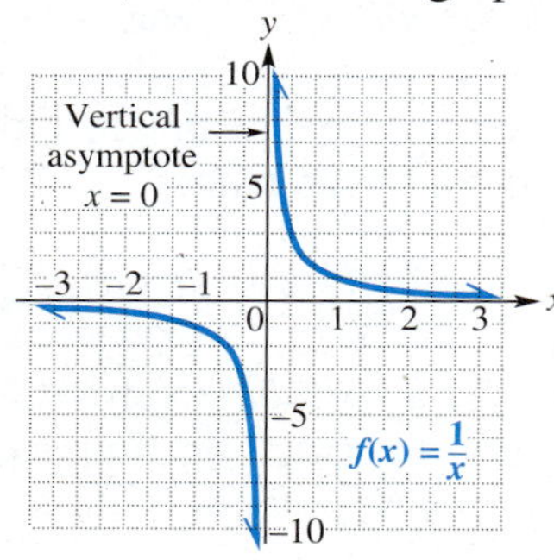

Figure 2

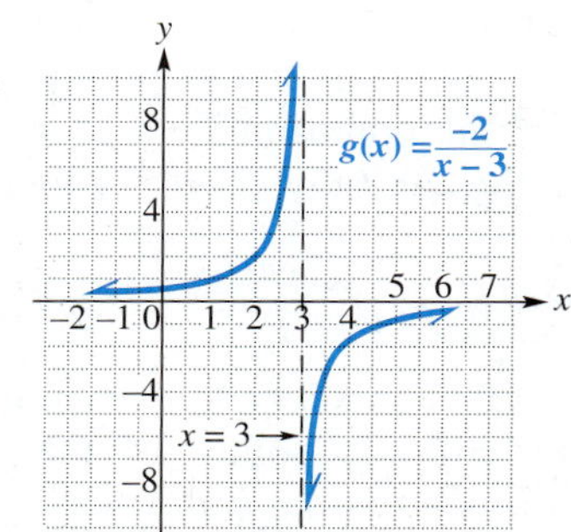

Figure 3

The graph of

$$g(x) = \frac{-2}{x-3}$$

is shown in Figure 3. Some ordered pairs are shown in the table.

x	-2	-1	0	1	2	2.5	2.75	3.25	3.5	4	5	6
y	$\frac{2}{5}$	$\frac{1}{2}$	$\frac{2}{3}$	1	2	4	8	-8	-4	-2	-1	$-\frac{2}{3}$

There is no point on the graph for $x = 3$ because 3 is excluded from the domain. The dashed line $x = 3$ represents the asymptote and is not part of the graph. As suggested by the points from the table, the graph gets closer to the vertical asymptote as the x-values get closer to 3.

Work Problem 1 at the Side.

The domain of a *rational expression* is the set of all possible values of the variable. Any value that makes the denominator 0 is excluded.

1 Graph each rational function, and give the equation of the vertical asymptote.

(a) $f(x) = -\frac{1}{x}$

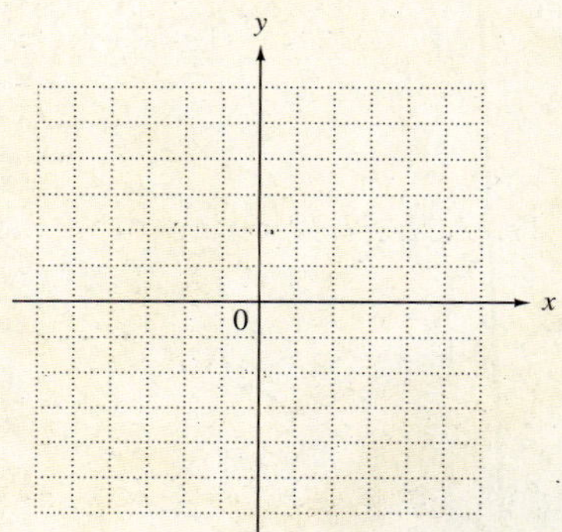

(b) $f(x) = \frac{2}{x+3}$

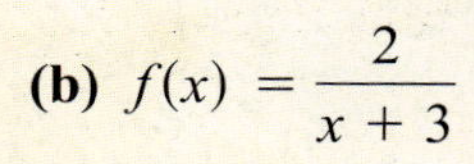

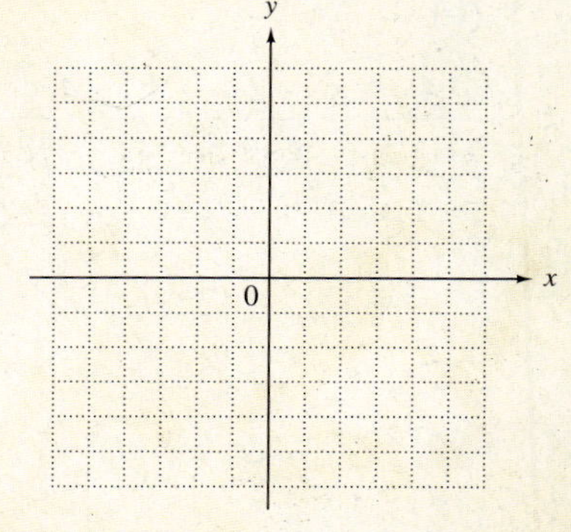

Answers

1. (a) asymptote: $x = 0$

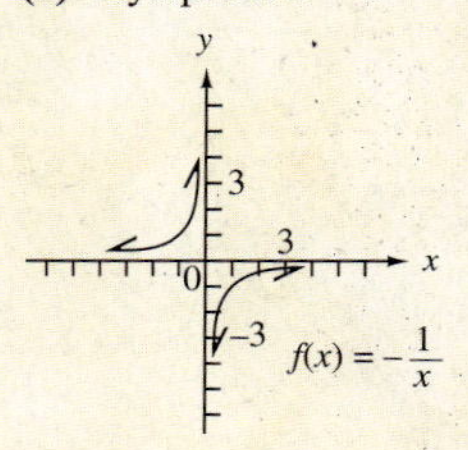

(b) asymptote: $x = -3$

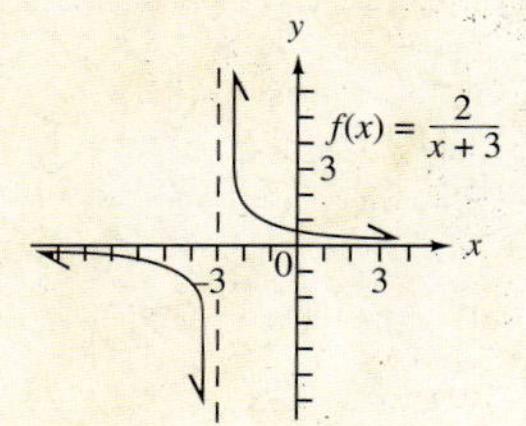

2 Find the domain of each equation.

(a) $\frac{3}{x} + \frac{1}{2} = \frac{5}{6x}$

(b) $\frac{4}{x-5} - \frac{2}{x+5} = \frac{1}{x^2-25}$

ANSWERS
2. **(a)** $\{x|x \neq 0\}$ **(b)** $\{x|x \neq \pm 5\}$

2 Determine the domain of a rational equation. The **domain of a rational equation** is the intersection (overlap) of the domains of the rational expressions in the equation.

Example 1 Determining the Domains of Rational Equations

Find the domain of each equation.

(a) $\frac{2}{x} - \frac{3}{2} = \frac{7}{2x}$

The domains of the three rational terms of the equation are, in order, $\{x|x \neq 0\}$, $(-\infty, \infty)$, and $\{x|x \neq 0\}$. The intersection of these three domains is all real numbers except 0, which may be written $\{x|x \neq 0\}$.

(b) $\frac{2}{x-3} - \frac{3}{x+3} = \frac{12}{x^2-9}$

The domains of these three terms are, respectively, $\{x|x \neq 3\}$, $\{x|x \neq -3\}$ and $\{x|x \neq \pm 3\}$. The domain of the equation is the intersection of the three domains, all real numbers except 3 and -3, written $\{x|x \neq \pm 3\}$.

Work Problem 2 at the Side.

3 Solve rational equations. The easiest way to solve most equations involving rational expressions is to multiply all terms in the equation by the least common denominator. This step will clear the equation of all denominators, as the next examples show. *We can do this only with equations, not expressions.*

Because the first step in solving a rational equation is to multiply each side of the equation by a common denominator, it is *necessary* to either check the solutions or verify that the solutions are in the domain.

CAUTION

When each side of an equation is multiplied by a *variable* expression, the resulting "solutions" may not satisfy the original equation. You *must* either determine and observe the domain or check all potential solutions in the original equation. *It is wise to do both.*

Example 2 Solving an Equation with Rational Expressions

Solve $\frac{2}{x} - \frac{3}{2} = \frac{7}{2x}$.

The domain, which excludes 0, was found in Example 1(a). Multiply each side of the equation by the LCD, $2x$.

$$2x\left(\frac{2}{x} - \frac{3}{2}\right) = 2x\left(\frac{7}{2x}\right)$$

$$2x\left(\frac{2}{x}\right) - 2x\left(\frac{3}{2}\right) = 2x\left(\frac{7}{2x}\right) \quad \text{Distributive property}$$

$$4 - 3x = 7 \quad \text{Multiply.}$$

$$-3x = 3 \quad \text{Subtract 4.}$$

$$x = -1 \quad \text{Divide by } -3.$$

Continued on Next Page

To check, replace x with -1 in the original equation.

$$\frac{2}{x} - \frac{3}{2} = \frac{7}{2x}$$

$$\frac{2}{-1} - \frac{3}{2} \stackrel{?}{=} \frac{7}{2(-1)} \quad \text{Let } x = -1.$$

$$-2 - \frac{3}{2} \stackrel{?}{=} -\frac{7}{2}$$

$$-\frac{7}{2} = -\frac{7}{2} \quad \text{True}$$

The solution set is $\{-1\}$.

Work Problem 3 at the Side.

Example 3 Solving an Equation with No Solution

Solve $\dfrac{2}{x-3} - \dfrac{3}{x+3} = \dfrac{12}{x^2-9}$.

Using the result from Example 1(b), we know that the domain excludes 3 and -3. Multiply each side by the LCD, $(x+3)(x-3)$.

$$(x+3)(x-3)\left(\frac{2}{x-3} - \frac{3}{x+3}\right) = (x+3)(x-3)\left(\frac{12}{x^2-9}\right)$$

$$2(x+3) - 3(x-3) = 12 \quad \text{Distributive property}$$

$$2x + 6 - 3x + 9 = 12 \quad \text{Distributive property}$$

$$-x + 15 = 12 \quad \text{Combine terms.}$$

$$-x = -3 \quad \text{Subtract 15.}$$

$$x = 3 \quad \text{Divide by } -1.$$

Since 3 is not in the domain, it cannot be a solution of the equation. Substitute 3 in the original equation.

$$\frac{2}{x-3} - \frac{3}{x+3} = \frac{12}{x^2-9}$$

$$\frac{2}{3-3} - \frac{3}{3+3} \stackrel{?}{=} \frac{12}{3^2-9} \quad \text{Let } x = 3.$$

$$\frac{2}{0} - \frac{3}{6} \stackrel{?}{=} \frac{12}{0}$$

Since division by 0 is undefined, the given equation has no solution, and the solution set is $\emptyset$.

Work Problem 4 at the Side.

Example 4 Solving an Equation with Rational Expressions

Solve $\dfrac{3}{p^2+p-2} - \dfrac{1}{p^2-1} = \dfrac{7}{2(p^2+3p+2)}$.

Factor each denominator to find the LCD, $2(p-1)(p+2)(p+1)$.

Continued on Next Page

3 Solve $-\dfrac{3}{20} + \dfrac{2}{x} = \dfrac{5}{4x}$.

4 Solve each equation.

(a) $\dfrac{3}{x+1} = \dfrac{1}{x-1} - \dfrac{2}{x^2-1}$

(b) $\dfrac{1}{x-3} + \dfrac{1}{x+3} = \dfrac{6}{x^2-9}$

ANSWERS

3. $\{5\}$

4. **(a)** $\emptyset$ **(b)** $\emptyset$

5 Solve

$$\frac{4}{x^2+x-6}-\frac{1}{x^2-4}=\frac{2}{x^2+5x+6}.$$

6 Solve

$$\frac{1}{x+4}+\frac{x}{x-4}=\frac{-8}{x^2-16}.$$

The domain excludes 1, −2, and −1. Multiply each side by the LCD.

$$2(p-1)(p+2)(p+1)\left(\frac{3}{(p+2)(p-1)}-\frac{1}{(p+1)(p-1)}\right)$$
$$=2(p-1)(p+2)(p+1)\left(\frac{7}{2(p+2)(p+1)}\right)$$

$$2\cdot 3(p+1)-2(p+2)=7(p-1) \quad \text{Distributive property}$$
$$6p+6-2p-4=7p-7 \quad \text{Distributive property}$$
$$4p+2=7p-7 \quad \text{Combine terms.}$$
$$9=3p$$
$$3=p$$

Note that 3 is in the domain; substitute 3 for p in the original equation to check that the solution set is $\{3\}$.

Work Problem 5 at the Side.

Example 5 Solving an Equation That Leads to a Quadratic Equation

Solve $\dfrac{2}{3x+1}=\dfrac{1}{x}-\dfrac{6x}{3x+1}$.

Since the denominator $3x+1$ cannot equal 0, $-\frac{1}{3}$ is excluded from the domain, as is 0. Multiply each side by the LCD, $x(3x+1)$.

$$x(3x+1)\left(\frac{2}{3x+1}\right)=x(3x+1)\left[\frac{1}{x}-\frac{6x}{3x+1}\right]$$
$$2x=3x+1-6x^2$$

Since this equation is quadratic, write it in standard form with 0 on the right side.

$$6x^2-3x+2x-1=0$$
$$6x^2-x-1=0 \quad \text{Standard form}$$
$$(3x+1)(2x-1)=0 \quad \text{Factor.}$$
$$3x+1=0 \quad \text{or} \quad 2x-1=0 \quad \text{Zero-factor property}$$
$$x=-\frac{1}{3} \quad \text{or} \quad x=\frac{1}{2}$$

Because $-\frac{1}{3}$ is not in the domain of the equation, it is not a solution. Check that the solution set is $\{\frac{1}{2}\}$.

Work Problem 6 at the Side.

ANSWERS
5. $\{-9\}$
6. $\{-1\}$

7.4 EXERCISES

FOR EXTRA HELP

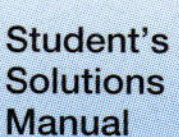
Student's Solutions Manual

MyMathLab.com

InterAct Math Tutorial Software

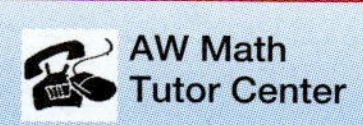
AW Math Tutor Center

www.mathxl.com

Digital Video Tutor CD 5 Videotape 12

Graph each rational function. Give the equation of the vertical asymptote. See Figures 1 and 2.

1. $f(x) = \dfrac{2}{x}$

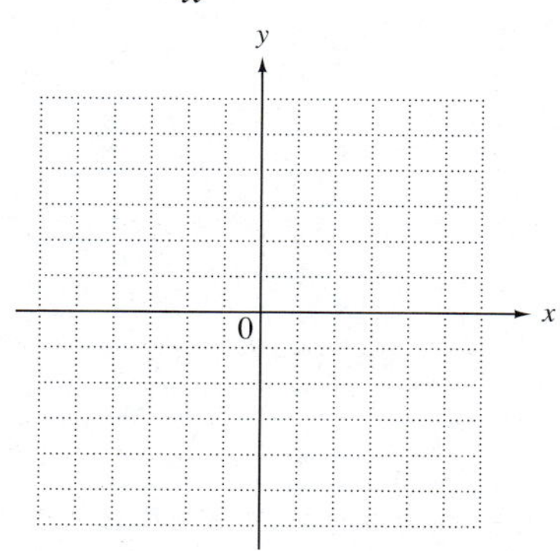

2. $f(x) = \dfrac{3}{x}$

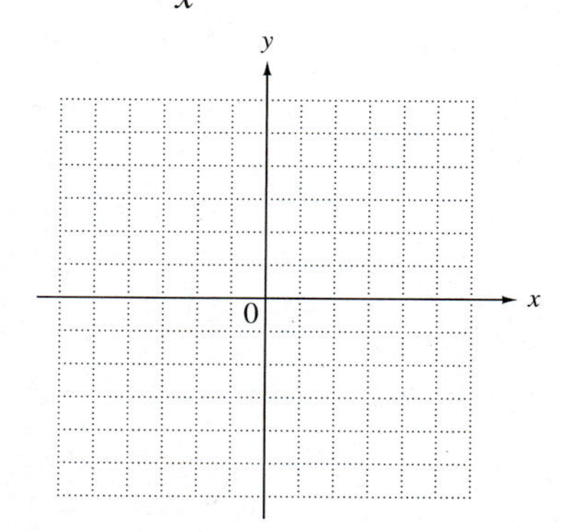

3. $f(x) = \dfrac{1}{x - 2}$

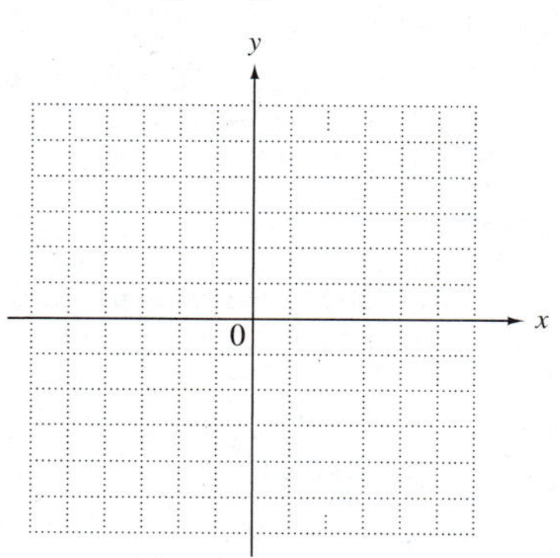

4. $f(x) = \dfrac{1}{x + 2}$

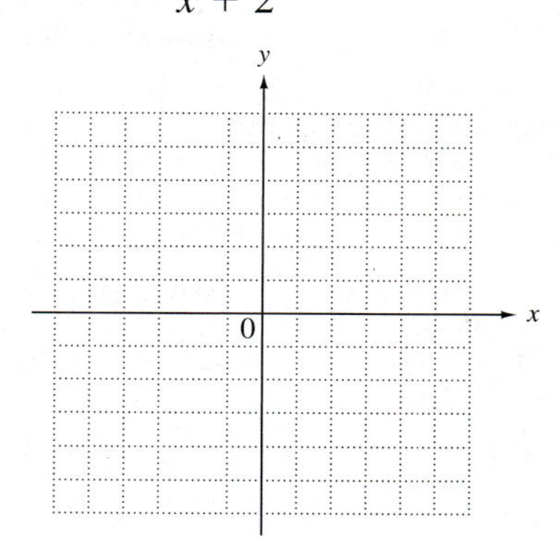

As explained in this section, any values that would cause a denominator to equal 0 *must be excluded from the domain and consequently as solutions of an equation that has variable expressions in the denominators.* ***(a)*** *Without actually solving the equation, list all possible numbers that would have to be rejected if they appeared as potential solutions.* ***(b)*** *Then give the domain using set notation. See Example 1.*

5. $\dfrac{1}{x + 1} - \dfrac{1}{x - 2} = 0$

6. $\dfrac{3}{x + 4} - \dfrac{2}{x - 9} = 0$

7. $\dfrac{5}{3x + 5} - \dfrac{1}{x} = \dfrac{1}{2x + 3}$

8. $\dfrac{6}{4x+7} - \dfrac{3}{x} = \dfrac{5}{6x-13}$

9. $\dfrac{1}{3x} + \dfrac{1}{2x} = \dfrac{x}{3}$

10. $\dfrac{5}{6x} - \dfrac{8}{2x} = \dfrac{x}{4}$

11. $\dfrac{3x+1}{x-4} = \dfrac{6x+5}{2x-7}$

12. $\dfrac{4x-1}{2x+3} = \dfrac{12x-25}{6x-2}$

13. $\dfrac{2}{x^2-x} + \dfrac{1}{x+3} = \dfrac{4}{x-2}$

14. Is it possible that any potential solutions to the equation

$$\frac{x+7}{4} - \frac{x+3}{3} = \frac{x}{12}$$

would have to be rejected? Explain.

Solve each equation. See Examples 2–5.

15. $\dfrac{-5}{2x} + \dfrac{3}{4x} = \dfrac{-7}{4}$

16. $\dfrac{6}{5x} - \dfrac{2}{3x} = \dfrac{-8}{45}$

17. $x - \dfrac{24}{x} = -2$

18. $p + \dfrac{15}{p} = -8$

19. $\dfrac{x-4}{x+6} = \dfrac{2x+3}{2x-1}$

20. $\dfrac{5x-8}{x+2} = \dfrac{5x-1}{x+3}$

21. $\dfrac{3x+1}{x-4} = \dfrac{6x+5}{2x-7}$

22. $\dfrac{4x-1}{2x+3} = \dfrac{12x-25}{6x-2}$

23. $\dfrac{1}{y-1} + \dfrac{5}{12} = \dfrac{-2}{3y-3}$

24. $\dfrac{4}{m+2} - \dfrac{11}{9} = \dfrac{1}{3m+6}$

25. $\dfrac{-2}{3t-6} - \dfrac{1}{36} = \dfrac{-3}{4t-8}$

26. $\dfrac{3}{4m+2} = \dfrac{17}{2} - \dfrac{7}{2m+1}$

27. $\dfrac{3}{k+2} - \dfrac{2}{k^2-4} = \dfrac{1}{k-2}$

28. $\dfrac{3}{x-2} + \dfrac{21}{x^2-4} = \dfrac{14}{x+2}$

29. $\dfrac{1}{y+2} + \dfrac{3}{y+7} = \dfrac{5}{y^2+9y+14}$

30. $\dfrac{1}{t+3} + \dfrac{4}{t+5} = \dfrac{2}{t^2+8t+15}$

31. $\dfrac{9}{x} + \dfrac{4}{6x-3} = \dfrac{2}{6x-3}$

32. $\dfrac{5}{n} + \dfrac{4}{6-3n} = \dfrac{2n}{6-3n}$

33. $\dfrac{6}{w+3} + \dfrac{-7}{w-5} = \dfrac{-48}{w^2-2w-15}$

34. $\dfrac{2}{r-5} + \dfrac{3}{2r+1} = \dfrac{22}{2r^2-9r-5}$

35. $\dfrac{x}{x-3} + \dfrac{4}{x+3} = \dfrac{18}{x^2-9}$

36. $\dfrac{2x}{x-3} + \dfrac{4}{x+3} = \dfrac{-24}{x^2-9}$

37. $\dfrac{6}{x-4} + \dfrac{5}{x} = \dfrac{-20}{x^2-4x}$

38. $\dfrac{7}{x-4} + \dfrac{3}{x} = \dfrac{-12}{x^2-4x}$

39. $\dfrac{2}{4x+7} + \dfrac{x}{3} = \dfrac{6}{12x+21}$

40. $\dfrac{5x+14}{x^2-9} = \dfrac{-2x^2-5x+2}{x^2-9} + \dfrac{2x+4}{x-3}$

41. $\dfrac{4x-7}{4x^2-9} = \dfrac{-2x^2+5x-4}{4x^2-9} + \dfrac{x+1}{2x+3}$

42. What is wrong with the following problem? "Solve $\dfrac{2x+1}{3x-4} + \dfrac{1}{2x+3}$."

Relating Concepts (Exercises 43–46) For Individual or Group Work

An equation of the form

$$\frac{A}{x+B} + \frac{x}{x-B} = \frac{C}{x^2 - B^2}$$

will have one rejected solution if the relationship $C = -2AB$ holds true. (This can be proved using methods not covered in intermediate algebra.) For example, if $A = 1$ and $B = 2$, then $C = -2AB = -2(1)(2) = -4$, and the equation becomes

$$\frac{1}{x+2} + \frac{x}{x-2} = \frac{-4}{x^2 - 4}.$$

This equation has solution set $\{-1\}$; the potential solution -2 must be rejected. To further understand this idea, ***work Exercises 43–46 in order.***

43. Show that the second equation does indeed have solution set $\{-1\}$ and -2 must be rejected.

44. Let $A = 2$ and let $B = 1$. What is the corresponding value of C? Solve the equation determined by A, B, and C. What is the solution set? What value must be rejected?

45. Let $A = 4$ and let $B = -3$. What is the corresponding value of C? Solve the equation determined by A, B, and C. What is the solution set? What value must be rejected?

46. Choose two numbers of your own, letting one be A and the other be B. Repeat the process described in Exercises 44 and 45.

Solve each problem.

47. The average number of vehicles waiting in line to enter a sports arena parking area is modeled by the rational function defined by

$$w(x) = \frac{x^2}{2(1-x)},$$

where x is a quantity between 0 and 1 known as the *traffic intensity.* (*Source:* Mannering, F. and W. Kilareski, *Principles of Highway Engineering and Traffic Control*, John Wiley and Sons, 1990.) To the nearest tenth, find the average number of vehicles waiting if the traffic intensity is

(a) .1 **(b)** .8 **(c)** .9.

(d) What happens to waiting time as traffic intensity increases?

48. The percent of deaths caused by smoking is modeled by the rational function defined by

$$p(x) = \frac{x-1}{x},$$

where x is the number of times a smoker is more likely to die of lung cancer than a nonsmoker. This is called the *incidence rate.* (*Source:* Walker, A., *Observation and Inference: An Introduction to the Methods of Epidemiology*, Epidemiology Resources Inc., 1991.) For example, $x = 10$ means that a smoker is 10 times more likely than a nonsmoker to die of lung cancer.

(a) Find $p(x)$ if x is 10.

(b) For what value of x is $p(x) = 80\%$? (*Hint:* Change 80% to a decimal.)

Summary Exercises on OPERATIONS AND EQUATIONS WITH RATIONAL EXPRESSIONS

A common student error is to confuse an equation, *such as* $\frac{x}{2} + \frac{x}{3} = -5$, *with an* operation, *such as* $\frac{x}{2} + \frac{x}{3}$. *Look for the equals sign to distinguish between them. Equations are solved for a numerical answer, while problems involving* operations *result in simplified* expressions, *as shown below.*

Solving an Equation

Solve: $\frac{x}{2} + \frac{x}{3} = -5$.

Multiply each side by the LCD, 6.

$$6\left(\frac{x}{2} + \frac{x}{3}\right) = 6(-5)$$
$$3x + 2x = -30$$
$$5x = -30$$
$$x = -6$$

Check that the solution set is $\{-6\}$.

Performing an Operation

Add: $\frac{x}{2} + \frac{x}{3}$.

Write both fractions with the LCD, 6.

$$\frac{x}{2} + \frac{x}{3} = \frac{x \cdot 3}{2 \cdot 3} + \frac{x \cdot 2}{3 \cdot 2}$$
$$= \frac{3x}{6} + \frac{2x}{6}$$
$$= \frac{3x + 2x}{6}$$
$$= \frac{5x}{6}$$

In each exercise, identify as an equation *or an* operation. *Then perform the indicated operation or solve the given equation, as appropriate.*

1. $\frac{x}{2} - \frac{x}{4} = 5$

2. $\frac{4x - 20}{x^2 - 25} \cdot \frac{(x + 5)^2}{10}$

3. $\frac{6}{7x} - \frac{4}{x}$

4. $\dfrac{\frac{1}{x} + \frac{1}{y}}{\frac{1}{x} - \frac{1}{y}}$

5. $\frac{5}{7t} = \frac{52}{7} - \frac{3}{t}$

6. $\frac{x - 5}{3} + \frac{1}{3} = \frac{x - 2}{5}$

7. $\frac{7}{6x} + \frac{5}{8x}$

8. $\frac{4}{x} - \frac{8}{x + 1} = 0$

9. $\dfrac{\frac{6}{x + 1} - \frac{1}{x}}{\frac{2}{x} - \frac{4}{x + 1}}$

10. $\frac{8}{r + 2} - \frac{7}{4r + 8}$

11. $\frac{x}{x + y} + \frac{2y}{x - y}$

12. $\frac{3p^2 - 6p}{p + 5} \div \frac{p^2 - 4}{8p + 40}$

13. $\dfrac{x-2}{9} \cdot \dfrac{5}{8-4x}$

14. $\dfrac{a-4}{3} + \dfrac{11}{6} = \dfrac{a+1}{2}$

15. $\dfrac{b^2+b-6}{b^2+2b-8} \cdot \dfrac{b^2+8b+16}{3b+12}$

16. $\dfrac{10z^2-5z}{3z^3-6z^2} \div \dfrac{2z^2+5z-3}{z^2+z-6}$

17. $\dfrac{5}{x^2-2x} - \dfrac{3}{x^2-4}$

18. $\dfrac{6}{t+1} + \dfrac{4}{5t+5} = \dfrac{34}{15}$

19. $\dfrac{\dfrac{5}{x}-\dfrac{3}{y}}{\dfrac{9x^2-25y^2}{x^2y}}$

20. $\dfrac{-2}{a^2+2a-3} - \dfrac{5}{3-3a} = \dfrac{4}{3a+9}$

21. $\dfrac{4y^2-13y+3}{2y^2-9y+9} \div \dfrac{4y^2+11y-3}{6y^2-5y-6}$

22. $\dfrac{8}{3k+9} - \dfrac{8}{15} = \dfrac{2}{5k+15}$

23. $\dfrac{3r}{r-2} = 1 + \dfrac{6}{r-2}$

24. $\dfrac{6z^2-5z-6}{6z^2+5z-6} \cdot \dfrac{12z^2-17z+6}{12z^2-z-6}$

25. $\dfrac{-1}{3-x} - \dfrac{2}{x-3}$

26. $\dfrac{\dfrac{t}{4}-\dfrac{1}{t}}{1+\dfrac{t+4}{t}}$

27. $\dfrac{2}{y+1} - \dfrac{3}{y^2-y-2} = \dfrac{3}{y-2}$

28. $\dfrac{7}{2x^2-8x} + \dfrac{3}{x^2-16}$

29. $\dfrac{3}{y-3} - \dfrac{3}{y^2-5y+6} = \dfrac{2}{y-2}$

30. $\dfrac{2k+\dfrac{5}{k-1}}{3k-\dfrac{2}{k}}$

7.5 Applications of Rational Expressions

1 Find the value of an unknown variable in a formula. Formulas may contain rational expressions, as does $t = \frac{d}{r}$. We now show how to work with formulas of this type.

Example 1 Finding the Value of a Variable in a Formula

In physics, the focal length, f, of a lens is given by the formula

$$\frac{1}{f} = \frac{1}{p} + \frac{1}{q},$$

where p is the distance from the object to the lens and q is the distance from the lens to the image. See Figure 4. Find q if $p = 20$ cm and $f = 10$ cm.

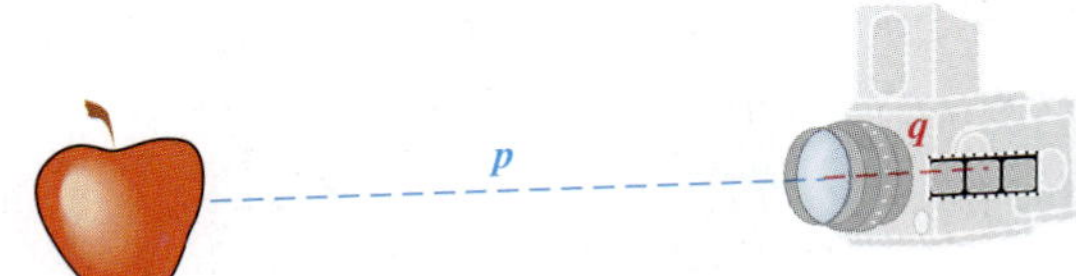

Focal Length of Camera Lens

Figure 4

Replace f with 10 and p with 20.

$$\frac{1}{f} = \frac{1}{p} + \frac{1}{q}$$

$$\frac{1}{10} = \frac{1}{20} + \frac{1}{q} \qquad \text{Let } f = 10,\ p = 20.$$

$$20q \cdot \frac{1}{10} = 20q\left(\frac{1}{20} + \frac{1}{q}\right) \qquad \text{Multiply by the LCD, } 20q.$$

$$2q = q + 20$$

$$q = 20$$

The distance from the lens to the image is 20 cm.

Work Problem 1 at the Side.

2 Solve a formula for a specified variable. The goal in solving for a specified variable is to isolate it on one side of the equals sign.

Example 2 Solving a Formula for a Specified Variable

Solve $\frac{1}{f} = \frac{1}{p} + \frac{1}{q}$ for p.

$$fpq \cdot \frac{1}{f} = fpq\left(\frac{1}{p} + \frac{1}{q}\right) \qquad \text{Multiply by the LCD, } fpq.$$

$$pq = fq + fp \qquad \text{Distributive property}$$

Continued on Next Page

Objectives

1. Find the value of an unknown variable in a formula.
2. Solve a formula for a specified variable.
3. Solve applications using proportions.
4. Solve applications about distance, rate, and time.
5. Solve applications about work rates.

1 Use the formula given in Example 1 to answer each part.

(a) Find p if $f = 15$ and $q = 25$.

(b) Find f if $p = 6$ and $q = 9$.

(c) Find q if $f = 12$ and $p = 16$.

Answers

1. (a) $\frac{75}{2}$ **(b)** $\frac{18}{5}$ **(c)** 48

2 Solve

$$\frac{3}{p}+\frac{3}{q}=\frac{5}{r}$$

for q.

3 Solve

$$A=\frac{Rr}{R+r}$$

for R.

Answers

2. $q=\dfrac{3rp}{5p-3r}$ or $q=\dfrac{-3rp}{3r-5p}$

3. $R=\dfrac{-Ar}{A-r}$ or $R=\dfrac{Ar}{r-A}$

Transform the equation so that the terms with p (the specified variable) are on the same side. One way to do this is to subtract fp from each side.

$$pq-fp=fq \qquad \text{Subtract } fp.$$
$$p(q-f)=fq \qquad \text{Factor out } p.$$
$$p=\frac{fq}{q-f} \qquad \text{Divide by } q-f.$$

Work Problem 2 at the Side.

Example 3 Solving a Formula for a Specified Variable

Solve $I=\dfrac{nE}{R+nr}$ for n.

$$(R+nr)I=(R+nr)\frac{nE}{R+nr} \qquad \text{Multiply by } R+nr.$$
$$RI+nrI=nE$$
$$RI=nE-nrI \qquad \text{Subtract } nrI.$$
$$RI=n(E-rI) \qquad \text{Factor out } n.$$
$$\frac{RI}{E-rI}=n \qquad \text{Divide by } E-rI.$$

CAUTION

Refer to the steps in Examples 2 and 3 that factor out the desired variable. This is a step that often gives students difficulty. Remember that the variable for which you are solving *must* be a factor on only one side of the equation, so each side can be divided by the remaining factor in the last step.

Work Problem 3 at the Side.

We can now solve problems that translate into equations with rational expressions. To do so, we continue to use the six-step problem-solving method from Chapter 2.

3 Solve applications using proportions. A **ratio** is a comparison of two quantities. The ratio of a to b may be written in any of the following ways:

$$a \text{ to } b, \quad a:b, \quad \text{or} \quad \frac{a}{b}.$$

Ratios are usually written as quotients in algebra. A **proportion** is a statement that two ratios are equal. Proportions are a useful and important type of rational equation.

Example 4 Solving a Proportion

In 1998, 16 of every 100 Americans had no health insurance coverage. The population at that time was about 227 million. How many million had no health insurance? (*Source*: U.S. Bureau of the Census.)

Step 1 **Read** the problem.

Step 2 **Assign a variable.** Let x = the number (in millions) who had no health insurance.

Continued on Next Page

Step 3 **Write an equation.** To get an equation, set up a proportion. The ratio x to 227 should equal the ratio 16 to 100. Write the proportion and solve the equation.

$$\frac{16}{100} = \frac{x}{227}$$

Step 4 **Solve.**

$$22{,}700\left(\frac{16}{100}\right) = 22{,}700\left(\frac{x}{227}\right) \quad \text{Multiply by a common denominator.}$$

$$3632 = 100x \quad \text{Simplify.}$$

$$x = 36.32$$

Step 5 **State the answer.** There were 36.32 million Americans with no health insurance in 1998.

Step 6 **Check** that the ratio of this number to 227 million is equivalent to $\frac{16}{100}$.

Work Problem 4 at the Side.

Example 5 Solving a Proportion Involving Rates

Marissa's car uses 10 gal of gas to travel 210 mi. She has 5 gal of gas in the car, and she wants to know how much more gas she will need to drive 640 mi. If we assume the car continues to use gas at the same rate, how many more gallons will she need?

Step 1 **Read** the problem.

Step 2 **Assign a variable.** Let x = the additional number of gallons of gas needed.

Step 3 **Write an equation.** To get an equation, set up a proportion.

$$\frac{\text{gallons}}{\text{miles}} \quad \frac{10}{210} = \frac{5 + x}{640} \quad \frac{\text{gallons}}{\text{miles}}$$

Step 4 **Solve.** The LCD is $10 \cdot 21 \cdot 64$.

$$\mathbf{10 \cdot 21 \cdot 64}\left(\frac{10}{210}\right) = \mathbf{10 \cdot 21 \cdot 64}\left(\frac{5 + x}{640}\right)$$

$$64 \cdot 10 = 21(5 + x)$$

$$640 = 105 + 21x \quad \text{Distributive property}$$

$$535 = 21x \quad \text{Subtract 105.}$$

$$25.5 \approx x \quad \text{Divide by 21; round to the nearest tenth.}$$

Step 5 **State the answer.** Marissa will need about 25.5 more gallons of gas.

Step 6 **Check** the answer in the words of the problem. The 25.5 gal plus the 5 gal equals 30.5 gal.

$$\frac{30.5}{640} \approx \mathbf{.0476} \quad \text{and} \quad \frac{10}{210} \approx \mathbf{.0476}$$

Since the rates are equal, the solution is correct.

Work Problem 5 at the Side.

4 **Solve applications about distance, rate, and time.** A familiar example of a rate is speed, which is the ratio of distance to time. The next examples use the distance formula $d = rt$ introduced in Chapter 2.

4 Solve the problem.
In 1998, approximately 15% of the 11,073,000 children in the United States had no health insurance. How many children were uninsured? (*Source:* U. S. Bureau of the Census.)

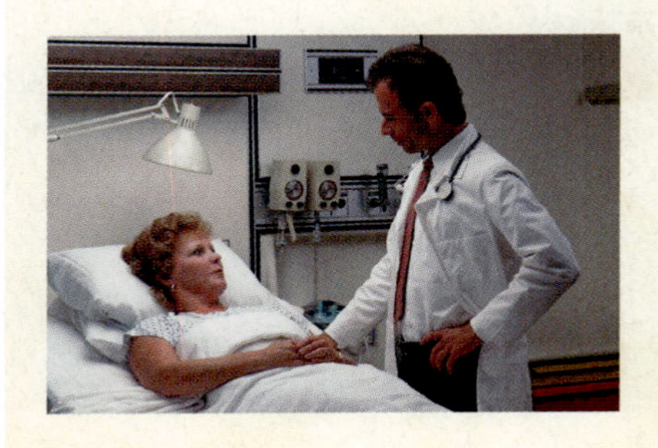

5 Solve the problem.
In 1997, the average American family spent 8.2 of every 100 dollars on health care. This amounted to \$3665 per family. To the nearest dollar, what was the average family income at that time? (*Source:* U.S. Health Care Financing Administration, U.S. Bureau of the Census.)

ANSWERS
4. 1,660,950
5. \$44,695

6 Solve the problem.

A plane travels 100 mi against the wind in the same time that it takes to travel 120 mi with the wind. The wind speed is 20 mph.

(a) Complete this table.

	d	r	t
Against Wind	100	$x - 20$	
With Wind	120	$x + 20$	

(b) Find the speed of the plane in still air.

Answers

6. (a) $\frac{100}{x-20}$; $\frac{120}{x+20}$ **(b)** 220 mph

Example 6 Solving a Problem about Distance, Rate, and Time

A tour boat goes 10 mi against the current in a small river in the same time that it goes 15 mi with the current. If the speed of the current is 3 mph, find the speed of the boat in still water.

Step 1 **Read** the problem. We must find the speed of the boat in still water.

Step 2 **Assign a variable.**

Let $x =$ the speed of the boat in still water; then

$x - 3 =$ the speed of the boat against the current;

$x + 3 =$ the speed of the boat with the current.

Because the time is the same going against the current as with the current, find time in terms of distance and rate (speed) for each situation.

Start with the distance formula, $d = rt$, and divide each side by r to get

$$t = \frac{d}{r}.$$

Going against the current, the distance is 10 mi and the rate is $x - 3$, giving

$$t = \frac{d}{r} = \frac{10}{x - 3}.$$

Going with the current, the distance is 15 mi and the rate is $x + 3$, so

$$t = \frac{d}{r} = \frac{15}{x + 3}.$$

This information is summarized in the following table.

	Distance	Rate	Time
Against Current	10	$x - 3$	$\frac{10}{x - 3}$
With Current	15	$x + 3$	$\frac{15}{x + 3}$

Times are equal.

Step 3 **Write an equation.** Because the times are equal,

$$\frac{10}{x - 3} = \frac{15}{x + 3}.$$

This is the equation to be solved.

Step 4 **Solve.** The LCD is $(x + 3)(x - 3)$.

$$(x + 3)(x - 3)\left(\frac{10}{x - 3}\right) = (x + 3)(x - 3)\left(\frac{15}{x + 3}\right) \quad \text{Multiply by the LCD.}$$

$$10(x + 3) = 15(x - 3)$$

$$10x + 30 = 15x - 45 \quad \text{Distributive property}$$

$$30 = 5x - 45 \quad \text{Subtract } 10x.$$

$$75 = 5x \quad \text{Add 45.}$$

$$15 = x \quad \text{Divide by 5.}$$

Step 5 **State the answer.** The speed of the boat in still water is 15 mph.

Step 6 **Check** the answer: $\frac{10}{15 - 3} = \frac{15}{15 + 3}$ is true.

Work Problem 6 at the Side.

Example 7 Solving a Problem about Distance, Rate, and Time

At O'Hare Airport, Cheryl and Bill are walking to the gate (at the same speed) to catch their flight to Akron, Ohio. Since Bill wants a window seat, he steps onto the moving sidewalk and continues to walk while Cheryl uses the stationary sidewalk. If the sidewalk moves at 1 m per sec and Bill saves 50 sec covering the 300-m distance, what is their walking speed?

Step 1 **Read** the problem. We must find their walking speed.

Step 2 **Assign a variable.** Let x represent their walking speed in meters per second. Thus Cheryl travels at x m per sec and Bill travels at $x + 1$ m per sec. Since Bill's time is 50 sec less than Cheryl's time, express their times in terms of the known distances and the variable rates. As in Example 6, start with $d = rt$ and divide each side by r to get

$$t = \frac{d}{r}.$$

For Cheryl, the distance is 300 m and the rate is x. Cheryl's time is

$$t = \frac{d}{r} = \frac{300}{x}.$$

Bill travels 300 m at a rate of $x + 1$, so his time is

$$t = \frac{d}{r} = \frac{300}{x + 1}.$$

This information is summarized in the following table.

	Distance	Rate	Time
Cheryl	300	x	$\frac{300}{x}$
Bill	300	$x + 1$	$\frac{300}{x+1}$

Step 3 **Write an equation** using the times from the table.

Bill's time is Cheryl's time less 50 seconds.

$$\frac{300}{x + 1} = \frac{300}{x} - 50$$

Step 4 **Solve.**

$$x(x + 1)\left(\frac{300}{x + 1}\right) = x(x + 1)\left(\frac{300}{x} - 50\right) \quad \text{Multiply by the LCD, } x(x + 1).$$

$$300x = 300(x + 1) - 50x(x + 1)$$

$$300x = 300x + 300 - 50x^2 - 50x \quad \text{Distributive property}$$

$$0 = 50x^2 + 50x - 300 \quad \text{Standard form}$$

$$0 = x^2 + x - 6 \quad \text{Divide by 50.}$$

$$0 = (x + 3)(x - 2) \quad \text{Factor.}$$

$$x + 3 = 0 \quad \text{or} \quad x - 2 = 0 \quad \text{Zero-factor property}$$

$$x = -3 \quad \text{or} \quad x = 2$$

Discard the negative answer, since speed cannot be negative.

Step 5 **State the answer.** Their walking speed is 2 m per sec.

Step 6 **Check** the solution in the words of the original problem.

Work Problem 7 at the Side.

7 Solve the problem.

Dona Kenly drove 300 mi north from San Antonio, mostly on the freeway. She usually averaged 55 mph, but an accident slowed her speed through Dallas to 15 mph. If her trip took 6 hr, how many miles did she drive at reduced speed?

	d	r	t
Normal Speed	$300 - x$	55	
Reduced Speed	x	15	

Answers

7. $11\frac{1}{4}$ mi

5 Solve applications about work rates. Problems about work are closely related to distance problems.

Problem Solving

People work at different rates. If the letters r, t, and A represent the rate at which the work is done, the time required, and the amount of work accomplished, respectively, then $A = rt$. Notice the similarity to the distance formula, $d = rt$. Amount of work can be measured in terms of jobs accomplished. Thus, if 1 job is completed, $A = 1$, and the formula gives the rate as

$$1 = rt$$
$$r = \frac{1}{t}.$$

Rate of Work

If a job can be accomplished in t units of time, then the rate of work is

$$\frac{1}{t} \text{ job per unit of time.}$$

To solve a work problem, we begin by using this fact to express all rates of work. See if you can identify the six steps used in the following example.

Example 8 Solving a Problem about Work

Letitia and Kareem are working on a neighborhood cleanup. Kareem can clean up all the trash in the area in 7 hr, while Letitia can do the same job in 5 hr. How long will it take them if they work together?

Let $x =$ the number of hours it will take the two people working together. Just as we made a table for the distance formula, $d = rt$, make a table here for $A = rt$, with $A = 1$. Since $A = 1$, the rate for each person will be $\frac{1}{t}$, where t is the time it takes the person to complete the job alone. For example, since Kareem can clean up all the trash in 7 hr, his rate is $\frac{1}{7}$ of the job per hour. Similarly, Letitia's rate is $\frac{1}{5}$ of the job per hour. Fill in the table as shown.

	Rate	Time Working Together	Fractional Part of the Job Done
Kareem	$\frac{1}{7}$	x	$\frac{1}{7}x$
Letitia	$\frac{1}{5}$	x	$\frac{1}{5}x$

Since together they complete 1 job, the sum of the fractional parts accomplished by them should equal 1.

Part done by Kareem	+	Part done by Letitia	is	1 whole job.
$\frac{1}{7}x$	+	$\frac{1}{5}x$	=	1

Continued on Next Page

Solve this equation. The LCD is 35.

$$35\left(\frac{1}{7}x + \frac{1}{5}x\right) = 35 \cdot 1$$
$$5x + 7x = 35$$
$$12x = 35$$
$$x = \frac{35}{12}$$

Working together, Kareem and Letitia can do the entire job in $\frac{35}{12}$ hr, or 2 hr and 55 min. Check this result in the original problem.

Work Problem 8 at the Side.

There is another way to approach problems about work. For instance, in Example 8, x represents the number of hours it will take the two people working together to complete the entire job. In one hour, $\frac{1}{x}$ of the entire job will be completed. Kareem completes $\frac{1}{7}$ of the job in one hour, and Letitia completes $\frac{1}{5}$ of the job, so the sum of their rates should equal $\frac{1}{x}$. This gives the equation

$$\frac{1}{7} + \frac{1}{5} = \frac{1}{x}.$$

When each side of this equation is multiplied by $35x$, the result is $5x + 7x = 35$. Notice that this is the same equation we got in Example 8 in the third line from the bottom. Thus the solution of the equation is the same using either approach.

8 Solve each problem.

(a) Stan needs 45 min to do the dishes, while Deb can do them in 30 min. How long will it take them if they work together?

	Rate	Time Working Together	Fractional Part of the Job Done
Stan	$\frac{1}{45}$	x	
Deb	$\frac{1}{30}$	x	

(b) Suppose it takes Stan 35 min to do the dishes, and together they can do them in 15 min. How long will it take Deb to do them alone?

ANSWERS

8. **(a)** 18 min **(b)** $26\frac{1}{4}$ min

Focus on Real-Data Applications

It Depends on What You Mean by "Average"

Finding an average seems to be a simple process. Don't we just add the values and divide by the number of values? Well, for rational expressions, it all depends on what you mean by "average."

- To find the average of two fractions, say $\frac{1}{3}$ and $\frac{3}{4}$, add the two fractions and divide by 2.

$$\frac{\frac{1}{3}+\frac{3}{4}}{2}=\frac{\left(\frac{1}{3}+\frac{3}{4}\right)\cdot 12}{2\cdot 12}=\frac{4+9}{24}=\frac{13}{24}$$

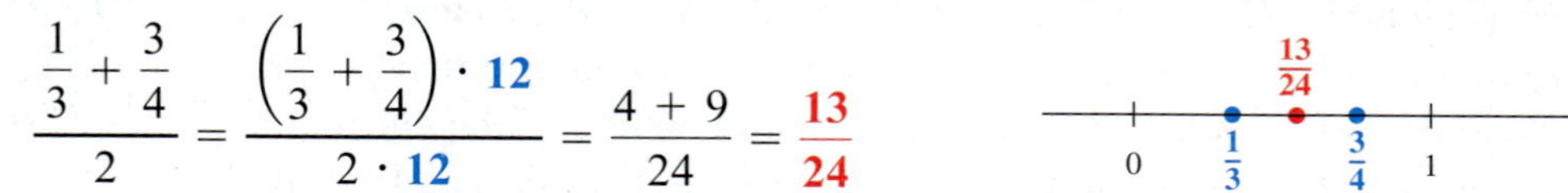

  On a number line, the fraction $\frac{13}{24}$ is the **arithmetic mean,** which is exactly halfway between the fractions $\frac{1}{3}$ and $\frac{3}{4}$.

- Suppose you travel one direction at 60 mph and return at 30 mph. To find your average rate, you have to calculate the total distance divided by the total time. Recall that $d = rt$, so the total distance is $2d$, the time going is $\frac{d}{60}$, and the time returning is $\frac{d}{30}$. Since $r = \frac{d}{t}$,

$$\frac{2d}{\frac{d}{60}+\frac{d}{30}}=\frac{2d\cdot 60}{\left(\frac{d}{60}+\frac{d}{30}\right)\cdot 60}=\frac{120d}{d+2d}=\frac{120d}{3d}=40 \text{ mph.}$$

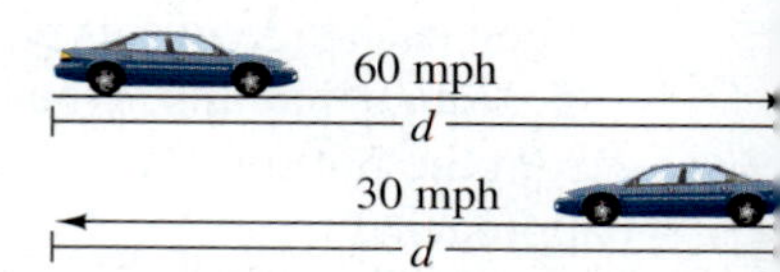

  The average rate is 40 mph. This is the *harmonic mean* of 60 and 30. The **harmonic mean** of two numbers a and b is defined as $\frac{2ab}{a+b}$. Note that

$$\frac{2\cdot 60\cdot 30}{60+30}=\frac{3600}{90}=40.$$

- To calculate a batting average, you find the **ratio** of the number of hits to the number of "at bats." Suppose a baseball player has 72 hits in 364 "at bats." His batting average would be $\frac{72}{364}\approx .198$. If the same player gets an additional 3 hits from 8 more "at bats" during the next week, then his revised batting average would be

$$\frac{72+3}{364+8}=\frac{75}{372}\approx .202.$$

For Group Discussion

A carpenter builds wine racks. For each situation, find the appropriate "average" quantity.

1. The carpenter told his helper to cut $\frac{1}{2}$ ft pieces from a dowel. The helper could not find a measuring tape, but he did recall that the distance from the tip of his middle finger to the tip of his thumb was approximately 6 in., so he estimated the lengths. When the carpenter checked his work, he found that the helper had actually cut two pieces that were $\frac{5}{12}$ and $\frac{1}{2}$ ft long. What was the average length of the two pieces?

2. Once the pieces are cut, the carpenter can assemble and finish a wine rack in 2 hr, working alone. His helper takes 4 hr to accomplish the same task, working alone. If the carpenter and the helper work together, what is their average time to assemble and finish a wine rack?

3. Of 115 wine racks built, 112 passed a quality control check. What was the acceptance rate? During the next week, the carpenter built 35 additional wine racks, of which 28 were acceptable. What was the revised acceptance rate? Round answers to the nearest thousandth.

7.5 EXERCISES

FOR EXTRA HELP

Student's Solutions Manual

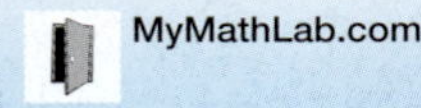

MyMathLab.com

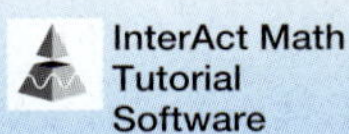

InterAct Math Tutorial Software

AW Math Tutor Center

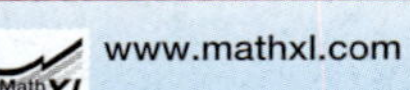

www.mathxl.com

Digital Video Tutor CD 6 Videotape 12

In Exercises 1–4, a familiar formula is given. Give the letter of the choice that is an equivalent form of the given formula.

1. $p = br$ (percent)

A. $b = \frac{p}{r}$ **B.** $r = \frac{b}{p}$

C. $b = \frac{r}{p}$ **D.** $p = \frac{r}{b}$

2. $V = LWH$ (geometry)

A. $H = \frac{LW}{V}$ **B.** $L = \frac{V}{WH}$

C. $L = \frac{WH}{V}$ **D.** $W = \frac{H}{VL}$

3. $m = \frac{F}{a}$ (physics)

A. $a = mF$ **B.** $F = \frac{m}{a}$

C. $F = \frac{a}{m}$ **D.** $F = ma$

4. $I = \frac{E}{R}$ (electricity)

A. $R = \frac{I}{E}$ **B.** $R = IE$

C. $E = \frac{I}{R}$ **D.** $E = RI$

Solve each problem. See Example 1.

5. A gas law in chemistry says that

$$\frac{PV}{T} = \frac{pv}{t}.$$

Suppose that $T = 300$, $t = 350$, $V = 9$, $P = 50$, and $v = 8$. Find p.

6. In work with electric circuits, the formula

$$\frac{1}{a} = \frac{1}{b} + \frac{1}{c}$$

occurs. Find b if $a = 8$ and $c = 12$.

7. A formula from anthropology says that

$$c = \frac{100b}{L}.$$

Find L if $c = 80$ and $b = 5$.

8. The gravitational force between two masses is given by

$$F = \frac{GMm}{d^2}.$$

Find M if $F = 10$, $G = 6.67 \times 10^{-11}$, $m = 1$, and $d = 3 \times 10^{-6}$.

Solve each formula for the specified variable. See Examples 2 and 3.

9. $F = \frac{GMm}{d^2}$ for G (physics)

10. $F = \frac{GMm}{d^2}$ for M (physics)

11. $\frac{1}{a} = \frac{1}{b} + \frac{1}{c}$ for a (electricity)

12. $\frac{1}{a} = \frac{1}{b} + \frac{1}{c}$ for b (electricity)

13. $\frac{PV}{T} = \frac{pv}{t}$ for v (chemistry)

14. $\frac{PV}{T} = \frac{pv}{t}$ for T (chemistry)

15. $I = \frac{nE}{R + nr}$ for r (engineering)

16. $a = \frac{V - v}{t}$ for V (physics)

17. $A = \frac{1}{2}h(B + b)$ for b (mathematics)

18. $S = \frac{n}{2}(a + \ell)d$ for n (mathematics)

19. $\frac{E}{e} = \frac{R + r}{r}$ for r (engineering)

20. $y = \frac{x + z}{a - x}$ for x

21. To solve the equation $m = \frac{ab}{a - b}$ for a, what is the first step?

22. Suppose you are asked to solve the equation

$$rp - rq = p + q$$

for r. What is the first step?

Solve each problem mentally. Use proportions in Exercises 23 and 24.

23. In a mathematics class, 3 of every 4 students are girls. If there are 20 students in the class, how many are girls? How many are boys?

24. In a certain southern state, sales tax on a purchase of \$1.50 is \$.12. What is the sales tax on a purchase of \$6.00?

25. If Marin can mow her yard in 2 hr, what is her rate (in job per hour)?

26. A van traveling from Atlanta to Detroit averages 50 mph and takes 14 hr to make the trip. How far is it from Atlanta to Detroit?

Use the bar graph to answer Exercises 27–30.

27. In which year was the ratio of truck accidents to car accidents the least?

28. In which year was the ratio of truck accidents to car accidents the greatest?

29. In which year was the ratio of car accidents to truck accidents closest to 3 to 1?

30. In which year was the ratio of car accidents to truck accidents less than 2 to 1?

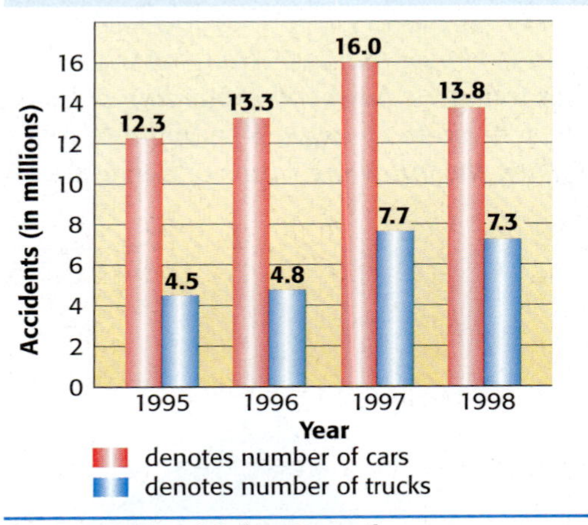

Source: National Safety Council.

Use a proportion to solve each problem. See Examples 4 and 5.

31. During the 1997–1998 academic year, the ratio of teachers to students in private high schools was approximately 1 to 24. If a private high school had 554 students, how many teachers would be at the school if this ratio was valid for that school? Round your answer to the nearest whole number. (*Source:* U.S. National Center for Education Statistics, *Private School Universe Survey,* 1997–98.)

32. During the 1998–1999 National Basketball Association season, Shaquille O'Neal of the Los Angeles Lakers played in 49 games for a total of 1705 min. If he had played in all 50 of the team's games, how many minutes would he have played, assuming that the ratio of games to minutes stayed the same? Round your answer to the nearest whole number. (*Source: Sports Illustrated 2000 Sports Almanac.*)

33. Biologists tagged 500 fish in a lake on January 1. On February 1 they returned and collected a random sample of 400 fish, 8 of which had been previously tagged. Approximately how many fish does the lake have based on this experiment?

34. Suppose that in the experiment of Exercise 33, 10 of the previously tagged fish were collected on February 1. What would be the estimate of the fish population?

35. In a recent year, 50 shares of common stock in Merck Company earned \$191.50. How much more would 75 shares of the stock have earned? (*Source:* Merck & Co., Inc., 1997 annual report.)

36. Seligman Communications and Information Fund, Inc. produced income of \$22,950 on an investment of \$100,000 in a recent year. If the investment had been increased to \$260,000, how much more income would have been produced? (*Source:* Seligman Communications and Information Fund, Inc.)

In geometry, it is shown that two triangles with corresponding angle measures equal, called similar triangles, *have corresponding sides proportional. For example, in the figure, angle A = angle D, angle B = angle E, and angle C = angle F, so the triangles are similar. Then the following ratios of corresponding sides are equal.*

$$\frac{4}{6} = \frac{6}{9} = \frac{2x + 1}{2x + 5}$$

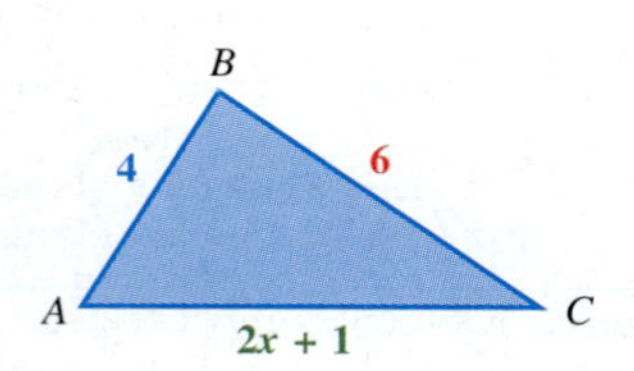

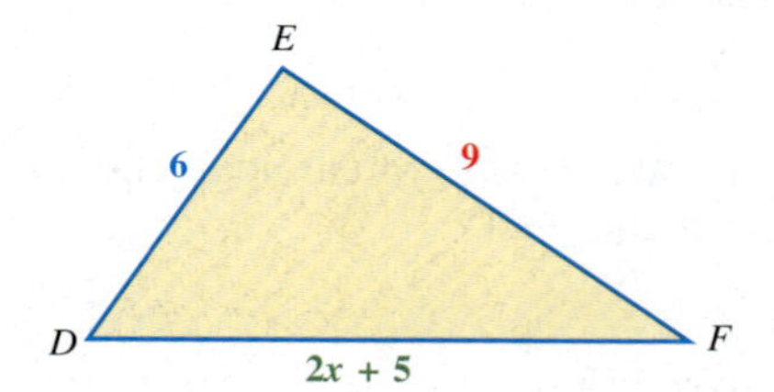

37. Solve for x using the given proportion to find the lengths of the third sides of the triangles.

38. Suppose the following triangles are similar. Find y and the lengths of the two longest sides of each triangle.

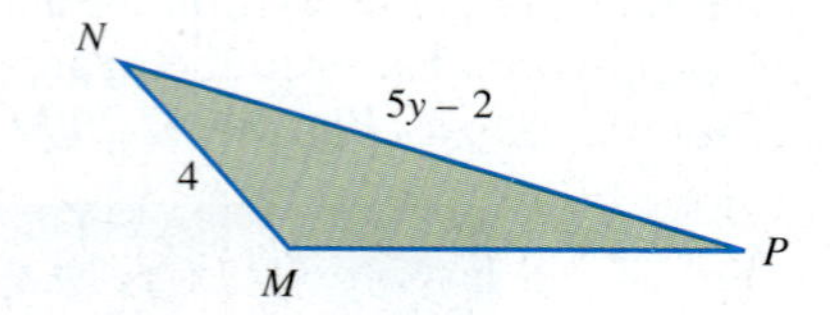

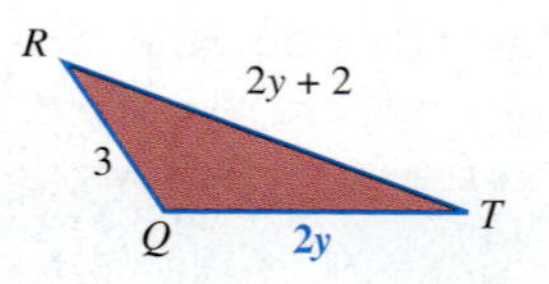

Solve each problem. See Examples 6 and 7.

39. Kellen's boat goes 12 mph. Find the rate of the current of the river if she can go 6 mi upstream in the same amount of time she can go 10 mi downstream.

	Distance	Rate	Time
Downstream	10	$12 + x$	
Upstream	6	$12 - x$	

40. Kasey can travel 8 mi upstream in the same time it takes her to go 12 mi downstream. Her boat goes 15 mph in still water. What is the rate of the current?

	Distance	Rate	Time
Downstream			
Upstream			

41. Driving from Tulsa to Detroit, Jeff averaged 50 mph. He figured that if he had averaged 60 mph, his driving time would have decreased 3 hr. How far is it from Tulsa to Detroit?

42. If Dr. Dawson rides his bike to his office, he averages 12 mph. If he drives his car, he averages 36 mph. His time driving is $\frac{1}{4}$ hr less than his time riding his bike. How far is his office from home?

43. A private plane traveled from San Francisco to a secret rendezvous. It averaged 200 mph. On the return trip, the average speed was 300 mph. If the total traveling time was 4 hr, how far from San Francisco was the secret rendezvous?

44. Johnny averages 30 mph when he drives on the old highway to his favorite fishing hole, and he averages 50 mph when most of his route is on the interstate. If both routes are the same length, and he saves 2 hr by traveling on the interstate, how far away is the fishing hole?

45. On the first part of a trip to Carmel traveling on the freeway, Marge averaged 60 mph. On the rest of the trip, which was 10 mi longer than the first part, she averaged 50 mph. Find the total distance to Carmel if the second part of the trip took 30 min more than the first part.

46. While on vacation, Jim and Annie decided to drive all day. During the first part of their trip on the highway, they averaged 60 mph. When they got to Houston, traffic caused them to average only 30 mph. The distance they drove in Houston was 100 mi less than their distance on the highway. What was their total driving distance if they spent 50 min more on the highway than they did in Houston?

Solve each problem. See Example 8.

47. Butch and Peggy want to pick up the mess that their grandson, Grant, has made in his playroom. Butch could do it in 15 min working alone. Peggy, working alone, could clean it in 12 min. How long will it take them if they work together?

	Rate	Time Working Together	Fractional Part of the Job Done
Butch	$\frac{1}{15}$	x	
Peggy	$\frac{1}{12}$	x	

48. Lou can groom Jay Beckenstein's dogs in 8 hr, but it takes his business partner, Janet, only 5 hr to groom the same dogs. How long will it take them to groom Jay's dogs if they work together?

	Rate	Time Working Together	Fractional Part of the Job Done
Lou	$\frac{1}{8}$	x	
Janet	$\frac{1}{5}$	x	

49. Ron Wood can paint a room in 6 hr working alone. If his son, Jason, helps him, the job takes 4 hr. How long would it take Jason to do the job if he worked alone?

50. Sandi and Cary Goldstein are refinishing a table. Working alone, Cary could do the job in 7 hr. If the two work together, the job takes 5 hr. How long will it take Sandi to refinish the table working alone?

51. If a vat of acid can be filled by an inlet pipe in 10 hr and emptied by an outlet pipe in 20 hr, how long will it take to fill the vat if both pipes are open?

52. A winery has a vat to hold chardonnay. An inlet pipe can fill the vat in 9 hr, while an outlet pipe can empty it in 12 hr. How long will it take to fill the vat if both the outlet and the inlet pipes are open?

53. Suppose that Hortense and Mort can clean their entire house in 7 hr, while their toddler, Mimi, just by being around, can completely mess it up in only 2 hr. If Hortense and Mort clean the house while Mimi is at her grandma's, and then start cleaning up after Mimi the minute she gets home, how long does it take from the time Mimi gets home until the whole place is a shambles?

54. An inlet pipe can fill an artificial lily pond in 60 min, while an outlet pipe can empty it in 80 min. Through an error, both pipes are left open. How long will it take for the pond to fill?

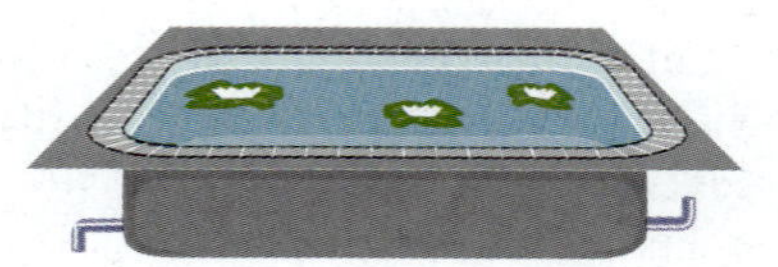

Chapter 7

SUMMARY

KEY TERMS

7.1	**rational expression**	A rational expression (algebraic fraction) is the quotient of two polynomials with denominator not 0.
	rational function	A rational function is a function that is defined by a rational expression in the form $$f(x) = \frac{P(x)}{Q(x)},$$ where $Q(x) \neq 0$.
7.2	**least common denominator (LCD)**	The least common denominator in a group of denominators is the product of all different factors from each denominator, with each factor raised to the greatest power that occurs in any denominator.
7.3	**complex fraction**	A complex fraction is an expression having a fraction in the numerator, denominator, or both.
7.4	**vertical asymptote**	A rational function of the form $f(x) = \frac{P(x)}{x - a}$ has the line $x = a$ as a vertical asymptote; the graph approaches the line on each side but does not intersect it.
	domain of a rational equation	The domain of a rational equation is the intersection (overlap) of the domains of the rational expressions in the equation.
7.5	**ratio**	A ratio is a comparison of two quantities using a quotient.
	proportion	A proportion is a statement that two ratios are equal.

TEST YOUR WORD POWER

See how well you have learned the vocabulary in this chapter. Answers follow the Quick Review.

1. A **rational expression** is
- **(a)** an algebraic expression made up of a term or the sum of a finite number of terms with real coefficients and integer exponents
- **(b)** a polynomial equation of degree 2
- **(c)** an expression with one or more fractions in the numerator, denominator, or both
- **(d)** the quotient of two polynomials with denominator not zero.

2. In a given set of fractions, the **least common denominator** is
- **(a)** the smallest denominator of all the denominators
- **(b)** the smallest expression that is divisible by all the denominators
- **(c)** the largest integer that evenly divides the numerator and denominator of all the fractions
- **(d)** the largest denominator of all the denominators.

3. A **complex fraction** is
- **(a)** an algebraic expression made up of a term or the sum of a finite number of terms with real coefficients and integer exponents
- **(b)** a polynomial equation of degree 2
- **(c)** an expression with one or more fractions in the numerator, denominator, or both
- **(d)** the quotient of two polynomials with denominator not zero.

4. A **ratio**
- **(a)** compares two quantities using a quotient
- **(b)** says that two quotients are equal
- **(c)** is a product of two quantities
- **(d)** is a difference between two quantities.

5. A **proportion**
- **(a)** compares two quantities using a quotient
- **(b)** says that two quotients are equal
- **(c)** is a product of two quantities
- **(d)** is a difference between two quantities.

Quick Review

Concepts	*Examples*
7.1 Rational Expressions and Functions; Multiplying and Dividing	
Fundamental Property of Rational Numbers If $\frac{a}{b}$ is a rational number and if c is any nonzero real number, then $$\frac{a}{b}=\frac{ac}{bc}.$$	$$\frac{3}{4}=\frac{3\cdot 5}{4\cdot 5}=\frac{15}{20}$$
Writing a Rational Expression in Lowest Terms Factor the numerator and the denominator completely. Then apply the fundamental property.	Write in lowest terms. $$\frac{2x+8}{x^2-16}=\frac{2(x+4)}{(x-4)(x+4)}=\frac{2}{x-4}$$
Multiplying Rational Expressions Factor numerators and denominators. Apply the fundamental property and replace all pairs of common factors in numerators and denominators by 1. Multiply the remaining factors in the numerator and in the denominator.	Multiply. $\frac{x^2+2x+1}{x^2-1}\cdot\frac{5}{3x+3}$ $$=\frac{(x+1)^2}{(x-1)(x+1)}\cdot\frac{5}{3(x+1)}$$ $$=\frac{5}{3(x-1)}$$
Dividing Rational Expressions Multiply the first rational expression by the reciprocal of the second.	Divide. $\frac{2x+5}{x-3}\div\frac{2x^2+3x-5}{x^2-9}$ $$=\frac{2x+5}{x-3}\cdot\frac{(x+3)(x-3)}{(2x+5)(x-1)}$$ $$=\frac{x+3}{x-1}$$
7.2 Adding and Subtracting Rational Expressions	
Adding or Subtracting Rational Expressions If the denominators are the same, add or subtract the numerators. Place the result over the common denominator. If the denominators are different, write all rational expressions with the LCD. Then add or subtract the numerators, and place the result over the common denominator. Be sure the answer is in lowest terms.	Subtract. $\frac{1}{x+6}-\frac{3}{x+2}$ $$=\frac{x+2}{(x+6)(x+2)}-\frac{3(x+6)}{(x+6)(x+2)}$$ $$=\frac{x+2-3(x+6)}{(x+6)(x+2)}$$ $$=\frac{x+2-3x-18}{(x+6)(x+2)}$$ $$=\frac{-2x-16}{(x+6)(x+2)}$$

Concepts	*Examples*
7.3 Complex Fractions	
Simplifying a Complex Fraction	Simplify the complex fraction.
Method 1 Simplify the numerator and denominator separately, as much as possible. Then multiply the numerator by the reciprocal of the denominator. Write the answer in lowest terms.	*Method 1* $$\frac{\frac{1}{x^2}-\frac{1}{y^2}}{\frac{1}{x}+\frac{1}{y}} = \frac{\frac{y^2}{x^2y^2}-\frac{x^2}{x^2y^2}}{\frac{y}{xy}+\frac{x}{xy}} = \frac{\frac{y^2-x^2}{x^2y^2}}{\frac{y+x}{xy}} = \frac{y^2-x^2}{x^2y^2} \div \frac{y+x}{xy} = \frac{(y+x)(y-x)}{x^2y^2} \cdot \frac{xy}{x+y} = \frac{y-x}{xy}$$
Method 2 Multiply the numerator and denominator of the complex fraction by the least common denominator of all fractions appearing in the complex fraction. Then simplify the result.	*Method 2* $$\frac{\frac{1}{x^2}-\frac{1}{y^2}}{\frac{1}{x}+\frac{1}{y}} = \frac{x^2y^2\left(\frac{1}{x^2}-\frac{1}{y^2}\right)}{x^2y^2\left(\frac{1}{x}+\frac{1}{y}\right)} = \frac{y^2-x^2}{xy^2+x^2y} = \frac{(y-x)(y+x)}{xy(y+x)} = \frac{y-x}{xy}$$
7.4 Graphs and Equations with Rational Expressions The graph of a simple rational function may have one or more breaks. At such points, the graph will approach an asymptote.	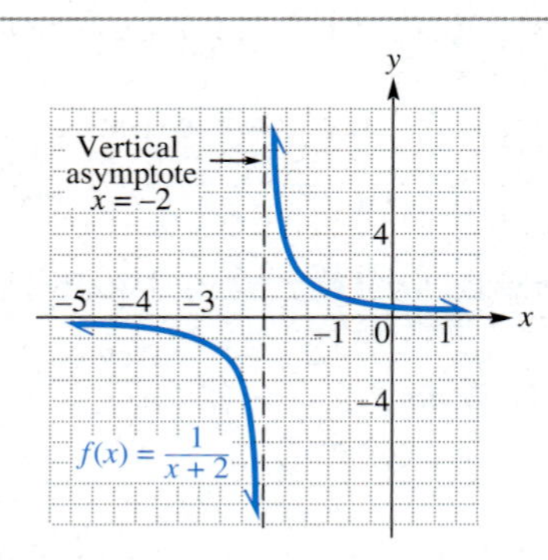
To solve an equation involving rational expressions, first determine the domain. Then multiply all the terms in the equation by the least common denominator. Solve the resulting equation. Each potential solution *must* be checked to see that it is in the domain of the equation.	Solve. $$\frac{1}{x} + x = \frac{26}{5}$$ Note that 0 is excluded from the domain. $$5 + 5x^2 = 26x \quad \text{Multiply by } 5x.$$ $$5x^2 - 26x + 5 = 0$$ $$(5x-1)(x-5) = 0$$ $$x = \frac{1}{5} \quad \text{or} \quad x = 5$$ Both check. The solution set is $\{\frac{1}{5}, 5\}$.

Concepts	Examples
7.5 Applications of Rational Expressions To solve a formula for a particular variable, isolate that variable on one side.	Solve for L. $c = \frac{100b}{L}$ $cL = 100b$ Multiply by L. $L = \frac{100b}{c}$ Divide by c.
To solve a motion problem, use the formula $d = rt$ or one of its equivalents, $t = \frac{d}{r}$ or $r = \frac{d}{t}$.	Solve. A canal has a current of 2 mph. Find the speed of Amy's boat in still water if it goes 11 mi downstream in the same time that it goes 8 mi upstream. Let x represent the speed of the boat in still water. (see table below) Because the times are the same, the equation is $\frac{11}{x+2} = \frac{8}{x-2}$. Use $t = \frac{d}{r}$. $11(x-2) = 8(x+2)$ Multiply by the LCD. $11x - 22 = 8x + 16$ Distributive property $3x = 38$ Subtract $8x$ and add 22. $x = 12\frac{2}{3}$ Divide by 3. The speed in still water is $12\frac{2}{3}$ mph.
To solve a work problem, use the fact that if a complete job is done in t units of time, the rate of work is $\frac{1}{t}$ job per unit of time.	

	Distance	Rate	Time
Downstream	11	$x + 2$	$\frac{11}{x+2}$
Upstream	8	$x - 2$	$\frac{8}{x-2}$

Answers to Test Your Word Power

1. **(d)** *Examples:* $-\frac{3}{4y^2}, \frac{5x^3}{x+2}, \frac{a+3}{a^2-4a-5}$ 2. **(b)** *Example:* The LCD of $\frac{1}{x}, \frac{2}{3}$, and $\frac{5}{x+1}$ is $3x(x+1)$.

3. **(c)** *Examples:* $\frac{\frac{2}{3}}{\frac{4}{7}}, \frac{x-\frac{1}{x}}{x+\frac{1}{y}}, \frac{\frac{2}{a+1}}{a^2-1}$ 4. **(a)** *Example:* $\frac{7 \text{ in.}}{12 \text{ in.}}$ compares two quantities.

5. **(b)** *Example:* The proportion $\frac{2}{3} = \frac{8}{12}$ states that the two ratios are equal.

Chapter 7 Review Exercises

[7.1] ***(a)*** *Find all real numbers that are excluded from the domain.* ***(b)*** *Give the domain using set notation.*

1. $f(x) = \dfrac{-7}{3x + 18}$

2. $f(x) = \dfrac{5x + 17}{x^2 - 7x + 10}$

3. $f(x) = \dfrac{9}{x^2 - 18x + 81}$

Write in lowest terms.

4. $\dfrac{12x^2 + 6x}{24x + 12}$

5. $\dfrac{25m^2 - n^2}{25m^2 - 10mn + n^2}$

6. $\dfrac{r - 2}{4 - r^2}$

7. What is meant by the reciprocal of a rational expression?

Multiply or divide. Write the answer in lowest terms.

8. $\dfrac{(2y + 3)^2}{5y} \cdot \dfrac{15y^3}{4y^2 - 9}$

9. $\dfrac{w^2 - 16}{w} \cdot \dfrac{3}{4 - w}$

10. $\dfrac{z^2 - z - 6}{z - 6} \cdot \dfrac{z^2 - 6z}{z^2 + 2z - 15}$

11. $\dfrac{m^3 - n^3}{m^2 - n^2} \div \dfrac{m^2 + mn + n^2}{m + n}$

[7.2] *Assume that each expression is the denominator of a rational expression. Find the least common denominator for each group.*

12. $32b^3$, $24b^5$

13. $9r^2$, $3r + 1$

14. $6x^2 + 13x - 5$, $9x^2 + 9x - 4$

Add or subtract as indicated.

15. $\dfrac{8}{z} - \dfrac{3}{2z^2}$

16. $\dfrac{5y + 13}{y + 1} - \dfrac{1 - 7y}{y + 1}$

17. $\dfrac{6}{5a + 10} + \dfrac{7}{6a + 12}$

18. $\dfrac{3r}{10r^2 - 3rs - s^2} + \dfrac{2r}{2r^2 + rs - s^2}$

[7.3] *Simplify each complex fraction.*

19. $\dfrac{\frac{3}{t}+2}{\frac{4}{t}-7}$

20. $\dfrac{\frac{2}{m-3n}}{\frac{1}{3n-m}}$

21. $\dfrac{\frac{3}{p}-\frac{2}{q}}{\frac{9q^2-4p^2}{qp}}$

22. $\dfrac{x^{-2}-y^{-2}}{x^{-1}-y^{-1}}$

[7.4]

23. Which is the graph of a rational function? What is the equation of its vertical asymptote?

A.

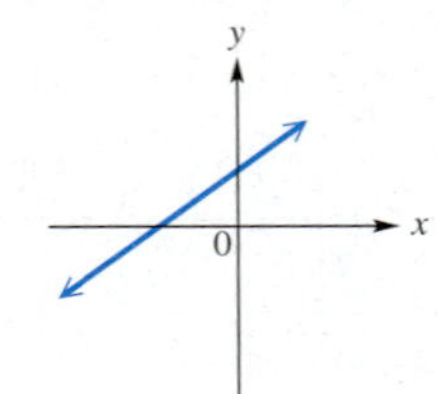

B.

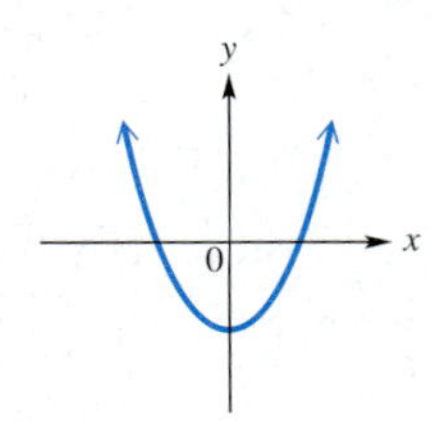

C.

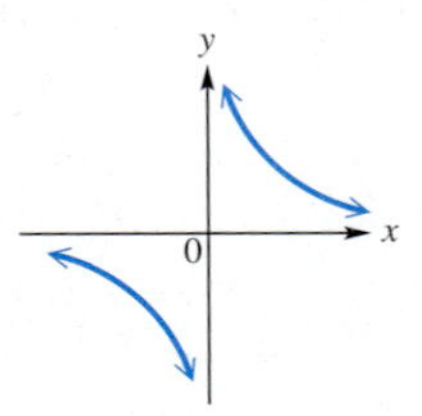

D.

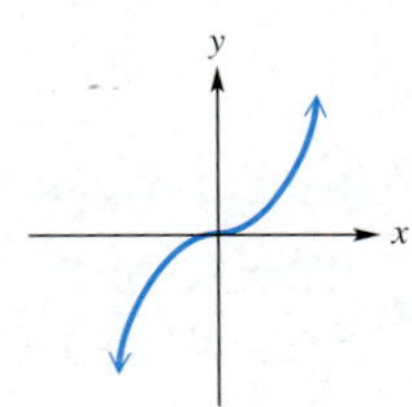

Solve each equation.

24. $\dfrac{1}{t+4}+\dfrac{1}{2}=\dfrac{3}{2t+8}$

25. $\dfrac{-5m}{m+1}+\dfrac{m}{3m+3}=\dfrac{56}{6m+6}$

26. $\dfrac{2}{k-1}-\dfrac{4k+1}{k^2-1}=\dfrac{-1}{k+1}$

27. $\dfrac{5}{x+2}+\dfrac{3}{x+3}=\dfrac{x}{x^2+5x+6}$

28. After solving the equation

$$\frac{3}{x-3}-\frac{2}{x-2}=\frac{3}{x^2-5x+6},$$

a student got $x = 3$ as her final step. She could not understand why the answer in the back of the book was "$\emptyset$," because she checked her algebra several times and was sure that all her algebraic work was correct. Was she wrong or was the answer in the back of the book wrong? Explain.

29. Explain the difference between simplifying the expression

$$\frac{4}{x}+\frac{1}{2}-\frac{1}{3}$$

and solving the equation

$$\frac{4}{x}+\frac{1}{2}=\frac{1}{3}.$$

[7.5]

30. According to a law from physics, $\dfrac{1}{A}=\dfrac{1}{B}+\dfrac{1}{C}$. Find A if $B = 30$ and $C = 10$.

Solve each formula for the specified variable.

31. $F = \dfrac{GMm}{d^2}$ for m (physics)

32. $\mu = \dfrac{Mv}{M + m}$ for M (electronics)

Solve each problem.

33. An article in *Scientific American* predicts that, in the year 2050, 23,200 of the 58,000 passenger-km per day in North America will be provided by high-speed trains. If the traffic volume in a typical region of North America is 15,000, how many passenger-kilometers per day will high-speed trains provide there? (*Source:* Schafer, Andreas and David Victor, "The Past and Future of Global Mobility," *Scientific American*, October, 1997.)

34. A river has a current of 4 km per hr. Find the speed of Lynn McTernan's boat in still water if it goes 40 km downstream in the same time that it takes to go 24 km upstream.

	d	r	t
Upstream	24	$x - 4$	
Downstream	40		

35. A sink can be filled by a cold-water tap in 8 min, and filled by the hot-water tap in 12 min. How long would it take to fill the sink with both taps open?

36. Jane Estrella and Jason Jordan need to sort a pile of bottles at the recycling center. Working alone, Jane could do the entire job in 9 hr, while Jason could do the entire job in 6 hr. How long will it take them if they work together?

Mixed Review Exercises

Write in lowest terms.

37. $\dfrac{x + 2y}{x^2 - 4y^2}$

38. $\dfrac{x^2 + 2x - 15}{x^2 - x - 6}$

Perform the indicated operations.

39. $\dfrac{2}{m} + \dfrac{5}{3m^2}$

40. $\dfrac{k^2 - 6k + 9}{1 - 216k^3} \cdot \dfrac{6k^2 + 17k - 3}{9 - k^2}$

41. $\dfrac{\dfrac{-3}{x}+\dfrac{x}{2}}{1+\dfrac{x+1}{x}}$

42. $\dfrac{9x^2+46x+5}{3x^2-2x-1} \div \dfrac{x^2+11x+30}{x^3+5x^2-6x}$

43. $\dfrac{\dfrac{3}{x}-5}{6+\dfrac{1}{x}}$

44. $\dfrac{9}{3-x}-\dfrac{2}{x-3}$

45. $\dfrac{4y+16}{30} \div \dfrac{2y+8}{5}$

46. $\dfrac{t^{-2}+s^{-2}}{t^{-1}-s^{-1}}$

47. $\dfrac{4a}{a^2-ab-2b^2}-\dfrac{6b-a}{a^2+4ab+3b^2}$

48. $\dfrac{a}{b}+\dfrac{b}{c}+\dfrac{c}{d}$

Solve.

49. $\dfrac{x+3}{x^2-5x+4}-\dfrac{1}{x}=\dfrac{2}{x^2-4x}$

50. $A=\dfrac{Rr}{R+r}$ for r

51. $1-\dfrac{5}{r}=\dfrac{-4}{r^2}$

52. $\dfrac{3x}{x-4}+\dfrac{2}{x}=\dfrac{48}{x^2-4x}$

53. The strength of a contact lens is given in units called diopters, and also in millimeters of arc. As the diopters increase, the millimeters of arc decrease. The rational function defined by

$$a=\frac{337}{d}$$

relates the arc measurement a to the diopter measurement d. (*Source:* Bausch and Lomb.)

(a) What arc measurement will correspond to 40.5-diopter lenses?

(b) A lens with an arc measurement of 7.51 will provide what diopter strength?

54. The hot-water tap can fill a tub in 20 min. The cold-water tap takes 15 min to fill the tub. How long would it take to fill the tub with both taps open?

55. At a certain gasoline station, 3 gal of unleaded gasoline cost \$4.86. How much would 13 gal of the same gasoline cost?

56. Three-fourths of a number is subtracted from seven-sixths of the number, giving 10. Find the number.

Chapter 7 TEST

1. Find all real numbers excluded from the domain of $f(x) = \dfrac{x + 3}{3x^2 + 2x - 8}$. Then give the domain using set notation.

2. Write $\dfrac{6x^2 - 13x - 5}{9x^3 - x}$ in lowest terms.

Multiply or divide.

3. $\dfrac{(x + 3)^2}{4} \cdot \dfrac{6}{2x + 6}$

4. $\dfrac{y^2 - 16}{y^2 - 25} \cdot \dfrac{y^2 + 2y - 15}{y^2 - 7y + 12}$

5. $\dfrac{x^2 - 9}{x^3 + 3x^2} \div \dfrac{x^2 + x - 12}{x^3 + 9x^2 + 20x}$

6. Find the least common denominator for the following group of denominators: $t^2 + t - 6, \quad t^2 + 3t, \quad t^2$.

Add or subtract as indicated.

7. $\dfrac{7}{6t^2} - \dfrac{1}{3t}$

8. $\dfrac{9}{x - 7} + \dfrac{4}{x + 7}$

9. $\dfrac{6}{x + 4} + \dfrac{1}{x + 2} - \dfrac{3x}{x^2 + 6x + 8}$

Simplify each complex fraction.

10. $\dfrac{\dfrac{12}{r + 4}}{\dfrac{11}{6r + 24}}$

11. $\dfrac{\dfrac{1}{a} - \dfrac{1}{b}}{\dfrac{a}{b} - \dfrac{b}{a}}$

12. $\dfrac{\dfrac{2}{x^2} + \dfrac{1}{y^2}}{\dfrac{1}{x} - \dfrac{1}{y}}$

13. Sketch the graph of the function defined by $f(x) = \dfrac{-2}{x + 1}$. Give the equation of its vertical asymptote.

1. ______________________

2. ______________________

3. ______________________

4. ______________________

5. ______________________

6. ______________________

7. ______________________

8. ______________________

9. ______________________

10. ______________________

11. ______________________

12. ______________________

13. ______________________

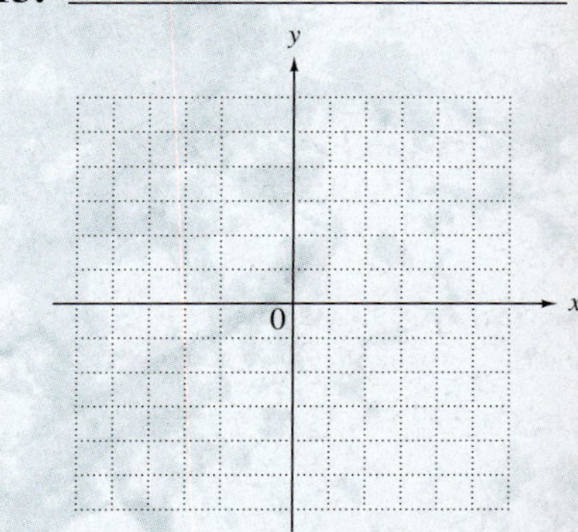

14. (a) ______________

(b) ______________

14. One of the following is an expression to be simplified by algebraic operations, and the other is an equation to be solved. Simplify the one that requires operations, and solve the one that is an equation.

(a) $\frac{2x}{3} + \frac{x}{4} - \frac{11}{2}$ **(b)** $\frac{2x}{3} + \frac{x}{4} = \frac{11}{2}$

Solve each equation.

15. ______________

16. ______________

15. $\frac{1}{x} - \frac{4}{3x} = \frac{1}{x - 2}$

16. $\frac{y}{y + 2} - \frac{1}{y - 2} = \frac{8}{y^2 - 4}$

17. ______________

17. Checking the solution(s) of an equation in Chapters 1–6 verified that the algebraic steps were performed correctly. When an equation includes a term with a variable denominator, what additional reason *requires* that the solutions be checked?

18. ______________

18. Solve for the variable ℓ in this formula from mathematics:

$$S = \frac{n}{2}(a + \ell).$$

Solve each problem.

19. ______________

19. Wayne can do the job in 9 hr, while Susan can do the same job in 5 hr. How long would it take them to do the job if they worked together?

20. ______________

20. The rate of the current in a stream is 3 mph. Nana's boat can go 36 mi downstream in the same time that it takes to go 24 mi upstream. Find the rate of her boat in still water.

21. ______________

21. Biologists collected a sample of 600 fish from Lake Linda on May 1 and tagged each of them. When they returned on June 1, a new sample of 800 fish was collected, and 10 of these had been previously tagged. Use this experiment to determine the approximate fish population of Lake Linda.

22. (a) ______________

(b) ______________

22. In biology, the function defined by

$$g(x) = \frac{5x}{2 + x}$$

gives the growth rate of a population for x units of available food. (*Source:* Smith, J. Maynard, *Models in Ecology*, Cambridge University Press, 1974.)

(a) What amount of food (in appropriate units) would produce a growth rate of 3 units of growth per unit of food?

(b) What is the growth rate if no food is available?

Cumulative Review Exercises CHAPTERS 1–7

Evaluate if $x = -4$, $y = 3$, and $z = 6$.

1. $|2x| + 3y - z^3$

2. $\dfrac{x(2x - 1)}{3y - z}$

Solve each equation.

3. $7(2x + 3) - 4(2x + 1) = 2(x + 1)$

4. $|6x - 8| - 4 = 0$

5. $ax + by = cx + d$ for x

Solve each inequality.

6. $\dfrac{2}{3}x + \dfrac{5}{12}x \leq 20$

7. $|3x + 2| \geq 4$

Solve each problem.

8. Otis Taylor invested some money at 4% interest and twice as much at 3% interest. His interest for the first year was $400. How much did he invest at each rate?

9. A triangle has an area of 42 m^2. The base is 14 m long. Find the height of the triangle.

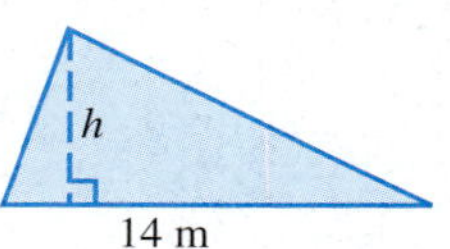

10. Graph $-4x + 2y = 8$ and give the intercepts.

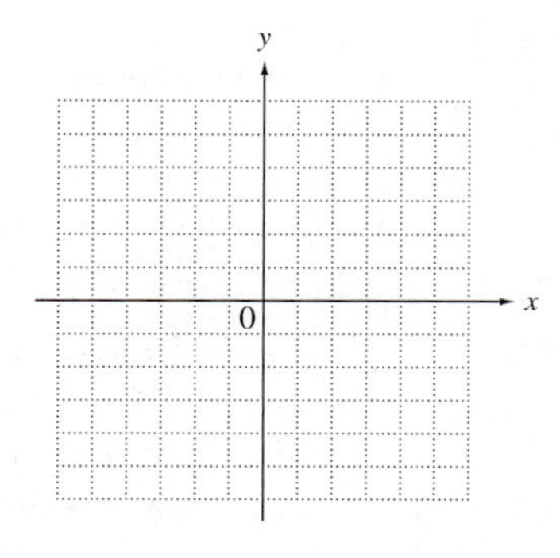

Find the slope of each line described in Exercises 11 and 12.

11. Through $(-5, 8)$ and $(-1, 2)$

12. Perpendicular to $4x - 3y = 12$

13. Write an equation of the line in Exercise 11 in the form $y = mx + b$.

Graph the solution set of each inequality.

14. $2x + 5y > 10$

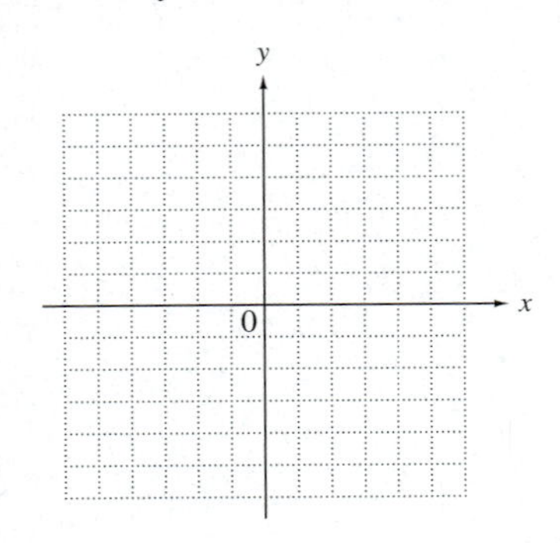

15. $x - y \geq 3$ and $3x + 4y \leq 12$

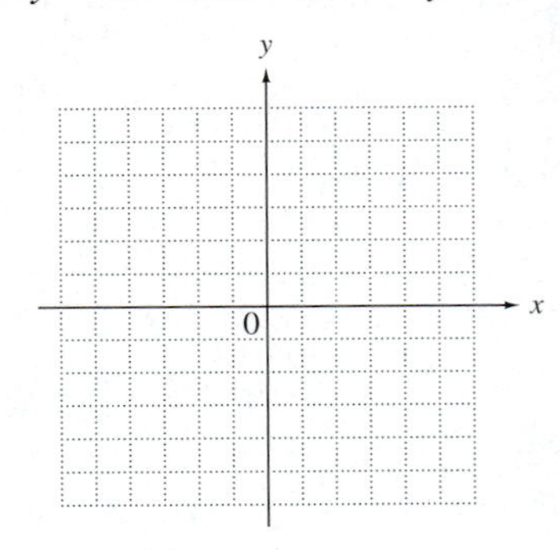

Decide whether each relation defined in Exercises 16–18 defines a function, and give its domain and range.

16. AVERAGE HOURLY WAGES IN MEXICO

Year	Wage (in dollars)
1990	1.25
1992	1.61
1994	1.80
1996	1.21
1998	1.94
2000	2.26

Source: John Christman, CIEMEX-WEFA.

17.

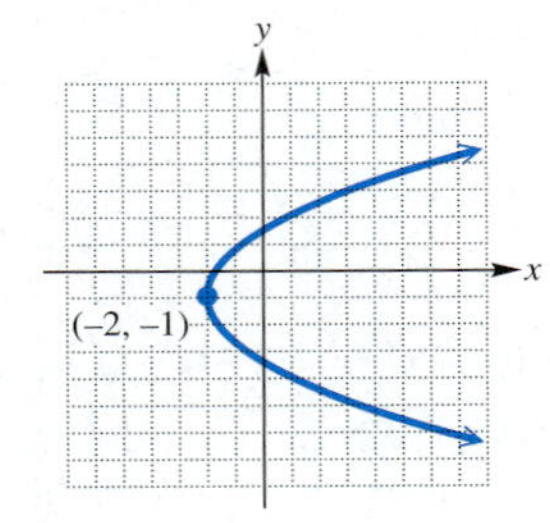

18. $y = -\sqrt{x + 2}$

19. Given the equation $5x - 3y = 8$,

(a) write it with function notation $f(x)$;

(b) find $f(1)$.

20. If $f(x) = 3x + 6$, what is $f(x + 3)$?

Solve each system.

21.
$$\begin{aligned} 4x - y &= -7 \\ 5x + 2y &= 1 \end{aligned}$$

22.
$$\begin{aligned} x + y - 2z &= -1 \\ 2x - y + z &= -6 \\ 3x + 2y - 3z &= -3 \end{aligned}$$

23.
$$\begin{aligned} x + 2y + z &= 5 \\ x - y + z &= 3 \\ 2x + 4y + 2z &= 11 \end{aligned}$$

24. Taking traffic into account, an automobile can travel on the average 7 km in the same time that an airplane can travel 100 km. The average speed of an airplane is 558 km per hr greater than that of an automobile. Find both speeds. (*Source:* Schafer, Andreas, and David Victor, "The Past and Future of Global Mobility," *Scientific American*, October 1997.)

Simplify. Write the answer with only positive exponents. Assume that all variables represent nonzero real numbers.

25. $\left(\dfrac{a^{-3}b^{4}}{a^{2}b^{-1}}\right)^{-2}$

26. $\left(\dfrac{m^{-4}n^{2}}{m^{2}n^{-3}}\right) \cdot \left(\dfrac{m^{5}n^{-1}}{m^{-2}n^{5}}\right)$

Perform the indicated operations.

27. $(3y^2 - 2y + 6) - (-y^2 + 5y + 12)$

28. $-6x^4(x^2 - 3x + 2)$

29. $(4f + 3)(3f - 1)$

30. $(7t^3 + 8)(7t^3 - 8)$

31. $\left(\dfrac{1}{4}x + 5\right)^2$

32. $(3x^3 + 13x^2 - 17x - 7) \div (3x + 1)$

33. Use synthetic division to divide $(2x^4 + 3x^3 - 8x^2 + x + 2)$ by $(x - 1)$.

Factor each polynomial completely.

34. $2x^2 - 13x - 45$

35. $100t^4 - 25$

36. $8p^3 + 125$

37. Solve the equation $3x^2 + 4x = 7$.

Write each rational expression in lowest terms.

38. $\dfrac{y^2 - 16}{y^2 - 8y + 16}$

39. $\dfrac{8x^2 - 18}{8x^2 + 4x - 12}$

Perform the indicated operations. Express the answer in lowest terms.

40. $\dfrac{2a^2}{a + b} \cdot \dfrac{a - b}{4a}$

41. $\dfrac{x + 4}{x - 2} + \dfrac{2x - 10}{x - 2}$

42. $\dfrac{2x}{2x - 1} + \dfrac{4}{2x + 1} + \dfrac{8}{4x^2 - 1}$

43. Solve the equation

$$\frac{-3x}{x + 1} + \frac{4x + 1}{x} = \frac{-3}{x^2 + x}.$$

44. Solve the formula

$$\frac{1}{f} = \frac{1}{p} + \frac{1}{q}$$

for q.

Solve each problem.

45. Lucinda can fly her plane 200 mi against the wind in the same time it takes her to fly 300 mi with the wind. The wind blows at 30 mph. Find the speed of her plane in still air.

46. Machine A can complete a certain job in 2 hr. To speed up the work, Machine B, which could complete the job alone in 3 hr, is brought in to help. How long will it take the two machines to complete the job working together?

Roots and Radicals

Many real-life situations are modeled by equations of the form $y = ax^2 + bx + c$. Some common examples are the height of an object thrown upward after a given amount of time and data that change at an increasing rate, such as the number of U.S. cell-phone subscribers discussed in the exercises for Section 8.6. In such cases, if y is known and we want to find the corresponding value of x, we must solve an equation with radicals. In this chapter we will learn to work with radical expressions and solve radical equations.

8.1 Radical Expressions and Graphs

Objectives

1. Find roots of numbers.
2. Find principal roots.
3. Graph functions defined by radical expressions.
4. Find *n*th roots of *n*th powers.
5. Use a calculator to find roots.

1 Simplify.

(a) $\sqrt[3]{8}$

(b) $\sqrt[3]{1000}$

(c) $\sqrt[4]{81}$

(d) $\sqrt[6]{64}$

Answers

1. (a) 2 (b) 10 (c) 3 (d) 2

1 Find roots of numbers. In Section 1.3 we found square roots of positive numbers such as

$$\sqrt{36} = 6, \text{ because } 6 \cdot 6 = 36 \quad \text{and} \quad \sqrt{144} = 12, \text{ because } 12 \cdot 12 = 144.$$

In this section we extend our discussion of roots to cube roots, fourth roots, and higher roots. In general, $\sqrt[n]{a}$ is a number whose nth power equals a. That is,

$$\sqrt[n]{a} = b \quad \textbf{means} \quad b^n = a.$$

The number a is the **radicand,** n is the **index** or **order,** and the expression $\sqrt[n]{a}$ is a **radical.**

Example 1 Simplifying Higher Roots

Simplify.

(a) $\sqrt[3]{27} = 3$, because $3^3 = 27$.

(b) $\sqrt[3]{125} = 5$, because $5^3 = 125$.

(c) $\sqrt[4]{16} = 2$, because $2^4 = 16$.

(d) $\sqrt[5]{32} = 2$, because $2^5 = 32$.

Work Problem 1 at the Side.

2 Find principal roots. If n is even, positive numbers have two nth roots. For example, both 4 and -4 are square roots of 16, and 2 and -2 are fourth roots of 16. In such cases, the notation $\sqrt[n]{a}$ represents the positive root, called the **principal root.**

***n*th Root**

If n is *even* and a is *positive* or 0, then

$\sqrt[n]{a}$ represents the principal nth root of a, and
$-\sqrt[n]{a}$ represents the negative nth root of a.

If n is *even* and a is *negative*, then

$\sqrt[n]{a}$ is not a real number.

If n is odd, then

there is exactly one nth root of a, written $\sqrt[n]{a}$.

The two nth roots of a are often written together as $\pm\sqrt[n]{a}$, with $\pm$ read "positive or negative."

Example 2 **Finding Roots**

Find each root.

(a) $\sqrt{100} = 10$
Because the radicand is positive, there are two square roots, 10 and -10. We want the principal root, which is 10.

(b) $-\sqrt{100} = -10$
Here, we want the negative square root, -10.

(c) $\sqrt[4]{81} = 3$

(d) $\sqrt[6]{-64}$
The index is even and the radicand is negative, so this is not a real number.

(e) $\sqrt[3]{-8} = -2$, because $(-2)^3 = -8$.

Work Problem **2** at the Side.

3 **Graph functions defined by radical expressions.** A **radical expression** is an algebraic expression that contains radicals. For example,

$$3 - \sqrt{x}, \quad \sqrt[3]{x}, \quad \text{and} \quad \sqrt{2x - 1}$$

are radical expressions.

In earlier chapters we graphed functions defined by polynomial and rational expressions. Now we examine the graphs of functions defined by the radical expressions $f(x) = \sqrt{x}$ and $f(x) = \sqrt[3]{x}$.

Figure 1 shows the graph of the **square root function** with a table of selected points.

x	$f(x) = \sqrt{x}$
0	0
1	1
4	2
9	3

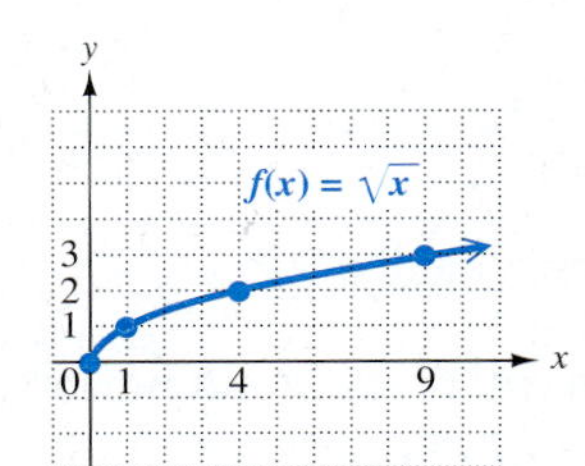

Figure 1

Only nonnegative values can be used for x, so the domain is $[0, \infty)$. Because $\sqrt{x}$ is the principal square root of x, it always has a nonnegative value, so the range is also $[0, \infty)$.

Figure 2 shows the graph of the **cube root function** and a table of selected points.

x	$f(x) = \sqrt[3]{x}$
-8	-2
-1	-1
0	0
1	1
8	2

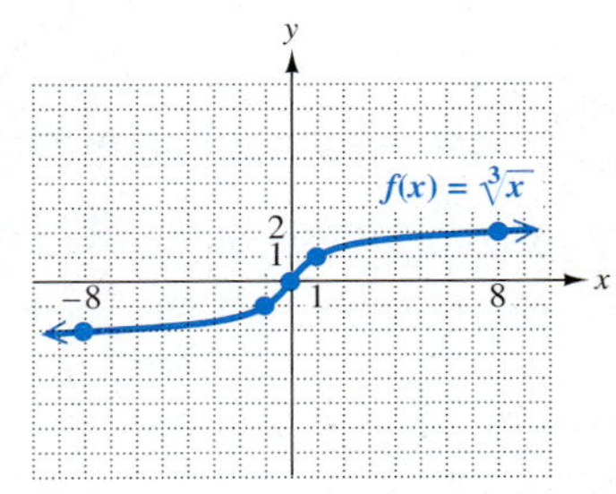

Figure 2

Since any real number (positive, negative, or 0) can be used for x in the cube root function, $\sqrt[3]{x}$ can be positive, negative, or 0. Thus both the domain and the range of the cube root function are $(-\infty, \infty)$.

2 Find each root.

(a) $\sqrt{4}$

(b) $\sqrt[3]{27}$

(c) $-\sqrt{36}$

(d) $\sqrt[4]{625}$

(e) $\sqrt[5]{-32}$

(f) $\sqrt[4]{-16}$

ANSWERS
2. **(a)** 2 **(b)** 3 **(c)** -6 **(d)** 5 **(e)** -2 **(f)** not a real number

3 Graph each function by creating a table of values. Give the domain and range.

(a) $f(x) = \sqrt{x} + 2$

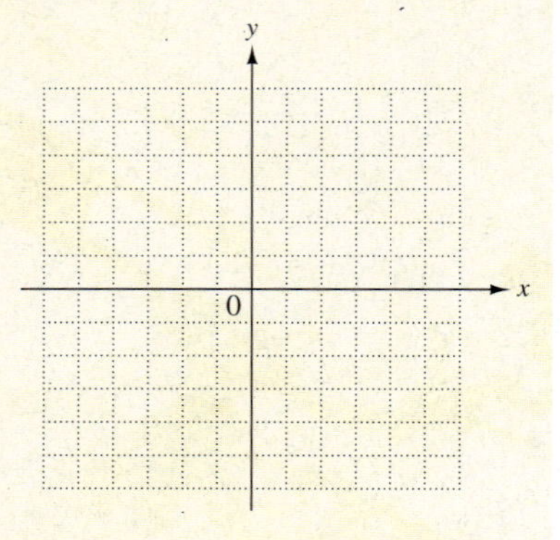

(b) $f(x) = \sqrt[3]{x - 1}$

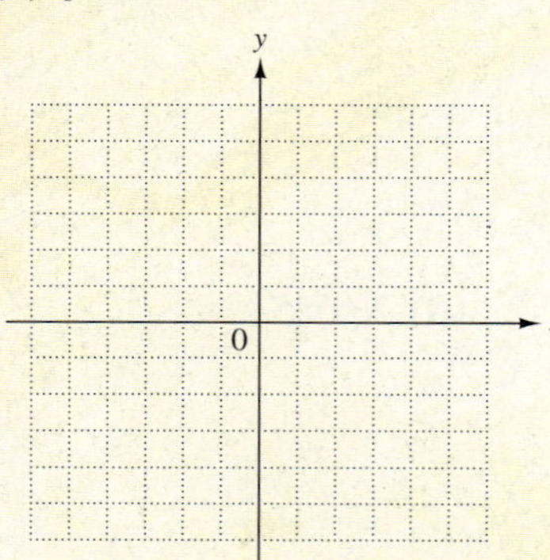

Answers

3. (a) domain: $[0, \infty)$; range: $[2, \infty)$

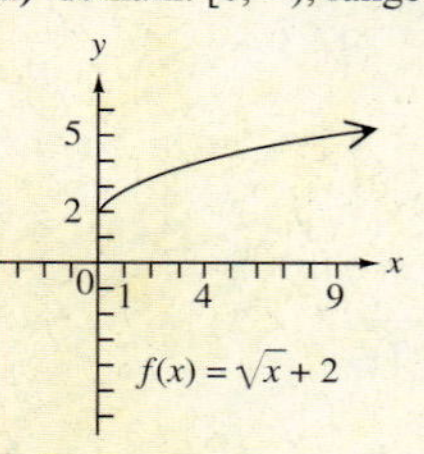

(b) domain: $(-\infty, \infty)$; range: $(-\infty, \infty)$

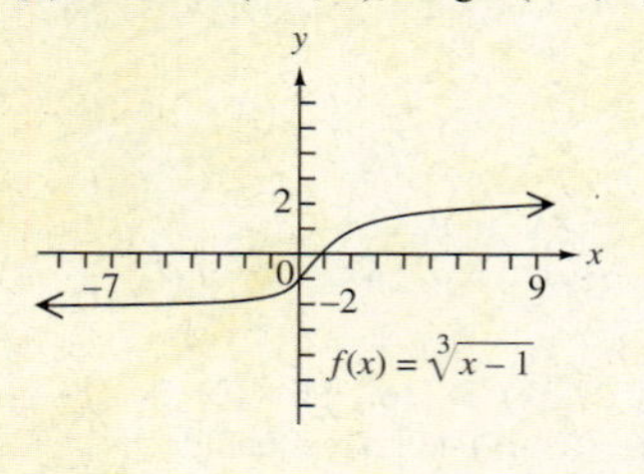

Example 3 Graphing Functions Defined with Radicals

Graph each function by creating a table of values. Give the domain and the range.

(a) $f(x) = \sqrt{x - 3}$

A table of values is shown. The x-values were chosen in such a way that the function values are all integers. For the radicand to be nonnegative, we must have $x - 3 \geq 0$, or $x \geq 3$. Therefore, the domain is $[3, \infty)$. Again, function values are positive or 0, so the range is $[0, \infty)$. The graph is shown in Figure 3.

x	$f(x) = \sqrt{x - 3}$
3	$\sqrt{3 - 3} = 0$
4	$\sqrt{4 - 3} = 1$
7	$\sqrt{7 - 3} = 2$

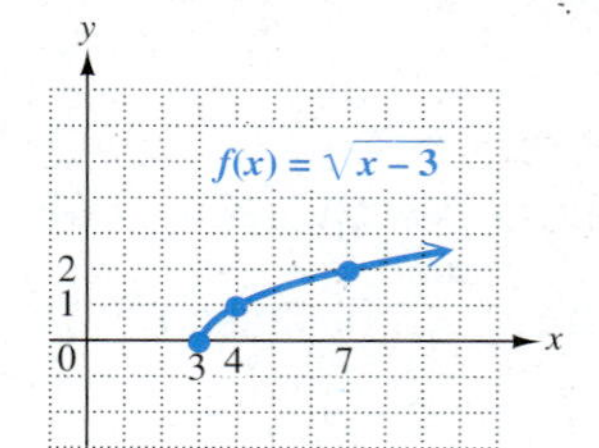

Figure 3

(b) $f(x) = \sqrt[3]{x} + 2$

See the table and Figure 4. Both the domain and the range are $(-\infty, \infty)$.

x	$f(x) = \sqrt[3]{x} + 2$
-8	$\sqrt[3]{-8} + 2 = 0$
-1	$\sqrt[3]{-1} + 2 = 1$
0	$\sqrt[3]{0} + 2 = 2$
1	$\sqrt[3]{1} + 2 = 3$
8	$\sqrt[3]{8} + 2 = 4$

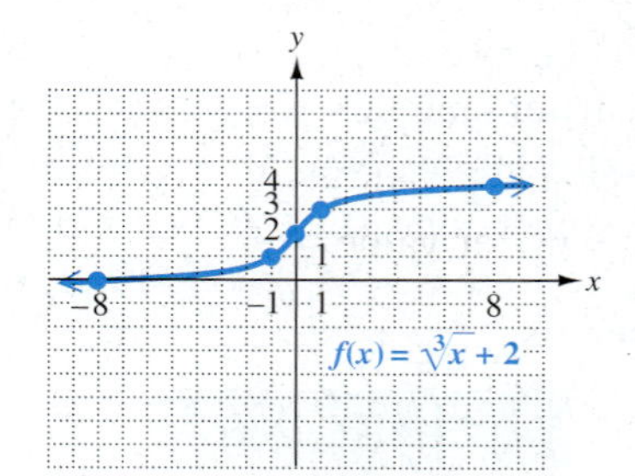

Figure 4

Work Problem 3 at the Side.

4 Find *n*th roots of *n*th powers. A square root of a^2 (where $a \neq 0$) is a number that can be squared to give a^2. This number is either a or $-a$. Since the symbol $\sqrt{a^2}$ represents the *nonnegative* square root, we must write $\sqrt{a^2}$ with absolute value bars, as $|a|$, because a may be a negative number.

For any real number a, $\sqrt{a^2} = |a|$.

Example 4 Simplifying Square Roots Using Absolute Value

Find each square root that is a real number.

(a) $\sqrt{7^2} = |7| = 7$

(b) $\sqrt{(-7)^2} = |-7| = 7$

(c) $\sqrt{k^2} = |k|$

(d) $\sqrt{(-k)^2} = |-k| = |k|$

Work Problem 4 at the Side.

We can generalize this idea to any nth root.

$\sqrt[n]{a^n}$

If n is an *even* positive integer, $\sqrt[n]{a^n} = |a|$,

and if n is an *odd* positive integer, $\sqrt[n]{a^n} = a$.

In words, use absolute value when n is even; do not use absolute value when n is odd.

Example 5 Simplifying Higher Roots Using Absolute Value

Simplify each root.

(a) $\sqrt[6]{(-3)^6} = |-3| = 3$ $\quad$ n is even; use absolute value.

(b) $\sqrt[5]{(-4)^5} = -4$ $\quad$ n is odd.

(c) $-\sqrt[4]{(-9)^4} = -|-9| = -9$

(d) $\sqrt[3]{\frac{8}{27}} = \sqrt[3]{\left(\frac{2}{3}\right)^3} = \frac{2}{3}$

(e) $-\sqrt{m^4} = -|m^2| = -m^2$

No absolute value bars are needed here because m^2 is nonnegative for any real number value of m.

(f) $\sqrt[3]{a^{12}} = a^4$, because $a^{12} = (a^4)^3$.

(g) $\sqrt[4]{x^{12}} = |x^3|$

We use absolute value bars to guarantee that the result is not negative (because x^3 can be either positive or negative, depending on x). If desired, $|x^3|$ can be written as $x^2 \cdot |x|$.

Work Problem 5 at the Side.

5 Use a calculator to find roots. While numbers such as $\sqrt{9}$ and $\sqrt[3]{-8}$ are rational, radicals are often irrational numbers. To find approximations of roots such as $\sqrt{15}$, $\sqrt[3]{10}$, and $\sqrt[4]{2}$, we usually use scientific or graphing calculators. Using a calculator, we find

$$\sqrt{15} \approx 3.872983346, \quad \sqrt[3]{10} \approx 2.15443469, \quad \text{and} \quad \sqrt[4]{2} \approx 1.189207115,$$

where the symbol $\approx$ means "is approximately equal to." In this book we will usually show approximations rounded to three decimal places. Thus, we would write

$$\sqrt{15} \approx 3.873, \quad \sqrt[3]{10} \approx 2.154, \quad \text{and} \quad \sqrt[4]{2} \approx 1.189.$$

Calculator Tip The methods for finding approximations differ among makes and models, and you should always consult your owner's manual for keystroke instructions. Be aware that graphing calculators often differ from scientific calculators in the order in which keystrokes are made.

4 Find each square root that is a real number.

(a) $\sqrt{49}$

(b) $-\sqrt{\frac{36}{25}}$

(c) $\sqrt{(-6)^2}$

(d) $\sqrt{r^2}$

5 Simplify.

(a) $\sqrt[6]{64}$

(b) $-\sqrt[4]{16}$

(c) $\sqrt[3]{\frac{216}{125}}$

(d) $\sqrt[5]{-243}$

(e) $\sqrt[6]{(-p)^6}$

(f) $-\sqrt[6]{y^{24}}$

Answers

4. **(a)** 7 **(b)** $-\frac{6}{5}$ **(c)** 6 **(d)** $|r|$

5. **(a)** 2 **(b)** -2 **(c)** $\frac{6}{5}$ **(d)** -3 **(e)** $|p|$ **(f)** $-y^4$

Figure 5 shows how the preceding approximations are displayed on a TI-83 graphing calculator. In Figure 5(a), eight or nine decimal places are shown, while in Figure 5(b), the number of decimal places is fixed at three.

√(15)
3.872983346
³√(10)
2.15443469
4 ˣ√2
1.189207115

(a)

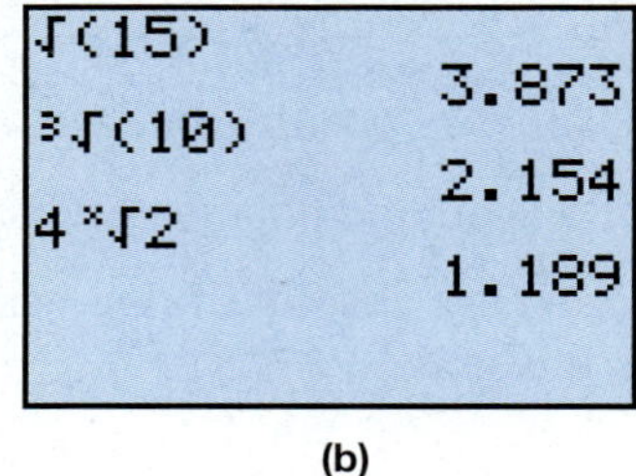

(b)

Figure 5

There is a simple way to check that a calculator approximation is "in the ballpark." Because 16 is a little larger than 15, $\sqrt{16} = 4$ should be a little larger than $\sqrt{15}$. Thus, 3.873 is a reasonable approximation for $\sqrt{15}$.

Example 6 Finding Approximations for Roots

Use a calculator to verify that each approximation is correct.

(a) $\sqrt{39} \approx 6.245$ **(b)** $-\sqrt{72} \approx -8.485$

(c) $\sqrt[3]{93} \approx 4.531$ **(d)** $\sqrt[4]{39} \approx 2.499$

Work Problem 6 at the Side.

6 Use a calculator to approximate each radical to three decimal places.

(a) $\sqrt{17}$

(b) $-\sqrt{362}$

(c) $\sqrt[3]{9482}$

(d) $\sqrt[4]{6825}$

Answers

6. **(a)** 4.123 **(b)** −19.026 **(c)** 21.166 **(d)** 9.089

8.1 EXERCISES

FOR EXTRA HELP

Student's Solutions Manual

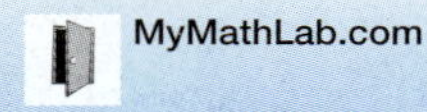
MyMathLab.com

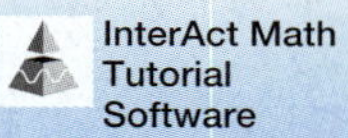
InterAct Math Tutorial Software

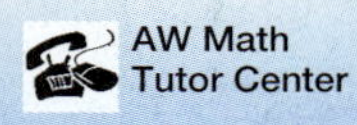
AW Math Tutor Center

www.mathxl.com

Digital Video Tutor CD 6 Videotape 13

Match each expression with the equivalent choice from A–F. Answers may be used more than once.

1. $-\sqrt{16}$ **2.** $\sqrt{-16}$ **3.** $\sqrt[3]{-27}$ **4.** $\sqrt[5]{-32}$ **5.** $\sqrt[4]{81}$ **6.** $\sqrt[3]{8}$

A. 3 **B.** -2 **C.** 2 **D.** -3 **E.** -4 **F.** Not a real number

Choose the closest approximation of each square root.

7. $\sqrt{123.5}$
A. 9 **B.** 10 **C.** 11 **D.** 12

8. $\sqrt{67.8}$
A. 7 **B.** 8 **C.** 9 **D.** 10

Refer to the figure to answer the questions in Exercises 9–10.

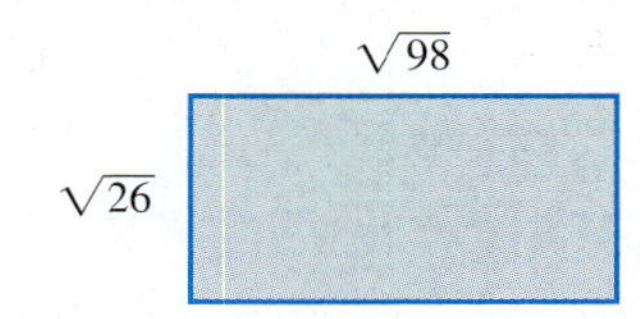

9. Which one of the following is the best estimate of its area?
A. 2500 **B.** 250 **C.** 50 **D.** 100

10. Which one of the following is the best estimate of its perimeter?
A. 15 **B.** 250 **C.** 100 **D.** 30

11. Consider the expression $-\sqrt{-a}$. Decide whether it is positive, negative, 0, or not a real number if
(a) $a > 0$, **(b)** $a < 0$, **(c)** $a = 0$.

12. If n is odd, under what conditions is $\sqrt[n]{a}$
(a) positive, **(b)** negative, **(c)** 0?

Find each root that is a real number. Use a calculator as necessary. See Examples 1 and 2.

13. $-\sqrt{81}$ **14.** $-\sqrt{121}$ **15.** $\sqrt[3]{216}$ **16.** $\sqrt[3]{343}$

17. $\sqrt[3]{-64}$ **18.** $\sqrt[3]{-125}$ **19.** $-\sqrt[3]{512}$ **20.** $-\sqrt[3]{1000}$

21. $\sqrt[4]{1296}$ **22.** $\sqrt[4]{625}$ **23.** $-\sqrt[4]{81}$ **24.** $-\sqrt[4]{256}$

25. $\sqrt[4]{-16}$ **26.** $\sqrt[4]{-81}$ **27.** $\sqrt[6]{(-2)^6}$ **28.** $\sqrt[6]{(-4)^6}$

29. $\sqrt[5]{(-9)^5}$ **30.** $\sqrt[5]{(-8)^5}$ **31.** $\sqrt{\frac{64}{81}}$ **32.** $\sqrt{\frac{100}{9}}$

33. $\sqrt[3]{\frac{8}{27}}$ **34.** $\sqrt[4]{\frac{81}{16}}$ **35.** $\sqrt[6]{\frac{1}{64}}$ **36.** $\sqrt[5]{\frac{1}{32}}$

Graph each function and give its domain and range. See Example 3.

37. $f(x) = \sqrt{x+3}$

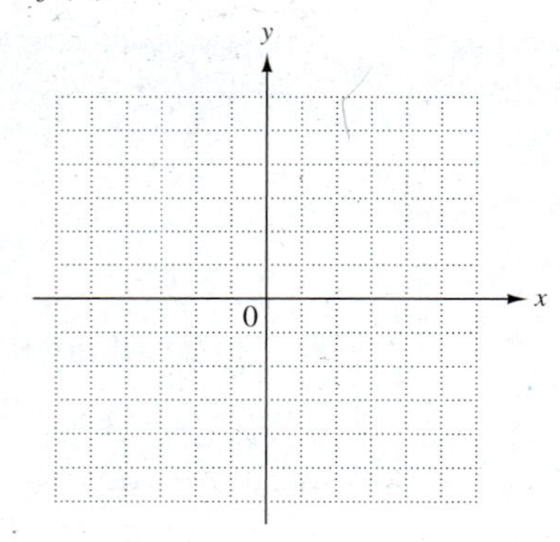

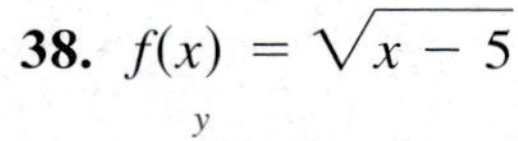

38. $f(x) = \sqrt{x-5}$

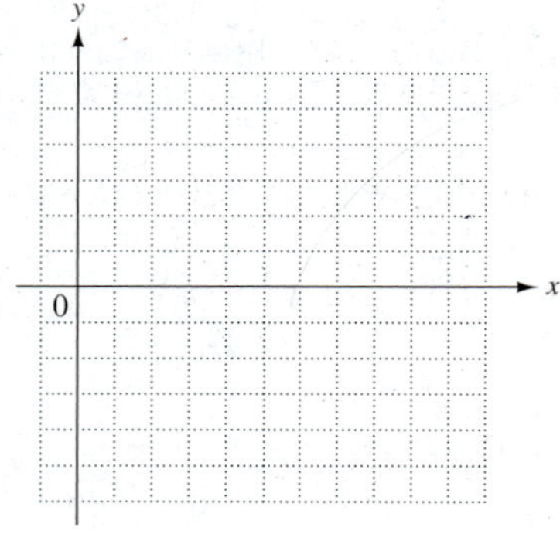

39. $f(x) = \sqrt{x} - 2$

40. $f(x) = \sqrt{x} + 4$

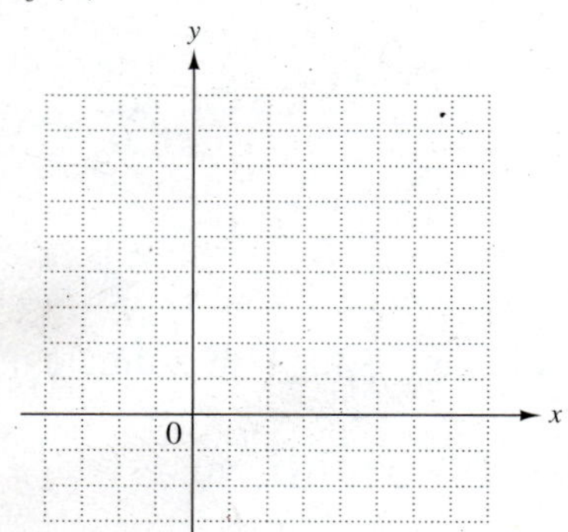

41. $f(x) = \sqrt[3]{x} - 3$

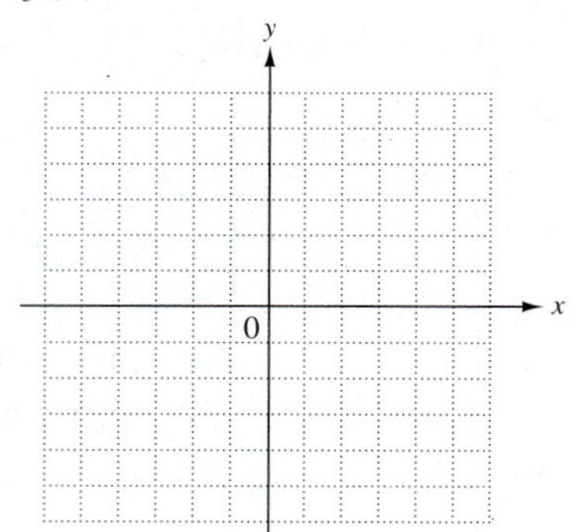

42. $f(x) = \sqrt[3]{x} + 1$

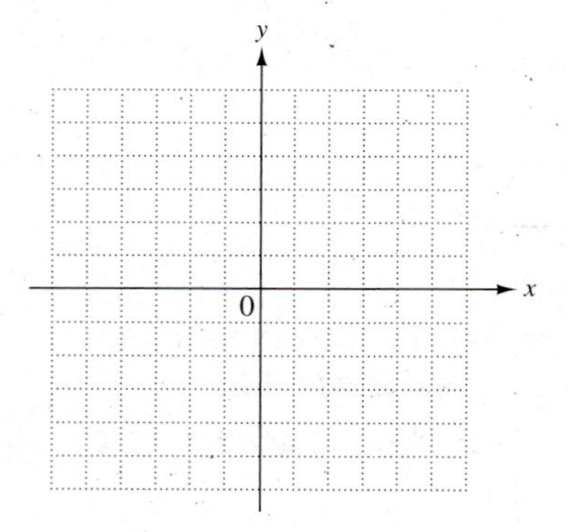

Simplify each root. See Examples 4 and 5.

43. $\sqrt{x^2}$ **44.** $-\sqrt{x^2}$ **45.** $\sqrt[3]{x^3}$ **46.** $-\sqrt[3]{x^3}$

47. $\sqrt[3]{x^{15}}$ **48.** $\sqrt[4]{k^{20}}$ **49.** $\sqrt{x^6}$ **50.** $\sqrt[4]{x^{12}}$

Use a calculator to find a decimal approximation for each radical. Round answers to three decimal places if necessary. See Example 6.

51. $\sqrt{9483}$ **52.** $\sqrt{6825}$ **53.** $\sqrt{284.361}$ **54.** $\sqrt{846.104}$

55. $\sqrt[4]{19.4481}$ **56.** $\sqrt[4]{39.0625}$ **57.** $\sqrt[3]{3.375}$ **58.** $\sqrt[3]{238.328}$

RELATING CONCEPTS (Exercises 59–64) FOR INDIVIDUAL OR GROUP WORK

Every positive number has two even nth roots, the principal (positive) root and a negative root. ***Work Exercises 59–64 in order,*** *to explore connections between these roots.*

59. Find the square roots of 16.

60. Find the principal square root of 16.

61. Find $\sqrt{16}$ and $-\sqrt{16}$.

62. What is the solution set of $x^2 = 16$?

63. Explain what is meant by $\pm\sqrt{16}$.

64. Explain why $\sqrt{x^2}$ is simplified as $|x|$.

8.2 Rational Exponents

1 Use exponential notation for *n*th roots. In mathematics we often formulate definitions so that previous rules remain valid. In Chapter 6 we defined 0 as an exponent in such a way that the rules for products, quotients, and powers would still be valid. Now we look at exponents that are rational numbers of the form $\frac{1}{n}$, where n is a natural number.

For the rules of exponents to remain valid, the product $(3^{1/2})^2 = 3^{1/2} \cdot 3^{1/2}$ should be found by adding exponents.

$$\begin{aligned}(3^{1/2})^2 &= 3^{1/2} \cdot 3^{1/2} \\ &= 3^{1/2+1/2} \\ &= 3^1 \\ &= 3\end{aligned}$$

However, by definition $(\sqrt{3})^2 = \sqrt{3} \cdot \sqrt{3} = 3$. Since both $(3^{1/2})^2$ and $(\sqrt{3})^2$ are equal to 3, we must have

$$3^{1/2} = \sqrt{3}.$$

This suggests the following generalization.

$a^{1/n}$

If $\sqrt[n]{a}$ is a real number, then

$$a^{1/n} = \sqrt[n]{a}.$$

Example 1 Evaluating Exponentials of the Form $a^{1/n}$

Evaluate each expression.

(a) $64^{1/3} = \sqrt[3]{64} = 4$

(b) $100^{1/2} = \sqrt{100} = 10$

(c) $-256^{1/4} = -\sqrt[4]{256} = -4$

(d) $(-256)^{1/4} = \sqrt[4]{-256}$ is not a real number because the radicand, -256, is negative and the index is even.

(e) $(-32)^{1/5} = \sqrt[5]{-32} = -2$

(f) $\left(\frac{1}{8}\right)^{1/3} = \sqrt[3]{\frac{1}{8}} = \frac{1}{2}$

CAUTION

Notice the difference between parts (c) and (d) in Example 1. The radical in part (c) is the *negative fourth root* of a positive number, while the radical in part (d) is the *principal fourth root of a negative number*, which is not a real number.

Work Problem 1 at the Side.

OBJECTIVES

1. Use exponential notation for *n*th roots.
2. Define $a^{m/n}$.
3. Convert between radicals and rational exponents.
4. Use the rules for exponents with rational exponents.

1 Evaluate each exponential.

(a) $8^{1/3}$

(b) $9^{1/2}$

(c) $-81^{1/4}$

(d) $(-16)^{1/4}$

(e) $64^{1/3}$

(f) $\left(\frac{1}{32}\right)^{1/5}$

ANSWERS

1. (a) 2 (b) 3 (c) -3 (d) not a real number (e) 4 (f) $\frac{1}{2}$

2 Evaluate each exponential.

(a) $64^{2/3}$

(b) $100^{3/2}$

(c) $-16^{3/4}$

(d) $(-16)^{3/4}$

ANSWERS
2. (a) 16 **(b)** 1000 **(c)** -8
(d) not a real number

2 **Define $a^{m/n}$.** How should we define a number like $8^{2/3}$? For past rules of exponents to be valid,

$$8^{2/3} = 8^{(1/3)2} = (8^{1/3})^2.$$

Since $8^{1/3} = \sqrt[3]{8}$,

$$8^{2/3} = (\sqrt[3]{8})^2 = 2^2 = 4.$$

Generalizing from this example, we define $a^{m/n}$ as follows.

$a^{m/n}$

If m and n are positive integers with m/n in lowest terms, then

$$a^{m/n} = (a^{1/n})^m,$$

provided that $a^{1/n}$ is a real number. If $a^{1/n}$ is not a real number, then $a^{m/n}$ is not a real number.

Example 2 Evaluating Exponentials of the Form $a^{m/n}$

Evaluate each exponential.

(a) $36^{3/2} = (36^{1/2})^3 = 6^3 = 216$

(b) $125^{2/3} = (125^{1/3})^2 = 5^2 = 25$

(c) $-4^{5/2} = -(4^{5/2}) = -(4^{1/2})^5 = -(2)^5 = -32$

(d) $(-27)^{2/3} = [(-27)^{1/3}]^2 = (-3)^2 = 9$

Notice how the $-$ sign is used in parts (c) and (d). In part (c), we first evaluate the exponential and then find its negative. In part (d), the $-$ sign is part of the base, -27.

(e) $(-100)^{3/2}$ is not a real number, since $(-100)^{1/2}$ is not a real number.

Work Problem 2 at the Side.

Example 3 Evaluating Exponentials with Negative Rational Exponents

Evaluate each exponential.

(a) $16^{-3/4}$

By the definition of a negative exponent,

$$16^{-3/4} = \frac{1}{16^{3/4}}.$$

Since $16^{3/4} = (\sqrt[4]{16})^3 = 2^3 = 8$,

$$16^{-3/4} = \frac{1}{16^{3/4}} = \frac{1}{8}.$$

(b) $25^{-3/2} = \frac{1}{25^{3/2}} = \frac{1}{(\sqrt{25})^3} = \frac{1}{5^3} = \frac{1}{125}$

(c) $\left(\frac{8}{27}\right)^{-2/3} = \frac{1}{\left(\frac{8}{27}\right)^{2/3}} = \frac{1}{\left(\sqrt[3]{\frac{8}{27}}\right)^2} = \frac{1}{\left(\frac{2}{3}\right)^2} = \frac{1}{\frac{4}{9}} = \frac{9}{4}$

Continued on Next Page

We could also use the rule $\left(\frac{b}{a}\right)^{-m} = \left(\frac{a}{b}\right)^{m}$ here, as follows.

$$\left(\frac{8}{27}\right)^{-2/3} = \left(\frac{27}{8}\right)^{2/3} = \left(\sqrt[3]{\frac{27}{8}}\right)^{2} = \left(\frac{3}{2}\right)^{2} = \frac{9}{4}$$

CAUTION

When using the rule in Example 3(c), we take the reciprocal only of the base, *not* the exponent. Also, be careful to distinguish between exponential expressions like $-16^{1/4}$, $16^{-1/4}$, and $-16^{-1/4}$.

$$-16^{1/4} = -2, \quad 16^{-1/4} = \frac{1}{2}, \quad \text{and} \quad -16^{-1/4} = -\frac{1}{2}.$$

Work Problem 3 at the Side.

We get an alternative definition of $a^{m/n}$ by using the power rule for exponents a little differently than in the earlier definition. If all indicated roots are real numbers,

$$a^{m/n} = a^{m(1/n)} = (a^m)^{1/n},$$

so

$$a^{m/n} = (a^m)^{1/n}.$$

$a^{m/n}$

If all indicated roots are real numbers, then

$$a^{m/n} = (a^{1/n})^m = (a^m)^{1/n}.$$

We can now evaluate an expression such as $27^{2/3}$ in two ways:

$$27^{2/3} = (27^{1/3})^2 = 3^2 = 9$$

or

$$27^{2/3} = (27^2)^{1/3} = 729^{1/3} = 9.$$

In most cases, it is easier to use $(a^{1/n})^m$.

This rule can also be expressed with radicals as follows.

Radical Form of $a^{m/n}$

If all indicated roots are real numbers, then

$$a^{m/n} = \sqrt[n]{a^m} = (\sqrt[n]{a})^m.$$

In words, we can raise to the power and then take the root, or take the root and then raise to the power.

For example,

$$8^{2/3} = \sqrt[3]{8^2} = \sqrt[3]{64} = 4, \quad \text{and} \quad 8^{2/3} = (\sqrt[3]{8})^2 = 2^2 = 4,$$

so

$$8^{2/3} = \sqrt[3]{8^2} = (\sqrt[3]{8})^2.$$

3 Convert between radicals and rational exponents. Using the definition of rational exponents, we can simplify many problems involving radicals by converting the radicals to numbers with rational exponents. After simplifying, we convert the answer back to radical form.

3 Evaluate each exponential.

(a) $36^{-3/2}$

(b) $32^{-4/5}$

(c) $\left(\frac{4}{9}\right)^{-5/2}$

ANSWERS

3. (a) $\frac{1}{216}$ (b) $\frac{1}{16}$ (c) $\frac{243}{32}$

4 Write each exponential as a radical. Assume all variables represent positive real numbers. Use the definition that takes the root first.

(a) $5^{2/3}$

(b) $4k^{3/5}$

(c) $(7r)^{4/3}$

(d) $(m^3 + n^3)^{1/3}$

5 Write each radical as an exponential and simplify. Assume all variables represent positive real numbers.

(a) $\sqrt{y^{10}}$

(b) $\sqrt[3]{27y^9}$

(c) $\sqrt[4]{t^4}$

Answers

4. **(a)** $(\sqrt[3]{5})^2$ **(b)** $4(\sqrt[5]{k})^3$
(c) $(\sqrt[3]{7r})^4$ **(d)** $\sqrt[3]{m^3 + n^3}$

5. **(a)** y^5 **(b)** $3y^3$ **(c)** t

Example 4 **Converting between Rational Exponents and Radicals**

Write each exponential as a radical. Assume that all variables represent positive real numbers. Use the definition that takes the root first.

(a) $13^{1/2} = \sqrt{13}$

(b) $6^{3/4} = (\sqrt[4]{6})^3$

(c) $9m^{5/8} = 9(\sqrt[8]{m})^5$

(d) $6x^{2/3} - (4x)^{3/5} = 6(\sqrt[3]{x})^2 - (\sqrt[5]{4x})^3$

(e) $r^{-2/3} = \dfrac{1}{r^{2/3}} = \dfrac{1}{(\sqrt[3]{r})^2}$

(f) $(a^2 + b^2)^{1/2} = \sqrt{a^2 + b^2}$ Note that $\sqrt{a^2 + b^2} \neq a + b$.

In (g)–(i), write each radical as an exponential. Simplify. Assume that all variables represent positive real numbers.

(g) $\sqrt{10} = 10^{1/2}$

(h) $\sqrt[4]{3^8} = 3^{8/4} = 3^2 = 9$

(i) $\sqrt[6]{z^6} = z$, since z is positive.

Work Problems 4 and 5 at the Side.

4 **Use the rules for exponents with rational exponents.** The definition of rational exponents allows us to apply the rules for exponents first introduced in Chapter 6.

Rules for Rational Exponents

Let r and s be rational numbers. For all real numbers a and b for which the indicated expressions exist:

$$a^r \cdot a^s = a^{r+s} \qquad a^{-r} = \frac{1}{a^r} \qquad \frac{a^r}{a^s} = a^{r-s} \qquad \left(\frac{a}{b}\right)^{-r} = \frac{b^r}{a^r}$$

$$(a^r)^s = a^{rs} \qquad (ab)^r = a^r b^r \qquad \left(\frac{a}{b}\right)^r = \frac{a^r}{b^r} \qquad a^{-r} = \left(\frac{1}{a}\right)^r.$$

Example 5 **Applying Rules for Rational Exponents**

Write with only positive exponents. Assume that all variables represent positive real numbers.

(a) $2^{1/2} \cdot 2^{1/4} = 2^{1/2+1/4} = 2^{3/4}$ Product rule

(b) $\dfrac{5^{2/3}}{5^{7/3}} = 5^{2/3-7/3} = 5^{-5/3} = \dfrac{1}{5^{5/3}}$ Quotient rule

(c) $\dfrac{(x^{1/2}y^{2/3})^4}{y} = \dfrac{(x^{1/2})^4(y^{2/3})^4}{y}$ Power rule

$= \dfrac{x^2y^{8/3}}{y^1}$ Power rule

$= x^2y^{8/3-1}$ Quotient rule

$= x^2y^{5/3}$

Continued on Next Page

(d) $m^{3/4}(m^{5/4} - m^{1/4}) = m^{3/4} \cdot m^{5/4} - m^{3/4} \cdot m^{1/4}$ Distributive property

$= m^{3/4+5/4} - m^{3/4+1/4}$ Product rule

$= m^{8/4} - m^{4/4}$

$= m^2 - m$

Do not make the common mistake of multiplying exponents in the first step.

Work Problem 6 at the Side.

CAUTION

Use the rules of exponents in problems like those in Example 5. Do not convert the expressions to radical form.

Example 6 Applying Rules for Rational Exponents

Rewrite all radicals as exponentials, and then apply the rules for rational exponents. Leave answers in exponential form. Assume that all variables represent positive real numbers.

(a) $\sqrt[3]{x^2} \cdot \sqrt[4]{x} = x^{2/3} \cdot x^{1/4}$ Convert to rational exponents.

$= x^{2/3+1/4}$ Product rule

$= x^{8/12+3/12}$ Write exponents with a common denominator.

$= x^{11/12}$

(b) $\dfrac{\sqrt{x^3}}{\sqrt[3]{x^2}} = \dfrac{x^{3/2}}{x^{2/3}} = x^{3/2-2/3} = x^{5/6}$

(c) $\sqrt{\sqrt[4]{z}} = \sqrt{z^{1/4}} = (z^{1/4})^{1/2} = z^{1/8}$

Work Problem 7 at the Side.

6 Write with only positive exponents. Assume that all variables represent positive real numbers.

(a) $11^{3/4} \cdot 11^{5/4}$

(b) $\dfrac{7^{3/4}}{7^{7/4}}$

(c) $\dfrac{9^{2/3}(x^{1/3})^4}{9^{-1/3}}$

(d) $a^{2/3}(a^{7/3} + a^{1/3})$

7 Simplify using the rules for rational exponents. Assume that all variables represent positive real numbers. Leave answers in exponential form.

(a) $\sqrt[5]{m^3} \cdot \sqrt{m}$

(b) $\dfrac{\sqrt[3]{p^5}}{\sqrt{p^3}}$

(c) $\sqrt[4]{\sqrt[3]{x}}$

ANSWERS

6. **(a)** 11^2 or 121 **(b)** $\dfrac{1}{7}$ **(c)** $9x^{4/3}$ **(d)** $a^3 + a$

7. **(a)** $m^{11/10}$ **(b)** $p^{1/6}$ **(c)** $x^{1/12}$

Focus on

Real-Data Applications

Windchill—A Radical Idea

When the wind blows, the air feels much colder than the actual temperature. The **windchill factor** measures the cooling effect that the wind has on one's skin. The formula that the National Weather Service uses to compute windchill is $T_{\text{wc}} = .0817(3.71\sqrt{V} + 5.81 - .25V)(T - 91.4) + 91.4$, where T_{wc} is windchill, V is wind speed in miles per hour (mph), and T is air temperature in degrees Fahrenheit. The windchill for various wind speeds and temperatures is shown in the table.

WINDCHILL FACTOR

Wind Speed (mph)	Air Temperature (°Fahrenheit)														
	35	30	25	20	15	10	5	0	−5	−10	−15	−20	−25	−30	−35
4	35	30	25	20	15	10	5	0	−5	−10	−15	−20	−25	−30	−35
5	32	27	22	16	11	6	0	−5	−10	−15	−21	−26	−31	−36	−42
10	22	16	10	3	−3	−9	−15	−22	−27	−34	−40	−46	−52	−58	−64
15	16	9	2	−5	−11	−18	−25	−31	−38	−45	−51	−58	−65	−72	−78
20	12	4	−3	−10	−17	−24	−31	−39	−46	−53	−60	−67	−74	−81	−88
25	8	1	−7	−15	−22	−29	−36	−44	−51	−59	−66	−74	−81	−88	−96
30	6	−2	−10	−18	−25	−33	−41	−49	−56	−64	−71	−79	−86	−93	−101
35	4	−4	−12	−20	−27	−35	−43	−52	−58	−67	−74	−82	−89	−97	−105
40	3	−5	−13	−21	−29	−37	−45	−53	−60	−69	−76	−84	−92	−100	−107
45	2	−6	−14	−22	−30	−38	−46	−54	−62	−70	−78	−85	−93	−102	−109

Source: USA Today.

If you consider the vertical columns of numbers in the table, the data represents the relationships of windchill versus wind speed for a constant air temperature. For example, if you choose one measure of air temperature to keep constant, such as 10°F, then the following data gives wind speed (V) as the *input* and windchill T_{wc} as the *output*. Wind speed is measured in miles per hour (mph).

V	4	5	10	15	20	25	30	35	40	45
T_{wc}	10	6	−9	−18	−24	−29	−33	−35	−37	−38

For Group Discussion

1. Choose a temperature of 10°F. Use the formula to calculate the windchill for wind speeds of 4, 10, 25, and 40 mph. Round the results to the nearest degree. Do your results match those in the tables?

2. On a sheet of graph paper, sketch a graph of windchill, T_{wc}, versus wind speed, V, data for the temperature 10°F. Describe the resulting graph. Is the graph a line or a parabola, for example?

3. For four representative pairs of points, calculate $\dfrac{\text{change in windchill}}{\text{change in wind speed}}$. For example,

using the ordered pairs (4, 10) and (5, 6), $\dfrac{\text{change in windchill}}{\text{change in wind speed}} = \dfrac{6 - 10}{5 - 4} = -4$.

Recall that if this ratio is constant, then the data is linearly related. Is it approximately constant?

8.2 EXERCISES

FOR EXTRA HELP

Student's Solutions Manual

MyMathLab.com

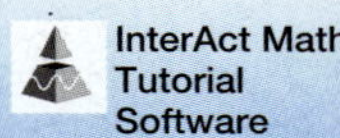
InterAct Math Tutorial Software

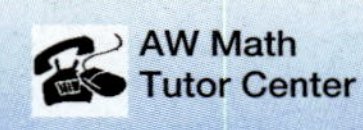
AW Math Tutor Center

www.mathxl.com

Digital Video Tutor CD 6 Videotape 13

Evaluate each exponential. See Example 1.

1. $169^{1/2}$ **2.** $121^{1/2}$ **3.** $729^{1/3}$ **4.** $512^{1/3}$

5. $16^{1/4}$ **6.** $625^{1/4}$ **7.** $-\left(\frac{64}{81}\right)^{1/2}$ **8.** $-\left(\frac{8}{27}\right)^{1/3}$

9. $(-27)^{1/3}$ **10.** $(-64)^{1/3}$ **11.** $(-100)^{1/2}$ **12.** $(-81)^{1/2}$

Match each expression with the equivalent choice from A–D.

13. $-6^{1/2}$ **14.** $-36^{.5}$ **15.** $\sqrt[4]{6^2}$ **16.** $\frac{3}{4}(4^{3/2})$

A. 6 **B.** -6 **C.** $\sqrt{6}$ **D.** $-\sqrt{6}$

17. Which one of the following is a positive number?

A. $(-27)^{2/3}$ **B.** $(-64)^{5/3}$ **C.** $(-100)^{1/2}$ **D.** $(-32)^{1/5}$

18. Explain why $(-64)^{1/2}$ is not a real number, while $-64^{1/2}$ is a real number.

Evaluate each exponential. See Examples 2 and 3.

19. $100^{5/2}$ **20.** $64^{3/2}$ **21.** $64^{4/3}$ **22.** $100^{7/2}$

23. $64^{-3/2}$ **24.** $81^{-3/2}$ **25.** $\left(\frac{625}{16}\right)^{-1/4}$ **26.** $\left(\frac{36}{25}\right)^{-3/2}$

27. $\left(-\frac{8}{27}\right)^{-2/3}$ **28.** $\left(-\frac{64}{125}\right)^{-2/3}$ **29.** $\left(-\frac{4}{9}\right)^{-1/2}$ **30.** $\left(-\frac{16}{25}\right)^{-1/2}$

Use a calculator to evaluate each exponential. Compare your answer to the answer found in the exercise or example noted in parentheses. (In Exercise 36, convert the answer in the example to a decimal to check.)

31. $169^{1/2}$ (Exercise 1) **32.** $121^{1/2}$ (Exercise 2) **33.** $100^{5/2}$ (Exercise 19)

34. $64^{3/2}$ (Exercise 20) **35.** $125^{2/3}$ (Example 2(b)) **36.** $16^{-3/4}$ (Example 3(a))

Write with radicals. Assume that all variables represent positive real numbers. See Examples 4(a)–(f).

37. $12^{1/2}$ **38.** $3^{1/2}$ **39.** $8^{3/4}$

40. $7^{2/3}$ **41.** $(9q)^{5/8} - (2x)^{2/3}$ **42.** $(3p)^{3/4} + (4x)^{1/3}$

43. $(2m)^{-3/2}$ **44.** $(5y)^{-3/5}$ **45.** $(2y + x)^{2/3}$

46. $(r + 2z)^{3/2}$ **47.** $(3m^4 + 2k^2)^{-2/3}$ **48.** $(5x^2 + 3z^3)^{-5/6}$

Simplify each radical by rewriting it with a rational exponent. Assume that all variables represent positive real numbers. See Examples 4(g)–(i).

49. $\sqrt{2^{12}}$ **50.** $\sqrt{5^{10}}$ **51.** $\sqrt[3]{4^9}$ **52.** $\sqrt[4]{6^8}$

53. $\sqrt{x^{20}}$ **54.** $\sqrt{r^{50}}$ **55.** $\sqrt[3]{a^{18}}$ **56.** $\sqrt[5]{k^{25}}$

Use the rules of exponents to simplify each expression. Write all answers with positive exponents. Assume that all variables represent positive real numbers. See Example 5.

57. $3^{1/2} \cdot 3^{3/2}$

58. $6^{4/3} \cdot 6^{2/3}$

59. $\dfrac{64^{5/3}}{64^{4/3}}$

60. $\dfrac{125^{7/3}}{125^{5/3}}$

61. $y^{7/3} \cdot y^{-4/3}$

62. $r^{-8/9} \cdot r^{17/9}$

63. $\dfrac{k^{1/3}}{k^{2/3} \cdot k^{-1}}$

64. $\dfrac{z^{3/4}}{z^{5/4} \cdot z^{-2}}$

65. $a^{5/6}a^{-1/3}$

66. $k^{-4/3}k^{2/5}$

67. $(2x^{-1/5}y^{3})^{-4}$

68. $(64a^{3/2}b^{6})^{-2/3}$

69. $(27x^{12}y^{15})^{2/3}$

70. $(64p^{4}q^{6})^{3/2}$

71. $\dfrac{(x^{2/3})^{2}}{(x^{2})^{7/3}}$

72. $\dfrac{(p^{3})^{1/4}}{(p^{5/4})^{2}}$

73. $\dfrac{m^{3/4}n^{-1/4}}{(m^{2}n)^{1/2}}$

74. $\dfrac{(a^{2}b^{5})^{-1/4}}{(a^{-3}b^{2})^{1/6}}$

75. $x^{1/2}(2x^{1/2} - x^{-1/2} - 2x)$

76. $m^{2/3}(3m^{1/3} + m^{-1/3} - 2m)$

Write each expression with rational exponents. Then apply the rules for exponents. Write answers in radical form. Assume that all variables represent positive real numbers. See Example 6.

77. $\sqrt[3]{x} \cdot \sqrt{x}$

78. $\sqrt[4]{y} \cdot \sqrt[5]{y^{2}}$

79. $\sqrt[4]{49y^{6}}$

80. $\sqrt[4]{100y^{10}}$

81. $\dfrac{\sqrt[3]{t^{4}}}{\sqrt[5]{t^{4}}}$

82. $\dfrac{\sqrt[4]{w^{3}}}{\sqrt[6]{w}}$

83. $\sqrt[4]{\sqrt{m}}$

84. $\sqrt[3]{\sqrt[3]{k}}$

RELATING CONCEPTS (Exercises 85–90) FOR INDIVIDUAL OR GROUP WORK

Earlier, we factored expressions like $x^4 - x^5$ by factoring out the greatest common factor to get $x^4 - x^5 = x^4(1 - x)$. We can adapt this approach to factor expressions with rational exponents. When one or more of the exponents is negative or a fraction, we use order on the number line discussed in Chapter 1 to decide on the common factor. In this type of factoring, we want the binomial factor to have only positive exponents, so we always factor out the variable with the least *exponent. A positive exponent is greater than a negative exponent, so in $7z^{5/8} + z^{-3/4}$, we factor out $z^{-3/4}$, because $-3/4$ is less than $5/8$.*

Work Exercises 85–90 in order.

Find the appropriate common factor in each expression.

85. $3x^{-1/2} - 4x^{1/2}$

86. $m^3 - 3m^{5/2}$

87. $9k^{-3/4} + 2k^{-1/4}$

Factor each expression.

88. $3x^{-1/2} - 4x^{1/2}$

89. $m^3 - 3m^{5/2}$

90. $9k^{-3/4} + 2k^{-1/4}$

Solve each problem.

91. Meteorologists can determine the duration of a storm by using the function defined by

$$T(D) = .07D^{3/2},$$

where D is the diameter of the storm in miles and T is the time in hours. Find the duration of a storm with a diameter of 16 mi. Round your answer to the nearest tenth of an hour.

92. The threshold weight T, in pounds, for a person is the weight above which the risk of death increases greatly. The threshold weight in pounds for men aged 40–49 is related to height in inches by the function defined by

$$h(T) = 12.3T^{1/3}.$$

What height corresponds to a threshold weight of 216 lb for a 43-yr-old man? Round your answer to the nearest inch, and then to the nearest tenth of a foot.

8.3 Simplifying Radical Expressions

Objectives

1. Use the product rule for radicals.
2. Use the quotient rule for radicals.
3. Simplify radicals.
4. Simplify products and quotients of radicals with different indexes.
5. Use the Pythagorean formula.
6. Use the distance formula.

1 Use the product rule for radicals. We now develop rules for multiplying and dividing radicals that have the same index. For example, is the product of two *n*th-root radicals equal to the *n*th root of the product of the radicands? Is $\sqrt{36 \cdot 4} = \sqrt{36} \cdot \sqrt{4}$? To find out, we simply do the computations:

$$\sqrt{36 \cdot 4} = \sqrt{144} = \mathbf{12}$$
$$\sqrt{36} \cdot \sqrt{4} = 6 \cdot 2 = \mathbf{12}.$$

Notice that in both cases the result is the same. This is an example of the **product rule for radicals.**

Product Rule for Radicals

If $\sqrt[n]{a}$ and $\sqrt[n]{b}$ are real numbers and n is a natural number,

$$\sqrt[n]{a} \cdot \sqrt[n]{b} = \sqrt[n]{ab}.$$

In words, the product of two radicals is the radical of the product.

We justify the product rule using the rules for rational exponents. Since $\sqrt[n]{a} = a^{1/n}$ and $\sqrt[n]{b} = b^{1/n}$,

$$\sqrt[n]{a} \cdot \sqrt[n]{b} = a^{1/n} \cdot b^{1/n} = (ab)^{1/n} = \sqrt[n]{ab}.$$

CAUTION

Use the product rule only when the radicals have the *same* indexes.

Example 1 Using the Product Rule

Multiply. Assume that all variables represent positive real numbers.

(a) $\sqrt{5} \cdot \sqrt{7} = \sqrt{5 \cdot 7} = \sqrt{35}$

(b) $\sqrt{2} \cdot \sqrt{19} = \sqrt{2 \cdot 19} = \sqrt{38}$

(c) $\sqrt{11} \cdot \sqrt{p} = \sqrt{11p}$

(d) $\sqrt{7} \cdot \sqrt{11xyz} = \sqrt{77xyz}$

Work Problem 1 at the Side.

Example 2 Using the Product Rule

Multiply. Assume that all variables represent positive real numbers.

(a) $\sqrt[3]{3} \cdot \sqrt[3]{12} = \sqrt[3]{3 \cdot 12} = \sqrt[3]{36}$

(b) $\sqrt[4]{8y} \cdot \sqrt[4]{3r^2} = \sqrt[4]{24yr^2}$

(c) $\sqrt[6]{10m^4} \cdot \sqrt[6]{5m} = \sqrt[6]{50m^5}$

(d) $\sqrt[4]{2} \cdot \sqrt[5]{2}$ cannot be simplified using the product rule for radicals, because the indexes (4 and 5) are different.

Work Problem 2 at the Side.

1 Multiply. Assume that all variables represent positive real numbers.

(a) $\sqrt{5} \cdot \sqrt{13}$

(b) $\sqrt{10y} \cdot \sqrt{3k}$

(c) $\sqrt{\frac{5}{a}} \cdot \sqrt{\frac{11}{z}}$

2 Multiply. Assume that all variables represent positive real numbers.

(a) $\sqrt[3]{2} \cdot \sqrt[3]{7}$

(b) $\sqrt[6]{8r^2} \cdot \sqrt[6]{2r^3}$

(c) $\sqrt[5]{9y^2x} \cdot \sqrt[5]{8xy^2}$

(d) $\sqrt{7} \cdot \sqrt[3]{5}$

Answers

1. (a) $\sqrt{65}$ (b) $\sqrt{30yk}$ (c) $\sqrt{\frac{55}{az}}$
2. (a) $\sqrt[3]{14}$ (b) $\sqrt[6]{16r^5}$ (c) $\sqrt[5]{72y^4x^2}$ (d) cannot be simplified using the product rule

3 Simplify. Assume that all variables represent positive real numbers.

(a) $\sqrt{\frac{100}{81}}$

(b) $\sqrt{\frac{11}{25}}$

(c) $\sqrt[3]{\frac{18}{125}}$

(d) $\sqrt{\frac{y^8}{16}}$

(e) $\sqrt[3]{\frac{x^2}{r^{12}}}$

ANSWERS

3. (a) $\frac{10}{9}$ (b) $\frac{\sqrt{11}}{5}$ (c) $\frac{\sqrt[3]{18}}{5}$ (d) $\frac{y^4}{4}$ (e) $\frac{\sqrt[3]{x^2}}{r^4}$

2 **Use the quotient rule for radicals.** The quotient rule for radicals is similar to the product rule.

Quotient Rule for Radicals

If $\sqrt[n]{a}$ and $\sqrt[n]{b}$ are real numbers, $b \neq 0$, and n is a natural number, then

$$\sqrt[n]{\frac{a}{b}} = \frac{\sqrt[n]{a}}{\sqrt[n]{b}}.$$

In words, the radical of a quotient is the quotient of the radicals.

Example 3 Using the Quotient Rule

Simplify. Assume that all variables represent positive real numbers.

(a) $\sqrt{\frac{16}{25}} = \frac{\sqrt{16}}{\sqrt{25}} = \frac{4}{5}$

(b) $\sqrt{\frac{7}{36}} = \frac{\sqrt{7}}{\sqrt{36}} = \frac{\sqrt{7}}{6}$

(c) $\sqrt[3]{-\frac{8}{125}} = \sqrt[3]{\frac{-8}{125}} = \frac{\sqrt[3]{-8}}{\sqrt[3]{125}} = \frac{-2}{5} = -\frac{2}{5}$

(d) $\sqrt[3]{\frac{7}{216}} = \frac{\sqrt[3]{7}}{\sqrt[3]{216}} = \frac{\sqrt[3]{7}}{6}$

(e) $\sqrt[5]{\frac{x}{32}} = \frac{\sqrt[5]{x}}{\sqrt[5]{32}} = \frac{\sqrt[5]{x}}{2}$

(f) $\sqrt[3]{\frac{m^6}{125}} = \frac{\sqrt[3]{m^6}}{\sqrt[3]{125}} = \frac{m^2}{5}$

Work Problem 3 at the Side.

3 **Simplify radicals.** We use the product and quotient rules to simplify radicals. A radical is **simplified** if the following four conditions are met.

Simplified Radical

1. The radicand has no factor raised to a power greater than or equal to the index.
2. The radicand has no fractions.
3. No denominator contains a radical.
4. Exponents in the radicand and the index of the radical have no common factor (except 1).

Example 4 Simplifying Roots of Numbers

Simplify.

(a) $\sqrt{24}$

Check to see whether 24 is divisible by a perfect square (the square of a natural number) such as 4, 9, ... Choose the largest perfect square that divides into 24. The largest such number is 4. Write 24 as the product of 4 and 6, and then use the product rule.

$$\sqrt{24} = \sqrt{4 \cdot 6} = \sqrt{4} \cdot \sqrt{6} = 2\sqrt{6}$$

Continued on Next Page

(b) $\sqrt{108}$

The number 108 is divisible by the perfect square 36: $\sqrt{108} = \sqrt{36 \cdot 3}$. If this is not obvious, try factoring 108 into its prime factors.

$$\begin{aligned} \sqrt{108} &= \sqrt{2^2 \cdot 3^3} \\ &= \sqrt{2^2 \cdot 3^2 \cdot 3} \\ &= 2 \cdot 3 \cdot \sqrt{3} \quad \text{Product rule} \\ &= 6\sqrt{3} \end{aligned}$$

(c) $\sqrt{10}$

No perfect square (other than 1) divides into 10, so $\sqrt{10}$ cannot be simplified further.

(d) $\sqrt[3]{16}$

Look for the largest perfect *cube* that divides into 16. The number 8 satisfies this condition, so write 16 as $8 \cdot 2$ (or factor 16 into prime factors).

$$\sqrt[3]{16} = \sqrt[3]{8 \cdot 2} = \sqrt[3]{8} \cdot \sqrt[3]{2} = 2\sqrt[3]{2}$$

(e)
$$\begin{aligned} \sqrt[4]{162} &= \sqrt[4]{81 \cdot 2} \quad \text{81 is a perfect 4th power.} \\ &= \sqrt[4]{81} \cdot \sqrt[4]{2} \quad \text{Product rule} \\ &= 3\sqrt[4]{2} \end{aligned}$$

CAUTION

In simplifying an expression like that in Example 4(b), be careful with which factors belong *outside* the radical sign and which belong *inside*. Note how $2 \cdot 3$ is written outside because $\sqrt{2^2} = 2$ and $\sqrt{3^2} = 3$, while the remaining 3 is left inside the radical.

Work Problem 4 at the Side.

Example 5 Simplifying Radicals Involving Variables

Simplify. Assume that all variables represent positive real numbers.

(a)
$$\begin{aligned} \sqrt{16m^3} &= \sqrt{16m^2 \cdot m} \\ &= \sqrt{16m^2} \cdot \sqrt{m} \\ &= 4m\sqrt{m} \end{aligned}$$

No absolute value bars are needed around the m in color because of the assumption that all the variables represent *positive* real numbers.

(b)
$$\begin{aligned} \sqrt{200k^7q^8} &= \sqrt{10^2 \cdot 2 \cdot (k^3)^2 \cdot k \cdot (q^4)^2} \quad \text{Factor.} \\ &= 10k^3q^4\sqrt{2k} \quad \text{Remove perfect square factors.} \end{aligned}$$

(c)
$$\begin{aligned} \sqrt[3]{8x^4y^5} &= \sqrt[3]{(8x^3y^3)(xy^2)} \quad 8x^3y^3 \text{ is the largest perfect cube that divides } 8x^4y^5. \\ &= \sqrt[3]{8x^3y^3} \cdot \sqrt[3]{xy^2} \\ &= 2xy\sqrt[3]{xy^2} \end{aligned}$$

(d)
$$\begin{aligned} \sqrt[4]{32y^9} &= \sqrt[4]{(16y^8)(2y)} \quad 16y^8 \text{ is the largest 4th power that divides } 32y^9. \\ &= \sqrt[4]{16y^8} \cdot \sqrt[4]{2y} \\ &= 2y^2\sqrt[4]{2y} \end{aligned}$$

4 Simplify.

(a) $\sqrt{32}$

(b) $\sqrt{45}$

(c) $\sqrt{300}$

(d) $\sqrt{35}$

(e) $\sqrt[3]{54}$

(f) $\sqrt[4]{243}$

ANSWERS

4. **(a)** $4\sqrt{2}$ **(b)** $3\sqrt{5}$ **(c)** $10\sqrt{3}$ **(d)** cannot be simplified further **(e)** $3\sqrt[3]{2}$ **(f)** $3\sqrt[4]{3}$

NOTE

From Example 5 we see that if a variable is raised to a power with an exponent divisible by 2, it is a perfect square. If it is raised to a power with an exponent divisible by 3, it is a perfect cube. In general, if it is raised to a power with an exponent divisible by n, it is a perfect nth power.

Work Problem 5 at the Side.

The conditions for a simplified radical given earlier state that an exponent in the radicand and the index of the radical should have no common factor (except 1). The next example shows how to simplify radicals with such common factors.

Example 6 Simplifying Radicals by Using Smaller Indexes

Simplify. Assume that all variables represent positive real numbers.

(a) $\sqrt[9]{5^6}$

We can write this radical using rational exponents and then write the exponent in lowest terms. We then express the answer as a radical.

$$\sqrt[9]{5^6} = 5^{6/9} = 5^{2/3} = \sqrt[3]{5^2} \quad \text{or} \quad \sqrt[3]{25}$$

(b) $\sqrt[4]{p^2} = p^{2/4} = p^{1/2} = \sqrt{p}$ (Recall the assumption that $p > 0$.)

These examples suggest the following rule.

If m is an integer, n and k are natural numbers, and all indicated roots exist,

$$\sqrt[kn]{a^{km}} = \sqrt[n]{a^m}.$$

Work Problem 6 at the Side.

4 **Simplify products and quotients of radicals with different indexes.** Since the product and quotient rules for radicals apply only when they have the same index, we multiply and divide radicals with different indexes by using rational exponents.

Example 7 Multiplying Radicals with Different Indexes

Simplify $\sqrt{7} \cdot \sqrt[3]{2}$.

Because the different indexes, 2 and 3, have a least common index of 6, use rational exponents to write each radical as a sixth root.

$$\sqrt{7} = 7^{1/2} = 7^{3/6} = \sqrt[6]{7^3} = \sqrt[6]{343}$$

$$\sqrt[3]{2} = 2^{1/3} = 2^{2/6} = \sqrt[6]{2^2} = \sqrt[6]{4}$$

Therefore,

$$\sqrt{7} \cdot \sqrt[3]{2} = \sqrt[6]{343} \cdot \sqrt[6]{4} = \sqrt[6]{1372}. \quad \text{Product rule}$$

Work Problem 7 at the Side.

5 Simplify. Assume that all variables represent positive real numbers.

(a) $\sqrt{25p^7}$

(b) $\sqrt{72y^3x}$

(c) $\sqrt[3]{y^7x^5z^6}$

(d) $\sqrt[4]{32a^5b^7}$

6 Simplify. Assume that all variables represent positive real numbers.

(a) $\sqrt[12]{2^3}$

(b) $\sqrt[6]{t^2}$

7 Simplify $\sqrt{5} \cdot \sqrt[3]{4}$.

Answers

5. **(a)** $5p^3\sqrt{p}$ **(b)** $6y\sqrt{2yx}$ **(c)** $y^2xz^2\sqrt[3]{yx^2}$ **(d)** $2ab\sqrt[4]{2ab^3}$

6. **(a)** $\sqrt[4]{2}$ **(b)** $\sqrt[3]{t}$

7. $\sqrt[6]{2000}$

5 **Use the Pythagorean formula.** The **Pythagorean formula** relates the lengths of the three sides of a right triangle.

Pythagorean Formula

If c is the length of the longest side of a right triangle and a and b are the lengths of the shorter sides, then

$$c^2 = a^2 + b^2.$$

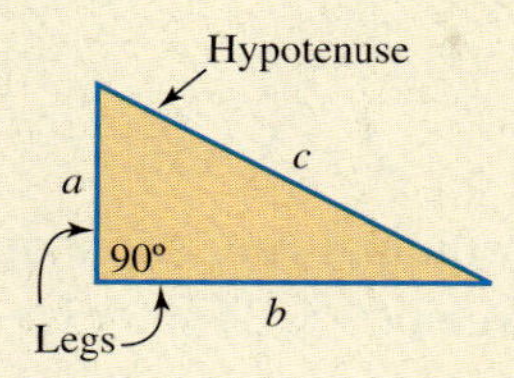

The longest side is the **hypotenuse** and the two shorter sides are the **legs** of the triangle. The hypotenuse is the side opposite the right angle.

Example 8 **Using the Pythagorean Formula**

Use the Pythagorean formula to find the length of the hypotenuse in the triangle in Figure 6.

To find the length of the hypotenuse c, let $a = 4$ and $b = 6$. Then, use the formula.

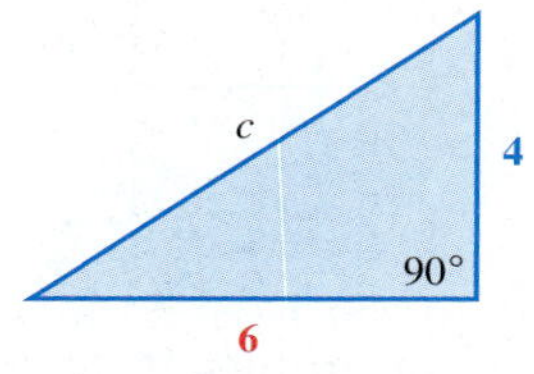

Figure 6

$$c^2 = a^2 + b^2$$
$$c^2 = 4^2 + 6^2 \quad \text{Let } a = 4 \text{ and } b = 6.$$
$$c^2 = 52$$
$$c = \sqrt{52} \quad \text{Choose the principal root.}$$
$$c = \sqrt{4 \cdot 13} \quad \text{Factor.}$$
$$c = \sqrt{4} \cdot \sqrt{13} \quad \text{Product rule}$$
$$c = 2\sqrt{13}$$

The length of the hypotenuse is $2\sqrt{13}$.

Work Problem 8 at the Side.

6 **Use the distance formula.** An important result in algebra is derived by using the Pythagorean formula. The **distance formula** allows us to find the distance between two points in the coordinate plane, or the length of the line segment joining those two points. Figure 7 on the next page shows the points $(3, -4)$ and $(-5, 3)$. The vertical line through $(-5, 3)$ and the horizontal line through $(3, -4)$ intersect at the point $(-5, -4)$. Thus, the point $(-5, -4)$ becomes the vertex of the right angle in a right triangle. By the Pythagorean formula, the square of the length of the hypotenuse, d, of the right triangle in Figure 7 is equal to the sum of the squares of the lengths of the two legs a and b:

$$d^2 = a^2 + b^2.$$

The length a is the difference between the y-coordinates of the endpoints. Since the x-coordinate of both points is -5, the side is vertical, and we can find a by finding the difference between the y-coordinates. We subtract -4 from 3 to get a positive value for a.

$$a = 3 - (-4) = 7$$

8 Find the length of the unknown side in each triangle.

(a)

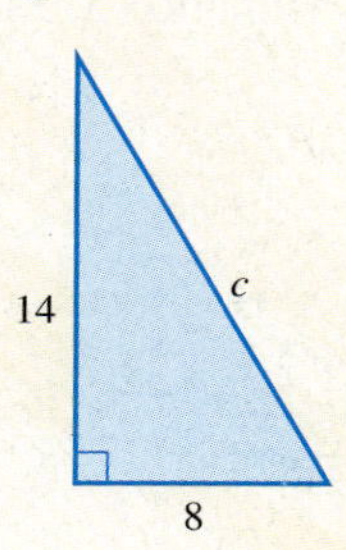

(b)

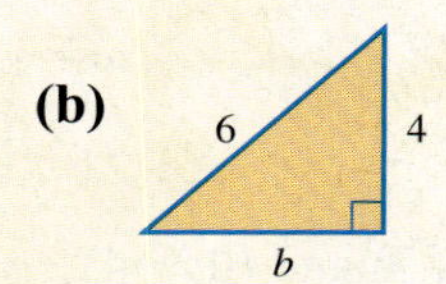

(*Hint:* Write the Pythagorean formula as $b^2 = c^2 - a^2$ here.)

Answers

8. **(a)** $2\sqrt{65}$ **(b)** $2\sqrt{5}$

9 Find the distance between each pair of points.

(a) (2, −1) and (5, 3)

(b) (−3, 2) and (0, −4)

Similarly, we find b by subtracting −5 from 3.

$$b = 3 - (-5) = 8$$

Substituting these values into the formula, we have

$$d^2 = a^2 + b^2$$
$$d^2 = 7^2 + 8^2 \quad \text{Let } a = 7 \text{ and } b = 8.$$
$$d^2 = 49 + 64$$
$$d^2 = 113$$
$$d = \sqrt{113}.$$

We choose the principal root since distance cannot be negative. Therefore, the distance between (−5, 3) and (3, −4) is $\sqrt{113}$.

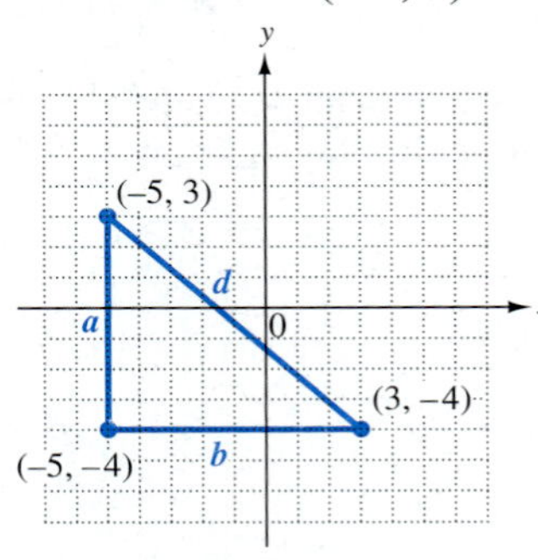

Figure 7

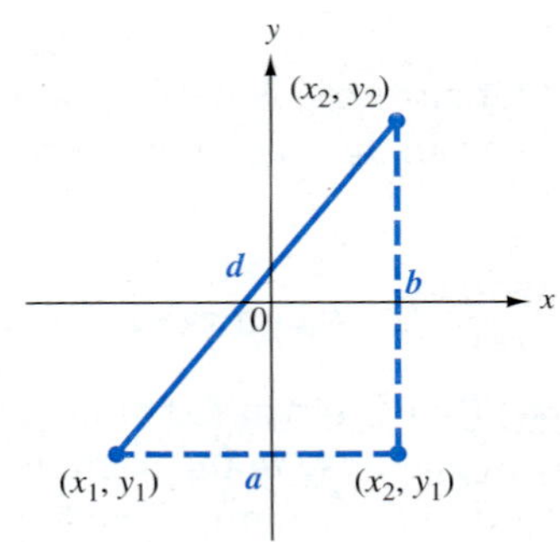

Figure 8

This result can be generalized. Figure 8 shows the two points (x_1, y_1) and (x_2, y_2). To find a formula for the distance d between these two points, notice that the distance between (x_1, y_1) and (x_2, y_1) is given by

$$a = x_2 - x_1,$$

and the distance between (x_2, y_2) and (x_2, y_1) is given by

$$b = y_2 - y_1.$$

From the Pythagorean formula,

$$d^2 = a^2 + b^2$$
$$= (x_2 - x_1)^2 + (y_2 - y_1)^2.$$

Choosing the principal square root gives the **distance formula.**

Distance Formula

The distance between the points (x_1, y_1) and (x_2, y_2) is

$$d = \sqrt{(x_2 - x_1)^2 + (y_2 - y_1)^2}.$$

Example 9 Using the Distance Formula

Find the distance between (−3, 5) and (6, 4).

When using the distance formula to find the distance between two points, designating the points as (x_1, y_1) and (x_2, y_2) is arbitrary. Let us choose $(x_1, y_1) = (-3, 5)$ and $(x_2, y_2) = (6, 4)$.

$$d = \sqrt{(x_2 - x_1)^2 + (y_2 - y_1)^2}$$
$$= \sqrt{(6 - (-3))^2 + (4 - 5)^2} \quad x_2 = 6, y_2 = 4, x_1 = -3, y_1 = 5$$
$$= \sqrt{9^2 + (-1)^2}$$
$$= \sqrt{82}$$

Work Problem 9 at the Side.

ANSWERS
9. **(a)** 5 **(b)** $\sqrt{45}$ or $3\sqrt{5}$

8.3 EXERCISES

FOR EXTRA HELP

Student's Solutions Manual

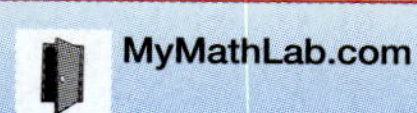
MyMathLab.com

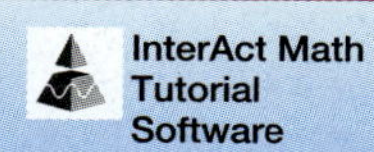
InterAct Math Tutorial Software

AW Math Tutor Center

www.mathxl.com

Digital Video Tutor CD 6 Videotape 13

Decide whether each statement is true *or* false *by using the product rule explained in this section. Then support your answer by finding a calculator approximation for each expression.*

1. $2\sqrt{12} = \sqrt{48}$

2. $\sqrt{72} = 2\sqrt{18}$

3. $3\sqrt{8} = 2\sqrt{18}$

4. $5\sqrt{72} = 6\sqrt{50}$

5. Explain why $\sqrt[3]{x} \cdot \sqrt[3]{x}$ is not equal to x. What is it equal to?

6. Explain why $\sqrt[4]{x} \cdot \sqrt[4]{x}$ is not equal to x, but *is* equal to $\sqrt{x}$, for $x \geq 0$.

Multiply. See Examples 1 and 2.

7. $\sqrt{5} \cdot \sqrt{6}$

8. $\sqrt{10} \cdot \sqrt{3}$

9. $\sqrt[3]{7x} \cdot \sqrt[3]{2y}$

10. $\sqrt[3]{9x} \cdot \sqrt[3]{4y}$

11. $\sqrt[4]{12} \cdot \sqrt[4]{3}$

12. $\sqrt[4]{6} \cdot \sqrt[4]{9}$

Simplify each radical. Assume that all variables represent positive real numbers. See Example 3.

13. $\sqrt{\dfrac{64}{121}}$

14. $\sqrt{\dfrac{16}{49}}$

15. $\sqrt{\dfrac{3}{25}}$

16. $\sqrt{\dfrac{13}{49}}$

17. $\sqrt{\dfrac{x}{25}}$

18. $\sqrt{\dfrac{k}{100}}$

19. $\sqrt{\dfrac{p^6}{81}}$

20. $\sqrt{\dfrac{w^{10}}{36}}$

21. $\sqrt[3]{\dfrac{27}{64}}$

22. $\sqrt[3]{\dfrac{216}{125}}$

23. $\sqrt[3]{-\dfrac{r^2}{8}}$

24. $\sqrt[3]{-\dfrac{t}{125}}$

Express each radical in simplified form. See Example 4.

25. $\sqrt{12}$ **26.** $\sqrt{18}$ **27.** $\sqrt{288}$ **28.** $\sqrt{72}$ **29.** $-\sqrt{32}$

30. $-\sqrt{48}$ **31.** $-\sqrt{28}$ **32.** $-\sqrt{24}$ **33.** $\sqrt{-300}$ **34.** $\sqrt{-150}$

35. $\sqrt[3]{128}$ **36.** $\sqrt[3]{24}$ **37.** $\sqrt[3]{-16}$ **38.** $\sqrt[3]{-250}$ **39.** $\sqrt[3]{40}$

40. $\sqrt[3]{375}$ **41.** $-\sqrt[4]{512}$ **42.** $-\sqrt[4]{1250}$ **43.** $\sqrt[5]{64}$ **44.** $\sqrt[5]{128}$

45. A student claimed that $\sqrt[3]{14}$ is not in simplified form, since $14 = 8 + 6$, and 8 is a perfect cube. Was his reasoning correct? Why or why not?

46. Explain in your own words why $\sqrt[3]{k^4}$ is not a simplified radical.

Express each radical in simplified form. Assume that all variables represent positive real numbers. See Example 5.

47. $\sqrt{72k^2}$ **48.** $\sqrt{18m^2}$ **49.** $\sqrt[3]{\frac{81}{64}}$ **50.** $\sqrt[3]{\frac{32}{216}}$

51. $\sqrt{121x^6}$ **52.** $\sqrt{256z^{12}}$ **53.** $-\sqrt[3]{27t^{12}}$ **54.** $-\sqrt[3]{64y^{18}}$

55. $-\sqrt{100m^8z^4}$ **56.** $-\sqrt{25t^6s^{20}}$ **57.** $-\sqrt[3]{-125a^6b^9c^{12}}$ **58.** $-\sqrt[3]{-216y^{15}x^6z^3}$

59. $\sqrt[4]{\frac{1}{16}r^8t^{20}}$ **60.** $\sqrt[4]{\frac{81}{256}t^{12}u^8}$ **61.** $\sqrt{50x^3}$ **62.** $\sqrt{300z^3}$

63. $-\sqrt{500r^{11}}$ **64.** $-\sqrt{200p^{13}}$ **65.** $\sqrt{13x^7y^8}$ **66.** $\sqrt{23k^9p^{14}}$

67. $\sqrt[3]{8z^6w^9}$ **68.** $\sqrt[3]{64a^{15}b^{12}}$ **69.** $\sqrt[3]{-16z^5t^7}$ **70.** $\sqrt[3]{-81m^4n^{10}}$

71. $\sqrt[4]{81x^{12}y^{16}}$ **72.** $\sqrt[4]{81t^8u^{28}}$ **73.** $-\sqrt[4]{162r^{15}s^{10}}$ **74.** $-\sqrt[4]{32k^5m^{10}}$

75. $\sqrt{\frac{y^{11}}{36}}$ **76.** $\sqrt{\frac{v^{13}}{49}}$ **77.** $\sqrt[3]{\frac{x^{16}}{27}}$ **78.** $\sqrt[3]{\frac{y^{17}}{125}}$

Simplify each radical. Assume that $x \geq 0$. See Example 6.

79. $\sqrt[4]{48^2}$ **80.** $\sqrt[4]{50^2}$ **81.** $\sqrt[10]{x^{25}}$ **82.** $\sqrt[12]{x^{44}}$

Simplify by first writing the radicals with the same index. Then multiply. See Example 7.

83. $\sqrt[3]{4} \cdot \sqrt{3}$ **84.** $\sqrt[3]{5} \cdot \sqrt{6}$ **85.** $\sqrt[4]{3} \cdot \sqrt[3]{4}$ **86.** $\sqrt[5]{7} \cdot \sqrt[7]{5}$

Find the unknown length in each right triangle. Simplify the answer if necessary. See Example 8.

87.

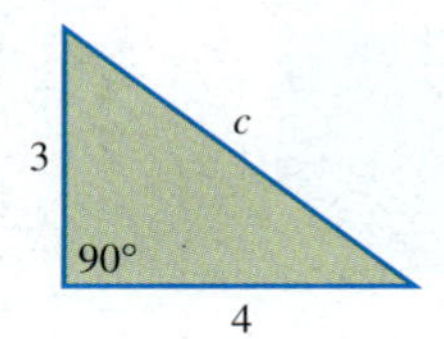

88.

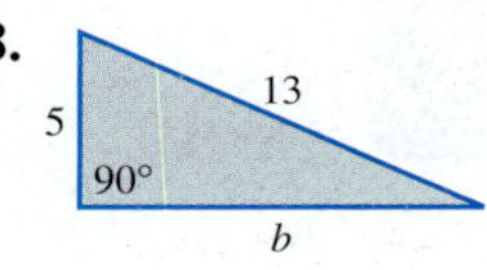

89.

90.

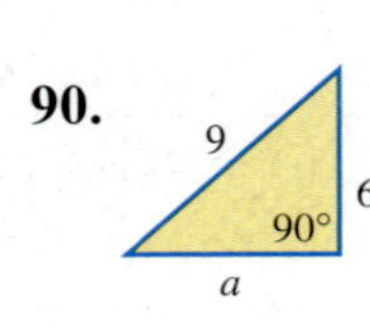

Find the distance between each pair of points. See Example 9.

91. $(5, 3)$ and $(-1, 2)$

92. $(-1, 4)$ and $(5, 3)$

93. $(-1, 5)$ and $(-7, 7)$

94. $(4, 5)$ and $(-8, 4)$

95. $(\sqrt{2}, \sqrt{6})$ and $(-2\sqrt{2}, 4\sqrt{6})$

96. $(\sqrt{7}, 9\sqrt{3})$ and $(-\sqrt{7}, 4\sqrt{3})$

97. $(x + y, y)$ and $(x - y, x)$

98. $(c, c - d)$ and $(d, c + d)$

Solve each problem.

99. A Sanyo color television, model AVM-2755, has a rectangular screen with a 21.7-in. width. Its height is 16 in. What is the diameter of the screen to the nearest tenth of an inch? (*Source:* Actual measurements of the author's television.)

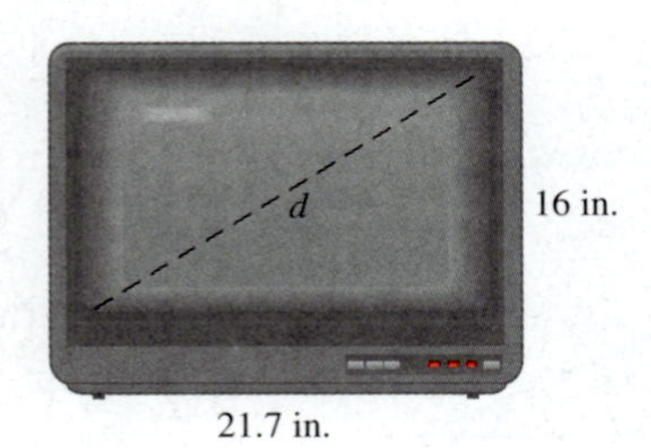

100. The length of the diagonal of a box is given by

$$D = \sqrt{L^2 + W^2 + H^2},$$

where L, W, and H are the length, width, and height of the box. Find the length of the diagonal, D, of a box that is 4 ft long, 3 ft high, and 2 ft wide. Give the exact value, then round to the nearest tenth of a foot.

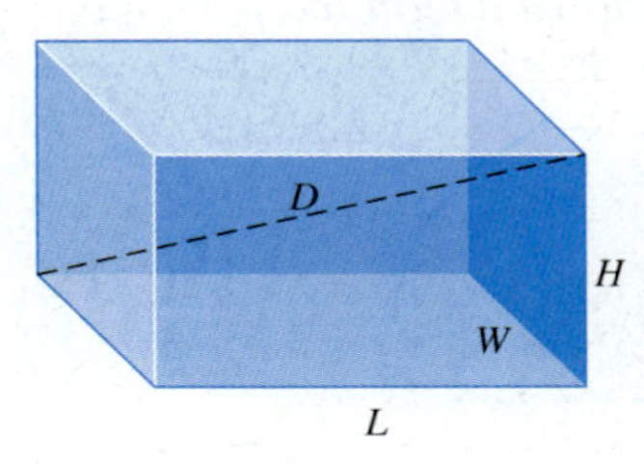

8.4 Adding and Subtracting Radical Expressions

The examples in the preceding section discussed simplifying radical expressions that involve multiplication and division. Now we show how to simplify radical expressions that involve addition and subtraction.

1 Simplify radical expressions involving addition and subtraction. An expression such as $4\sqrt{2} + 3\sqrt{2}$ can be simplified by using the distributive property.

$$4\sqrt{2} + 3\sqrt{2} = (4 + 3)\sqrt{2} = 7\sqrt{2}$$

As another example, $2\sqrt{3} - 5\sqrt{3} = (2 - 5)\sqrt{3} = -3\sqrt{3}$. This is similar to simplifying $2x + 3x$ to $5x$ or $5y - 8y$ to $-3y$.

CAUTION

Only radical expressions with the *same index* and the *same radicand* may be combined. Expressions such as $5\sqrt{3} + 2\sqrt{2}$ or $3\sqrt{3} + 2\sqrt[3]{3}$ cannot be simplified by combining terms.

Example 1 Adding and Subtracting Radicals

Add or subtract to simplify each radical expression.

(a) $3\sqrt{24} + \sqrt{54}$

Begin by simplifying each radical; then use the distributive property to combine terms.

$$\begin{aligned} 3\sqrt{24} + \sqrt{54} &= 3\sqrt{4} \cdot \sqrt{6} + \sqrt{9} \cdot \sqrt{6} && \text{Product rule} \\ &= 3 \cdot 2\sqrt{6} + 3\sqrt{6} \\ &= 6\sqrt{6} + 3\sqrt{6} \\ &= 9\sqrt{6} && \text{Combine terms.} \end{aligned}$$

(b) $$\begin{aligned} 2\sqrt{20x} - \sqrt{45x} &= 2\sqrt{4} \cdot \sqrt{5x} - \sqrt{9} \cdot \sqrt{5x} && \text{Product rule} \\ &= 2 \cdot 2\sqrt{5x} - 3\sqrt{5x} \\ &= 4\sqrt{5x} - 3\sqrt{5x} \\ &= \sqrt{5x} && \text{Combine terms.} \end{aligned}$$

Because the radicand is $5x$, we must have $x \geq 0$.

(c) $2\sqrt{3} - 4\sqrt{5}$

Here the radicals differ and are already simplified, so $2\sqrt{3} - 4\sqrt{5}$ cannot be simplified further.

Work Problem 1 at the Side.

CAUTION

Do not confuse the product rule with combining like terms. The root of a sum ***does not equal*** the sum of the roots. For example,

$$\sqrt{9 + 16} \neq \sqrt{9} + \sqrt{16}, \text{ since}$$

$$\sqrt{9 + 16} = \sqrt{25} = 5, \quad \text{but} \quad \sqrt{9} + \sqrt{16} = 3 + 4 = 7.$$

OBJECTIVES

1 Simplify radical expressions involving addition and subtraction.

1 Add or subtract to simplify each radical expression.

(a) $3\sqrt{5} + 7\sqrt{5}$

(b) $2\sqrt{11} - \sqrt{11} + 3\sqrt{44}$

(c) $5\sqrt{12y} + 6\sqrt{75y}, \quad y \geq 0$

(d) $3\sqrt{8} - 6\sqrt{50} + 2\sqrt{200}$

(e) $9\sqrt{5} - 4\sqrt{10}$

ANSWERS

1. (a) $10\sqrt{5}$ (b) $7\sqrt{11}$ (c) $40\sqrt{3y}$ (d) $-4\sqrt{2}$ (e) cannot be simplified further

2 Add or subtract to simplify each radical expression. Assume that all variables represent positive real numbers.

(a) $7\sqrt[3]{81} + 3\sqrt[3]{24}$

(b) $-2\sqrt[4]{32} - 7\sqrt[4]{162}$

(c) $\sqrt[3]{p^4q^7} - \sqrt[3]{64pq}$

3 Add. Assume that all variables represent positive real numbers.

$$\sqrt{\frac{80}{y^4}} + \sqrt{\frac{81}{y^{10}}}$$

ANSWERS

2. **(a)** $27\sqrt[3]{3}$ **(b)** $-25\sqrt[4]{2}$ **(c)** $(pq^2 - 4)\sqrt[3]{pq}$

3. $\frac{4y^3\sqrt{5} + 9}{y^5}$

Example 2 Adding and Subtracting Radicals with Higher Indexes

Add or subtract to simplify each radical expression. Assume that all variables represent positive real numbers.

(a)
$$\begin{aligned} 2\sqrt[3]{16} - 5\sqrt[3]{54} &= 2\sqrt[3]{8 \cdot 2} - 5\sqrt[3]{27 \cdot 2} && \text{Factor.} \\ &= 2\sqrt[3]{8} \cdot \sqrt[3]{2} - 5\sqrt[3]{27} \cdot \sqrt[3]{2} && \text{Product rule} \\ &= 2 \cdot 2 \cdot \sqrt[3]{2} - 5 \cdot 3 \cdot \sqrt[3]{2} \\ &= 4\sqrt[3]{2} - 15\sqrt[3]{2} \\ &= -11\sqrt[3]{2} && \text{Combine terms.} \end{aligned}$$

(b)
$$\begin{aligned} 2\sqrt[3]{x^2y} + \sqrt[3]{8x^5y^4} &= 2\sqrt[3]{x^2y} + \sqrt[3]{(8x^3y^3)x^2y} && \text{Factor.} \\ &= 2\sqrt[3]{x^2y} + 2xy\sqrt[3]{x^2y} && \text{Product rule} \\ &= (2 + 2xy)\sqrt[3]{x^2y} && \text{Distributive property} \end{aligned}$$

CAUTION

Remember to write the index when working with cube roots, fourth roots, and so on.

Work Problem 2 at the Side.

Example 3 Adding and Subtracting Radicals with Fractions

Perform the indicated operations. Assume that all variables represent positive real numbers.

(a)
$$\begin{aligned} 2\sqrt{\frac{75}{16}} + 4\frac{\sqrt{8}}{\sqrt{32}} &= 2\frac{\sqrt{25 \cdot 3}}{\sqrt{16}} + 4\frac{\sqrt{4 \cdot 2}}{\sqrt{16 \cdot 2}} && \text{Quotient rule} \\ &= 2\left(\frac{5\sqrt{3}}{4}\right) + 4\left(\frac{2\sqrt{2}}{4\sqrt{2}}\right) && \text{Product rule} \\ &= \frac{5\sqrt{3}}{2} + 2 && \text{Multiply; } \tfrac{\sqrt{2}}{\sqrt{2}} = 1. \\ &= \frac{5\sqrt{3}}{2} + \frac{4}{2} && \text{Write with a common denominator.} \\ &= \frac{5\sqrt{3} + 4}{2} \end{aligned}$$

(b)
$$\begin{aligned} 10\sqrt[3]{\frac{5}{x^6}} - 3\sqrt[3]{\frac{4}{x^9}} &= 10\frac{\sqrt[3]{5}}{\sqrt[3]{x^6}} - 3\frac{\sqrt[3]{4}}{\sqrt[3]{x^9}} && \text{Quotient rule} \\ &= \frac{10\sqrt[3]{5}}{x^2} - \frac{3\sqrt[3]{4}}{x^3} \\ &= \frac{10x\sqrt[3]{5}}{x^3} - \frac{3\sqrt[3]{4}}{x^3} && \text{Write with a common denominator.} \\ &= \frac{10x\sqrt[3]{5} - 3\sqrt[3]{4}}{x^3} \end{aligned}$$

Work Problem 3 at the Side.

8.4 EXERCISES

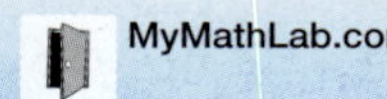

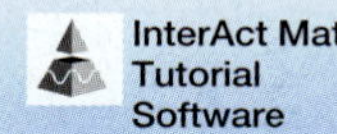

Digital Video Tutor CD 6
Videotape 13

1. Which one of the following sums could be simplified without first simplifying the individual radical expressions?

A. $\sqrt{50} + \sqrt{32}$ **B.** $3\sqrt{6} + 9\sqrt{6}$ **C.** $\sqrt[3]{32} - \sqrt[3]{108}$ **D.** $\sqrt[5]{6} - \sqrt[5]{192}$

2. Let $a = 1$ and $b = 64$.

(a) Evaluate $\sqrt{a} + \sqrt{b}$. Then find $\sqrt{a + b}$. Are they equal?

(b) Evaluate $\sqrt[3]{a} + \sqrt[3]{b}$. Then find $\sqrt[3]{a + b}$. Are they equal?

(c) Complete the following: In general, $\sqrt[n]{a} + \sqrt[n]{b} \neq$ ________, based on the observations in parts (a) and (b) of this exercise.

3. Even though the indexes of the terms are not equal, the sum $\sqrt{64} + \sqrt[3]{125} + \sqrt[4]{16}$ can be simplified quite easily. What is this sum? Why can these terms be combined so easily?

4. Explain why $28 - 4\sqrt{2}$ is not equal to $24\sqrt{2}$. (This error is a common one among algebra students.)

Add or subtract. Assume that all variables represent positive real numbers. See Examples 1–3.

5. $\sqrt{36} - \sqrt{100}$

6. $\sqrt{25} - \sqrt{81}$

7. $-2\sqrt{48} + 3\sqrt{75}$

8. $4\sqrt{32} - 2\sqrt{8}$

9. $6\sqrt{18} - \sqrt{32} + 2\sqrt{50}$

10. $5\sqrt{8} + 3\sqrt{72} - 3\sqrt{50}$

11. $-2\sqrt{63} + 2\sqrt{28} + 2\sqrt{7}$

12. $-\sqrt{27} + 2\sqrt{48} - \sqrt{75}$

13. $2\sqrt{5} + 3\sqrt{20} + 4\sqrt{45}$

14. $5\sqrt{54} - 2\sqrt{24} - 2\sqrt{96}$

15. $8\sqrt{2x} - \sqrt{8x} + \sqrt{72x}$

16. $4\sqrt{18k} - \sqrt{72k} + \sqrt{50k}$

17. $3\sqrt{72m^2} - 5\sqrt{32m^2} - 3\sqrt{18m^2}$

18. $9\sqrt{27p^2} - 14\sqrt{108p^2} + 2\sqrt{48p^2}$

19. $-\sqrt[3]{54} + 2\sqrt[3]{16}$

20. $15\sqrt[3]{81} - 4\sqrt[3]{24}$

21. $2\sqrt[3]{27x} - 2\sqrt[3]{8x}$

22. $6\sqrt[3]{128m} + 3\sqrt[3]{16m}$

23. $5\sqrt[4]{32} + 3\sqrt[4]{162}$

24. $2\sqrt[4]{512} + 4\sqrt[4]{32}$

25. $3\sqrt[4]{x^5y} - 2x\sqrt[4]{xy}$

26. $2\sqrt[4]{m^9p^6} - 3m^2p\sqrt[4]{mp^2}$

27. $\sqrt[3]{64xy^2} + \sqrt[3]{27x^4y^5}$

28. $\sqrt[4]{625s^3t} - \sqrt[4]{81s^7t^5}$

29. $\sqrt{\frac{8}{9}} + \sqrt{\frac{18}{36}}$

30. $\sqrt{\frac{12}{16}} + \sqrt{\frac{48}{64}}$

31. $\frac{\sqrt{32}}{3} + \frac{2\sqrt{2}}{3} - \frac{\sqrt{2}}{\sqrt{9}}$

32. $\frac{\sqrt{27}}{2} - \frac{3\sqrt{3}}{2} + \frac{\sqrt{3}}{\sqrt{4}}$

In Example 1(a) we show that $3\sqrt{24} + \sqrt{54} = 9\sqrt{6}$. To support this result, we can find a calculator approximation of $3\sqrt{24}$, then find a calculator approximation of $\sqrt{54}$, and add these two approximations. Then, we find a calculator approximation of $9\sqrt{6}$. It should correspond to the sum that we just found. (For this example, both approximations are 22.04540769. Due to rounding procedures, there may be a discrepancy in the final digit if you try to duplicate this work.) Follow this procedure to support the statements in Exercises 33–36.

33. $3\sqrt{32} - 2\sqrt{8} = 8\sqrt{2}$

34. $4\sqrt{12} - 7\sqrt{27} = -13\sqrt{3}$

35. $2\sqrt{40} + 6\sqrt{90} - 3\sqrt{160} = 10\sqrt{10}$

36. $5\sqrt{28} - 3\sqrt{63} + 2\sqrt{112} = 9\sqrt{7}$

37. A rectangular yard has a length of $\sqrt{192}$ m and a width of $\sqrt{48}$ m. Choose the best estimate of its dimensions. Then estimate the perimeter.

A. 14 m by 7 m **B.** 5 m by 7 m
C. 14 m by 8 m **D.** 15 m by 8 m

38. If the sides of a triangle are $\sqrt{65}$ in., $\sqrt{35}$ in., and $\sqrt{26}$ in., which one of the following is the best estimate of its perimeter?

A. 20 in. **B.** 26 in.
C. 19 in. **D.** 24 in.

Solve each problem. Give answers as simplified radical expressions.

39. Find the perimeter of the triangle.

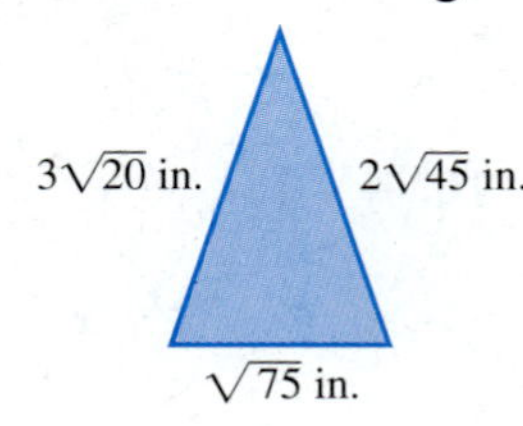

40. Find the perimeter of the rectangle.

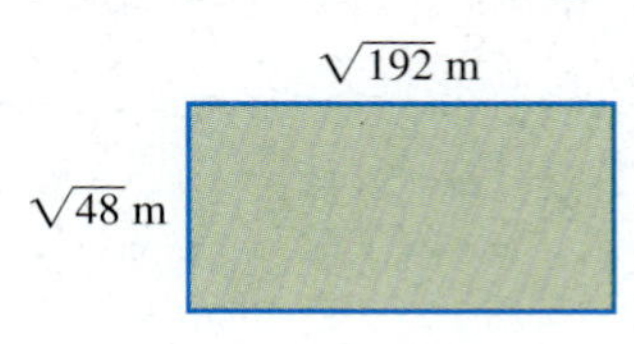

41. What is the perimeter of the computer graphic?

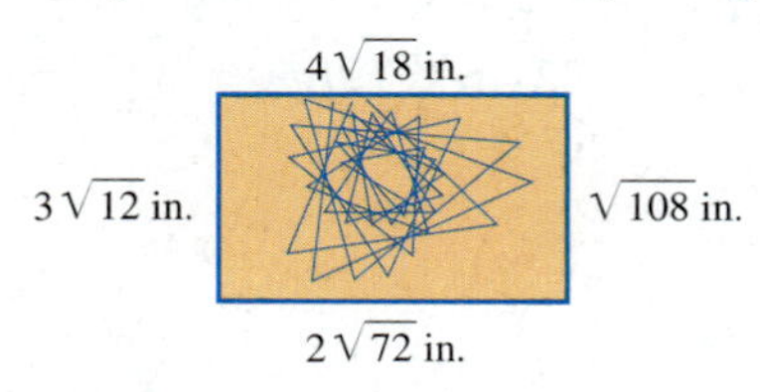

42. Find the area of the trapezoid.

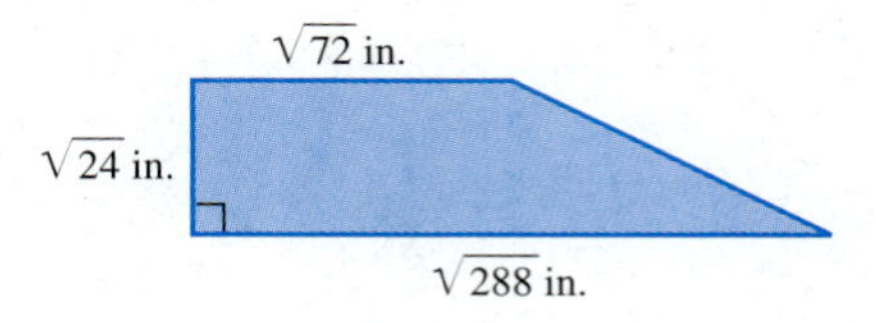

8.5 Multiplying and Dividing Radical Expressions

Objectives

1. Multiply radical expressions.
2. Rationalize denominators with one radical term.
3. Rationalize denominators with binomials involving radicals.
4. Write radical quotients in lowest terms.

1 Multiply radical expressions. We multiply binomial expressions involving radicals by using the FOIL (First, Outer, Inner, Last) method. For example, we find the product of the binomials $\sqrt{5} + 3$ and $\sqrt{6} + 1$ as follows.

$$(\sqrt{5} + 3)(\sqrt{6} + 1) = \underbrace{\sqrt{5} \cdot \sqrt{6}}_{\text{First}} + \underbrace{\sqrt{5} \cdot 1}_{\text{Outer}} + \underbrace{3 \cdot \sqrt{6}}_{\text{Inner}} + \underbrace{3 \cdot 1}_{\text{Last}}$$
$$= \sqrt{30} + \sqrt{5} + 3\sqrt{6} + 3$$

This result cannot be simplified further.

Example 1 **Multiplying Binomials Involving Radical Expressions**

Multiply using FOIL.

(a) $(7 - \sqrt{3})(\sqrt{5} + \sqrt{2}) = \overset{\text{F}}{7\sqrt{5}} + \overset{\text{O}}{7\sqrt{2}} - \overset{\text{I}}{\sqrt{3} \cdot \sqrt{5}} - \overset{\text{L}}{\sqrt{3} \cdot \sqrt{2}}$

$= 7\sqrt{5} + 7\sqrt{2} - \sqrt{15} - \sqrt{6}$

(b) $(\sqrt{10} + \sqrt{3})(\sqrt{10} - \sqrt{3})$

$= \sqrt{10} \cdot \sqrt{10} - \sqrt{10} \cdot \sqrt{3} + \sqrt{10} \cdot \sqrt{3} - \sqrt{3} \cdot \sqrt{3}$

$= 10 - 3$

$= 7$

Notice that this is the kind of product that results in the difference of squares:

$$(a + b)(a - b) = a^2 - b^2.$$

Here, $a = \sqrt{10}$ and $b = \sqrt{3}$.

(c) $(\sqrt{7} - 3)^2 = (\sqrt{7} - 3)(\sqrt{7} - 3)$

$= \sqrt{7} \cdot \sqrt{7} - 3\sqrt{7} - 3\sqrt{7} + 3 \cdot 3$

$= 7 - 6\sqrt{7} + 9$

$= 16 - 6\sqrt{7}$

(d) $(5 - \sqrt[3]{3})(5 + \sqrt[3]{3}) = 5 \cdot 5 + 5\sqrt[3]{3} - 5\sqrt[3]{3} - \sqrt[3]{3} \cdot \sqrt[3]{3}$

$= 25 - \sqrt[3]{3^2}$

$= 25 - \sqrt[3]{9}$

1 Multiply using FOIL.

(a) $(2 + \sqrt{3})(1 + \sqrt{5})$

(b) $(2\sqrt{3} + \sqrt{5})(\sqrt{6} - 3\sqrt{5})$

(c) $(4 + \sqrt{3})(4 - \sqrt{3})$

(d) $(\sqrt{6} - \sqrt{5})^2$

(e) $(4 + \sqrt[3]{7})(4 - \sqrt[3]{7})$

NOTE

In Example 1(c) we could have used the formula for the square of a binomial,

$$(a - b)^2 = a^2 - 2ab + b^2,$$

to get the same result.

$$\begin{aligned}(\sqrt{7} - 3)^2 &= (\sqrt{7})^2 - 2(\sqrt{7})(3) + 3^2\\ &= 7 - 6\sqrt{7} + 9\\ &= 16 - 6\sqrt{7}\end{aligned}$$

Work Problem 1 at the Side.

2 Rationalize denominators with one radical term. As defined earlier, a simplified radical expression will have no radical in the denominator. The origin of this agreement no doubt occurred before the days of high-speed calculation, when computation was a tedious process performed by hand. To see this, consider the radical expression $\frac{1}{\sqrt{2}}$. To find a decimal approximation by hand, it would be necessary to divide 1 by a decimal approximation for $\sqrt{2}$, such as 1.414. It would be much easier if the divisor were a whole number. This can be accomplished by multiplying $\frac{1}{\sqrt{2}}$ by 1 in the form $\frac{\sqrt{2}}{\sqrt{2}}$:

$$\frac{1}{\sqrt{2}} \cdot \frac{\sqrt{2}}{\sqrt{2}} = \frac{\sqrt{2}}{2}.$$

Now the computation would require dividing 1.414 by 2 to obtain .707, a much easier task.

With current technology, either form of this fraction can be approximated with the same number of keystrokes. See Figure 9, which shows how a calculator gives the same approximation for both forms of the expression.

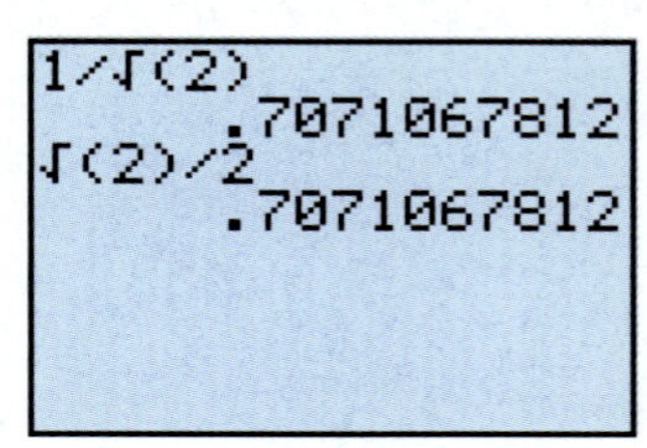

Figure 9

It is still important to be able to find equivalent forms of radical expressions. A common way of "standardizing" the form of a radical expression is to have the denominator contain no radicals. The process of removing radicals from a denominator so that the denominator contains only rational numbers is called **rationalizing the denominator.**

Example 2 Rationalizing Denominators with Square Roots

Rationalize each denominator.

(a) $\frac{3}{\sqrt{7}}$

Multiply the numerator and denominator by $\sqrt{7}$. This is, in effect, multiplying by 1.

Continued on Next Page

ANSWERS

1. **(a)** $2 + 2\sqrt{5} + \sqrt{3} + \sqrt{15}$
(b) $6\sqrt{2} - 6\sqrt{15} + \sqrt{30} - 15$
(c) 13 **(d)** $11 - 2\sqrt{30}$
(e) $16 - \sqrt[3]{49}$

$$\frac{3}{\sqrt{7}} = \frac{3 \cdot \sqrt{7}}{\sqrt{7} \cdot \sqrt{7}}$$

In the denominator, since $\sqrt{7} \cdot \sqrt{7} = \sqrt{7 \cdot 7} = \sqrt{49} = 7$,

$$\frac{3}{\sqrt{7}} = \frac{3\sqrt{7}}{7}.$$

The denominator is now a rational number.

(b) $\dfrac{5\sqrt{2}}{\sqrt{5}} = \dfrac{5\sqrt{2} \cdot \sqrt{5}}{\sqrt{5} \cdot \sqrt{5}} = \dfrac{5\sqrt{10}}{5} = \sqrt{10}$

(c) $\dfrac{6}{\sqrt{12}}$

Less work is involved if the radical in the denominator is simplified first.

$$\frac{6}{\sqrt{12}} = \frac{6}{\sqrt{4 \cdot 3}} = \frac{6}{2\sqrt{3}} = \frac{3}{\sqrt{3}}$$

Now rationalize the denominator by multiplying the numerator and denominator by $\sqrt{3}$.

$$\frac{3 \cdot \sqrt{3}}{\sqrt{3} \cdot \sqrt{3}} = \frac{3\sqrt{3}}{3} = \sqrt{3}$$

Work Problem 2 at the Side.

Example 3 Rationalizing Denominators in Roots of Fractions

Simplify each radical.

(a) $\sqrt{\dfrac{18}{125}}$

$$\sqrt{\frac{18}{125}} = \frac{\sqrt{18}}{\sqrt{125}} \quad \text{Quotient rule}$$

$$= \frac{\sqrt{9 \cdot 2}}{\sqrt{25 \cdot 5}} \quad \text{Factor.}$$

$$= \frac{3\sqrt{2}}{5\sqrt{5}} \quad \text{Product rule}$$

$$= \frac{3\sqrt{2} \cdot \sqrt{5}}{5\sqrt{5} \cdot \sqrt{5}} \quad \text{Multiply by } \frac{\sqrt{5}}{\sqrt{5}}.$$

$$= \frac{3\sqrt{10}}{5 \cdot 5} \quad \text{Product rule}$$

$$= \frac{3\sqrt{10}}{25}$$

Continued on Next Page

2 Rationalize each denominator.

(a) $\dfrac{8}{\sqrt{3}}$

(b) $\dfrac{\sqrt{3}}{\sqrt{7}}$

(c) $\dfrac{3}{\sqrt{48}}$

(d) $\dfrac{-16}{\sqrt{32}}$

ANSWERS

2. **(a)** $\dfrac{8\sqrt{3}}{3}$ **(b)** $\dfrac{\sqrt{21}}{7}$ **(c)** $\dfrac{\sqrt{3}}{4}$ **(d)** $-2\sqrt{2}$

3 Simplify. Assume that all variables represent positive real numbers.

(a) $\sqrt{\frac{8}{45}}$

(b) $\sqrt{\frac{72}{y}}$

(c) $\sqrt{\frac{200k^6}{y^7}}$

4 Simplify.

(a) $\sqrt[3]{\frac{15}{32}}$

(b) $\sqrt[3]{\frac{m^{12}}{n}}, \quad n \neq 0$

(c) $\sqrt[4]{\frac{6y}{w^2}}, \quad y \geq 0, w \neq 0$

Answers

3. **(a)** $\frac{2\sqrt{10}}{15}$ **(b)** $\frac{6\sqrt{2y}}{y}$ **(c)** $\frac{10k^3\sqrt{2y}}{y^4}$

4. **(a)** $\frac{\sqrt[3]{30}}{4}$ **(b)** $\frac{m^4\sqrt[3]{n^2}}{n}$ **(c)** $\frac{\sqrt[4]{6yw^2}}{w}$

(b) $\sqrt{\frac{50m^4}{p^5}}, \quad p > 0$

$$\sqrt{\frac{50m^4}{p^5}} = \frac{\sqrt{50m^4}}{\sqrt{p^5}} \quad \text{Quotient rule}$$

$$= \frac{5m^2\sqrt{2}}{p^2\sqrt{p}} \quad \text{Product rule}$$

$$= \frac{5m^2\sqrt{2} \cdot \sqrt{p}}{p^2\sqrt{p} \cdot \sqrt{p}} \quad \text{Multiply by } \frac{\sqrt{p}}{\sqrt{p}}.$$

$$= \frac{5m^2\sqrt{2p}}{p^2 \cdot p} \quad \text{Product rule}$$

$$= \frac{5m^2\sqrt{2p}}{p^3}$$

Work Problem **3** at the Side.

Example 4 Rationalizing Denominators with Cube Roots

Simplify.

(a) $\sqrt[3]{\frac{27}{16}}$

Use the quotient rule and simplify the numerator and denominator.

$$\sqrt[3]{\frac{27}{16}} = \frac{\sqrt[3]{27}}{\sqrt[3]{16}} = \frac{3}{\sqrt[3]{8} \cdot \sqrt[3]{2}} = \frac{3}{2\sqrt[3]{2}}$$

To get a rational denominator, multiply the numerator and denominator by a number that will result in a perfect cube in the radicand in the denominator. Since $2 \cdot 4 = 8$, a perfect cube, multiply the numerator and denominator by $\sqrt[3]{4}$.

$$\sqrt[3]{\frac{27}{16}} = \frac{3}{2\sqrt[3]{2}} = \frac{3 \cdot \sqrt[3]{4}}{2\sqrt[3]{2} \cdot \sqrt[3]{4}} = \frac{3\sqrt[3]{4}}{2\sqrt[3]{8}} = \frac{3\sqrt[3]{4}}{2 \cdot 2} = \frac{3\sqrt[3]{4}}{4}$$

(b) $\sqrt[4]{\frac{5x}{z}}, \quad x \geq 0, \; z > 0$

$$\sqrt[4]{\frac{5x}{z}} = \frac{\sqrt[4]{5x}}{\sqrt[4]{z}} \cdot \frac{\sqrt[4]{z^3}}{\sqrt[4]{z^3}} = \frac{\sqrt[4]{5xz^3}}{\sqrt[4]{z^4}} = \frac{\sqrt[4]{5xz^3}}{z}$$

CAUTION

It is easy to make mistakes in problems like the one in Example 4(a). A typical error is to multiply the numerator and denominator by $\sqrt[3]{2}$, forgetting that

$$\sqrt[3]{2} \cdot \sqrt[3]{2} \neq 2.$$

You need *three* factors of 2 to get 2^3 under the radical. As implied in Example 4(a),

$$\sqrt[3]{2} \cdot \sqrt[3]{2} \cdot \sqrt[3]{2} = 2.$$

Work Problem **4** at the Side.

3 Rationalize denominators with binomials involving radicals. Recall the special product

$$(a + b)(a - b) = a^2 - b^2.$$

To rationalize a denominator that contains a binomial expression (one that contains exactly two terms) involving radicals, such as

$$\frac{3}{1 + \sqrt{2}},$$

we must use conjugates. The conjugate of $1 + \sqrt{2}$ is $1 - \sqrt{2}$. In general, $a + b$ and $a - b$ are **conjugates.**

Rationalizing Binomial Denominators

Whenever a radical expression has a sum or difference with square root radicals in the denominator, we rationalize the denominator by multiplying both the numerator and denominator by the conjugate of the denominator.

For the expression $\frac{3}{1 + \sqrt{2}}$, we rationalize the denominator by multiplying both the numerator and denominator by $1 - \sqrt{2}$, the conjugate of the denominator.

$$\frac{3}{1 + \sqrt{2}} = \frac{3(1 - \sqrt{2})}{(1 + \sqrt{2})(1 - \sqrt{2})}$$

Then $(1 + \sqrt{2})(1 - \sqrt{2}) = 1^2 - (\sqrt{2})^2 = 1 - 2 = -1$. Placing -1 in the denominator gives

$$= \frac{3(1 - \sqrt{2})}{-1}$$

$$= \frac{3}{-1}(1 - \sqrt{2})$$

$$= -3(1 - \sqrt{2}) \quad \text{or} \quad -3 + 3\sqrt{2}.$$

Example 5 Rationalizing Binomial Denominators

Rationalize each denominator.

(a) $\frac{5}{4 - \sqrt{3}}$

To rationalize the denominator, multiply both the numerator and denominator by the conjugate of the denominator, $4 + \sqrt{3}$.

$$\frac{5}{4 - \sqrt{3}} = \frac{5(4 + \sqrt{3})}{(4 - \sqrt{3})(4 + \sqrt{3})}$$

$$= \frac{5(4 + \sqrt{3})}{16 - 3}$$

$$= \frac{5(4 + \sqrt{3})}{13}$$

Notice that the numerator is left in factored form. This makes it easier to determine whether the expression is written in lowest terms.

Continued on Next Page

5 Rationalize each denominator.

(a) $\dfrac{-4}{\sqrt{5}+2}$

(b) $\dfrac{15}{\sqrt{7}+\sqrt{2}}$

(c) $\dfrac{\sqrt{3}+\sqrt{5}}{\sqrt{2}-\sqrt{7}}$

(d) $\dfrac{2}{\sqrt{k}+\sqrt{z}}$
$(k \neq z, k > 0, z > 0)$

6 Write each quotient in lowest terms.

(a) $\dfrac{15-5\sqrt{3}}{5}$

(b) $\dfrac{24-36\sqrt{7}}{16}$

Answers

5. **(a)** $-4(\sqrt{5}-2)$ **(b)** $3(\sqrt{7}-\sqrt{2})$
(c) $\dfrac{-(\sqrt{6}+\sqrt{21}+\sqrt{10}+\sqrt{35})}{5}$
(d) $\dfrac{2(\sqrt{k}-\sqrt{z})}{k-z}$

6. **(a)** $3-\sqrt{3}$ **(b)** $\dfrac{6-9\sqrt{7}}{4}$

(b) $\dfrac{\sqrt{2}-\sqrt{3}}{\sqrt{5}+\sqrt{3}}$

Multiply the numerator and denominator by $\sqrt{5}-\sqrt{3}$ to rationalize the denominator.

$$\frac{\sqrt{2}-\sqrt{3}}{\sqrt{5}+\sqrt{3}} = \frac{(\sqrt{2}-\sqrt{3})(\sqrt{5}-\sqrt{3})}{(\sqrt{5}+\sqrt{3})(\sqrt{5}-\sqrt{3})}$$
$$= \frac{\sqrt{10}-\sqrt{6}-\sqrt{15}+3}{5-3}$$
$$= \frac{\sqrt{10}-\sqrt{6}-\sqrt{15}+3}{2}$$

(c) $$\frac{3}{\sqrt{5m}-\sqrt{p}} = \frac{3(\sqrt{5m}+\sqrt{p})}{(\sqrt{5m}-\sqrt{p})(\sqrt{5m}+\sqrt{p})}$$
$$= \frac{3(\sqrt{5m}+\sqrt{p})}{5m-p} \qquad (5m \neq p, m > 0, p > 0)$$

Work Problem **5** at the Side.

4 Write radical quotients in lowest terms.

Example 6 Writing Radical Quotients in Lowest Terms

Write each quotient in lowest terms.

(a) $\dfrac{6+2\sqrt{5}}{4}$

Factor the numerator and denominator, then write in lowest terms.

$$\frac{6+2\sqrt{5}}{4} = \frac{2(3+\sqrt{5})}{2 \cdot 2} = \frac{3+\sqrt{5}}{2}$$

Here is an alternative method for writing this expression in lowest terms.

$$\frac{6+2\sqrt{5}}{4} = \frac{6}{4} + \frac{2\sqrt{5}}{4} = \frac{3}{2} + \frac{\sqrt{5}}{2} = \frac{3+\sqrt{5}}{2}$$

(b) $$\frac{5y-\sqrt{8y^2}}{6y} = \frac{5y-2y\sqrt{2}}{6y}, y > 0 \qquad \text{Product rule}$$
$$= \frac{y(5-2\sqrt{2})}{6y} \qquad \text{Factor the numerator.}$$
$$= \frac{5-2\sqrt{2}}{6}$$

Note that the final fraction cannot be simplified further because there is no common factor of 2 in the numerator.

CAUTION

Be careful to factor *before* writing a quotient in lowest terms.

Work Problem **6** at the Side.

8.5 EXERCISES

FOR EXTRA HELP: Student's Solutions Manual | MyMathLab.com | InterAct Math Tutorial Software | AW Math Tutor Center | www.mathxl.com | Digital Video Tutor CD 6 Videotape 14

Match each part of a rule for a special product in Column I with the part it equals in Column II.

I	II
1. $(x + \sqrt{y})(x - \sqrt{y})$	**A.** $x - y$
2. $(\sqrt{x} + y)(\sqrt{x} - y)$	**B.** $x + 2y\sqrt{x} + y^2$
3. $(\sqrt{x} + \sqrt{y})(\sqrt{x} - \sqrt{y})$	**C.** $x - y^2$
4. $(\sqrt{x} + \sqrt{y})^2$	**D.** $x - 2\sqrt{xy} + y$
5. $(\sqrt{x} - \sqrt{y})^2$	**E.** $x^2 - y$
6. $(\sqrt{x} + y)^2$	**F.** $x + 2\sqrt{xy} + y$

Multiply, then simplify each product. Assume that all variables represent positive real numbers. See Example 1.

7. $\sqrt{3}(\sqrt{12} - 4)$

8. $\sqrt{5}(\sqrt{125} - 6)$

9. $\sqrt{2}(\sqrt{18} - \sqrt{3})$

10. $\sqrt{5}(\sqrt{15} + \sqrt{5})$

11. $(\sqrt{6} + 2)(\sqrt{6} - 2)$

12. $(\sqrt{7} + 8)(\sqrt{7} - 8)$

13. $(\sqrt{12} - \sqrt{3})(\sqrt{12} + \sqrt{3})$

14. $(\sqrt{18} + \sqrt{8})(\sqrt{18} - \sqrt{8})$

15. $(\sqrt{3} + 2)(\sqrt{6} - 5)$

16. $(\sqrt{7} + 1)(\sqrt{2} - 4)$

17. $(\sqrt{3x} + 2)(\sqrt{3x} - 2)$

18. $(\sqrt{6y} - 4)(\sqrt{6y} + 4)$

19. $(2\sqrt{x} + \sqrt{y})(2\sqrt{x} - \sqrt{y})$

20. $(\sqrt{p} + 5\sqrt{s})(\sqrt{p} - 5\sqrt{s})$

21. $(4\sqrt{x} + 3)^2$

22. $(5\sqrt{p} - 6)^2$

23. $(9 - \sqrt[3]{2})(9 + \sqrt[3]{2})$

24. $(7 + \sqrt[3]{6})(7 - \sqrt[3]{6})$

25. The correct answer to Exercise 7 is $6 - 4\sqrt{3}$. Explain why this is not equal to $2\sqrt{3}$.

26. When we rationalize the denominator in the radical expression $\frac{1}{\sqrt{2}}$, we multiply both the numerator and denominator by $\sqrt{2}$. What property of real numbers covered in Section 1.4 justifies this procedure?

Rationalize the denominator in each expression. Assume that all variables represent positive real numbers. See Example 2.

27. $\frac{7}{\sqrt{7}}$ **28.** $\frac{11}{\sqrt{11}}$ **29.** $\frac{15}{\sqrt{3}}$ **30.** $\frac{12}{\sqrt{6}}$ **31.** $\frac{\sqrt{3}}{\sqrt{2}}$

32. $\frac{\sqrt{7}}{\sqrt{6}}$ **33.** $\frac{9\sqrt{3}}{\sqrt{5}}$ **34.** $\frac{3\sqrt{2}}{\sqrt{11}}$ **35.** $\frac{-6}{\sqrt{18}}$ **36.** $\frac{-5}{\sqrt{24}}$

37. $\frac{-8\sqrt{3}}{\sqrt{k}}$ **38.** $\frac{-4\sqrt{13}}{\sqrt{m}}$ **39.** $\frac{6\sqrt{3y}}{\sqrt{y^3}}$ **40.** $\frac{-8\sqrt{5y}}{\sqrt{y^5}}$

41. Look again at the expression in Exercise 39. Start by multiplying both the numerator and the denominator by $\sqrt{y}$, to obtain the final answer. Then start over, multiplying both the numerator and denominator by $\sqrt{y^3}$, to obtain the same answer. Which method do you prefer? Why?

42. Explain why $\frac{1}{\sqrt[3]{2}}$ would not be written with the denominator rationalized if you begin by multiplying both the numerator and denominator by $\sqrt[3]{2}$. By what should you multiply them both to achieve the desired result?

Simplify. Assume that all variables represent positive real numbers. See Examples 3 and 4.

43. $\sqrt{\frac{7}{2}}$ **44.** $\sqrt{\frac{10}{3}}$ **45.** $-\sqrt{\frac{7}{50}}$ **46.** $-\sqrt{\frac{13}{75}}$ **47.** $\sqrt{\frac{24}{x}}$

48. $\sqrt{\frac{52}{y}}$ **49.** $-\sqrt{\frac{98r^3}{s}}$ **50.** $-\sqrt{\frac{150m^5}{n}}$ **51.** $\sqrt{\frac{288x^7}{y^9}}$ **52.** $\sqrt{\frac{242t^9}{u^{11}}}$

53. $\sqrt[3]{\frac{2}{3}}$ **54.** $\sqrt[3]{\frac{4}{5}}$ **55.** $\sqrt[3]{\frac{4}{9}}$ **56.** $\sqrt[3]{\frac{5}{16}}$

57. $-\sqrt[3]{\frac{2p}{r^2}}$ **58.** $-\sqrt[3]{\frac{6x}{y^2}}$ **59.** $\sqrt[4]{\frac{16}{x}}$ **60.** $\sqrt[4]{\frac{81}{y}}$

61. Explain the procedure you will use to rationalize the denominator of the expression in Exercise 63:

$$\frac{2}{4 + \sqrt{3}}.$$

62. Would multiplying both the numerator and denominator of $\frac{2}{4 + \sqrt{3}}$ by $4 + \sqrt{3}$ lead to a rationalized denominator? Why or why not?

Rationalize the denominator in each expression. Assume that all variables represent positive real numbers and that no denominators are 0. See Example 5.

63. $\frac{2}{4 + \sqrt{3}}$ **64.** $\frac{6}{5 + \sqrt{2}}$ **65.** $\frac{6}{\sqrt{5} + \sqrt{3}}$ **66.** $\frac{12}{\sqrt{6} + \sqrt{3}}$ **67.** $\frac{-4}{\sqrt{3} - \sqrt{7}}$

68. $\frac{-3}{\sqrt{2} + \sqrt{5}}$ **69.** $\frac{1 - \sqrt{2}}{\sqrt{7} + \sqrt{6}}$ **70.** $\frac{-1 - \sqrt{3}}{\sqrt{6} + \sqrt{5}}$ **71.** $\frac{4\sqrt{x}}{\sqrt{x} - 2\sqrt{y}}$

72. $\frac{5\sqrt{r}}{3\sqrt{r} + \sqrt{s}}$ **73.** $\frac{\sqrt{x} - \sqrt{y}}{\sqrt{2x} + \sqrt{3y}}$ **74.** $\frac{\sqrt{a} + \sqrt{b}}{\sqrt{5a} - \sqrt{2b}}$

75. If a and b are both positive numbers and $a^2 = b^2$, then $a = b$. Use this fact to show that

$$\frac{\sqrt{6} - \sqrt{2}}{4} = \frac{\sqrt{2 - \sqrt{3}}}{2}.$$

76. Use a calculator approximation to support the result in Exercise 75.

Write each quotient in lowest terms. Assume that all variables represent positive real numbers. See Example 6.

77. $\dfrac{25 + 10\sqrt{6}}{20}$

78. $\dfrac{12 - 6\sqrt{2}}{24}$

79. $\dfrac{16 + 4\sqrt{8}}{12}$

80. $\dfrac{12 + 9\sqrt{72}}{18}$

81. $\dfrac{6x + \sqrt{24x^3}}{3x}$

82. $\dfrac{11y + \sqrt{242y^5}}{22y}$

RELATING CONCEPTS (Exercises 83–86) FOR INDIVIDUAL OR GROUP WORK

Sometimes it is desirable to rationalize the numerator *in an expression. The procedure is similar to rationalizing the denominator. For example, to rationalize the numerator of*

$$\frac{6 - \sqrt{2}}{3},$$

we multiply both the numerator and denominator by the conjugate of the numerator, $6 + \sqrt{2}$.

$$\frac{6 - \sqrt{2}}{3} = \frac{(6 - \sqrt{2})(6 + \sqrt{2})}{3(6 + \sqrt{2})} = \frac{36 - 2}{3(6 + \sqrt{2})} = \frac{34}{3(6 + \sqrt{2})}$$

In the final expression, the numerator is rationalized. ***Work Exercises 83–86 in order.***

83. Rationalize the numerator of $\dfrac{8\sqrt{5} - 1}{6}$.

84. Rationalize the numerator of $\dfrac{3\sqrt{a} + \sqrt{b}}{\sqrt{b} - \sqrt{a}}$. Assume a and b are positive and $a \neq b$.

85. Rationalize the denominator of the expression in Exercise 84.

86. Describe the difference in the procedures used in Exercises 84 and 85.

8.6 EQUATIONS WITH RADICAL EXPRESSIONS

OBJECTIVES

1. Solve radical equations using the power rule.
2. Solve radical equations that require additional steps.
3. Solve radical equations with indexes greater than 2.

An equation that includes one or more radical expressions with a variable is called a **radical equation.** Some examples of radical equations are

$$\sqrt{x-4}=8, \quad \sqrt{5x+12}=3\sqrt{2x-1}, \quad \text{and} \quad \sqrt[3]{6+x}=27.$$

1 Solve radical equations using the power rule. The equation $x = 1$ has only one solution. Its solution set is $\{1\}$. If we square both sides of this equation, we get $x^2 = 1$. This new equation has two solutions: -1 and 1. Notice that the solution of the original equation is also a solution of the squared equation. However, the squared equation has another solution, -1, that is *not* a solution of the original equation. When solving equations with radicals, we use this idea of raising both sides to a power. It is an application of the *power rule.*

Power Rule for Solving Equations with Radicals

If both sides of an equation are raised to the same power, all solutions of the original equation are also solutions of the new equation.

Read the power rule carefully; it does *not* say that all solutions of the new equation are solutions of the original equation. They may or may not be. Solutions that do not satisfy the original equation are called **extraneous solutions;** they must be discarded.

CAUTION

When the power rule is used to solve an equation, ***every solution of the new equation* must *be checked in the original equation.***

Example 1 Using the Power Rule

Solve $\sqrt{3x+4}=8$.

Use the power rule and square both sides to get

$$(\sqrt{3x+4})^2 = 8^2$$
$$3x+4=64$$
$$3x=60$$
$$x=20.$$

To check, substitute the potential solution in the *original* equation.

$$\sqrt{3x+4}=8$$
$$\sqrt{3\cdot 20+4}=8 \quad ? \quad \text{Let } x=20.$$
$$\sqrt{64}=8 \quad ?$$
$$8=8 \quad \text{True}$$

Since 20 satisfies the *original* equation, the solution set is $\{20\}$.

Work Problem 1 at the Side.

The solution of the equation in Example 1 can be generalized to give a method for solving equations with radicals.

1 Solve.

(a) $\sqrt{r}=3$

(b) $\sqrt{5x+1}=4$

ANSWERS

1. (a) $\{9\}$ **(b)** $\{3\}$

2 Solve.

(a) $\sqrt{k} + 4 = -3$

(b) $\sqrt{x - 9} - 3 = 0$

Solving an Equation with Radicals

Step 1 **Isolate the radical.** Make sure that one radical term is alone on one side of the equation.

Step 2 **Apply the power rule.** Raise both sides of the equation to a power that is the same as the index of the radical.

Step 3 **Solve.** Solve the resulting equation; if it still contains a radical, repeat Steps 1 and 2.

Step 4 **Check.** It is essential that all potential solutions be checked in the original equation.

CAUTION

Remember Step 4 or you may get an incorrect solution set.

Example 2 **Using the Power Rule**

Solve $\sqrt{5q - 1} + 3 = 0$.

Step 1 To get the radical alone on one side, subtract 3 from each side.

$$\sqrt{5q - 1} = -3$$

Step 2 Now square both sides.

$$(\sqrt{5q - 1})^2 = (-3)^2$$

Step 3

$$5q - 1 = 9$$
$$5q = 10$$
$$q = 2$$

Step 4 Check the potential solution, 2, by substituting it in the original equation.

$$\sqrt{5q - 1} + 3 = 0$$
$$\sqrt{5 \cdot 2 - 1} + 3 = 0 \quad ? \quad \text{Let } q = 2.$$
$$3 + 3 = 0 \quad \text{False}$$

This false result shows that 2 is *not* a solution of the original equation; it is extraneous. The solution set is $\emptyset$.

NOTE

We could have determined after Step 1 that the equation in Example 2 has no solution because the expression on the left cannot be negative.

Work Problem 2 at the Side.

2 **Solve radical equations that require additional steps.** The next examples involve finding the square of a binomial. Recall that

$$(x + y)^2 = x^2 + 2xy + y^2.$$

ANSWERS

2. **(a)** $\emptyset$ **(b)** $\{18\}$

Example 3 Using the Power Rule; Squaring a Binomial

Solve $\sqrt{4-x} = x + 2$.

Step 1 The radical is alone on the left side of the equation.

Step 2 Square both sides; the square of $x + 2$ is $(x + 2)^2 = x^2 + 4x + 4$.

$$(\sqrt{4-x})^2 = (x+2)^2$$
$$4 - x = x^2 + 4x + 4$$

Twice the product of 2 and x

Step 3 The new equation is quadratic, so get 0 on one side.

$$0 = x^2 + 5x \quad \text{Subtract 4 and add } x.$$
$$0 = x(x+5) \quad \text{Factor.}$$
$$x = 0 \quad \text{or} \quad x + 5 = 0 \quad \text{Zero-factor property}$$
$$x = -5$$

Step 4 Check each potential solution in the original equation.

If $x = 0$, then

$$\sqrt{4-x} = x + 2$$
$$\sqrt{4-0} = 0 + 2 \quad ?$$
$$\sqrt{4} = 2 \quad ?$$
$$2 = 2. \quad \text{True}$$

If $x = -5$, then

$$\sqrt{4-x} = x + 2$$
$$\sqrt{4-(-5)} = -5 + 2 \quad ?$$
$$\sqrt{9} = -3 \quad ?$$
$$3 = -3. \quad \text{False}$$

The solution set is $\{0\}$. The other potential solution, -5, is extraneous.

CAUTION

When a radical equation requires squaring a binomial as in Example 3, remember to include the middle term.

$(x+2)^2 \neq x^2 + 4$	$(x+2)^2 = x^2 + 4x + 4$
INCORRECT	**CORRECT**

Work Problem 3 at the Side.

Example 4 Using the Power Rule; Squaring a Binomial

Solve $\sqrt{m^2 - 4m + 9} = m - 1$.

Squaring both sides gives $(m-1)^2 = m^2 - 2(m)(1) + 1^2$ on the right.

$$(\sqrt{m^2 - 4m + 9})^2 = (m-1)^2$$
$$m^2 - 4m + 9 = m^2 - 2m + 1$$

Twice the product of m and -1

Subtract m^2 and 1 from each side, then add $4m$ to each side to get

$$8 = 2m$$
$$4 = m.$$

Check this potential solution in the original equation.

$$\sqrt{m^2 - 4m + 9} = m - 1$$
$$\sqrt{4^2 - 4 \cdot 4 + 9} = 4 - 1 \quad ? \quad \text{Let } m = 4.$$
$$3 = 3 \quad \text{True}$$

The solution set of the original equation is $\{4\}$.

Work Problem 4 at the Side.

3 Solve.

(a) $\sqrt{3z - 5} = z - 1$

(b) $x + 1 = \sqrt{-2x - 2}$

4 Solve

$\sqrt{4a^2 + 2a - 3} = 2a + 7$.

Answers

3. **(a)** $\{2, 3\}$ **(b)** $\{-1\}$

4. $\{-2\}$

5 Solve
$\sqrt{p+1} - \sqrt{p-4} = 1.$

6 Solve each equation.

(a) $\sqrt[3]{p^2 + 3p + 12} = \sqrt[3]{p^2}$

(b) $\sqrt[4]{2k+5} + 1 = 0$

ANSWERS
5. {8}
6. (a) {−4} (b) ∅

Example 5 Using the Power Rule; Squaring Twice

Solve $\sqrt{5m+6} + \sqrt{3m+4} = 2$.

Start by getting one radical alone on one side of the equation by subtracting $\sqrt{3m+4}$ from each side.

$$\sqrt{5m+6} = 2 - \sqrt{3m+4}$$

Now square both sides.

$$(\sqrt{5m+6})^2 = (2 - \sqrt{3m+4})^2$$
$$5m + 6 = 4 - 4\sqrt{3m+4} + (3m+4)$$

Twice the product of 2 and $-\sqrt{3m+4}$

This equation still contains a radical, so square both sides again. Before doing this, isolate the radical term on the right.

$$5m + 6 = 8 + 3m - 4\sqrt{3m+4}$$
$$2m - 2 = -4\sqrt{3m+4} \quad \text{Subtract 8 and } 3m.$$
$$m - 1 = -2\sqrt{3m+4} \quad \text{Divide by 2.}$$
$$(m-1)^2 = (-2\sqrt{3m+4})^2 \quad \text{Square both sides again.}$$
$$m^2 - 2m + 1 = (-2)^2(\sqrt{3m+4})^2 \quad (ab)^2 = a^2b^2$$
$$m^2 - 2m + 1 = 4(3m+4)$$
$$m^2 - 2m + 1 = 12m + 16 \quad \text{Distributive property}$$
$$m^2 - 14m - 15 = 0 \quad \text{Standard form}$$
$$(m-15)(m+1) = 0 \quad \text{Factor.}$$
$$m - 15 = 0 \quad \text{or} \quad m + 1 = 0 \quad \text{Zero-factor property}$$
$$m = 15 \quad \text{or} \quad m = -1$$

Check each of these potential solutions in the original equation. Only -1 works, so the solution set, $\{-1\}$, has only one element.

Work Problem 5 at the Side.

3 Solve radical equations with indexes greater than 2. The power rule also works for powers greater than 2.

Example 6 Using the Power Rule for a Power Greater than 2

Solve $\sqrt[3]{z+5} = \sqrt[3]{2z-6}$.

Raise both sides to the third power.

$$(\sqrt[3]{z+5})^3 = (\sqrt[3]{2z-6})^3$$
$$z + 5 = 2z - 6$$
$$11 = z$$

Check this result in the original equation.

$$\sqrt[3]{z+5} = \sqrt[3]{2z-6}$$
$$\sqrt[3]{11+5} = \sqrt[3]{2 \cdot 11 - 6} \quad ? \quad \text{Let } z = 11.$$
$$\sqrt[3]{16} = \sqrt[3]{16} \quad \text{True}$$

The solution set is $\{11\}$.

Work Problem 6 at the Side.

8.6 Exercises

FOR EXTRA HELP

Student's Solutions Manual | MyMathLab.com | InterAct Math Tutorial Software | AW Math Tutor Center | www.mathxl.com | Digital Video Tutor CD 6 Videotape 14

1. Is 9 a solution of the equation $\sqrt{x} = -3$? If not, what is the solution of this equation?

2. Before even attempting to solve $\sqrt{3x + 18} = x$, how can you be sure that the equation cannot have a negative solution?

Solve each equation. See Examples 1 and 2.

3. $\sqrt{x - 3} = 4$

4. $\sqrt{y + 2} = 5$

5. $\sqrt{3k - 2} = 6$

6. $\sqrt{4t + 7} = 9$

7. $\sqrt{x} + 9 = 0$

8. $\sqrt{w} + 4 = 0$

9. $\sqrt{3x - 6} - 3 = 0$

10. $\sqrt{7y + 11} - 5 = 0$

11. $\sqrt{6x + 2} - \sqrt{5x + 3} = 0$

12. $\sqrt{3 + 5x} - \sqrt{x + 11} = 0$

13. $3\sqrt{x} = \sqrt{8x + 9}$

14. $6\sqrt{p} = \sqrt{30p + 24}$

15. Explain what is wrong with this step in the solution process for $\sqrt{3x + 4} = 8 - x$.

$$3x + 4 = 64 + x^2$$

16. Explain what is wrong with this step in the solution process for $\sqrt{5y + 6} = \sqrt{y + 3} - 3$.

$$5y + 6 = (y + 3) - 9$$

Solve each equation. See Examples 3 and 4.

17. $\sqrt{3x + 4} = 8 - x$

18. $\sqrt{5x + 1} = 2x - 2$

19. $\sqrt{13 + 4t} = t + 4$

20. $\sqrt{50 + 7k} = k + 8$

21. $\sqrt{r^2 - 15r + 15} + 5 = r$

22. $\sqrt{p^2 + 12p - 4} + 4 = p$

23. $\sqrt{3x + 7} - 3x = 5$

24. $\sqrt{4x + 13} - 2x = -1$

25. $\sqrt{4x + 2} - 4x = 0$

26. $\sqrt{4 - 2x} - 8 = 2x$

Solve each equation. See Example 5.

27. $\sqrt{r + 4} - \sqrt{r - 4} = 2$

28. $\sqrt{m + 1} - \sqrt{m - 2} = 1$

29. $\sqrt{11 + 2q} + 1 = \sqrt{5q + 1}$

30. $\sqrt{6 + 5y} - 3 = \sqrt{y + 3}$

31. $\sqrt{3 - 3p} - \sqrt{3p + 2} = 3$

32. $\sqrt{3x + 4} - \sqrt{2x - 4} = 2$

33. What is the smallest power to which you can raise both sides of the radical equation

$$\sqrt[3]{x+3} = \sqrt[3]{5+4x}$$

so that the radicals are eliminated?

34. What is the smallest power to which you can raise both sides of the radical equation

$$\sqrt{x+3} = \sqrt[3]{10x+14}$$

so that the radicals are eliminated?

Solve each equation. See Example 6.

35. $\sqrt[3]{2x^2+3x-7} = \sqrt[3]{2x^2+4x+6}$

36. $\sqrt[3]{3y^2-4y+6} = \sqrt[3]{3y^2-2y+8}$

37. $\sqrt[3]{1-2k} - \sqrt[3]{-k-13} = 0$

38. $\sqrt[3]{11-2t} - \sqrt[3]{-1-5t} = 0$

39. $\sqrt[4]{x-1} + 2 = 0$

40. $\sqrt[4]{2k+3} + 1 = 0$

41. $\sqrt[4]{x+7} = \sqrt[4]{2x}$

42. $\sqrt[4]{y+8} = \sqrt[4]{3y}$

For each equation, rewrite the expressions with rational exponents as radical expressions. Then solve using the procedures explained in this section.

43. $(5r-6)^{1/2} = 2 + (3r-6)^{1/2}$

44. $(3w+7)^{1/2} = 1 + (w+2)^{1/2}$

45. $(2w - 1)^{2/3} - w^{1/3} = 0$

46. $(x^2 - 2x)^{1/3} - x^{1/3} = 0$

If x represents the number of years since 1985, the equation $y = .4x^2$ approximates the number of U.S. cell-phone subscribers, in millions. For example, $x = 5$ represents 1990, $x = 10$ represents 1995, and so on.

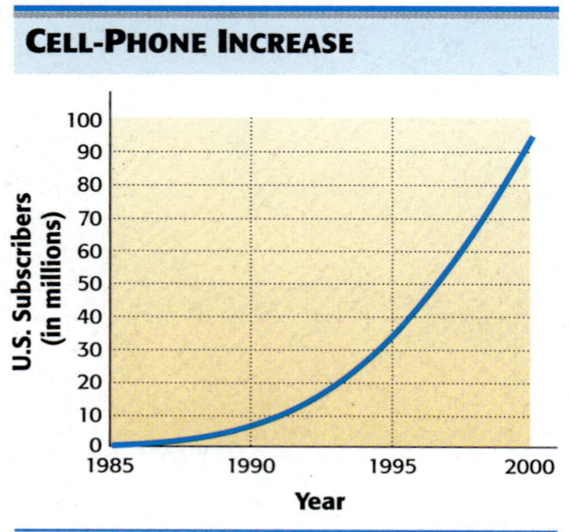

Source: Cellular Telecommunications Industry Association.

47. Replace x in the equation for each year shown in the figure, and give the value of y.

48. Use the figure to estimate the number of subscribers for each year shown.

49. Compare the values found from the equation with your estimates from the figure. Does the equation give a good approximation of the data from the figure? To the nearest million, in which year after 1985 is the approximation closest?

50. Use the equation to approximate the year when the number of cell-phone subscribers reached 70 million.

8.7 Complex Numbers

As we saw in Chapter 1, the set of real numbers includes many other number sets (the rational numbers, integers, and natural numbers, for example). In this section a new set of numbers is introduced that includes the set of real numbers, as well as numbers that are even roots of negative numbers, like $\sqrt{-2}$.

Objectives

1. Simplify numbers of the form $\sqrt{-b}$, where $b > 0$.
2. Recognize imaginary complex numbers.
3. Add and subtract complex numbers.
4. Multiply complex numbers.
5. Divide complex numbers.
6. Find powers of i.

1 Simplify numbers of the form $\sqrt{-b}$, where $b > 0$. The equation $x^2 + 1 = 0$ has no real number solution since any solution must be a number whose square is -1. In the set of real numbers, all squares are nonnegative numbers because the product of two positive numbers or two negative numbers is positive and $0^2 = 0$. To provide a solution for the equation $x^2 + 1 = 0$, a new number i is defined so that

$$i^2 = -1.$$

That is, i is a number whose square is -1, so $i = \sqrt{-1}$. This definition of i makes it possible to define any square root of a negative number as follows.

For any positive number b,

$$\sqrt{-b} = i\sqrt{b}.$$

Example 1 Simplifying Square Roots of Negative Numbers

Write each number as a product of a real number and i.

(a) $\sqrt{-100} = i\sqrt{100} = 10i$ **(b)** $-\sqrt{-36} = -i\sqrt{36} = -6i$

(c) $\sqrt{-2} = i\sqrt{2}$

CAUTION

It is easy to mistake $\sqrt{2}i$ for $\sqrt{2i}$, with the i under the radical. For this reason, we usually write $\sqrt{2}i$ as $i\sqrt{2}$, as in the definition of $\sqrt{-b}$.

Work Problem 1 at the Side.

When finding a product such as $\sqrt{-4} \cdot \sqrt{-9}$, we cannot use the product rule for radicals because it applies only to nonnegative radicands. For this reason, we change $\sqrt{-b}$ to the form $i\sqrt{b}$ before performing any multiplications or divisions. For example,

$$\begin{aligned} \sqrt{-4} \cdot \sqrt{-9} &= i\sqrt{4} \cdot i\sqrt{9} \\ &= i \cdot 2 \cdot i \cdot 3 \\ &= 6i^2 \\ &= 6(-1) \quad \text{Substitute: } i^2 = -1. \\ &= -6. \end{aligned}$$

1 Write each number as a product of a real number and i.

(a) $\sqrt{-16}$

(b) $-\sqrt{-81}$

(c) $\sqrt{-7}$

Answers

1. (a) $4i$ **(b)** $-9i$ **(c)** $i\sqrt{7}$

CAUTION

Using the product rule for radicals *before* using the definition of $\sqrt{-b}$ gives a *wrong* answer. The preceding example shows that

$$\sqrt{-4} \cdot \sqrt{-9} = -6, \text{ but}$$
$$\sqrt{-4(-9)} = \sqrt{36} = 6,$$

so

$$\sqrt{-4} \cdot \sqrt{-9} \neq \sqrt{-4(-9)}.$$

Example 2 **Multiplying Square Roots of Negative Numbers**

Multiply.

(a) $\sqrt{-3} \cdot \sqrt{-7} = i\sqrt{3} \cdot i\sqrt{7}$
$= i^2\sqrt{3 \cdot 7}$
$= (-1)\sqrt{21}$ Substitute: $i^2 = -1$.
$= -\sqrt{21}$

(b) $\sqrt{-2} \cdot \sqrt{-8} = i\sqrt{2} \cdot i\sqrt{8}$
$= i^2\sqrt{2 \cdot 8}$
$= (-1)\sqrt{16}$
$= (-1)4$
$= -4$

(c) $\sqrt{-5} \cdot \sqrt{6} = i\sqrt{5} \cdot \sqrt{6} = i\sqrt{30}$

Work Problem 2 at the Side.

The methods used to find products also apply to quotients.

Example 3 **Dividing Square Roots of Negative Numbers**

Divide.

(a) $\dfrac{\sqrt{-75}}{\sqrt{-3}} = \dfrac{i\sqrt{75}}{i\sqrt{3}} = \sqrt{\dfrac{75}{3}} = \sqrt{25} = 5$

(b) $\dfrac{\sqrt{-32}}{\sqrt{8}} = \dfrac{i\sqrt{32}}{\sqrt{8}} = i\sqrt{\dfrac{32}{8}} = i\sqrt{4} = 2i$

Work Problem 3 at the Side.

2 **Recognize imaginary complex numbers.** With the imaginary number i and the real numbers, a new set of numbers can be formed that includes the real numbers as a subset. The *complex numbers* are defined as follows.

Complex Number

If a and b are real numbers, then any number of the form $a + bi$ is called a **complex number.**

2 Multiply.

(a) $\sqrt{-7} \cdot \sqrt{-7}$

(b) $\sqrt{-5} \cdot \sqrt{-10}$

(c) $\sqrt{-15} \cdot \sqrt{2}$

3 Divide.

(a) $\dfrac{\sqrt{-32}}{\sqrt{-2}}$

(b) $\dfrac{\sqrt{-27}}{\sqrt{-3}}$

(c) $\dfrac{\sqrt{-40}}{\sqrt{10}}$

ANSWERS

2. (a) -7 **(b)** $-5\sqrt{2}$ **(c)** $i\sqrt{30}$
3. (a) 4 **(b)** 3 **(c)** $2i$

In the complex number $a + bi$, the number a is called the **real part** and b is called the **imaginary part.** When $b = 0$, $a + bi$ is a real number, so the real numbers are a subset of the complex numbers. Complex numbers with $b \neq 0$ are called **imaginary numbers.*** In spite of their name, imaginary numbers are very useful in applications, particularly in work with electricity.

The relationships among the various sets of numbers discussed in this book are shown in Figure 10.

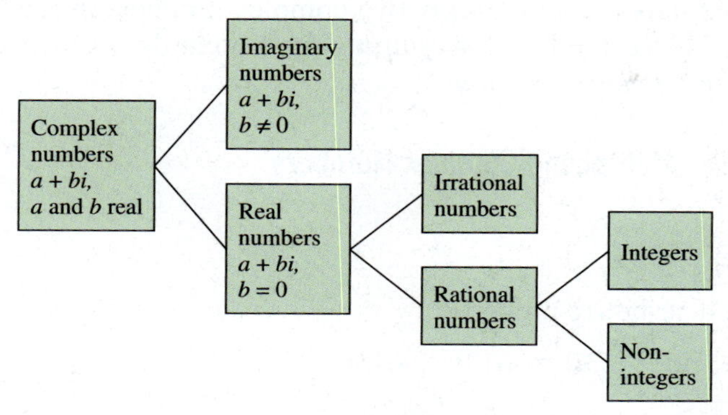

Figure 10

3 Add and subtract complex numbers. The commutative, associative, and distributive properties for real numbers are also valid for complex numbers. Thus, to add complex numbers, we add their real parts and add their imaginary parts.

Example 4 Adding Complex Numbers

Add.

(a) $(2 + 3i) + (6 + 4i)$

$= (2 + 6) + (3 + 4)i$ Commutative, associative, and distributive properties

$= 8 + 7i$

(b) $5 + (9 - 3i) = (5 + 9) - 3i$

$= 14 - 3i$

Work Problem 4 at the Side.

We subtract complex numbers by subtracting their real parts and subtracting their imaginary parts.

Example 5 Subtracting Complex Numbers

Subtract.

(a) $(6 + 5i) - (3 + 2i) = (6 - 3) + (5 - 2)i$

$= 3 + 3i$

(b) $(7 - 3i) - (8 - 6i) = (7 - 8) + [-3 - (-6)]i$

$= -1 + 3i$

(c) $(-9 + 4i) - (-9 + 8i) = (-9 + 9) + (4 - 8)i$

$= 0 - 4i$

$= -4i$

Work Problem 5 at the Side.

---

* Some texts define bi as the imaginary part of the complex number $a + bi$. Also, imaginary numbers are sometimes defined as complex numbers with $a = 0$ and $b \neq 0$.

4 Add.

(a) $(4 + 6i) + (-3 + 5i)$

(b) $(-1 + 8i) + (9 - 3i)$

5 Subtract.

(a) $(7 + 3i) - (4 + 2i)$

(b) $(-6 - i) - (-5 - 4i)$

(c) $8 - (3 - 2i)$

ANSWERS

4. (a) $1 + 11i$ **(b)** $8 + 5i$

5. (a) $3 + i$ **(b)** $-1 + 3i$ **(c)** $5 + 2i$

6 Multiply.

(a) $6i(4 + 3i)$

(b) $(6 - 4i)(2 + 4i)$

(c) $(3 - 2i)(3 + 2i)$

In Example 5(c), the answer was written as $0 - 4i$ and then as just $-4i$. A complex number written in the form $a + bi$, like $0 - 4i$, is in **standard form.** In this section, most answers will be given in standard form, but if a or b is 0, we consider answers such as a or bi to be in standard form.

4 Multiply complex numbers. We multiply complex numbers as we multiply polynomials. Complex numbers of the form $a + bi$ have the same form as binomials, so we multiply two complex numbers in standard form by using the FOIL method for multiplying binomials. (Recall that FOIL stands for *First, Outer, Inner, Last.*)

Example 6 Multiplying Complex Numbers

Multiply.

(a) $4i(2 + 3i)$

Use the distributive property.

$$\begin{aligned} 4i(2 + 3i) &= 4i(2) + 4i(3i) \\ &= 8i + 12i^2 \\ &= 8i + 12(-1) \quad \text{Substitute: } i^2 = -1. \\ &= -12 + 8i \end{aligned}$$

(b) $(3 + 5i)(4 - 2i)$

Use the FOIL method.

$$\begin{aligned} (3 + 5i)(4 - 2i) &= \underbrace{3(4)}_{\text{First}} + \underbrace{3(-2i)}_{\text{Outer}} + \underbrace{5i(4)}_{\text{Inner}} + \underbrace{5i(-2i)}_{\text{Last}} \\ &= 12 - 6i + 20i - 10i^2 \\ &= 12 + 14i - 10(-1) \quad \text{Substitute: } i^2 = -1. \\ &= 12 + 14i + 10 \\ &= 22 + 14i \end{aligned}$$

(c)
$$\begin{aligned} (2 + 3i)(1 - 5i) &= 2(1) + 2(-5i) + 3i(1) + 3i(-5i) \quad \text{FOIL} \\ &= 2 - 10i + 3i - 15i^2 \\ &= 2 - 7i - 15(-1) \\ &= 2 - 7i + 15 \\ &= 17 - 7i \end{aligned}$$

Work Problem 6 at the Side.

The two complex numbers $a + bi$ and $a - bi$ are called *conjugates* of each other. The product of a complex number and its conjugate is always a real number, as shown here.

$$\begin{aligned} (a + bi)(a - bi) &= a^2 - abi + abi - b^2i^2 \\ &= a^2 - b^2(-1) \end{aligned}$$

$$(a + bi)(a - bi) = a^2 + b^2$$

For example, $(3 + 7i)(3 - 7i) = 3^2 + 7^2 = 9 + 49 = 58$.

5 Divide complex numbers. The quotient of two complex numbers should be a complex number. To write the quotient as a complex number, we need to eliminate i in the denominator. We use conjugates to do this.

Answers

6. (a) $-18 + 24i$ **(b)** $28 + 16i$ **(c)** 13

Example 7 Dividing Complex Numbers

Find each quotient.

(a) $\dfrac{8 + 9i}{5 + 2i}$

Multiply both the numerator and denominator by the conjugate of the denominator. The conjugate of $5 + 2i$ is $5 - 2i$.

$$\frac{8 + 9i}{5 + 2i} = \frac{(8 + 9i)(5 - 2i)}{(5 + 2i)(5 - 2i)}$$

$$= \frac{40 - 16i + 45i - 18i^2}{5^2 + 2^2}$$

$$= \frac{58 + 29i}{29} \quad \text{Substitute: } i^2 = -1; \text{ combine terms.}$$

$$= \frac{29(2 + i)}{29} \quad \text{Factor the numerator.}$$

$$= 2 + i \quad \text{Lowest terms}$$

Notice that this is just like rationalizing a denominator. The final result is in standard form.

(b) $\dfrac{1 + i}{i}$

The conjugate of i is $-i$. Multiply both the numerator and denominator by $-i$.

$$\frac{1 + i}{i} = \frac{(1 + i)(-i)}{i(-i)}$$

$$= \frac{-i - i^2}{-i^2}$$

$$= \frac{-i - (-1)}{-(-1)} \quad \text{Substitute: } i^2 = -1.$$

$$= \frac{-i + 1}{1}$$

$$= 1 - i$$

Work Problem 7 at the Side.

Calculator Tip In Examples 4–7, we showed how complex numbers can be added, subtracted, multiplied, and divided using algebraic methods. Many current models of graphing calculators can perform these operations. Figure 11 shows how the computations in parts of Examples 4–7 are carried out by a TI-83 calculator. It is important to use parentheses as shown.

```
(2+3i)+(6+4i)
                 8+7i
(6+5i)-(3+2i)
                 3+3i
```

```
(3+5i)(4-2i)
               22+14i
(8+9i)/(5+2i)
                  2+i
```

Figure 11

7 Find each quotient.

(a) $\dfrac{2 + i}{3 - i}$

(b) $\dfrac{6 + 2i}{4 - 3i}$

(c) $\dfrac{5}{3 - 2i}$

(d) $\dfrac{5 - i}{i}$

ANSWERS

7. **(a)** $\frac{1}{2} + \frac{1}{2}i$ **(b)** $\frac{18}{25} + \frac{26}{25}i$ **(c)** $\frac{15}{13} + \frac{10}{13}i$ **(d)** $-1 - 5i$

8 Find each power of i.

(a) i^{21}

(b) i^{36}

(c) i^{50}

(d) i^{-9}

6 Find powers of *i*. Because i^2 is defined to be -1, we can find higher powers of i as shown in the following examples.

$$i^3 = i \cdot i^2 = i(-1) = -i \qquad i^6 = i^2 \cdot i^4 = (-1) \cdot 1 = -1$$
$$i^4 = i^2 \cdot i^2 = (-1)(-1) = 1 \qquad i^7 = i^3 \cdot i^4 = (-i) \cdot 1 = -i$$
$$i^5 = i \cdot i^4 = i \cdot 1 = i \qquad i^8 = i^4 \cdot i^4 = 1 \cdot 1 = 1$$

As these examples suggest, the powers of i rotate through the four numbers i, -1, $-i$, and 1. Larger powers of i can be simplified by using the fact that $i^4 = 1$. For example,

$$i^{75} = (i^4)^{18} \cdot i^3 = 1^{18} \cdot i^3 = 1 \cdot i^3 = i^3 = -i.$$

This example suggests a quick method for simplifying larger powers of i.

Example 8 Simplifying Powers of *i*

Find each power of i.

(a) $i^{12} = (i^4)^3 = 1^3 = 1$

(b) $i^{39} = i^{36} \cdot i^3$
$= (i^4)^9 \cdot i^3$
$= 1^9 \cdot (-i)$
$= -i$

(c) $i^{-2} = \dfrac{1}{i^2} = \dfrac{1}{-1} = -1$

(d) $i^{-1} = \dfrac{1}{i}$

To simplify this quotient, multiply both the numerator and denominator by $-i$, the conjugate of i.

$$\frac{1}{i} = \frac{1(-i)}{i(-i)} = \frac{-i}{-i^2} = \frac{-i}{-(-1)} = \frac{-i}{1} = -i$$

Work Problem 8 at the Side.

Answers

8. (a) i (b) 1 (c) -1 (d) $-i$

8.7 EXERCISES

FOR EXTRA HELP

 Student's Solutions Manual

 MyMathLab.com

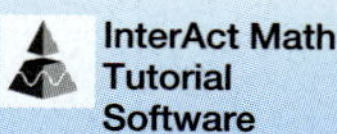 InterAct Math Tutorial Software

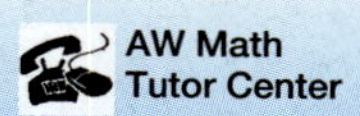 AW Math Tutor Center

 www.mathxl.com

Digital Video Tutor CD 7 Videotape 14

Decide whether each expression is equal to 1, -1, *i, or* $-i$.

1. $\sqrt{-1}$

2. $-i^2$

3. $\dfrac{1}{i}$

4. $(-i)^2$

5. Every real number is a complex number. Explain why this is so.

6. Not every complex number is a real number. Give an example of this, and explain why this statement is true.

Write each number as a product of a real number and i. Simplify all radical expressions. See Example 1.

7. $\sqrt{-169}$

8. $\sqrt{-225}$

9. $-\sqrt{-144}$

10. $-\sqrt{-196}$

11. $\sqrt{-5}$

12. $\sqrt{-21}$

13. $\sqrt{-48}$

14. $\sqrt{-96}$

Multiply or divide as indicated. See Examples 2 and 3.

15. $\sqrt{-15} \cdot \sqrt{-15}$

16. $\sqrt{-19} \cdot \sqrt{-19}$

17. $\sqrt{-4} \cdot \sqrt{-25}$

18. $\sqrt{-9} \cdot \sqrt{-81}$

19. $\dfrac{\sqrt{-300}}{\sqrt{-100}}$

20. $\dfrac{\sqrt{-40}}{\sqrt{-10}}$

21. $\dfrac{\sqrt{-75}}{\sqrt{3}}$

22. $\dfrac{\sqrt{-160}}{\sqrt{10}}$

Add or subtract as indicated. Write your answers in standard form. See Examples 4 and 5.

23. $(3 + 2i) + (-4 + 5i)$

24. $(7 + 15i) + (-11 + 14i)$

25. $(5 - i) + (-5 + i)$

26. $(-2 + 6i) + (2 - 6i)$

27. $(4 + i) - (-3 - 2i)$

28. $(9 + i) - (3 + 2i)$

29. $(-3 - 4i) - (-1 - 4i)$

30. $(-2 - 3i) - (-5 - 3i)$

31. $(-4 + 11i) + (-2 - 4i) + (7 + 6i)$

32. $(-1 + i) + (2 + 5i) + (3 + 2i)$

33. $[(7 + 3i) - (4 - 2i)] + (3 + i)$

34. $[(7 + 2i) + (-4 - i)] - (2 + 5i)$

35. Fill in the blank with the correct response: Because $(4 + 2i) - (3 + i) = 1 + i$, using the definition of subtraction we can check this to find that $(1 + i) + (3 + i) =$ ________.

36. Fill in the blank with the correct response: Because $\frac{-5}{2 - i} = -2 - i$, using the definition of division we can check this to find that $(-2 - i)(2 - i) =$ ________.

Multiply. See Example 6.

37. $(3i)(27i)$

38. $(5i)(125i)$

39. $(-8i)(-2i)$

40. $(-32i)(-2i)$

41. $5i(-6 + 2i)$

42. $3i(4 + 9i)$

43. $(4 + 3i)(1 - 2i)$

44. $(7 - 2i)(3 + i)$

45. $(4 + 5i)^2$

46. $(3 + 2i)^2$

47. $(12 + 3i)(12 - 3i)$

48. $(6 + 7i)(6 - 7i)$

49. **(a)** What is the conjugate of $a + bi$?

(b) If we multiply $a + bi$ by its conjugate, we get ________ + ________, which is always a real number.

50. Explain the procedure you would use to find the quotient

$$\frac{-1 + 5i}{3 + 2i}.$$

Write each quotient in the form $a + bi$. See Example 7.

51. $\dfrac{2}{1 - i}$

52. $\dfrac{29}{5 + 2i}$

53. $\dfrac{-7 + 4i}{3 + 2i}$

54. $\dfrac{-38 - 8i}{7 + 3i}$

55. $\dfrac{8i}{2 + 2i}$

56. $\dfrac{-8i}{1 + i}$

57. $\dfrac{2 - 3i}{2 + 3i}$

58. $\dfrac{-1 + 5i}{3 + 2i}$

RELATING CONCEPTS (Exercises 59–64) FOR INDIVIDUAL OR GROUP WORK

Consider these expressions:

Binomials	**Complex Numbers**
$x + 2, \quad 3x - 1$	$1 + 2i, \quad 3 - i.$

When we add, subtract, or multiply complex numbers in standard form, the rules are the same as those for the corresponding operations on binomials. That is, we add or subtract like terms, and we use FOIL to multiply. Division, however, is comparable to division by the sum or difference of radicals, where we multiply by the conjugate of the denominator to get a rational denominator. To express the quotient of two complex numbers in standard form, we also multiply by the conjugate of the denominator. ***Work Exercises 59–64 in order,*** *to better understand these ideas.*

59. (a) Add the two binomials.

(b) Add the two complex numbers.

60. (a) Subtract the second binomial from the first.

(b) Subtract the second complex number from the first.

61. (a) Multiply the two binomials.

(b) Multiply the two complex numbers.

62. (a) Rationalize the denominator: $\dfrac{\sqrt{3} - 1}{1 + \sqrt{2}}$.

(b) Write in standard form: $\dfrac{3 - i}{1 + 2i}$.

63. Explain why the answers for parts (a) and (b) in Exercise 61 do not correspond as the answers in Exercises 59 and 60 do.

64. Explain why the answers for parts (a) and (b) in Exercise 62 do not correspond as the answers in Exercises 59 and 60 do.

65. Recall that if $a \neq 0$, $\frac{1}{a}$ is called the reciprocal of a. Use this definition to express the reciprocal of $5 - 4i$ in the form $a + bi$.

66. Recall that if $a \neq 0$, a^{-1} is defined to be $\frac{1}{a}$. Use this definition to express $(4 - 3i)^{-1}$ in the form $a + bi$.

Find each power of i. See Example 8.

67. i^{18}

68. i^{26}

69. i^{89}

70. i^{45}

71. i^{96}

72. i^{48}

73. i^{-5}

74. i^{-17}

75. A student simplified i^{-18} as follows:

$$i^{-18} = i^{-18} \cdot i^{20} = i^{-18+20} = i^2 = -1.$$

Explain the mathematical justification for this correct work.

76. Explain why

$$(46 + 25i)(3 - 6i) \quad \text{and} \quad (46 + 25i)(3 - 6i)i^{12}$$

must be equal. (Do not actually perform the computation.)

Ohm's law for the current I in a circuit with voltage E, resistance R, capacitance reactance X_c, and inductive reactance X_L is

$$I = \frac{E}{R + (X_L - X_c)i}.$$

Use this law to work Exercises 77 and 78.

77. Find I if $E = 2 + 3i$, $R = 5$, $X_L = 4$, and $X_c = 3$.

78. Find E if $I = 1 - i$, $R = 2$, $X_L = 3$, and $X_c = 1$.

79. Show that $1 + 5i$ is a solution of

$$x^2 - 2x + 26 = 0.$$

80. Show that $3 + 2i$ is a solution of

$$x^2 - 6x + 13 = 0.$$

Chapter 8 Summary

Key Terms

	Term	Definition
8.1	**radicand, index**	In the expression $\sqrt[n]{a}$, a is the radicand and n is the index.
	radical	The expression $\sqrt[n]{a}$ is a radical.
	principal root	If a is positive and n is even, the principal nth root of a is the positive root.
	radical expression	A radical expression is an algebraic expression that contains radicals.
8.5	**rationalizing the denominator**	The process of removing radicals from the denominator so that the denominator contains only rational quantities is called rationalizing the denominator.
	conjugate	The conjugate of $a + b$ is $a - b$.
8.6	**radical equation**	A radical equation is an equation that includes one or more radical expressions with variables.
	extraneous solution	An extraneous solution of a radical equation is a solution of $x = a^2$ that is not a solution of $\sqrt{x} = a$.
8.7	**complex number**	A complex number is a number that can be written in the form $a + bi$, where a and b are real numbers.
	real part	The real part of $a + bi$ is a.
	imaginary part	The imaginary part of $a + bi$ is b.
	imaginary number	A complex number $a + bi$ with $b \neq 0$ is called an imaginary number.
	standard form (of a complex number)	A complex number is in standard form if it is written as $a + bi$.

New Symbols

Symbol	Meaning
$\sqrt{}$	radical sign
$\sqrt[n]{a}$	radical; principal nth root of a
$\pm$	positive or negative
$\approx$	is approximately equal to
$a^{1/n}$	a to the power $\frac{1}{n}$
$a^{m/n}$	a to the power $\frac{m}{n}$
i	a number whose square is -1

TEST YOUR WORD POWER

See how well you have learned the vocabulary in this chapter. Answers follow the Quick Review.

1. A **radicand** is
   (a) the index of a radical
   (b) the number or expression under the radical sign
   (c) the positive root of a number
   (d) the radical sign.

2. The **Pythagorean formula** states that, in a right triangle,
   (a) the sum of the measures of the angles is 180°
   (b) the sum of the lengths of the two shorter sides equals the length of the longest side
   (c) the longest side is opposite the right angle
   (d) the square of the length of the longest side equals the sum of the squares of the lengths of the two shorter sides.

3. A **hypotenuse** is
   (a) either of the two shorter sides of a triangle
   (b) the shortest side of a triangle
   (c) the side opposite the right angle in a triangle
   (d) the longest side in any triangle.

4. **Rationalizing the denominator** is the process of
   (a) eliminating fractions from a radical expression
   (b) changing the denominator of a fraction from a radical to a rational number
   (c) clearing a radical expression of radicals
   (d) multiplying radical expressions.

5. An **extraneous solution** is a solution
   (a) that does not satisfy the original equation
   (b) that makes an equation true
   (c) that makes an expression equal 0
   (d) that checks in the original equation.

6. A **complex number** is
   (a) a real number that includes a complex fraction
   (b) a zero multiple of i
   (c) a number of the form $a + bi$, where a and b are real numbers
   (d) the square root of -1.

QUICK REVIEW

Concepts	*Examples*
8.1 Radical Expressions and Graphs	
$\sqrt[n]{a} = b$ means $b^n = a$. $\sqrt[n]{a}$ is the principal nth root of a. $\sqrt[n]{a^n} = \lvert a \rvert$ if n is even. $\sqrt[n]{a^n} = a$ if n is odd.	The two square roots of 64 are $\sqrt{64} = 8$, the principal square root, and $-\sqrt{64} = -8$. $\sqrt[3]{-27} = -3$ $\quad$ $\sqrt[4]{(-2)^4} = \lvert -2 \rvert = 2$
Functions Defined by Radical Expressions The square root function with $f(x) = \sqrt{x}$ and the cube root function with $f(x) = \sqrt[3]{x}$ are two important functions defined by radical expressions.	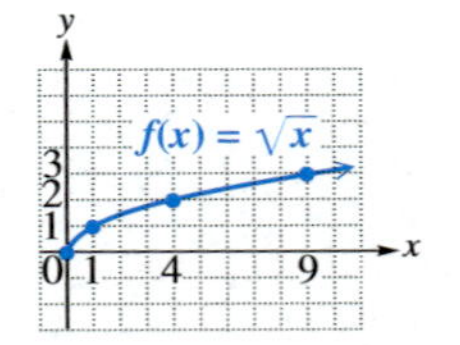 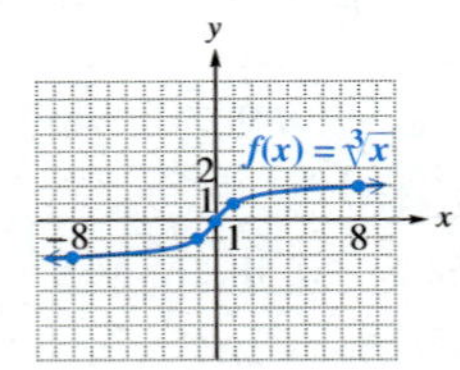
8.2 Rational Exponents $a^{1/n} = \sqrt[n]{a}$ whenever $\sqrt[n]{a}$ exists.	$81^{1/2} = \sqrt{81} = 9$ $\quad$ $-64^{1/3} = -\sqrt[3]{64} = -4$
If m and n are positive integers with m/n in lowest terms, then $a^{m/n} = (a^{1/n})^m$, provided that $a^{1/n}$ is a real number.	$8^{5/3} = (8^{1/3})^5 = 2^5 = 32$
All of the usual definitions and rules for exponents are valid for rational exponents.	$5^{-1/2} \cdot 5^{1/4} = 5^{-1/2+1/4} = 5^{-1/4} = \frac{1}{5^{1/4}}$ $\quad$ $(y^{2/5})^{10} = y^4$ $\frac{x^{-1/3}}{x^{-1/2}} = x^{-1/3-(-1/2)} = x^{-1/3+1/2} = x^{1/6}, \quad x > 0$

Concepts | *Examples*

8.3 Simplifying Radical Expressions

Product and Quotient Rules for Radicals

If $\sqrt[n]{a}$ and $\sqrt[n]{b}$ are real numbers and n is a natural number,

$$\sqrt[n]{a} \cdot \sqrt[n]{b} = \sqrt[n]{ab}$$

and

$$\sqrt[n]{\frac{a}{b}} = \frac{\sqrt[n]{a}}{\sqrt[n]{b}}, \quad b \neq 0.$$

$$\sqrt{3} \cdot \sqrt{7} = \sqrt{21}$$

$$\sqrt[5]{x^3y} \cdot \sqrt[5]{xy^2} = \sqrt[5]{x^4y^3}$$

$$\frac{\sqrt{x^5}}{\sqrt{x^4}} = \sqrt{\frac{x^5}{x^4}} = \sqrt{x}, \quad x > 0$$

Simplified Radical

1. The radicand has no factor raised to a power greater than or equal to the index.
2. The radicand has no fractions.
3. No denominator contains a radical.
4. Exponents in the radicand and the index of the radical have no common factors (except 1).

$$\sqrt{18} = \sqrt{9 \cdot 2} = 3\sqrt{2}$$

$$\sqrt[3]{54x^5y^3} = \sqrt[3]{27x^3y^3 \cdot 2x^2} = 3xy\sqrt[3]{2x^2}$$

$$\sqrt{\frac{7}{4}} = \frac{\sqrt{7}}{\sqrt{4}} = \frac{\sqrt{7}}{2}$$

$$\sqrt[9]{x^3} = x^{3/9} = x^{1/3} \quad \text{or} \quad \sqrt[3]{x}$$

Pythagorean Formula

If c is the length of the longest side of a right triangle and a and b are the lengths of the shorter sides, then $c^2 = a^2 + b^2$. The longest side is the hypotenuse and the two shorter sides are the legs of the triangle. The hypotenuse is opposite the right angle.

Find b for the triangle in the figure.

$$10^2 + b^2 = (2\sqrt{61})^2$$

$$b^2 = 4(61) - 100$$

$$b^2 = 144$$

$$b = 12$$

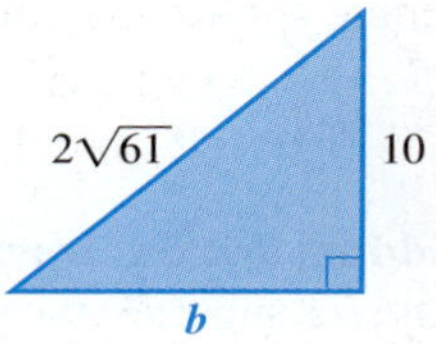

Distance Formula

The distance between (x_1, y_1) and (x_2, y_2) is

$$d = \sqrt{(x_2 - x_1)^2 + (y_2 - y_1)^2}.$$

The distance between $(3, -2)$ and $(-1, 1)$ is

$$\sqrt{(-1 - 3)^2 + [1 - (-2)]^2}$$

$$= \sqrt{(-4)^2 + 3^2} = \sqrt{16 + 9} = \sqrt{25} = 5.$$

8.4 Adding and Subtracting Radical Expressions

Only radical expressions with the same index and the same radicand may be combined.

$$3\sqrt{17} + 2\sqrt{17} - 8\sqrt{17} = (3 + 2 - 8)\sqrt{17}$$

$$= -3\sqrt{17}$$

$$\sqrt[3]{2} - \sqrt[3]{250} = \sqrt[3]{2} - 5\sqrt[3]{2}$$

$$= -4\sqrt[3]{2}$$

$$\left.\begin{array}{l}\sqrt{15} + \sqrt{30} \\ \sqrt{3} + \sqrt[3]{9}\end{array}\right\} \text{cannot be simplified further}$$

8.5 Multiplying and Dividing Radical Expressions

Multiply binomial radical expressions by using the FOIL method. Special products from Section 6.3 may apply.

$$(\sqrt{2} + \sqrt{7})(\sqrt{3} - \sqrt{6})$$

$$= \sqrt{6} - 2\sqrt{3} + \sqrt{21} - \sqrt{42} \qquad \sqrt{12} = 2\sqrt{3}$$

$$(\sqrt{5} - \sqrt{10})(\sqrt{5} + \sqrt{10}) = 5 - 10 = -5$$

$$(\sqrt{3} - \sqrt{2})^2 = 3 - 2\sqrt{3} \cdot \sqrt{2} + 2 = 5 - 2\sqrt{6}$$

Rationalize the denominator by multiplying both the numerator and denominator by the same expression.

$$\frac{\sqrt{7}}{\sqrt{5}} = \frac{\sqrt{7} \cdot \sqrt{5}}{\sqrt{5} \cdot \sqrt{5}} = \frac{\sqrt{35}}{5}$$

$$\frac{4}{\sqrt{5} - \sqrt{2}} = \frac{4(\sqrt{5} + \sqrt{2})}{(\sqrt{5} - \sqrt{2})(\sqrt{5} + \sqrt{2})}$$

$$= \frac{4(\sqrt{5} + \sqrt{2})}{5 - 2} = \frac{4(\sqrt{5} + \sqrt{2})}{3}$$

Concepts	*Examples*
8.6 Equations with Radical Expressions **Solving an Equation with Radicals** *Step 1* Isolate one radical on one side of the equation. *Step 2* Raise each side of the equation to a power that is the same as the index of the radical. *Step 3* Solve the resulting equation; if it still contains a radical, repeat Steps 1 and 2. *Step 4* Check all potential solutions in the *original* equation. Potential solutions that do not check are extraneous; they are not part of the solution set.	Solve $\sqrt{2x+3} - x = 0$. $\sqrt{2x+3} = x$ $(\sqrt{2x+3})^2 = x^2$ $2x + 3 = x^2$ $x^2 - 2x - 3 = 0$ $(x-3)(x+1) = 0$ $x - 3 = 0$ or $x + 1 = 0$ $x = 3$ or $x = -1$ A check shows that 3 is a solution, but -1 is extraneous. The solution set is $\{3\}$.
8.7 Complex Numbers $i^2 = -1$, so $i = \sqrt{-1}$. For any positive number b, $\sqrt{-b} = i\sqrt{b}$. To multiply radicals with negative radicands, first change each factor to the form $i\sqrt{b}$, then multiply. The same procedure applies to quotients.	$\sqrt{-25} = i\sqrt{25} = 5i$ $\sqrt{-3} \cdot \sqrt{-27} = i\sqrt{3} \cdot i\sqrt{27}$ $= i^2\sqrt{81}$ $= -1 \cdot 9$ $= -9$ $\dfrac{\sqrt{-18}}{\sqrt{-2}} = \dfrac{i\sqrt{18}}{i\sqrt{2}} = \sqrt{\dfrac{18}{2}} = \sqrt{9} = 3$
Adding and Subtracting Complex Numbers Add (or subtract) the real parts and add (or subtract) the imaginary parts.	$(5 + 3i) + (8 - 7i) = 13 - 4i$ $(5 + 3i) - (8 - 7i) = -3 + 10i$
Multiplying and Dividing Complex Numbers Multiply complex numbers by using the FOIL method.	$(2 + i)(5 - 3i) = 10 - 6i + 5i - 3i^2$ $= 10 - i - 3(-1)$ $= 10 - i + 3$ $= 13 - i$
Divide complex numbers by multiplying the numerator and the denominator by the conjugate of the denominator.	$\dfrac{2}{3+i} = \dfrac{2(3-i)}{(3+i)(3-i)} = \dfrac{2(3-i)}{9-i^2}$ $= \dfrac{2(3-i)}{10} = \dfrac{3-i}{5}$

ANSWERS TO TEST YOUR WORD POWER

1. (b) *Example:* In $\sqrt{3xy}$, $3xy$ is the radicand. **2. (d)** *Example:* In a right triangle where $a = 6$, $b = 8$, and $c = 10$, $6^2 + 8^2 = 10^2$. **3. (c)** *Example:* In a right triangle where the sides measure 9, 12, and 15 units, the hypotenuse is the side with measure 15 units. **4. (b)** *Example:* To rationalize the denominator of $\dfrac{5}{\sqrt{3}+1}$, multiply both the numerator and denominator by $\sqrt{3} - 1$ to get $\dfrac{5(\sqrt{3}-1)}{2}$. **5. (a)** *Example:* The potential solution 2 is extraneous in $\sqrt{5q - 1} + 3 = 0$. **6. (c)** *Examples:* -5 (or $-5 + 0i$), $7i$ (or $0 + 7i$), and $\sqrt{2} - 4i$.

Chapter 8 Review Exercises

[8.1] *Find each real number root. Use a calculator as necessary.*

1. $\sqrt{1764}$ **2.** $-\sqrt{289}$ **3.** $-\sqrt{-841}$ **4.** $\sqrt[3]{216}$

5. $\sqrt[5]{-32}$ **6.** $\sqrt{x^2}$ **7.** $\sqrt[3]{x^3}$ **8.** $\sqrt[4]{x^{20}}$

Graph each function. Give the domain and the range.

9. $f(x) = \sqrt{x-1}$

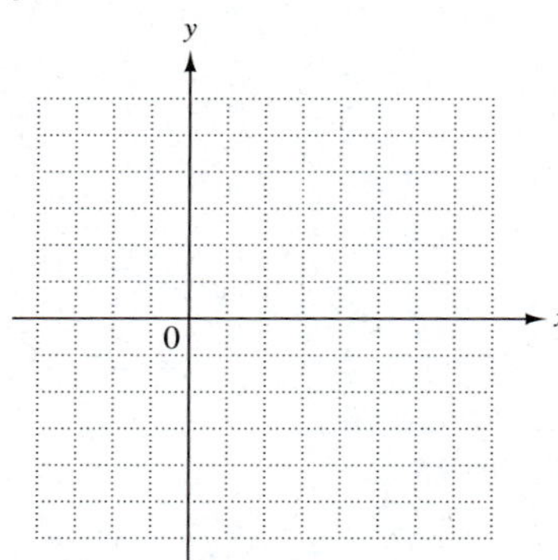

10. $f(x) = \sqrt[3]{x} + 4$

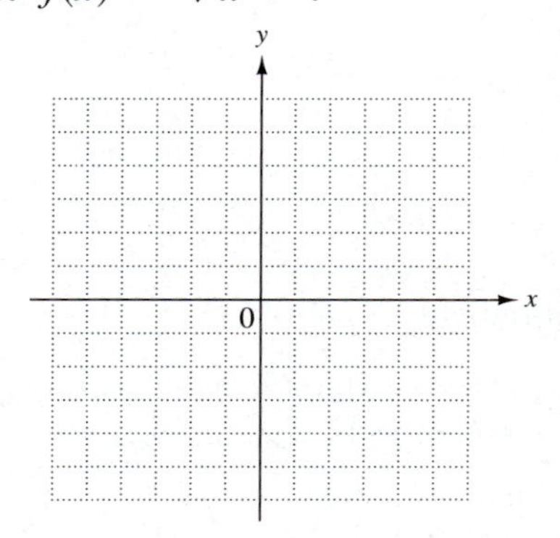

11. Under what conditions is $\sqrt[n]{a}$ not a real number?

12. If a is negative and n is even, what can be said about $a^{1/n}$?

Use a calculator to find a decimal approximation for each radical. Round to the nearest thousandth.

13. $\sqrt{40}$ **14.** $\sqrt{77}$ **15.** $\sqrt{310}$

16. According to an article in *The World Scanner Report* (August 1991), the distance D, in miles, to the horizon from an observer's point of view over water or "flat" earth is given by

$$D = \sqrt{2H},$$

where H is the height of the point of view, in feet. If a person whose eyes are 6 ft above ground level is standing at the top of a hill 44 ft above "flat" earth, approximately how far to the horizon will she be able to see?

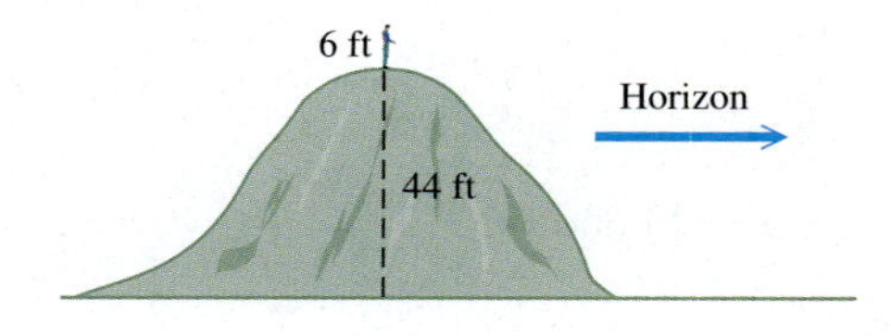

17. The time for one complete swing of a simple pendulum is given by

$$t = 2\pi\sqrt{\frac{L}{g}},$$

where t is time in seconds, L is the length of the pendulum in feet, and g, the force due to gravity, is about 32 ft per sec^2. Find the time of a complete swing of a 2-ft pendulum to the nearest tenth of a second.

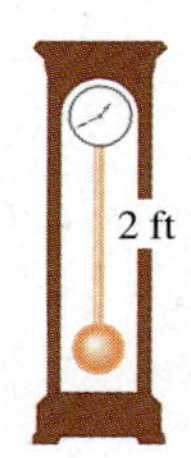

[8.2] *Find each real number root.*

18. $49^{1/2}$ **19.** $-8^{1/3}$ **20.** $(-16)^{1/4}$

21. Explain the relationship between the expressions $a^{m/n}$ and $\sqrt[n]{a^m}$.

Use the definitions and rules for exponents to simplify each expression. Assume that all variables represent positive real numbers.

22. $16^{5/4}$ **23.** $-8^{2/3}$ **24.** $-\left(\frac{36}{25}\right)^{3/2}$ **25.** $\left(-\frac{1}{8}\right)^{-5/3}$

26. $\left(\frac{81}{10,000}\right)^{-3/4}$ **27.** $7^{1/3} \cdot 7^{5/3}$ **28.** $\frac{96^{2/3}}{96^{-1/3}}$ **29.** $\frac{k^{2/3}k^{-1/2}k^{3/4}}{2(k^2)^{-1/4}}$

30. Write $2^{4/5}$ as a radical.

Use the rules of exponents to simplify each expression by first converting to rational exponents. Convert answers to radical form. Assume that all variables represent positive real numbers.

31. $\sqrt{3^{18}}$ **32.** $\sqrt{7^9}$ **33.** $\sqrt[3]{m^5} \cdot \sqrt[3]{m^8}$ **34.** $\sqrt[4]{k^2} \cdot \sqrt[4]{k^7}$

35. $\sqrt[3]{\sqrt{m}}$ **36.** $\sqrt[4]{16y^5}$ **37.** $\sqrt[5]{y} \cdot \sqrt[3]{y}$ **38.** $\frac{\sqrt[3]{y^2}}{\sqrt[4]{y}}$

[8.3] *Simplify each expression. Assume that all variables represent positive real numbers.*

39. $\sqrt{6} \cdot \sqrt{11}$ **40.** $\sqrt{5} \cdot \sqrt{r}$ **41.** $\sqrt[3]{6} \cdot \sqrt[3]{5}$ **42.** $\sqrt[4]{7} \cdot \sqrt[4]{3}$

43. $\sqrt{20}$ **44.** $-\sqrt{125}$ **45.** $\sqrt[3]{-108x^4y}$ **46.** $\sqrt[3]{64p^4q^6}$

47. $\sqrt{\frac{49}{81}}$ **48.** $\sqrt{\frac{y^3}{144}}$ **49.** $\sqrt[3]{\frac{m^{15}}{27}}$ **50.** $\sqrt[3]{\frac{r^2}{8}}$

51. $\frac{\sqrt[3]{2^4}}{\sqrt[4]{32}}$ **52.** $\frac{\sqrt{x}}{\sqrt[5]{x}}$

Find the distance between each pair of points.

53. (2, 7) and (−1, −4) **54.** (−3, −5) and (4, −3)

[8.4] *Perform the indicated operations. Assume that all variables represent positive real numbers.*

55. $2\sqrt{8} - 3\sqrt{50}$

56. $8\sqrt{80} - 3\sqrt{45}$

57. $-\sqrt{27y} + 2\sqrt{75y}$

58. $2\sqrt{54m^3} + 5\sqrt{96m^3}$

59. $3\sqrt[3]{54} + 5\sqrt[3]{16}$

60. $-6\sqrt[4]{32} + \sqrt[4]{512}$

[8.5] *Multiply, then simplify the products.*

61. $(\sqrt{3} + 1)(\sqrt{3} - 2)$

62. $(\sqrt{7} + \sqrt{5})(\sqrt{7} - \sqrt{5})$

63. $(3\sqrt{2} + 1)(2\sqrt{2} - 3)$

64. $(\sqrt{11} + 3\sqrt{5})(\sqrt{11} + 5\sqrt{5})$

65. $(\sqrt{13} - \sqrt{2})^2$

66. $(\sqrt{5} - \sqrt{7})^2$

Rationalize each denominator. Assume that all variables represent positive real numbers.

67. $\dfrac{-6\sqrt{3}}{\sqrt{2}}$

68. $\dfrac{3\sqrt{7p}}{\sqrt{y}}$

69. $-\sqrt[3]{\dfrac{9}{25}}$

70. $\sqrt[3]{\dfrac{108m^3}{n^5}}$

71. $\dfrac{1}{\sqrt{2} + \sqrt{7}}$

72. $\dfrac{-5}{\sqrt{6} - \sqrt{3}}$

[8.6] *Solve each equation.*

73. $\sqrt{8y + 9} = 5$

74. $\sqrt{2z - 3} - 3 = 0$

75. $\sqrt{3m + 1} = -1$

76. $\sqrt{7z + 1} = z + 1$

77. $3\sqrt{m} = \sqrt{10m - 9}$

78. $\sqrt{p^2 + 3p + 7} = p + 2$

79. $\sqrt{a + 2} - \sqrt{a - 3} = 1$

80. $\sqrt[3]{5m - 1} = \sqrt[3]{3m - 2}$

81. $\sqrt[4]{b + 6} = \sqrt[4]{2b}$

[8.7] *Write as a product of a real number and i.*

82. $\sqrt{-25}$

83. $\sqrt{-200}$

84. $\sqrt{-160}$

Perform the indicated operations. Write each imaginary number answer in standard form.

85. $(-2 + 5i) + (-8 - 7i)$

86. $(5 + 4i) - (-9 - 3i)$

87. $\sqrt{-5} \cdot \sqrt{-7}$

88. $\sqrt{-25} \cdot \sqrt{-81}$

89. $\dfrac{\sqrt{-72}}{\sqrt{-8}}$

90. $(2 + 3i)(1 - i)$

91. $(6 - 2i)^2$

92. $\dfrac{3 - i}{2 + i}$

93. $\dfrac{5 + 14i}{2 + 3i}$

Find each power of i.

94. i^{11}

95. i^{52}

96. i^{-13}

Mixed Review Exercises

Simplify. Assume that all variables represent positive real numbers.

97. $-\sqrt{169a^2b^4}$

98. $1000^{-2/3}$

99. $\dfrac{y^{-1/3} \cdot y^{5/6}}{y}$

100. $\dfrac{z^{-1/4}x^{1/2}}{z^{1/2}x^{-1/4}}$

101. $\sqrt[4]{k^{24}}$

102. $\sqrt[3]{54z^9t^8}$

103. $-5\sqrt{18} + 12\sqrt{72}$

104. $8\sqrt[3]{x^3y^2} - 2x\sqrt[3]{y^2}$

105. $(\sqrt{5} - \sqrt{3})(\sqrt{7} + \sqrt{3})$

106. $\dfrac{-1}{\sqrt{12}}$

107. $\sqrt[3]{\dfrac{12}{25}}$

108. $\dfrac{2\sqrt{z}}{\sqrt{z} - 2}$

109. $\sqrt{-49}$

110. $(4 - 9i) + (-1 + 2i)$

111. $\dfrac{\sqrt{50}}{\sqrt{-2}}$

Solve each equation.

112. $\sqrt{x + 4} = x - 2$

113. $\sqrt{6 + 2y} - 1 = \sqrt{7 - 2y}$

Solve each problem.

114. Carpenters stabilize wall frames with a diagonal brace as shown in the figure. The length of the brace is given by $L = \sqrt{H^2 + W^2}$. If the bottom of the brace is attached 9 ft from the corner and the brace is 12 ft long, how far up the corner post should it be nailed (to the nearest tenth of a foot)?

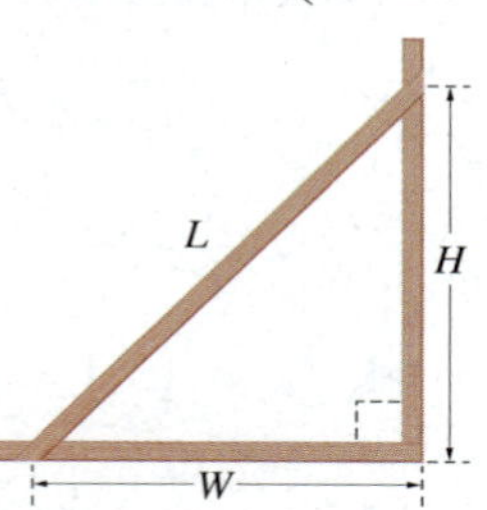

115. The sales in millions of dollars of Intel's flash memory chip are closely approximated by $f(x) = 11x^2 + 100$, where $x = 0$ represents the year 1995, $x = 5$ represents 2000, and so on. (*Source:* Cellular Telecommunications Industry Association, Intel Corp.)

(a) Use this function to approximate sales in 2001. Compare your answer with the company's estimate of \$490 million.

(b) According to the function, in what year were sales about \$270 million? Does this agree with the actual year, 1999?

Chapter 8 Test

Find each root. Use a calculator as necessary.

1. $-\sqrt{841}$

2. $125^{1/3}$

3. For $\sqrt{146.25}$, which choice gives the best estimate?

A. 10 **B.** 11 **C.** 12 **D.** 13

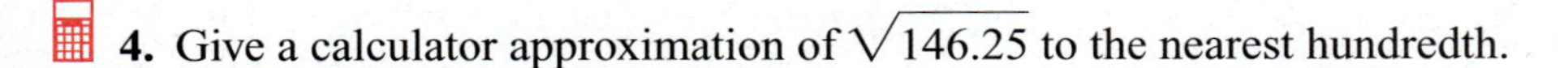

4. Give a calculator approximation of $\sqrt{146.25}$ to the nearest hundredth.

5. Graph the function defined by $f(x) = \sqrt{x + 6}$, and give the domain and the range.

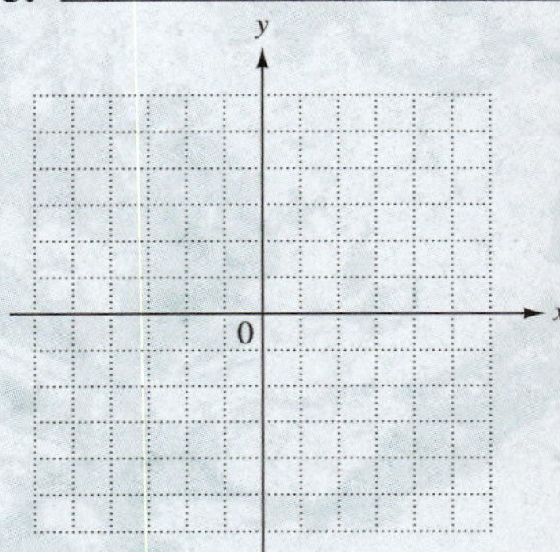

Simplify each expression. Assume that all variables represent positive real numbers.

6. $(-64)^{-4/3}$

7. $\dfrac{3^{2/5}x^{-1/4}y^{2/5}}{3^{-8/5}x^{7/4}y^{1/10}}$

8. $\sqrt{54x^5y^6}$

9. $\sqrt[4]{32a^7b^{13}}$

10. $\sqrt{2} \cdot \sqrt[3]{5}$

11. $3\sqrt{20} - 5\sqrt{80} + 4\sqrt{500}$

12. $(7\sqrt{5} + 4)(2\sqrt{5} - 1)$

1. ____________
2. ____________
3. ____________
4. ____________
5. ____________
6. ____________
7. ____________
8. ____________
9. ____________
10. ____________
11. ____________
12. ____________

13. ______________________

13. $\dfrac{-4}{\sqrt{7}+\sqrt{5}}$

14. ______________________

14. $\dfrac{-5}{\sqrt{40}}$

15. ______________________

15. $\dfrac{2}{\sqrt[3]{5}}$

16. ______________________

16. Find the distance between the points $(-3, 8)$ and $(2, 7)$.

17. ______________________

17. Use the Pythagorean formula to find the exact length of side b in the figure.

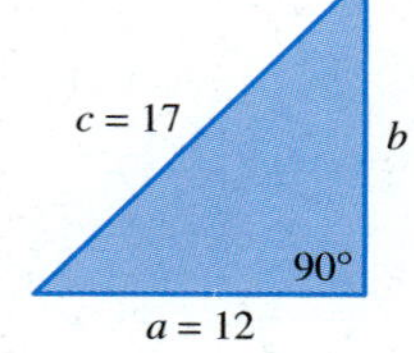

Solve each equation.

18. ______________________

18. $\sqrt[3]{5x} = \sqrt[3]{2x-3}$

19. ______________________

19. $\sqrt{7-x} + 5 = x$

Perform the indicated operations. Express answers in the form $a + bi$.

20. ______________________

20. $(-2 + 5i) - (3 + 6i) - 7i$

21. ______________________

21. $\dfrac{7+i}{1-i}$

22. ______________________

22. Simplify i^{35}.

Cumulative Review Exercises CHAPTERS 1–8

Solve each equation.

1. $7 - (4 + 3t) + 2t = -6(t - 2) - 5$

2. $|6x - 9| = |-4x + 2|$

Solve each inequality.

3. $-5 - 3(m - 2) < 11 - 2(m + 2)$

4. $1 + 4x > 5$ and $-2x > -6$

5. $-2 < 1 - 3y < 7$

6. Write an equation of the line through the points $(-4, 6)$ and $(7, -6)$.

7. The lines with equations $2x + 3y = 8$ and $6y = 4x + 16$ are
A. parallel, **B.** perpendicular, **C.** neither.

8. For the graph of $f(x) = -3x + 6$,
(a) what is the y-intercept?
(b) what is the x-intercept?

9. For many items, the cost per item to manufacture it varies inversely as the number made. Widgets are this type of item. It costs \$200 each to manufacture 1500 widgets. How much will it cost per widget to make 2500 widgets?

10. Graph the inequality $-2x + y < -6$.

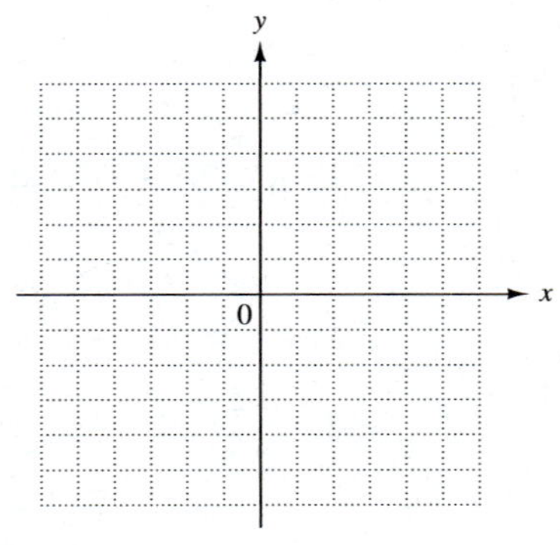

11. Find the measures of the marked angles.

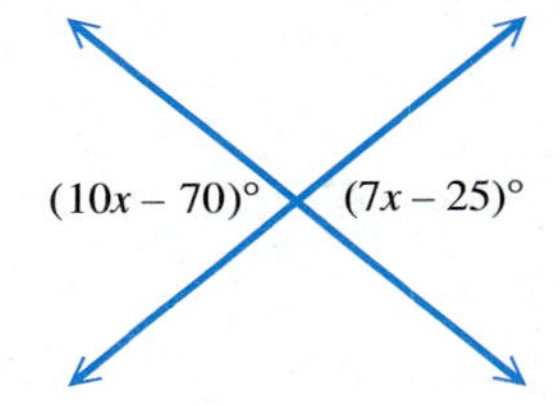

Solve each system.

12. $3x - y = 23$
$2x + 3y = 8$

13. $5x + 2y = 7$
$10x + 4y = 12$

14. $2x + y - z = 5$
$3x + 2y + z = 8$
$4x + 2y - 2z = 10$

15. In 1997, if you had sent five 2-oz letters and three 3-oz letters using first-class mail, it would have cost \$5.09. Sending three 2-oz letters and five 3-oz letters would have cost \$5.55. What was the 1997 postage rate for one 2-oz letter and for one 3-oz letter? (*Source:* U.S. Postal Service.)

Perform the indicated operations.

16. $(3k^3 - 5k^2 + 8k - 2) - (4k^3 + 11k + 7) + (2k^2 - 5k)$

17. $(8x - 7)(x + 3)$

18. $\dfrac{8z^3 - 16z^2 + 24z}{8z^2}$

19. $\dfrac{6y^4 - 3y^3 + 5y^2 + 6y - 9}{2y + 1}$

Factor each polynomial completely.

20. $2p^2 - 5pq + 3q^2$

21. $18k^4 + 9k^2 - 20$

22. $x^3 + 512$

Perform each operation and express answers in lowest terms.

23. $\dfrac{y^2 + y - 12}{y^3 + 9y^2 + 20y} \div \dfrac{y^2 - 9}{y^3 + 3y^2}$

24. $\dfrac{1}{x + y} + \dfrac{3}{x - y}$

Simplify each complex fraction.

25. $\dfrac{\dfrac{-6}{x - 2}}{\dfrac{8}{3x - 6}}$

26. $\dfrac{\dfrac{1}{a} - \dfrac{1}{b}}{\dfrac{a}{b} - \dfrac{b}{a}}$

Solve by factoring.

27. $2x^2 + 11x + 15 = 0$

28. $5t(t - 1) = 2(1 - t)$

Simplify.

29. $27^{-5/3}$

30. $\dfrac{x^{-2/3}}{x^{-3/4}}, \quad x \neq 0$

31. $8\sqrt{20} + 3\sqrt{80} - 2\sqrt{500}$

32. $\dfrac{-9}{\sqrt{80}}$

33. $\dfrac{4}{\sqrt{6} - \sqrt{5}}$

34. $\dfrac{12}{\sqrt[3]{2}}$

35. Find the distance between the points $(-4, 4)$ and $(-2, 9)$.

36. Solve $\sqrt{8x - 4} - \sqrt{7x + 2} = 0$.

Solve each problem.

37. The current of a river runs at 3 mph. Brent's boat can go 36 mi downstream in the same time that it takes to go 24 mi upstream. Find the speed of the boat in still water.

38. How many liters of pure alcohol must be mixed with 40 L of 18% alcohol to obtain a 22% alcohol solution?

39. A jar containing only dimes and quarters has 29 coins with a face value of $4.70. How many of each denomination are there?

40. Brenda rides her bike 4 mph faster than her husband, Chuck. If Brenda can ride 48 mi in the same time that Chuck can ride 24 mi, what are their speeds?

Quadratic Equations, Inequalities, and Graphs

In recent years, the number of U.S. companies filing for bankruptcy has been at its highest level since the recession of the early 1990s. One casualty of this trend is retailer Montgomery Ward. Started in 1872 as a mail-order catalog business, the company grew to include 250 stores in 30 states. After filing for Chapter 11 bankruptcy protection, the retailer closed for good in 2001. (*Source: USA Today,* December 29, 2000.) In Sections 9.4 and 9.5, we use *quadratic functions* to model the number of company bankruptcy filings.

9.1 The Square Root Property and Completing the Square

Objectives

1. Learn the square root property.
2. Solve quadratic equations of the form $(ax + b)^2 = c$ by using the square root property.
3. Solve quadratic equations by completing the square.
4. Solve quadratic equations with imaginary solutions.

We introduced quadratic equations in Section 6.9. Recall that a *quadratic equation* is defined as follows.

Quadratic Equation

An equation that can be written in the form

$$ax^2 + bx + c = 0,$$

where a, b, and c are real numbers, with $a \neq 0$, is a **quadratic equation.** The given form is called **standard form.**

A quadratic equation is a *second-degree equation*, that is, an equation with a squared term and no terms of higher degree. For example,

$$4m^2 + 4m - 5 = 0 \quad \text{and} \quad 3x^2 = 4x - 8$$

are quadratic equations, with the first equation in standard form.

Work Problem 1 at the Side.

In Section 6.9 we used factoring and the zero-factor property to solve quadratic equations.

Zero-Factor Property

If two numbers have a product of 0, then at least one of the numbers must be 0. That is, if $ab = 0$, then $a = 0$ or $b = 0$.

We solved a quadratic equation such as $3x^2 - 5x - 28 = 0$ using the zero-factor property as follows.

$$3x^2 - 5x - 28 = 0$$

$$(3x + 7)(x - 4) = 0 \quad \text{Factor.}$$

$$3x + 7 = 0 \quad \text{or} \quad x - 4 = 0 \quad \text{Zero-factor property}$$

$$3x = -7 \quad \text{or} \quad x = 4 \quad \text{Solve each equation.}$$

$$x = -\frac{7}{3}$$

The solution set is $\{-\frac{7}{3}, 4\}$.

Work Problem 2 at the Side.

1 Learn the square root property. Although factoring is the simplest way to solve quadratic equations, not every quadratic equation can be solved easily by factoring. In this section and the next, we develop other methods of solving quadratic equations based on the following property.

Square Root Property

If x and b are complex numbers and $x^2 = b$, then

$$x = \sqrt{b} \quad \text{or} \quad x = -\sqrt{b}.$$

1 (a) Which of the following are quadratic equations?

A. $x + 2y = 0$

B. $x^2 - 8x + 16 = 0$

C. $2t^2 - 5t = 3$

D. $x^3 + x^2 + 4 = 0$

(b) Which quadratic equation identified in part (a) is in standard form?

2 Solve each equation by factoring.

(a) $x^2 + 3x + 2 = 0$

(b) $3m^2 = 3 - 8m$

(*Hint:* Remember to write the equation in standard form first.)

Answers

1. (a) B, C (b) B

2. (a) $\{-2, -1\}$ (b) $\left\{-3, \frac{1}{3}\right\}$

The following steps justify the square root property.

$$x^2 = b$$

$$x^2 - b = 0 \quad \text{Subtract } b.$$

$$(x - \sqrt{b})(x + \sqrt{b}) = 0 \quad \text{Factor.}$$

$$x - \sqrt{b} = 0 \quad \text{or} \quad x + \sqrt{b} = 0 \quad \text{Zero-factor property}$$

$$x = \sqrt{b} \quad \text{or} \quad x = -\sqrt{b} \quad \text{Solve each equation.}$$

CAUTION

Remember that if $b \neq 0$, using the square root property always produces *two* square roots, one positive and one negative.

Example 1 Using the Square Root Property

Solve each equation.

(a) $r^2 = 5$

By the square root property,

$$r = \sqrt{5} \quad \text{or} \quad r = -\sqrt{5},$$

and the solution set is $\{\sqrt{5}, -\sqrt{5}\}$.

(b) $4x^2 - 48 = 0$

Solve for x^2.

$$4x^2 - 48 = 0$$

$$4x^2 = 48 \quad \text{Add 48.}$$

$$x^2 = 12 \quad \text{Divide by 4.}$$

$$x = \sqrt{12} \quad \text{or} \quad x = -\sqrt{12} \quad \text{Square root property}$$

$$x = 2\sqrt{3} \quad \text{or} \quad x = -2\sqrt{3} \quad \sqrt{12} = \sqrt{4} \cdot \sqrt{3} = 2\sqrt{3}$$

Check: $4x^2 - 48 = 0$ Original equation

$$4(2\sqrt{3})^2 - 48 = 0 \quad ? \qquad 4(-2\sqrt{3})^2 - 48 = 0 \quad ?$$

$$4(12) - 48 = 0 \quad ? \qquad 4(12) - 48 = 0 \quad ?$$

$$48 - 48 = 0 \quad ? \qquad 48 - 48 = 0 \quad ?$$

$$0 = 0 \quad \text{True} \qquad 0 = 0 \quad \text{True}$$

The solution set is $\{2\sqrt{3}, -2\sqrt{3}\}$.

Work Problem 3 at the Side.

NOTE

Recall that solutions such as those in Example 1 are sometimes abbreviated with the symbol $\pm$ (read "positive or negative"); with this symbol the solutions in Example 1 would be written $\pm\sqrt{5}$ and $\pm 2\sqrt{3}$.

3 Solve each equation.

(a) $m^2 = 64$

(b) $p^2 = 7$

(c) $3x^2 - 54 = 0$

ANSWERS

3. (a) $\{8, -8\}$ **(b)** $\{\sqrt{7}, -\sqrt{7}\}$ **(c)** $\{3\sqrt{2}, -3\sqrt{2}\}$

4 Solve the problem.

An expert marksman can hold a silver dollar at forehead level, drop it, draw his gun, and shoot the coin as it passes waist level. If the coin falls about 4 ft, use the formula in Example 2 to find the time that elapses between the dropping of the coin and the shot.

Example 2 **Using the Square Root Property in an Application**

Galileo Galilei (1564–1642) developed a formula for freely falling objects described by

$$d = 16t^2,$$

where d is the distance in feet that an object falls (disregarding air resistance) in t seconds, regardless of weight. Galileo dropped objects from the Leaning Tower of Pisa to develop this formula. If the Leaning Tower is about 180 ft tall, use Galileo's formula to determine how long it would take an object dropped from the tower to fall to the ground. (*Source:* Miller, Charles D., Heeren, Vern E., and Hornsby, John, *Mathematical Ideas, 9th Edition,* Addison-Wesley, 2001; *Microsoft Encarta Encyclopedia 2000.*)

We substitute 180 for d in Galileo's formula.

$$d = 16t^2$$

$$180 = 16t^2 \quad \text{Let } d = 180.$$

$$11.25 = t^2 \quad \text{Divide by 16.}$$

$$t = \sqrt{11.25} \quad \text{or} \quad t = -\sqrt{11.25} \quad \text{Square root property}$$

Since time cannot be negative, we discard the negative solution. In applied problems, we usually prefer approximations to exact values. Using a calculator, $\sqrt{11.25} \approx 3.4$ so $t \approx 3.4$. The object would fall to the ground in about 3.4 sec.

Work Problem 4 at the Side.

2 Solve quadratic equations of the form $(ax + b)^2 = c$ by using the square root property.

To solve more complicated equations using the square root property, such as

$$(x - 5)^2 = 36,$$

substitute $(x - 5)^2$ for x^2 and 36 for b, to get

$$x - 5 = \sqrt{36} \quad \text{or} \quad x - 5 = -\sqrt{36}$$

$$x - 5 = 6 \quad \text{or} \quad x - 5 = -6$$

$$x = 11 \quad \text{or} \quad x = -1.$$

Check: $(x - 5)^2 = 36$ Original equation

$$(11 - 5)^2 = 36 \ ? \qquad (-1 - 5)^2 = 36 \ ?$$

$$6^2 = 36 \ ? \qquad (-6)^2 = 36 \ ?$$

$$36 = 36 \quad \text{True} \qquad 36 = 36 \quad \text{True}$$

Example 3 **Using the Square Root Property**

Solve $(2x - 3)^2 = 18$.

By the square root property,

$$2x - 3 = \sqrt{18} \quad \text{or} \quad 2x - 3 = -\sqrt{18}$$

$$2x = 3 + \sqrt{18} \quad \text{or} \quad 2x = 3 - \sqrt{18}$$

$$x = \frac{3 + \sqrt{18}}{2} \quad \text{or} \quad x = \frac{3 - \sqrt{18}}{2}$$

Continued on Next Page

Answers

4. .5 sec

$$x = \frac{3 + 3\sqrt{2}}{2} \quad \text{or} \quad x = \frac{3 - 3\sqrt{2}}{2}. \qquad \sqrt{18} = \sqrt{9 \cdot 2} = 3\sqrt{2}$$

We show the check for the first solution. The check for the second solution is similar.

Check:

$$(2x - 3)^2 = 18 \qquad \text{Original equation}$$

$$\left[2\left(\frac{3 + 3\sqrt{2}}{2}\right) - 3\right]^2 = 18 \quad ?$$

$$(3 + 3\sqrt{2} - 3)^2 = 18 \quad ?$$

$$(3\sqrt{2})^2 = 18 \quad ?$$

$$18 = 18 \qquad \text{True}$$

The solution set is $\left\{\frac{3 + 3\sqrt{2}}{2}, \frac{3 - 3\sqrt{2}}{2}\right\}$.

Work Problem 5 at the Side.

5 Solve each equation.

(a) $(x - 3)^2 = 25$

(b) $(3k + 1)^2 = 2$

(c) $(2r + 3)^2 = 8$

3 Solve quadratic equations by completing the square. We can use the square root property to solve *any* quadratic equation by writing it in the form $(x + k)^2 = n$. That is, we must write the left side of the equation as a perfect square trinomial that can be factored as $(x + k)^2$, the square of a binomial, and the right side must be a constant. Rewriting a quadratic equation in this form is called **completing the square.**

Recall that the perfect square trinomial

$$x^2 + 10x + 25$$

can be factored as $(x + 5)^2$. In the trinomial, the coefficient of x (the first-degree term) is 10 and the constant term is 25. Notice that if we take half of 10 and square it, we get the constant term, 25.

Coefficient of x ↓ Constant ↓

$$\left[\frac{1}{2}(10)\right]^2 = 5^2 = 25$$

Similarly, in

$$x^2 + 12x + 36, \quad \left[\frac{1}{2}(12)\right]^2 = 6^2 = 36,$$

and in

$$m^2 - 6m + 9, \quad \left[\frac{1}{2}(-6)\right]^2 = (-3)^2 = 9.$$

This relationship is true in general and is the idea behind completing the square.

Example 4 Solving a Quadratic Equation by Completing the Square

Solve $x^2 + 8x + 10 = 0$.

This quadratic equation cannot be solved easily by factoring, and it is not in the correct form to solve using the square root property. To solve it by completing the square, we need a perfect square trinomial on the left side of the equation. To get this form, we first subtract 10 from each side.

Continued on Next Page

Answers

5. **(a)** $\{-2, 8\}$

(b) $\left\{\frac{-1 + \sqrt{2}}{3}, \frac{-1 - \sqrt{2}}{3}\right\}$

(c) $\left\{\frac{-3 + 2\sqrt{2}}{2}, \frac{-3 - 2\sqrt{2}}{2}\right\}$

6 Solve $n^2 + 6n + 4 = 0$ by completing the square.

$$x^2 + 8x + 10 = 0$$

$$x^2 + 8x = -10 \quad \text{Subtract 10.}$$

We must add a constant to get a perfect square trinomial on the left. To find this constant, we take half the coefficient of the first-degree term and square the result.

$$\left[\frac{1}{2}(8)\right]^2 = 4^2 = 16 \leftarrow \text{Desired constant}$$

Now we add 16 to *each* side of the equation. (Why?)

$$x^2 + 8x + 16 = -10 + 16$$

Next we factor on the left side and add on the right.

$$(x + 4)^2 = 6$$

We can now use the square root property.

$$x + 4 = \sqrt{6} \quad \text{or} \quad x + 4 = -\sqrt{6}$$

$$x = -4 + \sqrt{6} \quad \text{or} \quad x = -4 - \sqrt{6}$$

Check:

$$x^2 + 8x + 10 = 0 \quad \text{Original equation}$$

$$(-4 + \sqrt{6})^2 + 8(-4 + \sqrt{6}) + 10 = 0 \quad ? \quad \text{Let } x = -4 + \sqrt{6}$$

$$16 - 8\sqrt{6} + 6 - 32 + 8\sqrt{6} + 10 = 0 \quad ?$$

$$0 = 0 \quad \text{True}$$

The check for the second solution is similar. The solution set is

$$\{-4 + \sqrt{6}, -4 - \sqrt{6}\}.$$

Work Problem 6 at the Side.

The procedure from Example 4 can be generalized. Use the following steps to solve $ax^2 + bx + c = 0$ $(a \neq 0)$ by completing the square.

Completing the Square

Step 1 **Divide by *a*.** If $a \neq 1$, divide each side by a.

Step 2 **Rewrite the equation.** Rewrite the equation so that terms with variables are on one side of the equals sign, and the constant is on the other side.

Step 3 **Square half the coefficient of *x*.** Take half the coefficient of x (the first-degree term) and square it.

Step 4 **Add the square to each side.**

Step 5 **Factor the perfect square trinomial.** One side should now be a perfect square trinomial. Factor it and write it as the square of a binomial. Simplify the other side.

Step 6 **Use the square root property.** Use the square root property to complete the solution.

NOTE

Steps 1 and 2 can be done in either order. With some equations, it is more convenient to do Step 2 first.

ANSWERS

6. $\{-3 + \sqrt{5}, -3 - \sqrt{5}\}$

Example 5 **Solving a Quadratic Equation with $a = 1$ by Completing the Square**

Solve $k^2 + 5k - 1 = 0$.

Follow the steps in the box. Since the coefficient of the squared term is 1, begin with Step 2.

Step 2 $$k^2 + 5k = 1 \quad \text{Add 1 to each side.}$$

Step 3 Take half the coefficient of the first-degree term and square the result.

$$\left[\frac{1}{2}(5)\right]^2 = \left(\frac{5}{2}\right)^2 = \frac{25}{4}$$

Step 4 Add the square to each side of the equation to get

$$k^2 + 5k + \frac{25}{4} = 1 + \frac{25}{4}.$$

Step 5 $$\left(k + \frac{5}{2}\right)^2 = \frac{29}{4} \quad \text{Factor on the left; add on the right.}$$

Step 6 $$k + \frac{5}{2} = \sqrt{\frac{29}{4}} \quad \text{or} \quad k + \frac{5}{2} = -\sqrt{\frac{29}{4}} \quad \text{Square root property}$$

$$k + \frac{5}{2} = \frac{\sqrt{29}}{2} \quad \text{or} \quad k + \frac{5}{2} = -\frac{\sqrt{29}}{2}$$

$$k = -\frac{5}{2} + \frac{\sqrt{29}}{2} \quad \text{or} \quad k = -\frac{5}{2} - \frac{\sqrt{29}}{2}$$

$$k = \frac{-5 + \sqrt{29}}{2} \quad \text{or} \quad k = \frac{-5 - \sqrt{29}}{2}$$

Check that the solution set is $\left\{\frac{-5 + \sqrt{29}}{2}, \frac{-5 - \sqrt{29}}{2}\right\}$.

Work Problem 7 at the Side.

Example 6 **Solving a Quadratic Equation with $a \neq 1$ by Completing the Square**

Solve $2x^2 - 4x - 5 = 0$.

First divide each side of the equation by 2 to get 1 as the coefficient of the squared term.

$$x^2 - 2x - \frac{5}{2} = 0 \quad \text{Step 1}$$

$$x^2 - 2x = \frac{5}{2} \quad \text{Step 2}$$

$$\left[\frac{1}{2}(-2)\right]^2 = (-1)^2 = 1 \quad \text{Step 3}$$

$$x^2 - 2x + 1 = \frac{5}{2} + 1 \quad \text{Step 4}$$

$$(x - 1)^2 = \frac{7}{2} \quad \text{Step 5}$$

$$x - 1 = \sqrt{\frac{7}{2}} \quad \text{or} \quad x - 1 = -\sqrt{\frac{7}{2}} \quad \text{Step 6}$$

Continued on Next Page

7 Solve each equation by completing the square.

(a) $x^2 + 2x - 10 = 0$

(b) $r^2 + 3r - 1 = 0$

Answers

7. (a) $\{-1 + \sqrt{11}, -1 - \sqrt{11}\}$

(b) $\left\{\frac{-3 + \sqrt{13}}{2}, \frac{-3 - \sqrt{13}}{2}\right\}$

$$x = 1 + \sqrt{\frac{7}{2}} \quad \text{or} \quad x = 1 - \sqrt{\frac{7}{2}}$$

$$x = 1 + \frac{\sqrt{14}}{2} \quad \text{or} \quad x = 1 - \frac{\sqrt{14}}{2} \qquad \text{Rationalize denominators.}$$

Add the two terms in each solution as follows:

$$1 + \frac{\sqrt{14}}{2} = \frac{2}{2} + \frac{\sqrt{14}}{2} = \frac{2 + \sqrt{14}}{2}$$

$$1 - \frac{\sqrt{14}}{2} = \frac{2}{2} - \frac{\sqrt{14}}{2} = \frac{2 - \sqrt{14}}{2}.$$

Check that the solution set is $\left\{\frac{2 + \sqrt{14}}{2}, \frac{2 - \sqrt{14}}{2}\right\}$.

Work Problem 8 at the Side.

8 Solve each equation by completing the square.

(a) $2r^2 - 4r + 1 = 0$

(b) $3z^2 - 6z - 2 = 0$

(c) $8x^2 - 4x - 2 = 0$

4 Solve quadratic equations with imaginary solutions. So far, all the equations we have solved using the square root property have had two real solutions. In the equation $x^2 = b$, if $b < 0$, there will be two imaginary solutions.

Example 7 Solving Quadratic Equations with Imaginary Solutions

Solve each equation.

(a) $x^2 = -15$

$$x = \sqrt{-15} \quad \text{or} \quad x = -\sqrt{-15} \qquad \text{Square root property}$$

$$x = i\sqrt{15} \quad \text{or} \quad x = -i\sqrt{15} \qquad \sqrt{-1} = i$$

The solution set is $\{i\sqrt{15}, -i\sqrt{15}\}$.

(b) $(t + 2)^2 = -16$

$$t + 2 = \sqrt{-16} \quad \text{or} \quad t + 2 = -\sqrt{-16} \qquad \text{Square root property}$$

$$t + 2 = 4i \quad \text{or} \quad t + 2 = -4i \qquad \sqrt{-16} = 4i$$

$$t = -2 + 4i \quad \text{or} \quad t = -2 - 4i$$

The solution set is $\{-2 + 4i, -2 - 4i\}$.

(c) $x^2 + 2x + 7 = 0$

Solve by completing the square.

$$x^2 + 2x = -7 \qquad \text{Subtract 7.}$$

$$x^2 + 2x + 1 = -7 + 1 \qquad \left[\tfrac{1}{2}(2)\right]^2 = 1; \text{ add 1 to each side.}$$

$$(x + 1)^2 = -6 \qquad \text{Factor on the left; add on the right.}$$

$$x + 1 = \pm i\sqrt{6} \qquad \text{Square root property}$$

$$x = -1 \pm i\sqrt{6} \qquad \text{Subtract 1.}$$

The solution set is $\{-1 + i\sqrt{6}, -1 - i\sqrt{6}\}$.

Work Problem 9 at the Side.

9 Solve each equation.

(a) $x^2 = -17$

(b) $(k + 5)^2 = -100$

(c) $5t^2 - 15t + 12 = 0$

NOTE

The procedure for completing the square is also used in other areas of mathematics. For example, we will use it in Section 9.6 when we graph quadratic equations and again in Chapter 11 when we work with circles.

ANSWERS

8. (a) $\left\{\frac{2 + \sqrt{2}}{2}, \frac{2 - \sqrt{2}}{2}\right\}$
(b) $\left\{\frac{3 + \sqrt{15}}{3}, \frac{3 - \sqrt{15}}{3}\right\}$
(c) $\left\{\frac{1 + \sqrt{5}}{4}, \frac{1 - \sqrt{5}}{4}\right\}$

9. (a) $\{i\sqrt{17}, -i\sqrt{17}\}$
(b) $\{-5 + 10i, -5 - 10i\}$
(c) $\left\{\frac{15 + i\sqrt{15}}{10}, \frac{15 - i\sqrt{15}}{10}\right\}$

9.1 EXERCISES

FOR EXTRA HELP

Student's Solutions Manual

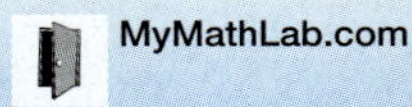
MyMathLab.com

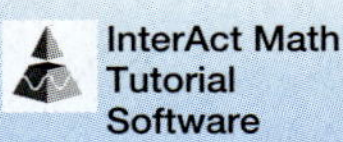
InterAct Math Tutorial Software

AW Math Tutor Center

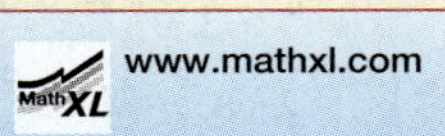
www.mathxl.com

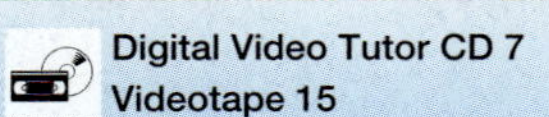
Digital Video Tutor CD 7 Videotape 15

1. A student was asked to solve the quadratic equation $x^2 = 16$ and did not get full credit for the solution set $\{4\}$. Why?

2. Why can't the zero-factor property be used to solve every quadratic equation?

3. Give a one-sentence description or explanation of each of the following.

(a) Quadratic equation in standard form

(b) Zero-factor property

(c) Square root property

4. What is wrong with the following "solution"?

$$x^2 - x - 2 = 5$$
$$(x - 2)(x + 1) = 5$$
$$x - 2 = 5 \quad \text{or} \quad x + 1 = 5 \qquad \text{Zero-factor property}$$
$$x = 7 \quad \text{or} \quad x = 4$$

Use the square root property to solve each equation. See Examples 1 and 3.

5. $x^2 = 81$

6. $z^2 = 225$

7. $t^2 = 17$

8. $k^2 = 19$

9. $m^2 = 32$

10. $x^2 = 54$

11. $t^2 - 20 = 0$

12. $p^2 - 50 = 0$

13. $3n^2 - 72 = 0$

14. $5z^2 - 200 = 0$

15. $(x + 2)^2 = 25$

16. $(t + 8)^2 = 9$

17. $(x - 4)^2 = 3$

18. $(x + 3)^2 = 11$

19. $(t + 5)^2 = 48$

20. $(m - 6)^2 = 27$

21. $(3k - 1)^2 = 7$

22. $(2x + 4)^2 = 10$

23. $(4p + 1)^2 = 24$

24. $(5k - 2)^2 = 12$

Solve Exercises 25 and 26 using Galileo's formula, $d = 16t^2$. Round answers to the nearest tenth. See Example 2.

25. The Gateway Arch in St. Louis, Missouri, is 630 ft tall. How long would it take an object dropped from the top of it to fall to the ground? (*Source: Home & Away*, November/December 2000.)

26. Mount Rushmore National Memorial in South Dakota features a sculpture of four of America's favorite presidents carved into the rim of the mountain, 500 ft above the valley floor. How long would it take a rock dropped from the top of the sculpture to fall to the ground? (*Source: Microsoft Encarta Encyclopedia 2000.*)

27. Of the two equations

$$(2x + 1)^2 = 5 \quad \text{and} \quad x^2 + 4x = 12,$$

one is more suitable for solving by the square root property, and the other is more suitable for solving by completing the square. Which method do you think most students would use for each equation?

28. Why would most students find the equation $x^2 + 4x = 20$ easier to solve by completing the square than the equation $5x^2 + 2x = 3$?

29. What would be the first step in solving $2x^2 + 8x = 9$ by completing the square?

30. *True* or *false:* Any quadratic equation can be solved by completing the square.

Determine the number that will complete the square to solve each equation after the constant term has been written on the right side. Do not actually solve. See Examples 4–6.

31. $x^2 + 4x - 2 = 0$

32. $t^2 + 2t - 1 = 0$

33. $x^2 + 10x + 18 = 0$

34. $x^2 + 8x + 11 = 0$

35. $3w^2 - w - 24 = 0$

36. $4z^2 - z - 39 = 0$

Solve each equation by completing the square. Use the results of Exercises 31–36 to solve Exercises 39–44. See Examples 4–6.

37. $x^2 - 2x - 24 = 0$

38. $m^2 - 4m - 32 = 0$

39. $x^2 + 4x - 2 = 0$

40. $t^2 + 2t - 1 = 0$

41. $x^2 + 10x + 18 = 0$

42. $x^2 + 8x + 11 = 0$

43. $3w^2 - w = 24$

44. $4z^2 - z = 39$

45. $2k^2 + 5k - 2 = 0$

46. $3r^2 + 2r - 2 = 0$

47. $5x^2 - 10x + 2 = 0$

48. $2x^2 - 16x + 25 = 0$

49. $9x^2 - 24x = -13$

50. $25n^2 - 20n = 1$

51. $z^2 - \frac{4}{3}z = -\frac{1}{9}$

52. $p^2 - \frac{8}{3}p = -1$

53. $.1x^2 - .2x - .1 = 0$
(*Hint:* First clear the decimals.)

54. $.1p^2 - .4p + .1 = 0$
(*Hint:* First clear the decimals.)

Find the imaginary solutions of each equation. See Example 7.

55. $x^2 = -12$

56. $y^2 = -18$

57. $(r - 5)^2 = -3$

58. $(t + 6)^2 = -5$

59. $(6k - 1)^2 = -8$

60. $(4m - 7)^2 = -27$

61. $m^2 + 4m + 13 = 0$

62. $t^2 + 6t + 10 = 0$

63. $3r^2 + 4r + 4 = 0$

64. $4x^2 + 5x + 5 = 0$

65. $-m^2 - 6m - 12 = 0$

66. $-k^2 - 5k - 10 = 0$

RELATING CONCEPTS (Exercises 67–72) FOR INDIVIDUAL OR GROUP WORK

The Greeks had a method of completing the square geometrically in which they literally changed a figure into a square. For example, to complete the square for $x^2 + 6x$, we begin with a square of side x, as in the figure. We add three rectangles of width 1 to the right side and the bottom to get a region with area $x^2 + 6x$. To fill in the corner (complete the square), we must add 9 1-by-1 squares as shown.

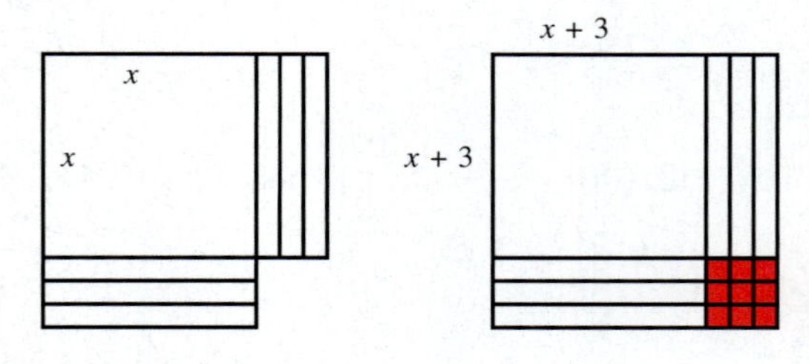

Work Exercises 67–72 in order.

67. What is the area of the original square?

68. What is the area of each strip?

69. What is the total area of the six strips?

70. What is the area of each small square in the corner of the second figure?

71. What is the total area of the small squares?

72. What is the area of the new, larger square?

9.2 The Quadratic Formula

Objectives

1. Solve quadratic equations using the quadratic formula.
2. Use the discriminant to determine the number and type of solutions.

The examples in the previous section showed that any quadratic equation can be solved by completing the square; however, completing the square can be tedious and time consuming. In this section, we complete the square to solve the general quadratic equation $ax^2 + bx + c = 0$, where a, b, and c are complex numbers and $a \neq 0$. The solution of this general equation gives a formula for finding the solution of any specific quadratic equation.

To solve $ax^2 + bx + c = 0$ by completing the square (assuming $a > 0$ for now), we follow the steps given in Section 9.1.

$$ax^2 + bx + c = 0$$

$$x^2 + \frac{b}{a}x + \frac{c}{a} = 0 \qquad \text{Divide by } a. \text{ (Step 1)}$$

$$x^2 + \frac{b}{a}x = -\frac{c}{a} \qquad \text{Subtract } \tfrac{c}{a}. \text{ (Step 2)}$$

$$\left[\frac{1}{2}\left(\frac{b}{a}\right)\right]^2 = \left(\frac{b}{2a}\right)^2 = \frac{b^2}{4a^2} \qquad \text{(Step 3)}$$

$$x^2 + \frac{b}{a}x + \frac{b^2}{4a^2} = -\frac{c}{a} + \frac{b^2}{4a^2} \qquad \text{Add } \tfrac{b^2}{4a^2} \text{ to each side. (Step 4)}$$

Write the left side as a perfect square, and rearrange the right side.

$$\left(x + \frac{b}{2a}\right)^2 = \frac{b^2}{4a^2} + \frac{-c}{a} \qquad \text{(Step 5)}$$

$$\left(x + \frac{b}{2a}\right)^2 = \frac{b^2}{4a^2} + \frac{-4ac}{4a^2} \qquad \text{Write with a common denominator.}$$

$$\left(x + \frac{b}{2a}\right)^2 = \frac{b^2 - 4ac}{4a^2} \qquad \text{Add fractions.}$$

$$x + \frac{b}{2a} = \sqrt{\frac{b^2 - 4ac}{4a^2}} \quad \text{or} \quad x + \frac{b}{2a} = -\sqrt{\frac{b^2 - 4ac}{4a^2}} \qquad \text{Square root property (Step 6)}$$

Since

$$\sqrt{\frac{b^2 - 4ac}{4a^2}} = \frac{\sqrt{b^2 - 4ac}}{\sqrt{4a^2}} = \frac{\sqrt{b^2 - 4ac}}{2a},$$

the result can be expressed as

$$x + \frac{b}{2a} = \frac{\sqrt{b^2 - 4ac}}{2a} \quad \text{or} \quad x + \frac{b}{2a} = \frac{-\sqrt{b^2 - 4ac}}{2a}$$

$$x = \frac{-b}{2a} + \frac{\sqrt{b^2 - 4ac}}{2a} \quad \text{or} \quad x = \frac{-b}{2a} - \frac{\sqrt{b^2 - 4ac}}{2a}$$

$$x = \frac{-b + \sqrt{b^2 - 4ac}}{2a} \quad \text{or} \quad x = \frac{-b - \sqrt{b^2 - 4ac}}{2a}.$$

If $a < 0$, the same two solutions are obtained. The result is the **quadratic formula,** often abbreviated as follows.

1 Identify the values of a, b, and c. (*Hint:* If necessary, write the equation in standard form first.) *Do not actually solve.*

(a) $-3q^2 + 9q - 4 = 0$

(b) $3x^2 = 6x + 2$

2 Solve $4x^2 - 11x - 3 = 0$ using the quadratic formula.

Quadratic Formula

The solutions of $ax^2 + bx + c = 0$ $(a \neq 0)$ are given by

$$x = \frac{-b \pm \sqrt{b^2 - 4ac}}{2a}.$$

CAUTION

In the quadratic formula, the square root is added to or subtracted from the value of $-b$ *before* dividing by $2a$.

1 **Solve quadratic equations using the quadratic formula.** To use the quadratic formula, first write the given equation in standard form $ax^2 + bx + c = 0$; then identify the values of a, b, and c and substitute them into the quadratic formula, as shown in the next examples.

Work Problem 1 at the Side.

Example 1 Using the Quadratic Formula (Rational Solutions)

Solve $6x^2 - 5x - 4 = 0$.

First, identify the values of a, b, and c of the general quadratic equation, $ax^2 + bx + c = 0$. Here a, the coefficient of the second-degree term, is 6, while b, the coefficient of the first-degree term, is -5, and the constant c is -4. Substitute these values into the quadratic formula.

$$x = \frac{-b \pm \sqrt{b^2 - 4ac}}{2a}$$

$$x = \frac{-(-5) \pm \sqrt{(-5)^2 - 4(6)(-4)}}{2(6)} \qquad a = 6, b = -5, c = -4$$

$$x = \frac{5 \pm \sqrt{25 + 96}}{12}$$

$$x = \frac{5 \pm \sqrt{121}}{12}$$

$$x = \frac{5 \pm 11}{12}$$

This last statement leads to two solutions, one from + and one from −.

$$x = \frac{5 + 11}{12} = \frac{16}{12} = \frac{4}{3} \quad \text{or} \quad x = \frac{5 - 11}{12} = \frac{-6}{12} = -\frac{1}{2}$$

Check each solution by substituting it in the original equation. The solution set is $\{-\frac{1}{2}, \frac{4}{3}\}$.

Work Problem 2 at the Side.

We could have used factoring to solve the equation in Example 1.

$$6x^2 - 5x - 4 = 0$$

$$(3x - 4)(2x + 1) = 0 \qquad \text{Factor.}$$

$$3x - 4 = 0 \quad \text{or} \quad 2x + 1 = 0 \qquad \text{Zero-factor property}$$

$$3x = 4 \quad \text{or} \quad 2x = -1 \qquad \text{Solve each equation.}$$

$$x = \frac{4}{3} \quad \text{or} \quad x = -\frac{1}{2} \qquad \text{Same solutions as in Example 1}$$

ANSWERS
1. (a) -3; 9; -4 (b) 3; -6; -2
2. $\{-\frac{1}{4}, 3\}$

When solving quadratic equations, it is a good idea to try factoring first. If the equation cannot be factored or if factoring is difficult, then use the quadratic formula. Later in this section, we will show a way to determine whether factoring can be used to solve a quadratic equation.

Example 2 **Using the Quadratic Formula (Irrational Solutions)**

Solve $4r^2 = 8r - 1$.

Rewrite the equation in standard form as

$$4r^2 - 8r + 1 = 0,$$

and identify $a = 4$, $b = -8$, and $c = 1$. Now use the quadratic formula.

$$r = \frac{-b \pm \sqrt{b^2 - 4ac}}{2a}$$

$$r = \frac{-(-8) \pm \sqrt{(-8)^2 - 4(4)(1)}}{2(4)} \qquad a = 4, b = -8, c = 1$$

$$= \frac{8 \pm \sqrt{64 - 16}}{8}$$

$$= \frac{8 \pm \sqrt{48}}{8}$$

$$= \frac{8 \pm 4\sqrt{3}}{8} \qquad \sqrt{48} = \sqrt{16} \cdot \sqrt{3} = 4\sqrt{3}$$

$$= \frac{4(2 \pm \sqrt{3})}{4(2)} \qquad \text{Factor.}$$

$$= \frac{2 \pm \sqrt{3}}{2} \qquad \text{Lowest terms}$$

The solution set is $\left\{\frac{2 + \sqrt{3}}{2}, \frac{2 - \sqrt{3}}{2}\right\}$.

CAUTION

When writing solutions in lowest terms, be sure to *factor first*; then divide out the common factor, as shown in the last two steps in Example 2.

Work Problem 3 at the Side.

Example 3 **Using the Quadratic Formula (Imaginary Solutions)**

Solve $(9q + 3)(q - 1) = -8$.

Every quadratic equation must be in standard form before we begin to solve it, whether we use factoring or the quadratic formula. To write this equation in standard form, we first multiply and collect all nonzero terms on the left.

$$(9q + 3)(q - 1) = -8$$

$$9q^2 - 6q - 3 = -8$$

$$9q^2 - 6q + 5 = 0 \qquad \text{Standard form}$$

Continued on Next Page

3 Solve each equation using the quadratic formula.

(a) $6x^2 + 4x - 1 = 0$

(b) $2k^2 + 19 = 14k$

ANSWERS

3. (a) $\left\{\frac{-2 + \sqrt{10}}{6}, \frac{-2 - \sqrt{10}}{6}\right\}$

(b) $\left\{\frac{7 + \sqrt{11}}{2}, \frac{7 - \sqrt{11}}{2}\right\}$

4 Solve each equation using the quadratic formula.

(a) $x^2 + x + 1 = 0$

(b) $(z + 2)(z - 6) = -17$

Now we identify $a = 9$, $b = -6$, and $c = 5$, and use the quadratic formula.

$$q = \frac{-(-6) \pm \sqrt{(-6)^2 - 4(9)(5)}}{2(9)}$$

$$= \frac{6 \pm \sqrt{-144}}{18}$$

$$= \frac{6 \pm 12i}{18} \qquad \sqrt{-144} = 12i$$

$$= \frac{6(1 \pm 2i)}{6(3)} \qquad \text{Factor.}$$

$$= \frac{1 \pm 2i}{3} \qquad \text{Lowest terms}$$

The solution set is $\left\{\frac{1 + 2i}{3}, \frac{1 - 2i}{3}\right\}$.

We could have written the solutions in Example 3 in the form $a + bi$, the standard form for complex numbers, as follows:

$$\frac{1 \pm 2i}{3} = \frac{1}{3} \pm \frac{2}{3}i.$$

Work Problem 4 at the Side.

2 Use the discriminant to determine the number and type of solutions. The solutions of the quadratic equation $ax^2 + bx + c = 0$ are given by

$$x = \frac{-b \pm \sqrt{b^2 - 4ac}}{2a}. \qquad \leftarrow \text{Discriminant}$$

If a, b, and c are integers, the type of solutions of a quadratic equation—that is, rational, irrational, or imaginary—is determined by the expression under the radical sign, $b^2 - 4ac$. Because it distinguishes among the three types of solutions, $b^2 - 4ac$ is called the **discriminant.** By calculating the discriminant before solving a quadratic equation, we can predict whether the solutions will be rational numbers, irrational numbers, or imaginary numbers. (This can be useful in an applied problem, for example, where irrational or imaginary solutions are not acceptable.)

Discriminant

The discriminant of $ax^2 + bx + c = 0$ is $b^2 - 4ac$. If a, b, and c are integers, then the number and type of solutions are determined as follows.

Discriminant	Number and Type of Solutions
Positive, and the square of an integer	Two rational solutions
Positive, but not the square of an integer	Two irrational solutions
Zero	One rational solution
Negative	Two imaginary solutions

ANSWERS

4. (a) $\left\{\frac{-1 + i\sqrt{3}}{2}, \frac{-1 - i\sqrt{3}}{2}\right\}$

(b) $\{2 + i, 2 - i\}$

Calculating the discriminant can also help you decide whether to solve a quadratic equation by factoring or by using the quadratic formula. If the discriminant is a perfect square (including 0), the equation can be solved by factoring. Otherwise, the quadratic formula should be used.

Example 4 Using the Discriminant

Find the discriminant. Use it to predict the number and type of solutions for each equation.

(a) $6x^2 - x - 15 = 0$

We find the discriminant by evaluating $b^2 - 4ac$.

$$b^2 - 4ac = (-1)^2 - 4(6)(-15) \quad a = 6, b = -1, c = -15$$
$$= 1 + 360$$
$$= 361$$

A calculator shows that $361 = 19^2$, a perfect square. Since a, b, and c are integers and the discriminant is a perfect square, there will be two rational solutions and the equation can be solved by factoring.

(b) $3m^2 - 4m = 5$

Rewrite the equation in standard form as $3m^2 - 4m - 5 = 0$ to find $a = 3$, $b = -4$, and $c = -5$.

$$b^2 - 4ac = (-4)^2 - 4(3)(-5)$$
$$= 16 + 60$$
$$= 76$$

Because 76 is positive but not the square of an integer and a, b, and c are integers, the equation will have two irrational solutions.

(c) $4x^2 + x + 1 = 0$

Since $a = 4$, $b = 1$, and $c = 1$, the discriminant is

$$1^2 - 4(4)(1) = -15.$$

Since the discriminant is negative and a, b, and c are integers, this quadratic equation will have two imaginary solutions.

(d) $4t^2 + 9 = 12t$

Rewrite the equation as $4t^2 - 12t + 9 = 0$ to find $a = 4$, $b = -12$, and $c = 9$. The discriminant is

$$b^2 - 4ac = (-12)^2 - 4(4)(9)$$
$$= 144 - 144$$
$$= 0.$$

Because the discriminant is 0, the quantity under the radical in the quadratic formula is 0, and there is only one rational solution. Again, the equation can be solved by factoring.

Work Problem 5 at the Side.

5 Find the discriminant. Use it to predict the number and type of solutions for each equation.

(a) $2x^2 + 3x = 4$

(b) $2x^2 + 3x + 4 = 0$

(c) $x^2 + 20x + 100 = 0$

(d) $15k^2 + 11k = 14$

(e) Which of the equations in parts (a)–(d) can be solved by factoring?

ANSWERS

5. (a) 41; two; irrational
(b) -23; two; imaginary
(c) 0; one; rational
(d) 961; two; rational **(e)** (c) and (d)

Focus on

Real-Data Applications

Almost, but Not Quite Right

Algebra is a precise language in which the correct order of operations must be followed and details must be watched. A slight change to a formula makes a big difference in the results.

- The Cadillac Bar in Houston, Texas, encourages patrons to write (tasteful) messages on their wall. Markers are provided, and the customers' creative juices flow. One person attempted to write the quadratic formula, which is shown in (1) below. Instead, that person wrote the formula shown in (2). It was not quite right.

(1) *Quadratic Formula:* $$\frac{-b \pm \sqrt{b^2 - 4ac}}{2a}$$

(2) *Formula on Wall of Cadillac Bar:* $$\frac{-b\sqrt{b^2 - 4ac}}{2a}$$

- An early version of Microsoft Word for Windows included the 1.0 edition of the Equation Editor. The documentation that explained how to use the Equation Editor used the following formula in the sample explanation. That author probably intended to use the quadratic formula, but again, it was not quite right.

(3) *Equation Editor Sample Formula:* $$-b \pm \frac{\sqrt{b^2 - 4ac}}{2a}$$

For Group Discussion

1. Evaluate the quadratic formula for $a = 2$, $b = 7$, and $c = -15$. Note that you have to evaluate the formula twice: first use the $+$ sign, then use the $-$ sign.

2. Using the order of operations, describe the steps to correctly evaluate the quadratic formula.

3. Explain how the Cadillac Bar formula used the order of operations differently than the correct quadratic formula.

4. Explain how the Microsoft Equation Editor formula used the order of operations differently than the correct quadratic formula.

9.2 EXERCISES

FOR EXTRA HELP

  Student's Solutions Manual

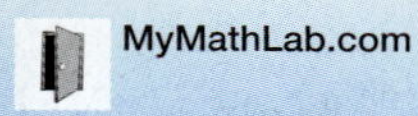 MyMathLab.com

 InterAct Math Tutorial Software

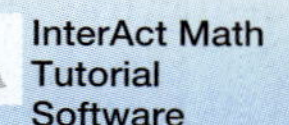 AW Math Tutor Center

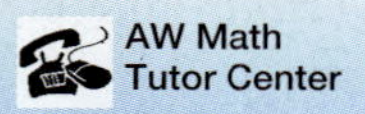 www.mathxl.com

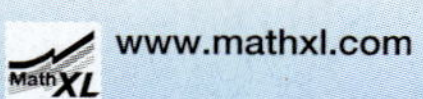 Digital Video Tutor CD 7 Videotape 15

1. A student wrote the following as the quadratic formula for solving $ax^2 + bx + c = 0, a \neq 0$:

$$x = -b \pm \frac{\sqrt{b^2 - 4ac}}{2a}.$$

Was this correct? Explain.

2. What is wrong with the following "solution" of $5x^2 - 5x + 1 = 0$?

$$x = \frac{5 \pm \sqrt{25 - 4(5)(1)}}{2(5)} \qquad a = 5, b = -5, c = 1$$

$$x = \frac{5 \pm \sqrt{5}}{10}$$

$$x = \frac{1}{2} \pm \sqrt{5}$$

Use the quadratic formula to solve each equation. (All solutions for these equations are real numbers.) See Examples 1 and 2.

3. $m^2 - 8m + 15 = 0$

4. $x^2 + 3x - 28 = 0$

5. $2k^2 + 4k + 1 = 0$

6. $2w^2 + 3w - 1 = 0$

7. $2x^2 - 2x = 1$

8. $9t^2 + 6t = 1$

9. $x^2 + 18 = 10x$

10. $x^2 - 4 = 2x$

11. $4k^2 + 4k - 1 = 0$

12. $4r^2 - 4r - 19 = 0$

13. $2 - 2x = 3x^2$

14. $26r - 2 = 3r^2$

15. $\frac{x^2}{4} - \frac{x}{2} = 1$
(*Hint:* First clear the fractions.)

16. $p^2 + \frac{p}{3} = \frac{1}{6}$
(*Hint:* First clear the fractions.)

17. $-2t(t + 2) = -3$

18. $-3x(x + 2) = -4$

19. $(r - 3)(r + 5) = 2$

20. $(k + 1)(k - 7) = 1$

Use the quadratic formula to solve each equation. (All solutions for these equations are imaginary numbers.) See Example 3.

21. $x^2 - 3x + 17 = 0$

22. $x^2 - 5x + 20 = 0$

23. $r^2 - 6r + 14 = 0$

24. $t^2 + 4t + 11 = 0$

25. $4x^2 - 4x = -7$

26. $9x^2 - 6x = -7$

27. $x(3x + 4) = -2$

28. $p(2p + 3) = -2$

Use the discriminant to determine whether the solutions for each equation are
A. *two rational numbers,* **B.** *one rational number,*
C. *two irrational numbers,* **D.** *two imaginary numbers.*
Do not actually solve. See Example 4.

29. $25x^2 + 70x + 49 = 0$

30. $4k^2 - 28k + 49 = 0$

31. $x^2 + 4x + 2 = 0$

32. $9x^2 - 12x - 1 = 0$

33. $3x^2 = 5x + 2$

34. $4x^2 = 4x + 3$

35. $3m^2 - 10m + 15 = 0$

36. $18x^2 + 60x + 82 = 0$

37. Using the discriminant, which equations in Exercises 29–36 can be solved by factoring?

38. Based on your answer in Exercise 37, solve the equation given in each exercise.
(a) Exercise 29 **(b)** Exercise 33

9.3 EQUATIONS QUADRATIC IN FORM

We have introduced four methods for solving quadratic equations written in standard form $ax^2 + bx + c = 0$. The following table lists some advantages and disadvantages of each method.

METHODS FOR SOLVING QUADRATIC EQUATIONS

Method	Advantages	Disadvantages
Factoring	This is usually the fastest method.	Not all polynomials are factorable; some factorable polynomials are hard to factor.
Square root property	This is the simplest method for solving equations of the form $(ax + b)^2 = c$.	Few equations are given in this form.
Completing the square	This method can always be used, although most people prefer the quadratic formula.	It requires more steps than other methods.
Quadratic formula	This method can always be used.	It is more difficult than factoring because of the square root, although calculators can simplify its use.

OBJECTIVES

1. Solve an equation with fractions by writing it in quadratic form.
2. Use quadratic equations to solve applied problems.
3. Solve an equation with radicals by writing it in quadratic form.
4. Solve an equation that is quadratic in form by substitution.

1 Solve an equation with fractions by writing it in quadratic form. A variety of nonquadratic equations can be written in the form of a quadratic equation and solved by using one of the methods in the table. As you solve the equations in this section, try to decide which method is best for each equation.

Example 1 Solving an Equation with Fractions that Leads to a Quadratic Equation

Solve $\frac{1}{x} + \frac{1}{x-1} = \frac{7}{12}$.

Clear fractions by multiplying each term by the least common denominator, $12x(x - 1)$. (Note that the domain must be restricted to $x \neq 0$ and $x \neq 1$.)

$$12x(x-1)\frac{1}{x} + 12x(x-1)\frac{1}{x-1} = 12x(x-1)\frac{7}{12}$$

$$12(x-1) + 12x = 7x(x-1)$$

$$12x - 12 + 12x = 7x^2 - 7x \quad \text{Distributive property}$$

$$24x - 12 = 7x^2 - 7x \quad \text{Combine terms.}$$

A quadratic equation must be in standard form before it can be solved by factoring or the quadratic formula. Combine and rearrange terms so that one side is 0. Then use factoring to solve the resulting equation.

$$0 = 7x^2 - 31x + 12 \quad \text{Standard form}$$

$$0 = (7x - 3)(x - 4) \quad \text{Factor.}$$

Using the zero-factor property gives the solutions $\frac{3}{7}$ and 4. Check by substituting these solutions in the original equation. The solution set is $\{\frac{3}{7}, 4\}$.

Work Problem 1 at the Side.

1 Solve each equation. Check your solutions.

(a) $\frac{5}{m} + \frac{12}{m^2} = 2$

(b) $\frac{2}{x} + \frac{1}{x-2} = \frac{5}{3}$

(c) $\frac{4}{m-1} + 9 = -\frac{7}{m}$

ANSWERS

1. (a) $\left\{-\frac{3}{2}, 4\right\}$ **(b)** $\left\{\frac{4}{5}, 3\right\}$ **(c)** $\left\{\frac{7}{9}, -1\right\}$

2 Use quadratic equations to solve applied problems. Earlier we solved distance-rate-time (or motion) problems that led to linear equations or rational equations. Now we can extend that work to motion problems that lead to quadratic equations. We continue to use the six-step problem-solving method from Chapter 2.

Example 2 Solving a Motion Problem

A riverboat for tourists averages 12 mph in still water. It takes the boat 1 hr, 4 min to go 6 mi upstream and return. Find the speed of the current. See Figure 1.

Figure 1

Step 1 **Read** the problem carefully.

Step 2 **Assign a variable.** Let $x =$ the speed of the current. The rate (or speed) upstream is the speed of the boat in still water less the speed of the current, or $12 - x$. Similarly, the speed downstream is $12 + x$. So,

$$12 - x = \text{the rate upstream;}$$
$$12 + x = \text{the rate downstream.}$$

Use the distance formula, $d = rt$, solved for time t.

$$t = \frac{d}{r}$$

This information can be used to complete a table.

	d	r	t
Upstream	6	$12 - x$	$\frac{6}{12 - x}$
Downstream	6	$12 + x$	$\frac{6}{12 + x}$

Times in hours

Step 3 **Write an equation.** The total time, 1 hr and 4 min, can be written as

$$1 + \frac{4}{60} = 1 + \frac{1}{15} = \frac{16}{15} \text{ hr.}$$

Because the time upstream plus the time downstream equals $\frac{16}{15}$ hr,

$$\underset{\text{Time upstream}}{\frac{6}{12 - x}} + \underset{\text{Time downstream}}{\frac{6}{12 + x}} = \underset{\text{Total time}}{\frac{16}{15}}.$$

Step 4 **Solve** the equation. Multiply each side by $15(12 - x)(12 + x)$, the LCD, and solve the resulting quadratic equation.

Continued on Next Page

$$15(12 + x)6 + 15(12 - x)6 = 16(12 - x)(12 + x)$$
$$90(12 + x) + 90(12 - x) = 16(144 - x^2)$$
$$1080 + 90x + 1080 - 90x = 2304 - 16x^2 \quad \text{Distributive property}$$
$$2160 = 2304 - 16x^2 \quad \text{Combine terms.}$$
$$16x^2 = 144$$
$$x^2 = 9 \quad \text{Divide by 16.}$$
$$x = 3 \quad \text{or} \quad x = -3 \quad \text{Square root property}$$

Step 5 **State the answer.** The speed of the current cannot be -3, so the answer is 3 mph.

Step 6 **Check** that this value satisfies the original problem.

CAUTION

As shown in Example 2, when a quadratic equation is used to solve an applied problem, sometimes only *one* answer satisfies the application. It is *always necessary* to check each answer in the words of the original problem.

Work Problem 2 at the Side.

In Chapter 7 we solved problems about work rates. Recall that a person's work rate is $\frac{1}{t}$ part of the job per hour, where t is the time in hours required to do the complete job. Thus, the part of the job the person will do in x hours is $\frac{1}{t}x$.

Example 3 Solving a Work Problem

It takes two carpet layers 4 hr to carpet a room. If each worked alone, one of them could do the job in 1 hr less time than the other. How long would it take each carpet layer to complete the job alone?

Step 1 **Read** the problem again. There will be two answers.

Step 2 **Assign a variable.** Let x represent the number of hours for the slower carpet layer to complete the job alone. Then the faster carpet layer could do the entire job in $x - 1$ hr. The slower person's rate is $\frac{1}{x}$, and the faster person's rate is $\frac{1}{x-1}$. Together, they do the job in 4 hr. Complete a table as shown.

	Rate	Time Working Together	Fractional Part of the Job Done
Slower Worker	$\frac{1}{x}$	4	$\frac{1}{x}(4)$
Faster Worker	$\frac{1}{x-1}$	4	$\frac{1}{x-1}(4)$

Sum is 1 whole job.

Continued on Next Page

2 Solve each problem.

(a) In 4 hr, Kerrie can go 15 mi upriver and come back. The speed of the current is 5 mph. Complete this table.

	d	*r*	*t*
Up			
Down			

(b) Find the speed of the boat from part (a) in still water.

(c) In $1\frac{3}{4}$ hr, Ken rows his boat 5 mi upriver and comes back. The speed of the current is 3 mph. How fast does Ken row?

Answers

2. (a) row 1: 15; $x - 5$; $\frac{15}{x-5}$;
row 2: 15; $x + 5$; $\frac{15}{x+5}$
(b) 10 mph **(c)** 7 mph

Step 3 **Write an equation.** The sum of the fractional parts done by the workers should equal 1 (the whole job).

$$\underset{\frac{4}{x}}{\text{Part done by slower worker}} + \underset{\frac{4}{x-1}}{\text{part done by faster worker}} = \underset{1}{\text{1 whole job.}}$$

Step 4 **Solve** the equation. Multiply each side by the LCD, $x(x-1)$.

$$4(x-1) + 4x = x(x-1)$$
$$4x - 4 + 4x = x^2 - x \quad \text{Distributive property}$$
$$0 = x^2 - 9x + 4 \quad \text{Standard form}$$

This equation cannot be solved by factoring, so use the quadratic formula.

$$x = \frac{9 \pm \sqrt{81-16}}{2} = \frac{9 \pm \sqrt{65}}{2} \quad a = 1, b = -9, c = 4$$

To the nearest tenth,

$$x = \frac{9+\sqrt{65}}{2} \approx 8.5 \quad \text{or} \quad x = \frac{9-\sqrt{65}}{2} \approx .5. \quad \text{Use a calculator.}$$

Step 5 **State the answer.** Only the solution 8.5 makes sense in the original problem. (Why?) Thus, the slower worker can do the job in about 8.5 hr and the faster in about $8.5 - 1 = 7.5$ hr.

Step 6 **Check** that these results satisfy the original problem.

Work Problem 3 at the Side.

3 Solve each problem. Round answers to the nearest tenth.

(a) Carlos can complete a certain lab test in 2 hr less time than Jaime can. If they can finish the job together in 2 hr, how long would it take each of them working alone?

	Rate	Time Working Together	Fractional Part of the Job Done
Carlos			
Jaime			

(b) Two chefs are preparing a banquet. One chef could prepare the banquet in 2 hr less time than the other. Together, they complete the job in 5 hr. How long would it take the faster chef working alone?

ANSWERS
3. (a) Jaime: 5.2 hr; Carlos: 3.2 hr **(b)** 9.1 hr

3 Solve an equation with radicals by writing it in quadratic form.

Example 4 Solving Radical Equations That Lead to Quadratic Equations

Solve each equation.

(a) $k = \sqrt{6k-8}$

This equation is not quadratic. However, squaring both sides of the equation gives a quadratic equation that can be solved by factoring.

$$k^2 = 6k - 8 \quad \text{Square both sides.}$$
$$k^2 - 6k + 8 = 0 \quad \text{Standard form}$$
$$(k-4)(k-2) = 0 \quad \text{Factor.}$$
$$k - 4 = 0 \quad \text{or} \quad k - 2 = 0 \quad \text{Zero-factor property}$$
$$k = 4 \quad \text{or} \quad k = 2 \quad \text{Potential solutions}$$

Recall from our work with radical equations in Section 8.6 that squaring both sides of an equation can introduce extraneous solutions that do not satisfy the original equation. Therefore, *all potential solutions must be checked in the original (not the squared) equation.*

Check: If $k = 4$, then

$$k = \sqrt{6k-8}$$
$$4 = \sqrt{6(4)-8} \quad ?$$
$$4 = \sqrt{16} \quad ?$$
$$4 = 4. \quad \text{True}$$

If $k = 2$, then

$$k = \sqrt{6k-8}$$
$$2 = \sqrt{6(2)-8} \quad ?$$
$$2 = \sqrt{4} \quad ?$$
$$2 = 2. \quad \text{True}$$

Both solutions check, so the solution set is $\{2, 4\}$.

Continued on Next Page

(b) $x + \sqrt{x} = 6$

$$\begin{aligned} \sqrt{x} &= 6 - x && \text{Get the radical alone on one side.} \\ x &= 36 - 12x + x^2 && \text{Square both sides.} \\ 0 &= x^2 - 13x + 36 && \text{Standard form} \\ 0 &= (x-4)(x-9) && \text{Factor.} \\ x - 4 = 0 \quad &\text{or} \quad x - 9 = 0 && \text{Zero-factor property} \\ x = 4 \quad &\text{or} \quad x = 9 && \text{Potential solutions} \end{aligned}$$

Check both potential solutions in the *original* equation.

If $x = 4$, then

$$x + \sqrt{x} = 6$$
$$4 + \sqrt{4} = 6 \quad ?$$
$$6 = 6. \quad \text{True}$$

If $x = 9$, then

$$x + \sqrt{x} = 6$$
$$9 + \sqrt{9} = 6 \quad ?$$
$$12 = 6. \quad \text{False}$$

Only the solution 4 checks, so the solution set is $\{4\}$.

Work Problem 4 at the Side.

4 Solve an equation that is quadratic in form by substitution. A nonquadratic equation that can be written in the form $au^2 + bu + c = 0$, for $a \neq 0$ and an algebraic expression u, is called **quadratic in form.**

Example 5 Solving Equations That Are Quadratic in Form

Solve each equation.

(a) $x^4 - 13x^2 + 36 = 0$

Because $x^4 = (x^2)^2$, we can write this equation in quadratic form with $u = x^2$ and $u^2 = x^4$.

$$\begin{aligned} x^4 - 13x^2 + 36 &= 0 \\ (x^2)^2 - 13x^2 + 36 &= 0 && x^4 = (x^2)^2 \\ u^2 - 13u + 36 &= 0 && \text{Let } u = x^2. \\ (u-4)(u-9) &= 0 && \text{Factor.} \\ u - 4 = 0 \quad \text{or} \quad u - 9 &= 0 && \text{Zero-factor property} \\ u = 4 \quad \text{or} \quad u &= 9 && \text{Solve.} \end{aligned}$$

To find x, we substitute x^2 for u.

$$x^2 = 4 \quad \text{or} \quad x^2 = 9$$
$$x = \pm 2 \quad \text{or} \quad x = \pm 3 \quad \text{Square root property}$$

The equation $x^4 - 13x^2 + 36 = 0$, a fourth-degree equation, has four solutions.* The solution set is $\{-3, -2, 2, 3\}$, which can be verified by substituting into the equation.

(b) $4x^4 + 1 = 5x^2$

Again, use the fact that $x^4 = (x^2)^2$ and let $u = x^2$ and $u^2 = x^4$.

$$4(x^2)^2 + 1 = 5x^2$$
$$4u^2 + 1 = 5u \quad \text{Let } u = x^2.$$

Continued on Next Page

* In general, an equation in which an nth-degree polynomial equals 0 has n solutions, although some of them may be repeated.

4 Solve each equation. Check your solutions.

(a) $x = \sqrt{7x - 10}$

(b) $2x = \sqrt{x} + 1$

ANSWERS

4. (a) $\{2, 5\}$ **(b)** $\{1\}$

5 Solve each equation. Check your solutions.

(a) $m^4 - 10m^2 + 9 = 0$

(b) $9k^4 - 37k^2 + 4 = 0$

$$4u^2 - 5u + 1 = 0 \quad \text{Standard form}$$
$$(4u - 1)(u - 1) = 0 \quad \text{Factor.}$$
$$4u - 1 = 0 \quad \text{or} \quad u - 1 = 0 \quad \text{Zero-factor property}$$
$$u = \frac{1}{4} \quad \text{or} \quad u = 1 \quad \text{Solve.}$$
$$x^2 = \frac{1}{4} \quad \text{or} \quad x^2 = 1 \quad \text{Substitute } x^2 \text{ for } u.$$
$$x = \pm\frac{1}{2} \quad \text{or} \quad x = \pm 1 \quad \text{Square root property}$$

Check that the solution set is $\{-1, -\frac{1}{2}, \frac{1}{2}, 1\}$.

NOTE

Some students prefer to solve equations like those in Example 5 by factoring directly. For example,

$$x^4 - 13x^2 + 36 = 0 \quad \text{Example 5(a) equation}$$
$$(x^2 - 9)(x^2 - 4) = 0 \quad \text{Factor.}$$
$$(x + 3)(x - 3)(x + 2)(x - 2) = 0. \quad \text{Factor again.}$$

Using the zero-factor property gives the same solutions obtained in Example 5(a).

Work Problem 5 at the Side.

Example 6 Solving Equations That Are Quadratic in Form

Solve each equation.

(a) $2(4m - 3)^2 + 7(4m - 3) + 5 = 0$

Because of the repeated quantity $4m - 3$, this equation is quadratic in form with $u = 4m - 3$. (Any letter except m could be used instead of u.)

$$2(4m - 3)^2 + 7(4m - 3) + 5 = 0$$
$$2u^2 + 7u + 5 = 0 \quad \text{Let } 4m - 3 = u.$$
$$(2u + 5)(u + 1) = 0 \quad \text{Factor.}$$
$$2u + 5 = 0 \quad \text{or} \quad u + 1 = 0 \quad \text{Zero-factor property}$$
$$u = -\frac{5}{2} \quad \text{or} \quad u = -1$$
$$4m - 3 = -\frac{5}{2} \quad \text{or} \quad 4m - 3 = -1 \quad \text{Substitute } 4m - 3 \text{ for } u.$$
$$4m = \frac{1}{2} \quad \text{or} \quad 4m = 2 \quad \text{Solve for } m.$$
$$m = \frac{1}{8} \quad \text{or} \quad m = \frac{1}{2}$$

Check that the solution set of the original equation is $\{\frac{1}{8}, \frac{1}{2}\}$.

Continued on Next Page

ANSWERS

5. (a) $\{-3, -1, 1, 3\}$ (b) $\left\{-2, -\frac{1}{3}, \frac{1}{3}, 2\right\}$

(b) $2a^{2/3} - 11a^{1/3} + 12 = 0$

Let $a^{1/3} = u$; then $a^{2/3} = (a^{1/3})^2 = u^2$. Substitute into the given equation.

$$2u^2 - 11u + 12 = 0 \qquad \text{Let } a^{1/3} = u;\ a^{2/3} = u^2.$$

$$(2u - 3)(u - 4) = 0 \qquad \text{Factor.}$$

$$2u - 3 = 0 \quad \text{or} \quad u - 4 = 0 \qquad \text{Zero-factor property}$$

$$u = \frac{3}{2} \quad \text{or} \quad u = 4$$

$$a^{1/3} = \frac{3}{2} \quad \text{or} \quad a^{1/3} = 4 \qquad u = a^{1/3}$$

$$(a^{1/3})^3 = \left(\frac{3}{2}\right)^3 \quad \text{or} \quad (a^{1/3})^3 = 4^3 \qquad \text{Cube each side.}$$

$$a = \frac{27}{8} \quad \text{or} \quad a = 64$$

Check that the solution set is $\{\frac{27}{8}, 64\}$.

CAUTION

A common error when solving problems like those in Examples 5 and 6 is to stop too soon. Once you have solved for u, remember to substitute and solve for the values of the *original* variable.

Work Problem 6 at the Side.

6 Solve each equation. Check your solutions.

(a) $5(r + 3)^2 + 9(r + 3) = 2$

(b) $4m^{2/3} = 3m^{1/3} + 1$

ANSWERS

6. (a) $\left\{-5, -\frac{14}{5}\right\}$ **(b)** $\left\{-\frac{1}{64}, 1\right\}$

Focus on Real-Data Applications

Smile! You're on Golden Ratio!

The Golden Ratio is the number phi, $\phi = \frac{1 + \sqrt{5}}{2}$. The Rhind Papyrus, dated 1600 B.C., referred to the **sacred ratio** used in building the Great Pyramids at Giza, Egypt. The ancient Greeks used ϕ in art and architecture, striving for a proportion that was the most pleasing to the eye. The Parthenon is the classic illustration of the use of ϕ in achieving that goal.

In a segment of length 1 that is divided into two parts, the Golden Ratio is defined as the proportion that equates the ratio of the whole segment to the larger segment and the ratio of the larger segment to the smaller segment. In the diagram, segment AC has length 1. Point B divides the segment so that AB has length x and BC has length $1 - x$. If x represents the length of the larger segment AB, then ϕ is the ratio $\frac{1}{x}$ and the Golden Proportion is

$$\frac{\text{whole}}{\text{larger}} = \frac{\text{larger}}{\text{smaller}} \quad \text{or} \quad \frac{1}{x} = \frac{x}{1 - x}.$$

x $1 - x$

A B C

The Golden Ratio is used in dentistry and medicine today. Eddy Levin, an English dentist, became interested in applications of the Golden Ratio, or Golden Proportion, to orthodontia and dentistry in 1978. His work is now a compulsory topic of study in U.S. dental schools. Viewed from the front, the "four front teeth, from central incisor to the premolar are the most significant part of the smile and they are in Golden Proportion to each other." He invented the Golden Mean Gauge, which is a tool that measures the Golden Proportion. (*Source:* www.goldenmeangauge.co.uk)

For Group Discussion

1. Write the Golden Proportion as a quadratic equation.

(a) Use the quadratic formula to solve this quadratic equation for x. Note that x must be a positive number since it is the length of AB.

(b) The Golden Ratio is $\phi = \frac{1}{x}$. Rationalize the denominator to write ϕ in exact form (using radicals).

(c) Write an approximate value for ϕ, rounded to 6 decimal places.

2. The Golden Ratio is a mathematically curious number. The reciprocal of ϕ is one less than ϕ, and the square of ϕ is one more than ϕ.

(a) Find $\frac{1}{\phi}$ and $\phi - 1$. (*Hint:* You have previously found the quantity $\frac{1}{\phi}$.)

(b) Find ϕ^2 and $\phi + 1$ in exact form.

9.3 EXERCISES

FOR EXTRA HELP

 Student's Solutions Manual
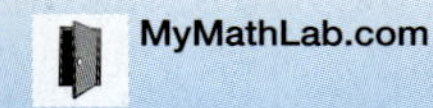 MyMathLab.com
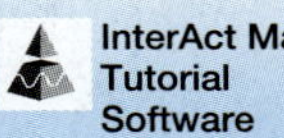 InterAct Math Tutorial Software
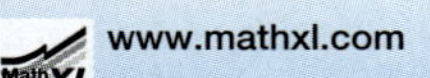 AW Math Tutor Center
www.mathxl.com
 Digital Video Tutor CD 7 Videotape 15

Based on the discussion and examples of this section, write a sentence describing the first step you would take to solve each equation. Do not actually solve.

1. $\frac{14}{x} = x - 5$

2. $\sqrt{1 + x} + x = 5$

3. $(r^2 + r)^2 - 8(r^2 + r) + 12 = 0$

4. $3t = \sqrt{16 - 10t}$

5. What is wrong with the following "solution"?

$$x = \sqrt{3x + 4}$$
$$x^2 = 3x + 4 \quad \text{Square both sides.}$$
$$x^2 - 3x - 4 = 0$$
$$(x - 4)(x + 1) = 0$$
$$x - 4 = 0 \quad \text{or} \quad x + 1 = 0$$
$$x = 4 \quad \text{or} \quad x = -1$$

Solution set: $\{4, -1\}$

6. What is wrong with the following "solution"?

$$2(m - 1)^2 - 3(m - 1) + 1 = 0$$
$$2u^2 - 3u + 1 = 0 \quad \text{Let } u = m - 1.$$
$$(2u - 1)(u - 1) = 0$$
$$2u - 1 = 0 \quad \text{or} \quad u - 1 = 0$$
$$u = \frac{1}{2} \quad \text{or} \quad u = 1$$

Solution set: $\left\{\frac{1}{2}, 1\right\}$

Solve each equation. Check your solutions. See Example 1.

7. $1 - \frac{3}{x} - \frac{28}{x^2} = 0$

8. $4 - \frac{7}{r} - \frac{2}{r^2} = 0$

9. $3 - \frac{1}{t} = \frac{2}{t^2}$

10. $1 + \frac{2}{k} = \frac{3}{k^2}$

11. $\frac{1}{x} + \frac{2}{x + 2} = \frac{17}{35}$

12. $\frac{2}{m} + \frac{3}{m + 9} = \frac{11}{4}$

13. $\frac{2}{x + 1} + \frac{3}{x + 2} = \frac{7}{2}$

14. $\frac{4}{3 - p} + \frac{2}{5 - p} = \frac{26}{15}$

15. $\frac{3}{2x} - \frac{1}{2(x + 2)} = 1$

16. $\frac{4}{3x} - \frac{1}{2(x + 1)} = 1$

ı hours to grade a set of papers, what is rate (in job per hour)?

18. A boat goes 20 mph in still water, and the rate of the current is t mph.

(a) What is the rate of the boat when it travels upstream?

(b) What is the rate of the boat when it travels downstream?

Solve each problem. See Examples 2 and 3.

19. On a windy day Yoshiaki found that he could go 16 mi downstream and then 4 mi back upstream at top speed in a total of 48 min. What was the top speed of Yoshiaki's boat if the current was 15 mph?

	d	r	t
Upstream	4	$x - 15$	
Downstream	16		

20. Lekesha flew her plane for 6 hr at a constant speed. She traveled 810 mi with the wind, then turned around and traveled 720 mi against the wind. The wind speed was a constant 15 mph. Find the speed of the plane.

	d	r	t
With Wind	810		
Against Wind	720		

21. In Canada, Medicine Hat and Cranbrook are 300 km apart. Harry rides his Honda 20 km per hr faster than Yoshi rides his Yamaha. Find Harry's average speed if he travels from Cranbrook to Medicine Hat in $1\frac{1}{4}$ hr less time than Yoshi. (*Source: State Farm Road Atlas.*)

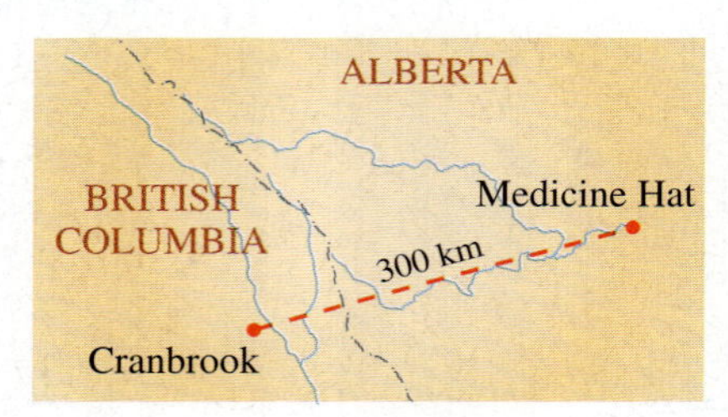

22. In California, the distance from Jackson to Lodi is about 40 mi, as is the distance from Lodi to Manteca. Rico drove from Jackson to Lodi during the rush hour, stopped in Lodi for a root beer, and then drove on to Manteca at 10 mph faster. Driving time for the entire trip was 88 min. Find his speed from Jackson to Lodi. (*Source: State Farm Road Atlas.*)

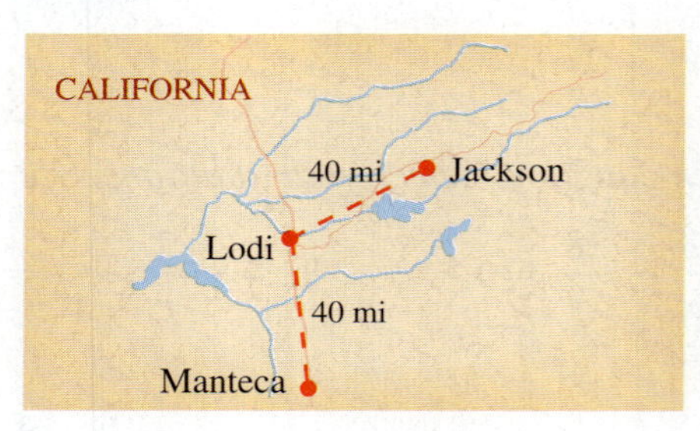

23. Working together, two people can cut a large lawn in 2 hr. One person can do the job alone in 1 hr less time than the other. How long (to the nearest tenth) would it take the faster person to do the job? (*Hint:* x is the time of the faster person.)

	Rate	Time Working Together	Fractional Part of the Job Done
Faster Worker	$\frac{1}{x}$	2	
Slower Worker		2	

24. A janitorial service provides two people to clean an office building. Working together, the two can clean the building in 5 hr. One person is new to the job and would take 2 hr longer than the other person to clean the building alone. How long (to the nearest tenth) would it take the new worker to clean the building alone?

	Rate	Time Working Together	Fractional Part of the Job Done
Faster Worker			
Slower Worker			

25. A washing machine can be filled in 6 min if both the hot and cold water taps are fully opened. Filling the washer with hot water alone takes 9 min longer than filling it with cold water alone. How long does it take to fill the washer with cold water?

26. Two pipes together can fill a large tank in 2 hr. One of the pipes, used alone, takes 3 hr longer than the other to fill the tank. How long would each pipe take to fill the tank alone?

Solve each equation. Check your solutions. See Example 4.

27. $2x = \sqrt{11x + 3}$

28. $4x = \sqrt{6x + 1}$

29. $3x = \sqrt{16 - 10x}$

30. $4t = \sqrt{8t + 3}$

31. $p - 2\sqrt{p} = 8$

32. $k + \sqrt{k} = 12$

33. $m = \sqrt{\dfrac{6 - 13m}{5}}$

34. $r = \sqrt{\dfrac{20 - 19r}{6}}$

Solve each equation. Check your solutions. See Examples 5 and 6.

35. $t^4 - 18t^2 + 81 = 0$

36. $x^4 - 8x^2 + 16 = 0$

37. $4k^4 - 13k^2 + 9 = 0$

38. $9x^4 - 25x^2 + 16 = 0$

39. $x^4 + 48 = 16x^2$

40. $z^4 = 17z^2 - 72$

41. $(x + 3)^2 + 5(x + 3) + 6 = 0$

42. $(k - 4)^2 + (k - 4) - 20 = 0$

43. $(t + 5)^2 + 6 = 7(t + 5)$

44. $3(m + 4)^2 - 8 = 2(m + 4)$

45. $2 + \dfrac{5}{3k - 1} = \dfrac{-2}{(3k - 1)^2}$

46. $3 - \dfrac{7}{2p + 2} = \dfrac{6}{(2p + 2)^2}$

47. $x^{2/3} + x^{1/3} - 2 = 0$

48. $x^{2/3} - 2x^{1/3} - 3 = 0$

49. $r^{2/3} + r^{1/3} - 12 = 0$

50. $3x^{2/3} - x^{1/3} - 24 = 0$

51. $2(1 + \sqrt{r})^2 = 13(1 + \sqrt{r}) - 6$

52. $(k^2 + k)^2 + 12 = 8(k^2 + k)$

Relating Concepts (Exercises 53–58) For Individual or Group Work

Consider the following equation, which contains variable expressions in the denominators.

$$\frac{x^2}{(x-3)^2} + \frac{3x}{x-3} - 4 = 0$$

Work Exercises 53–58 in order.

53. Why must 3 be excluded from the domain of this equation?

54. Multiply each side of the equation by the LCD, $(x - 3)^2$, and solve. There is only one solution—what is it?

55. Write the equation in a different manner so that it is quadratic in form using the expression $\frac{x}{x-3}$.

56. In your own words, explain why the expression $\frac{x}{x-3}$ cannot equal 1.

57. Solve the equation from Exercise 55 by making the substitution $t = \frac{x}{x-3}$. You should get two values for t. Why is one of them impossible for this equation?

58. Solve the equation $x^2(x - 3)^{-2} + 3x(x - 3)^{-1} - 4 = 0$ by letting $s = (x - 3)^{-1}$. You should get two values for s. Why is this impossible for this equation?

9.4 Formulas and Further Applications

Objectives

1. Solve formulas for variables involving squares and square roots.
2. Solve applied problems using the Pythagorean formula.
3. Solve applied problems using area formulas.
4. Solve applied problems using quadratic functions as models.

1 Solve formulas for variables involving squares and square roots. The methods presented earlier in this chapter and the previous one can be used to solve such formulas.

Example 1 Solving for Variables Involving Squares or Square Roots

Solve each formula for the given variable.

(a) $w = \dfrac{kFr}{v^2}$ for v

$$w = \frac{kFr}{v^2} \quad \text{Get } v \text{ alone on one side.}$$

$$v^2 w = kFr \qquad \text{Multiply by } v^2.$$

$$v^2 = \frac{kFr}{w} \qquad \text{Divide by } w.$$

$$v = \pm\sqrt{\frac{kFr}{w}} \qquad \text{Square root property}$$

$$v = \frac{\pm\sqrt{kFr}}{\sqrt{w}} \cdot \frac{\sqrt{w}}{\sqrt{w}} = \frac{\pm\sqrt{kFrw}}{w} \qquad \text{Rationalize the denominator.}$$

(b) $d = \sqrt{\dfrac{4A}{\pi}}$ for A

$$d = \sqrt{\frac{4A}{\pi}}$$

$$d^2 = \frac{4A}{\pi} \qquad \text{Square both sides.}$$

$$\pi d^2 = 4A \qquad \text{Multiply by } \pi.$$

$$\frac{\pi d^2}{4} = A \qquad \text{Divide by 4.}$$

Work Problem 1 at the Side.

1 Solve each formula for the given variable.

(a) $A = \pi r^2$ for r

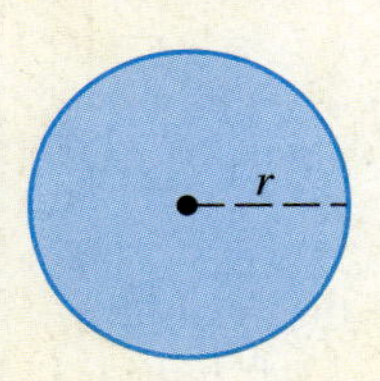

(b) $s = 30\sqrt{\dfrac{a}{p}}$ for a

NOTE

In many formulas like $v = \frac{\pm\sqrt{kFrw}}{w}$ in Example 1(a), we choose the positive value. In our work here, we will include both positive and negative values.

Example 2 Solving for a Squared Variable

Solve $s = 2t^2 + kt$ for t.

Since the equation has terms with t^2 and t, write it in standard form $ax^2 + bx + c = 0$, with t as the variable instead of x.

$$s = 2t^2 + kt$$

$$0 = 2t^2 + kt - s$$

Continued on Next Page

Answers

1. (a) $r = \dfrac{\pm\sqrt{A\pi}}{\pi}$ (b) $a = \dfrac{ps^2}{900}$

2 Solve $2t^2 - 5t + k = 0$ for t.

Now use the quadratic formula with $a = 2$, $b = k$, and $c = -s$.

$$t = \frac{-k \pm \sqrt{k^2 - 4(2)(-s)}}{2(2)} \qquad \text{Solve for } t.$$

$$t = \frac{-k \pm \sqrt{k^2 + 8s}}{4}$$

The solutions are $t = \dfrac{-k + \sqrt{k^2 + 8s}}{4}$ and $t = \dfrac{-k - \sqrt{k^2 + 8s}}{4}$.

Work Problem 2 at the Side.

Hypotenuse c

Leg a

$90°$

Leg b

$c^2 = a^2 + b^2$

Pythagorean Formula

2 Solve applied problems using the Pythagorean formula. The Pythagorean formula $a^2 + b^2 = c^2$, illustrated by the figure in the margin, was introduced in Chapter 8 and is used to solve applications involving right triangles. Such problems often require solving quadratic equations.

Example 3 Using the Pythagorean Formula

Two cars left an intersection at the same time, one heading due north, the other due west. Some time later, they were exactly 100 mi apart. The car headed north had gone 20 mi farther than the car headed west. How far had each car traveled?

Step 1 **Read** the problem carefully.

Step 2 **Assign a variable.** Let x be the distance traveled by the car headed west. Then $x + 20$ is the distance traveled by the car headed north. See Figure 2. The cars are 100 mi apart, so the hypotenuse of the right triangle equals 100.

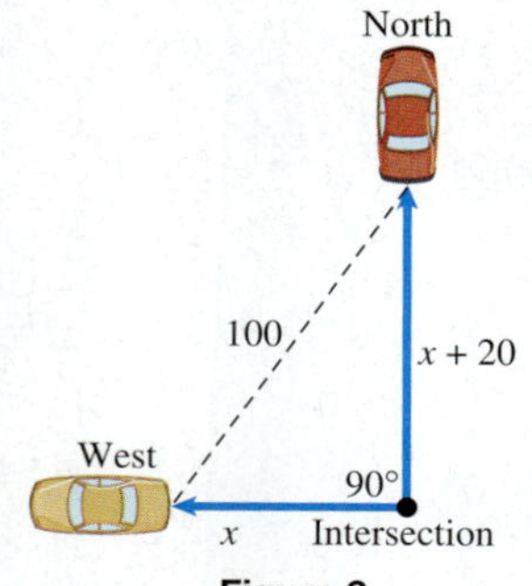

Figure 2

Step 3 **Write an equation.** Use the Pythagorean formula.

$$c^2 = a^2 + b^2$$

$$100^2 = x^2 + (x + 20)^2$$

Step 4 **Solve.**

$$10{,}000 = x^2 + x^2 + 40x + 400 \qquad \text{Square the binomial.}$$

$$0 = 2x^2 + 40x - 9600 \qquad \text{Standard form}$$

$$0 = x^2 + 20x - 4800 \qquad \text{Divide by 2.}$$

$$0 = (x + 80)(x - 60) \qquad \text{Factor.}$$

$$x + 80 = 0 \quad \text{or} \quad x - 60 = 0 \qquad \text{Zero-factor property}$$

$$x = -80 \quad \text{or} \quad x = 60$$

Step 5 **State the answer.** Since distance cannot be negative, discard the negative solution. The required distances are 60 mi and $60 + 20 = 80$ mi.

Step 6 **Check.** Since $60^2 + 80^2 = 100^2$, the answers are correct.

Work Problem 3 at the Side.

3 Solve the problem.

A 13-ft ladder is leaning against a house. The distance from the bottom of the ladder to the house is 7 ft less than the distance from the top of the ladder to the ground. How far is the bottom of the ladder from the house?

Answers

2. $t = \dfrac{5 + \sqrt{25 - 8k}}{4}$, $t = \dfrac{5 - \sqrt{25 - 8k}}{4}$

3. 5 ft

3 Solve applied problems using area formulas.

Example 4 Solving an Area Problem

A rectangular reflecting pool in a park is 20 ft wide and 30 ft long. The park gardener wants to plant a strip of grass of uniform width around the edge of the pool. She has enough seed to cover 336 ft². How wide will the strip be?

Step 1 **Read** the problem carefully.

Step 2 **Assign a variable.** The pool is shown in Figure 3. If x represents the unknown width of the grass strip, the width of the large rectangle is given by $20 + 2x$ (the width of the pool plus two grass strips), and the length is given by $30 + 2x$.

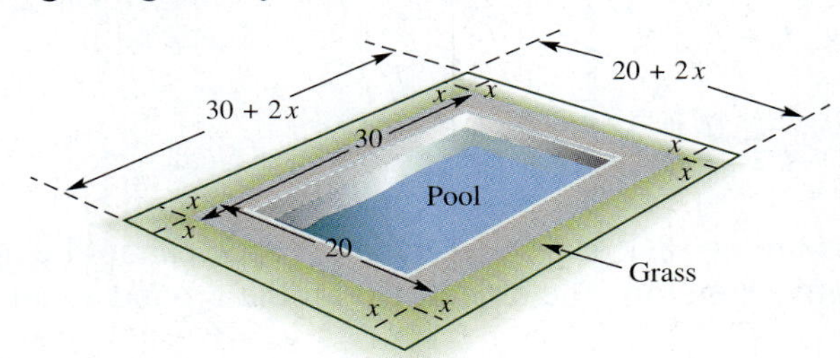

Figure 3

Step 3 **Write an equation.** The area of the large rectangle is given by the product of its length and width, $(30 + 2x)(20 + 2x)$. The area of the pool is $30 \cdot 20 = 600$ ft². The area of the large rectangle, minus the area of the pool, should equal the area of the grass strip. Since the area of the grass strip is to be 336 ft², the equation is

Area of rectangle − area of pool = area of grass.

$$(30 + 2x)(20 + 2x) - 600 = 336.$$

Step 4 **Solve.**

$$600 + 100x + 4x^2 - 600 = 336 \quad \text{Multiply.}$$
$$4x^2 + 100x - 336 = 0 \quad \text{Standard form}$$
$$x^2 + 25x - 84 = 0 \quad \text{Divide by 4.}$$
$$(x + 28)(x - 3) = 0 \quad \text{Factor.}$$
$$x = -28 \quad \text{or} \quad x = 3 \quad \text{Zero-factor property}$$

Step 5 **State the answer.** The width cannot be −28 ft, so the grass strip should be 3 ft wide.

Step 6 **Check.** If $x = 3$, then the area of the large rectangle (which includes the grass strip) is

$$(30 + 2 \cdot 3)(20 + 2 \cdot 3) = 36 \cdot 26 = 936 \text{ ft}^2. \quad \text{Area of pool and strip}$$

The area of the pool is $30 \cdot 20 = 600$ ft². So, the area of the grass strip is $936 - 600 = 336$ ft², which is the area the gardener had enough seed to cover. The answer is correct.

Work Problem 4 at the Side.

4 Solve the problem.
Suppose the pool in Example 4 is 20 ft by 40 ft and there is enough seed to cover 700 ft². How wide should the grass strip be?

4 Solve applied problems using quadratic functions as models.

Some applied problems can be modeled by *quadratic functions,* which can be written in the form

$$f(x) = ax^2 + bx + c,$$

for real numbers a, b, and c, $a \neq 0$.

Answers
4. 5 ft

5 Solve the problem.

A ball is propelled vertically upward from the ground. Its distance in feet from the ground at t seconds is

$$s(t) = -16t^2 + 64t.$$

At what times will the ball be 32 ft from the ground? Use a calculator and round answers to the nearest tenth. (*Hint:* There are two answers.)

Example 5 Solving an Applied Problem Using a Quadratic Function

If an object is propelled upward from the top of a 144-ft building at 112 ft per sec, its position (in feet above the ground) is given by

$$s(t) = -16t^2 + 112t + 144,$$

where t is time in seconds after it was thrown. When does it hit the ground?

When the object hits the ground, its distance above the ground is 0. We must find the value of t that makes $s(t) = 0$.

$$0 = -16t^2 + 112t + 144 \qquad \text{Let } s(t) = 0.$$

$$0 = t^2 - 7t - 9 \qquad \text{Divide by } -16.$$

$$t = \frac{7 \pm \sqrt{49 + 36}}{2} \qquad \text{Quadratic formula}$$

$$t = \frac{7 \pm \sqrt{85}}{2} \approx \frac{7 \pm 9.2}{2} \qquad \text{Use a calculator.}$$

The solutions are $t \approx 8.1$ or $t \approx -1.1$. Since time cannot be negative, discard the negative solution. The object will hit the ground about 8.1 sec after it is thrown.

Work Problem 5 at the Side.

Example 6 Using a Quadratic Function to Model Company Bankruptcy Filings

The number of companies filing for bankruptcy was high in the early 1990s due to an economic recession. The number then declined during the middle 1990s, and in recent years has increased again. The quadratic function defined by

$$f(x) = 2.84x^2 - 24.4x + 129$$

approximates the number of company bankruptcy filings during the years 1990–1999, where x is the number of years since 1990. (*Source:* www.BankruptcyData.com)

(a) Use the model to approximate the number of company bankruptcy filings in 1995.

For 1995, $x = 5$, so find $f(5)$.

$$f(5) = 2.84(5)^2 - 24.4(5) + 129 \qquad \text{Let } x = 5.$$

$$= 78$$

There were 78 company bankruptcy filings in 1995.

(b) In what year did company bankruptcy filings reach 140?

Find the value of x that makes $f(x) = 140$.

$$f(x) = 2.84x^2 - 24.4x + 129$$

$$140 = 2.84x^2 - 24.4x + 129 \qquad \text{Let } f(x) = 140.$$

$$0 = 2.84x^2 - 24.4x - 11 \qquad \text{Standard form}$$

Now use $a = 2.84$, $b = -24.4$, and $c = -11$ in the quadratic formula.

Work Problem 6 at the Side.

The positive solution is $x \approx 9$, so company bankruptcy filings reached 140 in $1990 + 9 = 1999$. (Reject the negative solution since the model is not valid for negative values of x.)

6 Use a calculator to evaluate

$$\frac{24.4 \pm \sqrt{(-24.4)^2 - 4(2.84)(-11)}}{2(2.84)}$$

for both solutions. Round to the nearest tenth. Which solution is valid for this problem?

Answers

5. at .6 sec and at 3.4 sec

6. 9.0, −.4; 9.0

9.4 EXERCISES

FOR EXTRA HELP

Student's Solutions Manual

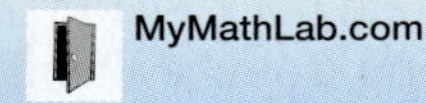
MyMathLab.com

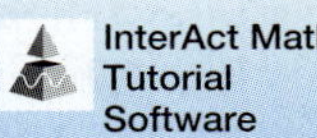
InterAct Math Tutorial Software

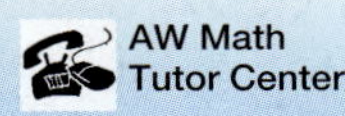
AW Math Tutor Center

www.mathxl.com

Digital Video Tutor CD 7 Videotape 15

1. What is the first step in solving a formula like $gw^2 = 2r$ for w?

2. What is the first step in solving a formula like $gw^2 = kw + 24$ for w?

In Exercises 3 and 4, solve for m in terms of the other variables ($m > 0$).

3.

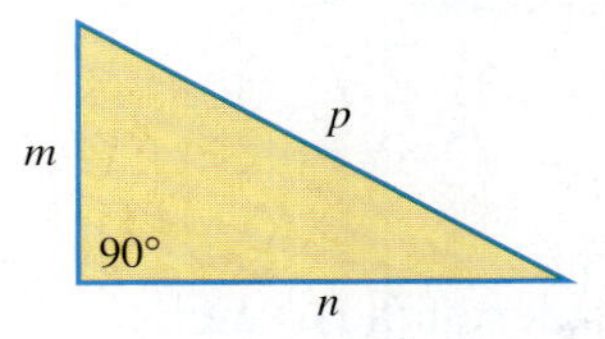

4.

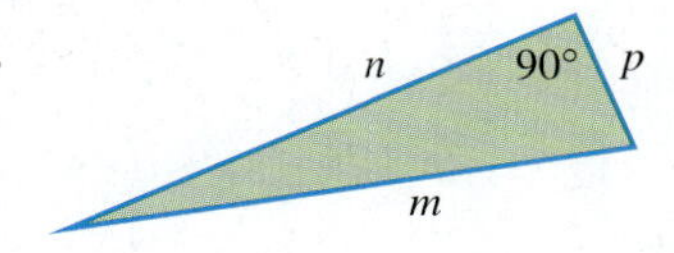

Solve each equation for the indicated variable. (Leave ± in your answers.) See Examples 1 and 2.

5. $d = kt^2$ for t

6. $s = kwd^2$ for d

7. $I = \dfrac{ks}{d^2}$ for d

8. $R = \dfrac{k}{d^2}$ for d

9. $F = \dfrac{kA}{v^2}$ for v

10. $L = \dfrac{kd^4}{h^2}$ for h

11. $V = \dfrac{1}{3}\pi r^2 h$ for r

12. $V = \pi(r^2 + R^2)h$ for r

13. $At^2 + Bt = -C$ for t

14. $S = 2\pi rh + \pi r^2$ for r

15. $D = \sqrt{kh}$ for h

16. $F = \dfrac{k}{\sqrt{d}}$ for d

17. $p = \sqrt{\dfrac{k\ell}{g}}$ for ℓ

18. $p = \sqrt{\dfrac{k\ell}{g}}$ for g

Solve each problem. When appropriate, round answers to the nearest tenth. See Example 3.

19. Two ships leave port at the same time, one heading due south and the other heading due east. Several hours later, they are 170 mi apart. If the ship traveling south traveled 70 mi farther than the other, how many miles did they each travel?

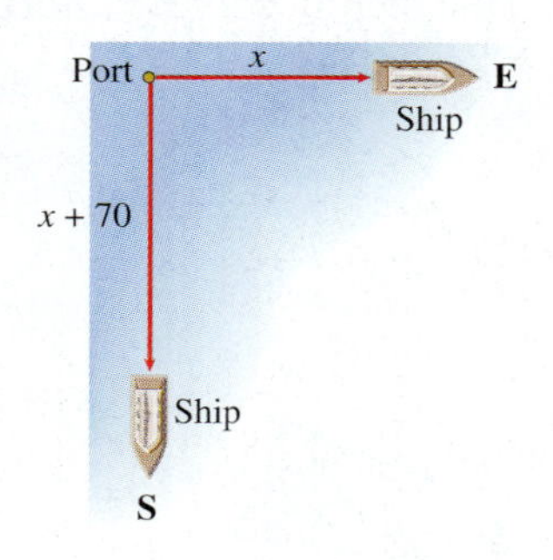

20. Allyson Pellissier is flying a kite that is 30 ft farther above her hand than its horizontal distance from her. The string from her hand to the kite is 150 ft long. How high is the kite?

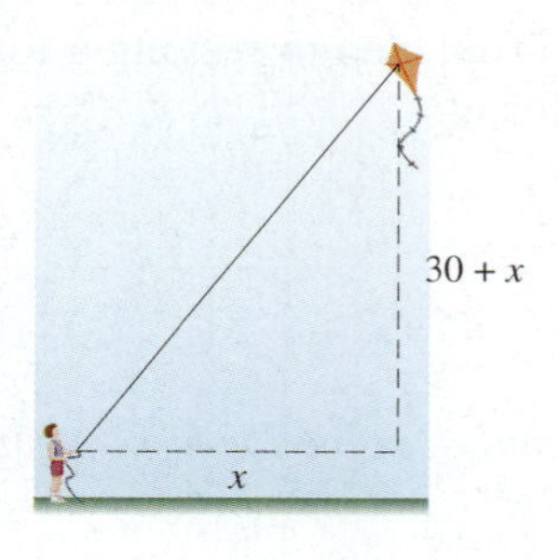

21. Find the lengths of the sides of the triangle.

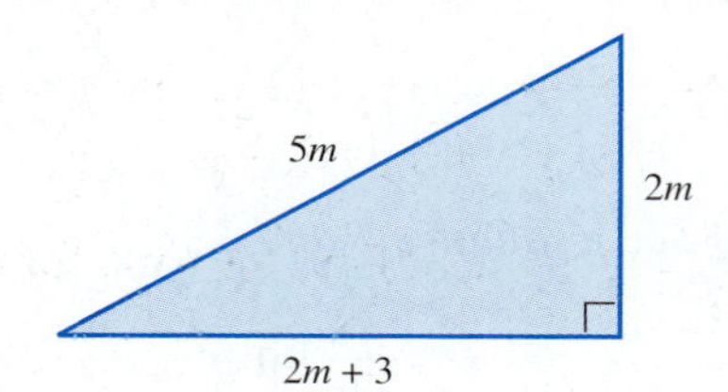

22. Find the lengths of the sides of the triangle.

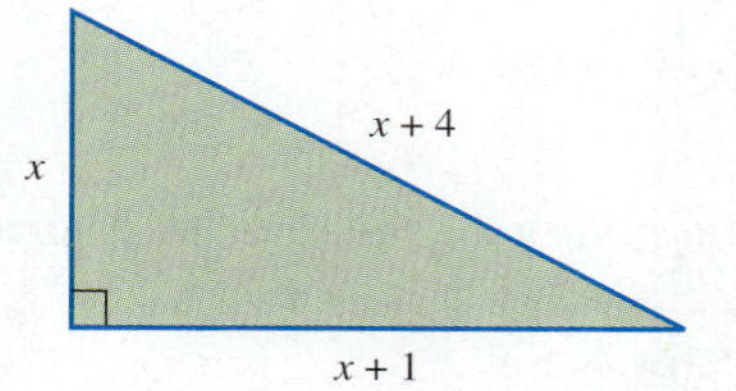

Solve each problem. See Example 4.

23. A couple wants to buy a rug for a room that is 20 ft long and 15 ft wide. They want to leave an even strip of flooring uncovered around the edges of the room. How wide a strip will they have if they buy a rug with an area of 234 ft^2?

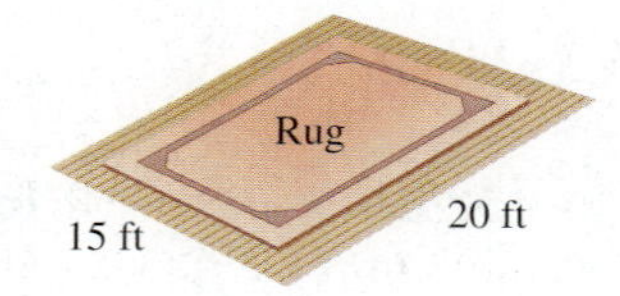

24. A club swimming pool is 30 ft wide and 40 ft long. The club members want an exposed aggregate border in a strip of uniform width around the pool. They have enough material for 296 ft^2. How wide can the strip be?

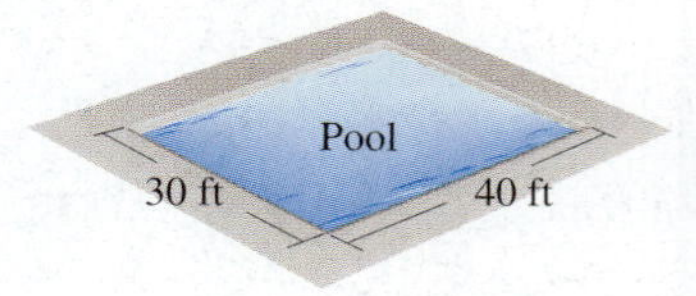

25. A rectangular piece of sheet metal has a length that is 4 in. less than twice the width. A square piece 2 in. on a side is cut from each corner. The sides are then turned up to form an uncovered box of volume 256 in.^3. Find the length and width of the original piece of metal.

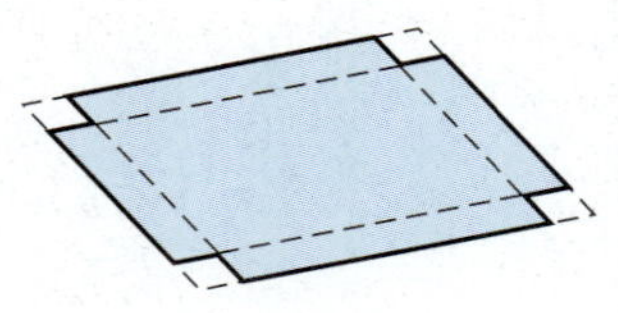

26. Another rectangular piece of sheet metal is 2 in. longer than it is wide. A square piece 3 in. on a side is cut from each corner. The sides are then turned up to form an uncovered box of volume 765 in.^3. Find the dimensions of the original piece of metal.

Solve each problem. Round answers to the nearest tenth. See Example 5.

27. A ball is projected upward from the ground. Its distance in feet from the ground in t seconds is given by

$$s(t) = -16t^2 + 128t.$$

At what times will the ball be 213 ft from the ground?

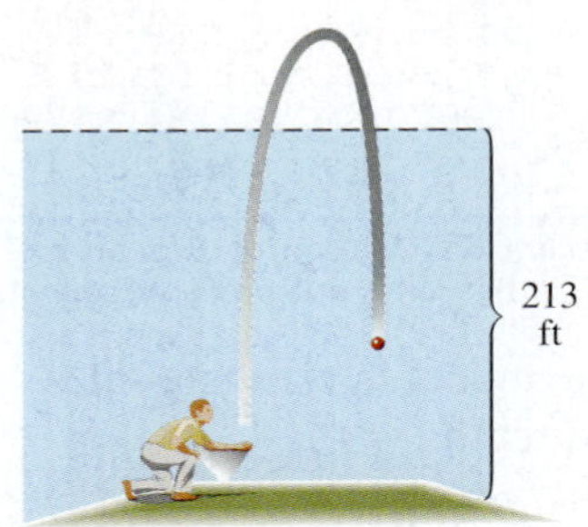

28. A toy rocket is launched from ground level. Its distance in feet from the ground in t seconds is given by

$$s(t) = -16t^2 + 208t.$$

At what times will the rocket be 550 ft from the ground?

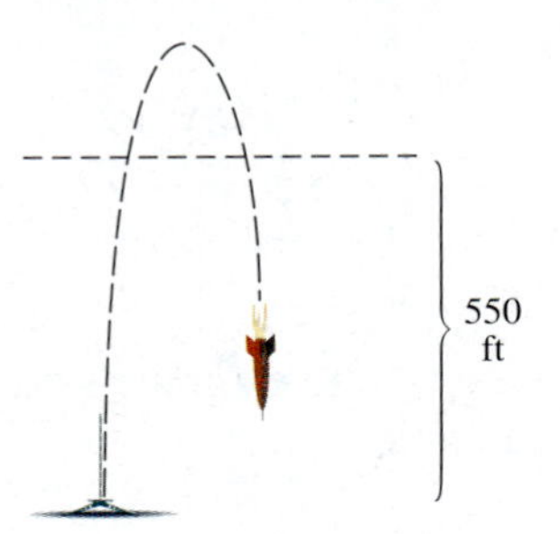

29. The function defined by

$$D(t) = 13t^2 - 100t$$

gives the distance in feet a car going approximately 68 mph will skid in t seconds. Find the time it would take for the car to skid 180 ft.

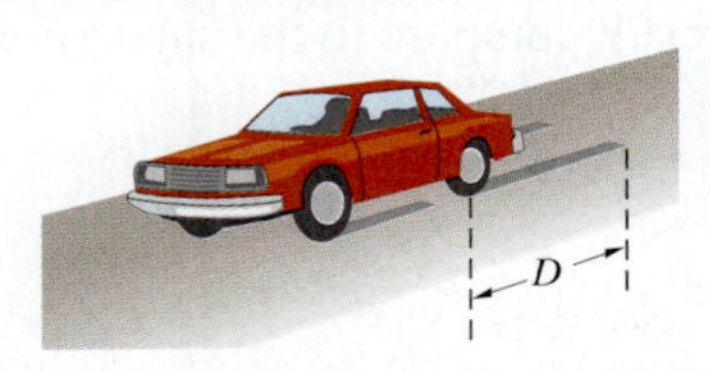

30. The function given in Exercise 29 becomes

$$D(t) = 13t^2 - 73t$$

for a car going 50 mph. Find the time for this car to skid 218 ft.

A rock is projected upward from ground level, and its distance in feet from the ground in t seconds is given by $s(t) = -16t^2 + 160t$. *Use algebra and a short explanation to answer Exercises 31 and 32.*

31. After how many seconds does it reach a height of 400 ft? How would you describe in words its position at this height?

32. After how many seconds does it reach a height of 425 ft? How would you interpret the mathematical result here?

Solve each problem using a quadratic equation.

33. A certain bakery has found that the daily demand for bran muffins is $\frac{3200}{p}$, where p is the price of a muffin in cents. The daily supply is $3p - 200$. Find the price at which supply and demand are equal.

34. In one area the demand for compact discs is $\frac{700}{P}$ per day, where P is the price in dollars per disc. The supply is $5P - 1$ per day. At what price does supply equal demand?

Sales of SUVs (sport utility vehicles) in the United States (in millions) for the years 1990–1999 are shown in the bar graph and can be modeled by the quadratic function defined by

$$f(x) = .016x^2 + .124x + .787.$$

Here, $x = 0$ represents 1990, $x = 1$ represents 1991, and so on. Use the graph and the model to work Exercises 35–38. See Example 6.

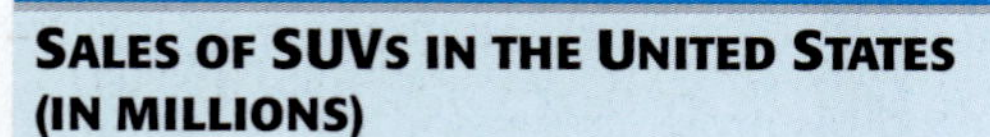

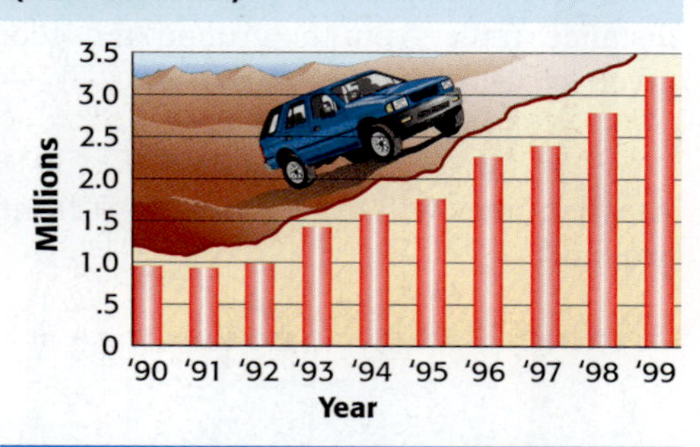

Source: CNW Marketing Research of Bandon, OR, based on automakers' reported sales.

35. (a) Use the graph to estimate sales in 1997 to the nearest tenth.

(b) Use the model to approximate sales in 1997 to the nearest tenth. How does this result compare to your estimate from part (a)?

36. (a) Use the model to estimate sales in 2000 to the nearest tenth.

(b) Sales through October 2000 were about 2.9 million. Based on this, is the sales estimate for 2000 from part (a) reasonable? Explain.

37. Based on the model, in what year did sales reach 2 million? (Round down to the nearest year.) How does this result compare to the sales shown in the graph?

38. Based on the model, in what year did sales reach 3 million? (Round down to the nearest year.) How does this result compare to the sales shown in the graph?

William Froude was a 19th century naval architect who used the expression

$$\frac{v^2}{g\ell}$$

in shipbuilding. This expression, known as the Froude number, was also used by R. McNeill Alexander in his research on dinosaurs. (Source: "How Dinosaurs Ran," Scientific American, *April 1991.) In Exercises 39 and 40, find the value of v (in meters per second), given that $g = 9.8$ m per sec^2.*

39. Rhinoceros: $\ell = 1.2$; Froude number $= 2.57$

40. Triceratops: $\ell = 2.8$; Froude number $= .16$

Recall from Chapter 7 that corresponding sides of similar triangles are proportional. Use this fact to find the lengths of the indicated sides of each pair of similar triangles. Check all possible solutions in both triangles. Sides of a triangle cannot be negative (and are not drawn to scale here).

41. Side AC

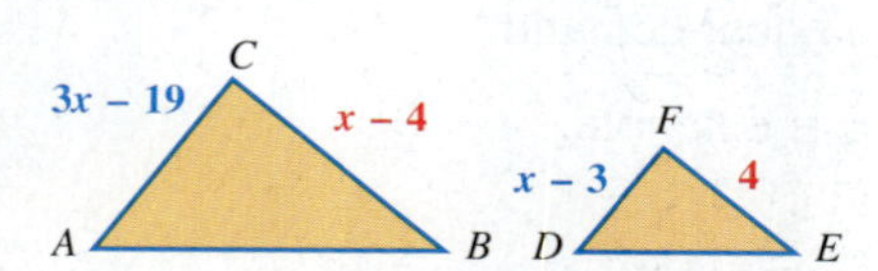

42. Side RQ

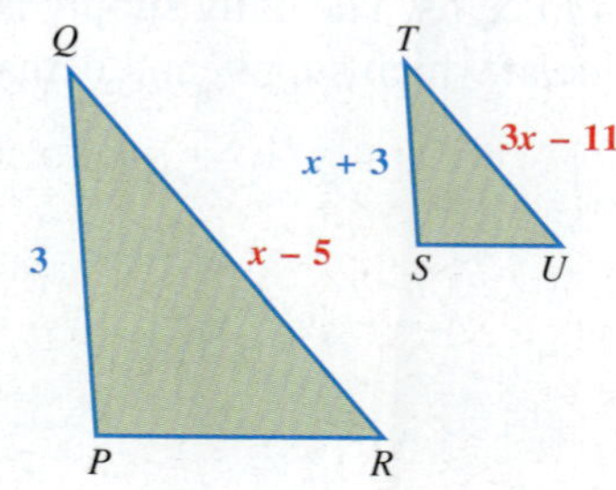

9.5 Graphs of Quadratic Functions

Objectives

1. Graph a quadratic function.
2. Graph parabolas with horizontal and vertical shifts.
3. Predict the shape and direction of a parabola from the coefficient of x^2.
4. Find a quadratic function to model data.

1 Graph a quadratic function. Polynomial functions were defined in Chapter 6, where we graphed a few simple second-degree polynomial functions by point-plotting. In Figure 4, we repeat a table of ordered pairs for the simplest quadratic function, defined by $y = x^2$, and the resulting graph.

x	y
−2	4
−1	1
0	0
1	1
2	4

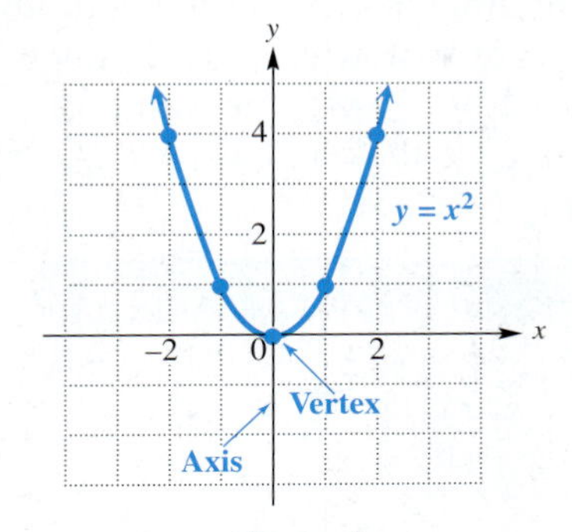

Figure 4

As mentioned in Chapter 6, this graph is called a **parabola.** The point (0, 0), the lowest point on the curve, is the **vertex** of this parabola. The vertical line through the vertex is the **axis** of the parabola, here $x = 0$. A parabola is **symmetric about its axis;** that is, if the graph were folded along the axis, the two portions of the curve would coincide. As Figure 4 suggests, x can be any real number, so the domain of the function defined by $y = x^2$ is $(-\infty, \infty)$. Since y is always nonnegative, the range is $[0, \infty)$.

In Section 9.4, we solved applications modeled by quadratic functions. In this section and the next, we consider graphs of more general quadratic functions as defined here.

Quadratic Function

A function that can be written in the form

$$f(x) = ax^2 + bx + c$$

for real numbers a, b, and c, with $a \neq 0$, is a **quadratic function.**

The graph of any quadratic function is a parabola with a vertical axis. We use the variable y and function notation $f(x)$ interchangeably when discussing parabolas. Although we use the letter f most often to name quadratic functions, other letters can be used. We use the capital letter F to distinguish between different parabolas graphed on the same coordinate axes.

Parabolas, which are a type of *conic section* (Chapter 11), have many applications. The large disks seen on the sidelines of televised football games, which are used by television crews to pick up the shouted signals of players on the field, have cross sections that are parabolas. Cross sections of radar dishes and automobile headlights also form parabolas. The cables that are used to support suspension bridges are shaped like parabolas.

2 Graph parabolas with horizontal and vertical shifts. Parabolas need not have their vertices at the origin, as does the graph of $f(x) = x^2$. For example, to graph a parabola of the form $F(x) = x^2 + k$, start by selecting sample values of x like those that were used to graph $f(x) = x^2$. The corresponding values of $F(x)$ in $F(x) = x^2 + k$ differ by k from those of $f(x) = x^2$. For this reason, the graph of $F(x) = x^2 + k$ is *shifted,* or *translated,* k units vertically compared with that of $f(x) = x^2$.

1 Graph each parabola. Give the vertex, domain, and range.

(a) $f(x) = x^2 + 3$

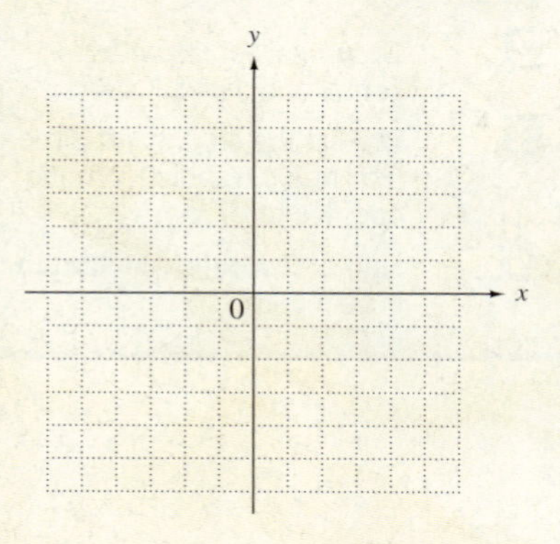

(b) $f(x) = x^2 - 1$

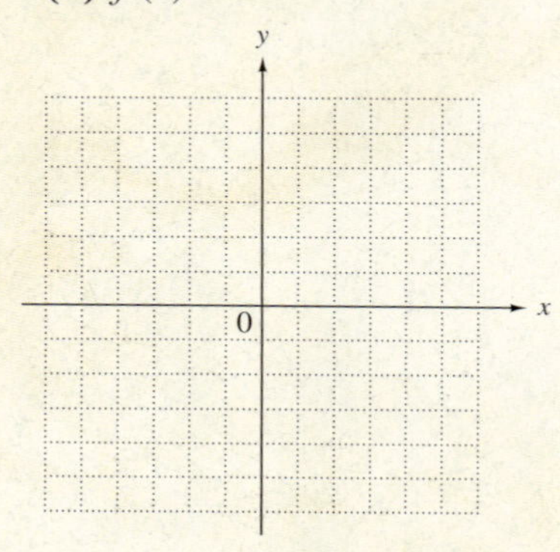

Example 1 Graphing a Parabola with a Vertical Shift

Graph $F(x) = x^2 - 2$.

This graph has the same shape as that of $f(x) = x^2$, but since k here is -2, the graph is shifted 2 units down, with vertex $(0, -2)$. Every function value is 2 less than the corresponding function value of $f(x) = x^2$. Plotting points on both sides of the vertex gives the graph in Figure 5. Notice that since the parabola is symmetric about its axis $x = 0$, the plotted points are "mirror images" of each other. Since x can be any real number, the domain is still $(-\infty, \infty)$; the value of y (or $F(x)$) is always greater than or equal to -2, so the range is $[-2, \infty)$. The graph of $f(x) = x^2$ is shown for comparison.

x	$f(x) = x^2$	$F(x) = x^2 - 2$
-2	4	2
-1	1	-1
0	0	-2
1	1	-1
2	4	2

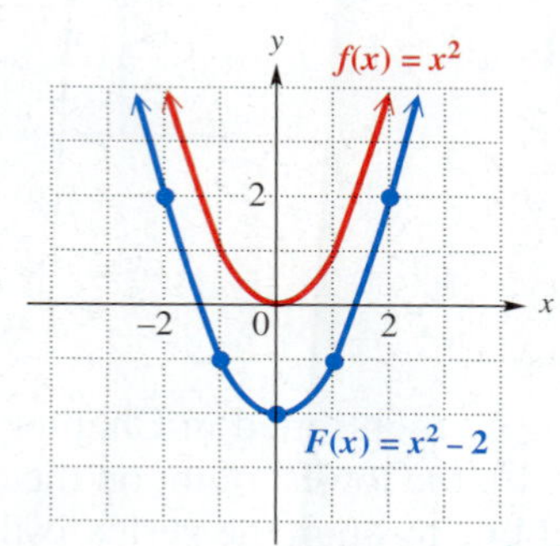

Figure 5

Vertical Shift

The graph of $F(x) = x^2 + k$ is a parabola with the same shape as the graph of $f(x) = x^2$. The parabola is shifted k units up if $k > 0$, and $|k|$ units down if $k < 0$. The vertex is $(0, k)$.

Work Problem 1 at the Side.

The graph of $F(x) = (x - h)^2$ is also a parabola with the same shape as that of $f(x) = x^2$. Because $(x - h)^2 \geq 0$ for all x, the vertex of $F(x) = (x - h)^2$ is the lowest point on the parabola. The lowest point occurs here when $F(x)$ is 0. To get $F(x)$ equal to 0, let $x = h$ so the vertex of $F(x) = (x - h)^2$ is $(h, 0)$. Based on this, the graph of $F(x) = (x - h)^2$ is shifted h units horizontally compared with that of $f(x) = x^2$.

Example 2 Graphing a Parabola with a Horizontal Shift

Graph $F(x) = (x - 2)^2$.

When $x = 2$, then $F(x) = 0$, giving the vertex $(2, 0)$. The graph of $F(x) = (x - 2)^2$ has the same shape as that of $f(x) = x^2$ but is shifted 2 units to the right. Plotting several points on one side of the vertex and using symmetry about the axis $x = 2$ to find corresponding points on the other side of the vertex gives the graph in Figure 6. Again, the domain is $(-\infty, \infty)$; the range is $[0, \infty)$.

Continued on Next Page

ANSWERS

1. (a)

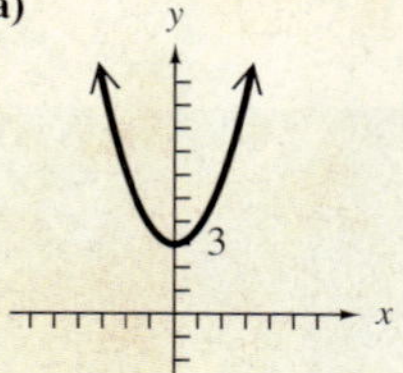

vertex: $(0, 3)$; domain: $(-\infty, \infty)$; range: $[3, \infty)$

(b)

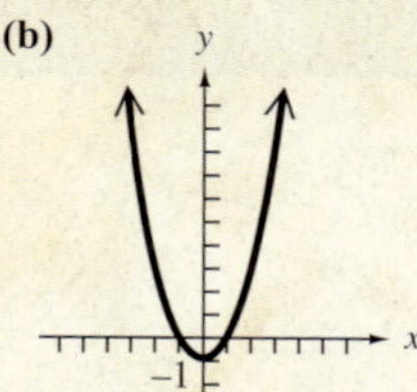

vertex: $(0, -1)$; domain: $(-\infty, \infty)$; range: $[-1, \infty)$

x	$F(x) = (x - 2)^2$
0	4
1	1
2	0
3	1
4	4

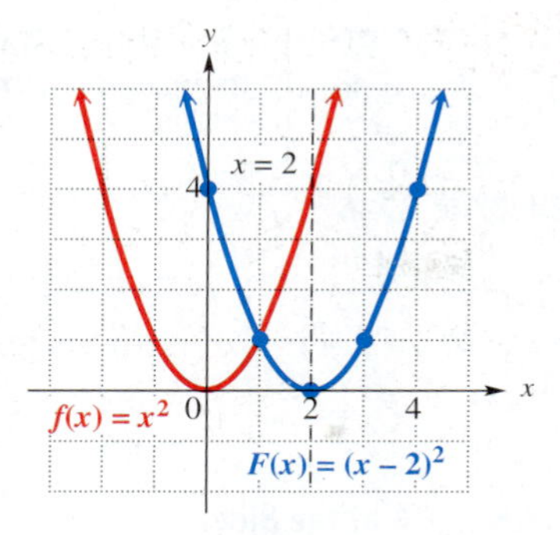

Figure 6

Horizontal Shift

The graph of $F(x) = (x - h)^2$ is a parabola with the same shape as the graph of $f(x) = x^2$. The parabola is shifted h units horizontally: h units to the right if $h > 0$, and $|h|$ units to the left if $h < 0$. The vertex is $(h, 0)$.

CAUTION

Errors frequently occur when horizontal shifts are involved. To determine the direction and magnitude of a horizontal shift, find the value that would cause the expression $x - h$ to equal 0. For example, the graph of $F(x) = (x - 5)^2$ would be shifted 5 units to the *right,* because $+5$ would cause $x - 5$ to equal 0. On the other hand, the graph of $F(x) = (x + 5)^2$ would be shifted 5 units to the *left,* because -5 would cause $x + 5$ to equal 0.

Work Problem 2 at the Side.

A parabola can have both horizontal and vertical shifts.

Example 3 Graphing a Parabola with Horizontal and Vertical Shifts

Graph $F(x) = (x + 3)^2 - 2$.

This graph has the same shape as that of $f(x) = x^2$, but is shifted 3 units to the left (since $x + 3 = 0$ if $x = -3$) and 2 units down (because of the -2). As shown in Figure 7, the vertex is $(-3, -2)$, with axis $x = -3$. This function has domain $(-\infty, \infty)$ and range $[-2, \infty)$.

x	F(x)
−5	2
−4	−1
−3	−2
−2	−1
−1	2

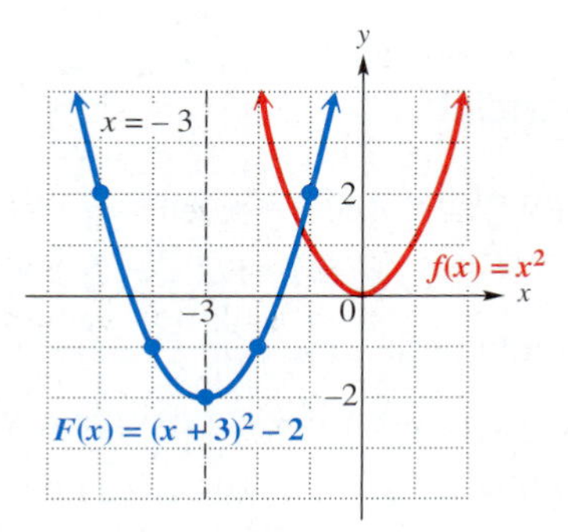

Figure 7

2 Graph each parabola. Give the vertex, axis, domain, and range.

(a) $f(x) = (x - 3)^2$

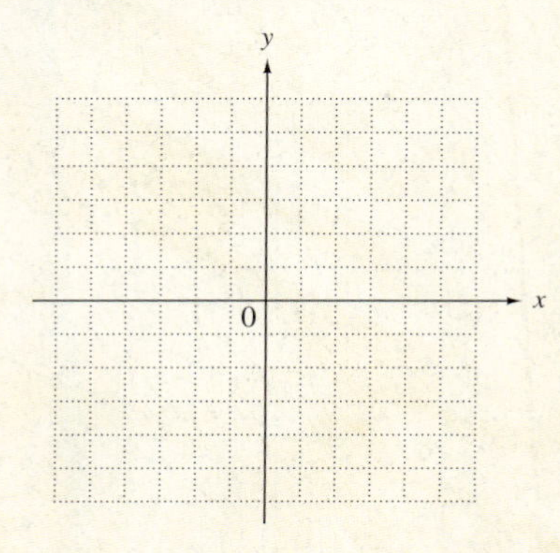

(b) $f(x) = (x + 2)^2$

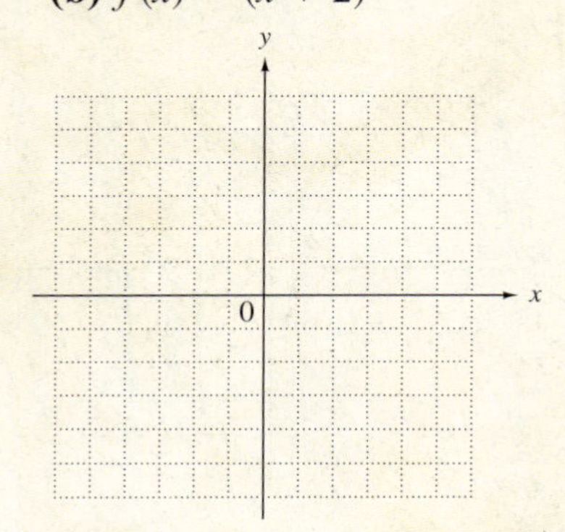

ANSWERS

2. (a)

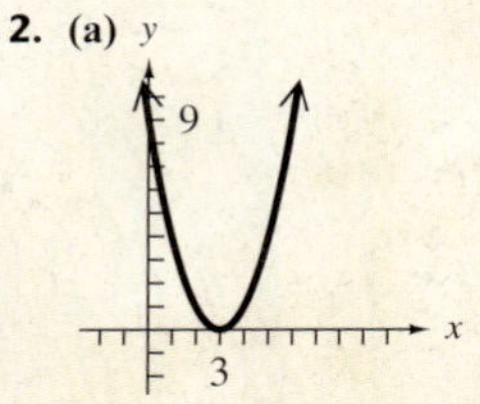

vertex: $(3, 0)$; axis: $x = 3$; domain: $(-\infty, \infty)$; range: $[0, \infty)$

(b)

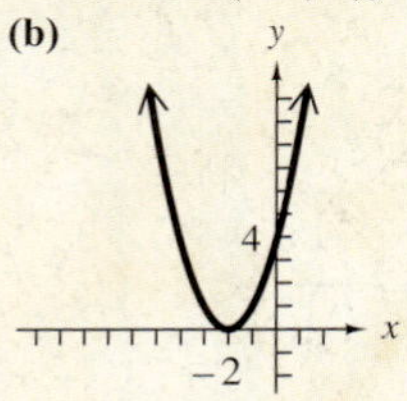

vertex: $(-2, 0)$; axis: $x = -2$; domain: $(-\infty, \infty)$; range: $[0, \infty)$

3 Graph each parabola. Give the vertex, axis, domain, and range.

(a) $f(x) = (x + 2)^2 - 1$

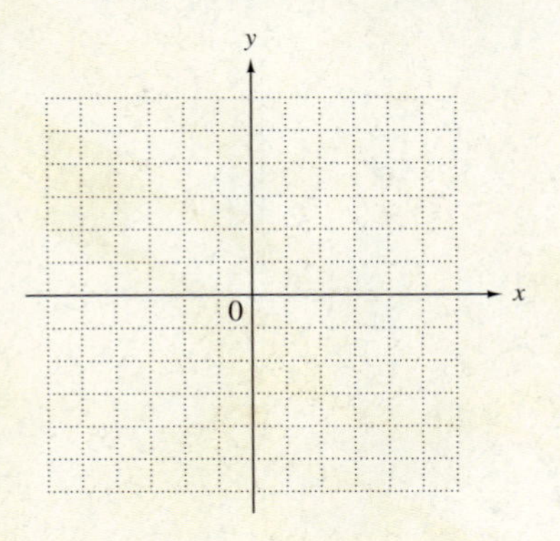

(b) $f(x) = (x - 2)^2 + 5$

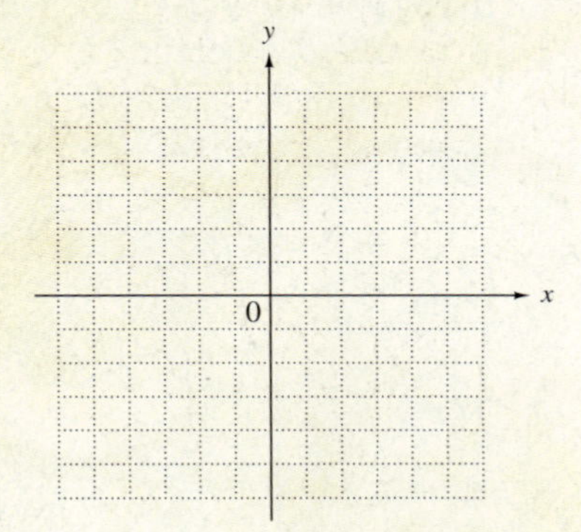

The characteristics of the graph of a parabola of the form $F(x) = (x - h)^2 + k$ are summarized as follows.

Vertex and Axis

The graph of $F(x) = (x - h)^2 + k$ is a parabola with the same shape as the graph of $f(x) = x^2$ with vertex (h, k). The axis is the vertical line $x = h$.

Work Problem 3 at the Side.

3 **Predict the shape and direction of a parabola from the coefficient of x^2.** Not all parabolas open up, and not all parabolas have the same shape as the graph of $f(x) = x^2$.

Example 4 Graphing a Parabola That Opens Down

Graph $f(x) = -\dfrac{1}{2}x^2$.

This parabola is shown in Figure 8. The coefficient $-\frac{1}{2}$ affects the shape of the graph; the $\frac{1}{2}$ makes the parabola wider (since the values of $\frac{1}{2}x^2$ increase more slowly than those of x^2), and the negative sign makes the parabola open down. The graph is not shifted in any direction; the vertex is still (0, 0). Unlike the parabolas graphed in Examples 1–3, the vertex here has the *largest* function value of any point on the graph. The domain is $(-\infty, \infty)$; the range is $(-\infty, 0]$.

x	$f(x)$
-2	-2
-1	$-\frac{1}{2}$
0	0
1	$-\frac{1}{2}$
2	-2

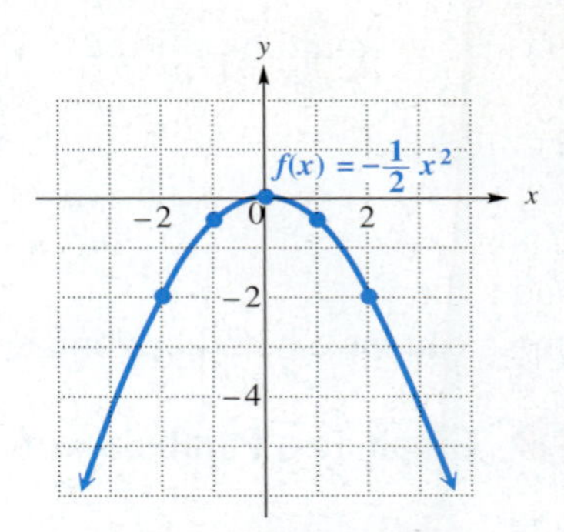

Figure 8

Some general principles concerning the graph of $F(x) = a(x - h)^2 + k$ are summarized as follows.

General Principles

1. The graph of the quadratic function defined by

$$F(x) = a(x - h)^2 + k, \quad a \neq 0$$

is a parabola with vertex (h, k) and the vertical line $x = h$ as axis.

2. The graph opens up if a is positive and down if a is negative.
3. The graph is wider than that of $f(x) = x^2$ if $0 < |a| < 1$. The graph is narrower than that of $f(x) = x^2$ if $|a| > 1$.

ANSWERS

3. (a)

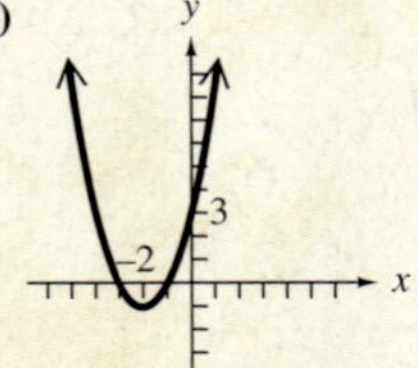

vertex: $(-2, -1)$; axis: $x = -2$; domain: $(-\infty, \infty)$; range: $[-1, \infty)$

(b)

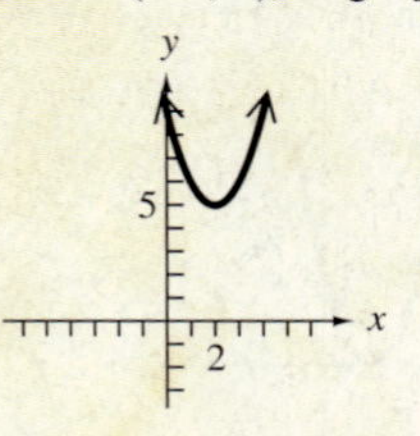

vertex: $(2, 5)$; axis: $x = 2$; domain: $(-\infty, \infty)$; range: $[5, \infty)$

Work Problems 4 and 5 at the Side.

Example 5 Using the General Principles to Graph a Parabola

Graph $F(x) = -2(x + 3)^2 + 4$.

The parabola opens down (because $a < 0$), and is narrower than the graph of $f(x) = x^2$, since $|-2| = 2 > 1$, causing values of $F(x)$ to decrease more quickly than those of $f(x) = -x^2$. This parabola has vertex $(-3, 4)$ as shown in Figure 9. To complete the graph, we plotted the ordered pairs $(-4, 2)$ and, by symmetry, $(-2, 2)$. Symmetry can be used to find additional ordered pairs that satisfy the equation, if desired.

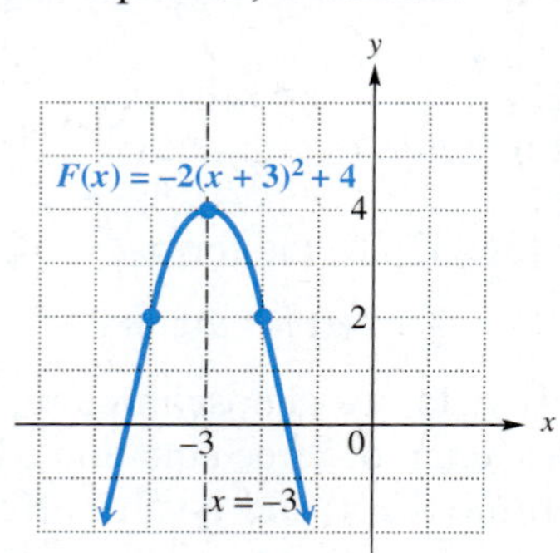

Figure 9

Work Problem 6 at the Side.

4 Find a quadratic function to model data.

Example 6 Finding a Quadratic Function to Model the Rise in Multiple Births

The number of higher-order multiple births in the United States is rising. Let x represent the number of years since 1970 and y represent the rate of higher-order multiples born per 100,000 births since 1971. The data are shown in the following table.

U.S. HIGHER-ORDER MULTIPLE BIRTHS

Year	x	y
1971	1	29.1
1976	6	35.0
1981	11	40.0
1986	16	47.0
1991	21	100.0
1996	26	152.6

Source: National Center for Health Statistics.

Find a quadratic function that models the data.

A scatter diagram of the ordered pairs (x, y) is shown in Figure 10 on the next page. Notice that the graphed points do not follow a linear pattern, so a linear function would not model the data very well. Instead, the general shape suggested by the scatter diagram indicates that a parabola should approximate these points, as shown by the dashed curve in the graph in Figure 11. The equation for such a parabola would have a positive coefficient for x^2 since the graph opens up.

Continued on Next Page

4 Decide whether each parabola opens up or down.

(a) $f(x) = -\frac{2}{3}x^2$

(b) $f(x) = \frac{3}{4}x^2 + 1$

(c) $f(x) = -2x^2 - 3$

(d) $f(x) = 3x^2 + 2$

5 Decide whether each parabola in Problem 4 is wider or narrower than the graph of $f(x) = x^2$.

6 Graph

$$f(x) = \frac{1}{2}(x - 2)^2 + 1.$$

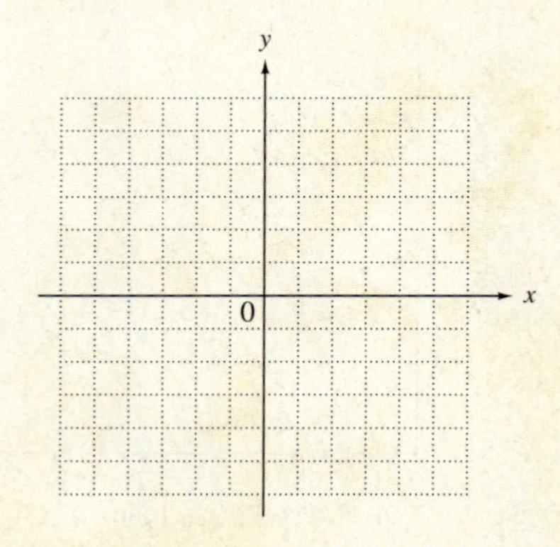

ANSWERS

4. **(a)** down **(b)** up **(c)** down **(d)** up

5. **(a)** wider **(b)** wider **(c)** narrower **(d)** narrower

6.

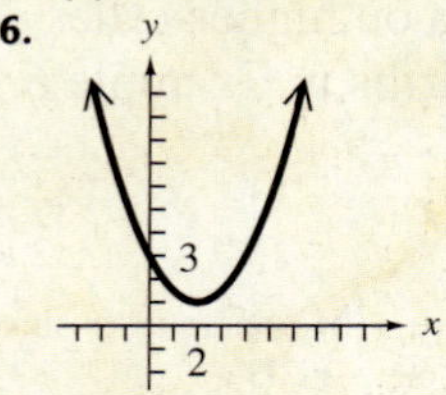

7 Tell whether a linear or quadratic function would be a more appropriate model for each set of graphed data. If linear, tell whether the slope should be positive or negative. If quadratic, tell whether the coefficient a of x^2 should be positive or negative.

(a)

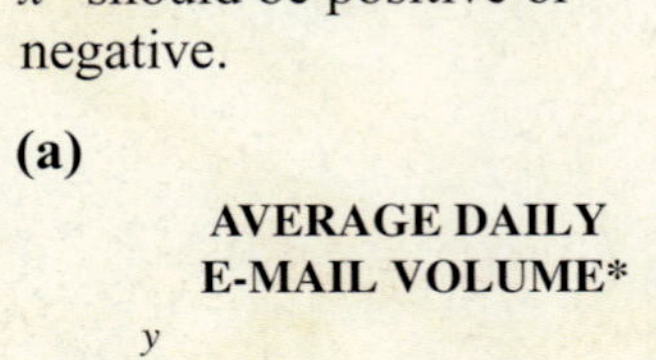

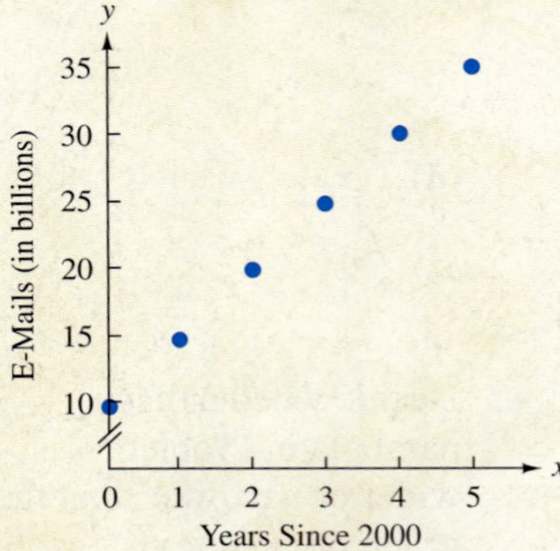

*Projected
Source: General Accounting Office.

(b)

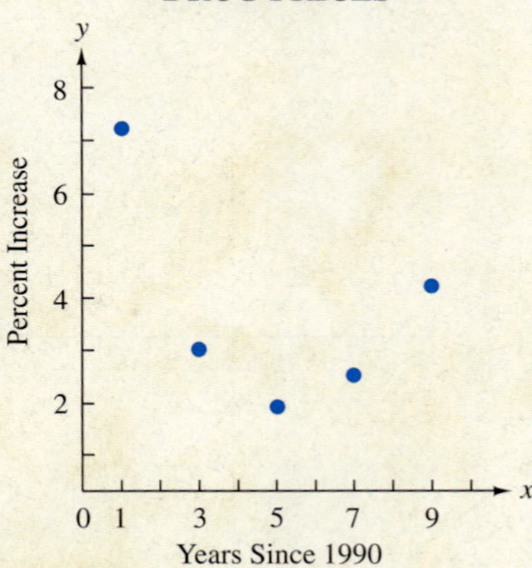

Source: *IMS Health*, Retail and Provider Perspective.

8 Using the points (1, 29.1), (6, 35), and (26, 152.6), find another quadratic model for the data on higher-order multiple births in Example 6.

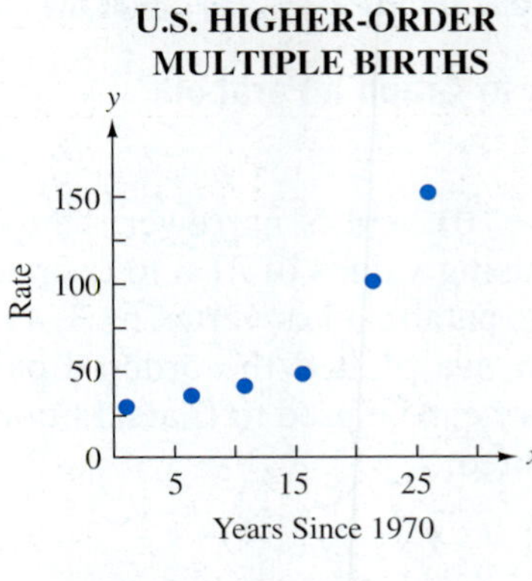

Figure 10

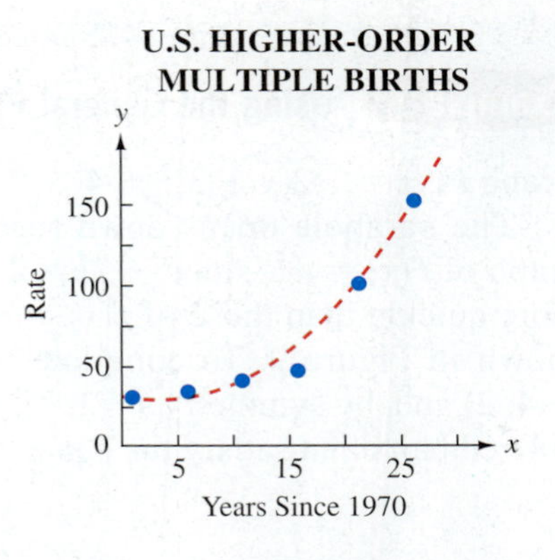

Figure 11

To find a quadratic function of the form

$$y = ax^2 + bx + c$$

that models, or *fits*, these data, we choose three representative ordered pairs and use them to write a system of three equations. Using (1, 29.1), (11, 40), and (21, 100), we substitute the x- and y-values from the ordered pairs into the quadratic form $y = ax^2 + bx + c$ to get the three equations

$$a(\mathbf{1})^2 + b(\mathbf{1}) + c = \mathbf{29.1} \quad \text{or} \quad a + b + c = 29.1 \quad (1)$$
$$a(\mathbf{11})^2 + b(\mathbf{11}) + c = \mathbf{40} \quad \text{or} \quad 121a + 11b + c = 40 \quad (2)$$
$$a(\mathbf{21})^2 + b(\mathbf{21}) + c = \mathbf{100} \quad \text{or} \quad 441a + 21b + c = 100. \quad (3)$$

We can find the values of a, b, and c by solving this system of three equations in three variables using the methods of Section 5.2. Multiplying equation (1) by -1 and adding the result to equation (2) gives

$$120a + 10b = 10.9. \quad (4)$$

Multiplying equation (2) by -1 and adding the result to equation (3) gives

$$320a + 10b = 60. \quad (5)$$

We can eliminate b from this system of equations in two variables by multiplying equation (4) by -1 and adding the result to equation (5) to get

$$200a = 49.1$$
$$a = .2455. \quad \text{Use a calculator.}$$

We substitute .2455 for a in equation (4) or (5) to find that $b = -1.856$. Substituting the values of a and b into equation (1) gives $c = 30.7105$. Using these values of a, b, and c, our model is defined by

$$y = .2455x^2 - 1.856x + 30.7105.$$

Work Problems 7 and 8 at the Side.

Calculator Tip The *quadratic regression* feature on a graphing calculator can be used to generate a quadratic model that fits given data. See your owner's manual for details on how to do this.

ANSWERS
7. (a) linear; positive **(b)** quadratic; positive
8. $y = .188x^2 - .136x + 29.05$

9.5 EXERCISES

FOR EXTRA HELP

Student's Solutions Manual

MyMathLab.com

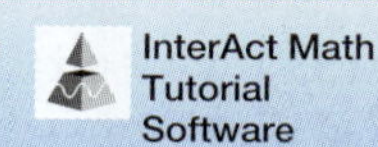
InterAct Math Tutorial Software

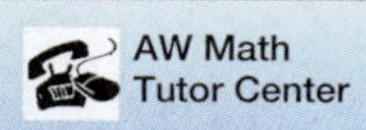
AW Math Tutor Center

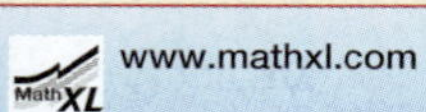
www.mathxl.com

Digital Video Tutor CD 7 Videotape 16

1. Match each quadratic function with its graph from choices A–D.

(a) $f(x) = (x + 2)^2 - 1$ **(b)** $f(x) = (x + 2)^2 + 1$ **(c)** $f(x) = (x - 2)^2 - 1$ **(d)** $f(x) = (x - 2)^2 + 1$

A.
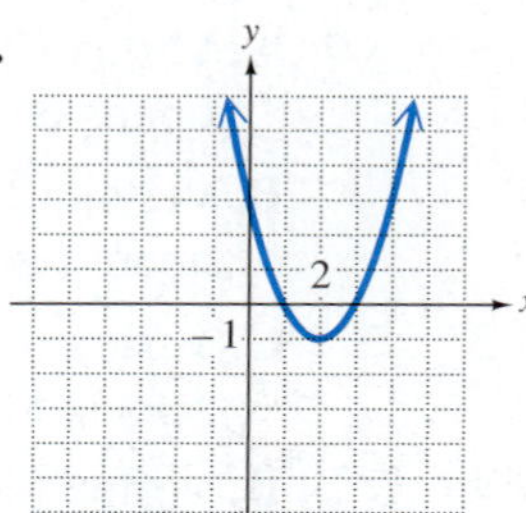

B.
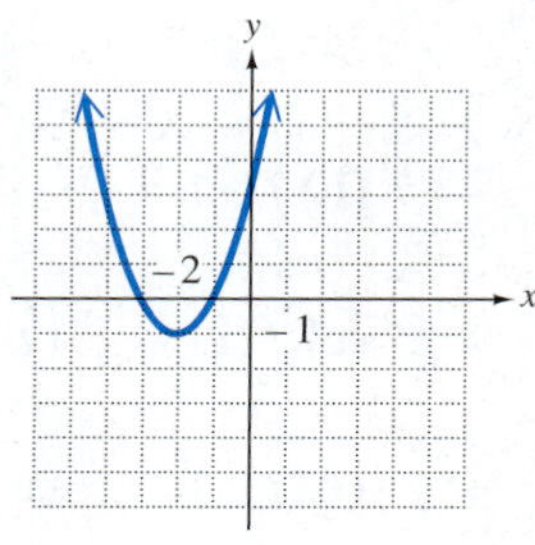

C.
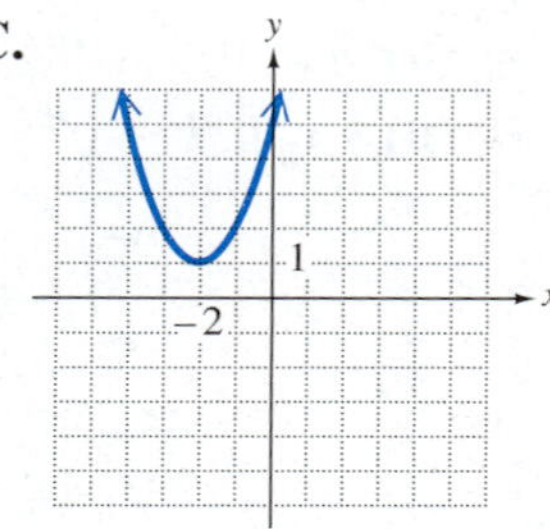

D.
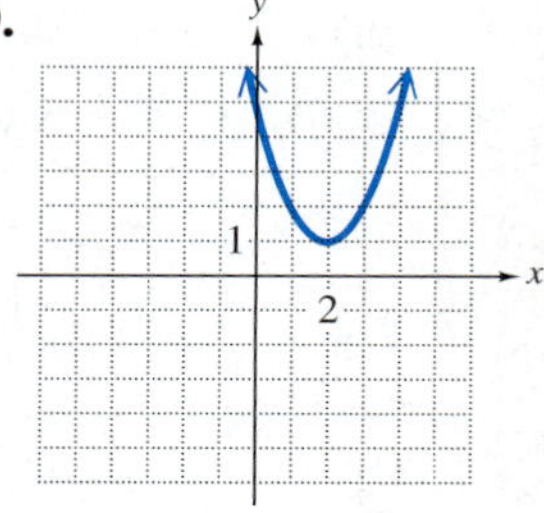

2. Match each quadratic function with its graph from choices A–D.

(a) $f(x) = -x^2 + 2$ **(b)** $f(x) = -x^2 - 2$ **(c)** $f(x) = -(x + 2)^2$ **(d)** $f(x) = -(x - 2)^2$

A.
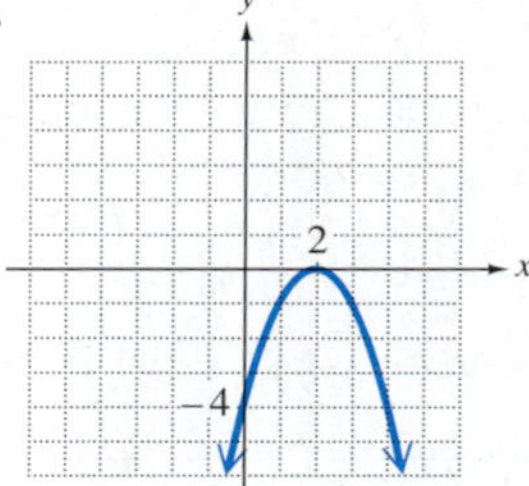

B.
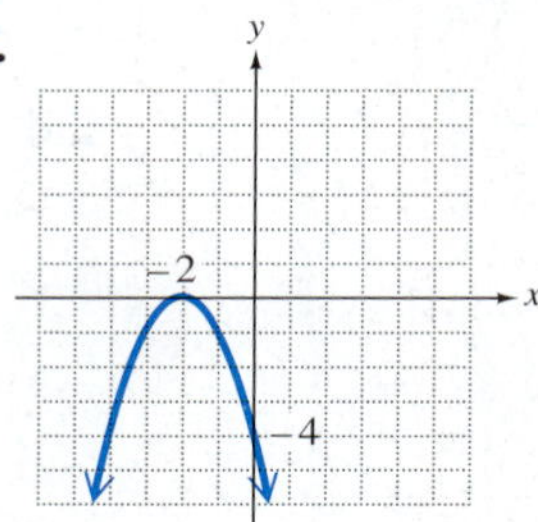

C.
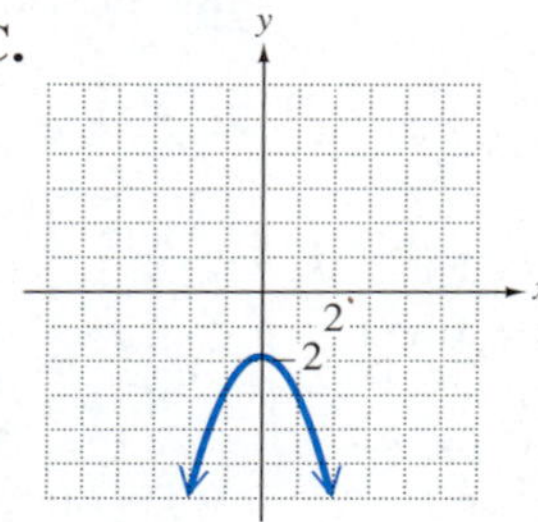

D.
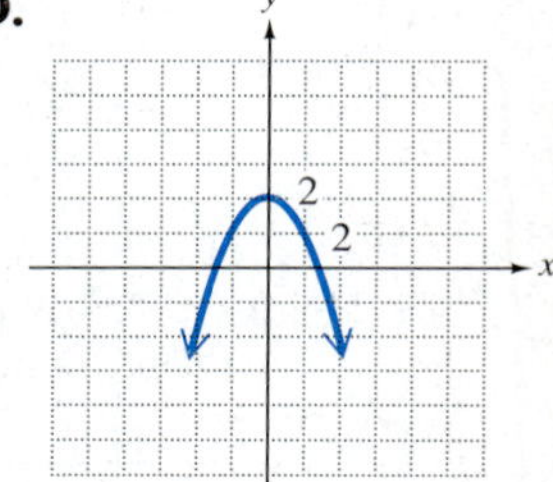

Identify the vertex of each parabola. See Examples 1–4.

3. $f(x) = -3x^2$

4. $f(x) = \frac{1}{2}x^2$

5. $f(x) = x^2 + 4$

6. $f(x) = x^2 - 4$

7. $f(x) = (x - 1)^2$

8. $f(x) = (x + 3)^2$

9. $f(x) = (x + 3)^2 - 4$

10. $f(x) = (x - 5)^2 - 8$

11. Describe how each of the parabolas in Exercises 9 and 10 is shifted compared to the graph of $f(x) = x^2$.

12. What does the value of a in $F(x) = a(x - h)^2 + k$ tell you about the graph of the equation compared to the graph of $f(x) = x^2$?

For each quadratic function, tell whether the graph opens up or down and whether the graph is wider, narrower, or the same shape as the graph of $f(x) = x^2$. See Examples 4 and 5.

13. $f(x) = -\frac{2}{5}x^2$

14. $f(x) = -2x^2$

15. $f(x) = 3x^2 + 1$

16. $f(x) = \frac{2}{3}x^2 - 4$

17. For $f(x) = a(x - h)^2 + k$, in what quadrant is the vertex if

(a) $h > 0, k > 0$; **(b)** $h > 0, k < 0$;

(c) $h < 0, k > 0$; **(d)** $h < 0, k < 0$?

18. Match each quadratic function with the description of the parabola that is its graph.

(a) $f(x) = (x - 4)^2 - 2$ **A.** Vertex $(2, -4)$, opens down

(b) $f(x) = (x - 2)^2 - 4$ **B.** Vertex $(2, -4)$, opens up

(c) $f(x) = -(x - 4)^2 - 2$ **C.** Vertex $(4, -2)$, opens down

(d) $f(x) = -(x - 2)^2 - 4$ **D.** Vertex $(4, -2)$, opens up

Sketch the graph of each parabola. Plot at least two points in addition to the vertex. In Exercises 25–32, give the axis, domain, and range of the parabola.

19. $f(x) = -2x^2$

20. $f(x) = \frac{1}{3}x^2$

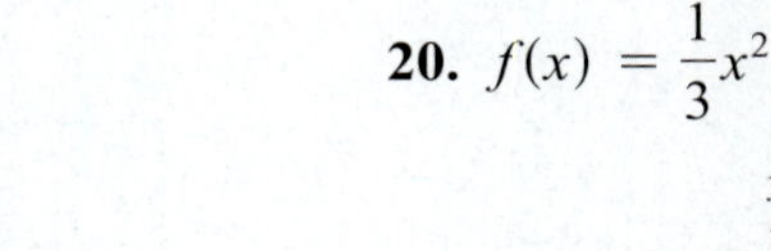

21. $f(x) = x^2 - 1$

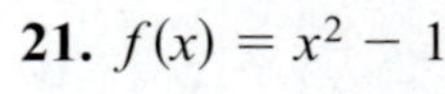

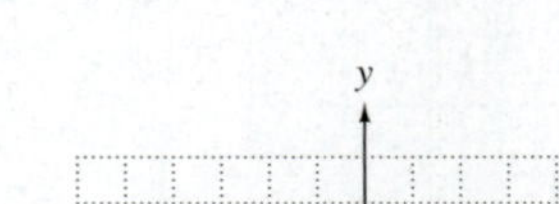

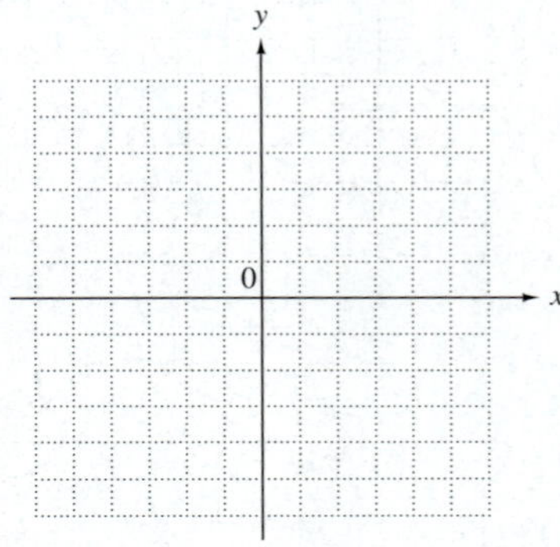

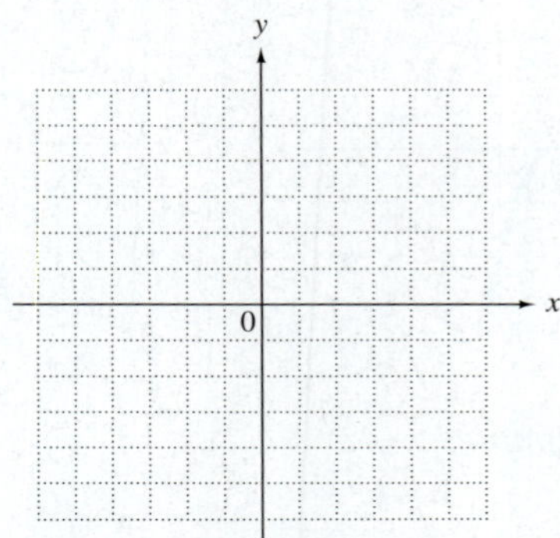

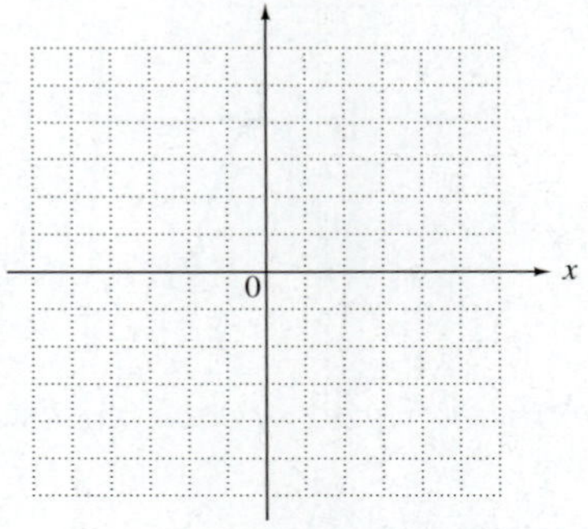

22. $f(x) = x^2 + 3$

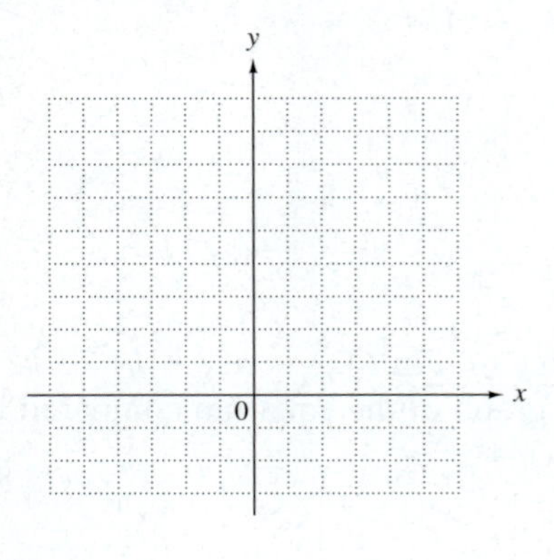

23. $f(x) = -x^2 + 2$

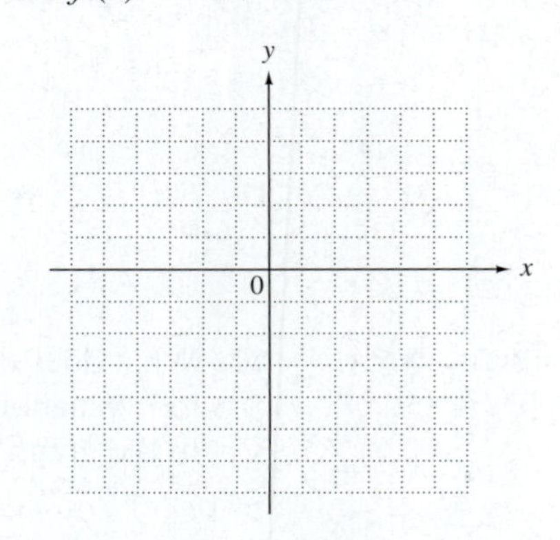

24. $f(x) = 2x^2 - 2$

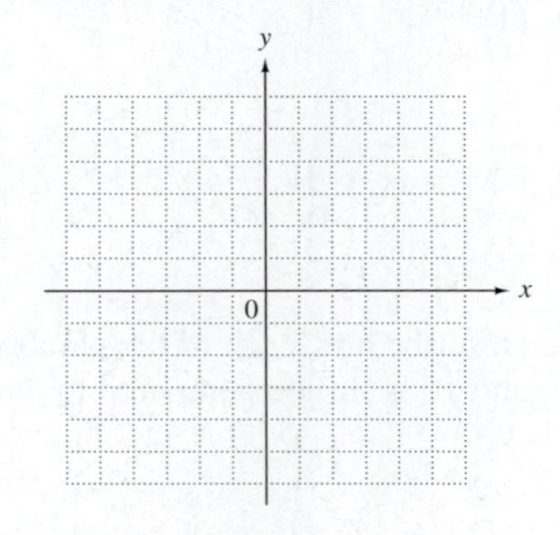

25. $f(x) = \frac{1}{2}(x - 4)^2$

axis:
domain:
range:

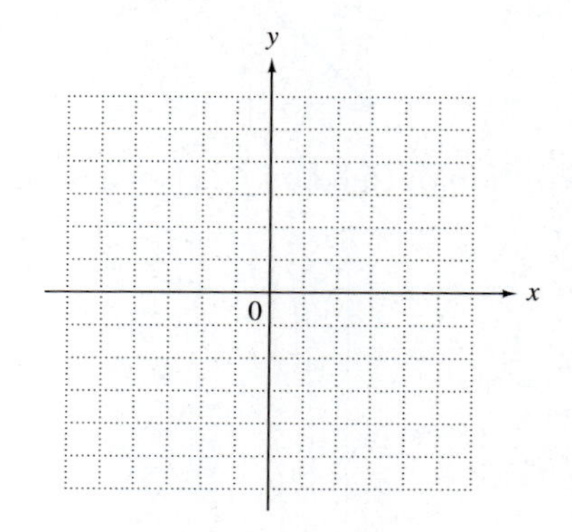

26. $f(x) = -2(x + 1)^2$

axis:
domain:
range:

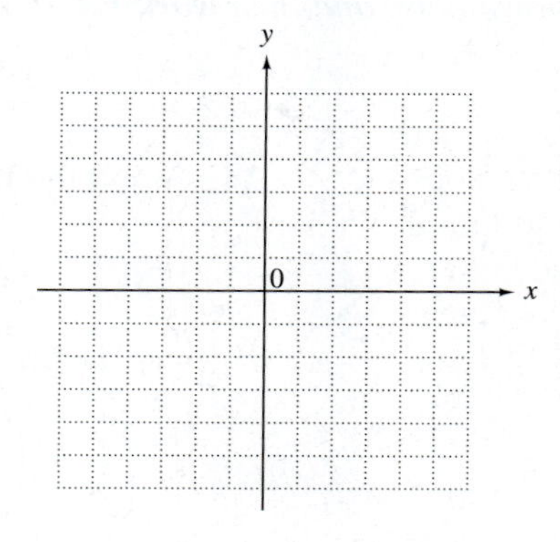

27. $f(x) = (x + 2)^2 - 1$

axis:
domain:
range:

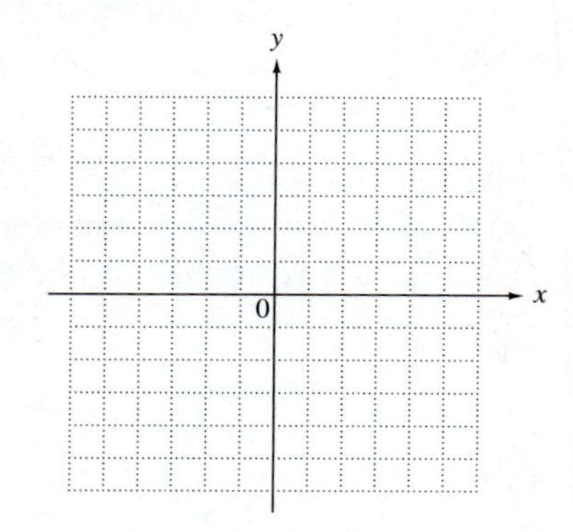

28. $f(x) = (x - 1)^2 + 2$

axis:
domain:
range:

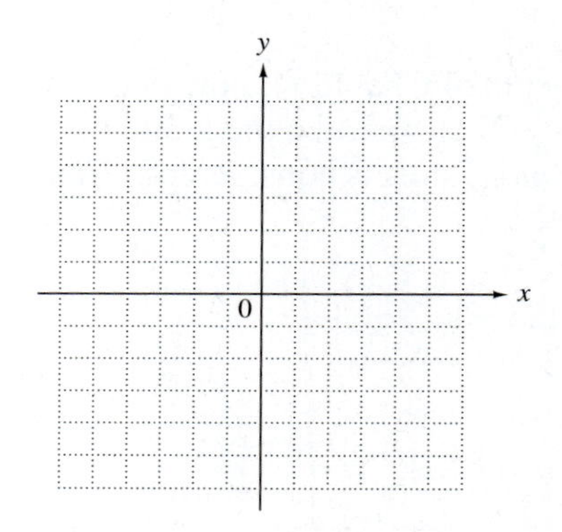

29. $f(x) = -2(x + 3)^2 + 4$

axis:
domain:
range:

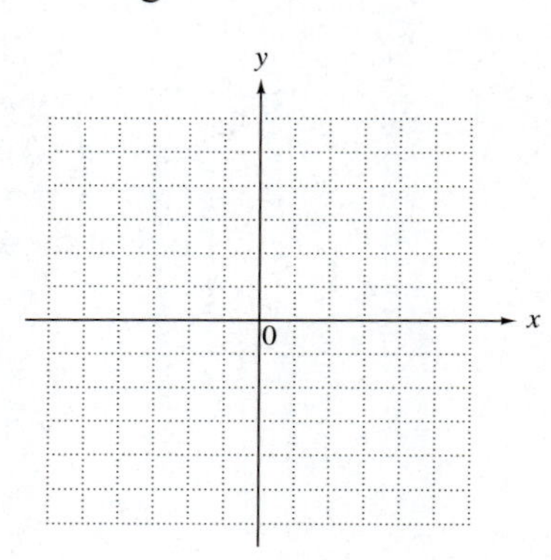

30. $f(x) = 2(x - 2)^2 - 3$

axis:
domain:
range:

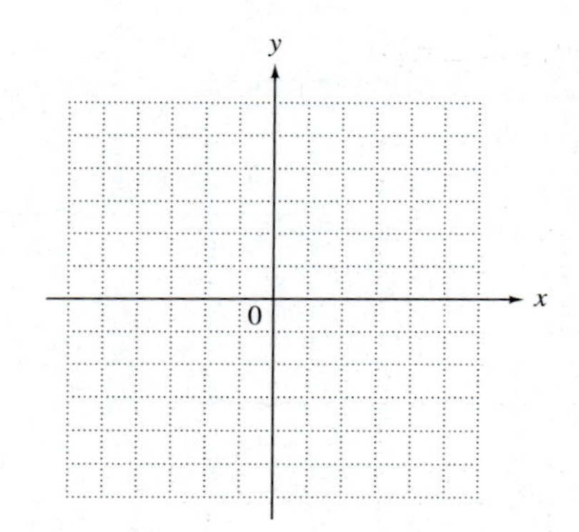

31. $f(x) = -\frac{2}{3}(x + 2)^2 + 1$

axis:
domain:
range:

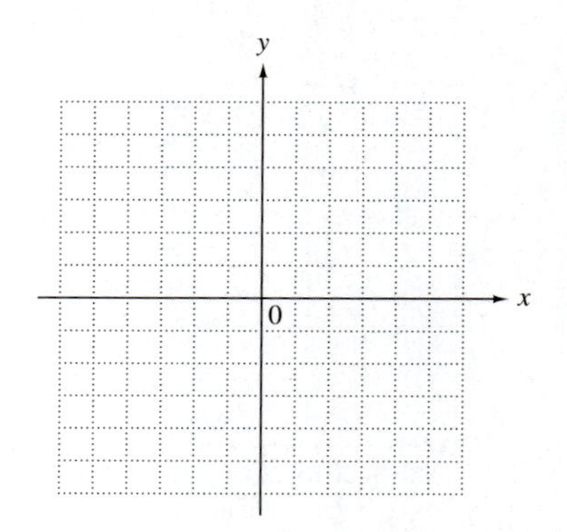

32. $f(x) = -\frac{1}{2}(x + 1)^2 + 2$

axis:
domain:
range:

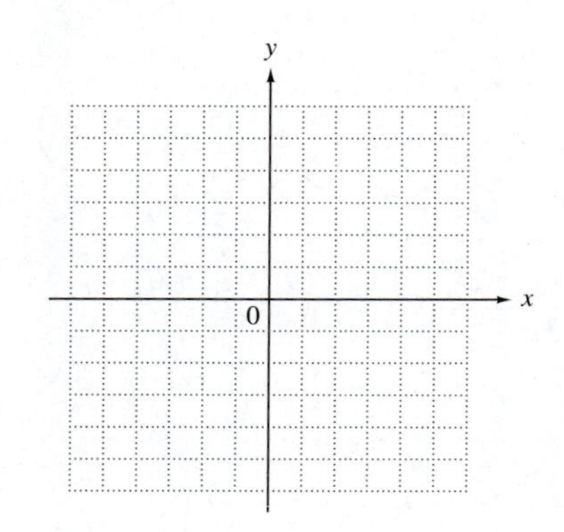

Relating Concepts (Exercises 33–38) For Individual or Group Work

The procedures described in this section that allow the graph of $f(x) = x^2$ to be shifted vertically and horizontally are applicable to other types of functions. In Section 4.5 we introduced linear functions of the form $g(x) = ax + b$. Consider the graph of the simplest linear function defined by $g(x) = x$, shown here, and then ***work Exercises 33–38 in order.***

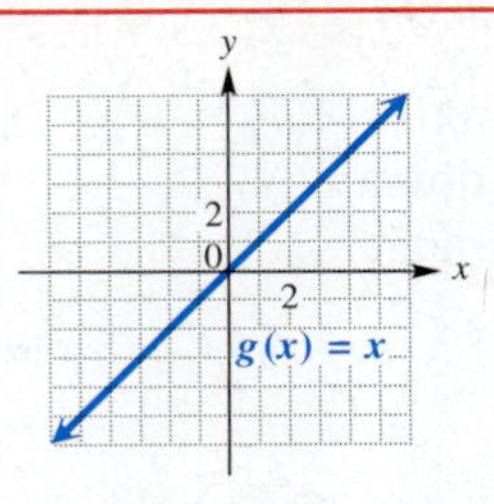

33. Based on the concepts of this section, how does the graph of $F(x) = x^2 + 6$ compare to the graph of $f(x) = x^2$ if a *vertical* shift is considered?

34. Graph the linear function defined by $G(x) = x + 6$.

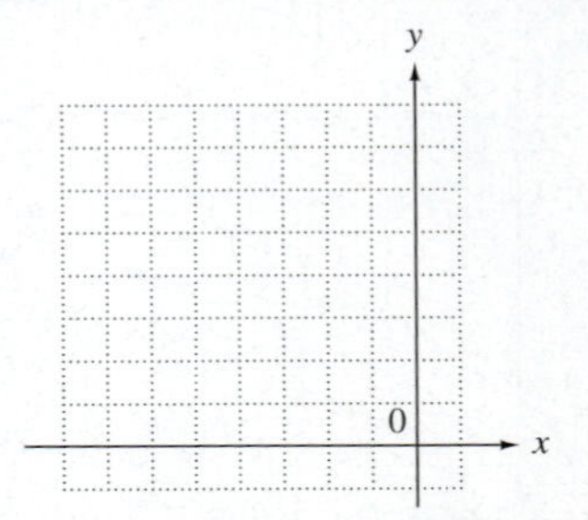

35. Based on the concepts of Chapter 4, how does the graph of $G(x) = x + 6$ compare to the graph of $g(x) = x$ if a *vertical* shift is considered? (*Hint:* Look at the y-intercept.)

36. Based on the concepts of this section, how does the graph of $F(x) = (x - 6)^2$ compare to the graph of $f(x) = x^2$ if a *horizontal* shift is considered?

37. Graph the linear function defined by $G(x) = x - 6$.

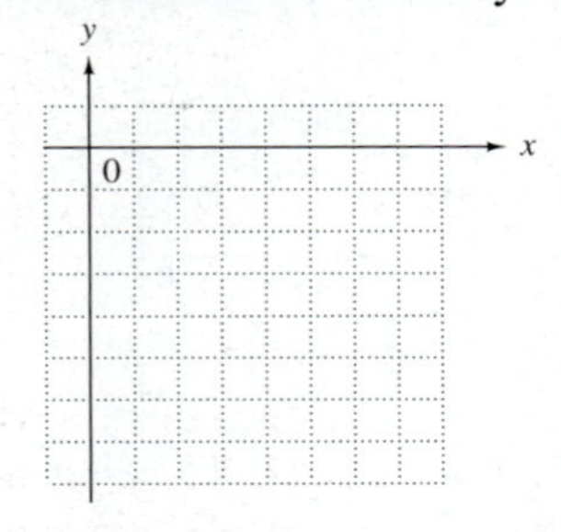

38. Based on the concepts of Chapter 4, how does the graph of $G(x) = x - 6$ compare to the graph of $g(x) = x$ if a *horizontal* shift is considered? (*Hint:* Look at the x-intercept.)

In Exercises 39–44, tell whether a linear or quadratic function would be a more appropriate model for each set of graphed data. If linear, tell whether the slope should be positive or negative. If quadratic, tell whether the coefficient a of x^2 should be positive or negative. See Example 6.

39.

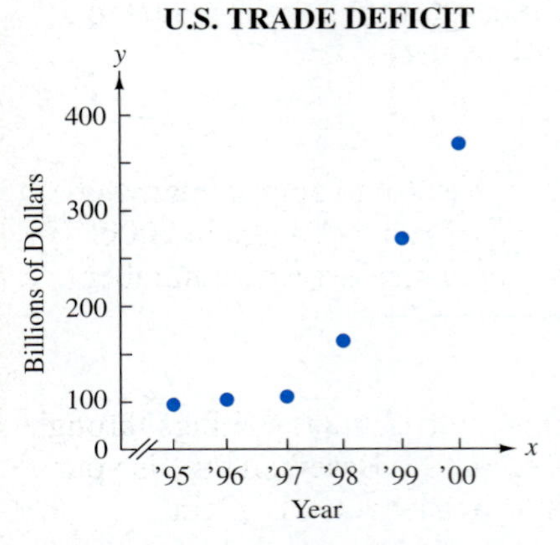

Source: U.S. Department of Commerce.

40.

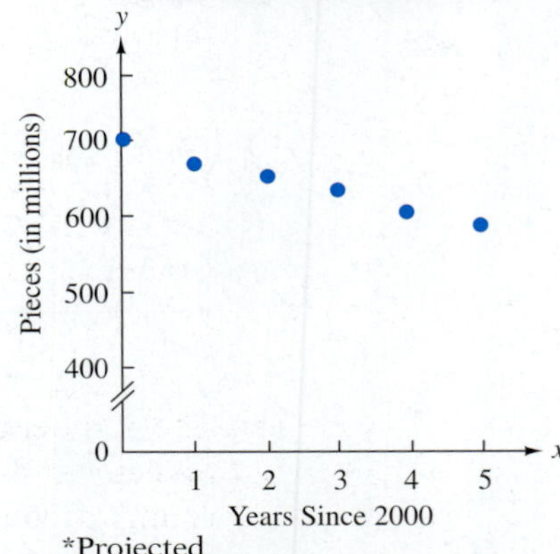

*Projected
Source: General Accounting Office.

41.

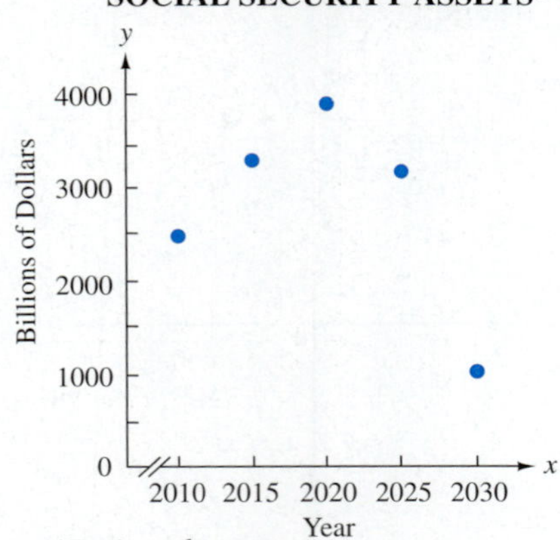

*Projected
Source: Social Security Administration.

42. **CEDAR RAPIDS SCHOOLS—GENERAL RESERVE FUND**

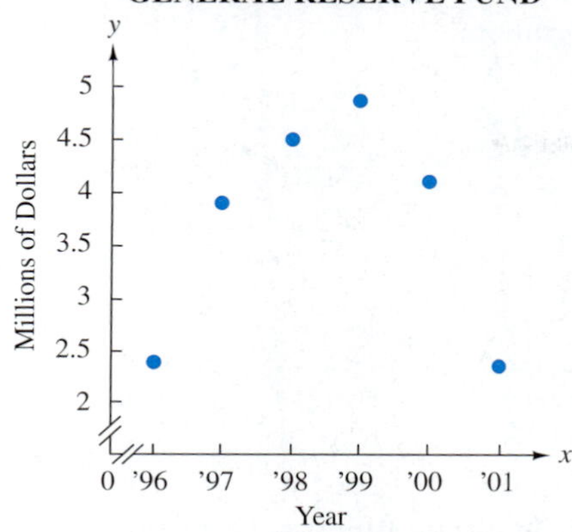

Source: Cedar Rapids School District.

43. **CONSUMER DEMAND FOR ELECTRICITY**

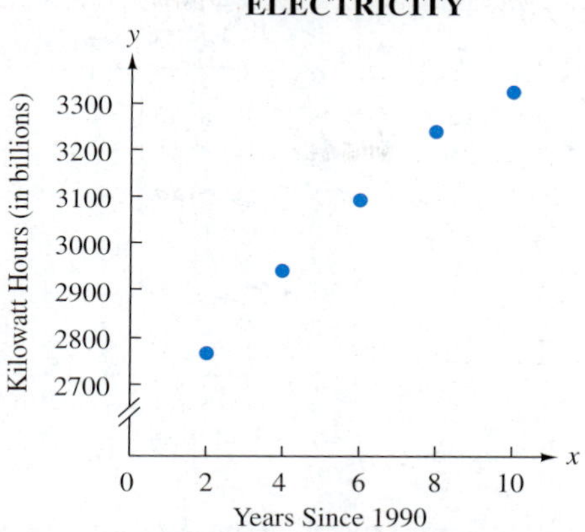

Source: U.S. Department of Energy.

44. **U.S. COMMERCIAL BANK FAILURES**

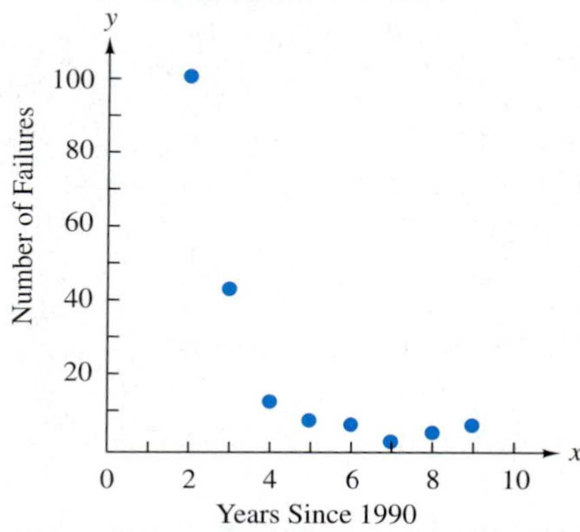

Source: www.ABA.com

Solve each problem. See Example 6.

45. The number of company bankruptcy filings for selected years between 1990 and 1999 are shown in the table. In the year column, 0 represents 1990, 2 represents 1992, and so on.

COMPANY BANKRUPTCY FILINGS

Year	Number of Bankruptcies
0	115
2	91
4	70
6	84
8	120
9	145

Source: www.BankruptcyData.com

(a) Use the ordered pairs (year, number of bankruptcies) to make a scatter diagram of the data.

COMPANY BANKRUPTCY FILINGS

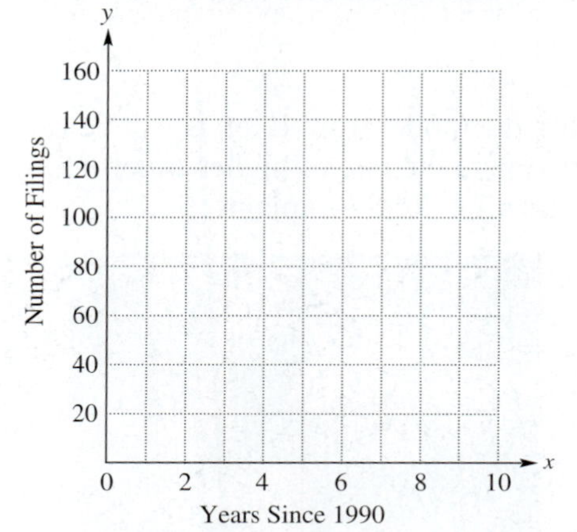

(b) Use the scatter diagram to decide whether a linear or quadratic function would better model the data. If quadratic, should the coefficient a of x^2 be positive or negative?

(c) Use the ordered pairs (0, 115), (4, 70), and (8, 120) to find a quadratic function that models the data. Round the values of a, b, and c in your model to three decimal places, as necessary.

(d) Use your model from part (c) to approximate the number of company bankruptcy filings in 2000. Round your answer to the nearest whole number.

(e) The number of company bankruptcy filings through September 8, 2000 was 124. Based on this, is your estimate from part (d) reasonable? Explain.

46. The number of new AIDS patients who survived the first year for the years from 1991 through 1997 are shown in the table. In the year column, 1 represents 1991, 2 represents 1992, and so on.

AIDS PATIENTS WHO SURVIVED THE FIRST YEAR

Year	Number of Patients
1	55
2	130
3	155
4	160
5	155
6	150
7	115

Source: HIV Health Services Planning Council.

(a) Use the ordered pairs (year, number of patients) to make a scatter diagram of the data.

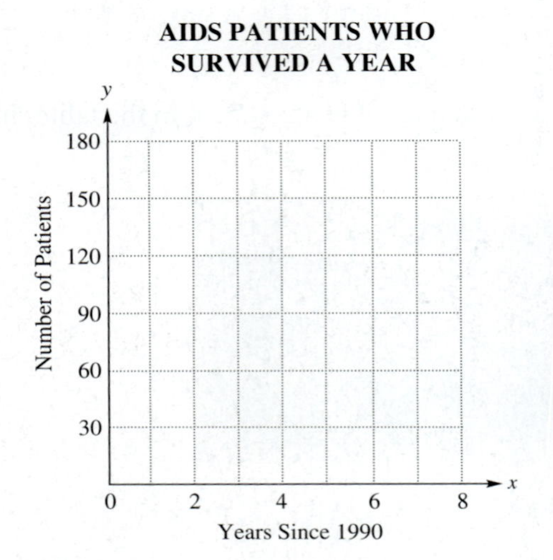

(b) Would a linear or quadratic function better model the data?

(c) Should the coefficient a of x^2 in a quadratic model be positive or negative?

(d) Use the ordered pairs (2, 130), (3, 155), and (7, 115) to find a quadratic function that models the data.

(e) Use your model from part (d) to approximate the number of AIDS patients who survived the first year in 1994 and 1996. How well does the model approximate the actual data from the table?

47. In Example 6, we determined that the quadratic function defined by

$$y = .2455x^2 - 1.856x + 30.7105$$

modeled the rate of higher-order multiple births, where x represents the number of years since 1970.

(a) Use this model to approximate the rate of higher-order births in 1998 to the nearest tenth.

(b) The actual rate of higher-order births in 1998 was 193.5. (*Source:* National Center for Health Statistics.) How does the approximation using the model compare to the actual rate for 1998?

48. Should the model from Exercise 47 be used to approximate the rate of higher-order multiple births in years after 1998? Explain.

9.6 More about Parabolas; Applications

Objectives

1. Find the vertex of a vertical parabola.
2. Graph a quadratic function.
3. Use the discriminant to find the number of x-intercepts of a vertical parabola.
4. Use quadratic functions to solve problems involving maximum or minimum value.
5. Graph horizontal parabolas.

1 Find the vertex of a vertical parabola. When the equation of a parabola is given in the form $f(x) = ax^2 + bx + c$, we need to locate the vertex in order to sketch an accurate graph. There are two ways to do this: complete the square as shown in Examples 1 and 2, or use a formula derived by completing the square.

Example 1 Completing the Square to Find the Vertex

Find the vertex of the graph of $f(x) = x^2 - 4x + 5$.

To find the vertex, we need to express $x^2 - 4x + 5$ in the form $(x - h)^2 + k$. We do this by completing the square on $x^2 - 4x$, as in Section 9.1. The process is a little different here because we want to keep $f(x)$ alone on one side of the equation. Instead of adding the appropriate number to each side, we *add and subtract* it on the right. This is equivalent to adding 0.

$$f(x) = x^2 - 4x + 5$$

$$= (x^2 - 4x \qquad) + 5 \qquad \text{Group the variable terms.}$$

$$\left[\frac{1}{2}(-4)\right]^2 = (-2)^2 = 4$$

$$= (x^2 - 4x + 4 - 4) + 5 \qquad \text{Add and subtract 4.}$$

$$= (x^2 - 4x + 4) - 4 + 5 \qquad \text{Bring } -4 \text{ outside the parentheses.}$$

$$f(x) = (x - 2)^2 + 1 \qquad \text{Factor; combine terms.}$$

The vertex of this parabola is (2, 1).

Work Problem 1 at the Side.

1 Find the vertex of each parabola.

(a) $f(x) = x^2 - 6x + 7$

(b) $f(x) = x^2 + 4x - 9$

Example 2 Completing the Square to Find the Vertex When $a \neq 1$

Find the vertex of the graph of $f(x) = -3x^2 + 6x - 1$.

We must complete the square on $-3x^2 + 6x$. Because the x^2-term has a coefficient other than 1, we factor that coefficient out of the first two terms and then proceed as in Example 1.

$$f(x) = -3x^2 + 6x - 1$$

$$= -3(x^2 - 2x) - 1 \qquad \text{Factor out } -3.$$

$$\left[\frac{1}{2}(-2)\right]^2 = (-1)^2 = 1$$

$$= -3(x^2 - 2x + 1 - 1) - 1 \qquad \text{Add and subtract 1.}$$

$$= -3(x^2 - 2x + 1) + (-3)(-1) - 1 \qquad \text{Distributive property}$$

$$= -3(x^2 - 2x + 1) + 3 - 1$$

$$f(x) = -3(x - 1)^2 + 2 \qquad \text{Factor; combine terms.}$$

The vertex is (1, 2).

Work Problem 2 at the Side.

2 Find the vertex of each parabola.

(a) $f(x) = 2x^2 - 4x + 1$

(b) $f(x) = -\frac{1}{2}x^2 + 2x - 3$

Answers

1. (a) $(3, -2)$ (b) $(-2, -13)$
2. (a) $(1, -1)$ (b) $(2, -1)$

3 Use the formula to find the vertex of the graph of each quadratic function.

(a) $f(x) = -2x^2 + 3x - 1$

(b) $f(x) = 4x^2 - x + 5$

To derive a formula for the vertex of the graph of the quadratic function $y = ax^2 + bx + c$, complete the square on the standard form of the equation.

$$f(x) = ax^2 + bx + c \quad (a \neq 0) \qquad \text{Standard form}$$

$$= a\left(x^2 + \frac{b}{a}x\right) + c \qquad \text{Factor } a \text{ from the first two terms.}$$

$$\left[\frac{1}{2}\left(\frac{b}{a}\right)\right]^2 = \left(\frac{b}{2a}\right)^2 = \frac{b^2}{4a^2}$$

$$= a\left(x^2 + \frac{b}{a}x + \frac{b^2}{4a^2} - \frac{b^2}{4a^2}\right) + c \qquad \text{Add and subtract } \tfrac{b^2}{4a^2}.$$

$$= a\left(x^2 + \frac{b}{a}x + \frac{b^2}{4a^2}\right) + a\left(-\frac{b^2}{4a^2}\right) + c \qquad \text{Distributive property}$$

$$= a\left(x^2 + \frac{b}{a}x + \frac{b^2}{4a^2}\right) - \frac{b^2}{4a} + c$$

$$= a\left(x + \frac{b}{2a}\right)^2 + \frac{4ac - b^2}{4a} \qquad \text{Factor; combine terms.}$$

$$f(x) = a\Bigg[x - \underbrace{\left(\frac{-b}{2a}\right)}_{h}\Bigg]^2 + \underbrace{\frac{4ac - b^2}{4a}}_{k} \qquad f(x) = (x - h)^2 + k$$

This equation shows that the vertex (h, k) can be expressed in terms of a, b, and c. However, it is not necessary to remember this expression for k, since it can be found by replacing x with $\frac{-b}{2a}$. Using function notation, if $y = f(x)$, the y-value of the vertex is $f(\frac{-b}{2a})$.

Vertex Formula

The graph of the quadratic function defined by $f(x) = ax^2 + bx + c$ has vertex

$$\left(\frac{-b}{2a}, f\left(\frac{-b}{2a}\right)\right),$$

and the axis of the parabola is the line

$$x = \frac{-b}{2a}.$$

Example 3 Using the Formula to Find the Vertex

Use the vertex formula to find the vertex of the graph of

$$f(x) = x^2 - x - 6.$$

For this function, $a = 1$, $b = -1$, and $c = -6$. The x-coordinate of the vertex of the parabola is given by

$$\frac{-b}{2a} = \frac{-(-1)}{2(1)} = \frac{1}{2}.$$

The y-coordinate is $f(\frac{-b}{2a}) = f(\frac{1}{2})$.

$$f\left(\frac{1}{2}\right) = \left(\frac{1}{2}\right)^2 - \frac{1}{2} - 6 = \frac{1}{4} - \frac{1}{2} - 6 = -\frac{25}{4}$$

The vertex is $(\frac{1}{2}, -\frac{25}{4})$.

Work Problem 3 at the Side.

ANSWERS

3. (a) $\left(\frac{3}{4}, \frac{1}{8}\right)$ (b) $\left(\frac{1}{8}, \frac{79}{16}\right)$

2 Graph a quadratic function. We give a general approach for graphing any quadratic function here.

Graphing a Quadratic Function f

Step 1 **Determine whether the graph opens up or down.** If $a > 0$, the parabola opens up; if $a < 0$, it opens down.

Step 2 **Find the vertex.** Use either the vertex formula or completing the square.

Step 3 **Find any intercepts.** To find the x-intercepts (if any), solve $f(x) = 0$. To find the y-intercept, evaluate $f(0)$.

Step 4 **Complete the graph.** Plot the points found so far. Find and plot additional points as needed, using symmetry about the axis.

Example 4 Using the Steps to Graph a Quadratic Function

Graph the quadratic function defined by

$$f(x) = x^2 - x - 6.$$

Step 1 From the equation, $a = 1$, so the graph of the function opens up.

Step 2 The vertex, $\left(\frac{1}{2}, -\frac{25}{4}\right)$, was found in Example 3 by substituting the values $a = 1$, $b = -1$, and $c = -6$ in the vertex formula.

Step 3 Now find any intercepts. Since the vertex, $\left(\frac{1}{2}, -\frac{25}{4}\right)$, is in quadrant IV and the graph opens up, there will be two x-intercepts. To find them, let $f(x) = 0$ and solve the equation.

$$\begin{aligned} f(x) &= x^2 - x - 6 \\ 0 &= x^2 - x - 6 && \text{Let } f(x) = 0. \\ 0 &= (x - 3)(x + 2) && \text{Factor.} \end{aligned}$$

$$x - 3 = 0 \quad \text{or} \quad x + 2 = 0 \qquad \text{Zero-factor property}$$

$$x = 3 \quad \text{or} \quad x = -2$$

The x-intercepts are $(3, 0)$ and $(-2, 0)$. To find the y-intercept, evaluate $f(0)$.

$$\begin{aligned} f(x) &= x^2 - x - 6 \\ f(0) &= 0^2 - 0 - 6 && \text{Let } x = 0. \\ f(0) &= -6 \end{aligned}$$

The y-intercept is $(0, -6)$.

Step 4 Plot the points found so far and additional points as needed using symmetry about the axis $x = \frac{1}{2}$. The graph is shown in Figure 12. The domain is $(-\infty, \infty)$, and the range is $\left[-\frac{25}{4}, \infty\right)$.

x	y
-2	0
-1	-4
0	-6
$\frac{1}{2}$	$-\frac{25}{4}$
2	-4
3	0

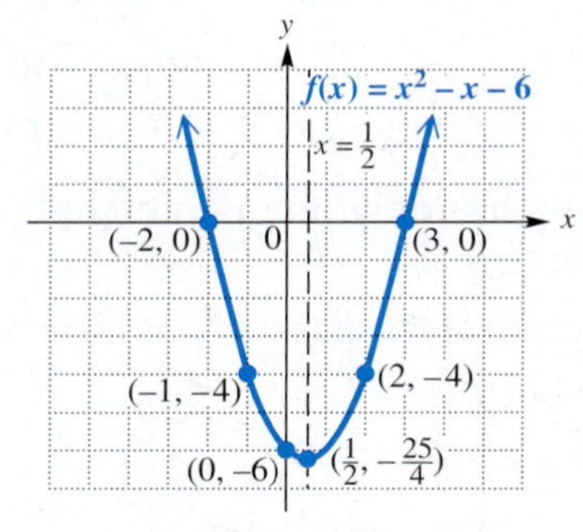

Figure 12

Work Problem 4 at the Side.

4 Graph the quadratic function defined by

$$f(x) = x^2 - 6x + 5.$$

Give the axis, domain, and range.

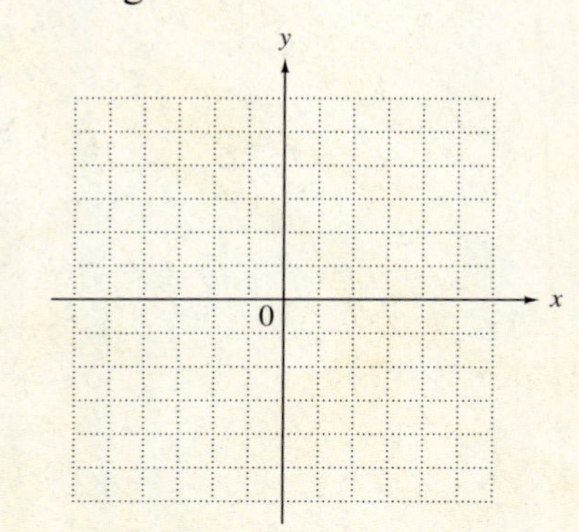

Answers

4.

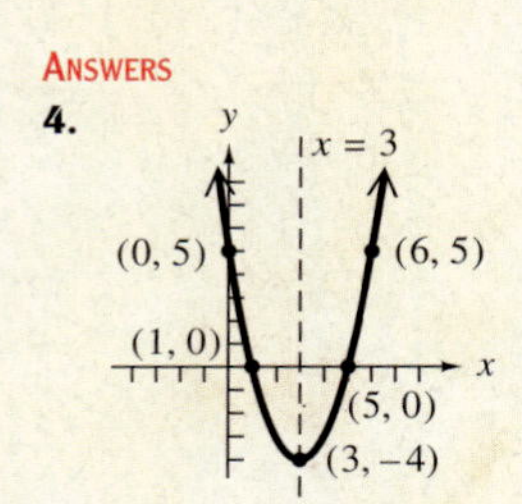

axis: $x = 3$; domain: $(-\infty, \infty)$; range: $[-4, \infty)$

5 Use the discriminant to determine the number of x-intercepts of the graph of each quadratic function.

(a) $f(x) = 4x^2 - 20x + 25$

(b) $f(x) = 2x^2 + 3x + 5$

(c) $f(x) = -3x^2 - x + 2$

3 Use the discriminant to find the number of x-intercepts of a vertical parabola. The graph of a quadratic function may have two x-intercepts, one x-intercept, or no x-intercepts, as shown in Figure 13. Recall from Section 9.2 that $b^2 - 4ac$ is called the *discriminant* of the quadratic equation $ax^2 + bx + c = 0$ and that we can use it to determine the number of real solutions of a quadratic equation.

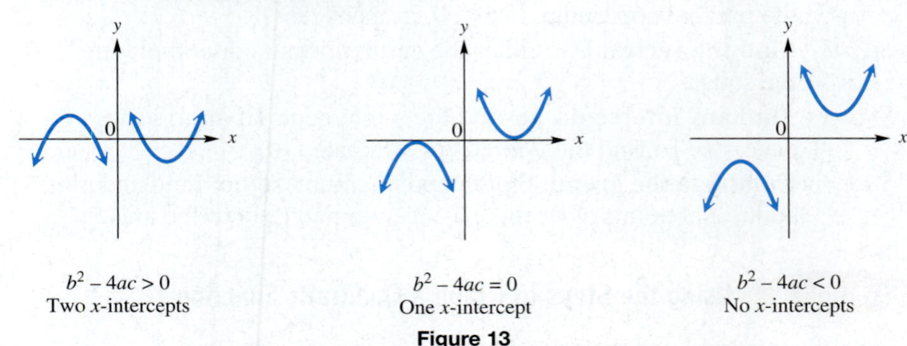

Figure 13

In a similar way, we can use the discriminant of a quadratic *function* to determine the number of x-intercepts of its graph. If the discriminant is positive, the parabola will have two x-intercepts. If the discriminant is 0, there will be only one x-intercept, and it will be the vertex of the parabola. If the discriminant is negative, the graph will have no x-intercepts.

Example 5 Using the Discriminant to Determine the Number of x-Intercepts

Use the discriminant to determine the number of x-intercepts of the graph of each quadratic function.

(a) $f(x) = 2x^2 + 3x - 5$

The discriminant is $b^2 - 4ac$. Here $a = 2$, $b = 3$, and $c = -5$, so

$$\begin{aligned} b^2 - 4ac &= 9 - 4(2)(-5) \\ &= 49. \end{aligned}$$

Since the discriminant is positive, the parabola has two x-intercepts.

(b) $f(x) = -3x^2 - 1$

In this equation, $a = -3$, $b = 0$, and $c = -1$. The discriminant is

$$\begin{aligned} b^2 - 4ac &= 0 - 4(-3)(-1) \\ &= -12. \end{aligned}$$

The discriminant is negative, so the graph has no x-intercepts.

(c) $f(x) = 9x^2 + 6x + 1$

Here, $a = 9$, $b = 6$, and $c = 1$. The discriminant is

$$\begin{aligned} b^2 - 4ac &= 36 - 4(9)(1) \\ &= 0. \end{aligned}$$

The parabola has only one x-intercept (its vertex) because the value of the discriminant is 0.

Work Problem 5 at the Side.

ANSWERS

5. (a) discriminant is 0; one x-intercept
(b) discriminant is -31; no x-intercepts
(c) discriminant is 25; two x-intercepts

4 Use quadratic functions to solve problems involving maximum or minimum value. The vertex of a parabola is either the highest or the lowest point on the parabola. The y-value of the vertex gives the maximum or minimum value of y, while the x-value tells where that maximum or minimum occurs.

In many applied problems we must find the largest or smallest value of some quantity. When we can express that quantity as a quadratic function, the value of k in the vertex gives that optimum value.

Example 6 Finding the Maximum Area of a Rectangular Region

A farmer has 120 ft of fencing. He wants to put a fence around a rectangular field next to a building. Find the maximum area he can enclose.

Figure 14

Figure 14 shows the field. Let x represent the width of the field. Since he has 120 ft of fencing,

$$x + x + \text{length} = 120 \quad \text{Sum of the sides is 120 ft.}$$
$$2x + \text{length} = 120 \quad \text{Combine terms.}$$
$$\text{length} = 120 - 2x. \quad \text{Subtract } 2x.$$

The area is given by the product of the width and length, so

$$A = x(120 - 2x)$$
$$= 120x - 2x^2.$$

To determine the maximum area, find the vertex of the parabola given by $A = 120x - 2x^2$ using the vertex formula. Writing the equation in standard form as $A = -2x^2 + 120x$ gives $a = -2$, $b = 120$, and $c = 0$, so

$$h = \frac{-b}{2a} = \frac{-120}{2(-2)} = \frac{-120}{-4} = 30;$$
$$f(30) = -2(30)^2 + 120(30) = -2(900) + 3600 = 1800.$$

The graph is a parabola that opens down, and its vertex is (30, 1800). Thus, the maximum area will be 1800 ft^2. This area will occur if x, the width of the field, is 30 ft.

6 Solve Example 6 if the farmer has only 100 ft of fencing.

7 Solve the problem.

A toy rocket is launched from the ground so that its distance in feet above the ground after t seconds is

$$s(t) = -16t^2 + 208t.$$

Find the maximum height it reaches and the number of seconds it takes to reach that height.

CAUTION

Be careful when interpreting the meanings of the coordinates of the vertex. The first coordinate, x, gives the value for which the *function value* is a maximum or a minimum. Be sure to read the problem carefully to determine whether you are asked to find the value of the independent variable, the function value, or both.

Work Problem 6 at the Side.

Example 7 Finding the Maximum Height Attained by a Projectile

If air resistance is neglected, a projectile on Earth shot straight upward with an initial velocity of 40 m per sec will be at a height s in meters given by

$$s(t) = -4.9t^2 + 40t,$$

where t is the number of seconds elapsed after projection. After how many seconds will it reach its maximum height, and what is this maximum height?

For this function, $a = -4.9$, $b = 40$, and $c = 0$. Use the vertex formula.

$$h = \frac{-b}{2a} = \frac{-40}{2(-4.9)} \approx \mathbf{4.1} \quad \text{Use a calculator.}$$

This indicates that the maximum height is attained at 4.1 sec. To find this maximum height, calculate $f(4.1)$.

$$f(\mathbf{4.1}) = -4.9(\mathbf{4.1})^2 + 40(\mathbf{4.1})$$

$$\approx \mathbf{81.6} \quad \text{Use a calculator.}$$

The projectile will attain a maximum height of approximately 81.6 m.

Work Problem 7 at the Side.

5 Graph horizontal parabolas. If x and y are interchanged in the equation $y = ax^2 + bx + c$, the equation becomes $x = ay^2 + by + c$. Because of the interchange of the roles of x and y, these parabolas are horizontal (with horizontal lines as axes), compared with the vertical ones graphed previously.

Graph of a Horizontal Parabola

The graph of

$$x = ay^2 + by + c \quad \text{or} \quad x = a(y - k)^2 + h$$

is a parabola with vertex (h, k) and the horizontal line $y = k$ as axis. The graph opens to the right if $a > 0$ and to the left if $a < 0$.

Example 8 Graphing a Horizontal Parabola

Graph $x = (y - 2)^2 - 3$.

This graph has its vertex at $(-3, 2)$, since the roles of x and y are reversed. It opens to the right, the positive x-direction, and has the same shape as $y = x^2$. Plotting a few additional points gives the graph shown in Figure 15. Note that the graph is symmetric about its axis, $y = 2$. The domain is $[-3, \infty)$, and the range is $(-\infty, \infty)$.

Continued on Next Page

ANSWERS
6. The field should be 25 ft by 50 ft with a maximum area of 1250 ft^2.
7. 676 ft; 6.5 sec

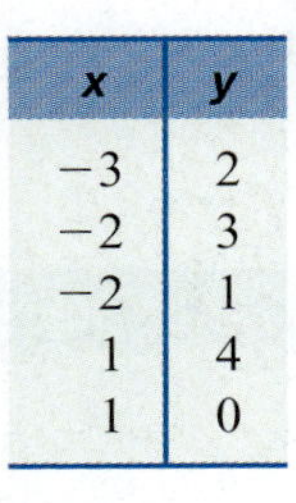

x	y
−3	2
−2	3
−2	1
1	4
1	0

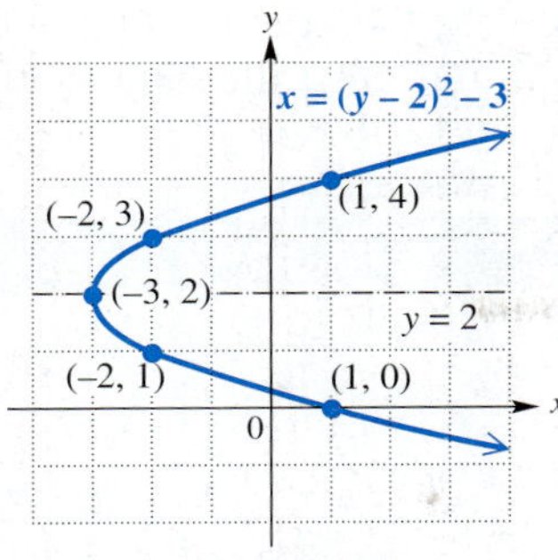

Figure 15

Work Problem 8 at the Side.

When a quadratic equation is given in the form $x = ay^2 + by + c$, completing the square on y will allow us to find the vertex.

Example 9 Completing the Square to Graph a Horizontal Parabola

Graph $x = -2y^2 + 4y - 3$. Give the domain and range of the relation.

$$x = -2y^2 + 4y - 3$$

$$= -2(y^2 - 2y) - 3 \qquad \text{Factor out } -2.$$

$$= -2(y^2 - 2y + 1 - 1) - 3 \qquad \text{Complete the square; add and subtract 1.}$$

$$= -2(y^2 - 2y + 1) + (-2)(-1) - 3 \qquad \text{Distributive property}$$

$$x = -2(y - 1)^2 - 1 \qquad \text{Factor; simplify.}$$

Because of the negative coefficient (-2), the graph opens to the left (the negative x-direction) and is narrower than the graph of $y = x^2$. As shown in Figure 16, the vertex is $(-1, 1)$. The domain is $(-\infty, -1]$, and the range is $(-\infty, \infty)$.

x	y
−3	2
−3	0
−1	1

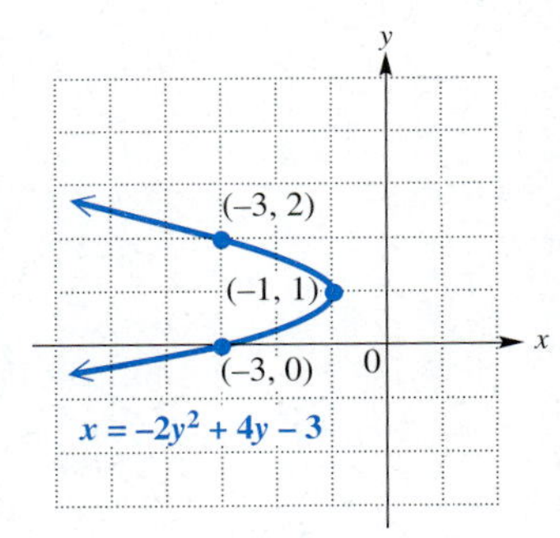

Figure 16

Work Problem 9 at the Side.

CAUTION

Only quadratic equations solved for y (whose graphs are vertical parabolas) are examples of functions. The horizontal parabolas in Examples 8 and 9 are *not* graphs of functions, because they do not satisfy the vertical line test. Furthermore, the vertex formula given earlier does not apply to parabolas with horizontal axes.

8 Graph $x = (y + 1)^2 - 4$. Give the axis, domain, and range.

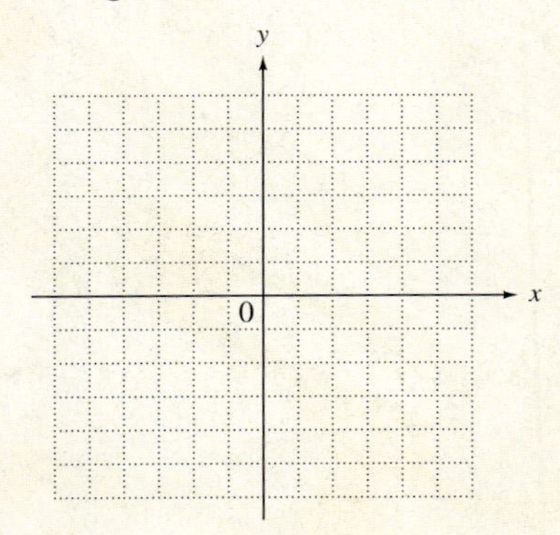

9 Find the vertex of each parabola. Tell whether the graph opens to the right or to the left. Give the domain and range.

(a) $x = 2y^2 - 6y + 5$

(b) $x = -y^2 + 2y + 5$

ANSWERS

8.

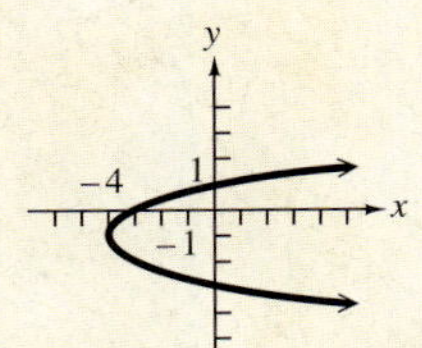

axis: $y = -1$; domain: $[-4, \infty)$; range: $(-\infty, \infty)$

9. (a) $\left(\frac{1}{2}, \frac{3}{2}\right)$; right; domain: $\left[\frac{1}{2}, \infty\right)$; range: $(-\infty, \infty)$

(b) $(6, 1)$; left; domain: $(-\infty, 6]$; range: $(-\infty, \infty)$

In summary, the graphs of parabolas studied in this section and the previous one fall into the following categories.

GRAPHS OF PARABOLAS

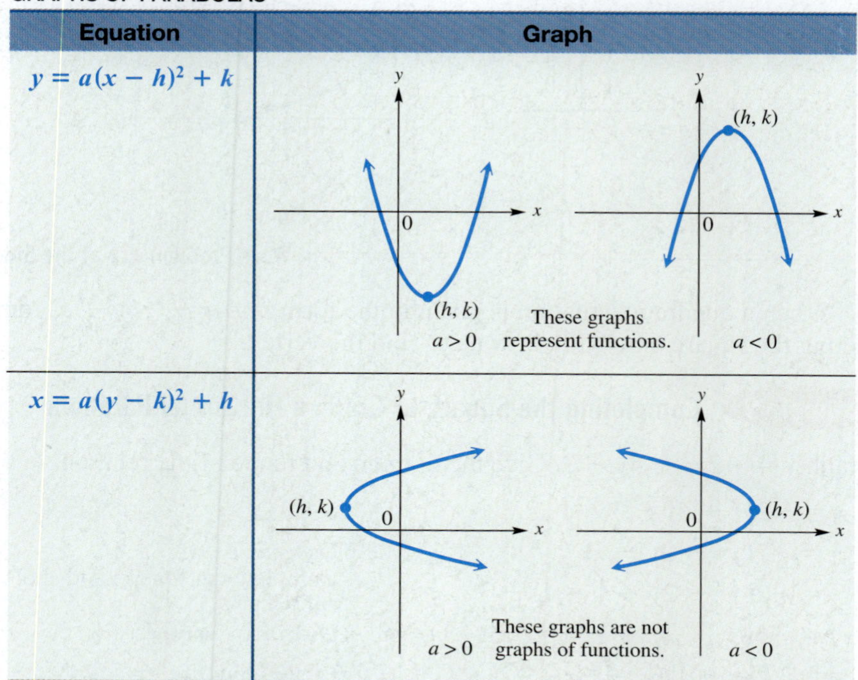

9.6 EXERCISES

FOR EXTRA HELP

Student's Solutions Manual | MyMathLab.com | InterAct Math Tutorial Software | AW Math Tutor Center |  www.mathxl.com |  Digital Video Tutor CD 8 Videotape 16

1. How can you determine just by looking at the equation of a parabola whether it has a vertical or a horizontal axis?

2. Why can't the graph of a quadratic function be a horizontal parabola?

3. How can you determine the number of x-intercepts of the graph of a quadratic function without graphing the function?

4. If the vertex of the graph of a quadratic function is $(1, -3)$ and the graph opens down, how many x-intercepts does the graph have?

Find the vertex of each parabola. For each equation, decide whether the graph opens up, down, to the left, or to the right, and whether it is wider, narrower, or the same shape as the graph of $y = x^2$. If it is a vertical parabola, use the discriminant to determine the number of x-intercepts. See Examples 1–3, 5, 8, and 9.

5. $y = 2x^2 + 4x + 5$

6. $y = 3x^2 - 6x + 4$

7. $y = -x^2 + 5x + 3$

8. $x = -y^2 + 7y - 2$

9. $x = \frac{1}{3}y^2 + 6y + 24$

10. $x = \frac{1}{2}y^2 + 10y - 5$

Graph each parabola using the techniques described in this section. Give the domain and range. See Examples 4, 8, and 9.

11. $f(x) = x^2 + 4x + 3$
domain:
range:

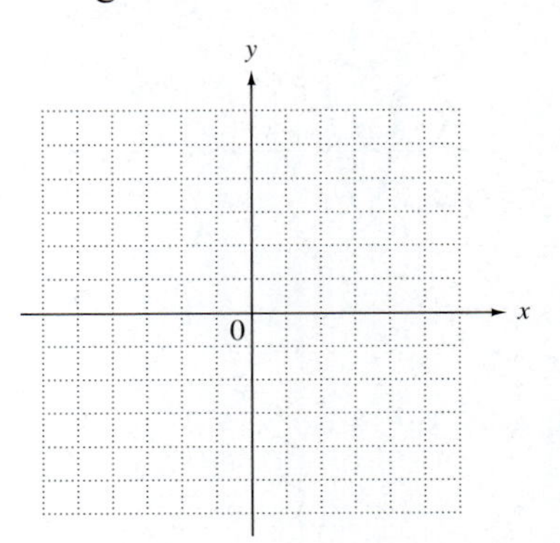

12. $f(x) = x^2 + 2x - 2$
domain:
range:

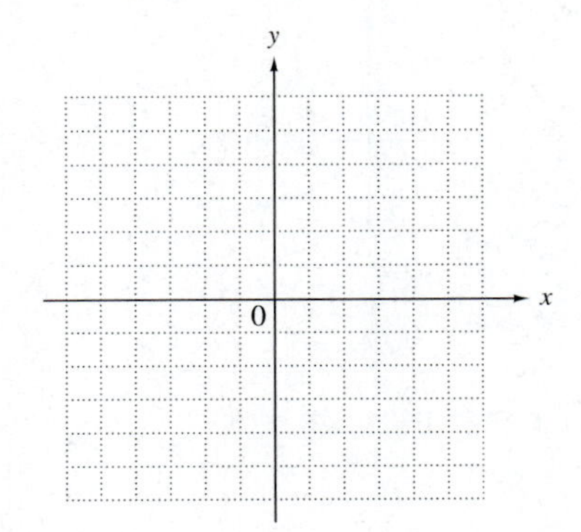

13. $f(x) = -2x^2 + 4x - 5$
domain:
range:

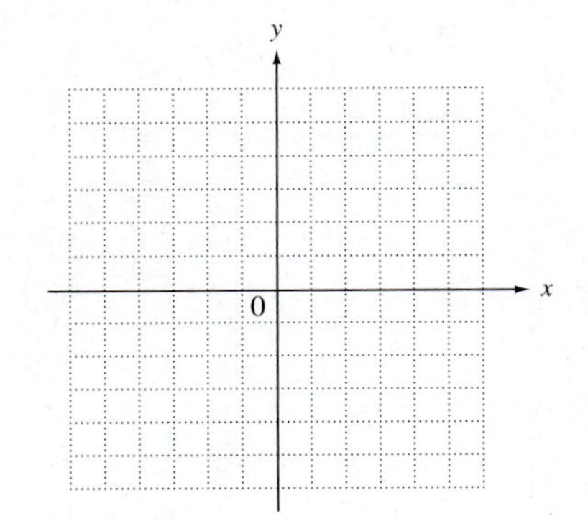

14. $f(x) = -3x^2 + 12x - 8$
domain:
range:

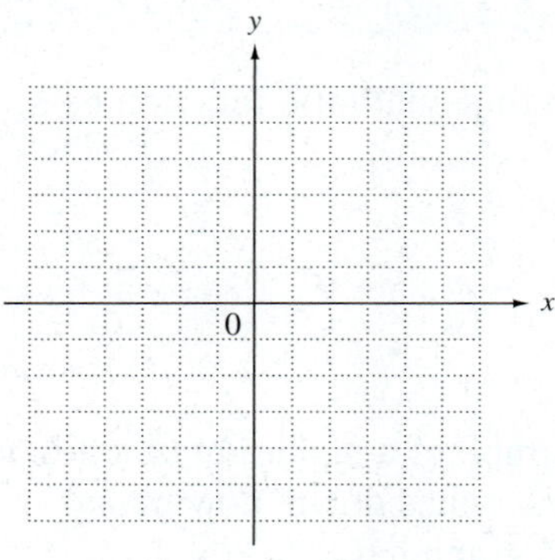

15. $x = -\frac{1}{5}y^2 + 2y - 4$
domain:
range:

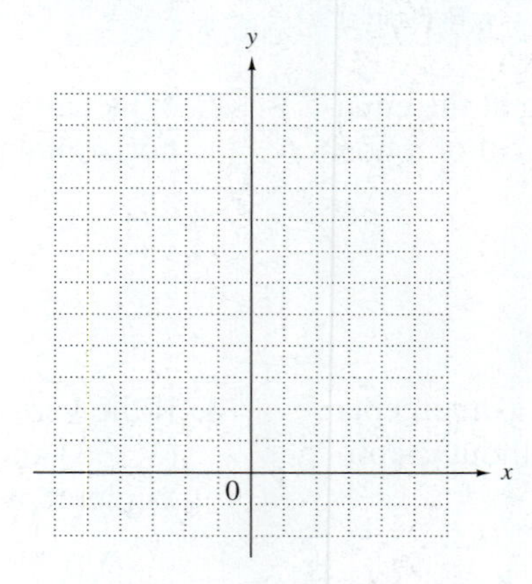

16. $x = -\frac{1}{2}y^2 - 4y - 6$
domain:
range:

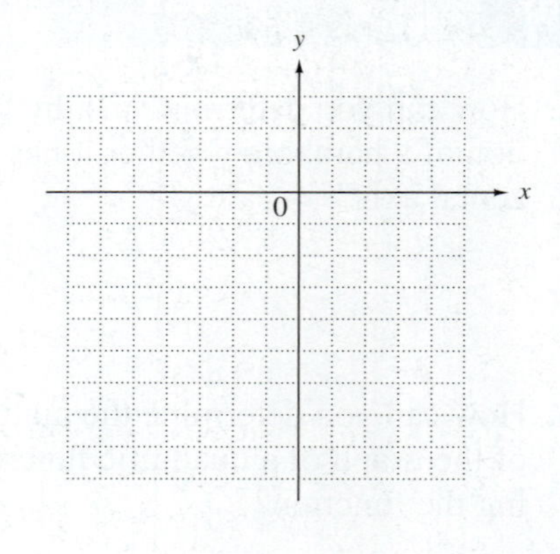

17. $x = 3y^2 + 12y + 5$
domain:
range:

18. $x = 4y^2 + 16y + 11$
domain:
range:

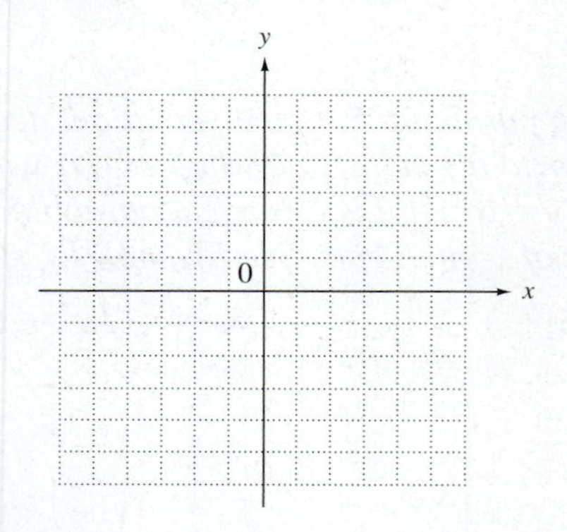

Use the concepts of this section to match each equation with its graph.

A.

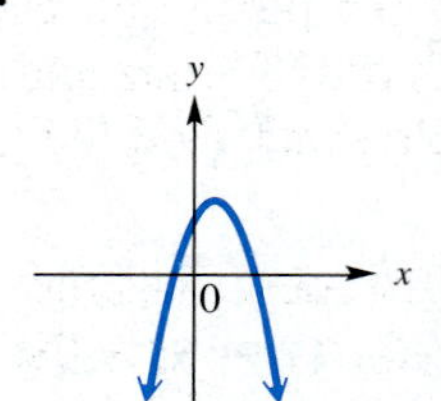

B.

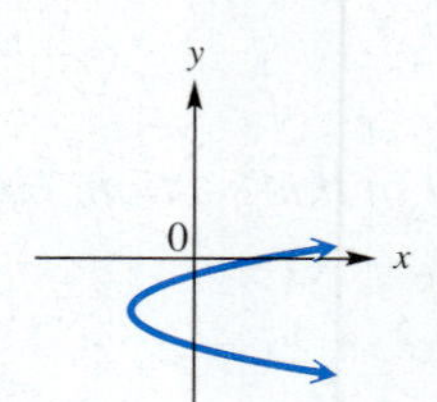

C.

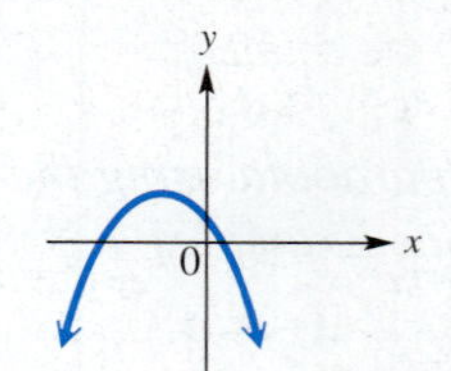

D.

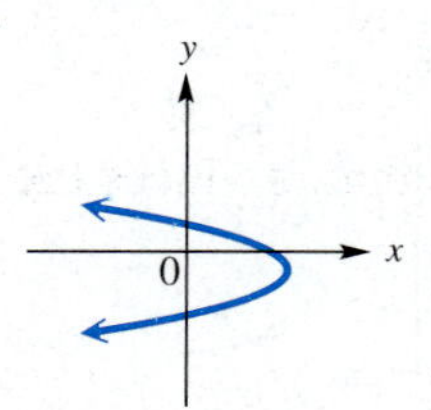

E.

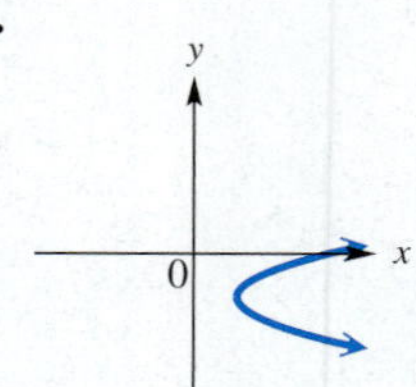

F.

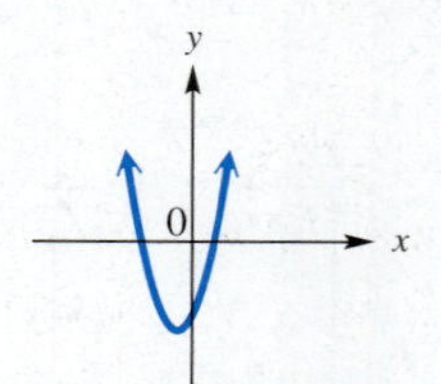

19. $y = 2x^2 + 4x - 3$

20. $y = -x^2 + 3x + 5$

21. $y = -\frac{1}{2}x^2 - x + 1$

22. $x = y^2 + 6y + 3$

23. $x = -y^2 - 2y + 4$

24. $x = 3y^2 + 6y + 5$

Solve each problem. See Examples 6 and 7.

25. Palo Alto College is planning to construct a rectangular parking lot on land bordered on one side by a highway. The plan is to use 640 ft of fencing to fence off the other three sides. What should the dimensions of the lot be if the enclosed area is to be a maximum?

26. Keisha Hughes has 100 m of fencing material to enclose a rectangular exercise run for her dog. What width will give the enclosure the maximum area?

27. Find the pair of numbers whose sum is 60 and whose product is a maximum. (*Hint:* Let x and $60 - x$ represent the two numbers.)

28. Find the pair of numbers whose sum is 10 and whose product is a maximum.

29. If an object on Earth is propelled upward with an initial velocity of 32 ft per sec, then its height (in feet) after t seconds is given by

$$h(t) = 32t - 16t^2.$$

Find the maximum height attained by the object and the number of seconds it takes to hit the ground.

30. A projectile on Earth is fired straight upward so that its distance (in feet) above the ground t seconds after firing is given by

$$s(t) = -16t^2 + 400t.$$

Find the maximum height it reaches and the number of seconds it takes to reach that height.

31. A charter flight charges a fare of $200 per person, plus $4 per person for each unsold seat on the plane. If the plane holds 100 passengers and if x represents the number of unsold seats, find the following.

(a) A function defined by $R(x)$ that describes the total revenue received for the flight (*Hint:* Multiply the number of people flying, $100 - x$, by the price per ticket, $200 + 4x$.)

(b) The number of unsold seats that will produce the maximum revenue

(c) The maximum revenue

32. For a trip to a resort, a charter bus company charges a fare of $48 per person, plus $2 per person for each unsold seat on the bus. If the bus has 42 seats and x represents the number of unsold seats, find the following.

(a) A function defined by $R(x)$ that describes the total revenue from the trip (*Hint*: Multiply the total number riding, $42 - x$, by the price per ticket, $48 + 2x$.)

(b) The number of unsold seats that produces the maximum revenue

(c) The maximum revenue

33. The annual percent increase in the amount pharmacies paid wholesalers for drugs in the years 1990–1999 can be modeled by the quadratic function with

$$f(x) = .228x^2 - 2.57x + 8.97,$$

where $x = 0$ represents 1990, $x = 1$ represents 1991, and so on. (*Source: IMS Health*, Retail and Provider Perspective.)

(a) Since the coefficient of x^2 in the model is positive, the graph of this quadratic function is a parabola that opens up. Will the y-value of the vertex of this graph be a maximum or minimum?

(b) In what year was the minimum percent increase? (Round down to the nearest year.) Use the actual x-value of the vertex, to the nearest tenth, to find this increase.

34. The U.S. domestic oyster catch (in millions) for the years 1990–1998 can be approximated by the quadratic function with

$$f(x) = -.566x^2 + 5.08x + 29.2,$$

where $x = 0$ represents 1990, $x = 1$ represents 1991, and so on. (*Source:* National Marine Fisheries Service.)

(a) Since the coefficient of x^2 in the model is negative, the graph of this quadratic function is a parabola that opens down. Will the y-value of the vertex of this graph be a maximum or minimum?

(b) In what year was the maximum domestic oyster catch? (Round down to the nearest year.) Use the actual x-value of the vertex, to the nearest tenth, to find this catch.

35. The graph shows how Social Security assets are expected to change as the number of retirees receiving benefits increases.

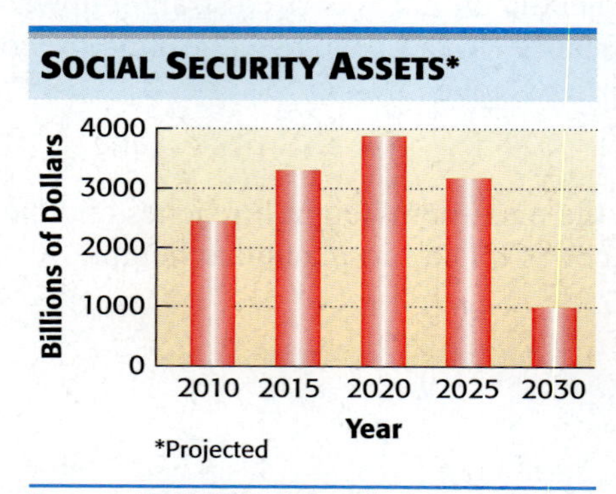

Source: Social Security Administration.

The graph suggests that a quadratic function would be a good fit to the data. The data are approximated by the function with

$$f(x) = -20.57x^2 + 758.9x - 3140.$$

In the model, $x = 10$ represents 2010, $x = 15$ represents 2015, and so on, and $f(x)$ is in billions of dollars.

(a) Explain why the coefficient of x^2 in the model is negative, based on the graph.

(b) Algebraically determine the vertex of the graph, with coordinates to four significant digits.

(c) Interpret the answer to part (b) as it applies to the application.

36. The graph shows the performance of investment portfolios with different mixtures of U.S. and foreign investments for the period January 1, 1971, to December 31, 1996.

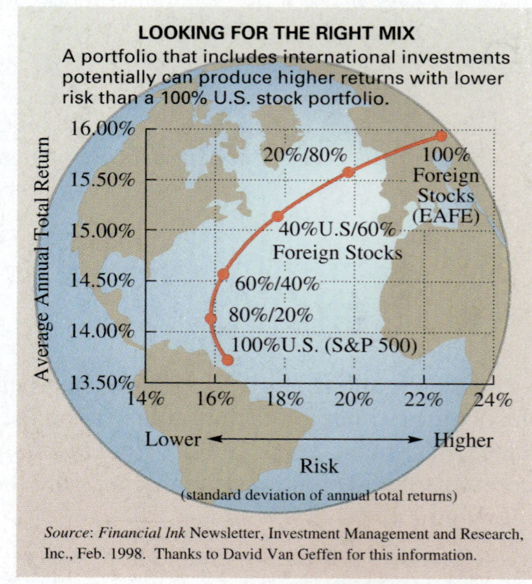

Source: Financial Ink Newsletter, Investment Management and Research, Inc., Feb. 1998. Thanks to David Van Geffen for this information.

(a) Is this the graph of a function? Explain.

(b) What investment mixture shown on the graph appears to represent the vertex? What relative amount of risk does this point represent? What return on investment does it provide?

(c) Which point on the graph represents the riskiest investment mixture? What return on investment does it provide?

9.7 Quadratic and Rational Inequalities

Objectives

1. Solve quadratic inequalities.
2. Solve polynomial inequalities of degree 3 or more.
3. Solve rational inequalities.

We discussed methods of solving linear inequalities in Chapter 3 and methods of solving quadratic equations in this chapter. Now we combine these ideas to solve *quadratic inequalities.*

Quadratic Inequality

A **quadratic inequality** can be written in the form

$$ax^2 + bx + c < 0 \quad \text{or} \quad ax^2 + bx + c > 0,$$

where a, b, and c are real numbers, with $a \neq 0$.

As before, $<$ and $>$ may be replaced with $\leq$ and $\geq$.

1 Solve quadratic inequalities. One method for solving a quadratic inequality is by graphing the related quadratic function.

Example 1 Solving Quadratic Inequalities by Graphing

Solve each inequality.

(a) $x^2 - x - 12 > 0$

To solve the inequality, we graph the related quadratic function defined by $f(x) = x^2 - x - 12$. We are particularly interested in the x-intercepts, which are found as in Section 9.6 by letting $f(x) = 0$ and solving the quadratic equation

$$x^2 - x - 12 = 0.$$

$$(x - 4)(x + 3) = 0 \qquad \text{Factor.}$$

$$x - 4 = 0 \quad \text{or} \quad x + 3 = 0 \qquad \text{Zero-factor property}$$

$$x = 4 \quad \text{or} \quad x = -3$$

Thus, the x-intercepts are (4, 0), and (−3, 0). The graph, which opens up since the coefficient of x^2 is positive, is shown in Figure 17(a). Notice from this graph that x-values less than −3 or greater than 4 result in y-values *greater than* 0. Therefore, the solution set of $x^2 - x - 12 > 0$, written in interval notation, is

$$(-\infty, -3) \cup (4, \infty).$$

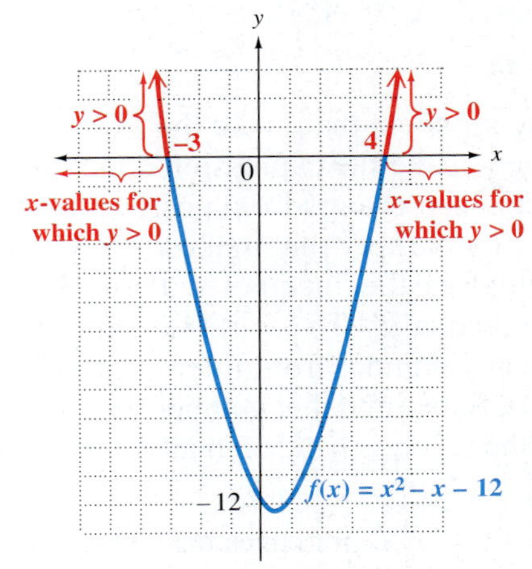

The graph is *above* the x-axis for $(-\infty, -3) \cup (4, \infty)$.

(a)

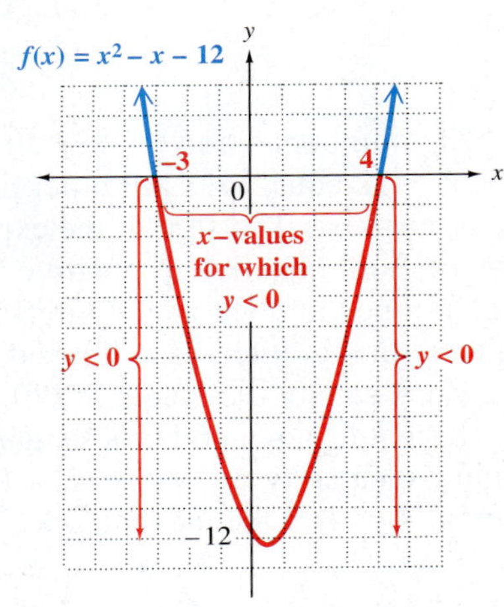

The graph is *below* the x-axis for $(-3, 4)$.

(b)

Figure 17

Continued on Next Page

1 Use the graph to solve each quadratic inequality.

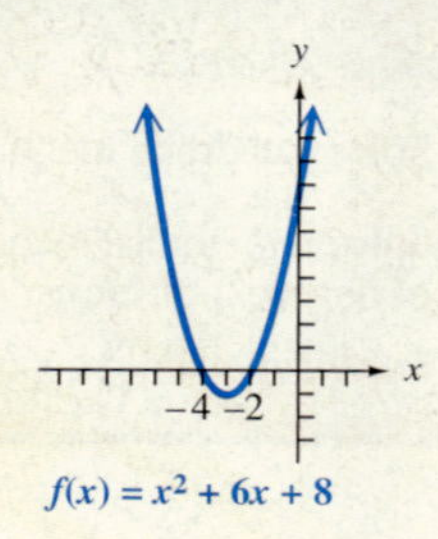

(a) $x^2 + 6x + 8 > 0$

(b) $x^2 + 6x + 8 < 0$

2 Solve each quadratic inequality by graphing.

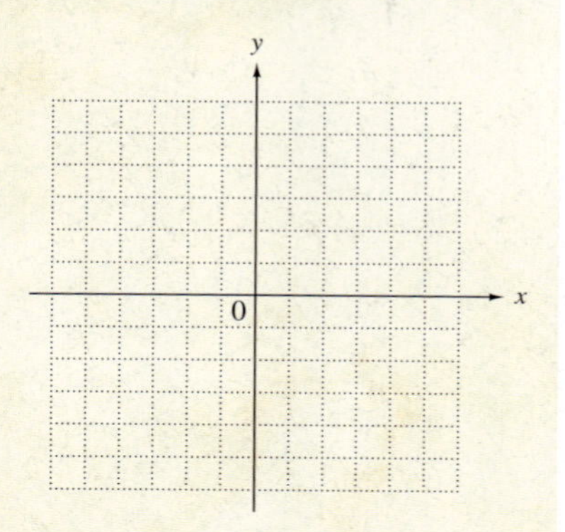

(a) $x^2 + 3x - 4 \geq 0$

(b) $x^2 + 3x - 4 \leq 0$

ANSWERS
1. (a) $(-\infty, -4) \cup (-2, \infty)$ (b) $(-4, -2)$
2. (a) $(-\infty, -4] \cup [1, \infty)$ (b) $[-4, 1]$

(b) $x^2 - x - 12 < 0$

Here we want values of y that are *less than* 0. Referring to Figure 17(b) on the previous page, we notice from the graph that x-values between -3 and 4 result in y-values less than 0. Therefore, the solution set of the inequality $x^2 - x - 12 < 0$, written in interval notation, is $(-3, 4)$.

NOTE

If the inequalities in Example 1 had used $\geq$ and $\leq$, the solution sets would have included the x-values of the intercepts and been written in interval notation as $(-\infty, -3] \cup [4, \infty)$ for Example 1(a) and $[-3, 4]$ for Example 1(b).

Work Problems 1 and 2 at the Side.

In Example 1, we used graphing to divide the x-axis into intervals. Then using the graphs in Figure 17, we determined which x-values resulted in y-values that were either greater than or less than 0. Another method for solving a quadratic inequality uses these basic ideas without actually graphing the related quadratic function.

Example 2 Solving a Quadratic Inequality Using Test Numbers

Solve $x^2 - x - 12 > 0$.

First solve the quadratic equation $x^2 - x - 12 = 0$ by factoring, as in Example 1(a).

$$(x - 4)(x + 3) = 0$$
$$x - 4 = 0 \quad \text{or} \quad x + 3 = 0$$
$$x = 4 \quad \text{or} \quad x = -3$$

The numbers 4 and -3 divide the number line into the three intervals shown in Figure 18. Be careful to put the smaller number on the left. (Notice the similarity between Figure 18 and the x-axis with intercepts $(-3, 0)$ and $(4, 0)$ in Figure 17(a).)

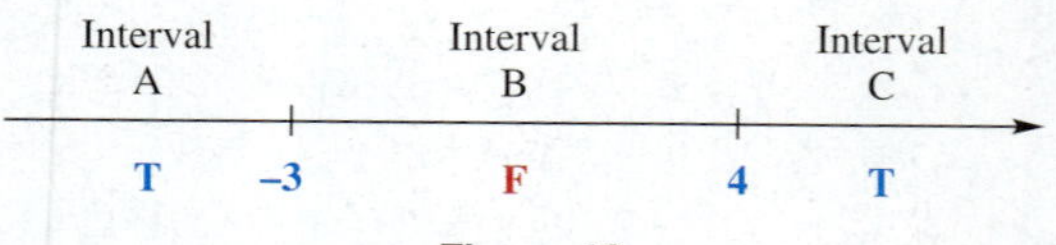

Figure 18

The numbers 4 and -3 are the only numbers that make the expression $x^2 - x - 12$ equal to 0. All other numbers make the expression either positive or negative. The sign of the expression can change from positive to negative or from negative to positive only at a number that makes it 0. Therefore, if one number in an interval satisfies the inequality, then all the numbers in that interval will satisfy the inequality. To see if the numbers in Interval A satisfy the inequality, choose any number from Interval A in Figure 18 (that is, any number less than -3). Substitute this number for x in the original inequality $x^2 - x - 12 > 0$. If the result is *true,* then all numbers in Interval A satisfy the inequality.

Continued on Next Page

We choose -5 from Interval A. Substitute -5 for x.

$$x^2 - x - 12 > 0 \quad \text{Original inequality}$$
$$(-5)^2 - (-5) - 12 > 0 \quad ?$$
$$25 + 5 - 12 > 0 \quad ?$$
$$18 > 0 \quad \text{True}$$

Because -5 from Interval A satisfies the inequality, all numbers from Interval A are solutions.

Try 0 from Interval B. If $x = 0$, then

$$0^2 - 0 - 12 > 0 \quad ?$$
$$-12 > 0. \quad \text{False}$$

The numbers in Interval B are *not* solutions.

Work Problem 3 at the Side.

In Problem 3 at the side, the number 5 satisfies the inequality, so the numbers in Interval C are also solutions.

Based on these results (shown by the colored letters in Figure 18), the solution set includes the numbers in Intervals A and C, as shown on the graph in Figure 19. The solution set is written in interval notation as

$$(-\infty, -3) \cup (4, \infty).$$

−3 0 4

Figure 19

This agrees with the solution set we found by graphing the related quadratic function in Example 1(a).

In summary, a quadratic inequality is solved by following these steps.

Solving a Quadratic Inequality

Step 1 **Write the inequality as an equation and solve it.**

Step 2 **Use the solutions from Step 1 to determine intervals.** Graph the numbers found in Step 1 on a number line. These numbers divide the number line into intervals.

Step 3 **Find the intervals that satisfy the inequality.** Substitute a number from each interval into the original inequality to determine the intervals that satisfy the inequality. All numbers in those intervals are in the solution set. A graph of the solution set will usually look like one of these. (Square brackets might be used instead of parentheses.)

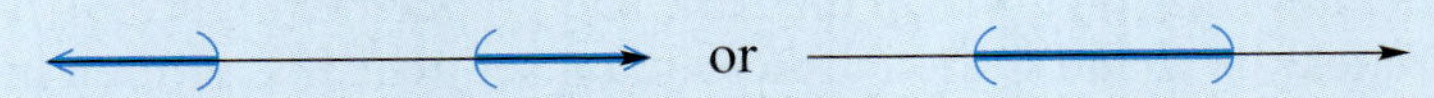

Step 4 **Consider the endpoints separately.** The numbers from Step 1 are included in the solution set if the inequality is $\leq$ or $\geq$; they are not included if it is $<$ or $>$.

Work Problem 4 at the Side.

3 Does the number 5 from Interval C satisfy $x^2 - x - 12 > 0$?

4 Solve each inequality, and graph the solution set.

(a) $x^2 + x - 6 > 0$

(b) $3m^2 - 13m - 10 \leq 0$

Answers

3. yes

4. (a) $(-\infty, -3) \cup (2, \infty)$

−3 0 2

(b) $\left[-\frac{2}{3}, 5\right]$

$-\frac{2}{3}$

−1 0 3 5 7

Special cases of quadratic inequalities may occur, as in the next example.

Example 3 Solving Special Cases

Solve $(2t - 3)^2 > -1$.

Because $(2t - 3)^2$ is never negative, it is always greater than -1. Thus, the solution is the set of all real numbers, $(-\infty, \infty)$. In the same way, there is no solution for $(2t - 3)^2 < -1$ and the solution set is $\emptyset$.

Work Problem 5 at the Side.

5 Solve each inequality.

(a) $(3k - 2)^2 > -2$

(b) $(5z + 3)^2 < -3$

2 Solve polynomial inequalities of degree 3 or more.

Higher-degree polynomial inequalities that can be factored are solved in the same way as quadratic inequalities.

Example 4 Solving a Third-Degree Polynomial Inequality

Solve $(x - 1)(x + 2)(x - 4) \leq 0$.

This is a *cubic* (third-degree) inequality rather than a quadratic inequality, but it can be solved using the method shown in the box by extending the zero-factor property to more than two factors. Begin by setting the factored polynomial *equal* to 0 and solving the equation. (Step 1)

$$(x - 1)(x + 2)(x - 4) = 0$$

$$x - 1 = 0 \quad \text{or} \quad x + 2 = 0 \quad \text{or} \quad x - 4 = 0$$

$$x = 1 \quad \text{or} \quad x = -2 \quad \text{or} \quad x = 4$$

Locate the numbers -2, 1, and 4 on a number line, as in Figure 20, to determine the Intervals A, B, C, and D. (Step 2)

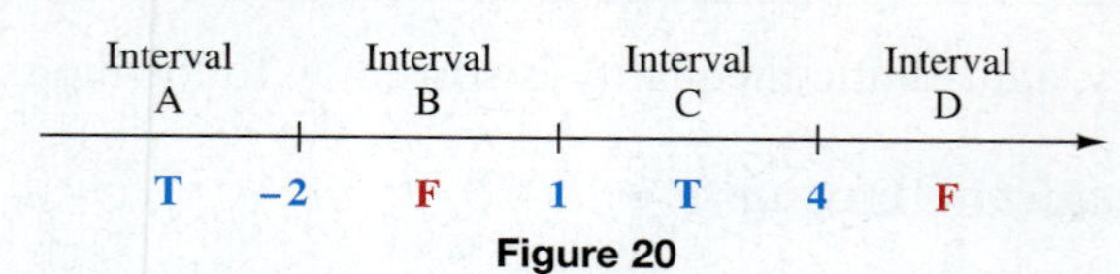

Figure 20

Substitute a number from each interval in the *original* inequality to determine which intervals satisfy the inequality. (Step 3) It is helpful to organize this information in a table.

Interval	Test Number	Test of Inequality	True or False?
A	-3	$-28 \leq 0$	T
B	0	$8 \leq 0$	F
C	2	$-8 \leq 0$	T
D	5	$28 \leq 0$	F

Verify the information given in the table and graphed in Figure 21. The numbers in Intervals A and C are in the solution set, which is written in interval notation as

$$(-\infty, -2] \cup [1, 4].$$

Notice that the three endpoints are included since the inequality symbol is $\leq$. (Step 4)

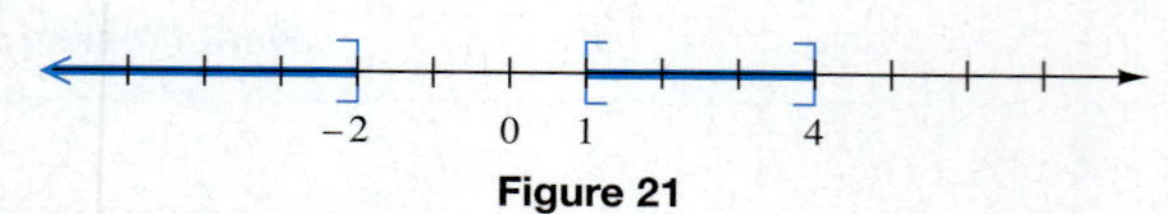

Figure 21

Work Problem 6 at the Side.

6 Solve each inequality, and graph the solution set.

(a) $(x - 3)(x + 2)(x + 1) > 0$

(b) $(k - 5)(k + 1)(k - 3) \leq 0$

Answers

5. (a) $(-\infty, \infty)$ **(b)** $\emptyset$

6. (a) $(-2, -1) \cup (3, \infty)$

−2 −1 0 1 2 3 4

(b) $(-\infty, -1] \cup [3, 5]$

3 Solve rational inequalities. Inequalities that involve rational expressions, called **rational inequalities,** are solved similarly using the following steps.

Solving a Rational Inequality

Step 1 **Write the inequality** so that 0 is on one side and there is a single fraction on the other side.

Step 2 **Determine the numbers that make the numerator and denominator equal to 0.**

Step 3 **Divide a number line into intervals.** Use the numbers from Step 2.

Step 4 **Find the intervals that satisfy the inequality.** Test a number from each interval by substituting it into the *original* inequality.

Step 5 **Consider the endpoints separately.** Exclude any values that make the denominator 0.

Example 5 Solving a Rational Inequality

Solve $\frac{-1}{p-3} > 1$.

Write the inequality so that 0 is on one side. (Step 1)

$$\frac{-1}{p-3} - 1 > 0 \quad \text{Subtract 1.}$$

$$\frac{-1}{p-3} - \frac{p-3}{p-3} > 0 \quad \text{Use } p-3 \text{ as the common denominator.}$$

$$\frac{-1 - p + 3}{p-3} > 0 \quad \text{Write the left side as a single fraction; Be careful with signs in the numerator.}$$

$$\frac{-p+2}{p-3} > 0 \quad \text{Combine terms.}$$

The sign of the rational expression $\frac{-p+2}{p-3}$ will change from positive to negative or negative to positive only at those numbers that make the numerator or denominator 0. The number 2 makes the numerator 0, and 3 makes the denominator 0. (Step 2) These two numbers, 2 and 3, divide a number line into three intervals. See Figure 22. (Step 3)

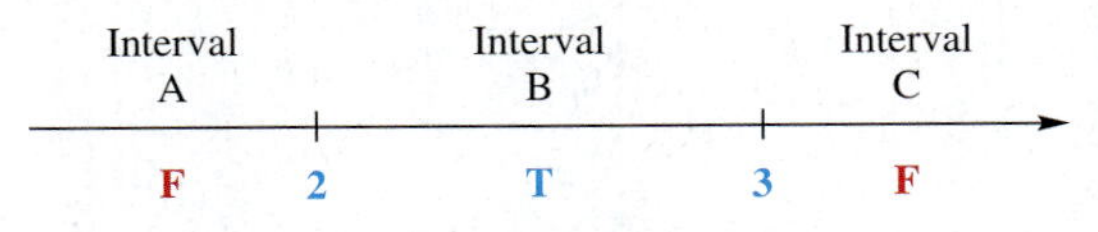

Figure 22

Testing a number from each interval in the *original* inequality, $\frac{-1}{p-3} > 1$, gives the results shown in the table. (Step 4)

Interval	Test Number	Test of Inequality	True or False?
A	0	$\frac{1}{3} > 1$	F
B	2.5	$2 > 1$	T
C	4	$-1 > 1$	F

Continued on Next Page

7 Solve each inequality, and graph the solution set.

(a) $\dfrac{2}{x-4} < 3$

(b) $\dfrac{5}{z+1} > 4$

The solution set is the interval $(2, 3)$. This interval does not include 3 since it would make the denominator of the original inequality 0; 2 is not included either since the inequality symbol is $>$. (Step 5) A graph of the solution set is given in Figure 23.

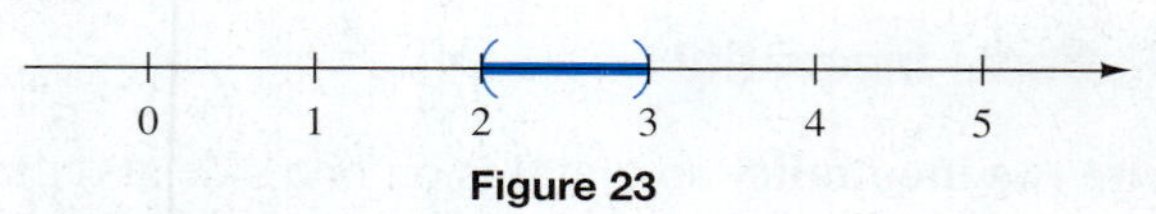

Figure 23

Work Problem 7 at the Side.

Example 6 Solving a Rational Inequality

Solve $\dfrac{m-2}{m+2} \leq 2.$

Write the inequality so that 0 is on one side. (Step 1)

$$\frac{m-2}{m+2} - 2 \leq 0 \qquad \text{Subtract 2.}$$

$$\frac{m-2}{m+2} - \frac{2(m+2)}{m+2} \leq 0 \qquad \text{Use } m+2 \text{ as the common denominator.}$$

$$\frac{m-2-2m-4}{m+2} \leq 0 \qquad \text{Write as a single fraction.}$$

$$\frac{-m-6}{m+2} \leq 0 \qquad \text{Combine terms.}$$

The number -6 makes the numerator 0, and -2 makes the denominator 0. (Step 2) These two numbers determine three intervals (Step 3). Test one number from each interval (Step 4) to see that the solution set is the interval

$$(-\infty, -6] \cup (-2, \infty).$$

The number -6 satisfies the original inequality, but -2 cannot be used as a solution since it makes the denominator 0 (Step 5). A graph of the solution set is shown in Figure 24.

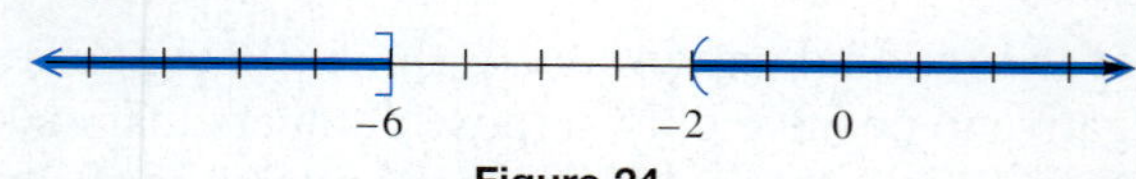

Figure 24

Work Problem 8 at the Side.

8 Solve $\dfrac{k+2}{k-1} \leq 5$, and graph the solution set.

ANSWERS

7. (a) $(-\infty, 4) \cup \left(\frac{14}{3}, \infty\right)$

$\frac{14}{3}$

0 1 2 3 4 5 6

(b) $\left(-1, \frac{1}{4}\right)$

$\frac{1}{4}$

−2 −1 0 1 2

8. $(-\infty, 1) \cup \left[\frac{7}{4}, \infty\right)$

0 1 $\frac{7}{4}$ 3

9.7 EXERCISES

FOR EXTRA HELP

Student's Solutions Manual

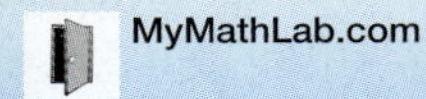
MyMathLab.com

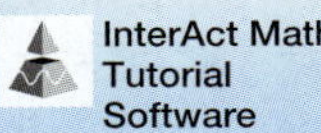
InterAct Math Tutorial Software

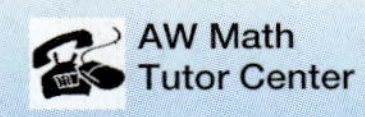
AW Math Tutor Center

www.mathxl.com

Digital Video Tutor CD 8
Videotape 16

In Example 1, we determined the solution sets of the quadratic inequalities $x^2 - x - 12 > 0$ and $x^2 - x - 12 < 0$ by graphing $f(x) = x^2 - x - 12$. The x-intercepts of this graph indicated the solutions of the equation $x^2 - x - 12 = 0$. The x-values of the points on the graph that were **above** the x-axis formed the solution set of $x^2 - x - 12 > 0$, and the x-values of the points on the graph that were **below** the x-axis formed the solution set of $x^2 - x - 12 < 0$.

In Exercises 1–4, the graph of a quadratic function f is given. Use the graph to find the solution set of each equation or inequality. See Example 1.

1. **(a)** $x^2 - 4x + 3 = 0$
(b) $x^2 - 4x + 3 > 0$
(c) $x^2 - 4x + 3 < 0$

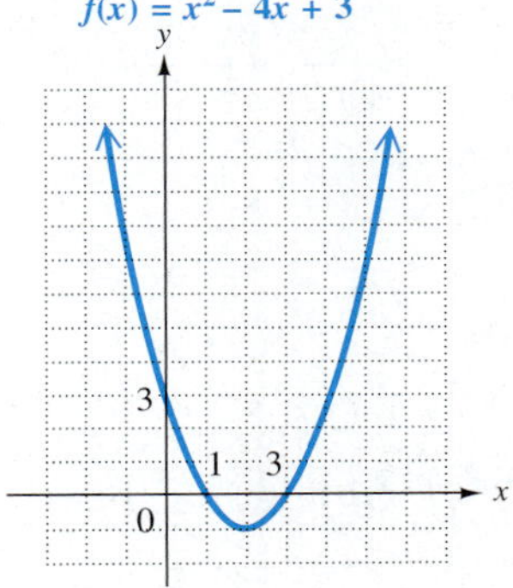

2. **(a)** $3x^2 + 10x - 8 = 0$
(b) $3x^2 + 10x - 8 \geq 0$
(c) $3x^2 + 10x - 8 < 0$

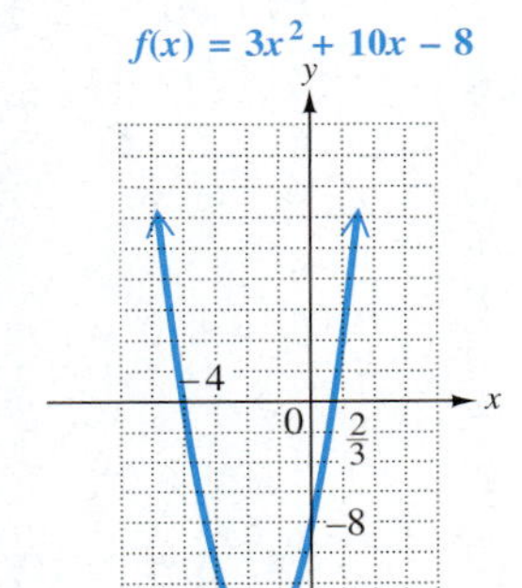

3. **(a)** $-2x^2 - x + 15 = 0$
(b) $-2x^2 - x + 15 \geq 0$
(c) $-2x^2 - x + 15 \leq 0$

$f(x) = -2x^2 - x + 15$

4. **(a)** $-x^2 + 3x + 10 = 0$
(b) $-x^2 + 3x + 10 \geq 0$
(c) $-x^2 + 3x + 10 \leq 0$

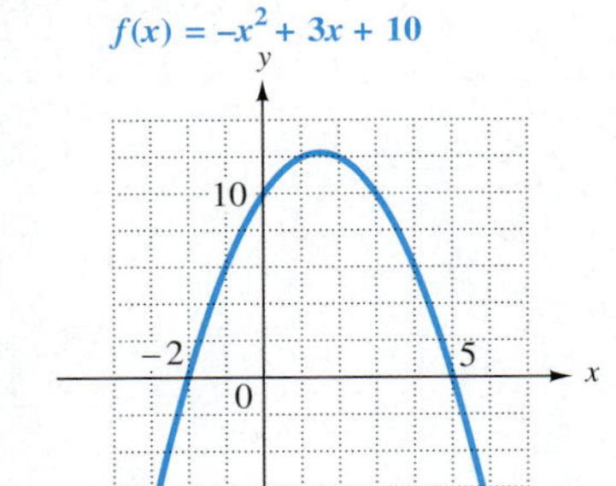

5. Explain how you determine whether to include or exclude endpoints when solving a quadratic or higher-degree inequality.

6. The solution set of the inequality $x^2 + x - 12 < 0$ is the interval $(-4, 3)$. Without actually performing any work, give the solution set of the inequality $x^2 + x - 12 \geq 0$.

Solve each inequality, and graph the solution set. See Example 2.

7. $(x + 1)(x - 5) > 0$

8. $(m + 6)(m - 2) > 0$

9. $(r + 4)(r - 6) < 0$

10. $(x + 4)(x - 8) < 0$

11. $x^2 - 4x + 3 \geq 0$

12. $m^2 - 3m - 10 \geq 0$

13. $10t^2 + 9t \geq 9$

14. $3r^2 + 10r \geq 8$

15. $9p^2 + 3p < 2$

16. $2x^2 + x < 15$

17. $6x^2 + x \geq 1$

18. $4m^2 + 7m \geq -3$

19. $x^2 - 6x + 6 \geq 0$
(*Hint:* Use the quadratic formula.)

20. $3k^2 - 6k + 2 \leq 0$
(*Hint:* Use the quadratic formula.)

Solve each inequality. See Example 3.

21. $(4 - 3x)^2 \geq -2$ **22.** $(6p + 7)^2 \geq -1$ **23.** $(3x + 5)^2 \leq -4$ **24.** $(8t + 5)^2 \leq -5$

Solve each inequality, and graph the solution set. See Example 4.

25. $(p - 1)(p - 2)(p - 4) < 0$

26. $(2r + 1)(3r - 2)(4r + 7) < 0$

27. $(x - 4)(2x + 3)(3x - 1) \geq 0$

28. $(z + 2)(4z - 3)(2z + 7) \geq 0$

Solve each inequality, and graph the solution set. See Examples 5 and 6.

29. $\dfrac{x - 1}{x - 4} > 0$ **30.** $\dfrac{x + 1}{x - 5} > 0$ **31.** $\dfrac{2n + 3}{n - 5} \leq 0$

32. $\dfrac{3t + 7}{t - 3} \leq 0$ **33.** $\dfrac{8}{x - 2} \geq 2$ **34.** $\dfrac{20}{x - 1} \geq 1$

35. $\frac{3}{2t - 1} < 2$

36. $\frac{6}{m - 1} < 1$

37. $\frac{w}{w + 2} \geq 2$

38. $\frac{m}{m + 5} \geq 2$

39. $\frac{4k}{2k - 1} < k$

40. $\frac{r}{r + 2} < 2r$

RELATING CONCEPTS (Exercises 41–44) FOR INDIVIDUAL OR GROUP WORK

A rock is projected vertically upward from the ground. Its distance s in feet above the ground after t seconds is given by the quadratic function defined by

$$s(t) = -16t^2 + 256t.$$

Work Exercises 41–44 in order, *to see how quadratic equations and inequalities are related.*

41. At what times will the rock be 624 ft above the ground? (*Hint:* Let $s(t) = 624$ and solve the quadratic *equation.*)

42. At what times will the rock be more than 624 ft above the ground? (*Hint:* Set $s(t) > 624$ and solve the quadratic *inequality.*)

43. At what times will the rock be at ground level? (*Hint:* Let $s(t) = 0$ and solve the quadratic *equation.*)

44. At what times will the rock be less than 624 ft above the ground? (*Hint:* Set $s(t) < 624$, solve the quadratic *inequality,* and observe the solutions in Exercises 42 and 43 to determine the smallest and largest possible values of t.)

Chapter 9

SUMMARY

KEY TERMS

9.1	**quadratic equation**	A quadratic equation is an equation that can be written in the form $ax^2 + bx + c = 0$, where a, b, and c are real numbers, with $a \neq 0$. This form is called standard form.
9.2	**quadratic formula**	The quadratic formula is a formula for solving quadratic equations.
	discriminant	The discriminant is the expression under the radical in the quadratic formula.
9.3	**quadratic in form**	A nonquadratic equation that can be written as a quadratic equation is called quadratic in form.
9.5	**parabola**	The graph of a quadratic function is a parabola.
	vertex	The point on a parabola that has the smallest y-value (if the parabola opens up) or the largest y-value (if the parabola opens down) is called the vertex of the parabola.
	axis	The vertical or horizontal line through the vertex of a parabola is its axis.
	quadratic function	A function defined by $f(x) = ax^2 + bx + c$, for real numbers a, b, and c, with $a \neq 0$, is a quadratic function.
9.7	**quadratic inequality**	A quadratic inequality is an inequality that can be written in the form $ax^2 + bx + c < 0$ or $ax^2 + bx + c > 0$, or with $\leq$ or $\geq$, where a, b, and c are real numbers, with $a \neq 0$.
	rational inequality	An inequality that involves a rational expression is a rational inequality.

y
4
2
$y = x^2$
x
−2
0
2
Vertex
Axis

TEST YOUR WORD POWER

See how well you have learned the vocabulary in this chapter. Answers follow the Quick Review.

1. The **quadratic formula** is
   - **(a)** a formula to find the number of solutions of a quadratic equation
   - **(b)** a formula to find the type of solutions of a quadratic equation
   - **(c)** the standard form of a quadratic equation
   - **(d)** a general formula for solving any quadratic equation.

2. A **quadratic function** is a function that can be written in the form
   - **(a)** $f(x) = mx + b$ for real numbers m and b
   - **(b)** $f(x) = \frac{P(x)}{Q(x)}$, where $Q(x) \neq 0$
   - **(c)** $f(x) = ax^2 + bx + c$ for real numbers a, b, and c $(a \neq 0)$
   - **(d)** $f(x) = \sqrt{x}$ for $x \geq 0$.

3. A **parabola** is the graph of
   - **(a)** any equation in two variables
   - **(b)** a linear equation
   - **(c)** an equation of degree 3
   - **(d)** a quadratic equation in 2 variables.

4. The **vertex** of a parabola is
   - **(a)** the point where the graph intersects the y-axis
   - **(b)** the point where the graph intersects the x-axis
   - **(c)** the lowest point on a parabola that opens up or the highest point on a parabola that opens down
   - **(d)** the origin.

5. The **axis** of a parabola is
   - **(a)** either the x-axis or the y-axis
   - **(b)** the vertical line (of a vertical parabola) or the horizontal line (of a horizontal parabola) through the vertex
   - **(c)** the lowest or highest point on the graph of a parabola
   - **(d)** a line through the origin.

6. A parabola is **symmetric about its axis** since
   - **(a)** its graph is near the axis
   - **(b)** its graph is identical on each side of the axis
   - **(c)** its graph looks different on each side of the axis
   - **(d)** its graph intersects the axis.

QUICK REVIEW

Concepts	*Examples*
9.1 The Square Root Property and Completing the Square	
Square Root Property If x and b are complex numbers and $x^2 = b$, then $x = \sqrt{b}$ or $x = -\sqrt{b}$.	Solve $(x-1)^2 = 8$. $x - 1 = \sqrt{8}$ or $x - 1 = -\sqrt{8}$ $x = 1 + 2\sqrt{2}$ or $x = 1 - 2\sqrt{2}$ Solution set: $\{1 + 2\sqrt{2}, 1 - 2\sqrt{2}\}$
Completing the Square To solve $ax^2 + bx + c = 0$ $(a \neq 0)$: *Step 1* If $a \neq 1$, divide each side by a. *Step 2* Write the equation with the variable terms on one side and the constant on the other. *Step 3* Take half the coefficient of x and square it. *Step 4* Add the square to each side. *Step 5* Factor the perfect square trinomial, and write it as the square of a binomial. Simplify the other side. *Step 6* Use the square root property to complete the solution.	Solve $2x^2 - 4x - 18 = 0$. $x^2 - 2x - 9 = 0$ Divide by 2. $x^2 - 2x = 9$ Add 9. $\left[\frac{1}{2}(-2)\right]^2 = (-1)^2 = 1$ $x^2 - 2x + 1 = 9 + 1$ $(x-1)^2 = 10$ $x - 1 = \sqrt{10}$ or $x - 1 = -\sqrt{10}$ $x = 1 + \sqrt{10}$ or $x = 1 - \sqrt{10}$ Solution set: $\{1 + \sqrt{10}, 1 - \sqrt{10}\}$
9.2 The Quadratic Formula	
Quadratic Formula The solutions of $ax^2 + bx + c = 0$ $(a \neq 0)$ are given by $x = \frac{-b \pm \sqrt{b^2 - 4ac}}{2a}$.	Solve $3x^2 + 5x + 2 = 0$. $x = \frac{-5 \pm \sqrt{5^2 - 4(3)(2)}}{2(3)} = \frac{-5 \pm 1}{6}$ $x = -1$ or $x = -\frac{2}{3}$ Solution set: $\{-1, -\frac{2}{3}\}$
The Discriminant If a, b, and c are integers, then the discriminant, $b^2 - 4ac$, of $ax^2 + bx + c = 0$ determines the number and type of solutions as follows.	For $x^2 + 3x - 10 = 0$, the discriminant is $3^2 - 4(1)(-10) = 49$. Two rational solutions For $4x^2 + x + 1 = 0$, the discriminant is $1^2 - 4(4)(1) = -15$. Two imaginary solutions

Discriminant	Number and Type of Solutions
Positive, the square of an integer	Two rational solutions
Positive, not the square of an integer	Two irrational solutions
Zero	One rational solution
Negative	Two imaginary solutions

Concepts	*Examples*
9.3 Equations Quadratic in Form	
A nonquadratic equation that can be written in the form $$au^2 + bu + c = 0,$$ for $a \neq 0$ and an algebraic expression u, is called quadratic in form. Substitute u for the expression, solve for u, and then solve for the variable in the expression.	Solve $3(x + 5)^2 + 7(x + 5) + 2 = 0$. $3u^2 + 7u + 2 = 0$ Let $u = x + 5$. $(3u + 1)(u + 2) = 0$ $u = -\frac{1}{3}$ or $u = -2$ $x + 5 = -\frac{1}{3}$ or $x + 5 = -2$ $x + 5 = u$ $x = -\frac{16}{3}$ or $x = -7$ Solution set: $\{-7, -\frac{16}{3}\}$
9.4 Formulas and Further Applications	
To solve a formula for a squared variable, proceed as follows. **(a) The variable appears only to the second power.** Isolate the squared variable on one side of the equation, then use the square root property.	Solve $A = \frac{2mp}{r^2}$ for r. $r^2A = 2mp$ Multiply by r^2. $r^2 = \frac{2mp}{A}$ Divide by A. $r = \pm\sqrt{\frac{2mp}{A}}$ Square root property $r = \frac{\pm\sqrt{2mpA}}{A}$ Rationalize the denominator.
(b) The variable appears to the first and second powers. Write the equation in standard form, then use the quadratic formula.	Solve $m^2 + rm = t$ for m. $m^2 + rm - t = 0$ Standard form $m = \frac{-r \pm \sqrt{r^2 - 4(1)(-t)}}{2(1)}$ $a = 1, b = r, c = -t$ $m = \frac{-r \pm \sqrt{r^2 + 4t}}{2}$
9.5 Graphs of Quadratic Functions	
1. The graph of the quadratic function with $F(x) = a(x - h)^2 + k, a \neq 0$, is a parabola with vertex at (h, k) and the vertical line $x = h$ as axis. 2. The graph opens up if a is positive and down if a is negative. 3. The graph is wider than the graph of $f(x) = x^2$ if $0 < \lvert a \rvert < 1$ and narrower if $\lvert a \rvert > 1$.	Graph $f(x) = -(x + 3)^2 + 1$. 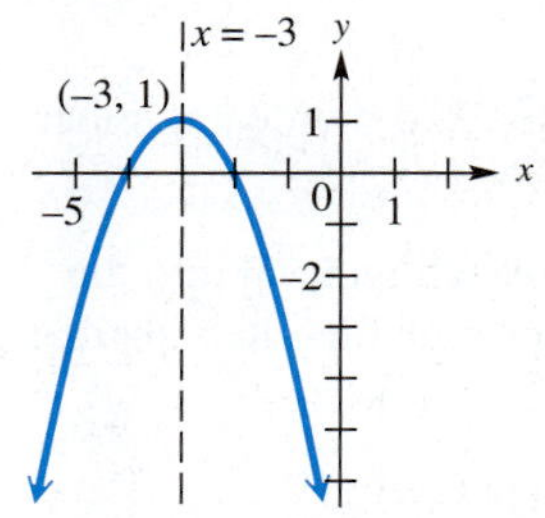 The graph opens down since $a < 0$. It is shifted 3 units left and 1 unit up, so the vertex is $(-3, 1)$, with axis $x = -3$.
9.6 More about Parabolas; Applications	
The vertex of the graph of $f(x) = ax^2 + bx + c, a \neq 0$, may be found by completing the square. The vertex has coordinates $$\left(\frac{-b}{2a}, f\left(\frac{-b}{2a}\right)\right).$$ **Graphing a Quadratic Function** *Step 1* Determine whether the graph opens up or down. *Step 2* Find the vertex. *Step 3* Find the x-intercepts (if any). Find the y-intercept. *Step 4* Find and plot additional points as needed.	Graph $f(x) = x^2 + 4x + 3$. 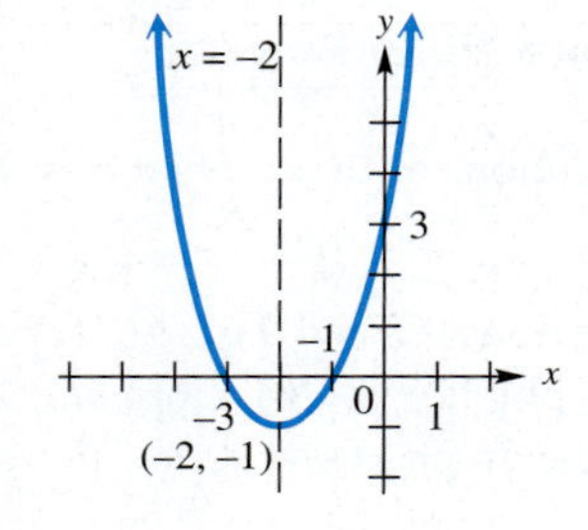 The graph opens up since $a > 0$. The vertex is $(-2, -1)$. The solutions of $x^2 + 4x + 3 = 0$ are -1 and -3, so the x-intercepts are $(-1, 0)$ and $(-3, 0)$. Since $f(0) = 3$, the y-intercept is $(0, 3)$.

Concepts	*Examples*
9.6 More about Parabolas; Applications (continued) The graph of $x = ay^2 + by + c$ is a horizontal parabola, opening to the right if $a > 0$ or to the left if $a < 0$. Horizontal parabolas do not represent functions.	Graph $x = 2y^2 + 6y + 5$. 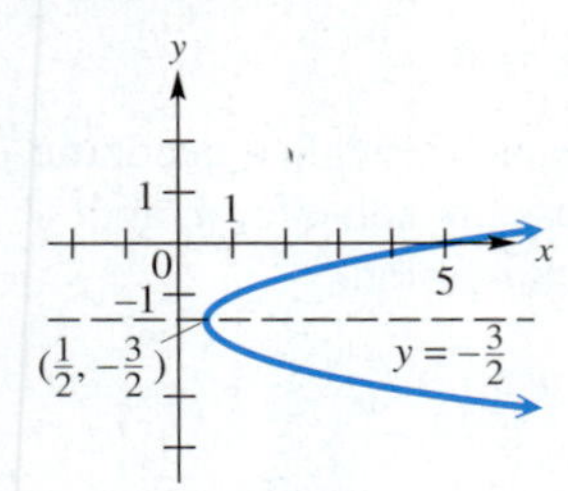
9.7 Quadratic and Rational Inequalities **Solving a Quadratic (or Higher-Degree Polynomial) Inequality**	Solve $2x^2 + 5x + 2 < 0$.
Step 1 Write the inequality as an equation and solve.	$2x^2 + 5x + 2 = 0$ $x = -\frac{1}{2}$ or $x = -2$
Step 2 Use the numbers found in Step 1 to divide a number line into intervals.	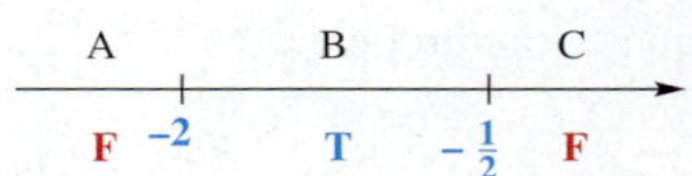
Step 3 Substitute a number from each interval into the original inequality to determine the intervals that belong in the solution set.	$x = -3$ makes the original inequality false; $x = -1$ makes it true; $x = 0$ makes it false.
Step 4 Consider the endpoints separately.	Solution set: $(-2, -\frac{1}{2})$
Solving a Rational Inequality	Solve $\frac{x}{x+2} \geq 4$.
Step 1 Write the inequality so that 0 is on one side and there is a single fraction on the other side.	$\frac{x}{x+2} - 4 \geq 0$ $\frac{x}{x+2} - \frac{4(x+2)}{x+2} \geq 0$ $\frac{-3x-8}{x+2} \geq 0$
Step 2 Determine the numbers that make the numerator and denominator 0.	$-\frac{8}{3}$ makes the numerator 0; -2 makes the denominator 0.
Step 3 Use the numbers from Step 2 to divide a number line into intervals.	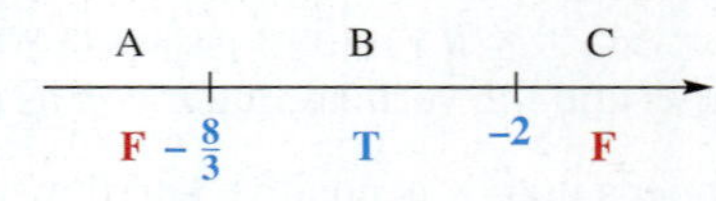
Step 4 Substitute a number from each interval into the original inequality to determine the intervals that belong in the solution set.	-4 makes the original inequality false; $-\frac{7}{3}$ makes it true; 0 makes it false.
Step 5 Consider the endpoints separately.	The solution set is $[-\frac{8}{3}, -2)$, since -2 makes the denominator 0.

Answers to Test Your Word Power

1. (d) *Example:* The solutions of $ax^2 + bx + c = 0$ $(a \neq 0)$ are given by $x = \frac{-b \pm \sqrt{b^2 - 4ac}}{2a}$.

2. (c) *Examples:* $f(x) = x^2 - 2$, $f(x) = (x + 4)^2 + 1$, $f(x) = x^2 - 4x + 5$ **3. (d)** *Examples:* See the figures in the Quick Review for Sections 9.5 and 9.6. **4. (c)** *Example:* The graph of $y = (x + 3)^2$ has vertex $(-3, 0)$, which is the lowest point on the graph. **5. (b)** *Example:* The axis of $y = (x + 3)^2$ is the vertical line $x = -3$.
6. (b) *Example:* Since the graph of $y = (x + 3)^2$ is symmetric about its axis $x = -3$, the points $(-2, 1)$ and $(-4, 1)$ are on the graph.

Chapter 9 Review Exercises

*Exercises marked * have imaginary number solutions.*

[9.1] *Solve each equation by using the square root property or completing the square.*

1. $t^2 = 121$

2. $p^2 = 3$

3. $(2x + 5)^2 = 100$

***4.** $(3k - 2)^2 = -25$

5. $x^2 + 4x = 15$

6. $2m^2 - 3m = -1$

7. A student gave the following "solution" to the equation $x^2 = 12$.

$$x^2 = 12$$
$$x = \sqrt{12} \quad \text{Square root property}$$
$$x = 2\sqrt{3}$$

What is wrong with this solution?

8. Navy Pier Center in Chicago, Illinois, features a 150-ft tall Ferris wheel. Use Galileo's formula $d = 16t^2$ to find how long it would take a wallet dropped from the top of the Ferris wheel to fall to the ground. Round your answer to the nearest tenth of a second. (*Source: Microsoft Encarta Encyclopedia 2000.*)

[9.2] *Solve each equation using the quadratic formula.*

9. $2x^2 + x - 21 = 0$

10. $k^2 + 5k = 7$

11. $(t + 3)(t - 4) = -2$

***12.** $2x^2 + 3x + 4 = 0$

***13.** $3p^2 = 2(2p - 1)$

14. $m(2m - 7) = 3m^2 + 3$

Use the discriminant to predict whether the solutions to each equation are

A. *two rational numbers;* **B.** *one rational number;*
C. *two irrational numbers;* **D.** *two imaginary numbers.*

15. $x^2 + 5x + 2 = 0$

16. $4t^2 = 3 - 4t$

17. $4x^2 = 6x - 8$

18. $9z^2 + 30z + 25 = 0$

[9.3] *Solve each equation.*

19. $\frac{15}{x} = 2x - 1$

20. $\frac{1}{n} + \frac{2}{n+1} = 2$

21. $-2r = \sqrt{\frac{48 - 20r}{2}}$

22. $8(3x + 5)^2 + 2(3x + 5) - 1 = 0$

23. $2x^{2/3} - x^{1/3} - 28 = 0$

24. $p^4 - 5p^2 + 4 = 0$

Solve each problem. Round answers to the nearest tenth, as necessary.

25. Phong paddled his canoe 20 mi upstream, then paddled back. If the speed of the current was 3 mph and the total trip took 7 hr, what was Phong's speed?

26. Maureen O'Connor drove 8 mi to pick up her friend Laurie, and then drove 11 mi to a mall at a speed 15 mph faster. If Maureen's total travel time was 24 min, what was her speed on the trip to pick up Laurie?

27. An old machine processes a batch of checks in 1 hr more time than a new one. How long would it take the old machine to process a batch of checks that the two machines together process in 2 hr?

28. Greg Tobin can process a stack of invoices 1 hr faster than Carter Fenton can. Working together, they take 1.5 hr. How long would it take each person working alone?

[9.4] *Solve each formula for the indicated variable. (Give answers with $\pm$.)*

29. $k = \frac{rF}{wv^2}$ for v

30. $p = \sqrt{\frac{yz}{6}}$ for y

31. $mt^2 = 3mt + 6$ for t

Solve each problem. Round answers to the nearest tenth, as necessary.

32. A large machine requires a part in the shape of a right triangle with a hypotenuse 9 ft less than twice the length of the longer leg. The shorter leg must be $\frac{3}{4}$ the length of the longer leg. Find the lengths of the three sides of the part.

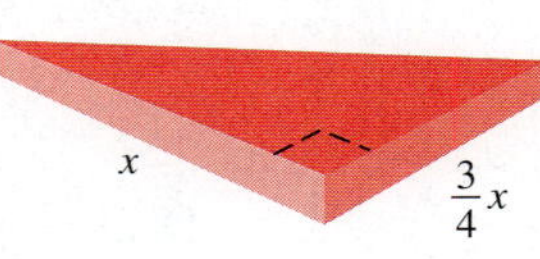

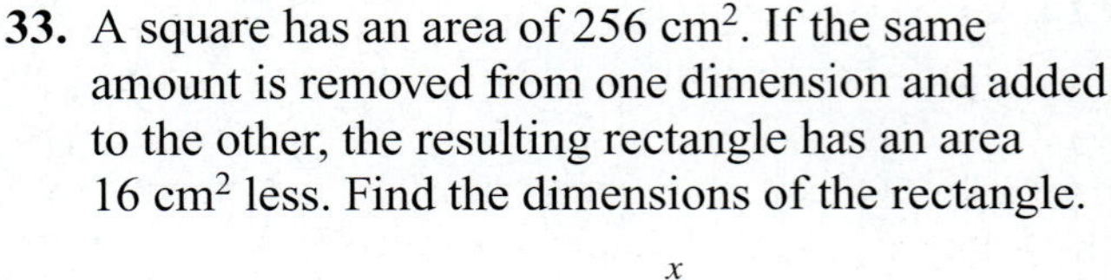

33. A square has an area of 256 cm^2. If the same amount is removed from one dimension and added to the other, the resulting rectangle has an area 16 cm^2 less. Find the dimensions of the rectangle.

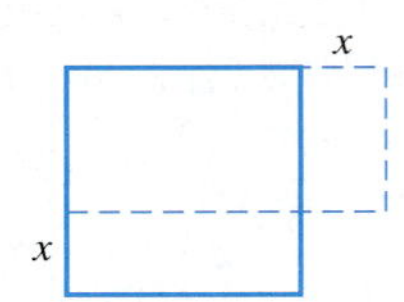

34. Nancy wants to buy a mat for a photograph that measures 14 in. by 20 in. She wants to have an even border around the picture when it is mounted on the mat. If the area of the mat she chooses is 352 in.^2, how wide will the border be?

35. A search light moves horizontally back and forth along a wall with the distance of the light from a starting point at t minutes given by the quadratic function defined by

$$f(t) = 100t^2 - 300t.$$

How long will it take before the light returns to the starting point?

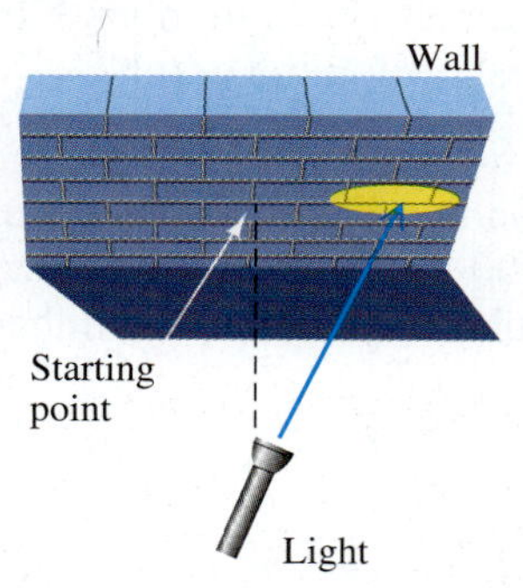

36. The Mart Hotel in Dallas, Texas, is 400 ft high. Suppose that a ball is projected upward from the top of the Mart, and its position in feet above the ground is given by the quadratic function defined by

$$f(t) = -16t^2 + 45t + 400,$$

where t is the number of seconds elapsed. How long will it take for the ball to reach a height of 200 ft above the ground? (*Source: World Almanac and Book of Facts,* 2000.)

37. The Toronto Dominion Center in Winnipeg, Manitoba, is 407 ft high. Suppose that a ball is projected upward from the top of the Center, and its position in feet above the ground is given by the quadratic function defined by

$$s(t) = -16t^2 + 75t + 407,$$

where t is the number of seconds elapsed. How long will it take for the ball to reach a height of 450 ft above the ground? (*Source: World Almanac and Book of Facts,* 2000.)

38. The manager of a fast-food outlet has determined that the demand for frozen yogurt is $\frac{25}{p}$ units per day, where p is the price (in dollars) per unit. The supply is $70p + 15$ units per day. Find the price at which supply and demand are equal.

39. Use the formula $A = P(1 + r)^2$ to find the interest rate r at which a principal P of \$10,000 will increase to \$10,920.25 in 2 yr.

40. The number of e-mail boxes in North America (in millions) for the years 1995–2001 are shown in the graph and can be modeled by the quadratic function defined by

$$f(x) = 3.29x^2 - 10.4x + 21.6.$$

In the model, $x = 5$ represents 1995, $x = 10$ represents 2000, and so on.

(a) Use the model to approximate the number of e-mail boxes in 2001 to the nearest whole number. How does this result compare to the number shown in the graph?

(b) Based on the model, in what year did the number of e-mail boxes reach 200 million? (Round down to the nearest year.) How does this result compare to the number shown in the graph?

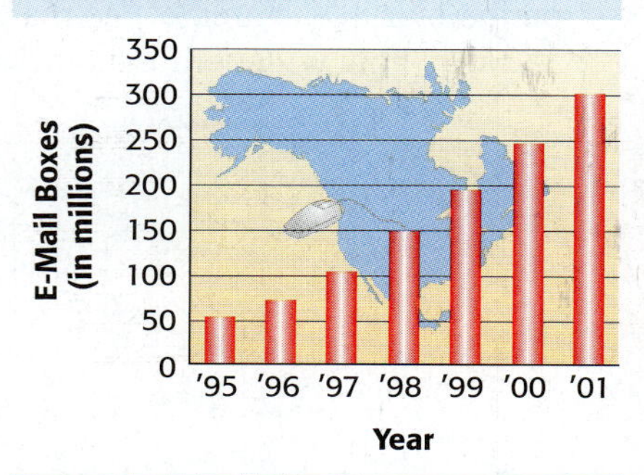

Source: IDC research.

[9.5–9.6] *Identify the vertex of each parabola.*

41. $f(x) = -(x - 1)^2$ **42.** $f(x) = (x - 3)^2 + 7$ **43.** $y = -3x^2 + 4x - 2$ **44.** $x = (y - 3)^2 - 4$

Graph each parabola. Give the domain and range.

45. $y = 2(x - 2)^2 - 3$
domain:
range:

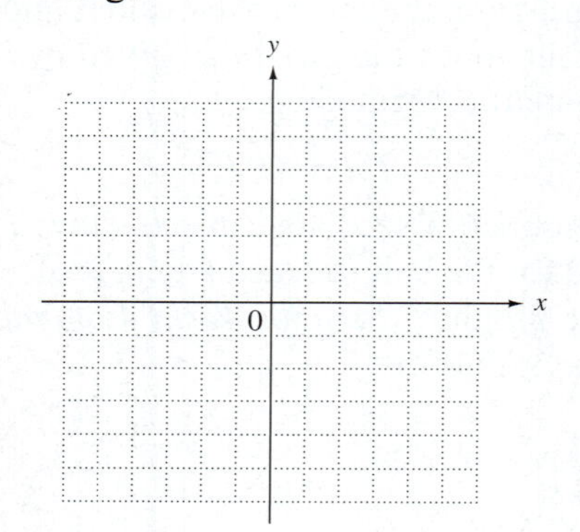

46. $f(x) = -2x^2 + 8x - 5$
domain:
range:

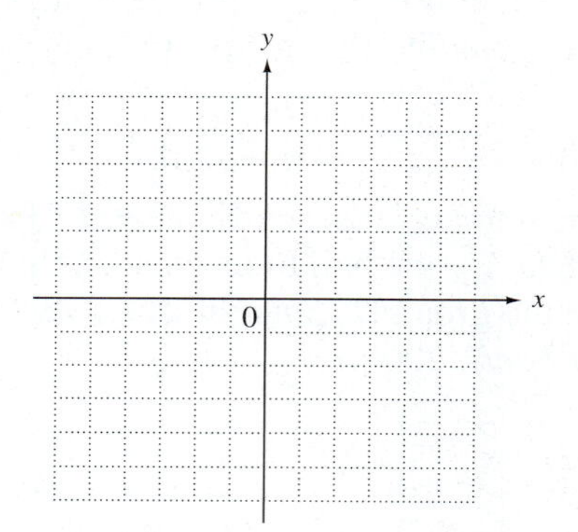

47. $x = 2(y + 3)^2 - 4$
domain:
range:

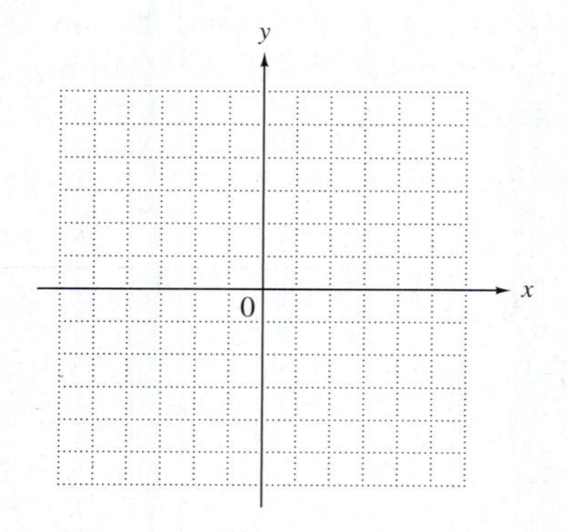

48. $x = -\frac{1}{2}y^2 + 6y - 14$
domain:
range:

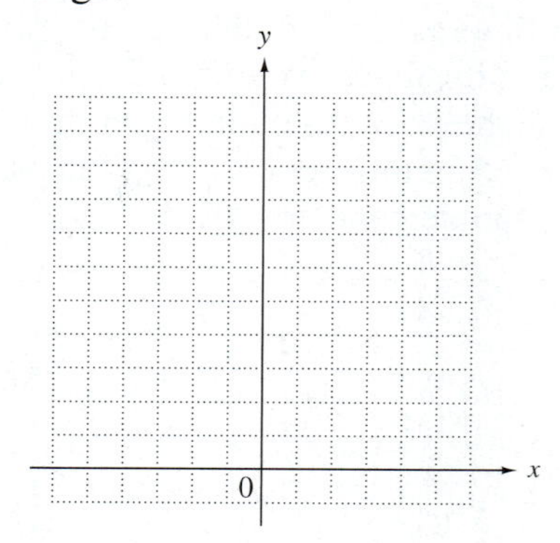

Solve each problem.

49. Consumer spending for home video games in dollars per person per year is given in the table. Let $x = 0$ represent 1990, $x = 2$ represent 1992, and so on.

(a) Use the data for 1990, 1994, and 1997 in the quadratic form $ax^2 + bx + c = y$ to write a system of three equations.

(b) Solve the system from part (a) to get a quadratic function f that models the data.

CONSUMER SPENDING FOR HOME VIDEO GAMES

Year	Dollars
1990	12.39
1992	13.08
1994	15.78
1996	19.43
1997	22.71
1998	24.14
1999	25.08

Source: Statistical Abstract of the United States.

(c) Use the model found in part (b) to approximate consumer spending for home video games in 1998 to the nearest cent. How does your answer compare to the actual data from the table?

50. The height (in feet) of a projectile t seconds after being fired from Earth into the air is given by

$$f(t) = -16t^2 + 160t.$$

Find the number of seconds required for the projectile to reach maximum height. What is the maximum height?

51. Find the length and width of a rectangle having a perimeter of 200 m if the area is to be a maximum.

[9.7] *Solve each inequality, and graph the solution set.*

52. $(x - 4)(2x + 3) > 0$

53. $x^2 + x \leq 12$

54. $(x + 2)(x - 3)(x + 5) \leq 0$

55. $(4m + 3)^2 \leq -4$

56. $\dfrac{6}{2z - 1} < 2$

57. $\dfrac{3t + 4}{t - 2} \leq 1$

Mixed Review Exercises

Solve.

58. $V = r^2 + R^2h$ for R

***59.** $3t^2 - 6t = -4$

60. $(x^2 - 2x)^2 = 11(x^2 - 2x) - 24$

61. $(r - 1)(2r + 3)(r + 6) < 0$

62. $(3k + 11)^2 = 7$

63. $S = \dfrac{Id^2}{k}$ for d

64. $2x - \sqrt{x} = 6$

65. $6 + \dfrac{15}{s^2} = -\dfrac{19}{s}$

66. $\dfrac{-2}{x + 5} \le -5$

67. Graph $f(x) = 4x^2 + 4x - 2$. Give the domain and range.

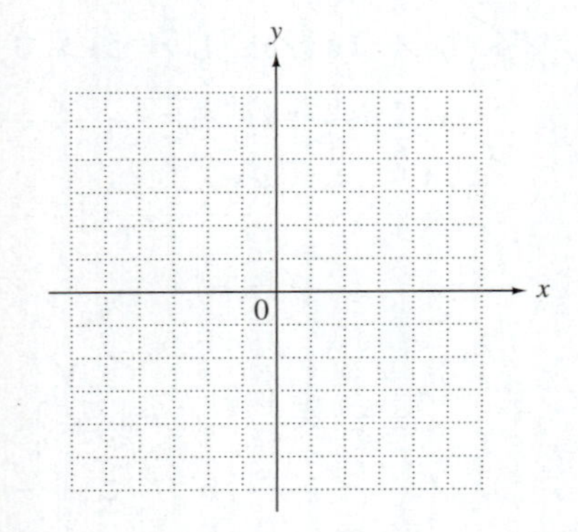

68. Natural gas use in the United States in trillions of cubic feet (ft^3) from 1970 through 1999 can be modeled by the quadratic function defined by

$$f(x) = .014x^2 - .396x + 21.2,$$

where $x = 0$ represents 1970, $x = 5$ represents 1975, and so on. (*Source:* Energy Information Administration.)

(a) Use the model to approximate natural gas use in 2000.

(b) Based on the model, in what year will natural gas use reach 25 trillion ft^3? (Round down to the nearest year.)

Chapter 9 Test

Study Skills Workbook
Activity 12

*Items marked * require knowledge of imaginary numbers.*

Solve by using either the square root property or completing the square.

1. $t^2 = 54$

1. ______________

2. $(7x + 3)^2 = 25$

2. ______________

3. $x^2 + 2x = 1$

3. ______________

Solve using the quadratic formula.

4. $2x^2 - 3x - 1 = 0$

4. ______________

***5.** $3t^2 - 4t = -5$

5. ______________

6. $3x = \sqrt{\dfrac{9x + 2}{2}}$

6. ______________

***7.** If k is a negative number, then which one of the following equations will have two imaginary solutions?

A. $x^2 = 4k$ **B.** $x^2 = -4k$

C. $(x + 2)^2 = -k$ **D.** $x^2 + k = 0$

7. ______________

8. What is the discriminant for $2x^2 - 8x - 3 = 0$? How many and what type of solutions does this equation have? (Do not actually solve.)

8. ______________

Solve by any method.

9. $3 - \dfrac{16}{x} - \dfrac{12}{x^2} = 0$

9. ______________

10. $4x^2 + 7x - 3 = 0$

10. ______________

11. ______________________

11. $9x^4 + 4 = 37x^2$

12. ______________________

12. $12 = (2n + 1)^2 + (2n + 1)$

13. ______________________

13. Solve for r: $S = 4\pi r^2$. (Leave $\pm$ in your answer.)

Solve each problem.

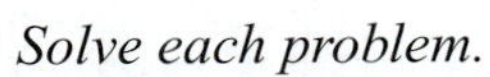

14. ______________________

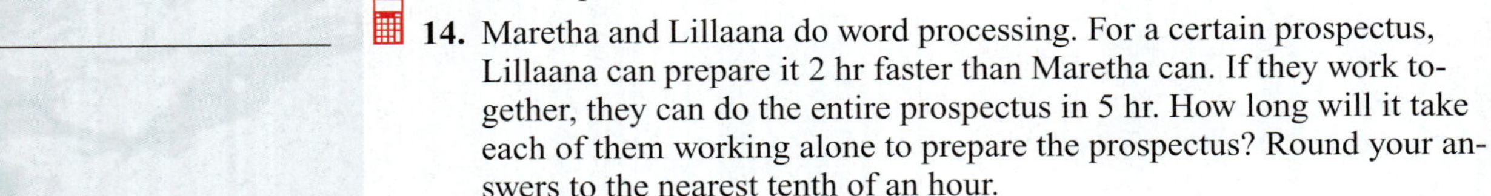

14. Maretha and Lillaana do word processing. For a certain prospectus, Lillaana can prepare it 2 hr faster than Maretha can. If they work together, they can do the entire prospectus in 5 hr. How long will it take each of them working alone to prepare the prospectus? Round your answers to the nearest tenth of an hour.

15. ______________________

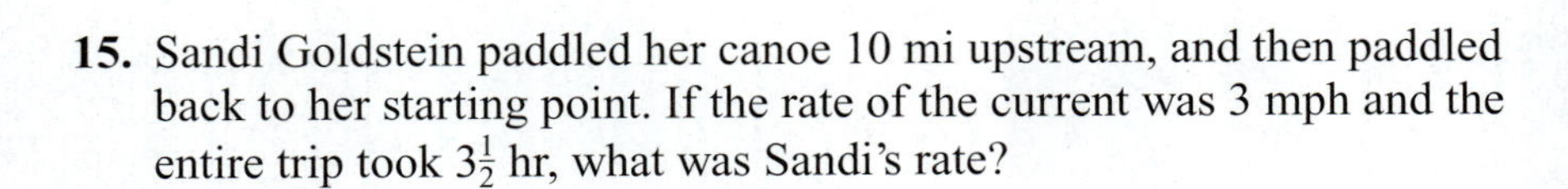

15. Sandi Goldstein paddled her canoe 10 mi upstream, and then paddled back to her starting point. If the rate of the current was 3 mph and the entire trip took $3\frac{1}{2}$ hr, what was Sandi's rate?

16. ______________________

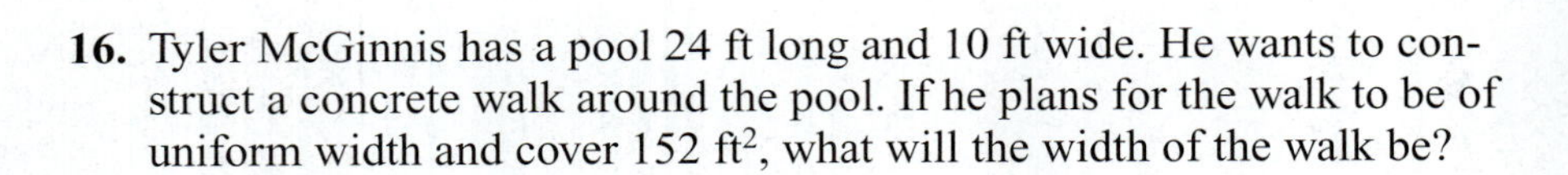

16. Tyler McGinnis has a pool 24 ft long and 10 ft wide. He wants to construct a concrete walk around the pool. If he plans for the walk to be of uniform width and cover 152 ft^2, what will the width of the walk be?

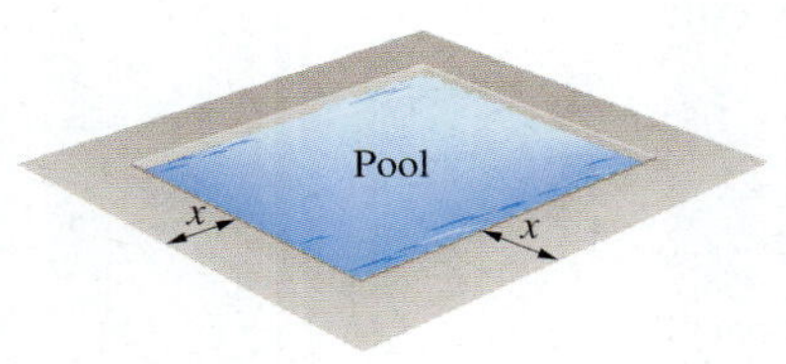

17. ______________________

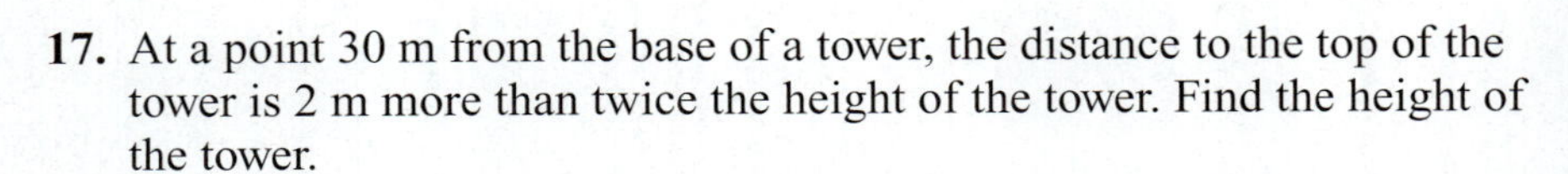

17. At a point 30 m from the base of a tower, the distance to the top of the tower is 2 m more than twice the height of the tower. Find the height of the tower.

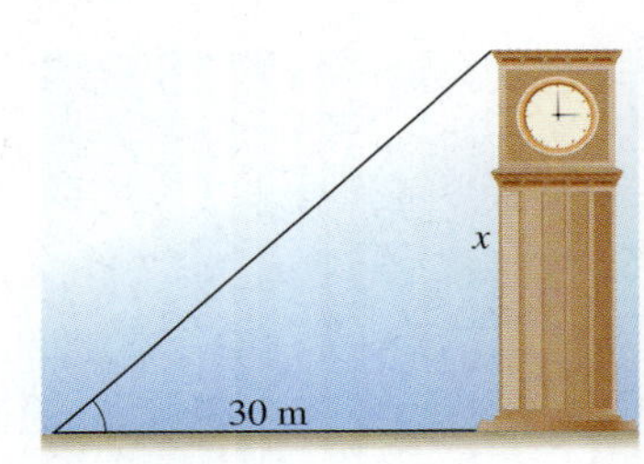

18. The percent increase for in-state tuition at Iowa public universities during the years 1992–2002 can be modeled by the quadratic function defined by

$$f(x) = .156x^2 - 2.05x + 10.2,$$

where $x = 2$ represents 1992, $x = 3$ represents 1993, and so on. (*Source:* Iowa Board of Regents.)

(a) Based on this model, by what percent (to the nearest tenth) did tuition increase in 2001?

(b) In what year was the minimum tuition increase? (Round down to the nearest year.) To the nearest tenth, by what percent did tuition increase that year?

18. (a) ______________________

(b) ______________________

19. Which one of the following most closely resembles the graph of $f(x) = a(x - h)^2 + k$ if $a < 0$, $h > 0$, and $k < 0$?

A.

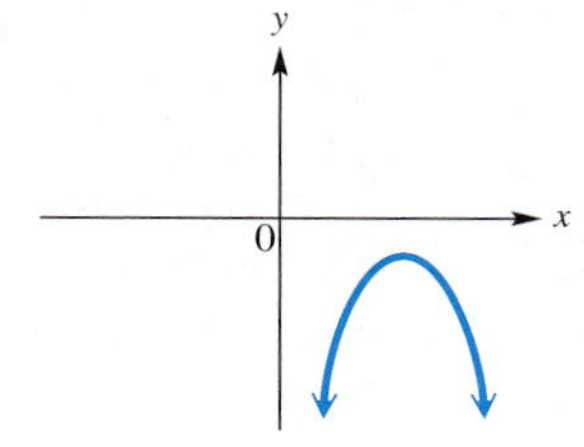

B.

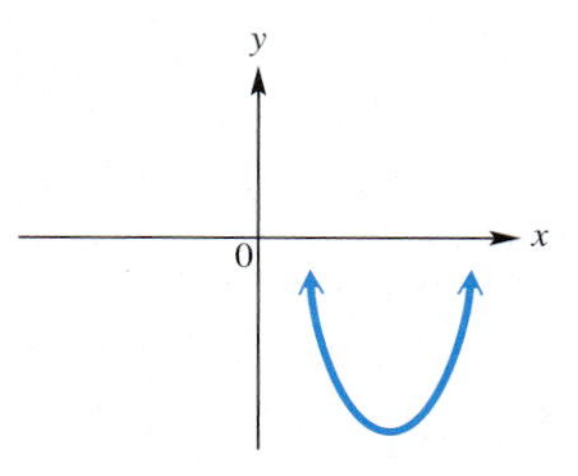

C.

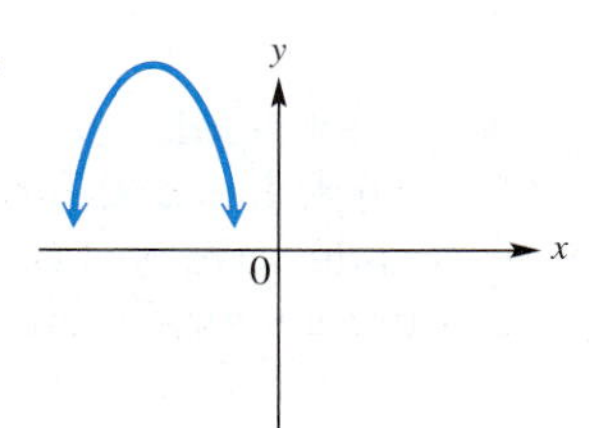

D.

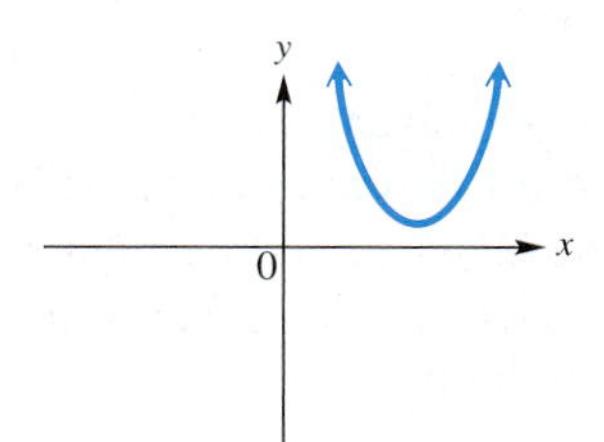

19. ______________________

Graph each parabola.

20. $f(x) = \frac{1}{2}x^2 - 2$

Give the vertex.

20. ______________________

21. ____________________

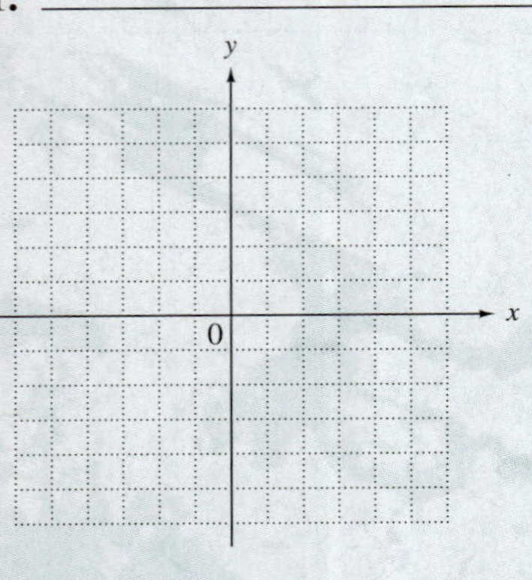

21. $f(x) = -x^2 + 4x - 1$

Give the vertex, domain, and range.

22. ____________________

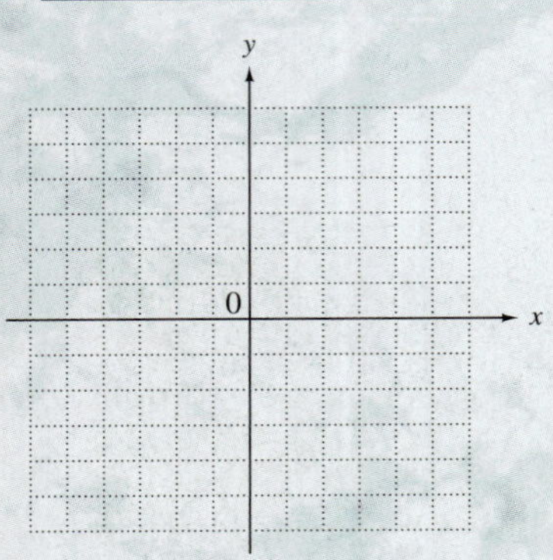

22. $x = 2y^2 + 8y + 3$

Give the vertex, domain, and range.

23. ____________________

23. Morgan's Department Store wants to construct a rectangular parking lot on land bordered on one side by a highway. The store has 280 ft of fencing that is to be used to fence off the other three sides. What should be the dimensions of the lot if the enclosed area is to be a maximum? What is the maximum area?

Solve. Graph each solution set.

24. ____________________

24. $2x^2 + 7x > 15$

25. ____________________

25. $\dfrac{5}{t - 4} \leq 1$

Cumulative Review Exercises CHAPTERS 1–9

1. Let $S = \{-\frac{7}{3}, -2, -\sqrt{3}, 0, .7, \sqrt{12}, \sqrt{-8}, 7, \frac{32}{3}\}$. List the elements of S that are elements of each set.

(a) Integers **(b)** Rational numbers **(c)** Real numbers **(d)** Complex numbers

Simplify each expression.

2. $|-3| + 8 - |-9| - (-7 + 3)$

3. $2(-3)^2 + (-8)(-5) + (-17)$

In this day of Automated Teller Machines (ATMs), people often find themselves doing what they have done for years when faced with a soft drink machine that won't respond: They talk to it. According to one report, the following are percentages of people in the United States, the United Kingdom (UK), and Germany who talk to ATMs and what they say.

	United States	UK	Germany
Thanking the ATM	22%	24%	14%
Cursing the ATM	31%	41%	53%
Telling the ATM to Hurry Up	47%	36%	33%

Source: BMRB International for NCR.

In a random sample of 3000 *people, how many would there be in each category?*

4. People in the United States who curse the ATM

5. People in the UK who thank the ATM

6. People in Germany who tell the ATM to hurry up

7. How many more German cursers would there be than United States thankers?

Solve each equation or inequality.

8. $-2x + 4 = 5(x - 4) + 17$

9. $-2x + 4 \leq -x + 3$

10. $|3x - 7| \leq 1$

11. Find the slope and y-intercept of the line with equation $2x - 4y = 7$.

12. Write the equation in standard form of the line through $(2, -1)$ and perpendicular to $-3x + y = 5$.

Graph each relation. Tell whether or not each is a function, and if it is, give its domain and range.

13. $4x - 5y = 15$

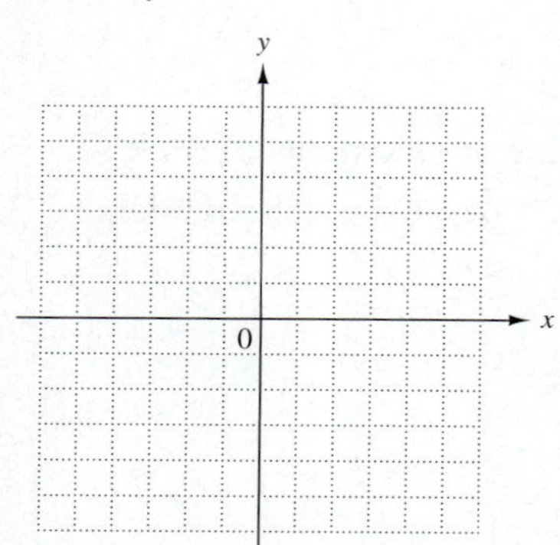

14. $4x - 5y < 15$

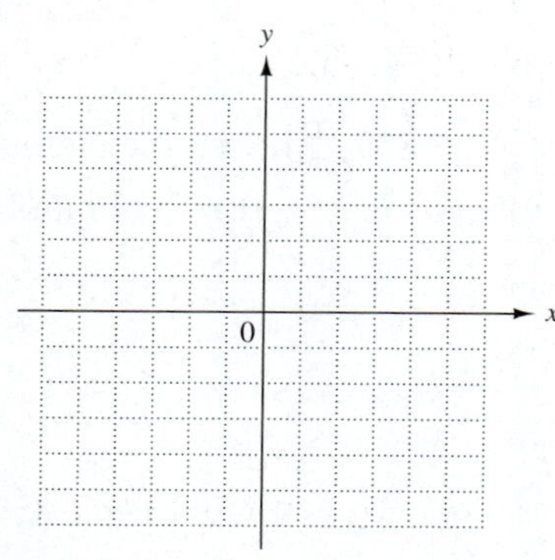

15. $f(x) = -2(x - 1)^2 + 3$

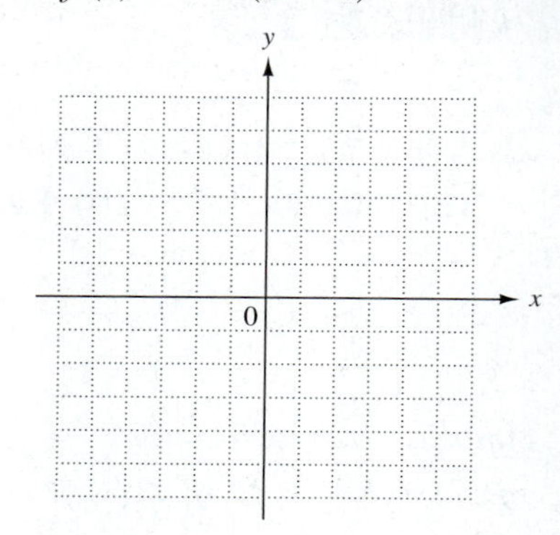

16. The record track-qualifying speeds at North Carolina Motor Speedway since Richard Petty captured the first pole in 1965 are given in the table and can be modeled by a linear equation. Let $x = 0$ represent 1965, $x = 10$ represent 1975, and so on.

(a) Use the ordered pairs (0, 116.26) and (20, 141.85) to write a linear equation that models these data.

(b) Use your model to approximate the record speed for 1998 to the nearest hundredth. How does it compare to the actual value from the table?

QUALIFYING RECORDS

Year	Speed (in mph)
1965	116.26
1975	132.02
1985	141.85
1995	155.38
1998	156.36

Source: NASCAR.

17. Does the relation $x = 5$ define a function? Explain why or why not.

Solve each system of equations.

18. $2x - 4y = 10$
$9x + 3y = 3$

19. $x + y + 2z = 3$
$-x + y + z = -5$
$2x + 3y - z = -8$

20. The recent merger of America Online and Time Warner is the largest in U.S. history. The two companies have combined sales of \$34.2 billion. Sales for AOL are \$.3 billion less than 4 times the sales of Time Warner. What are the sales for each company? (*Source:* Company reports.)

(a) Write a system of equations to solve the problem.

(b) Solve the problem.

Write with positive exponents only. Assume variables represent positive real numbers.

21. $\left(\dfrac{x^{-3}y^{2}}{x^{5}y^{-2}}\right)^{-1}$

22. $\dfrac{(4x^{-2})^{2}(2y^{3})}{8x^{-3}y^{5}}$

Perform the indicated operations.

23. $\left(\dfrac{2}{3}t + 9\right)^{2}$

24. $(3t^3 + 5t^2 - 8t + 7) - (6t^3 + 4t - 8)$

25. Divide $4x^3 + 2x^2 - x + 26$ by $x + 2$.

26. According to the Congressional Budget Office, the federal budget surplus is expected to total \$3,100,000,000,000 over the next decade. Write this amount using scientific notation. (*Source: The Gazette,* January 31, 2001.)

Factor completely.

27. $16x - x^3$

28. $24m^2 + 2m - 15$

29. $9x^2 - 30xy + 25y^2$

Perform the operations, and express answers in lowest terms. Assume denominators are nonzero.

30. $\dfrac{5t + 2}{-6} \div \dfrac{15t + 6}{5}$

31. $\dfrac{3}{2 - k} - \dfrac{5}{k} + \dfrac{6}{k^2 - 2k}$

32. $\dfrac{\dfrac{r}{s} - \dfrac{s}{r}}{\dfrac{r}{s} + 1}$

Simplify each radical expression.

33. $\sqrt[3]{\dfrac{27}{16}}$

34. $\dfrac{2}{\sqrt{7} - \sqrt{5}}$

Solve each equation.

35. $2x = \sqrt{\dfrac{5x + 2}{3}}$

36. $2x^2 - 4x - 3 = 0$

37. $z^2 - 2z = 15$

38. $\dfrac{3}{x - 3} - \dfrac{2}{x - 2} = \dfrac{3}{x^2 - 5x + 6}$

39. $p^4 - 10p^2 + 9 = 0$

40. Two cars left an intersection at the same time, one heading due south and the other due east. Later they were exactly 95 mi apart. The car heading east had gone 38 mi less than twice as far as the car heading south. How far had each car traveled?

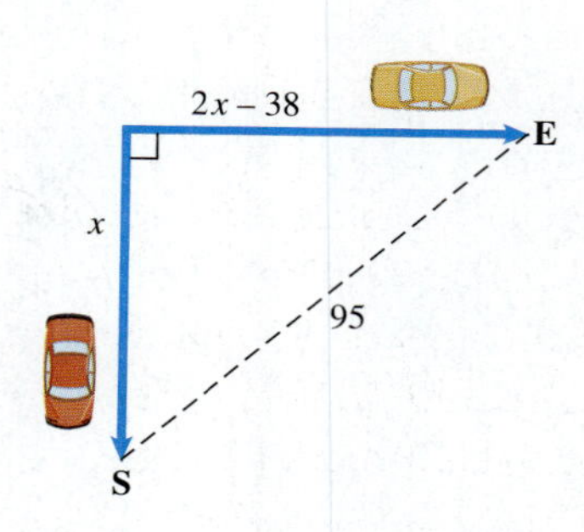

Exponential and Logarithmic Functions

The exponential and logarithmic functions introduced in this chapter are used to model a wide variety of situations, including environmental issues, compound interest, earthquake intensity, fossil dating, and sound levels. Recently, there has been concern about the level of sound Americans are subjected to daily. For example, action sequences in *Pearl Harbor, The Movie* reached 107 decibels, while the sound levels of *Lethal Weapon 4* often reached 100 decibels or more, compared to an average of 95 decibels for a motorcycle. In Section 10.5, Exercise 39, we give a logarithmic function to measure sound levels and to find the decibel levels of other recent movies. (*Source: World Almanac and Book of Facts,* 2001, www.lhh.org/noise/)

10.1 Inverse Functions

Objectives

1. Decide whether a function is one-to-one and, if it is, find its inverse.
2. Use the horizontal line test to determine whether a function is one-to-one.
3. Find the equation of the inverse of a function.
4. Graph f^{-1} from the graph of f.

In this chapter we will study two important types of functions, *exponential* and *logarithmic*. These functions are related in a special way: They are *inverses* of one another. We begin by discussing inverse functions in general.

Calculator Tip A calculator with the following keys will be essential in this chapter.

y^x, 10^x or LOG, e^x or ln x

We will explain how these keys are used at appropriate places in the chapter.

1 Decide whether a function is one-to-one and, if it is, find its inverse. Suppose we define the function

$$G = \{(-2, 2), (-1, 1), (0, 0), (1, 3), (2, 5)\}.$$

We can form another set of ordered pairs from G by interchanging the x- and y-values of each pair in G. Call this set F, with

$$F = \{(2, -2), (1, -1), (0, 0), (3, 1), (5, 2)\}.$$

To show that these two sets are related, F is called the *inverse* of G. For a function f to have an inverse, f must be *one-to-one.*

One-to-One Function

In a one-to-one function, each x-value corresponds to only one y-value, and each y-value corresponds to just one x-value.

The function shown in Figure 1(a) is not one-to-one because the y-value 7 corresponds to *two* x-values, 2 and 3. That is, the ordered pairs (2, 7) and (3, 7) both appear in the function. The function in Figure 1(b) is one-to-one.

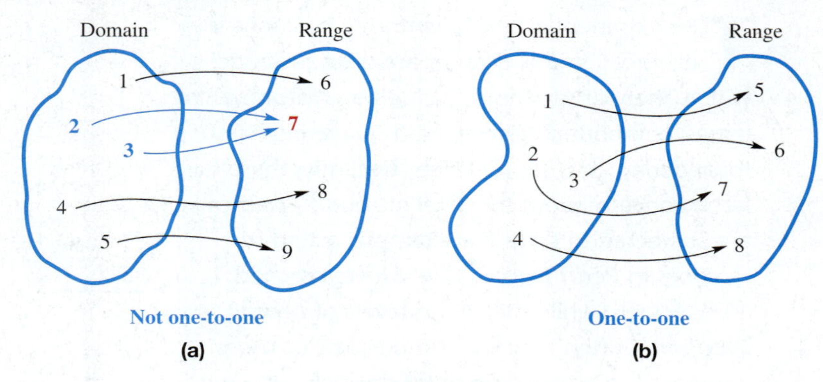

(a) (b)

Figure 1

The *inverse* of any one-to-one function f is found by interchanging the components of the ordered pairs of f. The inverse of f is written f^{-1}. Read f^{-1} as "the inverse of f" or "f-inverse."

CAUTION

The symbol $f^{-1}(x)$ does not represent $\dfrac{1}{f(x)}$.

The definition of the inverse of a function follows.

Inverse of a Function

The **inverse** of a one-to-one function f, written f^{-1}, is the set of all ordered pairs of the form (y, x), where (x, y) belongs to f. Since the inverse is formed by interchanging x and y, the domain of f becomes the range of f^{-1} and the range of f becomes the domain of f^{-1}.

For inverses f and f^{-1}, it follows that $f(f^{-1}(x)) = x$ and $f^{-1}(f(x)) = x$.

Example 1 Finding the Inverses of One-to-One Functions

Find the inverse of each one-to-one function.

(a) $F = \{(-2, 1), (-1, 0), (0, 1), (1, 2), (2, 2)\}$

Each x-value in F corresponds to just one y-value. However, the y-value 2 corresponds to two x-values, 1 and 2. Also, the y-value 1 corresponds to both -2 and 0. Because some y-values correspond to more than one x-value, F is not one-to-one and does not have an inverse.

(b) $G = \{(3, 1), (0, 2), (2, 3), (4, 0)\}$

Every x-value in G corresponds to only one y-value, and every y-value corresponds to only one x-value, so G is a one-to-one function. The inverse function is found by interchanging the x- and y-values in each ordered pair.

$$G^{-1} = \{(1, 3), (2, 0), (3, 2), (0, 4)\}$$

Notice how the domain and range of G become the range and domain, respectively, of G^{-1}.

(c) The U.S. Environmental Protection Agency has developed an indicator of air quality called the Pollutant Standard Index (PSI). If the PSI exceeds 100 on a particular day, that day is classified as unhealthy. The table shows the number of unhealthy days in Chicago for the years 1991–1997.

Year	Number of Unhealthy Days
1991	21
1992	4
1993	3
1994	8
1995	21
1996	6
1997	9

Source: U.S. Environmental Protection Agency.

Let f be the function defined in the table, with the years forming the domain and the numbers of unhealthy days forming the range. Then f is not one-to-one, because in two different years (1991 and 1995), the number of unhealthy days was the same, 21.

Work Problem 1 at the Side.

1 Find the inverse of each one-to-one function.

(a) $\{(1, 2), (2, 4), (3, 3), (4, 5)\}$

(b) $\{(0, 3), (-1, 2), (1, 3)\}$

(c) A Norwegian physiologist has developed a rule for predicting running times based on the time to run 5 km (5K). An example for one runner is shown here. (*Source:* Stephen Seiler, Agder College, Kristiansand, Norway.)

Distance	Time
1.5K	4:22
3K	9:18
5K	16:00
10K	33:40

Answers

1. (a) $\{(2, 1), (4, 2), (3, 3), (5, 4)\}$
(b) not a one-to-one function
(c)

Time	Distance
4:22	1.5K
9:18	3K
16:00	5K
33:40	10K

2 Use the horizontal line test to determine whether each graph is the graph of a one-to-one function.

(a)

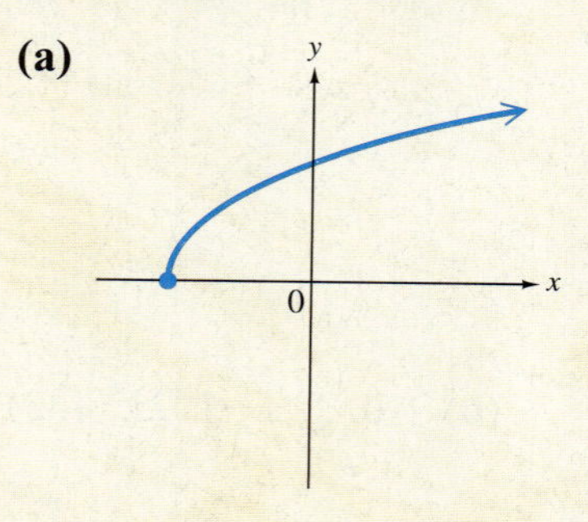

(b)

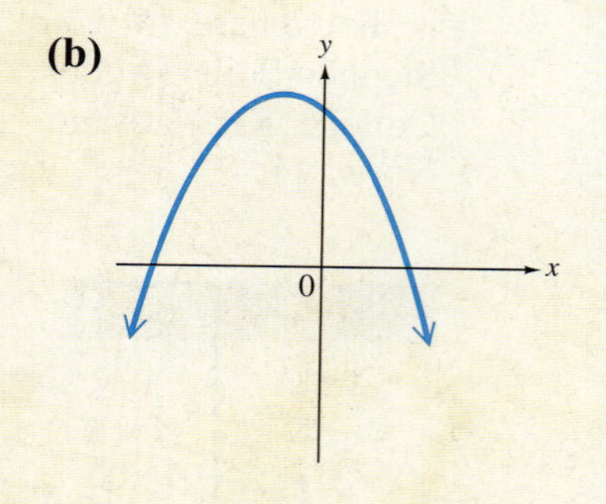

2 Use the horizontal line test to determine whether a function is one-to-one. It may be difficult to decide whether a function is one-to-one just by looking at the equation that defines the function. However, by graphing the function and observing the graph, we can use the *horizontal line test* to tell whether the function is one-to-one.

Horizontal Line Test

A function is one-to-one if every horizontal line intersects the graph of the function at most once.

The horizontal line test follows from the definition of a one-to-one function. Any two points that lie on the same horizontal line have the same y-coordinate. No two ordered pairs that belong to a one-to-one function may have the same y-coordinate, and therefore no horizontal line will intersect the graph of a one-to-one function more than once.

Example 2 Using the Horizontal Line Test

Use the horizontal line test to determine whether the graphs in Figures 2 and 3 are graphs of one-to-one functions.

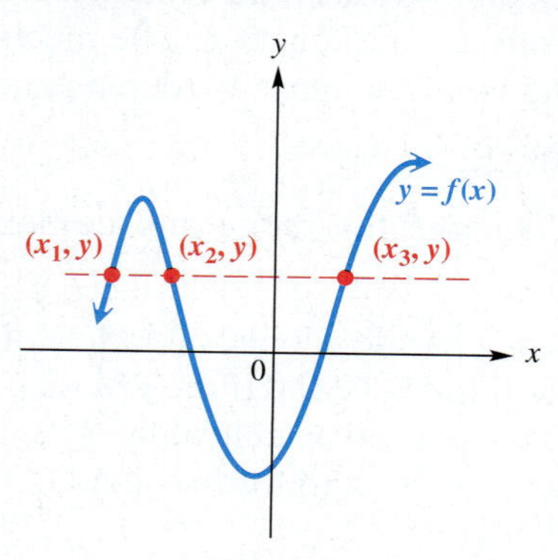

Figure 2

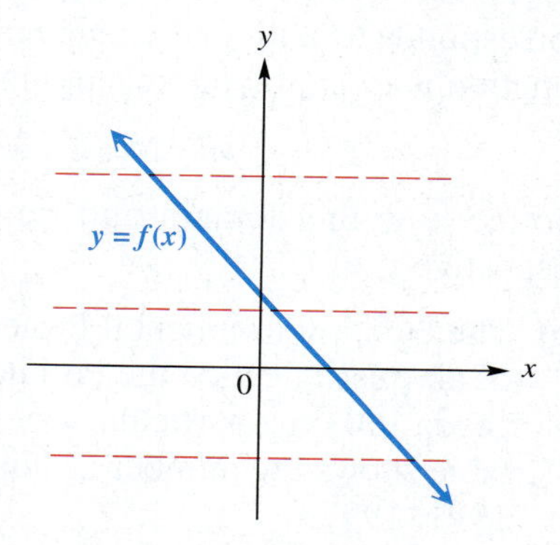

Figure 3

Because the horizontal line shown in Figure 2 intersects the graph in more than one point (actually three points), the function is not one-to-one.

Every horizontal line will intersect the graph in Figure 3 in exactly one point. This function is one-to-one.

Work Problem 2 at the Side.

3 Find the equation of the inverse of a function. By definition, the inverse of a function is found by interchanging the x- and y-values of each of its ordered pairs. The equation of the inverse of a function defined by $y = f(x)$ is found in the same way.

Finding the Equation of the Inverse of $y = f(x)$

For a one-to-one function f defined by an equation $y = f(x)$, find the defining equation of the inverse as follows.

Step 1 Interchange x and y.

Step 2 Solve for y.

Step 3 Replace y with $f^{-1}(x)$.

ANSWERS
2. (a) one-to-one (b) not one-to-one

Example 3 **Finding Equations of Inverses**

Decide whether each equation defines a one-to-one function. If so, find the equation of the inverse.

(a) $f(x) = 2x + 5$

The graph of $y = 2x + 5$ is a nonvertical line, so by the horizontal line test, f is a one-to-one function. To find the inverse, let $y = f(x)$ so that

$$y = 2x + 5$$

$$x = 2y + 5 \qquad \text{Interchange } x \text{ and } y. \text{ (Step 1)}$$

$$2y = x - 5 \qquad \text{Solve for } y. \text{ (Step 2)}$$

$$y = \frac{x - 5}{2}$$

$$f^{-1}(x) = \frac{x - 5}{2}. \qquad \text{(Step 3)}$$

Thus, f^{-1} is a linear function. In the function with $y = 2x + 5$, the value of y is found by starting with a value of x, multiplying by 2, and adding 5. The equation for the inverse has us *subtract* 5, and then *divide* by 2. This shows how an inverse is used to "undo" what a function does to the variable x.

(b) $y = x^2 + 2$

This equation has a vertical parabola as its graph, so some horizontal lines will intersect the graph at two points. For example, both $x = 3$ and $x = -3$ correspond to $y = 11$. Because of the x^2-term, there are many pairs of x-values that correspond to the same y-value. This means that the function defined by $y = x^2 + 2$ is not one-to-one and does not have an inverse.

If this is not noticed, following the steps for finding the equation of an inverse leads to

$$y = x^2 + 2$$

$$x = y^2 + 2 \qquad \text{Interchange } x \text{ and } y.$$

$$x - 2 = y^2 \qquad \text{Solve for } y.$$

$$\pm\sqrt{x - 2} = y \qquad \text{Square root property}$$

The last step shows that there are two y-values for each choice of $x > 2$, so the given function is not one-to-one and cannot have an inverse.

(c) $f(x) = (x - 2)^3$

Refer to Section 6.2 to see from its graph that a cubing function like this is a one-to-one function.

$$y = (x - 2)^3 \qquad \text{Replace } f(x) \text{ with } y.$$

$$x = (y - 2)^3 \qquad \text{Interchange } x \text{ and } y.$$

$$\sqrt[3]{x} = \sqrt[3]{(y - 2)^3} \qquad \text{Take the cube root on each side.}$$

$$\sqrt[3]{x} = y - 2$$

$$\sqrt[3]{x} + 2 = y$$

$$f^{-1}(x) = \sqrt[3]{x} + 2 \qquad \text{Replace } y \text{ with } f^{-1}(x).$$

Work Problem 3 at the Side.

3 Decide whether each equation defines a one-to-one function. If so, find the equation that defines the inverse.

(a) $f(x) = 3x - 4$

(b) $f(x) = x^3 + 1$

(c) $f(x) = (x - 3)^2$

Answers

3. (a) one-to-one function; $f^{-1}(x) = \frac{x + 4}{3}$

(b) one-to-one function; $f^{-1}(x) = \sqrt[3]{x - 1}$

(c) not a one-to-one function

4 Use the given graphs to graph each inverse.

(a)

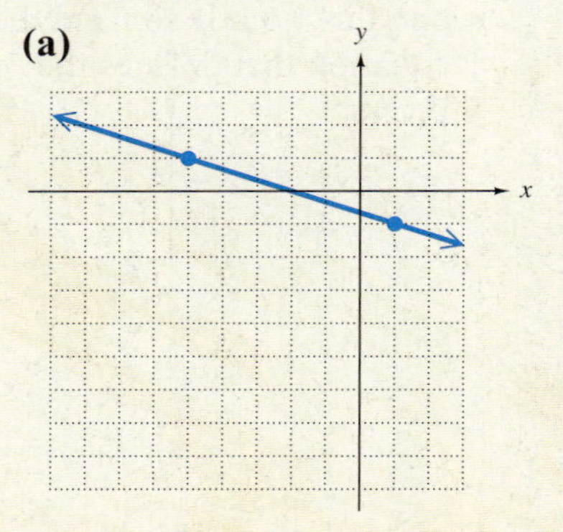

(b)

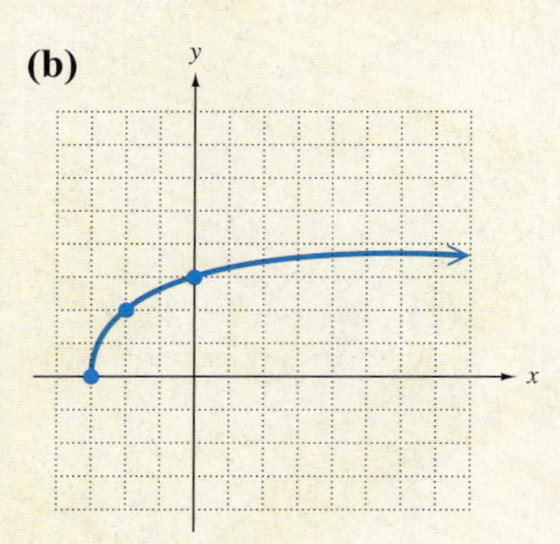

(c)

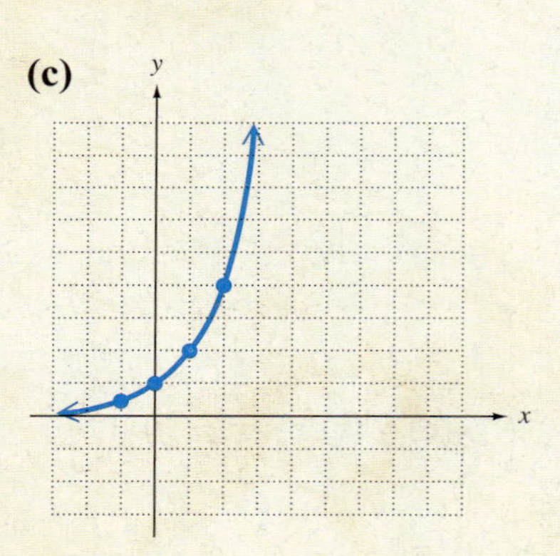

4 Graph f^{-1} from the graph of f. One way to graph the inverse of a function f whose equation is known is to find some ordered pairs that belong to f, interchange x and y to get ordered pairs that belong to f^{-1}, plot those points, and sketch the graph of f^{-1} through the points. A simpler way is to select points on the graph of f and use symmetry to find corresponding points on the graph of f^{-1}. For example, suppose the point (a, b) shown in Figure 4 belongs to a one-to-one function f. Then the point (b, a) belongs to f^{-1}. The line segment connecting (a, b) and (b, a) is perpendicular to, and cut in half by, the line $y = x$. The points (a, b) and (b, a) are "mirror images" of each other with respect to $y = x$. For this reason we can find the graph of f^{-1} from the graph of f by locating the mirror image of each point in f with respect to the line $y = x$.

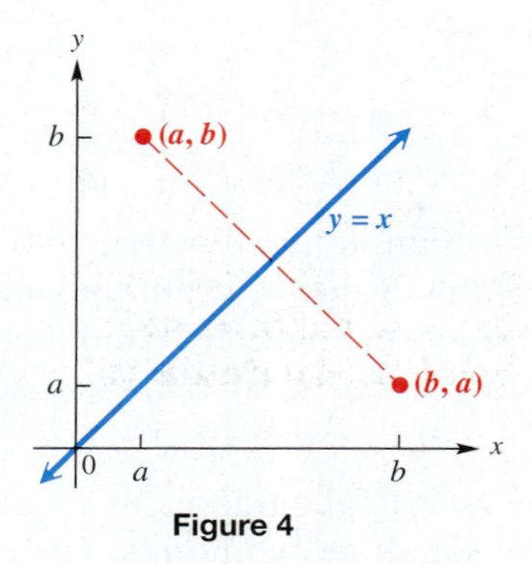

Figure 4

Example 4 Graphing the Inverse

Graph the inverses of the functions shown in Figure 5.

In Figure 5 the graphs of two functions are shown in blue. Their inverses are shown in red. In each case, the graph of f^{-1} is symmetric to the graph of f with respect to the line $y = x$.

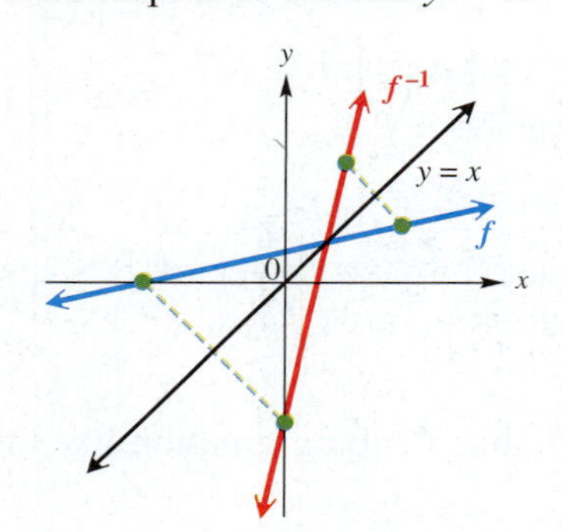

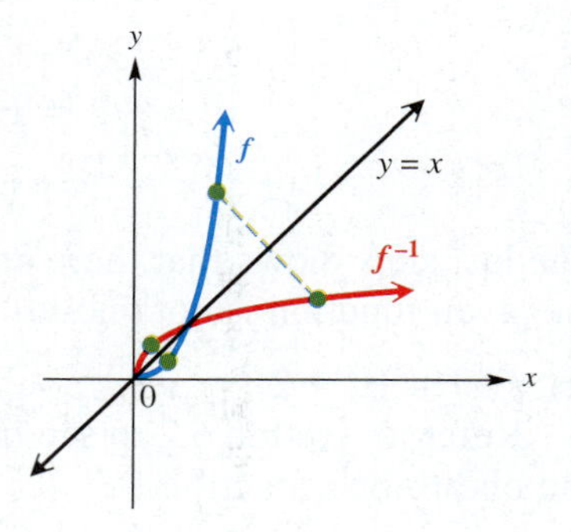

Figure 5

Work Problem 4 at the Side.

ANSWERS

4. (a) (b) (c)

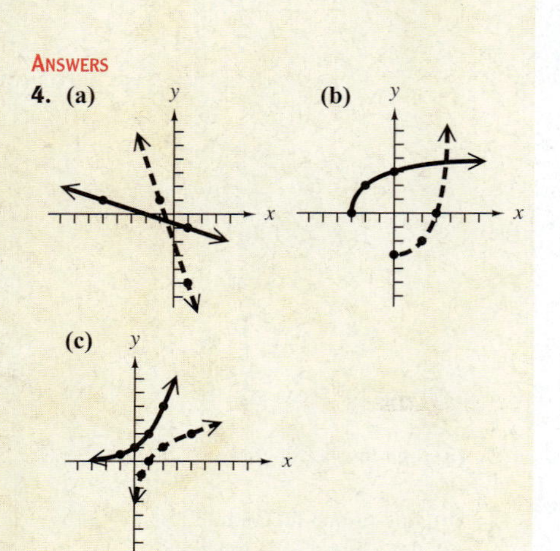

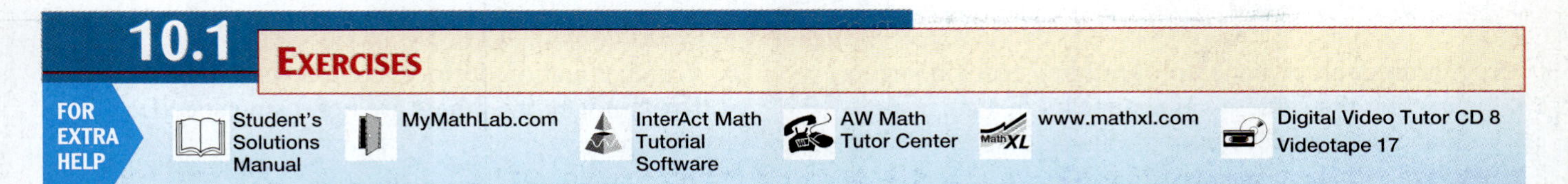

1. The table shows the number of uncontrolled hazardous waste sites that require further investigation to determine whether remedies are needed under the Superfund program. The seven states listed are ranked in the top ten in the United States.

If this correspondence is considered to be a function that pairs each state with its number of uncontrolled waste sites, is it one-to-one? If not, explain why.

State	Number of Sites
New Jersey	108
Pennsylvania	101
California	94
New York	79
Florida	53
Illinois	40
Wisconsin	40

Source: U.S. Environmental Protection Agency.

2. The table shows emissions of a major air pollutant, carbon monoxide, in the United States for the years 1991–1997.

If this correspondence is considered to be a function that pairs each year with its emissions amount, is it one-to-one? If not, explain why.

Year	Amount of Emissions (in thousand short tons)
1991	97,790
1992	94,400
1993	94,526
1994	98,854
1995	89,151
1996	90,611
1997	87,451

Source: U.S. Environmental Protection Agency.

3. Suppose you consider the set of ordered pairs (x, y) such that x represents a person in your mathematics class and y represents that person's mother. Explain how this function might not be a one-to-one function.

4. The road mileage between Denver, Colorado, and several selected U.S. cities is shown in the table below.

City	Distance to Denver (in miles)
Atlanta	1398
Dallas	781
Indianapolis	1058
Kansas City, MO	600
Los Angeles	1059
San Francisco	1235

If we consider this as a function that pairs each city with a distance, is it a one-to-one function? How could we change the answer to this question by adding 1 mile to one of the distances shown?

Choose the correct response from the given list.

5. If a function is made up of ordered pairs in such a way that the same y-value appears in a correspondence with two different x-values, then

A. the function is one-to-one
B. the function is not one-to-one
C. its graph does not pass the vertical line test
D. it has an inverse function associated with it.

6. Which equation defines a one-to-one function? Explain why the others are not, using specific examples.

A. $f(x) = x$ **B.** $f(x) = x^2$
C. $f(x) = |x|$ **D.** $f(x) = -x^2 + 2x - 1$

7. Only one of the graphs illustrates a one-to-one function. Which one is it?

A.

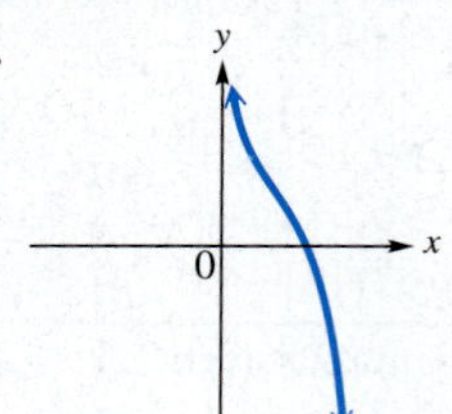

B.

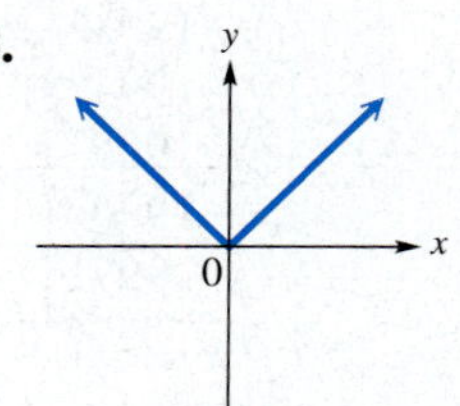

C.

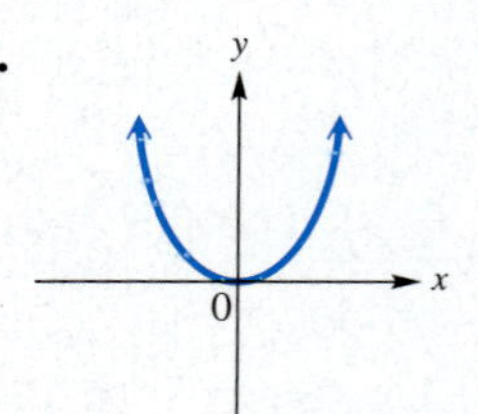

D.

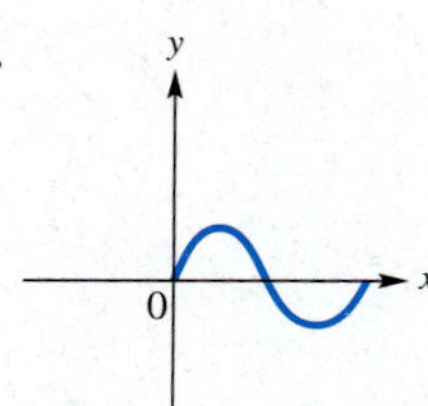

8. If a function f is one-to-one and the point (p, q) lies on the graph of f, then which point *must* lie on the graph of f^{-1}?

A. $(-p, q)$ **B.** $(-q, -p)$
C. $(p, -q)$ **D.** (q, p)

If the function is one-to-one, find its inverse. See Examples 1–3.

9. $\{(3, 6), (2, 10), (5, 12)\}$

10. $\{(-1, 3), (0, 5), (5, 0), (7, -\frac{1}{2})\}$

11. $\{(-1, 3), (2, 7), (4, 3), (5, 8)\}$

12. $\{(-8, 6), (-4, 3), (0, 6), (5, 10)\}$

13. $f(x) = 2x + 4$

14. $f(x) = 3x + 1$

15. $g(x) = \sqrt{x - 3}, x \geq 3$

16. $g(x) = \sqrt{x + 2}, x \geq -2$

17. $f(x) = 3x^2 + 2$

18. $f(x) = -4x^2 - 1$

19. $f(x) = x^3 - 4$

20. $f(x) = x^3 - 3$

Let $f(x) = 2^x$. We will see in the next section that the function f is one-to-one. Find each value, always working part **(a)** before *part* **(b)**.

21. **(a)** $f(3)$
(b) $f^{-1}(8)$

22. **(a)** $f(4)$
(b) $f^{-1}(16)$

23. **(a)** $f(0)$
(b) $f^{-1}(1)$

24. **(a)** $f(-2)$
(b) $f^{-1}(\frac{1}{4})$

The graphs of some functions are given in Exercises 25–30. ***(a)*** *Use the horizontal line test to determine whether each function is one-to-one.* ***(b)*** *If the function is one-to-one, graph the inverse of the function with a dashed line (or curve) on the same set of axes. (Remember that if f is one-to-one and $f(a) = b$, then $f^{-1}(b) = a$.) See Example 4.*

25.

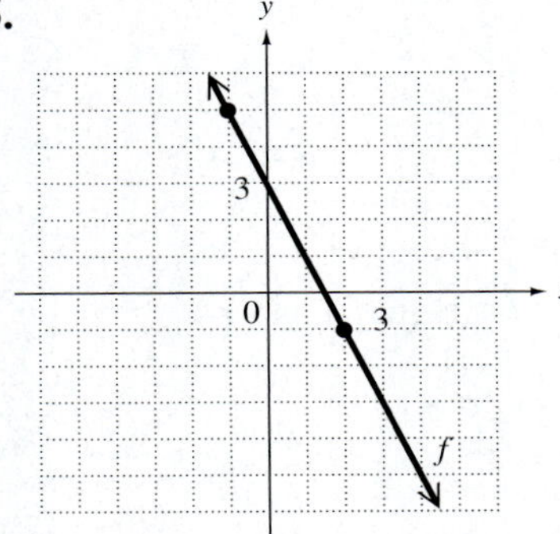

26.

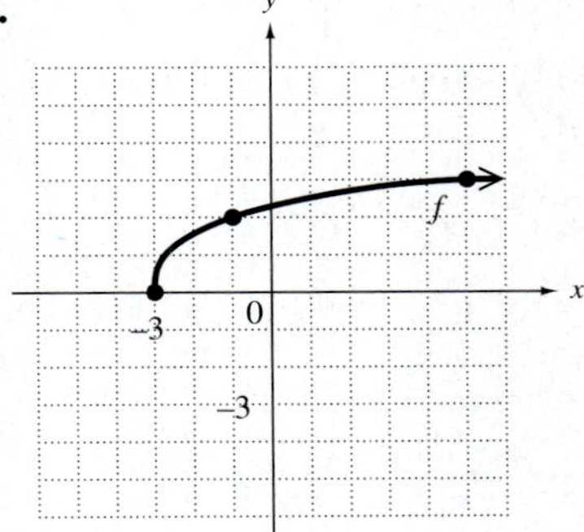

27.

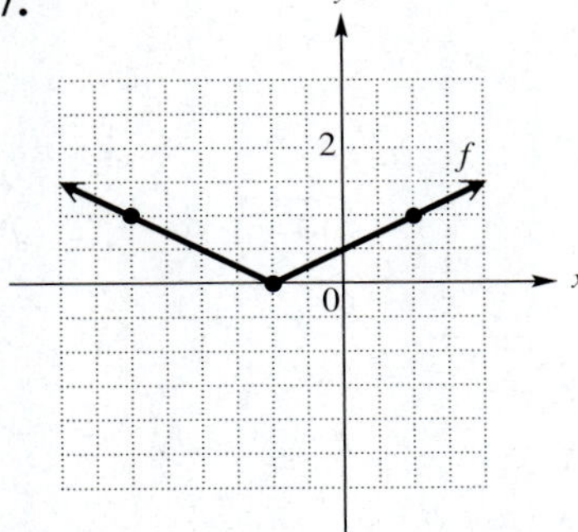

28.

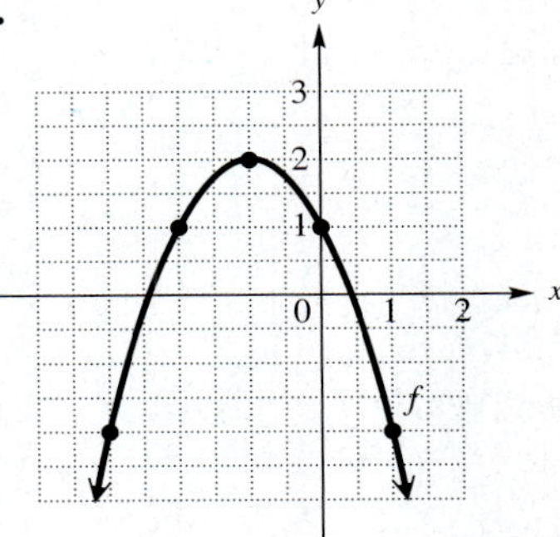

29.

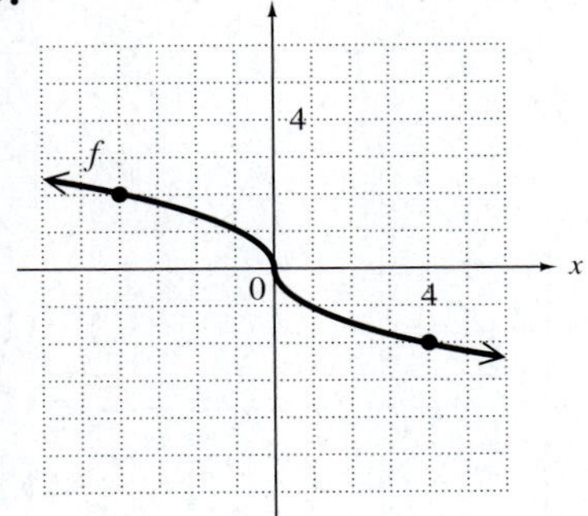

30.

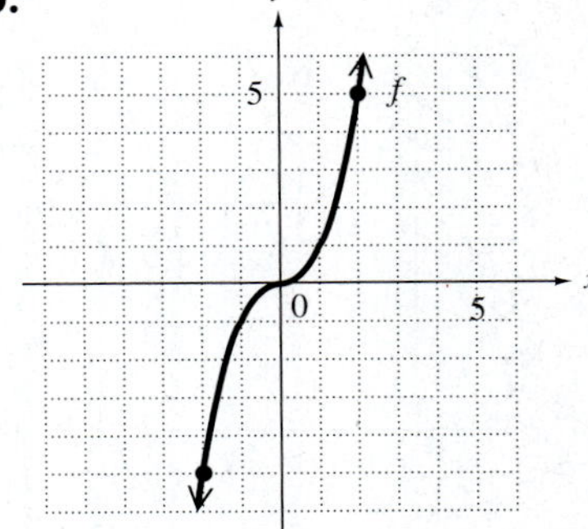

Each function defined in Exercises 31–38 is a one-to-one function. Graph the function as a solid line (or curve), and then graph its inverse on the same set of axes as a dashed line (or curve). In Exercises 37 and 38 you are given a table to complete so that graphing the function will be easier. See Example 4.

31. $f(x) = 2x - 1$

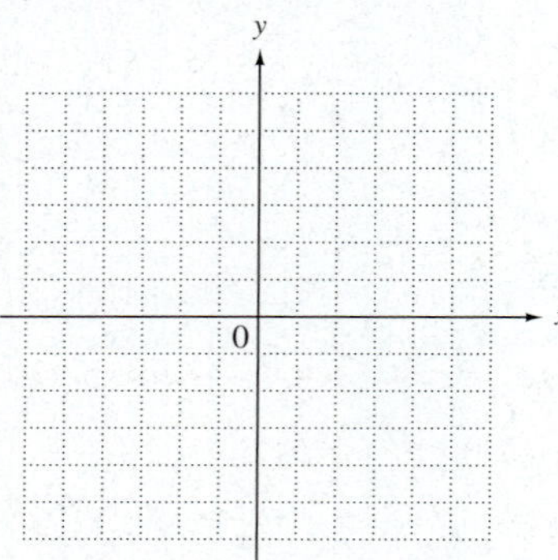

32. $f(x) = 2x + 3$

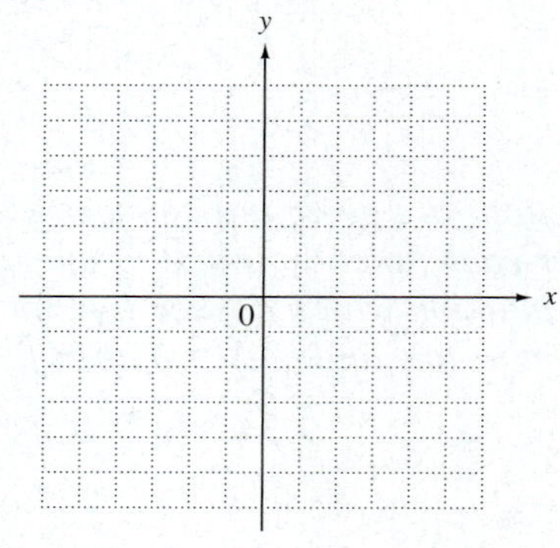

33. $g(x) = -4x$

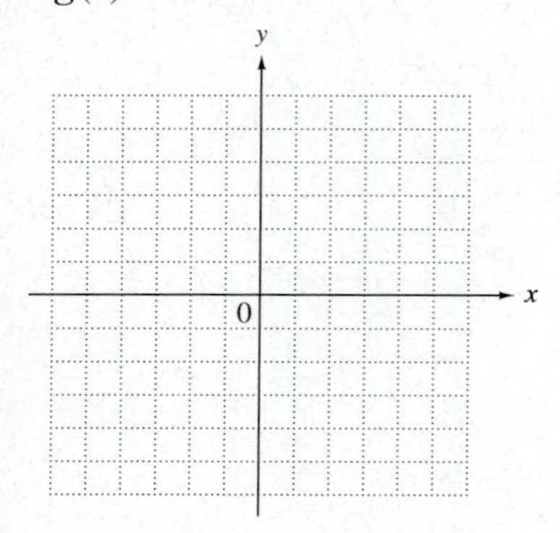

34. $g(x) = -2x$

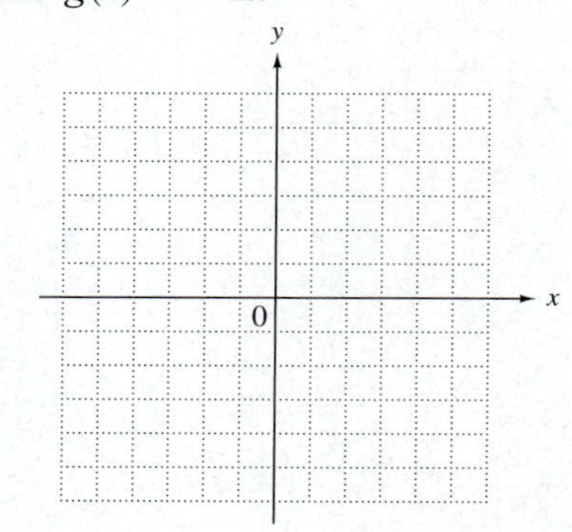

35. $f(x) = \sqrt{x}, x \geq 0$

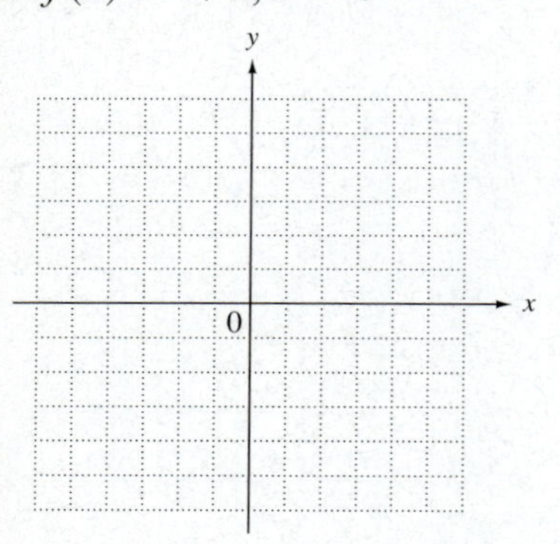

36. $f(x) = -\sqrt{x}, x \geq 0$

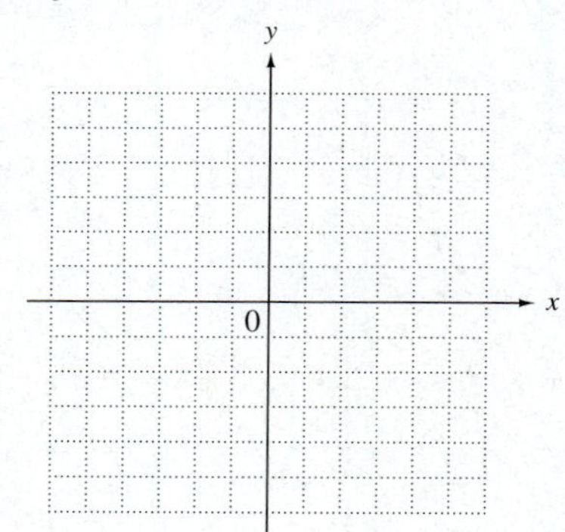

37. $y = x^3 - 2$

x	y
−1	
0	
1	
2	

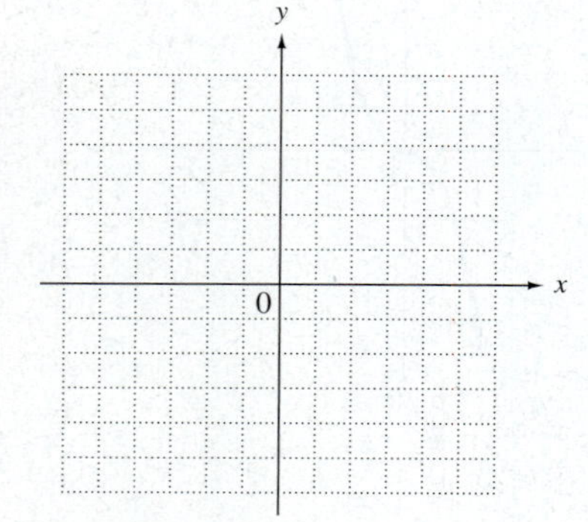

38. $y = x^3 + 3$

x	y
−2	
−1	
0	
1	

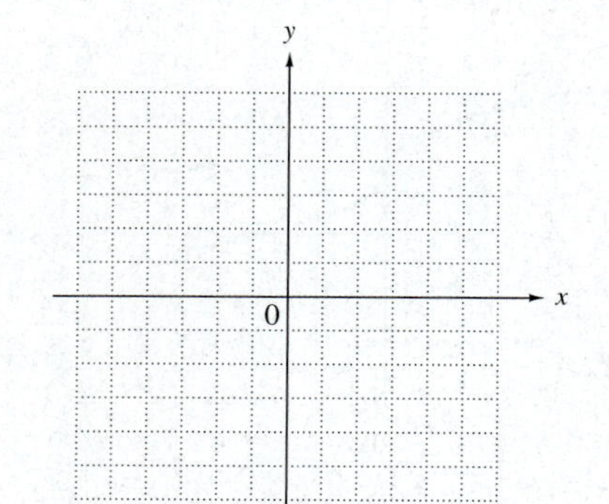

10.2 Exponential Functions

Objectives

1. Define exponential functions.
2. Graph exponential functions.
3. Solve exponential equations of the form $a^x = a^k$ for x.
4. Use exponential functions in applications involving growth or decay.

1 Define exponential functions. In Section 8.2 we showed how to evaluate 2^x for rational values of x. For example,

$$2^3 = 8, \qquad 2^{-1} = \frac{1}{2}, \qquad 2^{1/2} = \sqrt{2}, \qquad 2^{3/4} = \sqrt[4]{2^3} = \sqrt[4]{8}.$$

In more advanced courses it is shown that 2^x exists for all real number values of x, both rational and irrational. (Later in this chapter, we will see how to approximate the value of 2^x for irrational x.) The following definition of an exponential function assumes that a^x exists for all real numbers x.

Exponential Function

For $a > 0$, $a \neq 1$, and all real numbers x,

$$f(x) = a^x$$

defines an **exponential function.**

NOTE

The two restrictions on a in the definition of an exponential function are important. The restriction that a must be positive is necessary so that the function can be defined for all real numbers x. For example, letting a be negative ($a = -2$, for instance) and letting $x = \frac{1}{2}$ would give the expression $(-2)^{1/2}$, which is not real. The other restriction, $a \neq 1$, is necessary because 1 raised to any power is equal to 1, and the function would then be the linear function defined by $f(x) = 1$.

2 Graph exponential functions. We can graph an exponential function by finding several ordered pairs that belong to the function, plotting these points, and connecting them with a smooth curve.

CAUTION

Be sure to plot enough points to see how rapidly the graph rises.

Example 1 Graphing an Exponential Function with $a > 1$

Graph $f(x) = 2^x$.

Choose some values of x, and find the corresponding values of $f(x)$.

x	-3	-2	-1	0	1	2	3	4
$f(x) = 2^x$	$\frac{1}{8}$	$\frac{1}{4}$	$\frac{1}{2}$	1	2	4	8	16

Plotting these points and drawing a smooth curve through them gives the graph shown in Figure 6 on the next page. This graph is typical of the graphs of exponential functions of the form $F(x) = a^x$, where $a > 1$. The larger the value of a, the faster the graph rises. To see this, compare the graph of $F(x) = 5^x$ with the graph of $f(x) = 2^x$ in Figure 6.

Continued on Next Page

1 Graph.

(a) $f(x) = 10^x$

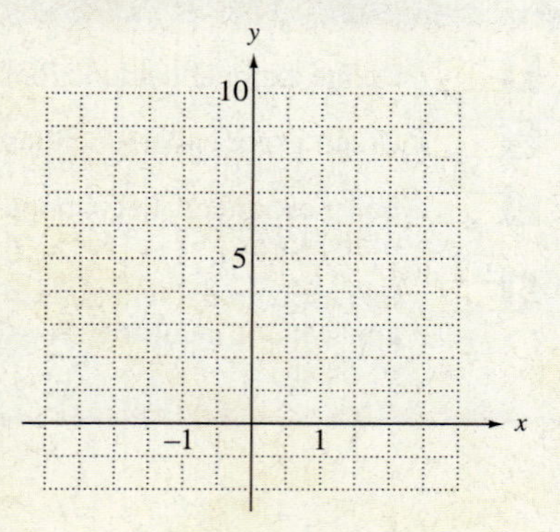

(b) $g(x) = \left(\frac{1}{4}\right)^x$

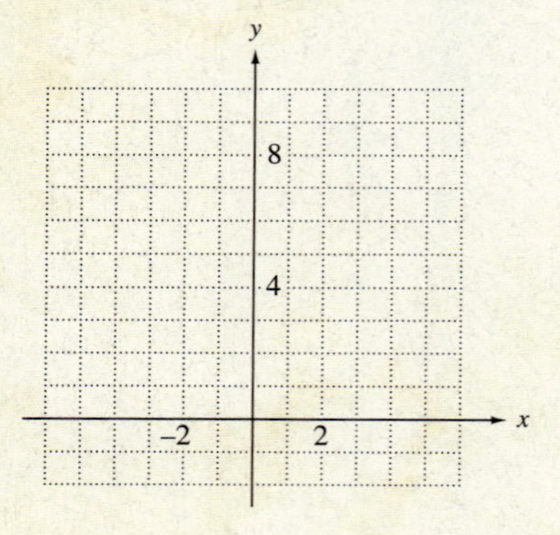

ANSWERS

1. (a)

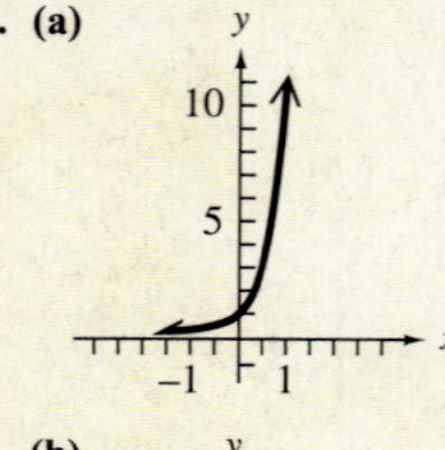

(b)

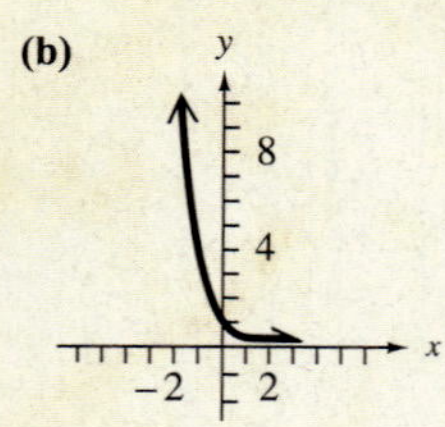

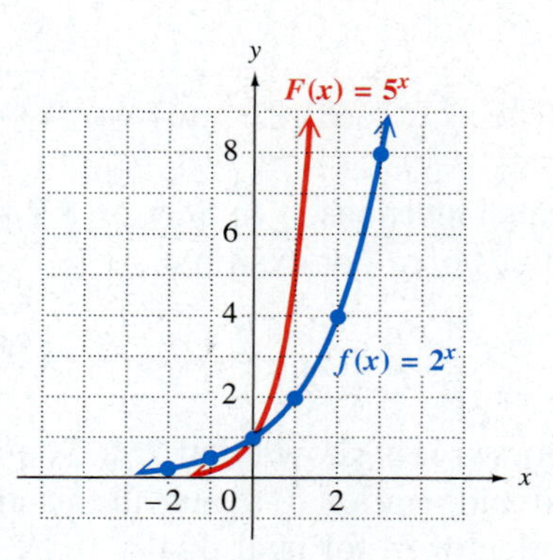

Figure 6

By the vertical line test, the graphs in Figure 6 represent functions. As these graphs suggest, the domain of an exponential function includes all real numbers. Because y is always positive, the range is $(0, \infty)$. Figure 6 also shows an important characteristic of exponential functions where $a > 1$: as x gets larger, y increases at a faster and faster rate.

Example 2 **Graphing an Exponential Function with $a < 1$**

Graph $g(x) = \left(\frac{1}{2}\right)^x$.

Again, find some points on the graph.

x	-3	-2	-1	0	1	2	3
$g(x) = (\frac{1}{2})^x$	8	4	2	1	$\frac{1}{2}$	$\frac{1}{4}$	$\frac{1}{8}$

The graph, shown in Figure 7, is very similar to that of $f(x) = 2^x$ (Figure 6) with the same domain and range, except that here as x gets larger, y *decreases*. This graph is typical of the graph of a function of the form $F(x) = a^x$, where $0 < a < 1$.

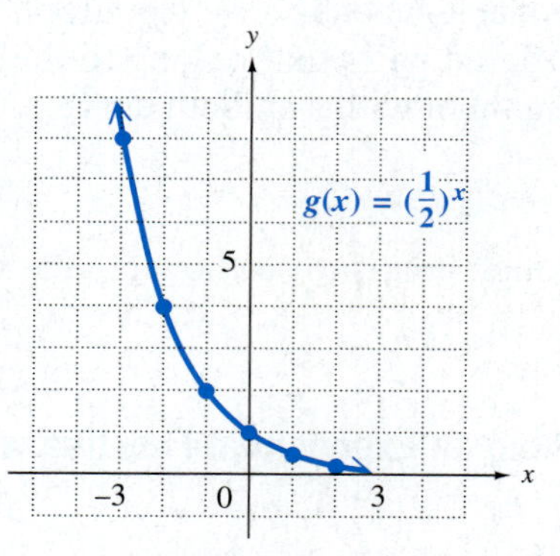

Figure 7

Work Problem 1 at the Side.

Based on Examples 1 and 2, we make the following generalizations about the graphs of exponential functions of the form $F(x) = a^x$.

Graph of $F(x) = a^x$

1. The graph will always contain the point (0, 1).
2. When $a > 1$, the graph will *rise* from left to right. When $0 < a < 1$, the graph will *fall* from left to right. In both cases, the graph goes from the second quadrant to the first.
3. The graph will approach the x-axis, but never touch it. (Recall from Chapter 7 that such a line is called an *asymptote*.)
4. The domain is $(-\infty, \infty)$, and the range is $(0, \infty)$.

Example 3 Graphing a More Complicated Exponential Function

Graph $f(x) = 3^{2x-4}$.

Find some ordered pairs.

$$\text{If } x = 0, \text{ then } y = 3^{2(0)-4} = 3^{-4} = \frac{1}{81}.$$

$$\text{If } x = 2, \text{ then } y = 3^{2(2)-4} = 3^0 = 1.$$

These ordered pairs, $(0, \frac{1}{81})$ and (2, 1), along with the other ordered pairs shown in the table, lead to the graph in Figure 8. The graph is similar to the graph of $f(x) = 3^x$ except that it is shifted to the right and rises more rapidly.

x	y
0	$\frac{1}{81}$
1	$\frac{1}{9}$
2	1
3	9

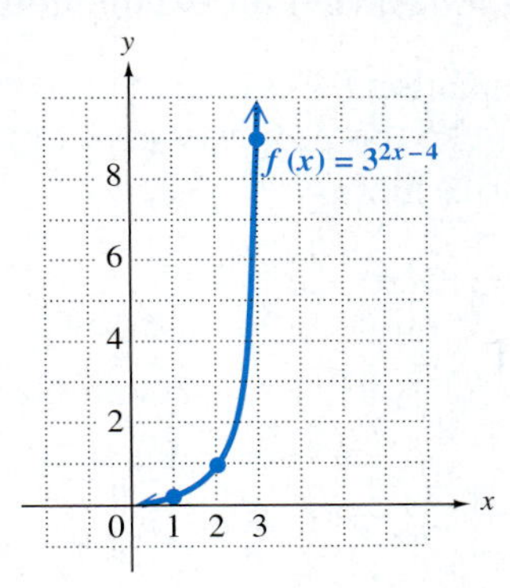

Figure 8

Work Problem 2 at the Side.

3 Solve exponential equations of the form $a^x = a^k$ for x. Until this chapter, we have solved only equations that had the variable as a base, like $x^2 = 8$; all exponents have been constants. An **exponential equation** is an equation that has a variable in an exponent, such as

$$9^x = 27.$$

By the horizontal line test, the exponential function defined by $F(x) = a^x$ is a one-to-one function, so we can use the following property to solve many exponential equations.

Property for Solving an Exponential Equation

For $a > 0$ and $a \neq 1$, if $a^x = a^y$ then $x = y$.

This property would not necessarily be true if $a = 1$.

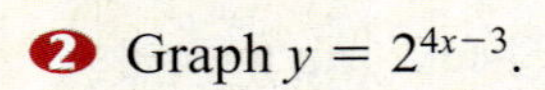

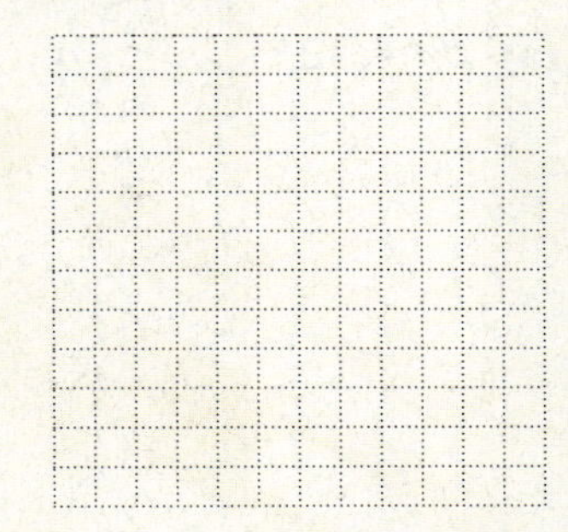

2.

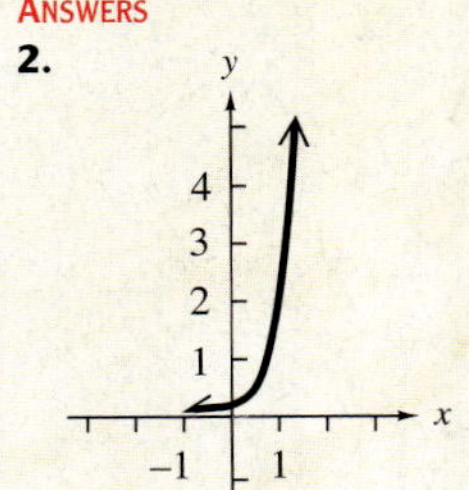

3 Solve. Check your answers.

(a) $25^x = 125$

(b) $4^x = 32$

(c) $81^p = 27$

To solve an exponential equation using this property, follow these steps.

Solving an Exponential Equation

Step 1 **Each side must have the same base.** If the two sides of the equation do not have the same base, express each as a power of the same base.

Step 2 **Simplify exponents.** If necessary, use the rules of exponents to simplify the exponents.

Step 3 **Set exponents equal.** Use the property given in this section to set the exponents equal.

Step 4 **Solve.** Solve the equation obtained in Step 3.

NOTE

These steps cannot be applied to an exponential equation like

$$3^x = 12$$

because Step 1 cannot easily be done. A method for solving such equations is given in Section 10.6.

Example 4 Solving an Exponential Equation

Solve the equation $9^x = 27$.

We can use the property given in the box if both sides are written with the same base. Since $9 = 3^2$ and $27 = 3^3$,

$$9^x = 27$$

$$(3^2)^x = 3^3 \quad \text{Write with the same base. (Step 1)}$$

$$3^{2x} = 3^3 \quad \text{Power rule for exponents (Step 2)}$$

$$2x = 3 \quad \text{If } a^x = a^y\text{, then } x = y. \text{ (Step 3)}$$

$$x = \frac{3}{2}. \quad \text{(Step 4)}$$

Check that the solution set is $\{\frac{3}{2}\}$ by substituting $\frac{3}{2}$ for x in the original equation.

Work Problem 3 at the Side.

Example 5 Solving Exponential Equations

Solve each equation.

(a) $4^{3x-1} = 16^{x+2}$

Since $4 = 2^2$ and $16 = 2^4$,

$$(2^2)^{3x-1} = (2^4)^{x+2} \quad \text{Write with the same base.}$$

$$2^{6x-2} = 2^{4x+8} \quad \text{Power rule for exponents}$$

$$6x - 2 = 4x + 8 \quad \text{Set exponents equal.}$$

$$2x = 10 \quad \text{Subtract } 4x\text{; add 2.}$$

$$x = 5. \quad \text{Divide by 2.}$$

Verify that the solution set is $\{5\}$.

Continued on Next Page

ANSWERS

3. (a) $\left\{\frac{3}{2}\right\}$ (b) $\left\{\frac{5}{2}\right\}$ (c) $\left\{\frac{3}{4}\right\}$

(b) $6^x = \frac{1}{216}$

$$6^x = \frac{1}{216}$$

$$6^x = \frac{1}{6^3} \qquad 216 = 6^3$$

$$6^x = 6^{-3} \qquad \text{Write with the same base; } \frac{1}{6^3} = 6^{-3}.$$

$$x = -3 \qquad \text{Set exponents equal.}$$

Verify that the solution set is $\{-3\}$.

(c) $\left(\frac{2}{3}\right)^x = \frac{9}{4}$

$$\left(\frac{2}{3}\right)^x = \left(\frac{4}{9}\right)^{-1} \qquad \frac{9}{4} = \left(\frac{4}{9}\right)^{-1}$$

$$\left(\frac{2}{3}\right)^x = \left[\left(\frac{2}{3}\right)^2\right]^{-1} \qquad \text{Write with the same base.}$$

$$\left(\frac{2}{3}\right)^x = \left(\frac{2}{3}\right)^{-2} \qquad \text{Power rule for exponents}$$

$$x = -2 \qquad \text{Set exponents equal.}$$

Check that the solution set is $\{-2\}$.

Work Problem 4 at the Side.

4 **Use exponential functions in applications involving growth or decay.**

Example 6 **Solving an Application Involving Exponential Growth**

One result of the rapidly increasing world population is an increase of carbon dioxide in the air, which scientists believe may be contributing to global warming. Both population and carbon dioxide in the air are increasing exponentially. This means that the growth rate is continually increasing. The graph in Figure 9 shows the concentration of carbon dioxide (in parts per million) in the air.

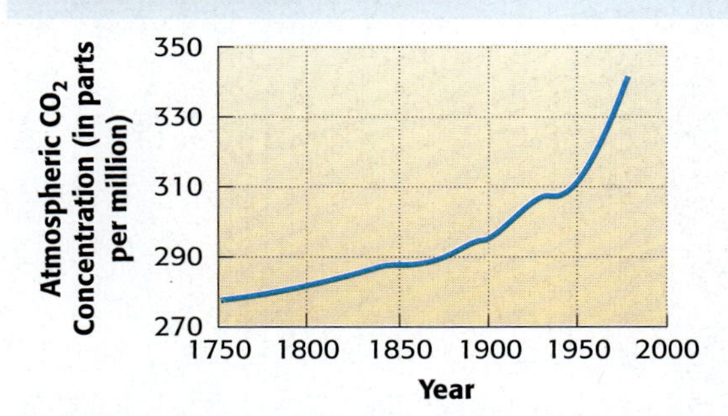

Source: Sacramento Bee, Monday, September 13, 1993.

Figure 9

Continued on Next Page

4 Solve each equation and check the solution.

(a) $25^{x-2} = 125^x$

(b) $4^x = \frac{1}{32}$

(c) $\left(\frac{3}{4}\right)^x = \frac{16}{9}$

ANSWERS

4. **(a)** $\{-4\}$ **(b)** $\left\{-\frac{5}{2}\right\}$ **(c)** $\{-2\}$

5 Solve each problem.

(a) Use the function in Example 6 to approximate the carbon dioxide concentration in 1925.

(b) Use the function in Example 7 to find the pressure at 8000 m.

The data are approximated by the function with

$$f(x) = 278(1.00084)^x,$$

where x is the number of years since 1750. Use this function and a calculator to approximate the concentration of carbon dioxide in parts per million for each year.

(a) 1900

Since x represents the number of years since 1750, in this case $x = 1900 - 1750 = 150$. Thus, evaluate $f(150)$.

$$f(150) = 278(1.00084)^{150} \quad \text{Let } x = 150.$$
$$\approx 315 \text{ parts per million} \quad \text{Use a calculator.}$$

(b) 1950

Use $x = 1950 - 1750 = 200$: $f(200) \approx 329$ parts per million.

Example 7 Applying an Exponential Decay Function

The atmospheric pressure (in millibars) at a given altitude x, in meters, can be approximated by the function defined by

$$f(x) = 1038(1.000134)^{-x},$$

for values of x between 0 and 10,000. Because the base is greater than 1 and the coefficient of x in the exponent is negative, the function values decrease as x increases. This means that as the altitude increases, the atmospheric pressure decreases. (*Source:* Miller, A. and J. Thompson, *Elements of Meteorology,* Fourth Edition, Charles E. Merrill Publishing Company, 1993.)

(a) According to this function, what is the pressure at ground level?

At ground level, $x = 0$, so

$$f(0) = 1038(1.000134)^{-0} = 1038(1) = 1038.$$

The pressure is 1038 millibars.

(b) What is the pressure at 5000 m?

Use a calculator to find $f(5000)$.

$$f(5000) = 1038(1.000134)^{-5000} \approx 531$$

The pressure is approximately 531 millibars.

Work Problem 5 at the Side.

ANSWERS

5. **(a)** 322 parts per million
(b) approximately 355 millibars

10.2 EXERCISES

FOR EXTRA HELP

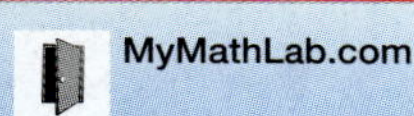

InterAct Math Tutorial Software

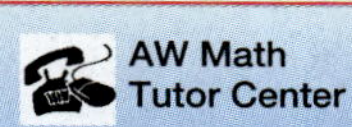

Digital Video Tutor CD 8 Videotape 17

Choose the correct response in Exercises 1–4.

1. Which point lies on the graph of $f(x) = 2^x$?

A. $(1, 0)$ **B.** $(2, 1)$

C. $(0, 1)$ **D.** $\left(\sqrt{2}, \frac{1}{2}\right)$

2. Which statement is true?

A. The y-intercept of the graph of $f(x) = 10^x$ is $(0, 10)$.

B. For any $a > 1$, the graph of $f(x) = a^x$ falls from left to right.

C. The point $(\frac{1}{2}, \sqrt{5})$ lies on the graph of $f(x) = 5^x$.

D. The graph of $y = 4^x$ rises at a faster rate than the graph of $y = 10^x$.

3. The asymptote of the graph of $F(x) = a^x$

A. is the x-axis. **B.** is the y-axis.

C. has equation $x = 1$. **D.** has equation $y = 1$.

4. Which equation is graphed here?

A. $y = 1000\left(\frac{1}{2}\right)^{.3x}$ **B.** $y = 1000\left(\frac{1}{2}\right)^{x}$

C. $y = 1000(2)^{.3x}$ **D.** $y = 1000^x$

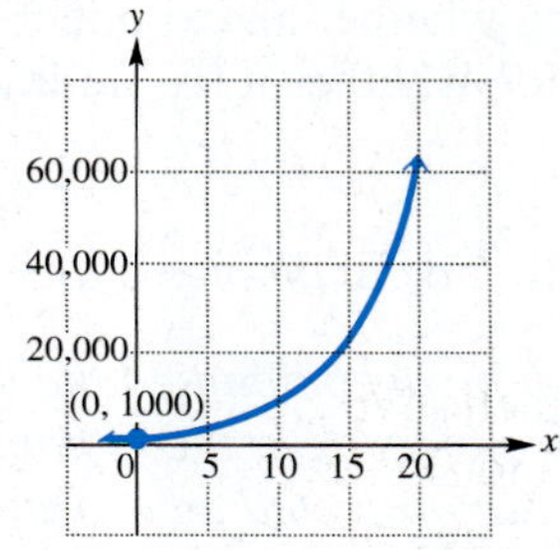

Graph each exponential function. See Examples 1–3.

5. $f(x) = 3^x$

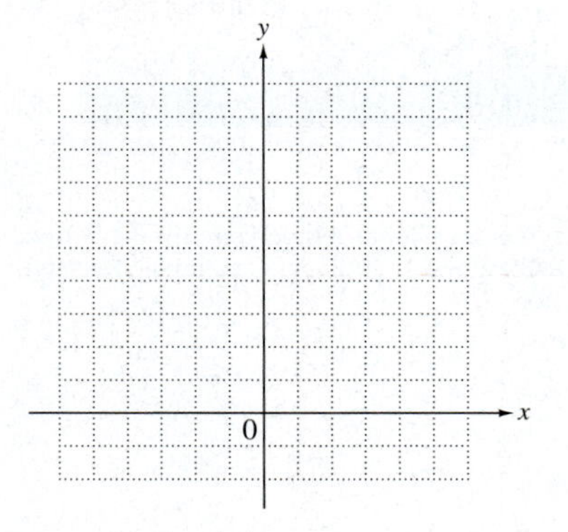

6. $f(x) = 5^x$

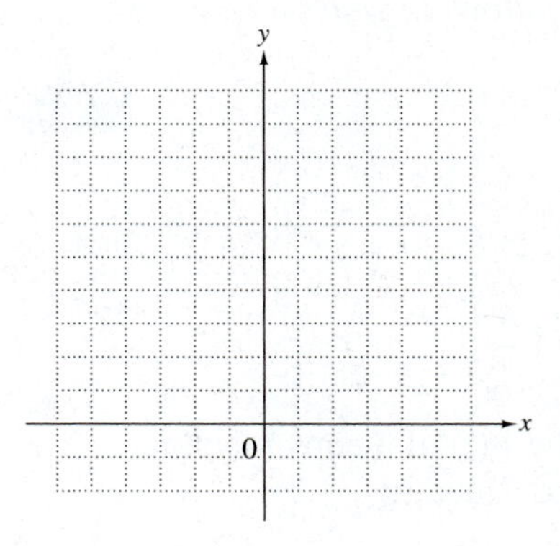

7. $g(x) = \left(\frac{1}{3}\right)^x$

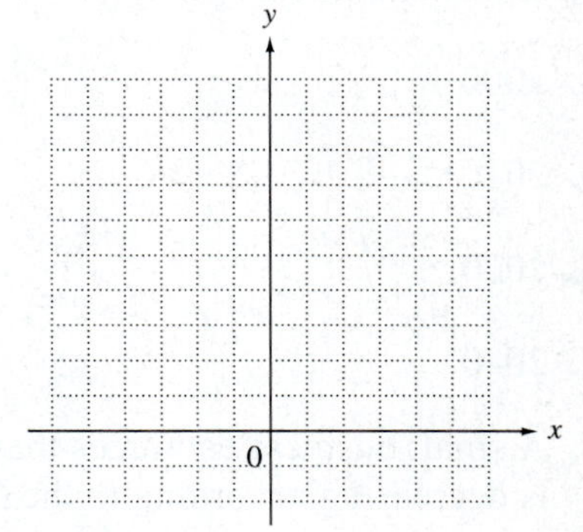

8. $g(x) = \left(\frac{1}{5}\right)^x$

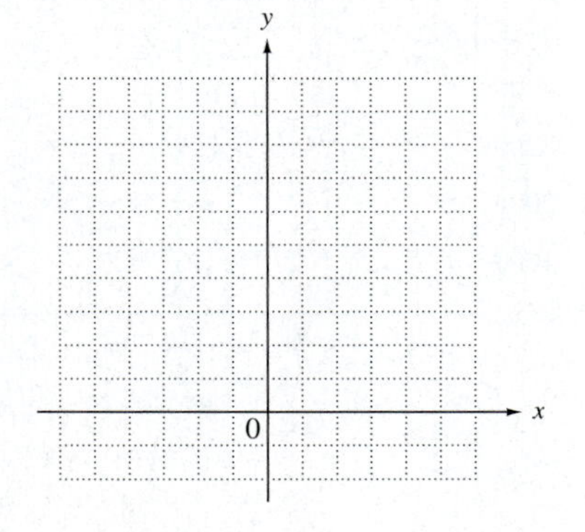

9. $y = 2^{2x-2}$

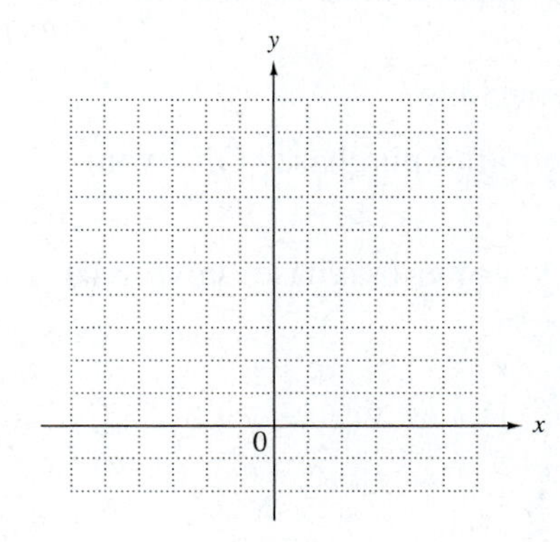

10. $y = 2^{2x+1}$

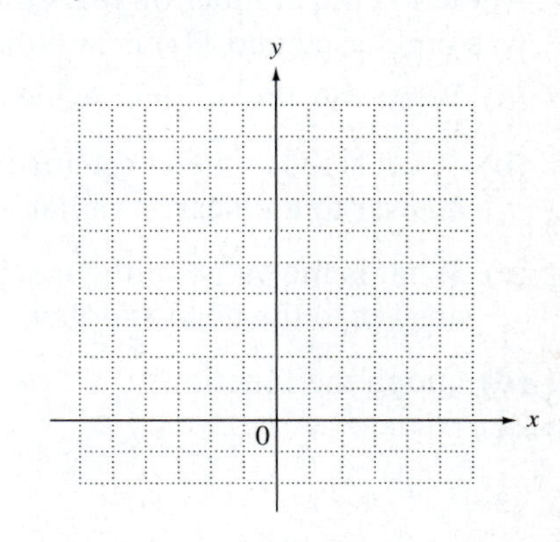

Solve each equation. See Examples 4 and 5.

11. $6^x = 36$

12. $8^x = 64$

13. $100^x = 1000$

14. $8^x = 4$

15. $16^{2x+1} = 64^{x+3}$

16. $9^{2x-8} = 27^{x-4}$

17. $5^x = \frac{1}{125}$

18. $3^x = \frac{1}{81}$

19. $5^x = .2$

20. $10^x = .1$

21. $\left(\frac{3}{2}\right)^x = \frac{8}{27}$

22. $\left(\frac{4}{3}\right)^x = \frac{27}{64}$

23. **(a)** For an exponential function defined by $f(x) = a^x$, if $a > 1$, the graph ________ (rises/falls) from left to right. If $0 < a < 1$, the graph ________ (rises/falls) from left to right.

(b) Based on your answers in part (a), make a conjecture (an educated guess) concerning whether an exponential function defined by $f(x) = a^x$ is one-to-one. Then decide whether it has an inverse based on the concepts of Section 10.1.

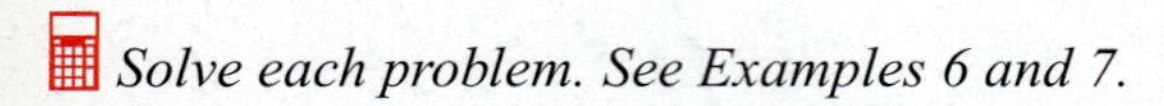
Solve each problem. See Examples 6 and 7.

The figure shown here accompanied the article "Is Our World Warming?" which appeared in the October 1990 issue of National Geographic. *It shows projected temperature increases using two graphs: one an exponential-type curve and the other linear. From the figure, approximate the increase* **(a)** *for the exponential curve, and* **(b)** *for the linear graph for each of the following years.*

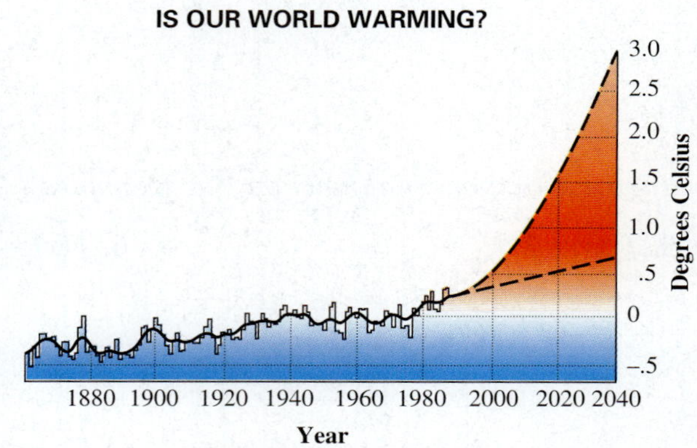

Graph, "Zero Equals Average Global Temperature for the Period 1950–1979." Dale D. Glasgow, © National Geographic Society. Reprinted by permission.

24. 2000

25. 2010

26. 2020

27. 2040

28. A small business estimates that the value $V(t)$ of a copy machine is decreasing according to the function defined by

$$V(t) = 5000(2)^{-.15t},$$

where t is the number of years that have elapsed since the machine was purchased and $V(t)$ is in dollars.

(a) What was the original value of the machine?

(b) What is the value of the machine 5 yr after purchase? Give your answer to the nearest dollar.

(c) What is the value of the machine 10 yr after purchase? Give your answer to the nearest dollar.

(d) Graph the function.

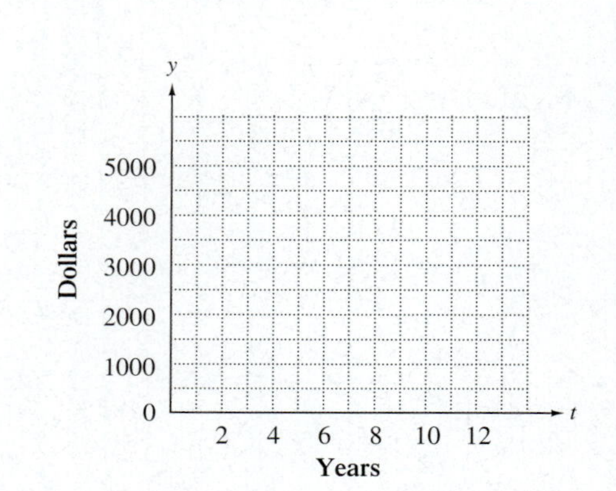

10.3 Logarithmic Functions

Objectives

1. Define a logarithm.
2. Convert between exponential and logarithmic forms.
3. Solve logarithmic equations of the form $\log_a b = k$ for a, b, or k.
4. Define and graph logarithmic functions.
5. Use logarithmic functions in applications of growth or decay.

The graph of $y = 2^x$ is the curve shown in blue in Figure 10. Because $y = 2^x$ defines a one-to-one function, it has an inverse. Interchanging x and y gives $x = 2^y$, the inverse of $y = 2^x$. As we saw in Section 10.1, the graph of the inverse is found by reflecting the graph of $y = 2^x$ about the line $y = x$. The graph of $x = 2^y$ is shown as a red curve in Figure 10.

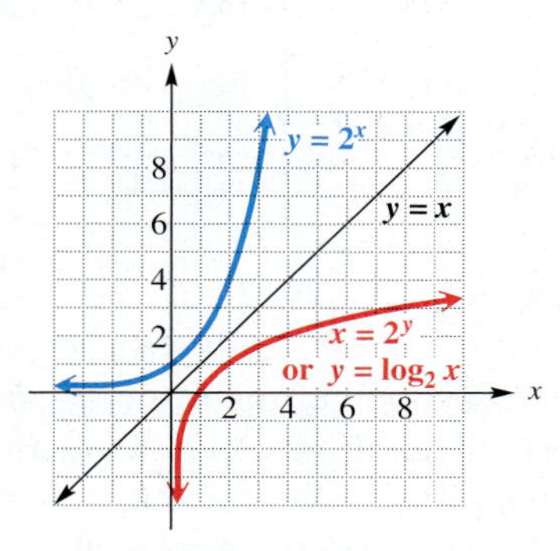

Figure 10

1 Define a logarithm. We cannot solve the equation $x = 2^y$ for the dependent variable y with the methods presented up to now. The following definition is used to solve $x = 2^y$ for y.

Logarithm

For all positive numbers a, $a \neq 1$, and all positive numbers x,

$$y = \log_a x \text{ means the same as } x = a^y.$$

This key statement should be memorized. The abbreviation **log** is used for **logarithm.** Read $\log_a x$ as "the logarithm of x to the base a." To remember the location of the base and the exponent in each form, refer to the following diagrams.

Logarithmic form: $y = \log_a x$ (Exponent: y; Base: a)

Exponential form: $x = a^y$ (Exponent: y; Base: a)

In working with logarithmic form and exponential form, remember the following.

Meaning of $\log_a x$

A logarithm is an exponent; $\log_a x$ is the exponent to which the base a must be raised to obtain x.

1 Complete the table.

Exponential Form	Logarithmic Form
$2^5 = 32$	______
$100^{1/2} = 10$	______
______	$\log_8 4 = \frac{2}{3}$
______	$\log_6 \frac{1}{1296} = -4$

2 **Convert between exponential and logarithmic forms.** We can use the definition of logarithm to write exponential statements in logarithmic form and logarithmic statements in exponential form. The following table shows several pairs of equivalent statements.

Exponential Form	Logarithmic Form
$3^2 = 9$	$\log_3 9 = 2$
$\left(\frac{1}{5}\right)^{-2} = 25$	$\log_{1/5} 25 = -2$
$10^5 = 100{,}000$	$\log_{10} 100{,}000 = 5$
$4^{-3} = \frac{1}{64}$	$\log_4 \frac{1}{64} = -3$

Work Problem 1 at the Side.

3 **Solve logarithmic equations of the form $\log_a b = k$ for a, b, or k.** A **logarithmic equation** is an equation with a logarithm in at least one term. We solve logarithmic equations of the form $\log_a b = k$ for any of the three variables by first writing the equation in exponential form.

Example 1 Solving Logarithmic Equations

Solve each equation.

(a) $\log_4 x = -2$

By the definition of logarithm, $\log_4 x = -2$ is equivalent to $x = 4^{-2}$. Solve this exponential equation.

$$x = 4^{-2} = \frac{1}{16}$$

The solution set is $\{\frac{1}{16}\}$.

(b) $\log_{1/2} (3x + 1) = 2$

$$3x + 1 = \left(\frac{1}{2}\right)^2 \quad \text{Write in exponential form.}$$

$$3x + 1 = \frac{1}{4}$$

$$12x + 4 = 1 \quad \text{Multiply by 4.}$$

$$12x = -3 \quad \text{Subtract 4.}$$

$$x = -\frac{1}{4} \quad \text{Divide by 12.}$$

The solution set is $\{-\frac{1}{4}\}$.

(c) $\log_x 3 = 2$

$$x^2 = 3 \quad \text{Write in exponential form.}$$

$$x = \pm\sqrt{3} \quad \text{Take square roots.}$$

Notice that only the principal square root satisfies the equation, since the base must be a positive number. The solution set is $\{\sqrt{3}\}$.

Continued on Next Page

ANSWERS

1. $\log_2 32 = 5$; $\log_{100} 10 = \frac{1}{2}$; $8^{2/3} = 4$; $6^{-4} = \frac{1}{1296}$

(d) $\log_{49} \sqrt[3]{7} = x$

$$49^x = \sqrt[3]{7} \quad \text{Write in exponential form.}$$
$$(7^2)^x = 7^{1/3}$$
$$7^{2x} = 7^{1/3} \quad \text{Write with the same base.}$$
$$2x = \frac{1}{3} \quad \text{Set exponents equal.}$$
$$x = \frac{1}{6} \quad \text{Divide by 2.}$$

The solution set is $\{\frac{1}{6}\}$.

Work Problem **2** at the Side.

For any real number b, we know that $b^1 = b$ and $b^0 = 1$. Writing these two statements in logarithmic form gives the following two properties of logarithms.

For any positive real number b, $b \neq 1$,

$$\log_b b = 1 \quad \text{and} \quad \log_b 1 = 0.$$

Example 2 **Using Properties of Logarithms**

Use the preceding two properties of logarithms to evaluate each logarithm.

(a) $\log_7 7 = 1$ **(b)** $\log_{\sqrt{2}} \sqrt{2} = 1$

(c) $\log_9 1 = 0$ **(d)** $\log_{.2} 1 = 0$

Work Problem **3** at the Side.

4 **Define and graph logarithmic functions.** Now we define the logarithmic function with base a.

Logarithmic Function

If a and x are positive numbers, with $a \neq 1$, then

$$G(x) = \log_a x$$

defines the **logarithmic function with base a.**

To graph a logarithmic function, it is helpful to write it in exponential form first. Then plot selected ordered pairs to determine the graph.

Example 3 **Graphing a Logarithmic Function**

Graph $y = \log_{1/2} x$.

By writing $y = \log_{1/2} x$ in exponential form as $x = (\frac{1}{2})^y$, we can identify ordered pairs that satisfy the equation. Here it is easier to choose values for y and find the corresponding values of x. See the table of ordered pairs.

Continued on Next Page

2 Solve each equation.

(a) $\log_3 27 = x$

(b) $\log_5 p = 2$

(c) $\log_m \frac{1}{16} = -4$

(d) $\log_x 12 = 3$

3 Evaluate each logarithm.

(a) $\log_{2/5} \frac{2}{5}$

(b) $\log_{.4} 1$

ANSWERS

2. **(a)** $\{3\}$ **(b)** $\{25\}$ **(c)** $\{2\}$ **(d)** $\sqrt[3]{12}$

3. **(a)** 1 **(b)** 0

4 Graph.

(a) $y = \log_3 x$

(b) $y = \log_{1/10} x$

x	y
$\frac{1}{4}$	2
$\frac{1}{2}$	1
1	0
2	-1
4	-2

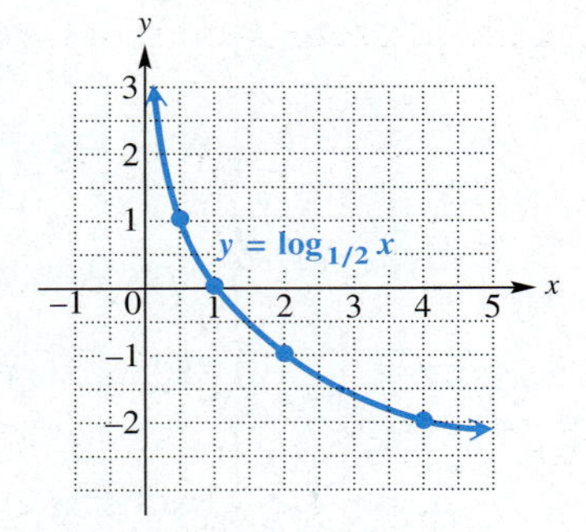

Figure 11

Plotting these points (be careful to get them in the right order) and connecting them with a smooth curve gives the graph in Figure 11. This graph is typical of logarithmic functions with $0 < a < 1$. The graph of $x = 2^y$ in Figure 10, which is equivalent to $y = \log_2 x$, is typical of graphs of logarithmic functions with base $a > 1$.

Work Problem 4 at the Side.

Based on the graphs of the functions defined by $y = \log_2 x$ in Figure 10 and $y = \log_{1/2} x$ in Figure 11, we make the following generalizations about the graphs of logarithmic functions of the form $G(x) = \log_a x$.

Graph of $G(x) = \log_a x$

1. The graph contains the point (1, 0).
2. When $a > 1$, the graph will *rise* from left to right, from the fourth quadrant to the first. When $0 < a < 1$, the graph will *fall* from left to right, from the first quadrant to the fourth.
3. The graph will approach the y-axis, but never touch it. (The y-axis is an asymptote.)
4. The domain is $(0, \infty)$, and the range is $(-\infty, \infty)$.

Compare these generalizations to the similar ones for exponential functions in Section 10.2.

5 Use logarithmic functions in applications of growth or decay. Logarithmic functions, like exponential functions, can be applied to growth or decay of real-world phenomena.

Example 4 Solving an Application of a Logarithmic Function

The function defined by

$$f(x) = 27 + 1.105 \log_{10}(x + 1)$$

approximates the barometric pressure in inches of mercury at a distance of x miles from the eye of a typical hurricane. (*Source:* Miller, A. and R. Anthes, *Meteorology,* Fifth Edition, Charles E. Merrill Publishing Company, 1985.)

Continued on Next Page

ANSWERS

4. (a)

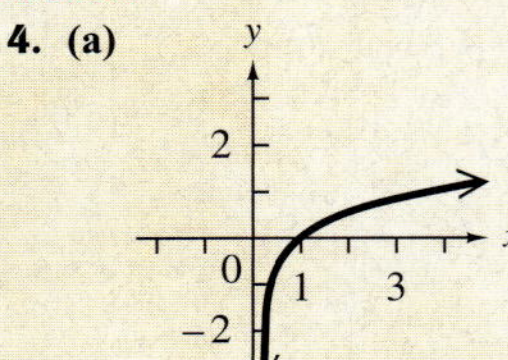

(b)

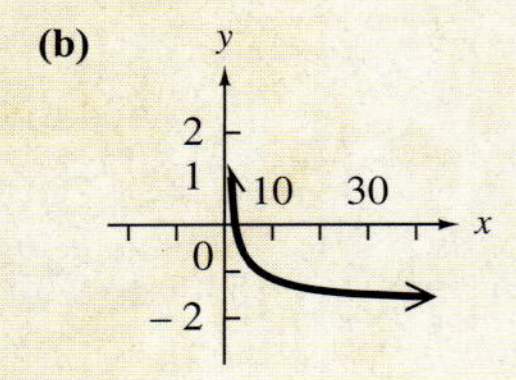

(a) Approximate the pressure 9 mi from the eye of the hurricane.
Let $x = 9$, and find $f(9)$.

$$
\begin{aligned}
f(9) &= 27 + 1.105 \log_{10}(9 + 1) && \text{Let } x = 9. \\
&= 27 + 1.105 \log_{10} 10 && \text{Add inside parentheses.} \\
&= 27 + 1.105(1) && \log_{10} 10 = 1 \\
&= 28.105 && \text{Add.}
\end{aligned}
$$

The pressure 9 mi from the eye of the hurricane is 28.105 in.

(b) Approximate the pressure 99 mi from the eye of the hurricane.

$$
\begin{aligned}
f(99) &= 27 + 1.105 \log_{10}(99 + 1) && \text{Let } x = 99. \\
&= 27 + 1.105 \log_{10} 100 && \text{Add inside parentheses.} \\
&= 27 + 1.105(2) && \log_{10} 100 = 2 \\
&= 29.21
\end{aligned}
$$

The pressure 99 mi from the eye of the hurricane is 29.21 in.

Work Problem 5 at the Side.

5 Solve the problem.

A population of mites in a laboratory is growing according to the function with

$$P(t) = 80 \log_{10}(t + 10),$$

where t is the number of days after a study is begun.

(a) Find the number of mites at the beginning of the study.

(b) Find the number present after 90 days.

(c) Find the number present after 990 days.

ANSWERS
5. (a) 80 **(b)** 160 **(c)** 240

Focus on **Real-Data Applications**

m&m's and Exponential Decay

Exponential functions are important for modeling decay patterns, including the life of a lightbulb and radioactive decaying elements, such as carbon-14. You can simulate an exponential decay problem with an m&m experiment.

Use a fun-size or small package of regular m&m's. Before you begin the simulation, check that each candy has the logo "m&m" stamped on one side—you may eat the candies with no logo. Place the m&m's in a cup, shake, and toss them onto a napkin. In the table, record the number of m&m's showing the logo. Discard (or eat) all the candies for which the m&m logo is not showing. Repeat until there are 1 or no candies left. The data from one such simulation using 64 m&m's are shown in the table.

EXPERIMENTAL DATA

Toss	Number of m&m Logos Showing	Your Results
0	64	
1	29	
2	17	
3	6	
4	4	
5	3	
6	2	
7	1	

For Group Discussion

1. In a perfect world, you might expect the number of candies left after each toss in our simulation using 64 m&m's to follow the pattern 64, 32, 16, 8, 4, 2, 1. An exponential model has an equation of the form $y = ab^x$. The constant a represents the initial quantity (value when $x = 0$), and the constant b represents the growth rate factor ($b > 1$) or decay rate factor ($0 < b < 1$). For our perfect-world data, what would be the values of a and b? What would be the model exponential equation?

2. On the grid, plot the points for the experimental data, your results, and the points for the theoretical (perfect-world) model. Use a different color to plot each set of data. How well do the graphs match?

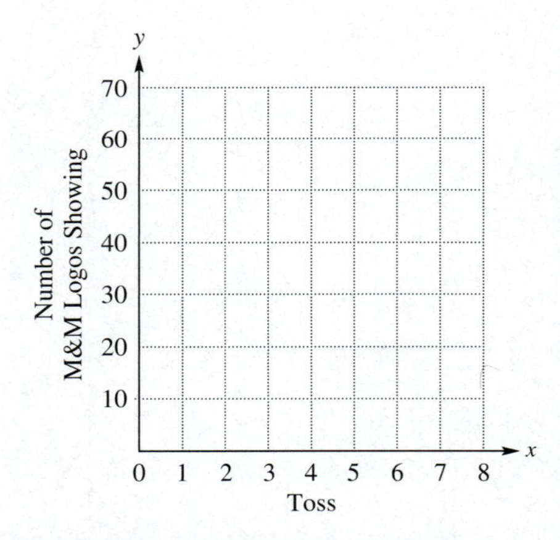

3. You can use a graphing calculator to find the statistical model for the sample experimental data, $y = 50.4(.56)^x$. Develop a table of values for $x = 0, 1, 2, \ldots, 7$ and superimpose the plot of the statistical model on your graph from Problem 2.

   (a) Does the theoretical or the statistical equation better model your experimental data?

   (b) Based on the statistical equation, give an estimate of the initial number of m&m's.

   (c) Based on the statistical equation, estimate the decay rate factor.

10.3 EXERCISES

FOR EXTRA HELP

Student's Solutions Manual

MyMathLab.com

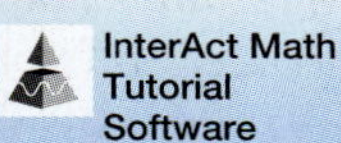

InterAct Math Tutorial Software

AW Math Tutor Center

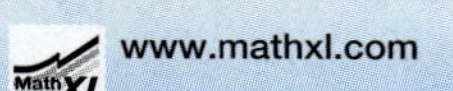

www.mathxl.com

Digital Video Tutor CD 8 Videotape 17

1. By definition, $\log_a x$ is the exponent to which the base a must be raised in order to obtain x. Use this definition to match the logarithm in Column I with its value in Column II. (*Example:* $\log_3 9$ is equal to 2 because 2 is the exponent to which 3 must be raised in order to obtain 9.)

I	II
(a) $\log_4 16$	**A.** -2
(b) $\log_3 81$	**B.** -1
(c) $\log_3\left(\frac{1}{3}\right)$	**C.** 2
(d) $\log_{10} .01$	**D.** 0
(e) $\log_5 \sqrt{5}$	**E.** $\frac{1}{2}$
(f) $\log_{13} 1$	**F.** 4

2. Match the logarithmic equation in Column I with the corresponding exponential equation from Column II.

I	II
(a) $\log_{1/3} 3 = -1$	**A.** $8^{1/3} = \sqrt[3]{8}$
(b) $\log_5 1 = 0$	**B.** $\left(\frac{1}{3}\right)^{-1} = 3$
(c) $\log_2 \sqrt{2} = \frac{1}{2}$	**C.** $4^1 = 4$
(d) $\log_{10} 1000 = 3$	**D.** $2^{1/2} = \sqrt{2}$
(e) $\log_8 \sqrt[3]{8} = \frac{1}{3}$	**E.** $5^0 = 1$
(f) $\log_4 4 = 1$	**F.** $10^3 = 1000$

Write in logarithmic form. See the table in Objective 2.

3. $4^5 = 1024$

4. $3^6 = 729$

5. $\left(\frac{1}{2}\right)^{-3} = 8$

6. $\left(\frac{1}{6}\right)^{-3} = 216$

7. $10^{-3} = .001$

8. $36^{1/2} = 6$

9. $\sqrt[4]{625} = 5$

10. $\sqrt[3]{343} = 7$

Write in exponential form. See the table in Objective 2.

11. $\log_4 64 = 3$

12. $\log_2 512 = 9$

13. $\log_{10} \frac{1}{10,000} = -4$

14. $\log_{100} 100 = 1$

15. $\log_6 1 = 0$

16. $\log_\pi 1 = 0$

17. $\log_9 3 = \frac{1}{2}$

18. $\log_{64} 2 = \frac{1}{6}$

19. When a student asked his teacher to explain to him how to evaluate $\log_9 3$ without showing any work, his teacher told him, "Think radically." Explain what the teacher meant by this hint.

20. A student told her teacher "I know that $\log_2 1$ is the exponent to which 2 must be raised in order to obtain 1, but I can't think of any such number." How would you explain to the student that the value of $\log_2 1$ is 0?

Solve each equation for x. See Examples 1 and 2.

21. $x = \log_{27} 3$

22. $x = \log_{125} 5$

23. $\log_x 9 = \frac{1}{2}$

24. $\log_x 5 = \frac{1}{2}$

25. $\log_x 125 = -3$

26. $\log_x 64 = -6$

27. $\log_{12} x = 0$

28. $\log_4 x = 0$

29. $\log_x x = 1$

30. $\log_x 1 = 0$

31. $\log_x \frac{1}{25} = -2$

32. $\log_x \frac{1}{10} = -1$

33. $\log_8 32 = x$

34. $\log_{81} 27 = x$

35. $\log_\pi \pi^4 = x$

36. $\log_{\sqrt{2}} \sqrt{2}^9 = x$

37. $\log_6 \sqrt{216} = x$

38. $\log_4 \sqrt{64} = x$

If the point (p, q) is on the graph of $f(x) = a^x$ (for $a > 0$ and $a \neq 1$), then the point (q, p) is on the graph of $f^{-1}(x) = \log_a x$. Use this fact and refer to the graphs required in Exercises 5–8 in Section 10.2 to graph each logarithmic function. See Example 3.

39. $y = \log_3 x$

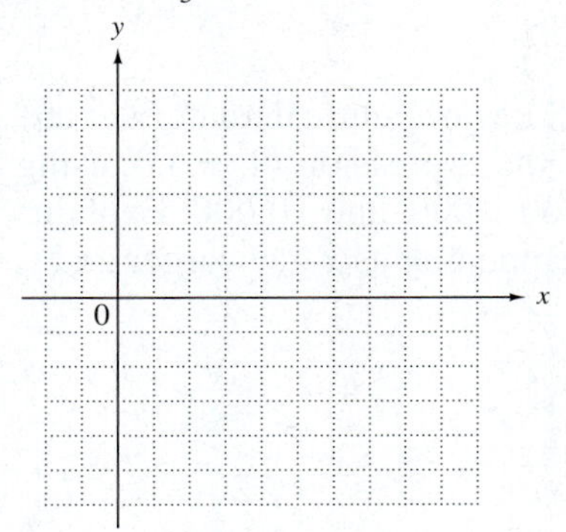

40. $y = \log_5 x$

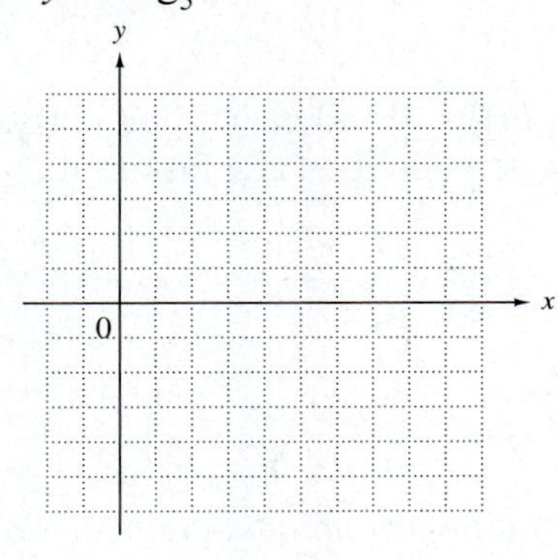

41. $y = \log_{1/3} x$

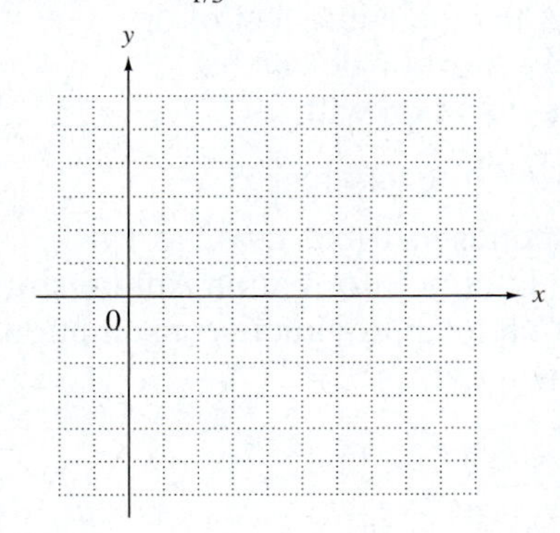

42. $y = \log_{1/5} x$

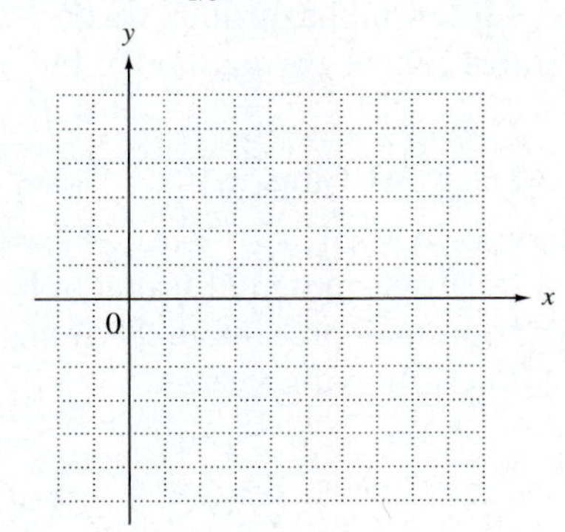

43. Compare the summary of facts about the graph of $F(x) = a^x$ in Section 10.2 with the similar summary of facts about the graph of $G(x) = \log_a x$ in this section. Make a list of the facts that reinforce the concept that F and G are inverse functions.

44. The domain of $F(x) = a^x$ is $(-\infty, \infty)$, while the range is $(0, \infty)$. Therefore, since $G(x) = \log_a x$ defines the inverse of F, the domain of G is ________, while the range of G is ________.

Use the graph to predict the value of $f(t)$ for each value of t.

45. $t = 0$

46. $t = 10$

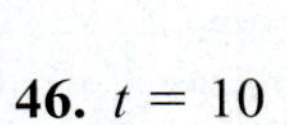

47. $t = 60$

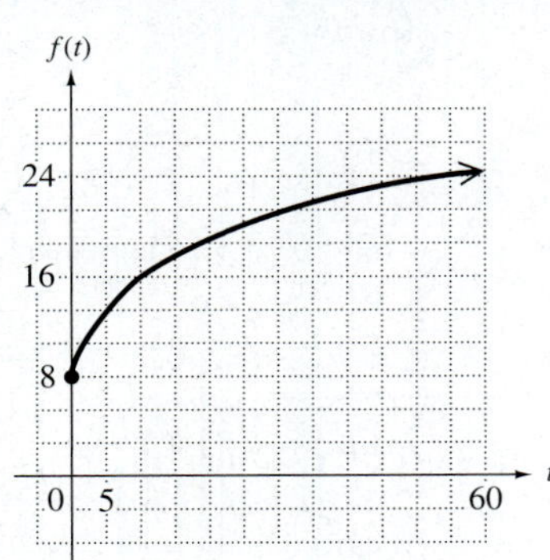

48. Show that the points determined in Exercises 45–47 lie on the graph of $f(t) = 8 \log_5(2t + 5)$.

49. Explain why 1 is not allowed as a base for a logarithmic function.

50. Explain why $\log_a 1$ is 0 for any value of a that is allowed as the base of a logarithm. Use a rule of exponents introduced earlier in your explanation.

51. The graphs of both $f(x) = 3^x$ and $g(x) = \log_3 x$ rise from left to right. Which one rises at a faster rate?

52. Use the exponential key of your calculator to find approximations for the expression $(1 + \frac{1}{x})^x$, using x values of 1, 10, 100, 1000, and 10,000. Explain what seems to be happening as x gets larger and larger.

Solve each application of a logarithmic function. See Example 4.

53. According to selected figures from 1981 through 1995, the number of Superfund hazardous waste sites in the United States can be approximated by the function with

$$f(x) = 11.34 + 317.01 \log_2 x,$$

where $x = 1$ corresponds to 1981, $x = 2$ to 1982, and so on. (*Source:* U.S. Environmental Protection Agency.) Use the function to approximate the number of sites in each of the following years.

(a) 1984

(b) 1988

(c) 1994

54. According to selected figures from 1980 through 1993, the number of trillion cubic feet of dry natural gas consumed worldwide can be approximated by the function with

$$f(x) = 51.47 + 6.044 \log_2 x,$$

where $x = 1$ corresponds to 1980, $x = 2$ to 1981, and so on. (*Source:* Energy Information Administration.) Use the function to approximate consumption in each of the following years.

(a) 1980

(b) 1987

(c) 1993

In the United States, the intensity of an earthquake is rated using the Richter scale. *The Richter scale rating of an earthquake of intensity x is given by*

$$R = \log_{10} \frac{x}{x_0},$$

where x_0 is the intensity of an earthquake of a certain (small) size. The figure shows Richter scale ratings for major southern California earthquakes since 1920. As the figure indicates, earthquakes "come in bunches" and the 1990s were an especially busy time.

55. The 1994 Northridge earthquake had a Richter scale rating of 6.7; the Landers earthquake had a rating of 7.3. How much more powerful was the Landers earthquake than the Northridge earthquake?

56. Compare the smallest rated earthquake in the figure (at 4.8) with the Landers quake. How much more powerful was the Landers quake?

MAJOR SOUTHERN CALIFORNIA EARTHQUAKES

Earthquakes with magnitudes greater than 4.8

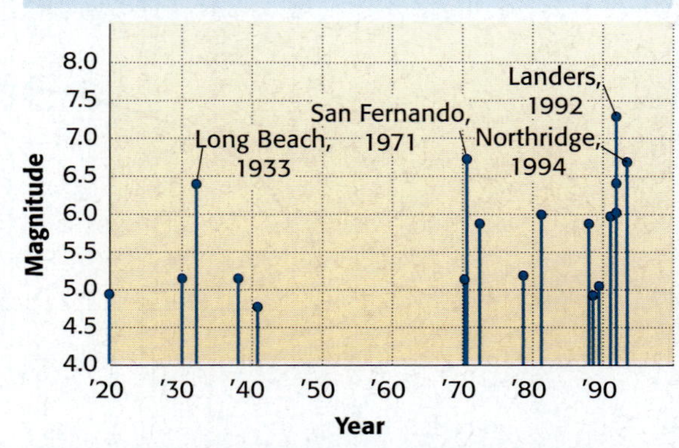

Source: Caltech; U.S. Geological Survey.

10.4 Properties of Logarithms

Logarithms have been used as an aid to numerical calculation for several hundred years. Today the widespread use of calculators has made the use of logarithms for calculation obsolete. However, logarithms are still very important in applications and in further work in mathematics.

Objectives

1. Use the product rule for logarithms.
2. Use the quotient rule for logarithms.
3. Use the power rule for logarithms.
4. Use properties to write alternative forms of logarithmic expressions.

1 Use the product rule for logarithms. One way in which logarithms simplify problems is by changing a problem of multiplication into one of addition. We know that $\log_2 4 = 2$, $\log_2 8 = 3$, and $\log_2 32 = 5$. Since $2 + 3 = 5$,

$$\log_2 32 = \log_2 4 + \log_2 8$$
$$\log_2(4 \cdot 8) = \log_2 4 + \log_2 8.$$

This is true in general.

Product Rule for Logarithms

If x, y, and b are positive real numbers, where $b \neq 1$, then

$$\log_b xy = \log_b x + \log_b y.$$

In words, the logarithm of a product is the sum of the logarithms of the factors.

NOTE

The word statement of the product rule can be restated by replacing "logarithm" with "exponent." The rule then becomes the familiar rule for multiplying exponential expressions: The *exponent* of a product is equal to the sum of the *exponents* of the factors.

To prove this rule, let $m = \log_b x$ and $n = \log_b y$, and recall that

$$\log_b x = m \quad \text{means} \quad b^m = x.$$
$$\log_b y = n \quad \text{means} \quad b^n = y.$$

Now consider the product xy.

$xy = b^m \cdot b^n$ — Substitution

$xy = b^{m+n}$ — Product rule for exponents

$\log_b xy = m + n$ — Convert to logarithmic form.

$\log_b xy = \log_b x + \log_b y$ — Substitution

The last statement is the result we wished to prove.

Example 1 Using the Product Rule

Use the product rule to rewrite each expression. Assume $x > 0$.

(a) $\log_5(6 \cdot 9)$
By the product rule,

$$\log_5(6 \cdot 9) = \log_5 6 + \log_5 9.$$

(b) $\log_7 8 + \log_7 12 = \log_7(8 \cdot 12) = \log_7 96$

Continued on Next Page

1 Use the product rule to rewrite each expression.

(a) $\log_6(5 \cdot 8)$

(b) $\log_4 3 + \log_4 7$

(c) $\log_8 8k, \quad k > 0$

(d) $\log_5 m^2, \quad m \neq 0$

(c) $\log_3(3x) = \log_3 3 + \log_3 x$

$= 1 + \log_3 x \qquad \log_3 3 = 1$

(d) $\log_4 x^3 = \log_4(x \cdot x \cdot x) \qquad x^3 = x \cdot x \cdot x$

$= \log_4 x + \log_4 x + \log_4 x \qquad$ Product rule

$= 3 \log_4 x$

Work Problem 1 at the Side.

2 **Use the quotient rule for logarithms.** The rule for division is similar to the rule for multiplication.

Quotient Rule for Logarithms

If x, y, and b are positive real numbers, where $b \neq 1$, then

$$\log_b \frac{x}{y} = \log_b x - \log_b y.$$

In words, the logarithm of a quotient is the difference between the logarithm of the numerator and the logarithm of the denominator.

The proof of this rule is very similar to the proof of the product rule.

Example 2 **Using the Quotient Rule**

Use the quotient rule to rewrite each logarithm.

(a) $\log_4 \frac{7}{9} = \log_4 7 - \log_4 9$

(b) $\log_5 6 - \log_5 x = \log_5 \frac{6}{x}$, for $x > 0$.

(c) $\log_3 \frac{27}{5} = \log_3 27 - \log_3 5$

$= 3 - \log_3 5 \qquad \log_3 27 = 3$

2 Use the quotient rule to rewrite each expression.

(a) $\log_7 \frac{9}{4}$

(b) $\log_3 p - \log_3 q$, $p > 0, \quad q > 0$

(c) $\log_4 \frac{3}{16}$

CAUTION

Remember that there is no property of logarithms to rewrite the logarithm of a *sum* or *difference*. For example, we *cannot* write $\log_b(x + y)$ in terms of $\log_b x$ and $\log_b y$. Also, $\log_b \frac{x}{y} \neq \frac{\log_b x}{\log_b y}$.

Work Problem 2 at the Side.

3 **Use the power rule for logarithms.** The next rule gives a method for evaluating powers and roots such as

$$2^{\sqrt{2}}, \quad (\sqrt{2})^{3/4}, \quad (.032)^{5/8}, \quad \text{and} \quad \sqrt[5]{12}.$$

ANSWERS

1. **(a)** $\log_6 5 + \log_6 8$ **(b)** $\log_4 21$ **(c)** $1 + \log_8 k$ **(d)** $2 \log_5 m, \quad m > 0$

2. **(a)** $\log_7 9 - \log_7 4$ **(b)** $\log_3 \frac{p}{q}$ **(c)** $\log_4 3 - 2$

This rule makes it possible to find approximations for numbers that could not be evaluated before. By the product rule for logarithms,

$$\begin{aligned}\log_5 2^3 &= \log_5(2 \cdot 2 \cdot 2)\\ &= \log_5 2 + \log_5 2 + \log_5 2\\ &= 3 \log_5 2.\end{aligned}$$

Also,

$$\begin{aligned}\log_2 7^4 &= \log_2(7 \cdot 7 \cdot 7 \cdot 7)\\ &= \log_2 7 + \log_2 7 + \log_2 7 + \log_2 7\\ &= 4 \log_2 7.\end{aligned}$$

Furthermore, we saw in Example 1(d) that $\log_4 x^3 = 3 \log_4 x$. These examples suggest the following rule.

Power Rule for Logarithms

If x and b are positive real numbers, where $b \neq 1$, and if r is any real number, then

$$\log_b x^r = r \log_b x.$$

In words, the logarithm of a number to a power equals the exponent times the logarithm of the number.

As examples of this result,

$$\log_b m^5 = 5 \log_b m \quad \text{and} \quad \log_3 5^4 = 4 \log_3 5.$$

To prove the power rule, let

$\log_b x = m.$	
$b^m = x$	Convert to exponential form.
$(b^m)^r = x^r$	Raise to the power r.
$b^{mr} = x^r$	Power rule for exponents
$\log_b x^r = mr$	Convert to logarithmic form.
$\log_b x^r = rm$	
$\log_b x^r = r \log_b x$	$m = \log_b x$

This is the statement to be proved.

As a special case of the power rule, let $r = \frac{1}{p}$, so

$$\log_b \sqrt[p]{x} = \log_b x^{1/p} = \frac{1}{p} \log_b x.$$

For example, using this result, with $x > 0$,

$$\log_b \sqrt[5]{x} = \log_b x^{1/5} = \frac{1}{5} \log_b x \quad \text{and} \quad \log_b \sqrt[3]{x^4} = \log_b x^{4/3} = \frac{4}{3} \log_b x.$$

Another special case is

$$\log_b \frac{1}{x} = \log_b x^{\ 1} = -\log_b x.$$

NOTE

For a review of rational exponents, refer to Section 8.2.

3 Use the power rule to rewrite each logarithm. Assume $a > 0, b > 0, x > 0, a \neq 1$, and $b \neq 1$.

(a) $\log_3 5^2$

(b) $\log_a x^4$

(c) $\log_b \sqrt{8}$

(d) $\log_2 \sqrt[3]{2}$

4 Find the value of each logarithmic expression.

(a) $\log_{10} 10^3$

(b) $\log_2 8$

(c) $5^{\log_5 3}$

ANSWERS

3. (a) $2 \log_3 5$ **(b)** $4 \log_a x$ **(c)** $\frac{1}{2}\log_b 8$ **(d)** $\frac{1}{3}$

4. (a) 3 **(b)** 3 **(c)** 3

Example 3 Using the Power Rule

Use the power rule to rewrite each logarithm. Assume $b > 0, x > 0$, and $b \neq 1$.

(a) $\log_5 4^2 = 2 \log_5 4$

(b) $\log_b x^5 = 5 \log_b x$

(c) $\log_b \sqrt{7}$

When using the power rule with logarithms of expressions involving radicals, begin by rewriting the radical expression with a rational exponent.

$$\log_b \sqrt{7} = \log_b 7^{1/2} \qquad \sqrt{x} = x^{1/2}$$

$$= \frac{1}{2} \log_b 7 \qquad \text{Power rule}$$

(d) $\log_2 \sqrt[5]{x^2} = \log_2 x^{2/5}$ $\quad \sqrt[5]{x^2} = x^{2/5}$

$$= \frac{2}{5} \log_2 x$$

Work Problem 3 at the Side.

Two special properties involving both exponential and logarithmic expressions come directly from the fact that logarithmic and exponential functions are inverses of each other.

Special Properties

If $b > 0$ and $b \neq 1$, then

$$b^{\log_b x} = x, \ x > 0 \quad \text{and} \quad \log_b b^x = x.$$

To prove the first statement, let

$$y = \log_b x.$$

$$b^y = x \qquad \text{Convert to exponential form.}$$

$$b^{\log_b x} = x \qquad \text{Replace } y \text{ with } \log_b x.$$

The proof of the second statement is similar.

Example 4 Using the Special Properties

Find the value of each logarithmic expression.

(a) $\log_5 5^4$

Since $\log_b b^x = x$,

$$\log_5 5^4 = 4.$$

(b) $\log_3 9 = \log_3 3^2 = 2$

(c) $4^{\log_4 10} = 10$

Work Problem 4 at the Side.

Here is a summary of the properties of logarithms.

Properties of Logarithms

If x, y, and b are positive real numbers, where $b \neq 1$, and r is any real number, then

Product Rule $\log_b xy = \log_b x + \log_b y$

Quotient Rule $\log_b \frac{x}{y} = \log_b x - \log_b y$

Power Rule $\log_b x^r = r \log_b x$

Special Properties $b^{\log_b x} = x$ and $\log_b b^x = x$.

4 **Use properties to write alternative forms of logarithmic expressions.** Applying the properties of logarithms is important for solving equations with logarithms and in calculus.

Example 5 Writing Logarithms in Alternative Forms

Use the properties of logarithms to rewrite each expression. Assume all variables represent positive real numbers.

(a) $\log_4 4x^3 = \log_4 4 + \log_4 x^3$ Product rule

$= 1 + 3 \log_4 x$ $\log_4 4 = 1$; Power rule

(b) $\log_7 \sqrt{\frac{m}{n}} = \log_7 \left(\frac{m}{n}\right)^{1/2}$

$= \frac{1}{2} \log_7 \frac{m}{n}$ Power rule

$= \frac{1}{2} (\log_7 m - \log_7 n)$ Quotient rule

(c) $\log_5 \frac{a^2}{bc} = \log_5 a^2 - \log_5 bc$ Quotient rule

$= 2 \log_5 a - \log_5 bc$ Power rule

$= 2 \log_5 a - (\log_5 b + \log_5 c)$ Product rule

$= 2 \log_5 a - \log_5 b - \log_5 c$

Notice the careful use of parentheses in the third step. Since we are subtracting the logarithm of a product and rewriting it as a sum of two terms, we must place parentheses around the sum.

(d) $4 \log_b m - \log_b n = \log_b m^4 - \log_b n$ Power rule

$= \log_b \frac{m^4}{n}$ Quotient rule

Continued on Next Page

(e) $\log_b(x + 1) + \log_b(2x - 1) - \frac{2}{3}\log_b x$

$$= \log_b(x + 1) + \log_b(2x - 1) - \log_b x^{2/3} \quad \text{Power rule}$$

$$= \log_b \frac{(x + 1)(2x - 1)}{x^{2/3}} \quad \text{Product and quotient rules}$$

$$= \log_b \frac{2x^2 + x - 1}{x^{2/3}}$$

(f) $\log_8(2p + 3r)$ cannot be rewritten by the properties of logarithms.

Work Problem 5 at the Side.

5 Use the properties of logarithms to rewrite each expression. Assume all variables represent positive real numbers.

(a) $\log_6 36m^5$

(b) $\log_2 \sqrt{9z}$

(c) $\log_q \frac{8r^2}{m - 1}, m \neq 1, q \neq 1$

(d) $2 \log_a x + 3 \log_a y, a \neq 1$

(e) $\log_4(3x + y)$

ANSWERS

5. **(a)** $2 + 5 \log_6 m$ **(b)** $\log_2 3 + \frac{1}{2}\log_2 z$
(c) $\log_q 8 + 2 \log_q r - \log_q(m - 1)$
(d) $\log_a x^2y^3$ **(e)** cannot be rewritten

10.4 EXERCISES

FOR EXTRA HELP: Student's Solutions Manual; MyMathLab.com; InterAct Math Tutorial Software; AW Math Tutor Center; www.mathxl.com; Digital Video Tutor CD 8 Videotape 18

Decide whether each statement of a logarithmic property is true *or* false. *If it is* false, *correct it by changing the right side of the equation.*

1. $\log_b x + \log_b y = \log_b(x + y)$

2. $\log_b \frac{x}{y} = \log_b x - \log_b y$

3. $\log_b b^x = x$

4. $\log_b x^r = \log_b rx$

Use the properties of logarithms introduced in this section to express each logarithm as a sum or difference of logarithms, or as a single number if possible. Assume that all variables represent positive real numbers. See Examples 1–5.

5. $\log_7 \frac{4}{5}$

6. $\log_8 \frac{9}{11}$

7. $\log_2 8^{1/4}$

8. $\log_3 9^{3/4}$

9. $\log_4 \frac{3\sqrt{x}}{y}$

10. $\log_5 \frac{6\sqrt{z}}{w}$

11. $\log_3 \frac{\sqrt[3]{4}}{x^2 y}$

12. $\log_7 \frac{\sqrt[3]{13}}{pq^2}$

13. $\log_3 \sqrt{\frac{xy}{5}}$

14. $\log_6 \sqrt{\frac{pq}{7}}$

15. $\log_2 \frac{\sqrt[3]{x} \cdot \sqrt[5]{y}}{r^2}$

16. $\log_4 \frac{\sqrt[4]{z} \cdot \sqrt[5]{w}}{s^2}$

17. A student erroneously wrote

$$\log_a(x + y) = \log_a x + \log_a y.$$

When his teacher explained that this was wrong, the student claimed he had used the distributive property. Write a few sentences explaining why the distributive property does not apply in this case.

18. Write a few sentences explaining how the rules for multiplying and dividing powers of the same base are similar to the rules for finding logarithms of products and quotients.

Use the properties of logarithms introduced in this section to rewrite each expression as a single logarithm. Assume all variables are defined in such a way that the variable expressions are positive, and bases are positive numbers not equal to 1. See Examples 1–5.

19. $\log_b x + \log_b y$

20. $\log_b 2 + \log_b z$

21. $3 \log_a m - \log_a n$

22. $5 \log_b x - \log_b y$

23. $(\log_a r - \log_a s) + 3 \log_a t$

24. $(\log_a p - \log_a q) + 2 \log_a r$

25. $3 \log_a 5 - 4 \log_a 3$

26. $3 \log_a 5 + \frac{1}{2} \log_a 9$

27. $\log_{10}(x + 3) + \log_{10}(x - 3)$

28. $\log_{10}(y + 4) + \log_{10}(y - 4)$

29. $3 \log_p x + \frac{1}{2} \log_p y - \frac{3}{2} \log_p z - 3 \log_p a$

30. $\frac{1}{3} \log_b x + \frac{2}{3} \log_b y - \frac{3}{4} \log_b s - \frac{2}{3} \log_b t$

31. Explain why the statement for the power rule for logarithms requires that x be a positive real number.

32. What is wrong with the following "proof" that $\log_2 16$ does not exist?

$$\begin{aligned} \log_2 16 &= \log_2(-4)(-4) \\ &= \log_2(-4) + \log_2(-4) \end{aligned}$$

Since the logarithm of a negative number is not defined, the final step cannot be evaluated, and so $\log_2 16$ does not exist.

RELATING CONCEPTS (Exercises 33–38) FOR INDIVIDUAL OR GROUP WORK

Work Exercises 33–38 in order.

33. Evaluate $\log_3 81$.

34. Write the *meaning* of the expression $\log_3 81$.

35. Evaluate $3^{\log_3 81}$.

36. Write the *meaning* of the expression $\log_2 19$.

37. Evaluate $2^{\log_2 19}$.

38. Keeping in mind that a logarithm is an exponent, and using the results from Exercises 33–37, what is the simplest form of the expression $k^{\log_k m}$?

10.5 Common and Natural Logarithms

As mentioned earlier, logarithms are important in many applications of mathematics to everyday problems, particularly in biology, engineering, economics, and social science. In this section we find numerical approximations for logarithms. Traditionally, base 10 logarithms were used most often because our number system is base 10. Logarithms to base 10 are called **common logarithms,** and $\log_{10} x$ is abbreviated as simply $\log x$, where the base is understood to be 10.

Objectives

1. Evaluate common logarithms using a calculator.
2. Use common logarithms in applications.
3. Evaluate natural logarithms using a calculator.
4. Use natural logarithms in applications.

1 Evaluate common logarithms using a calculator. We use calculators to evaluate common logarithms. In the next example we give the results of evaluating some common logarithms using a calculator with a LOG key. (This may be a second function key on some calculators.) For simple scientific calculators, just enter the number, then press the LOG key. For graphing calculators, these steps are reversed. We will give all logarithms to four decimal places.

Example 1 Evaluating Common Logarithms

Evaluate each logarithm using a calculator.

(a) $\log 327.1 \approx 2.5147$

(b) $\log 437{,}000 \approx 5.6405$

(c) $\log .0615 \approx -1.2111$

Notice that $\log .0615 \approx -1.2111$, a negative result. The common logarithm of a number between 0 and 1 is always negative because the logarithm is the exponent on 10 that produces the number. For example,

$$10^{-1.2111} \approx .0615.$$

If the exponent (the logarithm) were positive, the result would be greater than 1 because $10^0 = 1$. See Figure 12.

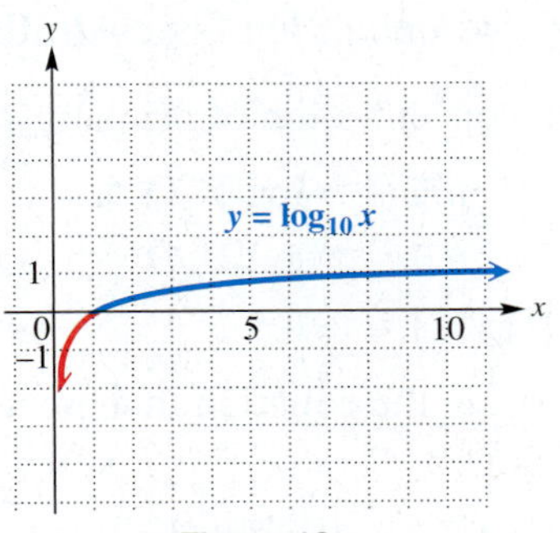

Figure 12

Work Problem 1 at the Side.

1 Evaluate each logarithm to four decimal places using a calculator.

(a) $\log 41{,}600$

(b) $\log 43.5$

(c) $\log .442$

2 Use common logarithms in applications. In chemistry, pH is a measure of the acidity or alkalinity of a solution; water, for example, has pH 7. In general, acids have pH numbers less than 7, and alkaline solutions have pH values greater than 7. The **pH** of a solution is defined as

$$\text{pH} = -\log[H_3O^+],$$

where $[H_3O^+]$ is the hydronium ion concentration in moles per liter. It is customary to round pH values to the nearest tenth.

Answers

1. (a) 4.6191 (b) 1.6385 (c) −.3546

2 Solve the problem.

Find the pH of water with a hydronium ion concentration of 1.2×10^{-3}. If this water had been taken from a wetland, is the wetland a rich fen, a poor fen, or a bog?

3 Find the hydronium ion concentrations of solutions with the following pH values.

(a) 4.6

(b) 7.5

Example 2 Using pH in an Application

Wetlands are classified as *bogs, fens, marshes,* and *swamps.* These classifications are based on pH values. A pH value between 6.0 and 7.5, such as that of Summerby Swamp in Michigan's Hiawatha National Forest, indicates that the wetland is a "rich fen." When the pH is between 4.0 and 6.0, the wetland is a "poor fen," and if the pH falls to 3.0 or less, it is a "bog." (*Source:* Mohlenbrock, R., "Summerby Swamp, Michigan," *Natural History,* March 1994.)

Suppose that the hydronium ion concentration of a sample of water from a wetland is 6.3×10^{-3}. How would this wetland be classified?

Use the definition of pH.

$$\begin{aligned} \text{pH} &= -\log(6.3 \times 10^{-3}) \\ &= -(\log 6.3 + \log 10^{-3}) && \text{Product rule} \\ &= -[.7993 - 3(1)] \\ &= -.7993 + 3 \\ &\approx 2.2 \end{aligned}$$

Since the pH is less than 3.0, the wetland is a bog.

Work Problem 2 at the Side.

Example 3 Finding Hydronium Ion Concentration

Find the hydronium ion concentration of drinking water with pH 6.5.

$$\begin{aligned} \text{pH} &= -\log[H_3O^+] \\ 6.5 &= -\log[H_3O^+] && \text{Let pH} = 6.5. \\ \log[H_3O^+] &= -6.5 && \text{Multiply by } -1. \end{aligned}$$

Solve for $[H_3O^+]$ by writing the equation in exponential form, remembering that the base is 10.

$$\begin{aligned} [H_3O^+] &= 10^{-6.5} \\ [H_3O^+] &\approx 3.2 \times 10^{-7} && \text{Use a calculator.} \end{aligned}$$

Work Problem 3 at the Side.

3 Evaluate natural logarithms using a calculator. The most important logarithms used in applications are **natural logarithms,** which have as base the number e. The number e is a fundamental number in our universe. For this reason e, like π, is called a *universal constant.* The letter e is used to honor Leonhard Euler, who published extensive results on the number in 1748. Since it is an irrational number, its decimal expansion never

ANSWERS

2. (a) 2.9; bog

3. (a) 2.5×10^{-5} **(b)** 3.2×10^{-8}

terminates and never repeats. The first few digits of the decimal value of e are 2.7182818285. A calculator key e^x or the two keys INV and ln x are used to approximate powers of e. For example, a calculator gives

$$e^2 \approx 7.389056099,$$
$$e^3 \approx 20.08553692,$$
and
$$e^{.6} \approx 1.8221188.$$

Logarithms to base e are called natural logarithms because they occur in biology and the social sciences in natural situations that involve growth or decay. The base e logarithm of x is written $\ln x$ (read "el en x"). A graph of $y = \ln x$, the equation that defines the natural logarithmic function, is given in Figure 13.

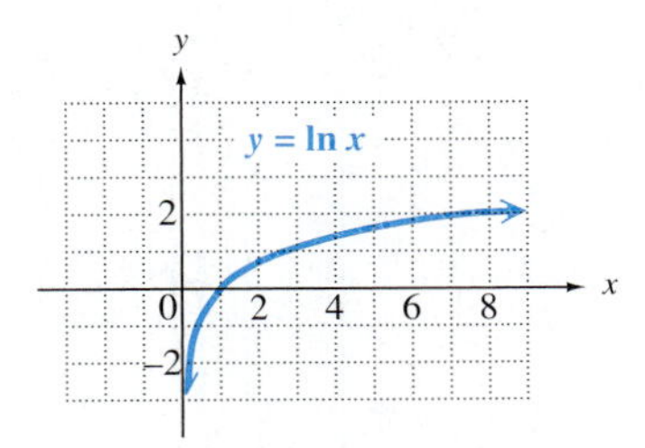

Figure 13

A calculator key labeled ln x is used to evaluate natural logarithms. If your calculator has an e^x key, but not a key labeled ln x, find natural logarithms by entering the number, pressing the INV key, and then pressing the e^x key. This works because $y = e^x$ defines the inverse function of $y = \ln x$ (or $y = \log_e x$).

Example 4 Finding Natural Logarithms

Find each logarithm to four decimal places.

(a) $\ln .5841 \approx -.5377$

As with common logarithms, a number between 0 and 1 has a negative natural logarithm.

(b) $\ln 192.7 \approx 5.2611$

(c) $\ln 10.84 \approx 2.3832$

Work Problem 4 at the Side.

4 Use natural logarithms in applications. A common application of natural logarithmic functions is to express growth or decay of a quantity, as in the next example.

Example 5 Applying Natural Logarithms

The altitude in meters that corresponds to an atmospheric pressure of x millibars is given by the logarithmic function with

$$f(x) = 51{,}600 - 7457 \ln x.$$

(*Source:* Miller, A. and J. Thompson, *Elements of Meteorology,* Fourth Edition, Charles E. Merrill Publishing Company, 1993.) Use this function to find the altitude when atmospheric pressure is 400 millibars.

Let $x = 400$ and substitute in the expression for $f(x)$.

$$f(400) = 51{,}600 - 7457 \ln 400$$
$$\approx 6900$$

Atmospheric pressure is 400 millibars at approximately 6900 m.

4 Find each logarithm to four decimal places.

(a) ln .01

(b) ln 27

(c) ln 529

Answers

4. (a) −4.6052 **(b)** 3.2958 **(c)** 6.2710

5 Use the logarithmic function in Example 5 to approximate the altitude at 700 millibars of pressure.

Calculator Tip In Example 5, the final answer was obtained using a calculator *without* rounding the intermediate values. In general, it is best to wait until the final step to round the answer; otherwise, a build-up of round-off error may cause the final answer to have an incorrect final decimal place digit.

Work Problem 5 at the Side.

ANSWERS

5. approximately 2700 m

10.5 EXERCISES

FOR EXTRA HELP: Student's Solutions Manual; MyMathLab.com; InterAct Math Tutorial Software; AW Math Tutor Center; MathXL www.mathxl.com; Digital Video Tutor CD 8 Videotape 18

Choose the correct response in Exercises 1–4.

1. What is the base in the expression $\log x$?
A. e **B.** 1 **C.** 10 **D.** x

2. What is the base in the expression $\ln x$?
A. e **B.** 1 **C.** 10 **D.** x

3. Since $10^0 = 1$ and $10^1 = 10$, between what two consecutive integers is the value of $\log 5.6$?
A. 5 and 6 **B.** 10 and 11 **C.** 0 and 1 **D.** -1 and 0

4. Since $e^1 \approx 2.718$ and $e^2 \approx 7.389$, between what two consecutive integers is the value of $\ln 5.6$?
A. 5 and 6 **B.** 2 and 3 **C.** 1 and 2 **D.** 0 and 1

5. Without using a calculator, give the value of $\log 10^{19.2}$.

6. Without using a calculator, give the value of $\ln e^{\sqrt{2}}$.

You will need a calculator for the remaining exercises in this set.

Find each logarithm. Give an approximation to four decimal places. See Examples 1 and 4.

7. $\log 43$

8. $\log 98$

9. $\log 328.4$

10. $\log 457.2$

11. $\log .0326$

12. $\log .1741$

13. $\log(4.76 \times 10^9)$

14. $\log(2.13 \times 10^4)$

15. $\ln 7.84$

16. $\ln 8.32$

17. $\ln .0556$

18. $\ln .0217$

19. $\ln 388.1$

20. $\ln 942.6$

21. $\ln(8.59 \times e^2)$

22. $\ln(7.46 \times e^3)$

23. $\ln 10$

24. $\log e$

25. Let m be the number of letters in your first name, and let n be the number of letters in your last name.

(a) In your own words, explain what $\log_m n$ means.

(b) Use your calculator to find $\log_m n$.

(c) Raise m to the power indicated by the number you found in part (b). What is your result?

26. Use your calculator to find approximations of the following logarithms.

(a) $\log 356.8$

(b) $\log 35.68$

(c) $\log 3.568$

(d) Observe your answers and make a conjecture concerning the decimal values of the common logarithms of numbers greater than 1 that have the same digits.

27. Try to find $\log(-1)$ using a calculator. (If you have a graphing calculator, it should be in real number mode.) What happens? Explain why this happens.

Refer to Example 2. In Exercises 28 and 29, suppose that water from a wetland area is sampled and found to have the given hydronium ion concentration. Determine whether the wetland is a rich fen, *a* poor fen, *or a* bog.

28. 2.5×10^{-5}

29. 2.5×10^{-2}

Use the formula $\text{pH} = -\log[H_3O^+]$ *to find the* pH *of the substance with the given hydronium ion concentration. See Example 2.*

30. Ammonia, 2.5×10^{-12}

31. Tuna, 1.3×10^{-6}

Use the formula for pH *to find the hydronium ion concentration of the substance with the given* pH. *See Example 3.*

32. Human blood plasma, 7.4

33. Human gastric contents, 2.0

34. Spinach, 5.4

35. Bananas, 4.6

Solve each problem. See Example 5.

36. The number of years, $N(r)$, since two independently evolving languages split off from a common ancestral language is approximated by

$$N(r) = -5000 \ln r,$$

where r is the percent of words (in decimal form) from the ancestral language common to both languages now. Find the number of years since the split for each percent of common words.

(a) 85% (or .85)

(b) 35% (or .35)

(c) 10% (or .10)

37. The time t in years for an amount increasing at a rate of r (in decimal form) to double is given by

$$t = \frac{\ln 2}{\ln(1 + r)}.$$

This is called *doubling time.* Find the doubling time to the nearest tenth for an investment at each interest rate.

(a) 2% = .02

(b) 5% = .05

(c) 8% = .08

38. The concentration of a drug injected into the bloodstream decreases with time. The intervals of time T when the drug should be administered are given by

$$T = \frac{1}{k} \ln \frac{C_2}{C_1},$$

where k is a constant determined by the drug in use, C_2 is the concentration at which the drug is harmful, and C_1 is the concentration below which the drug is ineffective. (*Source:* Horelick, Brindell and Sinan Koont, "Applications of Calculus to Medicine: Prescribing Safe and Effective Dosage," *UMAP Module 202,* 1977.) Thus, if $T = 4$, the drug should be administered every 4 hr. For a certain drug, $k = \frac{1}{3}$, $C_2 = 5$, and $C_1 = 2$. How often should the drug be administered? (*Hint:* Round down.)

39. The loudness of sounds is measured in a unit called a *decibel,* abbreviated dB. A very faint sound, called the *threshold sound*, is assigned an intensity I_0. If a particular sound has intensity I, then the decibel level of this louder sound is

$$D = 10 \log\left(\frac{I}{I_0}\right).$$

Find the average decibel level for each popular movie with the given intensity I. For comparison, a motorcycle or power saw has a decibel level of about 95 dB, and the sound of a jackhammer or helicopter is about 105 dB. (*Source: World Almanac and Book of Facts,* 2001; www.lhh.org/noise/)

(a) *Armageddon;* $5.012 \times 10^{10} I_0$

(b) *Godzilla;* $10^{10} I_0$

(c) *Saving Private Ryan;* $6{,}310{,}000{,}000\, I_0$

40. The growth of outpatient surgery as a percent of total surgeries at hospitals is approximated by

$$f(x) = -1317 + 304 \ln x,$$

where x represents the number of years since 1900. (*Source:* American Hospital Association.)

(a) What does this function predict for the percent of outpatient surgeries in 1998?

(b) When did outpatient surgeries reach 50%? (*Hint:* Substitute for y, then write the equation in exponential form to solve it.)

41. In the central Sierra Nevada of California, the percent of moisture p that falls as snow rather than rain is approximated reasonably well by

$$p = 86.3 \ln h - 680,$$

where h is the altitude in feet.

(a) What percent of the moisture at 5000 ft falls as snow?

(b) What percent at 7500 ft falls as snow?

42. The *cost-benefit equation*

$$T = -.642 - 189 \ln(1 - p)$$

describes the approximate tax T, in dollars per ton, that would result in a p% (in decimal form) reduction in carbon dioxide emissions.

(a) What tax will reduce emissions 25%?

(b) Explain why the equation is not valid for $p = 0$ or $p = 1$.

43. The age in years of a female blue whale is approximated by

$$t = -2.57 \ln\left(\frac{87 - L}{63}\right),$$

where L is its length in feet.

(a) How old is a female blue whale that measures 80 ft?

(b) The equation that defines t has domain $24 < L < 87$. Explain why.

10.6 Exponential and Logarithmic Equations; Further Applications

As mentioned earlier, exponential and logarithmic functions are important in many applications of mathematics. Using these functions in applications requires solving exponential and logarithmic equations. Some simple equations were solved in Sections 10.2 and 10.3. More general methods for solving these equations depend on the following properties.

Objectives

1. Solve equations involving variables in the exponents.
2. Solve equations involving logarithms.
3. Solve applications of compound interest.
4. Solve applications involving base *e* exponential growth and decay.
5. Use the change-of-base rule.

Properties for Solving Exponential and Logarithmic Equations

For all real numbers $b > 0$, $b \neq 1$, and any real numbers x and y:

1. If $x = y$, then $b^x = b^y$.
2. If $b^x = b^y$, then $x = y$.
3. If $x = y$, and $x > 0, y > 0$, then $\log_b x = \log_b y$.
4. If $x > 0, y > 0$, and $\log_b x = \log_b y$, then $x = y$.

We used Property 2 to solve exponential equations in Section 10.2.

1 Solve equations involving variables in the exponents. The first examples illustrate a general method for solving exponential equations using Property 3.

Example 1 Solving an Exponential Equation

Solve $3^m = 12$.

$$3^m = 12$$

$$\log 3^m = \log 12 \quad \text{Property 3}$$

$$m \log 3 = \log 12 \quad \text{Power rule}$$

$$m = \frac{\log 12}{\log 3} \quad \text{Divide by log 3.}$$

This quotient is the exact solution. To get a decimal approximation for the solution, use a calculator.

$$m \approx 2.262$$

The solution set is $\{2.262\}$. Check that $3^{2.262} \approx 12$.

CAUTION

Be careful: $\frac{\log 12}{\log 3}$ is *not* equal to $\log 4$ because $\log 4 \approx .6021$, but $\frac{\log 12}{\log 3} \approx 2.262$.

Work Problem 1 at the Side.

When an exponential equation has *e* as the base, it is easiest to use base *e* logarithms.

1 Solve each equation and give the decimal approximation to three places.

(a) $2^p = 9$

(b) $10^k = 4$

Answers

1. (a) $\{3.170\}$ (b) $\{.602\}$

2 Solve $e^{-.01t} = .38$.

3 Solve $\log_5 \sqrt{x - 7} = 1$.

Example 2 Solving an Exponential Equation with Base e

Solve $e^{.003x} = 40$.

Take base e logarithms on both sides.

$$\ln e^{.003x} = \ln 40$$

$$.003x \ln e = \ln 40 \quad \text{Power rule}$$

$$.003x = \ln 40 \quad \ln e = \ln e^1 = 1$$

$$x = \frac{\ln 40}{.003} \quad \text{Divide by .003.}$$

$$x \approx 1230 \quad \text{Use a calculator.}$$

The solution set is $\{1230\}$. Check that $e^{.003(1230)} \approx 40$.

Work Problem 2 at the Side.

General Method for Solving an Exponential Equation

Take logarithms to the same base on both sides and then use the power rule of logarithms or the special property $\log_b b^x = x$. (See Examples 1 and 2.)

As a special case, if both sides can be written as exponentials with the same base, do so, and set the exponents equal. (See Section 10.2.)

2 **Solve equations involving logarithms.** The properties of logarithms from Section 10.4 are useful here, as is using the definition of a logarithm to change the equation to exponential form.

Example 3 Solving a Logarithmic Equation

Solve $\log_2(x + 5)^3 = 4$.

$$(x + 5)^3 = 2^4 \quad \text{Convert to exponential form.}$$

$$(x + 5)^3 = 16$$

$$x + 5 = \sqrt[3]{16} \quad \text{Take the cube root on each side.}$$

$$x = -5 + \sqrt[3]{16}$$

$$x = -5 + 2\sqrt[3]{2} \quad \text{Simplify the radical.}$$

Verify that the solution satisfies the equation, so the solution set is $\{-5 + 2\sqrt[3]{2}\}$.

CAUTION

Recall that the domain of $y = \log_b x$ is $(0, \infty)$. For this reason, it is always necessary to check that the solution of an equation with logarithms yields only logarithms of positive numbers in the original equation.

Work Problem 3 at the Side.

ANSWERS
2. $\{96.8\}$
3. $\{32\}$

Example 4 Solving a Logarithmic Equation

Solve $\log_2(x + 1) - \log_2 x = \log_2 7$.

$$\log_2(x + 1) - \log_2 x = \log_2 7$$

$$\log_2 \frac{x + 1}{x} = \log_2 7 \quad \text{Quotient rule}$$

$$\frac{x + 1}{x} = 7 \quad \text{Property 4}$$

$$x + 1 = 7x \quad \text{Multiply by } x.$$

$$\frac{1}{6} = x \quad \text{Subtract } x\text{; divide by 6.}$$

Check this solution by substituting in the original equation. Here, both $x + 1$ and x must be positive. If $x = \frac{1}{6}$, this condition is satisfied, so the solution set is $\{\frac{1}{6}\}$.

Work Problem 4 at the Side.

4 Solve
$\log_8(2x + 5) + \log_8 3 = \log_8 33$.

Example 5 Solving a Logarithmic Equation

Solve $\log x + \log(x - 21) = 2$.

For this equation, write the left side as a single logarithm. Then write in exponential form and solve the equation.

$$\log x + \log(x - 21) = 2$$

$$\log x(x - 21) = 2 \quad \text{Product rule}$$

$$x(x - 21) = 10^2 \quad \text{Log } x = \log_{10} x\text{; write in exponential form.}$$

$$x^2 - 21x = 100$$

$$x^2 - 21x - 100 = 0 \quad \text{Standard form}$$

$$(x - 25)(x + 4) = 0 \quad \text{Factor.}$$

$$x - 25 = 0 \quad \text{or} \quad x + 4 = 0 \quad \text{Zero-factor property}$$

$$x = 25 \quad \text{or} \quad x = -4$$

The value -4 must be rejected as a solution since it leads to the logarithm of a negative number in the original equation:

$$\log(-4) + \log(-4 - 21) = 2. \quad \text{The left side is undefined.}$$

The only solution, therefore, is 25, and the solution set is $\{25\}$.

5 Solve
$\log_3 2x - \log_3(3x + 15) = -2$.

CAUTION

Do not reject a potential solution just because it is nonpositive. Reject any value that *leads to* the logarithm of a nonpositive number.

Work Problem 5 at the Side.

In summary, we use the following steps to solve a logarithmic equation.

ANSWERS
4. $\{3\}$
5. $\{1\}$

6 Find the value of \$2000 deposited at 5% compounded annually for 10 yr.

Solving a Logarithmic Equation

Step 1 **Get a single logarithm on one side.** Use the product rule or quotient rule of logarithms to do this.

Step 2 **(a) Use property 4.** If $\log_b x = \log_b y$, then $x = y$. (See Example 4.)
(b) Write the equation in exponential form. If $\log_b x = k$, then $x = b^k$. (See Examples 3 and 5.)

3 **Solve applications of compound interest.** So far in this book, problems involving applications of interest have been limited to simple interest using the formula $I = prt$. In most cases, interest paid or charged is compound interest (interest paid on both principal and interest). The formula for compound interest is an important application of exponential functions.

Compound Interest

If P dollars is deposited in an account paying an annual rate of interest r compounded (paid) n times per year, the account will contain

$$A = P\left(1 + \frac{r}{n}\right)^{nt}$$

dollars after t years.

In this formula, r is expressed as a decimal.

Example 6 **Solving a Compound Interest Problem for A**

How much money will there be in an account at the end of 5 yr if \$1000 is deposited at 6% compounded quarterly? (Assume no withdrawals are made.)

Because interest is compounded quarterly, $n = 4$. The other values given in the problem are $P = 1000$, $r = .06$ (because $6\% = .06$), and $t = 5$. Substitute into the compound interest formula to get the value of A.

$$A = 1000\left(1 + \frac{.06}{4}\right)^{4 \cdot 5}$$

$$A = 1000(1.015)^{20}$$

Now use the y^x key on a calculator, and round the answer to the nearest cent.

$$A = 1346.86$$

The account will contain \$1346.86. (The actual amount of interest earned is \$1346.86 − \$1000 = \$346.86. Why?)

Work Problem 6 at the Side.

Example 7 **Solving a Compound Interest Problem for t**

Suppose inflation is averaging 3% per year. How many years will it take for prices to double?

We want to find the number of years t for \$1 to grow to \$2 at a rate of 3% per year. In the compound interest formula, we let $A = 2$, $P = 1$, $r = .03$, and $n = 1$.

Continued on Next Page

ANSWERS
6. about \$3257.79

$$2 = 1\left(1 + \frac{.03}{1}\right)^{1t}$$

$$2 = (1.03)^t \quad \text{Simplify.}$$

$$\log 2 = \log(1.03)^t \quad \text{Property 3}$$

$$\log 2 = t \log 1.03 \quad \text{Power rule}$$

$$t = \frac{\log 2}{\log 1.03} \quad \text{Divide by } \log 1.03.$$

$$t \approx 23.45$$

Prices will double in about 23 yr. (This is called the *doubling time* of the money.) To check, verify that $1.03^{23.45} \approx 2$.

Work Problem 7 at the Side.

Banks sometimes compute interest based on **continuous compounding.** With this type of compounding, rather than paying interest a finite number of times per year, interest is earned at all times. As a result, the formula for compound interest cannot be applied because n is infinite. The formula used to determine the amount A in an account having initial principal P compounded continuously at an annual rate r for t years is

$$A = Pe^{rt}.$$

Example 8 Solving a Continuous Interest Problem

How much will $1000 grow to in 5 yr at an annual interest rate of 6% compounded continuously?

Use the formula.

$$A = Pe^{rt}$$

$$A = 1000e^{(.06)5} \quad \text{Let } P = 1000,\ r = .06, \text{ and } t = 5.$$

$$A \approx 1349.86 \quad \text{Use a calculator; round to two decimal places.}$$

The account will grow to $1349.86.

Work Problem 8 at the Side.

4 Solve applications involving base *e* exponential growth and decay. We saw some applications involving exponential growth and decay in Section 10.2. In many cases, quantities grow or decay according to a function defined by an exponential expression with base e. You have probably heard of the carbon-14 dating process used to determine the age of fossils. The method used is based on a base e exponential decay function.

7 Find the number of years it will take for $500 to increase to $750 in an account paying 4% interest compounded semiannually.

8 How much will $2500 grow to at 4% interest compounded continuously for 3 yr?

ANSWERS

7. about 10.24 yr

8. $2818.74

9 Radioactive strontium decays according to the function

$$y = y_0 e^{-.0239t},$$

where t is time in years.

(a) If an initial sample contains $y_0 = 12$ g of radioactive strontium, how many grams will be present after 35 yr?

(b) What is the half-life of radioactive strontium?

Example 9 Solving an Exponential Decay Application

Carbon-14 is a radioactive form of carbon that is found in all living plants and animals. After a plant or animal dies, the radioactive carbon-14 disintegrates according to the function with

$$y = y_0 e^{-.000121t},$$

where t is time in years, y is the amount of the sample at time t, and y_0 is the initial amount present at $t = 0$.

(a) If an initial sample contains $y_0 = 10$ g of carbon-14, how many grams will be present after 3000 yr?
Let $y_0 = 10$ and $t = 3000$ in the formula, and use a calculator.

$$y = 10e^{-.000121(3000)} \approx 6.96 \text{ g}$$

(b) How long would it take for the initial sample to decay to half of its original amount? (This is called the *half-life*.)
Let $y = \frac{1}{2}(10) = 5$, and solve for t.

$$5 = 10e^{-.000121t} \quad \text{Substitute.}$$

$$\frac{1}{2} = e^{-.000121t} \quad \text{Divide by 10.}$$

$$\ln \frac{1}{2} = -.000121t \quad \text{Take natural logarithms; } \ln e^k = k.$$

$$t = \frac{\ln \frac{1}{2}}{-.000121} \quad \text{Divide by } -.000121.$$

$$t \approx 5728 \quad \text{Use a calculator.}$$

The half-life is just over 5700 yr.

Work Problem 9 at the Side.

5 Use the change-of-base rule. In the previous section we used a calculator to approximate the values of common logarithms (base 10) or natural logarithms (base e). However, some applications involve logarithms to other bases. For example, for the years 1980–1996, the percentage of women who had a baby in the last year and returned to work is given by

$$y = 38.83 + 4.208 \log_2 x,$$

for year x. (*Source:* U.S. Bureau of the Census.) To use this function, we need to find a base 2 logarithm. The following rule is used to convert logarithms from one base to another.

Change-of-Base Rule

If $a > 0$, $a \neq 1$, $b > 0$, $b \neq 1$, and $x > 0$, then

$$\log_a x = \frac{\log_b x}{\log_b a}.$$

ANSWERS
9. **(a)** 5.20 g **(b)** 29 yr

NOTE

As an aid in remembering the change-of-base rule, notice that x is "above" a on both sides of the equation.

Any positive number other than 1 can be used for base b in the change-of-base rule, but usually the only practical bases are e and 10 because calculators give logarithms only for these two bases.

10 Find $\log_3 17$. (Use common logarithms.)

To derive the change-of-base rule, let $\log_a x = m$.

$$\log_a x = m$$

$$a^m = x \quad \text{Change to exponential form.}$$

$$\log_b(a^m) = \log_b x \quad \text{Property 3}$$

$$m \log_b a = \log_b x \quad \text{Power rule}$$

$$(\log_a x)(\log_b a) = \log_b x \quad \text{Substitute for } m.$$

$$\log_a x = \frac{\log_b x}{\log_b a} \quad \text{Divide by } \log_b a.$$

The last step gives the change-of-base rule.

Example 10 Using the Change-of-Base Rule

Find $\log_5 12$.

Use common logarithms and the change-of-base rule.

$$\log_5 12 = \frac{\log 12}{\log 5}$$

$$\approx 1.5440 \quad \text{Use a calculator.}$$

Work Problem **10** at the Side.

11 In Example 11, what percent of women returned to work after having a baby in 1990?

Example 11 Using the Change-of-Base Rule in an Application

Use natural logarithms in the change-of-base rule and the equation

$$y = 38.83 + 4.208 \log_2 x$$

(given earlier) to find the percent of women who returned to work after having a baby in 1995. In the equation, $x = 0$ represents 1980.

Substitute $1995 - 1980 = 15$ for x in the equation.

$$y = 38.83 + 4.208 \log_2 15$$

$$= 38.83 + 4.208\left(\frac{\ln 15}{\ln 2}\right) \quad \text{Change-of-base rule}$$

$$\approx 55.3\% \quad \text{Use a calculator.}$$

This is very close to the actual value of 55%.

Work Problem **11** at the Side.

ANSWERS

10. 2.5789

11. 52.8%; This is very close to the actual value of 53%.

Focus on **Real-Data Applications**

Evaluating Investments: The Rule of 72

The Rule of 72 gives an estimate of the doubling time of an investment. It is a useful tool in evaluating and comparing investments.

- The Rule of 72 is $\frac{72}{100r}$, where r is the annual interest rate. (Since r is the interest rate as a *decimal*, $100r$ is the interest rate as a *percent*.)
- The compound interest formula is $A = P(1 + \frac{r}{n})^{nt}$, for \$$P$ invested at interest rate r (in decimal form), compounded n times per year, that accumulates to \$$A$ after t years.
- The continuous interest formula is $A = Pe^{rt}$, for \$$P$ invested at interest rate r (in decimal form) that accumulates to \$$A$ after t years.

For Group Discussion

1. To investigate how the Rule of 72 works, we will use the Rule of 72 to estimate the doubling time for money invested at 10%.

(a) What is the estimated doubling time, to the nearest year, for the investment?

(b) If \$2000 is invested at 10% compounded quarterly, what is its accumulated value after the predicted doubling time? Did the Rule of 72 give a good estimate?

(c) If \$2000 is invested at 10% compounded continuously, what is its accumulated value after the predicted doubling time? Did the Rule of 72 give a good estimate?

2. If money is invested at 8%, the Rule of 72 predicts a 9-year doubling time. Sketch a graph to illustrate the doubling effect of an investment of \$2000 over time. The x-axis represents time in years with 0, 9, 18, 27, 36, and 45 representing five doubling-time periods. The y-axis represents the value of the investment in dollars.

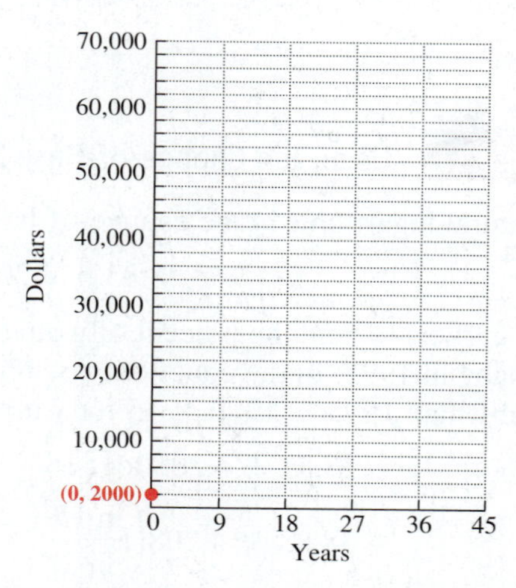

3. Now investigate why the Rule of 72 works. If an investment doubles in value, the continuous interest formula has the form $2P = Pe^{rt}$. Since P is not 0, divide each side of the equation by P to get $2 = e^{rt}$. To solve this equation for t, take the natural logarithm on each side, $\ln 2 = \ln e^{rt}$. Using the power rule, this simplifies to $\ln 2 = rt$. Therefore, $t = \frac{\ln 2}{r}$. Since $\ln 2 \approx .69$, this formula becomes $t = \frac{100 \ln 2}{100r} \approx \frac{69}{100r}$. The number 69 is less useful than 72, which has more factors (i.e., 2, 3, 4, 6, 8, 9, 12, 18, 24, 36), and the doubling time for compound interest will be slightly longer anyway. So, the Rule of 72 estimates this formula as $\frac{72}{100r}$. Does the Rule of 72 underestimate or overestimate the true doubling time for continuously compounded investments? Explain your answer.

10.6 EXERCISES

FOR EXTRA HELP

Student's Solutions Manual

MyMathLab.com

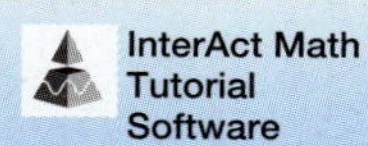
InterAct Math Tutorial Software

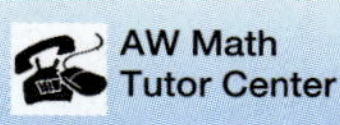
AW Math Tutor Center

www.mathxl.com

Digital Video Tutor CD 9 Videotape 18

RELATING CONCEPTS (Exercises 1–4) FOR INDIVIDUAL OR GROUP WORK

In Section 10.2 we solved an equation such as $5^x = 125$ by writing each side as a power of the same base, setting exponents equal, and then solving the resulting equation. The equation is solved as follows.

$$5^x = 125 \quad \text{Original equation}$$
$$5^x = 5^3 \quad 125 = 5^3$$
$$x = 3 \quad \text{Set exponents equal.}$$

Solution set: $\{3\}$

The method described in this section can also be used to solve this equation.

Work Exercises 1–4 in order, *to see how this is done.*

1. Take common logarithms on both sides, and write this equation.

2. Apply the power rule for logarithms on the left.

3. Get x alone on the left.

4. Use a calculator to find the decimal form of the solution. What is the solution set?

Many of the problems in the remaining exercises require a scientific calculator.

Solve each equation. Give solutions to three decimal places. See Example 1.

5. $7^x = 5$

6. $4^x = 3$

7. $9^{-x+2} = 13$

8. $6^{-t+1} = 22$

9. $3^{2x} = 14$

10. $5^{.3x} = 11$

11. $2^{y+3} = 5^y$

12. $6^{m+3} = 4^m$

Solve each equation. Use natural logarithms. Give solutions to three decimal places. See Example 2.

13. $e^{.006x} = 30$

14. $e^{.012x} = 23$

15. $e^{-.103x} = 7$

16. $e^{-.205x} = 9$

17. $\ln e^{x} = 4$

18. $\ln e^{3x} = 9$

19. $\ln e^{.04x} = \sqrt{3}$

20. $\ln e^{.45x} = \sqrt{7}$

21. Try solving one of the equations in Exercises 13–16 using common logarithms rather than natural logarithms. (You should get the same solution.) Explain why using natural logarithms is a better choice.

22. If you were asked to solve $10^{.0025x} = 75$, would natural or common logarithms be a better choice? Explain.

Solve each equation. Give the exact solution. See Example 3.

23. $\log_3(6x + 5) = 2$

24. $\log_5(12x - 8) = 3$

25. $\log_2(2x - 1) = 5$

26. $\log_6(4x + 2) = 2$

27. $\log_7(x + 1)^3 = 2$

28. $\log_4(y - 3)^3 = 4$

29. Suppose that in solving a logarithmic equation having the term $\log(x - 3)$ you obtain an apparent solution of 2. All algebraic work is correct. Explain why you must reject 2 as a solution of the equation.

30. Suppose that in solving a logarithmic equation having the term $\log(3 - x)$ you obtain an apparent solution of -4. All algebraic work is correct. Should you reject -4 as a solution of the equation? Explain why or why not.

Solve each equation. Give exact solutions. See Examples 4 and 5.

31. $\log(6x + 1) = \log 3$

32. $\log(7 - x) = \log 12$

33. $\log_5(3t + 2) - \log_5 t = \log_5 4$

34. $\log_2(x + 5) - \log_2(x - 1) = \log_2 3$

35. $\log 4x - \log(x - 3) = \log 2$

36. $\log(-x) + \log 3 = \log(2x - 15)$

37. $\log_2 x + \log_2(x - 7) = 3$

38. $\log(2x - 1) + \log 10x = \log 10$

39. $\log 5x - \log(2x - 1) = \log 4$

40. $\log_3 x + \log_3(2x + 5) = 1$

41. $\log_2 x + \log_2(x - 6) = 4$

42. $\log_2 x + \log_2(x + 4) = 5$

Solve each problem. See Examples 6–8.

43. **(a)** How much money will there be in an account at the end of 6 yr if $2000 is deposited at 4% compounded quarterly? (Assume no withdrawals are made.)

(b) To one decimal place, how long will it take for the account to grow to $3000?

44. **(a)** How much money will there be in an account at the end of 7 yr if $3000 is deposited at 3.5% compounded quarterly? (Assume no withdrawals are made.)

(b) To one decimal place, when will the account grow to $5000?

45. What will be the amount A in an account with initial principal $4000 if interest is compounded continuously at an annual rate of 3.5% for 6 yr?

46. Refer to Exercise 44. Does the money grow to a larger value under those conditions, or when invested for 7 yr at 3% compounded continuously?

47. How long would it take an initial principal P to double if it is invested at 4.5% compounded continuously?

48. How long would it take \$4000 to grow to \$6000 at 3.25% compounded continuously?

Solve each problem. See Example 9.

49. A sample of 400 g of lead-210 decays to polonium-210 according to the function with

$$A(t) = 400e^{-.032t},$$

where t is time in years. How much lead will be left in the sample after 25 yr?

50. How long will it take the initial sample of lead in Exercise 49 to decay to half of its original amount?

Use the change-of-base rule (with either common or natural logarithms) to find each logarithm. Give approximations to four decimal places. See Example 10.

51. $\log_6 13$

52. $\log_7 19$

53. $\log_{\sqrt{2}} \pi$

54. $\log_\pi \sqrt{2}$

55. $\log_{21} .7496$

56. $\log_{19} .8325$

Work each problem. See Example 11.

One measure of the diversity of the species in an ecological community is the index of diversity, *a logarithmic function defined by*

$$H(x) = -(p_1 \ln p_1 + p_2 \ln p_2 + \ldots + p_n \ln p_n),$$

where $p_1, p_2, \ldots, p_n$ are the proportions of a sample belonging to each of n species in the sample. (Source: Ludwig, John and James Reynolds, Statistical Ecology: A Primer on Methods and Computing, *New York, Wiley, 1988.) Find the index of diversity to three decimal places if a sample of* 100 *from a community produces the following numbers.*

57. 90 of one species, 10 of another

58. 60 of one species, 40 of another

Chapter 10

Summary

Key Terms

10.1 one-to-one function A one-to-one function is a function in which each x-value corresponds to just one y-value and each y-value corresponds to just one x-value.

inverse of a function f If f is a one-to-one function, the inverse of f is the set of all ordered pairs of the form (y, x), where (x, y) belongs to f.

Domain Range

One-to-one

10.2 exponential equation An equation involving an exponential, where the variable is in the exponent, is an exponential equation.

10.3 logarithm A logarithm is an exponent; $\log_a x$ is the exponent on the base a that gives the number x.

logarithmic equation A logarithmic equation is an equation with a logarithm in at least one term.

10.5 common logarithm A common logarithm is a logarithm to the base 10.

natural logarithm A natural logarithm is a logarithm to the base e.

New Symbols

f^{-1}	the inverse of f
$\log_a x$	the logarithm of x to the base a
$\log x$	common (base 10) logarithm of x
$\ln x$	natural (base e) logarithm of x
e	a constant, approximately 2.7182818285

Test Your Word Power

See how well you have learned the vocabulary in this chapter. Answers follow the Quick Review.

1. In a **one-to-one function**
- **(a)** each x-value corresponds to only one y-value
- **(b)** each x-value corresponds to one or more y-values
- **(c)** each x-value is the same as each y-value
- **(d)** each x-value corresponds to only one y-value and each y-value corresponds to only one x-value.

2. If f is a one-to-one function, then the **inverse** of f is
- **(a)** the set of all solutions of f
- **(b)** the set of all ordered pairs formed by interchanging the coordinates of the ordered pairs of f
- **(c)** an equation involving an exponential expression
- **(d)** the set of all ordered pairs that are the opposite (negative) of the coordinates of the ordered pairs of f.

3. An **exponential function** is a function defined by an expression of the form
- **(a)** $f(x) = ax^2 + bx + c$ for real numbers a, b, c $(a \neq 0)$
- **(b)** $f(x) = \log_a x$, for a and x positive numbers $(a \neq 1)$
- **(c)** $f(x) = a^x$ for all real numbers x $(a > 0, a \neq 1)$
- **(d)** $f(x) = \sqrt{x}$ for $x \geq 0$.

4. A **logarithm** is
- **(a)** an exponent
- **(b)** a base
- **(c)** an equation
- **(d)** a term.

5. A **logarithmic function** is a function that is defined by an expression of the form
- **(a)** $f(x) = ax^2 + bx + c$ for real numbers a, b, c $(a \neq 0)$
- **(b)** $f(x) = \log_a x$, for a and x positive numbers $(a \neq 1)$
- **(c)** $f(x) = a^x$ for all real numbers x $(a > 0, a \neq 1)$
- **(d)** $f(x) = \sqrt{x}$ for $x \geq 0$.

Quick Review

Concepts	*Examples*
10.1 Inverse Functions	
Horizontal Line Test If a horizontal line intersects the graph of a function in no more than one point, then the function is one-to-one.	Find f^{-1} if $f(x) = 2x - 3$. The graph of f is a straight line, so f is one-to-one by the horizontal line test.
Inverse Functions For a one-to-one function f defined by an equation $y = f(x)$, the equation that defines the inverse function f^{-1} is found by interchanging x and y, solving for y, and replacing y with $f^{-1}(x)$.	Interchange x and y in the equation $y = 2x - 3$. $x = 2y - 3$ Solve for y to get $y = \frac{x + 3}{2}$. Therefore, $f^{-1}(x) = \frac{x + 3}{2}$.
In general, the graph of f^{-1} is the mirror image of the graph of f with respect to the line $y = x$.	The graphs of a nonlinear function f and its inverse f^{-1} are shown here. 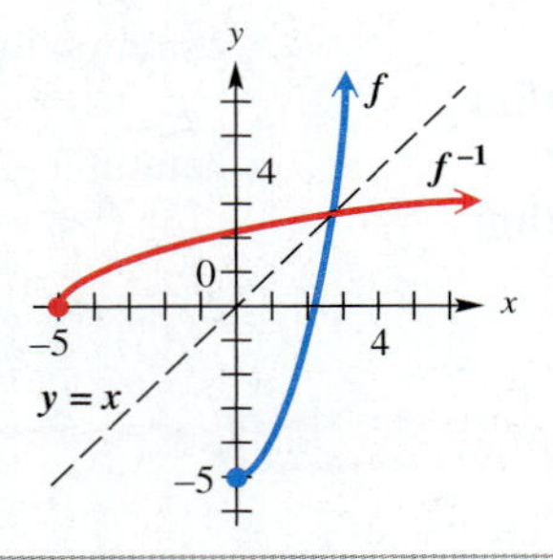
10.2 Exponential Functions For $a > 0$, $a \neq 1$, $f(x) = a^x$ defines an exponential function with base a.	$F(x) = 3^x$ defines an exponential function with base 3.
Graph of $F(x) = a^x$ **1.** The graph contains the point (0, 1). **2.** When $a > 1$, the graph rises from left to right. When $0 < a < 1$, the graph falls from left to right. **3.** The x-axis is an asymptote. **4.** The domain is $(-\infty, \infty)$; the range is $(0, \infty)$.	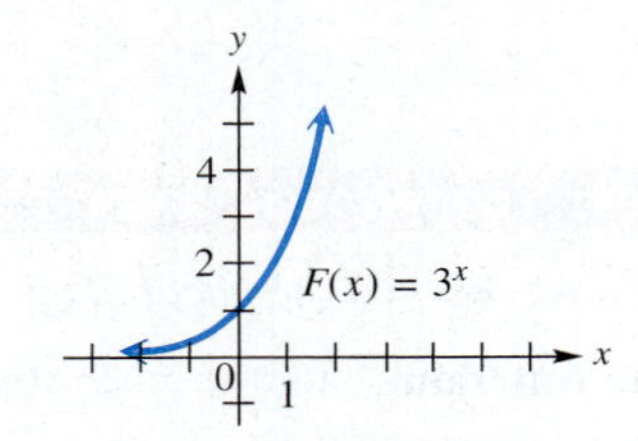
10.3 Logarithmic Functions $y = \log_a x$ means $x = a^y$.	$y = \log_2 x$ means $x = 2^y$.
For $b > 0$, $b \neq 1$, $\log_b b = 1$ and $\log_b 1 = 0$.	$\log_3 3 = 1$ $\log_5 1 = 0$
For $a > 0$, $a \neq 1$, $x > 0$, $G(x) = \log_a x$ defines the logarithmic function with base a.	$G(x) = \log_3 x$ defines the logarithmic function with base 3.
Graph of $G(x) = \log_a x$ **1.** The graph contains the point (1, 0). **2.** When $a > 1$, the graph rises from left to right. When $0 < a < 1$, the graph falls from left to right. **3.** The y-axis is an asymptote. **4.** The domain is $(0, \infty)$; the range is $(-\infty, \infty)$.	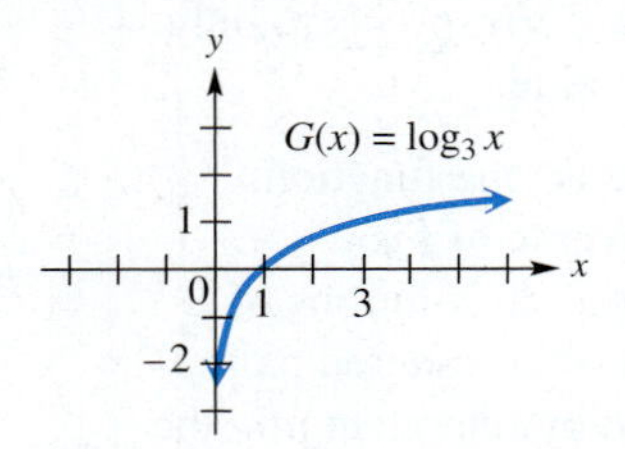

Concepts	*Examples*
10.4 Properties of Logarithms	
Product Rule $\log_a xy = \log_a x + \log_a y$	$\log_2 3m = \log_2 3 + \log_2 m$
Quotient Rule $\log_a \frac{x}{y} = \log_a x - \log_a y$	$\log_5 \frac{9}{4} = \log_5 9 - \log_5 4$
Power Rule $\log_a x^r = r \log_a x$	$\log_{10} 2^3 = 3 \log_{10} 2$
Special Properties $b^{\log_b x} = x$ and $\log_b b^x = x$	$6^{\log_6 10} = 10$ $\log_3 3^4 = 4$
10.5 Common and Natural Logarithms Common logarithms (base 10) are used in applications such as pH, sound level, and intensity of an earthquake. Use the LOG key of a calculator to evaluate common logarithms.	Use the formula $\text{pH} = -\log [H_3O^+]$ to find the pH (to one decimal place) of grapes with hydronium ion concentration 5.0×10^{-5}. $\text{pH} = -\log(5.0 \times 10^{-5})$ Substitute. $= -(\log 5.0 + \log 10^{-5})$ Property of logarithms ≈ 4.3 Evaluate.
Natural logarithms (base e) are most often used in applications of growth and decay, such as time for money invested to double, decay of chemical compounds, and biological growth. Use the ln x key or both the INV and e^x keys to evaluate natural logarithms.	Use the formula for doubling time (in years) $t = \frac{\ln 2}{\ln(1 + r)}$ to find the doubling time to the nearest tenth at an interest rate of 4%. $t = \frac{\ln 2}{\ln(1 + .04)}$ Substitute. ≈ 17.7 Evaluate. The doubling time is about 17.7 yr.
10.6 Exponential and Logarithmic Equations; Further Applications To solve exponential equations, use these properties ($b > 0$, $b \neq 1$).	
1. If $b^x = b^y$, then $x = y$.	Solve $2^{3x} = 2^5$. $3x = 5$ $x = \frac{5}{3}$ The solution set is $\{\frac{5}{3}\}$.
2. If $x = y$ $(x > 0, y > 0)$, then $\log_b x = \log_b y$.	Solve $5^m = 8$. $\log 5^m = \log 8$ $m \log 5 = \log 8$ $m = \frac{\log 8}{\log 5} \approx 1.2920$ The solution set is $\{1.2920\}$. *(continued)*

Concepts	*Examples*
10.6 Exponential and Logarithmic Equations; Further Applications (continued)	
To solve logarithmic equations, use these properties, where $b > 0$, $b \neq 1$, $x > 0$, $y > 0$. First use the properties of Section 10.4, if necessary, to get the equation in the proper form.	
1. If $\log_b x = \log_b y$, then $x = y$.	Solve $\log_3 2x = \log_3(x + 1)$. $2x = x + 1$ $x = 1$ The solution set is $\{1\}$.
2. If $\log_b x = y$, then $b^y = x$.	Solve $\log_2(3a - 1) = 4$. $3a - 1 = 2^4$ $3a - 1 = 16$ $3a = 17$ $a = \frac{17}{3}$ The solution set is $\{\frac{17}{3}\}$.
Change-of-Base Rule If $a > 0$, $a \neq 1$, $b > 0$, $b \neq 1$, $x > 0$, then $\log_a x = \dfrac{\log_b x}{\log_b a}$.	$\log_3 17 = \dfrac{\ln 17}{\ln 3} = \dfrac{\log 17}{\log 3} \approx 2.5789$

Answers to Test Your Word Power

1. (d) *Example:* The function $f = \{(0, 2), (1, -1), (3, 5), (-2, 3)\}$ is one-to-one.
2. (b) *Example:* The inverse of the one-to-one function f defined in Answer 1 is $f^{-1} = \{(2, 0), (-1, 1), (5, 3), (3, -2)\}$. **3. (c)** *Examples:* $f(x) = 4^x$, $g(x) = (\frac{1}{2})^x$, $h(x) = 2^{-x+3}$ **4. (a)** *Example:* $\log_a x$ is the exponent to which a must be raised to obtain x; $\log_3 9 = 2$ since $3^2 = 9$. **5. (b)** *Examples:* $y = \log_3 x$, $y = \log_{1/3} x$

Chapter 10 REVIEW EXERCISES

[10.1] *Determine whether each graph is the graph of a one-to-one function.*

1.

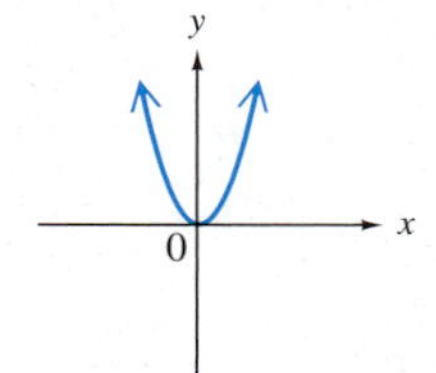

2.

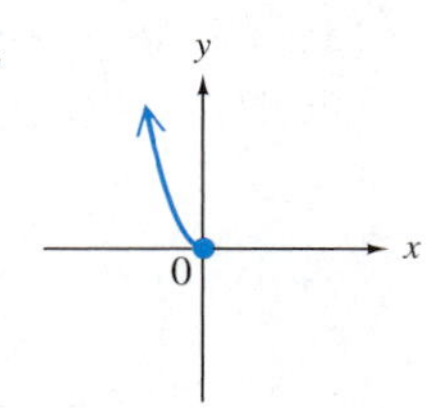

3. The table lists caffeine amounts in several popular 12-oz sodas. If the set of sodas is the domain and the set of caffeine amounts is the range of the function consisting of the six pairs listed, is it a one-to-one function? Why or why not?

Soda	Caffeine (mg)
Mountain Dew	55
Diet Coke	45
Dr. Pepper	41
Sunkist Orange Soda	41
Diet Pepsi-Cola	36
Coca-Cola Classic	34

Source: National Soft Drink Association.

Determine whether each function is one-to-one. If it is, find its inverse.

4. $f(x) = -3x + 7$

5. $f(x) = \sqrt[3]{6x - 4}$

6. $f(x) = -x^2 + 3$

Each function graphed is one-to-one. Graph its inverse.

7.

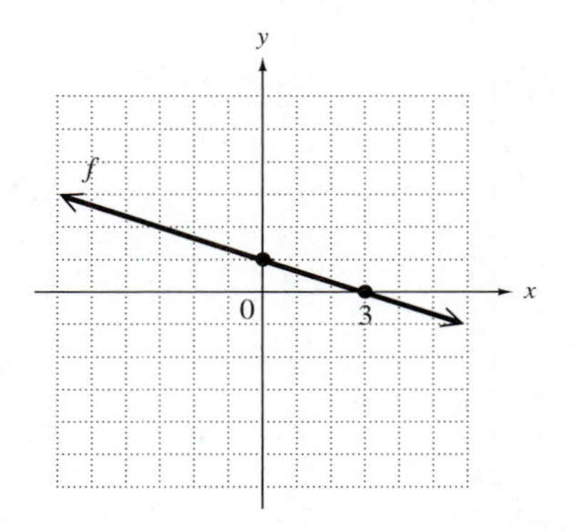

8.

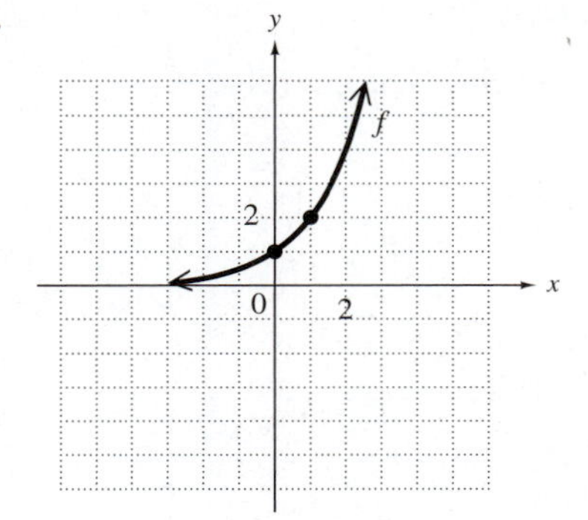

[10.2] *Graph each function.*

9. $f(x) = 3^x$

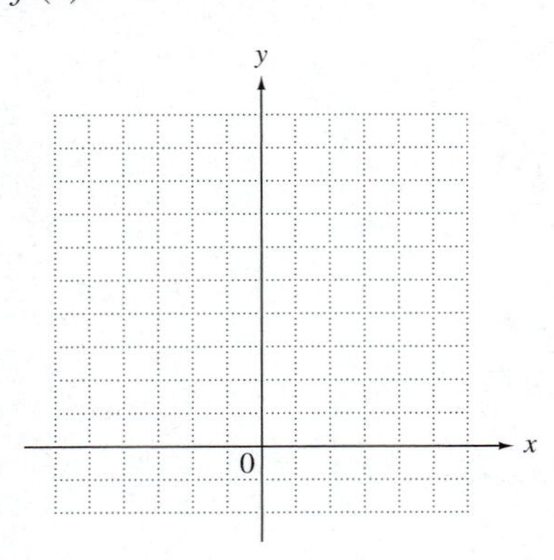

10. $f(x) = \left(\frac{1}{3}\right)^x$

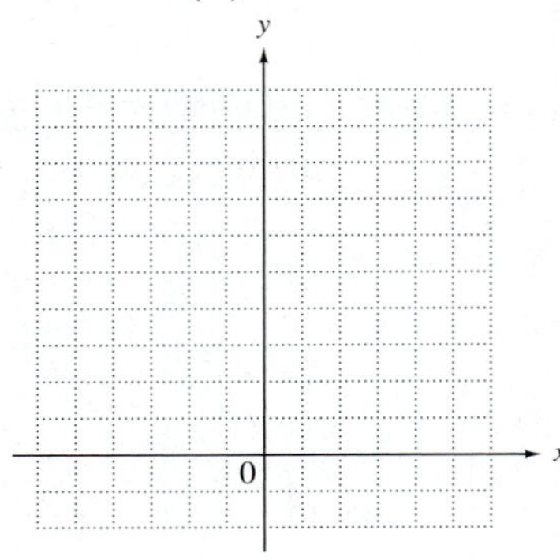

11. $y = 3^{x+1}$

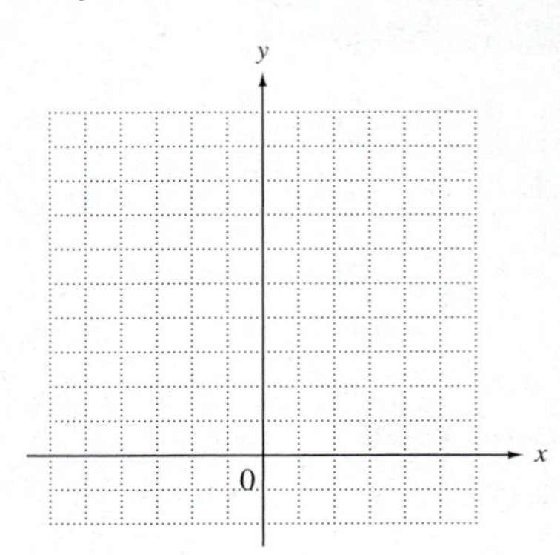

Solve each equation.

12. $4^{3x} = 8^{x+4}$

13. $\left(\frac{1}{27}\right)^{x-1} = 9^{2x}$

14. $5^x = 1$

In the remainder of the Chapter Review, many exercises will require a scientific calculator. We do not mark each such exercise.

15. The gross wastes generated in plastics, in millions of tons, from 1960 through 1990 can be approximated by the exponential function with

$$W(x) = .67(1.123)^x,$$

where $x = 0$ corresponds to 1960, $x = 5$ to 1965, and so on. Use this function to approximate the plastic waste amounts for the following years. (*Source:* U.S. Environmental Protection Agency, *Characterization of Municipal Solid Waste in the United States: 1994 Update,* 1995.)

(a) 1965 **(b)** 1975 **(c)** 1990

[10.3]

16. (a) Write in exponential form: $\log_5 625 = 4$.

(b) Write in logarithmic form: $5^{-2} = .04$.

17. (a) In your own words, explain the meaning of $\log_b a$.

(b) Based on the meaning of $\log_b a$, what is the simplest form of $b^{\log_b a}$?

Graph each function.

18. $g(x) = \log_3 x$ (*Hint:* See Exercise 9.)

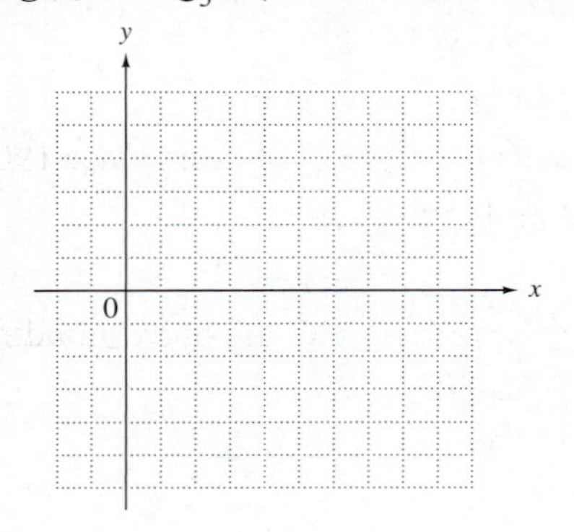

19. $g(x) = \log_{1/3} x$ (*Hint:* See Exercise 10.)

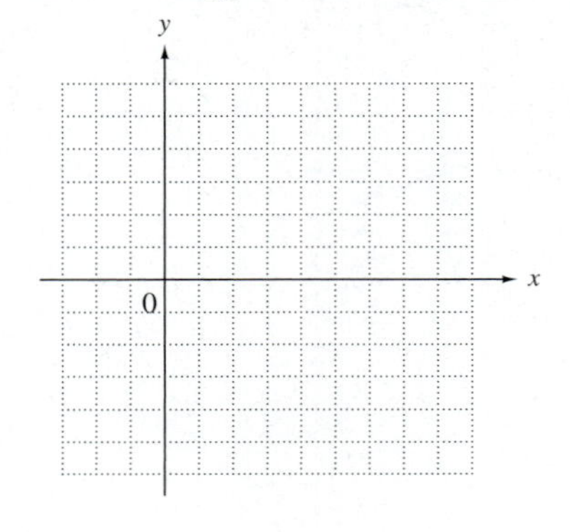

Solve each equation.

20. $\log_8 64 = x$

21. $\log_7\left(\frac{1}{49}\right) = x$

22. $\log_4 x = \frac{3}{2}$

23. $\log_b b^2 = 2$

[10.4] *Apply the properties of logarithms to express each logarithm as a sum or difference of logarithms. Assume that all variables represent positive real numbers.*

24. $\log_4 3x^2$

25. $\log_2 \frac{p^2 r}{\sqrt{z}}$

Use the properties of logarithms to write each expression as a single logarithm. Assume that all variables represent positive real numbers, $b \neq 1$.

26. $\log_b 3 + \log_b x - 2 \log_b y$

27. $\log_3(x + 7) - \log_3(4x + 6)$

[10.5] *Evaluate each logarithm. Give approximations to four decimal places.*

28. log 28.9

29. log .257

30. ln 28.9

31. ln .257

Use the formula pH $= -\log[H_3O^+]$ *to find the* pH *of each substance with the given hydronium ion concentration.*

32. Milk, 4.0×10^{-7}

33. Crackers, 3.8×10^{-9}

34. If orange juice has pH 4.6, what is its hydronium ion concentration?

Solve each problem.

35. Section 10.5 Exercise 37 introduced the *doubling function* defined by

$$t = \frac{\ln 2}{\ln(1 + r)},$$

that gives the number of years required to double your money when it is invested at interest rate r (in decimal form) compounded annually. How long does it take to double your money at each rate? Round answers to the nearest year.

(a) 4% **(b)** 6%

(c) 10% **(d)** 12%

(e) Compare each answer in parts (a)–(d) with these numbers:

$$\frac{72}{4}, \frac{72}{6}, \frac{72}{10}, \frac{72}{12}.$$

What do you find?

36. The graph shows the percent change in commercial rents in California from 1992 through 1999.

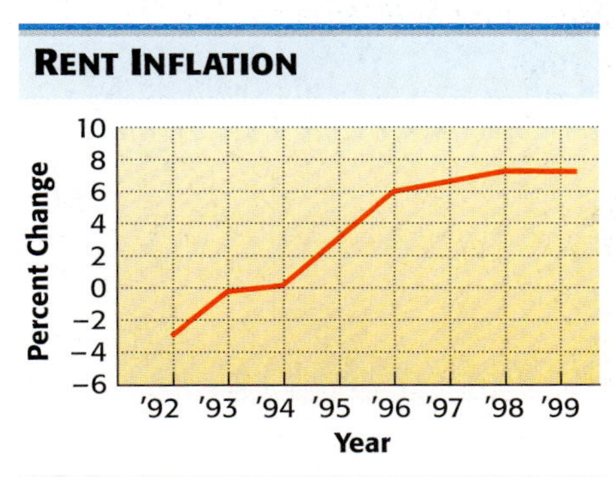

Source: CB Commercial/Torto Wheaton Research.

The percent change in rents is approximated by the logarithmic function with

$$g(x) = -650 + 143 \ln x,$$

where x represents the number of years since 1900.

(a) Find $g(92)$ and $g(99)$.

(b) Compare your results with the corresponding values in the graph.

[10.6] *Solve each equation. Give solutions to three decimal places.*

37. $3^x = 9.42$

38. $2^{x-1} = 15$

39. $e^{.06x} = 3$

Solve each equation. Give exact solutions.

40. $\log_3(9x + 8) = 2$

41. $\log_5(y + 6)^3 = 2$

42. $\log_3(p + 2) - \log_3 p = \log_3 2$

43. $\log(2x + 3) - \log x = 1$

44. $\log_4 x + \log_4(8 - x) = 2$

45. $\log_2 x + \log_2(x + 15) = 4$

Solve each problem.

46. How much would be in an account after 3 yr if \$6500.00 was invested at 3% annual interest, compounded daily (use $n = 365$)?

47. Which is a better plan?

Plan A: Invest \$1000.00 at 4% compounded quarterly for 3 yr

Plan B: Invest \$1000.00 at 3.9% compounded monthly for 3 yr

A machine purchased for business use depreciates, *or loses value, over a period of years. The value of the machine at the end of its useful life is called its* scrap value. *By one method of depreciation (where it is assumed a constant percentage of the value depreciates annually), the scrap value, S, is given by*

$$S = C(1 - r)^n,$$

where C is the original cost, n is the useful life in years, and r is the constant percent of depreciation.

48. Find the scrap value of a machine costing \$30,000, having a useful life of 12 yr and a constant annual rate of depreciation of 15%.

49. A machine has a "half-life" of 6 yr. Find the constant annual rate of depreciation.

Use the change-of-base rule (with either common or natural logarithms) to find each logarithm. Give approximations to four decimal places.

50. $\log_{16} 13$

51. $\log_4 12$

52. $\log_{\sqrt{6}} \sqrt{13}$

Mixed Review Exercises

Solve.

53. $\log_3(x + 9) = 4$

54. $\log_2 32 = x$

55. $\log_x \frac{1}{81} = 2$

56. $27^x = 81$

57. $2^{2x-3} = 8$

58. $\log_3(x + 1) - \log_3 x = 2$

59. $\log(3x - 1) = \log 10$

60. Find the value of n in the equation for Exercise 48 if the scrap value is \$10,000, the cost is \$30,000, and the depreciation rate is 15%.

Chapter 10 Test

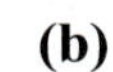

Study Skills Workbook
Activity 12

1. Decide whether each function is one-to-one.

(a) $f(x) = x^2 + 9$ **(b)**

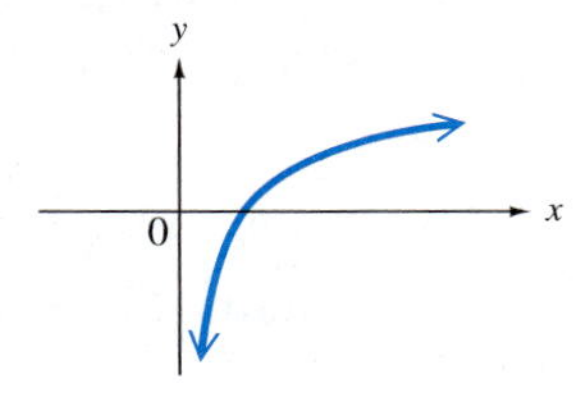

1. (a) ____________

(b) ____________

2. Find $f^{-1}(x)$ for the one-to-one function defined by $f(x) = \sqrt[3]{x + 7}$.

2. ____________

3. Graph the inverse of f, given the graph of f here.

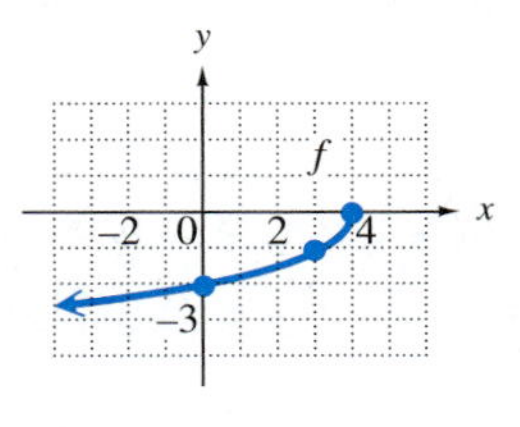

3.

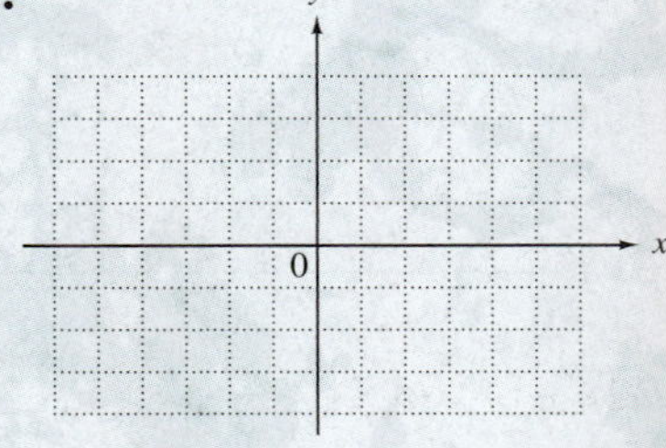

Graph each function.

4. $y = 6^x$

4.

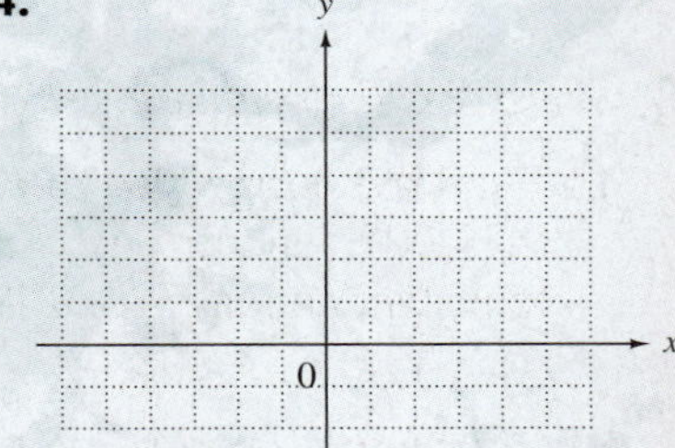

5. $y = \log_6 x$

5.

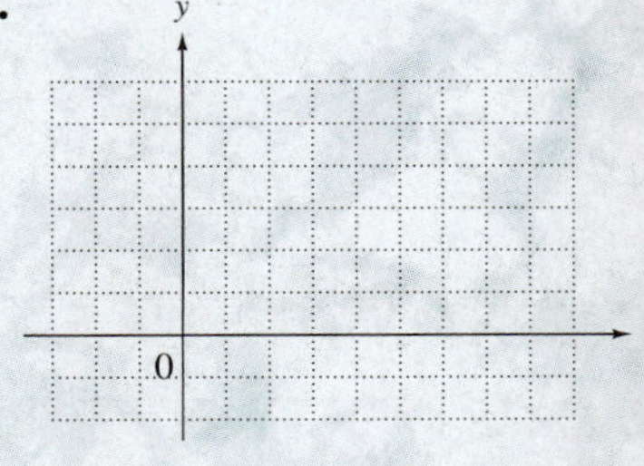

6. Explain how the graph of the function in Exercise 5 can be obtained from the graph of the function in Exercise 4.

6. ____________

Solve each equation. Give the exact solution.

7. $5^x = \dfrac{1}{625}$

8. $2^{3x-7} = 8^{2x+2}$

7. ____________

8. ____________

9. ______________

9. A recent report predicts that the U.S. Hispanic population will increase from 26.7 million in 1995 to 96.5 million in 2050. (*Source:* U.S. Bureau of the Census.) Assuming an exponential growth pattern, the population is approximated by

$$f(t) = 26.7e^{.023t},$$

where t represents the number of years since 1995. Use this function to estimate the population in 2000 and 2010.

10. ______________

10. Write in logarithmic form: $4^{-2} = .0625$.

11. ______________

11. Write in exponential form: $\log_7 49 = 2$.

Solve each equation.

12. ______________

13. ______________

14. ______________

12. $\log_{1/2} x = -5$

13. $x = \log_9 3$

14. $\log_x 16 = 4$

15. ______________

15. Use properties of logarithms to write $\log_3 x^2y$ as a sum or difference of logarithms. Assume the variables represent positive real numbers.

16. ______________

16. Use properties of logarithms to write $\frac{1}{4}\log_b r + 2\log_b s - \frac{2}{3}\log_b t$ as a single logarithm. Assume the variables represent positive real numbers, $b \neq 1$.

17. (a) ______________

(b) ______________

(c) ______________

17. Use a calculator to find an approximation to four decimal places for each logarithm.

(a) $\log 21.3$ **(b)** $\ln .43$ **(c)** $\log_6 45$

18. ______________

18. Solve $3^x = 78$, giving the solution to four decimal places.

19. ______________

19. Solve $\log_8(x + 5) + \log_8(x - 2) = \log_8 8$.

20. (a) ______________

(b) ______________

20. Suppose that \$10,000 is invested at 4.5% annual interest, compounded quarterly.

(a) How much will be in the account in 5 yr if no money is withdrawn?

(b) How long will it take for the initial principal to double?

Cumulative Review Exercises CHAPTERS 1–10

Let $S = \{-\frac{9}{4}, -2, -\sqrt{2}, 0, .6, \sqrt{11}, \sqrt{-8}, 6, \frac{30}{3}\}$. *List the elements of S that are elements of each set.*

1. Integers

2. Rational numbers

3. Irrational numbers

Simplify each expression.

4. $|-8| + 6 - |-2| - (-6 + 2)$

5. $2(-5) + (-8)(4) - (-3)$

Solve each equation or inequality.

6. $7 - (3 + 4a) + 2a = -5(a - 1) - 3$

7. $2m + 2 \leq 5m - 1$

8. $|2x - 5| = 9$

9. $|3p| - 4 = 12$

10. $|3k - 8| \leq 1$

11. $|4m + 2| > 10$

Graph each equation or inequality.

12. $y = -2.5x + 5$

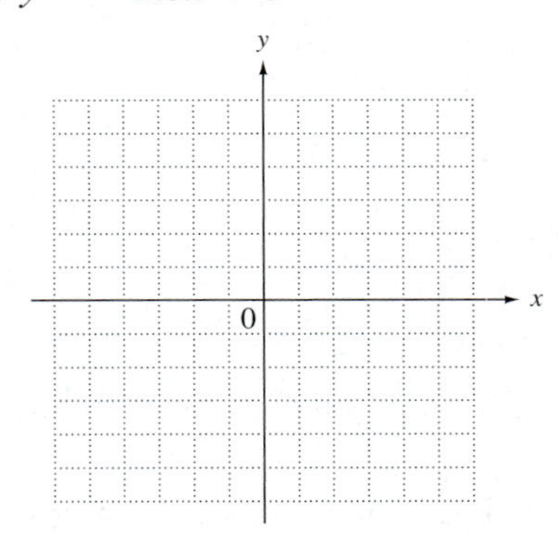

13. $-4x + y \leq 5$

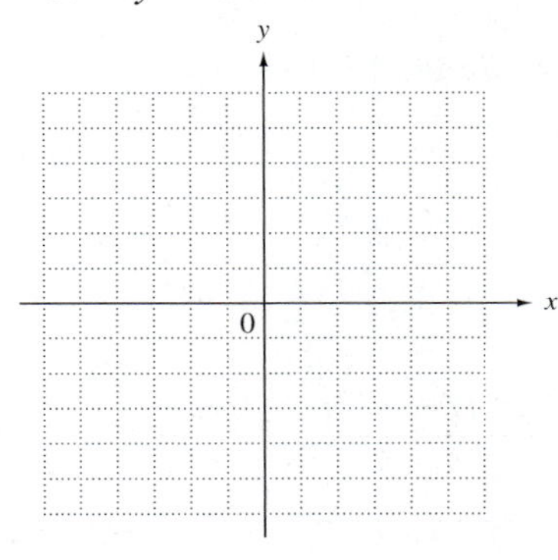

14. The graph indicates that timber harvests by Sierra Pacific Industries dropped from 17,716 acres in 1997 to 9733 acres in 1999.

(a) Is this the graph of a function?

(b) What is the slope of the line in the graph? Interpret the slope in the context of the timber harvests.

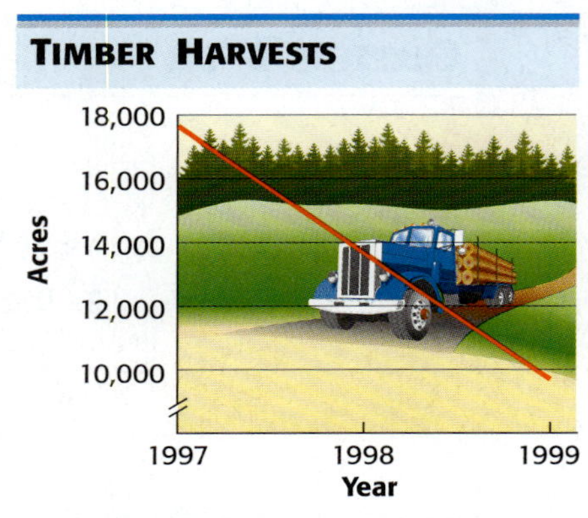

Source: Department of Forestry and Fire Protection.

15. Find the slope-intercept form of the equation of the line through $(5, -1)$ and parallel to the line with equation $3x - 4y = 12$.

Solve each system of equations.

16. $5x - 3y = 14$
$2x + 5y = 18$

17. $x + 2y + 3z = 11$
$3x - y + z = 8$
$2x + 2y - 3z = -12$

18. Candy worth \$1.00 per lb is to be mixed with candy worth \$1.96 per lb to get 16 lb of a mixture that will be sold for \$1.60 per lb. How many pounds of each candy should be used?

Price per Pound	Number of Pounds	Value
\$1.00	x	$1x$
	y	
\$1.60		

Perform the indicated operations.

19. $(2p + 3)(3p - 1)$

20. $(4k - 3)^2$

21. $(3m^3 + 2m^2 - 5m) - (8m^3 + 2m - 4)$

22. Divide $6t^4 + 17t^3 - 4t^2 + 9t + 4$ by $3t + 1$.

Factor completely.

23. $8x + x^3$

24. $24y^2 - 7y - 6$

25. $5z^3 - 19z^2 - 4z$

26. $16a^2 - 25b^4$

27. $8c^3 + d^3$

28. $16r^2 + 56rq + 49q^2$

Perform the indicated operations.

29. $\dfrac{(5p^3)^4(-3p^7)}{2p^2(4p^4)}$

30. $\dfrac{x^2 - 9}{x^2 + 7x + 12} \div \dfrac{x - 3}{x + 5}$

31. $\dfrac{2}{k + 3} - \dfrac{5}{k - 2}$

32. $\dfrac{3}{p^2 - 4p} - \dfrac{4}{p^2 + 2p}$

33. Solve $\dfrac{1}{x} - \dfrac{3}{2x} = \dfrac{1}{x + 1}$.

Simplify.

34. $\sqrt{288}$

35. $\dfrac{-8^{4/3}}{8^2}$

36. $2\sqrt{32} - 5\sqrt{98}$

37. Solve $\sqrt{2x + 1} - \sqrt{x} = 1$.

38. Multiply $(5 + 4i)(5 - 4i)$.

Solve each equation or inequality.

39. $3x^2 = x + 1$

40. $k^2 + 2k - 8 > 0$

41. $x^4 - 5x^2 + 4 = 0$

Graph.

42. $y = \frac{1}{3}(x - 1)^2 + 2$

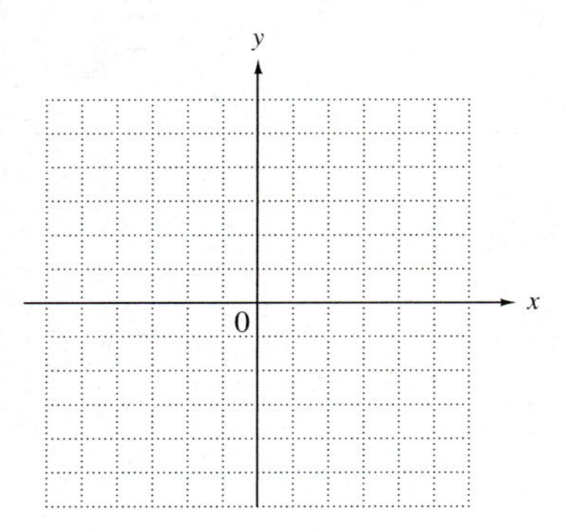

43. $f(x) = 2^x$

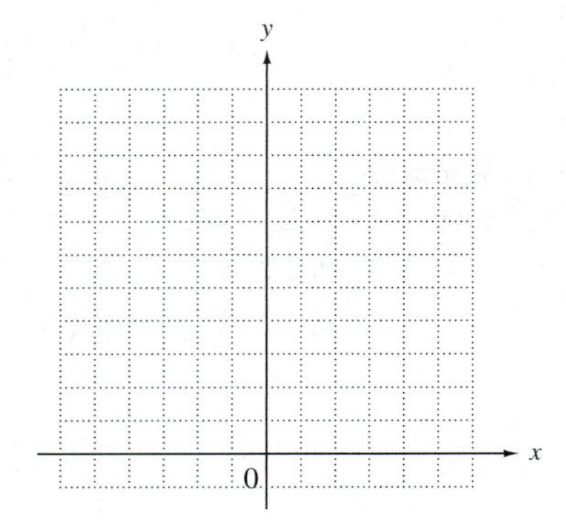

44. $f(x) = \log_3 x$

Solve.

45. $5^{x+3} = \left(\frac{1}{25}\right)^{3x+2}$

46. $\log_5 x + \log_5(x + 4) = 1$

47. Write $\log_5 125 = 3$ in exponential form.

48. Rewrite the following using the product, quotient, and power rules for logarithms:

$$\log \frac{x^3\sqrt{y}}{z}.$$

49. We used the formula for continuous compounding

$$A = Pe^{rt}$$

in Section 10.6. To three decimal places, what growth rate r will triple the value of P in 10 yr?

50. Let the number of bacteria present in a certain culture be given by

$$B(t) = 25{,}000e^{.2t},$$

where t is time measured in hours, and $t = 0$ corresponds to noon. Find, to the nearest hundred, the number of bacteria present at:

(a) noon **(b)** 2 P.M. **(c)** 5 P.M.

Nonlinear Functions, Conic Sections, and Nonlinear Systems

When a plane intersects an infinite cone at different angles, it produces curves called **conic sections.** In Chapter 9 we studied one example of conic sections, the *parabola.* In 1609, Johann Kepler (1571–1630) established the importance of another conic section, the *ellipse,* when he discovered that the orbits of the planets around the sun are elliptical, not circular. Exercises 37 and 38 of Section 11.2 involve the equations of the elliptical orbits formed by the planets Mars and Venus.

11.1 Additional Graphs of Functions; Composition

Objectives

1. Recognize the graphs of the elementary functions defined by $|x|$, $\frac{1}{x}$, and $\sqrt{x}$, and graph their translations.
2. Find the composition of functions.

In earlier chapters we introduced the function defined by $f(x) = x^2$, sometimes called the **squaring function.** This is one of the most important elementary functions in algebra.

1 Recognize the graphs of the elementary functions defined by $|x|$, $\frac{1}{x}$, and $\sqrt{x}$, and graph their translations. Another one of the elementary functions, defined by $f(x) = |x|$, is called the **absolute value function.** Its graph, along with a table of selected ordered pairs, is shown in Figure 1. Its domain is $(-\infty, \infty)$, and its range is $[0, \infty)$.

x	y
0	0
± 1	1
± 2	2
± 3	3

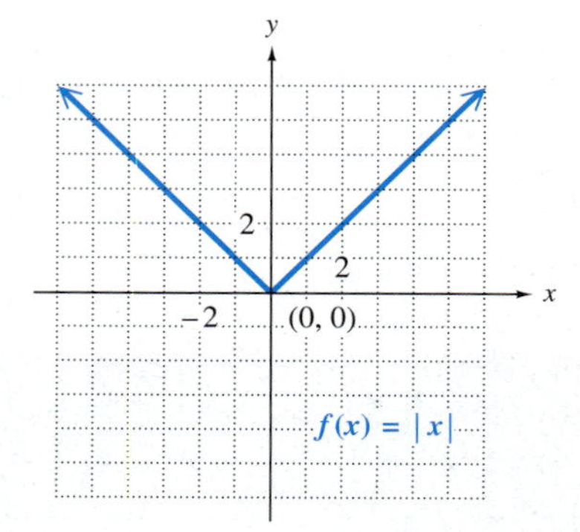

Figure 1

The **reciprocal function,** defined by $f(x) = \frac{1}{x}$, was introduced in Chapter 7. Its graph is shown in Figure 2, along with a table of selected ordered pairs. Notice that x can never equal 0 for this function, and as a result, as x gets closer and closer to 0, the graph approaches either ∞ or $-\infty$. Also, $\frac{1}{x}$ can never equal 0, and as x approaches ∞ or $-\infty$, $\frac{1}{x}$ approaches 0. The axes are called **asymptotes** for the function. (Asymptotes are studied in more detail in college algebra courses.) For the reciprocal function, the domain and the range are both $(-\infty, 0) \cup (0, \infty)$.

x	y
$\frac{1}{3}$	3
$\frac{1}{2}$	2
1	1
2	$\frac{1}{2}$
3	$\frac{1}{3}$

x	y
$-\frac{1}{3}$	-3
$-\frac{1}{2}$	-2
-1	-1
-2	$-\frac{1}{2}$
-3	$-\frac{1}{3}$

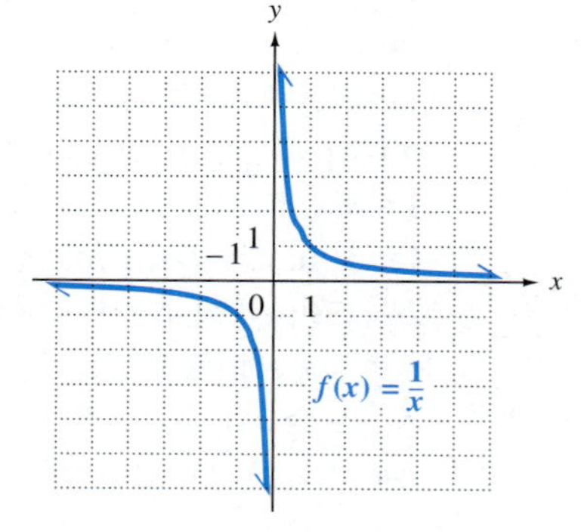

Figure 2

The **square root function,** defined by $f(x) = \sqrt{x}$, was introduced in Chapter 8. Its graph is shown in Figure 3. Notice that since we restrict function values to be real numbers, x cannot take on negative values. Thus, the domain of the square root function is $[0, \infty)$. Because the principal square root is always nonnegative, the range is also $[0, \infty)$. A table of values is shown along with the graph.

x	y
0	0
1	1
4	2

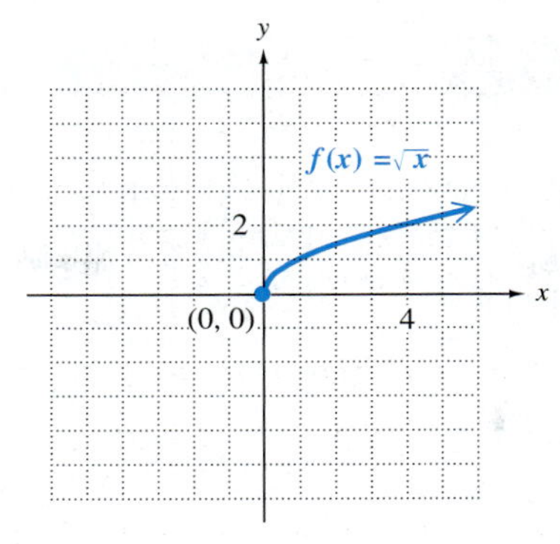

Figure 3

Just as the graph of $f(x) = x^2$ can be shifted, or translated, as we saw in Section 9.5, so can the graphs of these other elementary functions.

Example 1 Applying a Horizontal Shift

Graph $f(x) = |x - 2|$.

The graph of $y = (x - 2)^2$ is obtained by shifting the graph of $y = x^2$ two units to the right. In a similar manner, the graph of $f(x) = |x - 2|$ is found by shifting the graph of $y = |x|$ two units to the right, as shown in Figure 4. The table of ordered pairs accompanying the graph supports this, as you can see by comparing it to the table with Figure 1. The domain of this function is $(-\infty, \infty)$, and its range is $[0, \infty)$.

x	y
0	2
1	1
2	0
3	1
4	2

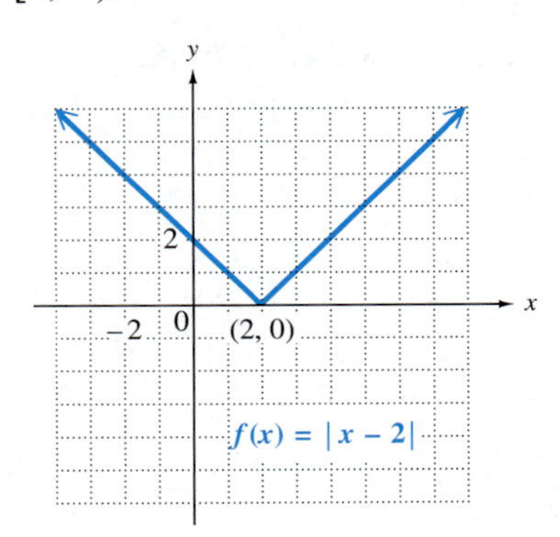

Figure 4

Work Problem 1 at the Side.

Example 2 Applying a Vertical Shift

Graph $f(x) = \dfrac{1}{x} + 3$.

The graph of this function is found by shifting the graph of $y = \frac{1}{x}$ three units up. See Figure 5 on the next page. The domain is $(-\infty, 0) \cup (0, \infty)$, and the range is $(-\infty, 3) \cup (3, \infty)$.

Continued on Next Page

1 Graph $f(x) = \sqrt{x + 4}$. Give the domain and range.

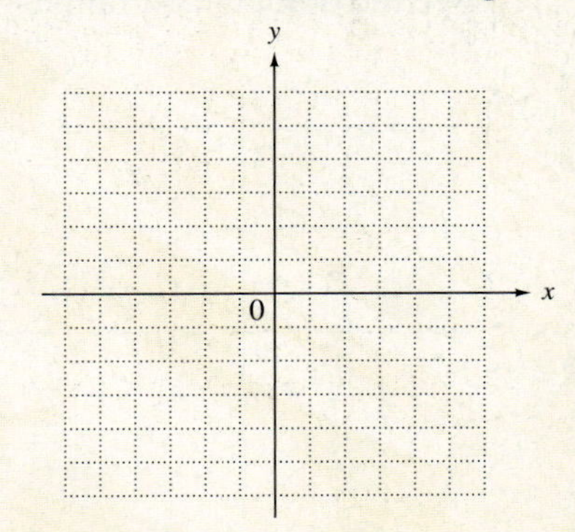

Answers

1. $[-4, \infty)$; $[0, \infty)$

2 Graph $f(x) = \frac{1}{x} - 2$.
Give the domain and range.

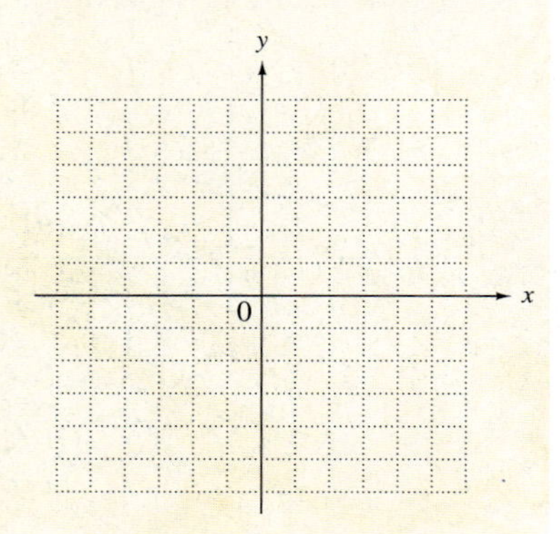

x	y
$\frac{1}{3}$	6
$\frac{1}{2}$	5
1	4
2	3.5

x	y
$-\frac{1}{3}$	0
$-\frac{1}{2}$	1
-1	2
-2	2.5

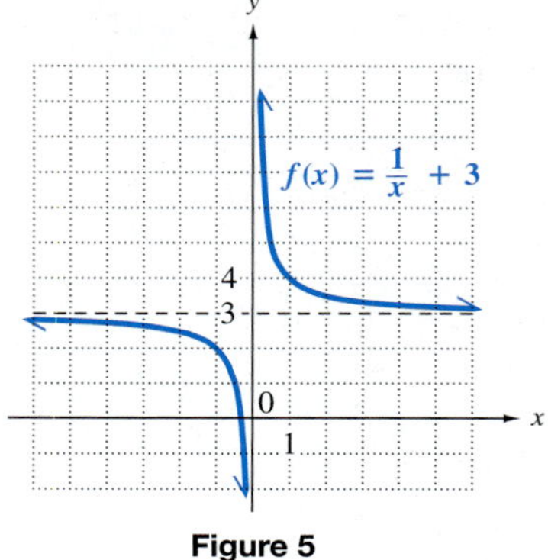

Figure 5

Work Problem **2** at the Side.

3 Graph $f(x) = |x + 2| + 1$.
Give the domain and range.

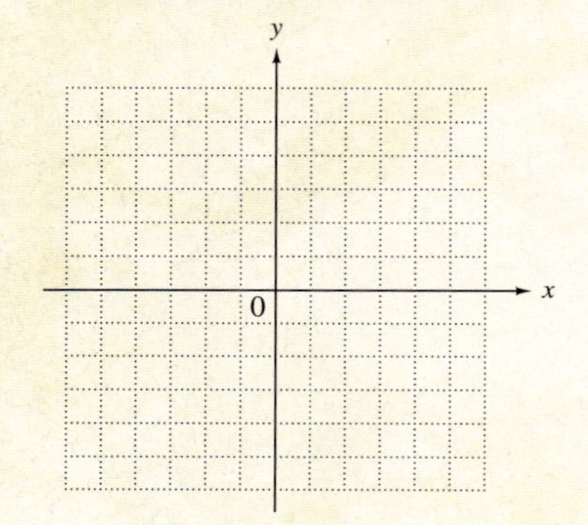

Example 3 Applying Both Horizontal and Vertical Shifts

Graph $f(x) = \sqrt{x + 1} - 4$.

The graph of $y = (x + 1)^2 - 4$ is obtained by shifting the graph of $y = x^2$ one unit to the left and four units down. Following this pattern here, we shift the graph of $y = \sqrt{x}$ one unit to the left and four units down to get the graph of $f(x) = \sqrt{x + 1} - 4$. See Figure 6. The domain is $[-1, \infty)$, and the range is $[-4, \infty)$.

x	y
-1	-4
0	-3
3	-2

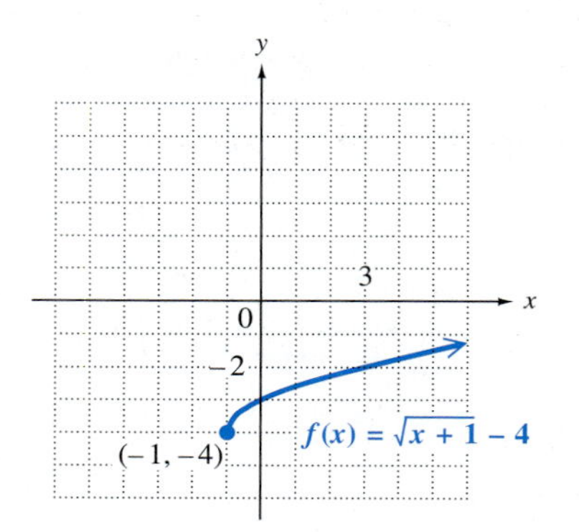

Figure 6

Work Problem **3** at the Side.

2 **Find the composition of functions.** The diagram in Figure 7 shows a function f that assigns to each element x of set X some element y of set Y. Suppose that a function g takes each element of set Y and assigns a value z of set Z. Using both f and g, then, an element x in X is assigned to an element z in Z. The result of this process is a new function h, which takes an element x in X and assigns an element z in Z.

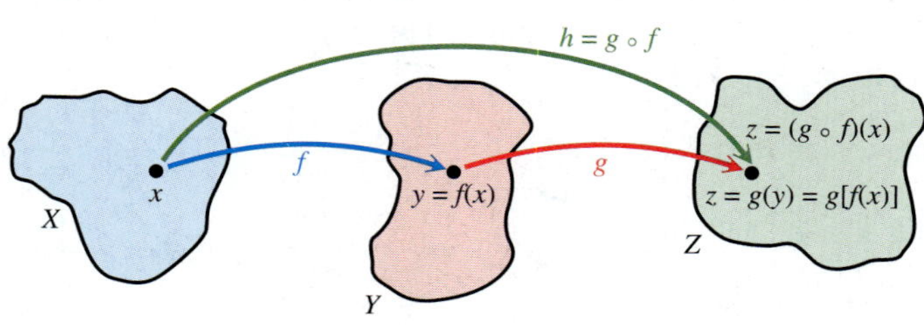

Figure 7

ANSWERS

2.

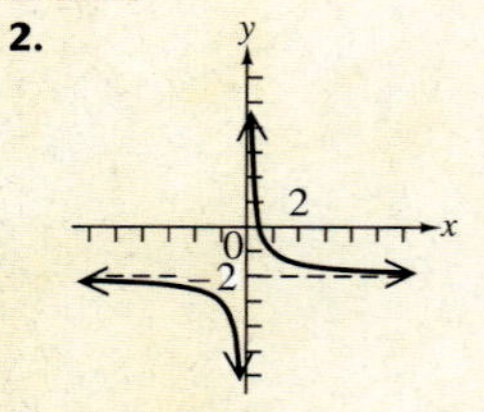

$(-\infty, 0) \cup (0, \infty)$; $(-\infty, -2) \cup (-2, \infty)$

3.

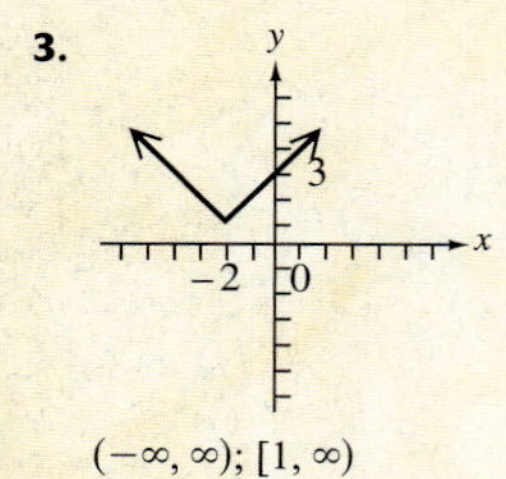

$(-\infty, \infty)$; $[1, \infty)$

This function h is called the *composition* of functions g and f, written $g \circ f$, and is defined as follows.

Composition of Functions

If f and g are functions, then the **composite function,** or **composition,** of g and f is defined by

$$(g \circ f)(x) = g[f(x)]$$

for all x in the domain of f such that $f(x)$ is in the domain of g.

Read $g \circ f$ as "g of f."

As a real-life example of how composite functions occur, suppose an oil well off the California coast is leaking, with the leak spreading oil in a circular layer over the surface. See Figure 8.

Figure 8

At any time t, in minutes, after the beginning of the leak, the radius of the circular oil slick is given by $r(t) = 5t$ ft. Since $A(r) = \pi r^2$ gives the area of a circle of radius r, the area can be expressed as a function of time by substituting $5t$ for r in $A(r) = \pi r^2$ to get

$$A(r) = \pi r^2$$
$$A[r(t)] = \pi(5t)^2 = 25\pi t^2.$$

The function $A[r(t)]$ is a composite function of the functions A and r.

Example 4 Finding a Composite Function

Let $f(x) = x^2$ and $g(x) = x + 3$. Find $(f \circ g)(4)$.

$$\begin{aligned}(f \circ g)(4) &= f[g(4)] && \text{Definition}\\ &= f(4 + 3) && \text{Use the rule for } g(x);\ g(4) = 4 + 3.\\ &= f(7) && \text{Add.}\\ &= 7^2 && \text{Use the rule for } f(x);\ f(7) = 7^2.\\ &= 49\end{aligned}$$

Notice in Example 4 that if we reverse the order of the functions, the composition of g and f is defined by $g[f(x)]$. Once again, letting $x = 4$, we have

$$\begin{aligned}(g \circ f)(4) &= g[f(4)] && \text{Definition}\\ &= g(4^2) && \text{Use the rule for } f(x);\ f(4) = 4^2.\\ &= g(16) && \text{Square 4.}\\ &= 16 + 3 && \text{Use the rule for } g(x);\ g(16) = 16 + 3.\\ &= 19.\end{aligned}$$

Here we see that $(f \circ g)(4) \neq (g \circ f)(4)$ because $49 \neq 19$. In general,

$$(f \circ g)(x) \neq (g \circ f)(x).$$

4 Let $f(x) = 3x + 6$ and $g(x) = x^3$. Find each of the following.

(a) $(f \circ g)(2)$

(b) $(g \circ f)(2)$

(c) $(f \circ g)(x)$

(d) $(g \circ f)(x)$

ANSWERS

4. (a) 30 **(b)** 1728 **(c)** $3x^3 + 6$ **(d)** $(3x + 6)^3$

Example 5 Finding Composite Functions

Let $f(x) = 4x - 1$ and $g(x) = x^2 + 5$. Find each of the following.

(a) $(f \circ g)(2)$

$$\begin{aligned}(f \circ g)(2) &= f[g(2)] \\ &= f(2^2 + 5) \\ &= f(9) \\ &= 4(9) - 1 \\ &= 35\end{aligned}$$

(b) $(f \circ g)(x)$

Here, use $g(x)$ as the input for the function f.

$$\begin{aligned}(f \circ g)(x) &= f[g(x)] \\ &= 4(g(x)) - 1 && \text{Use the rule for } f(x);\ f(x) = 4x - 1. \\ &= 4(x^2 + 5) - 1 && g(x) = x^2 + 5 \\ &= 4x^2 + 20 - 1 && \text{Distributive property} \\ &= 4x^2 + 19 && \text{Combine terms.}\end{aligned}$$

(c) Find $(f \circ g)(2)$ again, this time using the rule obtained in part (b).

$$\begin{aligned}(f \circ g)(x) &= 4x^2 + 19 && \text{From part (b)} \\ (f \circ g)(2) &= 4(2)^2 + 19 \\ &= 4(4) + 19 \\ &= 16 + 19 \\ &= 35\end{aligned}$$

The result, 35, is the same as the result in part (a).

Work Problem 4 at the Side.

11.1 EXERCISES

FOR EXTRA HELP

Student's Solutions Manual | MyMathLab.com | InterAct Math Tutorial Software | AW Math Tutor Center | 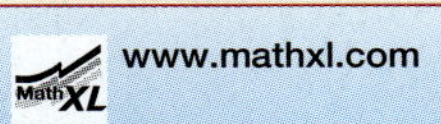 www.mathxl.com |  Digital Video Tutor CD 9 Videotape 19

Fill in each blank with the correct response.

1. For the reciprocal function defined by $f(x) = \frac{1}{x}$, ______ is the only real number not in the domain.

2. The range of the square root function, given by $f(x) = \sqrt{x}$, is ______.

3. The lowest point on the graph of $f(x) = |x|$ has coordinates (______, ______).

4. The range of $f(x) = x^2 + 4$, a translation of the squaring function, is ______.

Without actually plotting points, match each function defined by the absolute value expression with its graph. See Example 1.

5. $f(x) = |x - 2| + 2$

6. $f(x) = |x + 2| + 2$

7. $f(x) = |x - 2| - 2$

8. $f(x) = |x + 2| - 2$

A.

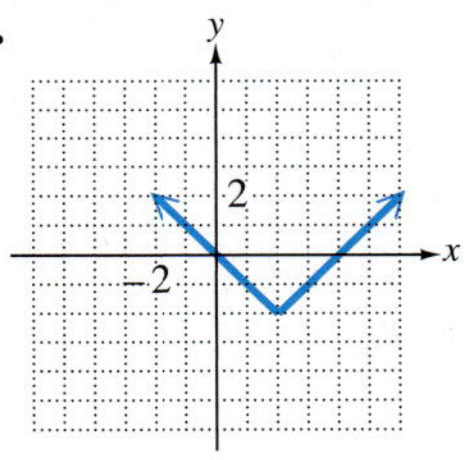

B.

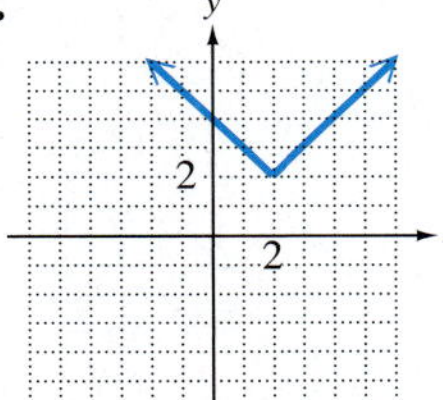

C.

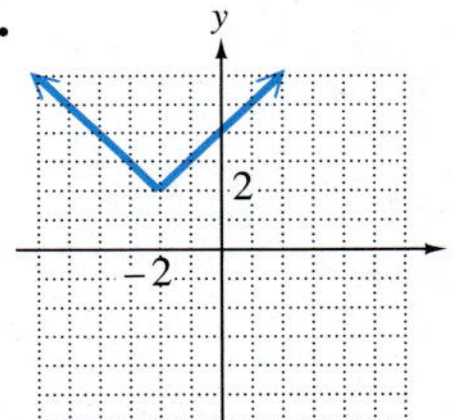

D.

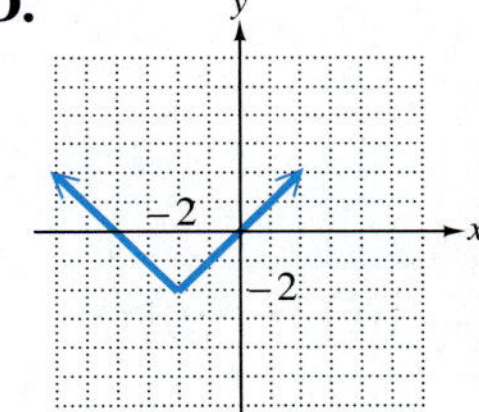

Graph each function. Give the domain and range. See Examples 1–3.

9. $f(x) = |x + 1|$

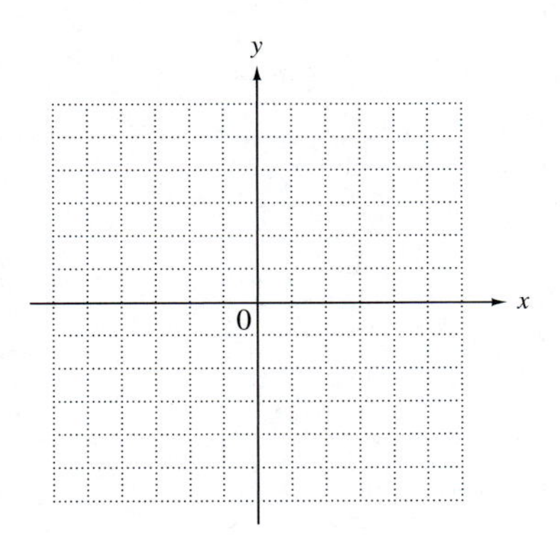

10. $f(x) = |x - 1|$

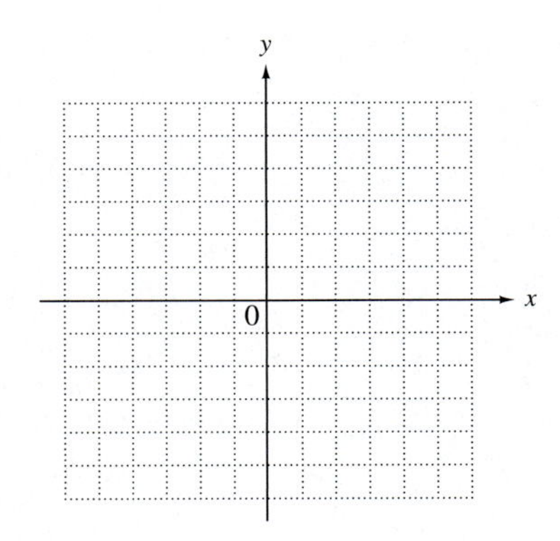

11. $f(x) = \frac{1}{x} + 1$

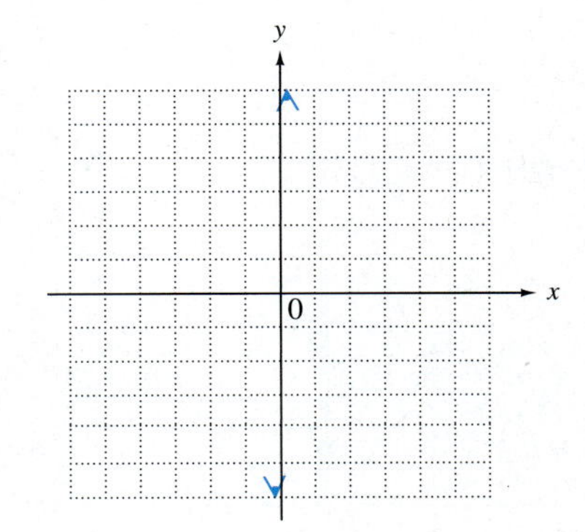

12. $f(x) = \frac{1}{x} - 1$

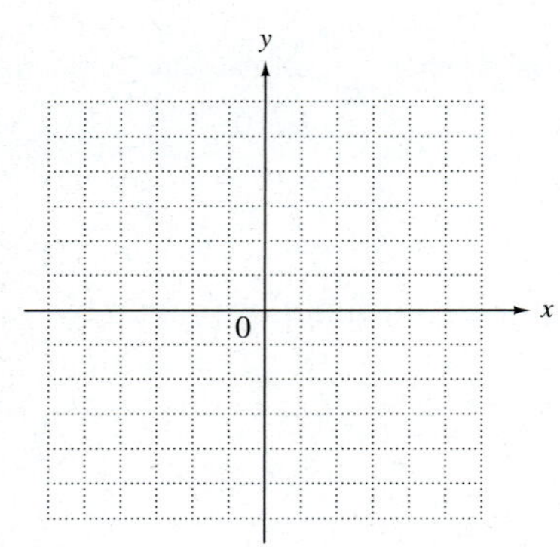

13. $f(x) = \sqrt{x - 2}$

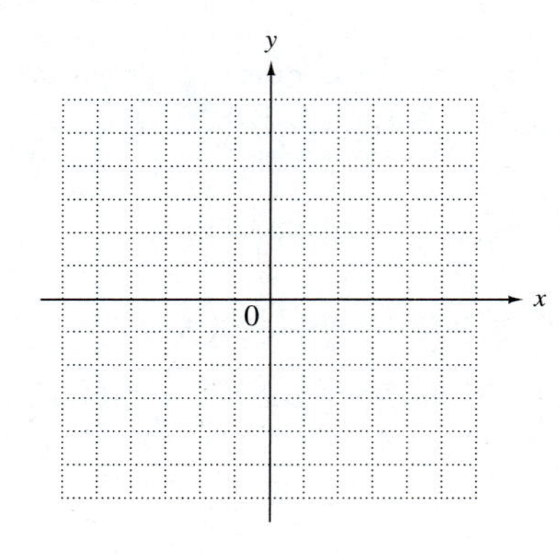

14. $f(x) = \sqrt{x + 5}$

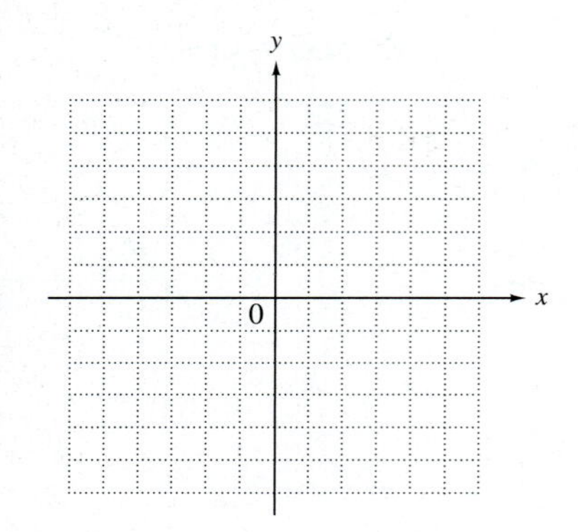

15. $f(x) = \frac{1}{x - 2}$

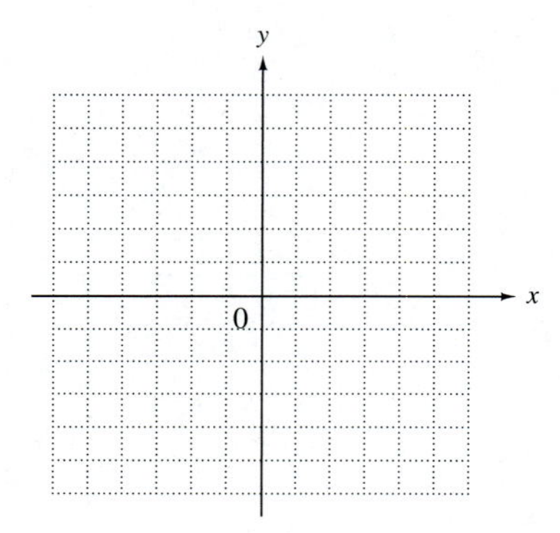

16. $f(x) = \frac{1}{x + 2}$

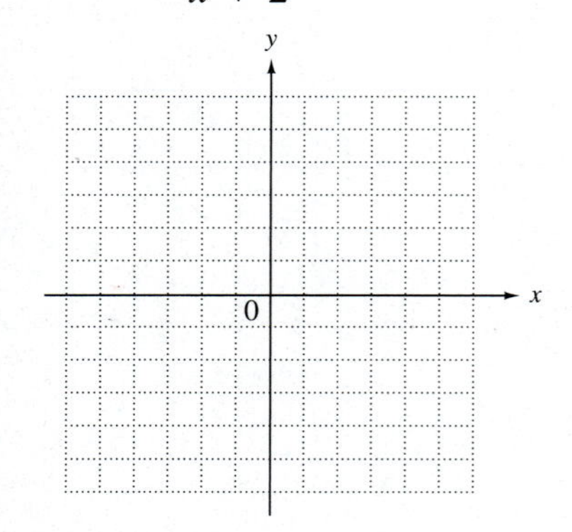

17. $f(x) = \sqrt{x + 3} - 3$

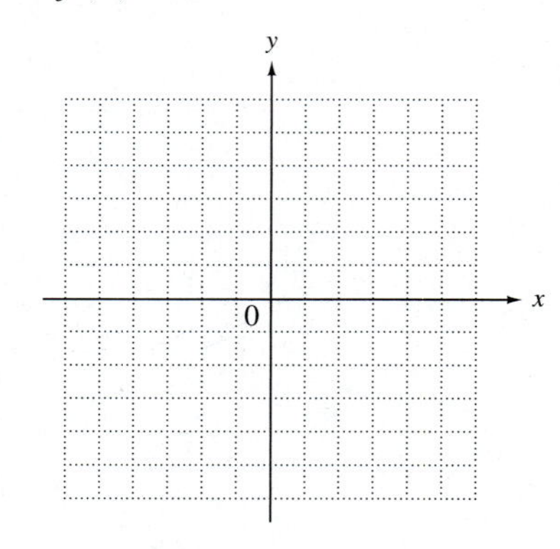

18. Explain how the graph of $f(x) = \frac{1}{x-3} + 2$ is obtained from the graph of $g(x) = \frac{1}{x}$.

Let $f(x) = x^2 + 4$, $g(x) = 2x + 3$, *and* $h(x) = x + 5$.
Find each value or expression. See Examples 4 and 5.

19. $(h \circ g)(4)$

20. $(f \circ g)(4)$

21. $(g \circ f)(6)$

22. $(h \circ f)(6)$

23. $(f \circ h)(-2)$

24. $(h \circ g)(-2)$

25. $(f \circ g)(x)$

26. $(g \circ h)(x)$

27. $(f \circ h)(x)$

28. $(g \circ f)(x)$

29. $(h \circ g)(x)$

30. $(h \circ f)(x)$

Solve each problem.

31. The function defined by $f(x) = 12x$ computes the number of inches in x ft and the function defined by $g(x) = 5280x$ computes the number of feet in x mi. What is $(f \circ g)(x)$ and what does it compute?

32. The perimeter x of a square with sides of length s is given by the formula $x = 4s$.

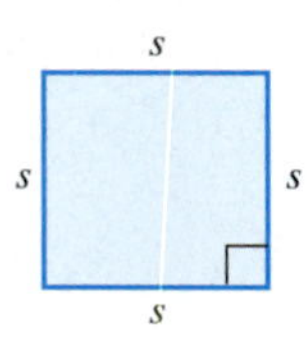

(a) Solve for s in terms of x.

(b) If y represents the area of this square, write y as a function of the perimeter x.

(c) Use the composite function of part (b) to find the area of a square with perimeter 6.

33. When a thermal inversion layer is over a city (as happens often in Los Angeles), pollutants cannot rise vertically but are trapped below the layer and must disperse horizontally. Assume that a factory smokestack begins emitting a pollutant at 8 A.M. Assume that the pollutant disperses horizontally over a circular area. Suppose that t represents the time, in hours, since the factory began emitting pollutants ($t = 0$ represents 8 A.M.), and assume that the radius of the circle of pollution is $r(t) = 2t$ mi. Let $A(r) = \pi r^2$ represent the area of a circle of radius r. Find and interpret $(A \circ r)(t)$.

34. An oil well off the Gulf Coast is leaking, with the leak spreading oil over the surface as a circle. At any time t, in minutes, after the beginning of the leak, the radius of the circular oil slick on the surface is $r(t) = 4t$ ft. Let $A(r) = \pi r^2$ represent the area of a circle of radius r. Find and interpret $(A \circ r)(t)$.

11.2 The Circle and the Ellipse

Objectives

1. Find the equation of a circle given the center and radius.
2. Determine the center and radius of a circle given its equation.
3. Recognize the equation of an ellipse.
4. Graph ellipses.

When an infinite cone is intersected by a plane, the resulting figure is called a *conic section.* The parabola is one example of a conic section; circles, ellipses, and hyperbolas may also result. See Figure 9.

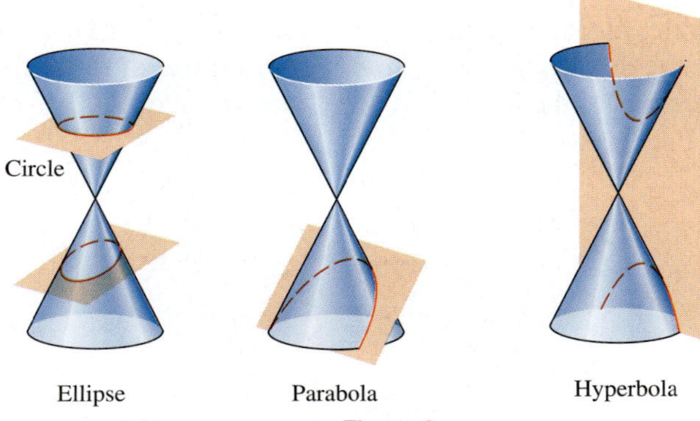

Figure 9

1 Find the equation of a circle given the center and radius. A **circle** is the set of all points in a plane that lie a fixed distance from a fixed point. The fixed point is called the **center,** and the fixed distance is called the **radius.** We use the distance formula to find an equation of a circle.

Example 1 Finding the Equation of a Circle and Graphing It

Find an equation of the circle with radius 3 and center at (0, 0), and graph it.

If the point (x, y) is on the circle, the distance from (x, y) to the center (0, 0) is 3. By the distance formula,

$$\sqrt{(x_2 - x_1)^2 + (y_2 - y_1)^2} = d$$

$$\sqrt{(x - 0)^2 + (y - 0)^2} = 3$$

$$x^2 + y^2 = 9. \quad \text{Square both sides.}$$

An equation of this circle is $x^2 + y^2 = 9$. The graph is shown in Figure 10.

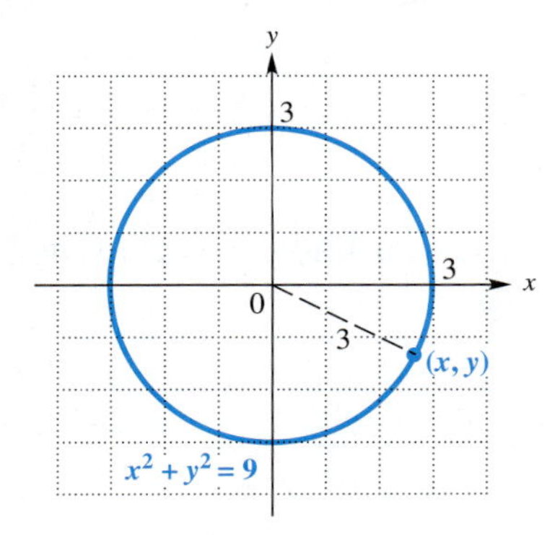

Figure 10

Work Problem 1 at the Side.

A circle may not be centered at the origin, as seen in the next example.

1 Find an equation of the circle with radius 4 and center (0, 0). Sketch its graph.

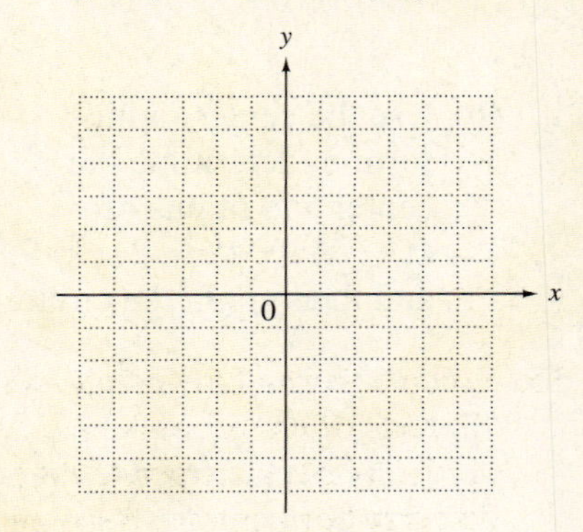

Answers

1. $x^2 + y^2 = 16$

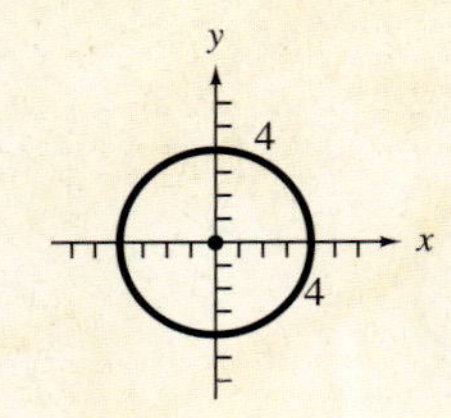

2 **(a)** Find an equation of the circle with center at $(3, -2)$ and radius 4. Graph the circle.

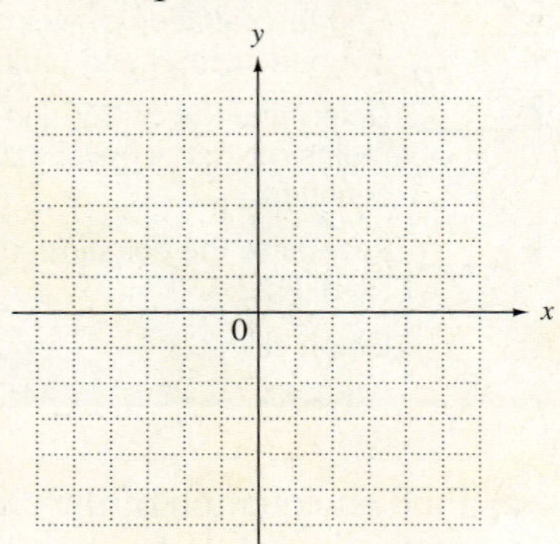

(b) Use the center-radius form to determine the center and radius of $(x - 5)^2 + (y + 2)^2 = 9$, and then graph the circle.

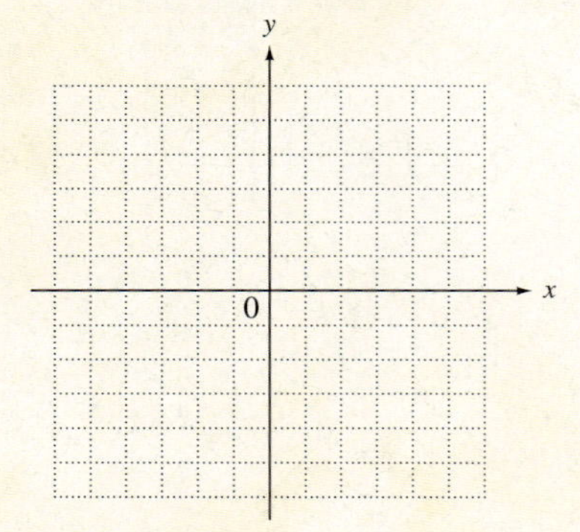

Answers

2. (a) $(x - 3)^2 + (y + 2)^2 = 16$

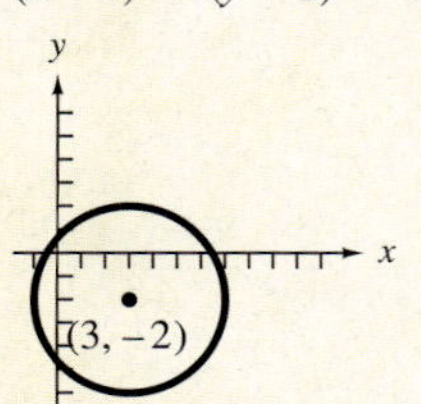

(b) center at $(5, -2)$; radius 3

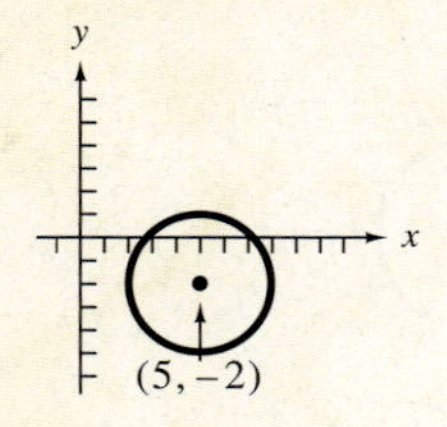

Example 2 Finding an Equation of a Circle and Graphing It

Find an equation of the circle with center at $(4, -3)$ and radius 5, and graph it.

Use the distance formula again.

$$\sqrt{(x - 4)^2 + [y - (-3)]^2} = 5$$

$$(x - 4)^2 + (y + 3)^2 = 25 \quad \text{Square both sides.}$$

To graph the circle, plot the center $(4, -3)$, then move 5 units right, left, up, and down from the center. Draw a smooth curve through these four points, sketching one quarter of the circle at a time. The graph of this circle is shown in Figure 11.

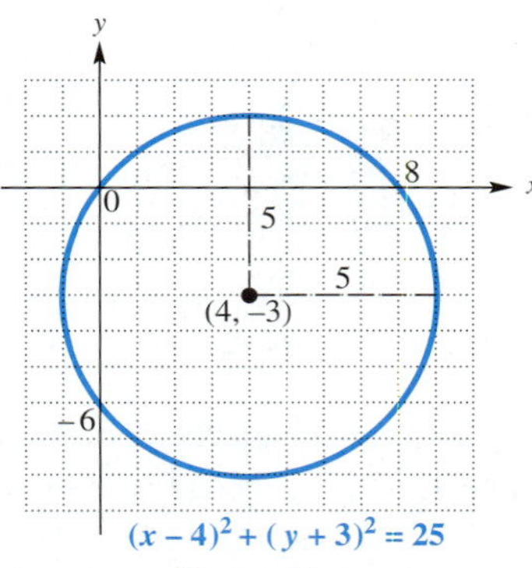

Figure 11

Examples 1 and 2 suggest the form of an equation of a circle with radius r and center at (h, k). If (x, y) is a point on the circle, the distance from the center (h, k) to the point (x, y) is r. Then by the distance formula,

$$\sqrt{(x - h)^2 + (y - k)^2} = r.$$

Squaring both sides gives us the following **center-radius form** of the equation of a circle.

Equation of a Circle (Center-Radius Form)

$$(x - h)^2 + (y - k)^2 = r^2$$

is an equation of the circle of radius r with center at (h, k).

Example 3 Using the Center-Radius Form of the Equation of a Circle

Find an equation of the circle with center at $(-1, 2)$ and radius 4.

Use the center-radius form, with $h = -1$, $k = 2$, and $r = 4$.

$$(x - h)^2 + (y - k)^2 = r^2$$

$$[x - (-1)]^2 + (y - 2)^2 = 4^2$$

$$(x + 1)^2 + (y - 2)^2 = 16$$

Work Problem 2 at the Side.

2 Determine the center and radius of a circle given its equation. In the equation found in Example 2, multiplying out $(x-4)^2$ and $(y+3)^2$ and then combining like terms gives

$$(x-4)^2 + (y+3)^2 = 25$$
$$x^2 - 8x + 16 + y^2 + 6y + 9 = 25$$
$$x^2 + y^2 - 8x + 6y = 0.$$

This general form suggests that an equation with both x^2- and y^2-terms with equal coefficients may represent a circle. The next example shows how to tell, by completing the square. This procedure was introduced in Chapter 9.

Example 4 **Completing the Square to Find the Center and Radius**

Graph $x^2 + y^2 + 2x + 6y - 15 = 0$.

Since the equation has x^2- and y^2-terms with equal coefficients, its graph might be that of a circle. To find the center and radius, complete the squares on x and y.

$$x^2 + y^2 + 2x + 6y = 15$$ Get the constant on the right.

$$(x^2 + 2x \qquad) + (y^2 + 6y \qquad) = 15$$ Rewrite in anticipation of completing the square.

$$\left[\frac{1}{2}(2)\right]^2 = 1 \qquad \left[\frac{1}{2}(6)\right]^2 = 9$$ Square half the coefficient of each middle term.

$$(x^2 + 2x + 1) + (y^2 + 6y + 9) = 15 + 1 + 9$$ Complete the squares on both x and y.

$$(x+1)^2 + (y+3)^2 = 25$$ Factor on the left; add on the right.

$$[x - (-1)]^2 + [y - (-3)]^2 = 5^2$$ Center-radius form

The last equation shows that the graph is a circle with center at $(-1, -3)$ and radius 5. The graph is shown in Figure 12.

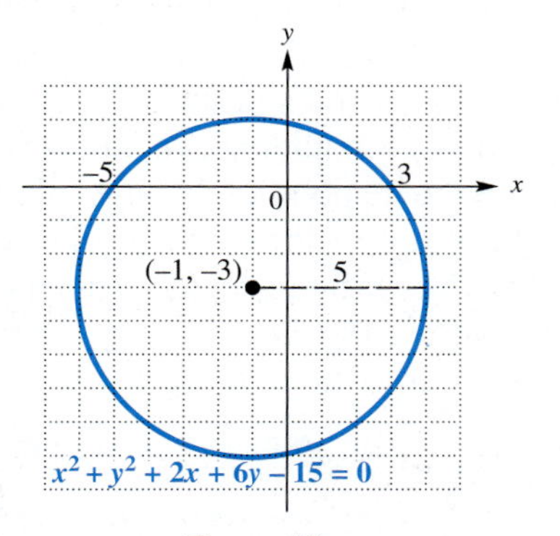

Figure 12

NOTE

If the procedure of Example 4 leads to an equation of the form $(x-h)^2 + (y-k)^2 = 0$, the graph is the single point (h, k). If the constant on the right side is negative, the equation has no graph.

Work Problem 3 at the Side.

3 Find the center and radius of the circle with equation

$$x^2 + y^2 - 6x + 8y - 4 = 0.$$

ANSWERS

3. center at $(3, -4)$; radius $\sqrt{29}$

3 Recognize the equation of an ellipse. An **ellipse** is the set of all points in a plane the *sum* of whose distances from two fixed points is constant. These fixed points are called **foci** (singular: *focus*). Figure 13 shows an ellipse whose foci are $(c, 0)$ and $(-c, 0)$, with x-intercepts $(a, 0)$ and $(-a, 0)$ and y-intercepts $(0, b)$ and $(0, -b)$. It can be shown that $c^2 = a^2 - b^2$ for an ellipse of this type. The origin is the **center** of the ellipse.

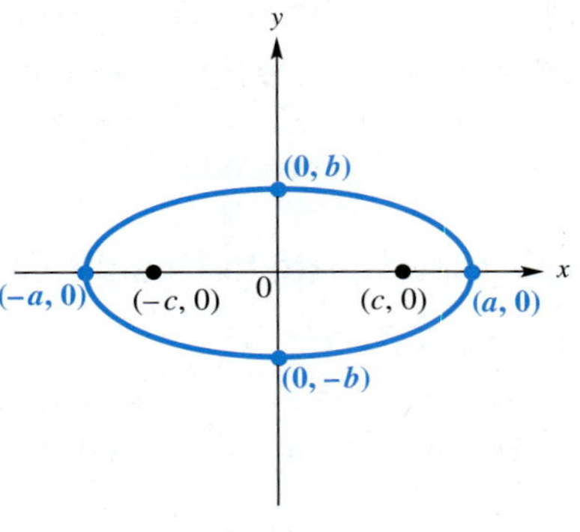

Figure 13

From the preceding definition, it can be shown by the distance formula that an ellipse has the following equation.

Equation of an Ellipse

The ellipse whose x-intercepts are $(a, 0)$ and $(-a, 0)$ and whose y-intercepts are $(0, b)$ and $(0, -b)$ has an equation of the form

$$\frac{x^2}{a^2} + \frac{y^2}{b^2} = 1.$$

A circle is a special case of an ellipse, where $a^2 = b^2$.

The paths of Earth and other planets around the sun are approximately ellipses; the sun is at one focus and a point in space is at the other. The orbits of communication satellites and other space vehicles are elliptical. Elliptical bicycle gears are designed to respond to the legs' natural strengths and weaknesses. At the top and bottom of the powerstroke, where the legs have the least leverage, the gear offers little resistance, but as the gear rotates, the resistance increases. This allows the legs to apply more power where it is most naturally available. See Figure 14.

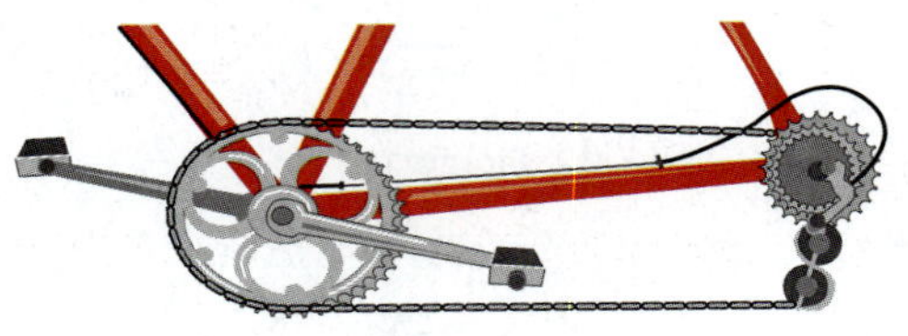

Figure 14

4 Graph ellipses. To graph an ellipse centered at the origin, we plot the four intercepts and then sketch the ellipse through those points.

Example 5 Graphing Ellipses

Graph each ellipse.

(a) $\dfrac{x^2}{49} + \dfrac{y^2}{36} = 1$

Here, $a^2 = 49$, so $a = \pm 7$, and the x-intercepts for this ellipse are $(7, 0)$ and $(-7, 0)$. Similarly, $b^2 = 36$, so $b = \pm 6$, and the y-intercepts are $(0, 6)$ and $(0, -6)$. Plotting the intercepts and sketching the ellipse through them gives the graph in Figure 15.

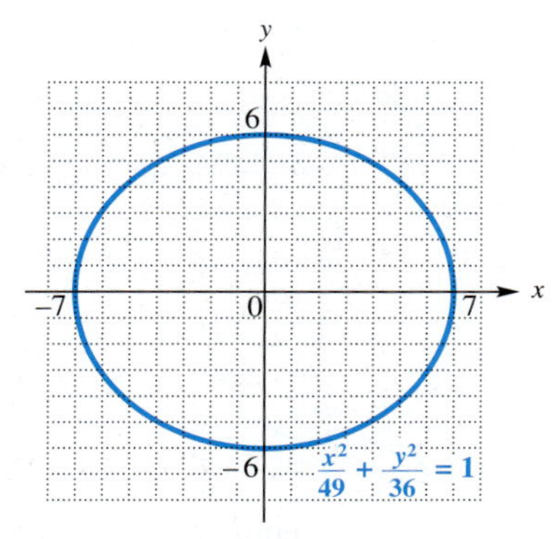

Figure 15

(b) $\dfrac{x^2}{36} + \dfrac{y^2}{121} = 1$

The x-intercepts for this ellipse are $(6, 0)$ and $(-6, 0)$, and the y-intercepts are $(0, 11)$ and $(0, -11)$. Join these with the smooth curve of an ellipse. The graph has been sketched in Figure 16.

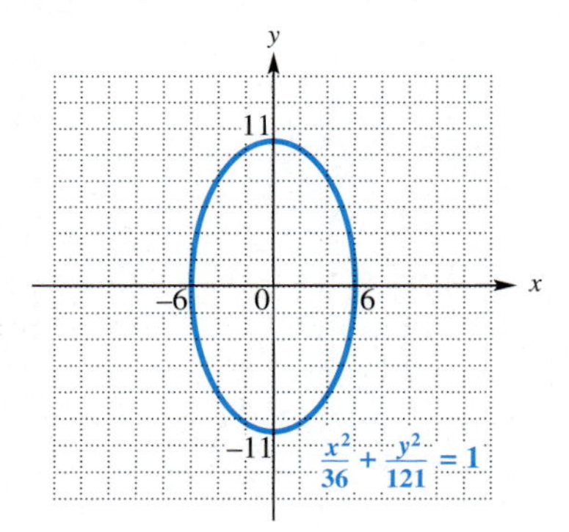

Figure 16

Work Problem 4 at the Side.

As with the graphs of parabolas and circles, the graph of an ellipse may be shifted horizontally and vertically, as in the next example.

4 Graph each ellipse.

(a) $\dfrac{x^2}{4} + \dfrac{y^2}{25} = 1$

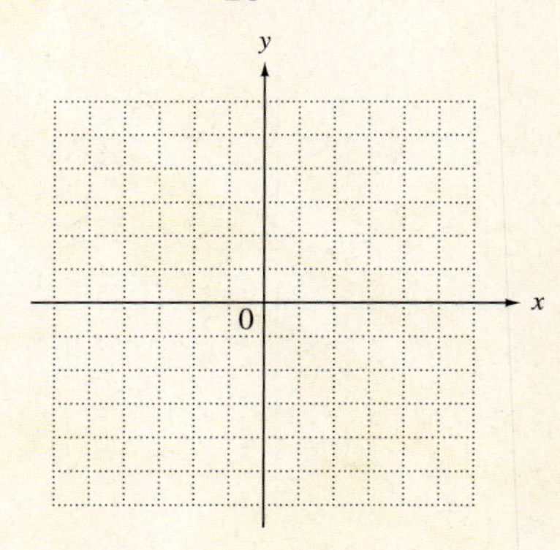

(b) $\dfrac{x^2}{64} + \dfrac{y^2}{49} = 1$

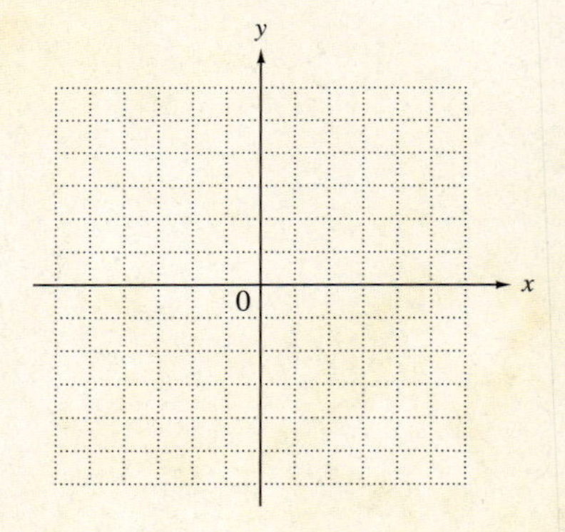

ANSWERS

4. (a) **(b)**

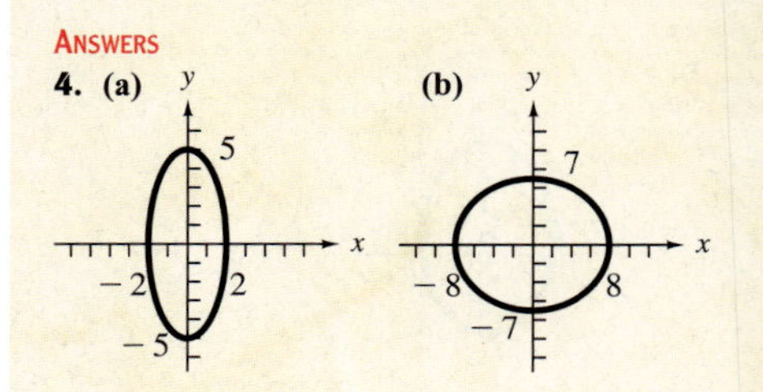

5 Graph

$$\frac{(x+4)^2}{16}+\frac{(y-1)^2}{36}=1.$$

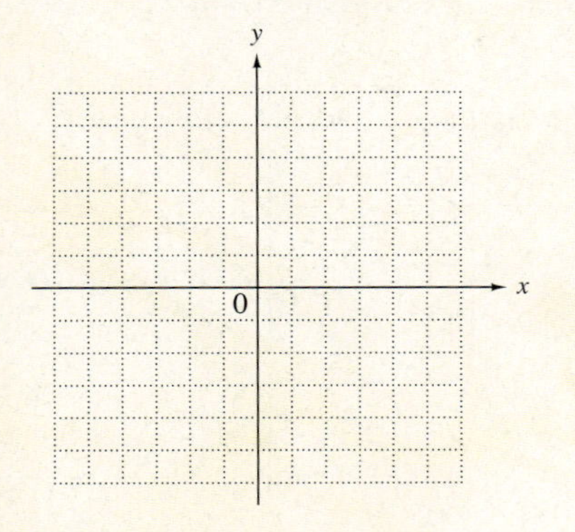

Example 6 Graphing an Ellipse Shifted Horizontally and Vertically

Graph $\dfrac{(x-2)^2}{25}+\dfrac{(y+3)^2}{49}=1.$

Just as $(x-2)^2$ and $(y+3)^2$ would indicate that the center of a circle would be $(2, -3)$, so it is with this ellipse. Figure 17 shows that the graph goes through the four points $(2, 4)$, $(7, -3)$, $(2, -10)$, and $(-3, -3)$. The x-values of these points are found by adding $\pm a = \pm 5$ to 2, and the y-values come from adding $\pm b = \pm 7$ to -3.

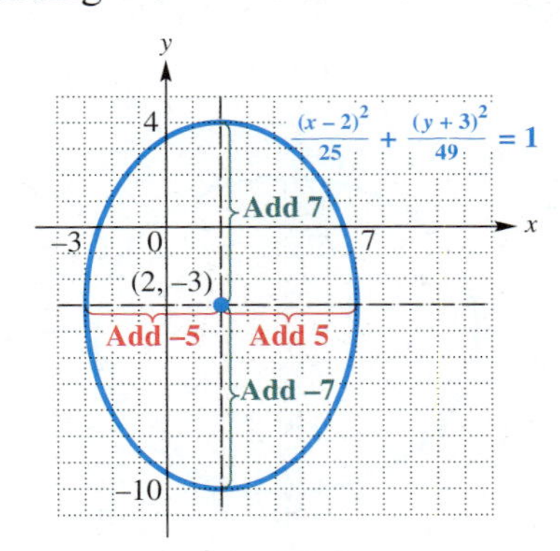

Figure 17

Work Problem 5 at the Side.

Notice that the graphs in this section are not graphs of functions. The only conic section whose graph is a function is the vertical parabola with equation $f(x) = ax^2 + bx + c$.

ANSWERS

5.

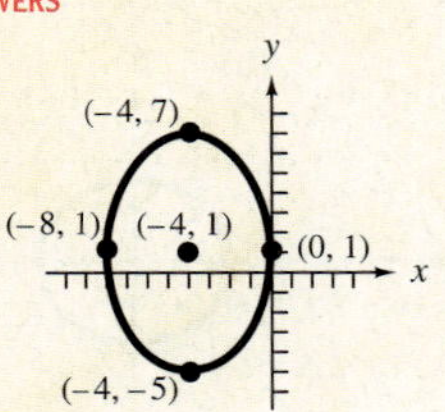

11.2 Exercises

FOR EXTRA HELP

Student's Solutions Manual

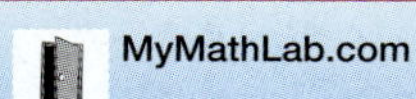
MyMathLab.com

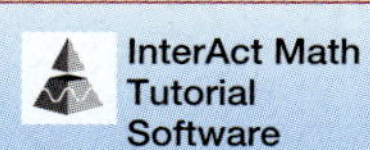
InterAct Math Tutorial Software

AW Math Tutor Center

www.mathxl.com

Digital Video Tutor CD 9
Videotape 19

1. See Example 1. Consider the circle whose equation is $x^2 + y^2 = 25$.

(a) What are the coordinates of its center?

(b) What is its radius?

(c) Sketch its graph.

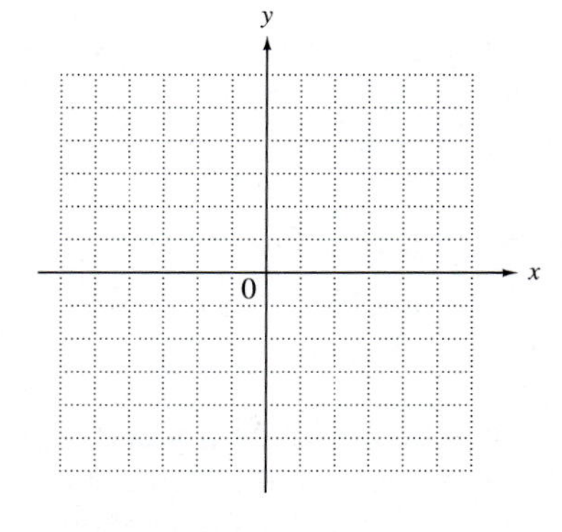

2. Explain why a set of points defined by a circle does not satisfy the definition of a function.

Match each equation with the correct graph. See Examples 1–3.

3. $(x - 3)^2 + (y - 2)^2 = 25$

4. $(x - 3)^2 + (y + 2)^2 = 25$

5. $(x + 3)^2 + (y - 2)^2 = 25$

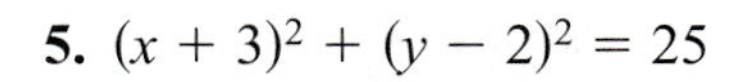

6. $(x + 3)^2 + (y + 2)^2 = 25$

A.

B.

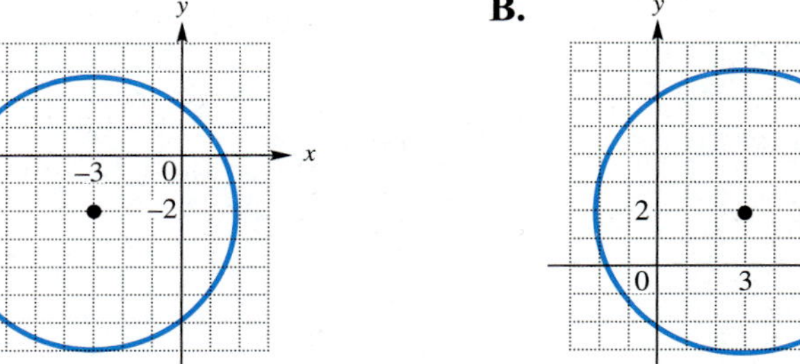

C.

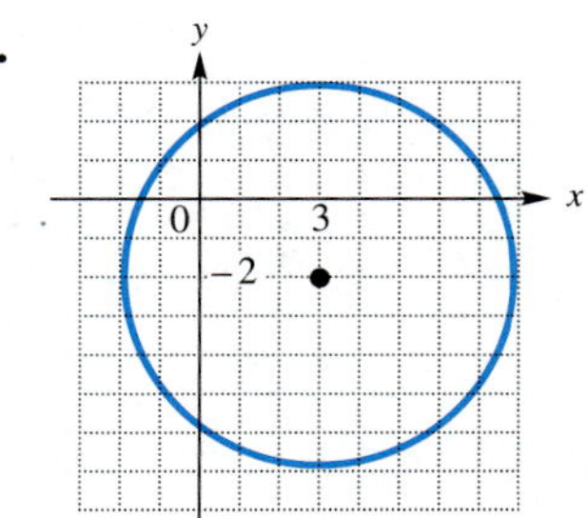

D.

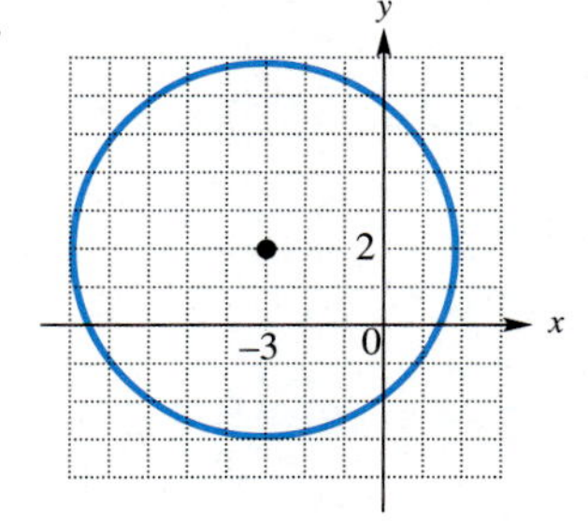

Find the equation of a circle satisfying the given conditions. See Examples 2 and 3.

7. Center: $(-4, 3)$; radius: 2

8. Center: $(5, -2)$; radius: 4

9. Center: $(-8, -5)$; radius: $\sqrt{5}$

10. Center: $(-12, 13)$; radius: $\sqrt{7}$

Find the center and radius of each circle. (Hint: In Exercises 15 and 16, divide each side by a common factor.) See Example 4.

11. $x^2 + y^2 + 4x + 6y + 9 = 0$

12. $x^2 + y^2 - 8x - 12y + 3 = 0$

13. $x^2 + y^2 + 10x - 14y - 7 = 0$

14. $x^2 + y^2 - 2x + 4y - 4 = 0$

15. $3x^2 + 3y^2 - 12x - 24y + 12 = 0$

16. $2x^2 + 2y^2 + 20x + 16y + 10 = 0$

17. A circle can be drawn on a piece of posterboard by fastening one end of a string with a thumbtack, pulling the string taut with a pencil, and tracing a curve, as shown in the figure. Explain why this method works.

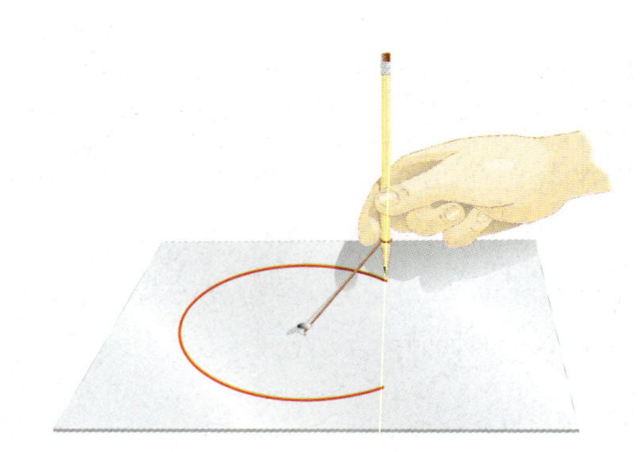

Graph each circle. See Examples 1–4.

18. $x^2 + y^2 = 9$

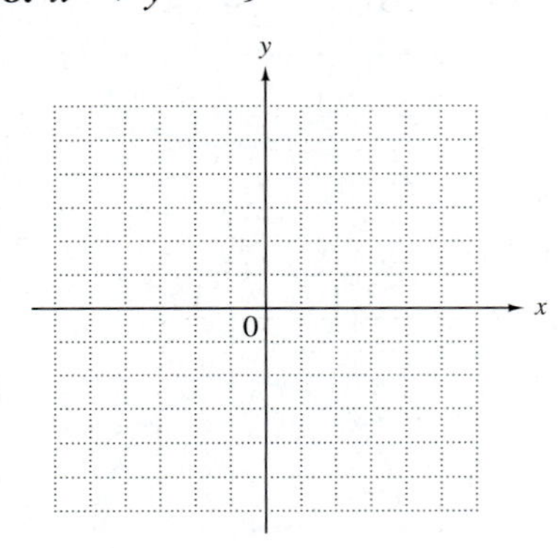

19. $x^2 + y^2 = 4$

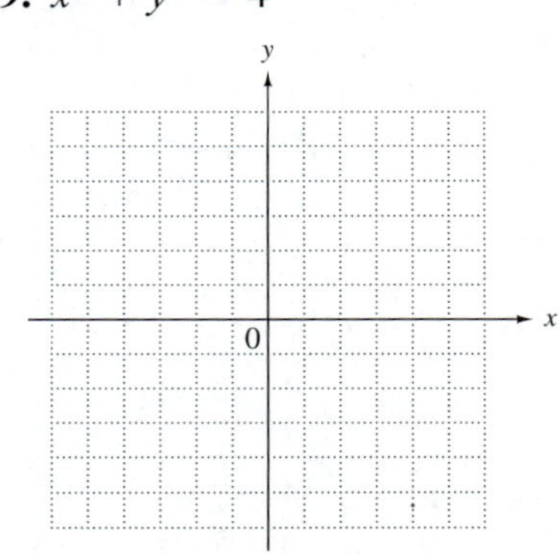

20. $2y^2 = 10 - 2x^2$

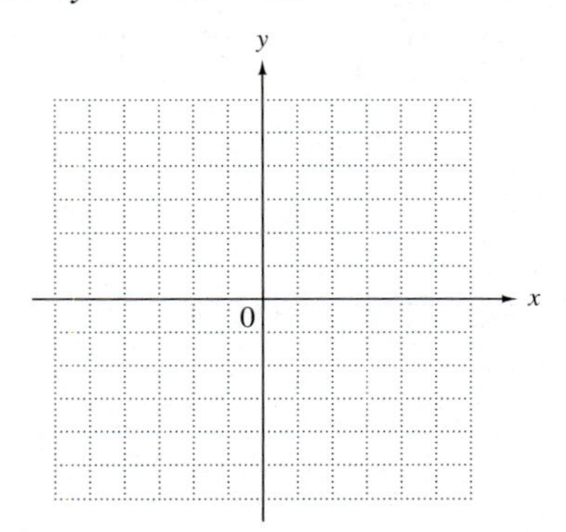

21. $3x^2 = 48 - 3y^2$

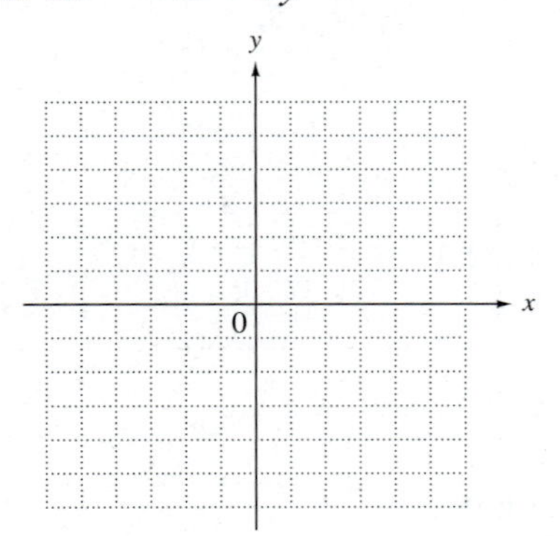

22. $(x + 3)^2 + (y - 2)^2 = 9$

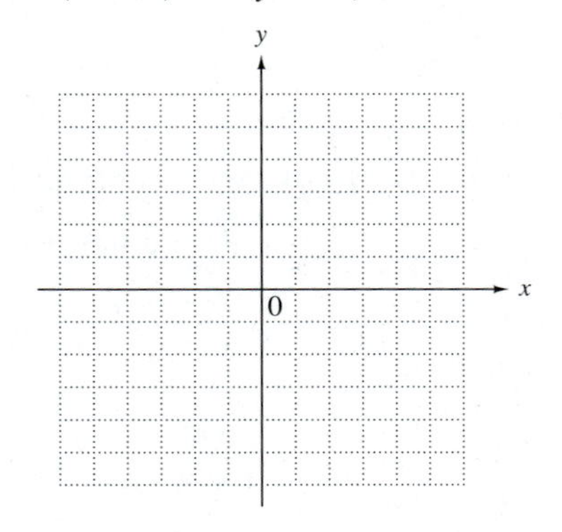

23. $(x - 1)^2 + (y + 3)^2 = 16$

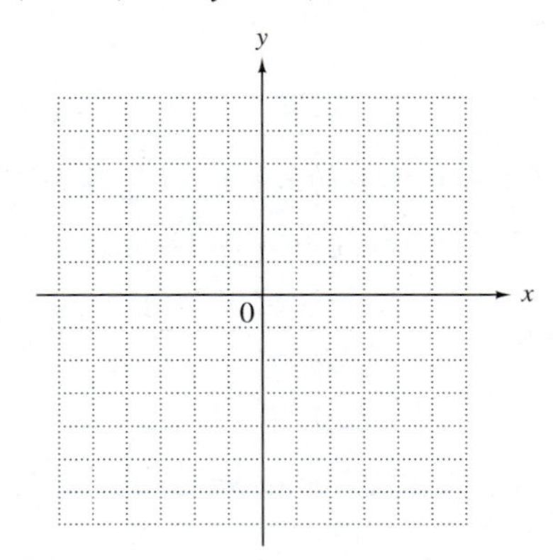

24. $x^2 + y^2 - 4x - 6y + 9 = 0$

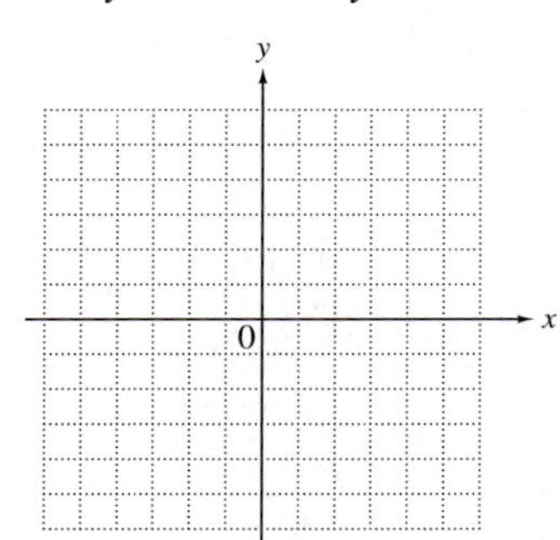

25. $x^2 + y^2 + 8x + 2y - 8 = 0$

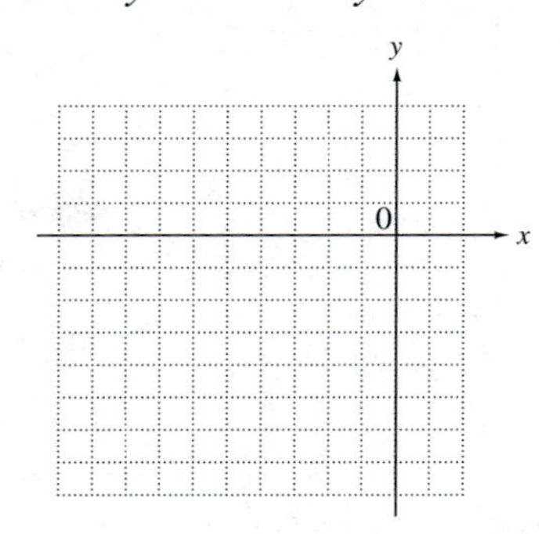

26. An ellipse can be drawn on a piece of posterboard by fastening two ends of a length of string with thumbtacks, pulling the string taut with a pencil, and tracing a curve, as shown in the figure. Explain why this method works.

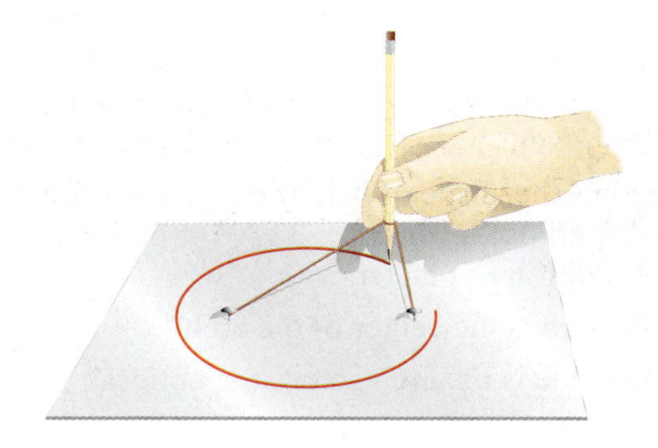

Graph each ellipse. See Examples 5 and 6.

27. $\frac{x^2}{9} + \frac{y^2}{25} = 1$

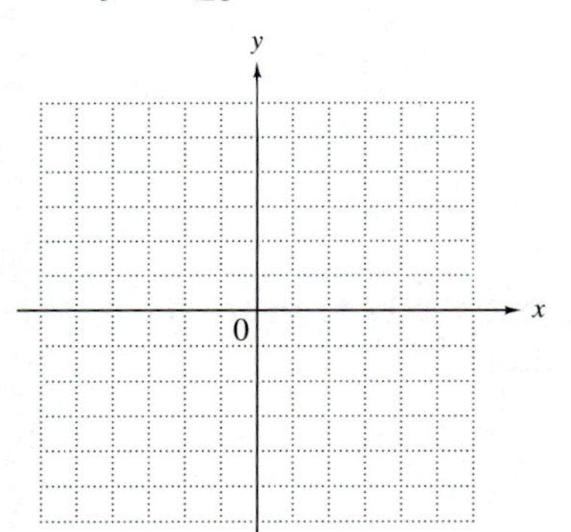

28. $\frac{x^2}{9} + \frac{y^2}{16} = 1$

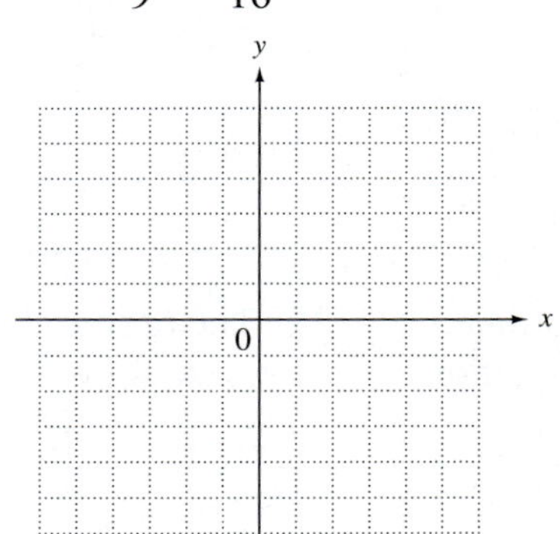

29. $\frac{x^2}{36} + \frac{y^2}{16} = 1$

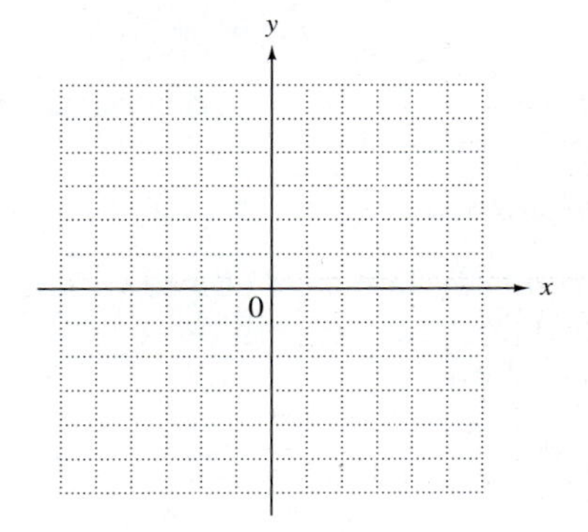

30. $\frac{x^2}{9} + \frac{y^2}{4} = 1$

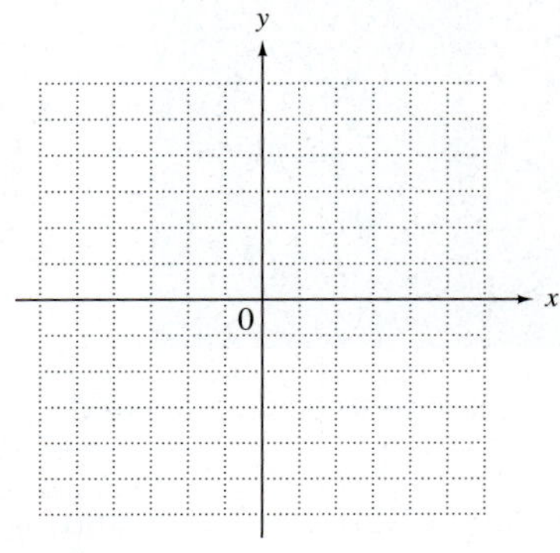

31. $\frac{x^2}{49} + \frac{y^2}{25} = 1$

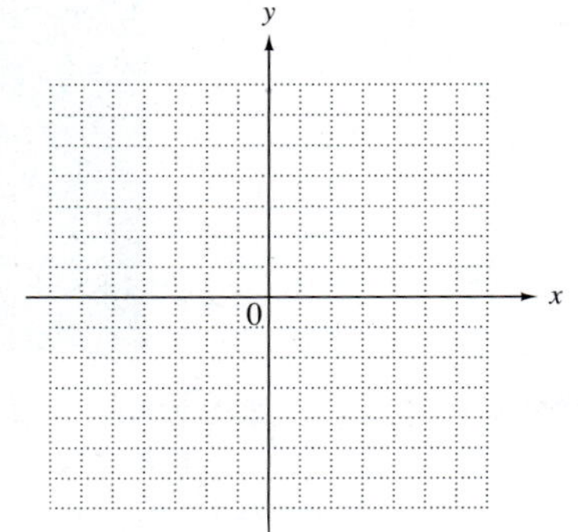

32. $\frac{x^2}{16} + \frac{y^2}{9} = 1$

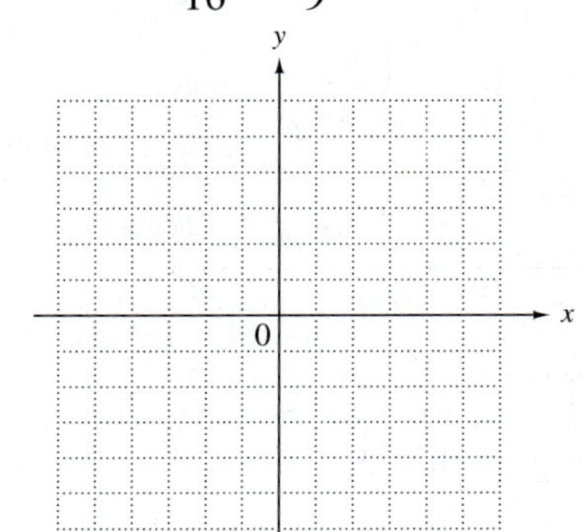

33. $\dfrac{(x-2)^2}{16} + \dfrac{(y-1)^2}{9} = 1$

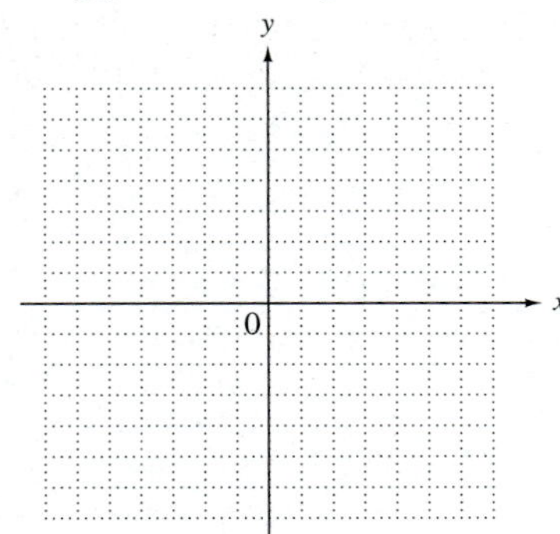

34. $\dfrac{(x-4)^2}{9} + \dfrac{(y+2)^2}{4} = 1$

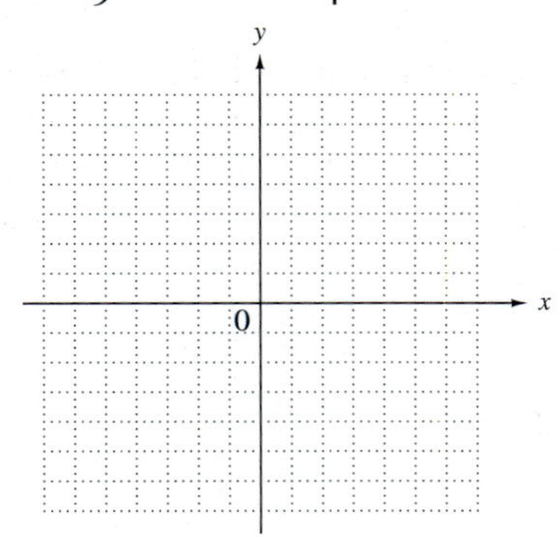

Solve each problem.

35. An arch has the shape of half an ellipse. The equation of the ellipse is $100x^2 + 324y^2 = 32{,}400$, where x and y are in meters.

(a) How high is the center of the arch?

(b) How wide is the arch across the bottom?

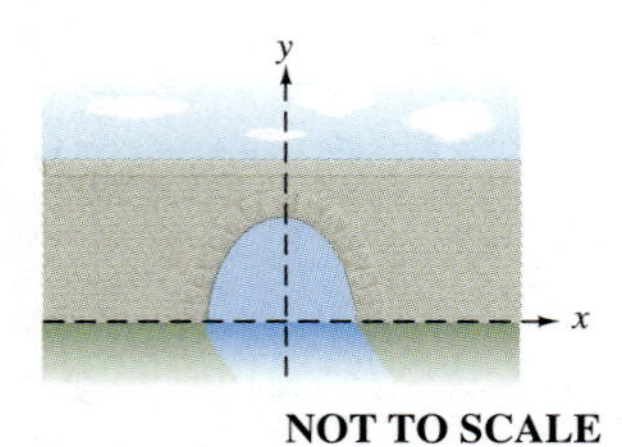

NOT TO SCALE

36. A one-way street passes under an overpass, which is in the form of the top half of an ellipse, as shown in the figure. Suppose that a truck 12 ft wide passes directly under the overpass. What is the maximum possible height of this truck?

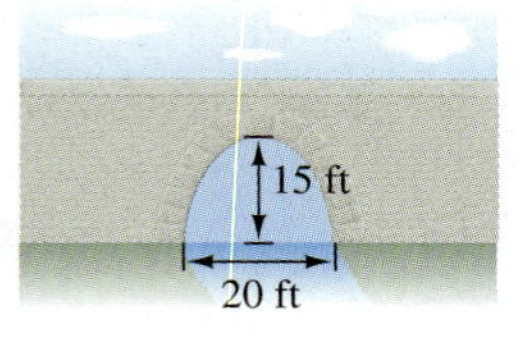

In Exercises 37 and 38, see Figure 13 and use the fact that $c^2 = a^2 - b^2$ where $a^2 > b^2$.

37. The orbit of Mars is an ellipse with the sun at one focus. For x and y in millions of miles, the equation of the orbit is

$$\frac{x^2}{141.7^2} + \frac{y^2}{141.1^2} = 1.$$

(*Source:* Kaler, James B., *Astronomy!*, Addison-Wesley, 1997.)

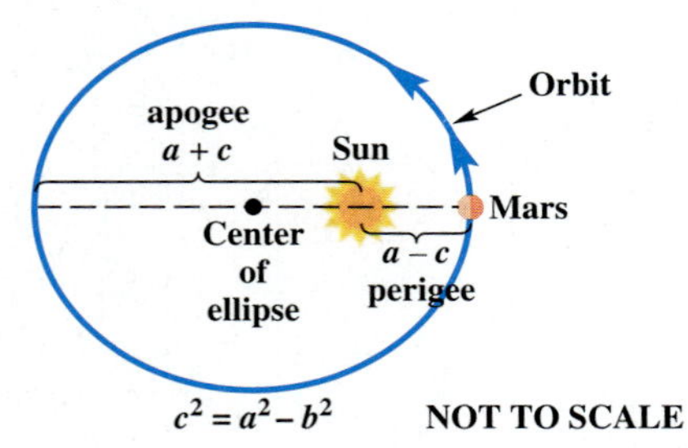

NOT TO SCALE

(a) Find the greatest distance (the *apogee*) from Mars to the sun.

(b) Find the smallest distance (the *perigee*) from Mars to the sun.

38. The orbit of Venus around the sun (one of the foci) is an ellipse with equation

$$\frac{x^2}{5013} + \frac{y^2}{4970} = 1,$$

where x and y are measured in millions of miles.

(*Source:* Kaler, James B., *Astronomy!*, Addison-Wesley, 1997.)

(a) Find the greatest distance between Venus and the sun.

(b) Find the smallest distance between Venus and the sun.

11.3 THE HYPERBOLA AND OTHER FUNCTIONS DEFINED BY RADICALS

OBJECTIVES

1. Recognize the equation of a hyperbola.
2. Graph hyperbolas by using asymptotes.
3. Identify conic sections by their equations.
4. Graph certain square root functions.

1 Recognize the equation of a hyperbola. A **hyperbola** is the set of all points in a plane such that the absolute value of the *difference* of the distances from two fixed points (called *foci*) is constant. Figure 18 shows a hyperbola; using the distance formula and the definition above, we can show that this hyperbola has equation $\frac{x^2}{16} - \frac{y^2}{12} = 1$.

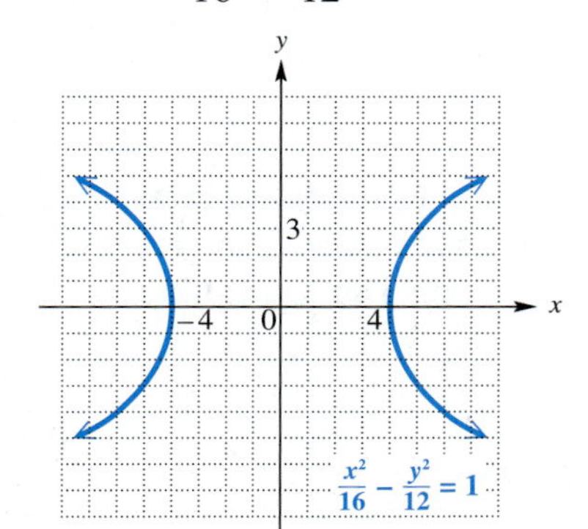

Figure 18

To graph hyperbolas centered at the origin, we need to find their intercepts. For the hyperbola in Figure 18, we proceed as follows.

***x*-Intercepts**

Let $y = 0$.

$$\frac{x^2}{16} - \frac{0^2}{12} = 1 \quad \text{Let } y = 0.$$

$$\frac{x^2}{16} = 1$$

$$x^2 = 16 \quad \text{Multiply by 16.}$$

$$x = \pm 4$$

The x-intercepts are $(4, 0)$ and $(-4, 0)$.

***y*-Intercepts**

Let $x = 0$.

$$\frac{0^2}{16} - \frac{y^2}{12} = 1 \quad \text{Let } x = 0.$$

$$-\frac{y^2}{12} = 1$$

$$y^2 = -12 \quad \text{Multiply by } -12.$$

Because there are no *real* solutions to $y^2 = -12$, the graph has no y-intercepts.

The graph of $\frac{x^2}{16} - \frac{y^2}{12} = 1$ has no y-intercepts. On the other hand, the hyperbola in Figure 19 has no x-intercepts. Its equation is

$$\frac{y^2}{25} - \frac{x^2}{9} = 1,$$

with y-intercepts $(0, 5)$ and $(0, -5)$.

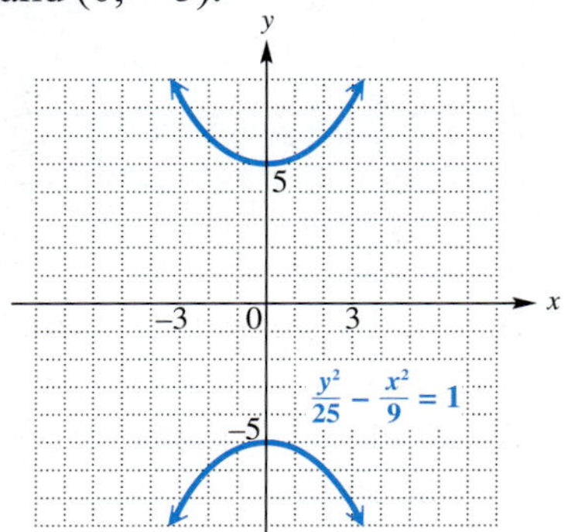

Figure 19

Equations of Hyperbolas

A hyperbola with x-intercepts $(a, 0)$ and $(-a, 0)$ has an equation of the form

$$\frac{x^2}{a^2} - \frac{y^2}{b^2} = 1,$$

and a hyperbola with y-intercepts $(0, b)$ and $(0, -b)$ has an equation of the form

$$\frac{y^2}{b^2} - \frac{x^2}{a^2} = 1.$$

2 Graph hyperbolas by using asymptotes. The two branches of the graph of a hyperbola approach a pair of intersecting straight lines, which are its asymptotes. See Figure 20. The asymptotes are useful for sketching the graph of the hyperbola.

Asymptotes of Hyperbolas

The extended diagonals of the rectangle with corners at the points (a, b), $(-a, b)$, $(-a, -b)$, and $(a, -b)$ are the **asymptotes** of the hyperbolas

$$\frac{x^2}{a^2} - \frac{y^2}{b^2} = 1 \quad \text{and} \quad \frac{y^2}{b^2} - \frac{x^2}{a^2} = 1.$$

This rectangle is called the **fundamental rectangle.** Using the methods of Chapter 4, we could show that the equations of these asymptotes are

$$y = \frac{b}{a}x \quad \text{and} \quad y = -\frac{b}{a}x.$$

To graph hyperbolas, follow these steps.

Graphing a Hyperbola

Step 1 **Find the intercepts.** Locate the intercepts at $(a, 0)$ and $(-a, 0)$ if the x^2-term has a positive coefficient, or at $(0, b)$ and $(0, -b)$ if the y^2-term has a positive coefficient.

Step 2 **Find the fundamental rectangle.** Locate the corners of the fundamental rectangle at (a, b), $(-a, b)$, $(-a, -b)$, and $(a, -b)$.

Step 3 **Sketch the asymptotes.** The extended diagonals of the rectangle are the asymptotes of the hyperbola, and they have equations $y = \pm \frac{b}{a}x$.

Step 4 **Draw the graph.** Sketch each branch of the hyperbola through an intercept and approaching (but not touching) the asymptotes.

Example 1 Graphing a Horizontal Hyperbola

Graph $\frac{x^2}{16} - \frac{y^2}{25} = 1$.

Step 1 Here $a = 4$ and $b = 5$. The x-intercepts are $(4, 0)$ and $(-4, 0)$.

Step 2 The four points $(4, 5)$, $(-4, 5)$, $(-4, -5)$, and $(4, -5)$ are the corners of the fundamental rectangle, as shown in Figure 20.

Steps 3 and 4 The equations of the asymptotes are $y = \pm\frac{5}{4}x$, and the hyperbola approaches these lines as x and y get larger and larger in absolute value.

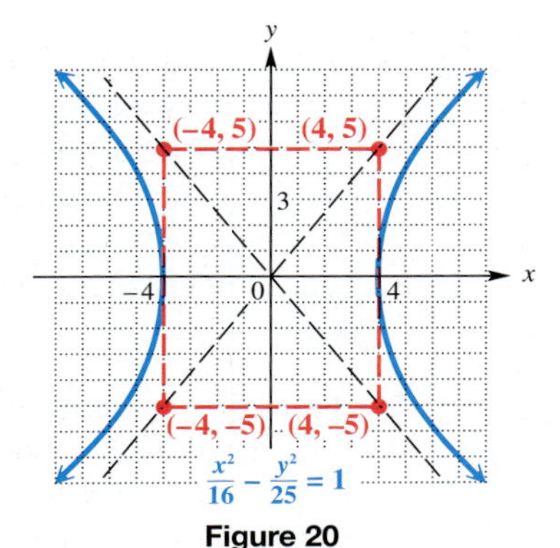

Figure 20

CAUTION

When sketching the graph of a hyperbola, be sure that the branches do not touch the asymptotes.

Work Problem 1 at the Side.

Example 2 Graphing a Vertical Hyperbola

Graph $\frac{y^2}{49} - \frac{x^2}{16} = 1$.

This hyperbola has y-intercepts $(0, 7)$ and $(0, -7)$. The asymptotes are the extended diagonals of the rectangle with corners at $(4, 7)$, $(-4, 7)$, $(-4, -7)$, and $(4, -7)$. Their equations are $y = \pm\frac{7}{4}x$. See Figure 21.

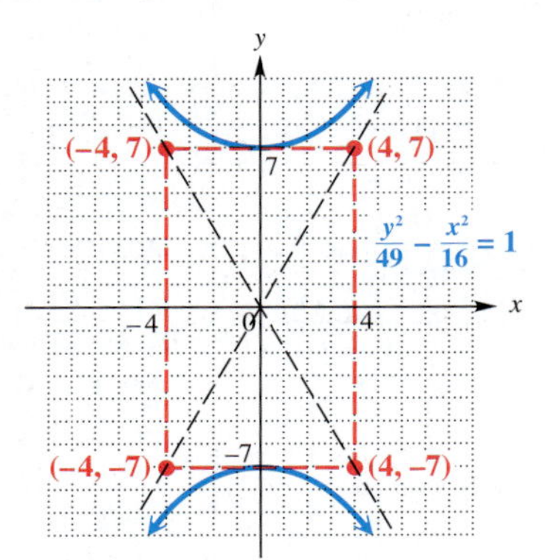

Figure 21

Work Problem 2 at the Side.

1 Graph $\frac{x^2}{4} - \frac{y^2}{25} = 1$.

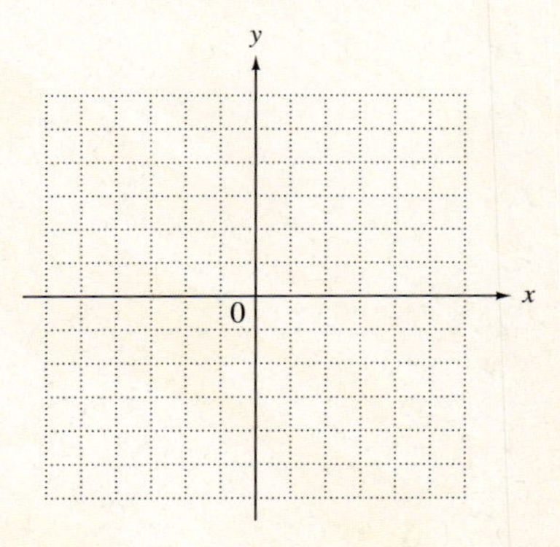

2 Graph $\frac{y^2}{81} - \frac{x^2}{64} = 1$.

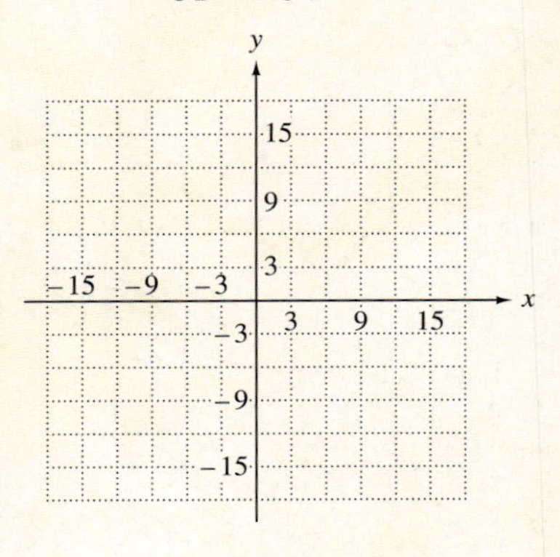

ANSWERS

1.

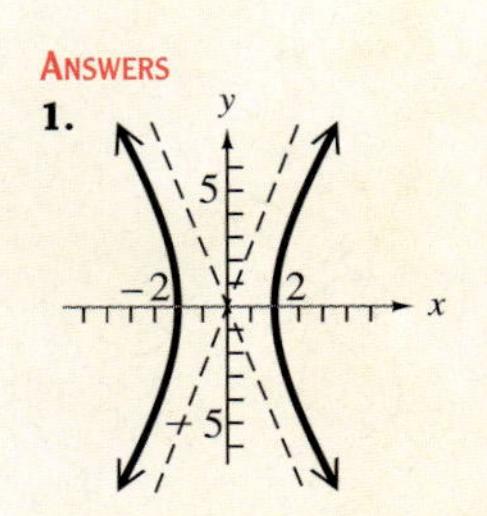

2.

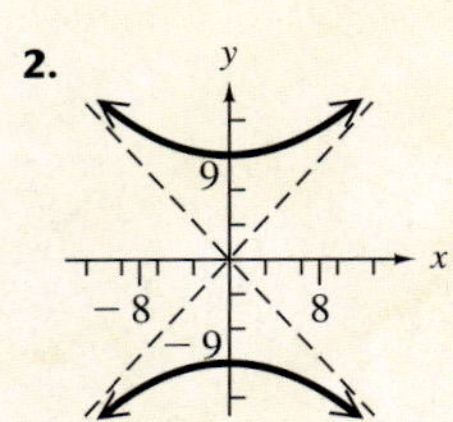

3 Identify conic sections by their equations. Rewriting a second-degree equation in one of the forms given for ellipses, hyperbolas, circles, or parabolas makes it possible to determine when the graph is one of these.

SUMMARY OF CONIC SECTIONS

Equation	Graph	Description	Identification
$y = a(x - h)^2 + k$	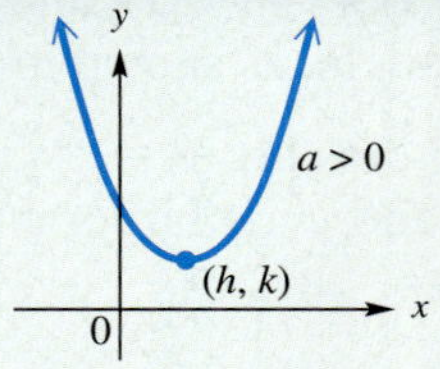 Parabola	It opens up if $a > 0$, down if $a < 0$. The vertex is (h, k).	It has an x^2-term. y is not squared.
$x = a(y - k)^2 + h$	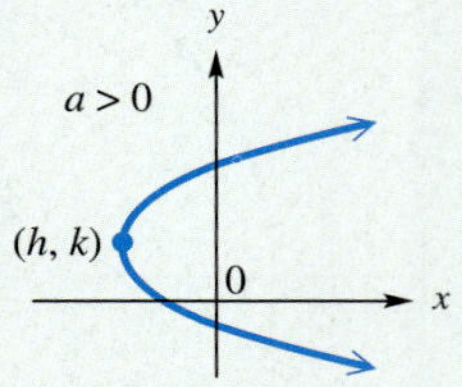 Parabola	It opens to the right if $a > 0$, to the left if $a < 0$. The vertex is (h, k).	It has a y^2-term. x is not squared.
$(x - h)^2 + (y - k)^2 = r^2$	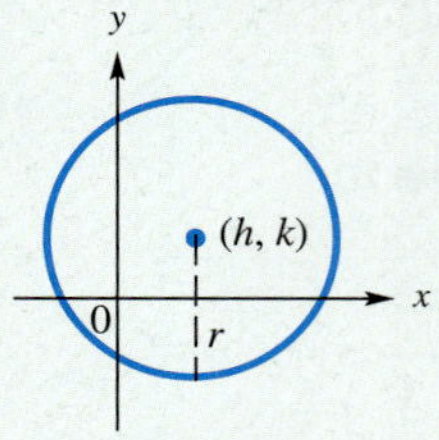 Circle	The center is (h, k), and the radius is r.	x^2- and y^2-terms have the same positive coefficient.
$\frac{x^2}{a^2} + \frac{y^2}{b^2} = 1$	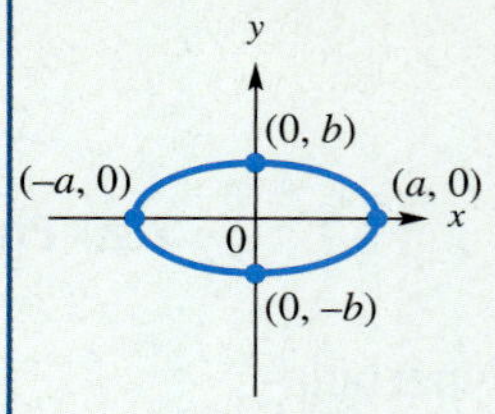 Ellipse	The x-intercepts are $(a, 0)$ and $(-a, 0)$. The y-intercepts are $(0, b)$ and $(0, -b)$.	x^2- and y^2-terms have different positive coefficients.
$\frac{x^2}{a^2} - \frac{y^2}{b^2} = 1$	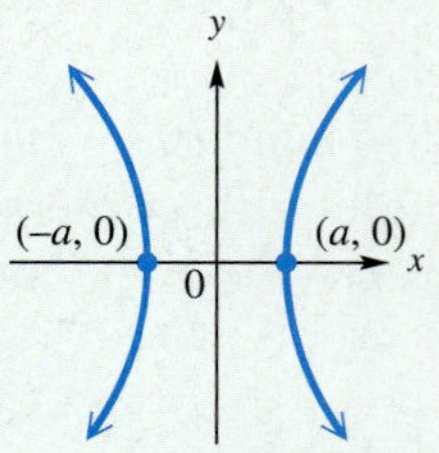 Hyperbola	The x-intercepts are $(a, 0)$ and $(-a, 0)$. The asymptotes are found from (a, b), $(a, -b)$, $(-a, -b)$, and $(-a, b)$.	x^2 has a positive coefficient. y^2 has a negative coefficient.
$\frac{y^2}{b^2} - \frac{x^2}{a^2} = 1$	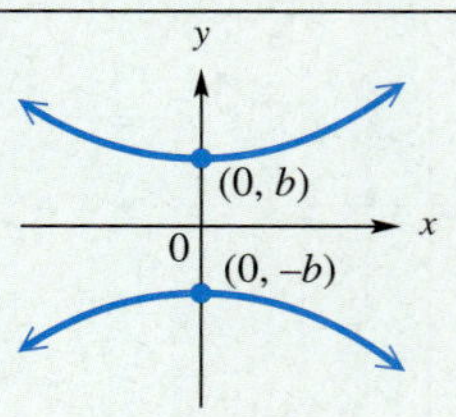 Hyperbola	The y-intercepts are $(0, b)$ and $(0, -b)$. The asymptotes are found from (a, b), $(a, -b)$, $(-a, -b)$, and $(-a, b)$.	y^2 has a positive coefficient. x^2 has a negative coefficient.

Example 3 **Identifying the Graphs of Equations**

Identify the graph of each equation.

(a) $9x^2 = 108 + 12y^2$

Both variables are squared, so the graph is either an ellipse or a hyperbola. (This situation also occurs for a circle, which is a special case of the ellipse.) To see which one it is, rewrite the equation so that the x^2- and y^2-terms are on one side of the equation and 1 is on the other.

$$9x^2 - 12y^2 = 108 \quad \text{Subtract } 12y^2.$$

$$\frac{x^2}{12} - \frac{y^2}{9} = 1 \quad \text{Divide by 108.}$$

Because of the minus sign, the graph of this equation is a hyperbola.

(b) $x^2 = y - 3$

Only one of the two variables, x, is squared, so this is the vertical parabola $y = x^2 + 3$.

(c) $x^2 = 9 - y^2$

Get the variable terms on the same side of the equation.

$$x^2 + y^2 = 9 \quad \text{Add } y^2.$$

The graph of this equation is a circle with center at the origin and radius 3.

Work Problem 3 at the Side.

3 Identify the graph of each equation.

(a) $3x^2 = 27 - 4y^2$

(b) $6x^2 = 100 + 2y^2$

(c) $3x^2 = 27 - 4y$

(d) $3x^2 = 27 - 3y^2$

4 **Graph certain square root functions.** Recall that no vertical line will intersect the graph of a function in more than one point. Thus, horizontal parabolas and all circles, ellipses, and the hyperbolas discussed in this chapter are examples of graphs that do not satisfy the conditions of a function. However, by considering only a part of the graph of each of these we have the graph of a function, as seen in Figure 22.

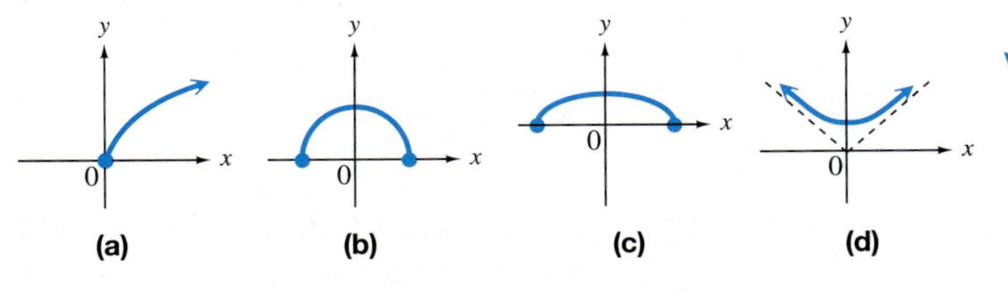

Figure 22

In parts (a), (b), (c), and (d) of Figure 22, the top portion of a conic section is shown (parabola, circle, ellipse, and hyperbola, respectively). In part (e), the top two portions of a hyperbola are shown. In each case, the graph is that of a function since the graph satisfies the conditions of the vertical line test.

In Sections 8.1 and 11.1 we observed the square root function defined by $f(x) = \sqrt{x}$. To find equations for the types of graphs shown in Figure 22, we extend its definition.

Square Root Function

A function of the form

$$f(x) = \sqrt{u}$$

for an algebraic expression u, with $u \geq 0$, is called a **square root function.**

ANSWERS

3. (a) ellipse (b) hyperbola (c) parabola (d) circle

4 Graph $f(x) = \sqrt{36 - x^2}$. Give the domain and range.

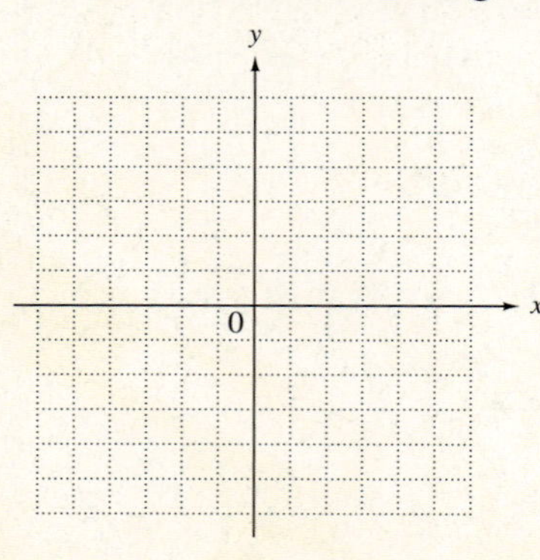

5 Graph

$$\frac{y}{3} = -\sqrt{1 - \frac{x^2}{4}}.$$

Give the domain and range.

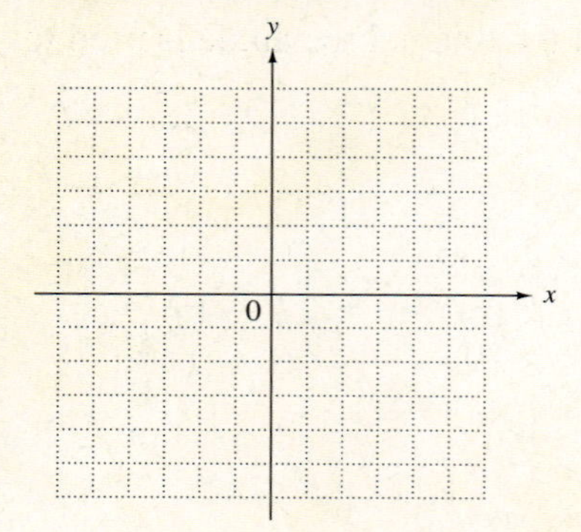

Answers

4.

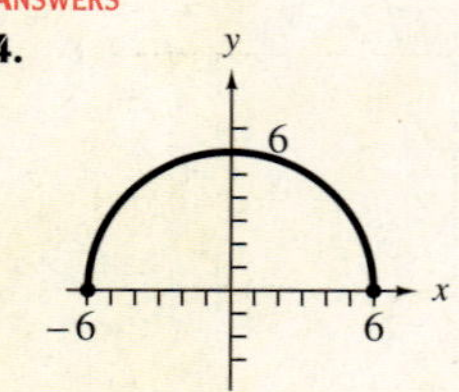

$[-6, 6]$; $[0, 6]$

5.

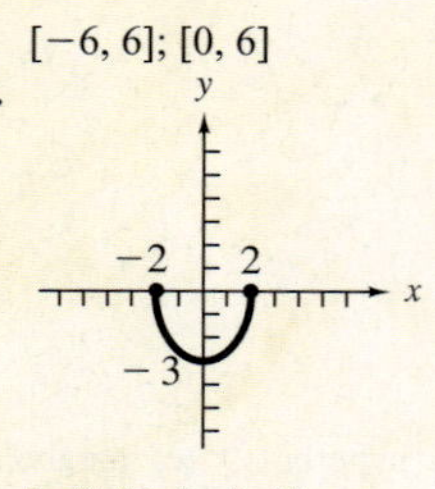

$[-2, 2]$; $[-3, 0]$

Example 4 Graphing a Semicircle

Graph $f(x) = \sqrt{25 - x^2}$. Give the domain and range.

Replace $f(x)$ with y and square both sides to get the equation

$$y^2 = 25 - x^2, \quad \text{or} \quad x^2 + y^2 = 25.$$

This is the graph of a circle with center at (0, 0) and radius 5. Since $f(x)$, or y, represents a principal square root in the original equation, $f(x)$ must be nonnegative. This restricts the graph to the upper half of the circle, as shown in Figure 23. Use the graph and the vertical line test to verify that it is indeed a function. The domain is $[-5, 5]$, and the range is $[0, 5]$.

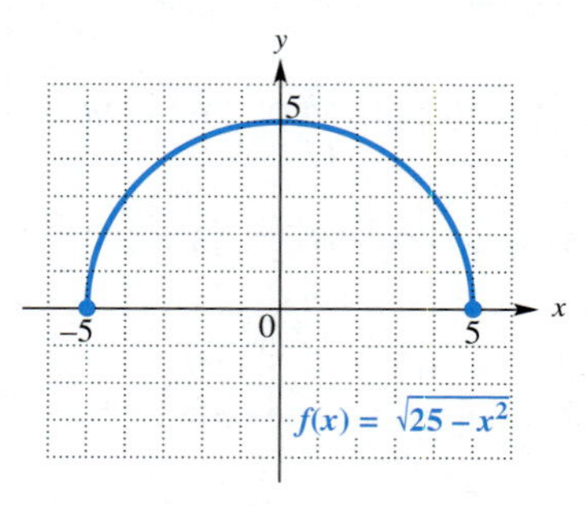

Figure 23

Work Problem **4** at the Side.

Example 5 Graphing a Portion of an Ellipse

Graph $\frac{y}{6} = -\sqrt{1 - \frac{x^2}{16}}$. Give the domain and range.

Square both sides to get an equation whose form is known.

$$\frac{y^2}{36} = 1 - \frac{x^2}{16}$$

$$\frac{x^2}{16} + \frac{y^2}{36} = 1 \qquad \text{Add } \tfrac{x^2}{16}.$$

This is the equation of an ellipse with x-intercepts (4, 0) and $(-4, 0)$ and y-intercepts (0, 6) and $(0, -6)$. Since $\frac{y}{6}$ equals a negative square root in the original equation, y must be nonpositive, restricting the graph to the lower half of the ellipse, as shown in Figure 24. Verify that this is the graph of a function, using the vertical line test. The domain is $[-4, 4]$, and the range is $[-6, 0]$.

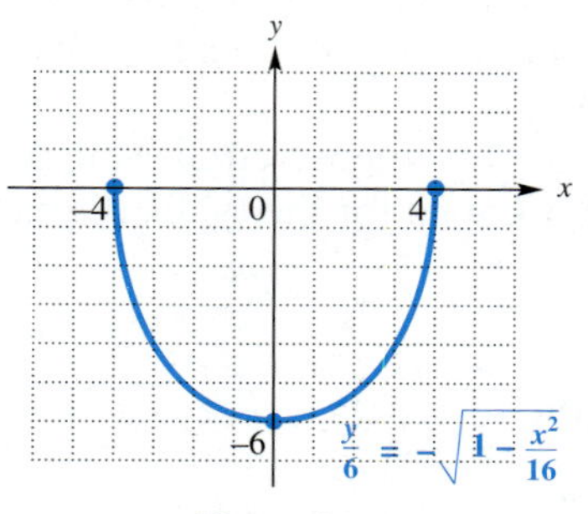

Figure 24

Work Problem **5** at the Side.

11.3 EXERCISES

FOR EXTRA HELP: Student's Solutions Manual; MyMathLab.com; InterAct Math Tutorial Software; AW Math Tutor Center; www.mathxl.com; Digital Video Tutor CD 9 Videotape 19

Based on the discussions of ellipses in the previous section and of hyperbolas in this section, match each equation with its graph.

1. $\dfrac{x^2}{25} + \dfrac{y^2}{9} = 1$

2. $\dfrac{x^2}{9} + \dfrac{y^2}{25} = 1$

3. $\dfrac{x^2}{9} - \dfrac{y^2}{25} = 1$

4. $\dfrac{x^2}{25} - \dfrac{y^2}{9} = 1$

A.

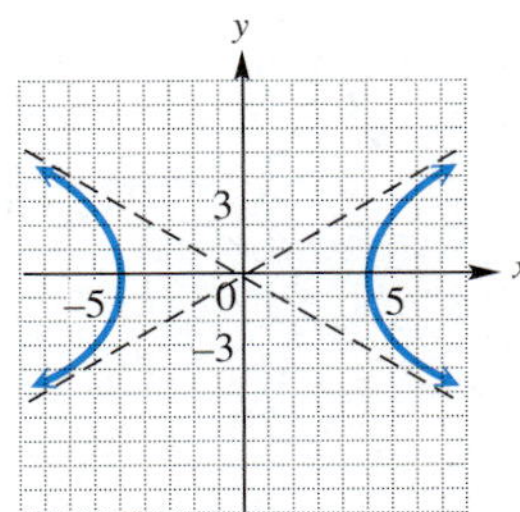

B.

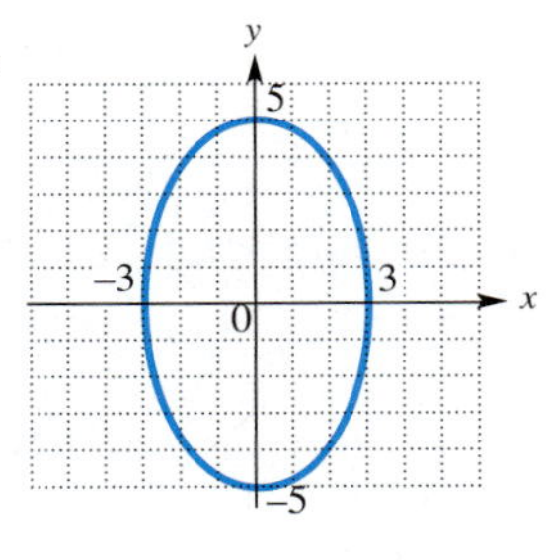

C.

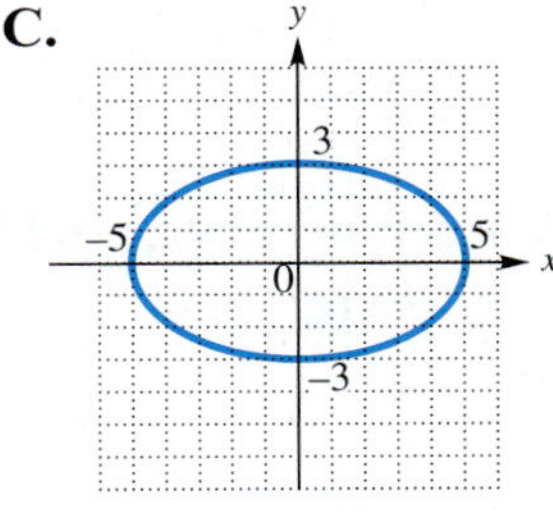

D.

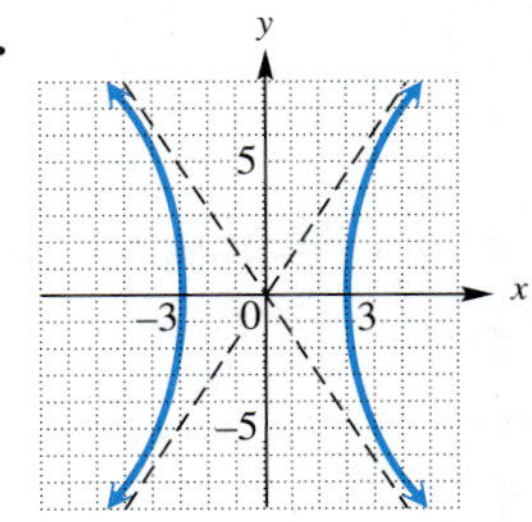

5. Write an explanation of how you can tell from the equation whether the branches of a hyperbola open up and down or left and right.

6. Describe how the fundamental rectangle is used to sketch a hyperbola.

Graph each hyperbola. See Examples 1 and 2.

7. $\dfrac{x^2}{16} - \dfrac{y^2}{9} = 1$

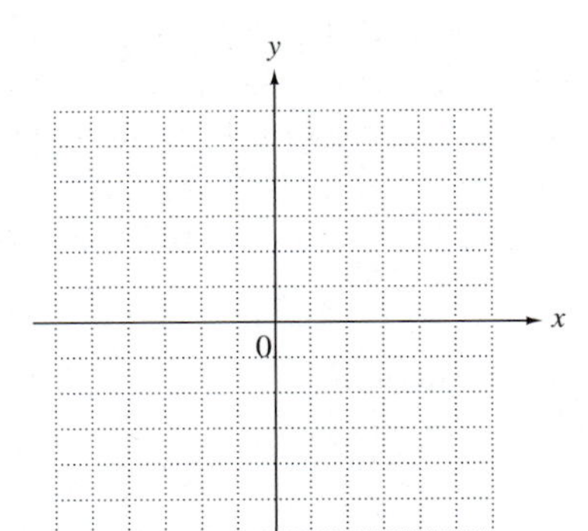

8. $\dfrac{y^2}{4} - \dfrac{x^2}{25} = 1$

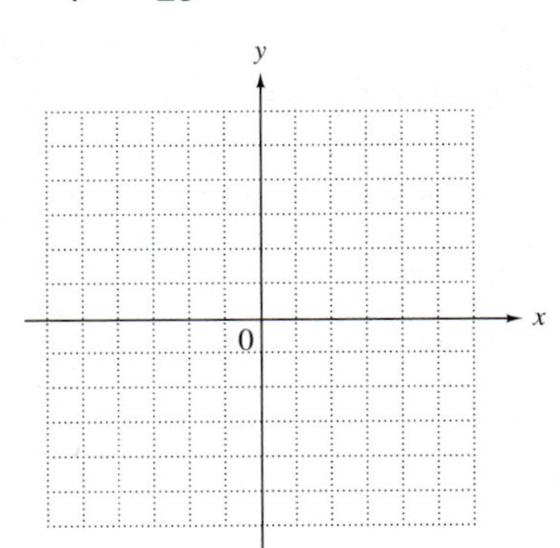

9. $\dfrac{y^2}{9} - \dfrac{x^2}{9} = 1$

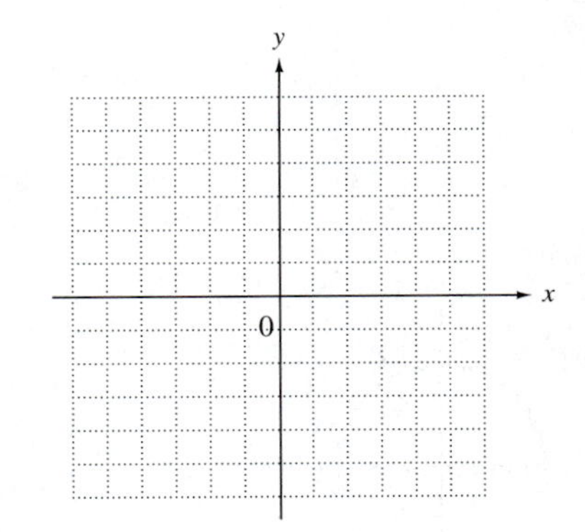

10. $\dfrac{x^2}{49} - \dfrac{y^2}{16} = 1$

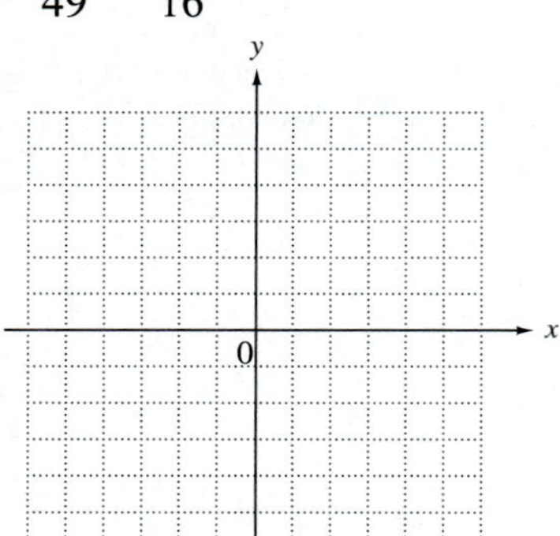

11. $\dfrac{x^2}{25} - \dfrac{y^2}{36} = 1$

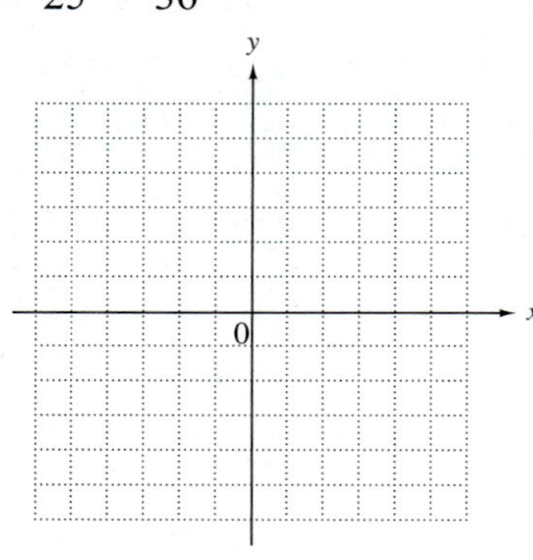

12. $\dfrac{y^2}{9} - \dfrac{x^2}{4} = 1$

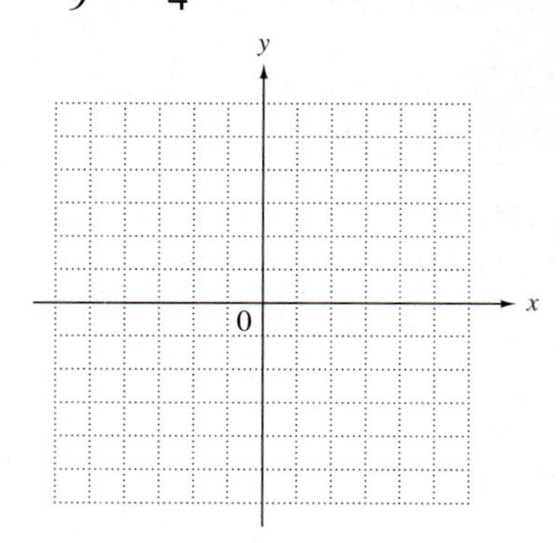

Identify the graph of each equation as a parabola, circle, ellipse, *or* hyperbola, *and sketch it. See Example 3.*

13. $x^2 - y^2 = 16$

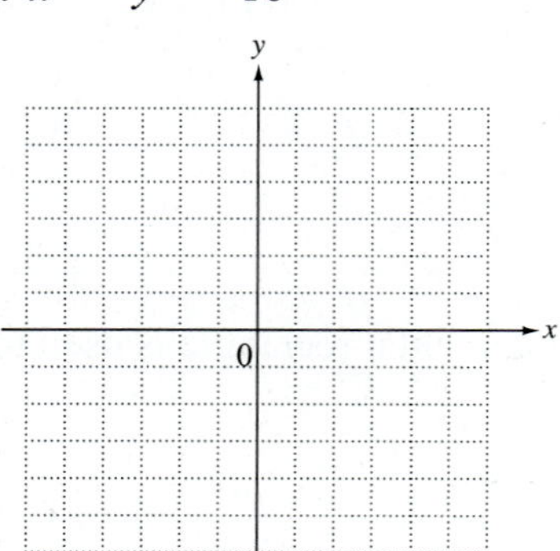

14. $x^2 + y^2 = 16$

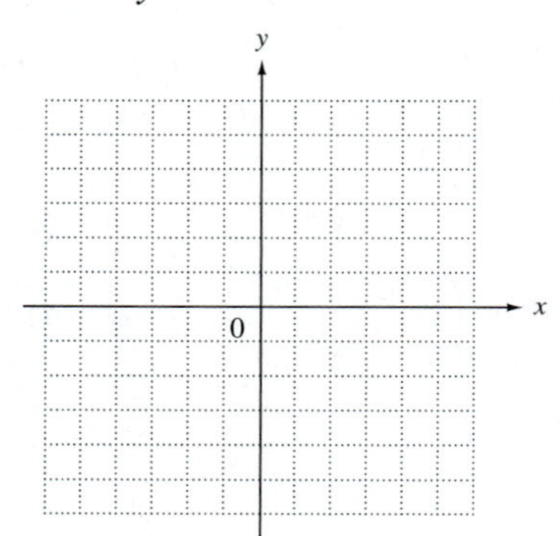

15. $4x^2 + y^2 = 16$

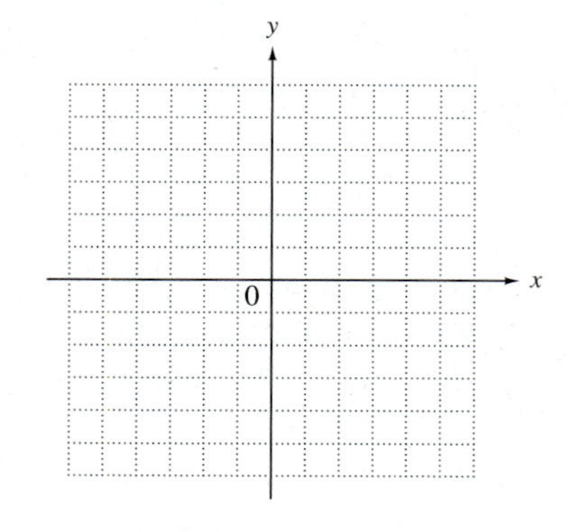

16. $x^2 - 2y = 0$

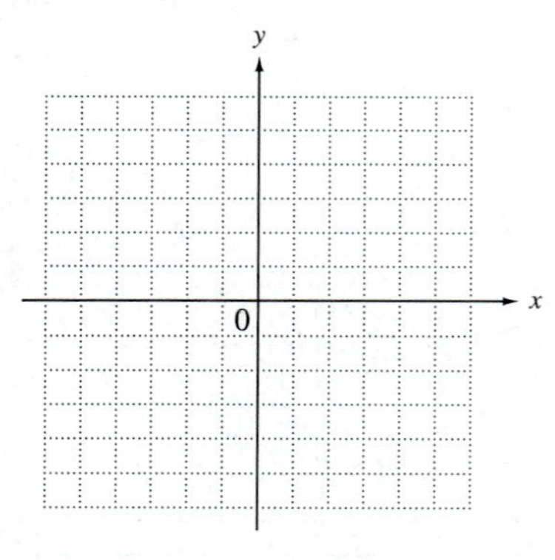

17. $y^2 = 36 - x^2$

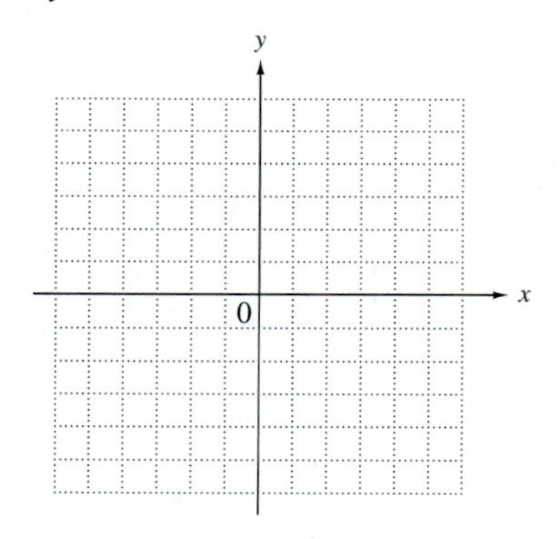

18. $9x^2 + 25y^2 = 225$

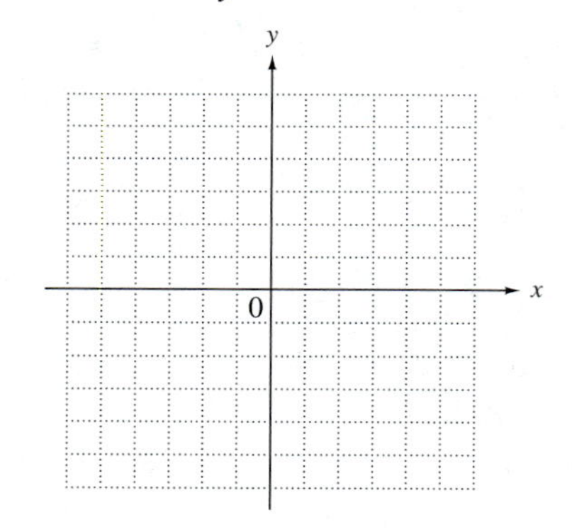

19. $9x^2 = 144 + 16y^2$

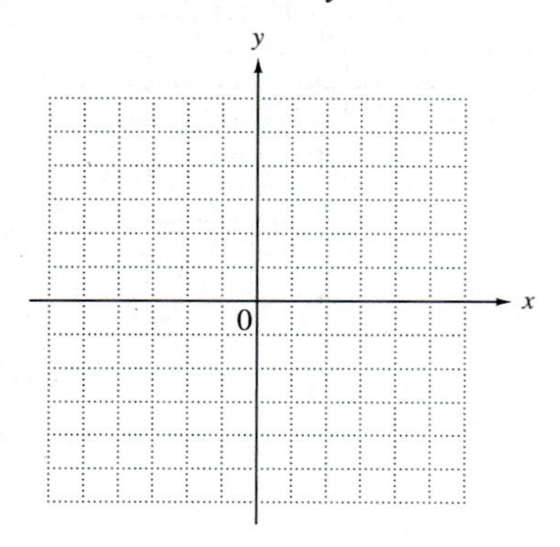

20. $y^2 = 4 + x^2$

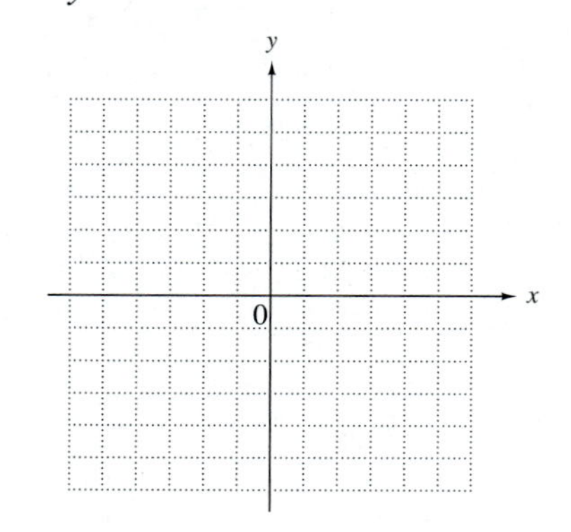

21. $x^2 + 9y^2 = 9$

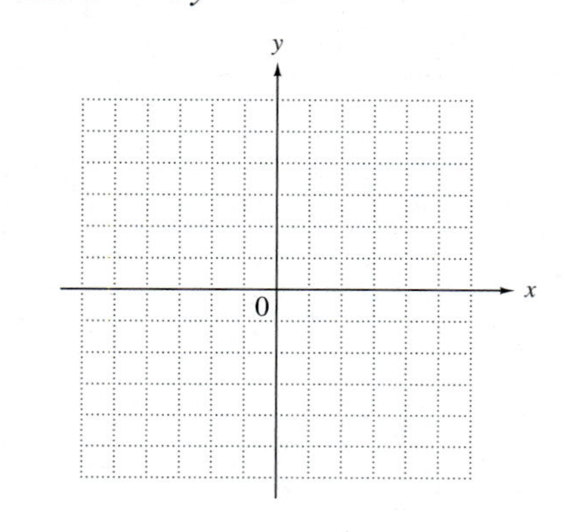

Graph each function defined by a radical expression. Give the domain and range. See Examples 4 and 5.

22. $f(x) = \sqrt{16 - x^2}$

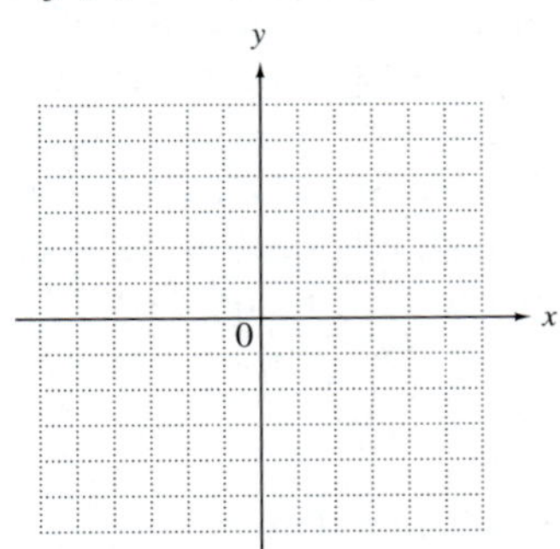

23. $f(x) = \sqrt{9 - x^2}$

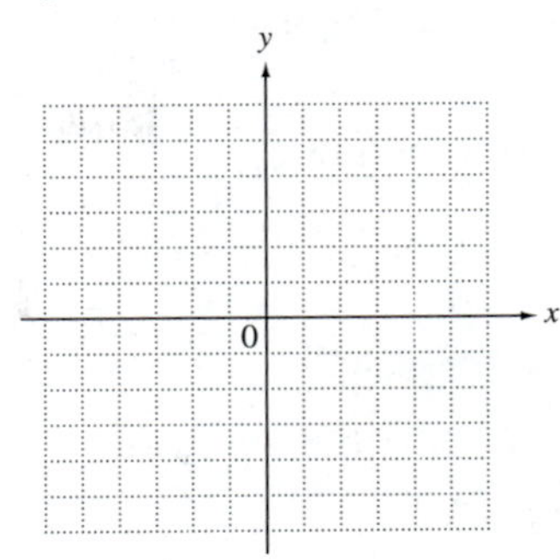

24. $f(x) = -\sqrt{36 - x^2}$

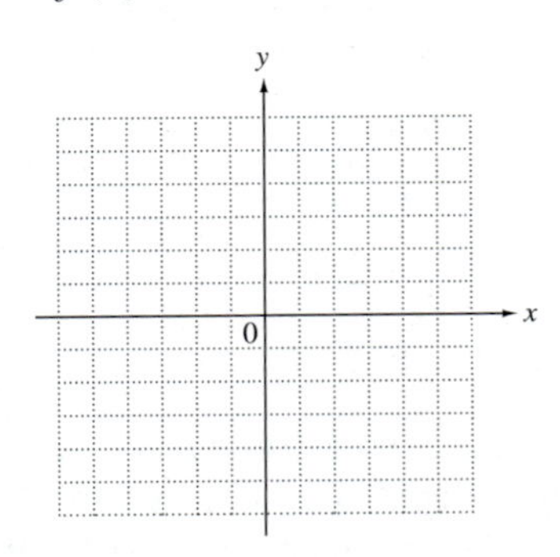

25. $f(x) = -\sqrt{25 - x^2}$

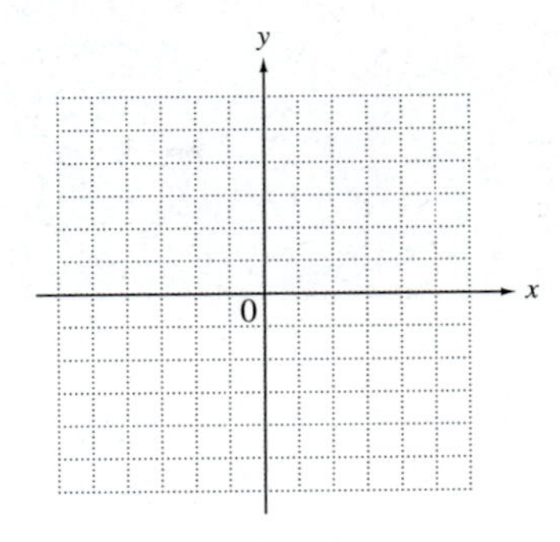

26. $\frac{y}{3} = \sqrt{1 + \frac{x^2}{9}}$

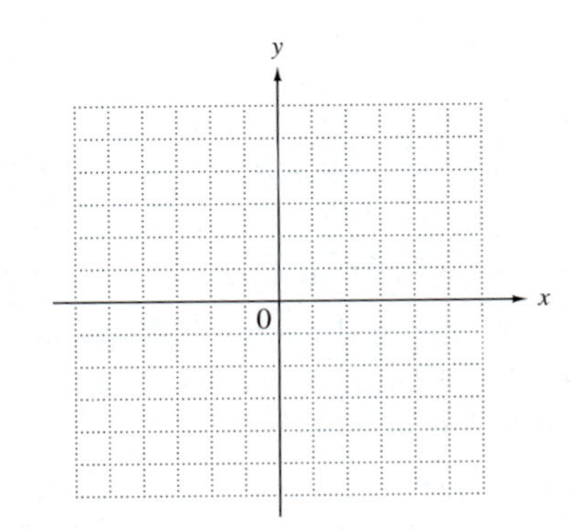

27. $y = \sqrt{\frac{x + 4}{2}}$

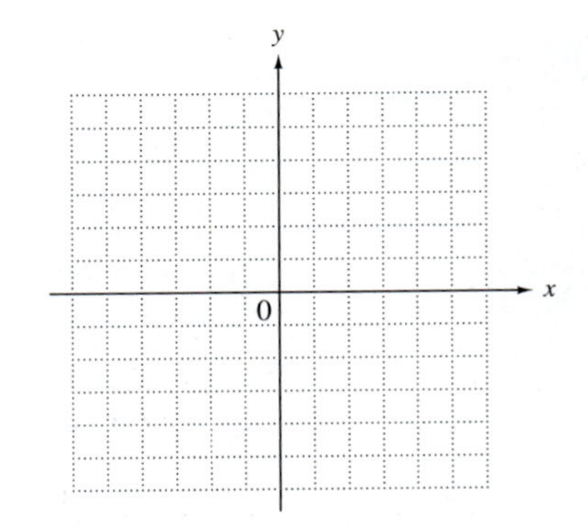

28. $y = -2\sqrt{\frac{9 - x^2}{9}}$

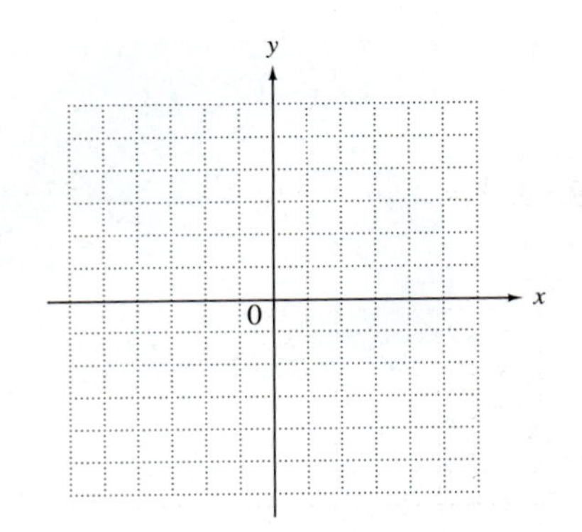

Solve each problem.

29. Two buildings in a sports complex are shaped and positioned like a portion of the branches of the hyperbola with equation

$$400x^2 - 625y^2 = 250{,}000,$$

where x and y are in meters.

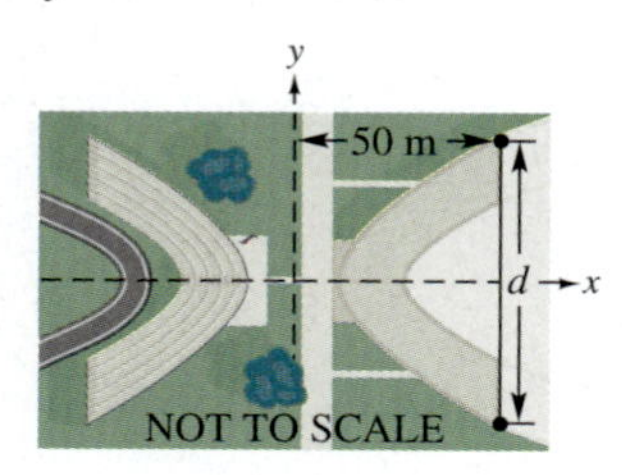

(a) How far apart are the buildings at their closest point?

(b) Find the distance d in the figure.

30. In rugby, after a *try* (similar to a touchdown in American football) the scoring team attempts a kick for extra points. The ball must be kicked from directly behind the point where the try was scored. The kicker can choose the distance but cannot move the ball sideways. It can be shown that the kicker's best choice is on the hyperbola with equation

$$\frac{x^2}{g^2} - \frac{y^2}{g^2} = 1,$$

where $2g$ is the distance between the goal posts. Since the hyperbola approaches its asymptotes, it is easier for the kicker to estimate points on the asymptotes instead of on the hyperbola. What are the asymptotes of this hyperbola? Why is it relatively easy to estimate them? (*Source:* Isaksen, Daniel C., "How to Kick a Field Goal," *The College Mathematics Journal,* September 1996.)

31. When a satellite is launched into orbit, the shape of its trajectory is determined by its velocity. The trajectory will be hyperbolic if the velocity V, in meters per second, satisfies the inequality

$$V > \frac{2.82 \times 10^7}{\sqrt{D}},$$

where D is the distance, in meters, from the center of Earth. For what values of V will the trajectory be hyperbolic if $D = 4.25 \times 10^7$ m? (*Source:* Kaler, James B., *Astronomy!,* Addison-Wesley, 1997.)

32. The percent of women in the work force has increased steadily for many years. The line graph shows the change for the period from 1975 to 1999, where $x = 75$ represents 1975, $x = 80$ represents 1980, and so on.

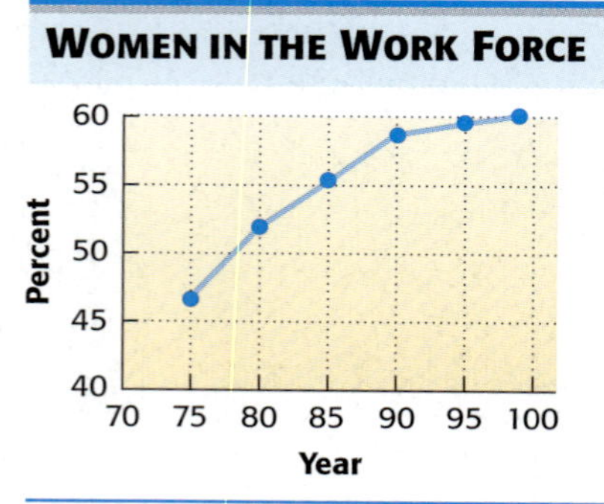

Source: U.S. Bureau of Labor Statistics.

The graph resembles the upper branch of a horizontal hyperbola. Using statistical methods, we found the corresponding square root equation

$$y = .607\sqrt{383.9 + x^2},$$

which closely approximates the line graph.

(a) According to the graph, what percent of women were in the work force in 1985?

(b) According to the equation, what percent of women worked in 1985? (Round to the nearest percent.)

11.4 Nonlinear Systems of Equations

Objectives

1. Solve a nonlinear system by substitution.
2. Use the elimination method to solve a system with two second-degree equations.
3. Solve a system that requires a combination of methods.

An equation in which some terms have more than one variable or a variable of degree 2 or greater is called a **nonlinear equation.** A **nonlinear system of equations** includes at least one nonlinear equation.

When solving a nonlinear system, it helps to visualize the types of graphs of the equations of the system to determine the possible number of points of intersection. For example, if a system includes two equations where the graph of one is a parabola and the graph of the other is a line, then there may be 0, 1, or 2 points of intersection, as illustrated in Figure 25.

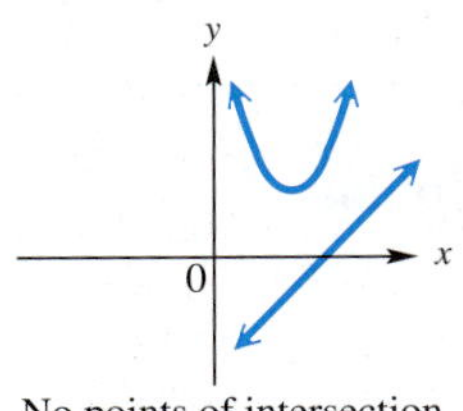

No points of intersection

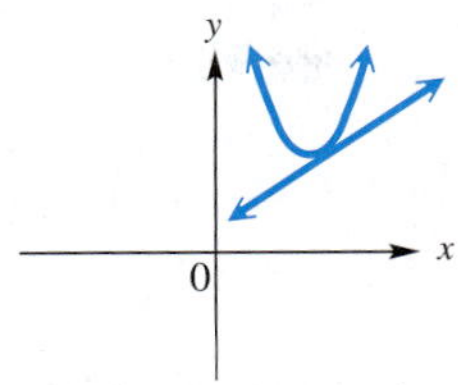

One point of intersection

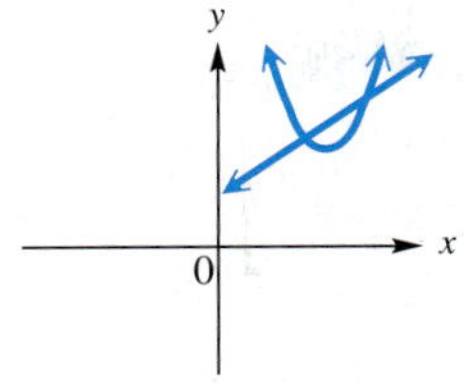

Two points of intersection

Figure 25

1 Solve a nonlinear system by substitution. We solve nonlinear systems by the elimination method, the substitution method, or a combination of the two. The substitution method is usually best when one of the equations is linear.

Example 1 Solving a Nonlinear System by Substitution

Solve the system.

$$x^2 + y^2 = 9 \quad (1)$$
$$2x - y = 3 \quad (2)$$

The graph of (1) is a circle and the graph of (2) is a line. Visualizing the possible ways the graphs could intersect indicates that there may be 0, 1, or 2 points of intersection. It is best to solve the linear equation first for one of the two variables; then substitute the resulting expression into the nonlinear equation to obtain an equation in one variable.

$$2x - y = 3 \quad (2)$$
$$y = 2x - 3 \quad (3)$$

Substitute $2x - 3$ for y in equation (1).

$$x^2 + (\mathbf{2x - 3})^2 = 9$$
$$x^2 + 4x^2 - 12x + 9 = 9$$
$$5x^2 - 12x = 0$$
$$x(5x - 12) = 0 \quad \text{GCF is } x.$$
$$x = 0 \quad \text{or} \quad x = \frac{12}{5} \quad \text{Zero-factor property}$$

Let $x = 0$ in equation (3) to get $y = -3$. If $x = \frac{12}{5}$, then $y = \frac{9}{5}$. The solution set of the system is $\{(0, -3), (\frac{12}{5}, \frac{9}{5})\}$. The graph in Figure 26 on the next page confirms the two points of intersection.

Continued on Next Page

1 Solve each system.

(a) $x^2 + y^2 = 10$
$x = y + 2$

(b) $x^2 - 2y^2 = 8$
$y + x = 6$

2 Solve each system.

(a) $xy = 8$
$x + y = 6$

(b) $xy + 10 = 0$
$4x + 9y = -2$

Answers

1. (a) $\{(3, 1), (-1, -3)\}$
(b) $\{(4, 2), (20, -14)\}$
2. (a) $\{(4, 2), (2, 4)\}$
(b) $\left\{(-5, 2), \left(\frac{9}{2}, -\frac{20}{9}\right)\right\}$

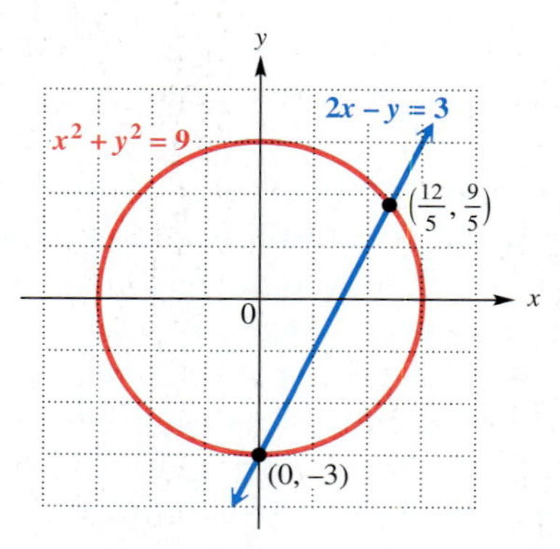

Figure 26

Work Problem 1 at the Side.

Example 2 Solving a Nonlinear System by Substitution

Solve the system.

$$6x - y = 5 \quad (1)$$
$$xy = 4 \quad (2)$$

The graph of (1) is a line. We have not specifically mentioned equations like (2); however, it can be shown by plotting points that its graph is a hyperbola. Visualizing a line and a hyperbola indicates that there may be 0, 1, or 2 points of intersection. Since neither equation has a squared term, we can solve either equation for one of the variables and then substitute the result into the other equation. Solving $xy = 4$ for x gives $x = \frac{4}{y}$. Substitute $\frac{4}{y}$ for x in equation (1).

$$6\left(\frac{4}{y}\right) - y = 5 \qquad \text{Let } x = \frac{4}{y}.$$
$$\frac{24}{y} - y = 5$$
$$24 - y^2 = 5y \qquad \text{Multiply by } y\ (y \neq 0).$$
$$0 = y^2 + 5y - 24$$
$$0 = (y - 3)(y + 8) \qquad \text{Factor.}$$
$$y = 3 \quad \text{or} \quad y = -8 \qquad \text{Zero-factor property}$$

We substitute these results into $x = \frac{4}{y}$ to obtain the corresponding values of x.

$$\text{If } y = 3, \text{ then } x = \frac{4}{3}. \qquad \text{If } y = -8, \text{ then } x = -\frac{1}{2}.$$

The solution set of the system is $\{(\frac{4}{3}, 3), (-\frac{1}{2}, -8)\}$. The graph in Figure 27 shows that there are two points of intersection.

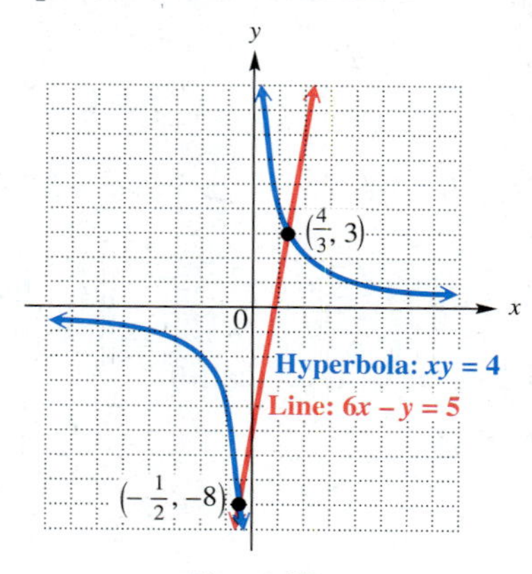

Figure 27

Work Problem 2 at the Side.

2 Use the elimination method to solve a system with two second-degree equations. The elimination method is often used when both equations are second degree.

Example 3 Solving a Nonlinear System by Elimination

Solve the system.

$$x^2 + y^2 = 9 \quad (1)$$
$$2x^2 - y^2 = -6 \quad (2)$$

The graph of (1) is a circle, while the graph of (2) is a hyperbola. By analyzing the possibilities we conclude that there may be 0, 1, 2, 3, or 4 points of intersection. Adding the two equations will eliminate y, leaving an equation that can be solved for x.

$$\begin{aligned} x^2 + y^2 &= 9 \\ 2x^2 - y^2 &= -6 \\ \hline 3x^2 \quad\quad &= 3 \end{aligned}$$
$$x^2 = 1$$
$$x = 1 \quad \text{or} \quad x = -1$$

Each value of x gives corresponding values for y when substituted into one of the original equations. Using equation (1) gives the following.

If $x = 1$, then
$$1^2 + y^2 = 9$$
$$y^2 = 8$$
$$y = \sqrt{8} \quad \text{or} \quad y = -\sqrt{8}$$
$$y = 2\sqrt{2} \quad \text{or} \quad y = -2\sqrt{2}.$$

If $x = -1$, then
$$(-1)^2 + y^2 = 9$$
$$y^2 = 8$$
$$y = 2\sqrt{2} \quad \text{or} \quad y = -2\sqrt{2}.$$

The solution set is $\{(1, 2\sqrt{2}), (1, -2\sqrt{2}), (-1, 2\sqrt{2}), (-1, -2\sqrt{2})\}$. Figure 28 shows the four points of intersection.

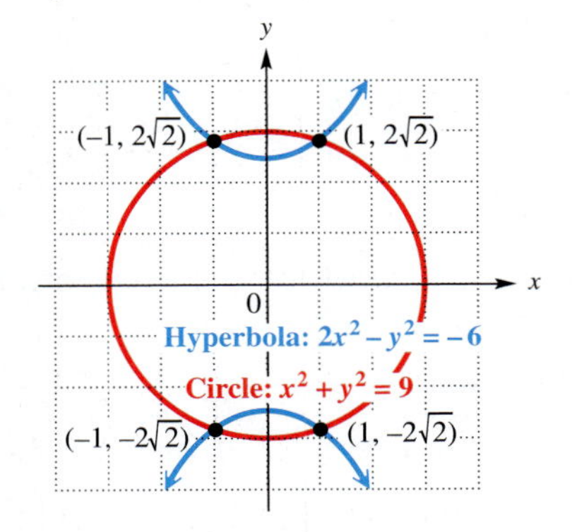

Figure 28

Work Problem 3 at the Side.

3 Solve a system that requires a combination of methods. Solving a system of second-degree equations may require a combination of methods.

---

3 Solve each system.

(a) $x^2 + y^2 = 41$
$x^2 - y^2 = 9$

(b) $x^2 + 3y^2 = 40$
$4x^2 - y^2 = 4$

Answers

3. (a) $\{(5, 4), (5, -4), (-5, 4), (-5, -4)\}$
(b) $\{(2, 2\sqrt{3}), (2, -2\sqrt{3}), (-2, 2\sqrt{3}), (-2, -2\sqrt{3})\}$

Example 4 **Solving a Nonlinear System by a Combination of Methods**

Solve the system.

$$x^2 + 2xy - y^2 = 7 \quad (1)$$
$$x^2 - y^2 = 3 \quad (2)$$

While we have not graphed equations like (1), its graph is a hyperbola. The graph of (2) is also a hyperbola. Two hyperbolas may have 0, 1, 2, 3, or 4 points of intersection. We use the elimination method here in combination with the substitution method. We begin by eliminating the squared terms by multiplying each side of equation (2) by -1 and then adding the result to equation (1).

$$\begin{aligned} x^2 + 2xy - y^2 &= 7 \\ -x^2 \qquad\quad + y^2 &= -3 \\ \hline 2xy \qquad\quad &= 4 \end{aligned}$$

Next, we solve $2xy = 4$ for y. (Either variable would do.)

$$2xy = 4$$
$$y = \frac{2}{x} \quad (3)$$

Now, we substitute $y = \frac{2}{x}$ into one of the original equations. It is easier to do this with equation (2).

$$\begin{aligned} x^2 - y^2 &= 3 && (2) \\ x^2 - \left(\frac{2}{x}\right)^2 &= 3 \\ x^2 - \frac{4}{x^2} &= 3 \\ x^4 - 4 &= 3x^2 && \text{Multiply by } x^2, x \neq 0. \\ x^4 - 3x^2 - 4 &= 0 && \text{Subtract } 3x^2. \\ (x^2 - 4)(x^2 + 1) &= 0 && \text{Factor.} \end{aligned}$$

$$x^2 - 4 = 0 \quad \text{or} \quad x^2 + 1 = 0$$
$$x^2 = 4 \quad \text{or} \quad x^2 = -1$$
$$x = 2 \quad \text{or} \quad x = -2 \qquad x = i \quad \text{or} \quad x = -i$$

Substituting these four values of x into equation (3) gives the corresponding values for y.

If $x = 2$, then $y = 1$. If $x = i$, then $y = -2i$.
If $x = -2$, then $y = -1$. If $x = -i$, then $y = 2i$.

Note that if we substitute the x-values we found into equation (1) or (2) instead of into equation (3), we get extraneous solutions. It is always wise to check all solutions in both of the given equations. There are four ordered pairs in the solution set, two with real values and two with imaginary values. The solution set is

$$\{(2, 1), (-2, -1), (i, -2i), (-i, 2i)\}.$$

Continued on Next Page

The graph of the system, shown in Figure 29, shows only the two real intersection points because the graph is in the real number plane. The two ordered pairs with imaginary components are solutions of the system, but do not appear on the graph.

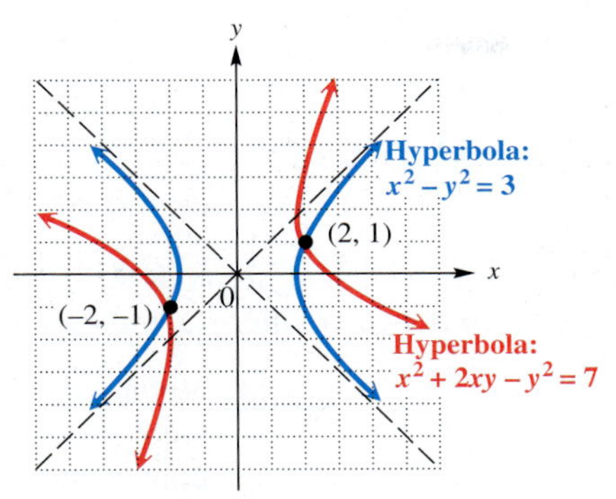

Figure 29

Work Problem 4 at the Side.

NOTE

In the examples of this section, we analyzed the possible number of points of intersection of the graphs in each system. However, in Examples 2 and 4, we worked with equations whose graphs had not been studied. Keep in mind that it is not absolutely essential to visualize the number of points of intersection in order to solve the system. Furthermore, as in Example 4, there are sometimes imaginary solutions to nonlinear systems which do not appear as points of intersection in the real plane. Visualizing the geometry of the graphs is only an aid to solving these systems.

4 Solve each system.

(a)
$$\begin{aligned} x^2 + xy + y^2 &= 3 \\ x^2 \qquad + y^2 &= 5 \end{aligned}$$

(b)
$$\begin{aligned} x^2 + 7xy - 2y^2 &= -8 \\ -2x^2 \qquad + 4y^2 &= 16 \end{aligned}$$

ANSWERS

4. (a) $\{(1, -2), (-1, 2), (2, -1), (-2, 1)\}$
(b) $\{(0, 2), (0, -2), (2i\sqrt{2}, 0), (-2i\sqrt{2}, 0)\}$

Focus on Real-Data Applications

Who Arrived First?

Suppose Vivian, Tommy, Carmen, and Manuel all leave home at 9:00 A.M. to drive to Atlanta, Georgia, along U.S. Interstate 75. The distance between Valdosta, Georgia and Atlanta is 245 mi.

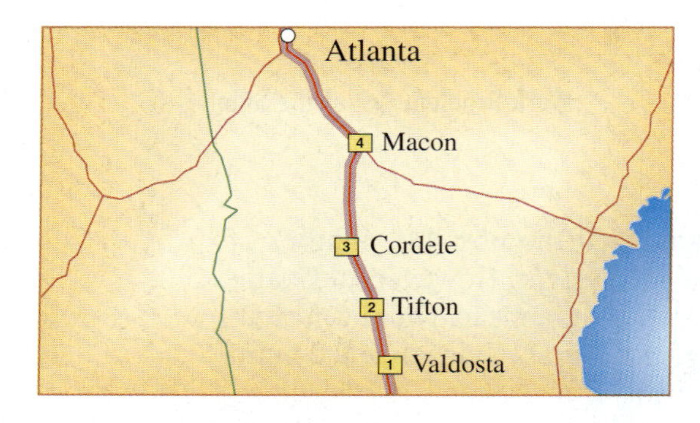

1. Vivian leaves from Valdosta, Georgia. She drives a Miata convertible and cruises at 85 mph.
2. Tommy leaves from Tifton, Georgia, which is 50 mi north of Valdosta. He drives a 1984 Toyota truck averaging 65 mph.
3. Carmen leaves from Cordele, Georgia, which is 40 mi north of Tifton. She drives a 1995 Honda Accord averaging 60 mph.
4. Manuel leaves from Macon, Georgia, which is 65 mi north of Cordele. He is riding a twenty-speed bike at 25 mph.

For Group Discussion

The independent variable is the number of hours traveled after 9:00 A.M. The dependent variable is distance, relative to Valdosta, after t hours. To compare the four trips, distances are measured from a common starting point in Valdosta. Recall that distance equals the product of rate and time. Tommy will have driven a distance of $65t$ mi after t hours, and since he starts 50 mi north of Valdosta, an equation that represents Tommy's distance relative to Valdosta is $d = 65t + 50$.

1. Write equations to represent the distances after t hr (relative to Valdosta) of Vivian, Carmen, and Manuel.

2. On a sheet of graph paper, sketch graphs of the four distance equations. Graph the horizontal line $d = 245$ to represent the distance between Valdosta and Atlanta. Based on your graphs, list the order in which the drivers reach Atlanta.

3. Based on your equations from Problem 1, at what time (rounded to the nearest minute) does each driver reach Atlanta? Are the results consistent with your conclusions based on your graphs?

4. Use a system of equations to find each time and location (distance from Valdosta). Round times to the nearest minute and distances to the nearest tenth of a mile, as necessary.
   (a) Find the time and location at which Vivian passes Tommy.
   (b) Find the time and location at which Carmen passes Manuel.
   (c) Does Carmen pass any other traveler before reaching Atlanta?

11.4 EXERCISES

FOR EXTRA HELP: Student's Solutions Manual | MyMathLab.com | InterAct Math Tutorial Software | AW Math Tutor Center | www.mathxl.com | Digital Video Tutor CD 9 Videotape 20

1. Write an explanation of the steps you would use to solve the system

$$x^2 + y^2 = 25$$
$$y = x - 1$$

by the substitution method. Why would the elimination method not be appropriate for this system?

2. Write an explanation of the steps you would use to solve the system

$$x^2 + y^2 = 12$$
$$x^2 - y^2 = 13$$

by the elimination method.

Each sketch represents the graphs of a pair of equations in a system. How many points are in each solution set?

3.

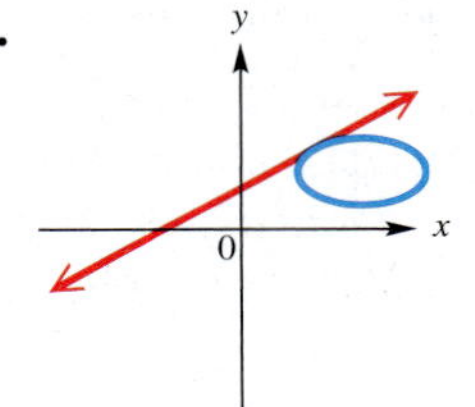

4.

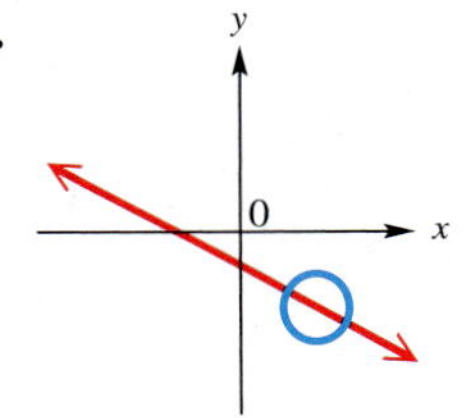

5.

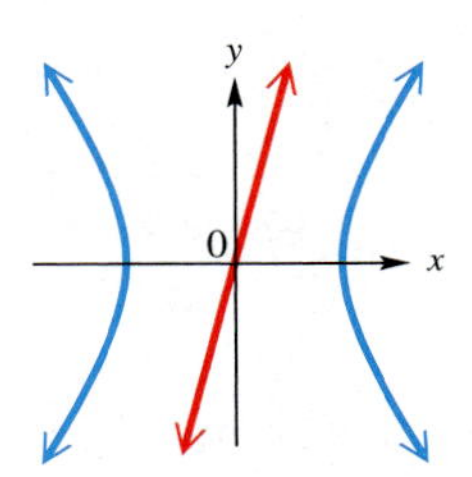

6.

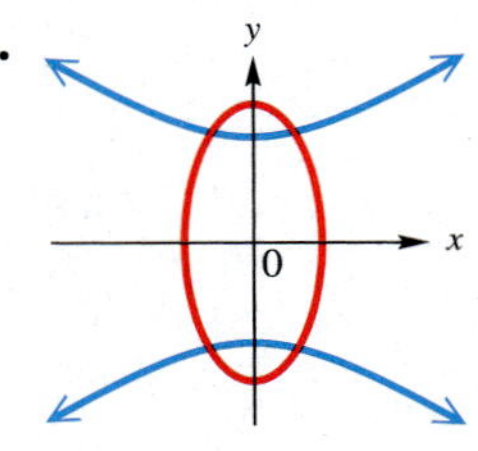

Suppose that a nonlinear system is composed of equations whose graphs are those described, and the number of points of intersection of the two graphs is as given. Make a sketch satisfying these conditions. (There may be more than one way to do this.)

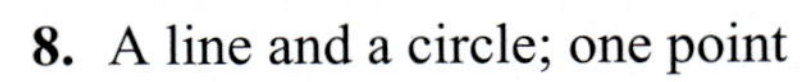

7. A line and a circle; no points

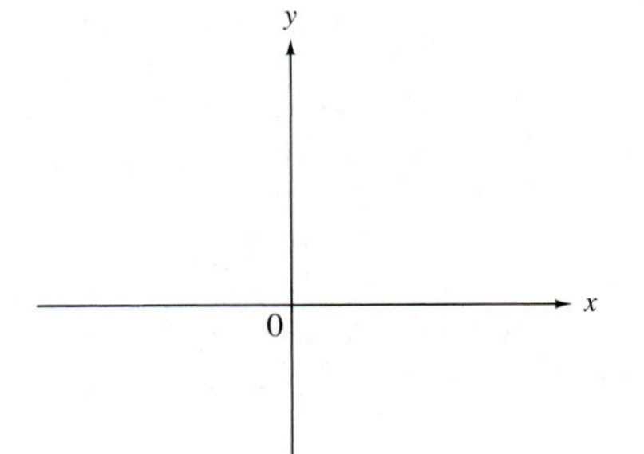

8. A line and a circle; one point

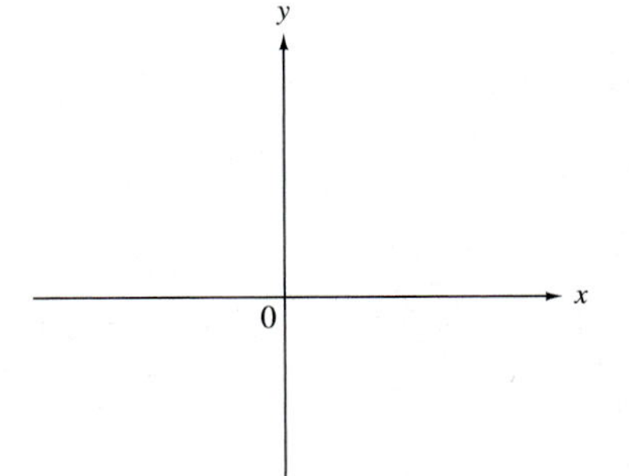

9. A line and an ellipse; two points

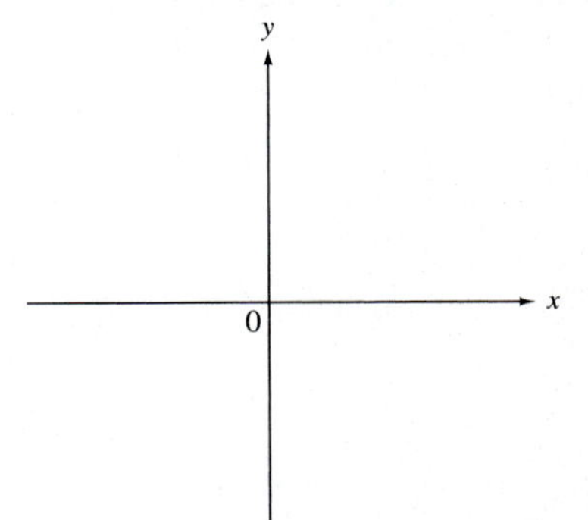

10. A line and a hyperbola; no points

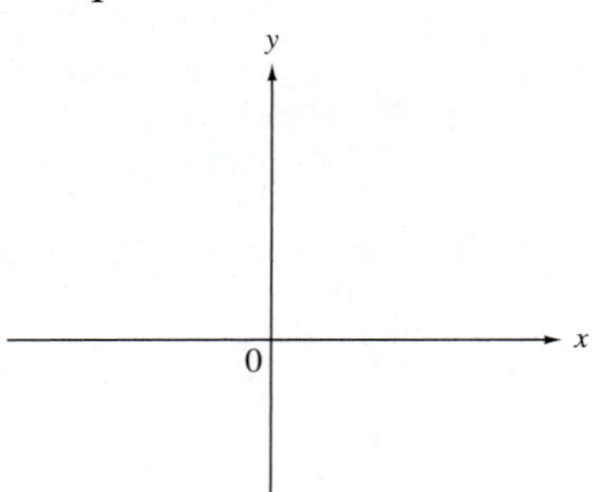

11. A circle and an ellipse; four points

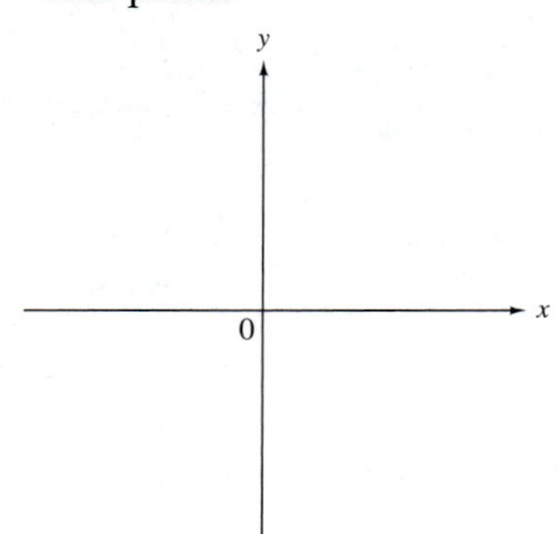

12. A parabola and an ellipse; one point

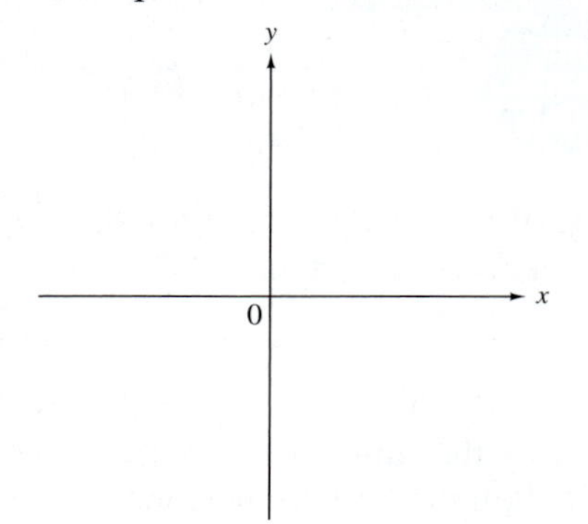

Solve each system by the substitution method. See Examples 1 and 2.

13. $y = 4x^2 - x$
$y = x$

14. $y = x^2 + 6x$
$3y = 12x$

15. $y = x^2 + 6x + 9$
$x + y = 3$

16. $y = x^2 + 8x + 16$
$x - y = -4$

17. $x^2 + y^2 = 2$
$2x + y = 1$

18. $2x^2 + 4y^2 = 4$
$x = 4y$

19. $xy = 4$
$3x + 2y = -10$

20. $xy = -5$
$2x + y = 3$

21. $xy = -3$
$x + y = -2$

22. $xy = 12$
$x + y = 8$

23. $y = 3x^2 + 6x$
$y = x^2 - x - 6$

24. $y = 2x^2 + 1$
$y = 5x^2 + 2x - 7$

25. $2x^2 - y^2 = 6$
$y = x^2 - 3$

26. $x^2 + y^2 = 4$
$y = x^2 - 2$

Solve each system using the elimination method or a combination of the elimination and substitution methods. See Examples 3 and 4.

27. $3x^2 + 2y^2 = 12$
$x^2 + 2y^2 = 4$

28. $2x^2 + y^2 = 28$
$4x^2 - 5y^2 = 28$

29. $xy = 6$
$3x^2 - y^2 = 12$

30. $xy = 5$
$2y^2 - x^2 = 5$

31. $2x^2 + 2y^2 = 8$
$3x^2 + 4y^2 = 24$

32. $5x^2 + 5y^2 = 20$
$x^2 + 2y^2 = 2$

33. $x^2 + xy + y^2 = 15$
$x^2 + y^2 = 10$

34. $2x^2 + 3xy + 2y^2 = 21$
$x^2 + y^2 = 6$

35. $3x^2 + 2xy - 3y^2 = 5$
$-x^2 - 3xy + y^2 = 3$

36. $-2x^2 + 7xy - 3y^2 = 4$
$2x^2 - 3xy + 3y^2 = 4$

Solve each problem by using a nonlinear system.

37. The area of a rectangular rug is 84 ft^2 and its perimeter is 38 ft. Find the length and width of the rug.

38. Find the length and width of a rectangular room whose perimeter is 50 m and whose area is 100 m^2.

39. A company has found that the price p (in dollars) of its scientific calculator is related to the supply x (in thousands) by the equation

$$px = 16.$$

The price is related to the demand x (in thousands) for the calculator by the equation

$$p = 10x + 12.$$

The *equilibrium price* is the value of p where demand equals supply. Find the equilibrium price and the supply/demand at that price by solving a system of equations. (*Hint:* Demand, price, and supply must all be positive.)

40. The calculator company in Exercise 39 has also determined that the cost y to make x (thousand) calculators is

$$y = 4x^2 + 36x + 20,$$

while the revenue y from the sale of x (thousand) calculators is

$$36x^2 - 3y = 0.$$

Find the *break-even point,* where cost equals revenue, by solving a system of equations.

41. Historically in the United States, the number of bachelor's degrees earned by men has been greater than the number earned by women. In the 1970s, however, this began to change as the number earned by men decreased. It stayed fairly constant in the 1980s, and then in the 1990s slowly began to increase again. Meanwhile, the number of bachelor's degrees earned by women has continued to rise steadily throughout this period. Functions that model the situation are defined by the following equations, where y is the number of degrees (in thousands) granted in year x, with $x = 0$ corresponding to 1970.

Men: $y = .138x^2 + .064x + 451$

Women: $y = 12.1x + 334$

Solve this system of equations to find the year when the same number of bachelor's degrees was awarded to men and women. How many bachelor's degrees were awarded in that year? Give answer to the nearest ten thousand. (*Source:* U.S. National Center for Education Statistics, *Digest of Education Statistics,* annual.)

42. Andy Grove, chairman of chip maker Intel Corp., recently noted that decreasing prices for computers and stable prices for Internet access implied that the trend lines for these costs either have crossed or soon will. He predicted that the time is not far away when computers, like cell phones, may be given away to sell on-line time. To see this, assume a price of \$1000 for a computer, and let x represent the number of months it will be used. (*Source:* Corcoran, Elizabeth, "Can Free Computers Be That Far Away?", *Washington Post,* from *Sacramento Bee,* February 3, 1999.)

(a) Write an equation for the monthly cost y of the computer over this period.

(b) The average monthly on-line cost is about \$20. Assume this will remain constant and write an equation to express this cost.

(c) Solve the system of equations from parts (a) and (b). Interpret your answer in relation to the situation.

11.5 SECOND-DEGREE INEQUALITIES AND SYSTEMS OF INEQUALITIES

OBJECTIVES

1. Graph second-degree inequalities.
2. Graph the solution set of a system of inequalities.

1 Graph second-degree inequalities. The linear inequality $3x + 2y \leq 5$ is graphed by first graphing the boundary line $3x + 2y = 5$. A **second-degree inequality** is an inequality with at least one variable of degree 2 and no variable with degree greater than 2. An example is $x^2 + y^2 \leq 36$. Such inequalities are graphed in the same way. The boundary of the inequality $x^2 + y^2 \leq 36$ is the graph of the equation $x^2 + y^2 = 36$, a circle with radius 6 and center at the origin, as shown in Figure 30. The inequality $x^2 + y^2 \leq 36$ will include either the points outside the circle or the points inside the circle, as well as the boundary. We decide which region to shade by substituting any test point not on the circle, such as (0, 0), into the original inequality. Since $0^2 + 0^2 \leq 36$ is a true statement, the original inequality includes the points inside the circle, the shaded region in Figure 30, and the boundary.

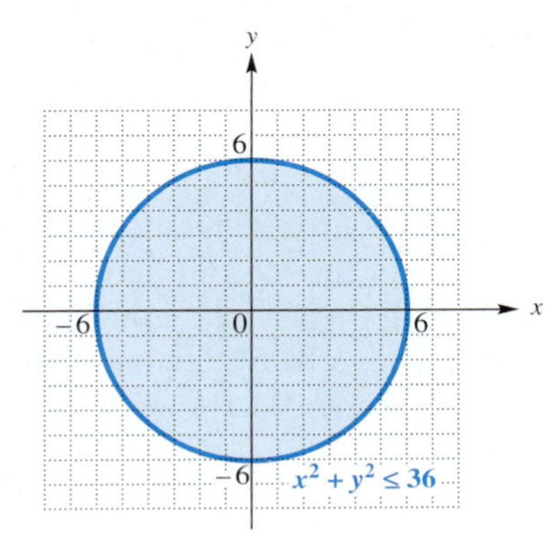

Figure 30

Example 1 Graphing a Second-Degree Inequality

Graph $y < -2(x - 4)^2 - 3$.

The boundary, $y = -2(x - 4)^2 - 3$, is a parabola that opens down with vertex at (4, −3). Using (0, 0) as a test point gives

$$0 < -2(0 - 4)^2 - 3 \quad ?$$
$$0 < -32 - 3 \quad ?$$
$$0 < -35. \quad \text{False}$$

Because the final inequality is a false statement, the points in the region containing (0, 0) do not satisfy the inequality. Figure 31 shows the final graph; the parabola is drawn as a dashed curve since the points of the parabola itself do not satisfy the inequality, and the region inside (or below) the parabola is shaded.

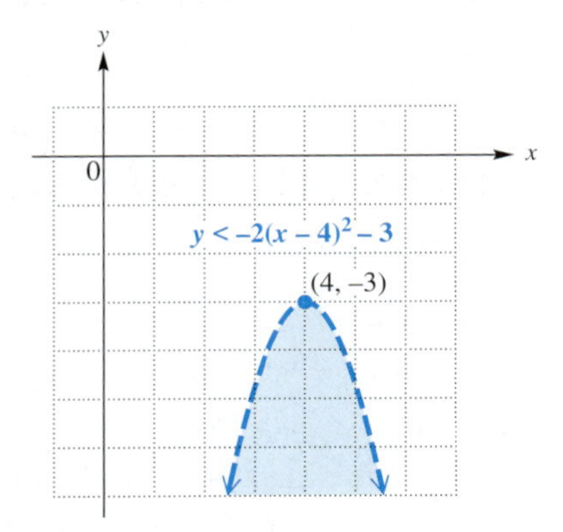

Figure 31

1 Graph $y \geq (x+1)^2 - 5$.

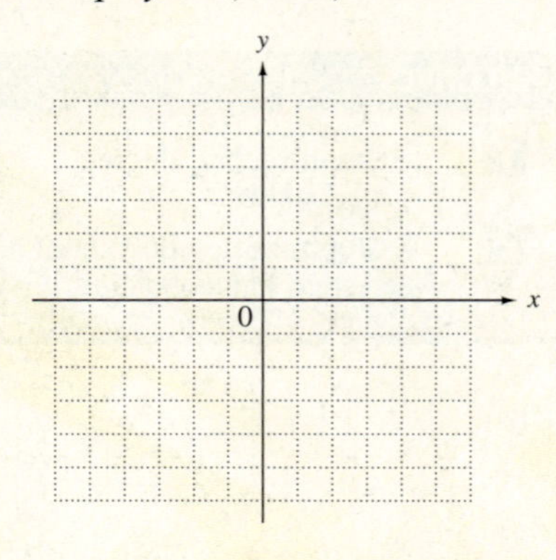

NOTE

Since the substitution is easy, the origin is the test point of choice unless the graph actually passes through (0, 0).

Work Problem 1 at the Side.

Example 2 Graphing a Second-Degree Inequality

Graph $16y^2 \leq 144 + 9x^2$.

First rewrite the inequality as follows.

$$16y^2 - 9x^2 \leq 144 \quad \text{Subtract } 9x^2.$$

$$\frac{y^2}{9} - \frac{x^2}{16} \leq 1 \quad \text{Divide by 144.}$$

This form shows that the boundary is the hyperbola given by

$$\frac{y^2}{9} - \frac{x^2}{16} = 1.$$

Since the graph is a vertical hyperbola, the desired region will be either the region between the branches or the regions above the top branch and below the bottom branch. Choose (0, 0) as a test point. Substituting into the original inequality leads to $0 \leq 144$, a true statement, so the region between the branches containing (0, 0) is shaded, as shown in Figure 32.

2 Graph $x^2 + 4y^2 > 36$.

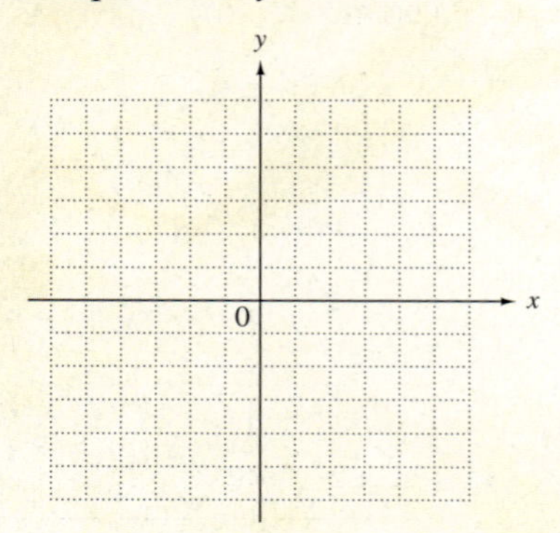

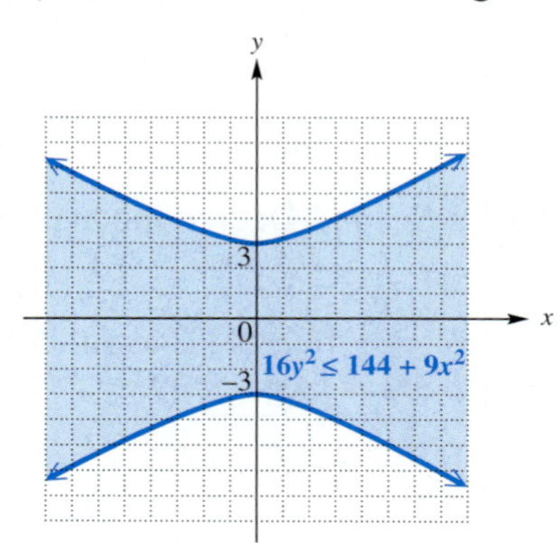

Figure 32

Work Problem 2 at the Side.

2 **Graph the solution set of a system of inequalities.** If two or more inequalities are considered at the same time, we have a **system of inequalities.** To find the solution set of the system, we find the intersection of the graphs (solution sets) of the inequalities in the system.

Example 3 Graphing a System of Two Inequalities

Graph the solution set of the system.

$$2x + 3y > 6$$
$$x^2 + y^2 < 16$$

Begin by graphing the solution set of $2x + 3y > 6$. The boundary line is the graph of $2x + 3y = 6$ and is a dashed line because of the symbol $>$. The test point (0, 0) leads to a false statement in the inequality $2x + 3y > 6$,

Continued on Next Page

ANSWERS

1.

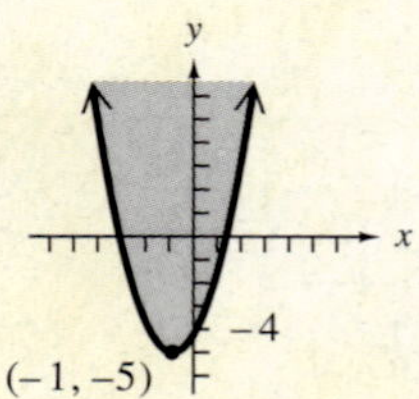

2.

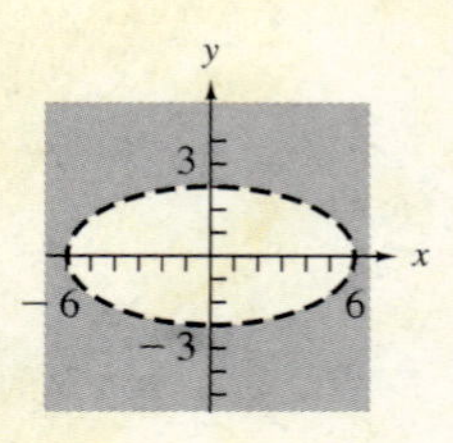

so shade the region above the line, as shown in Figure 33. The graph of $x^2 + y^2 < 16$ is the interior of a dashed circle centered at the origin with radius 4. This is shown in Figure 34.

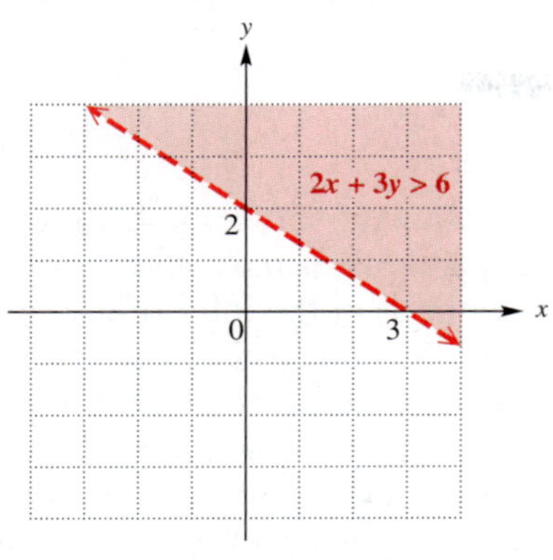

Figure 33

Figure 34

Finally, to get the graph of the solution set of the system, determine the intersection of the graphs of the two inequalities. The overlapping region in Figure 35 is the solution set.

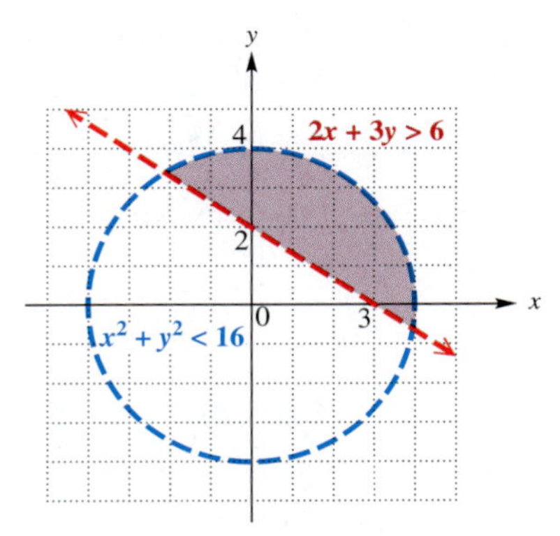

Figure 35

Work Problem **3** at the Side.

3 Graph the solution set of the system.

$$\begin{aligned} x^2 + y^2 &\le 25 \\ x + y &\le 3 \end{aligned}$$

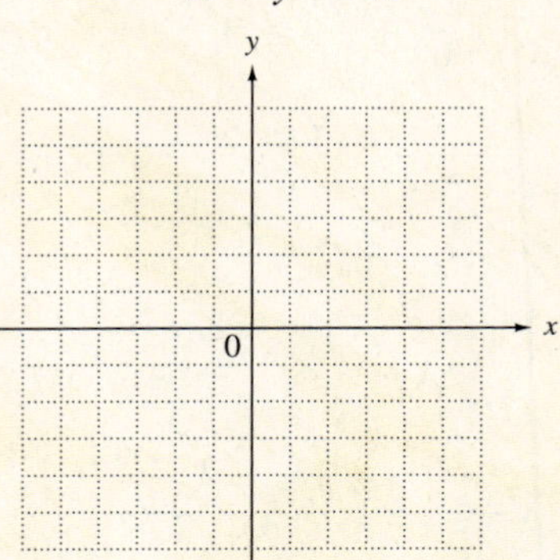

4 Graph the solution set of the system.

$$\begin{aligned} 3x - 4y &\ge 12 \\ x + 3y &\ge 6 \\ y &\le 2 \end{aligned}$$

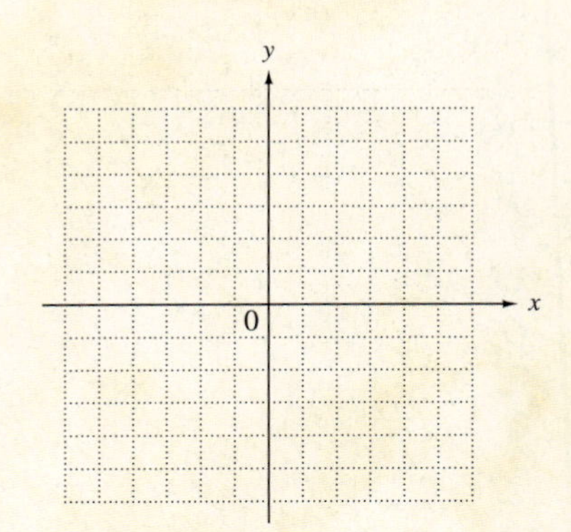

Example 4 Graphing a Linear System with Three Inequalities

Graph the solution set of the system.

$$\begin{aligned} x + y &< 1 \\ y &\le 2x + 3 \\ y &\ge -2 \end{aligned}$$

Graph each inequality separately, on the same axes. The graph of $x + y < 1$ consists of all points below the dashed line $x + y = 1$. The graph of $y \le 2x + 3$ is the region that lies below the solid line $y = 2x + 3$. Finally, the graph of $y \ge -2$ is the region above the solid horizontal line $y = -2$. The graph of the system, the intersection of these three graphs, is the triangular region enclosed by the three boundary lines in Figure 36, including two of its boundaries.

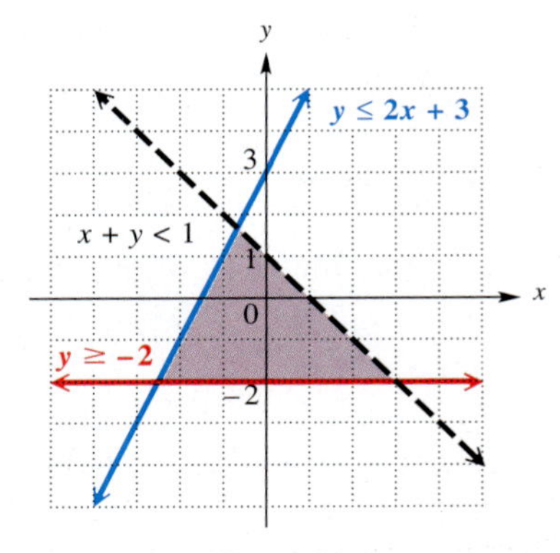

Figure 36

Work Problem **4** at the Side.

ANSWERS

3.

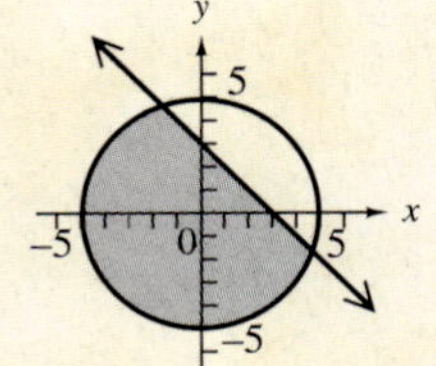

4.

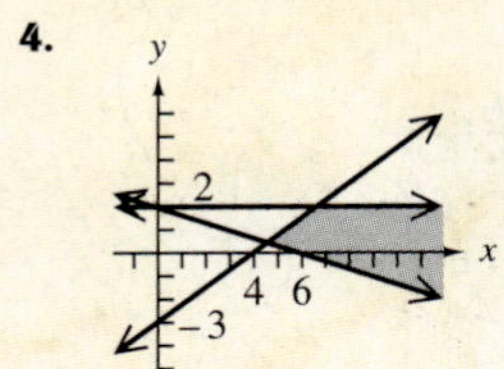

5 Graph the solution set of the system.

$$y \geq x^2 + 1$$
$$\frac{x^2}{9} + \frac{y^2}{4} \geq 1$$
$$y \leq 5$$

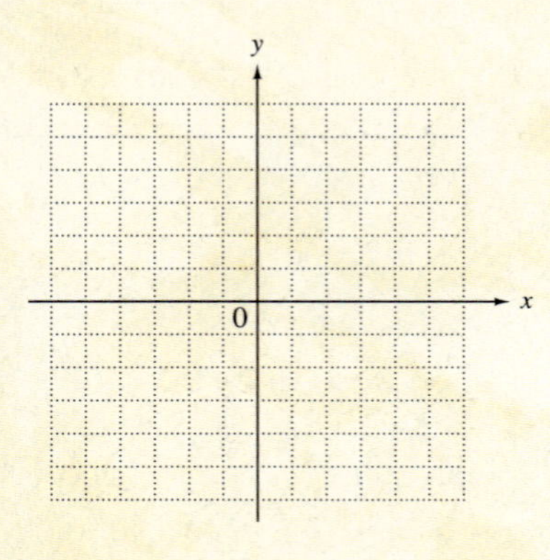

Example 5 Graphing a System with Three Inequalities

Graph the solution set of the system.

$$y \geq x^2 - 2x + 1$$
$$2x^2 + y^2 > 4$$
$$y < 4$$

The graph of $y = x^2 - 2x + 1$ is a parabola with vertex at $(1, 0)$. Those points above (or in the interior of) the parabola satisfy the condition $y > x^2 - 2x + 1$. Thus, points on the parabola or in the interior are in the solution set of $y \geq x^2 - 2x + 1$. The graph of the equation $2x^2 + y^2 = 4$ is an ellipse. We draw it as a dashed curve. To satisfy the inequality $2x^2 + y^2 > 4$, a point must lie outside the ellipse. The graph of $y < 4$ includes all points below the dashed line $y = 4$. Finally, the graph of the system is the shaded region in Figure 37 that lies outside the ellipse, inside or on the boundary of the parabola, and below the line $y = 4$.

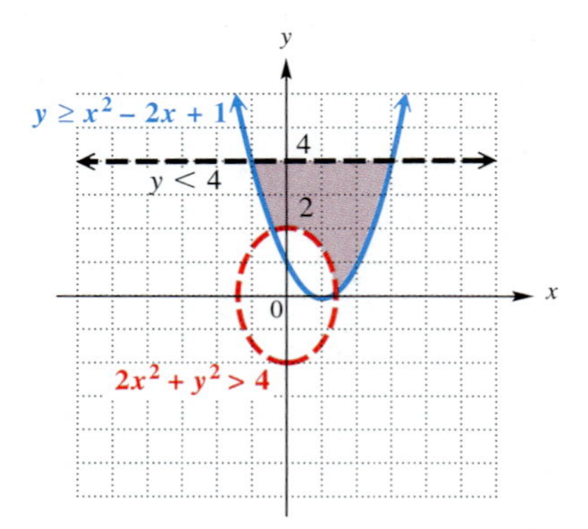

Figure 37

Work Problem 5 at the Side.

Answers

5.

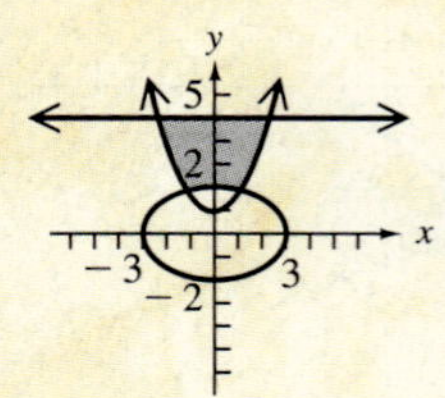

11.5 EXERCISES

FOR EXTRA HELP

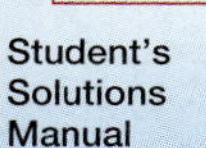
Student's Solutions Manual

MyMathLab.com

InterAct Math Tutorial Software
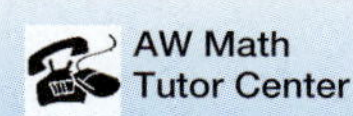
AW Math Tutor Center

www.mathxl.com
Digital Video Tutor CD 9 Videotape 20

1. Which one of the following is a description of the graph of the solution set of this system?

$$x^2 + y^2 < 25$$
$$y > -2$$

A. All points outside the circle $x^2 + y^2 = 25$ and above the line $y = -2$

B. All points outside the circle $x^2 + y^2 = 25$ and below the line $y = -2$

C. All points inside the circle $x^2 + y^2 = 25$ and above the line $y = -2$

D. All points inside the circle $x^2 + y^2 = 25$ and below the line $y = -2$

2. Fill in each blank with the appropriate response. The graph of the system

$$y > x^2 + 1$$
$$\frac{x^2}{9} + \frac{y^2}{4} > 1$$
$$y < 5$$

consists of all points ____________ (above/below) the parabola $y = x^2 + 1$, ____________ (inside/outside) the ellipse $\frac{x^2}{9} + \frac{y^2}{4} = 1$, and ____________ (above/below) the line $y = 5$.

3. Explain how to graph the solution set of a nonlinear inequality.

4. Explain how to graph the solution set of a system of inequalities.

Graph each inequality. See Examples 1 and 2.

5. $y > x^2 - 1$

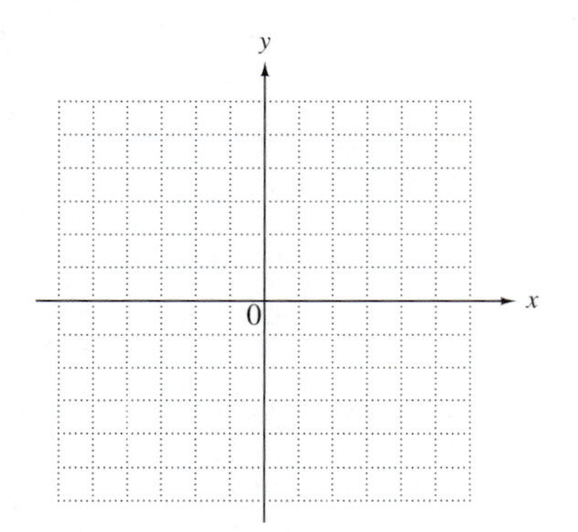

6. $y^2 > 4 + x^2$

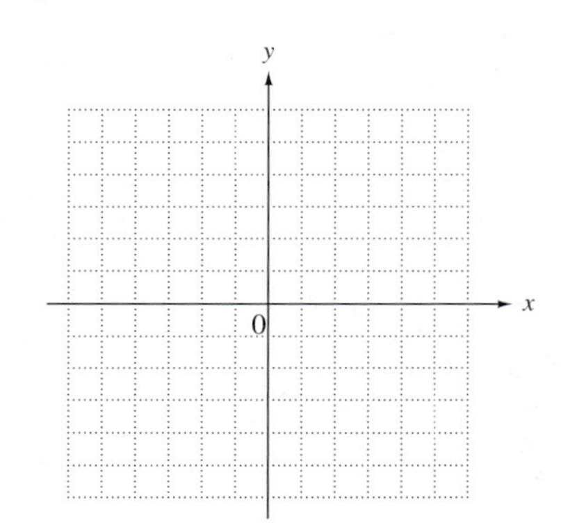

7. $y^2 \leq 4 - 2x^2$

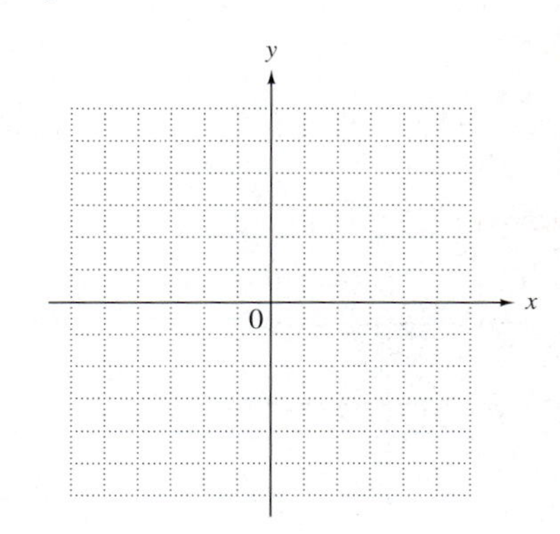

8. $y + 2 \geq x^2$

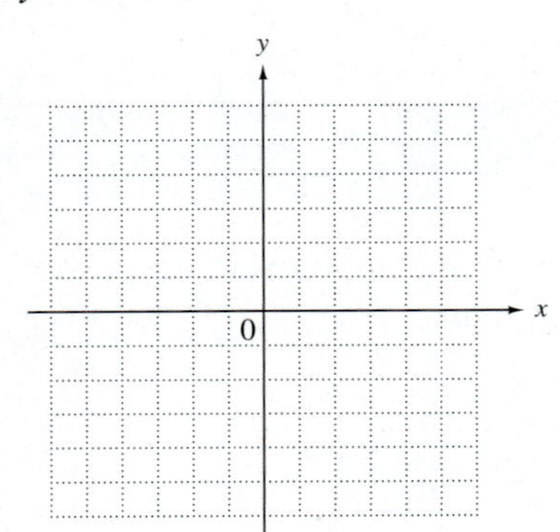

9. $x^2 \leq 16 - y^2$

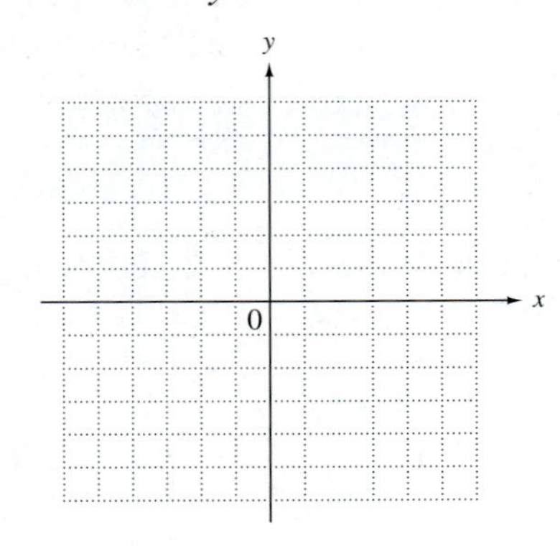

10. $2y^2 \geq 8 - x^2$

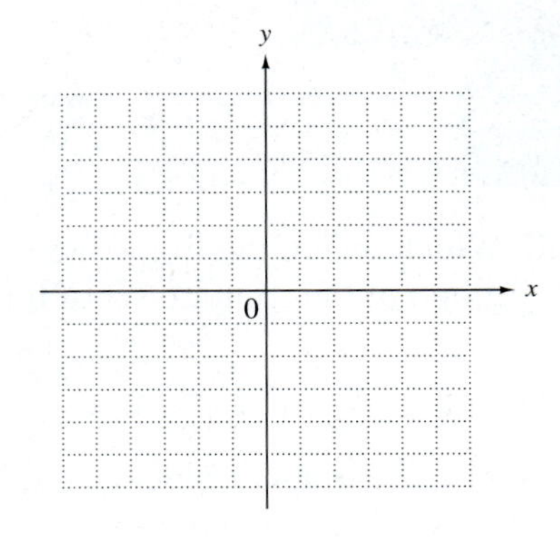

11. $x^2 \leq 16 + 4y^2$

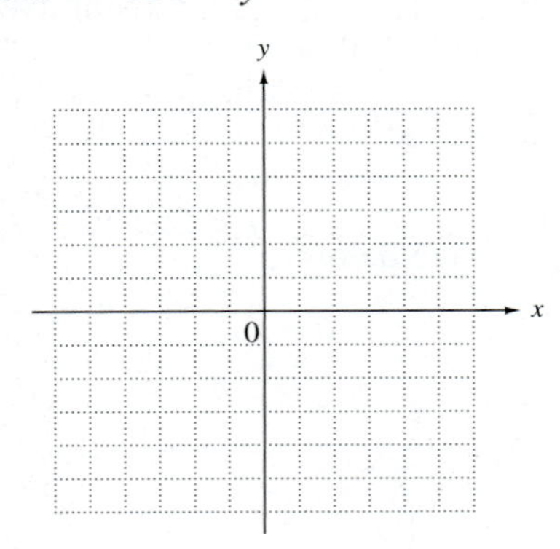

12. $y \leq x^2 + 4x + 2$

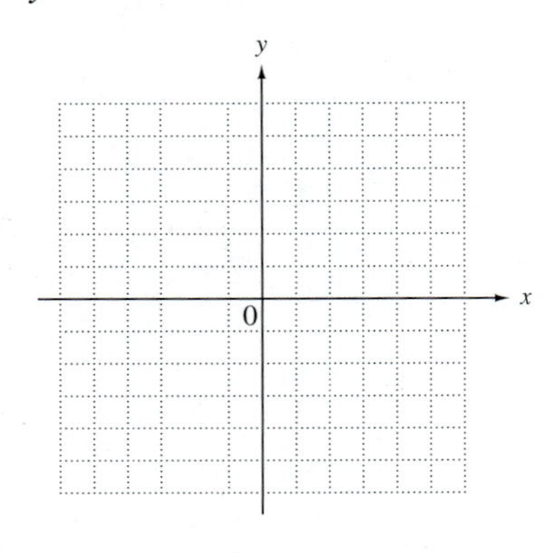

13. $9x^2 < 16y^2 - 144$

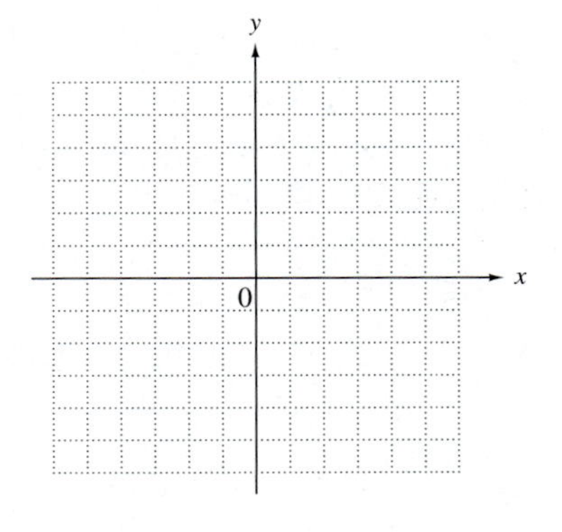

14. $9x^2 > 16y^2 + 144$

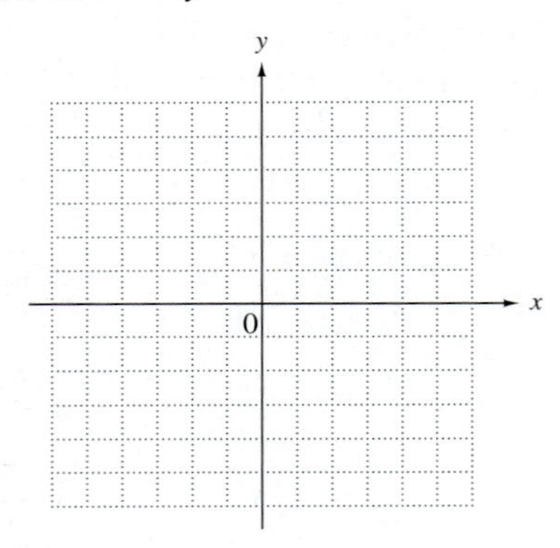

15. $4y^2 \leq 36 - 9x^2$

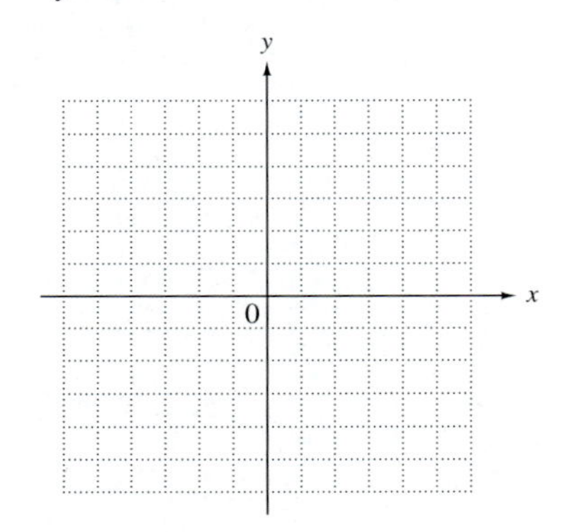

16. $x^2 - 4 \geq -4y^2$

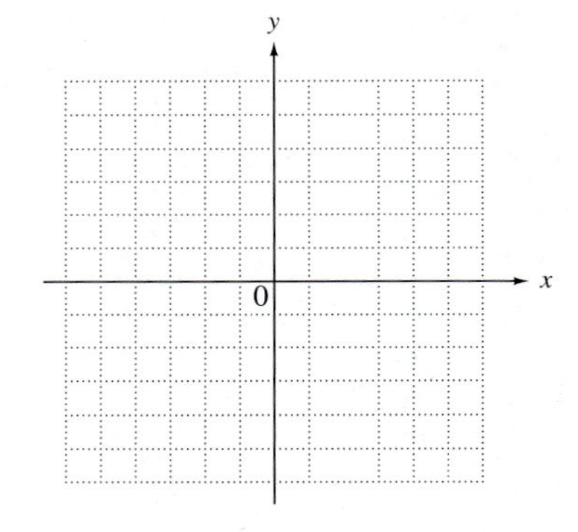

17. $x \geq y^2 - 8y + 14$

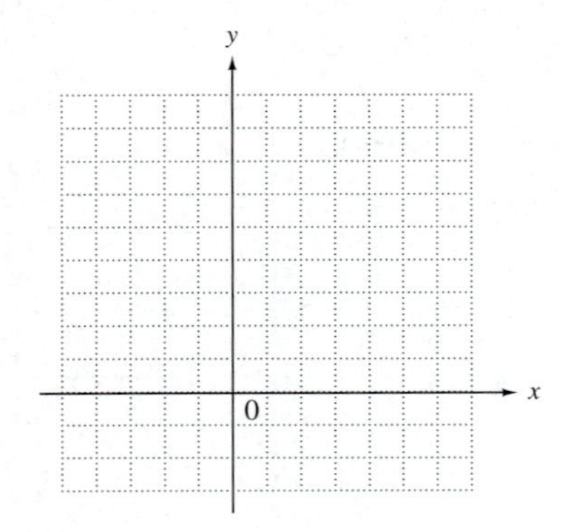

18. $x \leq -y^2 + 6y - 7$

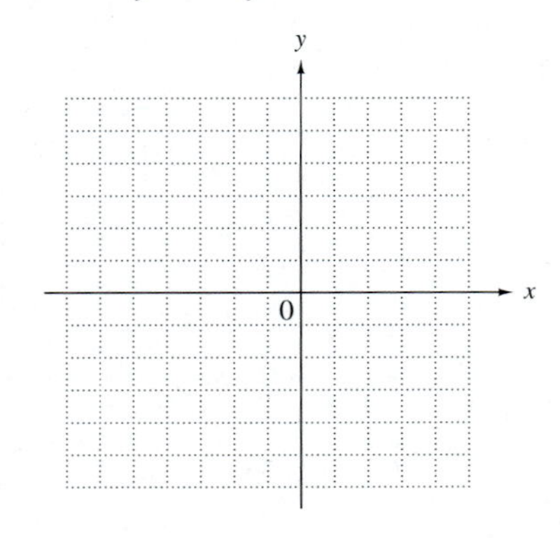

19. $25x^2 \leq 9y^2 + 225$

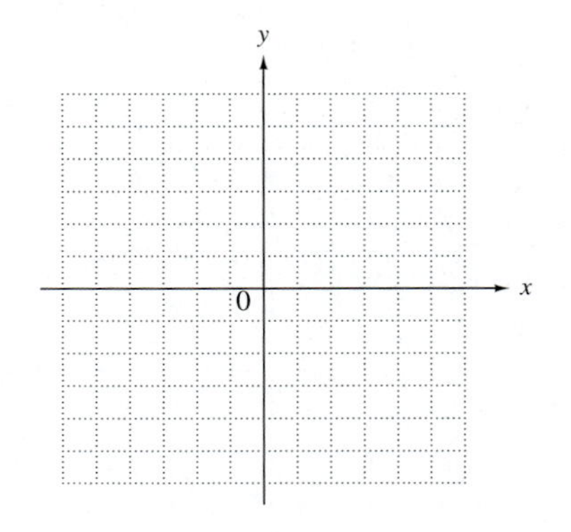

Graph each system of inequalities. See Examples 3–5.

20. $2x + 5y < 10$
$x - 2y < 4$

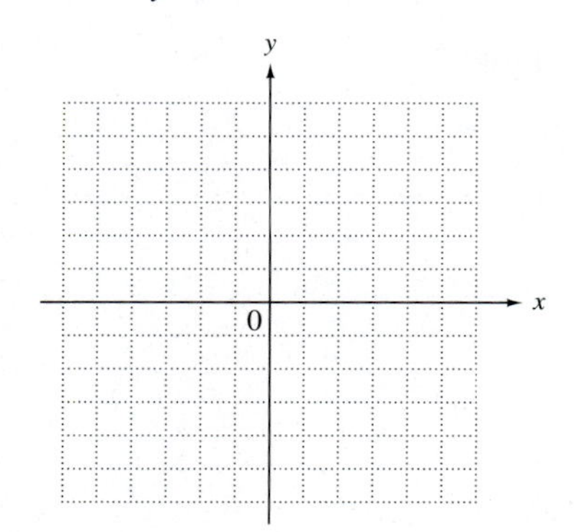

21. $3x - y > -6$
$4x + 3y > 12$

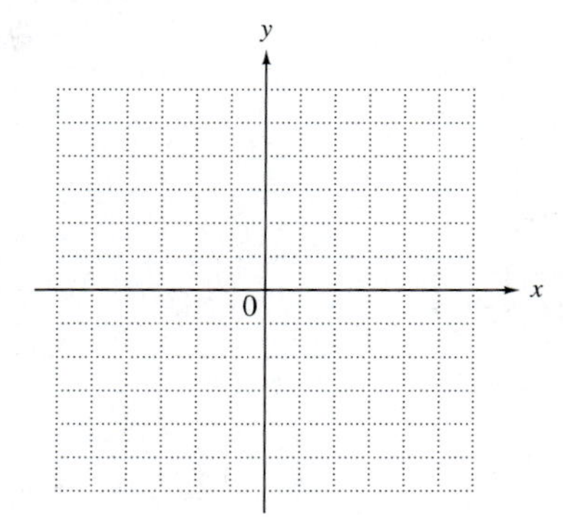

22. $5x - 3y \leq 15$
$4x + y \geq 4$

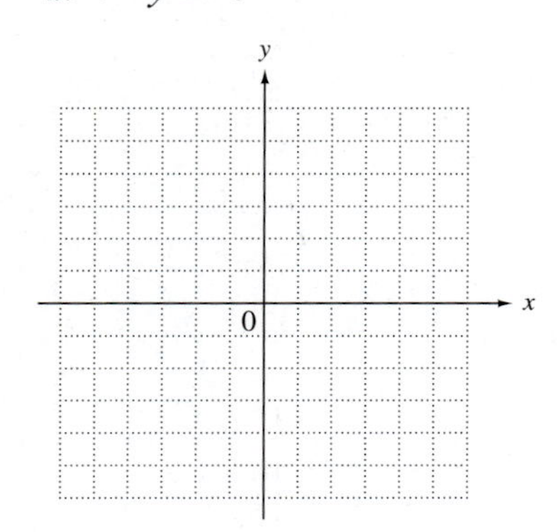

23. $4x - 3y \leq 0$
$x + y \leq 5$

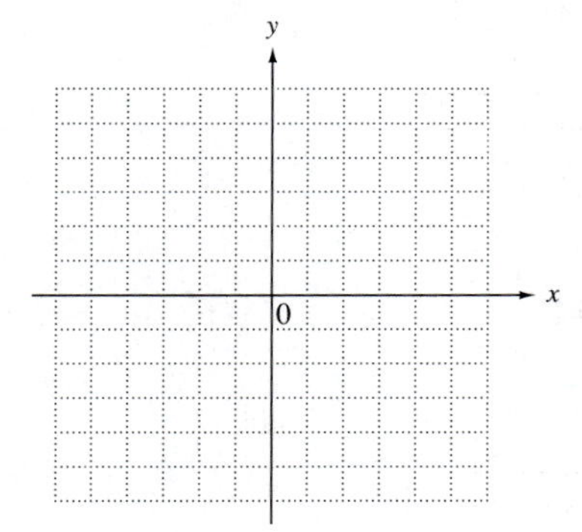

24. $x \leq 5$
$y \leq 4$

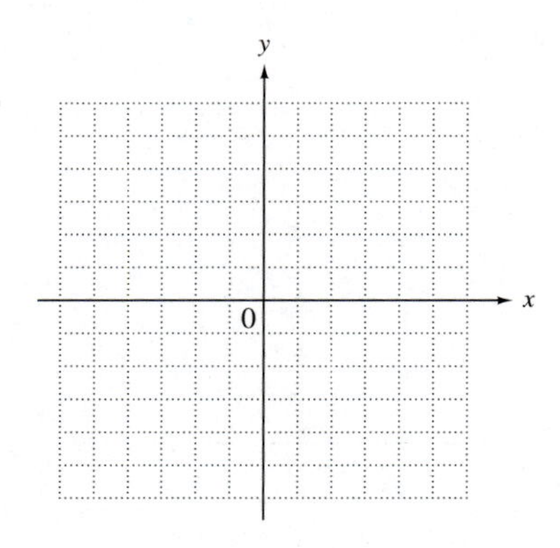

25. $x \geq -2$
$y \leq 4$

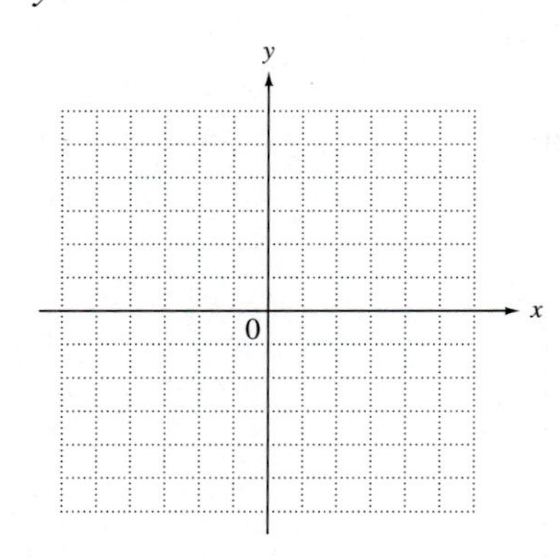

26. $y > x^2 - 4$
$y < -x^2 + 3$

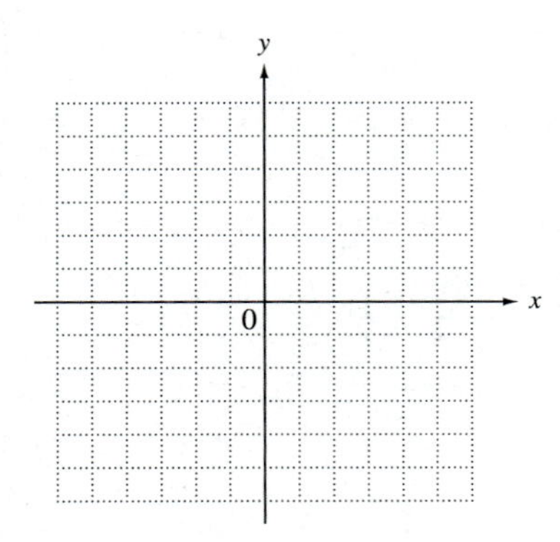

27. $x^2 - y^2 \geq 9$
$\dfrac{x^2}{16} + \dfrac{y^2}{9} \leq 1$

28. $y^2 - x^2 \geq 4$
$-5 \leq y \leq 5$

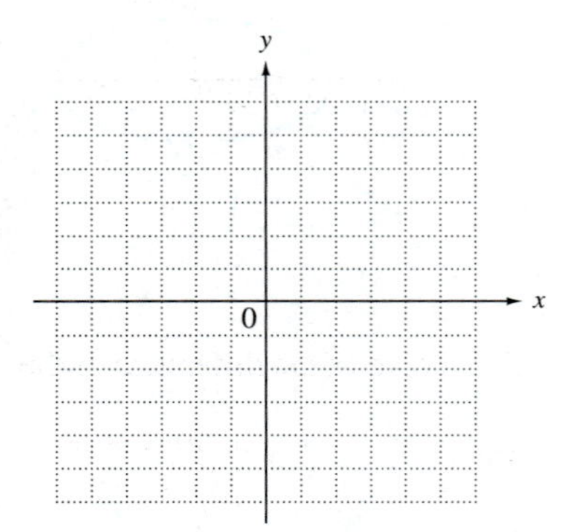

29. $x \geq 0$
$y \geq 0$
$x^2 + y^2 \geq 4$
$x + y \leq 5$

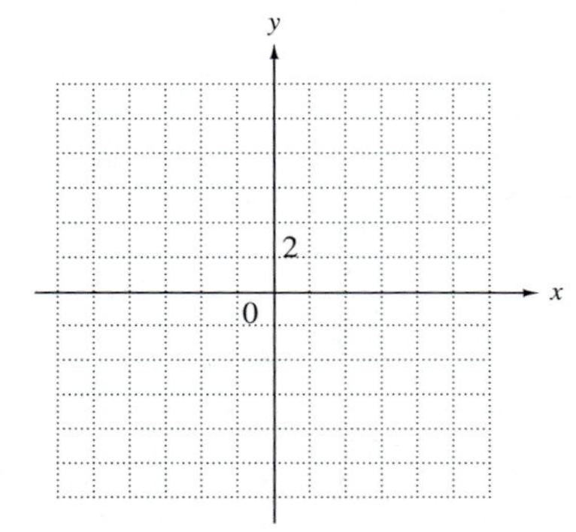

30. $y \leq -x^2$
$y \geq x - 3$
$y \leq -1$
$x < 1$

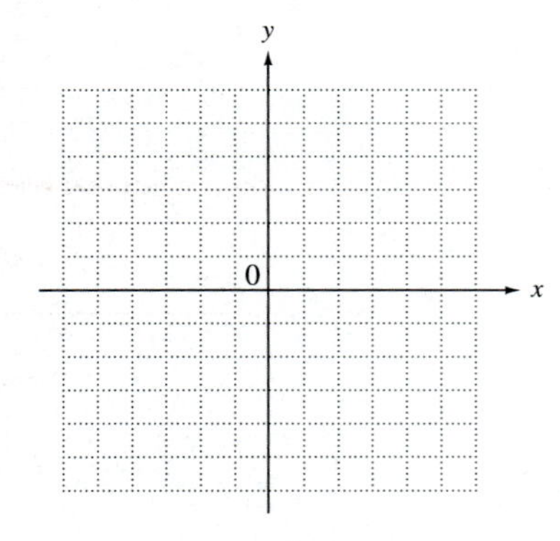

31. $y < x^2$
$y > -2$
$x + y < 3$
$3x - 2y > -6$

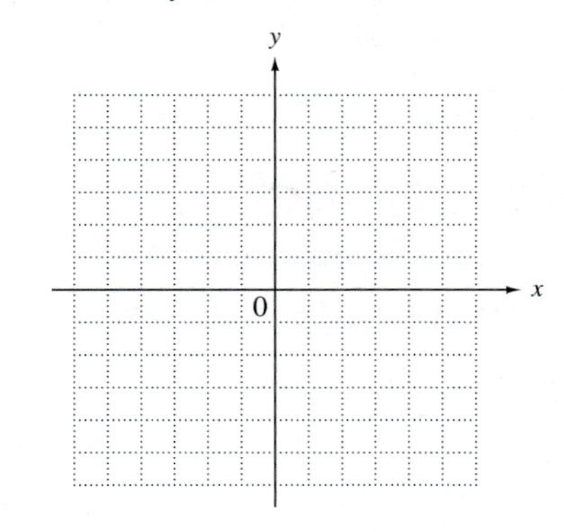

Chapter 11

Summary

Key Terms

11.1

asymptotes Lines that a graph approaches, such as the x- and y-axes for the graph of the reciprocal function, are called asymptotes of the graph.

composition If f and g are functions, then the composition of g and f is defined by $(g \circ f)(x) = g[f(x)]$ for all x in the domain of f such that $f(x)$ is in the domain of g.

11.2

circle A circle is the set of all points in a plane that lie a fixed distance from a fixed point.

center The fixed point discussed in the definition of a circle is the center of the circle.

radius The radius of a circle is the fixed distance between the center and any point on the circle.

$x^2 + y^2 = r^2$

Radius $= r$

Center is (0,0).

ellipse An ellipse is the set of all points in a plane the sum of whose distances from two fixed points is constant.

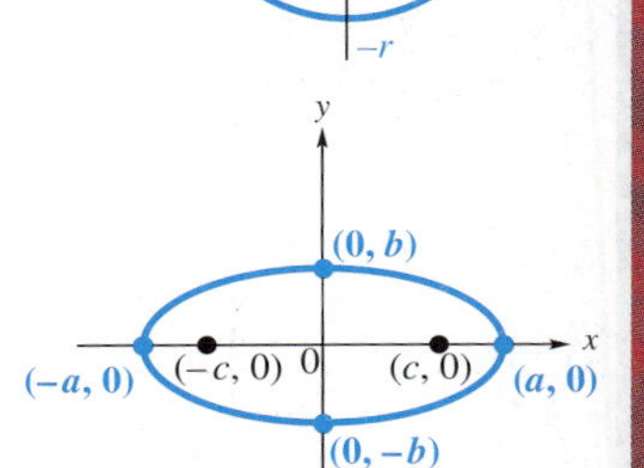

11.3

hyperbola A hyperbola is the set of all points in a plane such that the absolute value of the difference of the distances from two fixed points is constant.

asymptotes of a hyperbola The two intersecting lines that the branches of a hyperbola approach are called asymptotes of the hyperbola.

fundamental rectangle The asymptotes of a hyperbola are the extended diagonals of its fundamental rectangle.

11.4

nonlinear equation An equation in which some terms have more than one variable or a variable of degree 2 or greater is called a nonlinear equation.

nonlinear system of equations A nonlinear system of equations is a system with at least one nonlinear equation.

11.5

second-degree inequality A second-degree inequality is an inequality with at least one variable of degree 2 and no variable with degree greater than 2.

system of inequalities A system of inequalities consists of two or more inequalities to be solved at the same time.

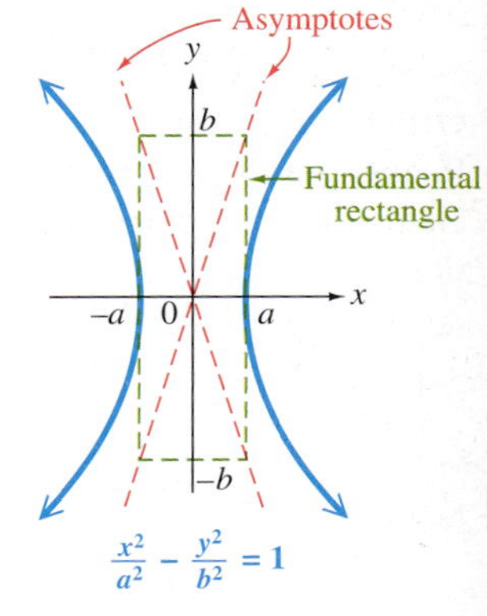

New Symbols

$(f \circ g)(x) = f[g(x)]$ composite function

Test Your Word Power

See how well you have learned the vocabulary in this chapter. Answers follow the Quick Review.

1. Conic sections are
- **(a)** graphs of first-degree equations
- **(b)** the result of two or more intersecting planes
- **(c)** graphs of first-degree inequalities
- **(d)** figures that result from the intersection of an infinite cone with a plane.

2. A **circle** is the set of all points in a plane
- **(a)** the difference of whose distances from two fixed points is constant
- **(b)** that lie a fixed distance from a fixed point
- **(c)** the sum of whose distances from two fixed points is constant
- **(d)** that make up the graph of any second-degree equation.

3. An **ellipse** is the set of all points in a plane
- **(a)** such that the absolute value of the difference of the distances from two fixed points is constant
- **(b)** that lie a fixed distance from a fixed point
- **(c)** the sum of whose distances from two fixed points is constant
- **(d)** that make up the graph of any second-degree equation.

4. A **hyperbola** is the set of all points in a plane
- **(a)** such that the absolute value of the difference of the distances from two fixed points is constant
- **(b)** that lie a fixed distance from a fixed point
- **(c)** the sum of whose distances from two fixed points is constant
- **(d)** that make up the graph of any second-degree equation.

5. A **nonlinear equation** is an equation
- **(a)** in which some terms have more than one variable or a variable of degree 2 or greater
- **(b)** in which the terms have only one variable
- **(c)** of degree 1
- **(d)** of a linear function.

6. A **nonlinear system of equations** is a system
- **(a)** with at least one linear equation
- **(b)** with two or more inequalities
- **(c)** with at least one nonlinear equation
- **(d)** with at least two linear equations.

Quick Review

Concepts	*Examples*
11.1 Additional Graphs of Functions; Composition **Other Functions** In addition to the squaring function, some other important elementary functions in algebra are the absolute value function, defined by $f(x) = \|x\|$; the reciprocal function, defined by $f(x) = \frac{1}{x}$; and the square root function, defined by $f(x) = \sqrt{x}$.	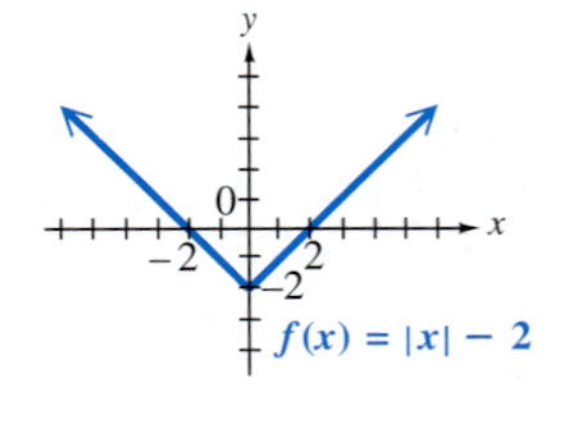 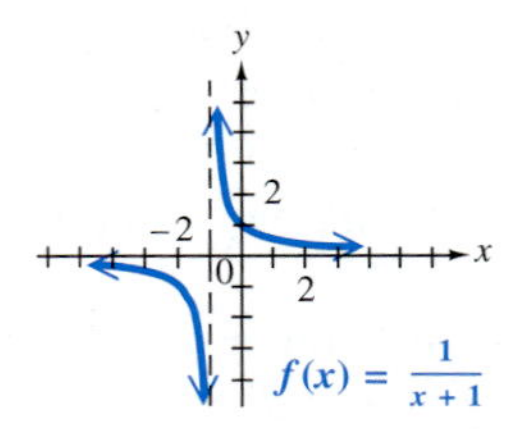 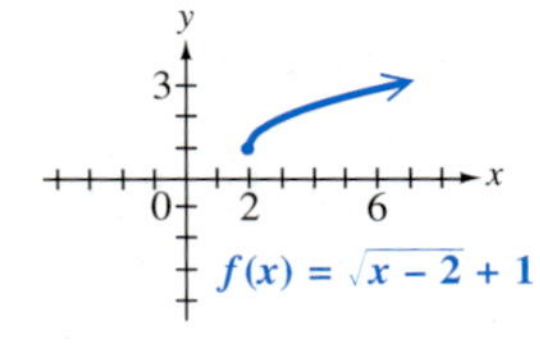
Composition of f and g $(f \circ g)(x) = f[g(x)]$	If $f(x) = x^2$ and $g(x) = 2x + 1$, then $(f \circ g)(x) = f[g(x)]$ $= (2x + 1)^2 = 4x^2 + 4x + 1$ and $(g \circ f)(x) = g[f(x)]$ $= 2x^2 + 1.$

Concepts	*Examples*
11.2 The Circle and The Ellipse **Circle** The circle with radius r and center at (h, k) has an equation of the form $$(x-h)^2+(y-k)^2=r^2.$$	The circle with equation $(x+2)^2+(y-3)^2=25$ has center $(-2, 3)$ and radius 5. 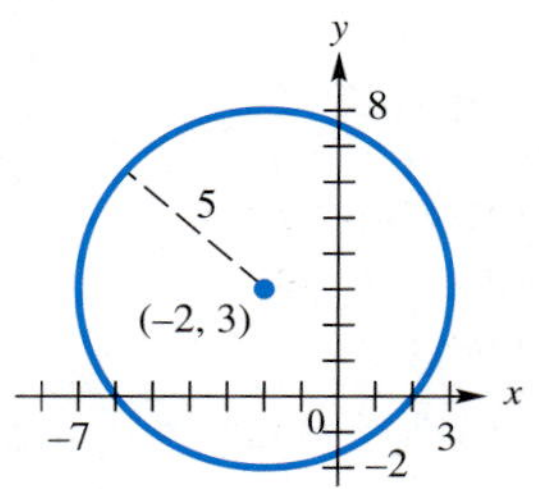
Ellipse The ellipse whose x-intercepts are $(a, 0)$ and $(-a, 0)$ and whose y-intercepts are $(0, b)$ and $(0, -b)$ has an equation of the form $$\frac{x^2}{a^2}+\frac{y^2}{b^2}=1.$$	Graph $\frac{x^2}{9}+\frac{y^2}{4}=1$. 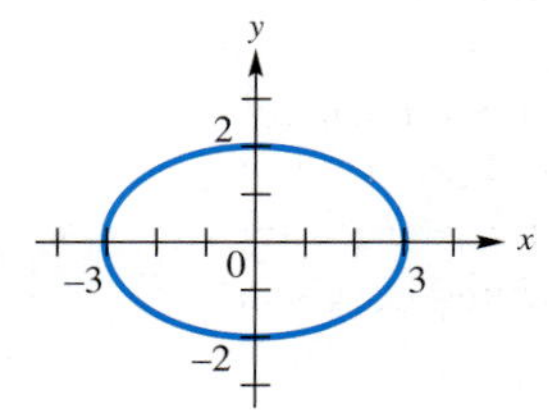
11.3 The Hyperbola and Other Functions Defined by Radicals **Hyperbola** A hyperbola with x-intercepts $(a, 0)$ and $(-a, 0)$ has an equation of the form $$\frac{x^2}{a^2}-\frac{y^2}{b^2}=1,$$ and a hyperbola with y-intercepts $(0, b)$ and $(0, -b)$ has an equation of the form $$\frac{y^2}{b^2}-\frac{x^2}{a^2}=1.$$	Graph $\frac{x^2}{4}-\frac{y^2}{4}=1$. The graph has x-intercepts $(2, 0)$ and $(-2, 0)$. 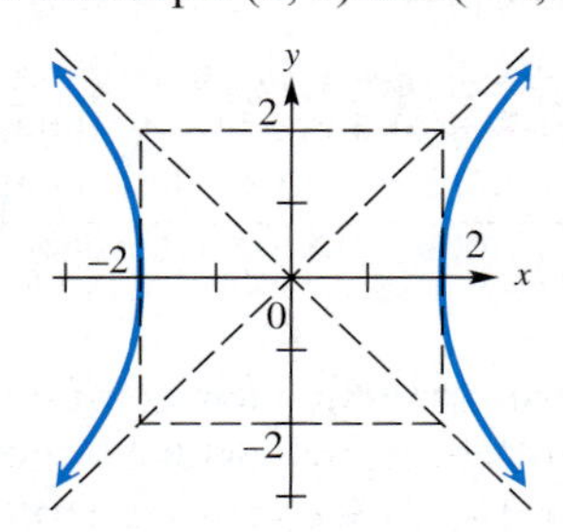
The extended diagonals of the fundamental rectangle with corners at the points (a, b), $(-a, b)$, $(-a, -b)$, and $(a, -b)$ are the asymptotes of these hyperbolas.	The fundamental rectangle has corners at $(2, 2)$, $(-2, 2)$, $(-2, -2)$, and $(2, -2)$.
Graphing a Square Root Function To graph a square root function, square both sides so that the equation can be easily recognized. Then graph only the part indicated by the original equation.	Graph $y=-\sqrt{4-x^2}$. Square both sides and rearrange terms to get $$x^2+y^2=4.$$ This equation has a circle as its graph. However, graph only the lower half of the circle, since the original equation indicates that y cannot be positive. 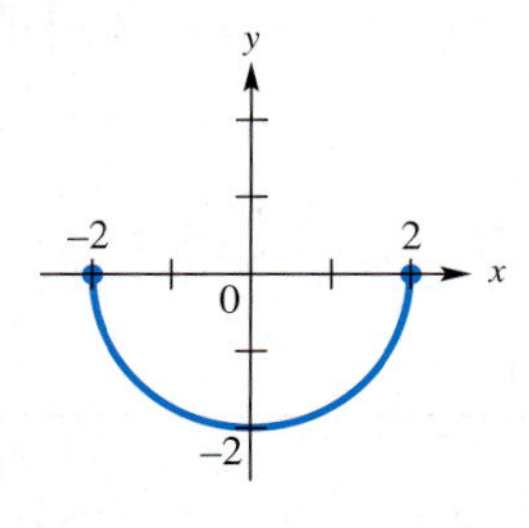

Concepts	*Examples*

11.4 Nonlinear Systems of Equations

Solving a Nonlinear System

A nonlinear system can be solved by the substitution method, the elimination method, or a combination of the two.

Solve the system.

$$x^2 + 2xy - y^2 = 14 \quad (1)$$
$$x^2 - y^2 = -16 \quad (2)$$

Multiply equation (2) by -1 and use elimination.

$$\begin{aligned} x^2 + 2xy - y^2 &= 14 \\ -x^2 \qquad + y^2 &= 16 \\ \hline 2xy \qquad &= 30 \\ xy &= 15 \end{aligned}$$

Solve for y to obtain $y = \frac{15}{x}$, and substitute into equation (2).

$$x^2 - \left(\frac{15}{x}\right)^2 = -16$$

$$x^2 - \frac{225}{x^2} = -16$$

$$x^4 + 16x^2 - 225 = 0 \qquad \text{Multiply by } x^2\text{; add } 16x^2.$$

$$(x^2 - 9)(x^2 + 25) = 0 \qquad \text{Factor.}$$

$$x = \pm 3 \quad \text{or} \quad x = \pm 5i \qquad \text{Zero-factor property}$$

Find corresponding y-values to get the solution set

$$\{(3, 5), (-3, -5), (5i, -3i), (-5i, 3i)\}.$$

11.5 Second-Degree Inequalities and Systems of Inequalities

Graphing a Second-Degree Inequality

To graph a second-degree inequality, graph the corresponding equation as a boundary and use test points to determine which region(s) form the solution set. Shade the appropriate region(s).

Graphing a System of Inequalities

The solution set of a system of inequalities is the intersection of the solution sets of the individual inequalities.

Graph $y \geq x^2 - 2x + 3$.

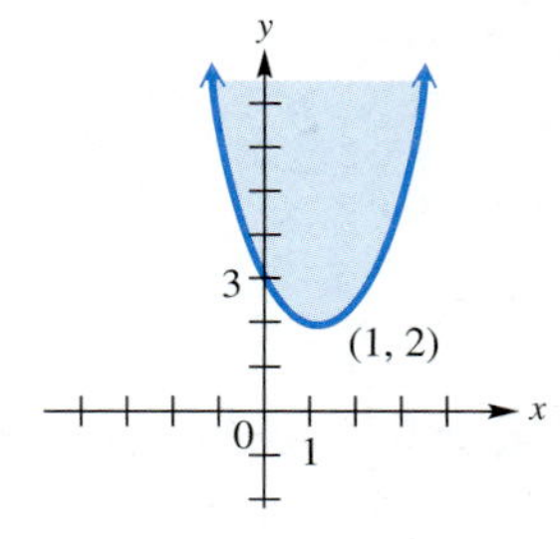

Graph the solution set of the system.

$$3x - 5y > -15$$
$$x^2 + y^2 \leq 25$$

Answers to Test Your Word Power

1. (d) *Example:* Parabolas, circles, ellipses, and hyperbolas are conic sections. **2. (b)** *Example:* See the graph of $x^2 + y^2 = 9$ in Figure 10 of Section 11.2. **3. (c)** *Example:* See the graph of $\frac{x^2}{49} + \frac{y^2}{36} = 1$ in Figure 15 of Section 11.2. **4. (a)** *Example:* See the graph of $\frac{x^2}{16} - \frac{y^2}{12} = 1$ in Figure 18 of Section 11.3. **5. (a)** *Examples:* $y = x^2 + 8x + 16$, $xy = 5$, $2x^2 - y^2 = 6$ **6. (c)** *Example:* $x^2 + y^2 = 2$, $2x + y = 1$

Chapter 11 Review Exercises

[11.1] *Graph each function.*

1. $f(x) = |x + 4|$

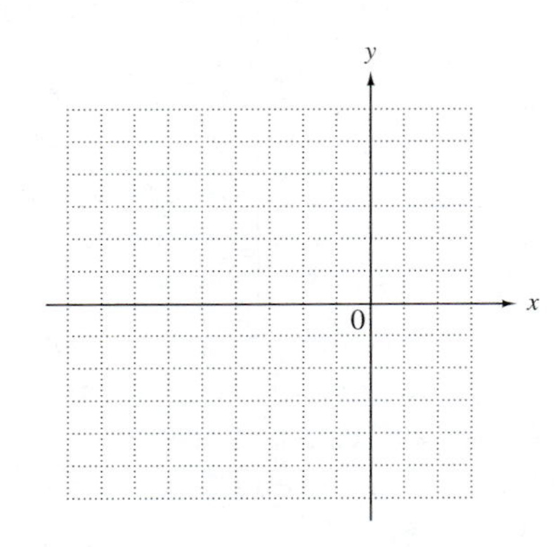

2. $f(x) = \dfrac{1}{x - 4}$

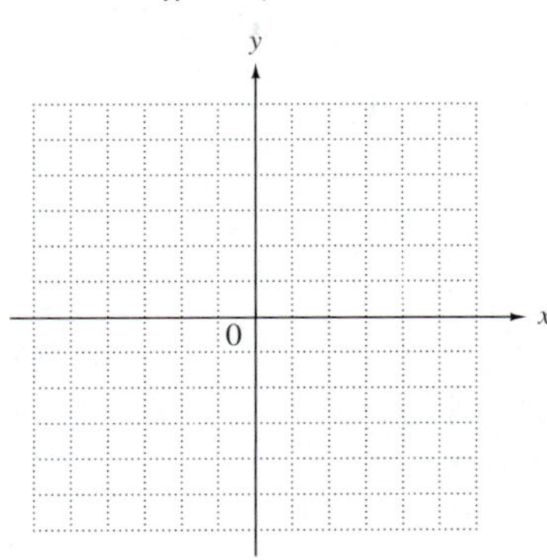

3. $f(x) = \sqrt{x} + 3$

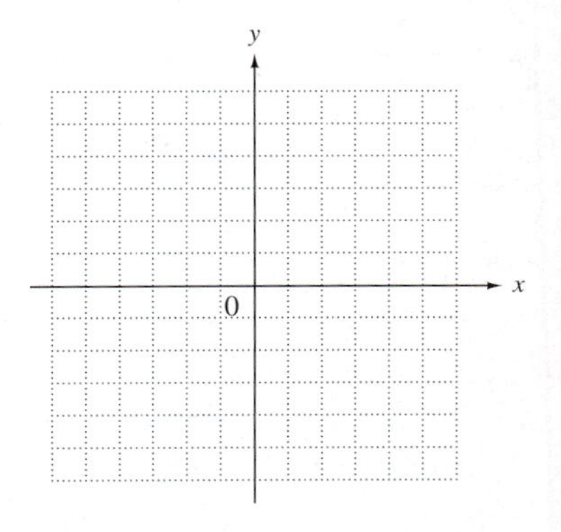

Let $f(x) = 3x^2 + 2x - 1$ *and* $g(x) = 5x + 7$. *Find each of the following.*

4. $(g \circ f)(3)$

5. $(f \circ g)(3)$

6. $(f \circ g)(-2)$

7. $(g \circ f)(-2)$

8. $(f \circ g)(x)$

9. $(g \circ f)(x)$

10. Based on your answers to Exercises 4–9, discuss whether composition of functions is a commutative operation.

[11.2] *Write an equation for each circle.*

11. Center $(-2, 4)$, $r = 3$

12. Center $(-1, -3)$, $r = 5$

13. Center $(4, 2)$, $r = 6$

Find the center and radius of each circle.

14. $x^2 + y^2 + 6x - 4y - 3 = 0$

15. $x^2 + y^2 - 8x - 2y + 13 = 0$

16. $2x^2 + 2y^2 + 4x + 20y = -34$

17. $4x^2 + 4y^2 - 24x + 16y = 48$

Graph each equation.

18. $x^2 + y^2 = 16$

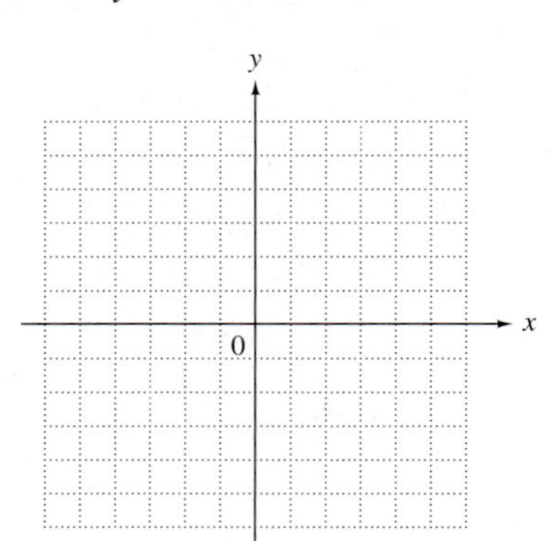

19. $\dfrac{x^2}{16} + \dfrac{y^2}{9} = 1$

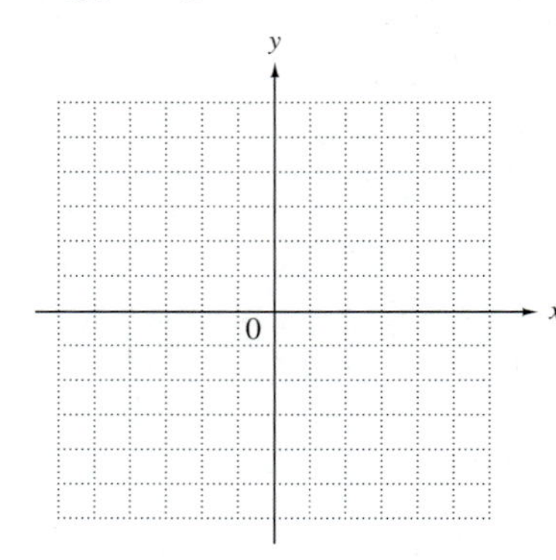

20. $\dfrac{x^2}{49} + \dfrac{y^2}{25} = 1$

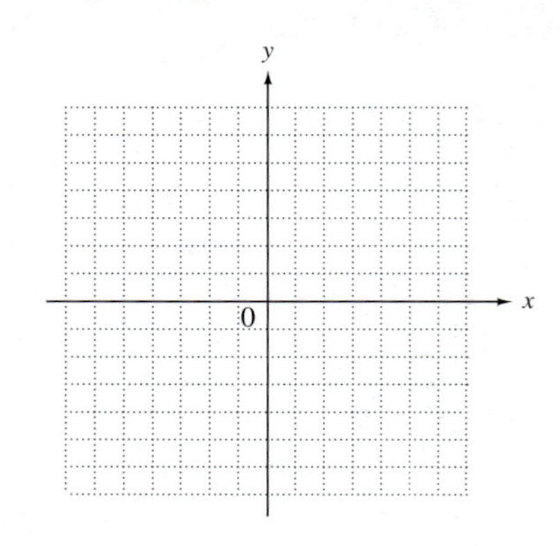

21. A satellite is in an elliptical orbit around Earth with perigee altitude of 160 km and apogee altitude of 16,000 km. See the figure. (*Source:* Kastner, Bernice, *Space Mathematics,* NASA, 1985.) Find the equation of the ellipse.

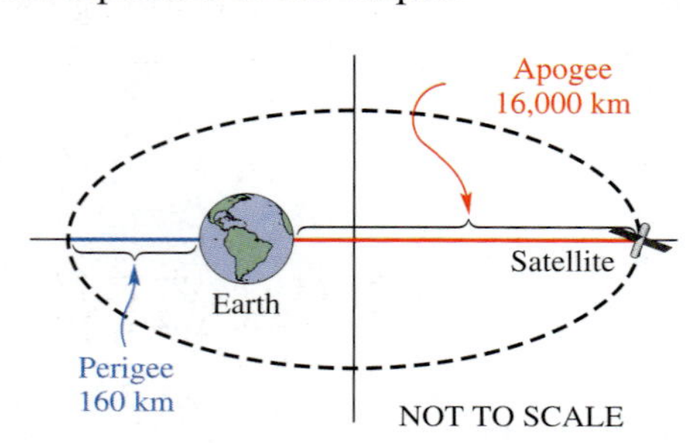

22. This figure illustrates how the crawfish race is conducted at the Crawfish Festival in Breaux Bridge, Louisiana. Explain why a circular "racetrack" is appropriate for such a race.

[11.3] *Graph each equation.*

23. $\dfrac{x^2}{16} - \dfrac{y^2}{25} = 1$

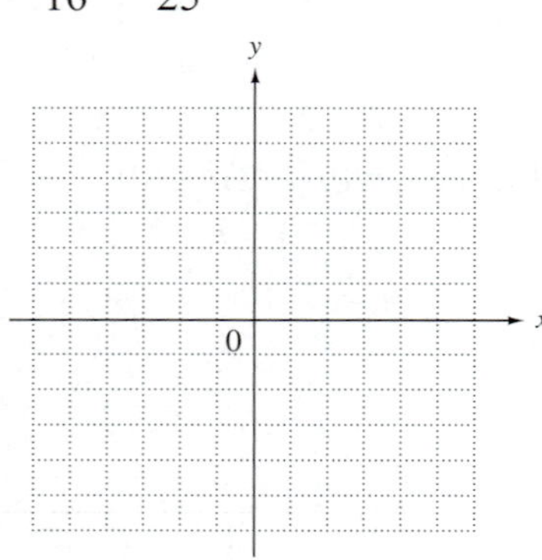

24. $\dfrac{y^2}{25} - \dfrac{x^2}{4} = 1$

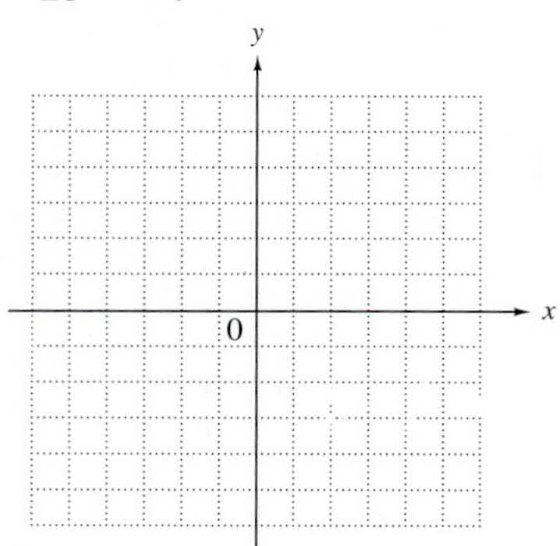

25. $f(x) = -\sqrt{16 - x^2}$

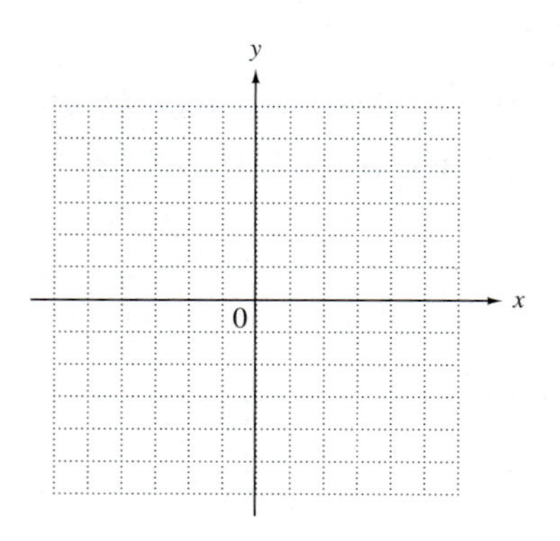

Identify the graph of each equation as a parabola, circle, ellipse, *or* hyperbola.

26. $x^2 + y^2 = 64$

27. $y = 2x^2 - 3$

28. $y^2 = 2x^2 - 8$

29. $y^2 = 8 - 2x^2$

30. $x = y^2 + 4$

31. $x^2 - y^2 = 64$

32. Ships and planes often use a location-finding system called LORAN. With this system, a radio transmitter at M sends out a series of pulses. (See the figure.) When each pulse is received at transmitter S, it then sends out a pulse. A ship at P receives pulses from both M and S. A receiver on the ship measures the difference in the arrival times of the pulses. A special map gives hyperbolas that correspond to the differences in arrival times (which give the distances d_1 and d_2 in the figure). The ship can then be located as lying on a branch of a particular hyperbola. Suppose $d_1 = 80$ mi and $d_2 = 30$ mi, and the distance between transmitters M and S is 100 mi. Use the definition to find an equation of the hyperbola the ship is located on.

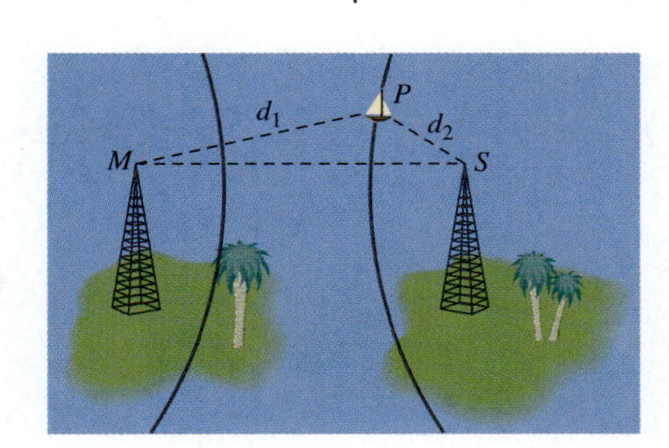

[11.4] *Solve each system.*

33. $2y = 3x - x^2$
$x + 2y = -12$

34. $y + 1 = x^2 + 2x$
$y + 2x = 4$

35. $x^2 + 3y^2 = 28$
$y - x = -2$

36. $xy = 8$
$x - 2y = 6$

37. $x^2 + y^2 = 6$
$x^2 - 2y^2 = -6$

38. $3x^2 - 2y^2 = 12$
$x^2 + 4y^2 = 18$

39. How many solutions are possible for a system of two equations whose graphs are a circle and a line?

40. How many solutions are possible for a system of two equations whose graphs are a parabola and a hyperbola?

[11.5] *Graph each inequality.*

41. $9x^2 \geq 16y^2 + 144$

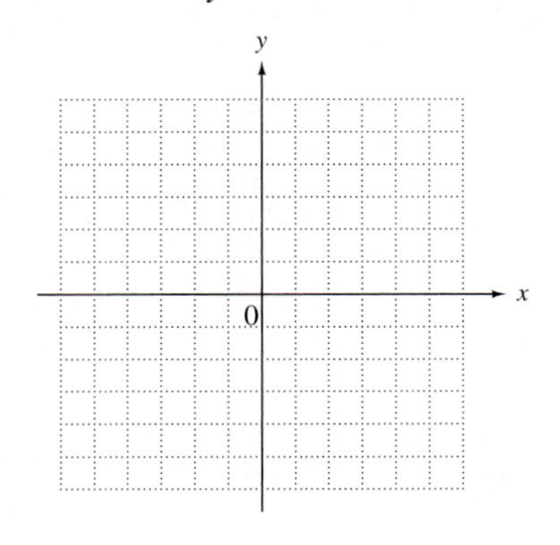

42. $4x^2 + y^2 \geq 16$

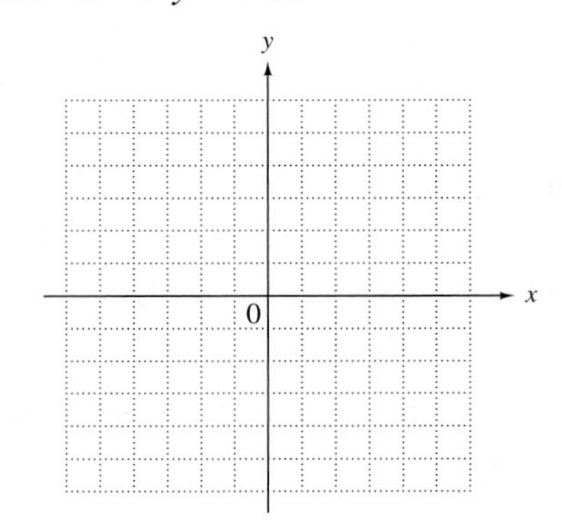

43. $y < -(x + 2)^2 + 1$

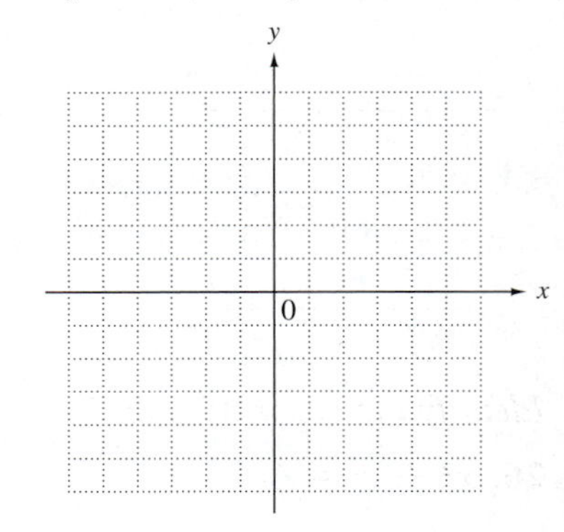

Graph each system of inequalities.

44. $2x + 5y \leq 10$
$3x - \quad y \leq 6$

45. $|x| \leq 2$
$|y| > 1$
$4x^2 + 9y^2 \leq 36$

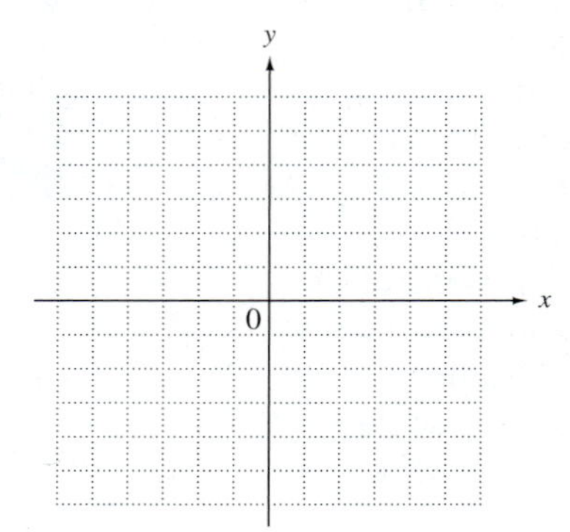

46. $9x^2 \leq 4y^2 + 36$
$x^2 + y^2 \leq 16$

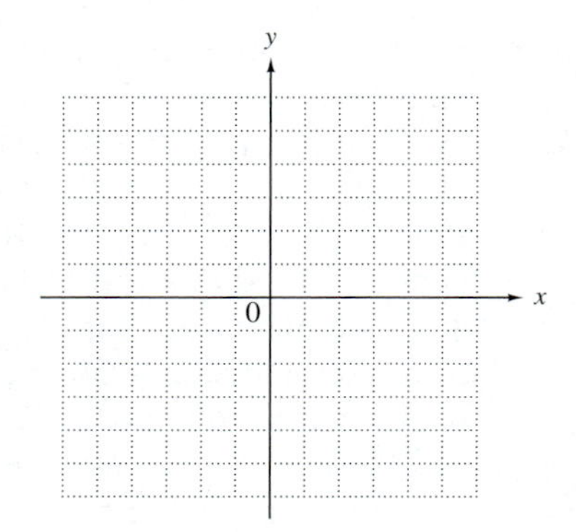

RELATING CONCEPTS (Exercises 47–51) FOR INDIVIDUAL OR GROUP WORK

In Chapter 5 we discussed several methods of solving systems of linear equations in three variables. Now these methods can be used to find an equation of a circle through three points in a plane that are not on the same line. The equation of a circle can be written in the form $x^2 + y^2 + ax + by + c = 0$ *for some values of a, b, and c.* ***Work Exercises 47–51 in order,*** *to find the equation of the circle through the points* (2, 4), (5, 1), *and* (−1, 1).

47. Determine one equation in a, b, and c by letting $x = 2$ and $y = 4$ in the general form given above. Write it with a, b, and c on the left and the constant on the right.

48. Repeat Exercise 47 for the point (5, 1).

49. Repeat Exercise 47 for the point (−1, 1).

50. Solve the system formed by the equations found in Exercises 47–49, and give the equation of the circle that satisfies these conditions.

51. Use the methods of this chapter to find the center and the radius of the circle in Exercise 50.

MIXED REVIEW EXERCISES

Graph.

52. $\dfrac{x^2}{64} + \dfrac{y^2}{25} = 1$

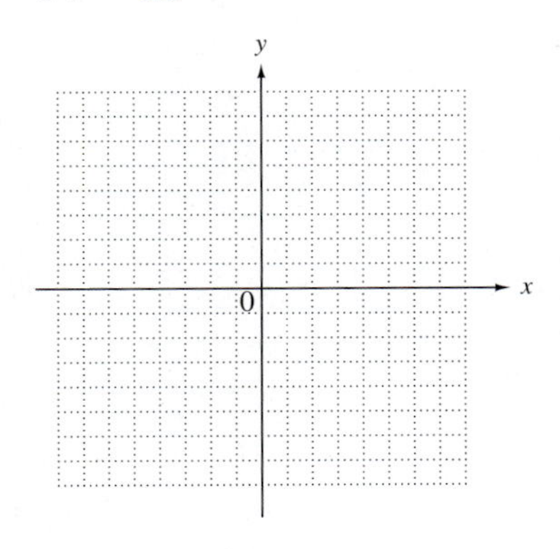

53. $\dfrac{y^2}{4} - 1 = \dfrac{x^2}{9}$

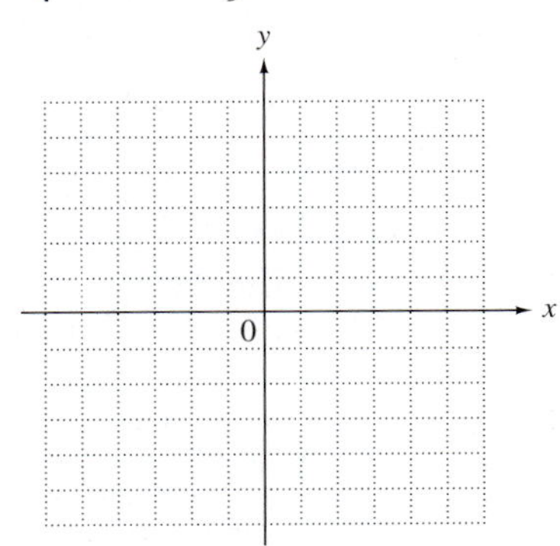

54. $x^2 + y^2 = 25$

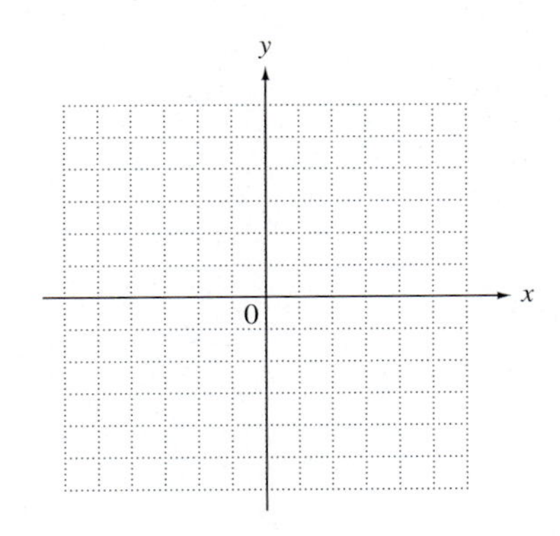

55. $x^2 + 9y^2 = 9$

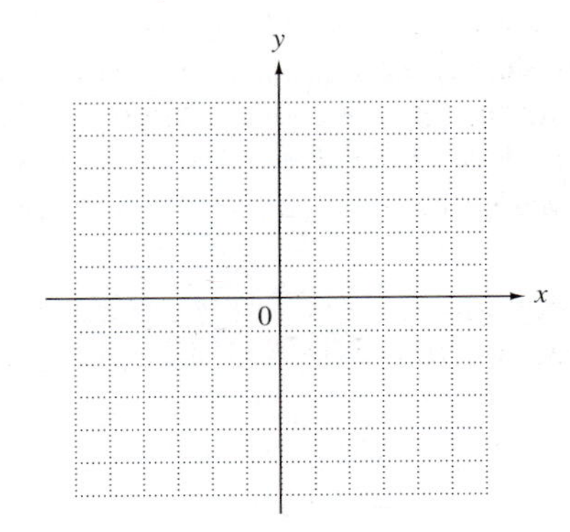

56. $x^2 - 9y^2 = 9$

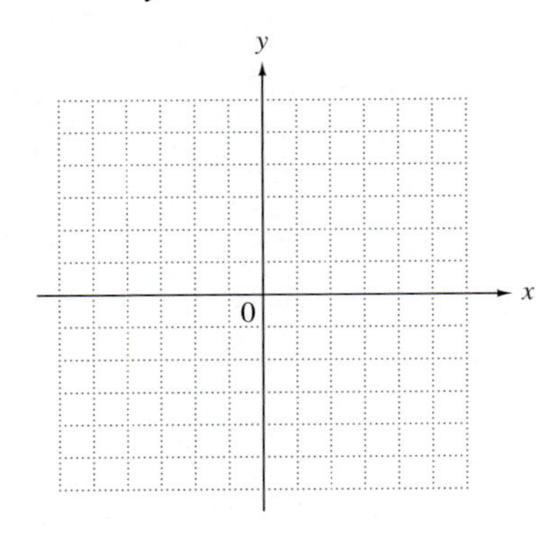

57. $f(x) = \sqrt{4 - x}$

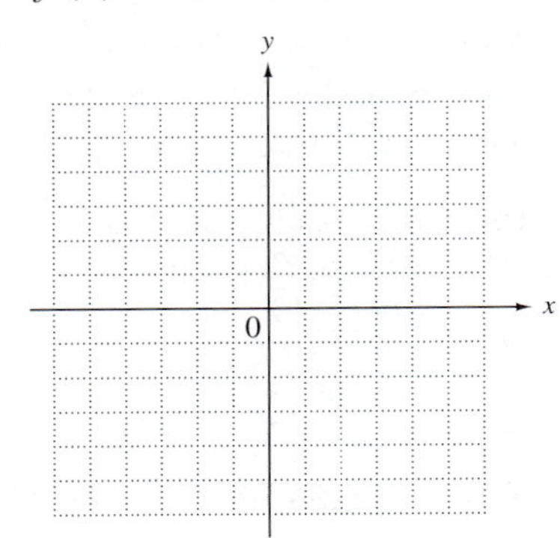

58. $3x + 2y \geq 0$
$y \leq 4$
$x \leq 4$

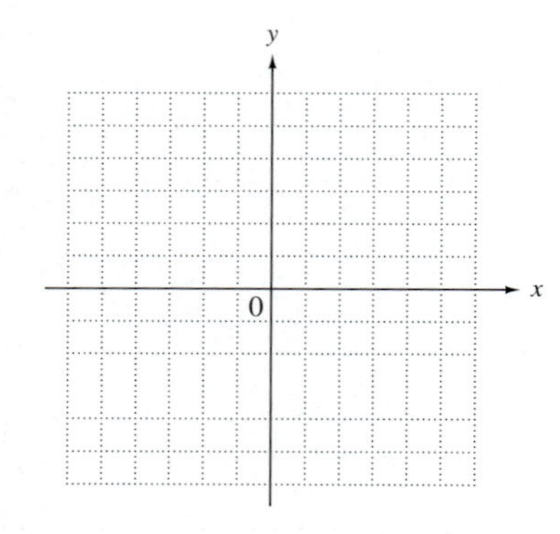

59. $4y > 3x - 12$
$x^2 < 16 - y^2$

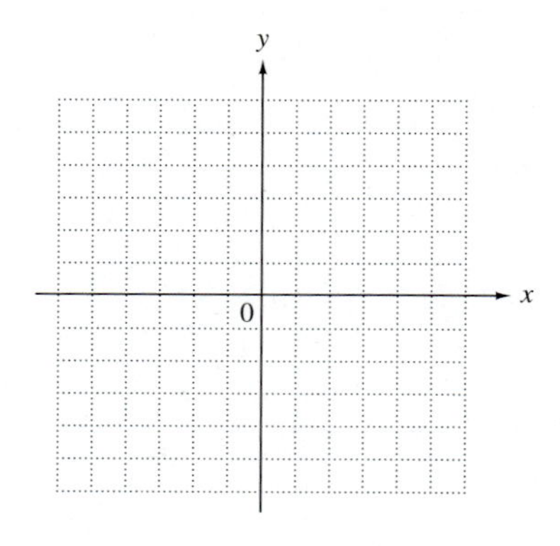

60. Explain why a set of points that form an ellipse does not satisfy the definition of a function.

The orbit of Mercury around the sun (a focus) is an ellipse with equation

$$\frac{x^2}{3352} + \frac{y^2}{3211} = 1,$$

where x and y are measured in million kilometers.

61. Find its apogee, its greatest distance from the sun. (*Hint:* Refer to Section 11.2, Exercise 37.)

62. Find its perigee, its smallest distance from the sun.

Chapter 11 Test

Study Skills — *Study Skills Workbook* **Activity 12**

Match each function with its graph from choices A, B, C, and D.

1. $f(x) = \sqrt{x - 2}$

2. $f(x) = \sqrt{x + 2}$

3. $f(x) = \sqrt{x} + 2$

4. $f(x) = \sqrt{x} - 2$

A.

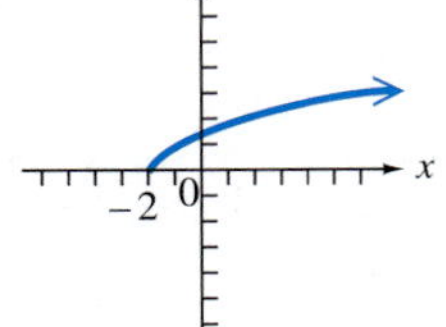

B.

y
x
0
−2

C.

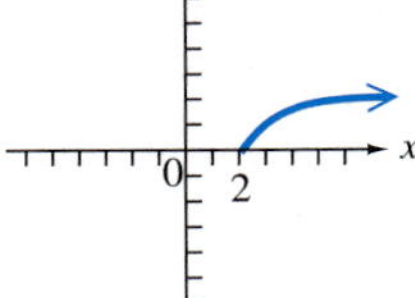

D.

y
x
2
0

1. ______________________

2. ______________________

3. ______________________

4. ______________________

5. Sketch the graph of $f(x) = |x - 3| + 4$.

5.

y
x
0

6. For $f(x) = 3x + 5$ and $g(x) = x^2 + 2$, find each of the following.

(a) $(f \circ g)(-2)$

(b) $(f \circ g)(x)$

(c) $(g \circ f)(x)$

6. (a) ______________________

(b) ______________________

(c) ______________________

7. Find the center and radius of the circle whose equation is $(x - 2)^2 + (y + 3)^2 = 16$. Sketch the graph.

7. ______________________

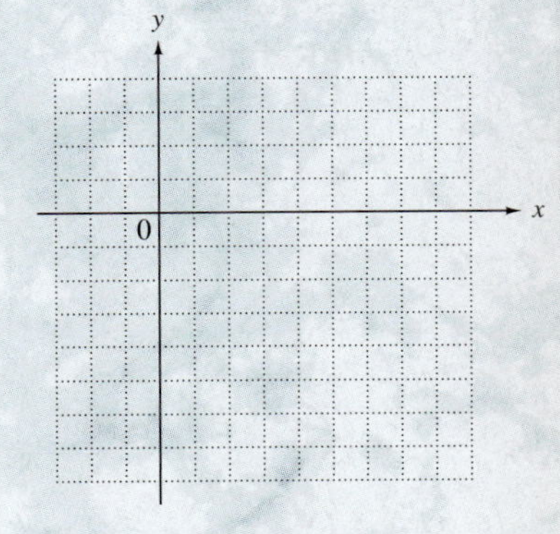

8. ____________________

8. Find the center and radius of the circle whose equation is $x^2 + y^2 + 8x - 2y = 8$.

Graph.

9.

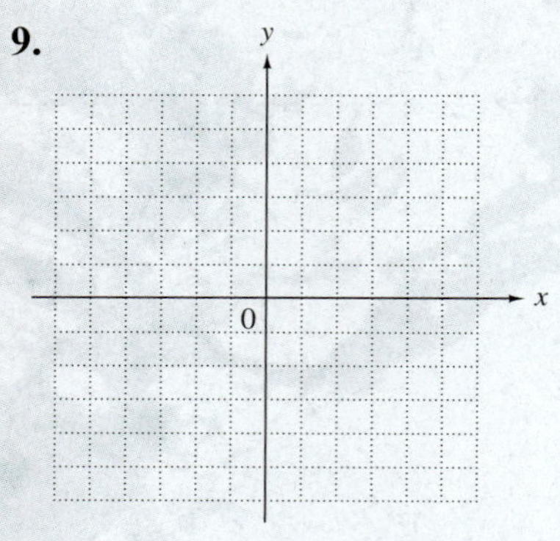

9. $f(x) = \sqrt{9 - x^2}$

10.

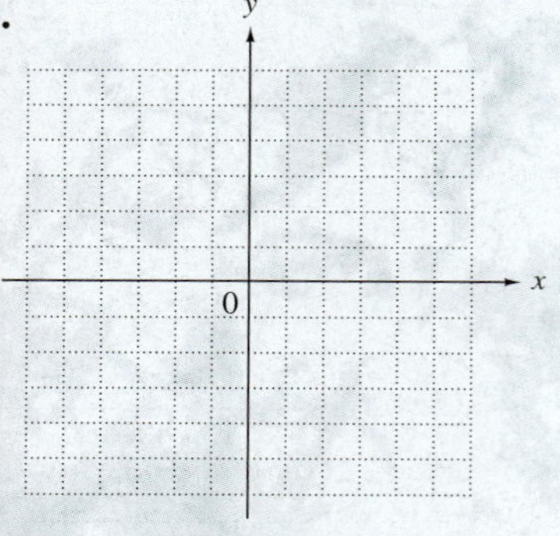

10. $4x^2 + 9y^2 = 36$

11.

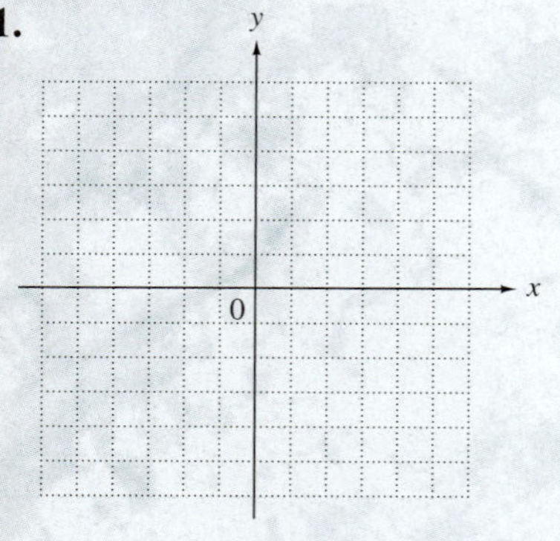

11. $16y^2 - 4x^2 = 64$

12. $\frac{y}{2} = -\sqrt{1 - \frac{x^2}{9}}$

12.

Identify the graph of each equation as a parabola, hyperbola, ellipse, *or* circle.

13. $6x^2 + 4y^2 = 12$

13. ______________________

14. $16x^2 = 144 + 9y^2$

14. ______________________

15. $4y^2 + 4x = 9$

15. ______________________

Solve each nonlinear system.

16. $2x - y = 9$
$xy = 5$

16. 

17. ______________________

17. $x - 4 = 3y$
$x^2 + y^2 = 8$

18. ______________________

18. $x^2 + y^2 = 25$
$x^2 - 2y^2 = 16$

19.

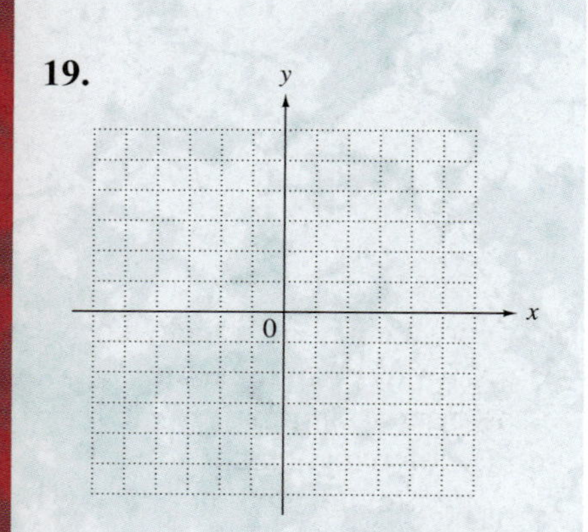

19. Graph the inequality $y < x^2 - 2$.

20.

y
x
0

20. Graph the system.

$$x^2 + 25y^2 \leq 25$$
$$x^2 + y^2 \leq 9$$

Cumulative Review Exercises CHAPTERS 1–11

1. Simplify $-10 + |-5| - |3| + 4$.

Solve.

2. $4 - (2x + 3) + x = 5x - 3$

3. $-4k + 7 \geq 6k + 1$

4. $|5m| - 6 = 14$

5. $|2p - 5| > 15$

6. Find the slope of the line through $(2, 5)$ and $(-4, 1)$.

7. Find the equation of the line through $(-3, -2)$ and perpendicular to the graph of $2x - 3y = 7$.

Solve each system.

8. $3x - y = 12$
   $2x + 3y = -3$

9. $x + y - 2z = 9$
   $2x + y + z = 7$
   $3x - y - z = 13$

10. $xy = -5$
    $2x + y = 3$

Solve each problem.

11. Al and Bev traveled from their apartment to a picnic 20 mi away. Al traveled on his bike while Bev, who left later, took her car. Al's average speed was half of Bev's average speed. The trip took Al $\frac{1}{2}$ hr longer than Bev. What was Bev's average speed?

12. The president of InstaTune, a chain of franchised automobile tune-up shops, reports that people who buy a franchise and open a shop pay a weekly fee (in dollars) to company headquarters, according to the linear function defined by

$$f(x) = .07x + 135,$$

where $f(x)$ is the fee and x is the total amount of money taken in during the week by the shop. Find the weekly fee if \$2000 is taken in for the week. (*Source: Business Week.*)

Perform the indicated operations.

13. $(5y - 3)^2$

14. $(2r + 7)(6r - 1)$

15. $\dfrac{8x^4 - 4x^3 + 2x^2 + 13x + 8}{2x + 1}$

Factor.

16. $12x^2 - 7x - 10$

17. $2y^4 + 5y^2 - 3$

18. $z^4 - 1$

19. $a^3 - 27b^3$

Perform each operation.

20. $\dfrac{5x - 15}{24} \cdot \dfrac{64}{3x - 9}$

21. $\dfrac{y^2 - 4}{y^2 - y - 6} \div \dfrac{y^2 - 2y}{y - 1}$

22. $\dfrac{5}{c + 5} - \dfrac{2}{c + 3}$

23. $\dfrac{p}{p^2 + p} + \dfrac{1}{p^2 + p}$

Solve.

24. Kareem and Jamal want to clean their office. Kareem can do the job alone in 3 hr, while Jamal can do it alone in 2 hr. How long will it take them if they work together?

Simplify. Assume all variables represent positive real numbers.

25. $\left(\dfrac{4}{3}\right)^{-1}$

26. $\dfrac{(2a)^{-2}a^4}{a^{-3}}$

27. $4\sqrt[3]{16} - 2\sqrt[3]{54}$

28. $\dfrac{3\sqrt{5x}}{\sqrt{2x}}$

29. $\dfrac{5 + 3i}{2 - i}$

Solve.

30. $2\sqrt{k} = \sqrt{5k + 3}$

31. $10q^2 + 13q = 3$

32. $(4x - 1)^2 = 8$

33. $3k^2 - 3k - 2 = 0$

34. $2(x^2 - 3)^2 - 5(x^2 - 3) = 12$

35. $F = \frac{kwv^2}{r}$ for v

36. If $f(x) = x^3 + 4$, find $f^{-1}(x)$.

37. Evaluate $3^{\log_3 4}$.

38. Evaluate $e^{\ln 7}$.

39. Use properties of logarithms to write

$$2 \log(3x + 7) - \log 4$$

as a single logarithm.

40. Solve $\log(x + 2) + \log(x - 1) = 1$.

41. If \$10,000 is invested at 5% for 4 yr, how much will there be in the account if interest is compounded

(a) quarterly,

(b) continuously?

The bar graph shows historic and projected annual on-line retail sales (in billions of dollars) over the Internet. A reasonable model for sales y in billions of dollars is the exponential function defined by

$$y = 1.38(1.65)^x.$$

The years are coded such that x is the number of years since 1995.

GROWTH IN ON-LINE SALES

Year	Billions of Dollars
1997	\$3
1998	\$7.1
1999	\$12
2000	\$18.4
2001	\$27.3
2002	\$41.1

Source: Jupiter Communications.

42. Use the model to estimate sales in the year 2000. (*Hint:* Let $x = 5$.)

43. Use the model to estimate sales in the year 2003.

44. If $f(x) = x^2 + 2x - 4$ and $g(x) = 3x + 2$, find

(a) $(g \circ f)(1)$ **(b)** $(f \circ g)(x)$.

Graph.

45. $f(x) = -3x + 5$

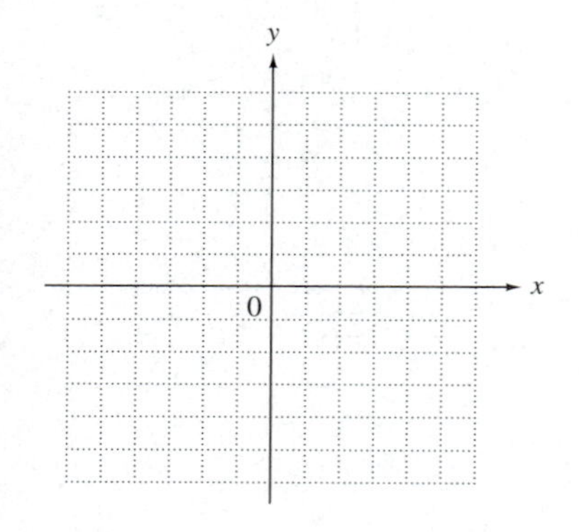

46. $f(x) = -2(x - 1)^2 + 3$

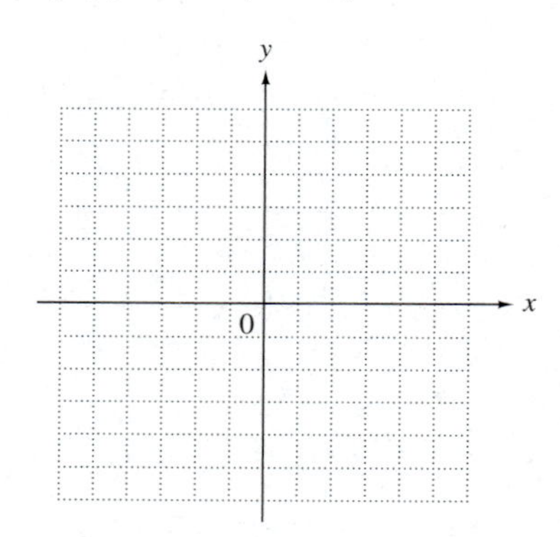

47. $\dfrac{x^2}{25} + \dfrac{y^2}{16} \le 1$

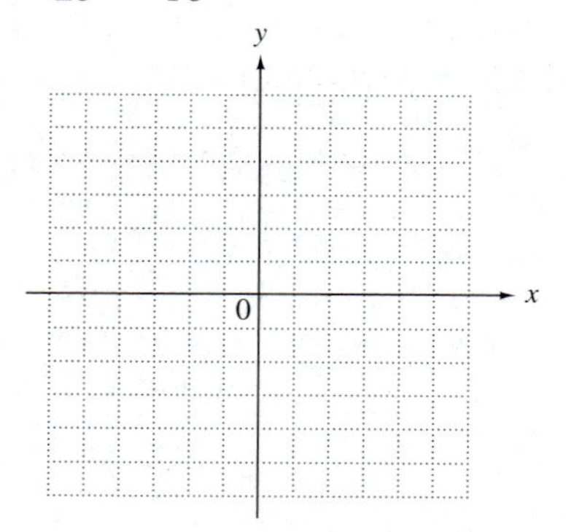

48. $f(x) = \sqrt{x - 2}$

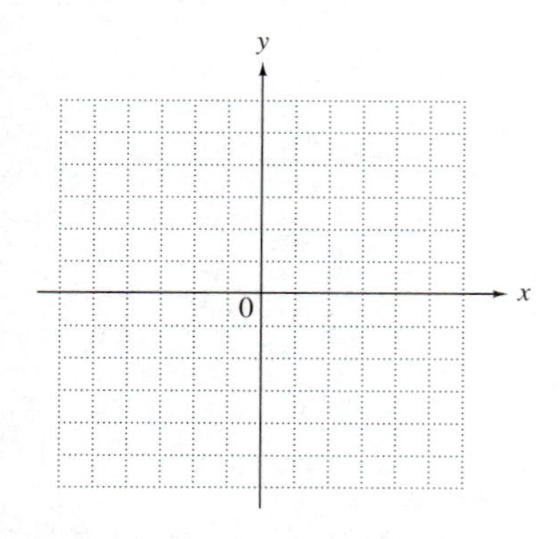

49. $\dfrac{x^2}{4} - \dfrac{y^2}{16} = 1$

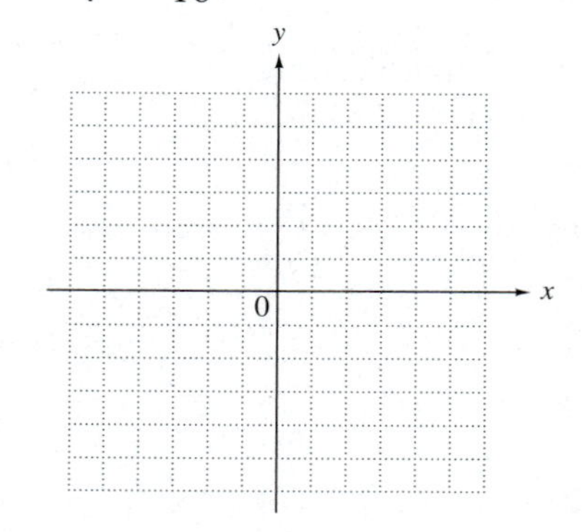

50. $f(x) = 3^x$

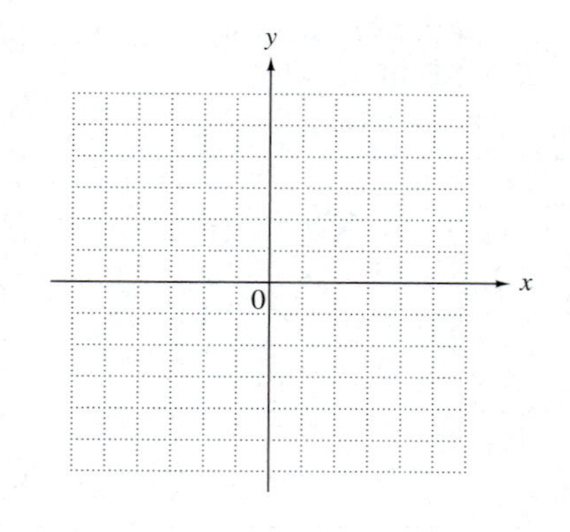

Appendix A
Review of Fractions

A Review of Fractions

Objectives

1. Identify prime numbers.
2. Write numbers in prime factored form.
3. Write fractions in lowest terms.
4. Multiply and divide fractions.
5. Add and subtract fractions.

Studying algebra requires good arithmetic skills. Most people do not get much practice using fractions, so we review the rules for fractions in this appendix.

The numbers used most often in everyday life are the **whole numbers,**

$$0, 1, 2, 3, 4, 5, \ldots$$

and **fractions,** such as

$$\frac{1}{3}, \quad \frac{5}{4}, \quad \text{and} \quad \frac{11}{12}.$$

The parts of a fraction are named as follows.

$$\text{Fraction bar} \rightarrow \frac{4 \leftarrow \text{Numerator}}{7 \leftarrow \text{Denominator}}$$

If the numerator of a fraction is smaller than the denominator, we call it a **proper fraction.** A proper fraction has a value less than 1. If the numerator is greater than the denominator, the fraction is an **improper fraction.** An improper fraction, which has a value greater than 1, is often written as a **mixed number.** For example, $\frac{12}{5}$ may be written as $2\frac{2}{5}$. In algebra, we prefer to use the improper form because it is easier to work with. In applications, we usually convert answers to mixed number form, which is more meaningful.

1 Identify prime numbers. In work with fractions, we will need to write the numerators and denominators as products. A **product** is the answer to a multiplication problem. When 12 is written as the product $2 \cdot 6$, for example, 2 and 6 are called **factors** of 12. Other factors of 12 are 1, 3, 4, and 12. A whole number is **prime** if it has exactly two different factors (itself and 1). The first dozen primes are listed here.

$$2, 3, 5, 7, 11, 13, 17, 19, 23, 29, 31, 37$$

A whole number greater than 1 that is not prime is called a **composite number.** For example, 4, 6, 8, 9, and 12 are composite numbers. The number 1 is neither prime nor composite.

Example 1 Distinguishing between Prime and Composite Numbers

Decide whether each number is prime or composite.

(a) 33
33 has factors of 3 and 11 as well as 1 and 33, so it is composite.

(b) 43
Since there are no numbers other than 1 and 43 itself that divide *evenly* into 43, the number 43 is prime.

(c) 9832
9832 can be divided by 2, giving $2 \cdot 4916$, so it is composite.

Work Problem 1 at the Side.

1 Tell whether each number is prime or composite.

(a) 12

(b) 13

(c) 27

(d) 59

(e) 1806

Answers
1. (a) composite **(b)** prime **(c)** composite **(d)** prime **(e)** composite

2 Write each number in prime factored form.

(a) 70

(b) 72

(c) 693

(d) 97

2 Write numbers in prime factored form. As mentioned earlier, to factor a number means to write it as the product of two or more numbers. Factoring is just the reverse of multiplying two numbers to get the product.

In algebra, a dot is used instead of the × symbol to indicate multiplication because × may be confused with the letter x. Each composite number can be written as the product of prime numbers in only one way (disregarding the order of the factors). A number written using factors that are all prime numbers is in **prime factored form.** We will write the factored form with the prime factors in order by size, although any ordering is correct.

Example 2 **Writing Numbers in Prime Factored Form**

Write each number in prime factored form.

(a) 35

Factor 35 as the product of the prime factors 5 and 7, or as

$$35 = 5 \cdot 7.$$

(b) 24

A handy way to keep track of the factors is to use a tree, as shown below. The prime factors are circled.

Divide by the smallest prime, 2, to get	$24 = 2 \cdot 12.$	24 → ② · 12
Now divide 12 by 2 to find factors of 12.	$24 = 2 \cdot 2 \cdot 6$	② · 6
Since 6 can be factored as $2 \cdot 3$, $24 = 2 \cdot 2 \cdot 2 \cdot 3$, where all factors are prime.	$24 = 2 \cdot 2 \cdot 2 \cdot 3$	② · ③

Work Problem 2 at the Side.

3 Write fractions in lowest terms. We use prime factors to write fractions in *lowest terms.* A fraction is in **lowest terms** when the numerator and denominator have no factors in common (other than 1). We write a fraction in this form by using the following facts.

Properties of 1

Any nonzero number divided by itself is equal to 1.

Any number multiplied by 1 remains the same.

For example,

$$\frac{3}{3} = 1, \quad \frac{8}{8} = 1, \quad \text{and} \quad 17 \cdot 1 = 17.$$

ANSWERS

2. (a) $2 \cdot 5 \cdot 7$ **(b)** $2 \cdot 2 \cdot 2 \cdot 3 \cdot 3$ **(c)** $3 \cdot 3 \cdot 7 \cdot 11$ **(d)** 97 is prime.

Writing a Fraction in Lowest Terms

Step 1 Write the numerator and denominator in prime factored form.

Step 2 Replace each pair of factors common to the numerator and denominator with 1.

Step 3 Multiply the remaining factors in the numerator and in the denominator.

(This procedure is sometimes called "simplifying the fraction.")

Example 3 **Writing Fractions in Lowest Terms**

Write each fraction in lowest terms.

(a) $\frac{10}{15} = \frac{2 \cdot 5}{3 \cdot 5} = \frac{2}{3} \cdot \frac{5}{5} = \frac{2}{3} \cdot 1 = \frac{2}{3}$

Since 5 is a common factor of 10 and 15, we use the first property of 1 to replace $\frac{5}{5}$ with 1.

(b) $\frac{15}{45} = \frac{3 \cdot 5}{3 \cdot 3 \cdot 5} = \frac{1 \cdot 3 \cdot 5}{3 \cdot 3 \cdot 5} = \frac{1}{3} \cdot \frac{3}{3} \cdot \frac{5}{5} = \frac{1}{3} \cdot 1 \cdot 1 = \frac{1}{3}$

Multiplying by 1 in the numerator does not change the value of the numerator and makes it possible to rewrite the expression as the product of three fractions in the next step.

(c) $\frac{150}{200}$

It is not always necessary to factor into *prime* factors in Step 1. Here, if you see that 50 is a common factor of the numerator and the denominator, factor as follows:

$$\frac{150}{200} = \frac{3 \cdot 50}{4 \cdot 50} = \frac{3}{4} \cdot 1 = \frac{3}{4}.$$

NOTE

When you are writing a fraction in lowest terms, look for the largest common factor in the numerator and the denominator. If none is obvious, factor the numerator and the denominator into prime factors. *Any* common factor can be used, and the fraction can be simplified in stages. For example,

$$\frac{150}{200} = \frac{15 \cdot 10}{20 \cdot 10} = \frac{3 \cdot 5 \cdot 10}{4 \cdot 5 \cdot 10} = \frac{3}{4}.$$

Work Problem 3 at the Side.

3 Write each fraction in lowest terms.

(a) $\frac{8}{14}$

(b) $\frac{35}{42}$

(c) $\frac{120}{72}$

ANSWERS

3. **(a)** $\frac{4}{7}$ **(b)** $\frac{5}{6}$ **(c)** $\frac{5}{3}$

4 Multiply and divide fractions.

Multiplying Fractions

To multiply two fractions, multiply the numerators to get the numerator of the product, and multiply the denominators to get the denominator of the product. The product must be written in lowest terms.

In practice, we will show the products of the numerator and the denominator in factored form to make it easier to write the product in lowest terms. We often simplify before performing the multiplication, as shown in the next example.

4 Find each product, and write it in lowest terms.

(a) $\frac{5}{8} \cdot \frac{2}{10}$

(b) $\frac{1}{10} \cdot \frac{12}{5}$

(c) $\frac{7}{9} \cdot \frac{12}{14}$

(d) $3\frac{1}{3} \cdot 1\frac{3}{4}$

Answers

4. **(a)** $\frac{1}{8}$ **(b)** $\frac{6}{25}$ **(c)** $\frac{2}{3}$ **(d)** $\frac{35}{6}$ or $5\frac{5}{6}$

Example 4 **Multiplying Fractions**

Find each product, and write it in lowest terms.

(a) $\frac{3}{8} \cdot \frac{4}{9} = \frac{3 \cdot 4}{8 \cdot 9}$ Multiply numerators. Multiply denominators.

$= \frac{3 \cdot 4}{2 \cdot 4 \cdot 3 \cdot 3}$ Factor.

$= \frac{1}{2 \cdot 3} = \frac{1}{6}$ Write in lowest terms.

(b) $2\frac{1}{3} \cdot 5\frac{1}{2} = \frac{7}{3} \cdot \frac{11}{2}$ Write as improper fractions.

$= \frac{77}{6}$ or $12\frac{5}{6}$ Multiply numerators and denominators.

Work Problem 4 at the Side.

Two fractions are **reciprocals** of each other if their product is 1. For example, $\frac{3}{4}$ and $\frac{4}{3}$ are reciprocals because

$$\frac{3}{4} \cdot \frac{4}{3} = 1.$$

The numbers $\frac{7}{11}$ and $\frac{11}{7}$ are reciprocals also. Other examples are $\frac{1}{5}$ and 5, $\frac{4}{9}$ and $\frac{9}{4}$, and 16 and $\frac{1}{16}$.

Because division is the opposite or inverse of multiplication, we use reciprocals to divide fractions.

Dividing Fractions

To divide two fractions, multiply the first fraction by the reciprocal of the second. The result is called the **quotient.**

Example 5 **Dividing Fractions**

Find each quotient, and write it in lowest terms.

(a) $\frac{3}{4} \div \frac{8}{5} = \frac{3}{4} \cdot \frac{5}{8} = \frac{3 \cdot 5}{4 \cdot 8} = \frac{15}{32}$

Multiply by the reciprocal of the second fraction.

(b) $\frac{3}{4} \div \frac{5}{8} = \frac{3}{4} \cdot \frac{8}{5} = \frac{3 \cdot 8}{4 \cdot 5} = \frac{3 \cdot 4 \cdot 2}{4 \cdot 5} = \frac{6}{5}$

(c) $\frac{5}{8} \div 10 = \frac{5}{8} \div \frac{10}{1} = \frac{5}{8} \cdot \frac{1}{10} = \frac{1}{16}$

Write 10 as $\frac{10}{1}$.

(d) $1\frac{2}{3} \div 4\frac{1}{2} = \frac{5}{3} \div \frac{9}{2}$ Write as improper fractions.

$= \frac{5}{3} \cdot \frac{2}{9}$ Multiply by the reciprocal of the second fraction.

$= \frac{10}{27}$

CAUTION

Notice that *only* the second fraction (the divisor) is replaced by its reciprocal in the multiplication.

Work Problem 5 at the Side.

5 **Add and subtract fractions.** The result of adding two numbers is called the **sum** of the numbers. For example, since $2 + 3 = 5$, the sum of 2 and 3 is 5.

Adding Fractions

To find the sum of two fractions with the *same* denominator, add their numerators and keep the *same* denominator.

Example 6 **Adding Fractions with the Same Denominator**

Add. Write sums in lowest terms.

(a) $\frac{3}{7} + \frac{2}{7} = \frac{3+2}{7} = \frac{5}{7}$ Denominator does not change.

(b) $\frac{2}{10} + \frac{3}{10} = \frac{2+3}{10} = \frac{5}{10} = \frac{1}{2}$ Write in lowest terms.

Work Problem 6 at the Side.

If the fractions to be added do not have the same denominator, the procedure above can still be used, but only *after* the fractions are rewritten with a common denominator. For example, to rewrite $\frac{3}{4}$ as a fraction with a denominator of 32,

$$\frac{3}{4} = \frac{?}{32},$$

we must find the number that can be multiplied by 4 to give 32. Since $4 \cdot 8 = 32$, we use the number 8. By the second property of 1, we can multiply the numerator and the denominator by 8.

$$\frac{3}{4} = \frac{3}{4} \cdot 1 = \frac{3}{4} \cdot \frac{8}{8} = \frac{3 \cdot 8}{4 \cdot 8} = \frac{24}{32}$$

Finding the Least Common Denominator (LCD)

Step 1 Factor all denominators to prime factored form.

Step 2 The LCD is the product of every (different) factor that appears in any of the factored denominators. If a factor is repeated, use the largest number of repeats as factors of the LCD.

Step 3 Write each fraction with the LCD as the denominator, using the second property of 1.

5 Find each quotient, and write it in lowest terms.

(a) $\frac{3}{10} \div \frac{2}{7}$

(b) $\frac{3}{4} \div \frac{7}{16}$

(c) $\frac{4}{3} \div 6$

(d) $3\frac{1}{4} \div 1\frac{2}{5}$

6 Add. Write sums in lowest terms.

(a) $\frac{3}{5} + \frac{4}{5}$

(b) $\frac{5}{14} + \frac{3}{14}$

ANSWERS

5. (a) $\frac{21}{20}$ **(b)** $\frac{12}{7}$ **(c)** $\frac{2}{9}$ **(d)** $\frac{65}{28}$ or $2\frac{9}{28}$

6. (a) $\frac{7}{5}$ **(b)** $\frac{4}{7}$

7 Add. Write sums in lowest terms.

(a) $\frac{7}{30} + \frac{2}{45}$

(b) $\frac{17}{10} + \frac{8}{27}$

(c) $2\frac{1}{8} + 1\frac{2}{3}$

(d) $132\frac{4}{5} + 28\frac{3}{4}$

Example 7 Adding Fractions with Different Denominators

Add. Write sums in lowest terms.

(a) $\frac{4}{15} + \frac{5}{9}$

Step 1 To find the LCD, we first factor both denominators to prime factored form.

$$15 = 5 \cdot \mathbf{3} \quad \text{and} \quad 9 = \mathbf{3} \cdot 3$$

3 is a factor of both denominators.

Step 2

$$\text{LCD} = \underbrace{5 \cdot 3}_{15} \cdot 3 = 45 \quad (9 = 3 \cdot 3)$$

In this example, the LCD needs one factor of 5 and two factors of 3 because the second denominator has two factors of 3.

Step 3 Now we can use the second property of 1 to write each fraction with 45 as the denominator.

$$\frac{4}{15} = \frac{4}{15} \cdot \frac{\mathbf{3}}{\mathbf{3}} = \frac{12}{45} \quad \text{and} \quad \frac{5}{9} = \frac{5}{9} \cdot \frac{\mathbf{5}}{\mathbf{5}} = \frac{25}{45}$$

Now add the two equivalent fractions to get the required sum.

$$\frac{4}{15} + \frac{5}{9} = \frac{12}{\mathbf{45}} + \frac{25}{\mathbf{45}} = \frac{37}{\mathbf{45}}$$

(b) $3\frac{1}{2} + 2\frac{3}{4} = \frac{7}{2} + \frac{11}{4}$ Change to improper fractions.

$= \frac{14}{4} + \frac{11}{4}$ Get a common denominator.

$= \frac{25}{4}$ or $6\frac{1}{4}$ Add.

(c) $45\frac{2}{3} + 73\frac{1}{2}$

We could use an alternative vertical method here, adding the whole numbers and the fractions separately.

$$\begin{array}{r} 45\frac{2}{3} = 45\frac{4}{6} \\ + \ 73\frac{1}{2} = 73\frac{3}{6} \\ \hline 118\frac{\mathbf{7}}{\mathbf{6}} = 118 + \left(\mathbf{1} + \frac{\mathbf{1}}{\mathbf{6}}\right) = 119\frac{1}{6} \end{array}$$

Work Problem 7 at the Side.

The **difference** between two numbers is found by subtracting the numbers. For example, $9 - 5 = 4$, so the difference between 9 and 5 is 4. We find the difference between two fractions as follows.

ANSWERS

7. (a) $\frac{5}{18}$ **(b)** $\frac{539}{270}$ **(c)** $\frac{91}{24}$ or $3\frac{19}{24}$ **(d)** $161\frac{11}{20}$

Subtracting Fractions

To find the difference between two fractions with the *same* denominator, subtract their numerators and keep the *same* denominator.

If the fractions have *different* denominators, write them with a common denominator first.

Example 8 Subtracting Fractions

Subtract. Write differences in lowest terms.

(a) $\frac{15}{8} - \frac{3}{8} = \frac{15-3}{8} = \frac{12}{8} = \frac{3}{2}$ Lowest terms

(b) $\frac{15}{16} - \frac{4}{9}$

Since $16 = 2 \cdot 2 \cdot 2 \cdot 2$ and $9 = 3 \cdot 3$ have no common factors, the LCD is $16 \cdot 9 = 144$.

$$\frac{15}{16} - \frac{4}{9} = \frac{15 \cdot 9}{16 \cdot 9} - \frac{4 \cdot 16}{9 \cdot 16}$$ Get a common denominator.

$$= \frac{135}{144} - \frac{64}{144}$$

$$= \frac{71}{144}$$ Subtract numerators; keep the same denominator.

(c) $2\frac{1}{2} - 1\frac{3}{4} = \frac{5}{2} - \frac{7}{4}$ Change to improper fractions.

$= \frac{10}{4} - \frac{7}{4}$ Get a common denominator.

$= \frac{3}{4}$ Subtract.

Work Problem 8 at the Side.

We often see mixed numbers used in applications of mathematics.

Example 9 Solving an Applied Problem Requiring Addition of Fractions

The given diagram appears in the book *Woodworker's 39 Sure-Fire Projects.* It is a view of a corner bookcase/desk. Add the fractions shown in the diagram to find the height of the bookcase/desk to the top of the writing surface.

We must find the following sum (" means inches).

$$\frac{3}{4} + 4\frac{1}{2} + 9\frac{1}{2} + \frac{3}{4} + 9\frac{1}{2} + \frac{3}{4} + 4\frac{1}{2}$$

$\frac{3}{4}$" $4\frac{1}{2}$" $9\frac{1}{2}$" $\frac{3}{4}$" $9\frac{1}{2}$" $\frac{3}{4}$" $4\frac{1}{2}$"

Writing surface

Cut 3 leg sections from ready-made turned leg.

Continued on Next Page

8 Subtract.

(a) $\frac{9}{11} - \frac{3}{11}$

(b) $\frac{13}{15} - \frac{5}{6}$

(c) $2\frac{3}{8} - 1\frac{1}{2}$

(d) $50\frac{1}{4} - 32\frac{2}{3}$

ANSWERS

8. **(a)** $\frac{6}{11}$ **(b)** $\frac{1}{30}$ **(c)** $\frac{7}{8}$ **(d)** $17\frac{7}{12}$

We change the mixed numbers to improper fractions.

$$\frac{3}{4} + \frac{9}{2} + \frac{19}{2} + \frac{3}{4} + \frac{19}{2} + \frac{3}{4} + \frac{9}{2}$$

The LCD is 4. Change all fractions to fourths.

$$\frac{3}{4} + \frac{18}{4} + \frac{38}{4} + \frac{3}{4} + \frac{38}{4} + \frac{3}{4} + \frac{18}{4}$$

Now we can add and simplify the answer.

$$\frac{3}{4} + \frac{18}{4} + \frac{38}{4} + \frac{3}{4} + \frac{38}{4} + \frac{3}{4} + \frac{18}{4} = \frac{121}{4} \text{ or } 30\frac{1}{4}$$

The height is $30\frac{1}{4}$ in.

Work Problem 9 at the Side.

9 Solve the problem.

To make a three-piece outfit from the same fabric, Wei Jen needs $1\frac{1}{4}$ yd for the blouse, $1\frac{2}{3}$ yd for the skirt, and $2\frac{1}{2}$ yd for the jacket. How much fabric does she need?

Example 10 Solving an Applied Problem Requiring Division of Fractions

An upholsterer needs $2\frac{1}{4}$ yd of fabric to recover a chair. How many chairs can be covered with $23\frac{2}{3}$ yd of fabric?

It helps to understand the problem if we replace the fractions with whole numbers. Suppose each chair requires 2 yd, and we have 24 yd of fabric. Dividing 24 by 2 gives the number of chairs (12) that can be recovered. To solve the original problem, we must divide $23\frac{2}{3}$ by $2\frac{1}{4}$.

$$\begin{aligned} 23\frac{2}{3} \div 2\frac{1}{4} &= \frac{71}{3} \div \frac{9}{4} \\ &= \frac{71}{3} \cdot \frac{4}{9} \\ &= \frac{284}{27} \text{ or } 10\frac{14}{27} \end{aligned}$$

Thus, 10 chairs can be recovered with some fabric left over.

Work Problem 10 at the Side.

10 Solve the problem.

A gallon of paint covers 500 ft^2. (ft^2 means square feet.) To paint his house, Tram needs enough paint to cover 4200 ft^2. How many gallons of paint should he buy?

Answers

9. $5\frac{5}{12}$ yd

10. $8\frac{2}{5}$ gal are needed, so he must buy 9 gal.

Appendix A

EXERCISES

FOR EXTRA HELP

Student's Solutions Manual

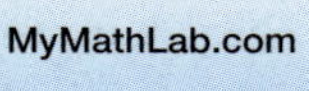
MyMathLab.com

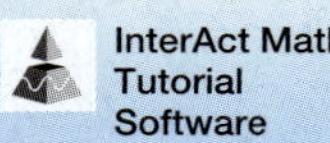
InterAct Math Tutorial Software

AW Math Tutor Center

MathXL www.mathxl.com

Decide whether each statement is true *or* false. *If it is* false, *say why.*

1. In the fraction $\frac{3}{7}$, 3 is the numerator and 7 is the denominator.

2. The mixed number equivalent of $\frac{41}{5}$ is $8\frac{1}{5}$.

3. The fraction $\frac{17}{51}$ is in lowest terms.

4. The reciprocal of $\frac{8}{2}$ is $\frac{4}{1}$.

5. The product of 8 and 2 is 10.

6. The difference between 12 and 2 is 6.

Identify each number as prime, composite, *or* neither. *See Example 1.*

7. 19 **8.** 29 **9.** 52 **10.** 99

11. 2468 **12.** 3125 **13.** 1 **14.** 14

Write each number in prime factored form. See Example 2.

15. 30 **16.** 40 **17.** 252 **18.** 168

19. 124 **20.** 165 **21.** 29 **22.** 31

Write each fraction in lowest terms. See Example 3.

23. $\frac{8}{16}$ **24.** $\frac{4}{12}$ **25.** $\frac{15}{18}$ **26.** $\frac{16}{20}$

27. $\frac{15}{75}$ **28.** $\frac{24}{64}$ **29.** $\frac{144}{120}$ **30.** $\frac{132}{77}$

31. For the fractions $\frac{p}{q}$ and $\frac{r}{s}$, which can serve as a common denominator?

A. $q \cdot s$

B. $q + s$

C. $p \cdot r$

D. $p + r$

32. Which is the correct way to write $\frac{16}{24}$ in lowest terms?

A. $\frac{16}{24} = \frac{8+8}{8+16} = \frac{8}{16} = \frac{1}{2}$

B. $\frac{16}{24} = \frac{4 \cdot 4}{4 \cdot 6} = \frac{4}{6}$

C. $\frac{16}{24} = \frac{8 \cdot 2}{8 \cdot 3} = \frac{2}{3}$

D. $\frac{16}{24} = \frac{14+2}{21+3} = \frac{2}{3} + \frac{2}{3} = \frac{4}{3}$

Find each product or quotient, and write it in lowest terms. See Examples 4 and 5.

33. $\frac{4}{5} \cdot \frac{6}{7}$

34. $\frac{5}{9} \cdot \frac{10}{7}$

35. $\frac{1}{10} \cdot \frac{12}{5}$

36. $\frac{6}{11} \cdot \frac{2}{3}$

37. $\frac{15}{4} \cdot \frac{8}{25}$

38. $\frac{4}{7} \cdot \frac{21}{8}$

39. $2\frac{2}{3} \cdot 5\frac{4}{5}$

40. $3\frac{3}{5} \cdot 7\frac{1}{6}$

41. $\frac{5}{4} \div \frac{3}{8}$

42. $\frac{7}{6} \div \frac{9}{10}$

43. $\frac{32}{5} \div \frac{8}{15}$

44. $\frac{24}{7} \div \frac{6}{21}$

45. $\frac{3}{4} \div 12$

46. $\frac{2}{5} \div 30$

47. $2\frac{5}{8} \div 1\frac{15}{32}$

48. $2\frac{3}{10} \div 7\frac{4}{5}$

49. In your own words, explain how to divide two fractions.

50. In your own words, explain how to add two fractions that have different denominators.

Find each sum or difference, and write it in lowest terms. See Examples 6–8.

51. $\frac{7}{12} + \frac{1}{12}$

52. $\frac{3}{16} + \frac{5}{16}$

53. $\frac{5}{9} + \frac{1}{3}$

54. $\frac{4}{15} + \frac{1}{5}$

55. $3\frac{1}{8} + \frac{1}{4}$

56. $5\frac{3}{4} + \frac{2}{3}$

57. $\frac{7}{12} - \frac{1}{9}$

58. $\frac{11}{16} - \frac{1}{12}$

59. $6\frac{1}{4} - 5\frac{1}{3}$

60. $8\frac{4}{5} - 7\frac{4}{9}$

61. $\frac{5}{3} + \frac{1}{6} - \frac{1}{2}$

62. $\frac{7}{15} + \frac{1}{6} - \frac{1}{10}$

Use the chart, which appears on a package of Quaker Quick Grits, to answer the questions in Exercises 63 and 64.

63. How many cups of water would be needed for eight microwave servings?

64. How many teaspoons of salt would be needed for five stove top servings? (*Hint:* 5 is halfway between 4 and 6.)

Microwave		**Stove Top**		
Servings	**1**	**1**	**4**	**6**
Water	$\frac{3}{4}$ cup	1 cup	3 cups	4 cups
Grits	3 Tbsp	3 Tbsp	$\frac{3}{4}$ cup	1 cup
Salt (optional)	Dash	Dash	$\frac{1}{4}$ tsp	$\frac{1}{2}$ tsp

Solve each applied problem. See Examples 9 and 10.

65. A motel owner has decided to expand his business by buying a piece of property next to the motel. The property has an irregular shape, with five sides as shown in the figure. Find the total distance around the piece of property. This is called the *perimeter* of the figure.

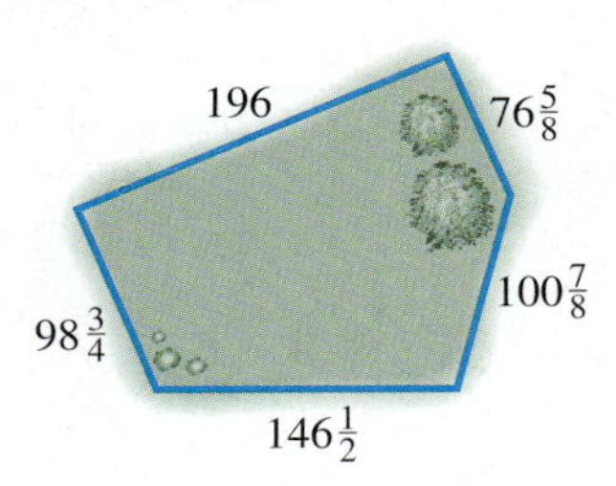

Measurements in feet

66. A triangle has sides of lengths $5\frac{1}{4}$ ft, $7\frac{1}{2}$ ft, and $10\frac{1}{8}$ ft. Find the perimeter of the triangle. See Exercise 65.

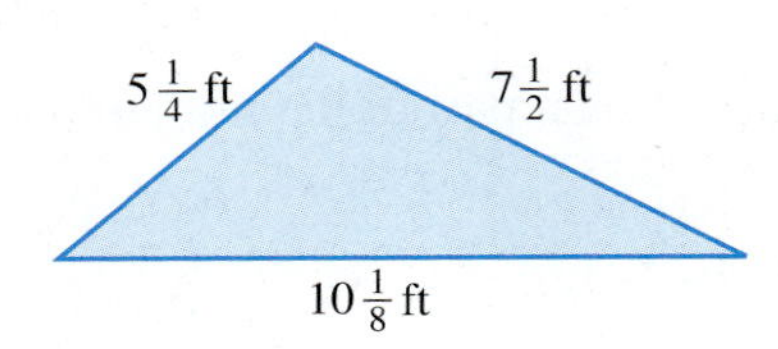

67. A hardware store sells a 40-piece socket wrench set. The measure of the largest socket is $\frac{3}{4}$ in., while the measure of the smallest socket is $\frac{3}{16}$ in. What is the difference between these measures?

68. Two sockets in a socket wrench set have measures of $\frac{9}{16}$ in. and $\frac{3}{8}$ in. What is the difference between these two measures?

69. Under existing standards, most of the holes in Swiss cheese must have diameters between $\frac{11}{16}$ and $\frac{13}{16}$ in. To accommodate new high-speed slicing machines, the USDA wants to reduce the minimum size to $\frac{3}{8}$ in. How much smaller is $\frac{3}{8}$ in. than $\frac{11}{16}$ in.? (*Source:* U.S. Department of Agriculture.)

70. Tex's favorite recipe for barbecue sauce calls for $2\frac{1}{3}$ cups of tomato sauce. The recipe makes enough barbecue sauce to serve 7 people. How much tomato sauce is needed for 1 serving?

More than 8 million immigrants were admitted to the United States between 1990 and 1997. The pie chart gives the fractional number from each region of birth for these immigrants. Use the chart to answer the following questions.

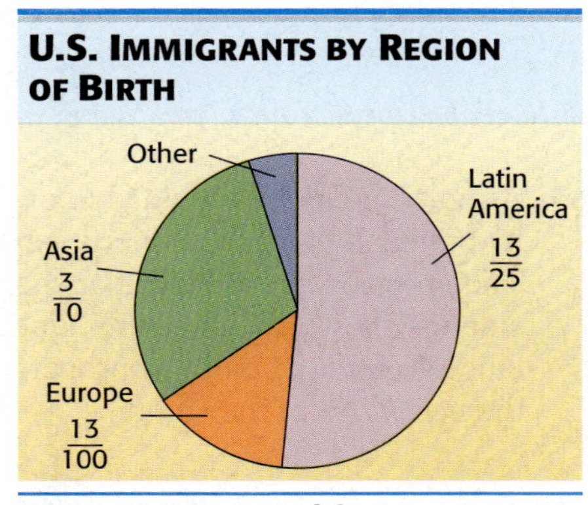

Source: U.S. Bureau of the Census.

71. What fractional part of the immigrants were from other regions?

72. What fractional part of the immigrants were from Latin America or Asia?

73. How many (in millions) were from Europe?

Appendix B Determinants and Cramer's Rule

B DETERMINANTS AND CRAMER'S RULE

OBJECTIVES

1. Evaluate 2 × 2 determinants.
2. Use expansion by minors to evaluate 3 × 3 determinants.
3. Understand the derivation of Cramer's rule.
4. Apply Cramer's rule to solve linear systems.

Recall from Section 5.4 that an ordered array of numbers within square brackets is called a *matrix* (plural *matrices*). Matrices are named according to the number of rows and columns they contain. A *square matrix* has the same number of rows and columns.

$$\text{Rows} \rightarrow \begin{bmatrix} 2 & 3 & 5 \\ 7 & 1 & 2 \end{bmatrix} \quad 2 \times 3 \text{ matrix} \qquad \begin{bmatrix} -1 & 0 \\ 1 & -2 \end{bmatrix} \quad 2 \times 2 \text{ square matrix}$$

(Columns indicated by arrows above the first matrix.)

Associated with every *square matrix* is a real number called the **determinant** of the matrix. A determinant is symbolized by the entries of the matrix placed between two vertical lines, such as

$$\begin{vmatrix} 2 & 3 \\ 7 & 1 \end{vmatrix} \quad 2 \times 2 \text{ determinant} \qquad \begin{vmatrix} 7 & 4 & 3 \\ 0 & 1 & 5 \\ 6 & 0 & 1 \end{vmatrix}. \quad 3 \times 3 \text{ determinant}$$

Like matrices, determinants are named according to the number of rows and columns they contain.

1 Evaluate 2 × 2 determinants. As mentioned above, the value of a determinant is a *real number*. The value of the 2 × 2 determinant

$$\begin{vmatrix} a & b \\ c & d \end{vmatrix}$$

is defined as follows.

Value of a 2 × 2 Determinant

$$\begin{vmatrix} a & b \\ c & d \end{vmatrix} = ad - bc$$

Example 1 Evaluating a 2 × 2 Determinant

Evaluate the determinant.

$$\begin{vmatrix} -1 & -3 \\ 4 & -2 \end{vmatrix}$$

Here $a = -1$, $b = -3$, $c = 4$, and $d = -2$, so

$$\begin{vmatrix} -1 & -3 \\ 4 & -2 \end{vmatrix} = -1(-2) - (-3)4 = 2 + 12 = 14.$$

Work Problem 1 at the Side.

1 Evaluate each determinant.

(a) $\begin{vmatrix} -4 & 6 \\ 2 & 3 \end{vmatrix}$

(b) $\begin{vmatrix} 3 & -1 \\ 0 & 2 \end{vmatrix}$

(c) $\begin{vmatrix} -2 & 5 \\ 1 & 5 \end{vmatrix}$

ANSWERS

1. (a) −24 (b) 6 (c) −15

A 3×3 determinant can be evaluated in a similar way.

Value of a 3 × 3 Determinant

$$\begin{vmatrix} a_1 & b_1 & c_1 \\ a_2 & b_2 & c_2 \\ a_3 & b_3 & c_3 \end{vmatrix} = (a_1b_2c_3 + b_1c_2a_3 + c_1a_2b_3) - (a_3b_2c_1 + b_3c_2a_1 + c_3a_2b_1)$$

This rule for evaluating a 3×3 determinant is hard to remember. A method for calculating a 3×3 determinant that is easier to use is based on the rule. Rearranging terms and using the distributive property gives

$$\begin{vmatrix} a_1 & b_1 & c_1 \\ a_2 & b_2 & c_2 \\ a_3 & b_3 & c_3 \end{vmatrix} = a_1(b_2c_3 - b_3c_2) - a_2(b_1c_3 - b_3c_1) + a_3(b_1c_2 - b_2c_1). \quad (1)$$

Each of the quantities in parentheses represents a 2×2 determinant that is the part of the 3×3 determinant remaining when the row and column of the multiplier are eliminated, as shown below.

$$a_1(b_2c_3 - b_3c_2) \qquad \begin{vmatrix} a_1 & b_1 & c_1 \\ a_2 & b_2 & c_2 \\ a_3 & b_3 & c_3 \end{vmatrix}$$

$$a_2(b_1c_3 - b_3c_1) \qquad \begin{vmatrix} a_1 & b_1 & c_1 \\ a_2 & b_2 & c_2 \\ a_3 & b_3 & c_3 \end{vmatrix}$$

$$a_3(b_1c_2 - b_2c_1) \qquad \begin{vmatrix} a_1 & b_1 & c_1 \\ a_2 & b_2 & c_2 \\ a_3 & b_3 & c_3 \end{vmatrix}$$

These 2×2 determinants are called **minors** of the elements in the 3×3 determinant. In the determinant above, the minors of a_1, a_2, and a_3 are, respectively,

$$\begin{vmatrix} b_2 & c_2 \\ b_3 & c_3 \end{vmatrix}, \quad \begin{vmatrix} b_1 & c_1 \\ b_3 & c_3 \end{vmatrix}, \quad \text{and} \quad \begin{vmatrix} b_1 & c_1 \\ b_2 & c_2 \end{vmatrix}.$$

2 Use expansion by minors to evaluate 3 × 3 determinants. A 3×3 determinant can be evaluated by multiplying each element in the first column by its minor and combining the products as indicated in equation (1). This is called **expansion of the determinant by minors** about the first column.

Example 2 Evaluating a 3 × 3 Determinant

Evaluate the determinant using expansion by minors about the first column.

$$\begin{vmatrix} 1 & 3 & -2 \\ -1 & -2 & -3 \\ 1 & 1 & 2 \end{vmatrix}$$

In this determinant, $a_1 = 1$, $a_2 = -1$, and $a_3 = 1$. Multiply each of these numbers by its minor, and combine the three terms using the definition. Notice that the second term in the definition is *subtracted.*

Continued on Next Page

$$\begin{vmatrix} 1 & 3 & -2 \\ -1 & -2 & -3 \\ 1 & 1 & 2 \end{vmatrix} = 1\begin{vmatrix} -2 & -3 \\ 1 & 2 \end{vmatrix} - (-1)\begin{vmatrix} 3 & -2 \\ 1 & 2 \end{vmatrix} + 1\begin{vmatrix} 3 & -2 \\ -2 & -3 \end{vmatrix}$$

$$= 1[-2(2) - (-3)1] + 1[3(2) - (-2)1] + 1[3(-3) - (-2)(-2)]$$

$$= 1(-1) + 1(8) + 1(-13)$$

$$= -1 + 8 - 13$$

$$= -6$$

Work Problem 2 at the Side.

To get equation (1) we could have rearranged terms in the definition of the determinant and used the distributive property to factor out the three elements of the second or third column or of any of the three rows. Therefore, expanding by minors about any row or any column results in the same value for a 3×3 determinant. To determine the correct signs for the terms of other expansions, the following **array of signs** is helpful.

Array of Signs for a 3 × 3 Determinant

$$\begin{matrix} + & - & + \\ - & + & - \\ + & - & + \end{matrix}$$

The signs alternate for each row and column beginning with a + in the first row, first column position. For example, if the expansion is to be about the second column, the first term would have a minus sign associated with it, the second term a plus sign, and the third term a minus sign.

Example 3 Evaluating a 3 × 3 Determinant

Evaluate the determinant of Example 2 using expansion by minors about the second column.

$$\begin{vmatrix} 1 & 3 & -2 \\ -1 & -2 & -3 \\ 1 & 1 & 2 \end{vmatrix} = -3\begin{vmatrix} -1 & -3 \\ 1 & 2 \end{vmatrix} + (-2)\begin{vmatrix} 1 & -2 \\ 1 & 2 \end{vmatrix} - 1\begin{vmatrix} 1 & -2 \\ -1 & -3 \end{vmatrix}$$

$$= -3(1) - 2(4) - 1(-5)$$

$$= -3 - 8 + 5$$

$$= -6$$

As expected, the result is the same as in Example 2.

Work Problem 3 at the Side.

2 Evaluate each determinant using expansion by minors about the first column.

(a) $\begin{vmatrix} 0 & -1 & 0 \\ 2 & 4 & 2 \\ 3 & 1 & 5 \end{vmatrix}$

(b) $\begin{vmatrix} 2 & 1 & 4 \\ -3 & 0 & 2 \\ -2 & 1 & 5 \end{vmatrix}$

3 Evaluate each determinant using expansion by minors about the second column.

(a) $\begin{vmatrix} 2 & 1 & 3 \\ -1 & 0 & 4 \\ 2 & 4 & 3 \end{vmatrix}$

(b) $\begin{vmatrix} 5 & -1 & 2 \\ 0 & 4 & 3 \\ -1 & 2 & 0 \end{vmatrix}$

ANSWERS

2. (a) 4 **(b)** -5

3. (a) -33 **(b)** -19

Calculator Tip The graphing calculator function det(A) assigns to each square matrix A one and only one real number, the determinant of A. For example, Figure 1 shows how a graphing calculator displays the correct value for the determinant in Example 1. Similarly, Figure 2 supports the results of Examples 2 and 3.

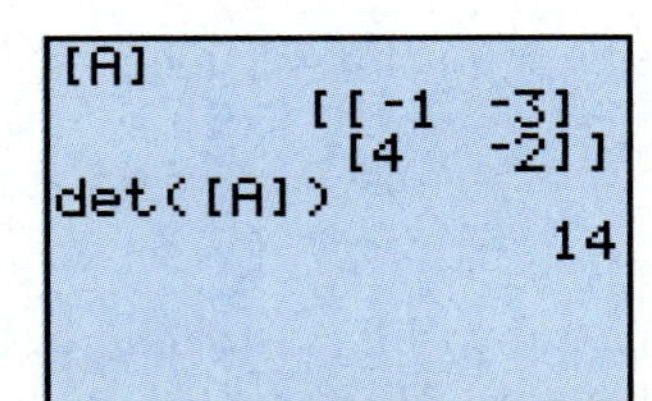

[A]
[[-1 -3]
[4 -2]]
det([A])
14
Figure 1

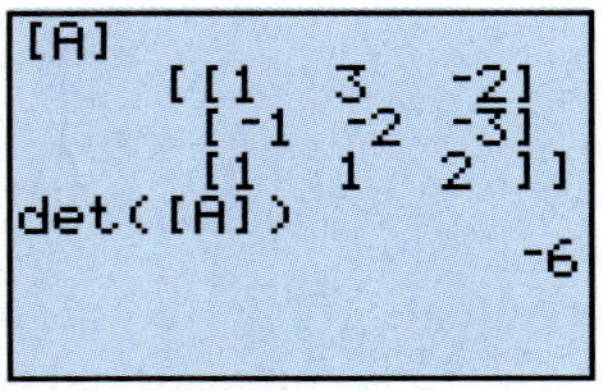

[A]
[[1 3 -2]
[-1 -2 -3]
[1 1 2]]
det([A])
-6
Figure 2

3 **Understand the derivation of Cramer's rule.** Determinants can be used to solve a system of the form

$$a_1x + b_1y = c_1 \qquad (1)$$
$$a_2x + b_2y = c_2. \qquad (2)$$

The result will be a formula that can be used to solve any system of two equations with two variables. To get this general solution, we eliminate y and solve for x by first multiplying each side of equation (1) by b_2 and each side of equation (2) by $-b_1$. Then we add these results and solve for x.

$$\begin{aligned} a_1b_2x + \mathbf{b_1b_2y} &= c_1b_2 && \text{Multiply equation (1) by } b_2.\\ -a_2b_1x - \mathbf{b_1b_2y} &= -c_2b_1 && \text{Multiply equation (2) by } -b_1.\\ \hline (a_1b_2 - a_2b_1)x &= c_1b_2 - c_2b_1 \\ x &= \frac{c_1b_2 - c_2b_1}{a_1b_2 - a_2b_1} && (\text{if } a_1b_2 - a_2b_1 \neq 0) \end{aligned}$$

To solve for y, we multiply each side of equation (1) by $-a_2$ and each side of equation (2) by a_1 and add.

$$\begin{aligned} \mathbf{-a_1a_2x} - a_2b_1y &= -a_2c_1 && \text{Multiply equation (1) by } -a_2.\\ \mathbf{a_1a_2x} + a_1b_2y &= a_1c_2 && \text{Multiply equation (2) by } a_1.\\ \hline (a_1b_2 - a_2b_1)y &= a_1c_2 - a_2c_1 \\ y &= \frac{a_1c_2 - a_2c_1}{a_1b_2 - a_2b_1} && (\text{if } a_1b_2 - a_2b_1 \neq 0) \end{aligned}$$

Both numerators and the common denominator of these values for x and y can be written as determinants because

$$a_1c_2 - a_2c_1 = \begin{vmatrix} a_1 & c_1 \\ a_2 & c_2 \end{vmatrix},$$

$$c_1b_2 - c_2b_1 = \begin{vmatrix} c_1 & b_1 \\ c_2 & b_2 \end{vmatrix},$$

and

$$a_1b_2 - a_2b_1 = \begin{vmatrix} a_1 & b_1 \\ a_2 & b_2 \end{vmatrix}.$$

Using these results, the solutions for x and y become

$$x = \frac{\begin{vmatrix} c_1 & b_1 \\ c_2 & b_2 \end{vmatrix}}{\begin{vmatrix} a_1 & b_1 \\ a_2 & b_2 \end{vmatrix}} \quad \text{and} \quad y = \frac{\begin{vmatrix} a_1 & c_1 \\ a_2 & c_2 \end{vmatrix}}{\begin{vmatrix} a_1 & b_1 \\ a_2 & b_2 \end{vmatrix}}, \quad \begin{vmatrix} a_1 & b_1 \\ a_2 & b_2 \end{vmatrix} \neq 0.$$

For convenience, denote the three determinants in the solution as

$$\begin{vmatrix} a_1 & b_1 \\ a_2 & b_2 \end{vmatrix} = D, \quad \begin{vmatrix} c_1 & b_1 \\ c_2 & b_2 \end{vmatrix} = D_x, \quad \text{and} \quad \begin{vmatrix} a_1 & c_1 \\ a_2 & c_2 \end{vmatrix} = D_y.$$

Notice that the elements of D are the four coefficients of the variables in the given system; the elements of D_x are obtained by replacing the coefficients of x by the respective constants; the elements of D_y are obtained by replacing the coefficients of y by the respective constants.

These results are summarized as **Cramer's rule.**

Cramer's Rule for 2 × 2 Systems

Given the system

$$\begin{aligned} a_1x + b_1y &= c_1 \\ a_2x + b_2y &= c_2 \quad \text{with} \quad a_1b_2 - a_2b_1 = D \neq 0, \end{aligned}$$

then

$$x = \frac{\begin{vmatrix} c_1 & b_1 \\ c_2 & b_2 \end{vmatrix}}{\begin{vmatrix} a_1 & b_1 \\ a_2 & b_2 \end{vmatrix}} = \frac{D_x}{D} \quad \text{and} \quad y = \frac{\begin{vmatrix} a_1 & c_1 \\ a_2 & c_2 \end{vmatrix}}{\begin{vmatrix} a_1 & b_1 \\ a_2 & b_2 \end{vmatrix}} = \frac{D_y}{D}.$$

NOTE

Swiss mathematician and physicist Gabriel Cramer (1704–1752) was looking for a method to determine the equation of a curve given several points on the curve. In 1750, he wrote down the general equation for a curve and then substituted each point for which he had two coordinates into the equation. For the resulting system of equations, he gave "a rule very convenient and general to solve any number of equations and unknowns which are of no more than first degree." This is the rule that now bears his name. (*Source:* Lial, Margaret L., Hornsby, John, and Schneider, David I., *College Algebra,* Eighth Edition, Addison-Wesley, 2001.)

4 **Apply Cramer's rule to solve linear systems.** To use Cramer's rule to solve a system of equations, find the three determinants, D, D_x, and D_y, and then write the necessary quotients for x and y.

CAUTION

As indicated in the box, Cramer's rule does not apply if $D = a_1b_2 - a_2b_1$ is 0. When $D = 0$, the system is inconsistent or has dependent equations. For this reason, it is a good idea to evaluate D first.

4 Solve each system using Cramer's rule.

(a) $x + y = 5$
$x - y = 1$

(b) $2x - 3y = -26$
$3x + 4y = 12$

(c) $4x - 5y = -8$
$3x + 7y = -6$

Example 4 Using Cramer's Rule to Solve a 2 × 2 System

Use Cramer's rule to solve the system.

$$5x + 7y = -1$$
$$6x + 8y = 1$$

By Cramer's rule, $x = \frac{D_x}{D}$ and $y = \frac{D_y}{D}$. As previously mentioned, it is a good idea to find D first since if $D = 0$, Cramer's rule does not apply. If $D \neq 0$, then find D_x and D_y.

$$D = \begin{vmatrix} 5 & 7 \\ 6 & 8 \end{vmatrix} = 5(8) - 7(6) = -2$$

$$D_x = \begin{vmatrix} -1 & 7 \\ 1 & 8 \end{vmatrix} = -1(8) - 7(1) = -15$$

$$D_y = \begin{vmatrix} 5 & -1 \\ 6 & 1 \end{vmatrix} = 5(1) - (-1)6 = 11$$

From Cramer's rule,

$$x = \frac{D_x}{D} = \frac{-15}{-2} = \frac{15}{2} \quad \text{and} \quad y = \frac{D_y}{D} = \frac{11}{-2} = -\frac{11}{2}.$$

The solution set is $\{(\frac{15}{2}, -\frac{11}{2})\}$, as can be verified by checking in the given system.

Work Problem 4 at the Side.

In a similar manner, Cramer's rule can be applied to systems of three equations with three variables.

Cramer's Rule for 3 × 3 Systems

Given the system

$$a_1x + b_1y + c_1z = d_1$$
$$a_2x + b_2y + c_2z = d_2$$
$$a_3x + b_3y + c_3z = d_3$$

with

$$D_x = \begin{vmatrix} d_1 & b_1 & c_1 \\ d_2 & b_2 & c_2 \\ d_3 & b_3 & c_3 \end{vmatrix}, \quad D_y = \begin{vmatrix} a_1 & d_1 & c_1 \\ a_2 & d_2 & c_2 \\ a_3 & d_3 & c_3 \end{vmatrix},$$

$$D_z = \begin{vmatrix} a_1 & b_1 & d_1 \\ a_2 & b_2 & d_2 \\ a_3 & b_3 & d_3 \end{vmatrix}, \quad D = \begin{vmatrix} a_1 & b_1 & c_1 \\ a_2 & b_2 & c_2 \\ a_3 & b_3 & c_3 \end{vmatrix} \neq 0,$$

then

$$x = \frac{D_x}{D}, \quad y = \frac{D_y}{D}, \quad \text{and} \quad z = \frac{D_z}{D}.$$

ANSWERS

4. (a) $\{(3, 2)\}$ **(b)** $\{(-4, 6)\}$ **(c)** $\{(-2, 0)\}$

Example 5 Using Cramer's Rule to Solve a 3 × 3 System

Use Cramer's rule to solve the system.

$$\begin{aligned} x + y - z + 2 &= 0 \\ 2x - y + z + 5 &= 0 \\ x - 2y + 3z - 4 &= 0 \end{aligned}$$

To use Cramer's rule, first rewrite the system in the form

$$\begin{aligned} x + y - z &= -2 \\ 2x - y + z &= -5 \\ x - 2y + 3z &= 4. \end{aligned}$$

Expand by minors about row 1 to find D.

$$D = \begin{vmatrix} \mathbf{1} & \mathbf{1} & \mathbf{-1} \\ 2 & -1 & 1 \\ 1 & -2 & 3 \end{vmatrix}$$

$$= \mathbf{1}\begin{vmatrix} -1 & 1 \\ -2 & 3 \end{vmatrix} - \mathbf{1}\begin{vmatrix} 2 & 1 \\ 1 & 3 \end{vmatrix} + (\mathbf{-1})\begin{vmatrix} 2 & -1 \\ 1 & -2 \end{vmatrix}$$

$$= 1(-1) - 1(5) - 1(-3)$$

$$= -3$$

Expanding D_x by minors about row 1 gives

$$D_x = \begin{vmatrix} \mathbf{-2} & \mathbf{1} & \mathbf{-1} \\ -5 & -1 & 1 \\ 4 & -2 & 3 \end{vmatrix}$$

$$= \mathbf{-2}\begin{vmatrix} -1 & 1 \\ -2 & 3 \end{vmatrix} - \mathbf{1}\begin{vmatrix} -5 & 1 \\ 4 & 3 \end{vmatrix} + (\mathbf{-1})\begin{vmatrix} -5 & -1 \\ 4 & -2 \end{vmatrix}$$

$$= -2(-1) - 1(-19) - 1(14)$$

$$= 7.$$

Work Problem 5 at the Side.

Using the results for D and D_x and the results from Problem 5 at the side, apply Cramer's rule to get

$$x = \frac{D_x}{D} = \frac{7}{-3} = -\frac{7}{3}, \quad y = \frac{D_y}{D} = \frac{-22}{-3} = \frac{22}{3}, \quad z = \frac{D_z}{D} = \frac{-21}{-3} = 7.$$

Check that the solution set is $\{(-\frac{7}{3}, \frac{22}{3}, 7)\}$.

Work Problem 6 at the Side.

As mentioned earlier, Cramer's rule does not apply when $D = 0$. The next example illustrates this case.

5 Find D_y and D_z for Example 5.

6 Solve each system using Cramer's rule.

(a) $$\begin{aligned} x + y + z &= 2 \\ 2x \quad - z &= -3 \\ y + 2z &= 4 \end{aligned}$$

(b) $$\begin{aligned} 3x - 2y + 4z &= 5 \\ 4x + y + z &= 14 \\ x - y - z &= 1 \end{aligned}$$

ANSWERS

5. $D_y = -22$ and $D_z = -21$

6. **(a)** $\{(-1, 2, 1)\}$ **(b)** $\{(3, 2, 0)\}$

7 Solve by Cramer's rule (if applicable).

$$\begin{aligned} x - y + z &= 6 \\ 3x + 2y + z &= 4 \\ 2x - 2y + 2z &= 14 \end{aligned}$$

Example 6 **Determining When Cramer's Rule Does Not Apply**

Use Cramer's rule to solve the system.

$$\begin{aligned} 2x - 3y + 4z &= 8 \\ 6x - 9y + 12z &= 24 \\ x + 2y - 3z &= 5 \end{aligned}$$

First, find D.

$$D = \begin{vmatrix} 2 & -3 & 4 \\ 6 & -9 & 12 \\ 1 & 2 & -3 \end{vmatrix}$$

$$= 2\begin{vmatrix} -9 & 12 \\ 2 & -3 \end{vmatrix} - 6\begin{vmatrix} -3 & 4 \\ 2 & -3 \end{vmatrix} + 1\begin{vmatrix} -3 & 4 \\ -9 & 12 \end{vmatrix}$$

$$= 2(3) - 6(1) + 1(0)$$

$$= 0$$

Since $D = 0$ here, Cramer's rule does not apply and we must use another method to solve the system. Multiplying each side of the first equation by 3 shows that the first two equations have the same solution set, so this system has dependent equations and an infinite solution set.

Work Problem 7 at the Side.

ANSWERS

7. Cramer's rule does not apply.

Appendix B Exercises

FOR EXTRA HELP

Student's Solutions Manual

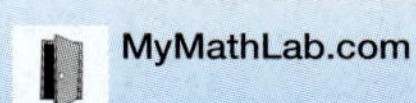
MyMathLab.com

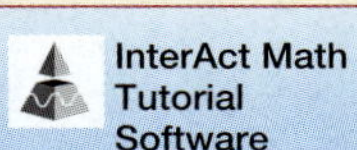
InterAct Math Tutorial Software

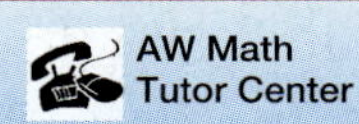
AW Math Tutor Center

www.mathxl.com

Decide whether each statement is true *or* false.

1. A matrix is an array of numbers, while a determinant is just a number.

2. A square matrix has the same number of rows as columns.

3. The determinant $\begin{vmatrix} a & b \\ c & d \end{vmatrix}$ is equal to $ad + bc$.

4. The value of $\begin{vmatrix} 0 & 0 \\ x & y \end{vmatrix}$ is 0 for any replacements for x and y.

Evaluate each determinant. See Example 1.

5. $\begin{vmatrix} -2 & 5 \\ -1 & 4 \end{vmatrix}$

6. $\begin{vmatrix} 3 & -6 \\ 2 & -2 \end{vmatrix}$

7. $\begin{vmatrix} 1 & -2 \\ 7 & 0 \end{vmatrix}$

8. $\begin{vmatrix} -5 & -1 \\ 1 & 0 \end{vmatrix}$

9. $\begin{vmatrix} 0 & 4 \\ 0 & 4 \end{vmatrix}$

10. $\begin{vmatrix} 8 & -3 \\ 0 & 0 \end{vmatrix}$

Evaluate each determinant using expansion by minors about the first column. See Example 2.

11. $\begin{vmatrix} -1 & 2 & 4 \\ -3 & -2 & -3 \\ 2 & -1 & 5 \end{vmatrix}$

12. $\begin{vmatrix} 2 & -3 & -5 \\ 1 & 2 & 2 \\ 5 & 3 & -1 \end{vmatrix}$

13. $\begin{vmatrix} 1 & 0 & -2 \\ 0 & 2 & 3 \\ 1 & 0 & 5 \end{vmatrix}$

14. $\begin{vmatrix} 2 & -1 & 0 \\ 0 & -1 & 1 \\ 1 & 2 & 0 \end{vmatrix}$

15. $\begin{vmatrix} 1 & 0 & 0 \\ 0 & 1 & 0 \\ 0 & 0 & 1 \end{vmatrix}$

16. $\begin{vmatrix} 0 & 0 & 1 \\ 0 & 1 & 0 \\ 1 & 0 & 0 \end{vmatrix}$

Evaluate each determinant by expansion about any row or column. (Hint: If possible, choose a row or column with 0s.) See Example 3.

17. $\begin{vmatrix} 4 & 4 & 2 \\ 1 & -1 & -2 \\ 1 & 0 & 2 \end{vmatrix}$

18. $\begin{vmatrix} 3 & -1 & 2 \\ 1 & 5 & -2 \\ 0 & 2 & 0 \end{vmatrix}$

19. $\begin{vmatrix} 2 & 0 & 1 \\ -1 & 0 & 2 \\ 5 & 0 & 4 \end{vmatrix}$

20. $\begin{vmatrix} 2 & -4 & 0 \\ 3 & -5 & 0 \\ 6 & -7 & 0 \end{vmatrix}$

21. $\begin{vmatrix} -6 & 3 & 5 \\ -3 & 2 & 2 \\ 0 & 0 & 0 \end{vmatrix}$

22. $\begin{vmatrix} 0 & 0 & 0 \\ 4 & 0 & -2 \\ 2 & -1 & 3 \end{vmatrix}$

23. $\begin{vmatrix} 3 & 5 & -2 \\ 1 & -4 & 1 \\ 3 & 1 & -2 \end{vmatrix}$

24. $\begin{vmatrix} 1 & 3 & 2 \\ 3 & -1 & -2 \\ 1 & 10 & 20 \end{vmatrix}$

25. For the system

$$\begin{aligned} 8x - 4y &= 8 \\ x + 3y &= 22, \end{aligned}$$

$D_x = 112$, $D_y = 168$, and $D = 28$. What is the solution set of the system?

26. For the system

$$\begin{aligned} x + 3y - 6z &= 7 \\ 2x - y + z &= 1 \\ x + 2y + 2z &= -1, \end{aligned}$$

the solution set is $\{(1, 0, -1)\}$ and $D = -43$. Find the values of D_x, D_y, and D_z.

Use Cramer's rule to solve each system. See Example 4.

27. $\begin{aligned} 3x + 5y &= -5 \\ -2x + 3y &= 16 \end{aligned}$

28. $\begin{aligned} 5x + 2y &= -3 \\ 4x - 3y &= -30 \end{aligned}$

29. $\begin{aligned} 3x + 2y &= 3 \\ 2x - 4y &= 2 \end{aligned}$

30. $\begin{aligned} 7x - 2y &= 6 \\ 4x - 5y &= 15 \end{aligned}$

31. $\begin{aligned} 8x + 3y &= 1 \\ 6x - 5y &= 2 \end{aligned}$

32. $\begin{aligned} 3x - y &= 9 \\ 2x + 5y &= 8 \end{aligned}$

Use Cramer's rule (where applicable) to solve each system. If Cramer's rule does not apply, say so. See Examples 5 and 6.

33. $\begin{aligned} 2x + 3y + 2z &= 15 \\ x - y + 2z &= 5 \\ x + 2y - 6z &= -26 \end{aligned}$

34. $\begin{aligned} x - y + 6z &= 19 \\ 3x + 3y - z &= 1 \\ x + 9y + 2z &= -19 \end{aligned}$

35. $\begin{aligned} 2x + 2y + z &= 10 \\ 4x - y + z &= 20 \\ -x + y - 2z &= -5 \end{aligned}$

36. $\begin{aligned} x + 3y - 4z &= -12 \\ 3x + y - z &= -5 \\ 5x - y + z &= -3 \end{aligned}$

37. $\begin{aligned} 2x - 3y + 4z &= 8 \\ 6x - 9y + 12z &= 24 \\ -4x + 6y - 8z &= -16 \end{aligned}$

38. $\begin{aligned} 7x + y - z &= 4 \\ 2x - 3y + z &= 2 \\ -6x + 9y - 3z &= -6 \end{aligned}$

39. $\begin{aligned} 3x + 5z &= 0 \\ 2x + 3y &= 1 \\ -y + 2z &= -11 \end{aligned}$

40. $\begin{aligned} -x + 2y &= 4 \\ 3x + y &= -5 \\ 2x + z &= -1 \end{aligned}$

41. $\begin{aligned} x - 3y &= 13 \\ 2y + z &= 5 \\ -x + z &= -7 \end{aligned}$

Answers to Selected Exercises

In this section we provide the answers that we think most students will obtain when they work the exercises using the methods explained in the text. If your answer does not look exactly like the one given here, it is not necessarily wrong. In many cases there are equivalent forms of the answer that are correct. For example, if the answer section shows $\frac{3}{4}$ and your answer is .75, you have obtained the correct answer but written it in a different (yet equivalent) form. Unless the directions specify otherwise, .75 is just as valid an answer as $\frac{3}{4}$.

In general, if your answer does not agree with the one given in the text, see whether it can be transformed into the other form. If it can, then it is the correct answer. If you still have doubts, talk with your instructor.

Diagnostic Pretest

(page xxix)

1. -3 **2.** 32,243 ft **3.** -9 **4.** $\{-5\}$ **5.** faster train: 67 mph; slower train: 54 mph **6.** 24°, 31°, 125°

7. $\left(-\infty, \frac{1}{2}\right]$

8. $(-4, 5]$

9. $\left\{-\frac{5}{8}, \frac{13}{2}\right\}$ **10.** $-\frac{11}{7}$

11. x-intercept: $(-5, 0)$; y-intercept: $(0, 3)$

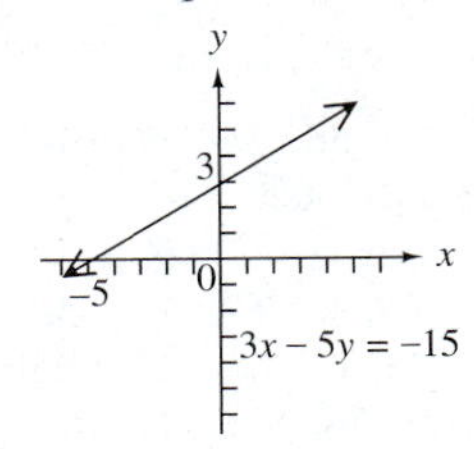

12.

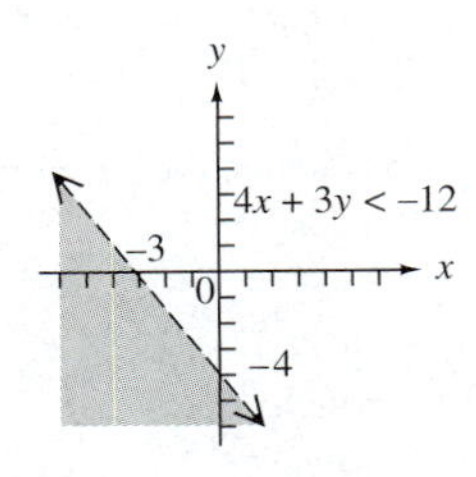

13. B; The set A includes two ordered pairs with the same first component (24.5 or 25.0) and different second components, so it is not a function. **14.** $\{(-8, -1)\}$ **15.** $\{(5, -3)\}$ **16.** $\{(1, -2, 3)\}$

17. 20 lb of nuts, 12 lb of raisins **18.** $\frac{49t^2}{s^8}$

19. $x^2 + 3x - 5 + \frac{2}{2x - 1}$ **20.** $(y + 3)(y + 2)(y - 2)$

21. $\left\{-5, \frac{3}{4}\right\}$ **22.** $\frac{z + 4}{z + 1}$ **23.** $\frac{5x^2 + 11x + 12}{(x + 3)(x - 3)}$ **24.** $4x + 1$

25. $\frac{16m^{10}}{n^6}$ **26.** $5y^2z^3\sqrt[3]{2yz^2}$ **27.** $\{7\}$ **28.** 89 **29.** $\left\{-5, -\frac{1}{3}\right\}$

30. $\left\{\frac{-5 + \sqrt{37}}{6}, \frac{-5 - \sqrt{37}}{6}\right\}$ **31.** east: 60 mi; south: 45 mi

32. vertex: $(3, 5)$
domain: $(-\infty, \infty)$
range: $(-\infty, 5]$

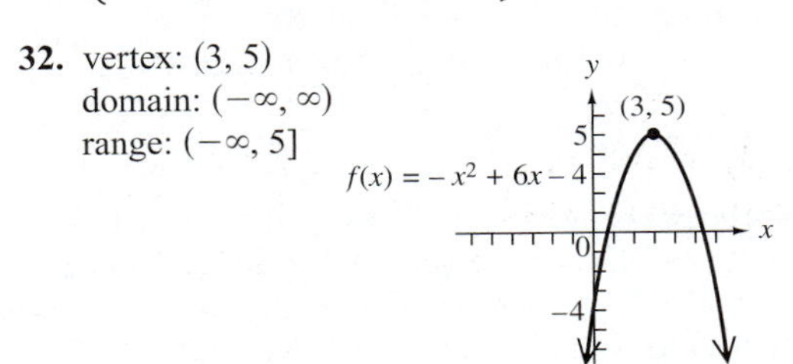

33. $f^{-1}(x) = \sqrt[3]{x + 8}$ **34.** $\{2\}$ **35.** $\{27\}$ **36.** $\{8\}$

37. **(a)** 12 **(b)** 13 **(c)** $4x^2 - 4x + 4$ **(d)** $2x^2 + 5$

38.

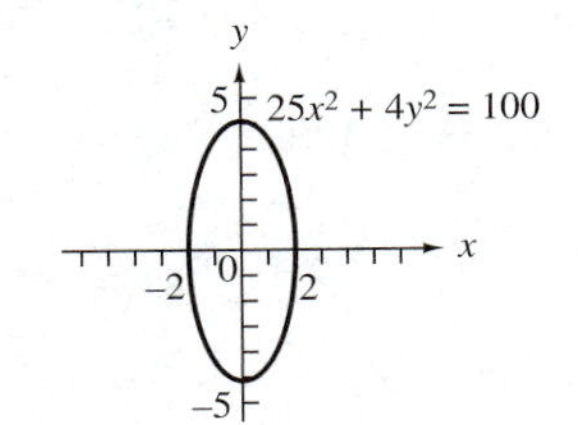

39.

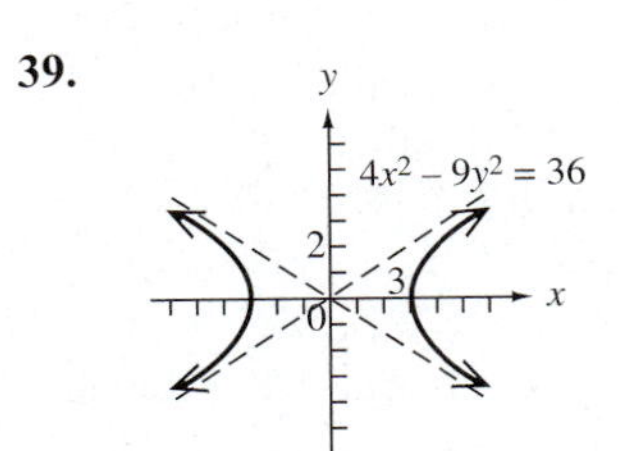

40. $\{(-1, -3), (3, 5)\}$

Chapter 1

Section 1.1 (page 11)

1. $\{1, 2, 3, 4, 5\}$ **3.** $\{5, 6, 7, 8, \ldots\}$ **5.** $\{10, 12, 14, 16, \ldots\}$ **7.** $\emptyset$ **9.** $\{-4, 4\}$ **11.** $\{x \mid x$ is an even natural number less than or equal to $8\}$ **13.** $\{x \mid x$ is a multiple of 4 greater than $0\}$

15.

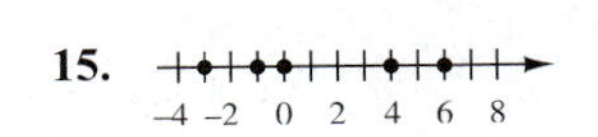

17.

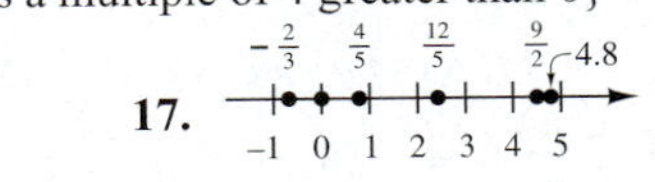

19. **(a)** $5, 17, \frac{40}{2}$ (or 20) **(b)** $0, 5, 17, \frac{40}{2}$ **(c)** $-8, 0, 5, 17, \frac{40}{2}$ **(d)** $-8, -.6, 0, \frac{3}{4}, 5, \frac{13}{2}, 17, \frac{40}{2}$ **(e)** $-\sqrt{5}, \sqrt{3}, \pi$ **(f)** All are real numbers. **21.** False; some are integers, but others, like $\frac{3}{4}$, are not. **23.** False; no irrational number is an integer. **25.** true **27.** true **29.** true **31.** **(a)** -6 **(b)** 6 **33.** **(a)** 12 **(b)** 12 **35.** **(a)** $-\frac{6}{5}$ **(b)** $\frac{6}{5}$ **37.** 8 **39.** $\frac{3}{2}$ **41.** -5 **43.** -2 **45.** -4.5 **47.** 5 **49.** 6 **51.** 0 **53.** **(a)** Philadelphia; The population declined 10.6%. **(b)** Chicago; The population increased .6%. **55.** Pacific Ocean, Indian Ocean, Caribbean Sea, South China Sea, Gulf of California **57.** true **59.** true **61.** false **63.** true **65.** true **67.** $7 > y$ **69.** $5 \geq 5$ **71.** $3t - 4 \leq 10$ **73.** $5x + 3 \neq 0$ **75.** $-6 < 10$; true **77.** $10 \geq 10$; true **79.** $-3 \geq -3$; true **81.** $-8 > -6$; false **83.** less than **85.** California (CA), Minnesota (MN), North Carolina (NC) **87.** $x = y$

Section 1.2 (page 21)

1. the numbers are additive inverses; $4 + (-4) = 0$ **3.** negative; $-7 + (-21) = -28$ **5.** the positive number has larger absolute value; $15 + (-2) = 13$ **7.** the number with smaller absolute value is subtracted from the one with larger absolute value; $-15 - (-3) = -12$ **9.** negative; $-5(15) = -75$ **11.** 9 **13.** -19 **15.** $-\frac{19}{12}$ **17.** -1.85 **19.** -11 **21.** 21 **23.** -13 **25.** -10.18 **27.** $\frac{67}{30}$ **29.** -6 **31.** -35 **33.** 40 **35.** 2 **37.** -12 **39.** $\frac{6}{5}$ **41.** 1 **43.** 5.88 **45.** -10.676 **47.** $\frac{1}{6}$ **49.** $-\frac{1}{7}$ **51.** $-\frac{3}{2}$ **53.** 5 **55.** 50 **57.** -1000 **59.** -7 **61.** 6 **63.** -4 **65.** 0 **67.** undefined **69.** $\frac{25}{102}$ **71.** $-\frac{9}{13}$ **73.** -2.1 **75.** 10,000 **77.** -11 **79.** 16 **81.** -4 **83.** -19 **85.** 112°F **87.** \$491 billion **89.** 2000: \$129 billion; 2010: \$206 billion; 2020: \$74 billion; 2030: $-$\$501 billion **91.** $-11{,}478$ **93.** C

Section 1.3 (page 31)

1. False; $-4^6 = -(4^6)$. **3.** true **5.** true **7.** true **9.** False; the base is 3. **11.** **(a)** 64 **(b)** -64 **(c)** 64 **(d)** -64 **13.** 8^3 **15.** $\left(\frac{1}{2}\right)^2$ **17.** $(-4)^4$ **19.** z^7 **21.** 16 **23.** .021952 **25.** $\frac{1}{125}$ **27.** $\frac{343}{1000}$ **29.** -125 **31.** 256 **33.** -729 **35.** -4096 **37.** exponent: 7; base: -4.1 **39.** exponent: 7; base: 4.1 **41.** 9 **43.** 13 **45.** -20 **47.** $\frac{10}{11}$ **49.** $-.7$ **51.** not a real number **53.** **(a)** B **(b)** C **(c)** A **55.** not a real number **57.** 24 **59.** 4 **61.** 14 **63.** 15 **65.** 55 **67.** -91 **69.** -8 **71.** -48 **73.** 8 **75.** -2 **77.** undefined **79.** -1 **81.** 17 **83.** -96 **85.** $-\frac{15}{238}$ **87.** \$1572 **89.** \$3296 **91.** .035 **93.** Decreased weight will result in higher BACs; .040; .053

Section 1.4 (page 41)

1. B **3.** A **5.** product; 0 **7.** grouping **9.** like **11.** $8k$ **13.** $-2r$ **15.** cannot be simplified **17.** $6a$ **19.** $2m + 2p$ **21.** $-10d + 5f$ **23.** 1900 **25.** 75 **27.** 431 **29.** $-6y + 3$ **31.** $p + 11$ **33.** $-2k + 15$ **35.** $m - 14$ **37.** -1 **39.** $2p + 7$ **41.** $-6z - 39$ **43.** $(5 + 8)x = 13x$ **45.** $(5 \cdot 9)r = 45r$ **47.** $9y + 5x$ **49.** 7 **51.** $8(-4) + 8x = -32 + 8x$ **53.** Answers will vary. One example is washing your face and brushing your teeth. **55.** associative property **56.** associative property **57.** commutative property **58.** associative property **59.** distributive property **60.** add

Chapter 1 Review Exercises (page 47)

1. 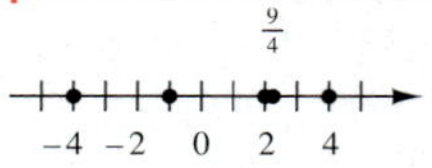

2. 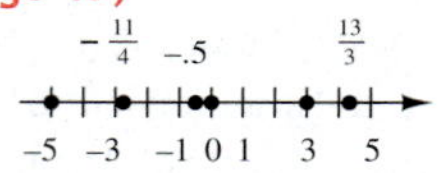

3. 16 **4.** 23 **5.** -4 **6.** 5 **7.** $0, \frac{12}{3}$ (or 4) **8.** $-9, -\sqrt{4}$ (or -2), $0, \frac{12}{3}$ (or 4) **9.** $-9, -\frac{4}{3}, -\sqrt{4}$ (or -2), $-.25, 0, .\overline{35}, \frac{5}{3}, \frac{12}{3}$ (or 4) **10.** All are real numbers except $\sqrt{-9}$. **11.** $\{4, 5, 6, 7, 8\}$ **12.** $\{0, 1, 2, 3\}$ **13.** true **14.** false **15.** true **16.** Hyundai; 50% **17.** General Motors; -5% **18.** false **19.** true **20.** $\frac{41}{24}$ **21.** $-\frac{1}{2}$ **22.** -3 **23.** -17.09 **24.** -39 **25.** -1 **26.** $\frac{23}{20}$ **27.** $-\frac{5}{18}$ **28.** -35 **29.** 11,331 ft **30.** -90 **31.** $\frac{2}{3}$ **32.** -11.408 **33.** -15 **34.** 3.21 **35.** $\frac{5}{7-7}$ **36.** 10,000 **37.** $\frac{27}{343}$ **38.** -125 **39.** -125 **40.** 2.89 **41.** 20 **42.** -14 **43.** $\frac{8}{11}$ **44.** $-.9$ **45.** not a real number **46.** -4 **47.** 44 **48.** -2 **49.** -30 **50.** -30 **51.** $-\frac{8}{51}$ **52.** **(a)** 24 **(b)** Answers will vary. **53.** $21q$ **54.** $-4z$ **55.** $5m$ **56.** $4p$ **57.** $-2k - 6$ **58.** $6r + 18$ **59.** $18m + 27n$ **60.** $-p - 3q$ **61.** $y + 1$ **62.** 0 **63.** $-18m$ **64.** $(2 + 3)x = 5x$ **65.** -4 **66.** $(2 \cdot 4)x = 8x$ **67.** $13 + (-3) = 10$ **68.** 0 **69.** $5x + 5z$ **70.** 7 **71.** 1 **72.** $(3 + 5 + 6)a = 14a$ **73.** 0 **74.** \$2.32 million (in the red) **75.** \$22.88 million (in the red) **76.** \$25.59 million (in the black) **77.** $\frac{256}{625}$ **78.** 25 **79.** 31 **80.** 9 **81.** 0 **82.** -5 **83.** $\frac{4}{3}$ **84.** -6.16 **85.** -9 **86.** 2 **87.** 2 **88.** not a real number **89.** -116 **90.** Work inside the parentheses first.

Chapter 1 Test (page 51)

1. 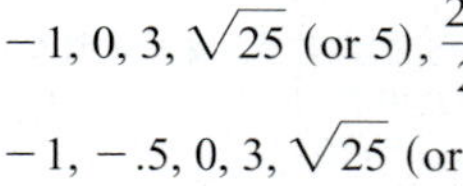

2. $0, 3, \sqrt{25}$ (or 5), $\frac{24}{2}$ (or 12) **3.** $-1, 0, 3, \sqrt{25}$ (or 5), $\frac{24}{2}$ (or 12) **4.** $-1, -.5, 0, 3, \sqrt{25}$ (or 5), $7.5, \frac{24}{2}$ (or 12) **5.** All are real numbers except $\sqrt{-4}$. **6.** 0 **7.** -26 **8.** 19 **9.** 1 **10.** $\frac{16}{7}$

11. $\frac{11}{23}$ **12.** 50,395 ft **13.** 37,486 ft **14.** 1345 ft **15.** 14 **16.** -15 **17.** not a real number **18. (a)** a must be positive. **(b)** a must be negative. **(c)** a must be 0. **19.** 2 **20.** $-\frac{6}{23}$ **21.** $10k - 10$ **22.** Both terms change sign and are added to $3r + 8$; $7r + 2$. **23.** B **24.** E **25.** D **26.** A **27.** F **28.** C **29.** C **30.** E

Chapter 2

Section 2.1 (page 61)

1. A and C **3.** Both sides are evaluated as 30, so 6 is a solution. **5.** Any number is a solution. For example, if the last name is Lincoln, $x = 7$. Both sides are evaluated as -48. **7. (a)** equation **(b)** expression **(c)** equation **(d)** expression **9.** $\{-1\}$ **11.** $\{-7\}$ **13.** $\{0\}$ **15.** $\left\{-\frac{5}{3}\right\}$ **17.** $\left\{-\frac{1}{2}\right\}$ **19.** $\{2\}$ **21.** $\{-2\}$ **23.** $\{7\}$ **25.** $\{-5\}$ **27.** $\{-8\}$ **29.** 12 **31.** Yes, you will get the correct solution. The coefficients will be larger, but in the end the solution will be the same. **33.** $\{4\}$ **35.** $\{0\}$ **37.** $\{0\}$ **39.** $\{2000\}$ **41.** $\{25\}$ **43.** $\{40\}$ **45.** A conditional equation has one solution, an identity has infinitely many solutions, and a contradiction has no solution. **47.** contradiction; $\emptyset$ **49.** conditional; $\{0\}$ **51.** identity; {all real numbers} **53.** solution set **55.** The solution set of the first equation is $\{0\}$, while the solution set of the second equation is $\emptyset$. They are not the same. **57.** The solution sets, $\{4\}$ for $k = 4$ and $\{-4, 4\}$ for $k^2 = 16$, are not the same. **59.** 1995–1996 and 1996–1997 **61.** 9.2 million; yes

Section 2.2 (page 71)

1. (a) $3x = 5x + 8$ **(b)** $ct = bt + k$ **2. (a)** $3x - 5x = 8$ **(b)** $ct - bt = k$ **3. (a)** $-2x = 8$; distributive property **(b)** $t(c - b) = k$; distributive property **4. (a)** $x = -4$ **(b)** $t = \frac{k}{c - b}$ **5.** $c \neq b$; If $c = b$, the denominator is 0. **6.** To solve an equation for a particular variable, such as solving the second equation for t, go through the same steps as you would in solving for x in the first equation. Treat all other variables as constants. **7.** $r = \frac{I}{pt}$ **9.** $L = \frac{P - 2W}{2}$ or $L = \frac{P}{2} - W$ **11.** $W = \frac{V}{LH}$ **13.** $r = \frac{C}{2\pi}$ **15.** $B = \frac{2A}{h} - b$ or $B = \frac{2A - bh}{h}$ **17.** $C = \frac{5}{9}(F - 32)$ **19.** D **21.** $r = \frac{-2k - 3y}{a - 1}$ or $r = \frac{2k + 3y}{1 - a}$ **23.** $y = \frac{-x}{w - 3}$ or $y = \frac{x}{3 - w}$ **25.** 4.388 hr **27.** 104°F **29.** 230 m **31.** radius: 240 in.; diameter: 480 in. **33.** 8 ft **35.** 75% water, 25% alcohol **37.** 3% **39.** \$10.51 **41.** \$45.66 **43. (a)** .600 **(b)** .542 **(c)** .490 **45.** 1500 **47.** 12,250 **49.** \$52,846

Section 2.3 (page 85)

1. (a) $x + 12$ **(b)** $12 > x$ **3. (a)** $x - 4$ **(b)** $4 < x$ **5.** D **7.** $2x - 13$ **9.** $12 + 3x$ **11.** $8(x - 12)$ **13.** $\frac{3x}{7}$ **15.** $x + 6 = -31$; -37 **17.** $x - (-4x) = x + 9$; $\frac{9}{4}$ **19.** $12 - \frac{2}{3}x = 10$; 3 **21.** expression **23.** equation **25.** expression **27.** Read the problem; Assign a variable; Write an equation; Solve the equation; State the answer; Check the answer. **29.** width: 165 ft; length: 265 ft **31.** 850 mi, 925 mi, 1300 mi **33.** General Motors: \$6.9 billion; General Electric: \$6.6 billion **35.** Clinton: 379 votes; Dole: 159 votes **37.** 1.9% **39.** 1,018,500 **41.** \$225 **43.** \$4000 at 3%; \$8000 at 4% **45.** \$10,000 at 4.5%; \$19,000 at 3% **47.** \$58,000 **49.** 5 L **51.** 4 L **53.** 1 gal **55.** 150 lb **57.** We cannot expect the final mixture to be worth more than either of the ingredients. **59. (a)** $800 - x$ **(b)** $800 - y$ **60. (a)** $.05x$; $.10(800 - x)$ **(b)** $.05y$; $.10(800 - y)$ **61. (a)** $.05x + .10(800 - x) = 800(.0875)$ **(b)** $.05y + .10(800 - y) = 800(.0875)$ **62. (a)** \$200 at 5%; \$600 at 10% **(b)** 200 L of 5% acid; 600 L of 10% acid **63.** The processes are the same. The amounts of money in Problem A correspond to the amounts of solution in Problem B.

Section 2.4 (page 95)

1. \$4.50 **3.** 52 mph **5.** The problem asks for the *distance* to the workplace. To find this distance, we must multiply the rate, 10 mph, by the time, $\frac{3}{4}$ hr. **7.** 17 pennies, 17 dimes, 10 quarters **9.** 26 quarters, 21 half-dollars **11.** 28 \$10 coins, 13 \$20 coins **13.** 450 floor tickets, 100 balcony tickets **15.** 7.91 m per sec **17.** 8.42 m per sec **19.** $2\frac{1}{2}$ hr **21.** 7:50 P.M. **23.** 15 mph **25.** $\frac{1}{2}$ hr **27.** 60°, 60°, 60° **29.** 40°, 45°, 95° **31.** 40°, 80° **32.** 120° **33.** The sum is equal to the measure of the angle found in Exercise 32. **34.** The sum of the measures of angles ① and ② is equal to the measure of angle ③. **35.** Both measure 122°. **37.** 64°, 26° **39.** 19, 20, 21 **41.** 61 years old

Summary Exercises on Solving Applied Problems (page 99)

1. length: 8 in.; width: 5 in. **2.** length: 60 m; width: 30 m **3.** \$425 **4.** \$8.95 **5.** \$800 at 4%; \$1600 at 5% **6.** \$12,000 at 3%; \$14,000 at 4% **7.** Roosevelt: 449; Willkie: 82 **8.** Eisner: \$40.1 million; Horrigan: \$21.7 million **9.** Tanui: 2.14 hr; Pippig: 2.44 hr **10.** 10 ft **11.** $1\frac{1}{2}$ cm **12.** 6 in., 12 in., 16 in. **13.** California: 960 pairs; Washington: 671 pairs **14.** London: 29,000,000; Frankfurt: 45,000,000 **15.** 31, 32, 33 **16.** 20°, 30°, 130°

Chapter 2 Review Exercises (page 105)

1. $\left\{-\frac{9}{5}\right\}$ **2.** $\left\{\frac{1}{3}\right\}$ **3.** $\{10\}$ **4.** $\left\{-\frac{7}{5}\right\}$ **5.** $\emptyset$ **6.** $\{0\}$ **7.** $\{16\}$ **8.** $\{300\}$ **9.** B **10.** Begin by subtracting 5 from each side. Then divide each side by -2. **11.** identity; {all real numbers} **12.** contradiction; $\emptyset$ **13.** conditional; $\{0\}$ **14.** $H = \frac{V}{LW}$ **15.** $h = \frac{2A}{B + b}$ **16.** $d = \frac{C}{\pi}$ **17.** 6 ft **18.** 17.4% **19.** 6.5% **20.** 25° **21.** approximately 17,415,099 **22.** 100 mm **23.** $9 - \frac{1}{3}x$ **24.** $\frac{4x}{x + 9}$ **25.** length: 13 m; width: 8 m **26.** 17 in., 17 in., 19 in. **27.** 12 kg **28.** 30 L **29.** 10 L **30.** \$10,000 at 6%; \$6000 at 4% **31.** A **32. (a)** 530 mi **(b)** 328 mi **33.** 2.2 hr **34.** 50 km per hr; 65 km per hr **35.** 1 hr **36.** 46 mph **37.** $\left\{\frac{7}{6}\right\}$ **38.** $\{0\}$ **39.** $\emptyset$ **40.** 12 in., 24 in., 32 in. **41.** $k = \frac{6t - bt}{a + s}$ or $k = \frac{bt - 6t}{-a - s}$ **42.** {all real numbers} **43.** 6 in. **44.** Gore: 266; Bush: 271 **45.** eastbound car: 3 hr; westbound car: 2 hr **46.** \$1300 at 4%; \$1800 at 5% **47.** $\{1500\}$ **48.** $W = \frac{P - 2L}{2}$ or $W = \frac{P}{2} - L$

Chapter 2 Test (page 109)

1. $\{-19\}$ **2.** $\{5\}$ **3.** $\{4\}$ **4.** contradiction; $\emptyset$
5. $v = \frac{S + 16t^2}{t}$ or $v = \frac{S}{t} + 16t$
6. $r = \frac{-2 - 6t}{a - 3}$ or $r = \frac{2 + 6t}{3 - a}$ **7.** 3.372 hr **8.** 6.25%
9. 73.3% **10.** \$8000 at 3%; \$20,000 at 5% **11.** faster car: 60 mph; slower car: 45 mph **12.** 40°, 40°, 100° **13.** 10%
14. 13.33% **15.** 1050

Cumulative Review Exercises: Chapters 1–2 (page 111)

1. $9, \sqrt{36}$ (or 6) **2.** $0, 9, \sqrt{36}$ (or 6) **3.** $-8, 0, 9, \sqrt{36}$ (or 6)
4. $-8, -\frac{2}{3}, 0, \frac{4}{5}, 9, \sqrt{36}$ (or 6) **5.** $-\sqrt{6}$
6. All are real numbers. **7.** $-\frac{22}{21}$ **8.** 7.9 **9.** 8 **10.** 0
11. -243 **12.** $\frac{216}{343}$ **13.** 4096 **14.** -4096 **15.** $\sqrt{-36}$
16. $\frac{4 + 4}{4 - 4}$ **17.** -16 **18.** -34 **19.** 184 **20.** $\frac{27}{16}$
21. $-20r + 17$ **22.** $13k + 42$ **23.** commutative property
24. distributive property **25.** inverse property **26.** $\{5\}$
27. $\{30\}$ **28.** $\{15\}$ **29.** $c = P - a - b$ **30.** $\emptyset$
31. {all real numbers} **32.** 2 L **33.** 9 pennies, 12 nickels, 8 quarters **34.** \$5000 at 5%; \$7000 at 6% **35.** $\frac{1}{8}$ hr **36.** 420
37. 25.7 **38.** 44 mg

Chapter 3

Section 3.1 (page 123)

1. D **3.** B **5.** F **7.** Use a parenthesis when an endpoint is not included; use a bracket when it is included.
9. $[5, \infty)$ 5
11. $(7, \infty)$ 7 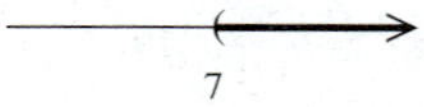
13. $(-4, \infty)$ 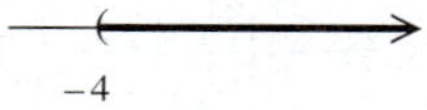 −4
15. $(-\infty, -40]$ −40 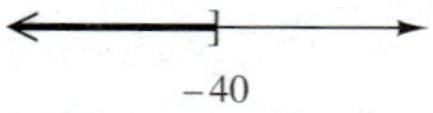
17. $(-\infty, 4]$ 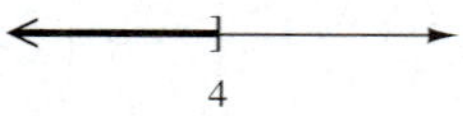 4
19. $\left(-\infty, -\frac{15}{2}\right)$ $-\frac{15}{2}$ 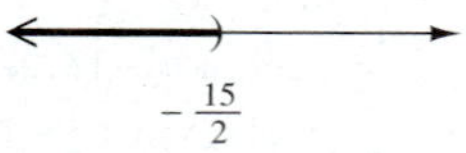
21. $\left[\frac{1}{2}, \infty\right)$ $\frac{1}{2}$
23. $(3, \infty)$ 3 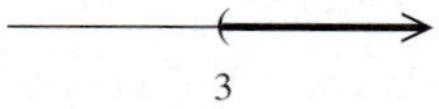
25. $(-\infty, 4)$ 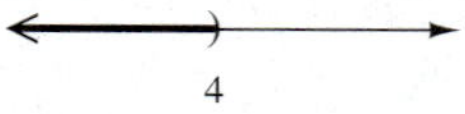 4
27. $\left(-\infty, \frac{23}{6}\right]$ $\frac{23}{6}$ 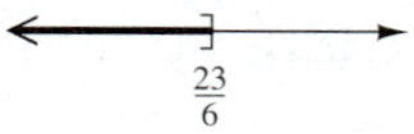
29. $\left(-\infty, \frac{76}{11}\right)$ $\frac{76}{11}$
31. $\{-9\}$ −9 
32. $(-9, \infty)$ 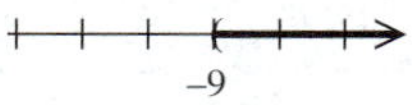 −9
33. $(-\infty, -9)$ −9 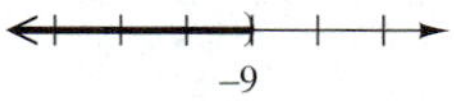
34. We obtain the set of all real numbers. −9
35. $(-\infty, -3)$
37. $(1, 11)$  1 11
39. $[-14, 10]$ −14 10 
41. $[-5, 6]$  −5 6
43. $(-6, -4)$ 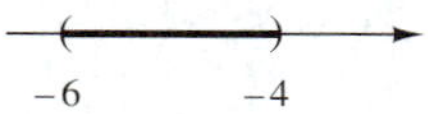 −6 −4
45. $\left[-\frac{13}{3}, \frac{11}{3}\right]$ $-\frac{13}{3}$ $\frac{11}{3}$ 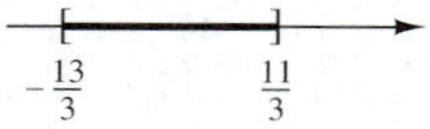
47. from about 8:00 A.M. to 10:15 A.M. and after about 9:00 P.M.
49. about 65°F–67°F **51.** at least 82 **53.** 628.6 mi **55.** 921 deliveries **57. (a)** 130 to 157 beats per min **(b)** Answers will vary.

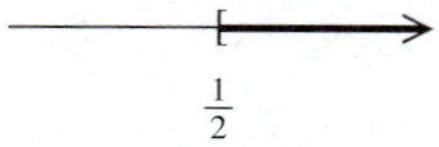

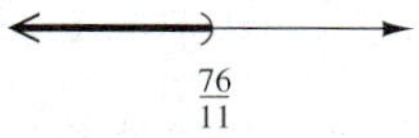

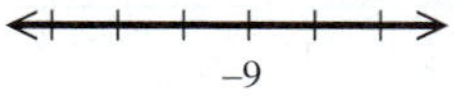

Section 3.2 (page 135)

1. true **3.** False; The union is $(-\infty, 6) \cup (6, \infty)$. **5.** $\{4\}$ or D
7. $\emptyset$ **9.** $\{1, 2, 3, 4, 5, 6\}$ or A **11.** $\{1, 3, 5, 6\}$
13. 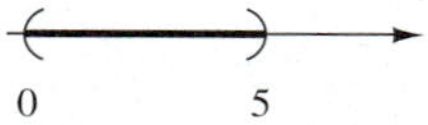 0 5
15. 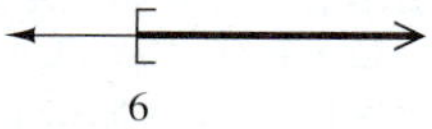 6
17. 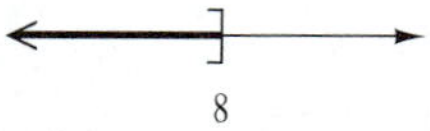 8
19. Answers will vary. One example is: The intersection of two streets is the region common to *both* streets.
21. $(-3, 2)$  −3 2
23. $(-\infty, 2]$ 2 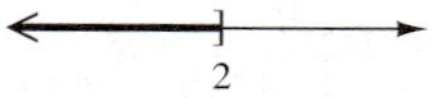
25. $\emptyset$ 
27. $[5, 9]$ 5 9
29. $(-\infty, 4]$ 4 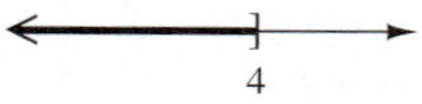
31. $(-\infty, 8]$  8
33. $[-2, \infty)$ −2

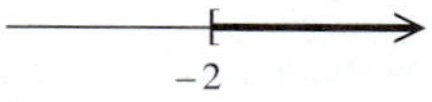

35. $(-\infty, \infty)$

0

37. $(-\infty, -5) \cup (5, \infty)$

−5 5

39. $(-\infty, 2) \cup (2, \infty)$

2

41. $[-4, -1]$ **43.** $[-9, -6]$ **45.** $(-\infty, 3)$ **47.** $[3, 9)$

49. intersection; $(-5, -1)$

−5 −1

51. union; $(-\infty, 4)$

4

53. intersection; $[4, 12]$

4 12

55. union; $(-\infty, 0] \cup [2, \infty)$

0 2

57. Mario, Joe **58.** none of them **59.** none of them **60.** Luigi, Than **61.** none

Section 3.3 (page 145)

1. E; C; D; B; A **3.** Use *or* for the equality statement and the > statement. Use *and* for the < statement. **5.** $\{-12, 12\}$ **7.** $\{-5, 5\}$ **9.** $\{-6, 12\}$ **11.** $\{-4, 3\}$ **13.** $\left\{-3, \frac{11}{2}\right\}$

15. $\left\{-\frac{19}{2}, \frac{9}{2}\right\}$ **17.** $\{-10, -2\}$ **19.** $\left\{-8, \frac{32}{3}\right\}$

21. $(-\infty, -3) \cup (3, \infty)$

−3 3

23. $(-\infty, -4] \cup [4, \infty)$

−4 4

25. $(-\infty, -12) \cup (8, \infty)$

−12 8

27. $\left(-\infty, -\frac{7}{3}\right] \cup [3, \infty)$

$-\frac{7}{3}$ 3

29. $(-\infty, -2) \cup (8, \infty)$

−2 8

31. **(a)**

−5 0 4

(b)

−5 0 4

33. $[-3, 3]$

−3 3

35. $(-4, 4)$

−4 4

37. $[-12, 8]$

−12 8

39. $\left(-\frac{7}{3}, 3\right)$

$-\frac{7}{3}$ 3

41. $[-2, 8]$

−2 8

43. $(-\infty, -5) \cup (13, \infty)$

−5 13

45. $\{-6, -1\}$

−6 −1

47. $\left[-\frac{10}{3}, 4\right]$

$-\frac{10}{3}$ 4

49. $\left[-\frac{7}{6}, -\frac{5}{6}\right]$

$-\frac{7}{6}$ $-\frac{5}{6}$

51. $\{-5, 5\}$ **53.** $\{-5, -3\}$ **55.** $(-\infty, -3) \cup (2, \infty)$ **57.** $[-10, 0]$ **59.** $\{-1, 3\}$ **61.** $\left\{-3, \frac{5}{3}\right\}$ **63.** $\left\{-\frac{1}{3}, -\frac{1}{15}\right\}$ **65.** $\left\{-\frac{5}{4}\right\}$ **67.** $\emptyset$ **69.** $\left\{-\frac{1}{4}\right\}$ **71.** $\emptyset$ **73.** $(-\infty, \infty)$ **75.** $\left\{-\frac{3}{7}\right\}$ **77.** $(-\infty, \infty)$ **79.** $\left(-\infty, -\frac{7}{10}\right) \cup \left(-\frac{7}{10}, \infty\right)$

81. $|x - 1000| \le 100$; $900 \le x \le 1100$

83. 472.9 ft **84.** 1201 Walnut, Fidelity Bank and Trust Building, City Hall, Kansas City Power and Light, Hyatt Regency

85. City Center Square, Commerce Tower, Federal Office Building, 1201 Walnut, Fidelity Bank and Trust Building, City Hall, Kansas City Power and Light, Hyatt Regency **86.** **(a)** $|x - 472.9| \ge 75$ **(b)** $x \ge 547.9$ or $x \le 397.9$ **(c)** AT&T Town Pavilion, One Kansas City Place **(d)** It makes sense because it includes all buildings *not* listed earlier.

Summary Exercises on Solving Linear and Absolute Value Equations and Inequalities (page 151)

1. $\{12\}$ **2.** $\{-5, 7\}$ **3.** $\{7\}$ **4.** $\left\{-\frac{2}{5}\right\}$ **5.** $\emptyset$ **6.** $(-\infty, -1]$ **7.** $\left[-\frac{2}{3}, \infty\right)$ **8.** $\{-1\}$ **9.** $\{-3\}$ **10.** $\left\{1, \frac{11}{3}\right\}$ **11.** $(-\infty, 5]$ **12.** $(-\infty, \infty)$ **13.** $\{2\}$ **14.** $(-\infty, -8] \cup [8, \infty)$ **15.** $\emptyset$ **16.** $(-\infty, \infty)$ **17.** $(-5.5, 5.5)$ **18.** $\left\{\frac{13}{3}\right\}$ **19.** $\left\{-\frac{96}{5}\right\}$ **20.** $(-\infty, 32]$ **21.** $(-\infty, -24)$ **22.** $\left\{\frac{3}{8}\right\}$ **23.** $\left\{\frac{7}{2}\right\}$ **24.** $(-6, 8)$ **25.** $(-\infty, \infty)$ **26.** $(-\infty, 5)$ **27.** $(-\infty, -4) \cup (7, \infty)$ **28.** $\{24\}$ **29.** $\left\{-\frac{1}{5}\right\}$ **30.** $\left(-\infty, -\frac{5}{2}\right]$ **31.** $\left[-\frac{1}{3}, 3\right]$ **32.** $[1, 7]$ **33.** $\left\{-\frac{1}{6}, 2\right\}$ **34.** $\{-3\}$ **35.** $(-\infty, -1] \cup \left[\frac{5}{3}, \infty\right)$ **36.** $\left[\frac{3}{4}, \frac{15}{8}\right]$ **37.** $\left\{-\frac{5}{2}\right\}$ **38.** $\{60\}$ **39.** $\left[-\frac{9}{2}, \frac{15}{2}\right]$ **40.** $(1, 9)$ **41.** $(-\infty, \infty)$ **42.** $\left\{\frac{1}{3}, 9\right\}$ **43.** $(-\infty, \infty)$ **44.** $\left\{-\frac{10}{9}\right\}$ **45.** $\{-2\}$ **46.** $\emptyset$ **47.** $(-\infty, -1) \cup (2, \infty)$ **48.** $[-3, -2]$

Chapter 3 Review Exercises (page 157)

1. $(-9, \infty)$

−9

2. $(-\infty, -3]$

−3

3. $\left(\frac{3}{2}, \infty\right)$

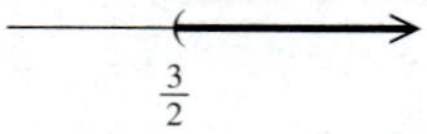

4. $\left(-\infty, -\frac{14}{9}\right)$

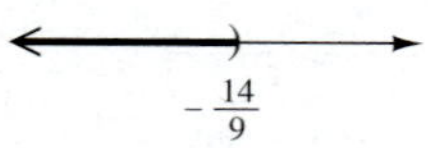

5. $[-3, \infty)$

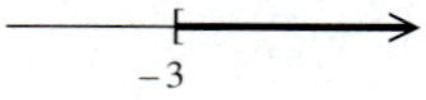

6. $[-3, 12]$

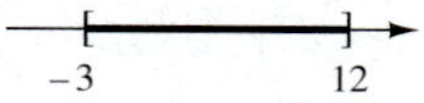

7. $[3, 5)$

8. $\left(-3, \frac{7}{2}\right)$

9. 38 m or less **10.** 99 tickets or less **11.** any grade greater than or equal to 61% **12.** Because the statement $-8 < -13$ is *false*, the inequality has no solution. **13.** {a, c} **14.** {a} **15.** {a, c, e, f, g} **16.** {a, b, c, d, e, f, g}

17. $(6, 9)$

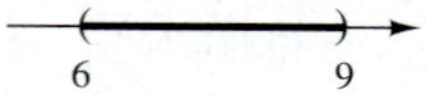

18. $(8, 14)$

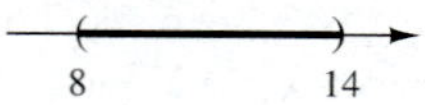

19. $(-\infty, -3] \cup (5, \infty)$

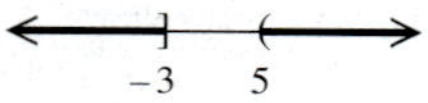

20. $(-\infty, \infty)$

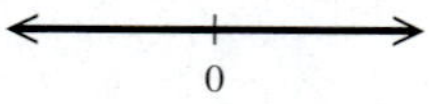

21. $\emptyset$

22. $(-\infty, -2] \cup [7, \infty)$

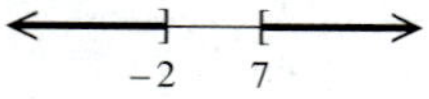

23. $(-3, 4)$ **24.** $(-\infty, 2)$ **25.** $(4, \infty)$ **26.** $(1, \infty)$ **27.** **(a)** managerial and professional specialty **(b)** managerial and professional specialty, mathematical and computer scientists **28.** $\{-7, 7\}$ **29.** $\{-11, 7\}$ **30.** $\left\{-\frac{1}{3}, 5\right\}$ **31.** $\emptyset$ **32.** $\{0, 7\}$ **33.** $\left\{-\frac{3}{2}, \frac{1}{2}\right\}$ **34.** $\left\{-\frac{3}{4}, \frac{1}{2}\right\}$ **35.** $\left\{-\frac{1}{2}\right\}$

36. $(-14, 14)$

37. $[-1, 13]$

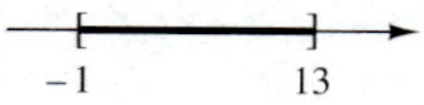

38. $[-3, -2]$

39. $(-\infty, \infty)$

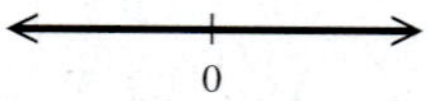

40. $\left(-\infty, -\frac{8}{5}\right) \cup (2, \infty)$

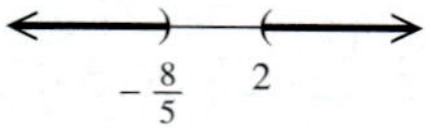

41. $(-\infty, \infty)$

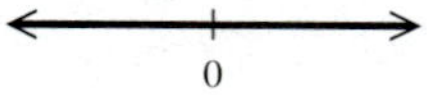

42. $\left(-\infty, \frac{7}{6}\right]$ **43.** $[-4, 5)$ **44.** $\left(-\infty, \frac{14}{17}\right)$ **45.** any amount greater than or equal to \$1100 **46.** $(-\infty, 2]$

47. $(-\infty, -1) \cup \left(\frac{11}{7}, \infty\right)$ **48.** $\{-5, 15\}$ **49.** $[-16, 10]$

50. $(-\infty, \infty)$ **51.** $\left\{-4, -\frac{2}{3}\right\}$

52.

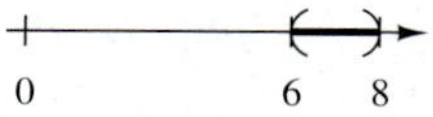

53.

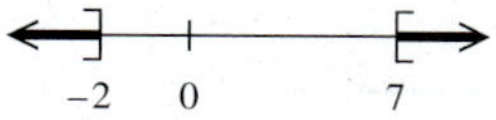

54. **(a)** $\emptyset$ **(b)** $(-\infty, \infty)$ **(c)** $\emptyset$

Chapter 3 Test (page 161)

1. Reverse the direction of the inequality symbol.

2. $[1, \infty)$

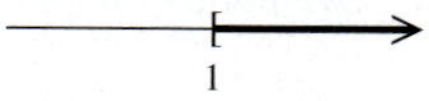

3. $(-\infty, 28)$

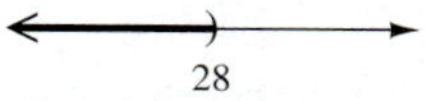

4. $[-3, 3]$

5. C **6.** **(a)** 1993–1998 **(b)** 1985–1989 **(c)** 1990, 1993–1995

7. 82% **8.** $[500, \infty)$ **9.** **(a)** $\{1, 5\}$ **(b)** $\{1, 2, 5, 7, 9, 12\}$

10. $\{2\}$

11. $[2, 9)$

12. $(-\infty, 3) \cup [6, \infty)$

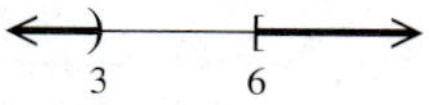

13. $\left[-\frac{5}{2}, 1\right]$

14. $\left(-\infty, -\frac{7}{6}\right) \cup \left(\frac{17}{6}, \infty\right)$

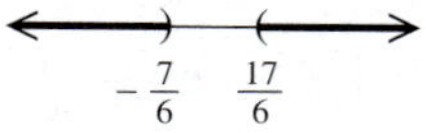

15. $\emptyset$ **16.** $\left\{-\frac{5}{3}, 3\right\}$ **17.** $\left\{-\frac{5}{7}, \frac{11}{3}\right\}$

Cumulative Review Exercises: Chapters 1–3 (page 163)

1. $\frac{3}{4}$ **2.** true **3.** $\frac{37}{60}$ **4.** $\frac{48}{5}$ **5.** 11 **6.** -8 **7.** -36

8. -125 **9.** $\frac{81}{16}$ **10.** -34 **11.** $\frac{3}{16}$ **12.** distributive property **13.** commutative property **14.** $2k - 11$ **15.** $\{-1\}$ **16.** $\{-12\}$

17. $\{26\}$ **18.** $\left\{\frac{3}{4}, \frac{7}{2}\right\}$ **19.** $y = \frac{24 - 3x}{4}$ **20.** $n = \frac{A - P}{iP}$

21. $[-14, \infty)$

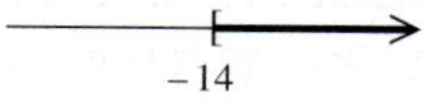

22. $\left[\frac{5}{3}, 3\right)$

23. $(-\infty, 0) \cup (2, \infty)$ 0 2

24. $\left(-\infty, -\frac{1}{7}\right] \cup [1, \infty)$ $-\frac{1}{7}$ 1

25. \$5000 **26.** $6\frac{1}{3}$ g **27.** 74 or greater **28.** 40 mph; 60 mph

29. **(a)** 122 **(b)** 7.6% **30.** 4 cm; 9 cm; 27 cm

Chapter 4

Section 4.1 (page 173)

1. **(a)** x represents the year; y represents the percent of women in math or computer science professions. **(b)** 1990–2000 **(c)** 1990 **(d)** 1980 **3.** You should choose a bar graph. **5.** origin **7.** y; x **9.** two **11.** **(a)** I **(b)** III **(c)** II **(d)** IV **(e)** no quadrant **13.** **(a)** I or III **(b)** II or IV **(c)** II or IV **(d)** I or III

15–24.

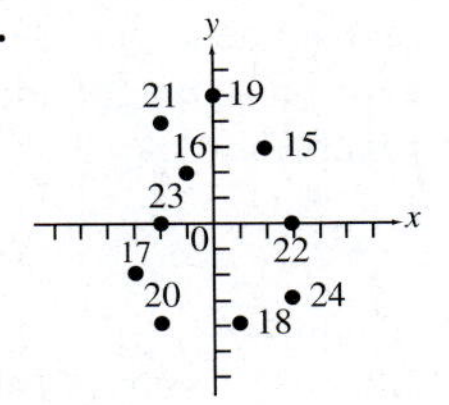

25. -3; 3; 2; -1

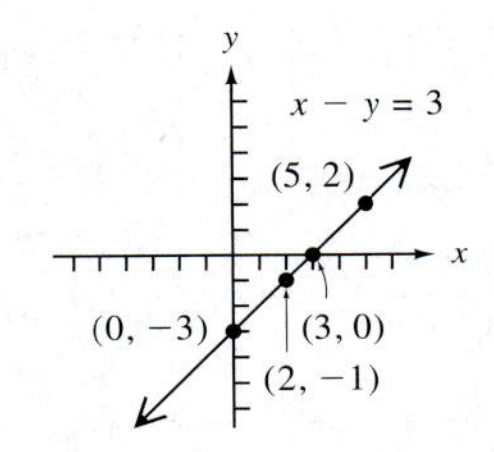

27. $\frac{5}{2}$; 5; $\frac{3}{2}$; 1

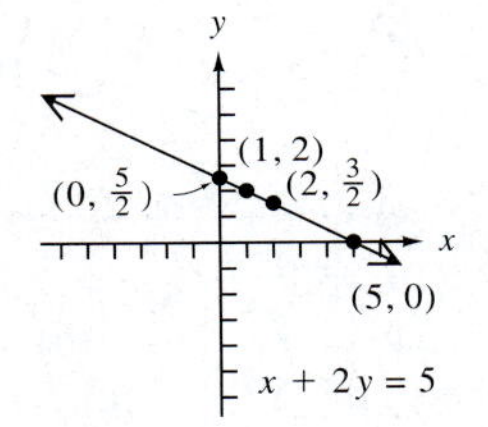

29. -4; 5; $-\frac{12}{5}$; $\frac{5}{4}$

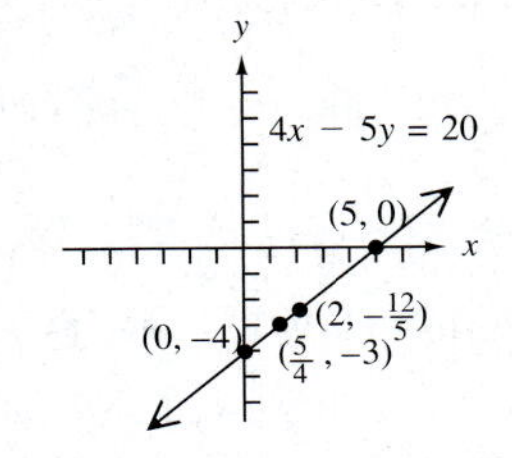

31. In quadrant III, both coordinates of the ordered pairs are negative. If $x + y = k$ and k is positive, then either x or y must be positive because the sum of two negative numbers is negative.

33. (6, 0); (0, 4)

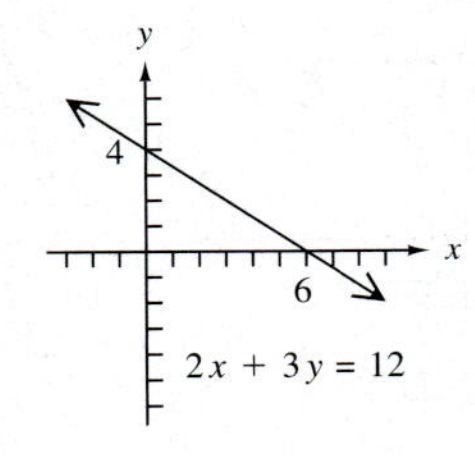

35. (6, 0); (0, -2)

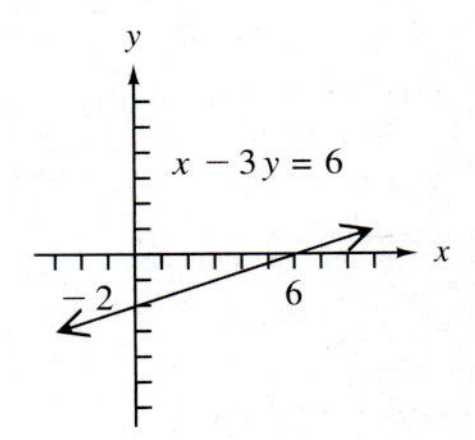

37. (3, 0); $\left(0, -\frac{9}{7}\right)$

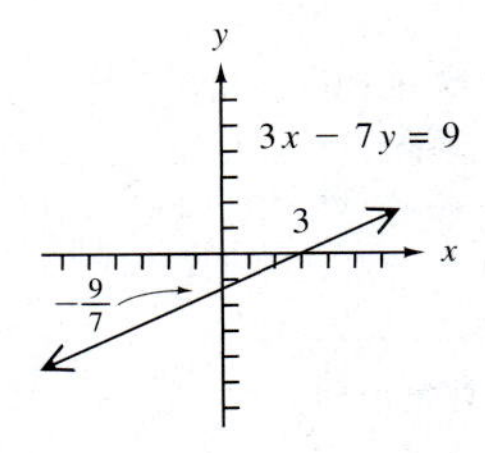

39. none; (0, 5)

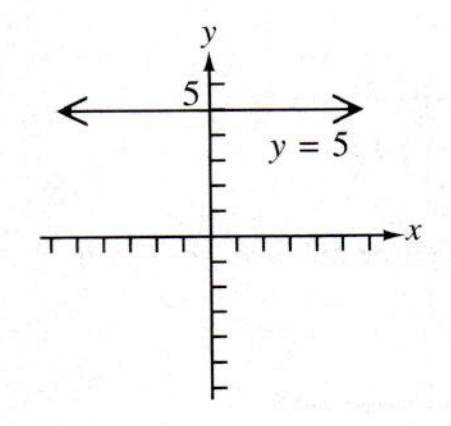

41. (2, 0); none

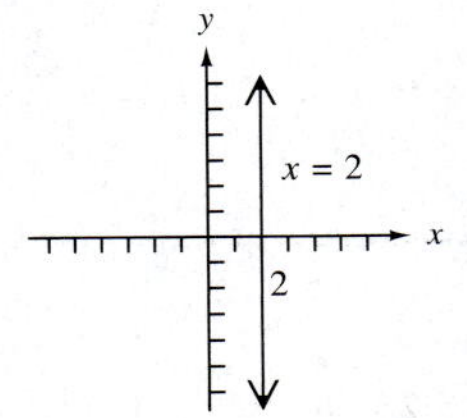

43. (0, 0); (0, 0)

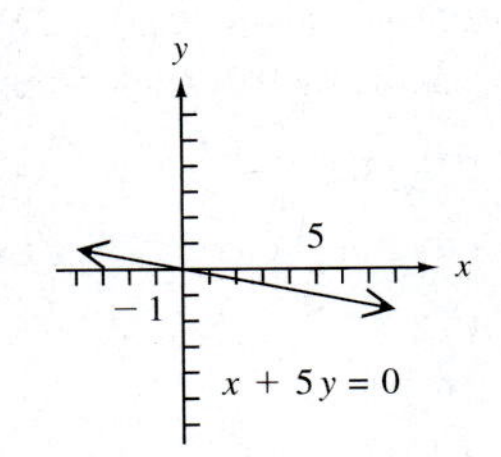

45. 154.6 mph **47.** (6, -2) **48.** (5, -2) **49.** (6, 0) **50.** (5, 0) **51.** 5; 0 **52.** The x-coordinate of M is the average of the x-coordinates of P and Q. The y-coordinate of M is the average of the y-coordinates of P and Q.

Section 4.2 (page 185)

1. A, B, D **3.** 2 **5.** undefined **7.** 2 **9.** $\frac{5}{2}$ **11.** 0 **13.** 8 **15.** $\frac{5}{6}$ **17.** 0 **19.** $-\frac{5}{2}$ **21.** undefined **23.** B **25.** A

27. $-\frac{1}{2}$

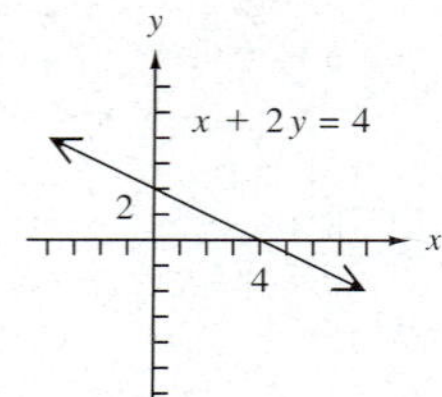

29. 1

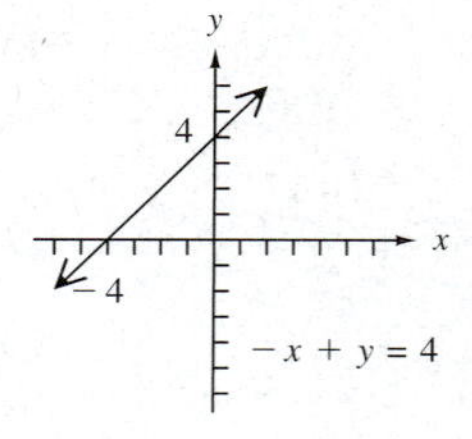

31. $-\frac{6}{5}$

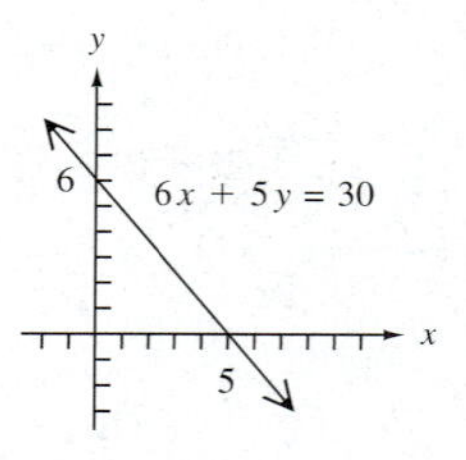

33. undefined

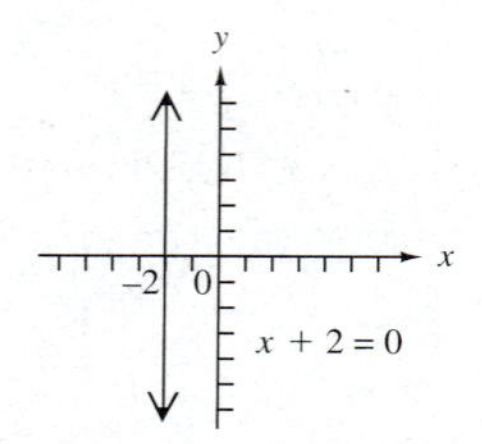

35. 4

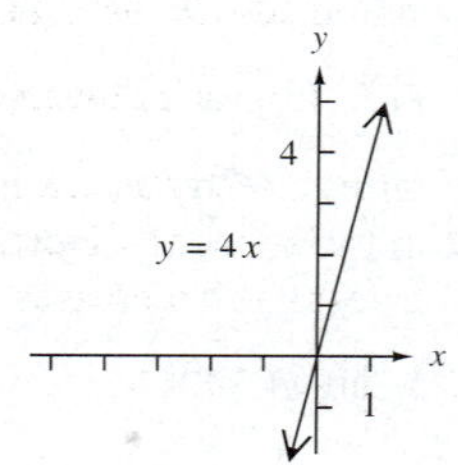

37. 0

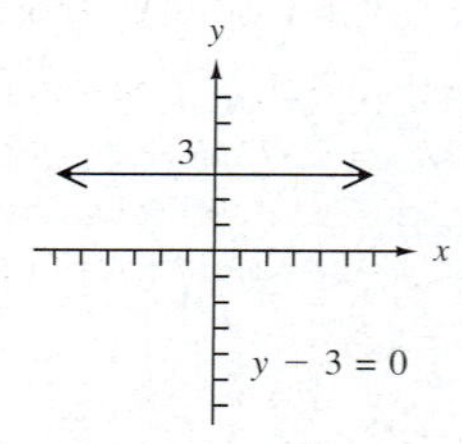

39.

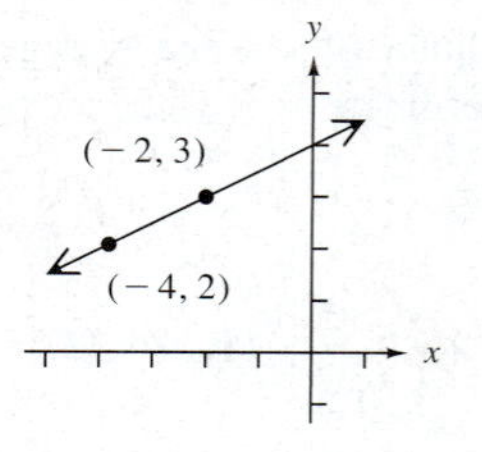

41.

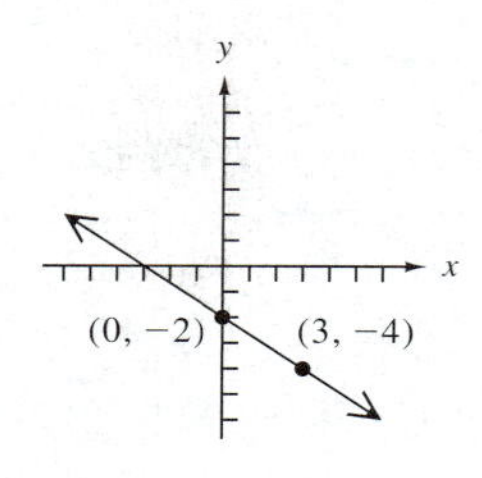

43.

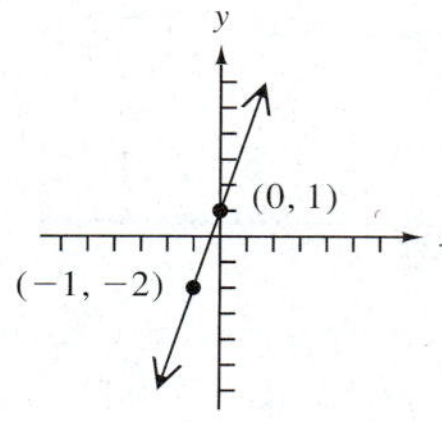

45. perpendicular **47.** parallel **49.** neither **51.** parallel **53.** neither **55.** perpendicular **57.** $\frac{7}{10}$ **59.** **(a)** \$200 million per yr **(b)** The positive slope means expenditures *increased* an average of \$200 million each year. **61.** **(a)** 1000 million/yr; 1000 million/yr; 1000 million/yr; The average rate of change is the same. This is true because the data points lie on a straight line. **(b)** 5.3%, 5.3%, 4.5%; They are all approximately 5%. **63.** $\frac{1}{3}$ **64.** $\frac{1}{3}$ **65.** $\frac{1}{3}$ **66.** $\frac{1}{3} = \frac{1}{3} = \frac{1}{3}$ is true. **67.** They are collinear. **68.** They are not collinear.

Section 4.3 (page 197)

1. A **3.** A **5.** $3x + y = 10$ **7.** A **9.** C **11.** H **13.** B **15.** $3x + 4y = 10$ **17.** $2x + y = 18$ **19.** $x - 2y = -13$ **21.** $y = 12$ **23.** $x = 9$ **25.** $y = .2$ **27.** $2x - y = 2$ **29.** $x + 2y = 8$ **31.** $2x - 13y = -6$ **33.** $y = 5$ **35.** $x = 7$ **37.** $y = 5x + 15$ **39.** $y = -\frac{2}{3}x + \frac{4}{5}$ **41.** $y = \frac{2}{5}x + 5$ **43.** $y = -\frac{5}{2}x + 10$; $-\frac{5}{2}$; $(0, 10)$ **45.** $y = \frac{2}{3}x - \frac{10}{3}$; $\frac{2}{3}$; $\left(0, -\frac{10}{3}\right)$ **47.** $y = 3x - 19$ **49.** $y = \frac{1}{2}x - 1$ **51.** $y = -\frac{1}{2}x + 9$ **53.** $y = 7$ **55.** $y = 45x$; (0, 0), (5, 225), (10, 450) **57.** $y = 1.50x$; (0, 0), (5, 7.50), (10, 15.00) **59.** **(a)** $y = 39x + 99$ **(b)** (5, 294); The cost of a 5-month membership is \$294. **(c)** \$567 **61.** **(a)** $y = .20x + 50$ **(b)** (5, 51); The charge for driving 5 mi is \$51. **(c)** 173 mi **63.** $y = -176.25x + 29{,}362$ **65.** **(a)** $y = 838.5x + 19{,}180.5$ **(b)** \$23,373; It is close to the actual value. **67.** 32; 212 **68.** (0, 32) and (100, 212) **69.** $\frac{9}{5}$ **70.** $F = \frac{9}{5}C + 32$ **71.** $C = \frac{5}{9}(F - 32)$ **72.** $-40°$ **73.** 60° **74.** 59°; They differ by 1°. **75.** 90°; 86°; They differ by 4°. **76.** Since $\frac{9}{5}$ is a little less than 2, and 32 is a little more than 30, $\frac{9}{5}C + 32 \approx 2C + 30$.

Section 4.4 (page 207)

1. solid; below **3.** dashed; above **5.** The graph of $Ax + By = C$ divides the plane into two regions. In one of these regions, the ordered pairs satisfy $Ax + By < C$; in the other, they satisfy $Ax + By > C$.

7.

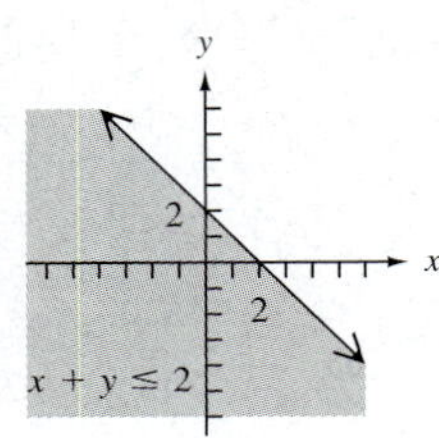

9.

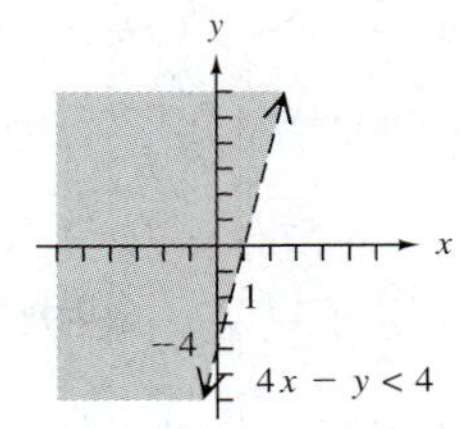

11.

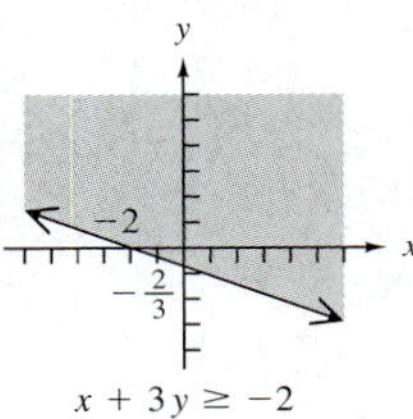

13.

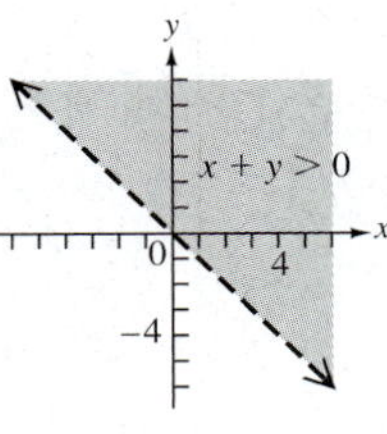

15.

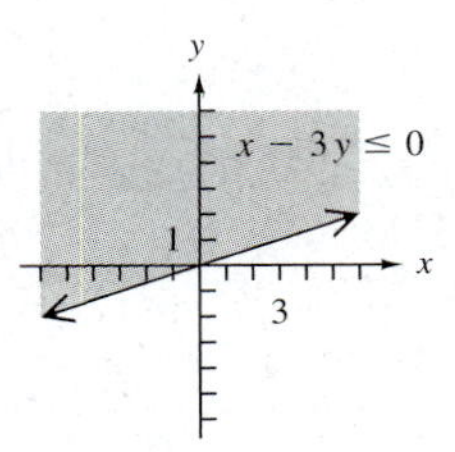

17.

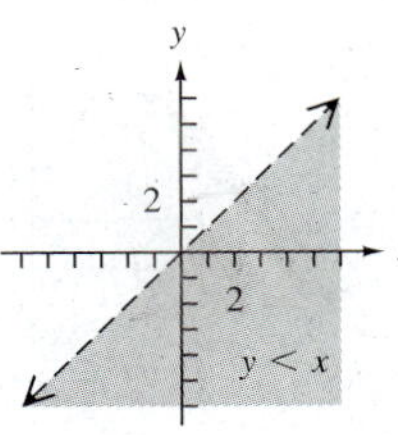

19.

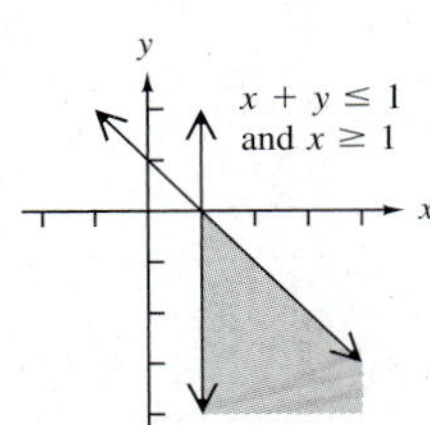

21.

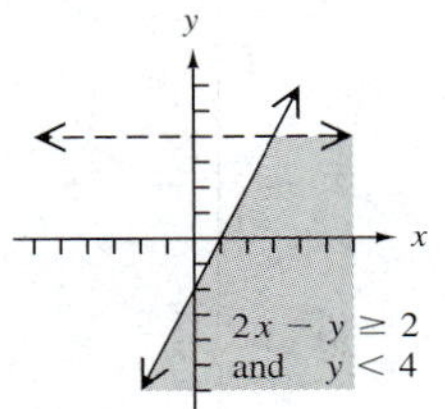

23.

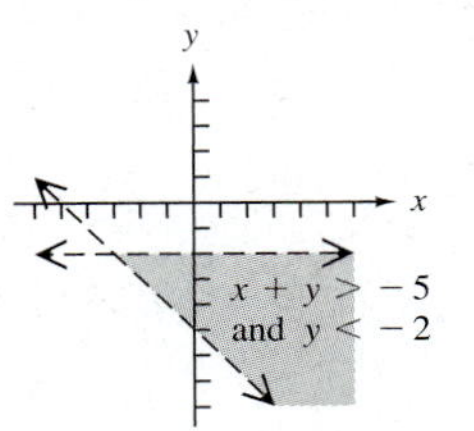

25.

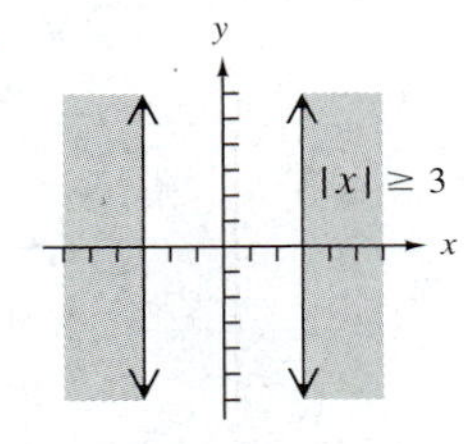

27.

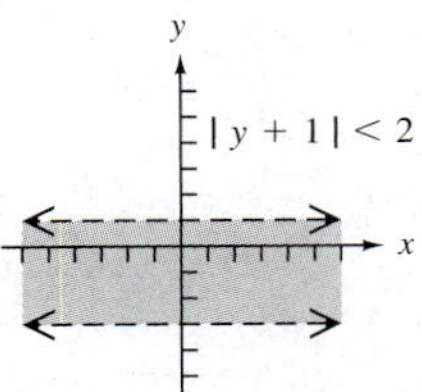

29.

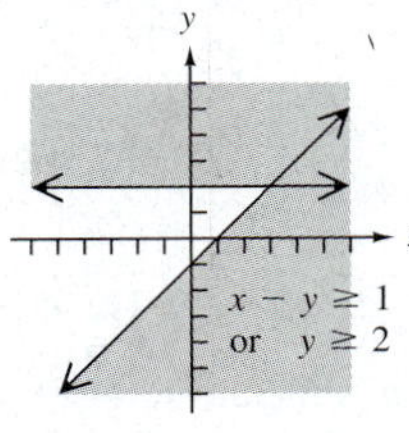

31.

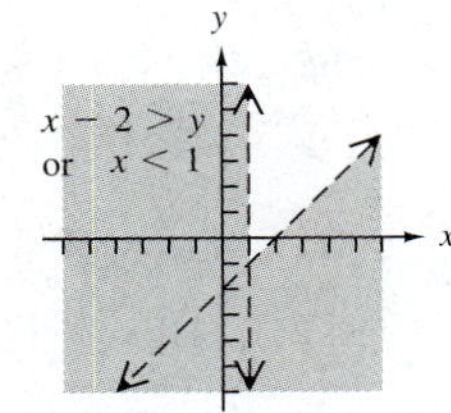

33.

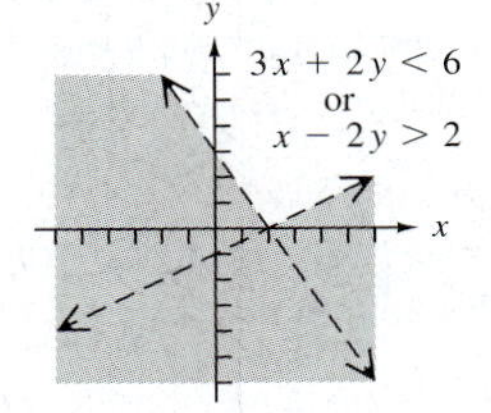

Section 4.5 (page 221)

1. independent variable **3. (a)** A relation is a set of ordered pairs. **(b)** The domain is the set of all first components (x-values). **(c)** The range is the set of all second components (y-values). **(d)** A function is a relation in which each domain element is paired with one and only one range element. **5.** function; domain: $\{8, 5, 9, 3\}$; range: $\{0, 4, 3, 9\}$ **7.** not a function; domain: $\{9, -3\}$; range: $\{-2, 5, 1\}$ **9.** not a function; domain: $(0, \infty)$; range: $(-\infty, 0) \cup (0, \infty)$ **11.** function; domain: $(-\infty, \infty)$; range: $(-\infty, 4]$ **13.** not a function; domain: $[3, \infty)$; range: $(-\infty, \infty)$ **15.** function; domain: $(-\infty, \infty)$ **17.** not a function; domain: $[0, \infty)$ **19.** not a function; domain: $(-\infty, \infty)$ **21.** function; domain: $[0, \infty)$ **23.** function; domain: $(-\infty, 0) \cup (0, \infty)$ **25.** function; domain: $(-\infty, \infty)$ **27.** function; domain: $\left[-\frac{1}{2}, \infty\right)$ **29.** function; domain: $(-\infty, 9) \cup (9, \infty)$ **31. (a)** $[0, 3000]$ **(b)** 25 hr; 25 hr **(c)** 2000 gal **(d)** $f(0) = 0$; The pool is empty at time 0. **33.** Here is one example. The cost of gasoline; number of gallons purchased; cost; number of gallons **35.** 4 **37.** -11 **39.** $-3p + 4$ **41.** $3x + 4$ **43.** $-3x - 2$ **45.** $-\frac{p^2}{9} + \frac{4p}{3} + 1$ **47.** line; -2; linear; $-2x + 4$; -2; 3; -2 **49. (a)** $f(x) = \frac{12 - x}{3}$ **(b)** 3 **51. (a)** $f(x) = 3 - 2x^2$ **(b)** -15 **53. (a)** $f(x) = \frac{8 - 4x}{-3}$ **(b)** $\frac{4}{3}$ **55. (a)** \$0; \$1.50; \$3.00; \$4.50 **(b)** $1.50x$ **(c)**

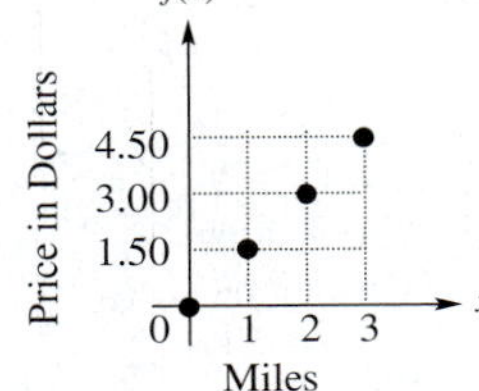

Section 4.6 (page 231)

1. inverse **3.** direct **5.** joint **7.** combined **9.** 36 **11.** .625 **13.** $222\frac{2}{9}$ **15.** increases; decreases **17.** If y varies inversely as x, x is in the denominator; however, if y varies directly as x, x is in the numerator. Also, for $k > 0$, with inverse variation, as x increases, y decreases. With direct variation, y increases as x increases. **19.** $\$1.69\frac{9}{10}$ **21.** about 450 cm^3 **23.** about \$9211 **25.** $21\frac{1}{3}$ foot-candles **27.** \$420 **29.** 448.1 lb **31.** approximately 68,600 calls **33.** 11.8 lb **35.** $(0, 0)$, $(1, 1.75)$ **36.** 1.75 **37.** $y = 1.75x + 0$ or $y = 1.75x$ **38.** $a = 1.75$, $b = 0$ **39.** It is the price per gallon and the slope of the line. **40.** It can be written in the form $y = kx$ (where $k = a$). The value of a is called the constant of variation.

Chapter 4 Review Exercises (page 239)

1. 3; 2; $\frac{10}{3}$

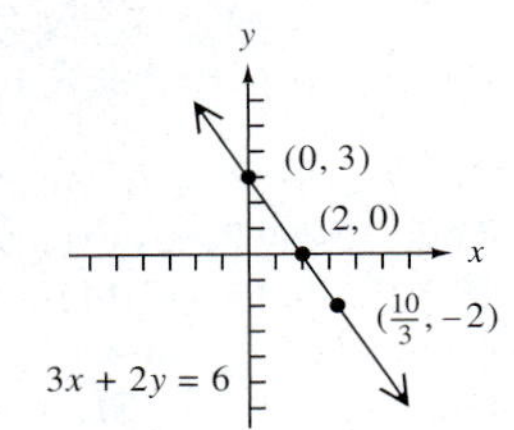

2. $-4, 3; -5; 4$

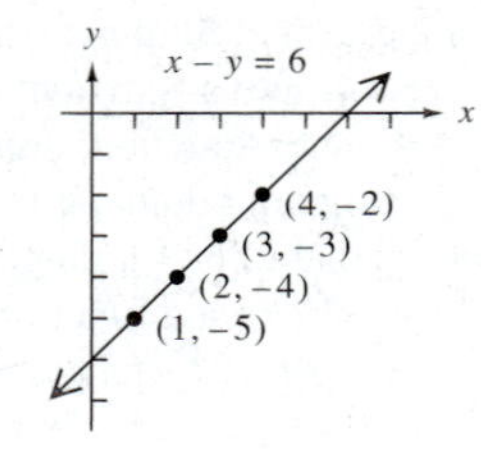

3. $(3, 0); (0, 4)$

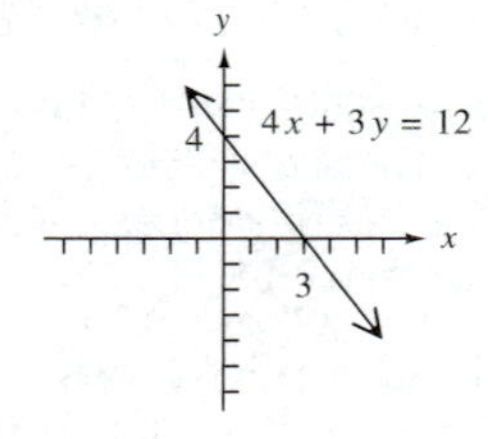

4. $(3, 0); \left(0, \frac{15}{7}\right)$

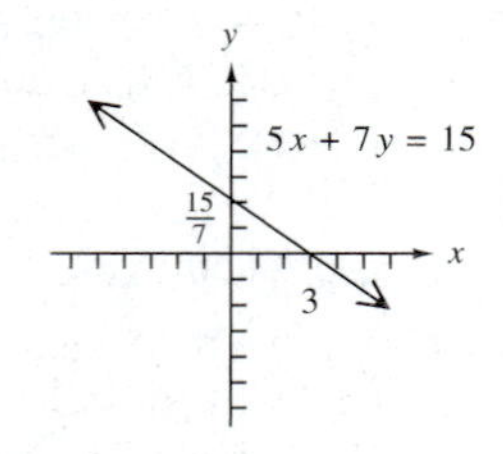

5. $-\frac{8}{5}$ **6.** 2 **7.** $\frac{3}{4}$ **8.** 0 **9.** $-\frac{2}{3}$ **10.** $-\frac{1}{3}$
11. positive slope **12.** negative slope **13.** 0 slope
14. undefined slope **15.** 12 ft **16.** \$1321 per yr
17. $y = \frac{3}{5}x - 8$ **18.** $y = -\frac{1}{3}x + 5$ **19.** $y = 12$ **20.** $x = 2$
21. $y = 4$ **22.** $x = .3$ **23.** $y = -9x + 13$ **24.** $y = \frac{7}{5}x + \frac{16}{5}$
25. $y = 4x - 26$ **26.** $y = -\frac{5}{2}x + 1$
27. (a) $y = 57x + 159$; \$843 **(b)** $y = 47x + 159$; \$723
28.

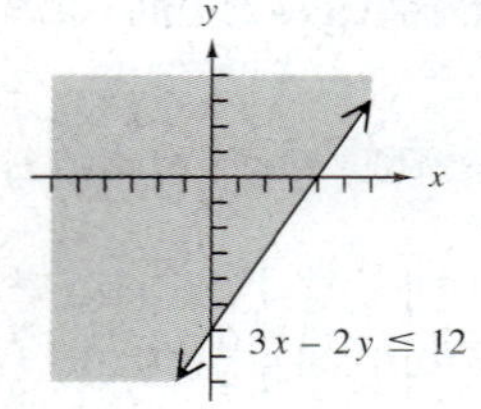

29.

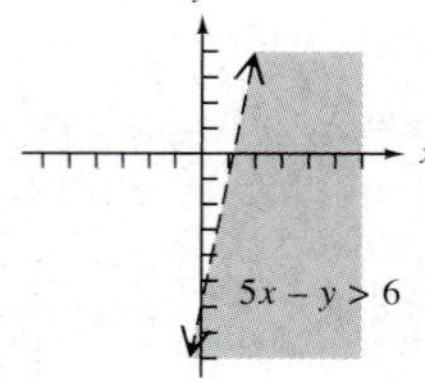

30.

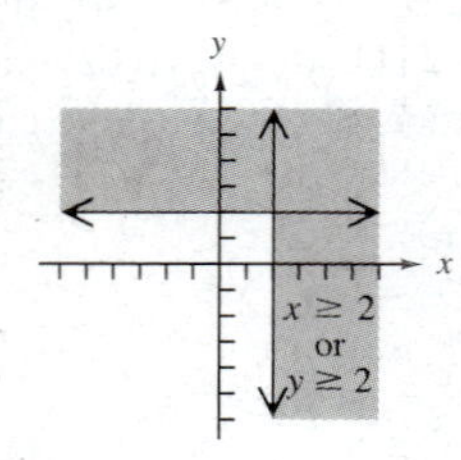

31.

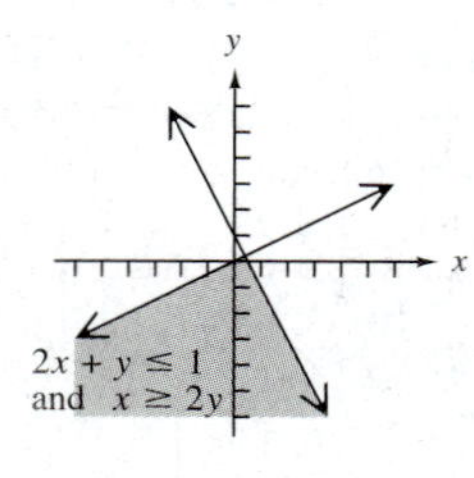

32. domain: $\{-4, 1\}$; range: $\{2, -2, 5, -5\}$; not a function
33. domain: {California, New York, Texas, Pennsylvania, Washington}; range: {71,266, 50,101, 48,010, 42,142, 38,240}; function **34.** domain: $[-4, 4]$; range: $[0, 2]$; function
35. function; linear function; domain: $(-\infty, \infty)$ **36.** not a function; domain: $(-\infty, \infty)$ **37.** function; domain: $(-\infty, \infty)$
38. function; domain: $\left[-\frac{7}{4}, \infty\right)$ **39.** not a function; domain: $[0, \infty)$
40. function; domain: $(-\infty, 36) \cup (36, \infty)$ **41.** If no vertical line intersects the graph in more than one point, then it is the graph of a function. **42.** -6 **43.** -15 **44.** $-2p^2 + 3p - 6$

45. $-2k^2 - 3k - 6$ **46.** $f(x) = 2x^2$; 18 **47.** C **48.** 800 gal
49. 430 mm **50.** $71\frac{1}{9}$ lb **51.** $.71\pi$ sec

Chapter 4 Test (page 243)

1. $\frac{1}{2}$ **2.** $\frac{3}{2}; \left(\frac{13}{3}, 0\right); \left(0, -\frac{13}{2}\right)$ **3.** 0; none; $(0, 5)$
4. The graph is a vertical line.
5. $(-3, 0); (0, 4)$

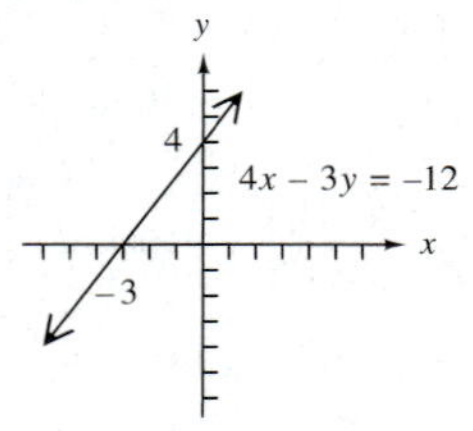

6. none; $(0, 2)$

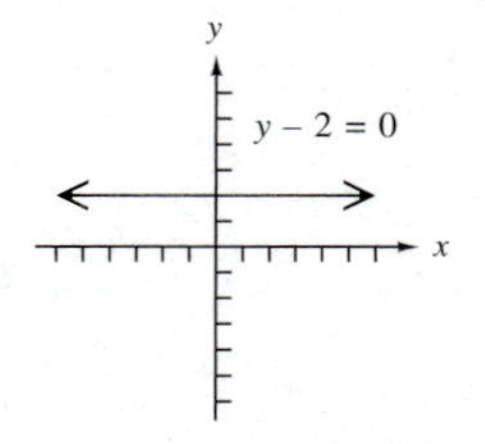

7. $(0, 0); (0, 0)$

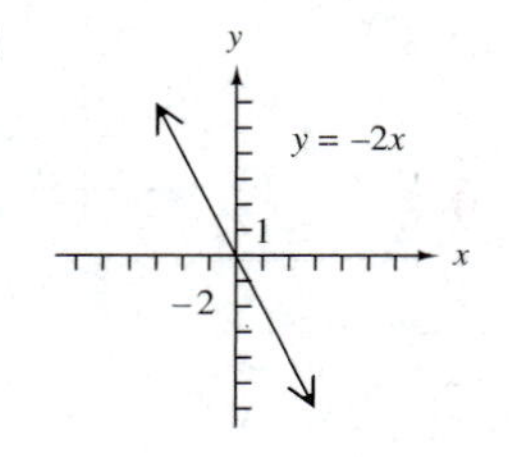

8.

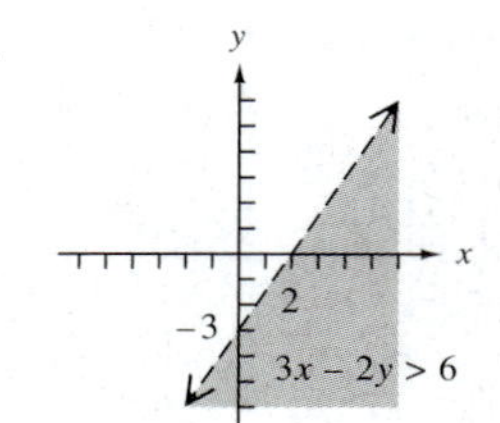

9. $y = 14$ **10.** $y = -5x + 19$ **11. (a)** $y = -\frac{3}{5}x - \frac{11}{5}$
(b) $y = -\frac{1}{2}x - \frac{3}{2}$ **12.** D **13. (a)** $[3, \infty)$ **(b)** 2 **14.** B;
The set in A includes, for example, the two ordered pairs (8.8, 1993) and (8.8, 1994). In a function the independent variable cannot correspond to more than one dependent variable. **15.** 256 ft
16. .8 lb

Cumulative Review Exercises: Chapters 1–4 (page 245)

1. always true **2.** always true **3.** never true **4.** sometimes true; for example, $3 + (-3) = 0$, but $3 + (-1) = 2 \neq 0$ **5.** 4
6. .64 **7.** not a real number **8.** $\frac{8}{5}$ **9.** $4m - 3$ **10.** $2x^2 + 5x + 4$
11. $-\frac{19}{2}$ **12.** $(-3, 5]$ **13.** no **14.** -24 **15.** 56
16. undefined **17.** $\left\{\frac{7}{6}\right\}$ **18.** $\{-1\}$ **19.** $h = \frac{3V}{\pi r^2}$ **20.** 2 hr
21. 4 white pills **22.** 6 in. **23.** The union of the three solution sets is $(-\infty, \infty)$. **24.** $\left(-\frac{1}{2}, \infty\right)$ **25.** $(2, 3)$ **26.** $(-\infty, 2) \cup (3, \infty)$

27. $\left\{-\frac{16}{5}, 2\right\}$ **28.** $(-11, 7)$ **29.** $(-\infty, -2] \cup [7, \infty)$

30. $(0, -3), (4, 0), \left(2, -\frac{3}{2}\right)$

31. x-intercept: $(-2, 0)$; y-intercept: $(0, 4)$

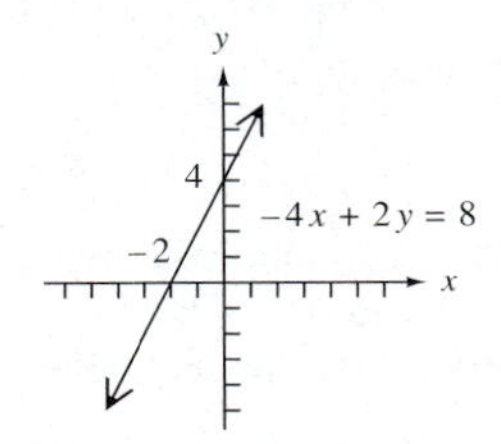

32. $-\frac{3}{2}$ **33.** $-\frac{1}{2}$ **34.** $-\frac{3}{4}$ **35.** $y = -\frac{3}{4}x - 1$ **36.** $y = -2$

37. $y = -\frac{4}{3}x + \frac{7}{3}$ **38. (a)** $(-\infty, \infty)$ **(b)** 22 **39.** 10.5

40. the segment for 1992 through 2000

Chapter 5

Section 5.1 (page 257)

1. 3; −6 **3.** $\emptyset$ **5.** 0 **7.** D; The ordered pair solution must be in quadrant IV, since that is where the graphs of the equations intersect.

9. (a) B **(b)** C **(c)** A **(d)** D

11. $\{(-2, -3)\}$ **13.** yes

15. no **17.** $\{(1, 2)\}$ **19.** $\left\{\left(\frac{22}{9}, \frac{22}{3}\right)\right\}$ **21.** $\{(2, 3)\}$

23. $\{(5, 4)\}$ **25.** $\left\{\left(-5, -\frac{10}{3}\right)\right\}$ **27.** $\{(2, 6)\}$ **29.** $\{(3, -1)\}$

31. $\{(2, -3)\}$ **33.** $\left\{\left(\frac{3}{2}, -\frac{3}{2}\right)\right\}$ **35.** $\{(x, y) \mid 7x + 2y = 6\}$; dependent equations **37.** $\{(2, -4)\}$ **39.** $\emptyset$; inconsistent system

41. $y = -\frac{3}{7}x + \frac{4}{7}$; $y = -\frac{3}{7}x + \frac{3}{14}$; 0

43. Both are $y = -\frac{2}{3}x + \frac{1}{3}$; infinitely many

45. (a) Use substitution since the second equation is solved for y. **(b)** Use elimination since the coefficients of the y-terms are opposites. **(c)** Use elimination since the equations are in standard form with no coefficients of 1 or −1. Solving by substitution would involve fractions. **47.** $\{(-3, 2)\}$ **49.** $\{(-4, 6)\}$

51. $\{(x, y) \mid 4x - y = -2\}$ **53.** $\left\{\left(1, \frac{1}{2}\right)\right\}$ **55. (a)** years 0 to 6 **(b)** year 6; about \$650 **57. (a)** 1989–1997 **(b)** 1997; NBC; 17% **(c)** 1989, share 20%; 1998, share 16% **(d)** NBC and ABC; (2000, 16) **(e)** Viewership has generally declined during these years. **59.** 1995–1997 **61.** $\{(2.3, 5.8)\}$ (Answers may vary slightly depending on how the system was solved.) **63.** $\{(1, 3)\}$

64. $f(x) = -3x + 6$; linear **65.** $g(x) = \frac{2}{3}x + \frac{7}{3}$; linear

66. one; 1; 3; 1; 3; 1; 3

Section 5.2 (page 269)

1. The statement means that when −1 is substituted for x, 2 is substituted for y, and 3 is substituted for z in the three equations, the resulting three statements are true. **3.** $\{(3, 2, 1)\}$

5. $\{(1, 4, -3)\}$ **7.** $\left\{\left(1, \frac{3}{10}, \frac{2}{5}\right)\right\}$ **9.** $\{(0, 2, -5)\}$

11. $\left\{\left(-\frac{7}{3}, \frac{22}{3}, 7\right)\right\}$ **13.** $\{(4, 5, 3)\}$ **15.** $\{(2, 2, 2)\}$

17. $\left\{\left(\frac{8}{3}, \frac{2}{3}, 3\right)\right\}$ **19.** Answers will vary. Some possible answers are **(a)** two perpendicular walls and the ceiling in a normal room, **(b)** the floors of three different levels of an office building, and **(c)** three pages of this book (since they intersect in the spine).

21. $\emptyset$ **23.** $\{(x, y, z) \mid x - y + 4z = 8\}$

25. $\{(x, y, z) \mid 2x + y - z = 6\}$ **27.** $\{(0, 0, 0)\}$

29. $128 = a + b + c$ **30.** $140 = 2.25a + 1.5b + c$

31. $80 = 9a + 3b + c$ **32.** $a + b + c = 128$; $2.25a + 1.5b + c = 140$; $9a + 3b + c = 80$; $\{(-32, 104, 56)\}$

33. $f(x) = -32x^2 + 104x + 56$ **34.** height; time **35.** 56 ft

36. 140.5 ft

Section 5.3 (page 281)

1. wins: 95; losses: 67 **3.** length: 78 ft; width: 36 ft **5.** Exxon-Mobil: \$214 billion; General Motors: \$185 billion **7.** $x = 40$ and $y = 50$, so the angles measure 40° and 50°. **9.** NHL: \$219.74; NBA: \$203.38 **11.** single: \$2.09; double: \$3.19 **13. (a)** 6 oz **(b)** 15 oz **(c)** 24 oz **(d)** 30 oz **15.** \$.99x **17.** 6 gal of 25%; 14 gal of 35% **19.** 6 L of pure acid; 48 L of 10% acid **21.** 14 kg of nuts; 16 kg of cereal **23.** \$1000 at 2%; \$2000 at 4% **25.** 25y

27. freight train: 50 km per hr; express train: 80 km per hr

29. boat: 21 mph; current: 3 mph **31.** Turner: \$80.4 million; 'N Sync: \$76.6 million **33.** 76 general admission; 108 with student ID **35.** 8 for a citron; 5 for a wood apple

37. $x + y + z = 180$; angle measures: 70°, 30°, 80°

39. first: 20°; second: 70°; third: 90° **41.** shortest: 12 cm; middle: 25 cm; longest: 33 cm **43.** Independent: 38; Democrat: 34; Republican: 28 **45.** \$10 tickets: 350; \$18 tickets: 250; \$30 tickets: 50 **47.** type A: 80; type B: 160; type C: 250

Section 5.4 (page 293)

1. (a) 0, 5, −3 **(b)** 1, −3, 8 **(c)** yes; The number of rows is the same as the number of columns (three).

(d) $\begin{bmatrix} 1 & 4 & 8 \\ 0 & 5 & -3 \\ -2 & 3 & 1 \end{bmatrix}$ **(e)** $\begin{bmatrix} 1 & -\frac{3}{2} & -\frac{1}{2} \\ 0 & 5 & -3 \\ 1 & 4 & 8 \end{bmatrix}$ **(f)** $\begin{bmatrix} 1 & 15 & 25 \\ 0 & 5 & -3 \\ 1 & 4 & 8 \end{bmatrix}$

3. $\left[\begin{array}{cc|c} 1 & 2 & 11 \\ 2 & -1 & -3 \end{array}\right]$; $\left[\begin{array}{cc|c} 1 & 2 & 11 \\ 0 & -5 & -25 \end{array}\right]$; $\left[\begin{array}{cc|c} 1 & 2 & 11 \\ 0 & 1 & 5 \end{array}\right]$; $x + 2y = 11$; $y = 5$; $\{(1, 5)\}$ **5.** $\{(4, 1)\}$ **7.** $\{(1, 1)\}$ **9.** $\{(-1, 4)\}$ **11.** $\emptyset$

13. $\left[\begin{array}{ccc|c} 1 & 1 & -1 & -3 \\ 0 & -1 & 3 & 10 \\ 0 & -6 & 7 & 38 \end{array}\right]$; $\left[\begin{array}{ccc|c} 1 & 1 & -1 & -3 \\ 0 & 1 & -3 & -10 \\ 0 & -6 & 7 & 38 \end{array}\right]$;

$\left[\begin{array}{ccc|c} 1 & 1 & -1 & -3 \\ 0 & 1 & -3 & -10 \\ 0 & 0 & -11 & -22 \end{array}\right]$; $\left[\begin{array}{ccc|c} 1 & 1 & -1 & -3 \\ 0 & 1 & -3 & -10 \\ 0 & 0 & 1 & 2 \end{array}\right]$; $x + y - z = -3$; $y - 3z = -10$; $z = 2$; $\{(3, -4, 2)\}$ **15.** $\{(4, 0, 1)\}$

17. $\{(-1, 23, 16)\}$ **19.** $\{(3, 2, -4)\}$

21. $\{(x, y) \mid x - 2y + z = 4\}$ **23.** \$3000 at 5%; \$1000 at 6%; \$6000 at 8%

Chapter 5 Review Exercises (page 291)

1. $\{(2, 2)\}$ **2. (a)** 1978 and 1982 **(b)** just less than 500,000

3. $\left\{\left(-\frac{8}{9}, -\frac{4}{3}\right)\right\}$ **4.** $\{(0, 4)\}$ **5.** $\{(2, 4)\}$ **6.** $\{(-1, 2)\}$

7. $\{(-6, 3)\}$ **8.** $\left\{\left(\frac{68}{13}, -\frac{31}{13}\right)\right\}$ **9.** $\{(x, y) \mid 3x - y = -6\}$
10. $\emptyset$ **11.** Answers will vary. **12.** Answers will vary.
13. Answers will vary. **14.** Because the lines have the same slope (3) but different y-intercepts ((0, 2) and (0, −4)), the lines do not intersect. Thus, the system has no solution. **15.** $\{(1, -5, 3)\}$
16. $\emptyset$ **17.** $\{(1, 2, 3)\}$ **18.** length: 200 ft; width: 85 ft
19. 3 weekend days; 3 weekdays **20.** plane: 300 mph; wind: 20 mph **21.** 30 lb of \$2 per lb nuts; 70 lb of \$1 per lb candy
22. 4 vats of green algae; 7 vats of brown algae **23.** 85°, 60°, 35°
24. 5 L of 8%; 3 L of 20%; none of 10% **25.** Mantle: 54; Maris: 61; Blanchard: 21 **26.** $\{(3, -2)\}$ **27.** $\{(-1, 5)\}$
28. $\{(0, 0, -1)\}$ **29.** $\{(12, 9)\}$ **30.** $\left\{\left(\frac{82}{23}, -\frac{4}{23}\right)\right\}$
31. $\{(3, -1)\}$ **32.** $\{(5, 3)\}$ **33.** $\{(0, 4)\}$ **34.** $\emptyset$ **35.** 20 L
36. U.S.: 97; Russia: 88; China: 59 **37.** $2a + b + c = -5$
38. $-a + c = -1$ **39.** $3a + 3b + c = -18$ **40.** $a = 1$, $b = -7$, $c = 0$; $x^2 + y^2 + x - 7y = 0$ **41.** The relation is not a function because a vertical line intersects its graph more than once.

Chapter 5 Test (page 303)

1. No; The graph for Ruth lies completely below the graph for Aaron. **2.** Aaron; Ruth
3. $\{(6, 1)\}$

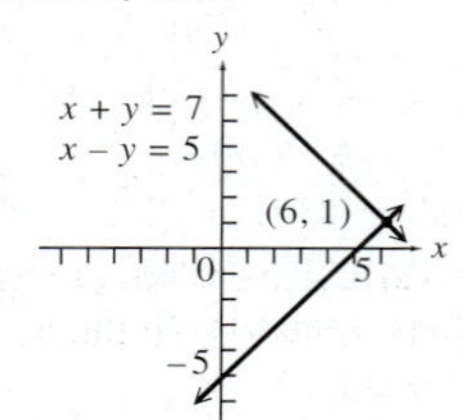

4. $\{(6, -4)\}$ **5.** $\{(x, y) \mid 12x - 5y = 8\}$ **6.** $\left\{\left(-\frac{9}{4}, \frac{5}{4}\right)\right\}$
7. $\{(3, 3)\}$ **8.** $\{(0, -2)\}$ **9.** $\emptyset$ **10.** $\left\{\left(-\frac{2}{3}, \frac{4}{5}, 0\right)\right\}$
11. $\{(3, -2, 1)\}$ **12.** *Pretty Woman:* \$178.4 million; *Runaway Bride:* \$152.3 million **13.** 45 mph, 75 mph **14.** 4 L of 20%; 8 L of 50% **15.** AC adaptor: \$8; rechargeable flashlight: \$15
16. 60 oz of Orange Pekoe; 30 oz of Irish Breakfast; 10 oz of Earl Grey **17.** $\left\{\left(\frac{2}{5}, \frac{7}{5}\right)\right\}$ **18.** $\{(-1, 2, 3)\}$

Cumulative Review Exercises: Chapters 1–5 (page 305)

1. 81 **2.** −81 **3.** −81 **4.** .7 **5.** −.7 **6.** not a real number
7. −199 **8.** 455 **9.** 14 **10.** $\left\{-\frac{15}{4}\right\}$ **11.** $\{11\}$
12. $x = \frac{d - by}{a - c}$ or $x = \frac{by - d}{c - a}$ **13.** $\left\{\frac{2}{3}, 2\right\}$ **14.** $\left(-\infty, \frac{240}{13}\right]$
15. $\left[-2, \frac{2}{3}\right]$ **16.** $(-\infty, \infty)$ **17.** 2010; 1813; 62.8%; 57.2%
18. not guilty: 105; guilty: 95 **19.** 46°, 46°, 88° **20.** $y = 6$
21. $x = 4$ **22.** $-\frac{4}{3}$ **23.** $\frac{3}{4}$ **24.** $4x + 3y = 10$
25. $f(x) = -\frac{4}{3}x + \frac{10}{3}$

26.

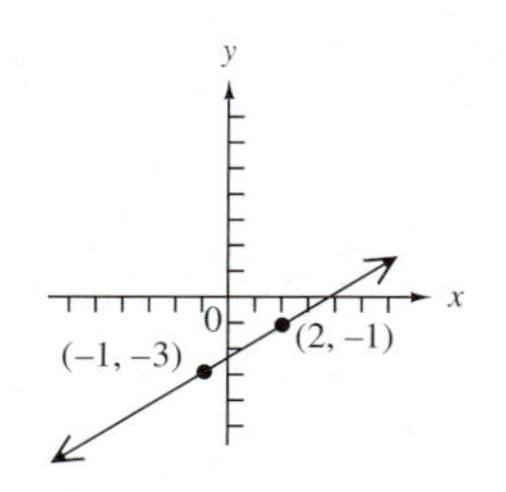

27.

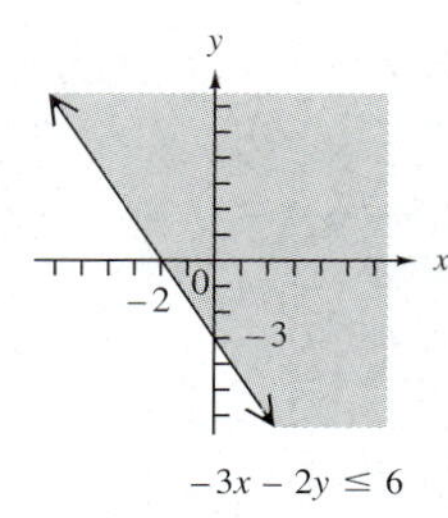

28. $\{(3, -3)\}$ **29.** $\{(5, 3, 2)\}$ **30.** Tickle Me Elmo: \$27.63; Snacktime Kid: \$36.26 **31.** 50 lb of \$1.20 candy; 30 lb of \$2.40 candy **32.** $x = 8$ or 800 parts; \$3000 **33.** about \$400

Chapter 6

Section 6.1 (page 317)

1. incorrect; $(ab)^2 = a^2b^2$ **3.** incorrect; $\left(\frac{4}{a}\right)^3 = \frac{4^3}{a^3}$ **5.** correct
7. 13^{12} **9.** x^{17} **11.** $-27w^8$ **13.** $18x^3y^8$ **15.** The product rule does not apply. **17.** 1 **19.** −1 **21.** 1 **23.** −2
25. $\frac{1}{5^4}$ or $\frac{1}{625}$ **27.** $\frac{1}{8}$ **29.** $\frac{1}{16x^2}$ **31.** $\frac{4}{x^2}$ **33.** $-\frac{1}{a^3}$ **35.** $\frac{1}{a^4}$
37. $\frac{11}{30}$ **39.** $-\frac{5}{24}$ **41.** 16 **43.** $\frac{27}{4}$ **45.** $\frac{27}{8}$ **47.** $\frac{25}{16}$
49. 4^2 or 16 **51.** x^4 **53.** $\frac{1}{r^3}$ **55.** 6^6 **57.** $\frac{1}{6^{10}}$ **59.** 7^2 or 49
61. r^3 **63.** The quotient rule does not apply. **65.** x^{18} **67.** $\frac{27}{125}$
69. $64t^3$ **71.** $-216x^6$ **73.** $-\frac{64m^6}{t^3}$ **75.** $\frac{1}{3}$ **77.** $\frac{1}{a^5}$
79. $\frac{1}{k^2}$ **81.** $-4r^6$ **83.** $\frac{625}{a^{10}}$ **85.** $\frac{z^4}{x^3}$ **87.** $\frac{1}{5p^{10}}$ **89.** $\frac{4}{a^2}$
91. $\frac{1}{6y^{13}}$ **93.** $\frac{2^2k^5}{m^2}$ or $\frac{4k^5}{m^2}$ **95.** 5.3×10^2
97. 8.3×10^{-1} **99.** 6.92×10^{-6} **101.** -3.85×10^4
103. 72,000 **105.** .00254 **107.** −60,000 **109.** .000012
111. .0000025 **113.** 200,000 **115.** $\$1.37574 \times 10^{10}$
117. \$40,045 **119.** approximately 5.87×10^{12} mi
121. (a) 20,000 hr (b) 833 days **123.** 1.23×10^5
125. 4.24×10^5 **127.** 4.4×10^5 **129.** $-a^n = (-a)^n$ when n is an odd number. When n is even, $-a^n \neq (-a)^n$. **131.** Write the fraction as its reciprocal raised to the opposite of the negative power.

Section 6.2 (page 331)

1. neither **3.** ascending **5.** descending **7.** 7; 1 **9.** −15; 2
11. 1; 4 **13.** −1; 6 **15.** monomial; 0 **17.** binomial; 1
19. trinomial; 3 **21.** none of these; 5 **23.** $8z^4$ **25.** $7m^3$
27. $5x$ **29.** already simplified **31.** $-3y^2 + 7y$ **33.** $8k^2 + 2k - 7$
35. $-2n^4 - n^3 + n^2$ **37.** $-9p^2 + 11p - 9$ **39.** $5a + 18$
41. $14m^2 - 13m + 6$ **43.** $13z^2 + 10z - 3$
45. $10y^3 - 7y^2 + 5y + 8$ **47.** $-5a^4 - 6a^3 + 9a^2 - 11$
49. $r + 13$ **51.** $8x^2 + x - 2$ **53.** $-2a^2 - 2a - 7$
55. $-3z^5 + z^2 + 7z$ **57.** (a) −10 (b) 8 **59.** (a) 8 (b) 2
61. (a) 8 (b) 74 **63.** (a) −11 (b) 4 **65.** (a) 11,280 (b) 16,437 (c) 18,369 **67.** (a) 33,000,000 (b) 34,100,000 (c) 35,980,000 **69.** (a) $8x - 3$ (b) $2x - 17$
71. (a) $-x^2 + 12x - 12$ (b) $9x^2 + 4x + 6$ **73.** Answers will vary. For example, let $f(x) = 2x^3 + 3x^2 + x + 4$ and $g(x) = 2x^4 + 3x^3 - 9x^2 + 2x - 4$. For these functions, $(f - g)(x) = -2x^4 - x^3 + 12x^2 - x + 8$, and $(g - f)(x) = 2x^4 + x^3 - 12x^2 + x - 8$. Because the two differences are not equal, subtraction of polynomial functions is not commutative.

75. domain: $(-\infty, \infty)$; range: $(-\infty, \infty)$

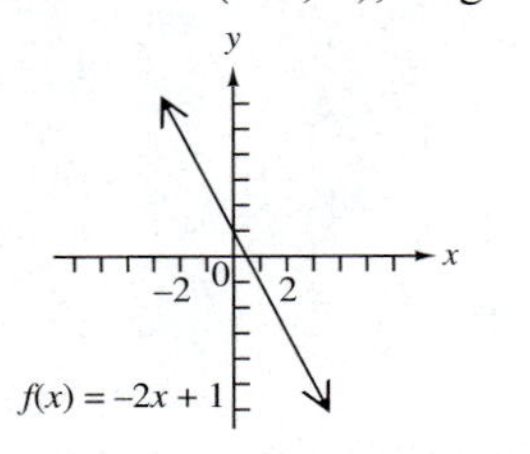

77. domain: $(-\infty, \infty)$; range: $(-\infty, 0]$

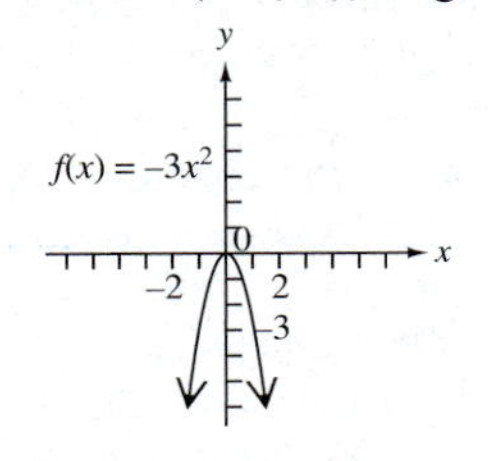

79. domain: $(-\infty, \infty)$; range: $(-\infty, \infty)$

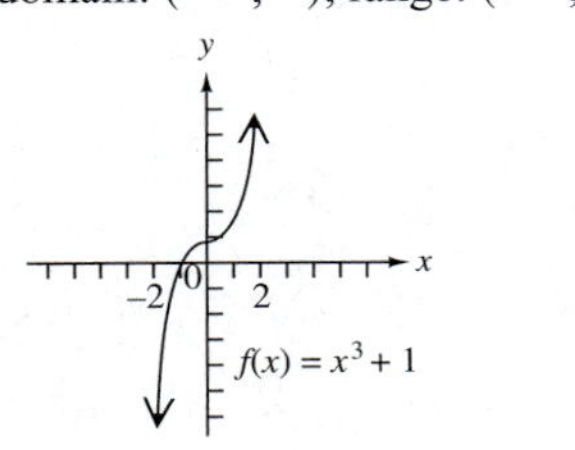

Section 6.3 (page 341)

1. $-24m^5$ **3.** $-6x^2 + 15x$ **5.** $-2q^3 - 3q^4$ **7.** $18k^4 + 12k^3 + 6k^2$ **9.** $6m^3 + m^2 - 14m - 3$ **11.** $4x^5 - 4x^4 - 24x^3$ **13.** $6y^2 + y - 12$ **15.** $-2b^3 + 2b^2 + 18b + 12$ **17.** $25m^2 - 9n^2$ **19.** $8z^4 - 14z^3 + 17z^2 + 20z - 3$ **21.** $6p^4 + p^3 + 4p^2 - 27p - 6$ **23.** $m^2 - 3m - 40$ **25.** $12k^2 + k - 6$ **27.** $3z^2 + zw - 4w^2$ **29.** $12c^2 + 16cd - 3d^2$ **31.** $.1x^2 + .63x - .13$ **33.** $3r^2 - \frac{23}{4}ry - \frac{1}{2}y^2$ **35.** The product of two binomials is the sum of the product of the first terms, the product of the outer terms, the product of the inner terms, and the product of the last terms. **37.** $4p^2 - 9$ **39.** $25m^2 - 1$ **41.** $9a^2 - 4c^2$ **43.** $16x^2 - \frac{4}{9}$ **45.** $16m^2 - 49n^4$ **47.** $25y^6 - 4$ **49.** $y^2 - 10y + 25$ **51.** $4p^2 + 28p + 49$ **53.** $16n^2 - 24nm + 9m^2$ **55.** $k^2 - \frac{10}{7}kp + \frac{25}{49}p^2$ **57.** $(x + y)^2 = x^2 + 2xy + y^2$, because it is a perfect square trinomial. Thus, it differs from $x^2 + y^2$ by $2xy$. **59.** $25x^2 + 10x + 1 + 60xy + 12y + 36y^2$ **61.** $4a^2 + 4ab + b^2 - 9$ **63.** $4h^2 - 4hk + k^2 - j^2$ **65.** $125r^3 - 75r^2s + 15rs^2 - s^3$ **67.** $m^4 - 4m^3p + 6m^2p^2 - 4mp^3 + p^4$ **69.** $a - b$ **70.** $A = s^2$; $(a - b)^2$ **71.** $(a - b)b$ or $ab - b^2$; $2ab - 2b^2$ **72.** b^2 **73.** a^2; a **74.** $a^2 - (2ab - 2b^2) - b^2 = a^2 - 2ab + b^2$ **75.** They must be equal to each other. **76.** $(a - b)^2 = a^2 - 2ab + b^2$; This reinforces the special product for the square of a binomial difference. **77.** $10x^2 - 2x$ **79.** $2x^2 - x - 3$ **81.** $8x^3 - 27$ **83.** $(2 + 3)^2 \neq 2^2 + 3^2$ because $25 \neq 13$; $(x + y)^2 = x^2 + 2xy + y^2$ **85.** $(2 + 3)^4 \neq 2^4 + 3^4$ because $625 \neq 97$; $(x + y)^4 = x^4 + 4x^3y + 6x^2y^2 + 4xy^3 + y^4$

Section 6.4 (page 349)

1. $3x^3 - 2x^2 + 1$ **3.** $3y + 4 - \frac{5}{y}$ **5.** $3m + 5 + \frac{6}{m}$ **7.** $n - \frac{3n^2}{2m} + 2$ **9.** $y - 3$ **11.** $t + 5$ **13.** $z^2 + 3$ **15.** $x^2 + 2x - 3 + \frac{6}{4x + 1}$ **17.** $2x - 5 + \frac{-4x + 5}{3x^2 - 2x + 4}$ **19.** $2k^2 + 3k - 1$ **21.** $9z^2 - 4z + 1 + \frac{-z + 6}{z^2 - z + 2}$ **23.** $\frac{2}{3}x - 1$ **25.** $\frac{3}{4}a - 2 + \frac{1}{4a + 3}$ **27.** $5x - 1$; 0 **29.** $2x - 3$; -1 **31.** $4x^2 + 6x + 9$; $\frac{3}{2}$ **33.** $2p + 7$ ft **35.** $t^3 + 6t^2 + 5t + 4$ **36.** $t + 4$ **37.** $t^2 + 2t - 3 + \frac{16}{t + 4}$ **38.** $118\frac{1}{7}$ **39.** $118\frac{1}{7}$ **40.** They are the same.

Section 6.5 (page 355)

1. C **3.** $x - 5$ **5.** $4m - 1$ **7.** $2a + 4 + \frac{5}{a + 2}$ **9.** $p - 4 + \frac{9}{p + 1}$ **11.** $4a^2 + a + 3$ **13.** $x^4 + 2x^3 + 2x^2 + 7x + 10 + \frac{18}{x - 2}$ **15.** $-4r^5 - 7r^4 - 10r^3 - 5r^2 - 11r - 8 + \frac{-8}{r - 1}$ **17.** $-3y^4 + 8y^3 - 21y^2 + 36y - 72 + \frac{143}{y + 2}$ **19.** $y^2 + y + 1 + \frac{2}{y - 1}$ **21.** 7 **23.** -2 **25.** 0 **27.** yes **29.** no **31.** no **33.** Since the variables are not present, a missing term will not be noticed in synthetic division, so the quotient will be wrong if placeholders are not inserted.

Section 6.6 (page 361)

1. To factor a polynomial means to write it as the product of two or more polynomials. **3.** $z^2(m + n)^4$ **5.** $3m$ **7.** $3(r + t)^2$ **9.** A **11.** $8k(k^2 + 3)$ **13.** $xy(3 - 5y)$ **15.** $-2p^2q^4(2p + q)$ **17.** $7x^3(3x^2 + 5x - 2)$ **19.** $5ac(3ac^2 - 5c + a)$ **21.** cannot be factored **23.** $(m - 4)(2m + 5)$ **25.** $11(2z - 1)$ **27.** $(2 - x)(10 - x - x^2)$ **29.** $r(-r^2 + 3r + 5)$; $-r(r^2 - 3r - 5)$ **31.** $12s^4(-s + 4)$; $-12s^4(s - 4)$ **33.** $2x^2(-x^3 + 3x + 2)$; $-2x^2(x^3 - 3x - 2)$ **35.** $(m + 3q)(x + y)$ **37.** $(5m + n)(2 + k)$ **39.** $(m - 3)(m + 5)$ **41.** $(p + q)(p - 4z)$ **43.** $(a + 5)(3a - 2)$ **45.** $(-3p + q)(5p + 2q)$ **47.** $(a^2 + b^2)(-3a + 2b)$ **49.** $(y - 2)(x - 2)$ **51.** $(3y - 2)(3y^3 - 4)$ **53.** $m^{-5}(3 + m^2)$ **55.** $p^{-3}(3 + 2p)$

Section 6.7 (page 369)

1. D **3.** B **5.** $(y - 3)(y + 10)$ **7.** $(p - 8)(p + 7)$ **9.** $-(m - 10)(m - 6)$ **11.** $(a + 5b)(a - 7b)$ **13.** prime **15.** $(xy + 9)(xy + 2)$ **17.** $-(6m - 5)(m + 3)$ **19.** $(5x - 6)(2x + 3)$ **21.** $(4k + 3)(5k + 8)$ **23.** $(3a - 2b)(5a - 4b)$ **25.** $(6m - 5)^2$ **27.** prime **29.** $(2xz - 1)(3xz + 4)$ **31.** $3(4x + 5)(2x + 1)$ **33.** $-5(a + 6)(3a - 4)$ **35.** $-11x(x - 6)(x - 4)$ **37.** $2xy^3(x - 12y)(x - 12y)$ **39.** $(5k + 4)(2k + 1)$ **41.** $(3m + 3p + 5)(m + p - 4)$ **43.** $(a + b)^2(a - 3b)(a + 2b)$ **45.** $(2x^2 + 3)(x^2 - 6)$ **47.** $(4x^2 + 3)(4x^2 + 1)$ **49.** $(6p^3 - r)(2p^3 - 5r)$ **51.** no **52.** 1, 3, 5, 9, 15, 45; no **53.** no **54.** $(5x + 2)(2x + 5)$; no **55.** Since k is odd, 2 is not a factor of $2x^2 + kx + 8$, and because 2 is a factor of $2x + 4$, the binomial $2x + 4$ cannot be a factor. **56.** $3y + 15$ cannot be a factor of $12y^2 - 11y - 15$ because 3 is a factor of $3y + 15$, but 3 is not a factor of $12y^2 - 11y - 15$.

Section 6.8 (page 375)

1. A, D **3.** B, C **5.** The sum of two squares can be factored only

if the binomial has a common factor. **7.** $(p + 4)(p - 4)$
9. $(5x + 2)(5x - 2)$ **11.** $2(3a + 7b)(3a - 7b)$
13. $4(4m^2 + y^2)(2m + y)(2m - y)$ **15.** $(y + z + 9)(y + z - 9)$
17. $(4 + x + 3y)(4 - x - 3y)$ **19.** $4pq$ **21.** $(k - 3)^2$ **23.** $(2z + w)^2$
25. $(4m - 1 + n)(4m - 1 - n)$ **27.** $(2r - 3 + s)(2r - 3 - s)$
29. $(x + y - 1)(x - y + 1)$ **31.** $2(7m + 3n)^2$ **33.** $(p + q + 1)^2$
35. $(a - b + 4)^2$ **37.** $(2x - y)(4x^2 + 2xy + y^2)$
39. $(4g + 3h)(16g^2 - 12gh + 9h^2)$
41. $3(2n + 3p)(4n^2 - 6np + 9p^2)$
43. $(y + z - 4)(y^2 + 2yz + z^2 + 4y + 4z + 16)$
45. $(m^2 - 5)(m^4 + 5m^2 + 25)$ **47.** $2b(3a^2 + b^2)$
48. $(x^3 - y^3)(x^3 + y^3)$; $(x - y)(x^2 + xy + y^2)(x + y)(x^2 - xy + y^2)$
49. $(x^2 + xy + y^2)(x^2 - xy + y^2)$ **50.** $(x^2 - y^2)(x^4 + x^2y^2 + y^4)$; $(x - y)(x + y)(x^4 + x^2y^2 + y^4)$ **51.** $x^4 + x^2y^2 + y^4$ **52.** The product must equal $x^4 + x^2y^2 + y^4$. Multiply $(x^2 + xy + y^2)(x^2 - xy + y^2)$ to verify this. **53.** Start by factoring as a difference of squares.

Summary Exercises on Factoring (page 377)

1. $(10a + 3b)(10a - 3b)$ **2.** $(5r - 1)(2r + 3)$
3. $6p^3(3p^2 - 4 + 2p^3)$ **4.** $5x(3x - 4)$ **5.** $(x + 7)(x - 5)$
6. $(3 + a - b)(3 - a + b)$ **7.** prime
8. $(x - 10)(x^2 + 10x + 100)$ **9.** $(6b + 1)(b - 3)$ **10.** prime
11. $3mn(3m + 2n)(2m - n)$ **12.** $(3t - 7u)(2t + 11u)$
13. $(2p + 5q)(p + 3q)$ **14.** $9m(m - 5 + 2m^2)$ **15.** $(2k + 7r)^2$
16. $2(3m - 10)(9m^2 + 30m + 100)$ **17.** $(m - 2)(n + 5)$
18. $(3m - 5n + p)(3m - 5n - p)$ **19.** $(x + 3)^2(x - 3)$
20. $7(2k - 5)(4k^2 + 10k + 25)$ **21.** prime
22. $(2p - 5)(4p^2 + 10p + 25)$ **23.** $(3k + 1)(2k - 1)$
24. $3(3m + 8n)^2$ **25.** $(x^2 + 25)(x + 5)(x - 5)$
26. $(5m^2 + 6)(25m^4 - 30m^2 + 36)$ **27.** $(a + 6)(b + c)$
28. $(p + 4)(p^2 - 4p + 16)$ **29.** $4y(y - 2)$ **30.** $(3a^2 + 2)(2a^2 - 5)$
31. $(7z + 2k)(2z - k)$ **32.** $6z(2z^2 - z + 3)$
33. $16(4b + 5c)(4b - 5c)$ **34.** prime
35. $8(5z + 4)(25z^2 - 20z + 16)$ **36.** $(8m + 5n)(8m - 5n)$
37. $(5r - s)(2r + 5s)$ **38.** $(3k - 5q)(4k + q)$
39. $8x^2(4 + 2x - 3x^3)$ **40.** $3(4k^2 + 9)(2k + 3)(2k - 3)$
41. $(7x + 5q)(2x - 5q)$ **42.** $5p(p - 2)$ **43.** $(y + 5)(y - 2)$
44. $(b - 9a)(b + 2a)$ **45.** $2a(a^2 + 3a - 2)$
46. $4rx(3m^2 + mn + 10n^2)$ **47.** $(9p - 5r)(2p + 7r)$
48. $(3a + b)(7a - 4b)$ **49.** $(x - 2y + 2)(x - 2y - 2)$
50. $(3m - n + 5)(3m - n - 5)$ **51.** $(5r + 2s - 3)^2$
52. $(p + 8q - 5)^2$ **53.** $(z + 2)(z - 2)(z^2 - 5)$
54. $(3m^2 - 5)(7m^2 + 1)$

Section 6.9 (page 385)

1. First rewrite the equation so one side is 0. Factor the other side and set each factor equal to 0. The solutions of these linear equations are solutions of the quadratic equation. **3.** $\{5, -10\}$
5. $\left\{-\frac{8}{3}, \frac{5}{2}\right\}$ **7.** $\{-2, 5\}$ **9.** $\{-6, -3\}$ **11.** $\left\{-\frac{1}{2}, 4\right\}$
13. $\left\{-\frac{1}{3}, \frac{4}{5}\right\}$ **15.** $\{-3, 4\}$ **17.** $\left\{-5, -\frac{1}{5}\right\}$
19. $\{0, 6\}$ **21.** $\{-3, 3\}$ **23.** $\{-2, 2\}$ **25.** $\{-4, 2\}$
27. $\left\{-\frac{1}{2}, 6\right\}$ **29.** $\{1, 6\}$ **31.** $\left\{-\frac{1}{2}, 0, 5\right\}$ **33.** $\left\{-\frac{4}{3}, 0, \frac{4}{3}\right\}$
35. $\left\{-\frac{5}{2}, -1, 1\right\}$ **37.** By dividing each side by a variable expression, she "lost" the solution 0. **39.** width: 16 ft; length: 20 ft **41.** base: 8 m; height: 11 m **43.** 50 ft by 100 ft
45. length: 15 in.; width: 9 in. **47.** 3 sec and 5 sec; 1 sec and 7 sec **49.** $6\frac{1}{4}$ sec

Chapter 6 Review Exercises (page 395)

1. 64 **2.** $\frac{1}{81}$ **3.** -125 **4.** 18 **5.** $\frac{81}{16}$ **6.** $\frac{16}{25}$ **7.** $\frac{11}{30}$ **8.** 0
9. $-12x^2y^8$ **10.** $-\frac{2n}{m^5}$ **11.** $\frac{10p^8}{q^7}$ **12.** $\frac{x^2}{y^2}$ **13.** $\frac{1}{3^8}$ **14.** x^8
15. $\frac{y^6}{x^2}$ **16.** $\frac{1}{z^{15}}$ **17.** $\frac{25}{m^{18}}$ **18.** $\frac{r^{17}}{9}$ **19.** $\frac{25}{z^4}$ **20.** $\frac{1}{96m^7}$
21. $\frac{2025}{8r^4}$ **22.** 1.345×10^4 **23.** 7.65×10^{-8} **24.** 1.38×10^{-1}
25. 1,210,000 **26.** .0058 **27.** $\$6.375 \times 10^{10}$ **28.** 2×10^{-4}; .0002 **29.** 1.5×10^3; 1500 **30.** 4.1×10^{-5}; .000041
31. 2.7×10^{-2}; .027 **32.** 14 **33.** -1 **34.** **(a)** $11k^3 - 3k^2 + 9k$ **(b)** trinomial **(c)** 3 **35.** **(a)** $9m^7 + 14m^6$ **(b)** binomial **(c)** 7
36. **(a)** $-7q^5r^3$ **(b)** monomial **(c)** 8 **37.** Answers will vary. An example is $x^5 + 2x^4 - x^2 + x + 2$. **38.** $-x^2 - 3x + 1$
39. $-5y^3 - 4y^2 + 6y - 12$ **40.** $6a^3 - 4a^2 - 16a + 15$
41. $8y^2 - 9y + 5$ **42.** $12x^2 + 8x + 5$ **43.** **(a)** -11 **(b)** 4
44. **(a)** $5x^2 - x + 5$ **(b)** $-5x^2 + 5x + 1$ **45.** 1994: 8992 thousand; 1995: 9257 thousand; 1996: 9566 thousand

46.

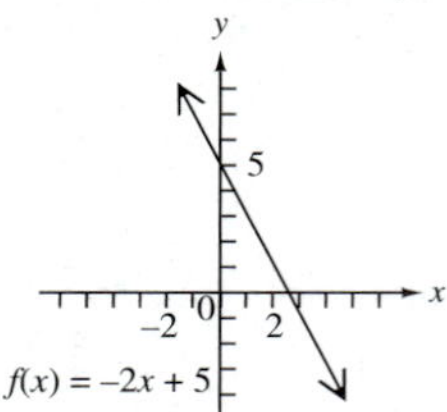

47.

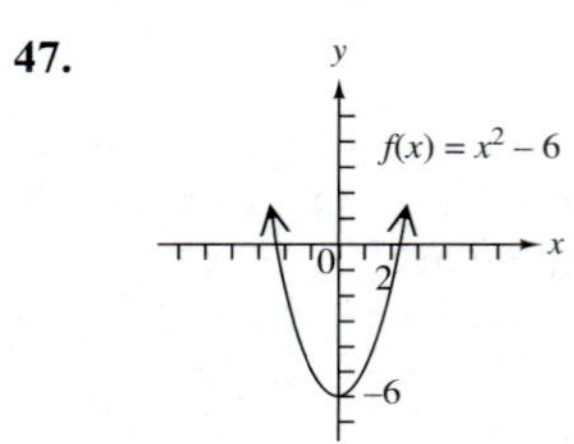

48.

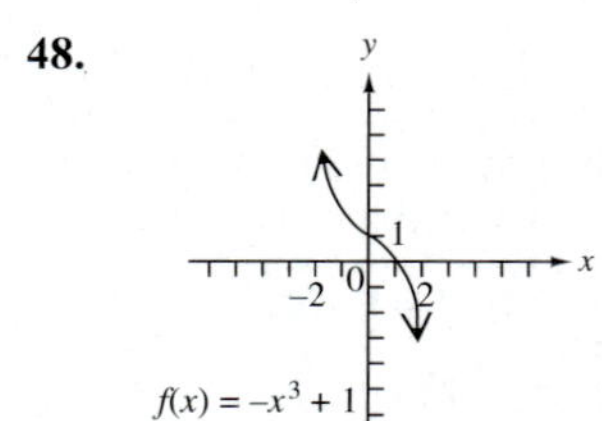

49. $-12k^3 - 42k$ **50.** $14y^2 + 5y - 24$ **51.** $6w^2 - 13wt + 6t^2$
52. $10p^4 + 30p^3 - 8p^2 - 24p$ **53.** $9z^4 - 12z^3 + 16z^2 - 11z + 2$
54. $36r^4 - 1$ **55.** $z^2 - \frac{9}{25}$ **56.** $16m^2 + 24m + 9$
57. $8x^3 + 60x^2 + 150x + 125$ **58.** $y^2 - 3y + \frac{5}{4}$
59. $p^2 + 6p + 9 + \frac{54}{2p - 3}$ **60.** $p^2 + 3p - 6$ **61.** $3p + 2$
62. $2k^2 + k + 3 + \frac{21}{k - 3}$ **63.** yes **64.** no **65.** -13
66. -5 **67.** $7y(3y + 5)$ **68.** $4qb(3q + 2b - 5q^2b)$
69. $(x + 3)(x - 3)$ **70.** $(z + 1)(3z - 1)$ **71.** $(m + q)(4 + n)$
72. $(x + y)(x + 5)$ **73.** $(m + 3)(2 - a)$ **74.** $(a - b)(2m - p)$
75. $(3p - 4)(p + 1)$ **76.** $(3r + 1)(4r - 3)$ **77.** $(2m + 5)(5m + 6)$
78. $(2k - h)(5k - 3h)$ **79.** prime **80.** $2x(4 + x)(3 - x)$
81. $(2k^2 + 1)(k^2 - 3)$ **82.** $(p + 2)^2(p + 3)(p - 2)$ **83.** It is not factored because it is a sum, not a *product*. The correct answer is $(y^2 - 6)(x^2 + 5)$. **84.** $p + 1$ **85.** $(4x + 5)(4x - 5)$
86. $(3t + 7)(3t - 7)$ **87.** $(x + 7)^2$ **88.** $(3k - 2)^2$

89. $(r + 3)(r^2 - 3r + 9)$ **90.** $(5x - 1)(25x^2 + 5x + 1)$
91. $(m + 1)(m^2 - m + 1)(m - 1)(m^2 + m + 1)$
92. $(x^4 + 1)(x^2 + 1)(x + 1)(x - 1)$ **93.** $(x + 3 + 5y)(x + 3 - 5y)$
94. $\left\{-1, -\frac{2}{5}\right\}$ **95.** $\{2, 3\}$ **96.** $\left\{-\frac{5}{2}, \frac{10}{3}\right\}$ **97.** $\left\{-\frac{3}{2}, \frac{1}{3}\right\}$
98. $\{-3, 3\}$ **99.** $\left\{-\frac{3}{2}, 0\right\}$ **100.** $\left\{\frac{1}{2}, 1\right\}$ **101.** $\{4\}$
102. $\left\{-\frac{7}{2}, 0, 4\right\}$ **103.** 3 ft **104.** length: 60 ft; width: 40 ft
105. after 16 sec **106.** after 1 sec and after 15 sec
107. The rock reaches a height of 240 ft once on its way up and once on its way down. **108.** $8x^2 - 10x - 3$ **109.** $\frac{y^4}{36}$
110. $\frac{1}{16y^{18}}$ **111.** $-14 + 16w - 8w^2$ **112.** $21p^9 + 7p^8 + 14p^7$
113. $-\frac{1}{5z^9}$ **114.** $x^2 + 2x - 3$ **115.** $-3k^2 + 4k - 7$
116. $a(6 - m)(5 + m)$ **117.** $(2 - a)(4 + 2a + a^2)$ **118.** prime
119. $5y^2(3y + 4)$ **120.** $\left\{-\frac{3}{5}, 4\right\}$ **121.** $\{-1, 0, 1\}$
122. width: 25 ft; length: 110 ft

Chapter 6 Test (page 401)

1. $\frac{4x^7}{9y^{10}}$ **2.** $\frac{6}{r^{14}}$ **3.** $\frac{16}{9p^{10}q^{28}}$ **4.** $\frac{16}{x^6y^{16}}$ **5. (a)** .00000091
(b) 3×10^{-4}; .0003 **6. (a)** -18 **(b)** $-2x^2 + 12x - 9$
(c) $-2x^2 - 2x - 3$
7.

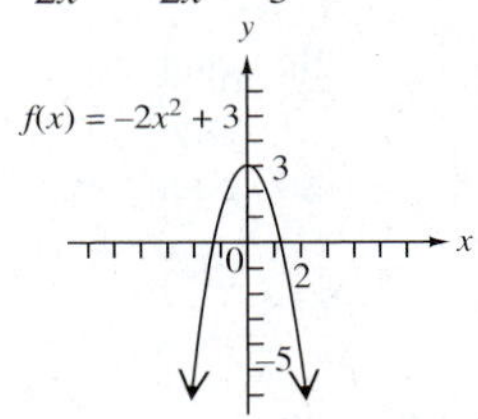

8. $x^3 - 2x^2 - 10x - 13$ **9.** $10x^2 - x - 3$
10. $6m^3 - 7m^2 - 30m + 25$ **11.** $36x^2 - y^2$ **12.** $9k^2 + 6kq + q^2$
13. $4y^2 - 9z^2 + 6zx - x^2$ **14.** $x^2 + 4x + 4$
15. (a) $2x + 7 + \frac{8}{x - 2}$ **(b)** 8 **16.** $11z(z - 4)$
17. $(x + y)(3 + b)$ **18.** $(4p - q)(p + q)$ **19.** $(4a + 5b)^2$
20. $(y - 6)(y^2 + 6y + 36)$ **21.** $(3k + 11j)(3k - 11j)$
22. $(2k^2 - 5)(3k^2 + 7)$ **23.** $(3x^2 + 1)(9x^4 - 3x^2 + 1)$
24. $-(x + 4)(x - 3)$ **25.** $(t^2 + 8)(t^2 + 2)$ **26.** It is not in factored form because there are two terms: $(x^2 + 2y)p$ and $3(x^2 + 2y)$. The common factor is $x^2 + 2y$, and the factored form is $(x^2 + 2y)(p + 3)$. **27.** $\left\{-2, -\frac{2}{3}\right\}$ **28.** $\left\{\frac{1}{5}, \frac{3}{2}\right\}$
29. length: 8 in.; width: 5 in. **30.** 2 sec and 4 sec

Cumulative Review Exercises: Chapters 1–6 (page 403)

1. $-2m + 6$ **2.** $4m - 3$ **3.** $2x^2 + 5x + 4$ **4.** -24 **5.** 204
6. undefined **7.** 10 **8.** $\left\{\frac{7}{6}\right\}$ **9.** $\{-1\}$ **10.** $\left(-\infty, \frac{15}{4}\right]$
11. $\left(-\frac{1}{2}, \infty\right)$ **12.** $(2, 3)$ **13.** $(-\infty, 2) \cup (3, \infty)$
14. $\left\{-\frac{16}{5}, 2\right\}$ **15.** $(-11, 7)$ **16.** $(-\infty, -2] \cup [7, \infty)$
17. $h = \frac{V}{lw}$ **18.** 2 hr
19.

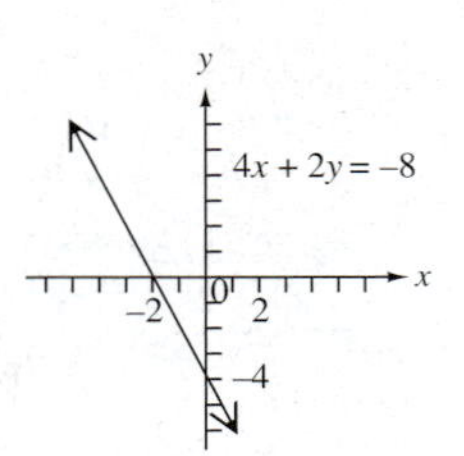

20. -1 **21.** 0 **22.** -1 **23.** $\left(-\frac{7}{2}, 0\right)$ **24.** $(0, 7)$ **25.** $\{(1, 5)\}$
26. $\{(1, 1, 0)\}$ **27.** $\frac{y}{18x}$ **28.** $\frac{5my^4}{3}$ **29.** $x^3 + 12x^2 - 3x - 7$
30. $49x^2 + 42xy + 9y^2$ **31.** $10p^3 + 7p^2 - 28p - 24$
32. $(2w + 7z)(8w - 3z)$ **33.** $(2x - 1 + y)(2x - 1 - y)$
34. $(2y - 9)^2$ **35.** $(10x^2 + 9)(10x^2 - 9)$
36. $(2p + 3)(4p^2 - 6p + 9)$ **37.** $\left\{-4, -\frac{3}{2}, 1\right\}$ **38.** $\left\{\frac{1}{3}\right\}$
39. 4 ft **40.** longer sides: 18 in.; distance between: 16 in.

Chapter 7

Section 7.1 (page 413)

1. C **3.** D **5.** E **7.** Replacing x with 2 makes the denominator 0 and the value of the expression undefined. To find the values excluded from the domain, set the denominator equal to 0 and solve the equation. All solutions of the equation are excluded from the domain. **9.** 7 **11.** $-\frac{1}{7}$ **13.** 0 **15.** $-2, \frac{3}{2}$ **17.** none
19. none **21. (a)** numerator: x^2, $4x$; denominator: x, 4 **(b)** First factor the numerator, getting $x(x + 4)$, then divide the numerator and denominator by the common factor of $x + 4$ to get $\frac{x}{1}$ or x.
23. B **25.** x **27.** $\frac{x - 3}{x + 5}$ **29.** $\frac{x + 3}{2x(x - 3)}$ **31.** already in lowest terms **33.** $\frac{6}{7}$ **35.** $\frac{z}{6}$ **37.** $\frac{2}{t - 3}$ **39.** $\frac{x - 3}{x + 1}$ **41.** $\frac{4x + 1}{4x + 3}$
43. $a^2 - ab + b^2$ **45.** $\frac{c + 6d}{c - d}$ **47.** $\frac{a + b}{a - b}$ **49.** -1 *In Exercises 51–55, there are other acceptable ways to express each answer.*
51. $-(x + y)$ **53.** $-\frac{x + y}{x - y}$ **55.** $-\frac{1}{2}$ **57.** already in lowest terms **59.** $\frac{x + 4}{x - 2}$ **61.** $\frac{2x + 3}{x + 2}$ **63.** $-\frac{35}{8}$ **65.** $\frac{7x}{6}$
67. $-\frac{p + 5}{2p}$ (There are other ways.) **69.** $\frac{-m(m + 7)}{m + 1}$ (There are other ways.) **71.** -2 **73.** $\frac{x + 4}{x - 4}$ **75.** $\frac{2x + 3y}{2x - 3y}$
77. $\frac{k + 5p}{2k + 5p}$ **79.** $(k - 1)(k - 2)$

Section 7.2 (page 423)

1. To add or subtract rational expressions that have a common denominator, first add or subtract the numerators. Then place the result over the common denominator. Write the answer in lowest terms. **3.** $\frac{9}{t}$ **5.** $\frac{2}{x}$ **7.** 1 **9.** $x - 5$ **11.** $\frac{1}{p + 3}$ **13.** $a - b$
15. $72x^4y^5$ **17.** $z(z - 2)$ **19.** $2(y + 4)$ **21.** $(x + 9)^2(x - 9)$
23. $(m + n)(m - n)$ **25.** $x(x - 4)(x + 1)$
27. $(t + 5)(t - 2)(2t - 3)$ **29.** $2y(y + 3)(y - 3)$
31. Yes, they could both be correct because the expressions are

equivalent. Multiplying $\frac{3}{5-y}$ by 1 in the form $\frac{-1}{-1}$ gives $\frac{-3}{y-5}$. **33.** $\frac{31}{3t}$ **35.** $\frac{5-22x}{12x^2y}$ **37.** $\frac{1}{x(x-1)}$ **39.** $\frac{5a^2-7a}{(a+1)(a-3)}$ **41.** 3 **43.** $\frac{3}{x-4}$ or $\frac{-3}{4-x}$ **45.** $\frac{w+z}{w-z}$ or $\frac{-w-z}{z-w}$ **47.** $\frac{-13}{12(3+x)}$ **49.** $\frac{2(2x-1)}{x-1}$ **51.** $\frac{7}{y}$ **53.** $\frac{6}{x-2}$ **55.** $\frac{3x-2}{x-1}$ **57.** $\frac{4x-7}{x^2-x+1}$ **59.** $\frac{2x+1}{x}$ **61.** $\frac{2x(x+12y)}{(x+2y)(x-y)(x+6y)}$ **63.** $c(x)=\frac{10x}{49(101-x)}$ **65.** $\frac{8}{9}$ **66.** $\frac{3}{7}+\frac{5}{9}-\frac{6}{63}$; They are the same. **67.** $\frac{8}{9}$; yes **68.** Answers will vary. Suppose the name is Bush, so that $x=4$. The problem is $\frac{3}{2}+\frac{5}{4}-\frac{6}{8}$. The predicted answer is $\frac{8}{4}=2$, which is correct. **69.** It causes $\frac{3}{x-2}$ and $\frac{6}{x^2-2x}$ to be undefined, since 0 appears in the denominators. **70.** 0

Section 7.3 (page 431)

1. Begin by simplifying the numerator. Then simplify the denominator. Write as a division problem, and proceed. **3.** $\frac{2x}{x-1}$ **5.** $\frac{2(k+1)}{3k-1}$ **7.** $\frac{5x^2}{9z^3}$ **9.** $\frac{1+x}{-1+x}$ **11.** $\frac{y+x}{y-x}$ **13.** $4x$ **15.** $x+4y$ **17.** $\frac{3y}{2}$ **19.** $\frac{x^2+5x+4}{x^2+5x+10}$ **21.** $\frac{m^2+6m-4}{m(m-1)}$ **22.** $\frac{m^2-m-2}{m(m-1)}$ **23.** $\frac{m^2+6m-4}{m^2-m-2}$ **24.** $m(m-1)$ **25.** $\frac{m^2+6m-4}{m^2-m-2}$ **26.** Method 1 involves simplifying the numerator and the denominator separately and then performing a division. Method 2 involves multiplying the fraction by a form of 1, the identity element for multiplication. (Preferences will vary.) **27.** $\frac{x^2y^2}{y^2+x^2}$ **29.** $\frac{y^2+x^2}{xy^2+x^2y}$ or $\frac{y^2+x^2}{xy(y+x)}$ **31.** $\frac{1}{2xy}$

Section 7.4 (page 437)

1. $x=0$ **3.** $x=2$

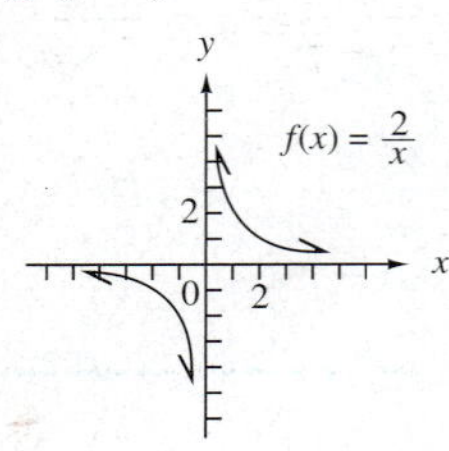

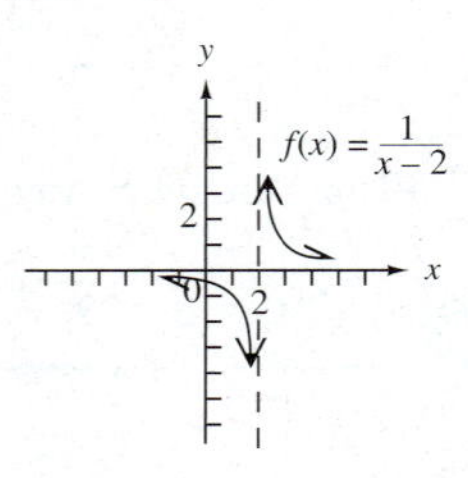

5. (a) $-1, 2$ **(b)** $\{x \mid x \neq -1, 2\}$ **7. (a)** $-\frac{5}{3}, 0, -\frac{3}{2}$ **(b)** $\left\{x \mid x \neq -\frac{5}{3}, 0, -\frac{3}{2}\right\}$ **9. (a)** 0 **(b)** $\{x \mid x \neq 0\}$ **11. (a)** $4, \frac{7}{2}$ **(b)** $\left\{x \mid x \neq 4, \frac{7}{2}\right\}$ **13. (a)** $0, 1, -3, 2$ **(b)** $\{x \mid x \neq 0, 1, -3, 2\}$ **15.** $\{1\}$ **17.** $\{-6, 4\}$ **19.** $\left\{-\frac{7}{12}\right\}$ **21.** $\emptyset$ **23.** $\{-3\}$ **25.** $\{5\}$ **27.** $\{5\}$ **29.** $\emptyset$ **31.** $\left\{\frac{27}{56}\right\}$ **33.** $\emptyset$ **35.** $\{-10\}$ **37.** $\emptyset$ **39.** $\{0\}$ **41.** $\left\{x \mid x \neq -\frac{3}{2}, \frac{3}{2}\right\}$ **43.** Substituting -1 for x gives a true statement, $\frac{4}{3}=\frac{4}{3}$. Substituting -2 for x leads to 0 in the first and third denominators. **44.** $C=-4$; $\{-2\}$; -1 is rejected. **45.** $C=24$; $\{-4\}$; 3 is rejected. **46.** Answers will vary. However, in every case, $-B$ will be the rejected solution, and $\{-A\}$ will be the solution set. **47. (a)** 0 **(b)** 1.6 **(c)** 4.1 **(d)** The waiting time also increases.

Summary Exercises on Operations and Equations with Rational Expressions (page 441)

1. equation; $\{20\}$ **2.** operation; $\frac{2(x+5)}{5}$ **3.** operation; $-\frac{22}{7x}$ **4.** operation; $\frac{y+x}{y-x}$ **5.** equation; $\left\{\frac{1}{2}\right\}$ **6.** equation; $\{7\}$ **7.** operation; $\frac{43}{24x}$ **8.** equation; $\{1\}$ **9.** operation; $\frac{5x-1}{-2x+2}$ or $\frac{5x-1}{-2(x-1)}$ **10.** operation; $\frac{25}{4(r+2)}$ **11.** operation; $\frac{x^2+xy+2y^2}{(x+y)(x-y)}$ **12.** operation; $\frac{24p}{p+2}$ **13.** operation; $-\frac{5}{36}$ **14.** equation; $\{0\}$ **15.** operation; $\frac{b+3}{3}$ **16.** operation; $\frac{5}{3z}$ **17.** operation; $\frac{2x+10}{x(x-2)(x+2)}$ **18.** equation; $\{2\}$ **19.** operation; $\frac{-x}{3x+5y}$ **20.** equation; $\{-13\}$ **21.** operation; $\frac{3y+2}{y+3}$ **22.** equation; $\left\{\frac{5}{4}\right\}$ **23.** equation; $\emptyset$ **24.** operation; $\frac{2z-3}{2z+3}$ **25.** operation; $\frac{-1}{x-3}$ or $\frac{1}{3-x}$ **26.** operation; $\frac{t-2}{8}$ **27.** equation; $\{-10\}$ **28.** operation; $\frac{13x+28}{2x(x+4)(x-4)}$ **29.** equation; $\emptyset$ **30.** operation; $\frac{k(2k^2-2k+5)}{(k-1)(3k^2-2)}$

Section 7.5 (page 451)

1. A **3.** D **5.** 65.625 **7.** $\frac{25}{4}$ **9.** $G=\frac{Fd^2}{Mm}$ **11.** $a=\frac{bc}{c+b}$ **13.** $v=\frac{PVt}{pT}$ **15.** $r=\frac{nE-IR}{In}$ **17.** $b=\frac{2A}{h}-B$ or $b=\frac{2A-Bh}{h}$ **19.** $r=\frac{eR}{E-e}$ **21.** Multiply each side by $a-b$. **23.** 15 girls, 5 boys **25.** $\frac{1}{2}$ job per hr **27.** 1996 **29.** 1996 **31.** 23 teachers **33.** 25,000 fish **35.** \$95.75 **37.** $x=\frac{7}{2}$; $AC=8$; $DF=12$ **39.** 3 mph **41.** 900 mi **43.** 480 mi **45.** 190 mi **47.** $6\frac{2}{3}$ min **49.** 12 hr **51.** 20 hr **53.** $2\frac{4}{5}$ hr

Chapter 7 Review Exercises (page 461)

1. (a) -6 **(b)** $\{x \mid x \neq -6\}$ **2. (a)** 2, 5 **(b)** $\{x \mid x \neq 2, 5\}$ **3. (a)** 9 **(b)** $\{x \mid x \neq 9\}$ **4.** $\frac{x}{2}$ **5.** $\frac{5m+n}{5m-n}$ **6.** $\frac{-1}{2+r}$ **7.** The reciprocal of a rational expression is another rational expression such that the two rational expressions have a product of 1. **8.** $\frac{3y^2(2y+3)}{2y-3}$ **9.** $\frac{-3(w+4)}{w}$ **10.** $\frac{z(z+2)}{z+5}$ **11.** 1 **12.** $96b^5$ **13.** $9r^2(3r+1)$ **14.** $(3x-1)(2x+5)(3x+4)$ **15.** $\frac{16z-3}{2z^2}$ **16.** 12 **17.** $\frac{71}{30(a+2)}$ **18.** $\frac{13r^2+5rs}{(5r+s)(2r-s)(r+s)}$ **19.** $\frac{3+2t}{4-7t}$ **20.** -2 **21.** $\frac{1}{3q+2p}$ **22.** $\frac{y+x}{xy}$ **23.** C; $x=0$ **24.** $\{-3\}$ **25.** $\{-2\}$ **26.** $\{0\}$ **27.** $\emptyset$ **28.** Although her algebra was correct, 3 is not a solution because it is not in the domain of the equation. Thus, $\emptyset$ is correct. **29.** In simplifying the expression, we are combining terms to get a single fraction with a denominator of $6x$, while in solving the equation, we are finding a value for x that makes the equation true. **30.** $\frac{15}{2}$ **31.** $m=\frac{Fd^2}{GM}$ **32.** $M=\frac{m\mu}{v-\mu}$ **33.** 6000 passenger-km per day **34.** 16 km per hr **35.** $4\frac{4}{5}$ min **36.** $3\frac{3}{5}$ hr **37.** $\frac{1}{x-2y}$ **38.** $\frac{x+5}{x+2}$ **39.** $\frac{6m+5}{3m^2}$ **40.** $\frac{k-3}{36k^2+6k+1}$ **41.** $\frac{x^2-6}{2(2x+1)}$ **42.** $\frac{x(9x+1)}{3x+1}$ **43.** $\frac{3-5x}{6x+1}$ **44.** $\frac{11}{3-x}$ or $\frac{-11}{x-3}$ **45.** $\frac{1}{3}$ **46.** $\frac{s^2+t^2}{st(s-t)}$ **47.** $\frac{5a^2+4ab+12b^2}{(a+3b)(a-2b)(a+b)}$ **48.** $\frac{acd+b^2d+bc^2}{bcd}$ **49.** $\left\{\frac{1}{3}\right\}$ **50.** $r=\frac{AR}{R-A}$ or $r=\frac{-AR}{A-R}$ **51.** $\{1,4\}$ **52.** $\left\{-\frac{14}{3}\right\}$ **53. (a)** 8.32 **(b)** 44.9 **54.** $8\frac{4}{7}$ min **55.** \$21.06 **56.** 24

Chapter 7 Test (page 465)

1. $-2, \frac{4}{3}$; $\left\{x \mid x \neq -2, \frac{4}{3}\right\}$ **2.** $\frac{2x-5}{x(3x-1)}$ **3.** $\frac{3(x+3)}{4}$ **4.** $\frac{y+4}{y-5}$ **5.** $\frac{x+5}{x}$ **6.** $t^2(t+3)(t-2)$ **7.** $\frac{7-2t}{6t^2}$ **8.** $\frac{13x+35}{(x-7)(x+7)}$ **9.** $\frac{4}{x+2}$ **10.** $\frac{72}{11}$ **11.** $-\frac{1}{a+b}$ **12.** $\frac{2y^2+x^2}{xy(y-x)}$ **13.** $x=-1$

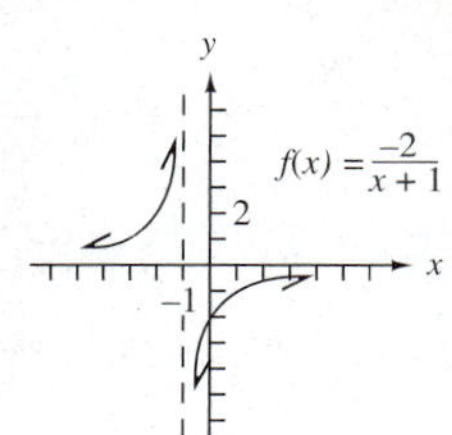

14. (a) operation; $\frac{11(x-6)}{12}$ **(b)** equation; $\{6\}$ **15.** $\left\{\frac{1}{2}\right\}$ **16.** $\{5\}$ **17.** A solution cannot make a denominator 0. **18.** $\ell=\frac{2S}{n}-a$ or $\ell=\frac{2S-na}{n}$ **19.** $3\frac{3}{14}$ hr **20.** 15 mph **21.** 48,000 fish **22. (a)** 3 units **(b)** 0

Cumulative Review Exercises: Chapters 1–7 (page 467)

1. -199 **2.** 12 **3.** $\left\{-\frac{15}{4}\right\}$ **4.** $\left\{\frac{2}{3}, 2\right\}$ **5.** $x=\frac{d-by}{a-c}$ or $x=\frac{by-d}{c-a}$ **6.** $\left(-\infty, \frac{240}{13}\right]$ **7.** $(-\infty,-2]\cup\left[\frac{2}{3},\infty\right)$ **8.** \$4000 at 4%; \$8000 at 3% **9.** 6 m **10.** x-intercept: $(-2, 0)$; y-intercept: $(0, 4)$

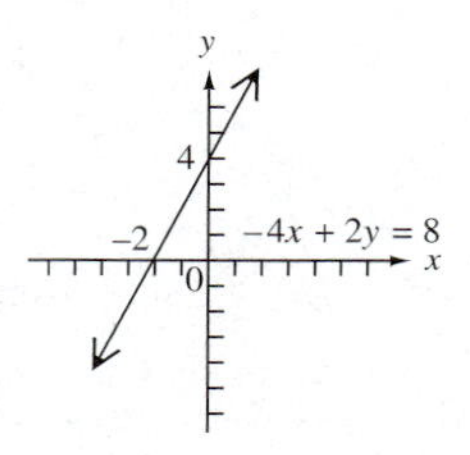

11. $-\frac{3}{2}$ **12.** $-\frac{3}{4}$ **13.** $y=-\frac{3}{2}x+\frac{1}{2}$

14.

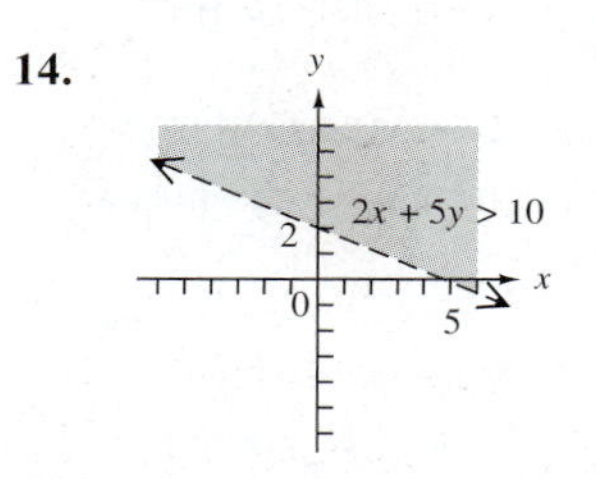

15.

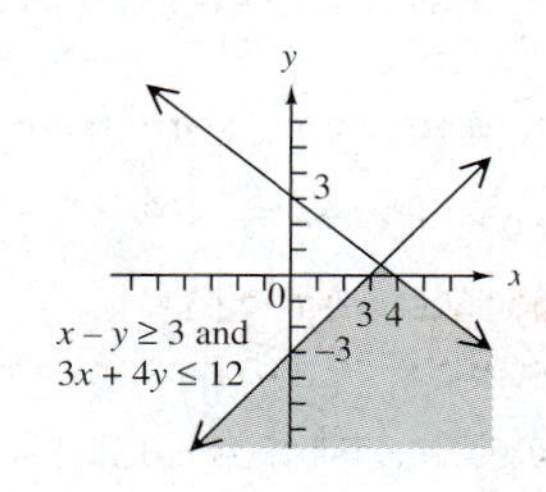

16. function; domain: $\{1990, 1992, 1994, 1996, 1998, 2000\}$; range: $\{1.25, 1.61, 1.80, 1.21, 1.94, 2.26\}$ **17.** not a function; domain: $[-2,\infty)$; range: $(-\infty,\infty)$ **18.** function; domain: $[-2,\infty)$; range: $(-\infty, 0]$ **19. (a)** $f(x)=\frac{5x-8}{3}$ or $f(x)=\frac{5}{3}x-\frac{8}{3}$ **(b)** -1 **20.** $3x+15$ **21.** $\{(-1,3)\}$ **22.** $\{(-2,3,1)\}$ **23.** $\emptyset$ **24.** automobile: 42 km per hr; airplane: 600 km per hr **25.** $\frac{a^{10}}{b^{10}}$ **26.** $\frac{m}{n}$ **27.** $4y^2-7y-6$ **28.** $-6x^6+18x^5-12x^4$ **29.** $12f^2+5f-3$ **30.** $49t^6-64$ **31.** $\frac{1}{16}x^2+\frac{5}{2}x+25$ **32.** x^2+4x-7 **33.** $2x^3+5x^2-3x-2$ **34.** $(2x+5)(x-9)$ **35.** $25(2t^2+1)(2t^2-1)$ **36.** $(2p+5)(4p^2-10p+25)$ **37.** $\left\{-\frac{7}{3}, 1\right\}$ **38.** $\frac{y+4}{y-4}$ **39.** $\frac{2x-3}{2(x-1)}$ **40.** $\frac{a(a-b)}{2(a+b)}$ **41.** 3 **42.** $\frac{2(x+2)}{2x-1}$ **43.** $\{-4\}$ **44.** $q=\frac{fp}{p-f}$ or $q=\frac{-fp}{f-p}$ **45.** 150 mph **46.** $1\frac{1}{5}$ hr

Chapter 8

Section 8.1 (page 477)

1. E **3.** D **5.** A **7.** C **9.** C **11. (a)** not a real number **(b)** negative **(c)** 0 **13.** -9 **15.** 6 **17.** -4 **19.** -8 **21.** 6 **23.** -3 **25.** not a real number **27.** 2 **29.** -9 **31.** $\frac{8}{9}$ **33.** $\frac{2}{3}$ **35.** $\frac{1}{2}$ **37.** $[-3,\infty)$; $[0,\infty)$

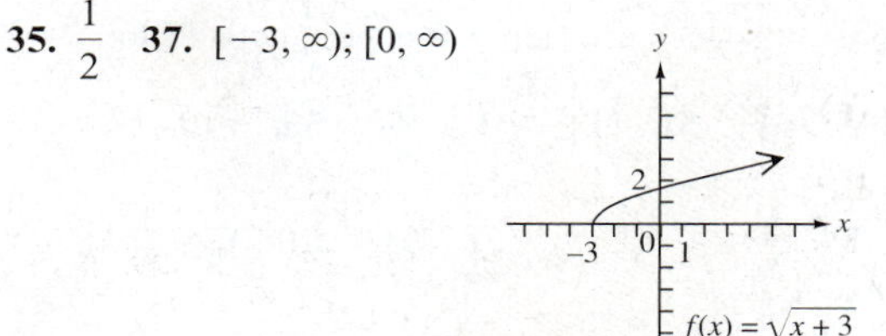

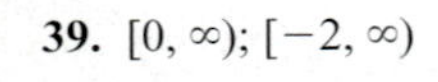

39. $[0, \infty)$; $[-2, \infty)$

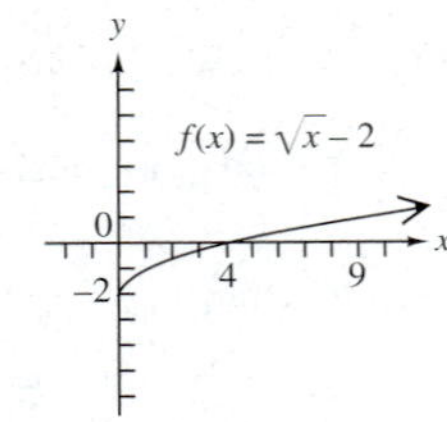

41. $(-\infty, \infty)$; $(-\infty, \infty)$

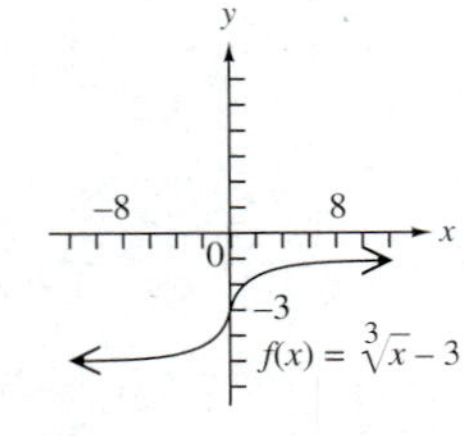

43. $|x|$ **45.** x **47.** x^5 **49.** $|x^3|$ **51.** 97.381 **53.** 16.863 **55.** 2.1 **57.** 1.5 **59.** -4 and 4 **60.** 4 **61.** 4, -4 **62.** $\{-4, 4\}$ **63.** $\pm\sqrt{16}$ represents the two numbers $\sqrt{16} = 4$ and $-\sqrt{16} = 4$. **64.** $\sqrt{x^2}$ is always nonnegative, so it must be simplified as $|x|$ because x may be negative.

Section 8.2 (page 485)

1. 13 **3.** 9 **5.** 2 **7.** $-\frac{8}{9}$ **9.** -3 **11.** not a real number **13.** D **15.** C **17.** A **19.** 100,000 **21.** 256 **23.** $\frac{1}{512}$ **25.** $\frac{2}{5}$ **27.** $\frac{9}{4}$ **29.** not a real number **31.** 13 **33.** 100,000 **35.** 25 **37.** $\sqrt{12}$ **39.** $(\sqrt[4]{8})^3$ **41.** $(\sqrt[8]{9q})^5 - (\sqrt[3]{2x})^2$ **43.** $\frac{1}{(\sqrt{2m})^3}$ **45.** $(\sqrt[3]{2y+x})^2$ **47.** $\frac{1}{(\sqrt[3]{3m^4+2k^2})^2}$ **49.** 64 **51.** 64 **53.** x^{10} **55.** a^6 **57.** 9 **59.** 4 **61.** y **63.** $k^{2/3}$ **65.** $a^{1/2}$ **67.** $\frac{x^{4/5}}{16y^{12}}$ **69.** $9x^8y^{10}$ **71.** $\frac{1}{x^{10/3}}$ **73.** $\frac{1}{m^{1/4}n^{3/4}}$ **75.** $2x - 1 - 2x^{3/2}$ **77.** $\sqrt[6]{x^5}$ **79.** $y\sqrt{7y}$ **81.** $\sqrt[15]{t^8}$ **83.** $\sqrt[8]{m}$ **85.** $x^{-1/2}$ **86.** $m^{5/2}$ **87.** $k^{-3/4}$ **88.** $x^{-1/2}(3 - 4x)$ **89.** $m^{5/2}(m^{1/2} - 3)$ **90.** $k^{-3/4}(9 + 2k^{1/2})$ **91.** 4.5 hr

Section 8.3 (page 495)

1. True; both are equal to $4\sqrt{3}$ and approximately 6.92820323. **3.** True; both are equal to $6\sqrt{2}$ and approximately 8.485281374. **5.** Because there are only two factors of $\sqrt[3]{x}$, $\sqrt[3]{x} \cdot \sqrt[3]{x} = (\sqrt[3]{x})^2$ or $\sqrt[3]{x^2}$. **7.** $\sqrt{30}$ **9.** $\sqrt[3]{14xy}$ **11.** $\sqrt[4]{36}$ **13.** $\frac{8}{11}$ **15.** $\frac{\sqrt{3}}{5}$ **17.** $\frac{\sqrt{x}}{5}$ **19.** $\frac{p^3}{9}$ **21.** $\frac{3}{4}$ **23.** $-\frac{\sqrt[3]{r^2}}{2}$ **25.** $2\sqrt{3}$ **27.** $12\sqrt{2}$ **29.** $-4\sqrt{2}$ **31.** $-2\sqrt{7}$ **33.** not a real number **35.** $4\sqrt[3]{2}$ **37.** $-2\sqrt[3]{2}$ **39.** $2\sqrt[3]{5}$ **41.** $-4\sqrt[4]{2}$ **43.** $2\sqrt[5]{2}$ **45.** His reasoning was incorrect. Here 8 is a term, not a factor. **47.** $6k\sqrt{2}$ **49.** $\frac{3\sqrt[3]{3}}{4}$ **51.** $11x^3$ **53.** $-3t^4$ **55.** $-10m^4z^2$ **57.** $5a^2b^3c^4$ **59.** $\frac{1}{2}r^2t^5$ **61.** $5x\sqrt{2x}$ **63.** $-10r^5\sqrt{5r}$ **65.** $x^3y^4\sqrt{13x}$ **67.** $2z^2w^3$ **69.** $-2zt^2\sqrt[3]{2z^2t}$ **71.** $3x^3y^4$ **73.** $-3r^3s^2\sqrt[4]{2r^3s^2}$ **75.** $\frac{y^5\sqrt{y}}{6}$ **77.** $\frac{x^5\sqrt[3]{x}}{3}$ **79.** $4\sqrt{3}$ **81.** $x^2\sqrt{x}$ **83.** $\sqrt[6]{432}$ **85.** $\sqrt[12]{6912}$ **87.** 5 **89.** $8\sqrt{2}$ **91.** $\sqrt{37}$ **93.** $2\sqrt{10}$ **95.** $6\sqrt{2}$ **97.** $\sqrt{5y^2 - 2xy + x^2}$ **99.** 27.0 in.

Section 8.4 (page 501)

1. B **3.** 15; each radicand is a whole number power corresponding to the index of the radical. **5.** -4 **7.** $7\sqrt{3}$ **9.** $24\sqrt{2}$ **11.** 0 **13.** $20\sqrt{5}$ **15.** $12\sqrt{2x}$ **17.** $-11m\sqrt{2}$ **19.** $\sqrt[3]{2}$ **21.** $2\sqrt[3]{x}$ **23.** $19\sqrt[4]{2}$ **25.** $x\sqrt[4]{xy}$ **27.** $(4 + 3xy)\sqrt[3]{xy^2}$ **29.** $\frac{7\sqrt{2}}{6}$ **31.** $\frac{5\sqrt{2}}{3}$ **33.** Both are approximately 11.3137085. **35.** Both are approximately 31.6227766. **37.** A; 42 m **39.** $12\sqrt{5} + 5\sqrt{3}$ in. **41.** $24\sqrt{2} + 12\sqrt{3}$ in.

Section 8.5 (page 509)

1. E **3.** A **5.** D **7.** $6 - 4\sqrt{3}$ **9.** $6 - \sqrt{6}$ **11.** 2 **13.** 9 **15.** $3\sqrt{2} - 5\sqrt{3} + 2\sqrt{6} - 10$ **17.** $3x - 4$ **19.** $4x - y$ **21.** $16x + 24\sqrt{x} + 9$ **23.** $81 - \sqrt[3]{4}$ **25.** $6 - 4\sqrt{3}$ is not equal to $2\sqrt{3}$ because 6 and $4\sqrt{3}$ are not like terms, so they cannot be combined. **27.** $\sqrt{7}$ **29.** $5\sqrt{3}$ **31.** $\frac{\sqrt{6}}{2}$ **33.** $\frac{9\sqrt{15}}{5}$ **35.** $-\sqrt{2}$ **37.** $\frac{-8\sqrt{3k}}{k}$ **39.** $\frac{6\sqrt{3}}{y}$ **41.** Both methods lead to the same result, $\frac{6\sqrt{3}}{y}$, but multiplying the numerator and denominator by $\sqrt{y}$ produces this result more directly, with less simplification required. **43.** $\frac{\sqrt{14}}{2}$ **45.** $-\frac{\sqrt{14}}{10}$ **47.** $\frac{2\sqrt{6x}}{x}$ **49.** $-\frac{7r\sqrt{2rs}}{s}$ **51.** $\frac{12x^3\sqrt{2xy}}{y^5}$ **53.** $\frac{\sqrt[3]{18}}{3}$ **55.** $\frac{\sqrt[3]{12}}{3}$ **57.** $-\frac{\sqrt[3]{2pr}}{r}$ **59.** $\frac{2\sqrt[4]{x^3}}{x}$ **61.** Multiply the numerator and denominator by $4 - \sqrt{3}$, so the denominator becomes $(4 + \sqrt{3})(4 - \sqrt{3}) = 16 - 3 = 13$, a rational number. **63.** $\frac{2(4 - \sqrt{3})}{13}$ **65.** $3(\sqrt{5} - \sqrt{3})$ **67.** $\sqrt{3} + \sqrt{7}$ **69.** $\sqrt{7} - \sqrt{6} - \sqrt{14} + 2\sqrt{3}$ **71.** $\frac{4\sqrt{x}(\sqrt{x} + 2\sqrt{y})}{x - 4y}$ **73.** $\frac{x\sqrt{2} - \sqrt{3xy} - \sqrt{2xy} + y\sqrt{3}}{2x - 3y}$ **75.** Square each side to show that each square is equal to $\frac{2 - \sqrt{3}}{4}$. **77.** $\frac{5 + 2\sqrt{6}}{4}$ **79.** $\frac{4 + 2\sqrt{2}}{3}$ **81.** $\frac{6 + 2\sqrt{6x}}{3}$ **83.** $\frac{319}{6(8\sqrt{5} + 1)}$ **84.** $\frac{9a - b}{(\sqrt{b} - \sqrt{a})(3\sqrt{a} - \sqrt{b})}$ **85.** $\frac{(3\sqrt{a} + \sqrt{b})(\sqrt{b} + \sqrt{a})}{b - a}$ **86.** In Exercise 84, we multiplied the numerator and denominator by the conjugate of the numerator, while in Exercise 85 we multiplied by the conjugate of the denominator.

Section 8.6 (page 517)

1. No; there is no solution. **3.** $\{19\}$ **5.** $\left\{\frac{38}{3}\right\}$ **7.** $\emptyset$ **9.** $\{5\}$

11. $\{1\}$ **13.** $\{9\}$ **15.** You cannot just square each term. The right-hand side should be $(8 - x)^2 = 64 - 16x + x^2$. **17.** $\{4\}$

19. $\{-3, -1\}$ **21.** $\emptyset$ **23.** $\{-1\}$ **25.** $\left\{\frac{1}{2}\right\}$ **27.** $\{5\}$

29. $\{7\}$ **31.** $\emptyset$ **33.** 3 **35.** $\{-13\}$ **37.** $\{14\}$ **39.** $\emptyset$

41. $\{7\}$ **43.** $\{2, 14\}$ **45.** $\left\{\frac{1}{4}, 1\right\}$ **47.** 1985: 0; 1990: 10 million; 1995: 40 million; 2000: 90 million

49. The approximation is reasonably good; 1990.

Section 8.7 (page 527)

1. i **3.** $-i$ **5.** $a + bi$ is a complex number if a and b are real numbers and i is the imaginary unit. Therefore, for every real number a, if $b = 0$, $a = a + 0i$ is a complex number. **7.** $13i$

9. $-12i$ **11.** $i\sqrt{5}$ **13.** $4i\sqrt{3}$ **15.** -15 **17.** -10 **19.** $\sqrt{3}$

21. $5i$ **23.** $-1 + 7i$ **25.** 0 **27.** $7 + 3i$ **29.** -2 **31.** $1 + 13i$

33. $6 + 6i$ **35.** $4 + 2i$ **37.** -81 **39.** -16 **41.** $-10 - 30i$

43. $10 - 5i$ **45.** $-9 + 40i$ **47.** 153 **49. (a)** $a - bi$ **(b)** a^2; b^2 **51.** $1 + i$ **53.** $-1 + 2i$ **55.** $2 + 2i$

57. $-\frac{5}{13} - \frac{12}{13}i$ **59. (a)** $4x + 1$ **(b)** $4 + i$ **60. (a)** $-2x + 3$ **(b)** $-2 + 3i$ **61. (a)** $3x^2 + 5x - 2$ **(b)** $5 + 5i$

62. (a) $-\sqrt{3} + \sqrt{6} + 1 - \sqrt{2}$ **(b)** $\frac{1}{5} - \frac{7}{5}i$

63. Because $i^2 = -1$, two pairs of like terms can be combined in Exercise 61(b). **64.** Because $i^2 = -1$, additional terms can be combined in the numerator and denominator. **65.** $\frac{5}{41} + \frac{4}{41}i$

67. -1 **69.** i **71.** 1 **73.** $-i$ **75.** Since $i^{20} = (i^4)^5 = 1^5 = 1$, the student multiplied by 1, which is justified by the identity property for multiplication. **77.** $\frac{1}{2} + \frac{1}{2}i$

79. $(1 + 5i)^2 - 2(1 + 5i) + 26$ will simplify to 0 when the operations are applied.

Chapter 8 Review Exercises (page 535)

1. 42 **2.** -17 **3.** not a real number **4.** 6 **5.** -2

6. $|x|$ **7.** x **8.** $|x|^5$ or $|x^5|$

9. domain: $[1, \infty)$; range: $[0, \infty)$

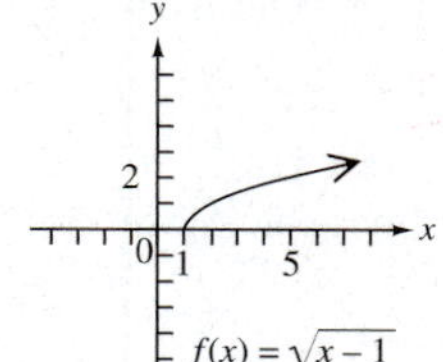

10. domain: $(-\infty, \infty)$; range: $(-\infty, \infty)$

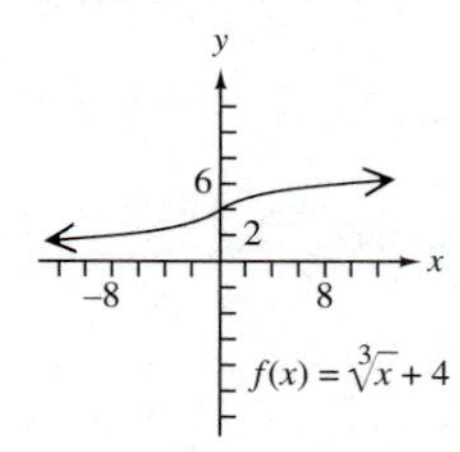

11. n must be even, and a must be negative. **12.** It is not a real number. **13.** 6.325 **14.** 8.775 **15.** 17.607 **16.** 10 mi

17. 1.6 sec **18.** 7 **19.** -2 **20.** not a real number

21. By a power rule for exponents and the definition of $x^{1/n}$, $a^{m/n} = (a^m)^{1/n} = \sqrt[n]{a^m}$. **22.** 32 **23.** -4 **24.** $-\frac{216}{125}$

25. -32 **26.** $\frac{1000}{27}$ **27.** 49 **28.** 96 **29.** $\frac{k^{17/12}}{2}$

30. $\sqrt[5]{2^4}$ or $\sqrt[5]{16}$ **31.** 3^9 **32.** $7^4\sqrt{7}$ **33.** $m^4\sqrt[3]{m}$ **34.** $k^2\sqrt[4]{k}$

35. $\sqrt[6]{m}$ **36.** $2y\sqrt[4]{y}$ **37.** $\sqrt[15]{y^8}$ **38.** $\sqrt[12]{y^5}$ **39.** $\sqrt{66}$

40. $\sqrt{5r}$ **41.** $\sqrt[3]{30}$ **42.** $\sqrt[4]{21}$ **43.** $2\sqrt{5}$ **44.** $-5\sqrt{5}$

45. $-3x\sqrt[3]{4xy}$ **46.** $4pq^2\sqrt[3]{p}$ **47.** $\frac{7}{9}$ **48.** $\frac{y\sqrt{y}}{12}$ **49.** $\frac{m^5}{3}$

50. $\frac{\sqrt[3]{r^2}}{2}$ **51.** $\sqrt[12]{2}$ **52.** $\sqrt[10]{x^3}$ **53.** $\sqrt{130}$ **54.** $\sqrt{53}$

55. $-11\sqrt{2}$ **56.** $23\sqrt{5}$ **57.** $7\sqrt{3y}$ **58.** $26m\sqrt{6m}$

59. $19\sqrt[3]{2}$ **60.** $-8\sqrt[4]{2}$ **61.** $1 - \sqrt{3}$ **62.** 2 **63.** $9 - 7\sqrt{2}$

64. $86 + 8\sqrt{55}$ **65.** $15 - 2\sqrt{26}$ **66.** $12 - 2\sqrt{35}$

67. $-3\sqrt{6}$ **68.** $\frac{3\sqrt{7py}}{y}$ **69.** $-\frac{\sqrt[3]{45}}{5}$ **70.** $\frac{3m\sqrt[3]{4n}}{n^2}$

71. $\frac{\sqrt{2} - \sqrt{7}}{-5}$ **72.** $\frac{-5(\sqrt{6} + \sqrt{3})}{3}$ **73.** $\{2\}$ **74.** $\{6\}$

75. $\emptyset$ **76.** $\{0, 5\}$ **77.** $\{9\}$ **78.** $\{3\}$ **79.** $\{7\}$ **80.** $\left\{-\frac{1}{2}\right\}$

81. $\{6\}$ **82.** $5i$ **83.** $10i\sqrt{2}$ **84.** $4i\sqrt{10}$ **85.** $-10 - 2i$

86. $14 + 7i$ **87.** $-\sqrt{35}$ **88.** -45 **89.** 3 **90.** $5 + i$

91. $32 - 24i$ **92.** $1 - i$ **93.** $4 + i$ **94.** $-i$ **95.** 1 **96.** $-i$

97. $-13ab^2$ **98.** $\frac{1}{100}$ **99.** $\frac{1}{y^{1/2}}$ **100.** $\frac{x^{3/4}}{z^{3/4}}$ **101.** k^6

102. $3z^3t^2\sqrt[3]{2t^2}$ **103.** $57\sqrt{2}$ **104.** $6x\sqrt[3]{y^2}$

105. $\sqrt{35} + \sqrt{15} - \sqrt{21} - 3$ **106.** $-\frac{\sqrt{3}}{6}$ **107.** $\frac{\sqrt[3]{60}}{5}$

108. $\frac{2\sqrt{z}(\sqrt{z} + 2)}{z - 4}$ **109.** $7i$ **110.** $3 - 7i$ **111.** $-5i$

112. $\{5\}$ **113.** $\left\{\frac{3}{2}\right\}$ **114.** 7.9 ft **115. (a)** \$496 million, which agrees closely with the estimate. **(b)** In year 3.9 or late in 1998, which agrees well with the actual year.

Chapter 8 Test (page 539)

1. -29 **2.** 5 **3.** C **4.** 12.09

5. domain: $[-6, \infty)$; range: $[0, \infty)$

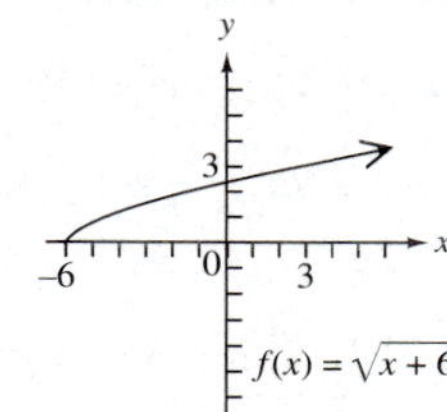

6. $\frac{1}{256}$ **7.** $\frac{9y^{3/10}}{x^2}$ **8.** $3x^2y^3\sqrt{6x}$ **9.** $2ab^3\sqrt[4]{2a^3b}$ **10.** $\sqrt[6]{200}$

11. $26\sqrt{5}$ **12.** $66 + \sqrt{5}$ **13.** $-2(\sqrt{7} - \sqrt{5})$ **14.** $\frac{-\sqrt{10}}{4}$

15. $\frac{2\sqrt[3]{25}}{5}$ **16.** $\sqrt{26}$ **17.** $\sqrt{145}$ **18.** $\{-1\}$ **19.** $\{6\}$

20. $-5 - 8i$ **21.** $3 + 4i$ **22.** $-i$

Cumulative Review Exercises: Chapters 1–8 (page 541)

1. $\left\{\frac{4}{5}\right\}$ **2.** $\left\{\frac{11}{10}, \frac{7}{2}\right\}$ **3.** $(-6, \infty)$ **4.** $(1, 3)$ **5.** $(-2, 1)$

6. $12x + 11y = 18$ **7.** C **8. (a)** $(0, 6)$ **(b)** $(2, 0)$ **9.** \$120

10.

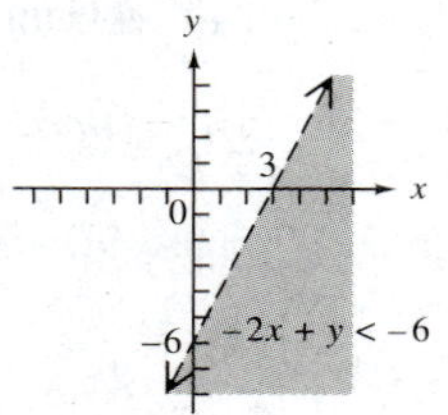

11. Both angles measure 80°. **12.** $\{(7, -2)\}$ **13.** $\emptyset$

14. infinite number of solutions **15.** 2-oz letter: \$.55; 3-oz letter: \$.78 **16.** $-k^3 - 3k^2 - 8k - 9$ **17.** $8x^2 + 17x - 21$ **18.** $z - 2 + \frac{3}{z}$ **19.** $3y^3 - 3y^2 + 4y + 1 + \frac{-10}{2y + 1}$ **20.** $(2p - 3q)(p - q)$ **21.** $(3k^2 + 4)(6k^2 - 5)$ **22.** $(x + 8)(x^2 - 8x + 64)$ **23.** $\frac{y}{y + 5}$ **24.** $\frac{4x + 2y}{(x + y)(x - y)}$ **25.** $-\frac{9}{4}$ **26.** $-\frac{1}{a + b}$ **27.** $\left\{-3, -\frac{5}{2}\right\}$ **28.** $\left\{-\frac{2}{5}, 1\right\}$ **29.** $\frac{1}{243}$ **30.** $x^{1/12}$ **31.** $8\sqrt{5}$ **32.** $\frac{-9\sqrt{5}}{20}$ **33.** $4(\sqrt{6} + \sqrt{5})$ **34.** $6\sqrt[3]{4}$ **35.** $\sqrt{29}$ **36.** $\{6\}$ **37.** 15 mph **38.** $\frac{80}{39}$ or $2\frac{2}{39}$ L **39.** 17 dimes and 12 quarters **40.** Brenda: 8 mph; Chuck: 4 mph

Chapter 9

Section 9.1 (page 551)

1. The equation is also true for $x = -4$. **3. (a)** A quadratic equation in standard form has a second-degree polynomial in decreasing powers equal to 0. **(b)** The zero-factor property states that if a product equals 0, then at least one of the factors equals 0. **(c)** The square root property states that if the square of a quantity equals a number, then the quantity equals the positive or negative square root of the number. **5.** $\{9, -9\}$ **7.** $\{\sqrt{17}, -\sqrt{17}\}$ **9.** $\{4\sqrt{2}, -4\sqrt{2}\}$ **11.** $\{2\sqrt{5}, -2\sqrt{5}\}$ **13.** $\{2\sqrt{6}, -2\sqrt{6}\}$ **15.** $\{-7, 3\}$ **17.** $\{4 + \sqrt{3}, 4 - \sqrt{3}\}$ **19.** $\{-5 + 4\sqrt{3}, -5 - 4\sqrt{3}\}$ **21.** $\left\{\frac{1 + \sqrt{7}}{3}, \frac{1 - \sqrt{7}}{3}\right\}$ **23.** $\left\{\frac{-1 + 2\sqrt{6}}{4}, \frac{-1 - 2\sqrt{6}}{4}\right\}$ **25.** 6.3 sec **27.** square root property for $(2x + 1)^2 = 5$; completing the square for $x^2 + 4x = 12$ **29.** Divide each side by 2. **31.** 4 **33.** 25 **35.** $\frac{1}{36}$ **37.** $\{-4, 6\}$ **39.** $\{-2 + \sqrt{6}, -2 - \sqrt{6}\}$ **41.** $\{-5 + \sqrt{7}, -5 - \sqrt{7}\}$ **43.** $\left\{-\frac{8}{3}, 3\right\}$ **45.** $\left\{\frac{-5 + \sqrt{41}}{4}, \frac{-5 - \sqrt{41}}{4}\right\}$ **47.** $\left\{\frac{5 + \sqrt{15}}{5}, \frac{5 - \sqrt{15}}{5}\right\}$ **49.** $\left\{\frac{4 + \sqrt{3}}{3}, \frac{4 - \sqrt{3}}{3}\right\}$ **51.** $\left\{\frac{2 + \sqrt{3}}{3}, \frac{2 - \sqrt{3}}{3}\right\}$ **53.** $\{1 + \sqrt{2}, 1 - \sqrt{2}\}$ **55.** $\{2i\sqrt{3}, -2i\sqrt{3}\}$ **57.** $\{5 + i\sqrt{3}, 5 - i\sqrt{3}\}$ **59.** $\left\{\frac{1 + 2i\sqrt{2}}{6}, \frac{1 - 2i\sqrt{2}}{6}\right\}$ **61.** $\{-2 + 3i, -2 - 3i\}$ **63.** $\left\{\frac{-2 + 2i\sqrt{2}}{3}, \frac{-2 - 2i\sqrt{2}}{3}\right\}$ **65.** $\{-3 + i\sqrt{3}, -3 - i\sqrt{3}\}$ **67.** x^2 **68.** x **69.** $6x$ **70.** 1 **71.** 9 **72.** $(x + 3)^2$ or $x^2 + 6x + 9$

Section 9.2 (page 561)

1. The student was incorrect, since the fraction bar should extend under the term $-b$. **3.** $\{3, 5\}$ **5.** $\left\{\frac{-2 + \sqrt{2}}{2}, \frac{-2 - \sqrt{2}}{2}\right\}$ **7.** $\left\{\frac{1 + \sqrt{3}}{2}, \frac{1 - \sqrt{3}}{2}\right\}$ **9.** $\{5 + \sqrt{7}, 5 - \sqrt{7}\}$ **11.** $\left\{\frac{-1 + \sqrt{2}}{2}, \frac{-1 - \sqrt{2}}{2}\right\}$ **13.** $\left\{\frac{-1 + \sqrt{7}}{3}, \frac{-1 - \sqrt{7}}{3}\right\}$ **15.** $\{1 + \sqrt{5}, 1 - \sqrt{5}\}$ **17.** $\left\{\frac{-2 + \sqrt{10}}{2}, \frac{-2 - \sqrt{10}}{2}\right\}$ **19.** $\{-1 + 3\sqrt{2}, -1 - 3\sqrt{2}\}$ **21.** $\left\{\frac{3 + i\sqrt{59}}{2}, \frac{3 - i\sqrt{59}}{2}\right\}$ **23.** $\{3 + i\sqrt{5}, 3 - i\sqrt{5}\}$ **25.** $\left\{\frac{1 + i\sqrt{6}}{2}, \frac{1 - i\sqrt{6}}{2}\right\}$ **27.** $\left\{\frac{-2 + i\sqrt{2}}{3}, \frac{-2 - i\sqrt{2}}{3}\right\}$ **29.** B **31.** C **33.** A **35.** D **37.** The equations in Exercises 29, 30, 33, and 34 can be solved by factoring.

Section 9.3 (page 571)

1. Multiply by the LCD, x. **3.** Substitute a variable for $r^2 + r$. **5.** The potential solution -1 does not check. The solution set is $\{4\}$. **7.** $\{-4, 7\}$ **9.** $\left\{-\frac{2}{3}, 1\right\}$ **11.** $\left\{-\frac{14}{17}, 5\right\}$ **13.** $\left\{-\frac{11}{7}, 0\right\}$ **15.** $\left\{\frac{-1 + \sqrt{13}}{2}, \frac{-1 - \sqrt{13}}{2}\right\}$ **17.** $\frac{1}{m}$ job per hr **19.** 25 mph **21.** 80 km per hr **23.** 3.6 hr **25.** 9 min **27.** $\{3\}$ **29.** $\left\{\frac{8}{9}\right\}$ **31.** $\{16\}$ **33.** $\left\{\frac{2}{5}\right\}$ **35.** $\{-3, 3\}$ **37.** $\left\{-\frac{3}{2}, -1, 1, \frac{3}{2}\right\}$ **39.** $\{-2\sqrt{3}, -2, 2, 2\sqrt{3}\}$ **41.** $\{-6, -5\}$ **43.** $\{-4, 1\}$ **45.** $\left\{-\frac{1}{3}, \frac{1}{6}\right\}$ **47.** $\{-8, 1\}$ **49.** $\{-64, 27\}$ **51.** $\{25\}$ **53.** It would cause both denominators to be 0, and division by 0 is undefined. **54.** $\frac{12}{5}$ **55.** $\left(\frac{x}{x - 3}\right)^2 + 3\left(\frac{x}{x - 3}\right) - 4 = 0$ **56.** The numerator can never equal the denominator, since the denominator is 3 less than the numerator. **57.** $\left\{\frac{12}{5}\right\}$; The values for t are -4 and 1. The value 1 is impossible because it leads to a contradiction $\left(\text{since } \frac{x}{x - 3} \text{ is never equal to } 1\right)$. **58.** $\left\{\frac{12}{5}\right\}$; The values for s are $\frac{1}{x}$ and $\frac{-4}{x}$. The value $\frac{1}{x}$ is impossible, since $\frac{1}{x} \neq \frac{1}{x - 3}$ for all x.

Section 9.4 (page 579)

1. Solve for w^2 by dividing each side by g. **3.** $m = \sqrt{p^2 - n^2}$ **5.** $t = \frac{\pm\sqrt{dk}}{k}$ **7.** $d = \frac{\pm\sqrt{skI}}{I}$ **9.** $v = \frac{\pm\sqrt{kAF}}{F}$ **11.** $r = \frac{\pm\sqrt{3\pi Vh}}{\pi h}$ **13.** $t = \frac{-B \pm \sqrt{B^2 - 4AC}}{2A}$ **15.** $h = \frac{D^2}{k}$ **17.** $\ell = \frac{p^2 g}{k}$ **19.** eastbound ship: 80 mi; southbound ship: 150 mi **21.** 2.3, 5.3, 5.8 **23.** 1 ft **25.** 20 in. by 12 in. **27.** 2.4 sec and 5.6 sec **29.** 9.2 sec **31.** It reaches its *maximum* height at 5 sec because this is the only time it reaches 400 ft. **33.** \$.80 **35. (a)** 2.4 million **(b)** 2.4 million; They are the same. **37.** 1995; The graph indicates that sales reached 2 million in 1996. **39.** 5.5 m per sec **41.** 5 or 14

Section 9.5 (page 589)

1. (a) B **(b)** C **(c)** A **(d)** D **3.** (0, 0) **5.** (0, 4) **7.** (1, 0) **9.** (−3, −4) **11.** In Exercise 9, the parabola is shifted 3 units to the left and 4 units down. The parabola in Exercise 10 is shifted 5 units to the right and 8 units down. **13.** down; wider **15.** up; narrower **17. (a)** I **(b)** IV **(c)** II **(d)** III

19.

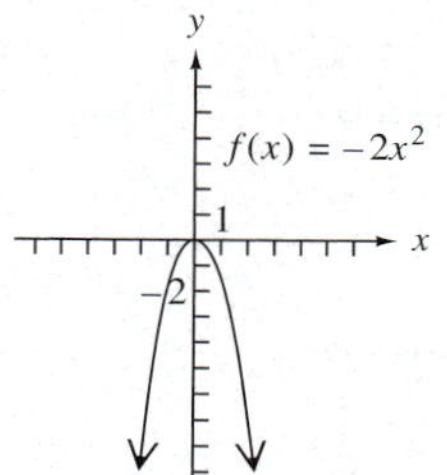

21.

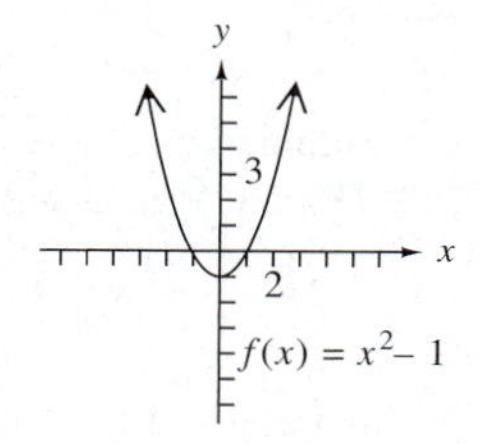

23.

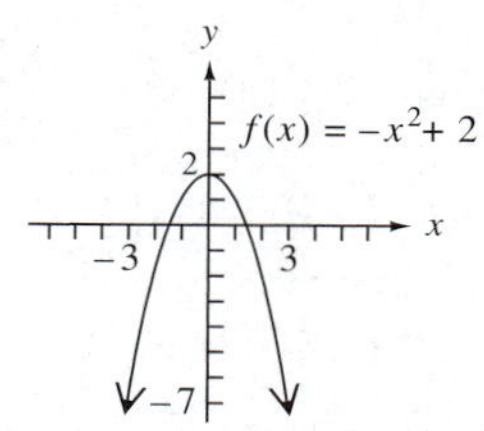

25. axis: $x = 4$; domain: $(-\infty, \infty)$; range: $[0, \infty)$

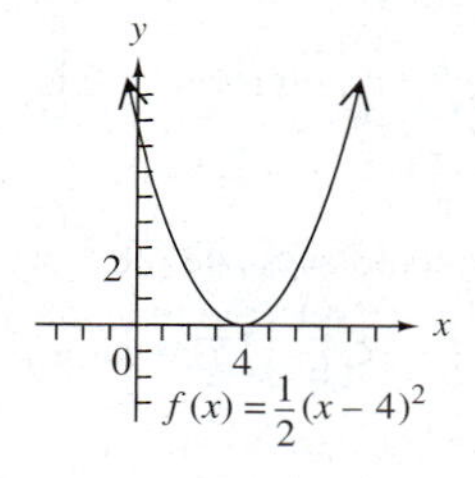

27. axis: $x = -2$; domain: $(-\infty, \infty)$; range: $[-1, \infty)$

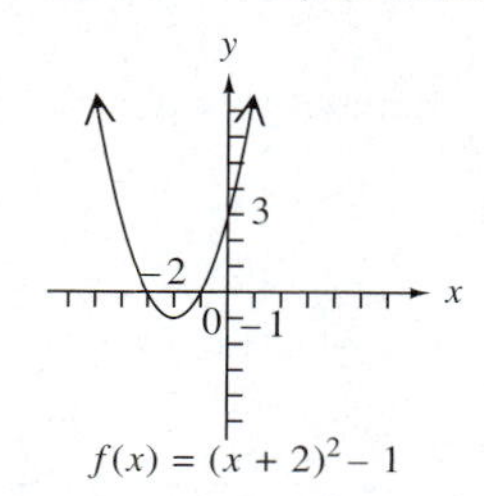

29. axis: $x = -3$; domain: $(-\infty, \infty)$; range: $(-\infty, 4]$

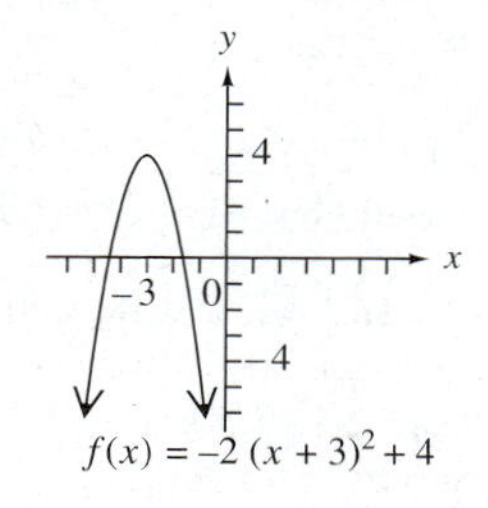

31. axis: $x = -2$; domain: $(-\infty, \infty)$; range: $(-\infty, 1]$

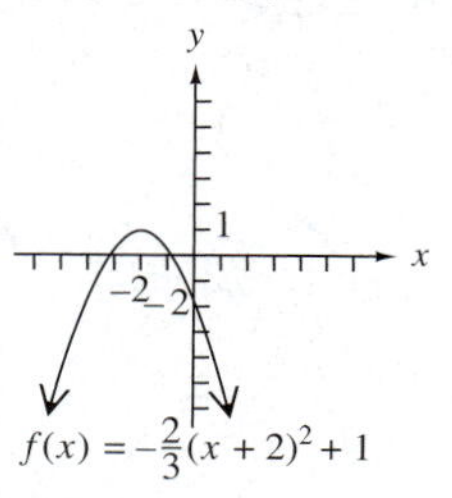

33. It is shifted 6 units up.

34.

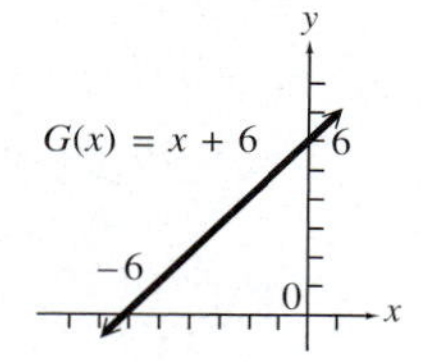

35. It is shifted 6 units up. **36.** It is shifted 6 units to the right.

37.

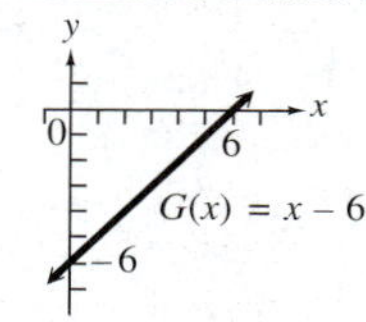

38. It is shifted 6 units to the right. **39.** quadratic; positive **41.** quadratic; negative **43.** linear; positive

45. (a)

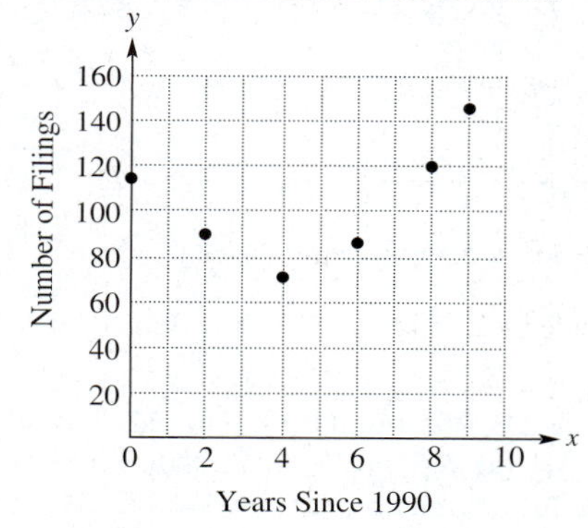

(b) quadratic; positive **(c)** $y = 2.969x^2 - 23.125x + 115$ **(d)** 181 **(e)** Yes. About 15 companies filed for bankruptcy each month, so at this rate, filings for 2000 would be about 180. **47. (a)** 171.2 **(b)** The approximation using the model is low.

Section 9.6 (page 603)

1. If x is squared, it has a vertical axis; if y is squared, it has a horizontal axis. **3.** Use the discriminant of the corresponding quadratic equation. If it is positive, there are two x-intercepts. If it is 0, there is just one x-intercept (the vertex), and if it is negative, there are no x-intercepts. **5.** (−1, 3); up; narrower; no x-intercepts **7.** $\left(\frac{5}{2}, \frac{37}{4}\right)$; down; same; two x-intercepts **9.** (−3, −9); to the right; wider

11. domain: $(-\infty, \infty)$; range: $[-1, \infty)$

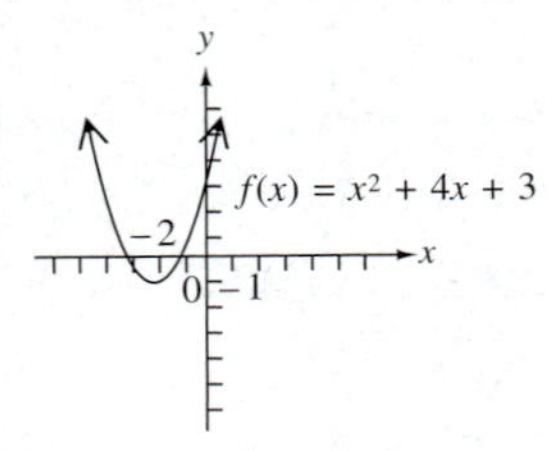

13. domain: $(-\infty, \infty)$; range: $(-\infty, -3]$

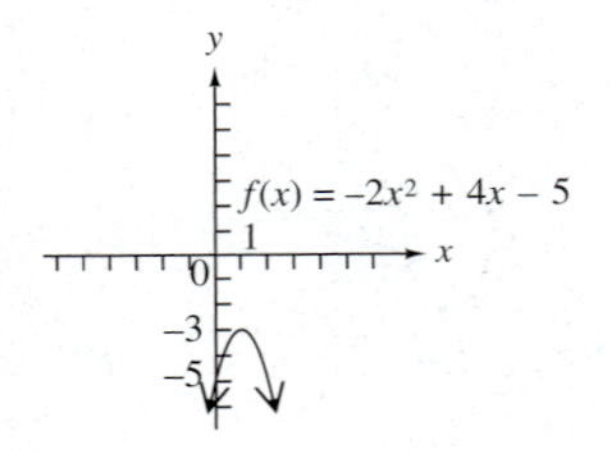

15. domain: $(-\infty, 1]$; range: $(-\infty, \infty)$

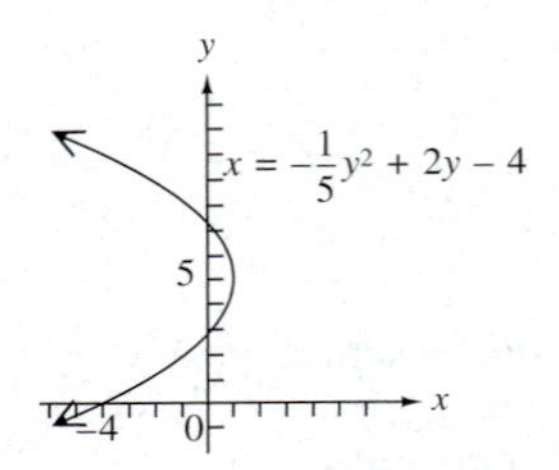

17. domain: $[-7, \infty)$; range: $(-\infty, \infty)$

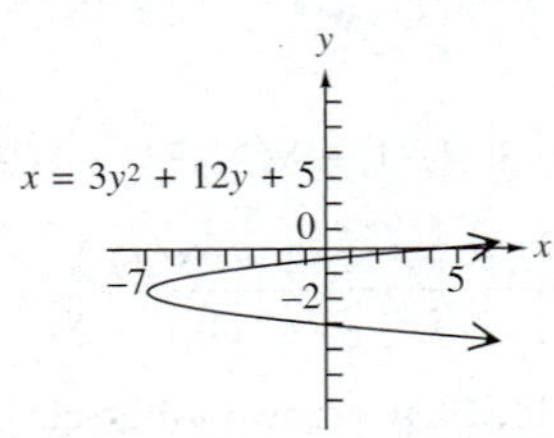

19. F **21.** C **23.** D **25.** 160 ft by 320 ft **27.** 30 and 30 **29.** 16 ft; 2 sec **31. (a)** $R(x) = 20{,}000 + 200x - 4x^2$ **(b)** 25 **(c)** \$22,500 **33. (a)** minimum **(b)** 1995; 1.7% **35. (a)** The coefficient of x^2 is negative because the parabola opens down. **(b)** (18.45, 3860) **(c)** In 2018 Social Security assets will reach their maximum value of \$3860 billion.

Section 9.7 (page 613)

1. (a) $\{1, 3\}$ **(b)** $(-\infty, 1) \cup (3, \infty)$ **(c)** $(1, 3)$

3. (a) $\left\{-3, \frac{5}{2}\right\}$ **(b)** $\left[-3, \frac{5}{2}\right]$ **(c)** $(-\infty, -3] \cup \left[\frac{5}{2}, \infty\right)$

5. Include the endpoints if the symbol is $\geq$ or $\leq$. Exclude the endpoints if the symbol is $>$ or $<$.

7. $(-\infty, -1) \cup (5, \infty)$

9. $(-4, 6)$

11. $(-\infty, 1] \cup [3, \infty)$

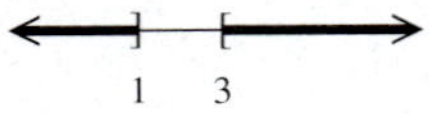

13. $\left(-\infty, -\frac{3}{2}\right] \cup \left[\frac{3}{5}, \infty\right)$

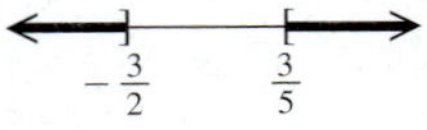

15. $\left(-\frac{2}{3}, \frac{1}{3}\right)$

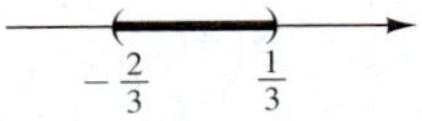

17. $\left(-\infty, -\frac{1}{2}\right] \cup \left[\frac{1}{3}, \infty\right)$

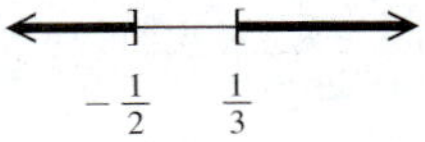

19. $(-\infty, 3 - \sqrt{3}] \cup [3 + \sqrt{3}, \infty)$

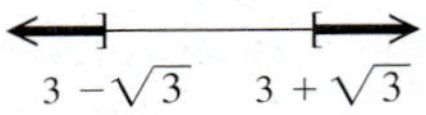

21. $(-\infty, \infty)$ **23.** $\emptyset$

25. $(-\infty, 1) \cup (2, 4)$

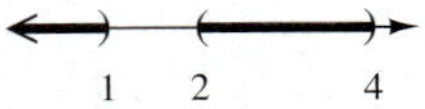

27. $\left[-\frac{3}{2}, \frac{1}{3}\right] \cup [4, \infty)$

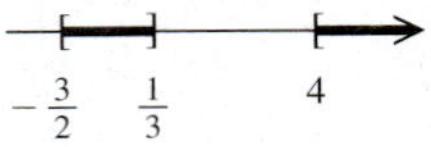

29. $(-\infty, 1) \cup (4, \infty)$

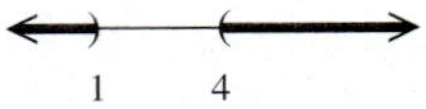

31. $\left[-\frac{3}{2}, 5\right)$

33. $(2, 6]$

35. $\left(-\infty, \frac{1}{2}\right) \cup \left(\frac{5}{4}, \infty\right)$

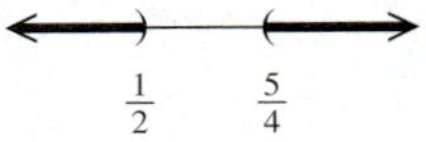

37. $[-4, -2)$

39. $\left(0, \frac{1}{2}\right) \cup \left(\frac{5}{2}, \infty\right)$

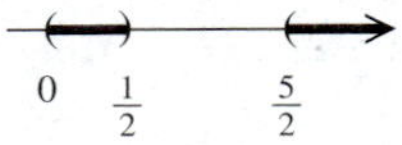

41. 3 sec and 13 sec **42.** between 3 sec and 13 sec **43.** at 0 sec (the time when it is initially projected) and at 16 sec (the time when it hits the ground) **44.** between 0 and 3 sec and between 13 and 16 sec

Chapter 9 Review Exercises (page 621)

1. $\{11, -11\}$ **2.** $\{\sqrt{3}, -\sqrt{3}\}$ **3.** $\left\{-\frac{15}{2}, \frac{5}{2}\right\}$

4. $\left\{\frac{2 + 5i}{3}, \frac{2 - 5i}{3}\right\}$ **5.** $\{-2 + \sqrt{19}, -2 - \sqrt{19}\}$

6. $\left\{\frac{1}{2}, 1\right\}$ **7.** By the square root property, $x = \sqrt{12}$ or $x = -\sqrt{12}$. **8.** 3.1 sec **9.** $\left\{-\frac{7}{2}, 3\right\}$

10. $\left\{\frac{-5 + \sqrt{53}}{2}, \frac{-5 - \sqrt{53}}{2}\right\}$ **11.** $\left\{\frac{1 + \sqrt{41}}{2}, \frac{1 - \sqrt{41}}{2}\right\}$

12. $\left\{\frac{-3 + i\sqrt{23}}{4}, \frac{-3 - i\sqrt{23}}{4}\right\}$ **13.** $\left\{\frac{2 + i\sqrt{2}}{3}, \frac{2 - i\sqrt{2}}{3}\right\}$

14. $\left\{\frac{-7 + \sqrt{37}}{2}, \frac{-7 - \sqrt{37}}{2}\right\}$ **15.** C **16.** A **17.** D

18. B **19.** $\left\{-\frac{5}{2}, 3\right\}$ **20.** $\left\{-\frac{1}{2}, 1\right\}$ **21.** $\{-4\}$

22. $\left\{-\frac{11}{6}, -\frac{19}{12}\right\}$ **23.** $\left\{-\frac{343}{8}, 64\right\}$ **24.** $\{-2, -1, 1, 2\}$

25. 7 mph **26.** 40 mph **27.** 4.6 hr **28.** Greg: 2.6 hr; Carter: 3.6 hr **29.** $v = \frac{\pm\sqrt{rFkw}}{kw}$ **30.** $y = \frac{6p^2}{z}$

31. $t = \frac{3m \pm \sqrt{9m^2 + 24m}}{2m}$ **32.** 9 ft, 12 ft, 15 ft

33. 12 cm by 20 cm **34.** 1 in. **35.** 3 min **36.** 5.2 sec **37.** .7 sec and 4.0 sec **38.** \$.50 **39.** 4.5% **40.** **(a)** 305; It is close to the number shown in the graph. **(b)** $x \approx 9$, which represents 1999; Based on the graph, the number of e-mail boxes did not quite reach 200 million in 1999. **41.** (1, 0) **42.** (3, 7)

43. $\left(\frac{2}{3}, -\frac{2}{3}\right)$ **44.** (−4, 3)

45. domain: $(-\infty, \infty)$; range: $[-3, \infty)$

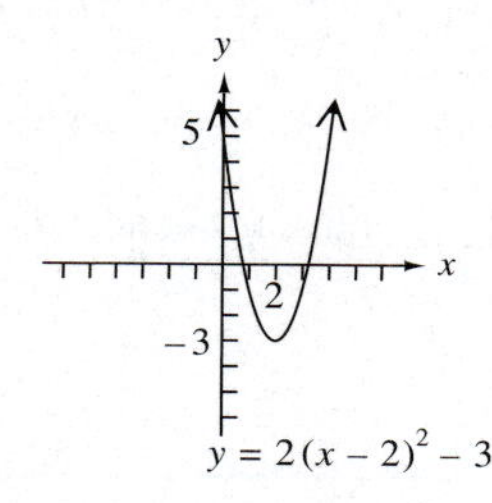

46. domain: $(-\infty, \infty)$; range: $(-\infty, 3]$

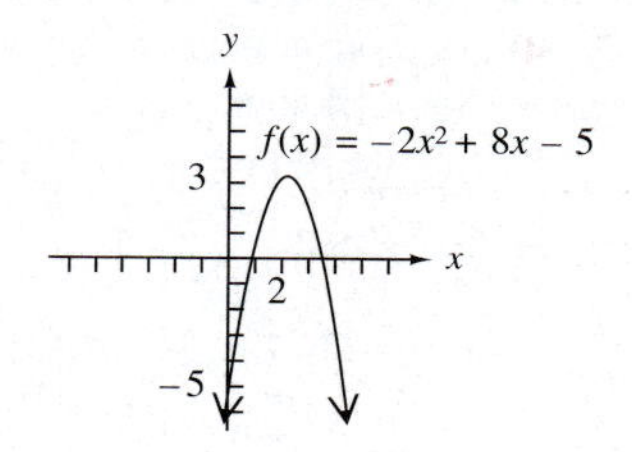

47. domain: $[-4, \infty)$; range: $(-\infty, \infty)$

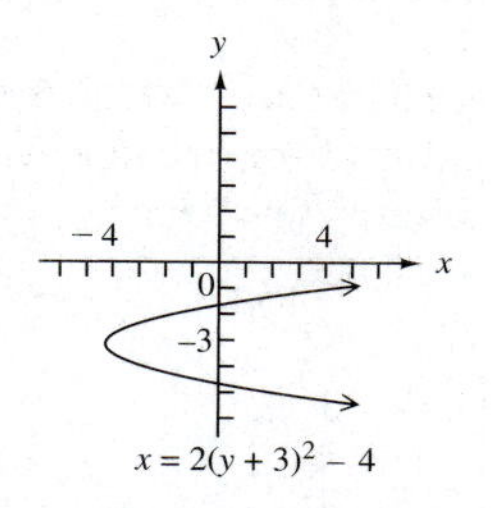

48. domain: $(-\infty, 4]$; range: $(-\infty, \infty)$

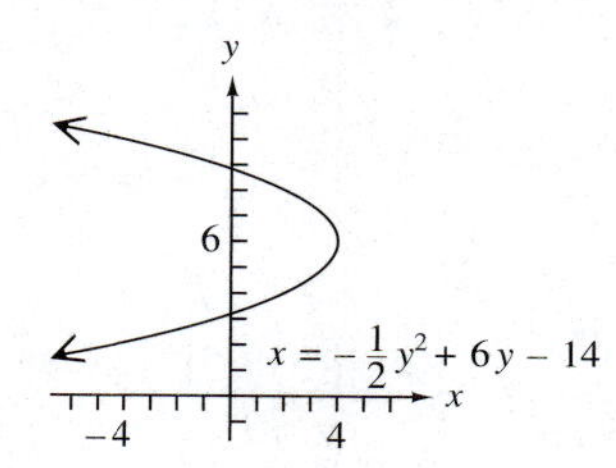

49. **(a)** $c = 12.39$, $16a + 4b + c = 15.78$, $49a + 7b + c = 22.71$ **(b)** $f(x) = .2089x^2 + .0118x + 12.39$ **(c)** \$25.85; The result using the model is a little high. **50.** 5 sec; 400 ft **51.** length: 50 m; width: 50 m

52. $\left(-\infty, -\frac{3}{2}\right) \cup (4, \infty)$

53. $[-4, 3]$

54. $(-\infty, -5] \cup [-2, 3]$

55. $\emptyset$ **56.** $\left(-\infty, \frac{1}{2}\right) \cup (2, \infty)$

57. $[-3, 2)$

58. $R = \frac{\pm\sqrt{Vh - r^2h}}{h}$ **59.** $\left\{\frac{3 + i\sqrt{3}}{3}, \frac{3 - i\sqrt{3}}{3}\right\}$

60. $\{-2, -1, 3, 4\}$ **61.** $(-\infty, -6) \cup \left(-\frac{3}{2}, 1\right)$

62. $\left\{\frac{-11 + \sqrt{7}}{3}, \frac{-11 - \sqrt{7}}{3}\right\}$ **63.** $d = \frac{\pm\sqrt{SkI}}{I}$

64. $\{4\}$ **65.** $\left\{-\frac{5}{3}, -\frac{3}{2}\right\}$ **66.** $\left(-5, -\frac{23}{5}\right]$

67. domain: $(-\infty, \infty)$; range: $[-3, \infty)$

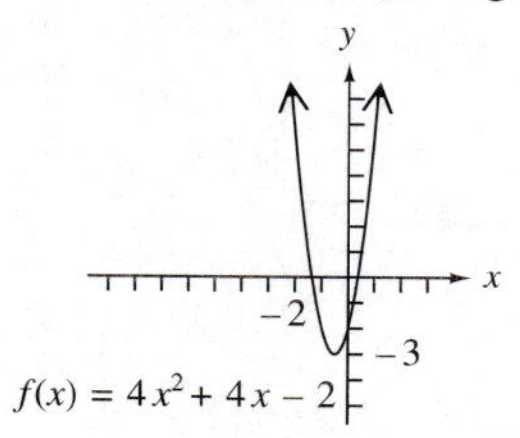

68. **(a)** 21.92 trillion ft³ **(b)** 2005

Chapter 9 Test (page 627)

1. $\{3\sqrt{6}, -3\sqrt{6}\}$ **2.** $\left\{-\frac{8}{7}, \frac{2}{7}\right\}$ **3.** $\{-1 + \sqrt{2}, -1 - \sqrt{2}\}$

4. $\left\{\frac{3 + \sqrt{17}}{4}, \frac{3 - \sqrt{17}}{4}\right\}$ **5.** $\left\{\frac{2 + i\sqrt{11}}{3}, \frac{2 - i\sqrt{11}}{3}\right\}$

6. $\left\{\frac{2}{3}\right\}$ **7.** A **8.** discriminant: 88; There are two irrational solutions. **9.** $\left\{-\frac{2}{3}, 6\right\}$ **10.** $\left\{\frac{-7 + \sqrt{97}}{8}, \frac{-7 - \sqrt{97}}{8}\right\}$

11. $\left\{-2, -\frac{1}{3}, \frac{1}{3}, 2\right\}$ **12.** $\left\{-\frac{5}{2}, 1\right\}$ **13.** $r = \frac{\pm\sqrt{\pi S}}{2\pi}$

14. Maretha: 11.1 hr; Lillaana: 9.1 hr **15.** 7 mph **16.** 2 ft **17.** 16 m **18.** **(a)** 6.5% **(b)** 1996; 3.5% **19.** A **20.** (0, −2)

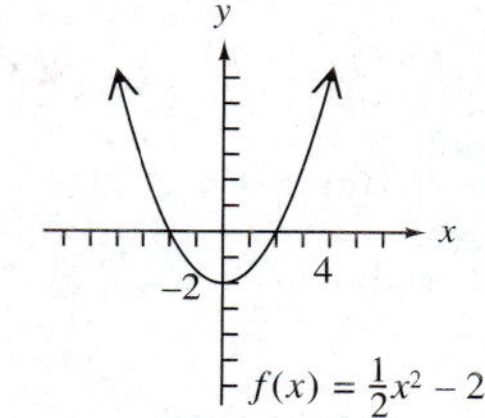

21. vertex: (2, 3); domain: $(-\infty, \infty)$; range: $(-\infty, 3]$

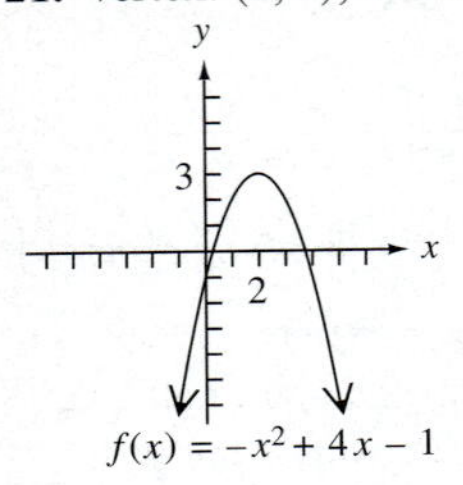

22. vertex: $(-5, -2)$; domain: $[-5, \infty)$; range: $(-\infty, \infty)$

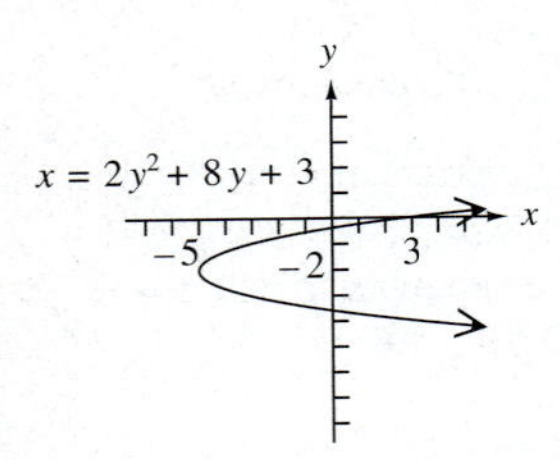

23. 140 ft by 70 ft; 9800 ft^2

24. $(-\infty, -5) \cup \left(\frac{3}{2}, \infty\right)$

25. $(-\infty, 4) \cup [9, \infty)$

Cumulative Review Exercises: Chapters 1–9 (page 631)

1. **(a)** $-2, 0, 7$ **(b)** $-\frac{7}{3}, -2, 0, .7, 7, \frac{32}{3}$ **(c)** all except $\sqrt{-8}$ **(d)** All are complex numbers. **2.** 6 **3.** 41 **4.** 930 **5.** 720 **6.** 990 **7.** 930 **8.** $\{1\}$ **9.** $[1, \infty)$ **10.** $\left[2, \frac{8}{3}\right]$

11. slope: $\frac{1}{2}$; y-intercept: $\left(0, -\frac{7}{4}\right)$ **12.** $x + 3y = -1$

13. function; domain: $(-\infty, \infty)$; range: $(-\infty, \infty)$

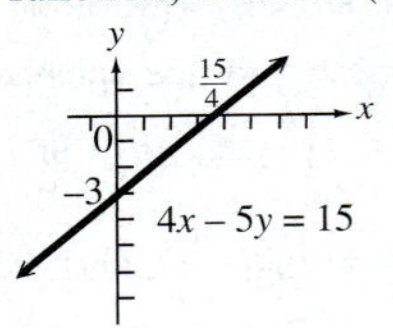

14. not a function

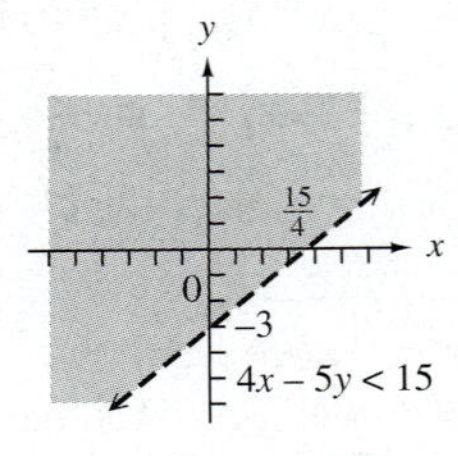

15. function; domain: $(-\infty, \infty)$; range: $(-\infty, 3]$

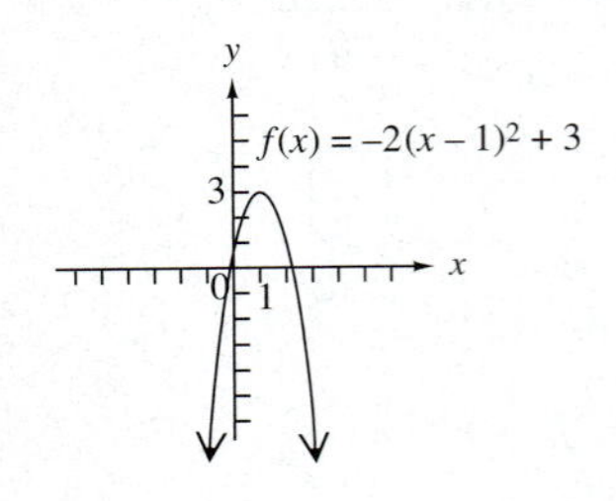

16. **(a)** $y = 1.2795x + 116.26$ **(b)** 158.48; It is a little too high. **17.** No, because the graph is a vertical line, which is not the graph of a function by the vertical line test. **18.** $\{(1, -2)\}$ **19.** $\{(3, -4, 2)\}$ **20.** **(a)** $x + y = 34.2$; $x = 4y - .3$ **(b)** AOL: \$27.3 billion; Time Warner: \$6.9 billion **21.** $\frac{x^8}{y^4}$

22. $\frac{4}{xy^2}$ **23.** $\frac{4}{9}t^2 + 12t + 81$ **24.** $-3t^3 + 5t^2 - 12t + 15$

25. $4x^2 - 6x + 11 + \frac{4}{x+2}$ **26.** $\$3.1 \times 10^{12}$

27. $x(4 + x)(4 - x)$ **28.** $(4m - 3)(6m + 5)$ **29.** $(3x - 5y)^2$

30. $-\frac{5}{18}$ **31.** $-\frac{8}{k}$ **32.** $\frac{r-s}{r}$ **33.** $\frac{3\sqrt[3]{4}}{4}$ **34.** $\sqrt{7} + \sqrt{5}$

35. $\left\{\frac{2}{3}\right\}$ **36.** $\left\{\frac{2+\sqrt{10}}{2}, \frac{2-\sqrt{10}}{2}\right\}$ **37.** $\{-3, 5\}$ **38.** $\emptyset$

39. $\{-3, -1, 1, 3\}$ **40.** southbound car: 57 mi; eastbound car: 76 mi

Chapter 10

Section 10.1 (page 641)

1. It is not one-to-one because both Illinois and Wisconsin are paired with the same range element, 40. **3.** Two or more siblings might be in the class. They would be paired with the same mother. **5.** B **7.** A **9.** $\{(6, 3), (10, 2), (12, 5)\}$ **11.** not one-to-one

13. $f^{-1}(x) = \frac{x-4}{2}$ **15.** $g^{-1}(x) = x^2 + 3, x \geq 0$

17. not one-to-one **19.** $f^{-1}(x) = \sqrt[3]{x+4}$ **21.** **(a)** 8 **(b)** 3 **23.** **(a)** 1 **(b)** 0

25. **(a)** one-to-one **27.** **(a)** not one-to-one

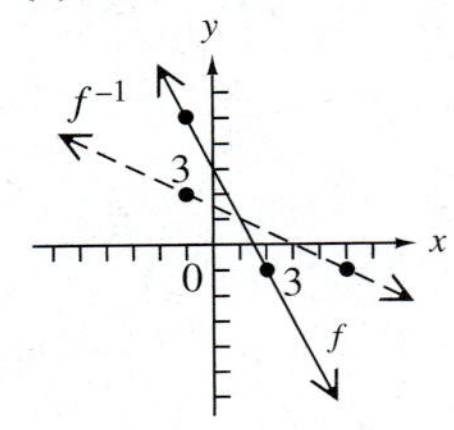

29. **(a)** one-to-one

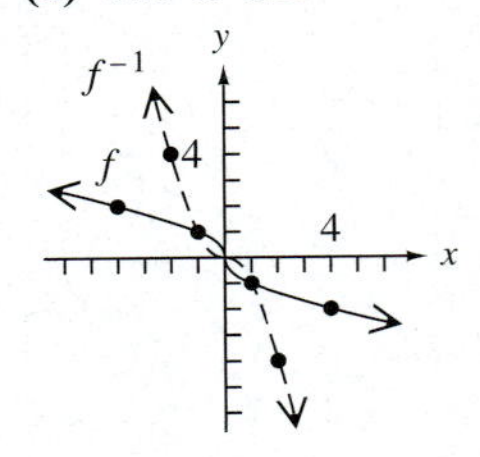

31.

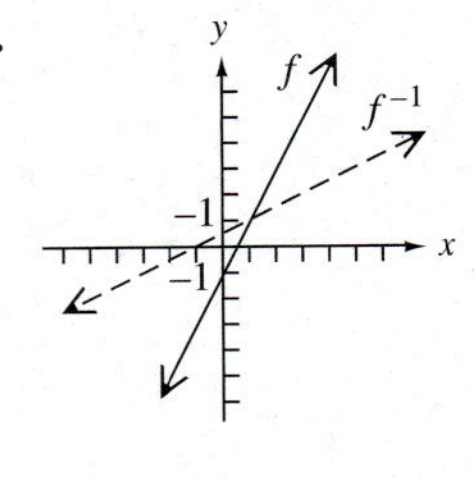

33.

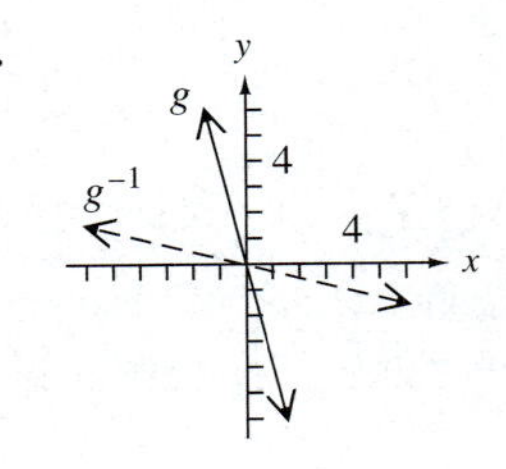

35.

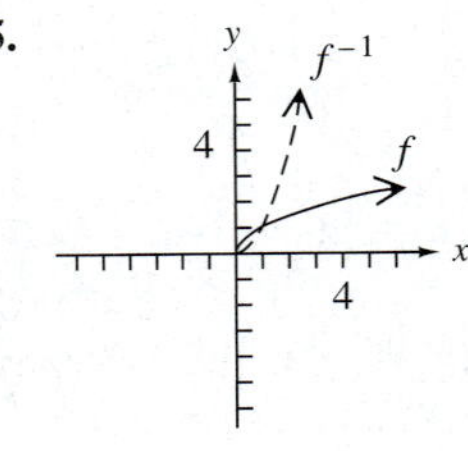

37.

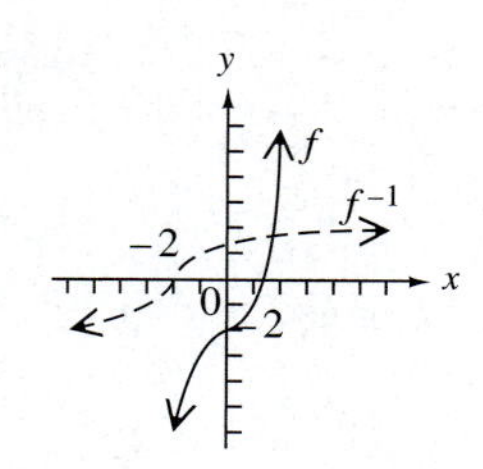

Section 10.2 (page 651)

1. C **3.** A

5.

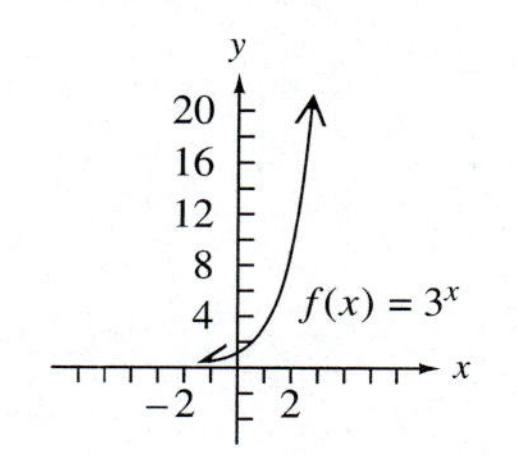

7.

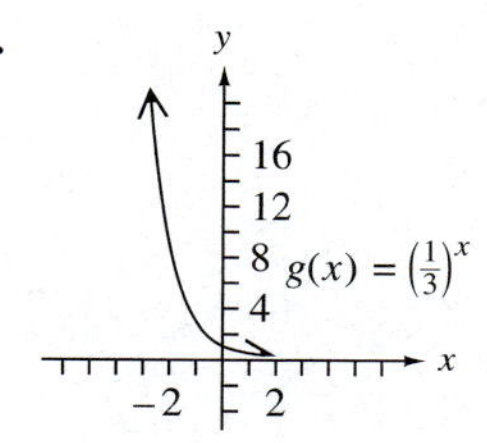

9.

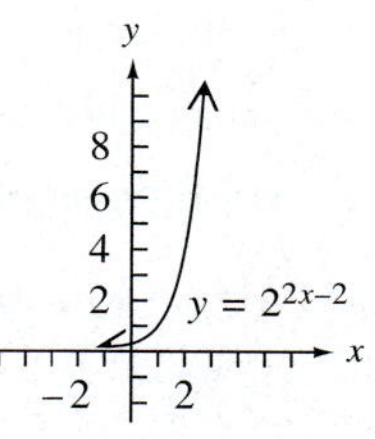

11. $\{2\}$ **13.** $\left\{\frac{3}{2}\right\}$ **15.** $\{7\}$ **17.** $\{-3\}$ **19.** $\{-1\}$ **21.** $\{-3\}$ **23. (a)** rises; falls **(b)** It is one-to-one and thus has an inverse. **25. (a)** 1.0°C **(b)** .4°C **27. (a)** 3.0°C **(b)** .7°C

Section 10.3 (page 659)

1. (a) C **(b)** F **(c)** B **(d)** A **(e)** E **(f)** D **3.** $\log_4 1024 = 5$ **5.** $\log_{1/2} 8 = -3$ **7.** $\log_{10} .001 = -3$ **9.** $\log_{625} 5 = \frac{1}{4}$ **11.** $4^3 = 64$ **13.** $10^{-4} = \frac{1}{10{,}000}$ **15.** $6^0 = 1$ **17.** $9^{1/2} = 3$ **19.** Since the radical $\sqrt{9} = 9^{1/2} = 3$, the exponent to which 9 must be raised is 1/2. **21.** $\left\{\frac{1}{3}\right\}$ **23.** $\{81\}$ **25.** $\left\{\frac{1}{5}\right\}$ **27.** $\{1\}$ **29.** $\{x \mid x > 0, x \neq 1\}$ **31.** $\{5\}$ **33.** $\left\{\frac{5}{3}\right\}$ **35.** $\{4\}$ **37.** $\left\{\frac{3}{2}\right\}$

39.

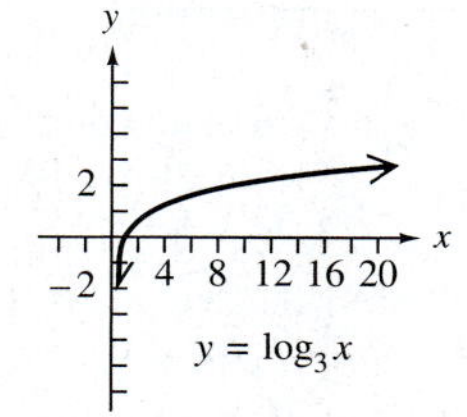

41.

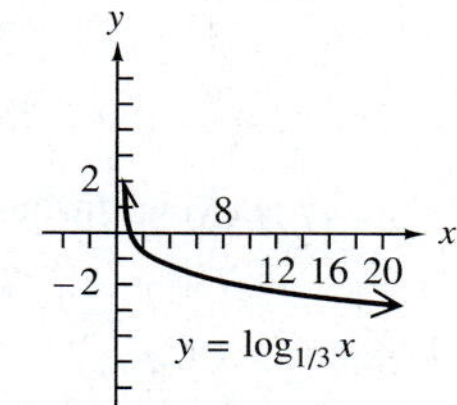

43. Answers will vary. **45.** 8 **47.** 24 **49.** Since every real number power of 1 equals 1, if $y = \log_1 x$, then $x = 1^y$ and so $x = 1$ for every y. This contradicts the definition of a function. **51.** $f(x) = 3^x$ **53. (a)** 645 sites **(b)** 962 sites **(c)** 1218 sites **55.** about 4 times as powerful

Section 10.4 (page 669)

1. false; $\log_b x + \log_b y = \log_b xy$ **3.** true **5.** $\log_7 4 - \log_7 5$ **7.** $\frac{1}{4}\log_2 8$ or $\frac{3}{4}$ **9.** $\log_4 3 + \frac{1}{2}\log_4 x - \log_4 y$ **11.** $\frac{1}{3}\log_3 4 - 2\log_3 x - \log_3 y$ **13.** $\frac{1}{2}\log_3 x + \frac{1}{2}\log_3 y - \frac{1}{2}\log_3 5$ **15.** $\frac{1}{3}\log_2 x + \frac{1}{5}\log_2 y - 2\log_2 r$ **17.** The distributive property tells us that the *product* $a(x + y)$ equals the sum $ax + ay$. In the notation $\log_a(x + y)$, the parentheses do not indicate multiplication. They indicate that $x + y$ is the result of raising a to some power. **19.** $\log_b xy$ **21.** $\log_a \frac{m^3}{n}$ **23.** $\log_a \frac{rt^3}{s}$ **25.** $\log_a \frac{125}{81}$ **27.** $\log_{10}(x^2 - 9)$ **29.** $\log_p \frac{x^3 y^{1/2}}{z^{3/2} a^3}$ **31.** For the power rule $\log_b x^r = r \log_b x$ to be true, x must be in the domain of $g(x) = \log_b x$, so $x > 0$. **33.** 4 **34.** It is the exponent to which 3 must be raised in order to obtain 81. **35.** 81 **36.** It is the exponent to which 2 must be raised in order to obtain 19. **37.** 19 **38.** m

Section 10.5 (page 675)

1. C **3.** C **5.** 19.2 **7.** 1.6335 **9.** 2.5164 **11.** −1.4868 **13.** 9.6776 **15.** 2.0592 **17.** −2.8896 **19.** 5.9613 **21.** 4.1506 **23.** 2.3026 **25.** Answers will vary. Suppose the name is Paul Bunyan, with $m = 4$ and $n = 6$. **(a)** $\log_4 6$ is the exponent to which 4 must be raised in order to obtain 6. **(b)** 1.29248125 **(c)** 6 (the value of n) **27.** An error message appears because we cannot find the common logarithm of a negative number. **29.** bog **31.** 5.9 **33.** 1.0×10^{-2} **35.** 2.5×10^{-5} **37. (a)** 35.0 yr **(b)** 14.2 yr **(c)** 9.0 yr **39. (a)** 107 dB **(b)** 100 dB **(c)** 98 dB **41. (a)** 55% **(b)** 90% **43. (a)** 5.6 yr **(b)** $t > 0$ and $\frac{87 - L}{63}$ is positive and in the domain of the function only if $24 < L < 87$.

Section 10.6 (page 687)

1. $\log 5^x = \log 125$ **2.** $x \log 5 = \log 125$ **3.** $x = \frac{\log 125}{\log 5}$ **4.** $\frac{\log 125}{\log 5} = 3$; $\{3\}$ **5.** $\{.827\}$ **7.** $\{.833\}$ **9.** $\{1.201\}$ **11.** $\{2.269\}$ **13.** $\{566.866\}$ **15.** $\{-18.892\}$ **17.** $\{4\}$ **19.** $\{43.301\}$ **21.** Natural logarithms are a better choice because e is the base. **23.** $\left\{\frac{2}{3}\right\}$ **25.** $\left\{\frac{33}{2}\right\}$ **27.** $\{-1 + \sqrt[3]{49}\}$ **29.** 2 cannot be a solution because $\log(2 - 3) = \log(-1)$, and -1 is not in the domain of $\log x$. **31.** $\left\{\frac{1}{3}\right\}$ **33.** $\{2\}$ **35.** $\emptyset$ **37.** $\{8\}$ **39.** $\left\{\frac{4}{3}\right\}$ **41.** $\{8\}$ **43. (a)** \$2539.47 **(b)** 10.2 yr **45.** \$4934.71 **47.** 15.4 yr **49.** about 180 g **51.** 1.4315 **53.** 3.3030 **55.** −.0947 **57.** .325

Chapter 10 Review Exercises (page 695)

1. not one-to-one **2.** one-to-one **3.** This function is not one-to-one because two sodas in the list have 41 mg of caffeine. **4.** $f^{-1}(x) = \frac{x - 7}{-3}$ or $\frac{7 - x}{3}$ **5.** $f^{-1}(x) = \frac{x^3 + 4}{6}$ **6.** not one-to-one

7.

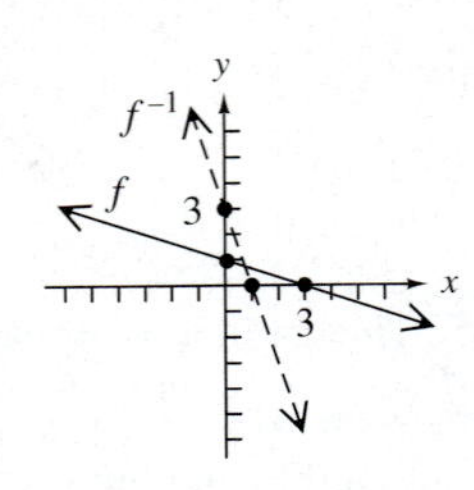

8.

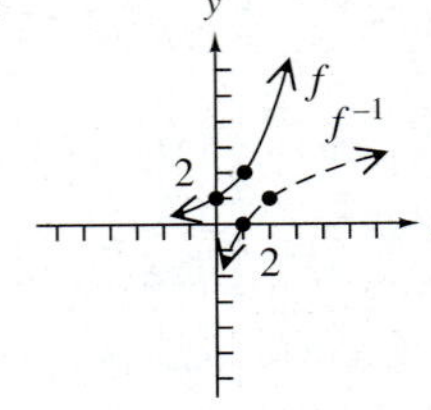

9.

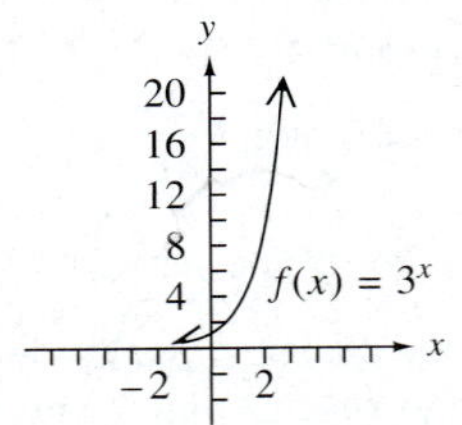

10.

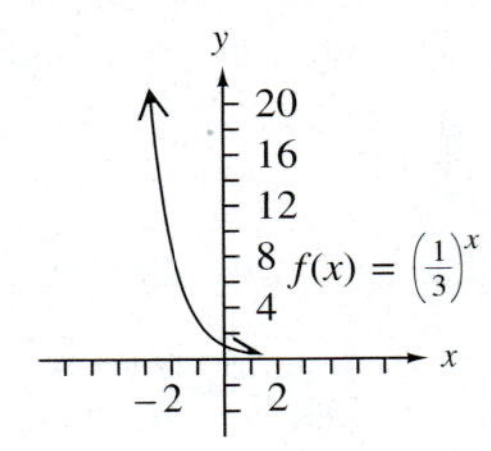

11.

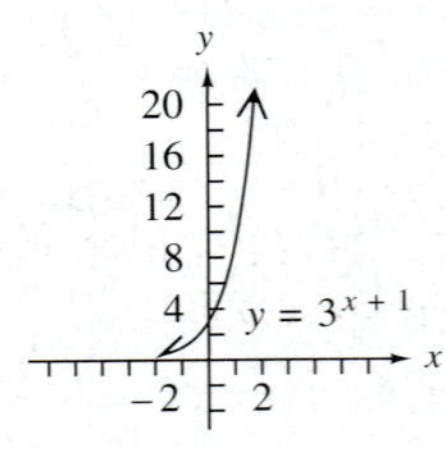

12. $\{4\}$ **13.** $\left\{\frac{3}{7}\right\}$ **14.** $\{0\}$ **15. (a)** 1.2 million tons **(b)** 3.8 million tons **(c)** 21.8 million tons **16. (a)** $5^4 = 625$ **(b)** $\log_5 .04 = -2$ **17. (a)** $\log_b a$ represents the exponent on b that equals a. **(b)** a

18.

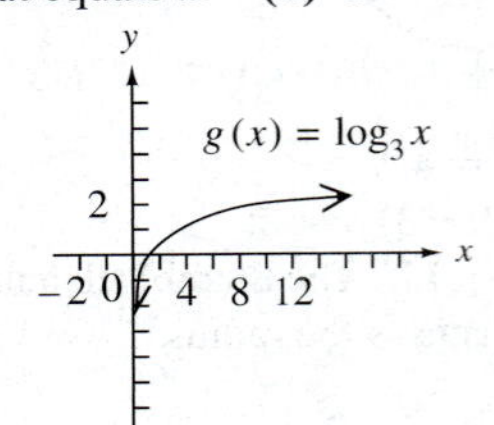

19.

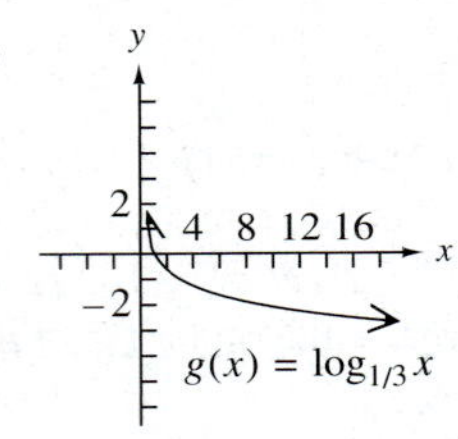

20. $\{2\}$ **21.** $\{-2\}$ **22.** $\{8\}$ **23.** $\{b \mid b > 0, b \neq 1\}$ **24.** $\log_4 3 + 2\log_4 x$ **25.** $2\log_2 p + \log_2 r - \frac{1}{2}\log_2 z$ **26.** $\log_b \frac{3x}{y^2}$ **27.** $\log_3 \frac{x+7}{4x+6}$ **28.** 1.4609 **29.** $-.5901$ **30.** 3.3638 **31.** -1.3587 **32.** 6.4 **33.** 8.4 **34.** 2.5×10^{-5} **35. (a)** 18 yr **(b)** 12 yr **(c)** 7 yr **(d)** 6 yr **(e)** Each comparison shows approximately the same number. For example, in part (a) the doubling time is 18 yr (rounded) and $\frac{72}{4} = 18$. Thus, the formula $t = \frac{72}{100r}$ (called the *rule of 72*) is an excellent approximation of the doubling time formula. (It is used by bankers for that purpose.) **36. (a)** $g(92) = -3.4$; $g(99) = 7.1$ **(b)** $g(92)$ agrees closely with the y-value from the graph of about -3; $g(99)$ also is reasonably close to the y-value from the graph of about 7.5. **37.** $\{2.042\}$ **38.** $\{4.907\}$ **39.** $\{18.310\}$ **40.** $\left\{\frac{1}{9}\right\}$ **41.** $\{-6 + \sqrt[3]{25}\}$ **42.** $\{2\}$ **43.** $\left\{\frac{3}{8}\right\}$ **44.** $\{4\}$ **45.** $\{1\}$ **46.** \$7112.11 **47.** Plan A; it would pay \$2.92 more. **48.** \$4267 **49.** about 11% **50.** .9251 **51.** 1.7925 **52.** 1.4315 **53.** $\{72\}$ **54.** $\{5\}$ **55.** $\left\{\frac{1}{9}\right\}$ **56.** $\left\{\frac{4}{3}\right\}$ **57.** $\{3\}$ **58.** $\left\{\frac{1}{8}\right\}$ **59.** $\left\{\frac{11}{3}\right\}$ **60.** 6.8 yr

Chapter 10 Test (page 699)

1. (a) not one-to-one **(b)** one-to-one **2.** $f^{-1}(x) = x^3 - 7$

3.

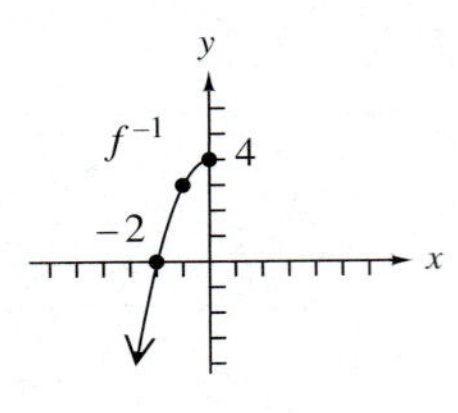

4.

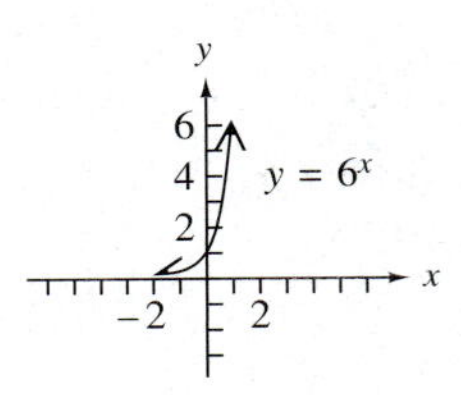

5.

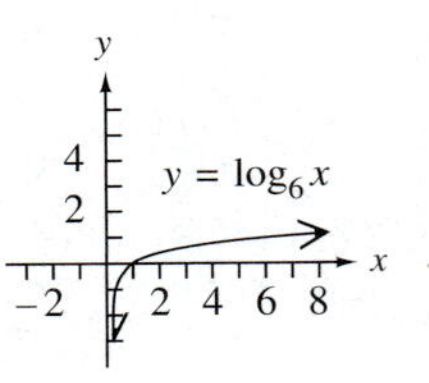

6. Interchange the x- and y-values of the ordered pairs, because the functions are inverses. **7.** $\{-4\}$ **8.** $\left\{-\frac{13}{3}\right\}$ **9.** 30.0 million; 37.7 million **10.** $\log_4 .0625 = -2$ **11.** $7^2 = 49$ **12.** $\{32\}$ **13.** $\left\{\frac{1}{2}\right\}$ **14.** $\{2\}$ **15.** $2\log_3 x + \log_3 y$ **16.** $\log_b \frac{r^{1/4}s^2}{t^{2/3}}$ **17. (a)** 1.3284 **(b)** $-.8440$ **(c)** 2.1245 **18.** $\{3.9656\}$ **19.** $\{3\}$ **20. (a)** \$12,507.51 **(b)** 15.5 yr

Cumulative Review Exercises: Chapters 1–10 (page 701)

1. $-2, 0, 6, \frac{30}{3}$ (or 10) **2.** $-\frac{9}{4}, -2, 0, .6, 6, \frac{30}{3}$ (or 10) **3.** $-\sqrt{2}, \sqrt{11}$ **4.** 16 **5.** -39 **6.** $\left\{-\frac{2}{3}\right\}$ **7.** $[1, \infty)$ **8.** $\{-2, 7\}$ **9.** $\left\{-\frac{16}{3}, \frac{16}{3}\right\}$ **10.** $\left[\frac{7}{3}, 3\right]$ **11.** $(-\infty, -3) \cup (2, \infty)$

12.

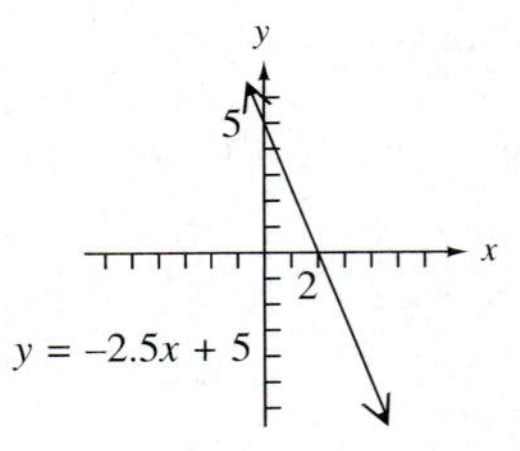

13.

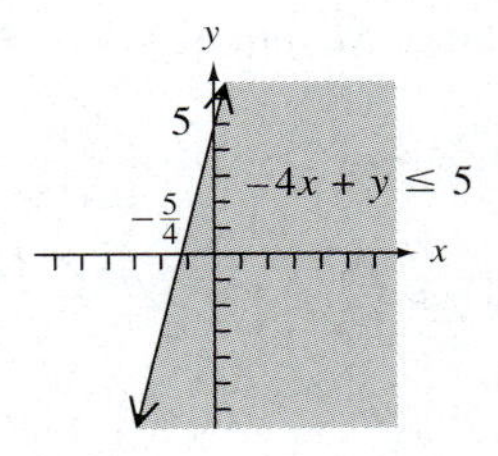

14. (a) yes **(b)** -4000; The number of acres harvested decreased by 4000 acres per year during 1997–1999. **15.** $y = \frac{3}{4}x - \frac{19}{4}$ **16.** $\{(4, 2)\}$ **17.** $\{(1, -1, 4)\}$ **18.** 6 lb of \$1.00 candy and 10 lb of \$1.96 candy **19.** $6p^2 + 7p - 3$ **20.** $16k^2 - 24k + 9$ **21.** $-5m^3 + 2m^2 - 7m + 4$ **22.** $2t^3 + 5t^2 - 3t + 4$ **23.** $x(8 + x^2)$ **24.** $(3y - 2)(8y + 3)$ **25.** $z(5z + 1)(z - 4)$ **26.** $(4a + 5b^2)(4a - 5b^2)$ **27.** $(2c + d)(4c^2 - 2cd + d^2)$ **28.** $(4r + 7q)^2$ **29.** $-\frac{1875p^{13}}{8}$ **30.** $\frac{x+5}{x+4}$ **31.** $\frac{-3k - 19}{(k+3)(k-2)}$ **32.** $\frac{22 - p}{p(p-4)(p+2)}$ **33.** $\left\{-\frac{1}{3}\right\}$ **34.** $12\sqrt{2}$ **35.** $-\frac{1}{4}$ **36.** $-27\sqrt{2}$ **37.** $\{0, 4\}$ **38.** 41

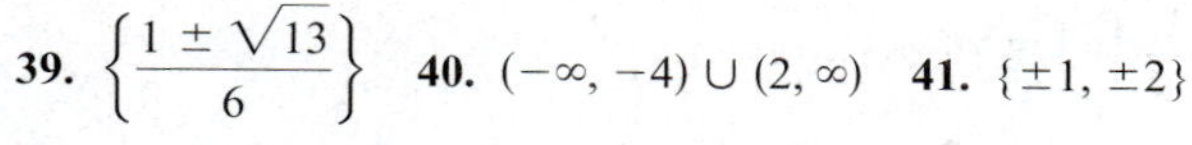
39. $\left\{\frac{1 \pm \sqrt{13}}{6}\right\}$ **40.** $(-\infty, -4) \cup (2, \infty)$ **41.** $\{\pm 1, \pm 2\}$

42.

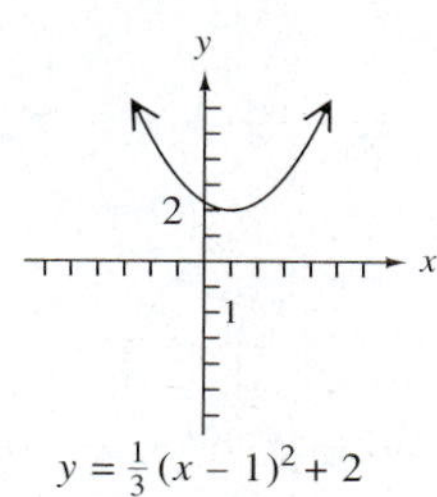

43.

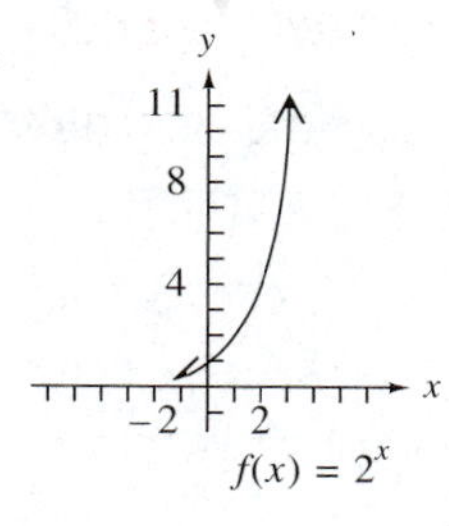

44.

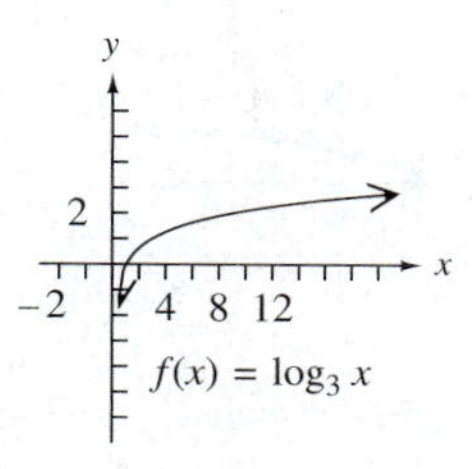

45. $\{-1\}$ **46.** $\{1\}$ **47.** $5^3 = 125$

48. $3 \log x + \frac{1}{2} \log y - \log z$ **49.** .110 or 11%

50. **(a)** 25,000 **(b)** 37,300 **(c)** 68,000

Chapter 11

Section 11.1 (page 711)

1. 0 **3.** 0; 0 **5.** B **7.** A

9. $(-\infty, \infty)$; $[0, \infty)$

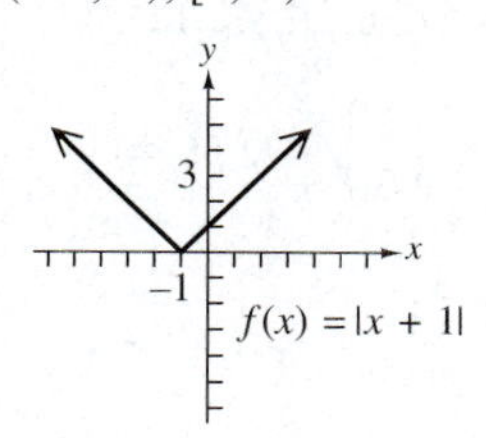

11. $(-\infty, 0) \cup (0, \infty)$; $(-\infty, 1) \cup (1, \infty)$

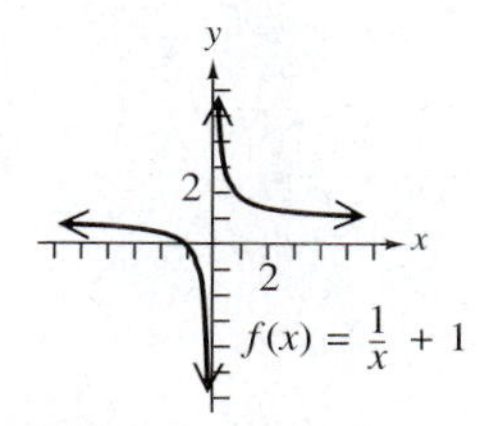

13. $[2, \infty)$; $[0, \infty)$

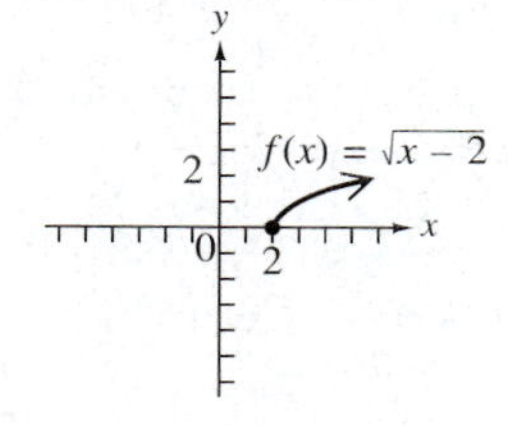

15. $(-\infty, 2) \cup (2, \infty)$; $(-\infty, 0) \cup (0, \infty)$

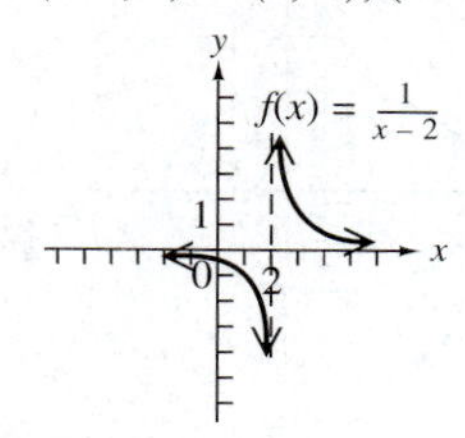

17. $[-3, \infty)$; $[-3, \infty)$

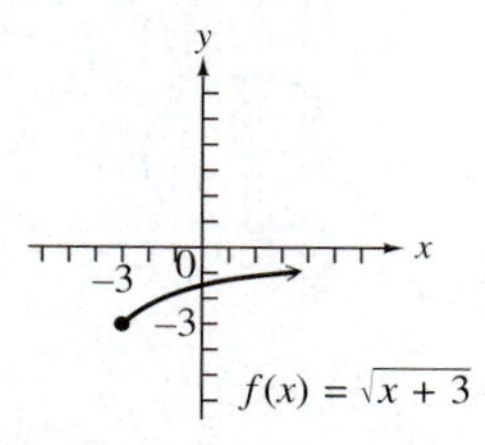

19. 16 **21.** 83 **23.** 13 **25.** $4x^2 + 12x + 13$
27. $x^2 + 10x + 29$ **29.** $2x + 8$ **31.** $(f \circ g)(x) = 63{,}360x$; It computes the number of inches in x mi. **33.** $(A \circ r)(t) = 4\pi t^2$; This is the area of the circular layer as a function of time.

Section 11.2 (page 721)

1. **(a)** $(0, 0)$ **(b)** 5 **(c)**

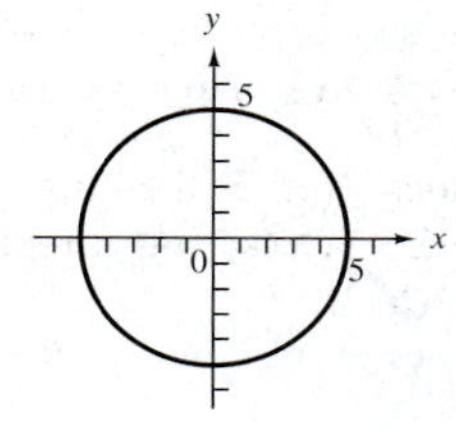

3. B **5.** D **7.** $(x + 4)^2 + (y - 3)^2 = 4$
9. $(x + 8)^2 + (y + 5)^2 = 5$ **11.** $(-2, -3)$; $r = 2$
13. $(-5, 7)$; $r = 9$ **15.** $(2, 4)$; $r = 4$ **17.** The thumbtack acts as the center and the length of the string acts as the radius.

19.

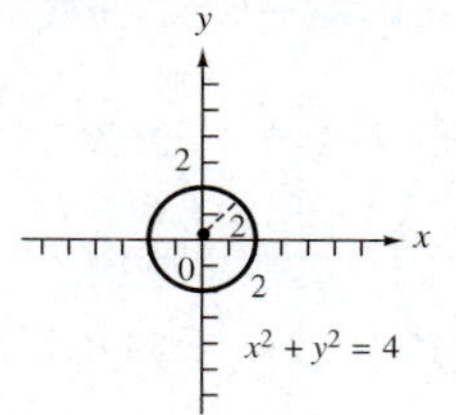

21.

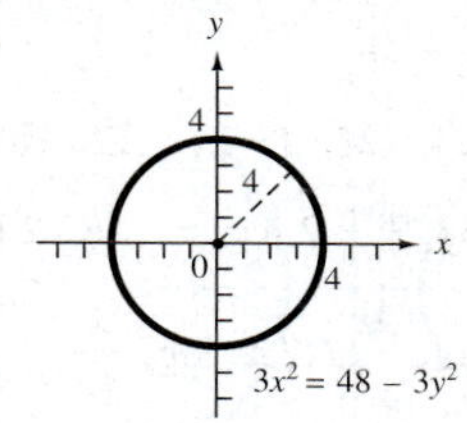

23.

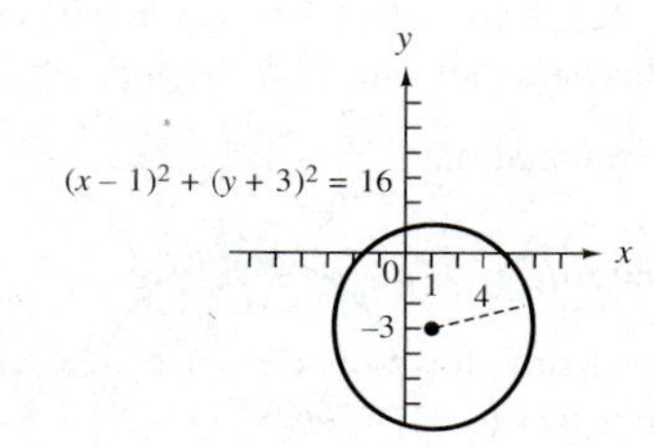

25.

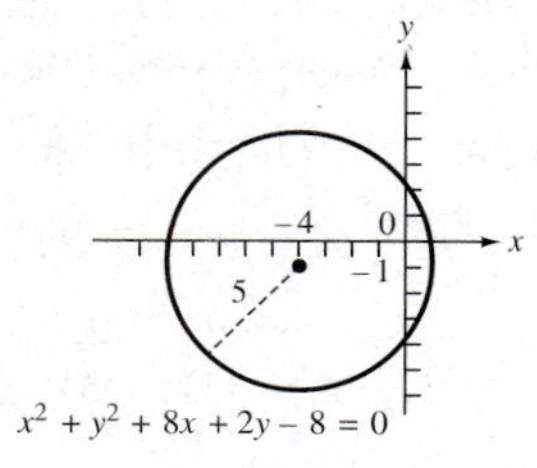

27.

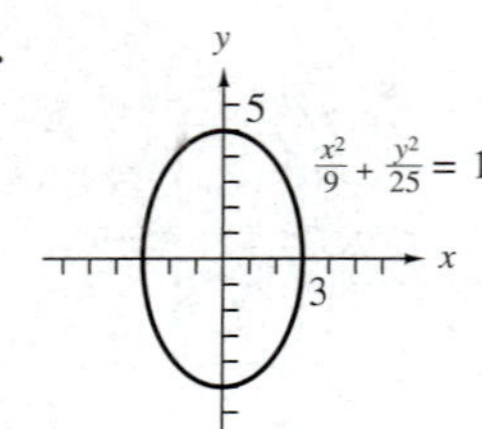

29.

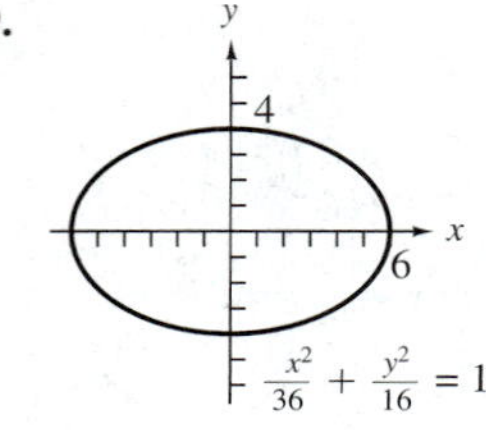

31.

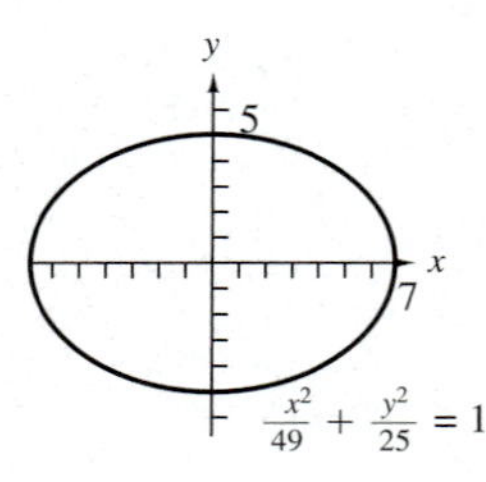

33.

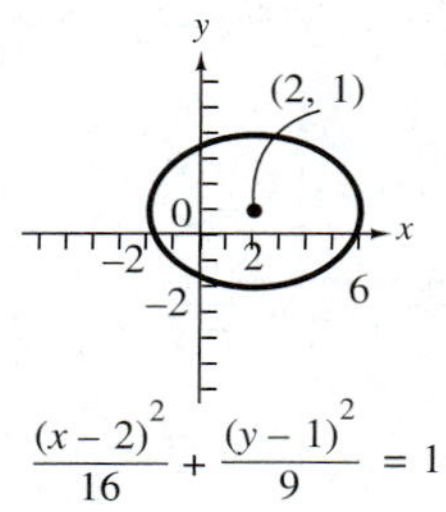

35. **(a)** 10 m **(b)** 36 m **37.** **(a)** 154.7 million mi **(b)** 128.7 million mi (Answers are rounded.)

Section 11.3 (page 731)

1. C **3.** D **5.** When written in one of the forms given in the box titled "Equations of Hyperbolas" in this section, it will open up and down if the $-$ sign precedes the x^2-term; it will open left and right if the $-$ sign precedes the y^2-term.

7.

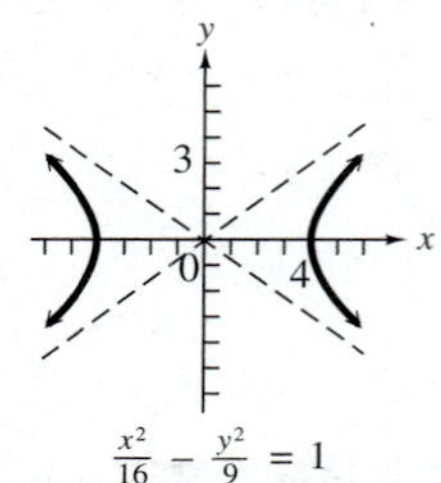

9.

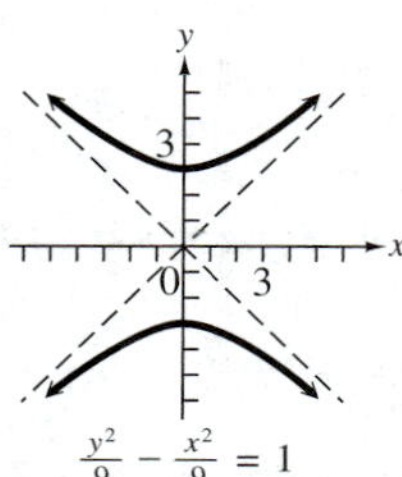

11.

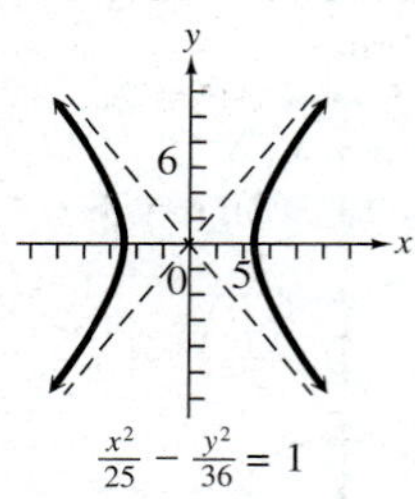

13. hyperbola

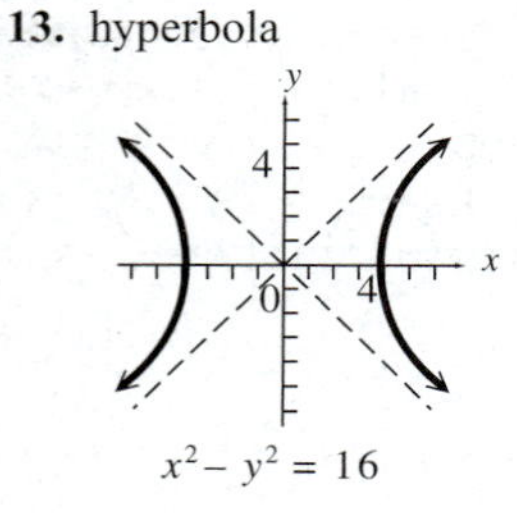

15. ellipse

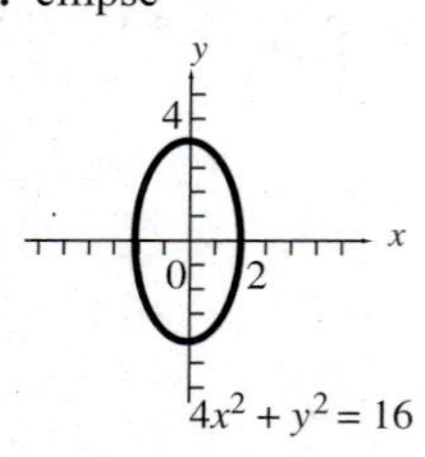

17. circle

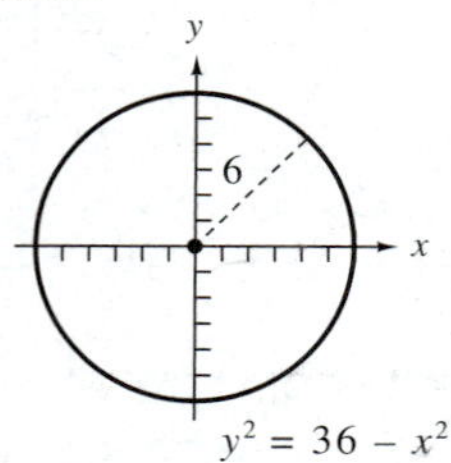

19. hyperbola

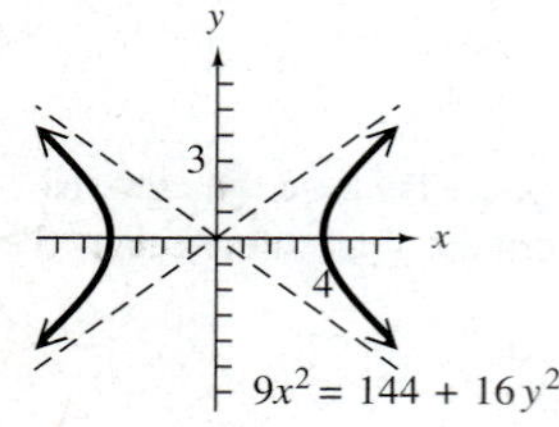

21. ellipse

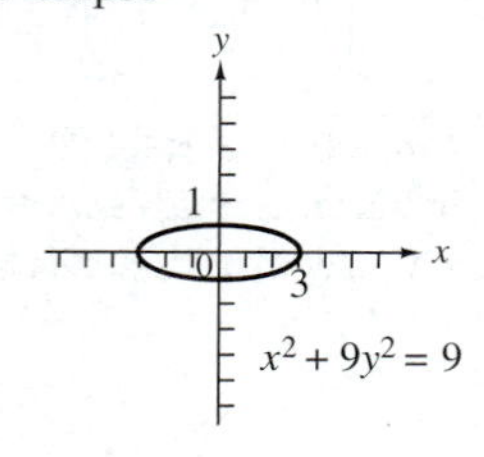

23. $[-3, 3]$; $[0, 3]$

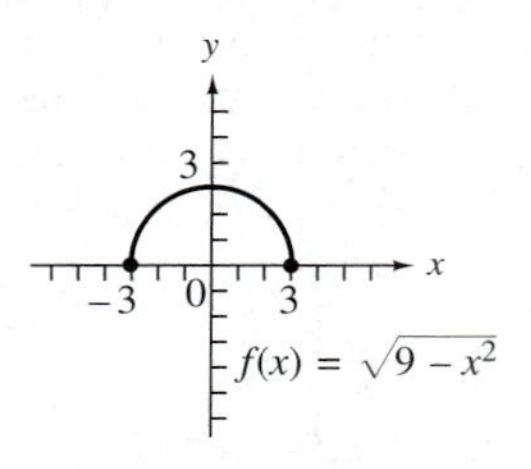

25. $[-5, 5]$; $[-5, 0]$

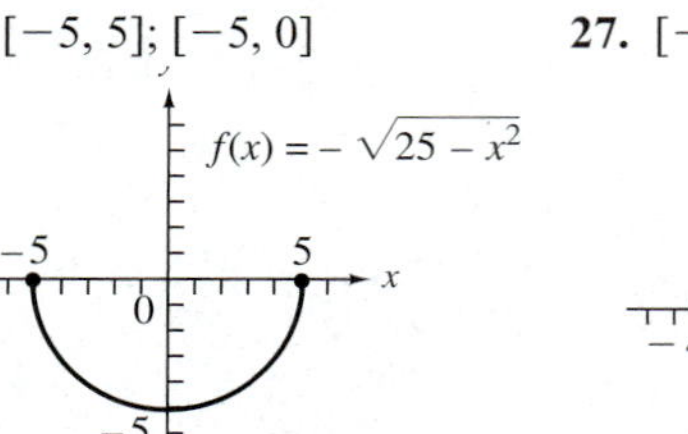

27. $[-4, \infty)$; $[0, \infty)$

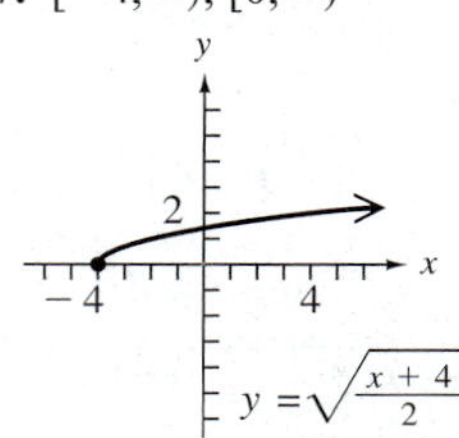

29. **(a)** 50 m **(b)** 69.3 m **31.** for V greater than 4325.68 m per sec

Section 11.4 (page 741)

1. Substitute $x - 1$ for y in the first equation. Then solve for x. Find the corresponding y-values by substituting back into $y = x - 1$. In the first equation, both variables are squared and in the second, both variables are to the first power, so the elimination method is not appropriate. **3.** one **5.** none

7.

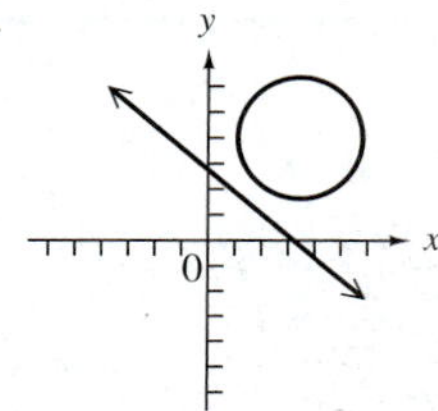

9.

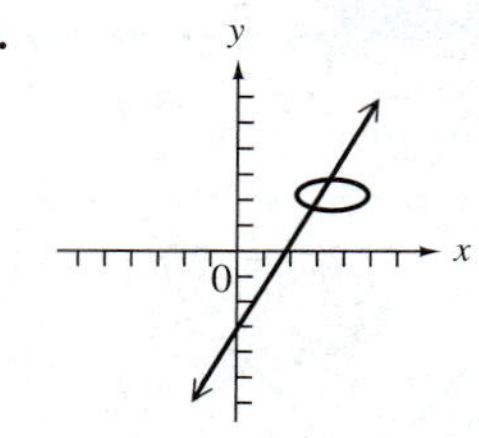

11.

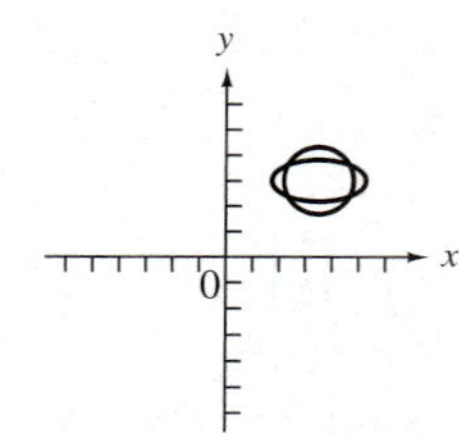

13. $\left\{(0, 0), \left(\frac{1}{2}, \frac{1}{2}\right)\right\}$ **15.** $\{(-6, 9), (-1, 4)\}$

17. $\left\{\left(-\frac{1}{5}, \frac{7}{5}\right), (1, -1)\right\}$ **19.** $\left\{(-2, -2), \left(-\frac{4}{3}, -3\right)\right\}$

21. $\{(-3, 1), (1, -3)\}$ **23.** $\left\{\left(-\frac{3}{2}, -\frac{9}{4}\right), (-2, 0)\right\}$

25. $\{(-\sqrt{3}, 0), (\sqrt{3}, 0), (-\sqrt{5}, 2), (\sqrt{5}, 2)\}$

27. $\{(-2, 0), (2, 0)\}$ **29.** $\{(i\sqrt{2}, -3i\sqrt{2}), (-i\sqrt{2}, 3i\sqrt{2}), (-\sqrt{6}, -\sqrt{6}), (\sqrt{6}, \sqrt{6})\}$ **31.** $\{(-2i\sqrt{2}, -2\sqrt{3}), (-2i\sqrt{2}, 2\sqrt{3}), (2i\sqrt{2}, -2\sqrt{3}), (2i\sqrt{2}, 2\sqrt{3})\}$

33. $\{(-\sqrt{5}, -\sqrt{5}), (\sqrt{5}, \sqrt{5})\}$ **35.** $\{(i, 2i), (-i, -2i), (2, -1), (-2, 1)\}$ **37.** length: 12 ft; width: 7 ft **39.** \$20; 800 calculators **41.** 1981; 470 thousand

Section 11.5 (page 749)

1. C **3.** Graph the corresponding equation as a solid curve if the inequality is $\leq$ or $\geq$, or as a dashed curve if the inequality is $<$ or $>$. Use a test point to decide which side of the boundary satisfies the inequality, and shade it. The shaded region is the solution set.

5.

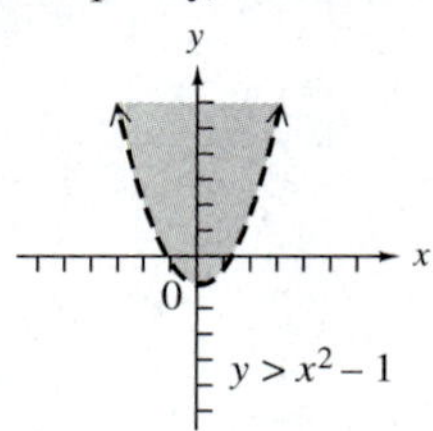

7.

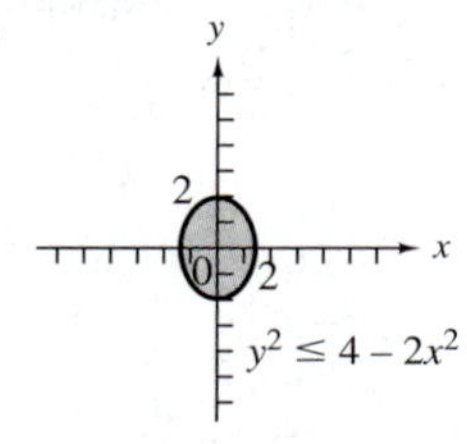

9.

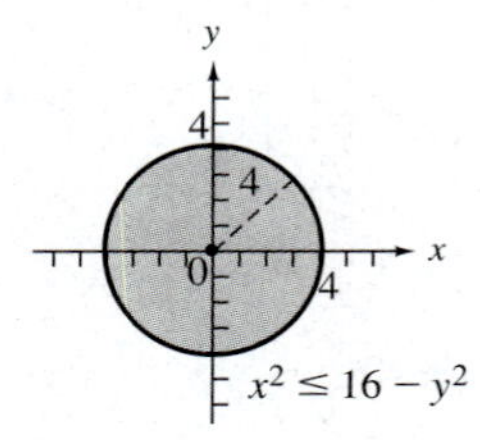

11.

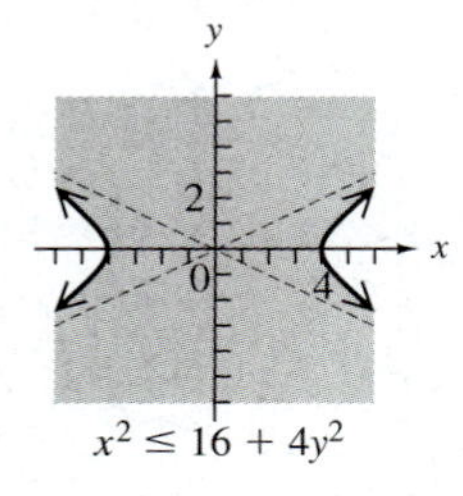

13.

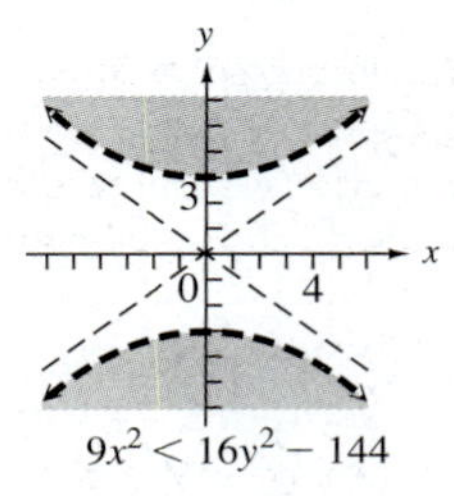

15.

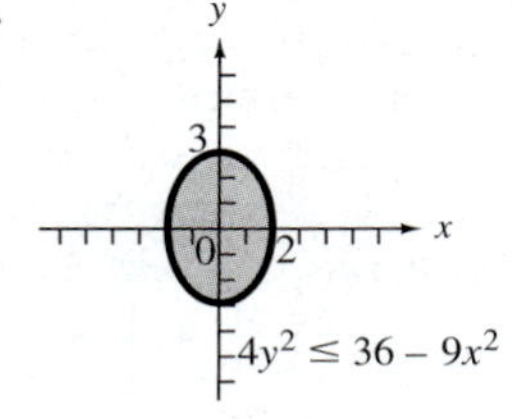

17.

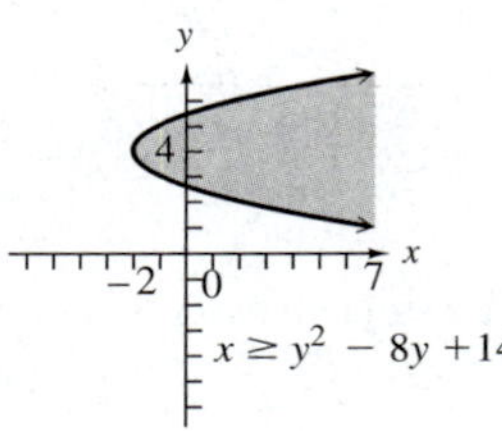

19.

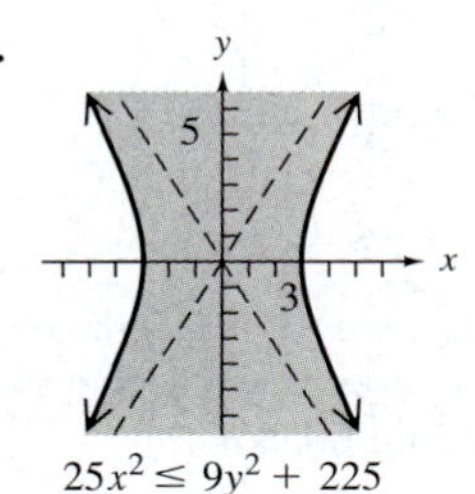

21.

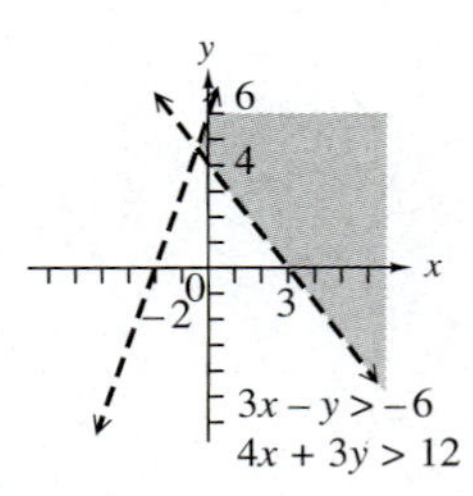

23.

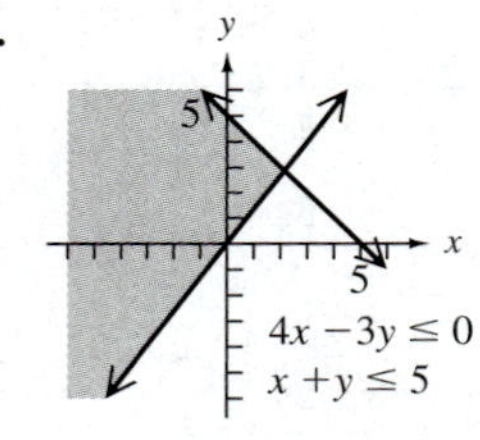

25.

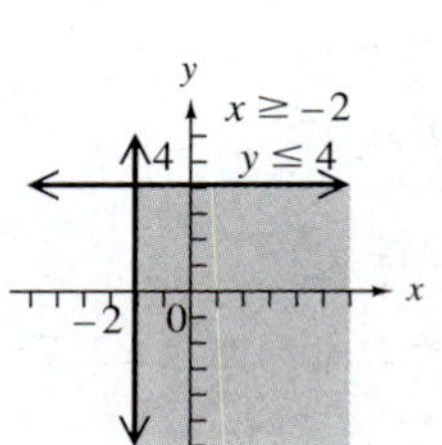

27.

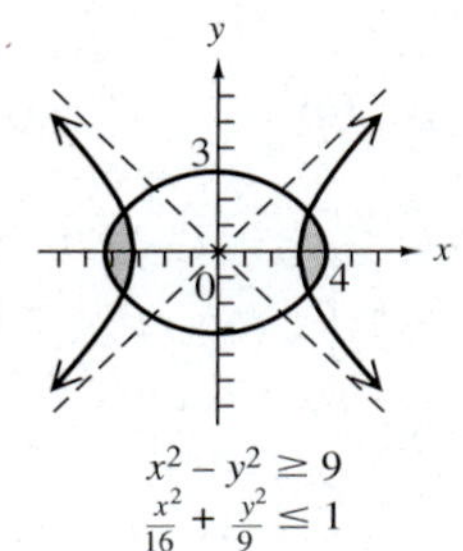

29.

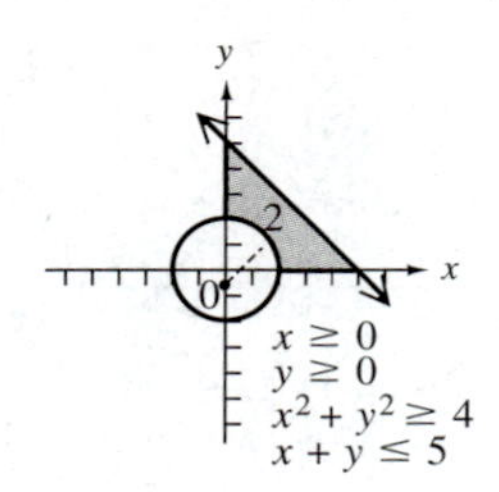

31.

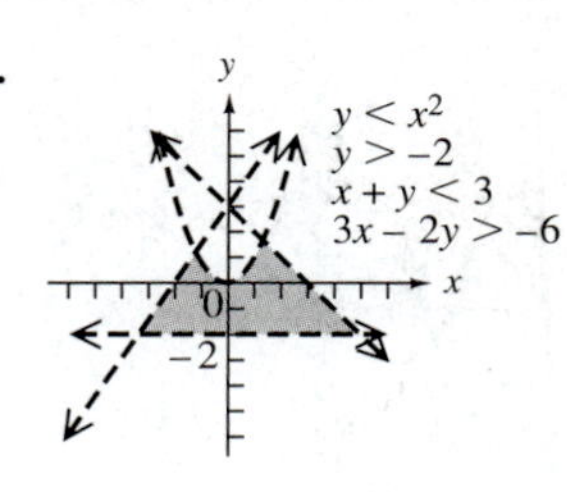

Chapter 11 Review Exercises (page 757)

1.

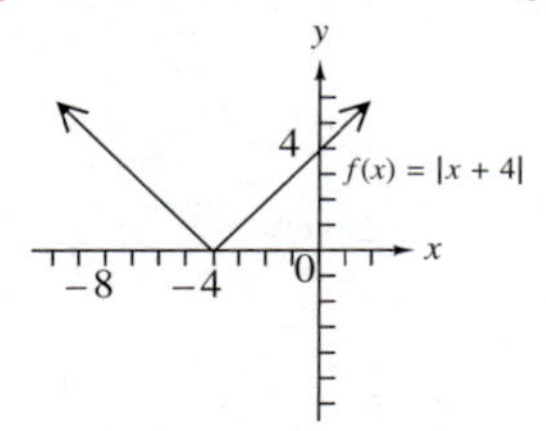

2.

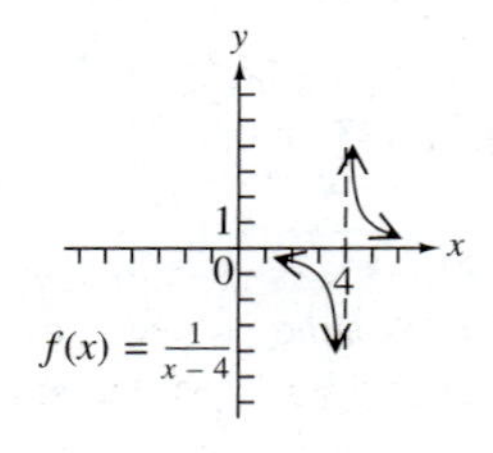

3.

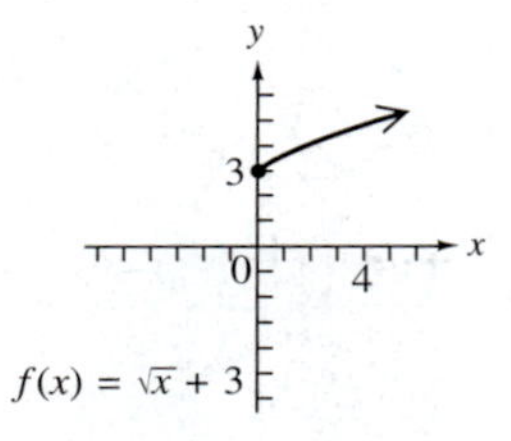

4. 167 **5.** 1495 **6.** 20 **7.** 42 **8.** $75x^2 + 220x + 160$ **9.** $15x^2 + 10x + 2$ **10.** No, composition of functions is not a commutative operation. For example, the results of Exercises 8 and 9 show that $(f \circ g)(x) \neq (g \circ f)(x)$ in this case. **11.** $(x + 2)^2 + (y - 4)^2 = 9$ **12.** $(x + 1)^2 + (y + 3)^2 = 25$ **13.** $(x - 4)^2 + (y - 2)^2 = 36$ **14.** $(-3, 2)$, $r = 4$ **15.** $(4, 1)$, $r = 2$ **16.** $(-1, -5)$, $r = 3$ **17.** $(3, -2)$, $r = 5$

18.

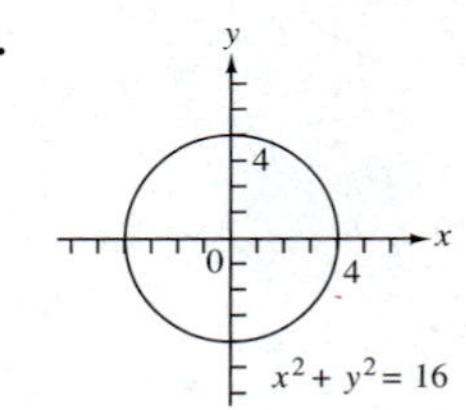

19.

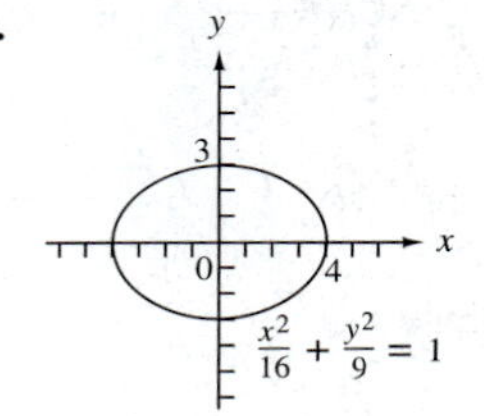

20.

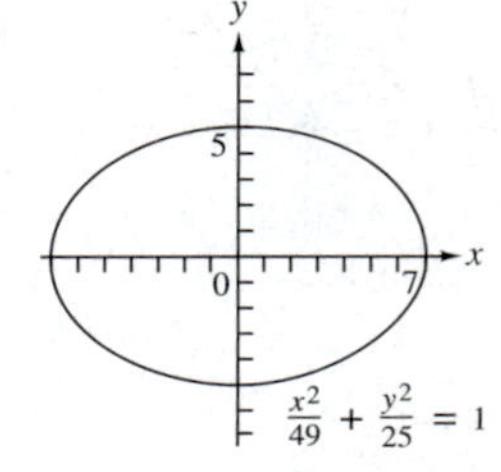

21. $\dfrac{x^2}{65{,}286{,}400} + \dfrac{y^2}{2{,}560{,}000} = 1$

22. A circular racetrack is most appropriate because the crawfish can move in any direction. Distance from the center determines the winner.

23.

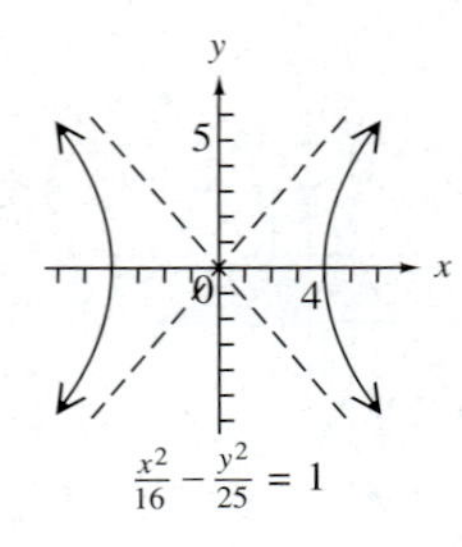

24.

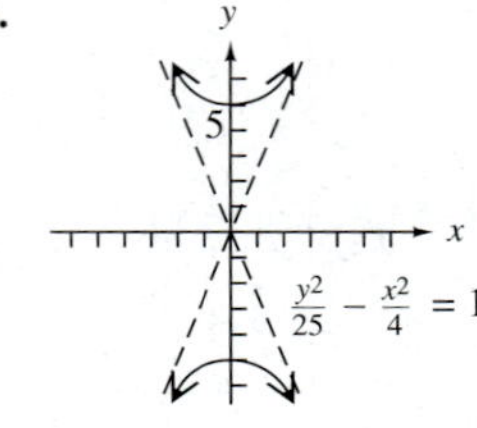

25.

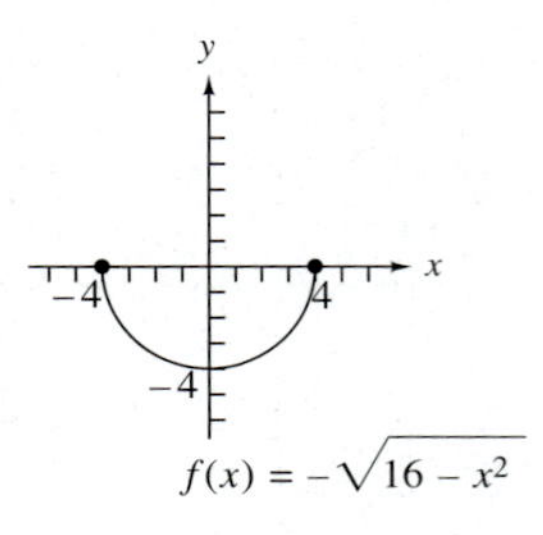

26. circle **27.** parabola **28.** hyperbola **29.** ellipse
30. parabola **31.** hyperbola **32.** $\frac{x^2}{625} - \frac{y^2}{1875} = 1$
33. $\{(6, -9), (-2, -5)\}$ **34.** $\{(1, 2), (-5, 14)\}$
35. $\{(4, 2), (-1, -3)\}$ **36.** $\{(-2, -4), (8, 1)\}$
37. $\{(-\sqrt{2}, 2), (-\sqrt{2}, -2), (\sqrt{2}, -2), (\sqrt{2}, 2)\}$
38. $\{(-\sqrt{6}, -\sqrt{3}), (-\sqrt{6}, \sqrt{3}), (\sqrt{6}, -\sqrt{3}), (\sqrt{6}, \sqrt{3})\}$
39. 0, 1, or 2 **40.** 0, 1, 2, 3, or 4

41.

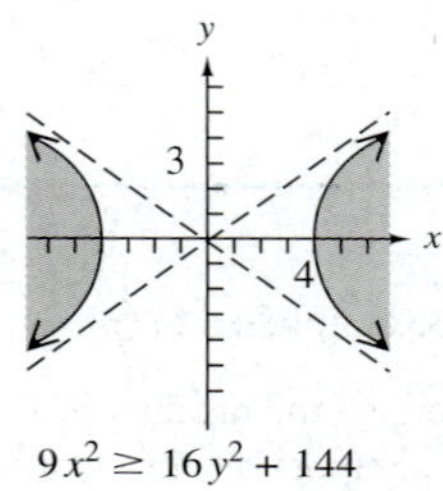

42.

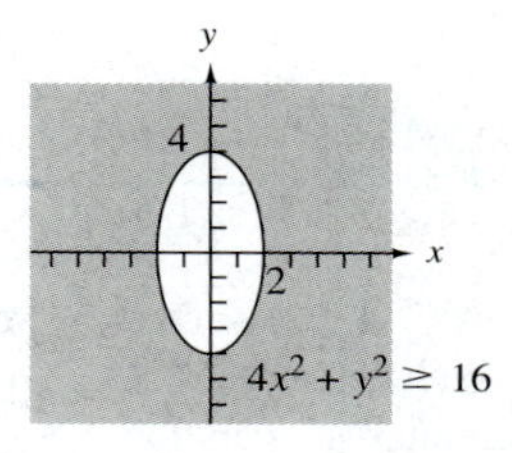

43.

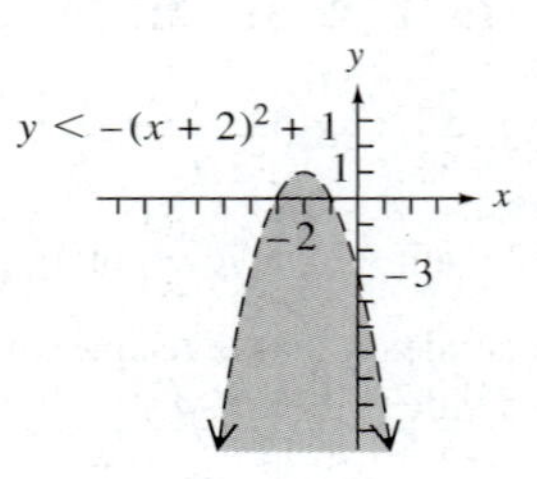

44.

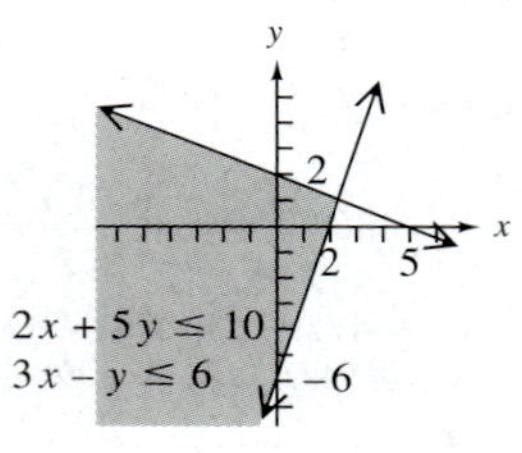

45.

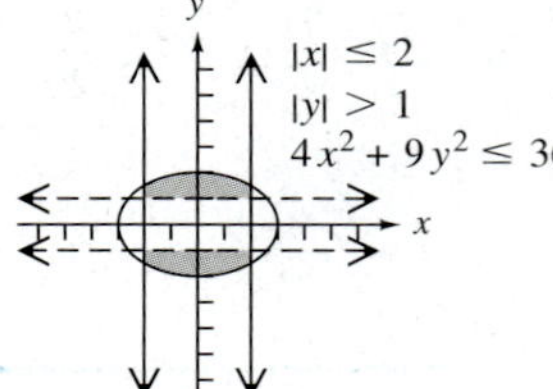

46.

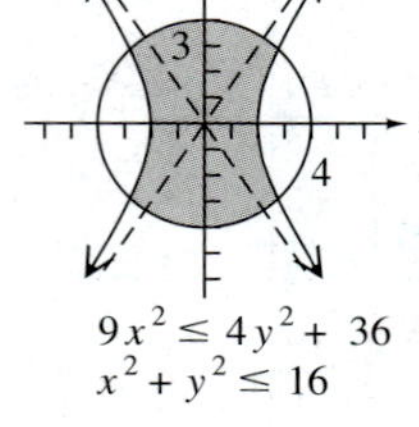

47. $2a + 4b + c = -20$ **48.** $5a + b + c = -26$
49. $-a + b + c = -2$ **50.** $\{(-4, -2, -4)\}$;
$x^2 + y^2 - 4x - 2y - 4 = 0$ **51.** center: (2, 1); radius: 3

52.

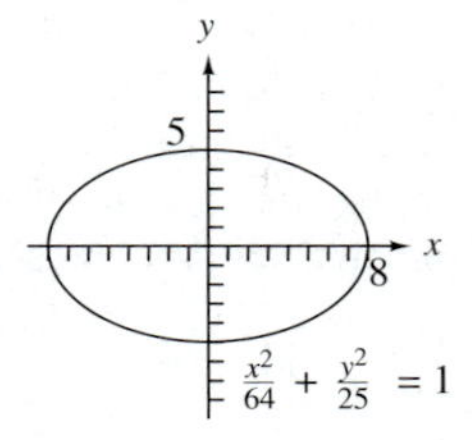

53.

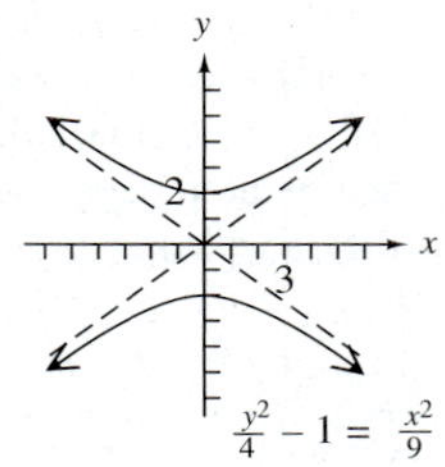

54.

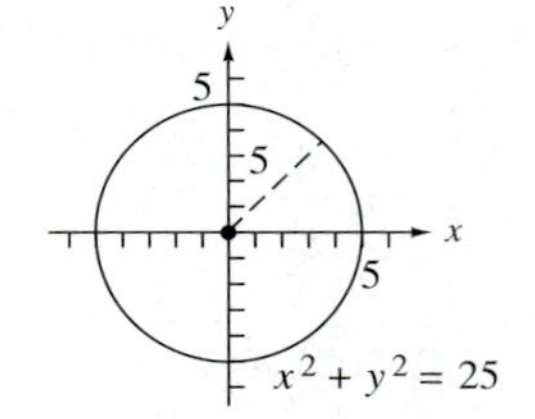

55.

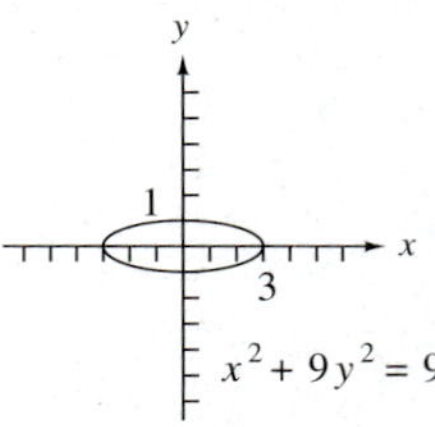

56.

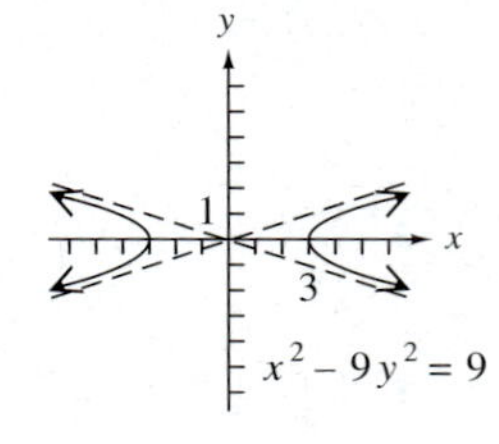

57.

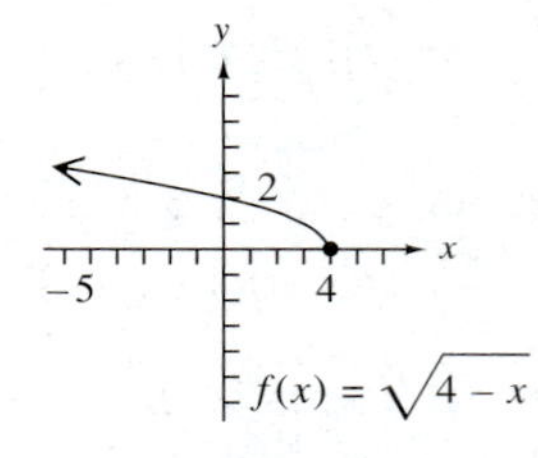

58.

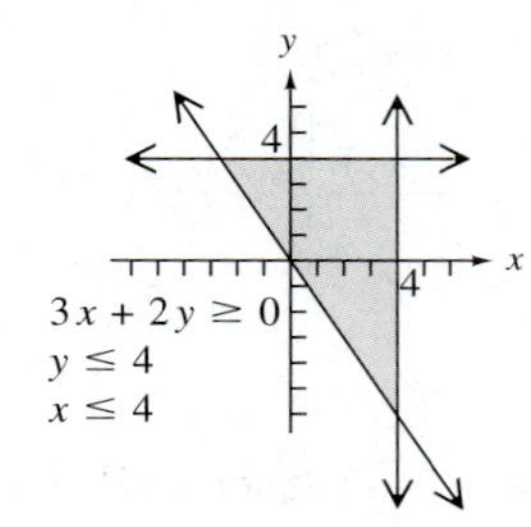

59.

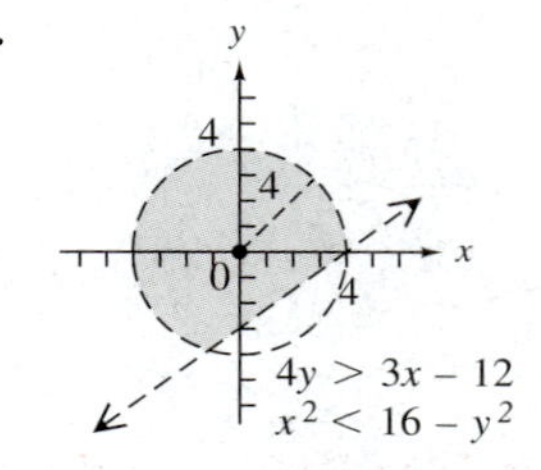

60. There are cases where one x-value will yield two y-values. In a function, every x yields one and only one y. **61.** 69.8 million km
62. 46.0 million km

Chapter 11 Test (page 763)

1. C **2.** A **3.** D **4.** B
5.

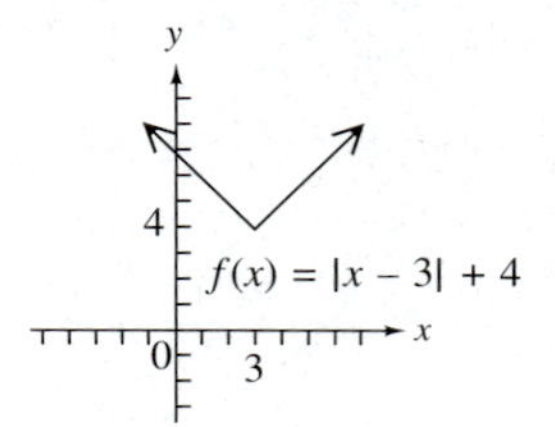

6. **(a)** 23 **(b)** $3x^2 + 11$ **(c)** $9x^2 + 30x + 27$
7. center: $(2, -3)$; radius: 4

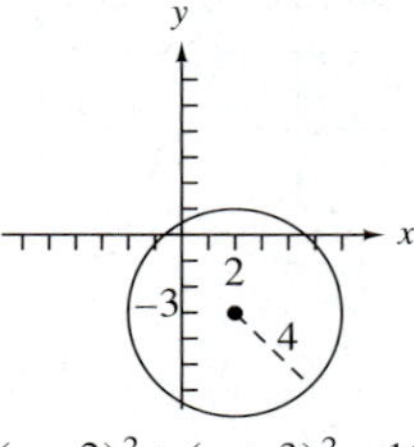

8. center: $(-4, 1)$; radius: 5

9.

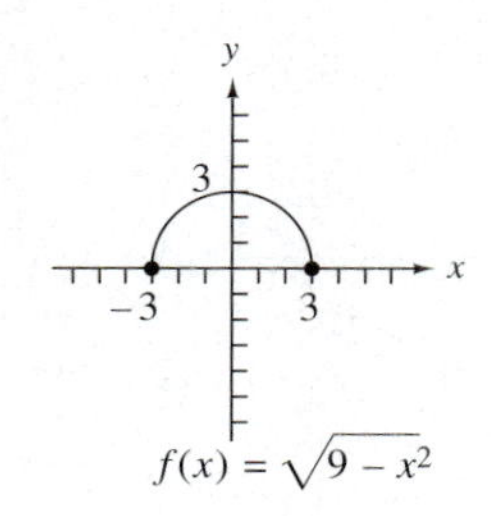

10.

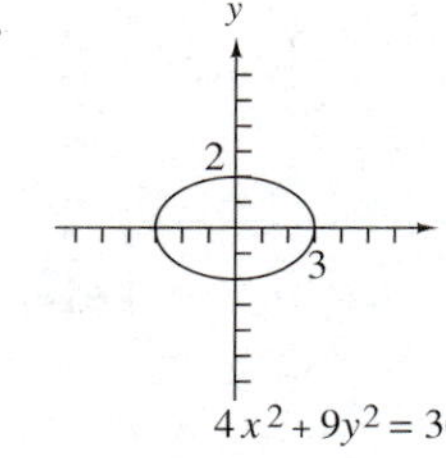

11.

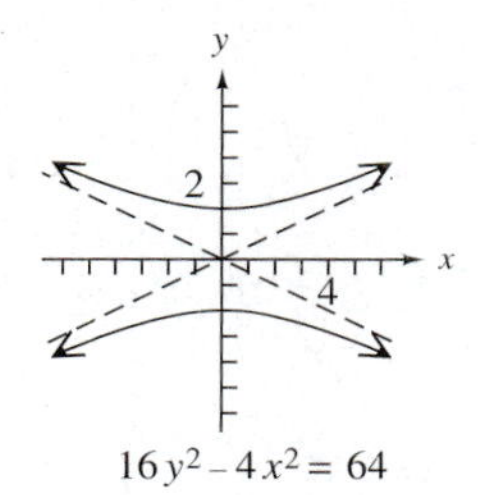

12.

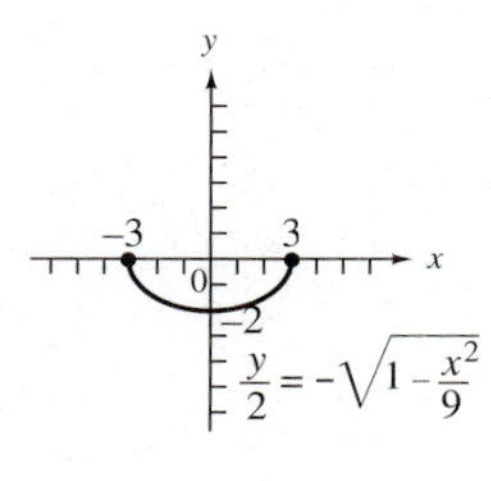

13. ellipse **14.** hyperbola **15.** parabola

16. $\left\{\left(-\frac{1}{2}, -10\right), (5, 1)\right\}$ **17.** $\left\{(-2, -2), \left(\frac{14}{5}, -\frac{2}{5}\right)\right\}$

18. $\{(-\sqrt{22}, -\sqrt{3}), (-\sqrt{22}, \sqrt{3}), (\sqrt{22}, -\sqrt{3}), (\sqrt{22}, \sqrt{3})\}$

19.

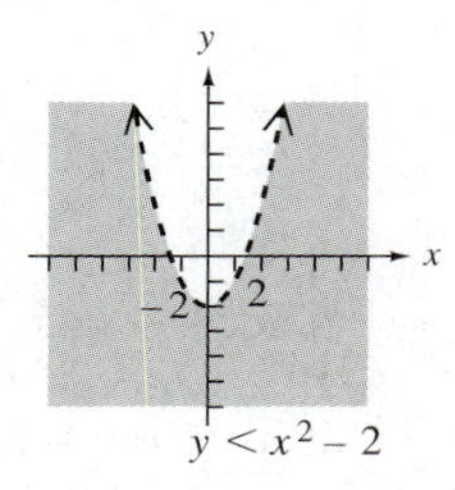

20.

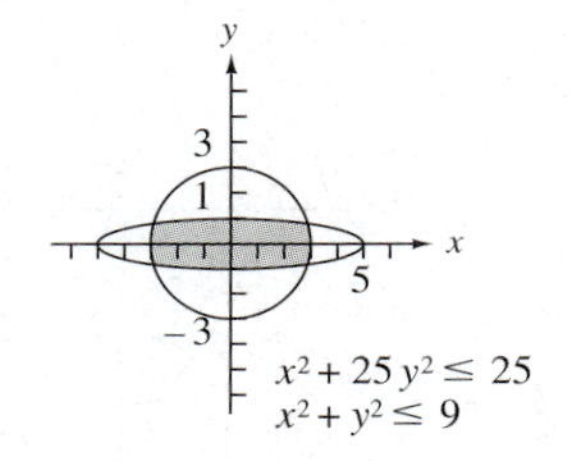

Cumulative Review Exercises: Chapters 1–11 (page 767)

1. -4 **2.** $\left\{\frac{2}{3}\right\}$ **3.** $\left(-\infty, \frac{3}{5}\right]$ **4.** $\{-4, 4\}$

5. $(-\infty, -5) \cup (10, \infty)$ **6.** $\frac{2}{3}$ **7.** $3x + 2y = -13$

8. $\{(3, -3)\}$ **9.** $\{(4, 1, -2)\}$ **10.** $\left\{(-1, 5), \left(\frac{5}{2}, -2\right)\right\}$

11. 40 mph **12.** \$275 **13.** $25y^2 - 30y + 9$ **14.** $12r^2 + 40r - 7$

15. $4x^3 - 4x^2 + 3x + 5 + \frac{3}{2x + 1}$ **16.** $(3x + 2)(4x - 5)$

17. $(2y^2 - 1)(y^2 + 3)$ **18.** $(z^2 + 1)(z + 1)(z - 1)$

19. $(a - 3b)(a^2 + 3ab + 9b^2)$ **20.** $\frac{40}{9}$ **21.** $\frac{y - 1}{y(y - 3)}$

22. $\frac{3c + 5}{(c + 5)(c + 3)}$ **23.** $\frac{1}{p}$ **24.** $1\frac{1}{5}$ hr **25.** $\frac{3}{4}$ **26.** $\frac{a^5}{4}$

27. $2\sqrt[3]{2}$ **28.** $\frac{3\sqrt{10}}{2}$ **29.** $\frac{7}{5} + \frac{11}{5}i$ **30.** $\emptyset$ **31.** $\left\{\frac{1}{5}, -\frac{3}{2}\right\}$

32. $\left\{\frac{1 + 2\sqrt{2}}{4}, \frac{1 - 2\sqrt{2}}{4}\right\}$ **33.** $\left\{\frac{3 + \sqrt{33}}{6}, \frac{3 - \sqrt{33}}{6}\right\}$

34. $\left\{-\frac{\sqrt{6}}{2}, \frac{\sqrt{6}}{2}, -\sqrt{7}, \sqrt{7}\right\}$ **35.** $v = \frac{\pm\sqrt{rFkw}}{kw}$

36. $f^{-1}(x) = \sqrt[3]{x - 4}$ **37.** 4 **38.** 7 **39.** $\log \frac{(3x + 7)^2}{4}$

40. $\{3\}$ **41.** **(a)** \$12,198.90 **(b)** \$12,214.03 **42.** \$16.9 billion
43. \$75.8 billion **44.** **(a)** -1 **(b)** $9x^2 + 18x + 4$

45.

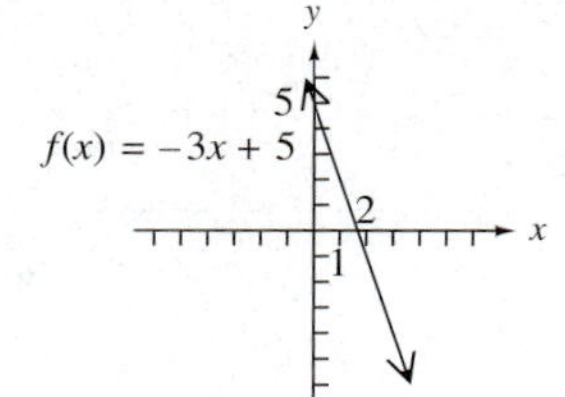

46.

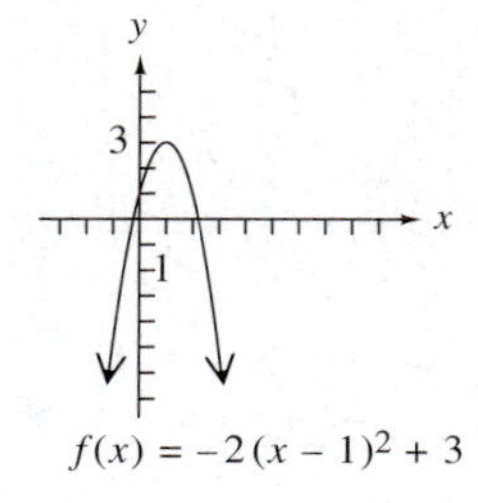

47.

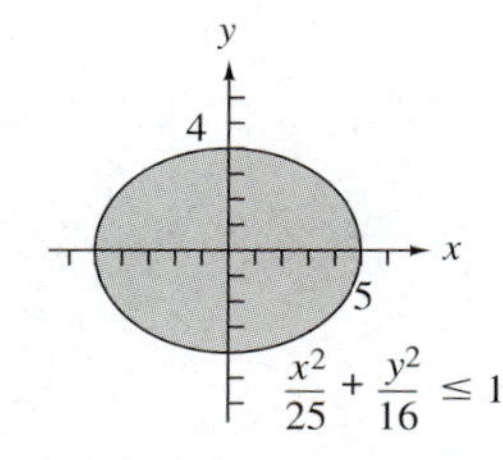

48.

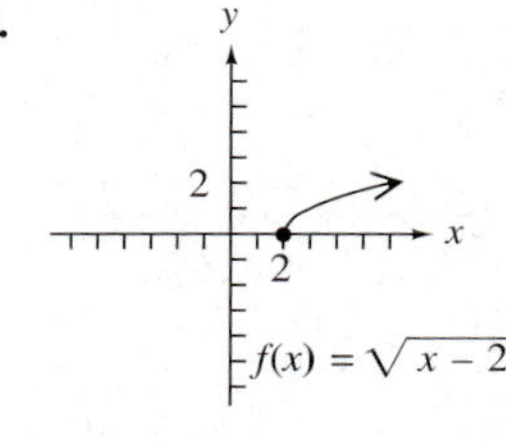

49.

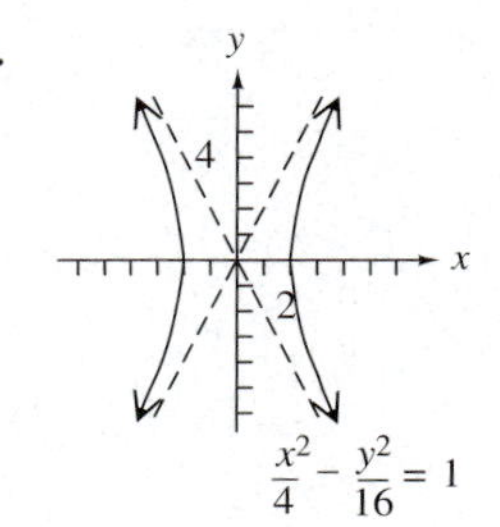

50.

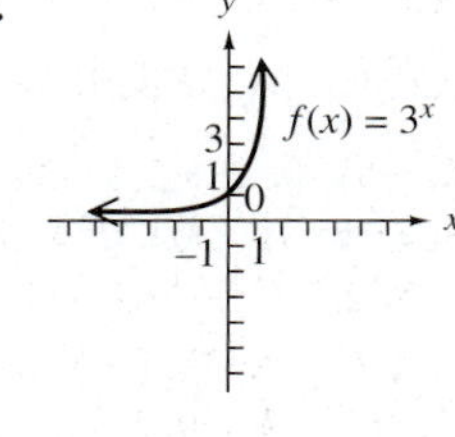

Appendix A

(page A–9)

1. true **3.** False; the fraction $\frac{17}{51}$ can be simplified to $\frac{1}{3}$.

5. False; *product* indicates multiplication, so the product of 8 and 2 is 16. **7.** prime **9.** composite **11.** composite **13.** neither

15. $2 \cdot 3 \cdot 5$ **17.** $2 \cdot 2 \cdot 3 \cdot 3 \cdot 7$ **19.** $2 \cdot 2 \cdot 31$ **21.** 29 **23.** $\frac{1}{2}$

25. $\frac{5}{6}$ **27.** $\frac{1}{5}$ **29.** $\frac{6}{5}$ **31.** A **33.** $\frac{24}{35}$ **35.** $\frac{6}{25}$ **37.** $\frac{6}{5}$

39. $\frac{232}{15}$ or $15\frac{7}{15}$ **41.** $\frac{10}{3}$ **43.** 12 **45.** $\frac{1}{16}$ **47.** $\frac{84}{47}$ or $1\frac{37}{47}$

49. Multiply the first fraction (the dividend) by the reciprocal of the second fraction (the divisor) to divide two fractions.

51. $\frac{2}{3}$ **53.** $\frac{8}{9}$ **55.** $\frac{27}{8}$ or $3\frac{3}{8}$ **57.** $\frac{17}{36}$ **59.** $\frac{11}{12}$ **61.** $\frac{4}{3}$

63. 6 cups **65.** $618\frac{3}{4}$ ft **67.** $\frac{9}{16}$ in. **69.** $\frac{5}{16}$ in. **71.** $\frac{1}{20}$

73. more than $1\frac{1}{25}$ million

Appendix B

(page A–21)

1. true **3.** false **5.** -3 **7.** 14 **9.** 0 **11.** 59 **13.** 14
15. 1 **17.** -22 **19.** 0 **21.** 0 **23.** 20 **25.** $\{(4, 6)\}$

27. $\{(-5, 2)\}$ **29.** $\{(1, 0)\}$ **31.** $\left\{\left(\frac{11}{58}, -\frac{5}{29}\right)\right\}$

33. $\{(-2, 3, 5)\}$ **35.** $\{(5, 0, 0)\}$ **37.** Cramer's rule does not apply. **39.** $\{(20, -13, -12)\}$ **41.** $\left\{\left(\frac{62}{5}, -\frac{1}{5}, \frac{27}{5}\right)\right\}$

Index

A

B

C

G

H

I

O

P

Q

R

S

Videotape Index

The purpose of this index is to show those exercises fromhe text that are used in the Real to Reel videotape series that accompanies *Intermed:te Algebra*, Seventh Edition.

Section	Exercises
1.1	41, 49
1.2	51
1.3	75
1.4	39
2.1	19, 41
2.2	21
2.3	19, 49
2.4	24
3.1	25, 53
3.2	5
3.3	29, 55
4.1	35
4.2	41
4.4	19, 27
4.5	15, 19
4.6	22
5.1	29
5.2	17
5.3	19
5.4	7, 17
6.1	45, 47
6.2	71
6.3	51, 55
6.4	9, 19
6.5	13, 23
6.6	27, 35, 48
6.7	15, 31
6.8	43
6.9	11, 17

Section	Exercises
7.1	11, 35
7.2	11, 41, 43, 47, 49
7.3	5, 11, 13, 29
7.4	7, 17, 27
7.5	9, 45
8.1	37
8.2	23
8.3	83
8.4	39
8.5	19, 35
8.6	19, 29
8.7	13, 19, 43, 53, 77
9.1	59
9.2	23
9.3	15, 31, 47
9.4	29
9.5	27
9.6	7, 29
9.7	11
10.1	19, 21
10.3	31
10.4	33, 37
10.5	15, 17, 23
10.6	7, 47
11.1	19, 27
11.2	13
11.3	13, 17, 21
11.4	35, 37
11.5	26